Geometric figures

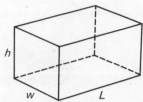

Rectangle
Area = wh
Length of diagonal
$= \sqrt{w^2 + h^2}$
(Pythagorean theorem)

Rectangular box
Volume = wLH
Surface area = $2Lw + 2Lh + 2wh$
Length of diagonal = $\sqrt{L^2 + w^2 + h^2}$
(Pythagorean theorem)

Triangle
Area = $\frac{1}{2}bh$

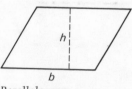

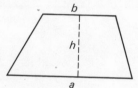

Parallelogram
Area = bh

Trapezoid
Area = $\frac{1}{2}(a + b)h$

Circle
Area = πr^2
Circumference = $2\pi r$

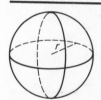

Sphere
Volume = $\frac{4}{3}\pi r^3$
Surface area = $4\pi r^2$

Right circular cone
Volume = $\frac{1}{3}\pi r^2 h$
Lateral surface area
$= \pi r \sqrt{r^2 + h^2}$

Volume
Lateral surface area = $2\pi rh$

The Greek alphabet

A α	Alpha	I ι	Iota	P ρ	Rho
B β	Beta	K κ	Kappa	Σ σ s	Sigma
Γ γ	Gamma	Λ λ	Lambda	T τ	Tau
Δ δ	Delta	M μ	Mu	Υ υ	Upsilon
E ε	Epsilon	N ν	Nu	Φ φ	Phi
Z ζ	Zeta	Ξ ξ	Xi	X χ	Chi
H η	Eta	O o	Omicron	Ψ ψ	Psi
Θ θ ϑ	Theta	Π π	Pi	Ω ω	Omega

SECOND EDITION

CALCULUS AND ANALYTIC GEOMETRY

AL SHENK

University of California
San Diego

Goodyear Publishing Company, Inc.
Santa Monica, California

To Genie

Library of Congress Cataloging in Publication Data
Shenk, Al
    Calculus and analytic geometry.
    Bibliography: p. 998
    1. Calculus.   2. Geometry, Analytic.   I. Title
QA303.S53   515'.15   L.C. 76–17206
ISBN 0–87620–195–8

Current printing (last digit)
10 9 8 7 6 5 4 3 2 1

Y–1958–1

Printed in the United States of America.
Aquiring Editor: Jack Pritchard
Managing Editor: Gerald P. Rafferty
Cover model and photograph by Tom Gould, based on the surface $z = \dfrac{\cos x \sin y}{1 + x^2}$

CONTENTS

PREFACE TO THE SECOND EDITION   ix

TO THE INSTRUCTOR   xi

TO THE STUDENT   xv

CHAPTER 1 COORDINATE PLANES AND GRAPHS OF EQUATIONS

1-1   Coordinate lines and inequalities   1
1-2   Solving inequalities   5
1-3   Coordinate planes   7
1-4   Graphs of equations   12
1-5   The slope of a line   17
1-6   Equations of lines   21
1-7   A brief catalog of curves   25
1-8   Historical notes (optional)   28
Appendix The algebra of real numbers   30
Miscellaneous exercises   36

CHAPTER 2 FUNCTIONS AND LIMITS

2-1   Functions   39
2-2   Graphs of functions   44
2-3   Finite limits   48
2-4   Limit theorems   53
2-5   Infinite limits   57
2-6   Limits of polynomials and ratio
        $x \to \pm \infty$ 62
2-7   Continuous functions   65
2-8   The Intermediate Value and Extreme Value
        Theorems   70
2-9   The $\epsilon\delta$-definitions of limits (optional)   74
Miscellaneous exercises   83

CHAPTER 3 THE DERIVATIVE

3-1   Derivatives and tangent lines   87
3-2   Leibniz notation and the differentiation
         operator d/dx 94
3-3   The derivative of x^n, derivatives of linear
         combinations   99
3-4   The product and quotient rules   104
3-5   Average velocity and velocity   107
3-6   Other rates of change   112
3-7   Composite functions and the chain rule   118
3-8   Angles, sines, and cosines   127
3-9   The derivatives of $\sin x$ and $\cos x$ 135
3-10   Remarks on the use of mathematical models
         (optional)   142
3-11   Historical notes (optional)   143
Miscellaneous exercises   148

CHAPTER 4 APPLICATIONS OF DIFFERENTIATION

4-1   Sketching the graphs of functions: The first
         derivative test   152
4-2   Higher order derivatives and the second
         derivative test   156
4-3   Sketching graphs of rational functions   160
4-4   Sketching graphs of other functions   164
4-5   Implicitly defined functions and implicit
         differentiation   168
4-6   Inverse functions   173
4-7   Related rates   179
4-8   Maximum-minimum problems in finite
         closed intervals   186
4-9   Other maximum-minimum problems   190
4-10   Tangent line approximations and differentials   194
4-11   Newton's method (optional)   199
4-12   The Mean Value Theorem for derivatives   202
4-13   Antiderivatives   206
Miscellaneous exercises   208

CHAPTER 5 THE INTEGRAL

Introduction   212
5-1   The geometric definitions of the integral   213
5-2   The analytic definition of the integral   220
5-3   The Fundamental Theorem of Calculus   229
5-4   Indefinite integrals and integrals of x^n ($n \neq -1$),
         $\sin x$, $\cos x$ 235
5-5   The algebra of integrals   239
5-6   Integration by substitution   242
5-7   Historical notes (optional)   247
Appendix The existence of definite integrals
   (optional)   250
Miscellaneous exercises   256

CHAPTER 6 APPLICATIONS OF INTEGRATION

6-1   Areas of plane regions   261
6-2   Computing volumes by slicing   266
6-3   Volumes of solids of revolution: The methods of disks and washers   269
6-4   Volumes of solids of revolution: The method of cylinders   276
6-5   Lengths of curves   282
6-6   Areas of surfaces of revolution   284
6-7   Integrals of rates of change   288
6-8   Work   295
6-9   Hydrostatic pressure   301
6-10   Density, weight, and centers of gravity   305
6-11   Average values of functions and the Mean-Value Theorem for integrals   312
Miscellaneous exercises   316

CHAPTER 7 TRANSCENDENTAL FUNCTIONS

7-1   The algebra of exponential functions and logarithms   322
7-2   The number e and the natural logarithm   329
7-3   The integral of x^{-1} 337
7-4   Logarithmic differentiation (optional)   341
7-5   The exponential function   342
7-6   The differential equation $dy/dt = ry$ and applications   349
7-7   Compound interest (optional)   354
7-8   The trigonometric functions   357
7-9   The inverse trigonometric functions   366
7-10   The hyperbolic functions   373
7-11   Historical notes (optional)   379
Miscellaneous exercises   382

CHAPTER 8 TECHNIQUES OF INTEGRATION

8-1   Review of integration by substitution   390
8-2   Changing variables in definite integrals (optional)   394
8-3   Integration by parts   397
8-4   Integrals of powers and products of trigonometric functions   402
8-5   Inverse trigonometric substitutions   407
8-6   Integration by partial fractions   411
8-7   Integrals of rational functions of sines and cosines (optional)   419
8-8   The use of integral tables (optional)   421
8-9   Approximate integration (optional)   423
Miscellaneous exercises   426

CHAPTER 9 SEPARABLE DIFFERENTIAL EQUATIONS

(Chapter 9 may be covered any time after Chapter 8.)

9-1   Separable differential equations   430
9-2   Applications of separable differential
      equations   437
Miscellaneous exercises   448

CHAPTER 10 CONIC SECTIONS AND POLAR COORDINATES

Introduction   452
10-1   Parabolas   453
10-2   Ellipses   458
10-3   Hyperbolas   463
10-4   Rotation of axes   470
10-5   Sections of cones; directrices and
      eccentricity   477
10-6   Polar coordinates   481
10-7   Graphs of equations in polar coordinates   484
10-8   Area in polar coordinates   491
Miscellaneous exercises   493

CHAPTER 11 CURVES AND VECTORS IN THE PLANE

11-1   Vectors   496
11-2   The dot product   504
11-3   Parametric equations of curves, arclength,
      speed   509
11-4   Vector representations of curves; velocity
      vectors   516
11-5   Curvature   524
11-6   Acceleration vectors and Newton's law   529
11-7   Polar forms of vectors, planetary motion
      (optional)   536
Miscellaneous exercises   542

CHAPTER 12 VECTORS IN SPACE AND SOLID ANALYTIC GEOMETRY

12-1   Rectangular coordinates and vectors in
      space   546
12-2   The cross product of two vectors in space   555
12-3   Equations of lines and planes   562
12-4   Curves in space   569
12-5   A brief catalog of surfaces   573
Appendix. Cramer's rule (optional)   580
Miscellaneous exercises   583

CHAPTER 13 PARTIAL DERIVATIVES

13-1   Functions of two variables   586
13-2   Partial derivatives   603
13-3   Limits, continuity, and the chain rule   612
13-4   Directional derivatives and gradients   625
13-5   Tangent planes   637
13-6   Functions of three variables   643
Miscellaneous exercises   651

CHAPTER 14 MAXIMA AND MINIMA WITH TWO OR THREE VARIABLES

14-1   The first derivative test   656
14-2   The second derivative test   666
14-3   Lagrange multipliers   674
Miscellaneous exercises   683

CHAPTER 15 MULTIPLE INTEGRALS

15-1   Double integrals   686
15-2   Applications of double integrals   695
15-3   Double integrals in polar coordinates   703
15-4   Triple integrals   707
15-5   Cylindrical and spherical coordinates   714
Miscellaneous exercises   726

CHAPTER 16 VECTOR ANALYSIS

16-1   Line integrals   729
16-2   Vector fields and path-independent line integrals   738
16-3   Green's Theorem, the Divergence Theorem, and Stokes' Theorem in the plane   749
Miscellaneous exercises   758

CHAPTER 17 TAYLOR'S THEOREM, L'HOPITAL'S RULE, AND IMPROPER INTEGRALS

(Chapters 17 and 18 may be covered any time after Chapter 8.)

17-1   Taylor polynomials and Taylor's Theorem   764
17-2   L'Hopital's rule   771
17-3   Improper integrals   777
Miscellaneous exercises   782

CHAPTER 18 INFINITE SEQUENCES AND SERIES

18-1   Infinite sequences   785

18-2   The ϵN-definition of the limit of a
          sequence   788

18-3   Finite geometric series   794

18-4   Infinite series and the comparison test for
          series with nonnegative terms   796

18-5   The Integral Test and the Ratio Test   806

18-6   Absolute convergence and the Alternating
          Series Test   812

18-7   Power series and Taylor series   819

18-8   Operations with power series   823

Miscellaneous exercises   830

CHAPTER 19 FURTHER TOPICS IN DIFFERENTIAL EQUATIONS

(Chapter 19 may be covered any time after Chapter 9.)

19-1   First-order linear differential equations   836

19-2   Exact differential equations   840

19-3   Second-order linear equations with constant
          coefficients   842

19-4   Applications of second-order equations   850

19-5   Power series solutions   858

Miscellaneous exercises   862

APPENDIX
          Reference tables   865
          Answers and outlines of solutions   882

REFERENCES   998

INDEX   1000

PREFACE TO THE SECOND EDITION

The Second Edition of *Calculus and Analytic Geometry* is an outgrowth of users' experience with the First Edition during its initial two years of publication. Instructors and students communicated suggestions for improving the book, and reviewers made section-by-section critiques. This information along with the author's experience in his own classes guided the revision. New material has been added and certain discussions have been modified or reorganized to make them easier to understand and to teach from, but the basic coverage, level, teaching philosophy, organization, and format of the book have not been changed.

Over 100 new examples have been added in 47 sections and over 750 routine exercises have been added in 79 sections. Metric units are used in many of the new exercises. Of the 114 new illustrations, 47 are in the answer section.

An optional discussion of mathematical induction has been provided in the Appendix to Chapter 1. The review of trigonometry in Chapters 3 and 7 has been expanded to include more on trigonometric identities, on evaluating trigonometric functions, and on solving equations that involve them. The discussion of sines and cosines in Chapter 3 has been divided into two sections, the first dealing with trigonometry and the second with derivatives. Section 4.6 on inverse functions has been entirely rewritten to clarify the geometric interpretations.

Chapter 5 has been reorganized to emphasize the relationship between the informal geometric definition of the integral in terms of areas and the formal analytic definition as a limit of Riemann sums. The analytic definition is now presented in Section 5.2 before the Fundamental Theorem of Calculus in Section 5.3. Background material on summation notation and the use of approximations by rectangles to compute areas is now given in the introductory Section 5.1.

New material on the algebra of logarithms and exponential functions and on solving equations that involve them has been added to Chapter 7. The informal discussion which serves as background for the definition of the natural logarithm as an integral has been moved from Section 7.2 to Section 7.1 and the presentation of the underlying theory has been revised.

A brief catalog of curves given by equations in polar coordinates has been added to Section 10.7, along with a discussion of tangent lines at the origin. Formal definitions of vectors as ordered pairs and triples of numbers have been included with the geometric definitions in Chapters 11 and 12.

The treatment of limits and continuity for functions of two and three variables in Section 13.3 has been revised to parallel that for functions of one variable in Chapter 2. Brief discussions of the Extreme Value Theorem for functions of two or three variables and of the existence of double and triple integrals have been added to Chapters 14 and 15.

Chapter 18 on infinite sequences and series has been reorganized so the Comparison Test, the Integral Test, and the Ratio Test are first given for series with nonnegative terms. Absolute convergence and the convergence tests for general series are studied in a later section along with the Alternating Series Test. An entirely new section on operations with power series has been added at the end of the chapter.

The author wishes to thank the reviewers and other users of the First Edition for their contributions to the revision.

Reviewers for the Second Edition   Dennis Berkey, Boston University; Gerald Bradley, Claremont Men's College; Ray Edwards, Chabot College; Paul Gormley, Villanova University; John Kenally, Clemson University; Josephs Krebs, Boston College; Stanley Lukawecki, Clemson University; Richard Marshall, Eastern Michigan University; Isaac Metts, Jr., The Citadel; Burton Rodin, University of California, San Diego; and David Sprows, Villanova University.

TO THE INSTRUCTOR

This book covers analytic geometry and differential and integral calculus of one, two, and three variables. It includes introductions to vector analysis, infinite sequences and series, and differential equations. It is designed primarily for calculus courses for math and science majors but can be used in shorter courses by limiting the coverage of some of the topics (see the course outlines on p. xiii).

Why this book was written

Calculus can and should be a clear, interesting subject. The concepts of limit, velocity, area, and volume are familiar in one form or another to everyone; their development into the basic ideas of calculus can be easily understood. Students can learn the theory if the purpose of the definitions and theorems is explained and if the geometric meanings are illustrated. They can work all types of problems if they are given enough practice and are taught step-by-step procedures to guide their reasoning. They can appreciate the significance of calculus if they see a variety of examples and exercises that indicate how it is used in other fields.

Distinctive features

The theory is presented concisely. Whereas a theorem-proof format is used throughout for key results, related proofs are given as exercises. Some results that would require unimportant technical details if proved as theorems are given as Rules.

The level of a course can be adjusted to meet the needs of its students. The Miscellaneous Exercises at the ends of the chapters are generally more difficult than the exercises at the ends of the sections, and certain theoretical topics such as the $\epsilon\delta$-definitions of limits and continuity are given in optional sections.

Geometric intuition is emphasized. Curve drawing is covered before other applications, and students are taught to use sketches in problem solving. There are over 900 illustrations in the text and over 400 more in the answer section at the back of the book. Computer-generated pictures are used to enable students to work with curves and surfaces that are too complex for them to draw. Graphical problems that require little or no algebra help teach the geometric meanings of results and procedures.

Applications are stressed. Examples and exercises have been collected to illustrate the role of calculus in fields such as physics, biology, chemistry, medicine, and economics.

Outlines of solutions of selected exercises (designated by ‡) are provided to help students learn step-by-step procedures of problem solving. These are given in the back of the book, along with the answers to approximately one-half of the remaining exercises (designated by †). A *Solutions Manual,* containing detailed solutions of selected exercises, is also available to students. The sets of "Additional Drill Exercises" given in certain sections provide extra practice for students who have difficulty mastering algebraic techniques.

Exercises to be worked by computer or electronic calculator are included to illustrate the limit process numerically in various contexts, including the definitions of the derivative and of the integral, Newton's method, and Simpson's rule.

Historical Notes survey the origins and development of calculus.

Organizational features

This book has been designed to be as flexible as possible. Instructors should take note of the following features when planning its use.

Much of the material in Chapter 1 (Coordinate Planes and Graphs of Equations) and Chapter 2 (Functions and Limits) is covered in precalculus courses in high schools and colleges and may be skipped or reviewed quickly in many calculus courses.

In Chapter 2 (Functions and Limits) we first discuss limits and continuity from an intuitive, geometric point of view. Those who choose to cover the final section of the chapter on the $\epsilon\delta$-definitions have the option of incorporating it with the earlier sections.

Differentiation and integration of the sine and cosine functions are studied in Chapter 3 (The Derivative) and Chapter 5 (The Integral), so that applications involving these functions may be included in Chapter 4 (Applications of Differentiation) and Chapter 6 (Applications of Integration). The calculus of the other trigonometric functions is presented in Chapter 7 (Transcendental Functions).

An optional Section 8.8 in Chapter 8 (Techniques of Integration) deals with the use of integral tables and may be covered in place of some or all of Sections 8.2 through 8.7 on special techniques of integration.

The material on separable differential equations is placed in a short Chapter 9, so that students can see more applications of transcendental functions and techniques of integration immediately after studying them. This chapter and Chapter 19 (Further Topics in Differential Equations) may be covered any time after Chapter 8 (Techniques of Integration).

Chapter 17 (Taylor's Theorem, l'Hopital's Rule, and Improper Integrals) and Chapter 18 (Infinite Sequences and Series) may also be covered any time after Chapter 8.

Instructors who wish to assign exercises for which solutions are not given in the *Solutions Manual* should choose them from the sets of Miscellaneous Exercises or from those not marked with † or ‡ at the ends of the sections.

Possible course outlines

For math and science majors (infinite series and differential equations covered last):

Four quarters

1st quarter: Chapters 1–5
2nd quarter: Chapters 6–8, 10, 11
3rd quarter: Chapters 12–16
4th quarter: Chapters 17, 18, 9, 19

Three semesters

1st semester: Chapters 1–7
2nd semester: Chapters 8, 10–14
3rd semester: Chapters 15–18, 9, 19

For math and science majors (multiple variable calculus covered last):

Four quarters

1st quarter: Chapters 1–5
2nd quarter: Chapters 6–8, 10, 11
3rd quarter: Chapters 17, 18, 9, 19
4th quarter: Chapters 12–16

Three semesters

1st semester: Chapters 1–7
2nd semester: Chapters 8, 9, 17–19
3rd semester: Chapters 10–16

For short calculus courses:

Three quarters

First quarter: Chapters 1–5
Second quarter: Sections 6.1, 6.7, 6.8, Chapter 7 (except Section 7.10), Sections 8.1 and 8.8, Chapter 9, Sections 19.3 and 19.4, Sections 17.1 and 17.2
Third quarter: Sections 11.1 through 11.4, Chapters 12, 13, 15

Two semesters:

First semester: Chapters 1–5, Sections 6.1, 6.7, and 6.8, Chapter 7 (except Section 7.10)
Second semester: Sections 8.1 and 8.8, Chapter 9, Sections 19.3 and 19.4, Sections 17.1 and 17.2, Sections 11.1 through 11.4, Chapters 12, 13, 15

Acknowledgements

This book owes a great deal to all those who contributed to its preparation and production, especially to its editor Jack Pritchard for his vision and support; to Genie Shenk, who helped revise the manuscript at each stage of its development; to Curt Cowan for his skillful handling of the challenges of production; to Jim and Tauri Rapp and the other graduate students who worked on the exercises; to Professor C. Truesdell and the late Carl B. Boyer for their suggestions concerning the Historical Notes; to the staff of the U.C.S.D. Computer Center for their assistance in the preparation of computer-generated pictures; to the reviewers for their thoughtful and valuable advice; and to many others among the author's friends and colleagues who gave him the benefit of their teaching experience in suggesting how topics might be presented and who contributed material for exercises.

Reviewers   Gerald Ball, Gerald Bradley, James Diederich, Albert Froderberg, William Fuller, D. W. Hall, Carol Holben, Joseph Krebs, David Ledbetter, Patrick Ledden, Burton Rodin, Lawrence Runyon, Dan Saltz, Alan Schoenfeld, C. H. Scott, Richard Thompson, Dale Walston, David Whitman, Jack Williamson.

Other contributors   Allen Altman, Diana Axelsen, George Backus, Dwight Bean, Ed Bender, Reginald Bickford, Rick Bradley, Jim Brom, Rod Canfield, Tom Carew, Leo Cashman, Tom Chino, Kathryn Christensen, Barsha Coleman, Chuck Cox, Guy Dimonte, Michael Dixon, Ara Djamboulian, Doug Draper, John Evans, Romeo Favreau, Jay Fillmore, Jerry Fitzsimmons, Ted Frankel, Michael Gilpin, Bill Goff, Janet Goff, David Golber, Joe Glover, Bill Gragg, Bill Helton, David Horowitz, Hugh Hudson, David Lesley, Jayne Lesley, Jon Luke, Alfred Manaster, Yue Pok Mack, Roger Marchand, Jerry Matthews, Joanna Mitro, Diego Muñoz, Rick Oberndorf, Roy H. Ogawa, Frank Oldham, Dan Olfe, Chris Parrish, Charles Perrin, Elizabeth Ralston, Jim Ralston, Richard Resco, John Rice, Paul Roudebush, Lea Rudee, Paul Savage, Alan Shapiro, Michael Sharpe, Lance Small, Dennis Smallwood, Don Smith, Michael Soule, George Stephenson, Ted Sweetser, Audrey Terras Melissa Tyson, Chris Veendorp, Peter Wagner, Dan Warner, John Wavrik, Gil Williamson, Christopher Wills, Clint Winant, Paul Zuzelo.

TO THE STUDENT

Calculus is a tool for solving problems that cannot be solved by other means. Arithmetic and algebra are all you need to determine the motion of a car down a straight road if the forces acting on it are constant, but if the road is curved and the forces vary, you need calculus. Geometry will suffice to compute the area of a circle or the volume of a cone, but to find the area of a region bounded by a parabola or the volume of a nonspherical lens, you need calculus.

Calculus is the foundation for much of higher mathematics, and is fundamental in all branches of the physical sciences. It has many applications, including the study of planetary motion, aerodynamics, fluid flow, heat transfer, light and sound, electricity and magnetism, biological growth and decay, chemical reactions, and numerous topics in medicine, economics, and the social sciences.

The basic operations of calculus, *differentiation* and *integration*, are not difficult to understand because they are based on the familiar concepts of velocity and area. With this textbook you will learn how these familiar concepts lead to the rules of calculus; you will develop your skill at working with these rules; and you will learn how to apply calculus to practical problems.

How to use this book

Study (or review) all of the precalculus material in Chapter 1, including Section 1.7 (A Brief Catalogue of Curves) and the appendix (The Algebra of Real Numbers), even if it is not covered in your course. Use the exercises in the appendix to test your ability to work with fractions and exponents.

The most frequently employed techniques from algebra for handling narrative problems are illustrated in the exercises in Section 2.1 and in the Miscellaneous Exercise section at the end of Chapter 2. These techniques should be studied before you begin work on the calculus narrative problems in Sections 4.7, 4.8, and 4.9 of Chapter 4.

You will probably find that your greatest resource for learning and using calculus is your geometric intuition. Learn the geometric meanings of results and procedures and develop the ability to use the underlying geometry to guide your reasoning when you work exercises. Draw lots of pictures.

Although you can learn to work routine problems mechanically, you need to understand the theory to use calculus in new types of problems. Study the definitions, the theorems, and their proofs. They may seem difficult to understand at first, but their meaning and purpose will become clear as you apply them and as your understanding of the subject increases. The small solid triangles (◀) in the margin indicate formulas that you will have to use and should learn.

You will master calculus by doing the exercises. Learn to write out solutions in an orderly, step-by-step fashion, so that you can concentrate on one detail at a time. To allow you to check your work, the answers to exercises marked with single daggers (†) are provided in the back of the book. Answers and outlines of solutions, which give the results of intermediate steps in working the problems, are given for exercises that are marked with double daggers (‡). In many cases the outlines of solutions also indicate the techniques required to carry out the various steps. Sets of "Additional Drill Exercises" are included in certain sections to provide more practice with algebraic techniques.

A *Solutions Manual* containing detailed solutions of selected exercises is also available.

The more challenging (and in many cases more interesting) Miscellaneous Exercises at the ends of the chapters are designed to consolidate your understanding of the material in the chapter and to show more ways in which it can be applied.

If you have access to a computer or electronic calculator, you will find that the exercise sets that call for their use will help you to understand the various types of limit processes that are fundamental to calculus.

The theory of calculus as presented in modern courses is so concise and formal that it may lead you to ask "Where did this come from and when did it all begin?" Calculus was not invented all at once but has evolved over the past 300 years. The optional Historical Notes are included to give you a better understanding of its origins and development.

CHAPTER 1

COORDINATE PLANES AND GRAPHS OF EQUATIONS

SUMMARY: This chapter covers the portion of analytic geometry that we will use to show the geometric meaning of calculus. In Section 1.1 we discuss the representation of numbers on coordinate lines. Section 1.2 deals with inequalities and Section 1.3 with coordinate planes. The concept of the graph of an equation is introduced in Section 1.4. Lines and their equations are studied in Sections 1.5 and 1.6. In Section 1.7 we present routines for sketching, without the use of calculus, four additional types of curves that will be used in examples and exercises of Chapters 2 and 3. Section 1.8 contains optional historical notes on algebra and analytic geometry. The laws of the algebra of real numbers are reviewed in an appendix to the chapter.

1-1 COORDINATE LINES AND INEQUALITIES

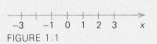

FIGURE 1.1

The REAL NUMBERS are the positive numbers, the negative numbers, and zero. In analytic geometry the real numbers are associated with the points on a COORDINATE LINE, which is a line with a scale marked off on it, such as the line sketched in Figure 1.1. The number 0 is associated with the point labeled "0" on the coordinate line. The arrow indicates that the positive numbers are associated with the points to the right of the point "0." The distance between the points labeled "0" and "1" is the UNIT DISTANCE. The letter "x" indicates that x is being used to refer to points on the coordinate line and to the real numbers that they represent.

Each positive number x is associated with the point x units (i.e., x times the unit distance) to the right of the point "0," and its negative $-x$ is associated with the point x units to the left of "0."

Conceptually, the coordinate line is infinite in both directions. In each instance we sketch whatever portion of it suits our purposes, and we use whatever scale (choice of unit distance) we find convenient in our sketch. The coordinate line in Figure 1.1 is called an x-COORDINATE LINE or an x-AXIS. We will use a variety of other letters, as well as "x," to label coordinate lines.

Because we deal almost exclusively with real numbers in introductory calculus, when we say "number" we mean "real number." Because of the association of each real number with a point on a coordinate line, we use the terms "the number x" and "the point x" interchangeably.

Inequalities

If one number x_1 on a coordinate line, as in Figure 1.1, is to the left of another number x_2, we say that x_1 is *less* than x_2 and that x_2 is *greater* than x_1. We write $x_1 < x_2$ or $x_2 > x_1$. The expression $x_1 \leq x_2$ means that x_1 is *less than or equal to* x_2. The expression $x_2 \geq x_1$ means that x_2 is *greater than or equal to* x_1.

Sketching sets of numbers on coordinate lines

The next two examples illustrate the conventions we will use in sketching sets (collections) of numbers that can be described by one or more inequalities.

EXAMPLE 1   Sketch on a coordinate line the set of numbers x that satisfy the condition $-1 < x \leq 3$.

FIGURE 1.2

Solution.   The sketch is shown in Figure 1.2. The dot at the point 3 indicates that the number 3 is in the set. This is the case because the number 3 is greater than -1 and is equal to 3. The circle at the point -1 indicates that the number -1 is not in the set; it is less than 3 but is not greater than -1. The heavy line between the points -1 and 3 indicates that all numbers greater than -1 and less than 3 are in the set.   □

EXAMPLE 2   Sketch on a coordinate line the set of numbers t that satisfy the condition, $t > -1$ or $t < -4$.

FIGURE 1.3

Solution.   The set is sketched in Figure 1.3. The circles at the points -1 and -4 indicate that the numbers -1 and -4 are not in the set. The heavy line to the left of the point -4 indicates that all numbers t less than -4 satisfy the condition and are in the set. The heavy line to the right of the point -1 indicates that all numbers t greater than -1 are in the set. The set extends infinitely far to the left and to the right.   □

Intervals

The sets of numbers most frequently encountered in calculus are INTERVALS, and there are several types. A FINITE CLOSED INTERVAL is the set of numbers satisfying a condition of the form $a \leq x \leq b$ for constants a and b with $a < b$ (Figure 1.4). A FINITE OPEN INTERVAL is the set of numbers x satisfying a condition of the form $a < x < b$

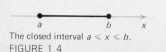

The closed interval $a \leq x \leq b$.
FIGURE 1.4

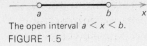

The open interval $a < x < b$.

FIGURE 1.5

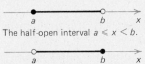

The half-open interval $a \leq x < b$.

The half-open interval $a < x \leq b$.

FIGURE 1.6

(Figure 1.5). The numbers a and b are called the END POINTS of these intervals. The interval $a \leq x \leq b$ is called closed because it includes its end points. The interval $a < x < b$ is called open because it does not include its end points. Sets described by conditions of the form $a \leq x < b$ or of the form $a < x \leq b$ are called HALF-OPEN INTERVALS (Figure 1.6).

These are several types of infinite intervals. The intervals $x \geq a$ and $x \leq b$ are called CLOSED HALF LINES (Figure 1.7). The intervals, $x > a$ and $x < b$, are called OPEN HALF LINES (Figure 1.8). The set of all real numbers is also called an interval. It is considered to be both open and closed. (It has no end points to include or exclude.)

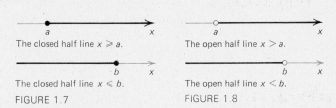

The closed half line $x \geq a$.

The open half line $x > a$.

The closed half line $x \leq b$.

The open half line $x < b$.

FIGURE 1.7

FIGURE 1.8

The INTERIOR of a nonopen interval is the open interval obtained by removing its end points; the interior of an open interval is the interval itself. Thus, the interiors of the intervals $1 \leq x < 5, x \leq -1$, and $0 < x < 1$ are the intervals $1 < x < 5, x < -1$, and $0 < x < 1$, respectively.

Absolute values

The ABSOLUTE VALUE of a real number x is denoted $|x|$ and is defined by the rule

$$|x| = \begin{cases} x & \text{if } x \text{ is positive or zero} \\ -x & \text{if } x \text{ is negative.} \end{cases}$$

Thus, $|3| = 3$ and $|-3| = -(-3) = 3$. This definition may also be expressed in the form $|x| = \sqrt{x^2}$.

The operation of taking absolute values obeys the laws

$$|ab| = |a|\,|b| \qquad \left|\frac{a}{b}\right| = \frac{|a|}{|b|} \quad \text{and} \quad |a^r| = |a|^r.$$

EXAMPLE 3   Solve the equation $|2x - 3| = 4$ for x.

Solution.   The equation $|2x - 3| = 4$ is equivalent to the statement

$$2x - 3 = 4 \quad \text{or} \quad 2x - 3 = -4.$$

Adding 3 to both sides of the two equations gives the statement

$$2x = 7 \quad \text{or} \quad 2x = -1.$$

Then we divide by 2 to obtain the solutions $x = \frac{7}{2}$ and $x = -\frac{1}{2}$.   □

Distance between two points on a coordinate line

If the number a is greater than the number b, then the distance between the points a and b on a coordinate line is the positive number $a - b$ (Figure 1.9). If a is less than b, the distance between the points a and b is the positive number $b - a$ (Figure 1.10). In any case, the distance between the two points is $|a - b|$:

$$\begin{bmatrix} \text{The distance between} \\ \text{the points } a \text{ and } b \end{bmatrix} = |a - b| = \begin{cases} a - b & \text{if } a \geq b \\ b - a & \text{if } b > a. \end{cases} \qquad \blacktriangleleft$$

FIGURE 1.9   FIGURE 1.10

Exercises

†*Answer provided.*
‡*Outline of solution provided.*

In Exercises 1 through 12 sketch the sets of numbers described by the given conditions.

1.† $x < -2$ and $x \geq -\frac{5}{2}$ **2.**† $x > -2$ or $x \leq -\frac{5}{2}$

3.† $-3 < t \leq 1$ and $-2 \leq t < 3$

4.† $-3 < t \leq 1$ or $-2 \leq t < 3$

(The "or" here is the nonexclusive "or" always used in mathematics. The statement, "A or B," means "either A or B or both.")

5.† $-1 < w < 3$ or $w \leq 0$ **6.**† $-1 < w < 3$ and $w \leq 0$

7. $p \leq -3$ or $p > 0$ **8.** $p \leq -3$ and $p > 0$

9. $-5 < u \leq 1$ and $-1 < u \leq 5$

10. $-5 < u \leq -1$ or $-1 < u \leq 5$

11.† $|-5 - t| > 3$ **12.** $|t - 3| \leq 2$

13.† Give an expression that does not involve absolute value signs for the distance on a coordinate line between the points -4 and x, where x is a number greater than -4.

14. Give an expression that does not involve absolute value signs for the distance on a coordinate line between the points 3 and y, where y is a number less than 3.

In each of Exercises 15 through 17 use a single inequality involving absolute value signs to describe the set of numbers that satisfy the given condition.

15.‡ $4 < x < 10$ **16.**‡ $t \leq -1$ or $t \geq 7$

17. $0 \leq y \leq 5$

18. Describe the interiors of the following intervals.
  a.† $-10 \leq x \leq 5$ **b.** $x \leq 0$ **c.**† $x > -50$
  d. $3 < x \leq 3.976$ **e.**† $-2 < x \leq 2$ **f.** $|x| \leq 2$

In Exercises 19 through 24 sketch on coordinate lines the sets of numbers described by the given inequalities.

19.[†] $0 < x \le 1$ or $x > 2$ **20.**[†] $-2 \le x \le 3$ and $x < 1$

21.[†] $|y - 2| < 1$ or $|y + 2| \le 2$ **22.**[†] $0 < t \le 3$ and $t^2 + 1 > 0$

23.[†] $z^2 \ge 4$ **24.**[†] $-1 < v < 2$ or $1 < v \le 3$

25.[†] Compute   **a.** $|1 + (-2)^3|$ and **b.** $|\frac{12}{5} - \frac{11}{3}|$

In Exercises 26 through 31 solve the equations for x.

26.[†] $|2x - 4| = 3$ **27.**[†] $x^2 = 4|x|$

28.[†] $|x^2 - 6| = |x|$ **29.**[†] $|x + 2| = |3x - 4|$

30.[†] $x^3 = |x|^3$ **31.**[†] $x^2 + 1 = 2|x|$

1-2 SOLVING INEQUALITIES

Inequalities are almost as important in calculus as equations, and to perform algebraic operations on them, we will frequently have to apply one or more of the following three rules.

Rule 1.1   Adding a number to or subtracting a number from both sides of an inequality gives an equivalent inequality.

Rule 1.2   Multiplying or dividing both sides of an inequality by a positive number gives an equivalent inequality.

Rule 1.3   Multiplying or dividing both sides of an inequality by a negative number and reversing the direction of the inequality gives an equivalent inequality.

The examples and exercises of this section call for "solving" inequalities for an "unknown" variable. To do this, the given inequalities in which the variable appears in an algebraic expression are replaced by equivalent inequalities in which the variable stands alone.

EXAMPLE 1   Solve the inequalities $1 \le -2x + 1 < 3$ for x and sketch the solution set.

Solution.   The given condition may be written

$$1 \le -2x + 1 \quad \text{and} \quad -2x + 1 < 3.$$

We solve these inequalities by performing the same operations on both of them.

First, we subtract 1 from each side of each inequality to obtain the equivalent condition

$$0 \le -2x \quad \text{and} \quad -2x < 2.$$

Then we divide each side of each inequality by the negative number -2 and reverse the inequalities to obtain

$$0 \ge x \quad \text{and} \quad x > -1$$

FIGURE 1.11

where the inequalities have been solved for x. We can also write the last condition in the form $-1 < x \le 0$. The solution set is the half-open interval sketched in Figure 1.11. □

EXAMPLE 2   Sketch the set of numbers t that satisfy the condition

$$2t - 6 > 4 \quad \text{or} \quad 2t - 6 \le -2$$

after solving the two inequalities for t.

Solution.   We first add 6 to each side of each inequality to obtain the equivalent condition

$$2t > 10 \quad \text{or} \quad 2t \le 4.$$

Then we divide each side of each equation by the positive number 2 to obtain

$$t > 5 \quad \text{or} \quad t \le 2$$

FIGURE 1.12

where the inequalities have been solved for t. The set is sketched in Figure 1.12. □

EXAMPLE 3   Solve the inequality $\left| -\frac{1}{2}u + 3 \right| \le 5$ for u and sketch the solution set.

Solution.   The given inequality is equivalent to the simultaneous inequalities $-5 \le -\frac{1}{2}u + 3 \le 5$. We subtract 3 from each of the three expressions to obtain $-8 \le -\frac{1}{2}u \le 2$. Then we multiply the resulting three expressions by the negative number -2 and reverse the inequalities to obtain the condition $16 \ge u \ge -4$, in which the inequalities are solved for u. The solution set is sketched in Figure 1.13. □

FIGURE 1.13

Exercises

†*Answer provided.*
‡*Outline of solution provided.*

In Exercises 1 through 21 solve the given inequalities and sketch the solution sets.

1.† $-5 \le 3x + 4 < 7$

2. $|2x + 7| < 15$

3.† $|w - 4| \le 16$

4.‡ $|5 - 2z| \ge 7$

5.† $-3y + 14 < 10$ or
      $-3y + 14 \ge 20$

6. $7t - 1 < -1$ or $7t - 1 \ge 5$

7.† $x + 3 < 6x + 10$

8. $-6 \le w + 14 < 13$

9.† $|6x - 7| > 10$

10. $|3 - t| \ge 15$

11.† $v - 7 \ge -5$ or
        $v - 7 \le -6$

12. $t - 3 \ge 7 - 6t$

13.‡ $0 < 3x + 1 \le 4x - 6$

14.† $t < 6t - 10$ or $t \ge 2t + 5$

15. $x + 12 < 3x < 4x - 1$

16.† $1 \le 2 - x < 3$

17.† $|2y - 3| > 4$

18.† $-6 < 3y + 3 \le 3$

19.† $2t - 1 > 1$ or $t + 3 < 4$

20.† $t + 4 \ge 2t - 5$

21.† $1 < x - 2 < 6 - x$

1-3   COORDINATE PLANES

Analytic geometry involves algebraic descriptions of geometric objects and concepts. It serves two main roles in this textbook: it allows us to apply algebra and calculus to geometric problems, and it enables us to study the geometric meaning of topics in algebra and calculus. In this section we illustrate how geometric problems can be solved with analytic geometry and algebra.

Coordinate planes

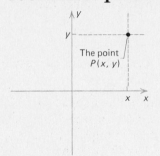

FIGURE 1.14

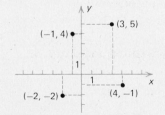

FIGURE 1.15

Plane analytic geometry is based on the use of RECTANGULAR COORDINATES to represent the points in a plane. The coordinates are shown on a horizontal and a vertical coordinate axis, perpendicular to each other and intersecting at their zero points. The axes are chosen so that the positive numbers are on the right of the horizontal axis and on the top of the vertical axis. A plane with such coordinate axes is called a COORDINATE PLANE. The coordinate plane shown in Figure 1.14, which has a horizontal x-axis and a vertical y-axis, is called an xy-COORDINATE PLANE or simply an xy-PLANE.

To find the coordinates of a point P in the xy-plane of Figure 1.14, we draw horizontal and vertical lines through the point. The value of x at the intersection of the vertical line with the x-axis is the x-COORDINATE (or ABSCISSA) of the point, and the value of y at the intersection of the horizontal line with the y-axis is the y-COORDINATE (or ORDINATE) of the point. The point P is represented by the symbol (x, y) or $P(x, y)$ with the x-coordinate to the left of the comma and the y-coordinate to the right. Figure 1.15 shows four points in the xy-plane labeled by their coordinates.

The expressions (x, y), which give the coordinates of the points in an xy-plane, are called ORDERED PAIRS of numbers. The term "ordered" refers to the fact that the order in which the numbers appear is taken into account. The expressions $(-1, 4)$ and $(4, -1)$, for example,

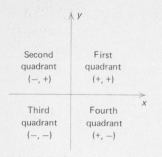

Second quadrant $(-, +)$ First quadrant $(+, +)$

Third quadrant $(-, -)$ Fourth quadrant $(+, -)$

FIGURE 1.16

represent different points, as is shown in Figure 1.15. *The choice of coordinate axes associates each point in the plane with an ordered pair of numbers and associates each ordered pair of numbers with a point in the plane.* As the variables x and y range over all real numbers, the corresponding point (x, y) ranges over the entire coordinate plane.

The points on the x-axis have y-coordinates equal to zero, and the points on the y-axis have x-coordinates equal to zero. The point where the axes cross has coordinates $(0, 0)$ and is called the ORIGIN of the coordinate plane. The coordinate axes divide the rest of the xy-plane into four QUADRANTS. These are distinguished by the signs of the coordinates in them (Figure 1.16).

The distance formula   We use the symbol $\overline{PQ}$ for the (nonnegative) distance between points $P(x_0, y_0)$ and $Q(x_1, y_1)$ in a coordinate plane. The Pythagorean theorem (Figure 1.17) enables us to express this distance in terms of the coordinates of the points:

(1) $$\overline{PQ} = \sqrt{(x_1 - x_0)^2 + (y_1 - y_0)^2}.$$ ◄

We assume here that the scales on the coordinate axes are equal.

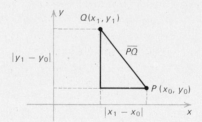

FIGURE 1.17

EXAMPLE 1   Plot the points $P(-3, 1)$ and $Q(3, 4)$ in an xy-plane. Compute the distance between them.

Solution.   The points are plotted in Figure 1.18. According to the distance formula (1) with (x_0, y_0) equal to $(-3, 1)$ and (x_1, y_1) equal to $(3, 4)$, the distance between the points is

$$\overline{PQ} = \sqrt{(3 - (-3))^2 + (4 - 1)^2} = \sqrt{6^2 + 3^2} = \sqrt{45}. \quad \square$$

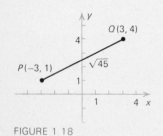

FIGURE 1.18

In the next example we use the following converse of the Pythagorean theorem.

Rule 1.4   If the square of the length of one side of a triangle is equal to the sum of the squares of the lengths of the other sides, then the triangle is a right triangle.

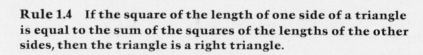

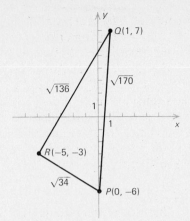

$\sqrt{136}$

$\sqrt{170}$

$R(-5, -3)$

$\sqrt{34}$

$P(0, -6)$

FIGURE 1.19

EXAMPLE 2   Sketch the triangle with vertices $P(0, -6)$, $Q(1, 7)$, and $R(-5, -3)$ in an xy-plane. Is it a right triangle? If so, at which vertex is the right angle?

Solution.   The triangle is sketched in Figure 1.19. The length of the side between the vertices $P(0, -6)$ and $Q(1, 7)$ is

(2) $\overline{PQ} = \sqrt{(1 - 0)^2 + (7 - (-6))^2} = \sqrt{1^2 + 13^2} = \sqrt{170}.$

The length of the side between $P(0, -6)$ and $R(-5, -3)$ is

(3) $\overline{PR} = \sqrt{(-5 - 0)^2 + (-3 - (-6))^2} = \sqrt{(-5)^2 + 3^2} = \sqrt{34}.$

The length of the side between $Q(1, 7)$ and $R(-5, -3)$ is

(4) $\overline{QR} = \sqrt{(-5 - 1)^2 + (-3 - 7)^2} = \sqrt{(-6)^2 + (-10)^2}$
      $= \sqrt{136}.$

The square of the length in equation (2) is the sum of the squares of the lengths in equations (3) and (4). Therefore, the triangle is a right triangle. The right angle is at the point R opposite the longest side.   □

The midpoint formula

The average $\frac{1}{2}(a + b)$ of two numbers a and b is the point midway between the points a and b on a coordinate line (Figure 1.20). Consequently, if we average the x-coordinates of two points (x_0, y_0) and (x_1, y_1), we obtain the x-coordinate of the midpoint of the line segment between them (Figure 1.21). If we average the y-coordinates of the two points, we obtain the y-coordinate of the midpoint. This yields the MIDPOINT FORMULA:

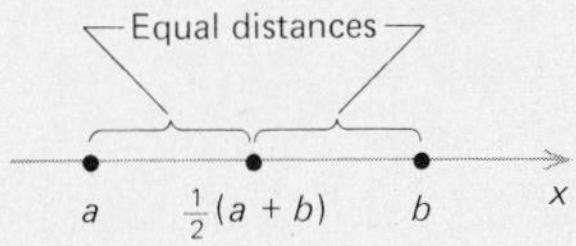

Equal distances

a $\frac{1}{2}(a + b)$ b x

FIGURE 1.20

(5) $\begin{bmatrix} \text{The midpoint of the} \\ \text{line segment between} \\ (x_0, y_0) \text{ and } (x_1, y_1) \end{bmatrix} = \left(\frac{1}{2}(x_0 + x_1), \frac{1}{2}(y_0 + y_1)\right).$ ◀

EXAMPLE 3   Find the midpoint of the line segment between $(4, 5)$ and $(-6, 2)$.

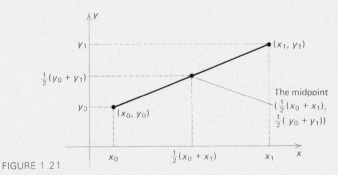

y_1 (x_1, y_1)

$\frac{1}{2}(y_0 + y_1)$

The midpoint
$(\frac{1}{2}(x_0 + x_1),$
$\frac{1}{2}(y_0 + y_1))$

y_0 (x_0, y_0)

x_0 $\frac{1}{2}(x_0 + x_1)$ x_1 x

FIGURE 1.21

Solution.   We apply formula (5) with (x_0, y_0) equal to $(4, 5)$ and (x_1, y_1) equal to $(-6, 2)$. The midpoint is

$$\left(\tfrac{1}{2}(4 + (-6)), \tfrac{1}{2}(5 + 2)\right) = \left(\tfrac{1}{2}(-2), \tfrac{1}{2}(7)\right) = \left(-1, \tfrac{7}{2}\right). \quad □$$

Labeling coordinate axes

We frequently use letters other than x and y to label coordinate axes. Figures 1.22 and 1.23 show a uv-plane and a tR-plane. In each instance the first letter in the name of the coordinate plane is the label on the horizontal axis and the second letter is the label on the vertical axis.

FIGURE 1.22   A uv-plane

FIGURE 1.23   A tR-plane

Exercises

†*Answer provided.*
‡*Outline of solution provided.*

In Exercises 1 through 4 plot the points and compute the distances between them.

1.† $(5, 7)$ and $(-4, 3)$ in an xy-plane

2. $(0, 6)$ and $(-3, -6)$ in an xy-plane

3.† $(4, 5)$ and $(-4, -5)$ in a uv-plane

4. $(-7, 14)$ and $(-5, 10)$ in an rs-plane

In Exercises 5 and 6 plot the points and compute the coordinates of the midpoints of the line segments between them.

5.† $(-3, 4)$ and $(6, -7)$ in a pq-plane

6. $(0, 8)$ and $(8, 0)$ in an xy-plane

7.† The point P has coordinates $(3, 4)$ in an xy-plane. Find the coordinates of the point Q such that the x-axis is the perpendicular bisector of the line segment PQ.

8. The point R has coordinates $(4, -5)$ in a uv-plane. Find the coordinates of the point S such that the v-axis is the perpendicular bisector of the line segment RS.

In the remaining exercises use xy-planes.

9.† Give the coordinates of the points that are a distance of 5 units from the origin and have x-coordinate equal to 4.

10. Give the coordinates of the points that are a distance of 13 units from the point $(0, 2)$ and have y-coordinate equal to 5.

11.† Sketch the triangle with vertices $P(-5, 6)$, $Q(4, -3)$, and $R(6, 8)$. Compute the lengths of its sides. Is it an isosceles triangle (does it have two equal sides)? Is it an equilateral triangle (are all of its sides equal)?

12. Sketch the quadrilateral $ABCD$ with vertices $A(0, 0)$, $B(4, 1)$, $C(2, 3)$, and $D(-2, 2)$. Compute the lengths of its sides. Is it a parallelogram (are its opposite sides equal)? Is it a rhombus (are all of its sides equal)?

13.† Sketch the triangle with vertices $A(2, -3)$, $B(-3, -3)$, and $C(3, 2)$. Compute the lengths of its sides. Is it a right triangle?

14.‡ Determine a number k such that the triangle with vertices $A(0, 0)$, $B(3, 1)$, and $C(2, k)$ is a right triangle with its right angle at B.

15.† Find a number k such that the quadrilateral $ABCD$ with vertices $A(-3, 0)$, $B(1, -2)$, $C(5, 0)$, and $D(1, k)$ is a parallelogram. Is it a rhombus? Is it a rectangle? Is it a square?

16. Show that for any nonzero number k the triangle with vertices $A(0, 0)$, $B(0, 4)$, and $C(k, 2)$ is isosceles. Find two values of k for which the triangle is an isosceles right triangle. Find two values of k for which it is equilateral.

17.† Use the distance formula to show that the points $A(1, 1)$, $B(2, 3)$, and $C(4, 7)$ are on a line.

18.† Use similar triangles to show that the points $A(1, 1)$, $B(2, 3)$, and $C(4, 7)$ are on a line.

19.† Prove that the midpoint of the hypotenuse of any right triangle is equidistant to the three vertices. (Introduce coordinate axes so that the triangle has vertices $A(0, 0)$, $B(a, 0)$, and $C(0, b)$, where a and b are arbitrary, nonzero numbers.)

20. The *medians* of a triangle are the line segments from the vertices to the midpoints of the opposite sides. Prove that the medians from the equal angles of an isosceles triangle are of equal length. (Introduce coordinate axes so that the triangle has vertices $A(-a, 0)$, $B(a, 0)$ and $C(0, b)$, where a and b are arbitrary positive numbers.)

21. Sketch the parallelogram $ABCD$ with vertices $A(1, 4)$, $B(7, 5)$, $C(10, 2)$, and $D(4, 1)$. Demonstrate that the diagonals of the parallelogram intersect at their midpoints by showing that the midpoints of the two diagonals coincide.

22. **a.** Show that the quadrilateral $ABCD$ with vertices $A(0, 0)$, $B(a, 0)$, $C(a + c, d)$, and $D(c, d)$ is a parallelogram by computing the lengths of its sides. **b.** Show that its diagonals bisect each other. **c.** Show that the sum of the squares of the lengths of its diagonals is equal to the sum of the squares of the lengths of its sides.

In Exercises 23 through 25 show that the points P, Q, and R form a right triangle PQR and give the coordinates of the midpoint of its hypotenuse.

23.† $P(-1, 4)$, $Q(4, 3)$, $R(1, 1)$ **24.**† $P(3, 12)$, $Q(5, 0)$, $R(-1, -1)$

25.† $P(0, 0)$, $Q(9, 3)$, $R(-1, 3)$

1-4   GRAPHS OF EQUATIONS

We use analytic geometry as a tool in calculus by studying GRAPHS of equations.

Definition 1.1   If an equation involves the variables x and y, then the graph of the equation in an xy-plane is the set of points whose coordinates (x, y) satisfy the equation.*

EXAMPLE 1   Sketch the graph of the equation

(1) $x^2 + y^2 = 4$

in an xy-plane.

Solution.   The left side of equation (1) is the square of the distance $\sqrt{x^2 + y^2}$ from the point (x, y) to the origin. The coordinates of the point (x, y) satisfy equation (1) if and only if the distance from the point to the origin is the square root of 4. The graph of equation (1) is the circle of radius 2 with its center at the origin (Figure 1.24).   □

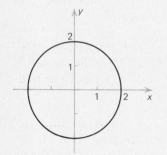

FIGURE 1.24   The graph of the equation $x^2 + y^2 = 4$

Equation (1) is called an EQUATION OF THE CIRCLE in Figure 1.24. We will frequently use an equation of a curve as a *name* for the curve. Thus, we will speak of "the circle" $x^2 + y^2 = 4$.

The equation of Example 1 is exceptionally easy to analyze. Sketching the graphs of most equations requires more involved procedures which will be discussed in Chapter 4. Meanwhile, we can sketch the graph of an equation without calculus procedures by *plotting a number of points on the graph*. We pick convenient values of one of the variables x or y and use the equation to compute the values of the other coordinate. This procedure is illustrated in the next examples.

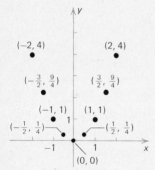

FIGURE 1.25

EXAMPLE 2   Sketch the graph of the equation $y = x^2$ in an xy-plane.

Solution.   Since y is given in terms of x in the equation, we pick values of x and compute the corresponding values of y. The table below gives the values of y for the points on the graph with x equal to 0, $\pm\frac{1}{2}$, ± 1, $\pm\frac{3}{2}$, and ± 2.

x	0	$\frac{1}{2}$	$-\frac{1}{2}$	1	-1	$\frac{3}{2}$	$-\frac{3}{2}$	2	-2
$y = x^2$	0	$\frac{1}{4}$	$\frac{1}{4}$	1	1	$\frac{9}{4}$	$\frac{9}{4}$	4	4

The nine points with coordinates given in this table are plotted in Figure 1.25. The graph of the equation $y = x^2$, on which the points lie, is sketched in Figure 1.26.   □

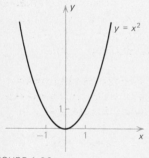

FIGURE 1.26

---

*Contrast this technical use of the term "graph," meaning a curve or other set of points, with everyday English, where a "graph" is an entire sketch or diagram.

Note.   Not all equations in two variables have graphs that we would call "curves." The graph of the equation $x^2 + y^2 = 0$, for example, consists of the single point $(0, 0)$ since the only real values of x and y that satisfy the equation are $x = 0$ and $y = 0$. The graph of the equation $x^2 + y^2 = -1$ is *empty*. There are no points in it because there are no real values of x and y that satisfy the equation. In contrast, the graph of the equation $(x + y)^2 = x^2 + 2xy + y^2$ consists of the entire xy-plane since this equation is satisfied for all real numbers x and y.   □

Symmetry of curves

Various types of SYMMETRY occur frequently in nature and in mathematics. Recognizing the symmetry of a mathematical object is important because it simplifies the study of that object. The most important types of symmetry of curves are symmetry about a point and symmetry about a line.

Symmetry about a point

Definition 1.2   Two points P and Q are symmetric about a third point R if R is the midpoint of the line segment PQ (Figure 1.27). A curve is symmetric about the point R if for any point P on the curve, the point Q which is symmetric to P about R is also on the curve (Figure 1.28).

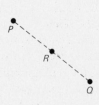

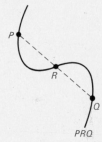

FIGURE 1.27   Points P and Q symmetric about the point R

FIGURE 1.28   A curve symmetric about the point R

A test for symmetry about the origin

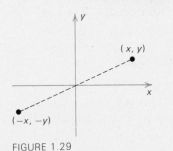

FIGURE 1.29

For any values of x and y, the points (x, y) and $(-x, -y)$ are symmetric about the origin (Figure 1.29). Therefore, if the graph of an equation is symmetric about the origin and the coordinates of a point (x, y) satisfy that equation, then the coordinates of the symmetric point $(-x, -y)$ must also satisfy the equation. This leads to the following rule.

Rule 1.5   The graph of an equation is symmetric about the origin if replacing x by −x and y by −y in the equation gives an equivalent equation (one with the same solutions).

EXAMPLE 3   Is the curve $y = x^3$ symmetric about the origin?

Solution.   Replacing x by $-x$ and y by $-y$ in the equation gives the equation $-y = (-x)^3$; multiplying out the cube gives the equation $-y = -x^3$; and then multiplying both sides by -1 gives the original equation. The new equation $-y = (-x)^3$ is equivalent to the original equation, and the curve is symmetric about the origin. The curve is sketched in Figure 1.30. □

FIGURE 1.30

Symmetry about a line

Definition 1.3   Two points P and Q are symmetric about a line if the line is the perpendicular bisector of the line segment PQ (Figure 1.31). A curve is symmetric about the line if for any point P on the curve, the point Q which is symmetric to P about the line is also on the curve (Figure 1.32).

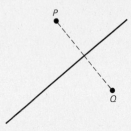

FIGURE 1.31   Points P and Q symmetric about a line

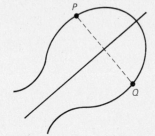

FIGURE 1.32   A curve symmetric about a line

Tests for symmetry about the x-axis and y-axis

For any values of x and y, the points (x, y) and $(x, -y)$ are symmetric about the x-axis, and the points (x, y) and $(-x, y)$ are symmetric about the y-axis (Figure 1.33). This leads to the following rules.

Rule 1.6   The graph of an equation is symmetric about the x-axis if replacing y by $-y$ and leaving x unchanged in the equation gives an equivalent equation (one with the same solutions).

Rule 1.7   The graph of an equation is symmetric about the y-axis if replacing x by $-x$ and leaving y unchanged in the equation gives an equivalent equation (one with the same solutions).

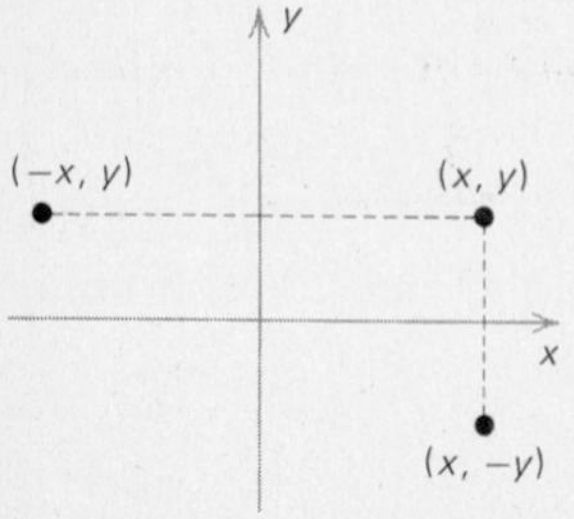

FIGURE 1.33

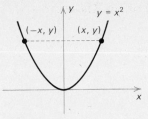

FIGURE 1.34

EXAMPLE 4   Is the graph of the equation $y = x^2$ symmetric about the y-axis? Is it symmetric about the x-axis?

Solution.   Replacing x by $-x$ and leaving y unchanged gives the equation $y = (-x)^2$, which is equivalent to the original equation $y = x^2$ because $(-x)^2 = x^2$. The graph of the equation is symmetric about the y-axis (Figure 1.34).

Replacing y by $-y$ and leaving x unchanged, on the other hand, gives the equation $-y = x^2$, which is not equivalent to the original equation, so the graph is not symmetric about the x-axis.   □

Intercepts

Definition 1.4   **The values of x where a curve intersects the x-axis are the x-intercepts of the curve. The values of y where the curve intersects the y-axis are its y-intercepts.**

The x-intercepts are found by setting y equal to zero in an equation for the curve and solving for x. The y-intercepts are found by setting x equal to zero and solving for y.

EXAMPLE 5   Find the x- and y-intercepts of the graph of the equation $x = 4 - y^2$.

Solution.   Setting y equal to zero in the equation $x = 4 - y^2$ gives the x-intercept $x = 4$. Setting x equal to zero gives the equation $0 = 4 - y^2$. Solving this equation gives the y-intercepts $y = \pm 2$. The curve is sketched in Figure 1.35.   □

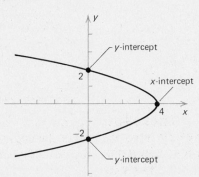

FIGURE 1.35   The graph of $x = 4 - y^2$

Exercises

In Exercises 1 through 12 sketch the graphs of the equations.

1.† $(x - 1)^2 + (y - 1)^2 = 1$ in an xy-plane

2. $(x + 2)^2 + y^2 = 9$ in an xy-plane

3.‡ $y = \sqrt{16 - x^2}$ in an xy-plane

4.† $y = -\sqrt{16 - x^2}$ in an xy-plane

5.† $x + \sqrt{16 - y^2} = 0$ in an xy-plane

6. $x - \sqrt{16 - y^2} = 4$ in an xy-plane

7.† $v = 4$ in a uv-plane   **8.** $t = -5$ in a tq-plane

9.† $u + v = 1$ in a uv-plane   **10.** $v = 3u$ in a uv-plane

11.† $x^2 = y^2$ in a xy-plane   **12.** $|r| = |s|$ in an rs-plane

In Exercises 13 through 18 find the x- and y-intercepts of the graphs of the equations and determine whether the graphs are symmetric about the origin, about the y-axis, and about the x-axis. Then sketch the graphs.

13.† $yx^2 = 4$ **14.** $xy = -3$

15.† $2x + 3y = 0$ **16.** $4x - y = 4$

17.† $4y - x^2 = 4$ **18.** $y^2 = 9 - x^2$

19.   Prove that if the graph of an equation is symmetric about the x-axis and about the y-axis, then it is symmetric about the origin.

20.†   An object is dropped from 1600 feet in the air at time $t = 0$ seconds. As we will see in Chapter 6, it follows that the object is then $s = 1600 - 16t^2$ feet above the ground at time t until $t = 10$, when it strikes the ground. Sketch the portion of the graph of the equation $s = 1600 - 16t^2$ for $0 \le t \le 10$. (Use a ts-plane with scales so that one second on the t-axis is equal to 200 feet on the s-axis.)

21.   When a man is x feet from a certain lamp post, his shadow is $y = 3x$ feet long. Sketch the portion of the graph of the equation $y = 3x$ for $x \ge 0$.

22.   The pressure p and the volume V of a certain container of gas satisfy the equation $pV = 3000$. The pressure is measured in atmospheres, and the volume is measured in cubic centimeters. Sketch the portion of the graph of the equation $pV = 3000$ for $p > 0$. (Use a pV-plane with scales so that one atmosphere on the p-axis is equal to 1000 cubic centimeters on the V-axis.)

23.   Find an equation for the locus (set) of points $P(x, y)$ such that the sum of the squares of the distances from P to $Q(a, 0)$ and $R(-a, 0)$ is $b(b \ge 2a^2)$. What is the curve?

In Exercises 24 through 28 find the x- and y-intercepts of the graphs and determine whether the graphs are symmetric about the x-axis, about the y-axis, and about the origin.

24.† $4x^2 + xy + y^2 = 4$ **25.**† $x^3y^2 + xy^4 + y^2 - x = 9$

26.† $\sqrt{1 - x^2} + \sqrt{4 - y^2} = 2$ **27.**† $(x^3 + 1)(y^2 - 1) = 7$

28.† $x^4 + x^2y^2 - y^4 = 16$

1-5   THE SLOPE OF A LINE

In this and the following section we study lines in a coordinate plane. This material will be used in Chapter 3, when we formulate the definition of a derivative, and for applications of calculus to geometry and to other fields where quantities are related by equations whose graphs are lines.

The simplest way to study lines analytically is to use the concept of the SLOPE of a line. The slope of a line is a *number* that indicates whether the line points up or down to the right and gives a measure of how steep it is.

Definition 1.5 **The slope m of a nonvertical line is the ratio $m = \dfrac{\text{rise}}{\text{run}}$ of the change in y (the rise) and the change in x (the run) as we go from one point to a second point on the line (Figure 1.36). (Vertical lines do not have slopes.)***

If the points on the line are $P(x_0, y_0)$ and $Q(x_1, y_1)$, then the rise as we go from P to Q is $y_1 - y_0$ and the run is $x_1 - x_0$. Hence,

(1)   The slope of the line = Slope $PQ = \dfrac{\text{Rise}}{\text{Run}} = \dfrac{y_1 - y_0}{x_1 - x_0}$.   ◀

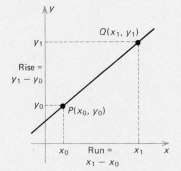

FIGURE 1.36   Line of slope
$m = \dfrac{\text{Rise}}{\text{Run}} = \dfrac{y_1 - y_0}{x_1 - x_0}$

EXAMPLE 1   Compute the slope of the line through the points $P(2, 5)$ and $Q(-6, 3)$.

Solution.   We apply formula (1) with (x_0, y_0) equal to $(2, 5)$ and (x_1, y_1) equal to $(-6, 3)$. We obtain

$$\text{Slope } PQ = \frac{3 - 5}{-6 - 2} = \frac{-2}{-8} = \frac{1}{4}. \quad \square$$

A basic property of the slope of a line is that it is the same number no matter which two points (x_0, y_0) and (x_1, y_1) on the line are used in formula (1) to compute it. Reversing the roles of the two points gives the same slope because that procedure multiplies both the numerator and denominator of (1) by -1. An argument using similar triangles (Figure 1.37) shows that we get the same slope if we use two different points (x_2, y_2) and (x_3, y_3) on the line in place of (x_0, y_0) and (x_1, y_1).

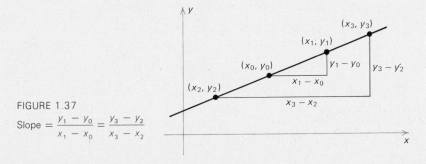

FIGURE 1.37

Slope $= \dfrac{y_1 - y_0}{x_1 - x_0} = \dfrac{y_3 - y_2}{x_3 - x_2}$

---

*Contrast this technical use of the term "slope" with everyday English, where a "slope" is a hillside or an incline rather than a number.

Lines that point *up to the right,* such as the lines of Figures 1.36 and 1.37, have *positive* slopes. If the run as we go from one point to another on such a line is positive, then the corresponding rise is also positive. Lines that point *down to the right,* such as the line of Figure 1.38, have *negative* slopes. If the run as we go from one point to another is positive, then the corresponding rise is negative.

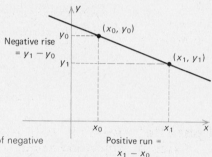

FIGURE 1.38   A line of negative slope

Horizontal lines have *zero* slope. All points on a horizontal line have the same *y*-coordinate, so if $P(x_0, y_0)$ and $Q(x_1, y_1)$ are two points on the line (Figure 1.39), then $y_0 = y_1$ and

$$\text{Slope } PQ = \frac{y_1 - y_0}{x_1 - x_0} = \frac{0}{x_1 - x_0} = 0.$$

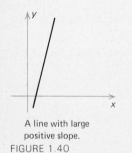

FIGURE 1.39   Line of zero slope

For pairs of points on steep lines, the rise is large relative to the run. *Steep* lines therefore have *large* (positive or negative) slopes. Lines that are *close to being horizontal* have *small* (positive or negative) slopes (Figure 1.40).

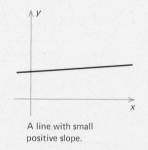 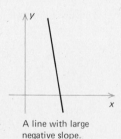 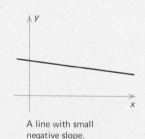

A line with large positive slope.

A line with small positive slope.

A line with large negative slope.

A line with small negative slope.

FIGURE 1.40

Angle of inclination

Definition 1.6   The angle of inclination of a nonhorizontal line is the angle ϕ (*phi*) from the positive side of the x-axis to the line (Figure 1.41). Horizontal lines have zero angle of inclination.

Note.   If the scales on the coordinate axes are equal, then the slope of a nonvertical angle is the *tangent* of its angle of inclination. To demonstrate this statement for a line that points up to the right, as in

FIGURE 1.41   Angle of inclination of a line

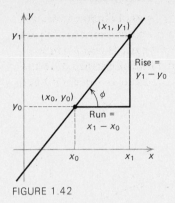

FIGURE 1.42

Figure 1.42, we pick two points (x_0, y_0) and (x_1, y_1) on the line. The rise as we go from (x_0, y_0) to (x_1, y_1) is the length of the side opposite the angle ϕ in the right triangle. The corresponding run is the length of the side adjacent to ϕ. Therefore,

$$\left[\begin{array}{c}\text{The tangent}\\ \text{of the angle } \phi\end{array}\right] = \frac{\text{Opposite}}{\text{Adjacent}} = \frac{\text{Rise}}{\text{Run}} = \left[\begin{array}{c}\text{The slope}\\ \text{of the line}\end{array}\right].$$

Analogous arguments show that the slopes of lines that point down to the right or are horizontal are also equal to the tangents of the angles of inclination. □

Slopes of parallel lines

Rule 1.8   Nonvertical lines are parallel if and only if their slopes are equal.

This rule follows from the fact that parallel lines have equal angles of inclination.

EXAMPLE 2   Use slopes to show that the quadrilateral $ABCD$ with vertices $A(1, 3)$, $B(6, 5)$, $C(10, 4)$, and $D(5, 2)$ is a parallelogram.

Solution.   The quadrilateral is shown in Figure 1.43. The slopes

$$\text{Slope } AB = \frac{5 - 3}{6 - 1} = \frac{2}{5} \quad \text{and} \quad \text{Slope } CD = \frac{2 - 4}{5 - 10} = \frac{2}{5}$$

are equal, so the sides AB and CD are parallel. The slopes

$$\text{Slope } AD = \frac{2 - 3}{5 - 1} = -\frac{1}{4} \quad \text{and} \quad \text{Slope } BC = \frac{4 - 5}{10 - 6} = -\frac{1}{4}$$

are equal, so the sides AD and BC are also parallel. □

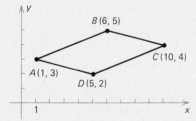

FIGURE 1.43

Slopes of perpendicular lines

Rule 1.9   Two nonvertical lines are perpendicular if and only if their slopes are negative reciprocals of each other. If m is the slope of one of the lines, then $-1/m$ is the slope of the other.

If the first line, like one of the lines in Figure 1.44, points steeply up to the right and has large positive slope m, then the line perpendicular to it points slightly down to the right and has small negative slope $-1/m$. Conversely, the slope m of the first line is the negative reciprocal of the slope $-1/m$ of the second line.

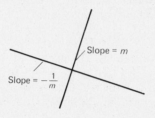

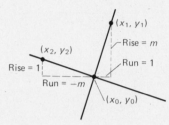

FIGURE 1.44   FIGURE 1.45

Proof.   If the lines are perpendicular, one has positive slope m. We let (x_0, y_0) be the intersection of the two lines, as in Figure 1.45. We let (x_1, y_1) denote the point one unit to the right of the intersection on the line of slope m and let (x_2, y_2) be the point one unit above the intersection on the perpendicular line. The run as we go from the point (x_0, y_0) to (x_1, y_1) is 1 and the line the points are on has slope m. Therefore, the rise as we go from (x_0, y_0) to (x_1, y_1) is m. Because the two lines are perpendicular, the two right triangles in Figure 1.45 are congruent. Therefore, since the rise as we go from (x_0, y_0) to (x_2, y_2) is 1, the run is $-m$. The slope of that line is $\dfrac{\text{rise}}{\text{run}} = \dfrac{1}{-m} = -\dfrac{1}{m}$.

Conversely, if the slopes are m and $-1/m$, the lines are perpendicular because the triangles in Figure 1.45 are congruent. This gives Rule 1.9.   Q.E.D.

In our next example we rework Example 2 of Section 1.3 by using slopes rather than by using the converse of the Pythagorean theorem.

EXAMPLE 3   By using slopes, show that the triangle with vertices $P(0, -6)$, $Q(1, 7)$, and $R(-5, -3)$ is a right triangle.

Solution.   The triangle is sketched in Figure 1.46. We have

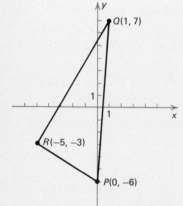

FIGURE 1.46

(2)   Slope $RQ = \dfrac{7 - (-3)}{1 - (-5)} = \dfrac{7 + 3}{1 + 5} = \dfrac{10}{6} = \dfrac{5}{3}$

(3)   Slope $RP = \dfrac{-6 - (-3)}{0 - (-5)} = \dfrac{-6 + 3}{5} = -\dfrac{3}{5}$.

The slopes (2) and (3) are the negative reciprocals of each other, so the triangle is a right triangle. $\square$

Exercises

In Exercises 1 through 4 compute the slopes of the lines through the given points.

1.† $(2, 5)$ and $(6, -7)$

2. $(3, -4)$ and $(-1, 4)$

3.† $(-6, -1)$ and $(-5, 11)$

4. $(0, 10)$ and $(10, 0)$

In each of Exercises 5 through 8 use slopes to determine if the three points lie on a line.

5.‡ $P(-3, 6), Q(3, 2), R(6, 0)$

6. $P(0, 1), Q(4, 0), R(-2, 2)$

7.† $P(11, -3), Q(25, 23), R(4, 7)$

8. $P(15, 10), Q(10, 16), R(25, -2)$

In each of Exercises 9 through 12 use slopes to determine if the three points are the vertices of a right triangle. If they are, sketch the triangle and compute its area.

9.‡ $P(-1, -1), Q(-6, 6), R(0, 5)$

10. $P(0, -1), Q(-4, -3), R(-3, 3)$

11.† $P(1, 1), Q(4, 5), R(12, -1)$

12. $P(0, 0), Q(-5, -2), R(-4, 10)$

13. Use slopes to determine a constant k such that the triangle with vertices $P(-2, -1), Q(2, 2), R(-1, k)$ is a right triangle with its right angle at the point Q.

14.† Use slopes to show that for any constant k the points $A(0, 0), B(2, 1), C(k + 2, 6)$ and $D(k, 5)$ are the vertices of a parallelogram $ABCD$. Determine a value of k for which it is a rectangle.

15. Use slopes to find a constant k such that the points $A(-3, 0), B(-1, -1), C(4, 4),$ and $D(k, 5)$ are the vertices of a parallelogram $ABCD$.

16.† The point $(-2, -4)$ lies on the circle $x^2 + y^2 = 20$. The tangent line to the circle at that point is perpendicular to the radius. Find the slope of the tangent line.

17. Show that the quadrilateral $ABCD$ with vertices $A(0, 0), B(5, 5), C(-2, 6),$ and $D(-7, 1)$ is a rhombus. Then show that its diagonals are perpendicular.

18.† The graph of the equation $3x - y = 6$ is a line. Find the coordinates of the points on that line at $x = 0$ and at $x = 1$. Use these points to compute the slope of the line.

19. The equation $4x + 6y = 9$ is an equation of a line. Find the coordinates of two points on that line and use them to compute its slope.

1-6 EQUATIONS OF LINES

In the last section we discussed the formula

(1)   Slope $= m = \dfrac{y_1 - y_0}{x_1 - x_0}$

for the slope of the nonvertical line through the points (x_0, y_0) and (x_1, y_1). In this section we use (1) to derive formulas that will enable us to obtain equations of lines from verbal descriptions of them.

The point-slope equation of a nonvertical line

Suppose that (x_0, y_0) is a point on a nonvertical line of slope m. If (x, y) is any other point on the line, then by formula (1) with (x_1, y_1) replaced by (x, y) we have

$$m = \frac{y - y_0}{x - x_0}.$$

When we multiply this equation by $x - x_0$, we obtain the equation

$$(2) \quad y - y_0 = m(x - x_0) \qquad \blacktriangleleft$$

which is also satisfied for $(x, y) = (x_0, y_0)$. Equation (2) is known as the POINT-SLOPE EQUATION of the line through (x_0, y_0) with slope m (Figure 1.47).

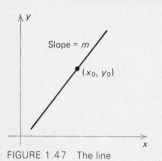

FIGURE 1.47   The line $y - y_0 = m(x - x_0)$

EXAMPLE 1   Give an equation for the line through the point $(6, -5)$ and with slope -3.

Solution.   We use formula (2) with $x_0 = 6$, $y_0 = -5$, and $m = -3$. This gives the equation $y - (-5) = -3(x - 6)$ or $y = -3x + 13$.   □

Horizontal lines

If the slope m is zero, then equation (2) reads $y - y_0 = 0$ or

$$(3) \quad y = y_0. \qquad \blacktriangleleft$$

This is an equation of the horizontal line through the point (x_0, y_0) (Figure 1.48).

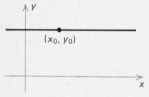

FIGURE 1.48   The line $y = y_0$

The two-point equation

If we substitute formula (1) for the slope into the point-slope equation (2), we obtain the equation

$$(4) \quad y - y_0 = \frac{y_1 - y_0}{x_1 - x_0}(x - x_0). \qquad \blacktriangleleft$$

This expression is known as the TWO-POINT EQUATION for the line through the points (x_0, y_0) and (x_1, y_1) (Figure 1.49).

Note.   The two-point equation (4) does not have to be learned as a separate formula. If the coordinates of two points on a line are given, formula (1) can be used to compute the slope of the line, and the point-slope equation (2) can be used to obtain an equation for the line.

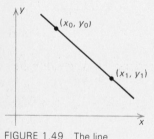

FIGURE 1.49   The line $y - y_0 = \frac{y_1 - y_0}{x_1 - x_0}(x - x_0)$

EXAMPLE 2   Give an equation for the line through the points $(-5, 0)$ and $(6, 10)$.

Solution.   The slope of the line is $\dfrac{\text{rise}}{\text{run}} = \dfrac{10 - 0}{6 - (-5)} = \dfrac{10}{11}$. Using this slope and the point $(-5, 0)$ in the point-slope equation (2) gives

$$y - 0 = \frac{10}{11}(x - (-5)) \text{ or } y = \frac{10}{11}(x + 5)$$ as an equation of the line.   □

The slope-intercept equation

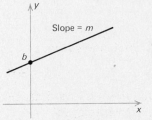

FIGURE 1.50   The line $y = mx + b$

If a nonvertical line has slope m and y-intercept b, then the point $(0, b)$ is on the line, and the point-slope formula (2) with $x_0 = 0$ and $y_0 = b$ gives the equation $y - b = m(x - 0)$ or

(5) $y = mx + b$ ◀

for the line. This is the SLOPE-INTERCEPT EQUATION (Figure 1.50).

EXAMPLE 3   Sketch the line $x + 2y = 6$ in an xy-plane.

Solution.   We first solve for y to put the equation in the form $y = -\frac{1}{2}x + 3$. With the equation in slope-intercept form, we can see that the line has slope $m = -\frac{1}{2}$ and y-intercept $b = 3$. A point on it falls one unit for every two units it moves to the right, because the slope is $-\frac{1}{2}$ (Figure 1.51).   □

Note.   We could also sketch the line of Example 3 by computing its x- and y-intercepts instead of its y-intercept and slope. Setting $x = 0$ in the equation $x + 2y = 6$ gives $y = 3$ as the y-intercept. Setting $y = 0$ gives $x = 6$ as the x-intercept (Figure 1.52).   □

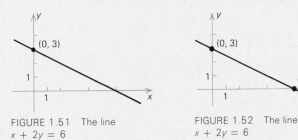

FIGURE 1.51   The line $x + 2y = 6$

FIGURE 1.52   The line $x + 2y = 6$

The two-intercept equation

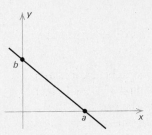

FIGURE 1.53   The line $\dfrac{x}{a} + \dfrac{y}{b} = 1$

If we know that the x-intercept of a nonvertical, nonhorizontal line is a and that its y-intercept is b, then an application of the two-point formula (4) with $(x_0, y_0) = (a, 0)$ and $(x_1, y_1) = (0, b)$ gives the equation

(6) $\dfrac{x}{a} + \dfrac{y}{b} = 1$ ◀

for the line (Figure 1.53). This is the TWO-INTERCEPT EQUATION for the line.

Vertical lines

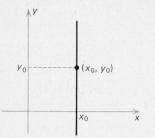

FIGURE 1.54   The line $x = x_0$

The formulas based on slope do not apply to vertical lines. The vertical line through the point (x_0, y_0) consists of all points with x-coordinate x_0 (Figure 1.54) and, consequently, has the equation

(7) $x = x_0$.   ◄

Note. Any equation of the form $Ax + By = C$ with at least one of the constants A and B not zero is the equation of a line. If the constant A is zero, the line is the horizontal line $y = C/B$. If the constant B is zero, the line is the vertical line $x = C/A$. If neither A nor B is zero, it is the oblique line $y = -(A/B)x + (C/B)$ with slope $-A/B$.

If both A and B are zero, the equation $Ax + By = C$ reads $0 = C$ and its graph is not a line. Either the constant C is zero and the graph is the entire xy-plane or C is not zero and the graph is empty.   □

Exercises

†*Answer provided.*
‡*Outline of solution provided.*

In Exercises 1 through 6 give equations for the lines and then sketch them.

1.† The line through the point $(3, 5)$ and with slope -4

2. The line through the point $(-4, 7)$ and with slope $\frac{1}{2}$

3.† The line through the points $(5, 1)$ and $(-5, 2)$

4. The line through the points $(-3, -5)$ and $(2, -2)$

5.† The line through the points $(7, 3)$ and $(-7, 3)$

6. The line through the points $(3, 7)$ and $(3, -7)$

In Exercises 7 through 10 find the slopes of the lines by putting the equations in slope-intercept form. Then sketch the lines.

7.† $5x + 7y = 35$ **8.** $x = 3y + 2$

9.† $y + 5 = 0$ **10.** $3x - 7y + 21 = 0$

11.† Give an equation for the line that is parallel to the line $6x - 4y = 7$ and passes through the point $(1, -1)$.

12. Give an equation for the line that is perpendicular to the line $3x + y = 0$ and passes through the point $(4, 7)$.

13.† Give an equation for the line with x-intercept 5 and y-intercept -6. What is its slope?

14. Give an equation for the perpendicular bisector of the line segment between the points $(4, 6)$ and $(6, 0)$.

15.† Give equations of the lines through the point $(-4, 7)$ and parallel to
a. the x-axis and   **b.** the y-axis.

16. Give an equation for the line through the origin that passes through the second and fourth quadrants and bisects the angles between the coordinate axes.

17.‡ Find the coordinates of the intersection of the lines $2x + y = 1$ and $x + y = 4$.

18.† **a.** Find the coordinates of the vertices of the triangle formed by the lines $y = \frac{1}{2}x$, $y = -2x$, and $y = -\frac{1}{3}x + 5$. **b.** Show it is a right triangle and **c.** compute its area.

19. Find the coordinates of the vertices of the triangle formed by the lines $x = 9y$, $y = x$, and $x - 3y = 6$.

20. The temperature in degrees Farenheit T_F and the temperature in degrees Centigrade T_C satisfy a linear relation, that is, they satisfy an equation whose graph is a line. Find the equation and sketch its graph in a $T_F T_C$-plane. (Water freezes at 0°C and 32°F and boils at 100°C and 212°F.)

21.† The water pressure in the ocean is proportional to the depth beneath the surface. The pressure increases 62.4 pounds per square foot with each 1 foot increase in depth. Give an equation relating the pressure P in pounds per square foot and the depth h in feet and sketch its graph in an hP-plane.

22. It costs the Barnhills fifty cents to produce each gallon of their product. They also have to pay ten dollars a week to preserve the secrecy of their operation. Express their total cost (C dollars) for a week's production in terms of the amount (M gallons) they produce. Sketch the graph of this equation in an MC-plane.

1-7   A BRIEF CATALOG OF CURVES

In Chapter 4 we will study methods, based on techniques of calculus, for making accurate sketches of a wide variety of equations. Meanwhile in Chapters 2 and 3 we will illustrate the geometric aspects of the material by using the four types of curves

$$(1) \quad y = ax^2 + b \qquad y = ax^3 + b \qquad y = \frac{a}{x} + b \qquad y = \frac{a}{x^2} + b.$$

In this section we show how to sketch these curves without using calculus and with a minimum of point plotting. We determine the general shape of the curve, and once this is done, an adequate sketch of the curve is readily obtained by plotting a few points on it.

The curves $y = ax^2 + b$

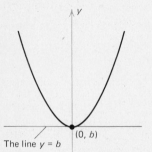

The line $y = b$

FIGURE 1.55   The parabola $y = ax^2 + b$ with $a > 0$

The quantity x^2 is zero for $x = 0$ and positive for all $x \neq 0$. It is very small when x is small (when $|x|$ is substantially less than 1), and it is very large when x is large (when $|x|$ is substantially larger than 1). The graph of $y = ax^2 + b$ has a U-like shape as in Figure 1.55 or 1.56 and is called a PARABOLA. (We will study the geometric definitions and properties of parabolas in Chapter 10.) The x-axis is not shown in the figures because its location depends on the value of the constant b. The curves $y = ax^2 + b$ are symmetric about the y-axis.

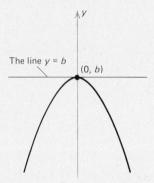

The line $y = b$

FIGURE 1.56   The parabola $y = ax^2 + b$ with $a < 0$

The curves $y = ax^3 + b$

A curve with an equation of the form $y = ax^3 + b$ is called a CUBIC, because its equation involves the cube of the variable x. Like the parabola $y = ax^2 + b$, the cubic $y = ax^3 + b$ has y-intercept b, is very close to the horizontal line $y = b$ for x close to 0, and is very far from that line for x far from 0. However, because x^3 is positive when x is positive and negative when x is negative, the cubic is on one side of the line $y = b$ for positive x and on the other side for negative x (Figures 1.57 and 1.58). The cubic $y = ax^3 + b$ is symmetric about the point $(0, b)$.

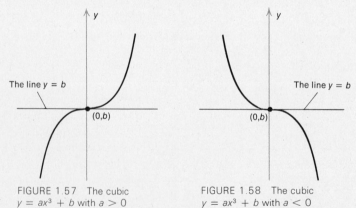

FIGURE 1.57   The cubic $y = ax^3 + b$ with $a > 0$

FIGURE 1.58   The cubic $y = ax^3 + b$ with $a < 0$

The curves $y = \dfrac{a}{x} + b$

Curves with equations of the form $y = (a/x) + b$ are a type of HYPERBOLA. To determine the shape of the curve $y = (a/x) + b$, we notice that the expression $1/x$ is small when x is large and large when x is small. Therefore, the curve is close to the line $y = b$ when x is far from zero and far from that line when x is close to zero (Figures 1.59 and 1.60). The curve is in two pieces, separated by the y-axis and by the horizontal line $y = b$. The curve is symmetric about the point $(0, b)$, where the line $y = b$ intersects the y-axis. (We will study the geometric properties of hyperbolas along with those of parabolas in Chapter 10.)

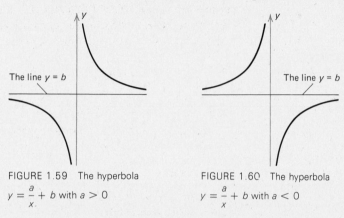

FIGURE 1.59   The hyperbola
$y = \dfrac{a}{x} + b$ with $a > 0$

FIGURE 1.60   The hyperbola
$y = \dfrac{a}{x} + b$ with $a < 0$

The curves
$$y = \frac{a}{x^2} + b$$

The curves with equations of the form $y = (a/x^2) + b$ do not have a common name. They are of special interest because of the inverse square laws of physics (see Exercises 10 and 11).

   The curve $y = (a/x^2) + b$, like the hyperbola $y = (a/x) + b$, has no y-intercept and is close to the line $y = b$ for x far from zero and far from that line for x close to zero. Because of the square of x in the equation $y = (a/x^2) + b$, however, its entire graph lies on one side or the other of the line $y = b$ (Figures 1.61 and 1.62). In either case, the curve is symmetric about the y-axis.

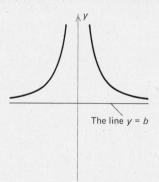

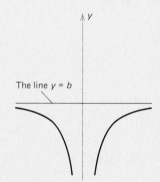

FIGURE 1.61   The curve
$y = \dfrac{a}{x^2} + b$ with $a > 0$

FIGURE 1.62   The curve
$y = \dfrac{a}{x^2} + b$ with $a < 0$

Exercises

In these exercises we use other letters in addition to x and y as variables, but the curves considered are of the types discussed in this section.
In Exercises 1 through 8 sketch the curves.

1.‡ $y = x^2 - 3$ in an xy-plane

2.† $y = 3 - \frac{1}{4}x^3$ in an xy-plane

3. $y = 3 - \frac{1}{4}x^2$ in an xy-plane

4.‡ $v = 4u^{-2}$ in a uv-plane

5.† $y = -(6/x)$ in an xy-plane

6. $q = (2/p) - 2$ in a pq-plane

7.† $R = 3 + t^{-2}$ in a tR-plane

8. $R = 6 - t^{-1}$ in a tR-plane

9.‡ A boy throws a ball straight up in the air at time $t = -1$ (seconds) and catches it at time $t = 1$. For each value of t in the interval $-1 \leq t \leq 1$, the ball is $s = 16 - 16t^2$ feet above the boy's hand. The portion of the parabola $s = 16 - 16t^2$ for $-1 \leq t \leq 1$ shows the relationship between the height of the ball above the boy's hand and the time. Sketch that portion of the parabola. Use a ts-plane and scales so that $\frac{1}{4}$ second on the t-axis equals 2 feet on the s-axis.

10.† According to Coulomb's law, the force of attraction between one particle with a positive charge of 3 statcoulombs and a second particle with a negative charge of 3 statcoulombs is $F = 9r^{-2}$ dynes when the charges are r centimeters apart. Sketch the portion of the curve $F = 9r^{-2}$ for $r > 0$ in an rF-plane.

11. According to Newton's law of gravity, an object that weighs one pound on the surface of the earth weighs $w = 16r^{-2}$ pounds when it is r thousand miles from the center of the earth. (The radius of the earth is 4 thousand miles.) Sketch the portion of the curve $w = 16r^{-2}$ for $r > 0$ in an rw-plane.

12. It costs a perfume factory \$5 to manufacture each pint of Nature-All perfume, plus an overhead of \$100 on each batch produced. The cost on a batch of p pints of the perfume is therefore $100 + 5p$ dollars, and the average cost per pint on a batch of p pints is

$$A = \frac{100 + 5p}{p} = \frac{100}{p} + 5 \text{ dollars per pint.}$$

Sketch the portion of the curve $A = \dfrac{100}{p} + 5$ for $p > 0$ in a pA-plane.

1-8   HISTORICAL NOTES   (Optional)

The development of calculus during the second half of the seventeenth century was facilitated by improvements in algebraic notation and by the development of the rudiments of analytic geometry during the first half of the century.

The concept of an unknown in an algebraic equation was employed in the sixteenth century, but the notation used was somewhat clumsy.

FIGURE 1.63   Gerolamo Cardano
(1501–1576)
The Bettmann Archive, Inc.

FIGURE 1.64   François Viète
(1540–1603)
Culver Pictures, Inc.

For example, the Italian physician Gerolamo Cardano (1501–1576), who did important work in algebra, used the Latin sentence *Cubus $\overline{p}$ 6 rebus aequalis* 20 for the equation that would read $x^3 + 6x = 20$ in modern notation. In Cardano's sentence, the word "cubus" denotes the cube of the unknown, the symbol "$\overline{p}$" stands for "plus," the word "rebus" denotes the unknown, and "aequalis" means "equals." Somewhat later, a French lawyer and mathematician François Viète (1540–1603) used the letter "C" for the cube of the unknown, the letter "Q" for the square of the unknown, and the letter "N" for the unknown itself. He wrote, for example, 1C − 8Q + 16N *aequalis* 40, which would read $x^3 - 8x^2 + 16x = 40$ in modern notation.

The use of exponents for positive integer powers and of a single letter for the unknown evolved during the end of the sixteenth century and the beginning of the seventeenth. Algebraic notation that was very similar to what we use today was then popularized by a treatise called "La Géométrie," written in 1637 by the French philosopher René Descartes (1596–1650) as an appendix to a work on the philosophy of science.

Descartes and a French lawyer Pierre Fermat (1601–1665) are considered the inventors of analytic geometry. Greek geometers, including Menaechmus (ca. 360 B.C.), Euclid (ca. 300 B.C.), Archimedes (287–212 B.C.), and Apollonius (ca. 225 B.C.), used the equivalent of coordinate systems with their theory of proportions for studying geometric figures. In the sixteenth century, Viète and others used algebra to solve geometric problems. It was Descartes and Fermat, however, who introduced the point of view that is characteristic of analytic geometry. They considered curves defined by equations as well as by geometric properties, and they made extensive use of the close association between the algebra of the equations and the geometry of the curves.

FIGURE 1.65   René Descartes
(1596–1650)
Culver Pictures, Inc.

FIGURE 1.66   Pierre Fermat
(1601–1665)
The Bettmann Archive, Inc.

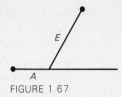

FIGURE 1.67

Descartes and Fermat used only what corresponds to the positive half of a modern horizontal coordinate axis and did not use an explicit vertical axis. To plot a point corresponding to positive solutions of an equation in two unknowns A and E, Fermat measured the distance A along a horizontal line from a fixed point. The fixed point corresponded to the origin in a modern coordinate plane. He then measured the distance E along a line at a fixed angle, but not necessarily a right angle, to the horizontal line (Figure 1.67).

Descartes' procedure for associating solutions of equations in two unknowns to points in a plane was similar to Fermat's. He also used occasionally what would correspond to the fourth quadrant of a modern coordinate plane, in that he allowed negative values of the second unknown. Negative coordinates were first used systematically by Isaac Newton (1642–1727) in his *Enumeration of Curves of Third Degree* (1676).

Descartes introduced his techniques of analytic geometry in his treatise "La Géométrie" (1637). Fermat's independent work was not published until more than forty years later, after his death, so Descartes often receives more credit for inventing what is now known as analytic or "cartesian" geometry.

FURTHER READING

D. STRUIK, [1] *A Source Book in Mathematics* (Cardano, pp. 62–69; Fermat, pp. 143–150; Descartes, p. 155–157).

C. BOYER, [2] *History of Analytic Geometry*, pp. 74–102.

G. CARDANO, [1] *The Book of My Life*.

A. WYKES, [1] *Doctor Cardano, Physician Extraordinary*.

E. BELL, [1] *Men of Mathematics* (Descartes, pp. 35–55; Fermat, pp. 56–72).

E. HALDANE, [1] *Descartes: His Life and Times*.

APPENDIX TO CHAPTER 1.   THE ALGEBRA OF REAL NUMBERS

Two real numbers can be added or multiplied together, and one real number can be divided by another nonzero real number to yield other real numbers. These operations obey the following laws of arithmetic.

Commutative laws   For any real numbers a and b, $a + b = b + a$ and $ab = ba$.

Associative laws   For any real numbers a, b, and c, $a + (b + c) = (a + b) + c$ and $a(bc) = (ab)c$.

Distributive law   For any real numbers a, b, and c, $a(b + c) = ab + ac$.

Existence of 0 For any real number a, $a + 0 = a$ and $a + (-a) = a - a = 0$.

Existence of 1 For any real number a, $a(1) = a$, and for any nonzero real number a, $a(1/a) = a/a = 1$.

Rational and irrational numbers

A real number is called RATIONAL if it can be expressed as the ratio p/q of two integers p and q, with q not zero. The integers themselves are rational numbers, because we can express any integer p in the form $p/1$. Any positive or negative fraction is also a rational number. A real number that cannot be expressed as the ratio of two integers is called IRRATIONAL. It can be shown that the numbers $\sqrt{2}$ and π, for example, are irrational numbers.

Exponentiation

The symbol a^n, where n is a positive integer and a is any real number, denotes the product of n a's.

For nonzero a and a positive integer n, the symbol a^{-n} denotes the reciprocal of a^n: $a^{-n} = 1/a^n$.

If q is an *odd* integer greater than 1 ($q = 3, 5, 7$, etc.), then every real number a has one real qth root, which is denoted $\sqrt[q]{a}$ or $a^{1/q}$. The root is positive if a is positive, zero if a is zero, and negative if a is negative. We have, for example, $\sqrt[3]{8} = 8^{1/3} = 2$ and $\sqrt[3]{-8} = (-8)^{1/3} = -2$.

If q is an *even* positive integer ($q = 2, 4, 6$, etc.), then every *positive* number a has two real qth roots, one positive and one negative. In this case the symbols $\sqrt[q]{a}$ and $a^{1/q}$ are used to denote the *positive* root. We have, for example, $\sqrt{9} = 9^{1/2} = 3$. The negative qth root of the positive number a is denoted $-\sqrt[q]{a}$ or $-a^{1/q}$; thus, $-\sqrt[4]{16} = -2$.

If q is *even* and a is *negative*, then a *does not have* a real qth root. In this case we consider the symbols $\sqrt[q]{a}$ and $a^{1/q}$ to be undefined.

The expression $a^{p/q}$ denotes the pth power of the qth root of a or the qth root of the pth power of a. The symbol a^0 denotes the number 1 for any nonzero real number a. If a is not zero and a^r is defined with r a positive rational number, then we write a^{-r} for $1/a^r$.

Negative, fractional, and zero exponents are introduced so that rules from algebra that deal with powers, reciprocals, and roots can be expressed concisely. We have

$$a^r a^s = a^{r+s} \qquad [a^r]^s = a^{rs} \qquad a^r b^r = [ab]^r$$ ◄

$$a^{-r} = \frac{1}{a^r} \quad \text{and} \quad \frac{a^r}{a^s} = a^{r-s}$$

for any real numbers a and b and any rational numbers r and s for which the expressions involved are defined.

Note.   The symbol $-a^r$ stands for $-(a^r)$ and not for $(-a)^r$. We have, for example, $-2^2 = -4$ and $-9^{1/2} = -3$.   □

The order relation $\leq$ has the following basic properties for all real numbers a, b, and c.

Reflexivity $a \leq a$.

Anti-symmetry   If $a \leq b$ and $b \leq a$, then $a = b$.

Transitivity   If $a \leq b$ and $b \leq c$, then $a \leq c$.

These statements are also valid with $\leq$ replaced by $\geq$.

Decimal representations of real numbers

We may consider real numbers to be given by their DECIMAL EXPANSIONS, which are representations of the numbers as infinite decimals, with minus signs in front of those expansions representing negative numbers.

If a nonzero number has a decimal expansion that contains an infinite final string of zeros, then that number has a second decimal expansion. This second decimal expansion is obtained from the first expansion by lowering its last nonzero digit by 1 and replacing the infinite string of zeros by an infinite string of 9's. The number $\frac{1}{4}$, for example, has the two decimal expansions 0.25000 . . . and 0.24999 . . . .

It can be shown that the decimal expansions of rational numbers are the *repeating* decimals (see Exercise 12). Each of these consists, from a certain point on, of a finite string of digits repeated an infinite number of times.

The correspondence between rational numbers and repeating decimals implies that the irrational numbers are the numbers with nonrepeating decimals. The decimal 0.101001000100001 . . . , for example, in which the 1's are separated by an increasing number of 0's, does not repeat and, therefore, is the decimal expansion of an irrational number.

Mathematical induction (optional)

Often when we want to prove that a statement or formula involving an integer n is valid for all integers $n \geq 1$, we can use the principal of MATHEMATICAL INDUCTION. We let P_n denote the statement or formula that involves the integer n and perform the following two steps.

Step 1   We show that P_1 (the statement or formula obtained by setting $n = 1$) is valid.

Step 2   We show that if P_k (the statement or formula with $n = k$) is valid for any positive integer k, then P_{k+1} (the statement or formula with $n = k + 1$) is also valid.

This shows that P_n is valid for all positive integers n.

EXAMPLE 1   When we study integrals in Chapter 5, we will use the formula

(1) $1 + 2 + 3 + \cdots + n = \frac{1}{2}n(n + 1)$

for the sum of the first n positive integers. Use mathematical induction to prove (1) for all positive integers n.

Solution.   We let P_n denote the statement (1).

Step 1   When we set $n = 1$ in (1) we obtain $1 = \frac{1}{2}(1)(1 + 1)$, which is P_1 and is correct.

Step 2   If P_k is valid, where k is an arbitrary, unspecified positive integer, then we have

(2) $1 + 2 + 3 + \cdots + k = \frac{1}{2}k(k + 1)$.

We add $k + 1$ to each side of (2) and then factor it from the right side to obtain

$$1 + 2 + 3 + \cdots + k + (k + 1) = \frac{1}{2}k(k + 1) + (k + 1)$$

$$= (k + 1)(\tfrac{1}{2}k + 1) = \tfrac{1}{2}(k + 1)(k + 2).$$

This is statement P_{k+1}, so the proof of (1) by mathematical induction is complete.   □

If we are to prove a statement or formula P_n for all integers $n \geq N$ with N an integer other than 1, then Step 1 of a proof by mathematical induction consists of verifying statement P_N and in Step 2 we consider all $k \geq N$. This is the case in the next example with $N = 2$.

EXAMPLE 2   When we study derivatives in Chapter 3, we will need the formula

$$(3) \qquad (a + b)^n = a^n + na^{n-1}b + \begin{bmatrix} \text{Terms involving } b^2 \\ \text{and higher powers of } b \end{bmatrix}$$

where a and b are arbitrary numbers and n is an arbitrary integer ≥ 2. Verify this formula by mathematical induction.

Solution.   We let P_n denote statement (3).

Step 1   We set $n = 2$ in (3) to obtain P_2 which reads

$$(4) \qquad (a + b)^2 = a^2 + 2ab + \begin{bmatrix} \text{Terms involving } b^2 \text{ and} \\ \text{higher powers of } b \end{bmatrix}.$$

This is correct because $(a + b)^2 = a^2 + 2ab + b^2$.

Step 2   If P_k is true with k an integer ≥ 2, then we have

$$(5) \qquad (a + b)^k = a^k + ka^{k-1}b + \begin{bmatrix} \text{Terms involving } b^2 \text{ and} \\ \text{higher powers of } b \end{bmatrix}.$$

Multiplying both sides of (5) by $(a + b)$ yields

$$(a + b)^{k+1} = (a + b)\left\{ a^k + ka^{k-1}b + \begin{bmatrix} \text{Terms involving } b^2 \text{ and} \\ \text{higher powers of } b \end{bmatrix} \right\}$$

$$= (a + b)a^k + (a + b)ka^{k-1}b + (a + b)\begin{bmatrix} \text{Terms involving } b^2 \\ \text{and higher powers of } b \end{bmatrix}$$

$$= (a^{k+1} + a^k b) + (ka^k b + ka^{k-1}b^2) + \begin{bmatrix} \text{Terms involving } b^2 \text{ and} \\ \text{higher powers of } b \end{bmatrix}$$

$$= a^{k+1} + (k + 1)a^k b + \begin{bmatrix} \text{Terms involving } b^2 \text{ and} \\ \text{higher powers of } b \end{bmatrix}.$$

This is P_{k+1}, so the proof is complete. In the second step of this calculation we used the fact that $(a + b)$ multiplied by terms involving b^2 and higher powers of b yields other terms involving b^2 and higher powers of b.   □

Exercises

†Answer provided.
‡Outline of solution provided.

1.† Calculate for $n = 0, -2$, and 3:

a. -3^n **b.** $(-3)^n$ **c.** $-n + 1^n$

d. $-(n + 1)^n$ **e.** $(-n + 1)^n$

2. Calculate for $n = 1, -1, 2$, and -2:

a. $-n^2$ **b.** $(-n)^2$ **c.** $-n^n$

d. $(-n)^n$ **e.** n^{1+n}

3. Put each of the following expressions in the form x^n with n a rational number. (Here x is positive.)

a.† $\left[\dfrac{x^2}{\sqrt[5]{x^3}}\right]^3$ **b.†** $\sqrt{\dfrac{x^{1/2}}{x^{2/3}}}$ **c.†** $\sqrt[3]{\dfrac{(x^{1/5})(x^{3/4})}{x^{10}}}$

d.† $\left[\dfrac{x^3}{\sqrt[3]{x^2}}\right]^{-5}$ **e.** $\dfrac{\sqrt[3]{x^4}}{\sqrt[4]{x^3}}$ **f.** $[\sqrt{x}\ \sqrt[3]{x}\ \sqrt[4]{x}]^{12}$

4.† Which of the following expressions equal $\dfrac{5x + 1}{3x + 1}$ for all values of x for which no denominator is zero?

a. $\dfrac{1 + (5/x)}{1 + (3/x)}$ **b.** $\dfrac{5 + (1/x)}{3 + (1/x)}$ **c.** $(5/3)x + 1$

d. $\dfrac{1}{1 + (2x/(3x + 1))}$ **e.** $1 - \dfrac{8x}{3x - 1}$ **f.** $\dfrac{8x}{3x + 1} - 1$

g. $5x + \dfrac{1}{3x + 1}$ **h.** $5x + \dfrac{1 - 15x^2}{3x + 1}$

5.† Which of the following expressions equal $\dfrac{y - x}{x + y}$ for all x and y for which no denominator is zero?

a. $\dfrac{y}{x} - \dfrac{x}{y}$ **b.** $\dfrac{(1/x) - (1/y)}{(1/x) + (1/y)}$ **c.** $\dfrac{1 - (x/y)}{1 + (x/y)}$

d. $\dfrac{(1/x) + (1/y)}{(1/x) - (1/y)}$ **e.** $\dfrac{(x/y) - 1}{(x/y) + 1}$ **f.** $\dfrac{(y/x) - 1}{(x/y) + 1}$

g. $1 - \dfrac{2x}{x + y}$ **h.** $\dfrac{(y^2/x) - y}{y^2 + xy}$

6. Which of the following expressions equal xy for all x and y for which no denominator is zero?

a. $\dfrac{(1/y) - (1/x)}{x - y}$ **b.** $\dfrac{y - x}{(1/y) - (1/x)}$ **c.** $\dfrac{y - x}{(1/x) - (1/y)}$

d. $\dfrac{(3/x) - (4/y)}{3y - 4x}$ **e.** $\dfrac{5/x}{5y}$ **f.** $\dfrac{5x}{1/5y}$

g. $\dfrac{5x}{5/y}$ **h.** $\dfrac{xy + x}{(1/y) + (1/y^2)}$

7.[†] Which of the following expressions define real numbers?

    a. 0^{-5} **b.** $(-5)^0$ **c.** $(-7)^{643/645}$ **d.** $7^{-643/644}$

    e. $(-7)^{643/644}$ **f.** $\dfrac{(-2)^{-1/3}}{(-3)^{-1/2}}$ **g.** $(1-1)^{1+1}$ **h.** $(1+1)^{1-1}$

8. Which of the following expressions define real numbers?

    a. $(-16)^0$ **b.** 0^{-16} **c.** $(-16)^{3/8}$

    d. $(16)^{-3/8}$ **e.** $(-16)^{8/3}$ **f.** $(-3y)^{-7/4}$ with $y < 0$

    g. $(4+4)^{6-6}$

9. Compute for $n = 1, -1, 3,$ and -3:

    a. $|1 - 3n|$ **b.** $|4n - n^2|$ **c.** $|n^n|$ **d.** $|n|^{|n|}$

10. Compute for $n = 2, -2, 4,$ and -4:

    a. $\left|\dfrac{n-2}{n+3}\right|$ **b.** $|5 - 2n|$ **c.** $n + \dfrac{1}{n}$ **d.** $\dfrac{n}{|n|}$

11. Express each of the following as a sum of constants times powers of x.

    a.[‡] $(x^2 + x^{1/2} + 1)(x^3 - x^{-3})$ **b.**[†] $(x^3 + 1)^2$

    c.[†] $(x - x^{-2})^3$ **d.** $x[x - (x+1)^2]$

    e. $(x^{3/4} + 2x)(x^{1/4} - 10)$ **f.** $\sqrt{x}(x+1)(x-3)$

    g.[†] $(2x+1)(3x-1)(4x+5)$ **h.** $(\sqrt{x} - \sqrt[3]{x})(\sqrt{x} + \sqrt[5]{x} - \sqrt[6]{x})$

12.[†] **a.** The decimal expansion of a rational number p/q may be found by long division. By analyzing the long division process, show that the decimal expansion of p/q must repeat at least every q digits after the decimal point.   **b.** Show that any repeating decimal is the decimal representation of a rational number. (Use the fact that the fraction $1/(999 \ldots 9)$ with n 9's has the repeating decimal expansion $0.000 \ldots 01000 \ldots 01 \ldots$, where each string of $n - 1$ zeros is followed by a 1.)

Use mathematical induction to verify the formulas in Exercises 13 through 20 for all positive integers n.

13.[†] $1 + 3 + 5 + \cdots + (2n - 1) = n^2$

14.[†] $1^2 + 2^2 + 3^2 + \cdots + n^2 = \frac{1}{6}n(n+1)(2n+1)$

15.[†] $1 \cdot 2 + 2 \cdot 3 + 3 \cdot 4 + \cdots + n(n+1) = \frac{1}{3}n(n+1)(n+2)$

16.[†] $\dfrac{1}{1 \cdot 2} + \dfrac{1}{2 \cdot 3} + \dfrac{1}{3 \cdot 4} + \cdots + \dfrac{1}{n(n+1)} = \dfrac{n}{n+1}$

17. $1 + 5 + 9 + \cdots + (4n - 3) = n(2n - 1)$

18. $\dfrac{1}{1 \cdot 3} + \dfrac{1}{3 \cdot 5} + \dfrac{1}{5 \cdot 7} + \cdots + \dfrac{1}{(2n-1)(2n+1)} = \dfrac{n}{2n+1}$

19. $1^3 + 2^3 + 3^3 + \cdots + n^3 = \frac{1}{4}n^2(n+1)^2$

20. $2 \cdot 2 + 3 \cdot 2^2 + 4 \cdot 2^3 + \cdots + (n+1)2^n = n2^{n+1}$

MISCELLANEOUS EXERCISES

†*Answer provided.*
‡*Outline of solution provided.*

1.† Compute for $n = 1, 2, 4, -1, -2$, and -4:

   a. $(1 + (1/n))^n$ **b.** $16^{(1+1/n)}$

2.† Compute n^{n^n} for $n = 1, 2, -1$

3.† Suppose p and q are unspecified positive integers. Express the number

   $\left(1 - \dfrac{1}{p} - \dfrac{1}{q}\right)\left(1 + \dfrac{1}{p} + \dfrac{1}{q}\right)^{-1}$ as the ratio $\dfrac{m}{n}$ of two integers.

4.† Compute $-[x^2 - x^3]$ for $x = 1, 2, -1$, and -2.

5.‡ Solve $|2 - 3x| = 4$ for x.

6.† Solve $\left|\dfrac{t}{t + 2}\right| = 4$ for t.

7. Solve $|u^2 - 17| = 8$ for u.

8.‡ Solve $|x + 1| = |2x - 1|$ for x.

9. Solve $|2y| = |3y + 4|$ for y.

10.† Solve $|1 - z| = |1 + z|$ for z.

11. Derive the "triangle inequality," $|x + y| \leq |x| + |y|$. (Show that $|x + y| = |x| + |y|$, if x and y are both positive or both negative or if either one is zero, and that $|x + y| < |x| + |y|$ if x and y are of opposite signs.)

12.† Use the triangle inequality of Exercise 11 to show that $\big||x| - |y|\big| \leq |x - y|$ for all numbers x and y.

In Exercises 13 through 21 solve the given inequalities and sketch the solution sets.

13.† $-3 \leq 2 - x < 4$ **14.**† $|5y + 6| > 0$

15. $|2z - 3| \leq 5$ **16.**‡ $\dfrac{3 + x}{x} < 2$

17.† $\dfrac{1 + 2x}{x - 1} \geq 1$ **18.** $\dfrac{y}{6 - y} > 2$

19.‡ $x^2 - x - 2 < 0$ **20.**† $u^3 \geq u^2$

21. $z^3(z + 1) > 0$

22.† Sketch the triangle with vertices $A(1, 1)$, $B(3, 4)$, and $C(-5, 5)$ in an xy-plane. Compute the lengths of its sides. Is it a right triangle?

23. Find numbers h and k so that the points $A(h, k)$, $B(0, 0)$, $C(3, 3)$, and $D(0, 6)$ are the corners of a square in a coordinate plane.

24.† Find a number k so that the points $A(-1, k)$, $B(3, 2)$, and $C(7, 3)$ are on a line.

25. There are three parallelograms that have the points $A(0,0)$, $B(2,1)$, and $C(4,0)$ as vertices. Find the fourth vertex of each of them.

26. The point two-thirds of the way along the line segment from (x_0, y_0) to (x_1, y_1) has coordinates $(\frac{1}{3}x_0 + \frac{2}{3}x_1, \frac{1}{3}y_0 + \frac{2}{3}y_1)$. Use this fact to show that the three medians of any triangle meet at a point two-thirds of the way along each median from the corresponding vertex of the triangle. (Introduce coordinate axes, so that the triangle has vertices $A(0,0)$, $B(a,0)$, and $C(b,c)$. Show that the point two-thirds of the way along the median from the vertex B coincides with the point two-thirds of the way along the median from the vertex C.)

In Exercises 27 through 30 sketch the graphs of the equations in xy-planes.

27.[†] $(2x + 3y)^2 = 36$ **28.** $y^2 = (1/16)x^4$

29.[†] $\sqrt{y-1} = 3$ **30.** $x^2 - x = 2$

In Exercises 31 through 34 determine whether the graphs of the equations are symmetric about the x-axis, about the y-axis, and about the origin. Do not sketch the graphs.

31.[†] $x^2 = y^3$ **32.** $x^3 + y^4 - xy^2 = 4$

33.[†] $y^4 - 3xy + x^4 = 1$ **34.** $x^4y^2 + x^2y^4 = 1$

35.[†] For what values of m does the line $y = mx$ pass through the acute angle ABC determined by the points $A(2,5)$, $B(0,0)$, $C(5,2)$?

36. Are the four points $A(3,4)$, $B(6,8)$, $C(-9,-12)$, and $D(12,-16)$ on the same line?

37.[†] Show that any triangle inscribed in the semicircle $y = \sqrt{a^2 - x^2}$ $(a > 0)$ and with one side along the diameter is a right triangle.

38. Show that the quadrilateral $ABCD$, with vertices $A(0,0)$, $B(a,0)$, $C(a+b,c)$, and $D(b,c)$ is a parallelogram. Show that it is a rhombus, if and only if its diagonals are perpendicular.

39.[†] Give equations for the lines that form the sides of the parallelogram $ABCD$ with vertices $A(-2,-2)$, $B(1,1)$, $C(1,5)$, and $D(-2,2)$.

40. Give equations for the lines that form the sides of the triangle with vertices $A(10,2)$, $B(-15,-4)$, and $C(3,50)$.

41.[‡] Find the two right triangles whose legs are formed by the x- and y-axes, whose hypotenuses pass through the point $(2,4)$, and that have areas equal to 18.

42.[†] Give equations for the lines that have their x-intercepts equal to their y-intercepts and pass through the point $(-4,-2)$.

43. Give equations of the two tangent lines to the circle $x^2 + y^2 = 25$ that have slopes equal to $\frac{3}{4}$.

44.[‡] The line segment between the points $A(1,3)$ and $B(3,1)$ is a chord of two circles of radius $\sqrt{10}$. Give the coordinates of the centers of the two circles.

45.[†] An *altitude* of a triangle is the line through one vertex that is perpendicular to the opposite side.   **a.**   Give equations for the altitudes of the triangle with vertices $A(0, 0)$, $B(a, 0)$, and $C(b, c)$. **b.**   Show that the three altitudes intersect at a single point.

In Exercises 46 through 49 sketch the curves.

46.[†] $x = y^2$

47. $xy^2 = 9$

48.[†] $4x = y^3$

49. $4y = (x - 2)^2$

CHAPTER 2
FUNCTIONS AND LIMITS

SUMMARY: In this chapter we study the concepts of function and limit, which will be needed in defining the basic operations of differentiation and integration in later chapters. We discuss functions in Section 2.1 and their graphs in Section 2.2. Sections 2.3 through 2.8 deal with limits and the related concept of continuity. The $\epsilon\delta$-definitions of limits and continuity are presented in an optional Section 2.9.

2-1 FUNCTIONS

TABLE 2.1 Rainfall in Dateland, Arizona, 1928–1943

YEAR	INCHES OF RAIN
1928	1.35
1929	0.54
1930	2.38
1931	10.97
1932	5.62
1933	3.32
1934	2.17
1935	6.66
1936	2.12
1937	2.57
1938	3.90
1939	7.19
1940	6.20
1941	14.23
1942	4.52
1943	0.55

(C. Green, [1] *Arizona Climate*, p. 273)

In the first half of this textbook we will study REAL-VALUED FUNCTIONS OF ONE REAL VARIABLE, which are rules that describe how one quantity is determined by another quantity. The formula πr^2, for example, gives the area of a circle as a *function* of its radius r; and Table 2.1 gives the number of inches of rainfall in a town as a *function* of the year.

In the formula for the area of a circle, the radius r is called the VARIABLE of the function and the area πr^2 is called the corresponding VALUE OF THE FUNCTION. In the rainfall table the variable is the year and the values of the function are the numbers of inches of rain.

The DOMAIN of a function is the set of values of the variable for which it is defined. The domain of the function that gives the areas of circles is the set of all positive numbers r, since circles have all possible positive radii. The domain of the function given by Table 2.1 is the set of integers from 1928 to 1943.

Functions are frequently given letter names, such as "f," "g," and "P." If the function is f and its variable is x, then the value of f at x is denoted by the symbol "$f(x)$." This notation is read "f at x" or "f of x." With it we can make the following general definition:

Definition 2.1   A real-valued function f of a real variable x is a rule that assigns a single real number $f(x)$ to each x in the domain of the function. The domain of the function is a set of real numbers.

We usually abbreviate the term "real-valued function of a real variable" to "function."

EXAMPLE 1   Let F be the function of the variable Y whose value $F(Y)$ is the number of inches of rainfall in Dateland, Arizona in year Y for $1928 \le Y \le 1943$. Find $F(1931)$ and $F(1943)$.

Solution. $F(1931)$ and $F(1943)$ are the amounts of rain that fell in the years 1931 and 1943, respectively. Table 2.1 gives the values $F(1931) = 10.97$ inches and $F(1943) = 0.55$ inches.   □

EXAMPLE 2   Let g be the function defined for $r > 0$ whose value $g(r)$ is the area of a circle of radius r. Compute $g(4)$.

Solution.   The values of the function g are given by the formula $g(r) = \pi r^2$. We use square brackets around this algebraic expression with the equation $r = 4$ as a subscript to denote the value of that expression at $r = 4$. We obtain

$$g(4) = [\pi r^2]_{r=4} = \pi(4)^2 = 16\pi. \quad □$$

Algebraic expressions as names of functions

Frequently we use an *algebraic expression* involving a variable as the *name* of a function, rather than introducing a letter name for the function such as "f" and "g." In this case we adopt the following convention: *the values of the function are those given by the algebraic expression, and the domain is assumed to consist of all values of the variable for which the algebraic expression is defined.*

For example, the expression πr^2 is defined for all real numbers r. Therefore, if we speak of "the function" πr^2, we mean the function whose domain consists of all real numbers r and whose value at r is πr^2. This function is considered to be different from the function g of Example 2, even though the values of the two functions are given by the same algebraic expression, because they have different domains. We specified the domain of the function g to consist of only the positive numbers r.

EXAMPLE 3   What is the domain of the function $\sqrt{5 - x}$ and what is its value at $x = 1$?

Solution.   The expression $\sqrt{5 - x}$ is defined for all x such that $5 - x$ is nonnegative. Solving the inequality $5 - x \ge 0$ for x gives the inequality $x \le 5$. The domain of the function $\sqrt{5 - x}$ is the closed half line $x \le 5$. The value of the function at $x = 1$ is $[\sqrt{5 - x}]_{x=1} = \sqrt{5 - 1} = 2.$ □

Deriving formulas for functions

In order to apply calculus to a problem, we often have to determine formulas for the functions relevant to the problem. Some techniques for deriving such formulas are reviewed in the following examples, in a number of exercises at the end of this section, and in the Miscellaneous Exercises at the end of the chapter.

EXAMPLE 4   A woman wants to design a rectangular garden to have an area of 400 square feet. She needs to know how much fencing would be required for various dimensions of the garden. Let $P(w)$ denote the perimeter of a garden of width w (feet). Give a formula for the values of the function P.

Solution.   We let L denote the length (feet) of the rectangular garden (Figure 2.1). Since the area of the garden is to be 400 square feet, we have $wL = 400$. Solving this for L in terms of w gives $L = (400/w)$. The perimeter is twice the width plus twice the length, so we obtain $P(w) = 2w + 2L = 2w + (800/w)$.   □

FIGURE 2.1   Area $= wL = 400$
Perimeter $= 2w + 2L$

EXAMPLE 5   A potter has 200 pots to sell during the Christmas season. In previous years he found that he could sell 50 pots for $10 each. He anticipates that for each dollar that he lowers the price, he could sell 15 more. Let $R(x)$ be the total revenue he would receive from the pots if he lowers the price x dollars. Give a formula for the values of the function R. What is its domain?

Solution.   The domain of the function consists of the integers x with $0 \le x \le 10$. If he lowers the price x dollars, he will sell $50 + 15x$ pots for $10 - x$ dollars each and his revenue will be

$$R(x) = [50 + 15x \text{ pots}]\left[10 - x\,\frac{\text{dollars}}{\text{pot}}\right]$$
$$= (50 + 15x)(10 - x) \text{ dollars.} \quad □$$

Exercises

1.† The domain of a function g is the set of all numbers x and its values are given by $g(x) = x/(x^2 + 4)$. What are $g(-10), g(-3), g(0), g(3)$, and $g(10)$?

2. The domain of a function h is the interval $0 < x < 4$, and its values are given by $h(x) = (4x - x^2)^{-3/2}$. What are the values of the function h at $x = 1, x = 2$, and $x = 3$?

3.† A function f is defined for all x and its values are given by

$$f(x) = \begin{cases} \sqrt{3 - x} & \text{for } x < 0 \\ \sqrt{3 + x} & \text{for } x \ge 0. \end{cases}$$

Compute $f(-33), f(0)$, and $f(78)$.

4.   A function K is defined for $t > -3$ by

$$K(t) = \begin{cases} \dfrac{1}{t} & \text{for } -3 < t < 0 \\ 5 & \text{for} \qquad t = 0 \\ t^3 & \text{for} \qquad t > 0. \end{cases}$$

What are $K(-2)$, $K(-1)$, $K(0)$, $K(1)$ and $K(100)$?

5.† a.   What is the domain of the function

$$\frac{(x - 3)(x + 4)}{(x + 2)(x - 1)(x + 5)}$$

and what are its values at   **b.** $x = -3$ **c.** $x = 0$ **d.** $x = 3$?

6. a.   What is the domain of the function

$$\frac{x^2}{\sqrt{5 - 2x}}$$

and what are its values at   **b.** $x = 0$ **c.** $x = -2$
d. $x = -22$ **e.** $x = -100$?

7.†   The balance, $B(r)$ dollars, after one year on an initial deposit of $1000 that earns $r\%$ interest per year, compounded semiannually, is

$$B(r) = 1000 + 10r + \frac{1}{40} r^2.$$

a. What is the balance after one year if the interest rate is 4% per year?
b. 5% per year?

8.   According to the *Handbook of Chemistry and Physics* ([1], p. 2205) the density $D(T)$ of dry air at a pressure of 76 centimeters of mercury and at a temperature of T degrees centigrade is

$$D(T) = \frac{0.001293}{1 + (0.00367)T} \text{ grams per milliliter.}$$

a.   What is the density at 10°C?   **b.** At 50°C?   **c.** Does the density increase or decrease as the temperature increases?

9.‡   Two corners of a trapezoid are at the points $(-2, 0)$ and $(2, 0)$ in an xy-plane. The other corners are at the points with y-coordinate y on the parabola $y = x^2$.   **a.** Express the area of the trapezoid as a function of y.
b. What is the domain of this function?

10.†   Express the area of a rectangle inscribed in a circle of radius 10 as a function of its width.

11.‡   Equal squares are cut from the corners of a rectangular piece of material five feet wide and seven feet long. The sides are then folded up to form a topless box.   **a.** Express the volume of the box as a function of the width of the squares that are cut out.   **b.** What is the domain of this function?

12.[†] A farmer uses 100 feet of fencing to make three sides of a rectangular pen. He uses the straight bank of a river for the fourth side.   **a.** Express the area of the pen as a function of the length of the side opposite the river. **b.** What is the domain of this function?

13.[†] Express the distance from the point $(-2, 3)$ to a point on the parabola $y = 3x^2 - 4$ as a function of the x-coordinate of the point on the curve.

14.[†] A certain right triangle has a hypotenuse that is 11 feet long.   **a.** Express the length of one of the legs of the triangle as a function of the length of the other leg.   **b.** What is the domain of that function?

15.[‡] A man six feet tall is standing on a level street near a fifteen-foot lamppost. Express the length of his shadow as a function of his distance from the lamppost.

16.[†] A water tank has the shape of a right circular cone with its vertex pointing downward. The radius of the top is 30 feet, and the height of the tank is 90 feet. Express the volume of the water in the tank as a function of its depth.

17.[‡] A *right triangular prism* is a solid whose ends are congruent triangles and whose three rectangular sides are perpendicular to the ends. A right triangular prism with isosceles right triangles for ends is ten inches long. Express its total surface area as a function of the lengths of the legs of the triangles.

18.[‡] A pile of sand is in the form of a right circular cone with a 60° angle at its vertex.   **a.** Express the volume of the pile as a function of its height. **b.** Express the height of the pile as a function of its volume.

19.[‡] A man is in a rowboat 2 miles east of his house, which is on the shore of a lake. The shore is straight and runs north–south. He wants to go to a store that is 3 miles south of his house on the shore of the lake. He can row 3/2 miles per hour and can run 6 miles per hour. How long will it take him to get to the store if he rows directly to a point on the shore x miles south of his house ($0 \le x \le 3$) and then runs along the shore to the store?

20.[†] Two men start walking from the same point at the same time along perpendicular straight paths. One walks 2 miles an hour, and the other walks 3 miles an hour. Express the distance between them as a function of the time they have spent walking.

21.[‡] A rectangular box with a square base is to be built to have a volume of 1000 cubic inches. The four sides of the box are to be made of a material that costs 2 cents per square inch. The base and top are to be made of a material that costs 3 cents per square inch. Express the costs of the material for the box as a function of its width.

22.[†] It costs a toy company $1000 + 20x^{2/3}$ dollars to manufacture x Big Bertha dolls. Each doll sells for $5.   **a.** Express the profit on x dolls as a function of x.   **b.** Express the average profit on x dolls as a function of x. **c.** What are the domains of these functions?

Describe the domains of the functions in Exercises 23 through 30.

23.[†] $\sqrt{1 + x^3}$

24.[†] $\sqrt{1 + x^2}$

25.[†] $(1 + x^5)^{-1/3}$

26.[†] $\dfrac{x}{x + 3} - \dfrac{x^2}{x - 2} + \dfrac{1}{x + 4}$

27.[†] $\dfrac{1}{x^2 - 4x}$

28.[†] $\sqrt{x + 2} - \sqrt{3 - x}$

29.[†] $f(x) = \begin{cases} x^2 & \text{for } x < 2 \\ 4x & \text{for } 2 < x \le 10 \end{cases}$

30.[†] $(2 - x)^{-1/2}$

2-2 GRAPHS OF FUNCTIONS

A geometric picture of a function is given by its GRAPH in a coordinate plane. If the letter "x" is being used as the variable of the function, usually an xy-coordinate plane is used. We consider the domain of a function f to be a portion (or all) of the x-axis, and we indicate the values $f(x)$ of the function on the y-axis.

Definition 2.2   The graph of the function f in an xy-plane is the set of points (x, y) where x is in the domain of f and y is the corresponding value $f(x)$ of f.

Figure 2.2 shows the graph of a function f. In this sketch the point labeled "x" on the x-axis represents a typical value of the variable, and the point $(x, f(x))$ with x-coordinate x and y-coordinate $f(x)$ is the corresponding point on the graph of f. To find the value of f at x, we move vertically from the point x on the x-axis to the corresponding point on the graph of f. Then we move horizontally to the y-axis and read the value $f(x)$ of the function off that axis.

TABLE 2.1

Y	R
1928	1.35
1929	0.54
1930	2.38
1931	10.97
1932	5.62
1933	3.32
1934	2.17
1935	6.66
1936	2.12
1937	2.57
1938	3.90
1939	7.19
1940	6.20
1941	14.23
1942	4.52
1943	0.55

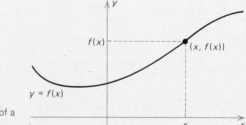

FIGURE 2.2   The graph of a function f

Just as we can use letters other than "x" for the variables of functions, we can use letters other than "y" to label the vertical axes when sketching the graphs of functions. In the next example we use a YR-plane to sketch the graph of the function given by the rainfall table of Section 2.1, repeated here as Table 2.1.

EXAMPLE 1   Let $F(Y)$ be the number of inches of rainfall in Dateland, Arizona, in year Y, as shown in Table 2.1. Sketch the graph of the function F in a YR-plane.

Solution.   We use the letter Y to suggest "year" and the letter R to suggest "rainfall." The graph shown in Figure 2.3 consists of sixteen points, represented by the sixteen dots in the sketch, one point for each of the sixteen years between 1928 and 1943. The Y-coordinate of each dot is the corresponding year and the R-coordinate is the number of inches of rain in that year.   □

We can refer to the graph of a function f in an xy-plane as *the graph of the equation* $y = f(x)$, because the graph of the equation is the set of points (x, y) whose coordinates satisfy the equation. For this to be the case, x must be in the domain of f and y must be the value of f at x.

Thus, the graph of the function f in Figure 2.2 is labeled with the equation $y = f(x)$, and the graph of the function F in Figure 2.3 with the equation $R = F(Y)$.

EXAMPLE 2   Let g be the function, defined for $r > 0$, whose value $g(r) = \pi r^2$ is the area of a circle of radius r. Sketch the graph of g in an rA-plane.

Solution.   We use r to suggest "radius" and A to suggest "area." The domain of the function is the open half line $r > 0$. The graph is the portion of the parabola $A = \pi r^2$ for these values of r (Figure 2.4).   □

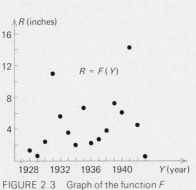

FIGURE 2.3   Graph of the function F (Rainfall in Dateland, Arizona)

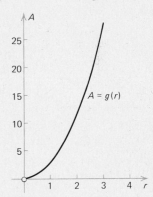

FIGURE 2.4   The graph of the function g

EXAMPLE 3   The domain of a function K is the closed half line $x \leq 6$. The values of K are given by the formula

(1) $$K(x) = \begin{cases} -x & \text{for} & x < 0 \\ -3 & \text{for} & x = 0 \\ \dfrac{3}{x} & \text{for } 0 < x \leq 6. \end{cases}$$

Sketch the graph of K in an xy-plane.

Solution.   In definition (1) the domain of the function is divided into the open half line $x < 0$, the single point $x = 0$, and the half open interval $0 < x \leq 6$. The graph of the function is, accordingly, in three pieces.

The graph is shown in Figure 2.5. The first piece of the graph, on the left, is the portion of the line $y = -x$ for $x < 0$. The second piece is the single point at $(0, -3)$. The third piece, on the right, is the portion of the hyperbola $y = 3/x$ for $0 < x \leq 6$.

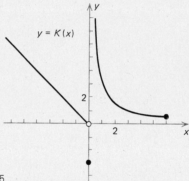

FIGURE 2.5

The circle at the origin indicates that the origin is not on the graph. The number $x = 0$ satisfies the second and not the first condition in the definition (1). The point on the graph at $x = 0$ is the point indicated by the dot at $(0, -3)$.

The dot at the point $(6, \tfrac{1}{2})$ shows that the point $(6, \tfrac{1}{2})$ on the hyperbola is on the graph of K, and that the graph does not extend to the right beyond that point. The graph extends infinitely far to the left and extends up infinitely far just to the right of the y-axis.   □

Curves that are not the graphs of functions

We defined a "function" to be a rule that assigns a single number to each x in its domain. *The graph of a function, therefore, consists of exactly one point for each value of x in its domain.* The y-coordinate of that single point is the value of the function at x.

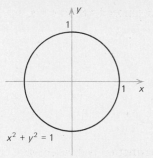

FIGURE 2.6   Not the graph of a function

Because of this property of graphs of functions, *a set of points in an xy-plane that contains two distinct points with the same x-coordinate is not the graph of a function.* The circle $x^2 + y^2 = 1$ in Figure 2.6 contains two points for each value of x with $-1 < x < 1$. It is not the graph of a function.

When we solve the equation $x^2 + y^2 = 1$ for y, we obtain $y = \pm\sqrt{1 - x^2}$, which gives formulas for two functions whose graphs are parts of the circle in Figure 2.6. The graph of the function $\sqrt{1 - x^2}$ is the upper half of the circle and the graph of the function $-\sqrt{1 - x^2}$ is the lower half.

Exercises

†*Answer provided.*
‡*Outline of solution provided.*

In Exercises 1 through 10 sketch the graphs of the given functions in the indicated coordinate planes.

1.† $\frac{1}{9}x^2 + 2$ in an xy-plane

2. $2x + 1$ in an xy-plane

3.† $(6/x) - 3$ in an xy-plane

4. $3 + (9/x^2)$ in an xy-plane

5.† $8 - t^2$ in a ts-plane

6. $4 - (8/u)$ in a uv-plane

7.† $R(x) = \begin{cases} 2 - x^2 & \text{for } x < 0 \\ 0 & \text{for } x = 0 \\ -2 + x^2 & \text{for } x > 0 \end{cases}$

in an xy-plane

8. $V(x) = \begin{cases} \dfrac{4}{x} & \text{for} & x < -2 \\ x & \text{for } -2 \le x \le 2 \\ \dfrac{4}{x} & \text{for} & x > 2 \end{cases}$

in an xy-plane

9.† $K(u) = \begin{cases} -1 + \dfrac{2}{u} & \text{for } -4 \le u < 0 \\ 0 & \text{for} & u = 0 \\ 1 + \dfrac{2}{u} & \text{for } \ 0 < u \le 4 \end{cases}$

in a uv-plane

10. $W(p) = \begin{cases} -p - 2 & \text{for } -6 \le p < -2 \\ p + 2 & \text{for } -2 \le p < 0 \\ -p + 2 & \text{for } \ \ 0 \le p < 2 \\ p - 2 & \text{for } \ \ 2 \le p \le 6 \end{cases}$

in a pq-plane

11.†   Ten minutes after he arrives at a park a hot dog vendor sells one hot dog. Fifteen minutes later he sells four hot dogs. Another half hour later he sells three hot dogs. In the next thirty-five minutes he doesn't sell any, so he leaves. Let $H(t)$ be the number of hot dogs he sells during the first t minutes he is at the park. The function H is defined for $0 \le t \le 90$. Give a formula for the values of H and sketch its graph.

12. The country of Bocabajo has a regressive tax structure. Every resident pays 20% tax on the first 100,000 donas of his taxable income, 10% on the next 100,000, and 5% on any taxable income over 200,000 donas. Give a formula for the tax $F(x)$ (donas) on a taxable income of x (donas) and sketch the graph of the function F.

13.[†] A boy sends up his model rocket ship at time $t = -3$ (seconds). At time $t = -1$, it runs out of fuel, and at time $t = \sqrt{5}$, it hits the ground. Its height above the ground $h(t)$ (feet) at time t (seconds) for $-1 \le t \le \sqrt{5}$ is given by the formula

$$h(t) = \begin{cases} 96 + 32t & \text{for } -3 \le t \le -1 \\ 80 - 16t^2 & \text{for } -1 < t \le \sqrt{5}. \end{cases}$$

Sketch the graph of the function h in a ts-plane. How high does the rocket go?

2-3   FINITE LIMITS

This section and Section 2.4 contain background material for the study of the derivative in Chapter 3. We will see there that the derivative is defined as a LIMIT. We discuss the intuitive and geometric meaning of this type of limit in this section and present techniques for evaluating such limits in Section 2.4.* We begin by defining some notation.

Suppose that f is a function whose domain includes an open interval $a < x < x_0$ to the left of x_0 and an open interval $x_0 < x < b$ to the right of x_0. If $f(x)$ approaches a number L as x approaches x_0, then we say that L is the limit of $f(x)$ as x approaches (or tends to) x_0, and we write either

$$\lim_{x \to x_0} f(x) = L$$

or

$$f(x) \to L \quad \text{as} \quad x \to x_0.$$

TABLE 2.2

x	x^2
1.9	3.61
1.99	3.9601
1.999	3.996001
1.9999	3.9960001
2.1	4.41
2.01	4.0401
2.001	4.004001
2.0001	4.00040001

EXAMPLE 1   Find the limit of x^2 as x tends to 2.

Solution.   If x is slightly greater than 2 or slightly less than 2, the value x^2 of the function is close to the number 4 (see Table 2.2). Furthermore, as x approaches 2 the corresponding number x^2 approaches 4. Therefore, the limit of x^2 as x tends to 2 is 4.   □

The next example is typical of the type of limits that we will encounter when we study the derivative.

*In Section 9 of this chapter we describe the $\epsilon\delta$-definitions of this and other types of limits.

EXAMPLE 2   Guess the value of the limit

$$\lim_{x \to 1} \frac{\sqrt{x} - 1}{x - 1}$$

by computing the values of the function for several values of x close to 1.

Solution.   The function

(1) $$\frac{\sqrt{x} - 1}{x - 1}$$

TABLE 2.3

x	$\dfrac{\sqrt{x} - 1}{x - 1}$
0.9	0.5131670 . . .
0.99	0.5012562 . . .
0.999	0.5001250 . . .
0.9999	0.5000125 . . .
1.1	0.4880848 . . .
1.01	0.4987562 . . .
1.001	0.4998750 . . .
1.0001	0.4999875 . . .

is not defined at $x = 1$. If we set $x = 1$ in formula (1), we obtain the meaningless expression 0/0. The values of the function in Table 2.3, however, suggest that

$$\lim_{x \to 1} \frac{\sqrt{x} - 1}{x - 1} = \frac{1}{2}.$$

In Section 2.4, when techniques for determining exact values of such limits are introduced, we will see that this is the case.   □

An essential feature of the concept of the limit

$$\lim_{x \to x_0} f(x) = L$$

is that in determining the limit *no reference is made to the value of the function at the limiting value x_0 of its variable.* The limit L might be the function's value at the limiting value of x, as in Example 1, where $f(x) = x^2$ and

$$\lim_{x \to 2} f(x) = \lim_{x \to 2} x^2 = 4 = [x^2]_{x=2} = f(2).$$

The function might be undefined at the limiting value of x, as with the limit

$$\lim_{x \to 1} \frac{\sqrt{x} - 1}{x - 1} = \frac{1}{2}$$

of Example 2. Or, the function's value at the limiting value of x might be different from the function's limit, as in the next example.

EXAMPLE 3   Sketch the graph of the function g defined by

$$g(x) = \begin{cases} 3 - x^2 & \text{for } x < 1 \\ 1 & \text{for } x = 1 \\ x + 1 & \text{for } x > 1. \end{cases}$$

Then determine the limit of $g(x)$ as x tends to 1.

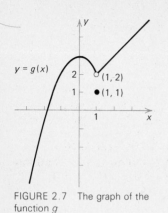

FIGURE 2.7   The graph of the function g

Solution.   The graph of the function g is sketched in Figure 2.7. It consists of the portion of the parabola $y = 3 - x^2$ for $x < 1$, the point $(1, 1)$, and the portion of the line $y = x + 1$ for $x > 1$. The portion of the parabola and the portion of the line meet at the point $(1, 2)$, which is marked with a circle in the figure. That point is not on the graph of the function g, because $g(1) = 1$ and the point $(1, 1)$ is the point at $x = 1$ on the graph.

Figure 2.8 shows what happens as x approaches 1 from the left, that is, as x gets closer and closer to the number 1 while still remaining less than 1. Figure 2.9 shows what happens as x approaches 1 from the right, that is, as x gets closer and closer to the number 1 while remaining greater than 1. In either case the point $(x, g(x))$ on the graph of g approaches the circle at $(1, 2)$, and consequently $g(x) \to 2$ as $x \to 1$. $\square$

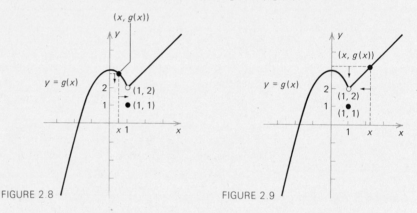

FIGURE 2.8

FIGURE 2.9

One-sided limits   The type of limit described above is a TWO-SIDED LIMIT because it deals with the variable x as it approaches its limiting value x_0 from both sides. We obtain a ONE-SIDED LIMIT by considering x only as it approaches x_0 from the left or from the right.

If the domain of a function f includes an open interval $a < x < x_0$ to the left of x_0, and if $f(x)$ approaches a number L as x approaches x_0 from the left, we write

(2) $\lim_{x \to x_0^-} f(x) = L$

or

(3) $f(x) \to L$ as $x \to x_0^-.$

If the domain of f includes an open interval $x_0 < x < b$ to the right of x_0 and if $f(x)$ approaches the number L as x approaches x_0 from the right, we write

(4) $\lim_{x \to x_0^+} f(x) = L$

or

(5) $f(x) \to L$ as $x \to x_0^+.$

EXAMPLE 4   Sketch the graph of the function h given by

$$h(x) = \begin{cases} 3 & \text{for} & x \le -2 \\ -\tfrac{1}{2}x^2 & \text{for} & -2 < x < 2 \\ 3 & \text{for} & x \ge 2. \end{cases}$$

Then determine the limits of $h(x)$ (a) as $x \to -2^-$,   (b) as $x \to -2^+$, (c) as $x \to 2^-$, and (d) as $x \to 2^+$.

Solution.   The graph of the function h is shown in Figure 2.10. It consists of the portions of the horizontal line $y = 3$ for $x \le -2$ and for $x \ge 2$ and the portion of the parabola $y = -\tfrac{1}{2}x^2$ for $-2 < x < 2$.

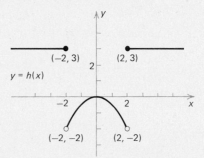

FIGURE 2.10   The graph of the function h

(a) As x approaches -2 from the left, the point $(x, h(x))$ on the graph is on the line $y = 3$ and $h(x)$ is equal to the constant 3. Therefore, $h(x) \to 3$ as $x \to -2^-$.

(b) For x just to the right of -2 (specifically for $-2 < x < 2$), the corresponding point on the graph of h is on the parabola $y = -\tfrac{1}{2}x^2$. As x approaches -2 from the right, the point $(x, h(x))$ on the graph approaches the circle at $(-2, -2)$. Hence, $h(x)$ tends to the y-coordinate of that circle, and $h(x) \to -2$ as $x \to -2^+$.

(c) Similarly, as x approaches 2 from the left, the point $(x, h(x))$ approaches the circle at $(2, -2)$, so $h(x) \to -2$ as $x \to 2^-$.

(d) Finally, for $x > 2$, the point $(x, h(x))$ is on the line $y = 3$, and $h(x)$ has the constant value 3. Therefore, $h(x) \to 3$ as $x \to 2^+$.   □

Nonexisting two-sided limits   If the one-sided limits

$$\lim_{x \to x_0^-} f(x) \quad \text{and} \quad \lim_{x \to x_0^+} f(x)$$

are different numbers, then we say that the two-sided limit

$$\lim_{x \to x_0} f(x)$$

does not exist. The two-sided limits

$$\lim_{x \to -2} h(x) \quad \text{and} \quad \lim_{x \to 2} h(x)$$

do not exist for the function h of Example 4.

Exercises

†Answer provided.
‡Outline of solution provided.

In Exercises 1 through 6 find the limits.

1.† $\lim\limits_{x \to 1} [3x^2 - 2x + 1]$

2. $\lim\limits_{x \to -1} [x^3 + 2]$

3.† $\lim\limits_{x \to 9} \sqrt{x}$

4. $\lim\limits_{x \to 0} \sqrt{\dfrac{x + 20}{2x + 5}}$

5.† $\lim\limits_{t \to -2} \left[\dfrac{t^2 + t}{3t + 4}\right]$

6. $\lim\limits_{y \to 5} \left[\dfrac{4y^2}{y - 1}\right]$

In Exercises 7 through 12 sketch the graphs of the functions in xy-planes and find the indicated limits.

7.† $f(x) = \begin{cases} -x & \text{for } x < -1 \\ -2 & \text{for } x = -1 \\ x^2 & \text{for } x > -1 \end{cases}$

    a. $\lim\limits_{x \to -1} f(x)$ **b.** $\lim\limits_{x \to 0} f(x)$

8. $g(x) = \begin{cases} 1 + x & \text{for } x < 0 \\ 2 & \text{for } x = 0 \\ 1 - x & \text{for } x > 0 \end{cases}$

    a. $\lim\limits_{x \to -1} g(x)$ **b.** $\lim\limits_{x \to 0} g(x)$ **c.** $\lim\limits_{x \to 1} g(x)$

9.† $h(x) = \begin{cases} -\dfrac{4}{x} & \text{for } -4 \le x < -1 \\ 4 & \text{for } -1 < x < 1 \\ \dfrac{4}{x} & \text{for } 1 < x \le 4 \end{cases}$

    a. $\lim\limits_{x \to -1} h(x)$ **b.** $\lim\limits_{x \to 0} h(x)$ **c.** $\lim\limits_{x \to 1} h(x)$

10. $F(x) = \begin{cases} -6 - x & \text{for} & x < -3 \\ x & \text{for } -3 < x < 3 \\ 6 - x & \text{for} & x > 3 \end{cases}$

    a. $\lim\limits_{x \to -6} F(x)$ **b.** $\lim\limits_{x \to -3} F(x)$ **c.** $\lim\limits_{x \to 3} F(x)$ **d.** $\lim\limits_{x \to 6} F(x)$

11.† $G(x) = \begin{cases} x + 4 & \text{for } x < 0 \\ x^2 + 1 & \text{for } x > 0 \end{cases}$

    a. $\lim\limits_{x \to 0^-} G(x)$ **b.** $\lim\limits_{x \to 0^+} G(x)$ **c.** $\lim\limits_{x \to 0} G(x)$

12. $P(x) = \begin{cases} x & \text{for } x < 1 \\ x - 2 & \text{for } x \ge 1 \end{cases}$

    a. $\lim\limits_{x \to 1^-} P(x)$ **b.** $\lim\limits_{x \to 1^+} P(x)$ **c.** $\lim\limits_{x \to 1} P(x)$

13. If you have access to an electronic calculator or computer, guess the limit of each of the following expressions as x tends to 1 by making a list of its values at $x = 0.9, 1.1, 0.99, 1.01, 0.999, 1.001, 0.9999,$ and 1.0001.

    a.‡ $\dfrac{x^2 - 1}{x - 1}$ **b.**‡ $\dfrac{x^3 - 1}{1 - x^4}$ **c.**‡ $\dfrac{x^2 + x - 2}{\sqrt{x} - 1}$

    d.‡ $\dfrac{x^5 - 1}{\sqrt{2x + 7} - 3}$ **e.**† $\dfrac{\sqrt[3]{x} - 1}{x^3 - 1}$ **f.**† $\sqrt[3]{\dfrac{2 - 2x}{x^2 - 1}}$

    g. $\dfrac{x^{17} - 1}{x - 1}$ **h.** $\dfrac{x^3 - 2x + 1}{4 - 3x^2 - x^4}$ **i.** $\dfrac{(x + 1)^3 - 8}{(x - 3)^3 + 8}$

In Exercises 14 through 18 find the limits of $f(x)$ **a.** as $x \to 0^-$,   **b.** as $x \to 0^+$, **c.** as $x \to 0$,   **d.** as $x \to -1^-$, and   **e.** as $x \to 2^+$.

14.[†] $f(x) = \dfrac{x}{|x|}$

15.[†] $f(x) = \dfrac{x}{|x|} + \dfrac{x-1}{|x-1|}$

16.[†] $f(x) = \begin{cases} \sqrt{5-x} & \text{for } x < 0 \\ 4 & \text{for } x = 0 \\ x^3 - 2x^2 + 1 & \text{for } x > 0 \end{cases}$

17.[†] $f(x) = \begin{cases} 4x^3 - 3x^2 & \text{for } x < 1 \\ x^5 & \text{for } 1 < x < 2 \\ 18x & \text{for } x > 2 \end{cases}$

18.[†] $f(x) = \sqrt{x^2 + x + 2}$

2-4 LIMIT THEOREMS

Many functions encountered in calculus are constructed by taking sums, products, and quotients of simpler functions. For example, the function

(1) $\dfrac{x^2}{3x + 4}$

is constructed in this way from the function x and the constant functions, 3 and 4.

The study of the limits of such functions occasionally can be simplified by the application of the following LIMIT THEOREM, which deals with limits of sums, products, and quotients of functions.

Theorem 2.1   If both of the limits $\lim\limits_{x \to x_0} f(x) = L$ and $\lim\limits_{x \to x_0} g(x) = M$ exist, then

(2) $\lim\limits_{x \to x_0} [f(x) + g(x)] = \left[\lim\limits_{x \to x_0} f(x) \right] + \left[\lim\limits_{x \to x_0} g(x) \right] = L + M$

and

(3) $\lim\limits_{x \to x_0} [f(x)g(x)] = \left[\lim\limits_{x \to x_0} f(x) \right] \left[\lim\limits_{x \to x_0} g(x) \right] = LM.$

If the limit M of $g(x)$ is not zero, then we also have

(4) $\lim\limits_{x \to x_0} \dfrac{f(x)}{g(x)} = \dfrac{\lim\limits_{x \to x_0} f(x)}{\lim\limits_{x \to x_0} g(x)} = \dfrac{L}{M}.$

Theorem 2.1 follows from a common sense understanding of limits. (Its formal proof requires the $\epsilon\delta$-definition of limits and is presented in Section 2.9.) There are corresponding results for one-sided limits.

In the next example we apply Theorem 2.1 to the function (1). To do this we need to use the facts that for any number x_0 and any constant C,

(5) $\qquad \lim_{x \to x_0} [x] = x_0$

and

(6) $\qquad \lim_{x \to x_0} [C] = C.$

EXAMPLE 1   Use statements (5) and (6) and Theorem 2.1 to show that

$$\lim_{x \to 5} \frac{x^2}{3x + 4} = \frac{25}{19}.$$

Solution.   Equations (3) and (5) show that

$$\lim_{x \to 5} [x^2] = \left[\lim_{x \to 5} x\right]^2 = 5^2 = 25.$$

From (2), (3), (5), and (6) we have

(7) $\qquad \lim_{x \to 5} [3x + 4] = \left[\lim_{x \to 5} 3\right]\left[\lim_{x \to 5} x\right] + \left[\lim_{x \to 5} 4\right]$
$$= 3(5) + 4 = 19.$$

Because the limit (7) is not zero, it follows from (4) that

$$\lim_{x \to 5} \left[\frac{x^2}{3x + 4}\right] = \frac{\lim\limits_{x \to 5} [x^2]}{\lim\limits_{x \to 5} [3x + 4]} = \frac{25}{19}. \quad \square$$

In practice we do not often have to refer explicitly to limit theorems when computing limits, just as we do not have to refer explicitly to the laws of algebra when making arithmetic computations. We refer to the limit theorems in theoretical discussions or to clarify our thinking when we are dealing with complex or subtle questions involving limits.

When the limit theorems do not apply   In elementary calculus the most important feature of Theorem 2.1 is that rule (4) for finding the limit of a quotient of two functions *does not apply* when the limit of the denominator is zero. This is important because, as we will see in Chapter 3, the derivative is defined to be such a limit.

In such instances, where the limit theorems do not apply, further work has to be done to determine the limit, as illustrated in the next example.

EXAMPLE 2   Determine the limit

(8) $\qquad \lim_{x \to 3} \frac{x^2 - 9}{x - 3}.$

Solution.   We cannot apply rule (4) because the limit of the denominator in (8) is zero: $x - 3 \to 0$ as $x \to 3$. In fact, the numerator also tends to zero: $x^2 - 9 \to 0$ as $x \to 3$. If we tried to apply rule (4), we would obtain the meaningless expression $0/0$ as the limit.

To find the limit (8), we factor the numerator $x^2 - 9$ into the product of the functions $x - 3$ and $x + 3$: $x^2 - 9 = (x - 3)(x + 3)$. We make this substitution and then cancel the denominator with the $x - 3$ in the numerator to obtain the equation

(9) $$\frac{x^2 - 9}{x - 3} = \frac{(x - 3)(x + 3)}{x - 3} = x + 3.$$

Thus, the function on the left side of equation (9) is equal to the function $x + 3$ for all $x \neq 3$. The limit theorems (and common sense) tell us that the limit of $x + 3$ is 6, so we obtain

$$\lim_{x \to 3} \frac{x^2 - 9}{x - 3} = \lim_{x \to 3} [x + 3] = 3 + 3 = 6. \quad \square$$

Rationalization

In Example 2 of the last section we guessed the limit

(10) $$\lim_{x \to 1} \frac{\sqrt{x} - 1}{x - 1} = \frac{1}{2}$$

by computing values of the function $\dfrac{\sqrt{x} - 1}{x - 1}$ for values of x close to 1.

In the next example we show that our guess was correct. The presence of the square root calls for a procedure known as RATIONALIZATION.

EXAMPLE 3   Verify limit (10).

Solution.   To apply the rationalization procedure, we view $\sqrt{x} - 1$ in (10) as the difference of square roots $\sqrt{x} - \sqrt{1}$ and multiply and divide by the sum $\sqrt{x} + \sqrt{1}$ of those square roots. This gives for nonnegative $x \neq 1$

$$\frac{\sqrt{x} - \sqrt{1}}{x - 1} = \frac{(\sqrt{x} - \sqrt{1})(\sqrt{x} + \sqrt{1})}{(x - 1)(\sqrt{x} + \sqrt{1})} = \frac{(\sqrt{x})^2 - (\sqrt{1})^2}{(x - 1)(\sqrt{x} + 1)}$$
$$= \frac{x - 1}{(x - 1)(\sqrt{x} + 1)}.$$

The new numerator $x - 1$ cancels with the $x - 1$ in the denominator to leave

$$\frac{\sqrt{x} - 1}{x - 1} = \frac{1}{\sqrt{x} + 1}.$$

The denominator of the expression on the right does not tend to zero as x tends to 1, so we can determine its limit by inspection. We obtain

$$\lim_{x \to 1} \frac{\sqrt{x} - 1}{x - 1} = \lim_{x \to 1} \frac{1}{\sqrt{x} + 1} = \frac{1}{\sqrt{1} + 1} = \frac{1}{2}. \quad \square$$

In the next example we have to simplify a fraction before factoring and cancelling to find the limit.

EXAMPLE 4   What is the limit of

$$(11) \quad \frac{\dfrac{6}{x} - 3}{x - 2}$$

as x tends to 2?

Solution.   We write for $x \neq 2$

$$\frac{\dfrac{6}{x} - 3}{x - 2} = \frac{1}{x - 2}\left(\frac{6}{x} - 3\right) = \frac{1}{x - 2}\left(\frac{6 - 3x}{x}\right) = \frac{3(2 - x)}{(x - 2)x} = -\frac{3}{x}.$$

Hence

$$\lim_{x \to 2}\left[\frac{\dfrac{6}{x} - 3}{x - 2}\right] = \lim_{x \to 2}\left(-\frac{3}{x}\right) = -\frac{3}{2}. \quad \square$$

Exercises

In Exercises 1 through 4 use equations (1) to (5) and the analogous statements for one-sided limits.

1.† Show that $\lim\limits_{x \to 2} \dfrac{x + (4/x)}{x - 3} = -4$.

2. Show that $\lim\limits_{t \to 1} (t^2 - t^3 + 5t)/(t^2 - 6) = -1$.

3.† Show that if $\lim\limits_{x \to 10^-} f(x) = -5$ and $\lim\limits_{x \to 10^-} g(x) = 4$, then
$\lim\limits_{x \to 10^-} [f(x)(3g(x) - x)] = -10$.

4. Show that $\lim\limits_{x \to 0^-} \dfrac{K(x)^2}{x - 3K(x)} = \frac{2}{3}$ if $\lim\limits_{x \to 0^-} K(x) = -2$.

In Exercises 5 through 16 find the indicated limits.

5.† $\lim\limits_{x \to 100} \dfrac{F(x) - 5G(x)^2}{F(x) - 100}$, where $\lim\limits_{x \to 100} F(x) = 350$ and $\lim\limits_{x \to 100} G(x) = -4$.

6. $\lim\limits_{x \to 5^-} [5p(x)^4 - 4p(x)^3 + 3p(x)^2 - 2p(x) + 1]$, where $\lim\limits_{x \to 5^-} p(x) = -1$.

7.† $\lim\limits_{x \to -2} \dfrac{2 + x}{4 - x^2}$

8. $\lim\limits_{x \to 1} \dfrac{x^2 - 1}{3x - 3}$

9.† $\lim\limits_{x \to 9} \dfrac{\sqrt{x} - 3}{x^2 - 9x}$

10. $\lim\limits_{x \to 25} \dfrac{x - 25}{5 - \sqrt{x}}$

11.‡ $\lim\limits_{x \to -1} \dfrac{1 + (1/x)}{x^2 - 1}$

12.† $\lim\limits_{x \to 5} \dfrac{25 - 5x}{1 - (25/x^2)}$

13. $\lim\limits_{x \to 2} \dfrac{x^2 - 4}{(1/x) - \frac{1}{2}}$

14.‡ $\lim\limits_{h \to 0} \dfrac{(3 + h)^2 - 9}{h}$

15.† $\lim\limits_{y \to 0} \dfrac{1 - \sqrt{1 + y}}{7y}$

16. $\lim\limits_{t \to 0} \dfrac{4 - (t + 2)^2}{9 - (t + 3)^2}$

17.† Evaluate the limit $\lim\limits_{x \to -1} (x^2 + 3x + 2)/(x + 1)$, and then sketch the graph of the function $(x^2 + 3x + 2)/(x + 1)$.

18. **a.** Sketch the graph of $(x^3 - 6x)/3x$ and find its limits as
b. $x \to -2$ **c.** $x \to 0$ **d.** $x \to 1$.

Additional drill exercises

Find the limits in Exercises 19–28.

19.† $\lim\limits_{x \to 2} \dfrac{(1/x) - (1/2)}{x^2 - 4}$

20.† $\lim\limits_{x \to 3} \dfrac{(9/x^2) - 1}{12 - 4x}$

21.† $\lim\limits_{x \to 4+} \dfrac{x - 4}{\sqrt{x} - 2}$

22.† $\lim\limits_{x \to 0+} \dfrac{3x^2 - 2\sqrt{x}}{\sqrt{x}}$

23.† $\lim\limits_{x \to 0} \dfrac{(1 + x)^3 - 1}{(2 + x)^2 - 4}$

24.† $\lim\limits_{x \to 0} \dfrac{(3x + 1)^2 - 1}{x^3 - 3x}$

25.† $\lim\limits_{x \to 1} \dfrac{x^2 - 1}{x^2 + 1}$

26.† $\lim\limits_{x \to -1} \dfrac{x + 1}{1 - x^3}$

27.† $\lim\limits_{x \to 1} \dfrac{(x - 1)^2}{x^2 - 1}$

28.† $\lim\limits_{x \to 0} \dfrac{x^2 + 4x}{6x - x^3}$

2-5 INFINITE LIMITS

We say that a variable x IS TENDING TO INFINITY and we write $x \to \infty$ or $x \to +\infty$ if x is becoming a larger and larger positive number, increasing without bound. If x is negative and decreasing without bound, so that $|x|$ is tending to ∞, we say that x is TENDING TO MINUS INFINITY, and we write $x \to -\infty$.

In this section we discuss a number of types of INFINITE LIMITS where either the variable or the value of the function tends to infinity or to minus infinity. We will describe the various types of infinite limits by examining a few examples, rather than by making a list of formal definitions.

Limits as x tends to ∞ or to −∞ and infinite one-sided limits

EXAMPLE 1   Find the limits of $1 - 1/x$ (a) as $x \to \infty$, (b) as $x \to -\infty$, (c) as $x \to 0^+$, and (d) as $x \to 0^-$.

Solution.   The graph of $1 - 1/x$ is the hyperbola $y = 1 - 1/x$ shown in Figure 2.11.

(a) If x is a large positive number, then $1/x$ is small, and as x tends to infinity, $1/x$ tends to zero. Therefore, $1 - 1/x \to 1$ as $x \to \infty$ and the graph in Figure 2.11 approaches the horizontal line $y = 1$ as it moves off to the right.

(b) The quantity $1/x$ also tends to zero as x tends to minus infinity, so $1 - 1/x \rightarrow 1$ as $x \rightarrow -\infty$ and the graph approaches the line $y = 1$ as it moves off to the left.

(c) If x is a very small positive number, then $-1/x$ and $1 - 1/x$ are very large negative numbers. Therefore, $1 - 1/x \rightarrow -\infty$ as $x \rightarrow 0^+$ and the graph goes down just to the right of the y-axis.

(d) If x is a very small negative number, then $-1/x$ and $1 - 1/x$ are very large positive numbers. Hence, $1 - 1/x \rightarrow \infty$ as $x \rightarrow 0^-$ and the graph goes up just to the left of the y-axis.   □

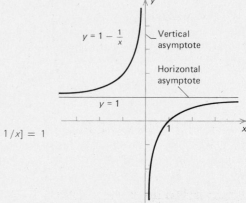

FIGURE 2.11 $\lim\limits_{x \to \infty} [1 - 1/x] = 1$

$\lim\limits_{x \to -\infty} [1 - 1/x] = 1$

$\lim\limits_{x \to 0^+} [1 - 1/x] = -\infty$

$\lim\limits_{x \to 0^-} [1 - 1/x] = \infty$

Horizontal and vertical asymptotes

Definition 2.3   A line $y = y_0$ is a horizontal asymptote of the graph of a function f, if $f(x)$ tends to y_0 as x tends to ∞ or as x tends to $-\infty$. A line $x = x_0$ is a vertical asymptote of the graph of f, if $f(x)$ tends to ∞ or to $-\infty$ as x tends to x_0 from the right or from the left.

The line $y = 1$ is a horizontal asymptote of the graph of the function $1 - 1/x$ in Figure 2.11, and the line $x = 0$ (the y-axis) is a vertical asymptote.

Infinite two-sided limits

If the limit of $f(x)$ as x tends to x_0 from the left and from the right is ∞, we say that the (two-sided) limit of $f(x)$ as x tends to x_0 is ∞ and we write $\lim\limits_{x \to x_0} f(x) = \infty$. Similarly, the statement $\lim\limits_{x \to x_0} f(x) = -\infty$ means that $f(x) \rightarrow -\infty$ as x tends to x_0 from the left and from the right.

EXAMPLE 2   Find the limit

$$\lim_{x \to 0} \left[\frac{2}{x^2} - 2 \right].$$

Solution.   If x is a small positive or negative number, then $(2/x^2) - 2$ is a large positive number. We have

$$\lim_{x \to 0} \left[\frac{2}{x^2} - 2 \right] = \infty. \quad □$$

The geometric interpretation of the result of Example 2 is shown in Figure 2.12. The graph of the function $(2/x^2) - 2$ goes up on either side of the y-axis. We also have $\lim\limits_{x \to \infty} [(2/x^2) - 2] = -2$ and $\lim\limits_{x \to -\infty} [(2/x^2) - 2] = -2$. The y-axis is a vertical asymptote and the line $y = -2$ is a horizontal asymptote of the graph in Figure 2.12.

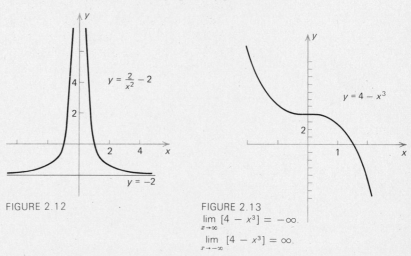

FIGURE 2.12

FIGURE 2.13
$\lim\limits_{x \to \infty} [4 - x^3] = -\infty.$

$\lim\limits_{x \to -\infty} [4 - x^3] = \infty.$

Infinite limits as x tends to ∞ or to −∞

EXAMPLE 3   Find the limits $\lim\limits_{x \to \infty} [4 - x^3]$ and $\lim\limits_{x \to -\infty} [4 - x^3]$.

Solution.   As x becomes a larger and larger positive number, $-x^3$ becomes a larger and larger negative number, and $4 - x^3 \to -\infty$ as $x \to \infty$. As x becomes a larger and larger negative number, $-x^3$ becomes a larger and larger positive number, and $4 - x^3 \to \infty$ as $x \to -\infty$. We write

$$\lim_{x \to \infty} [4 - x^3] = -\infty \quad \text{and} \quad \lim_{x \to -\infty} [4 - x^3] = \infty. \quad \square$$

The geometric interpretations of these limits can be seen in Figure 2.13, where we show the graph $y = 4 - x^3$ of the function $4 - x^3$. The graph drops down infinitely far as it moves off to the right and it rises up infinitely far as it moves off to the left.

Interpreting infinite limits

EXAMPLE 4   It costs a small chemical company $3 for each pound of gunpowder they produce, plus $120 overhead each day. Let $A(x)$ denote the average cost of the gunpowder, measured in dollars per pound, in a day when x pounds are produced. (a) Give a formula for $A(x)$ and determine the limits $\lim\limits_{x \to 0^+} A(x)$ and $\lim\limits_{x \to \infty} A(x)$. (b) Sketch the graph of the function A and describe what the results from part (a) mean to the chemical company.

Solution.   (a) It costs the company $120 + 3x$ dollars to produce x pounds of gunpowder in one day. The average cost is

$$(1) \qquad A(x) = \frac{[120 + 3x] \text{ dollars}}{x \text{ pounds}} = \frac{120}{x} + 3 \frac{\text{dollars}}{\text{pound}}.$$

We have

$$(2) \qquad \lim_{x \to 0^+} A(x) = \lim_{x \to 0^+} \left[\frac{120}{x} + 3 \right] = \infty$$

and

$$(3) \qquad \lim_{x \to \infty} A(x) = \lim_{x \to \infty} \left[\frac{120}{x} + 3 \right] = 3.$$

(b) The function A is defined for $x > 0$. Its graph is the right half of the hyperbola $y = (120/x) + 3$, which is sketched in Figure 2.14. It has the y-axis as a vertical asymptote and the line $y = 3$ as a horizontal asymptote.

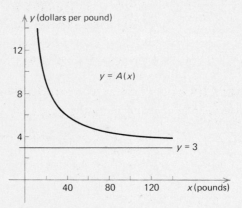

FIGURE 2.14

We notice that the term $120/x$ in formula (1) for the average cost is due to the fixed overhead, while the term 3 is due to the other costs of production. If the company produces a large amount of gunpowder in one day (i.e., if x is large), then the contribution $120/x$ to the average cost is small, because the overhead is spread out over a large number of pounds of gunpowder. The average cost tends to 3 as x tends to infinity, as stated in (3), because the more gunpowder produced in one day, the closer the average cost comes to the \$3 per pound it would be with no overhead.

The meaning of result (2) to the company is that their average cost per pound will be very large if they produce a very small amount in a day. If they produce only one half pound, for example, their average cost will be

$$A\left(\frac{1}{2}\right) = \frac{120}{1/2} + 3 = 240 + 3 = 243 \frac{\text{dollars}}{\text{pound}}.$$

The average cost per pound tends to infinity as the number of pounds they produce tends to zero because their overhead is the same no matter how little gunpowder is made. □

On the existence of limits (optional)

Whereas we have seen many examples in which a two-sided limit does not exist because the one-sided limits are different, all one-sided limits exist for the functions that are normally encountered in calculus. The one-sided limits are either numbers or ∞ or $-\infty$.

It is easy, however, to construct a function for which a one-sided limit does not exist. We define $f(x)$ to be 1 for $\frac{1}{2} < x \le 1$, to be -1 for $\frac{1}{4} < x \le \frac{1}{2}$, to be 1 for $\frac{1}{8} < x \le \frac{1}{4}$, to be -1 for $\frac{1}{16} < x \le \frac{1}{8}$, and so forth. To describe this definition in one statement, we write

$$f(x) = (-1)^n \quad \text{for} \quad \frac{1}{2^{n+1}} < x \le \frac{1}{2^n}, \quad n = 0, 1, 2, 3, \ldots$$

This function is then defined for $0 < x \le 1$. The portion of its graph for $1/16 < x \le 1$ is sketched in Figure 2.15.

The one-sided limit $\lim_{x \to 0^+} f(x)$ does not exist because as x approaches 0 from the right $f(x)$ flips back and forth between 1 and -1 infinitely often. The limit must be a single number and cannot be both 1 and -1.

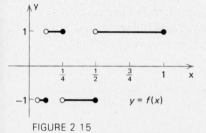

FIGURE 2.15

Exercises

†Answer provided.
‡Outline of solution provided.

In Exercises 1 through 10 find the limits of the functions

a. as $x \to -\infty$ **b.** as $x \to 0^-$ **c.** as $x \to 0^+$ **d.** as $x \to 0$ **e.** as $x \to \infty$.

Sketch the graphs of the functions and show any vertical or horizontal asymptotes.

1.† $2 + (1/x^2)$ ⠀⠀⠀⠀⠀⠀⠀⠀⠀ **2.** $-3 - (9/x^2)$

3.† $5 - x^2$ ⠀⠀⠀⠀⠀⠀⠀⠀⠀⠀⠀⠀ **4.** $x^3 - 10$

5.† $(3/x) - 4$ ⠀⠀⠀⠀⠀⠀⠀⠀⠀ **6.** $5 - (4/x)$

7.† $V(x) = \begin{cases} \dfrac{3}{x} + 3 & \text{for } x < 0 \\ 0 & \text{for } x = 0 \\ \dfrac{3}{x} - 3 & \text{for } x > 0 \end{cases}$ ⠀⠀ **8.** $W(x) = \begin{cases} -4 + \frac{1}{4}x^2 & \text{for } x < 0 \\ 4 - \frac{1}{4}x^2 & \text{for } x > 0 \end{cases}$

9.† $g(x) = \begin{cases} x + 6 & \text{for } & x \le -3 \\ \frac{1}{9}x^3 & \text{for } -3 < x < 3 \\ x - 6 & \text{for } & x \ge 3 \end{cases}$

10. $h(x) = \begin{cases} x & \text{for } & x < -2 \\ \dfrac{4}{x} & \text{for } -2 \le x < 0 \\ \dfrac{4}{x} & \text{for } \ 0 < x \le 2 \\ x & \text{for } & x > 2 \end{cases}$

2-6 LIMITS OF POLYNOMIALS AND RATIONAL FUNCTIONS AS $x \to \pm\infty$

A POLYNOMIAL in the variable x is an expression of the form

$$a_n x^n + a_{n-1} x^{n-1} + \cdots + a_2 x^2 + a_1 x + a_0$$

where n is a nonnegative integer and the symbols $a_n, a_{n-1}, \ldots, a_0$ denote real constants. The functions $x^7 - 3x^5 + 2$, $-x^4$, and 15 are polynomials.

A RATIONAL FUNCTION is a function formed from polynomials by taking sums, products, and quotients. The functions

$$\frac{x^2 + x - 1}{x - 5} \qquad 1 + x + \frac{3}{x} \qquad \frac{x(x + 1)}{x^{10} + 2}$$

are rational functions.

In this section we describe rules for determining the limits of polynomials and rational functions as their variables tend to ∞ or to $-\infty$.

Limits of polynomials as x tends to ∞ or to $-\infty$

Rule 2.1   The limit of a polynomial as its variable x tends to ∞ or to $-\infty$ is the same as the limit of the term in it that involves the highest power of x.

This rule can be verified in any particular case by factoring the highest power of x from the polynomial. The other factor is not a polynomial but a function whose limits as x tends to ∞ and to $-\infty$ are easy to recognize.

EXAMPLE 1   Use Rule 2.1 to determine the limits of the polynomial $2x^4 - 3x^2 + x$, as x tends to ∞ and to $-\infty$. Then verify your conclusions by factoring out the highest power of x from the polynomial.

Solution.   The highest power term in the polynomial $2x^4 - 3x^2 + x$ is $2x^4$. So according to Rule 2.1,

(1) $\displaystyle \lim_{x \to \infty} [2x^4 - 3x^2 + x] = \lim_{x \to \infty} [2x^4] = \infty$

and

(2) $\displaystyle \lim_{x \to -\infty} [2x^4 - 3x^2 + x] = \lim_{x \to -\infty} [2x^4] = \infty.$

To justify Rule 2.1 in this case, we factor x^4 from the polynomial. We write for $x \neq 0$

$$x^2 = x^4 \left(\frac{1}{x^2} \right) \quad \text{and} \quad x = x^4 \left(\frac{1}{x^3} \right).$$

Then we have

$$(3) \qquad 2x^4 - 3x^2 + x = 2x^4 - 3x^4\left(\frac{1}{x^2}\right) + x^4\left(\frac{1}{x^3}\right) = x^4\left(2 - \frac{3}{x^2} + \frac{1}{x^3}\right).$$

The second factor $2 - (3/x^2) + (1/x^3)$ on the right side of equation (3) tends to 2 as x tends to ∞ or to $-\infty$. Consequently, for large (positive or negative) x, $2x^4 - 3x^2 + x$ is equal to x^4 multiplied by a positive number close to 2. Since x^4 tends to ∞ as x tends to ∞ or to $-\infty$, the polynomial $2x^4 - 3x^2 + x$ has the same limits, as stated in the rule.   □

Limits of rational functions as x tends to ∞ or to −∞

Rule 2.2   The limit of a quotient of polynomials as x tends to ∞ or to −∞ is the same as the limit of the quotient of the highest power terms in the numerator and denominator.

This rule can be justified in any particular instance by factoring the highest power of x from the numerator and the highest power of x from the denominator.

EXAMPLE 2   Use Rule 2.2 to determine the limits of the rational function

$$(4) \qquad \frac{5x^3 + 2x^2 - 1}{x - 6x^4}$$

as x tends to ∞ and as x tends to $-\infty$, then verify your conclusions by factoring the highest powers of x from the numerator and denominator of the function.

Solution.   The highest power term in the numerator is $5x^3$ and the highest power term in the denominator is $-6x^4$. Therefore, according to Rule 2.2, the limits of the function (4) as x tends to ∞ and to $-\infty$ are the same as the limits of the function

$$(5) \qquad \frac{5x^3}{-6x^4} = -\frac{5}{6x}.$$

This function tends to zero as x tends to ∞ or to $-\infty$; therefore, the function (4) also tends to zero as x tends to ∞ or to $-\infty$.

To justify this application of Rule 2.2, we factor x^3 from the numerator and x^4 from the denominator; then we cancel the x^3 in the numerator with x^3 in the denominator. For $x \neq 0$ we have

$$(6) \qquad \frac{5x^3 + 2x^2 - 1}{x - 6x^4} = \frac{x^3\left(5 + \dfrac{2}{x} - \dfrac{1}{x^3}\right)}{x^4\left(\dfrac{1}{x^3} - 6\right)} = \frac{1}{x}\left[\frac{5 + \dfrac{2}{x} - \dfrac{1}{x^3}}{\dfrac{1}{x^3} - 6}\right].$$

The expression in square brackets on the right side of equation (6) tends to $5/(-6) = -5/6$ as x tends to ∞ or to $-\infty$. Consequently, for large x (positive or negative), the function (6) is equal to $1/x$ multiplied by a number close to $-5/6$. The function (6) tends to zero as x tends to ∞ or to $-\infty$, as was indicated by the rule. $\square$

EXAMPLE 3   Use Rule 2.2 to determine the limits of the function

$$(7) \qquad \frac{2 + x - 6x^2}{2x^2 + 10}$$

as x tends to ∞ and to $-\infty$.

Solution.   The expression (7) has the same limits as x tends to ∞ or to $-\infty$ as the function

$$\frac{-6x^2}{2x^2} = -3.$$

The limits are both -3. $\square$

Exercises

In Exercises 1 through 7 use Rules 2.1 and 2.2 to determine the limits of the functions   **a.** as x tends to ∞ **b.** as x tends to $-\infty$ **c.** Verify your conclusions by factoring the highest powers of x from the polynomials.

1.† $\ 1 - 2x^2 - 4x^3$

2. $\ x - 5x^3 - 6x^4$

3.† $\dfrac{2x - 3}{4 - 5x}$

4. $\dfrac{x}{x^2 + 17}$

5.‡ $(1 - x)(2 - x)(3 - x)$

6.‡ $\dfrac{(x^2 + 1)(3x - 4)}{1 + 2x + 3x^2}$

7. $\dfrac{(1 - x)(2 - 3x)}{1 - 10x^2}$

In Exercises 8 through 14 use Rules 2.1 and 2.2 to determine the limits of the functions   **a.** as x tends to ∞ **b.** as x tends to $-\infty$.

8.† $\ 5x^3 - 3x^2 + 2x - 100$

9. $\ 1 + x - x^2 + x^3 - x^4$

10.† $\dfrac{x^2(5 - 6x)}{5x^3 + x^2 + 1}$

11. $\dfrac{3x}{4x^2 + 2x - 1}$

12.‡ $\dfrac{x^2}{x + 1} - \dfrac{x^2}{x + 3}$

13.† $\ x + \dfrac{2x^2 - 1}{3 - 2x}$

14. $\dfrac{x^3}{x - 5} - x^2$

Additional drill exercises   In Exercises 15 through 21 find the limits of the functions   **a.** as $x \to -\infty$ and   **b.** as $x \to \infty$.

15.† $x - 3x^2 + 5x^3 - 6x^4 + 7x^5$ **16.**† $\dfrac{3x^2 + 2x - 1}{5 - 4x + 2x^2}$

17.† $(x - 1)(1 - x)(x - 2)(2 - x)$ **18.**† $\dfrac{(2x + 1)(3x - 1)}{8x^2 + 7}$

19.† $\dfrac{1}{x} - \dfrac{1}{x + 1}$ **20.**† $x - \dfrac{1}{x} + x^2 - \dfrac{1}{x^2}$

21.† $\dfrac{1}{x} + \dfrac{2x}{x + 1}$

22.† Find the limits of $x - (x - 2)^{-2}$ **a.** as $x \to 2^-$,   **b.** as $x \to 2^+$, and **c.** as $x \to -2^+$.

23.† Find the limits of $x/(x - 3)$ **a.** as $x \to -3^+$,   **b.** as $x \to 3^-$, and **c.** as $x \to 4$.

24.† Find the limits of $x(x - 1)^{-1}(x - 2)^{-2}(x - 3)^{-3}$ **a.** as $x \to 1^-$, **b.** as $x \to 1^+$,   **c.** as $x \to 2$, and   **d.** as $x \to 3$.

25.† Find the limits of $(8 - x^3)^{-1/3}$ **a.** as $x \to 0^+$,   **b.** as $x \to 0^-$, **c.** as $x \to 2^+$,   **d.** as $x \to 2^-$,   **e.** as $x \to -2^+$, and   **f.** as $x \to -2^-$.

2-7   CONTINUOUS FUNCTIONS

Most natural phenomena are described by CONTINUOUS FUNCTIONS. These are functions whose graphs are not "broken." If we plot atmospheric pressure as a function of the altitude above sea level, we obtain an unbroken curve (Figure 2.16). We also obtain unbroken curves if we plot the length of a fish as a function of its age (Figure 2.17) or the height of a ball thrown into the air as a function of time (Figure 2.18).

Figure 2.19 shows the graph of a DISCONTINUOUS FUNCTION. The function gives a bank balance as a function of time. The graph is "broken" because the value of the function "jumps" whenever a deposit is made or a check is paid.

As we will see in the next section, continuous functions have special properties that make them important in a discussion of calculus.

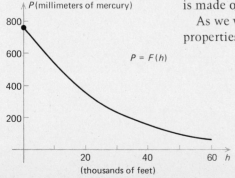

FIGURE 2.16 $F(h) =$ The atmospheric pressure at an altitude h above sea level. (Adapted from W. G. Brombacher and M. R. Houseman, [1] *The Stratosphere Flight of 1935 in the Balloon "Explorer II,"* p. 227.)

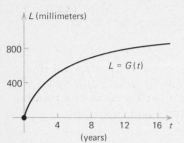

FIGURE 2.17 $G(t)$ = Length (millimeters) of a summer flounder at age t (years). (Adapted from C. E. Richards, [1] "Analog simulation in fish population studies," p. 204.)

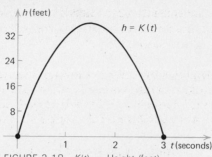

FIGURE 2.18 $K(t)$ = Height (feet) of a ball above the ground at time t (seconds).

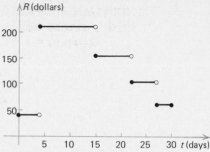

FIGURE 2.19 $f(t)$ = Bank balance t days after the beginning of the month.

The mathematical definition of continuity

The notion that a continuous function is one whose graph is "unbroken" is not precise enough to serve as a mathematical definition. The mathematical definition of continuity is based on the concept of limit.

Definition 2.4   A function f defined in an open interval containing the point x_0 is continuous at x_0 if

(1) $$\lim_{x \to x_0} f(x) = f(x_0).$$

Condition (1) means that the one-sided limits of $f(x)$ as x approaches x_0 from the right and from the left $\lim_{x \to x_0^+} f(x)$ and $\lim_{x \to x_0^-} f(x)$ both exist, are finite, and are equal to the number $f(x_0)$. This condition is what makes the graph of f "fit together" and not be "broken" at x_0.

EXAMPLE 1   A function f is given by

$$f(x) = \begin{cases} -1 & \text{for} & x < -4 \\ 2 & \text{for} & x = -4 \\ \dfrac{4}{x} & \text{for } -4 < x < 0 \\ x & \text{for} & x \geq 0. \end{cases}$$

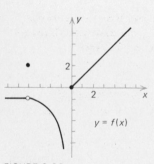

FIGURE 2.20

Sketch the graph of f and determine the values of x at which it is discontinuous.

Solution.   The graph of f is shown in Figure 2.20. The function is discontinuous at $x = -4$ and at $x = 0$, where its graph is "broken." The function is discontinuous at $x = -4$ because the two numbers $\lim_{x \to -4} f(x) = -1$ and $f(-4) = 2$ are different. It is discontinuous at $x = 0$ because the limit $\lim_{x \to 0} f(x)$ does not exist. This two-sided limit does not exist because $\lim_{x \to 0^-} f(x) = -\infty$ and $\lim_{x \to 0^+} f(x) = 0$.   □

Continuity at end points of domains

If we are considering the domain of a function to be an interval, we test for the continuity of the function at an endpoint of the interval by using a one-sided limit.

Definition 2.5   If the domain of a function is an interval that includes its left end point a, we say that f is continuous at a if $f(x) \to f(a)$ as $x \to a^+$. If the interval includes its right end point b, we say that f is continuous at b if $f(x) \to f(b)$ as $x \to b^-$.

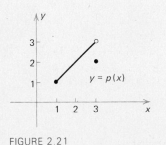

FIGURE 2.21

This definition is illustrated by the function

$$p(x) = \begin{cases} x & \text{for } 1 \le x < 3 \\ 2 & \text{for } \quad x = 3 \end{cases}$$

whose graph is shown in Figure 2.21. The function is continuous at $x = 1$ and discontinuous at $x = 3$.

The algebra of functions

The SUM OF TWO FUNCTIONS f and g is denoted $f + g$. It is the function whose value at x is the sum of the values of f and g: $(f + g)(x) = f(x) + g(x)$. The PRODUCT of these two functions is denoted fg. Its value at x is the product of the values of f and g: $(fg)(x) = f(x)g(x)$. The function f/g is a QUOTIENT of the two functions. Its value at x is given by the formula $(f/g)(x) = f(x)/g(x)$. A LINEAR COMBINATION of the functions f and g is a function of the form $Af + Bg$ where A and B are constants.

The domain of the sum, product, and linear combinations of the functions consists of all numbers x that are in the domain of both functions. The domain of the quotient (f/g) consists of those x numbers in the domains of both functions such that $g(x) \ne 0$.

Continuity theorems

When studying the continuity of functions that are formed from other functions by taking sums, products, and quotients, we can use the following CONTINUITY THEOREM.

Theorem 2.2   If the functions f and g are both continuous at x_0, then their sum $f + g$ and their product fg are also continuous at x_0. If $g(x_0)$ is not zero, then the quotient f/g is also continuous at x_0.

Theorem 2.2 follows from the limit theorem (Theorem 2.1) of Section 2.4 and from the corresponding results for one-sided limits.

Continuity of polynomials and rational functions

The statements $\lim_{x \to x_0} x = x_0$ and $\lim_{x \to x_0} C = C$ are valid for any number x_0 and any constant C. Therefore, the function x and all constant functions are continuous for all x. These facts along with Theorem 2.2 give the following result.

Theorem 2.3   All polynomials are continuous for all x. A rational function is continuous wherever it is defined, that is, at all points except where one or more of its denominators is zero.

Proof.   Polynomials are continuous for all x because they can be constructed by taking products and sums of the function x and constant functions. A rational function is continuous except where one of its denominators is zero because it is formed from polynomials by taking sums and quotients.   Q.E.D.

EXAMPLE 2   At what values of x are the functions (a) $x^3 - 6x$, (b) $\dfrac{1}{x} + \dfrac{1}{(x-2)(x-3)}$, and (c) $\dfrac{x^5 - 6x^3 + 10}{x^2 + 1}$ continuous?

Solution.   The polynomial (a) is continuous for all x. The rational function (b) is continuous except at the points $x = 0, x = 2$, and $x = 3$ where its denominators are zero. The denominator of the rational function (c) is never zero, so the function is continuous for all x.   □

Exercises

†*Answer provided.*
‡*Outline of solution provided.*

In each of Exercises 1 through 4 sketch the graph of the function and list the values of x in its domain where it is discontinuous.

1.† $g(x) = \begin{cases} 5 - x^2 & \text{for } -2 \le x \le 2 \\ 2 & \text{for } \quad 2 < x < 5 \\ 1 & \text{for } \qquad x = 5 \end{cases}$

2. $G(x) = \begin{cases} x + 4 & \text{for } -6 \le x < -2 \\ x & \text{for } -2 \le x < 2 \\ x - 4 & \text{for } \quad 2 \le x \le 6 \end{cases}$

3.† $Z(x) = \begin{cases} -\frac{1}{8}x^2 & \text{for } \qquad x \le -2 \\ \frac{1}{4}x & \text{for } -2 < x < 2 \\ \frac{1}{8}x^2 & \text{for } \qquad x \ge 2 \end{cases}$

4. $V(x) = \begin{cases} x + 2 & \text{for } \qquad x \le -1 \\ x^3 & \text{for } -1 < x \le 1 \\ x & \text{for } \qquad x > 1 \end{cases}$

In Exercises 5 through 10 describe where the functions are continuous.

5.† $\dfrac{1 + 2/x}{1 - 2/x}$

6. $\dfrac{x^2 - 25}{x^2 + 25}$

7.† $1 + \dfrac{1}{x} + \dfrac{1}{x + 1} - \dfrac{1}{x - 1}$

8. $\dfrac{x^2 + 25}{x^2 - 25}$

9.† $1 + 2x^2 + 3x^3 + 4x^4 + 5x^5 + 6x^6 + 7x^7$

10. $\dfrac{x + 2}{(x + 1)(x - 2)}$

11.‡ Find a constant k such that the function

$$f(x) = \begin{cases} \dfrac{x^2 - 9}{x - 3} & \text{for } x \neq 3 \\ k & \text{for } x = 3 \end{cases}$$

is continuous for all x.

12.† Find a constant k such that the function

$$H(x) = \begin{cases} \dfrac{1 - \sqrt{x}}{x - 1} & \text{for } 0 \leq x < 1 \\ k & \text{for } \quad x = 1 \end{cases}$$

is continuous for $0 \leq x \leq 1$.

13. Find a constant k such that the function

$$T(x) = \begin{cases} \dfrac{x^3 + 2x^2 + 3x}{x} & \text{for } x \neq 0 \\ k & \text{for } x = 0 \end{cases}$$

is continuous for all x.

14.† Find constants a and b such that the function

$$g(x) = \begin{cases} x^3 & \text{for} & x < -1 \\ ax + b & \text{for } -1 \leq x < 1 \\ x^2 + 2 & \text{for} & x \geq 1 \end{cases}$$

is continuous for all x.

15. Find constants a and b such that the function

$$W(x) = \begin{cases} -x & \text{for } -3 \leq x \leq -2 \\ ax^2 + b & \text{for } -2 < x < 0 \\ 6 & \text{for} & x = 0 \end{cases}$$

is continuous for $-3 \leq x \leq 0$.

In Exercises 16 through 19 describe where the functions are continuous.

16.† $f(x) = \begin{cases} x^3 - x^4 + 2 & \text{for } x \leq -1 \\ 2x^5 + x + 3 & \text{for } -1 < x < 1 \\ 7 & \text{for } x \geq 1 \end{cases}$

17.† $f(x) = \begin{cases} 1/x & \text{for } x < 0 \\ 6/(x + 1) & \text{for } 0 \leq x \leq 2 \\ 2 & \text{for } x > 2 \end{cases}$

18.† $f(x) = \dfrac{x^2 - 1}{|x^2 - 1|}$

19.† $f(x) = \begin{cases} x^3/|x| & \text{for } x < 0 \\ x^6 + 6x & \text{for } 0 \leq x < 1 \\ 7x & \text{for } x \geq 1 \end{cases}$

2-8 THE INTERMEDIATE VALUE AND EXTREME VALUE THEOREMS

The Intermediate Value Theorem

We occasionally want to determine whether an equation of the form

(1) $f(x) = L$

has a solution x in a given interval. The INTERMEDIATE VALUE THEOREM implies that there is such a solution, provided the function f is continuous in the interval and the number L lies between two values of the function in the interval:

Theorem 2.4   If a function f is continuous in an interval that includes the points a and b ($a < b$), then for each number L between $f(a)$ and $f(b)$ there is at least one number x in the interval $a \leq x \leq b$ such that $f(x) = L$.

The proof of the Intermediate Value Theorem is presented in advanced courses that deal with the logical foundations of calculus (see R. Courant, [1] *Differential and Integral Calculus*). It involves the $\epsilon\delta$-definition of continuity, which we discuss in Section 2.9, as well as the axioms of the real numbers, which we do not study in this textbook.*

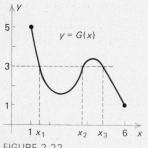

FIGURE 2.22

The idea behind the Intermediate Value Theorem is that the continuous function f cannot "jump" the value L as its value changes from $f(a)$ at $x = a$ to $f(b)$ at $x = b$. The function G, whose graph is shown in Figure 2.22, for example, is continuous in the closed interval $1 \leq x \leq 6$. The number 3 is between the values $G(1) = 5$ and $G(6) = 1$ of G at 1 and 6. According to the Intermediate Value Theorem, there should be at least one value of x in the interval such that $G(x) = 3$. The graph in Figure 2.22 shows that there are three such values of x, labeled x_1, x_2, and x_3.

The Intermediate Value Theorem is not valid if we drop the condition in the hypothesis that the function be continuous. This is shown by the function H of Figure 2.23. The function H has the same values at 1 and at 6 as the function G of Figure 2.22, but there are no numbers x in the interval $1 \leq x \leq 6$ that are solutions of the equation $H(x) = 3$. The discontinuity of the function H allows it to "jump" the value 3 as it goes from the value $H(1) = 5$ to the value $H(6) = 1$.

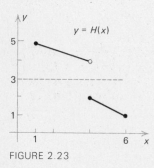

FIGURE 2.23

EXAMPLE 1   Use the Intermediate Value Theorem to show that there must be at least one value of x in the interval $0 \leq x \leq 1$ that is a solution of the equation

(2) $x^5 + 4x^2 - x = 3.$

Except in Chapter 18 where we use the least upper bound axiom.

Solution.   By Theorem 2.3 the polynomial $x^5 + 4x^2 - x$ is continuous for all x. Its value at $x = 0$ is 0, and its value at $x = 1$ is 4. The number 3 is between 0 and 4, so by the Intermediate Value Theorem, there is a solution of equation (2) in the interval $0 \leq x \leq 1$.   □

In the next example we apply the Intermediate Value Theorem to a narrative problem.

EXAMPLE 2   A boy weighed $\frac{1}{2}$ stone (7 pounds) at birth and 5 stones (70 pounds) when he was 8 years old. Was there a time in his life when his weight in stones was equal to his age in years?

Solution.   We let $f(t)$ denote the boy's weight in stones when he was t years old. We consider f to be a continuous function of the time t. Figure 2.24 shows the graph of a possible function f along with the graph of the function t, which is the line $w = t$.

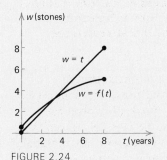

FIGURE 2.24

Because the boy weighed $\frac{1}{2}$ stone at birth, $f(0)$ is $\frac{1}{2}$ and the graph of f is above the line $w = t$ at $t = 0$. Because the boy weighed 5 stones at age 8, $f(8)$ is 5 and the graph of f is below the line $w = t$ at $t = 8$.

The boy's weight in stones was equal to his age in years at any time t where the graph of f intersects the line $w = t$. Our geometric intuition tells us that this must occur at some time between $t = 0$ and $t = 8$ because the functions f and t are continuous and their graphs cross and are not "broken." To use the Intermediate Value Theorem, we set $g(t) = f(t) - t$. The function g is continuous and $g(0) = f(0) - 0 = \frac{1}{2}$, whereas $g(8) = f(8) - 8 = 5 - 8 = -3$. The number 0 is between these values of f, so there is a t between 0 and 8 such that $g(t) = 0$. At this t, $f(t) = t$ and the boy's weight was equal to his age.   □

EXAMPLE 3   In Example 2, was the boy's weight in pounds necessarily ever equal to his age in years?

Solution.   No. His weight in pounds was greater than his age at birth and at age 8. Consequently, it was possible (and, in fact, likely) that his weight in pounds was always greater than his age in years.   □

The Extreme Value Theorem

Definition 2.6   A value $f(x_0)$ of a function f is the maximum value of f in an interval if x_0 is in the interval and $f(x) \leq f(x_0)$ for all other numbers x in that interval. The value $f(x_0)$ is the minimum value of f in the interval if $f(x) \geq f(x_0)$ for all other numbers x in the interval.

The next result is known as the EXTREME VALUE THEOREM. It gives conditions under which we can be assured that a function has a maximum and minimum value in an interval.

Theorem 2.5   **If a function f is continuous in a finite closed interval $a \leq x \leq b$, then it has a maximum and a minimum value in that interval.**

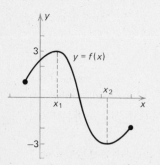

FIGURE 2.25   The maximum value of f for $-1 \leq x \leq 6$ is $f(x_1) = 3$. The minimum value is $f(x_2) = -3$.

The proof of the Extreme Value Theorem, like the proof of the Intermediate Value Theorem, is dealt with in advanced courses.

The function f whose graph is sketched in Figure 2.25 is continuous in the finite closed interval $-1 \leq x \leq 6$. According to the Extreme Value Theorem, it has a maximum and a minimum value in that interval. The maximum value of the function is 3 and occurs at the point labeled x_1 in the sketch. The minimum value of the function is -3 and occurs at the point labeled x_2.

If the hypotheses of the Extreme Value Theorem are not satisfied, that is, if the function is not continuous in the entire interval or if the interval is not closed or is infinite, then the function may or may not have a maximum or a minimum value in the interval, as illustrated in the next examples.

EXAMPLE 4   Does the function $1/x$ have a maximum and a minimum value in the interval $0 < x \leq 3$? If not, explain which of the hypotheses of the Extreme Value Theorem are not satisfied.

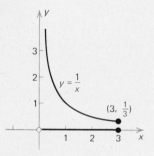

FIGURE 2.26

Solution.   The portion of the graph of the function $1/x$ for $0 < x \leq 3$ is sketched in Figure 2.26, along with the half-open interval $0 < x \leq 3$ on the x-axis.

The minimum value of $1/x$ for $0 < x \leq 3$ is $\frac{1}{3}$. It is the y-coordinate of the lowest point $(3, \frac{1}{3})$ on that portion of the graph.

The portion of the graph for $0 < x \leq 3$ has no highest point and the function $1/x$ has *no maximum value* in that interval. In fact, $(1/x) \to \infty$ as $x \to 0^+$. The function $1/x$ is continuous in the interval $0 < x \leq 3$, and that interval is finite. The hypotheses of the Extreme Value Theorem are not satisfied because that interval is not closed.   □

EXAMPLE 5   Does the function k defined by

$$k(x) = \begin{cases} 4 - \frac{1}{2}x^2 & \text{for } -2 \leq x < 4 \\ 2 & \text{for } 4 \leq x \leq 7 \end{cases}$$

have a maximum and a minimum value in the interval $-2 \leq x \leq 7$? If not, describe which of the hypotheses in the Extreme Value Theorem are not satisfied.

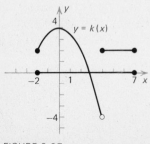

FIGURE 2.27

Solution.   The graph of the function k and the closed interval $-2 \leq x \leq 7$ are shown in Figure 2.27. The highest point on that portion of the graph is the point at $y = 4$ on the y-axis. Therefore, the maximum value of k in the interval is $k(0) = 4$.

A quick glance might suggest that the minimum value of the function is the y-coordinate -4 of the circle at $(4, -4)$. The number -4, however, *is not a value of the function.* The limit of $k(x)$ as x

approaches 4 from the left is -4, but the value of k at $x = 4$ is 2. The function k *has no minimum value* in the interval $-2 \leq x \leq 7$.

The interval $-2 \leq x \leq 7$ is closed and finite. The hypotheses of the Extreme Value Theorem are not satisfied, because the function k is discontinuous at the point $x = 4$ in the interval. $\square$

Exercises

†*Answer provided.*
‡*Outline of solution provided.*

1.† Use the Intermediate Value Theorem to show that there is at least one solution x of the equation $(x^3 - 2x)/(x + 3) = -3$ with $-2 \leq x \leq 2$.

2. Use the Intermediate Value Theorem to show that the equation $x^3 + x^2 - 4x = 15$ has a solution x with $-3 \leq x \leq 3$.

3.† A sprinter starts from rest and runs around a circular track. He stops when he gets back to the starting blocks. Show that at least once during that lap he must have had the same speed at diametrically opposite points.

4. A rock climber starts to climb a mountain at 0800 on Saturday and gets to the top at 1600 that afternoon. He camps on top and climbs back down on Sunday, starting at 0800 and getting back to his original starting point at 1600. Show that at some time of day on Sunday he was at the same elevation as he was at that time on Saturday.

5.† Use the Intermediate Value Theorem to show that the equation $ax^3 + bx^2 + cx + d = 0$ has at least one solution x for any choice of the constants $a, b, c,$ and d with $a \neq 0$.

In each of Exercises 6 through 9 sketch the graph of the function. Then give its maximum and minimum values, if they exist. If they do not, explain which hypotheses of the Extreme Value Theorem are not satisfied.

6.† $Q(x) = \begin{cases} 1 - x & \text{for } -4 \leq x \leq 0 \\ \dfrac{1}{x} & \text{for } \quad 0 < x \leq 4 \end{cases}$

7. $R(x) = \begin{cases} 2 & \text{for } -3 \leq x < 1 \\ \dfrac{6}{x} - 4 & \text{for } \quad\quad x \geq 1 \end{cases}$

8.† $S(x) = 5 - \dfrac{1}{2}x^2 \quad \text{for } -4 < x < 4$

9. $T(x) = \begin{cases} -\dfrac{3}{x} & \text{for } -3 \leq x < -1 \\ x & \text{for } -1 \leq x \leq 2 \end{cases}$

In Exercises 10 through 14 use a graph of the function to find   **a.** the maximum value and   **b.** the minimum value of $f(x)$ for all x in its domain (if the maximum and minimum exist).

10.† $f(x) = \begin{cases} x^3 & \text{for } -2 \leq x < 0 \\ 5 - x^2 & \text{for } \quad 0 \leq x \leq 2 \end{cases}$

11.† $f(x) = \begin{cases} 2/x & \text{for } -10 \leq x \leq -1 \\ 2 & \text{for } -1 < x \leq 0 \\ x & \text{for } \quad 0 < x \leq 5 \end{cases}$ 12.† $f(x) = 4 + x^2$

13.[†] $f(x) = \begin{cases} 2 - x^{-2} & \text{for} \quad 0 < |x| \le 1 \\ 1 & \text{for} \quad x = 0 \end{cases}$

14.[†] $f(x) = \begin{cases} -1/x & \text{for} \quad -2 \le x < -1 \\ -x & \text{for} \quad -1 \le x \le 1 \\ 1/x & \text{for} \quad 1 < x \le 2 \end{cases}$

2-9 THE $\epsilon\delta$-DEFINITIONS OF LIMITS (Optional)

In this section we present the definitions of the various types of limits whose intuitive meanings were described in Section 2.3 and 2.5. These definitions are important, not only because they provide calculus with a logically more rigorous foundation but also because they can be used to answer questions dealing with limits that cannot be resolved with less precise definitions.

Finite limits   The definition we will give for the statement

$$\lim_{x \to x_0} f(x) = L$$

can be paraphrased as follows: *The limit of $f(x)$ as x tends to x_0 is the number L if $f(x)$ can be made arbitrarily close to L by taking x sufficiently close to x_0.*

To make this statement precise, the phrase "$f(x)$ can be made arbitrarily close to L" is replaced with the phrase "for each positive number ϵ (epsilon), no matter how small, the distance $|f(x) - L|$ between $f(x)$ and L can be made less than ϵ" and the phrase "by taking x sufficiently close to x_0" is replaced with the phrase "by restricting x to satisfy the condition $0 < |x - x_0| < \delta$ for some positive number δ (delta)." The definition then requires that there be a suitable δ for each ϵ.

Definition 2.7   **Consider a function f defined in an open interval $a < x < x_0$ to the left of x_0 and in an open interval $x_0 < x < b$ to the right of x_0. The limit of $f(x)$ as x tends to x_0 is the number L if for each $\epsilon > 0$ there is a $\delta > 0$ such that**

(1) $|f(x) - L| < \epsilon$

for all x with

(2) $0 < |x - x_0| < \delta.$

One-sided finite limits have analogous definitions:

Definition 2.8   **Consider a function f defined in an open interval $a < x < x_0$ to the left of x_0. The limit of $f(x)$ as x tends to x_0 from the left is the number L if for each $\epsilon > 0$ there is a $\delta > 0$ such that**

(1) $|f(x) - L| < \epsilon$

for all x with

(3) $x_0 - \delta < x < x_0.$

Definition 2.9   Consider a function f defined in an open interval $x_0 < x < b$ to the right of x_0. The limit of $f(x)$ as x tends to x_0 from the right is the number L if for each $\epsilon > 0$ there is a $\delta > 0$ such that

(1) $|f(x) - L| < \epsilon$

for all x with

(4) $x_0 < x < x_0 + \delta.$

Verifying a definition of limit

To show that Definition 2.7, 2.8, or 2.9 is verified in a particular instance, we need to show that there is a suitable δ for each positive ϵ. The most common way of doing this is to give a *rule* that prescribes a suitable δ for each ϵ.

EXAMPLE 1   Show that, according to Definition 2.7, the limit of $2x$ as x tends to 3 is 6.

Solution.   We need a rule that provides for each $\epsilon > 0$ a number $\delta > 0$ such that

(5) $|2x - 6| < \epsilon$

for all x with

(6) $0 < |x - 3| < \delta.$

We will determine a suitable rule algebraically, and then give its geometric interpretation.

We have $|2x - 6| = |2(x - 3)| = 2|x - 3|$, which leads us to choose the δ's by the rule

(7) $\delta = \dfrac{\epsilon}{2}.$

To show that this rule provides suitable δ's, we let ϵ denote an arbitrary positive number and define δ by (7). Then for any x that satisfies condition (6), we have

$$|2x - 6| = 2|x - 3| < 2\delta = \epsilon$$

so that condition (5) is satisfied and the definition of the limit is verified. $\square$

The geometric interpretation of these considerations is indicated in Figure 2.28. We show there the graph of the function $2x$ and the δ that corresponds to a particular choice of ϵ.

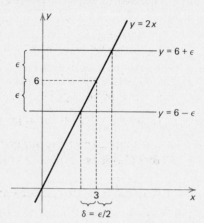

FIGURE 2.28

For the given $\epsilon > 0$ we have to find a $\delta > 0$, so that the portion of the graph with $0 < |x - 3| < \delta$ lies between the horizontal lines $y = 6 - \epsilon$ and $y = 6 + \epsilon$ at a distance of ϵ on either side of $y = 6$. As Figure 2.28 shows, choosing δ by rule (7) gives a suitable δ.

Rule (7) for choosing the δ in the solution of Example 1 gives the largest possible δ for each ϵ, but this is not necessary. Any rule that prescribes a smaller δ for each ϵ would also work. For example, we could prescribe the δ's by the rule $\delta = \epsilon/10$ in the solution of Example 1.

In most instances we can take advantage of this fact. Usually it is somewhat easier to find a rule that provides *a suitable* δ for each ϵ than it is to find *the largest possible* δ for each ϵ. This is illustrated in the next example.

EXAMPLE 2   Show that, according to Definition 2.7,

(8) $\lim_{x \to 2} (x^2 - 1) = 3.$

Solution.   We need a rule that gives for each $\epsilon > 0$ a $\delta > 0$ such that

(9) $|(x^2 - 1) - 3| < \epsilon$

for all x with

(10) $0 < |x - 2| < \delta.$

We write

(11) $|(x^2 - 1) - 3| = |x^2 - 4| = |x + 2| \, |x - 2|.$

The rule prescribing a δ for each ϵ cannot involve x, so we will employ an inequality of the form

(12) $|x + 2| \le M$

in conjunction with equation (11).

There is no constant M such that (12) is valid for all x, since $|x + 2|$ tends to ∞ as $|x|$ tends to ∞. However, we are considering the limit as x tends to 2, and we may restrict our attention to x in the interval

(13) $1 < x < 3$

centered at $x = 2$. For x in this interval we have

(14) $|x + 2| < 5$

and, therefore, by (11)

(15) $|(x^2 - 1) - 3| = |x + 2||x - 2| < 5|x - 2|.$

To make this quantity on the right of (15) less than ϵ, and to assure that (13) is satisfied so that (15) is valid, we define δ by the rule

(16) $\delta =$ the smaller of 1 and $\dfrac{\epsilon}{5}$.

To show that (16) gives a suitable δ for each ϵ, we let ϵ denote an arbitrary positive number and we define δ by (16). Then for numbers x satisfying (10) we have $|x - 2| < 1$, so that inequalities (13), (14), and (15) are valid. We also have $|x - 2| < \dfrac{\epsilon}{5}$, so that by (15)

$$|(x^2 - 1) - 3| < 5|x - 2| < 5\left(\frac{\epsilon}{5}\right) = \epsilon$$

which shows that rule (16) provides a suitable δ.

Figure 2.29 shows the situation for $\epsilon = \frac{1}{2}$. We need to choose the corresponding δ, so that the portion of the graph of $x^2 - 1$ for $0 < |x - 2| < \delta$ lies between the lines $y = \frac{5}{2}$ and $y = \frac{7}{2}$, one half unit above and below $y = 3$. Rule (16) gives $\delta = \frac{1}{10}$ for this ϵ, and as Figure 2.29 shows this is a slightly smaller δ than is necessary (see also Exercise 33). $\square$

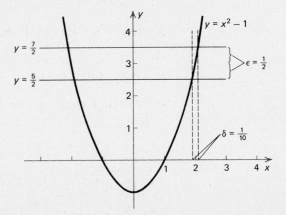

FIGURE 2.29

Infinite limits

The definition that we will give for the statement $\lim\limits_{x \to x_0} f(x) = \infty$ may be paraphrased as follows: *The limit of $f(x)$ as x tends to x_0 is ∞ if $f(x)$ can be made arbitrarily large (positive) by taking x sufficiently close to x_0.*

To obtain the definition, we replace the phrase, "$f(x)$ can be made arbitrarily large," by the phrase, "for each positive number Y, no matter how large, $f(x)$ can be made greater than Y." We then obtain the following definition.

Definition 2.10   **The limit of $f(x)$ as x tends to x_0 is ∞ if for each positive number Y there is a positive number δ such that**

(17) $f(x) > Y$

for all x with

(18) $0 < |x - x_0| < \delta.$

EXAMPLE 3   Show that, according to Definition 2.10,

(19) $\lim\limits_{x \to 0} \dfrac{1}{x^2} = \infty.$

Solution.   We need a rule that prescribes for each $Y > 0$ a $\delta > 0$ such that

(20) $\dfrac{1}{x^2} > Y$

for all x with

(21) $0 < |x| < \delta.$

Solving inequality (20) for x gives first $x^2 < (1/Y)$ and then $|x| < \sqrt{1/Y}$; therefore, we may take

(22) $\delta = \sqrt{1/Y}.$

To show that this choice of δ is satisfactory, we let Y denote an arbitrary positive number; we define δ by (22). Then condition (21) reads $0 < |x| < \sqrt{1/Y}$, which implies that $x^2 < (1/Y)$ and hence that (20) is satisfied. $\square$

There are a variety of other types of infinite limits. We will state the definition of one here and leave the formulations of the others as exercises.

Definition 2.11   **The limit of $f(x)$ is $-\infty$ as x tends to ∞ if for each positive number Y there is a positive number X such that**

(23) $f(x) < -Y$

for all x with

(24) $x > X.$

EXAMPLE 4   Show that, according to Definition 2.11, $\lim_{x \to \infty} [-x^2 + x]$ $= -\infty$.

Solution.   We need a rule that prescribes for each $Y > 0$ a number $X > 0$ such that

(25) $-x^2 + x < -Y$

for all x with

(26) $x > X$.

We have for $x \neq 0$

(27) $-x^2 + x = -x^2\left(1 - \dfrac{1}{x}\right).$

The factor $1 - (1/x)$ in the last expression tends to 1 as x tends to ∞ and has no effect on the limit; to show this, we notice that for $x > 2$, we have $1/x < 1/2$ and $1 - (1/x) > 1/2$. Therefore,

(28) $x^2\left(1 - \dfrac{1}{x}\right) > \dfrac{1}{2}x^2$ for $x > 2$.

Solving the inequality $\frac{1}{2}x^2 > Y$ for (positive) x gives the inequality $x > \sqrt{2Y}$. We are led to define X by the rule

(29) $X =$ The larger of 2 and $\sqrt{2Y}$.

To show that this rule provides a suitable X for each Y, we let Y denote an arbitrary positive number. We define X by (29). Then for $x > X$, we have $x > 2$ so that inequality (28) is valid. Furthermore, we have $x > \sqrt{2Y}$ so that $\frac{1}{2}x^2 > Y$ and

$$x^2 - x = x^2\left(1 - \dfrac{1}{x}\right) > \dfrac{1}{2}x^2 > Y$$

which gives us inequality (25) and verifies Definition 2.11.   □

Continuity   Definition 2.7, for a two-sided finite limit, leads to the following definition of continuity of a function.

Definition 2.12   **Suppose that the domain of a function f includes an open interval containing the point x_0. The function f is continuous at x_0 if for each $\epsilon > 0$ there is a $\delta > 0$ such that**

(30) $|f(x) - f(x_0)| < \epsilon$

for all x with

(31) $|x - x_0| < \delta.$

Limit theorems   We can now use Definition 2.7 for a finite two-sided limit to establish Theorem 2.1 of Section 2.4. We restate that theorem here.

Theorem 2.1   **If both of the limits $\lim\limits_{x \to x_0} f(x) = L$ and $\lim\limits_{x \to x_0} g(x) = M$ exist, then**

(32) $\lim\limits_{x \to x_0} [f(x) + g(x)] = \left[\lim\limits_{x \to x_0} f(x)\right] + \left[\lim\limits_{x \to x_0} g(x)\right] = L + M$

and

(33) $\lim\limits_{x \to x_0} [f(x)g(x)] = \left[\lim\limits_{x \to x_0} f(x)\right]\left[\lim\limits_{x \to x_0} g(x)\right] = LM.$

If the limit M of $g(x)$ is not zero, then we also have

(34) $\lim\limits_{x \to x_0} \dfrac{f(x)}{g(x)} = \dfrac{\lim\limits_{x \to x_0} f(x)}{\lim\limits_{x \to x_0} g(x)} = \dfrac{L}{M}.$

Proof of (32).   Let ϵ be an arbitrary positive number. Because the limit of $f(x)$ is L and the limit of $g(x)$ is M, there are positive numbers δ_1 and δ_2 such that

(35) $|f(x) - L| < \dfrac{\epsilon}{2}$ for $0 < |x - x_0| < \delta_1$

(36) $|g(x) - M| < \dfrac{\epsilon}{2}$ for $0 < |x - x_0| < \delta_2.$

Let δ be the smaller of δ_1 and δ_2. Then for $0 < |x - x_0| < \delta$ we have by (35) and (36)

$$|[f(x) + g(x)] - [L + M]| = |[f(x) - L] + [g(x) - M]|$$
$$\leq |f(x) - L| + |g(x) - M| < \dfrac{\epsilon}{2} + \dfrac{\epsilon}{2} = \epsilon$$

which gives us statement (32).   Q.E.D.

Proof of (33).   Because the limit of $f(x)$ is L, there is a positive number δ_1 such that $|f(x) - L| < 1$ for $0 < |x - x_0| < \delta_1$, which implies that

(37) $|f(x)| \leq |L| + 1$ for $0 < |x - x_0| < \delta_1.$

Set $C = |L| + 1 + |M|$. Let ϵ be an arbitrary positive number. Because the limit of $f(x)$ is L and the limit of $g(x)$ is M, there are positive numbers δ_2 and δ_3 such that

(38) $|f(x) - L| < \dfrac{\epsilon}{2C}$ for $0 < |x - x_0| < \delta_2$

(39) $|g(x) - M| < \dfrac{\epsilon}{2C}$ for $0 < |x - x_0| < \delta_3.$

Let δ be the smallest of the numbers δ_1, δ_2, and δ_3. Then for $0 < |x - x_0| < \delta$ we have by (37), (38), and (39)

$$\begin{aligned} |f(x)g(x) - LM| &= |f(x)g(x) - f(x)M + f(x)M - LM| \\ &= |f(x)[g(x) - M] + [f(x) - L]M| \\ &\leq |f(x)||g(x) - M| + |f(x) - L||M| \\ &\leq C|g(x) - M| + C|f(x) - L| \\ &< C\left(\frac{\epsilon}{2C}\right) + C\left(\frac{\epsilon}{2C}\right) = \epsilon \end{aligned}$$

which gives us statement (33).   Q.E.D.

We leave the proof of statement (34) as Exercise 27.

Exercises

†*Answer provided.*
‡*Outline of solution provided.*

In Exercises 1 through 6 formulate definitions of the indicated limits.

1.† $\displaystyle\lim_{x \to x_0^+} f(x) = \infty$

2. $\displaystyle\lim_{x \to x_0^-} f(x) = \infty$

3.† $\displaystyle\lim_{x \to x_0} f(x) = -\infty$

4. $\displaystyle\lim_{x \to -\infty} f(x) = \infty$

5.† $\displaystyle\lim_{x \to \infty} f(x) = L$

6. $\displaystyle\lim_{x \to -\infty} f(x) = L$

In Exercises 7 through 17 verify the appropriate definitions of limit.

7.† $\displaystyle\lim_{x \to 4} \sqrt{x} = 2$

8.† $\displaystyle\lim_{x \to 5} \frac{10}{x} = 2$

9. $\displaystyle\lim_{x \to 0} \frac{6}{x + 2} = 3$

10. $\displaystyle\lim_{x \to 0} x^{11} = 0$

11.† $\displaystyle\lim_{x \to 3} \frac{x^2 - 9}{x - 3} = 6$

12. $\displaystyle\lim_{x \to 1} \frac{\sqrt{x} - 1}{x - 1} = \frac{1}{2}$

13.† $\displaystyle\lim_{x \to 0^+} \frac{1}{\sqrt{x}} = \infty$

14. $\displaystyle\lim_{x \to 0^-} \frac{1}{x} = -\infty$

15.† $\displaystyle\lim_{x \to \infty} \frac{4}{x + 7} = 0$

16. $\displaystyle\lim_{x \to -\infty} x^3 = -\infty$

17. $\displaystyle\lim_{x \to \infty} \frac{6x}{2x - 1} = 3$

18. Why does the first inequality appear in the condition $0 < |x - x_0| < \delta$ in Definition 2.7 of a limit but does not appear in the condition $|x - x_0| < \delta$ in Definition 2.12 of continuity?

19.† Show that according to Definition 2.12, the function x^2 is continuous at $x = -4$.

20. Show that according to Definition 2.12, the function $(x - 1)/(x + 1)$ is continuous at $x = 1$.

21.‡ Figure 2.30 shows the graph of a function f such that $\lim_{x \to 2^-} f(x) = 1$.

   a. Which of the following values of δ are suitable for $\epsilon = \frac{1}{2}$ in the definition of this limit: $\frac{1}{4}, \frac{1}{2}, \frac{3}{4}, 1, 2, 3$?

   b. For which of the following values of ϵ is $\delta = 2$ suitable in the definition of the limit: $\frac{1}{4}, \frac{1}{2}, 1, \frac{3}{2}, 2$?

22. Figure 2.31 shows the graph of a function g such that $\lim_{x \to -1/2} g(x) = \frac{1}{2}$.

   a. Which of the following values of δ are suitable for $\epsilon = \frac{1}{8}$ in the definition of this limit: $\frac{1}{64}, \frac{1}{8}, \frac{1}{4}, \frac{1}{2}, \frac{3}{4}$?

   b. For which of the following values of ϵ is $\delta = \frac{1}{4}$ suitable in this definition: $\frac{1}{64}, \frac{1}{8}, \frac{1}{4}, \frac{1}{2}, \frac{3}{4}$?

23.‡ Figure 2.32 shows the graph of a function h such that the limit of $h(x)$ as x tends to 1 does not exist, which means that for each potential limit L, there is an $\epsilon > 0$ for which no suitable δ can be found. Give a value of $\epsilon > 0$ that has this property for all numbers L.

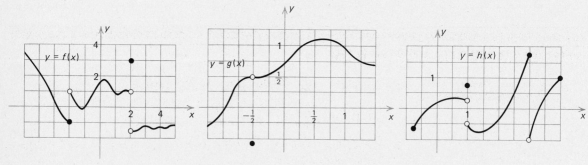

FIGURE 2.30   FIGURE 2.31   FIGURE 2.32

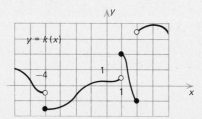

FIGURE 2.33

24. Figure 2.33 shows the graph of a function k that is discontinuous at $x = -4$. Give a value of $\epsilon > 0$ for which there is no suitable $\delta > 0$ in the definition of the statement, "k is continuous at $x = -4$."

25.† Verify the definition of the limit $\lim_{x \to 2^-} f(x) = 5$ for the function

$$f(x) = \begin{cases} 3x - 1 & \text{for } 0 < x < 2 \\ 100 & \text{for } \quad x = 2 \\ -x & \text{for } \quad x > 2. \end{cases}$$

26. Verify the definition of the limit $\lim_{x \to 1} g(x) = 7$ for the function

$$g(x) = \begin{cases} 6x + 1 & \text{for } -10 \le x < 1 \\ 0 & \text{for } \quad x = 1 \\ -x + 8 & \text{for } \quad 1 < x \le 5. \end{cases}$$

27. Prove statement (34).

28. Prove that, according to the definitions of this section, if $f(x)$ has a finite limit as x tends to x_0 while $g(x)$ tends to ∞, then $\dfrac{f(x)}{g(x)}$ tends to zero.

29. Prove statement (33) for limits as x tends to ∞.

30.[†] Show that, according to Definition 2.9, the statement $\lim\limits_{x\to 3^+} x^2 = 9.00001$ is false.

31.[†] Show that, according to Definition 2.7, the function $x/|x|$ does not have a limit as x tends to zero.

32. Show that, according to Definition 2.9, the function $\sin(1/x)$ does not have a limit as x tends to zero from the right.

33.[†] What is the largest possible δ for $\epsilon = \frac{1}{2}$ in Example 2?

MISCELLANEOUS EXERCISES

[†]*Answer provided.*
[‡]*Outline of solution provided.*

1.[†] Compute   **a.** $W(30)$ **b.** $W(-30)$ where
$$W(x) = \begin{cases} \sqrt[3]{2x-4} & \text{for } x < 0 \\ \sqrt[3]{4+2x} & \text{for } x \geq 0. \end{cases}$$

2. A function Q is defined for $-1 < x < 1$ by $Q(x) = (1-x^2)^{-1/2}$. What is the value of Q at $x = 3/5$?

In Exercises 3 through 8 describe the domains of the functions.

3.[†] $\dfrac{1 + (1/x)}{1 - (1/x)}$

4. $\sqrt{1 + 5x^2}$

5.[†] $\sqrt{x^2 - 9}$

6. $(x - 4)^{-1/2}$

7.[†] $(9 - x^2)^{-1/3}$

8. $\dfrac{x(x+3)(x-4)}{(x-5)(x+6)}$

9.[†] **a.** Express the distance from the point $(5, -3)$ to a point on the hyperbola $xy = 6$ as a function of the x-coordinate of the point on the graph.   **b.** What is the domain of that function?

10. Give the length D of the diagonals of a square as a function of its area A.

11.[†] Express the area A of an isosceles right triangle as a function of its perimeter P.

12. Give the area of a $30°-60°$ right triangle as a function of the length h of its hypotenuse.

13.[†] Express the surface area A of a sphere as a function of its volume V.

14. The surface area of a box with a square base and top is 1600 square inches. Express its volume V as a function of its width w.

15.[†] It costs 5 cents to manufacture one toy soldier, and a shipment of N soldiers ($0 < N \le 1000$) sells for $10 - \dfrac{N}{200}$ cents per soldier. Express the profit P on a shipment as a function of N.

16.[†] A child crawls at the rate of 3 miles per hour for x miles and at the rate of 2 miles per hour for $2x$ miles. How long does the crawl take?

17.[†] A baseball player runs around the bases at the constant speed of 30 feet per second. He starts from home plate at time $t = 0$ (seconds). Express his distance D from home plate as a function of the time t. The baseball diamond is a 90-foot square.

In Exercises 18 through 19 sketch the graphs of the functions.

18.[†] $S(x) = \begin{cases} \dfrac{3}{x^2} & \text{for} & x \le -1 \\[2mm] x^2 & \text{for} -1 < x < 1 \\[2mm] \dfrac{3}{x^2} & \text{for} & x \ge 1 \end{cases}$

19. $T(x) = \begin{cases} 4 - x^2 & \text{for} & x < -1 \\ -x^2 & \text{for} -1 \le x \le 1 \\ 4 - x^2 & \text{for} & x > 1 \end{cases}$

20.[†] It takes 1 calorie of heat to raise the temperature of 1 gram of water by $1°C$, and $\frac{1}{2}$ calorie of heat to raise the temperature of 1 gram of ice or steam by $1°C$. It takes 80 calories to melt 1 gram of ice at $0°C$ into 1 gram of water at $0°C$ and 539 calories to boil 1 gram of water at $100°C$ into 1 gram of steam at $100°C$.

Suppose that you start with 1 gram of ice at $-100°C$ and heat it. Let $W(Q)$ be the temperature of that gram of H_2O after it has absorbed Q calories of heat. Give a formula for the values of the function W and sketch its graph.

21. When we say that an object weighs 1 pound, we mean that the force of the earth's gravity on the object is 1 pound when it is on the surface of the earth. If we let $P(r)$ (pounds) be the force of the earth's gravity on such an object when it is r miles from the center of the earth, then

$$P(r) = \begin{cases} \dfrac{r}{4000} & \text{for } 0 \le r \le 4000 \\[3mm] \left[\dfrac{4000}{r}\right]^2 & \text{for } \quad r \ge 4000. \end{cases}$$

(Since the radius of the earth is 4000 miles, this formula gives the correct value $P(4000) = 1$ for the weight of the object on the surface of the earth.)   **a.** What is the force of the earth's gravity on the object if it is on the moon, 240,000 miles from the center of the earth?   **b.** What is the force of the earth's gravity on the object if it is 2000 miles from the center of the earth? How high *above* the earth must the object be to be subjected to that same force?   **c.** Sketch the graph of the function P in an rF-plane.

In Exercises 22 through 47 find the indicated limits.

22.[‡] **a.** $\lim\limits_{x \to 1^-} M(x)$ **b.** $\lim\limits_{x \to 1^+} M(x)$ **c.** $\lim\limits_{x \to 1} M(x)$ for

$$M(x) = \begin{cases} 3x & \text{for } x \le 1 \\ x^2 & \text{for } x > 1 \end{cases}$$

23.[†] **a.** $\lim\limits_{x \to -10^+} N(x)$ **b.** $\lim\limits_{x \to -10^-} N(x)$ **c.** $\lim\limits_{x \to -10} N(x)$ for

$$N(x) = \begin{cases} x^2 & \text{for} & x < -10 \\ -10x & \text{for} -10 < x < 15 \\ 7 & \text{for} & x > 15 \end{cases}$$

24. **a.** $\lim\limits_{x \to 0} L(x)$ **b.** $\lim\limits_{x \to 10} L(x)$ **c.** $\lim\limits_{x \to 100} L(x)$ for

$$L(x) = \begin{cases} x^2 + x + 1 & \text{for } x \le 10 \\ 11x & \text{for } x > 10 \end{cases}$$

25. **a.** $\lim\limits_{x \to -1} W(x)$ **b.** $\lim\limits_{x \to 0} W(x)$ **c.** $\lim\limits_{x \to 1} W(x)$ for

$$W(x) = \begin{cases} x^3 & \text{for } -10 < x < -1 \\ x^4 & \text{for } -1 < x < 1 \\ x^5 & \text{for } 1 < x < 10 \end{cases}$$

26.[†] $\lim\limits_{x \to 2} \dfrac{1 - (2/x)}{x^2 - 4}$ **27.** $\lim\limits_{x \to -2} \dfrac{1 + (2/x)}{x^2 + 4}$

28.[†] $\lim\limits_{t \to 0} \dfrac{(2 + t)^2 - 4}{t^3 + 3t}$ **29.** $\lim\limits_{y \to 1} \dfrac{y^2 - 2y + 1}{1 - y^2}$

30.[†] $\lim\limits_{y \to 4^+} \dfrac{\sqrt{y} - 2}{16 - y^2}$ **31.** $\lim\limits_{t \to 0} \dfrac{(t + 1)^2 - 1}{\sqrt{t + 4} - 2}$

32.[‡] $\lim\limits_{t \to 2} \dfrac{t^2 + t - 6}{t^2 + 3t - 10}$ **33.** $\lim\limits_{z \to 1} \dfrac{z^3 - 1}{z^2 - 1}$

34.[†] $\lim\limits_{x \to 2} \dfrac{x^2 - 4}{x^3 - 3x - 2}$ **35.** $\lim\limits_{x \to 2} \dfrac{x^2 - 4x + 4}{x^2 + 3x - 10}$

36.[‡] $\lim\limits_{x \to 1} \dfrac{x^4 + x^2 - 2}{x - 1}$ **37.** $\lim\limits_{x \to 2} \dfrac{x^4 - 16}{x - 2}$

38.[‡] $\lim\limits_{x \to 0} \dfrac{x^2 + 4}{x^2}$ **39.** $\lim\limits_{x \to 0^+} \dfrac{x^3 - 5}{2x}$

40.[‡] $\lim\limits_{x \to 2^-} \dfrac{x}{x - 2}$ **41.** $\lim\limits_{x \to 2^+} \dfrac{x^2}{x - 2}$

42.[†] $\lim\limits_{x \to \infty} [x^5 - x^4]$ **43.** $\lim\limits_{x \to -\infty} [x^5 - x^4]$

44.[†] $\lim\limits_{x \to \infty} \left[\dfrac{x^2 - (x + 1)^2}{x} \right]$ **45.** $\lim\limits_{x \to \infty} \left[\dfrac{x + (1/x)}{x^2} \right]$

46.[†] $\lim\limits_{x \to \infty} \dfrac{(2x + 3)(4x + 5)}{(6x + 7)(8x + 9)}$ **47.** $\lim\limits_{x \to -\infty} \dfrac{(3x + 1)^3}{(1 - 2x)^3}$

48.[†] Sketch the graph of $\dfrac{(x-1)(x-2)(x+1)(x+2)}{x^2-1}$.

49. Sketch the graph of $\dfrac{x^2-1}{x^4-x^2}$.

50.[†] List the points of discontinuity of the function

$$f(x) = \begin{cases} x^3 - 6x & \text{for} & x < 1 \\ -4 - x^2 & \text{for } 1 \le x \le 10 \\ 6x^2 + 46 & \text{for} & x > 10. \end{cases}$$

51. List the points of discontinuity of the function

$$g(x) = \begin{cases} \dfrac{x}{x+1} & \text{for} & x < -1 \\ x^5 & \text{for } -1 \le x < 3 \\ 9x^3 & \text{for} & x \ge 3. \end{cases}$$

52.[†] Figure 2.34 shows the graph of the function that gives the velocity, v (centimeters per second), of small water waves as a function of the distance, L (centimeters) between wave crests. What are the approximate values of v and L for the slowest waves? (The tiny waves are propelled primarily by surface tension. The smaller they are, the faster they move. The larger waves are propelled primarily by gravity. The larger they are, the faster they travel.)

53. Figure 2.35 shows the graph of the function that gives the percentage of chlorophyl production in a leaf from June to December as a function of the number of months, t, after June 1. When does the leaf have full chlorophyl production?

54. Figure 2.36 shows the graph of the function that gives the probable average weight W (kilograms) of the monsters in Loch Ness as a function of their average number P. If there are fifteen monsters in Loch Ness, what is their probable average weight?

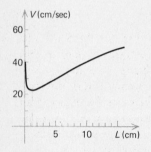

FIGURE 2.34   (Adapted from Blair Kinsman, [1] *Wind Waves*.)

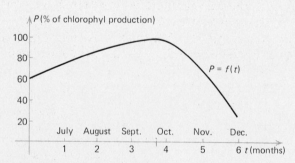

FIGURE 2.35   (Adapted from P. W. Spencer, [1] "The Turning of the Leaves," p. 63.)

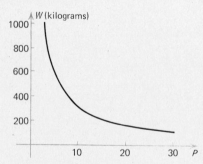

FIGURE 2.36   (Adapted from R. W. Shelton and S. R. Kerr, [1] "The Population Density of Monsters in Loch Ness," p. 549.)

CHAPTER 3
THE DERIVATIVE

SUMMARY: The derivative was developed for studying tangent lines and rates of change. In Sections 3.1 and 3.2 of this chapter we use the concept of tangent lines to introduce derivatives. The rules for differentiating x^n and for differentiating linear combinations, products, and quotients of functions are derived in Sections 3.3 and 3.4. We discuss the derivative as a rate of change in Sections 3.5 and 3.6 and the chain rule for differentiating composite functions in Section 3.7. The definitions of the sine and cosine and basic results from trigonometry dealing with these functions are given in Section 3.8. The formulas for the derivatives of $\sin x$ and $\cos x$ are derived in Section 3.9. In an optional Section 3.10 we discuss the role of mathematical models in applications of the derivative to physical problems. Section 3.11, which is also optional, contains historical notes on the derivative.

3-1 DERIVATIVES AND TANGENT LINES

We begin by describing the geometric meaning of the derivative and then giving the formal definition. As we observed in Chapter 1, the steepness of a line is determined by computing its *slope*. To measure the steepness of the graph of a function f at a point $(x_0, f(x_0))$ on it, we approximate the graph by its TANGENT LINE at that point (Figure 3.1).

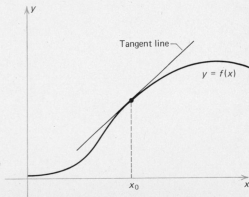

FIGURE 3.1   The derivative is the slope of the tangent line.

Because the tangent line closely approximates the graph near that point, its slope tells how steep the graph is there. *The slope of the tangent line is the derivative of f at x_0.* It is denoted $f'(x_0)$ (read "f prime at x_0").

We cannot compute the slope of the tangent line directly because we know the coordinates of only the one point $(x_0, f(x_0))$ on it. Instead, we consider the SECANT LINE through $(x_0, f(x_0))$ and through a second point $(x, f(x))$ on the graph of f (Figure 3.2).* For x close to x_0 the secant line is close to the tangent line, and as x varies, the secant line turns about the fixed point $(x_0, f(x_0))$. We find the tangent line as the *limiting position of the secant line* when x tends to x_0 (Figure 3.3).

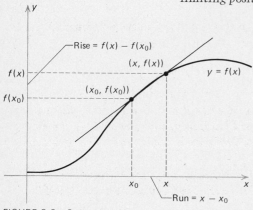

FIGURE 3.2   Secant line of slope
$$\frac{f(x) - f(x_0)}{x - x_0}$$

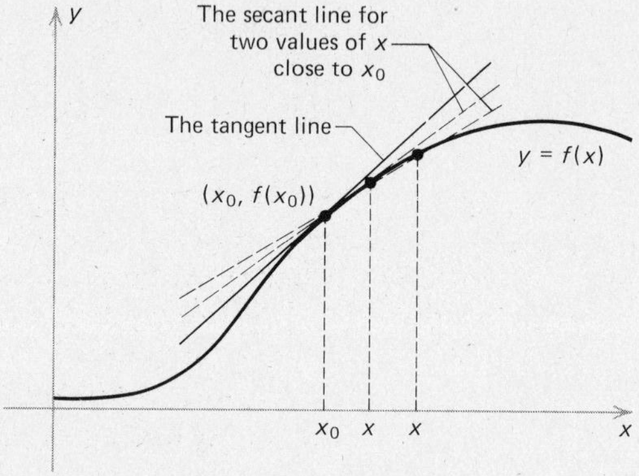

FIGURE 3.3

The rise as we go from the point $(x_0, f(x_0))$ on the secant line to the point $(x, f(x))$ is $f(x) - f(x_0)$ and the run is $x - x_0$. Therefore,

(1)   The slope of the secant line $= \dfrac{f(x) - f(x_0)}{x - x_0}$ for $x \neq x_0$.   ◄

To have the derivative be the slope of the tangent line, we define it as the limit of the slope (1) of the secant line:

Definition 3.1   For a function f defined in an open interval containing the point x_0, the derivative of f at x_0 is the limit

(2) $f'(x_0) = \lim\limits_{x \to x_0} \dfrac{f(x) - f(x_0)}{x - x_0}$ ◄

provided this limit exists and is finite. The tangent line to the graph of f at $(x_0, f(x_0))$ is the line through that point with slope $f'(x_0)$.

---

*A secant line is a line through two points on a curve. The use of the terms *tangent* and *secant* in this context comes directly from the etymology of the words (Latin: *tangere*, to touch, and *secare*, to cut) and not from any reference to trigonometry.

If the limit (2) does not exist or is infinite, we say that *f does not have a derivative at* x_0.

EXAMPLE 1   Compute the derivative of the function x^2 at $x = 1$.

Solution.   We set $f(x) = x^2$. Putting $x_0 = 1$ in definition (2) of the derivative gives us

$$(3) \quad f'(1) = \lim_{x \to 1} \frac{f(x) - f(1)}{x - 1}.$$

In the case of $f(x) = x^2$ we have $f(1) = 1^2 = 1$ and definition (3) reads

$$(4) \quad f'(1) = \lim_{x \to 1} \frac{x^2 - 1}{x - 1}.$$

Equation (4) expresses the derivative as the limit of the quotient $(x^2 - 1)/(x - 1)$. The numerator and denominator of this quotient both tend to zero as x tends to 1. Therefore, *the derivative is the limit of a quotient of quantities, both of which tend to zero.* This is the case for any derivative.

We saw in Section 2.4 that limits such as (4) can be found by factoring the numerator and canceling one of the factors with the denominator. We have

$$(5) \quad \frac{x^2 - 1}{x - 1} = \frac{(x + 1)(x - 1)}{x - 1} = x + 1.$$

Therefore,

$$f'(1) = \lim_{x \to 1} \frac{x^2 - 1}{x - 1} = \lim_{x \to 1} [x + 1] = 2. \quad \square$$

The geometric interpretation of the solution of Example 1 is indicated in Figure 3.4. The graph of the function x^2 is the parabola $y = x^2$. The derivative $f'(1) = 2$, which we computed in Example 1, is the slope of

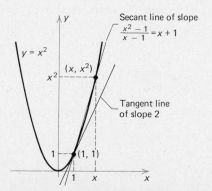

FIGURE 3.4

the tangent line to the graph at the point $(1, 1)$. Figure 3.4 shows the tangent line and the secant line through the point $(1, 1)$ and a second point (x, x^2) on the graph. Calculation (5) shows that the secant line has slope $x + 1$, and the limit of that slope as x tends to 1 is the slope 2 of the tangent line.

In the next example we have to simplify a fraction and then use a rationalization procedure to determine the limit that gives the derivative.

EXAMPLE 2   Compute $g'(5)$ where $g(x) = 1/\sqrt{x}$.

Solution.   Definition 3.1 with $x_0 = 5$ and with f replaced by g reads

$$g'(5) = \lim_{x \to 5} \frac{g(x) - g(5)}{x - 5}.$$

For positive $x \neq 5$ we have

$$\frac{g(x) - g(5)}{x - 5} = \frac{\dfrac{1}{\sqrt{x}} - \dfrac{1}{\sqrt{5}}}{x - 5} = \frac{1}{x - 5}\left(\frac{1}{\sqrt{x}} - \frac{1}{\sqrt{5}}\right)$$

$$= \frac{1}{x - 5}\left(\frac{\sqrt{5} - \sqrt{x}}{\sqrt{5x}}\right).$$

To "rationalize" the difference $\sqrt{5} - \sqrt{x}$ of square roots we multiply and divide by the sum $\sqrt{5} + \sqrt{x}$ of the square roots:

$$\frac{g(x) - g(5)}{x - 5} = \frac{1}{x - 5}\left(\frac{\sqrt{5} - \sqrt{x}}{\sqrt{5x}}\right)\left(\frac{\sqrt{5} + \sqrt{x}}{\sqrt{5} + \sqrt{x}}\right)$$

$$= \frac{(\sqrt{5})^2 - (\sqrt{x})^2}{(x - 5)(\sqrt{5x})(\sqrt{5} + \sqrt{x})}$$

$$= \frac{5 - x}{(x - 5)(\sqrt{5x})(\sqrt{5} + \sqrt{x})} = \frac{-1}{\sqrt{5x}(\sqrt{5} + \sqrt{x})}.$$

In the last step we obtained the -1 by dividing $x - 5$ into $5 - x$. Therefore,

$$g'(5) = \lim_{x \to 5} \frac{g(x) - g(5)}{x - 5} = \lim_{x \to 5} \frac{-1}{\sqrt{5x}(\sqrt{5} + \sqrt{x})}$$

$$= \frac{-1}{\sqrt{25}(\sqrt{5} + \sqrt{5})}$$

$$= \frac{-1}{10\sqrt{5}}. \quad \square$$

Equations of tangent lines

Once we can compute the derivative $f'(x_0)$, we can give an equation for the tangent line at $(x_0, f(x_0))$ on the graph of f. Its slope is the derivative, so by the point-slope formula (formula (2) of Section 1.6) the tangent line has the equation

$$(6) \quad y - f(x_0) = f'(x_0)(x - x_0). \quad \blacktriangleleft$$

EXAMPLE 3   Give an equation for the tangent line to the parabola $y = x^2$ at $x = 1$.

Solution.   When we speak of the tangent line at $x = 1$, we mean the tangent line at the point on the curve where $x = 1$. The parabola $y = x^2$ is the graph of the function $f(x) = x^2$, and we saw in Example 1 that for this function $f(1) = 1$ and $f'(1) = 2$. Therefore, formula (6) with $x_0 = 1$ gives the equation $y - 1 = 2(x - 1)$ or $y = 2x - 1$ for the tangent line.   □

Differentiable functions are continuous

We can expect that the graph of a function will not have a tangent line at a point where the function is discontinuous and the graph is "broken" (Figure 3.5). In other words a function must be continuous at any point where it has a derivative. We state this result in the following theorem.

Theorem 3.1   If the derivative $f'(x_0)$ exists, then the function f is continuous at x_0.

Proof.   For $x \neq x_0$ we write

$$(7) \quad f(x) - f(x_0) = \left[\frac{f(x) - f(x_0)}{x - x_0} \right] (x - x_0).$$

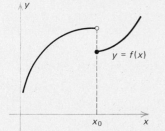

FIGURE 3.5   The graph of f does not have a tangent line at $x = x_0$. The function f does not have a derivative at x_0.

If the derivative $f'(x_0)$ exists, then the first factor on the right of (7) has that finite limit as x tends to x_0 whereas the second factor $x - x_0$ tends to zero. By a limit theorem (Theorem 2.1) the quantity (7) tends to zero and $f(x)$ tends to $f(x_0)$ as x tends to x_0. Hence, f is continuous at x_0.   Q.E.D.

Continuous functions that do not have derivatives

It is possible for a function f not to have a derivative at a point x_0 even though it is continuous there. This usually occurs in one of two ways: The slope of the secant line

$$(1) \quad \frac{f(x) - f(x_0)}{x - x_0}$$

might tend to ∞ or to $-\infty$ as x tends to x_0, so that the graph of f has a vertical tangent line at $x = x_0$ (Figure 3.6); or the slope (1) of the secant line might have different limits as x tends to x_0 from the left and from the right, so that the graph has a "corner" at $x = x_0$ (Figure 3.7).

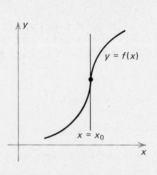

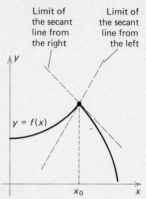

FIGURE 3.6 $f'(x_0)$ does not exist because the graph has a vertical tangent line at $x = x_0$.

FIGURE 3.7 $f'(x_0)$ does not exist because the graph has a "corner" at $x = x_0$.

Exercises

†*Answer provided.*
‡*Outline of solution provided.*

In Exercises 1 through 8 compute the derivatives by using Definition 3.1.

1.† $f'(3)$, where $f(x) = -2x^2$

2. $F'(1)$, where $F(x) = 6x + 3$

3.† $g'(2)$, where $g(x) = 4$

4.‡ $K'(2)$, where $K(t) = t^3 + 5$

5.‡ $h'(25)$, where $h(t) = \sqrt{t}$

6.‡ $G'(4)$, where $G(u) = 1/(5 - 2u)$

7. $P'(0)$, where $P(x) = (x + 4)/(x + 2)$

8. $f'(9)$, where $f(x) = x^{-1/2}$

In each of Exercises 9 through 11 compute the slope of the indicated secant line, then sketch the curve and the secant line.

9.‡ The secant line through the points at $x = 1$ and $x = 2$ on the curve $y = x^3$

10.† The secant line through the points at $x = 1$ and $x = 2$ on the curve $y = 2/x$

11. The secant line through the points at $x = -2$ and $x = -1$ on the curve $y = 4 - x^2$

In each of Exercises 12 through 17 use Definition 3.1 of the derivative to find an equation of the tangent line at the indicated values of x. Then sketch the curve and the tangent line.

12.‡ At $x = 2$ on the curve $y = 4x^{-2}$

13.† At $x = 3$ on the hyperbola $y = 2 - 6/x$

14. At $x = -1$ on the parabola $y = 4 - x^2$

15. At $x = 1$ on the cubic $y = x^3 + 4$

16.† At $x = 0$ on the cubic $y = -\frac{1}{4}x^3 - 2$

17. At $x = -2$ on the line $y = -3x + 4$

18. (Normal lines) The line through the point $(x_0, f(x_0))$ and perpendicular to the tangent line to the graph of f at that point is called the *normal line* to the graph at that point. (The term *normal* is used for *perpendicular*, because it comes from the Latin word for a carpenter's square, *norma*.) Show that the normal line to the graph of f at $(x_0, f(x_0))$ has the equation

$$y - f(x_0) = -\frac{1}{f'(x_0)}(x - x_0)$$ ◀

if $f'(x_0)$ is not zero, and that it has the equation $x = x_0$ if $f'(x_0)$ is zero.

19.† Give an equation for the normal line to the curve $y = x^3$ at $x = -1$.

20. Give an equation for the normal line to the curve $y = x^2 + 4$ at $x = 0$.

In Exercises 21 through 24 use Definition 3.1 to find equations for the indicated lines.

21.† The tangent line to the graph of $1/(x^2 + 1)$ at $x = -2$

22. The tangent line to the graph of $x^2 - 3x + 4$ at $x = 2$

23.† The normal line to the graph of $\sqrt{x + 5}$ at $x = 4$

24. The normal line to the graph of $x/(x + 7)$ at $x = 0$

25. If you have access to a computer or electronic calculator, guess the derivatives of the following functions at the indicated values of x_0 by computing the slope (1) of the secant line with $x - x_0$ equal to 0.1, -0.1, 0.001, -0.001, 0.0001, and -0.0001.

    a.† $f(x) = x^4, x_0 = 2$ **b.** $f(x) = (x^2 + 9)^3, x_0 = 1$
    c.† $f(x) = x^{4/3}, x_0 = 1$ **d.** $f(x) = x^{1/10}, x_0 = 1$
    e. $f(x) = (3x + 5)^{1/3}, x_0 = 1$ **f.** $f(x) = (5x - 4)^5, x_0 = 1$
    g.† $f(x) = (x^2 - 1)^{-1/3}, x_0 = 3$ **h.** $f(x) = x/(x - 1), x_0 = 2$
    i. $f(x) = (x^2 - 8)^{1/4}, x_0 = 3$ **j.**† $f(x) = (3 + x^{1/2})^{1/2}, x_0 = 1$

Additional drill exercises

In Exercises 26 through 35 compute the derivatives by using Definition 3.1.

26.† $f'(4)$ where $f(x) = 2/(3 + 2x)$

27.† $g'(1)$ where $g(x) = x^2 + 2x$

28.† $h'(5)$ where $h(t) = t^{-2}$

29.† $k'(0)$ where $k(y) = (5y + 1)/(2y + 1)$

30.† $F'(0)$ where $F(v) = v^5 + 4v$

31.† $G'(1)$ where $G(x) = \sqrt{4x + 1}$

32. $f'(3)$ where $f(x) = x^{-1} + 3x$

33. $g'(1)$ where $g(x) = x/(6x + 1)$

34. $h'(0)$ where $h(x) = \sqrt{3 - x}$

35. $F'(0)$ where $F(x) = x^{4/3}$

3-2   LEIBNIZ NOTATION AND THE DIFFERENTIATION OPERATOR d/dx

Isaac Newton and Gottfried Leibniz, who are considered to be the founders of calculus, each introduced notation for the derivative. The prime notation $f'(x_0)$ is similar to that used by Newton. The notation that Leibniz used has evolved into what is known today as *Leibniz notation*. It is important to be able to work with both types of notation because both are used extensively.

In Leibniz notation the derivative of f at x_0 is denoted by the symbol $\dfrac{df}{dx}(x_0)$. Here the symbol $\dfrac{df}{dx}$, which replaces the f' in the prime notation, is only a symbolic fraction and must be treated as a unit.

The Δx-formulation of the definition of the derivative

With Leibniz notation the definition of the derivative reads

(1) $$\frac{df}{dx}(x_0) = \lim_{x \to x_0} \frac{f(x) - f(x_0)}{x - x_0}.$$

Often, this definition is given a different formulation by using different notation for the slope

(2) $$\frac{f(x) - f(x_0)}{x - x_0}$$

of the secant line.

We set

(3) $$\Delta x = x - x_0.$$

The symbol "Δx" here denotes a single number and *not* the product of a "Δ" and an "x." (The Δ is the Greek capital letter delta.) Solving equation (3) for x gives

(4) $$x = x_0 + \Delta x.$$

We write Δf for the change $f(x_0 + \Delta x) - f(x_0)$ in the value of f that occurs when we change x by the amount Δx from x_0 to $x_0 + \Delta x$ (Figure 3.8). With this notation and equations (3) and (4), formula (2) for the slope of the secant line takes the form

(5) $$\frac{f(x_0 + \Delta x) - f(x_0)}{\Delta x} = \frac{\Delta f}{\Delta x}.$$

We can have $x = x_0 + \Delta x$ tend to x_0 by having Δx tend to zero; so, with this notation the definition of the derivative reads

(6) $$\frac{df}{dx}(x_0) = \lim_{\Delta x \to 0} \frac{\Delta f}{\Delta x} = \lim_{\Delta x \to 0} \frac{f(x_0 + \Delta x) - f(x_0)}{\Delta x}.$$ ◀

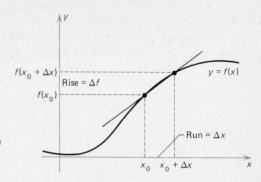

FIGURE 3.8   Secant line of slope
$$\frac{\Delta f}{\Delta x} = \frac{f(x_0 + \Delta x) - f(x_0)}{\Delta x}$$

EXAMPLE 1   Use formulation (6) of the definition of the derivative to compute $\dfrac{df}{dx}(3)$ where $f(x) = x^2$.

Solution.   Formula (5) for the slope of the secant line with $x_0 = 3$ and $f(x) = x^2$ gives

$$\frac{f(3 + \Delta x) - f(3)}{\Delta x} = \frac{(3 + \Delta x)^2 - 3^2}{\Delta x} = \frac{9 + 6\Delta x + (\Delta x)^2 - 9}{\Delta x}$$

$$= \frac{6\Delta x + (\Delta x)^2}{\Delta x} = 6 + \Delta x.$$

Therefore,

$$\frac{df}{dx}(3) = \lim_{\Delta x \to 0} \frac{f(3 + \Delta x) - f(3)}{\Delta x} = \lim_{\Delta x \to 0} (6 + \Delta x) = 6. \quad \Box$$

The derivative as a function   Up to now we have considered derivatives at specific single values of x. If we consider the derivative at a variable x, we obtain THE DERIVATIVE FUNCTION $\dfrac{df}{dx}$ whose value $\dfrac{df}{dx}(x)$ is the slope of the tangent line to the graph of f at the point $(x, f(x))$.

The Δx-formulation of the definition of the derivative is especially suitable for computing derivative functions. We use the formula

$$(7) \qquad \frac{df}{dx}(x) = \lim_{\Delta x \to 0} \frac{f(x + \Delta x) - f(x)}{\Delta x} \qquad \blacktriangleleft$$

obtained from (6) by replacing x_0 with x.

EXAMPLE 2   Use formulation (7) of the definition of the derivative to compute the derivative of the function $1/x$.

Solution.   We set $f(x) = 1/x$ for $x \neq 0$ and obtain

$$\frac{f(x + \Delta x) - f(x)}{\Delta x} = \frac{(1/(x + \Delta x)) - (1/x)}{\Delta x} = \frac{1}{\Delta x}\left(\frac{1}{x + \Delta x} - \frac{1}{x}\right)$$

$$= \frac{1}{\Delta x}\left[\frac{x - (x + \Delta x)}{(x + \Delta x)x}\right] = \frac{-\Delta x}{\Delta x(x + \Delta x)x}$$

$$= \frac{-1}{(x + \Delta x)x}.$$

Having simplified the difference quotient to a form whose limit as Δx tends to zero, we can recognize, we obtain

$$\frac{df}{dx}(x) = \lim_{\Delta x \to 0} \frac{f(x + \Delta x) - f(x)}{\Delta x} = \lim_{\Delta x \to 0}\left[\frac{-1}{(x + \Delta x)x}\right]$$

$$= \frac{-1}{(x + 0)x} = -\frac{1}{x^2}.$$

Thus, the derivative of the function $1/x$ is the function $-1/x^2$.   □

The graph of the function $1/x$ is shown in Figure 3.9 and the graph of its derivative in Figure 3.10. All of the graph of the derivative is below the x-axis because all of the tangent lines to the graph of $1/x$ have negative slopes. The value of the derivative is a large negative number for x near zero and a small negative number for x far from zero because the graph of the function $1/x$ has tangent lines that are very steep for x near zero and are close to horizontal for x far from zero.

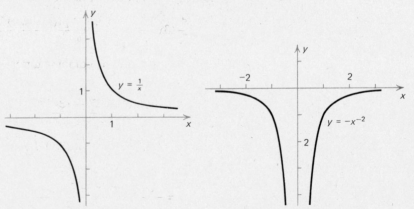

FIGURE 3.9   The graph of the function $\dfrac{1}{x}$

FIGURE 3.10   The graph of the derivative $\dfrac{d}{dx}\left(\dfrac{1}{x}\right) = -\dfrac{1}{x^2}$

The differentiation operator $\dfrac{d}{dx}$

We have introduced the two symbols f' and $\dfrac{df}{dx}$ for the derivative of a function with the letter name f and the symbols $f'(x)$ and $\dfrac{df}{dx}(x)$ for the value of the derivative at x. If a function is given by an algebraic expression or by any means other than a single letter name, we denote its derivative by using the DIFFERENTIATION OPERATOR $\dfrac{d}{dx}$ which transforms a function into its derivative.* Thus, in Example 2 we found that the derivative of $1/x$ is $-1/x^2$, so we write

$$(8) \quad \frac{d}{dx}\left(\frac{1}{x}\right) = \frac{-1}{x^2}.$$

Derivatives of powers

When we use the differentiation operator to express derivatives of powers as in the expressions $\dfrac{d}{dx}x^3$ and $\dfrac{d}{dx}x^{-1}$, we adopt the convention that the powers are to be computed before the derivatives. Accordingly, $\dfrac{d}{dx}x^3$ denotes $\dfrac{d}{dx}(x^3)$ and $\dfrac{d}{dx}x^{-1}$ denotes $\dfrac{d}{dx}(x^{-1})$ and we can express formula (8) in the form

$$\frac{d}{dx}x^{-1} = -x^{-2}.$$

EXAMPLE 3   Find the derivative of the function $3x^2 - 4$.

Solution.   The value of $3x^2 - 4$ at $x + \Delta x$ is $3(x + \Delta x)^2 - 4$. By formulation (7) of the definition of the derivative, we have

$$\frac{d}{dx}(3x^2 - 4) = \lim_{\Delta x \to 0} \frac{[3(x + \Delta x)^2 - 4] - [3x^2 - 4]}{\Delta x}.$$

We compute the square, combine terms, and then divide by Δx to obtain for $\Delta x \neq 0$

$$\frac{[3(x + \Delta x)^2 - 4] - [3x^2 - 4]}{\Delta x}$$

$$= \frac{3[x^2 + 2x\Delta x + (\Delta x)^2] - 4 - 3x^2 + 4}{\Delta x}$$

$$= \frac{3x^2 + 6x\Delta x + 3(\Delta x)^2 - 3x^2}{\Delta x} = \frac{6x\Delta x + 3(\Delta x)^2}{\Delta x}$$

$$= 6x + 3\Delta x.$$

---

*Some books use the symbol D or D_x for this differentiation operator.

Hence

(9) $\dfrac{d}{dx}(3x^2 - 4) = \lim_{\Delta x \to 0} (6x + 3\Delta x) = 6x.$ □

When we use the differentiation operator to denote a derivative, we use square brackets with the subscript "$x = x_0$" to designate the value of the derivative at x_0. For example, to compute the derivative of $3x^2 - 4$ at 5 from formula (9), we write

$$\left[\frac{d}{dx}(3x^2 - 4)\right]_{x=5} = [6x]_{x=5} = 6(5) = 30.$$

Exercises

In Exercises 1 through 4 use formulation (6) of the definition of the derivative to compute the derivatives.

1.† $\dfrac{df}{dx}(3)$, where $f(x) = 2x^2 - x$ **2.** $g'(5)$, where $g(x) = \dfrac{1}{x + 2}$

3.† $\dfrac{dh}{dt}(4)$, where $h(t) = \sqrt{t + 5}$ **4.** $\dfrac{dF}{dv}(16)$, where $F(v) = v^{-1/2}$

In Exercises 5 through 10 use formulation (7) of the definition of the derivative to compute the derivatives of the functions, then sketch the graphs of the functions and of their derivatives.

5.† $-\frac{1}{4}x^2$ **6.** $6 - 8/x$

7.† $5x - 3$ **8.** $\frac{1}{9}x^2 - 3$

9.† $\frac{1}{8}x^3$ **10.** $2 - \frac{1}{4}x^3$

In Exercises 11 through 14 use formulation (7) of the definition of the derivative to compute the derivatives of the functions.

11.† $2x^2 + 3x - 4$ **12.** $x^3 - x^2$

13.† $\sqrt{2x + 3}$ **14.** $1/x^2$

In Exercises 15 through 22 find the derivatives.

15.† $\dfrac{d}{dx}(6x + 4)$ **16.†** $\dfrac{d}{dx}(x + x^{-1})$

17.† $\dfrac{d}{dx}\left(\dfrac{1}{3x + 1}\right)$ **18.†** $\dfrac{d}{dx}x^{-1/2}$

19.† $\left[\dfrac{d}{dx}x^{1/2}\right]_{x=1}$ **20.†** $\left[\dfrac{d}{dx}(x^2 + 3x)\right]_{x=5}$

21.† $\left[\dfrac{d}{dx}(2 - x)^{-1}\right]_{x=0}$ **22.†** $\dfrac{d}{dx}\left(\dfrac{x}{x + 1}\right)$

3-3   THE DERIVATIVE OF x^n, DERIVATIVES OF LINEAR COMBINATIONS

In Sections 3.1 and 3.2 we saw how the derivatives of functions given by very simple algebraic expressions can be computed by using the definition of the derivative. This procedure would require a great deal of effort if it were applied to a function given by a complicated algebraic expression. The effectiveness of calculus as a mathematical tool rests in part on the fact that derivatives of many functions can be computed without referring back to the definition of the derivative; they can be computed by the application of RULES OF DIFFERENTIATION.

In this section we discuss the rules for differentiating the function x^n and for expressing the derivative of a linear combination of functions in terms of the derivatives of the functions themselves. These rules, along with the product, quotient, and chain rules, which will be discussed in Sections 3.4 and 3.7, will enable us to compute, with relatively little effort, the derivative of any function that is given by an algebraic expression.

The derivative of x^n

Theorem 3.2   **For any rational constant n the derivative of the function x^n is**

(1) $$\frac{d}{dx}x^n = nx^{n-1}.$$ ◄

Here x must be nonzero if $n - 1$ is negative and must be positive if n is a fraction, $n = p/q$, with p an integer and q an even integer.* In the case of $n = 0$ we interpret nx^{n-1} as the zero function.

Rule (1) is not valid for $x = 0$ if $n - 1$ is negative because then it would involve dividing by zero. It is not valid for $x \leq 0$ if n is a fraction (positive or negative) with an even denominator because then it would involve even roots of the negative number x.

We give here the proof of Theorem 3.2 first for $n = 0$ and 1, then for integers $n \geq 2$, and finally for arbitrary positive rational numbers n. The proof for negative rational n is left as Exercise 27 at the end of this section.

Proof of Theorem 3.2 for $n = 0$.   In the case of $n = 0$ the function x^n denotes the constant function 1, and formula (1) reads

(2) $$\frac{d}{dx}1 = 0.$$ ◄

The graph of the constant function 1 is the horizontal line $y = 1$ (Figure 3.11); its tangent line at any point is the line itself and has zero

FIGURE 3.11   The graph of the function 1

---

*In Chapter 7 we will see that (1) is also valid for positive x and irrational n.

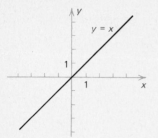

FIGURE 3.12   The graph of the function x

slope. The derivative of the constant function 1, therefore, is zero for all x, as stated in formula (2).   Q.E.D.

Proof of Theorem 3.2 for $n = 1$.   In this case the function x^n is the function $x^1 = x$, and formula (1) reads

(3) $\dfrac{d}{dx}x = 1.$ ◄

The graph of the function x is the line $y = x$ (Figure 3.12); its tangent line at any point is the line itself and has slope 1. Therefore, the derivative of the function x is 1 as is stated in formula (3).   Q.E.D.

Difference quotients   The slope of the secant line $\dfrac{f(x + \Delta x) - f(x)}{\Delta x}$, whose limit is the derivative $f'(x)$, is also known as a DIFFERENCE QUOTIENT because the denominator is the difference $\Delta x = (x + \Delta x) - x$ of values of the variable and the numerator is the difference $f(x + \Delta x) - f(x)$ of the corresponding values of the function.

Proof of Theorem 3.2 for integers $n \geq 2$.   The graph of x^n for integers $n \geq 2$ is not a line, and we have to compute its derivative as the limit of a difference quotient. We have

(4) $\dfrac{d}{dx}x^n = \lim_{\Delta x \to 0} \dfrac{(x + \Delta x)^n - x^n}{\Delta x}.$

To express the difference quotient in (4) in a form whose limit we can recognize, we use the formula from algebra

(5) $(x + \Delta x)^n = x^n + nx^{n-1}(\Delta x) + \left[\begin{array}{l}\text{Terms involving } (\Delta x)^2 \\ \text{and higher powers of } \Delta x\end{array}\right].$

Formula (5) is valid for any integer $n \geq 2$. For $n = 2$, for example, we have $(x + \Delta x)^2 = x^2 + 2x(\Delta x) + (\Delta x)^2$, and for $n = 3$, $(x + \Delta x)^3 = x^3 + 3x^2(\Delta x) + 3x(\Delta x)^2 + (\Delta x)^3$. The proof of (5) for arbitrary integers $n \geq 2$ is given in Example 2 of the Appendix to Chapter 1.

If we subtract x^n from both sides of equation (5) and divide both sides by Δx, we obtain

$\dfrac{(x + \Delta x)^n - x^n}{\Delta x} = nx^{n-1} + \left[\begin{array}{l}\text{Terms involving } \Delta x \\ \text{and powers of } \Delta x\end{array}\right].$

Because the terms involving Δx and powers of Δx tend to zero as Δx tends to zero, this equation yields

$\dfrac{d}{dx}x^n = \lim_{\Delta x \to 0} \dfrac{(x + \Delta x)^n - x^n}{\Delta x} = nx^{n-1}.$ Q.E.D.

EXAMPLE 1   Compute the derivative $\left[\dfrac{d}{dx}x^5\right]_{x=2}$.

Solution.   Formula (1) with $n = 5$ reads

$$\frac{d}{dx}x^5 = 5x^{5-1} = 5x^4.$$

Therefore,

$$\left[\frac{d}{dx}x^5\right]_{x=2} = [5x^4]_{x=2} = 5(2^4) = 80. \quad \square$$

Proof of Theorem 3.2 for positive rational n. Suppose that $n = p/q$ with p and q positive integers and that the numbers x and a are positive if q is even. Set $y = x^{1/q}$ and $b = a^{1/q}$, so that $x = y^q$ and $a = b^q$. Then for $x \neq a$ we have

$$\frac{x^n - a^n}{x - a} = \frac{(x^{1/q})^p - (a^{1/q})^p}{x - a} = \frac{y^p - b^p}{y^q - b^q} = \frac{\dfrac{y^p - b^p}{y - b}}{\dfrac{y^q - b^q}{y - b}}.$$

As x tends to a, $y = x^{1/q}$ tends to $b = a^{1/q}$, so we have

$$\left[\frac{d}{dx}x^n\right]_{x=a} = \lim_{x \to a}\frac{x^n - a^n}{x - a} = \frac{\displaystyle\lim_{y \to b}\frac{y^p - b^p}{y - b}}{\displaystyle\lim_{y \to b}\frac{y^q - b^q}{y - b}} = \frac{\left[\dfrac{d}{dy}y^p\right]_{y=b}}{\left[\dfrac{d}{dy}y^q\right]_{y=b}}.$$

We have established the differentiation formula for the positive integer powers p and q. Therefore, the last equation gives

$$\left[\frac{d}{dx}x^n\right]_{x=a} = \frac{pb^{p-1}}{qb^{q-1}} = \left(\frac{p}{q}\right)b^{p-q} = \left(\frac{p}{q}\right)(a^{1/q})^{p-q}$$

$$= \left(\frac{p}{q}\right)a^{(p/q)-1} = na^{n-1}.$$

This is differentiation formula (1) at the arbitrary value $x = a$. Q.E.D.

EXAMPLE 2   Compute the derivative $F'(8)$, where $F(x) = x^{4/3}$.

Solution.   By formula (1) with $n = \frac{4}{3}$ and $n - 1 = \frac{4}{3} - 1 = \frac{1}{3}$ we have

$$F'(8) = \left[\frac{d}{dx}x^{4/3}\right]_{x=8} = \left[\frac{4}{3}x^{1/3}\right]_{x=8} = \frac{4}{3}(8)^{1/3} = \frac{4}{3}(2) = \frac{8}{3}. \quad \square$$

Derivatives of linear combinations of functions

The next theorem, used in conjunction with rule (1) for differentiating powers, enables us to differentiate any polynomial or other linear combination of powers of x.

Theorem 3.3   If the functions f and g have derivatives at x, then so does the linear combination $Af + Bg$ for any constants A and B. The derivative of the linear combination of the functions is the same linear combination of their derivatives:

$$(6) \quad \frac{d}{dx}[Af(x) + Bg(x)] = A\frac{df}{dx}(x) + B\frac{dg}{dx}(x). \qquad \blacktriangleleft$$

Proof.   To prove this theorem, we notice that a difference quotient for the function $Af + Bg$ is equal to A times the difference quotient for f plus B times the difference quotient for g:

$$(7) \quad \frac{[Af(x + \Delta x) + Bg(x + \Delta x)] - [Af(x) + Bg(x)]}{\Delta x}$$
$$= A\left[\frac{f(x + \Delta x) - f(x)}{\Delta x}\right] + B\left[\frac{g(x + \Delta x) - g(x)}{\Delta x}\right].$$

When Δx tends to zero, equation (7) becomes equation (6).   Q.E.D.

EXAMPLE 3   Compute $\dfrac{d}{dx}(6x^{2/3} - 4x^{-2} + 5)$. For what values of x does this derivative exist?

Solution.   By the analog of rule (6) for linear combinations of three functions and by rule (1) with $n = \frac{2}{3}, n = -2$, and $n = 0$ we have

$$\frac{d}{dx}(6x^{2/3} - 4x^{-2} + 5) = 6\frac{d}{dx}x^{2/3} - 4\frac{d}{dx}x^{-2} + 5\frac{d}{dx}x^0$$
$$= 6(\tfrac{2}{3})x^{-1/3} - 4(-2)x^{-3} + 5(0)$$
$$= 4x^{-1/3} + 8x^{-3}.$$

This derivative exists for all $x \neq 0$. $\square$

Exercises

In Exercises 1 through 10 compute the derivatives and describe the values of the variables for which they exist.

1.† $\dfrac{d}{dx}(5x^3 - 6x^2 + 7x)$

2. $\dfrac{d}{dx}(x^7 + 3x^2 - 15)$

3.† $\dfrac{d}{dx}(6x^{-3} - 4x^{-1})$

4. $\dfrac{d}{dx}(7x^{-8} - 8x^{-7} + 3)$

5.† $\dfrac{d}{dt}(t^{3/2} - t^{2/3})$

6. $\dfrac{d}{dt}(4t^{-1/3} - 3t^{1/4})$

7.‡ $\dfrac{d}{dv}(\sqrt{v} - 4\sqrt[3]{v})$

8.† $\dfrac{d}{ds}(1 - s + \sqrt{s^5})$

9.† $\dfrac{d}{dx}\left(x - \dfrac{3}{\sqrt{x}}\right)$

10. $\dfrac{d}{dx}\left(\dfrac{1}{x} - \dfrac{2}{x^2} + \dfrac{3}{x^3}\right)$

In Exercises 11 through 16 compute the derivatives.

11.[†] $F'(2)$, where $F(x) = 6x^3 - x^{-1}$

12. $W'(3)$, where $W(t) = t^{-1} - 3t^2 + 4$

13.[†] $\left. \dfrac{d}{dx}(4x^4 - 4x^{-4}) \right|_{x=1}$ **14.** $\left. \dfrac{d}{dw}(3w^{4/3} - 6w^{-1/3}) \right|_{w=8}$

15.[†] $f'(10)$, where $f(x) = 3g(x) - 4h(x)$ and $g'(10) = 7$, $h'(10) = -3$

16. $\left. \dfrac{d}{dt}[5V(t) + 6W(t) - 7Z(t)] \right|_{t=0}$, where $V'(0) = -1$, $W'(0) = 4$, and $Z'(0) = 3$

In Exercises 17 through 20 give equations for the indicated lines.

17.[†] The tangent line to $y = x^3 - 4x$ at $x = 2$

18. The tangent line to $y = \sqrt[3]{x}$ at $x = 27$

19.[†] The normal line to $y = 53 + 7x^{10}$ at $x = 1$

20. The normal line to $y = x^{3/2}$ at $x = 9$

21.[†] Two curves are *tangent* at a point where they intersect if they have the same tangent line at that point. Find a constant k such that the curves $y = 1 + kx - kx^2$ and $y = x^4$ are tangent at the point $(1, 1)$.

22. Two curves are *perpendicular* at a point where they intersect if their tangent lines at that point are perpendicular. Find a constant n such that the curves $y = \frac{1}{2}x^{-2}$ and $y = \frac{1}{2}x^n$ are perpendicular at the point $(1, \frac{1}{2})$, then sketch the two curves in one coordinate plane.

In Exercises 23 through 26 use the definition of the derivative to derive the indicated formula.

23.[†] Formula (2)   **24.** Formula (3)

25.[†] Formula (1) for $n = \frac{1}{2}$ **26.** Formula (1) for $n = -1$

27.[†] Prove formula (1) for negative rational numbers n.

Compute the derivatives in Exercises 28 through 37 by expressing the functions as linear combinations of powers.

28.[†] $\dfrac{d}{dx}\left(\dfrac{\sqrt{x} - 3}{x} \right)$ **29.**[†] $\dfrac{d}{dx}(x + 3)^2$ (Compute the square)

30.[†] $\dfrac{d}{du}[(1 - u^2)/\sqrt{u}]$ **31.**[†] $\dfrac{d}{dw}(\sqrt[5]{w^3} + \sqrt[3]{w^5})$

32.[†] $\dfrac{d}{dy}[y^2(2y^3 + 3y^2 - 4)]$ **33.**[†] $\dfrac{d}{dx}[(x + 2)(3x - 4)]$

34.[†] $\dfrac{d}{dx}\left(\dfrac{2 + 3x^{30} - 5x^{40}}{x^{50}} \right)$ **35.**[†] $\dfrac{d}{dx}[x^{1/7}(x^{1/6} - x^{1/5})]$

36.[†] $\dfrac{d}{dt}\left(\dfrac{t - 1}{t} \right)^2$ **37.**[†] $\dfrac{d}{dx}\sqrt{\dfrac{x^{3/4}}{x^{2/3}}}$

38.[†] Find numbers a and b so that the curves $y = \sqrt{x}$ and $y = ax^2 + b$ are tangent at $(1, 1)$.

39.[†] Find constants a and n so that the curves $y = x^3 + x^2 + x$ and $y = ax^n$ are perpendicular at $(1, 3)$.

3-4   THE PRODUCT AND QUOTIENT RULES

Many functions may be expressed as products or quotients of simpler functions, and to compute their derivatives, we use the PRODUCT and QUOTIENT RULES, which we discuss in this section.

Theorem 3.4   (The Product Rule)   If the functions f and g have derivatives at x, then so does their product. The derivative of the product is

$$(1) \quad \frac{d}{dx}[f(x)g(x)] = f(x)\frac{dg}{dx}(x) + g(x)\frac{df}{dx}(x). \qquad \blacktriangleleft$$

Formula (1) states that the derivative of the product is equal to the first function multiplied by the derivative of the second, plus the second multiplied by the derivative of the first.

Proof.   To prove formula (1), we add and subtract the term $f(x + \Delta x)g(x)$ in the numerator of the difference quotient for the product and rewrite the result. We obtain

$$(2) \quad \begin{aligned} &\frac{f(x + \Delta x)g(x + \Delta x) - f(x)g(x)}{\Delta x} \\ &= \frac{f(x + \Delta x)g(x + \Delta x) - f(x + \Delta x)g(x)}{\Delta x} \\ &\quad + \frac{f(x + \Delta x)g(x) - f(x)g(x)}{\Delta x} \\ &= f(x + \Delta x)\left[\frac{g(x + \Delta x) - g(x)}{\Delta x}\right] + g(x)\left[\frac{f(x + \Delta x) - f(x)}{\Delta x}\right]. \end{aligned}$$

Because f has a derivative at x, it is continuous there and $f(x + \Delta x)$ tends to $f(x)$ as Δx tends to zero. The expressions in square brackets on the right side of equation (2) tend to the derivatives of g and f, while the left side of the equation tends to the derivative of their product. Equation (2), therefore, becomes equation (1) as Δx tends to zero.   Q.E.D.

EXAMPLE 1   Compute the derivative $\dfrac{d}{dx}[(x^2 + 3x - 1)(4x^{1/2} - 6)]$.

Solution.   According to the product rule (1) with $f(x) = x^2 + 3x - 1$ and $g(x) = 4x^{1/2} - 6$, we have

$$\frac{d}{dx}[(x^2 + 3x - 1)(4x^{1/2} - 6)]$$

$$= (x^2 + 3x - 1)\frac{d}{dx}(4x^{1/2} - 6) + (4x^{1/2} - 6)\frac{d}{dx}(x^2 + 3x - 1)$$

$$= (x^2 + 3x - 1)(2x^{-1/2}) + (4x^{1/2} - 6)(2x + 3). \quad \square$$

EXAMPLE 2   The functions H and K satisfy $H(3) = 4$, $H'(3) = -7$, $K(3) = 10$, and $K'(3) = 2$. Compute the derivative $\left[\dfrac{d}{dx}(H(x)K(x))\right]_{x=3}$.

Solution.   By the product rule

$$\frac{d}{dx}(H(x)K(x)) = H(x)\frac{d}{dx}K(x) + K(x)\frac{d}{dx}H(x)$$

$$= H(x)K'(x) + K(x)H'(x).$$

At $x = 3$ this expression reads

$$\left[\frac{d}{dx}(H(x)K(x))\right]_{x=3} = H(3)K'(3) + K(3)H'(3)$$

$$= 4(2) + 10(-7) = -62. \quad \square$$

Theorem 3.5   (The Quotient Rule)   If the functions f and g have derivatives at x and if $g(x)$ is not zero, then the quotient f/g also has a derivative at x. The derivative is

(3) $$\frac{d}{dx}\left[\frac{f(x)}{g(x)}\right] = \frac{g(x)\dfrac{df}{dx}(x) - f(x)\dfrac{dg}{dx}(x)}{g(x)^2}.$$ ◄

Formula (3) states that the derivative of the quotient is equal to the denominator multiplied by the derivative of the numerator, minus the numerator multiplied by the derivative of the denominator, all divided by the square of the denominator.

Proof.   We will deal here with the case where $f(x)$ is the constant function 1 and formula (3) gives the derivative of the reciprocal function $g(x)^{-1}$. The derivation of the general case then follows from the product rule and is left as Exercise 16.

Because g has a derivative at x, it is continuous there and $g(x + \Delta x)$ is not zero for sufficiently small $\Delta x \neq 0$. For such values of Δx we have

$$\frac{\dfrac{1}{g(x + \Delta x)} - \dfrac{1}{g(x)}}{\Delta x} = \frac{g(x) - g(x + \Delta x)}{\Delta x g(x + \Delta x)g(x)}$$

$$= \frac{-1}{g(x + \Delta x)g(x)}\left[\frac{g(x + \Delta x) - g(x)}{\Delta x}\right].$$

As Δx tends to 0, $g(x + \Delta x)$ tends to $g(x)$ and the expression in square brackets on the right side of the last equation tends to $\dfrac{dg}{dx}(x)$. Hence

$$\frac{d}{dx}\left[\frac{1}{g(x)}\right] = \lim_{\Delta x \to 0} \frac{\dfrac{1}{g(x + \Delta x)} - \dfrac{1}{g(x)}}{\Delta x} = \frac{-\dfrac{dg}{dx}(x)}{g(x)^2}.$$

This is formula (3) if f is the constant function 1 because then df/dx is the zero function. Q.E.D.

EXAMPLE 3   Compute the derivative $\dfrac{d}{dx}\left[\dfrac{x^2}{2x - 1}\right]$.

Solution.   According to the quotient rule (3) with $f(x) = x^2$ and $g(x) = 2x - 1$ we have

$$\frac{d}{dx}\left[\frac{x^2}{2x - 1}\right] = \frac{(2x - 1)\dfrac{d}{dx}(x^2) - x^2\dfrac{d}{dx}(2x - 1)}{(2x - 1)^2}$$

$$= \frac{(2x - 1)(2x) - x^2(2)}{(2x - 1)^2} = \frac{4x^2 - 2x - 2x^2}{(2x - 1)^2}$$

$$= \frac{2x^2 - 2x}{(2x - 1)^2}. \quad \square$$

Exercises

†Answer provided.
‡Outline of solution provided.

In Exercises 1 through 6 compute the derivatives.

1.‡ $\dfrac{d}{dx}[(1 + 3x - x^2)(x^2 - 5)]$ **2.†** $\dfrac{d}{dx}[(x - x^2)(1 + x^{-1} + x^{-2})]$

3. $\dfrac{d}{dx}[(4x^2 + 5)(6x^{-2} - 3)]$ **4.‡** $\dfrac{d}{dt}[(\sqrt{t} + \sqrt[3]{t})(\sqrt[4]{t} + \sqrt[5]{t})]$

5.† $\dfrac{d}{dw}\left[\left(w^2 + \dfrac{1}{w^2}\right)(2 + 3w)\right]$ **6.** $\dfrac{d}{dz}[(1 - \sqrt{z})(1 + \sqrt{z} - z)]$

7.‡ Compute $F'(-2)$ where $F(x) = x^2 G(x)$, $G(-2) = 3$, and $G'(-2) = 5$.

8.† Compute $\left[\dfrac{d}{dx}\left(\dfrac{P(x)}{x}\right)\right]\Big|_{x=5}$, where $P(5) = 10$ and $P'(5) = 2$

9. Compute $\left[\dfrac{d}{dt}R(t)\right]\Big|_{t=6}$, where $R(t) = P(t)Q(t)$, $P(6) = -1$, $Q(6) = 5$, $\dfrac{dP}{dt}(6) = 4$, and $\dfrac{dQ}{dt}(6) = 7$.

10.‡ Compute $\left[\dfrac{d}{dw}\left(\dfrac{f(w) + g(w)}{h(w)}\right)\right]\Big|_{w=0}$, where $f(0) = 1$, $g(0) = 2$, $h(0) = 3$, $f'(0) = 4$, $g'(0) = 5$, and $h'(0) = 6$.

11.† Compute $\dfrac{dV}{dy}(7)$, where $V(y) = \dfrac{y^2}{y - T(y)}$ and $T(7) = -3, \dfrac{dT}{dy}(7) = -5$.

12. Compute $K'(1)$, where $K(z) = \dfrac{z + f(z)}{z - f(z)}$ and $f(1) = 4, f'(1) = 2$.

In Exercises 13 through 15 give equations for the tangent lines.

13.‡ The tangent line to $y = \dfrac{2x + 3}{4x + 5}$ at $x = 0$

14.† The tangent line to $y = \dfrac{2}{2 - x}$ at $x = -2$

15. The tangent line to $y = \dfrac{1 + \sqrt{x}}{1 - \sqrt{x}}$ at $x = 9$

16. Use the product rule and the quotient rule for the special case of $f = 1$ to derive the quotient rule for general f.

Compute the derivatives of the functions in Exercises 17 through 25.

17.† $\dfrac{x^2 + 1}{x^3 - 1}$ **18.**† $(\sqrt{x} - 1)(\sqrt{x} + 1)$

19.† $(x^{-1} + 1 + x)(x^{-2} + 2 + x^2)$ **20.**† $\dfrac{\sqrt{x}}{3x - 4}$

21.† $\dfrac{1 + x + x^2}{1 + x^4}$ **22.**† $\dfrac{x + x^{-1}}{x - x^{-1}}$

23.† $(x^{1/3} + 1)(x^{2/3} - 1)$ **24.**† $\dfrac{x + 3}{x - 4}$

25.† $x^3 - \dfrac{3x}{x^2 + 1}$

26.† Derive the formula

$$\frac{d}{dx}[f(x)g(x)h(x)] = \frac{df}{dx}(x)g(x)h(x) + f(x)\frac{dg}{dx}(x)h(x) + f(x)g(x)\frac{dh}{dx}(x)$$

for the derivative of the product of three functions. Use the formula in Exercise 26 to compute the derivatives of the functions in Exercises 27 and 28.

27.† $(x^2 + 1)(x^3 + 2)(x^4 + 3)$ **28.**† $(x + 1)(x^{1/2} + x)(x^{-1} + x - 3)$

3-5 AVERAGE VELOCITY AND VELOCITY

We can say that a car has the constant velocity of 50 miles per hour if in every time interval of length T it travels (50 miles/hour)(T hours) $= 50T$ miles. This calculation requires only arithmetic. If, however, the car is accelerating or decelerating, then its velocity *varies*, and our calculations require the use of a derivative.

FIGURE 3.13

To see why a derivative is used, let us consider a car's motion down a straight road. We indicate the car's position on an s-axis as in Figure 3.13 with s measured in miles and the time t in hours. We let $f(t)$ denote the car's s-coordinate at time t. For illustration we will use the function f whose graph is shown in Figure 3.14.

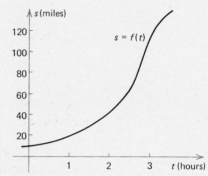

FIGURE 3.14 $f(t)$ is the car's s-coordinate at time t.

Average velocity

To find the car's velocity at time t_0, we first consider its AVERAGE VELOCITY during time intervals beginning and ending at t_0. The average velocity during a time interval is the distance traveled in the positive s-direction divided by the time taken to cover that distance.

During the time interval of length $t - t_0$ hours from time t_0 to time $t > t_0$ the car travels $f(t) - f(t_0)$ miles in the positive direction. During the time interval of length $t_0 - t$ hours from time $t < t_0$ to time t_0 the car travels $f(t_0) - f(t)$ miles in the positive direction. In either case its average velocity during the time interval is given by the following definition.

Definition 3.2 If an object is at $s = f(t)$ (miles) on an s-axis at time t (hours), then its average velocity during the time interval between times t_0 and t ($t \neq t_0$) is

(1) $$\frac{\text{Average}}{\text{velocity}} = \frac{\text{The distance traveled in the positive direction}}{\text{The time taken}}.$$

$$= \frac{f(t) - f(t_0)}{t - t_0} \frac{\text{miles}}{\text{hour}}.$$

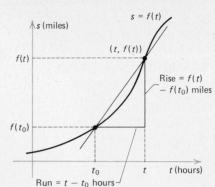

FIGURE 3.15   Secant line of slope equal to the average velocity

$$\frac{f(t) - f(t_0)}{t - t_0} \frac{\text{miles}}{\text{hour}}$$

As indicated in Figure 3.15, the average velocity (1) is the slope of the secant line through the points $(t_0, f(t_0))$ and $(t, f(t))$ on the graph of f.

EXAMPLE 1   A motorcycle traveling due south is $16t^3$ miles south of a gas station at time t (hours) for $t \geq 0$. What is the motorcycle's average velocity during the time interval $\frac{1}{2} \leq t \leq 1$?

Solution.   We imagine an s-axis with its scale measured in miles, its positive side pointing south, and its origin at the gas station (Figure 3.16). Then the motorcycle is at $s = 16t^3$ at time t. At time $t = \frac{1}{2}$ the motorcycle is at $s = [16t^3]_{t=1/2} = 16(\frac{1}{2})^3 = \frac{16}{8} = 2$ miles. At time $t = 1$ it is at $s = [16t^3]_{t=1} = 16(1)^3 = 16$ miles. During the time interval $\frac{1}{2} \leq t \leq 1$ the motorcycle travels $16 - 2 = 14$ miles, and the time interval is one half hour long; therefore, the average velocity is

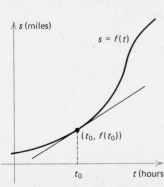

FIGURE 3.16

(2) $$\frac{\text{The distance traveled}}{\text{The time taken}} = \frac{14 \text{ miles}}{\frac{1}{2} \text{ hour}} = 28 \frac{\text{miles}}{\text{hour}}. \quad \square$$

Velocity as a derivative

We define an object's VELOCITY at t_0 to be the *limit* as t tends to t_0 of its average velocity between times t_0 and t. Since the average velocity is the difference quotient

(1) $$\frac{f(t) - f(t_0)}{t - t_0} \frac{\text{miles}}{\text{hour}}$$

the velocity at t_0 is the derivative $f'(t_0) = \dfrac{df}{dt}(t_0)$:

FIGURE 3.17   Tangent line of slope equal to the instantaneous velocity $\dfrac{df}{dt}(t_0) \dfrac{\text{miles}}{\text{hour}}$

Definition 3.3   Suppose that an object is at $s = f(t)$ miles on an s-axis at time t hours. Then its velocity in the positive direction at time t_0 is $\dfrac{df}{dt}(t_0)$ miles per hour.

Occasionally, the term *instantaneous velocity* is used in place of *velocity* to distinguish it from *average velocity*. The instantaneous velocity $\dfrac{df}{dt}(t_0)$ is the slope of the tangent line to the graph $s = f(t)$ at $t = t_0$ (Figure 3.17).

EXAMPLE 2   Compute the (instantaneous) velocity of the motorcycle of Example 1 at time $t = 1$.

Solution.   The motorcycle's velocity at time $t = 1$ is the derivative

$$\left[\frac{d}{dt}(16t^3)\right]_{t=1} = [16(3)t^2]_{t=1} = [48t^2]_{t=1} = 48\frac{\text{miles}}{\text{hour}}. \quad \square$$

Summary.   When $f(t)$ is an object's position on a coordinate axis at time t, we have three ways to describe the statement

$$\frac{df}{dt}(t_0) = \lim_{t \to t_0} \frac{f(t) - f(t_0)}{t - t_0}$$

verbally. The quantity

$$\frac{f(t) - f(t_0)}{t - t_0}$$

can be thought of (a) as a difference quotient, (b) as the slope of a secant line, or (c) as an average velocity. According to how we view this quantity, we view its limit $\frac{df}{dt}(t_0)$ (a) as a derivative, (b) as the slope of a tangent line, or (c) as an instantaneous velocity (Figure 3.18).

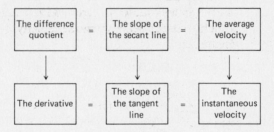

FIGURE 3.18

Note.   We can use either prime notation $f'(t_0)$ or Leibniz notation $\frac{df}{dt}(t_0)$ to denote velocity. One advantage of Leibniz notation is that it helps keep track of the units. If the object's coordinate $f(t)$ is measured in miles and the time t in hours, we assign the units of miles to the symbol "df" and the units of hours to the symbol "dt." Then the derivative $\frac{df \text{ (miles)}}{dt \text{ (hours)}}$ has the proper units of miles per hour. $\square$

Positive and negative velocity   In many situations the object we are studying moves in the negative direction along the coordinate axis we have chosen, or moves in the positive direction part of the time and in the negative direction part of the time. If the object moves in the negative direction during the time interval from t_0 to a later time t, then $f(t)$ is less than $f(t_0)$ and the average velocity $[f(t) - f(t_0)]/(t - t_0)$ is negative. Similarly, if the object's (instantaneous) velocity $f'(t_0)$ is negative, then the tangent line to the graph of f at t_0 has negative slope and points down to the right. Consequently, the object is moving in the negative direction at time t_0. Thus, we can determine when an object is moving in the positive and negative directions by determining when its velocity is positive and when it is negative. *The object is moving in the positive direction on the s-axis when its velocity is positive and in the negative direction when its velocity is negative.* This is illustrated in the next example.

EXAMPLE 3   A pellet is shot into the air at time $t = -3$ (seconds) and is $144 - 16t^2$ feet above the ground at time t until $t = 3$ when it hits the ground. For what values of t with $-3 \leq t \leq 3$ is the pellet's upward velocity positive? Zero? Negative?

Solution.   The pellet's upward velocity at time t is the derivative $\dfrac{d}{dt}(144 - 16t^2) = -32t \dfrac{\text{feet}}{\text{second}}$. It is positive for $-3 \leq t < 0$, zero at $t = 0$, and negative for $0 < t \leq 3$.   □

The graph of the function $144 - 16t^2$, which gives the pellet's height, is the parabola shown in Figure 3.19. The graph of the function $-32t$, which gives the pellet's upward velocity, is the line shown in Figure 3.20.

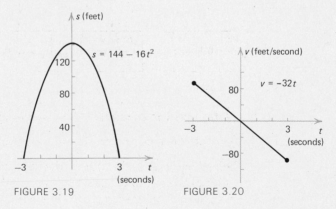

FIGURE 3.19   FIGURE 3.20

Exercises

†*Answers provided.*
‡*Outline of solution provided.*

In Exercises 1 through 6 an object's position on an *s*-axis is given as a function of the time *t*. Compute the indicated average and instantaneous velocities. Sketch the graphs of the functions and the corresponding secant and tangent lines.

1.† $s = 3 - t^2$ (feet) at time t (seconds)   **a.** average velocity for $0 \leq t \leq 1$
       b. instantaneous velocity at $t = 1$

2. $s = 100 - 20/t$ (miles) at time t (minutes)   **a.** average velocity for $\frac{1}{5} \le t \le 2$ **b.** instantaneous velocity at $t = 2$

3.[†] $s = 37t$ (miles) at time t (hours)   **a.** average velocity for $0 \le t \le 10$ **b.** instantaneous velocity at $t = 5$

4. $s = 10t^2$ (meters) at time t (hours)   **a.** average velocity for $-4 \le t \le 4$ **b.** instantaneous velocity at $t = -4$

5.[†] $s = 5t^3$ (furlongs) at time t (fortnights)   **a.** average velocity for $-1 \le t \le 0$ **b.** instantaneous velocity at $t = 0$

6. $s = 5t^2$ (inches) at time t (seconds)   **a.** average velocity for $-3 \le t \le 2$ **b.** instantaneous velocity at $t = -3$

7.[†]   A ball is $48 + 32t - 16t^2$ feet above the ground at time t (seconds) for $-1 \le t \le 3$.   **a.** What is its upward velocity at time $t = 0$?   **b.** When is its upward velocity zero? Positive? Negative?   **c.** How high does the ball go?

8.[†]   A car is $10t^{3/2} - 15t + 10$ miles east of a rest stop at time t (hours).   **a.** What is its velocity at time $t = \frac{1}{4}$ and which direction is it traveling?   **b.** Where is the car when its velocity is zero?

9.   A boy rolls a ball down a hill. After t seconds the ball has rolled $3t + \frac{1}{2}t^2$ feet.   **a.** How fast is the ball rolling after 5 seconds?   **b.** How far does the ball have to roll before it is rolling 11 feet per second?

3-6   OTHER RATES OF CHANGE

RATES OF CHANGE other than velocities are important in a variety of situations, and each of them is given by a derivative. The corresponding difference quotients are AVERAGE RATES OF CHANGE.

Definition 3.4   The difference quotient

(1) $\dfrac{f(x_1) - f(x_0)}{x_1 - x_0}$ $(x_1 \ne x_0)$

is the average rate of change of $f(x)$ with respect to x in the interval between x_0 and x_1. The derivative

(2) $f'(x_0) = \dfrac{df}{dx}(x_0)$

is the (instantaneous) rate of change of $f(x)$ with respect to x at x_0.

The instantaneous rate of change (2) is the limit of the average rate of change (1) as x_1 tends to x_0.

EXAMPLE 1   The heat produced by a certain type of precision heater is controlled by varying the current supplied to it. The heater produces $Q(I) = 150\ I^2$ calories of heat in 1 second when the current is I amperes.

What is the rate of change of the heat $Q(I)$ with respect to the current I when the current is 2 amperes?

Solution.   The desired rate of change is the derivative

$$\frac{dQ}{dI}(2) = \left[\frac{d}{dI}(150\,I^2)\right]_{I=2} = [300\,I]_{I=2} = 600\,\frac{\text{calories}}{\text{ampere}}. \quad \square$$

Abbreviated notation   It is frequently convenient to simplify notation by having one letter play two or more roles in a single discussion. In Example 1 we could write $Q = 150\,I^2$ in place of $Q(I) = 150\,I^2$. Here we would be using the letter "Q" as the name of the function and to denote its value at I. The letter "Q" is used in a third way in Figure 3.21, which shows the graph of the function Q; there "Q" is also used as the label on the vertical coordinate axis. When such simplifications of notation are made, the meanings of the letters must be given verbal descriptions or determined from the context in which they appear.

Often we will also find it convenient to use a symbol such as $Q(I)$ as the *name* of a function as well as to denote its value at I.

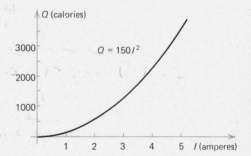

FIGURE 3.21   The graph of the function $Q = 150\,I^2$.

Estimating derivatives from graphs of functions   Sometimes a function is given by a sketch of its graph rather than by a precise formula for its values. This is frequently the case when the function is determined experimentally. Although we cannot determine the exact values of the function or of its derivative, we can find approximate values of the function by estimating the vertical coordinates of points on its graph and approximate values of the derivative by sketching plausible tangent lines to the graph and estimating their slopes. This is illustrated in the next example.

EXAMPLE 2   Figure 3.22 shows the graph of the function T, which gives the temperature (degrees centigrade) as a function of the altitude h (feet) above sea level during a subtropical inversion over a portion of Los Angeles County. (The temperature distribution indicated in the figure is called a temperature "inversion" because the air is warmer at 3000 feet than at 1000 feet, although the air is normally cooler at higher altitudes.) (a) What was the approximate temperature at 500 feet? At

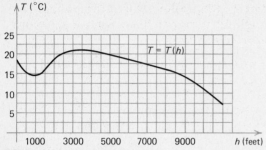

FIGURE 3.22 Air temperature as a function of altitude during a subtropical inversion over Santa Monica, July 1, 1957. (Adapted from M. Neiburger, [1] "The Role of Meteorology in the Study and Control of Air Pollution," pp. 951–965.)

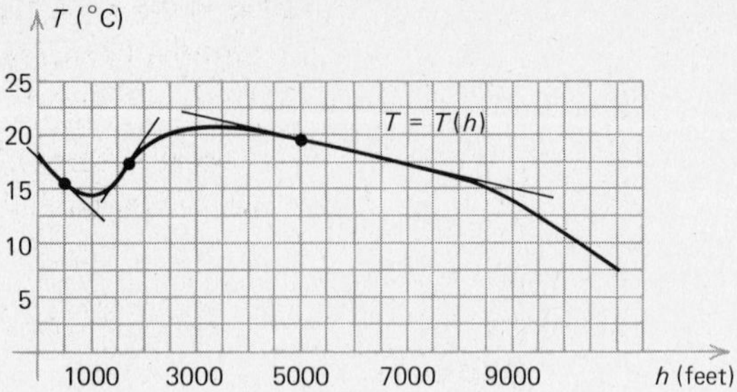

FIGURE 3.23

1700 feet? At 5000 feet? (b) At approximately what rate was the temperature rising or falling with increasing altitude at 500 feet? At 1700 feet? At 5000 feet?

Solution. (a) The temperature at an altitude of 500 feet is $T(500)$. From the sketch of the graph of the function in Figure 3.23, we obtain $T(500) \approx 16°C$. The temperatures at altitudes of 1700 and 5000 feet are $T(1700)$ and $T(5000)$. The sketch gives $T(1700) \approx 17\frac{1}{2}°C$ and $T(5000) \approx 19°C$.

(b) We have drawn plausible tangent lines to the graph at $h = 500$, $h = 1700$, and $h = 5000$ in Figure 3.23.

The slope of the tangent line at $h = 500$ is about $-1/200$, so

$$\frac{dT}{dh}(500) \approx -\frac{1}{200}$$

and the temperature is falling at the approximate rate of 1/200 degree per foot at an altitude of 500 feet. The slopes of the tangent lines at $h = 1700$ and at $h = 5000$ are approximately 1/120 and $-1/1000$, respectively. Therefore,

$$\frac{dT}{dh}(1700) \approx \frac{1}{120} \quad \text{and} \quad \frac{dT}{dh}(5000) \approx -\frac{1}{1000}.$$

The temperature is rising at the approximate rate of 1/120 degrees per foot at an altitude of 1700 feet and is dropping at the approximate rate of 1/1000 degrees per foot at an altitude of 5000 feet. □

Acceleration

The rate of change of an object's velocity with respect to time is its ACCELERATION. If the object's velocity is $v(t)$ (feet per second) at time t (seconds), then its acceleration at time t is

$$a(t) = \frac{dv}{dt}(t) \frac{\text{feet/second}}{\text{second}} = \frac{dv}{dt}(t) \frac{\text{feet}}{\text{second}^2}.$$

◀

EXAMPLE 3   A bullet fired straight up into the air at time $t = 0$ (seconds) has an upward velocity of $v(t) = 800 - 32t$ feet per second at time t. What is its upward acceleration?

Solution.   The bullet's upward acceleration is the derivative of its upward velocity with respect to t:

$$a(t) = \frac{dv}{dt}(t) = \frac{d}{dt}(800 - 32t) = -32\,\frac{\text{feet/second}}{\text{second}}$$

$$= -32\,\frac{\text{feet}}{\text{second}^2}.$$

The bullet's upward acceleration is negative because its upward velocity decreases with time.   □

Note.   The bullet's *downward* acceleration in Example 3 is 32 feet per second². The *downward acceleration caused by gravity* is often denoted by the letter "g" and is the same for all objects regardless of their weight.

A more precise value of g depends on the altitude above sea level and on the distance from the equator. At sea level, g is 32.0878 feet per second² at the equator and 32.2577 feet per second² at the North or South Pole. In the metric system the value of g is approximately 9.8 meters per second².   □

The term *marginal* in economics

Economists use the word MARGINAL for *derivative* or *rate of change*. Thus, if $C(x)$ is the cost of manufacturing x items, the derivative $C'(x)$ is the MARGINAL COST WITH RESPECT TO x of producing x items. MARGINAL REVENUE, MARGINAL PROFIT, and so forth, are defined similarly.

EXAMPLE 4   It costs a toy company $C(x) = 100 + \frac{1}{2}x - \frac{1}{1000}x^2$ dollars to manufacture x Baby Bonnie dolls in one day ($0 \leq x \leq 200$). What is the marginal cost of the production of 100 of the dolls?

Solution.   The marginal cost of manufacturing x dolls is the derivative

$$C'(x) = \frac{d}{dx}\left(100 + \frac{1}{2}x - \frac{1}{1000}x^2\right) = \frac{1}{2} - \frac{1}{500}x\,\frac{\text{dollars}}{\text{doll}}.$$

The marginal cost of manufacturing 100 dolls is the value of this derivative at $x = 100$ or

$$C'(100) = \left[\frac{1}{2} - \frac{1}{500}x\right]_{x=100} = \frac{1}{2} - \frac{1}{5} = 0.30\,\frac{\text{dollars}}{\text{doll}}.   □$$

In a strict mathematical model of the toy factory's production costs in the last example, the cost function $C(x)$ would be defined only for nonnegative integers x since the factory manufactures whole dolls only. We allow the variable x to take any values in the interval $0 \leq x \leq 200$ so that we can use derivatives to study the function. Then we can think of the marginal cost $C'(100)$ of manufacturing 100 dolls as being approximately equal to the cost of manufacturing the 101st doll because the cost of that doll is the difference quotient

$$C(101) - C(100) = \frac{C(101) - C(100)}{101 - 100}$$

which is approximately equal to the derivative $C'(100)$.

Exercises

†*Answer provided.*
‡*Outline of solution provided.*

1.† **a.** Express the width w (meters) of a cube as a function of its volume V (cubic meters).   **b.** What is the average rate of change of the width with respect to the volume for $1 \leq V \leq 27$?   **c.** What is the instantaneous rate of change of the width with respect to the volume at $V = 8$?

2.   **a.** What is the average rate of change of the volume $V = \frac{4}{3}\pi r^3$ (cubic feet) of a sphere of radius r (feet) with respect to its radius for $2 \leq r \leq 4$? **b.** Show that the instantaneous rate of change of the sphere's volume with respect to its radius is equal to the surface area of the sphere.

3.† Two positive charges, of 1 electrostatic unit each, exert a repulsive force of $F = 1/r^2$ dynes on each other when they are r centimeters apart. **a.** What is the average rate of change of the force with respect to the distance for $5 \leq r \leq 100$?   **b.** What is the instantaneous rate of change of the force with respect to the distance when the distance is $r = 10$ centimeters?

4.   The fundamental period of the vibration of a certain stretched string is $P = \frac{1}{20}T^{-1/2}$ seconds when the tension in the string is T pounds. **a.** What is the average rate of change of the period with respect to the tension for $1 \leq T \leq 9$?   **b.** What is the instantaneous rate of change of the period with respect to the tension at $T = 4$ pounds?

5.† An object is at $s = t^3 - 12t^2 + 45t$ (feet) on an s-axis at time t (seconds). **a.** What is its velocity at time t?   **b.** What is its average acceleration for $0 \leq t \leq 3$?   **c.** What is its acceleration at time $t = 2$?   **d.** At time $t = 4$?

6.   An object is at $s = t^4 + t^2 + 10t - 1$ (meters) on an s-axis at time t (hours).   **a.** What is its average acceleration for $0 \leq t \leq 2$?   **b.** What is its acceleration at time $t = 1$?

7.† The value $A(t)$ of the function in Figure 3.24 is the percentage of alcohol in a person's blood t hours after he has drunk 8 fluid ounces of whiskey. **a.** What is the approximate percentage of alcohol in the person's blood after $\frac{1}{2}$ hour?   **b.** After 6 hours?   **c.** At what approximate rate is the percentage of alcohol increasing after $\frac{1}{2}$ hour?   **d.** After 6 hours?

8.   The function L of Figure 3.25 gives the upward force (lift) of an airplane flying at a constant speed as a function of the angle x (degrees) that the wings make with the horizontal.   **a.** What is the approximate lift at an angle of 4 degrees?   **b.** At an angle of 13 degrees?   **c.** What is the approximate rate of change of the lift with respect to the angle at an angle of 4 degrees?   **d.** At an angle of 13 degrees?

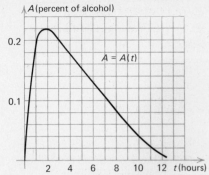

FIGURE 3.24   (Adapted from *Encyclopedia Britannica*. Vol. 1, p. 548.)

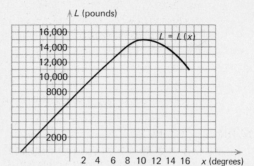

FIGURE 3.25   (Adapted from H. Schlichting,[1] *Boundary Layer Theory*, p. 21.)

9. The function shown in Figure 3.26 gives the upward velocity $v(t)$ (feet/second) of a V-2 rocket t seconds after blast-off. Give approximate values for the rocket's upward velocity and acceleration 30 and 55 seconds after blast-off. (The acceleration increases as the rocket shoots up because the rocket becomes lighter as its fuel supply is consumed and because the air resistance drops as the rocket's altitude increases.)

10.[†] It costs $C(x) = 50 + \frac{1}{4}x - \frac{1}{30}x^2$ dollars to make x dozen donuts in one day $(0 \le x \le 40)$ and the donuts sell for 75 cents a dozen.   **a.** What are the revenue $R(x)$ (gross income) and profit $P(x)$ on x dozen donuts?   **b.** What are the average and marginal profits on 30 dozen donuts?

11.[†] Suppose that a store will sell $x(p) = 1000 - 300p + 30p^2 - p^3$ pencils in one week if it sells them for p cents each $(0 \le p \le 10)$.   **a.** What will the revenue $R(p)$ be if the price is set at p cents each?   **b.** What will the marginal revenue dR/dp with respect to price be if the price is set at 5 cents each? (The function $x(p)$ is called a *demand function*.)

12. A small company's average cost, $A(x)$ dollars/pound, to produce x pounds of piñon incense is given by the function of Figure 3.27. Give approximate values for the company's cost and marginal cost in the production of 15 pounds of incense.

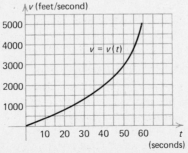

FIGURE 3.26   (Adapted from A. Feodosiev and A. Siniarev, [1] *Introduction to Rocket Technology*, p. 254.)

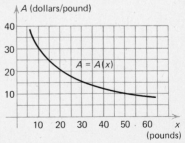

FIGURE 3.27

In each of Exercises 13 through 16 compute the acceleration of the object whose coordinate on an s-axis is given as a function of time.

13.[†] $s = t^4 + 2t - 1$ (meters) at time t (seconds)

14.[†] $s = \dfrac{t + 1}{t + 4}$ (centimeters) at time t (seconds)

15.[†] $s = t^{10} - 9\sqrt{t}$ (kilometers) at time t (hours)

16.[†] $s = t^{11/3}$ (kilometers) at time t (hours)

3-7 COMPOSITE FUNCTIONS AND THE CHAIN RULE

An important procedure for constructing functions is the *composition* of other functions. Suppose that we have a function $G = G(s)$ of the variable s and that we consider s to be a function $s = s(t)$ of the variable t. Then the function $G(s(t))$ of the variable t is a COMPOSITE FUNCTION; it is the composition of the functions G and s. The domain of $G(s(t))$ consists of all numbers t in the domain of the function s such that $s(t)$ is in the domain of the function G.

EXAMPLE 1   What is the value of the composite function $G(s(t))$ at $t = 5$ if $s(5) = 20$ and $G(20) = 8$?

Solution.   At $t = 5$ we have $G(s(5)) = G(20) = 8$.   □

EXAMPLE 2   The function s is defined by the formula $s(t) = 2t - 1$ and the function G by $G(s) = 4\sqrt{s}$. Give a formula for the composite function $G(s(t))$ in terms of its variable t. What is its domain?

Solution.   Replacing s by $s(t) = 2t - 1$ in the formula for $G(s)$ gives $G(s(t)) = 4\sqrt{s(t)} = 4\sqrt{2t - 1}$. The domain of $G(s) = 4\sqrt{s}$ is the closed half line $s \geq 0$, so the domain of the composite function consists of all t such that $s(t) = 2t - 1$ is nonnegative. The domain is the closed half line $t \geq \frac{1}{2}$. We could also determine this by examining the formula $4\sqrt{2t - 1}$ for the composite function.   □

The chain rule   The CHAIN RULE tells us how to compute the derivative of a composite function from the derivatives of the functions used to form it.

Theorem 3.6   (The Chain Rule)   The derivative of the composition $G(s(t))$ of the functions $G(s)$ and $s(t)$ is

(1) $$\frac{d}{dt} G(s(t)) = \frac{dG}{ds}(s(t)) \frac{ds}{dt}(t).$$

Here we assume that the function $s(t)$ has a derivative at t and that the function $G(s)$ has a derivative at $s = s(t)$.

Proof. We will derive (1) at $t = t_0$. We set $s_0 = s(t_0)$. Because we assume that $G(s)$ has a derivative at s_0, we have

$$(2) \qquad \lim_{s \to s_0} \frac{G(s) - G(s_0)}{s - s_0} = \frac{dG}{ds}(s_0).$$

We cannot set $s - s_0 = s(t) - s(t_0)$ in the denominator of the difference quotient in (2) because even for $t \neq t_0$, $s(t)$ might be equal to $s(t_0)$ and $s(t) - s(t_0)$ might be zero. Therefore, we need to reformulate definition (2) of the derivative so that it does not involve $s - s_0$ in the denominator.

To do this, we let $E(s)$ ($s \neq s_0$) denote the difference between the slope of the tangent line at $(s_0, G(s_0))$ and the slope of the secant line through the points $(s_0, G(s_0))$ and $(s, G(s))$. Then

$$(3) \qquad \frac{G(s) - G(s_0)}{s - s_0} = \frac{dG}{ds}(s_0) + E(s)$$

and statement (2) implies that $E(s)$ tends to zero as s tends to s_0.

Multiplying equation (3) by $s - s_0$ gives

$$(4) \qquad G(s) - G(s_0) = \left[\frac{dG}{ds}(s_0) + E(s) \right](s - s_0).$$

We set $E(s_0) = 0$ so that equation (4) is also valid at $s = s_0$. Equation (4) with the statement that $E(s)$ tends to zero is the reformulation of definition (2) that we need.

Making the substitutions $s = s(t)$ and $s_0 = s(t_0)$ in equation (4) and dividing by $t - t_0$ ($t \neq t_0$) yields

$$(5) \qquad \frac{G(s(t)) - G(s(t_0))}{t - t_0} = \left[\frac{dG}{ds}(s_0) + E(s(t)) \right]\left[\frac{s(t) - s(t_0)}{t - t_0} \right].$$

Because the derivative

$$(6) \qquad \frac{ds}{dt}(t_0) = \lim_{t \to t_0} \frac{s(t) - s(t_0)}{t - t_0}$$

exists, $s = s(t)$ tends to $s_0 = s(t_0)$ and $E(s(t)) = E(s)$ tends to zero as t tends to t_0. Hence equations (2), (5), and (6) give

$$\left[\frac{d}{dt}G(s(t)) \right]_{t=t_0} = \lim_{t \to t_0} \frac{G(s(t)) - G(s(t_0))}{t - t_0}$$

$$= \frac{dG}{ds}(s(t_0)) \frac{ds}{dt}(t_0)$$

which is formula (1) at $t = t_0$.   Q.E.D.

EXAMPLE 3   Suppose that $s(5) = 20$, $\dfrac{ds}{dt}(5) = 6$, and $\dfrac{dG}{ds}(20) = -4$.

What is $\dfrac{d}{dt}G(s(t))$ at $t = 5$?

Solution. The chain rule (1) at $t = 5$ reads

$$\left[\frac{d}{dt} G(s(t))\right]_{t=5} = \frac{dG}{ds}(s(5))\frac{ds}{dt}(5).$$

Making the substitution $s(5) = 20$, we obtain

$$\left[\frac{d}{dt} G(s(t))\right]_{t=5} = \frac{dG}{ds}(20)\frac{ds}{dt}(5) = (-4)(6) = -24. \quad \square$$

The chain rule in abbreviated Leibniz notation

The form of the chain rule (1) can be emphasized by writing it in abbreviated notation. We use the letter "G" to denote both the function $G(s)$ of s and the composite function $G(s(t))$ of t, and we drop the references to where the derivatives are evaluated. Then equation (1) reads

$$\frac{dG}{dt} = \frac{dG}{ds}\frac{ds}{dt}. \qquad \blacktriangleleft$$

This abbreviated formulation of the chain rule is useful because it is easy to remember. Even though the symbols representing the derivatives are *not* actual fractions, the ds terms appear to cancel giving the equation:

$$\frac{dG}{dt} = \frac{dG}{d\!\!\!/s}\frac{d\!\!\!/s}{dt}.$$

Interpreting the chain rule

Although the notation can make the general statement of the chain rule somewhat difficult to understand, in situations where the functions and their derivatives have everyday meaning, it follows from a common sense understanding of rates of change.

Let us examine the meaning of the chain rule (1) in the case where $s(t)$ is the distance (miles) a car has traveled along a route after t hours, and $G(s)$ is the amount (gallons) of gas it takes the car to travel s miles along the route. Then $G(s(t))$ is the amount of gas it takes the car up to time t, when it is at $s = s(t)$. The chain rule (1) reads

$$(7) \quad \frac{d}{dt} G(s(t)) \frac{\text{gallons}}{\text{hour}} = \left[\frac{dG}{ds}(s(t))\frac{\text{gallons}}{\text{mile}}\right]\left[\frac{ds}{dt}(t)\frac{\text{miles}}{\text{hour}}\right]$$

and states that the car's rate of consumption in gallons per hour is equal to its rate of consumption in gallons per mile, multiplied by its velocity in miles per hour.

Notice that in equation (7) the words "mile" and "miles," which give the units in the symbolic fractions, seem to cancel:

$$\frac{\text{gallons}}{\text{mile}}\frac{\text{miles}}{\text{hour}} = \frac{\text{gallons}}{\text{hour}}.$$

Symbolic calculations such as this are called DIMENSION ANALYSIS. They make it easier to remember equations such as (7).

The geometric meaning of the chain rule

Suppose that the functions $s(t)$ and $G(s)$ are those whose graphs are shown in Figures 3.28 and 3.29. The graph of the composite function $G(s(t))$ is then shown in Figure 3.30. The chain rule

$$(1) \quad \frac{d}{dt} G(s(t)) = \frac{dG}{ds}(s(t)) \frac{ds}{dt}(t)$$

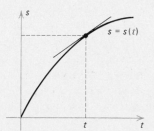

FIGURE 3.28   Graph of the function $s(t)$ and the tangent line of slope $\frac{ds}{dt}(t)$

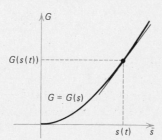

FIGURE 3.29   Graph of the function $G(s)$ and the tangent line of slope $\frac{dG}{ds}(s(t))$

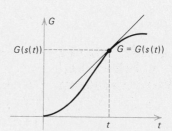

FIGURE 3.30   Graph of the composite function $G(s(t))$ and the tangent line of slope $\frac{d}{dt}G(s(t))$

$$= \frac{dG}{dt}(s(t)) \frac{ds}{dt}(t)$$

states that the slope $\frac{d}{dt} G(s(t))$ of the tangent line to the graph of the composite function is equal to the *product* of the slopes $\frac{dG}{ds}(s(t))$ and $\frac{ds}{dt}(t)$ of the tangent lines to the graphs of $G(s)$ and $s(t)$ at the corresponding points.

The use of prime notation in the chain rule

It is sometimes preferable to use the chain rule with some or all of the derivatives expressed in prime notation. To do this, we follow the convention that *the prime always denotes the function's derivative with respect to its own variable.*

   The variable of the function G in the chain rule (1) is s. Therefore, G' stands for $\frac{dG}{ds}$, and we can write (1) in the form

$$(8) \quad \frac{d}{dt} G(s(t)) = G'(s(t)) \frac{ds}{dt}(t). \qquad \blacktriangleleft$$

   The variable of the function s is t, so s' stands for $\frac{ds}{dt}$ and we can also express the chain rule in the form

$$(9) \quad \frac{d}{dt} G(s(t)) = G'(s(t)) \, s'(t). \qquad \blacktriangleleft$$

EXAMPLE 4   Suppose that $G'(8) = -7$. What is $\dfrac{d}{dt}G(t^3)$ at $t = 2$?

Solution.   Formulation (9) of the chain rule gives

$$\frac{d}{dt}G(t^3) = G'(t^3)\frac{d}{dt}t^3 = 3t^2\,G'(t^3).$$

At $t = 2$, this equation reads

$$\left[\frac{d}{dt}G(t^3)\right]_{t=2} = [3t^2\,G'(t^3)]_{t=2} = 12\overset{\bullet}{G'}(8) = 12(-7) = -84. \quad \square$$

The derivative of $s(t)^n$ If we take G in the chain rule (1) to be the function $G(s) = s^n$, where n is a rational constant, we obtain

$$\frac{dG}{ds}(s) = \frac{d}{ds}s^n = ns^{n-1}.$$

Since in this case $G(s(t)) = s(t)^n$, the chain rule gives

(10) $\dfrac{d}{dt}s(t)^n = ns(t)^{n-1}\dfrac{ds}{dt}(t).$ ◄

Note that in expressions such as on the left side of equation (10), as in expressions of the form $\dfrac{d}{dt}t^n$, we adopt the convention that the power is to be computed before the derivative. Thus, $\dfrac{d}{dt}s(t)^n$ stands for $\dfrac{d}{dt}[s(t)^n]$ (see the discussion of notation at the end of Section 3.2).

EXAMPLE 5   Suppose that $s(5) = 4$ and $s'(5) = -1$. What is $\dfrac{d}{dt}s(t)^3$ at $t = 5$?

Solution.   Applying rule (10) with $n = 3$ yields

$$\frac{d}{dt}s(t)^3 = 3s(t)^{3-1}\frac{d}{dt}s(t) = 3s(t)^2s'(t).$$

Then, setting $t = 5$ and using the conditions $s(5) = 4$ and $s'(5) = -1$, we obtain

$$\left[\frac{d}{dt}s(t)^3\right]_{t=5} = [3s(t)^2s'(t)]_{t=5} = 3[s(5)]^2\,s'(5)$$
$$= 3[4]^2(-1) = -48. \quad \square$$

EXAMPLE 6   Compute the derivative with respect to t of $\sqrt{t^2 + 4}$.

Solution.   We apply rule (10) with $s(t) = t^2 + 4$ and $n = \frac{1}{2}$ to obtain

$$\frac{d}{dt}\sqrt{t^2 + 4} = \frac{d}{dt}(t^2 + 4)^{1/2} = \frac{1}{2}(t^2 + 4)^{-1/2}\frac{d}{dt}(t^2 + 4)$$

$$= \frac{1}{2}(t^2 + 4)^{-1/2}(2t) = t(t^2 + 4)^{-1/2}. \quad \square$$

The differentiation of a complex expression may require the product rule or the quotient rule and one or more applications of the chain rule. We can determine the order in which to apply these operations by noting the order of the steps used in calculating values of the function. The differentiation proceeds in the reverse order. This is illustrated in the following three examples.

EXAMPLE 7   Find the derivative of $\left(\dfrac{t^2 + 1}{t - 3}\right)^5$.

Solution.   To compute $\left(\dfrac{t^2 + 1}{t - 3}\right)^5$ for a particular value of t, we first compute $t^2 + 1$ and $t - 3$, then we divide, and then we take the 5th power. Accordingly, to find the derivative we first apply the chain rule (10) with $n = 5$ and $s(t) = (t^2 + 1)/(t - 3)$; then we use the quotient rule; and finally we differentiate $t^2 + 1$ and $t - 3$:

$$\frac{d}{dt}\left(\frac{t^2 + 1}{t - 3}\right)^5 = 5\left(\frac{t^2 + 1}{t - 3}\right)^4 \frac{d}{dt}\left(\frac{t^2 + 1}{t - 3}\right)$$

$$= 5\left(\frac{t^2 + 1}{t - 3}\right)^4 \left[\frac{(t - 3)\dfrac{d}{dt}(t^2 + 1) - (t^2 + 1)\dfrac{d}{dt}(t - 3)}{(t - 3)^2}\right]$$

$$= 5\left(\frac{t^2 + 1}{t - 3}\right)^4 \left[\frac{(t - 3)(2t) - (t^2 + 1)(1)}{(t - 3)^2}\right]$$

$$= 5\left(\frac{t^2 + 1}{t - 3}\right)^4 \left[\frac{t^2 - 6t - 1}{(t - 3)^2}\right]. \quad \square$$

EXAMPLE 8   What is the derivative of $x^2(x^3 + 2x)^{10}$?

Solution.   To compute a value of $x^2(x^3 + 2x)^{10}$, we first compute $x^3 + 2x$, then the 10th power, then x^2, and finally the product. Therefore, to find the derivative, we first use the product rule; then we differentiate x^2 and use the chain rule (10) with x replacing t and $s(x) = x^3 + 2x$; and finally we differentiate $x^3 + 2x$:

$$\frac{d}{dx}[x^2(x^3 + 2x)^{10}] = \left(\frac{d}{dx}x^2\right)(x^3 + 2x)^{10} + x^2\left[\frac{d}{dx}(x^3 + 2x)^{10}\right]$$

$$= 2x(x^3 + 2x)^{10} + x^2\left[10(x^3 + 2x)^9\frac{d}{dx}(x^3 + 2x)\right]$$

$$= 2x(x^3 + 2x)^{10} + 10x^2(x^3 + 2x)^9(3x^2 + 2)$$

$$= 2x(x^3 + 2x)^{10} + (30x^4 + 20x^2)(x^3 + 2x)^9. \quad \square$$

EXAMPLE 9 Find the derivative of $[x^2 + (5x + 1)^3]^7$.

Solution. To compute $[x^2 + (5x + 1)^3]^7$ from x, we first compute $5x + 1$, then its third power and x^2, then the sum, and finally the seventh power. Accordingly, to find the derivative, we first apply the chain rule (10) with $s(x) = x^2 + (5x + 1)^3$; then we differentiate x^2 and use (10) with $s(x) = 5x + 1$; and finally, we differentiate $5x + 1$:

$$\frac{d}{dx}[x^2 + (5x + 1)^3]^7 = 7[x^2 + (5x + 1)^3]^6\frac{d}{dx}[x^2 + (5x + 1)^3]$$

$$= 7[x^2 + (5x + 1)^3]^6\left[2x + 3(5x + 1)^2\frac{d}{dx}(5x + 1)\right]$$

$$= 7[x^2 + (5x + 1)^3]^6[2x + 3(5x + 1)^2(5)]$$

$$= 7[x^2 + (5x + 1)^3]^6[2x + 15(5x + 1)^2]. \quad \square$$

Exercises

†*Answer provided.*
‡*Outline of solution provided.*

In Exercises 1 through 17 compute the derivatives.

1.‡ $\dfrac{d}{dx}(1 - x^3)^5$ **2.†** $f'(t)$, where $f(t) = (t^3 + 2)^{1/2}$

3. $\dfrac{d}{dx}(5x^4 - 4x^3 + 2)^{11}$ **4.‡** $\dfrac{d}{dx}\dfrac{1}{\sqrt{3x - 4}}$ **5.** $\dfrac{d}{du}\sqrt{u^3 + u^5}$

6.‡ $\left[\dfrac{d}{dx}\sqrt{f(x)}\right]_{x=1}$, where $f(1) = 4$ and $f'(1) = -5$

7.† $\dfrac{dW}{dt}(0)$, where $W(t) = v(t)^5$ and $v(0) = 2, \dfrac{dv}{dt}(0) = -4$

8. $p'(5)$, where $p(x) = g(x)^{-1}$ and $g(5) = 10, g'(5) = 4$

9.‡ $\left[\dfrac{d}{dx}G(x^3)\right]_{x=2}$, where $G'(8) = 5$

10.† $W'(9)$, where $W(t) = S(\sqrt{t})$ and $S'(3) = -4$

11. $\left[\dfrac{d}{dx}F(x^3 + 2x)\right]_{x=-1}$, where $F'(-3) = 10$

12.‡ $\left[\dfrac{dy}{dt}\right]_{t=1}$, where $y = y(x(t))$, and $x(1) = 3, \dfrac{dx}{dt}(1) = 5, \dfrac{dy}{dx}(3) = 6$

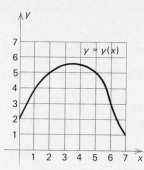

FIGURE 3.31

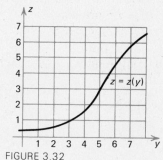

FIGURE 3.32

13.† $\left[\dfrac{dw}{dx}\right]_{x=0}$, where $w = w(v(x))$, and $v(0) = -1, \dfrac{dv}{dx}(0) = 2,$

$\dfrac{dw}{dv}(-1) = -3$

14. $\left[\dfrac{dF}{dt}\right]_{t=5}$, where $F = F(v(t))$, and $v(5) = 300, \dfrac{dv}{dt}(5) = 40, \dfrac{dF}{dv}(300) = 25$

15.‡ $\left[\dfrac{d}{dx}g(f(x))\right]_{x=3}$, where $f(3) = 5, f(6) = 4, f'(3) = 7, f'(6) = 10,$

$g(3) = 6, g(5) = 11, g'(3) = 10,$ and $g'(5) = -1$

16.† $C'(-1)$, where $C(x) = B(A(x))$, and $A(-1) = -3, A'(-1) = -4,$

$B(-3) = -6, B'(-3) = 9$

17. $\dfrac{dF}{dx}(3)$, where $F(x) = G(H(x))$, and $H(3) = 4, H'(3) = 7, G(4) = 5,$

$G'(4) = 6, G'(7) = 10, G'(3) = -2$

18.† The graphs of differentiable functions $y = y(x)$ and $z = z(y)$ are sketched in Figures 3.31 and 3.32. Give approximate values for $z(y(x))$ at $x = 2$ and $x = 5$ and for the derivative dz/dx at those two values of x.

19. The functions whose graphs are sketched in Figures 3.33 and 3.34 give an ice skater's velocity $v(t)$ (feet/minute) t minutes after he has begun skating and his oxygen consumption $A(v)$ (liters/minute) when he is skating v feet/minute. At what approximate rate is his oxygen consumption increasing 2 minutes after he starts skating?

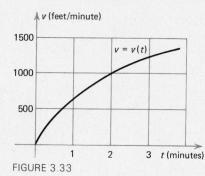

FIGURE 3.33

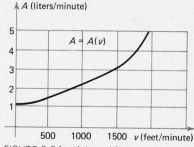

FIGURE 3.34   (Adapted from R. Shephard, [1] *The Physiology of Physical Activity*, p. 537.)

20.† The force of air resistance (drag) on a certain car is $D = \frac{1}{30}v^2$ pounds when the velocity is v miles per hour. The car is accelerating at the constant rate of 2 miles per hour every second. At what rate is the drag increasing when the car is going 50 miles per hour? (Adapted from S. Hoerner, [1] *Fluid Dynamic Drag*, p. 12.7.)

21. The function $D(v)$ of Figure 3.35 gives the air resistance on a certain blunt object as a function of its velocity. (Notice how sharply the curve rises near $v = 740$ miles per hour, the speed of sound. This represents the "sound barrier.") When the object is traveling 725 miles per hour, it is accelerating at the constant rate of 10 miles per hour every second. At approximately what rate is the drag increasing at that moment?

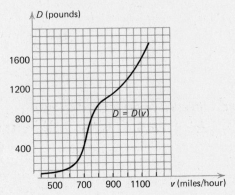

FIGURE 3.35   (Adapted from
S. Hoerner, [1] *Fluid Dynamic Drag*,
p. 16.26.)

Compute the derivatives of the functions in Exercises 22 through 36.

22.[†] $\sqrt{6x^2 + 3}$

23.[†] $(6\sqrt{x} + 3)^3$

24.[†] $\left(\dfrac{x + 1}{x - 1}\right)^5$

25.[†] $(8x)^{1/3} + (8x + 2)^{1/3}$

26.[†] $\dfrac{\sqrt{3x + 2}}{\sqrt{5x - 2}}$

27.[†] $\sqrt{\dfrac{3x + 2}{5x - 2}}$

28.[†] $x(5x + 4)^{1/4}$

29.[†] $[1 - \sqrt{2x + 1}]^{1/2}$

30.[†] $\dfrac{x + \sqrt{5x - 2}}{x^2 + 9}$

31.[†] $(x^2 + 3)^{1/3}(x^3 + 2)^{1/2}$

32. $\dfrac{x^2 - \sqrt{1 + x^2}}{x}$

33. $\dfrac{(2x + 1)^{10} - (3x + 1)^9}{(4x + 1)^9}$

34. $(3x^4 + 1)(3x + 1)^4$

35. $(x^{1/2} + 1)^2(x^{1/3} + 1)^3$

36. $(x^2 + 1)^2(x^3 + 1)^3(x^4 + 1)^4$

In each of Exercises 37 through 40 find   **a.** $\dfrac{dG}{du}(u)$ and   **b.** $\dfrac{du}{dx}(x)$.   **c.** Use the results of parts (a) and (b) to find $\dfrac{d}{dx}G(u(x))$.   **d.** Then give a formula for $G(u(x))$ in terms of x.   **e.** Use that formula to compute $\dfrac{d}{dx}G(u(x))$ directly.

37.[†] $G(u) = \sqrt{u}; u(x) = x^3 + 1$

38.[†] $G(u) = u^3 + 1; u(x) = \sqrt{x}$

39. $G(u) = \dfrac{1}{u} - u; \; u(x) = x^{1/3} + x$

40. $G(u) = u^{1/3} + u; \; u(x) = \dfrac{1}{x} - x$

41.[†] What is the domain of $G(x^3)$ if the domain of $G(u)$ is the interval $1 \le u < 8$?

42.[†] What is the domain of $G\left(\dfrac{1}{x}\right)$ if the domain of $G(u)$ is the interval $-1 \le u \le \tfrac{1}{4}$?

3-8   ANGLES, SINES, AND COSINES

How does the length of a shadow vary with the changing angle of the sun? How does the vertical velocity of a person on a ferris wheel vary as the ferris wheel turns? Questions such as these can be answered by employing calculus in conjunction with the TRIGONOMETRIC FUNCTIONS. In this section we give the definitions of the two basic trigonometric functions, the SINE function and the COSINE function and discuss trigonometric identities that relate them. We will derive formulas for their derivatives in the next section. The other four trigonometric functions, the TANGENT, COTANGENT, SECANT, and COSECANT functions, will be studied in Chapter 7.

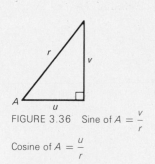

FIGURE 3.36   Sine of $A = \dfrac{v}{r}$

Cosine of $A = \dfrac{u}{r}$

The sine and cosine functions are introduced in trigonometry to relate the angles in right triangles to the lengths of their sides (Figure 3.36). In analytic geometry and calculus these functions are used to relate angles to the coordinates of points in planes. For this purpose it is convenient to give angles counterclockwise or clockwise orientations and to consider angles that are larger than those in triangles. It is also convenient in calculus to measure angles in RADIANS rather than in degrees, as is the custom in trigonometry.

Angles and their radian measure

We think of an angle as being formed by rotating one of its sides, the TERMINAL SIDE, from the position of its other side, the INITIAL SIDE. If this is a counterclockwise rotation, the angle is considered to be positive, and if it is a clockwise rotation, the angle is considered to be negative. The radian measure of an angle is then defined as follows:

Definition 3.5   Consider a circle of radius 1 with its center at the vertex of an angle. If the angle has a counterclockwise orientation, its radian measure is the length of the arc it subtends on the circle (Figure 3.37). If the angle has a clockwise orientation, its radian measure is the negative of the length of the arc it subtends.

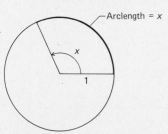

FIGURE 3.37   A positive angle of x radians subtends an arc of length x on a circle of radius 1.

Because the circumference of a circle of radius 1 is 2π, a full counterclockwise revolution is 2π radians, half a counterclockwise revolution is π radians, and a counterclockwise oriented right angle is $\pi/2$ radians. The radian measures of the corresponding clockwise oriented angles are the negatives of these amounts (Figure 3.38).

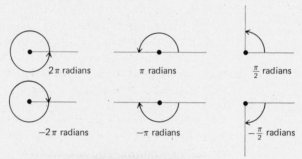

FIGURE 3.38

Because π radians corresponds to 180 degrees, radians are converted into degrees by multiplying by the ratio $\dfrac{180 \text{ degrees}}{\pi \text{ radians}}$. Degrees are converted into radians by multiplying by the inverse ratio $\dfrac{\pi \text{ radians}}{180 \text{ degrees}}$. Thus, one radian is $180/\pi$ degrees or about 57.3 degrees (an angle that subtends an arc with length equal to the radius of the circle). Thirty degrees is $\pi/6$ radians; forty-five degrees is $\pi/4$ radians; sixty degrees is $\pi/3$ radians; and ninety degrees is $\pi/2$ radians.

An angle of x radians ($0 \le x \le 2\pi$) with its vertex at the center of a circle of radius r subtends an arc of length rx on the circumference of the circle and forms a region of area $\frac{1}{2}r^2x$ inside the circle (Figure 3.38a). This is the case because the entire circle has circumference $2\pi r$ and area πr^2 and the angle forms the fraction $x/(2\pi)$ of the circle.

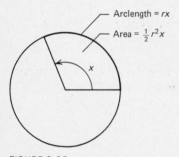

FIGURE 3.38a

Sines and cosines of arbitrary angles

To define the sine and cosine of an arbitrary angle x, we place the vertex of the angle at the origin in a uv-plane with the initial side of the angle along the positive u-axis. Then we draw the UNIT CIRCLE in the uv-plane, which is the circle of radius 1 with its vertex at the origin (Figure 3.39). (We use a uv-plane because we are using the letter "x" to denote angles.)

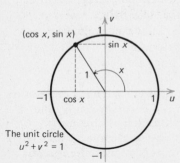

FIGURE 3.39

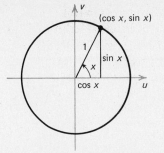

FIGURE 3.40 $\sin x = \dfrac{\text{Opposite}}{\text{Hypotenuse}}$

$\cos x = \dfrac{\text{Adjacent}}{\text{Hypotenuse}}$

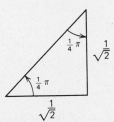

FIGURE 3.40a

$\sin\left(\dfrac{\pi}{4}\right) = \cos\left(\dfrac{\pi}{4}\right) = \dfrac{1}{\sqrt{2}}$

FIGURE 3.40b

$\cos\left(\dfrac{\pi}{3}\right) = \sin\left(\dfrac{\pi}{6}\right) = \dfrac{1}{2}$

$\sin\left(\dfrac{\pi}{3}\right) = \cos\left(\dfrac{\pi}{6}\right) = \dfrac{\sqrt{3}}{2}$

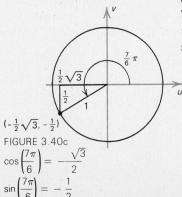

FIGURE 3.40c

$\cos\left(\dfrac{7\pi}{6}\right) = -\dfrac{\sqrt{3}}{2}$

$\sin\left(\dfrac{7\pi}{6}\right) = -\dfrac{1}{2}$

Definition 3.6   **If the initial side of an angle x is the positive u-axis in a uv-plane, then $\cos x$ and $\sin x$ are the coordinates $(\cos x, \sin x)$ of the point of intersection of the terminal side of the angle with the unit circle $u^2 + v^2 = 1$ (Figure 3.39).**

Figure 3.40 shows why Definition 3.6 agrees with the definitions of the sine and cosine from trigonometry for angles x between zero and $\pi/2$. In this case we can consider x to be an angle in a right triangle with hypotenuse of length 1. The numbers $\sin x$ and $\cos x$ are the lengths of the legs of the triangle opposite and adjacent to the angle x, respectively. These are equal to the length of the opposite and adjacent sides divided by the length of the hypotenuse, because the hypotenuse has length 1.

The sine and cosine of $\pi/4$, $\pi/3$, and $\pi/6$ radians can be determined by using the Pythagorean theorem with an isosceles right triangle or a $30°$-$60°$-$90°$ triangle. The isosceles right triangle has two angles of $\pi/4$ radians (Figure 3.40a). If its hypotenuse has length 1 and its legs length s, then $s^2 + s^2 = 1^2$ and consequently $s = 1/\sqrt{2}$. This shows that $\sin(\pi/4)$ and $\cos(\pi/4)$ are both $1/\sqrt{2}$.

The $30°$-$60°$-$90°$ triangle has angles of $\pi/6$ and $\pi/3$ radians and is half of an equilateral triangle (Figure 3.40b). If the hypotenuse of the right triangle has length 1, then the side adjacent to the angle of $\pi/3$ radians has length $\frac{1}{2}$ and the opposite side has length s where $s^2 + (\frac{1}{2})^2 = 1^2$. Solving this equation shows that $s = \frac{1}{2}\sqrt{3}$. Therefore, $\cos(\pi/3)$ and $\sin(\pi/6)$ are $\frac{1}{2}$; $\cos(\pi/6)$ and $\sin(\pi/3)$ are $\frac{1}{2}\sqrt{3}$.

The sine and cosine of integer multiples of $\pi/4$, $\pi/3$, and $\pi/6$ radians can be found by sketching the angle in a uv-plane as in Definition 3.6 and by examining an appropriate isosceles right or $30°$-$60°$-$90°$ triangle. This is illustrated in the next example.

EXAMPLE 1   What are $\sin(7\pi/6)$ and $\cos(7\pi/6)$?

Solution.   We draw the angle with a unit circle in a uv-plane (Figure 3.40c). Because $7\pi/6$ is $\pi/6$ more than π, the triangle shown in the figure is a $30°$-$60°$-$90°$ triangle with hypotenuse of length 1. The length of its vertical leg is $\frac{1}{2}$ and the length of its horizontal leg is $\frac{1}{2}\sqrt{3}$. Therefore, the point where the terminal side of the angle $7\pi/6$ intersects the unit circle has coordinates $(-\frac{1}{2}\sqrt{3}, -\frac{1}{2})$ and $\cos(7\pi/6) = -\frac{1}{2}\sqrt{3}$, $\sin(7\pi/6) = -\frac{1}{2}$. $\square$

The graph of sin x   The graph of the function $\sin x$ in an xy-plane is shown in Figure 3.41. To see why it has the shape it does, imagine the angle x in Figure 3.39 to vary. As the angle x increases, the point $(\cos x, \sin x)$ moves in a counterclockwise direction around the unit circle, and the v-coordinate of that point is $\sin x$. It is zero at $x = 0$, increases to 1 as x increases to $\pi/2$, decreases back to zero as x increases to π, decreases to -1 as x increases to $3\pi/2$, and then increases back to zero as x increases to 2π.

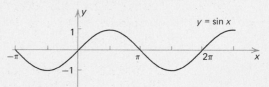

FIGURE 3.41

The shape of the graph of $\sin x$ for $0 \leq x \leq 2\pi$ is repeated an infinite number of times over intervals of length 2π because

✳ (1) $\sin(x + 2n\pi) = \sin x$ ◀

for all x and any integer n. Property (1) follows from the fact that the angles x and $x + 2n\pi$ have the same terminal sides. It is described by saying that $\sin x$ is PERIODIC with period 2π.

The graph of cos x   Similar considerations show that the graph of $\cos x$ has the shape shown in Figure 3.42. The cosine function is also periodic with period 2π: For any x and any integer n we have

✳ (2) $\cos(x + 2n\pi) = \cos x.$ ◀

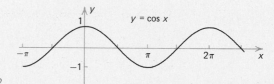

FIGURE 3.42

Trigonometric identities   Because $u = \cos x$ and $v = \sin x$ are defined as coordinates of a point on the unit circle $u^2 + v^2 = 1$ in a uv-plane, the functions $\cos x$ and $\sin x$ satisfy the PYTHAGOREAN IDENTITY

✳ (3) $\sin^2 x + \cos^2 x = 1.$ ◀

Other trigonometric identities also follow readily from the definitions of the sine and cosine. Because the terminal sides of angles x and $-x$ are symmetric about the u-axis (Figure 3.42a), we have for all x

✳ (4) $\cos(-x) = \cos x$ and $\sin(-x) = -\sin x.$ ◀

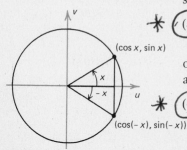

FIGURE 3.42a
$\cos(-x) = \cos x$
$\sin(-x) = -\sin x$

✳ - need to know for test

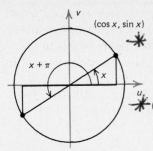

FIGURE 3.42b
$\cos(x + \pi) = -\cos x$
$\sin(x + \pi) = -\sin x$

$(\cos(x + \tfrac{1}{2}\pi), \sin(x + \tfrac{1}{2}\pi))$

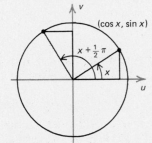

FIGURE 3.42c
$\cos(x + \tfrac{1}{2}\pi) = -\sin x$
$\sin(x + \tfrac{1}{2}\pi) = \cos x$

The identities

(5) $\cos(x + \pi) = -\cos x$ and $\sin(x + \pi) = -\sin x$ ◄

hold because the terminal sides of angles x and $x + \pi$ are symmetric about the origin (Figure 3.42b). We obtain the identities

(6) $\cos(x + \tfrac{1}{2}\pi) = -\sin x$ and $\sin(x + \tfrac{1}{2}\pi) = \cos x$ ◄

because the terminal sides of angles x and $x + \tfrac{1}{2}\pi$ form the hypotenuses of congruent right triangles as in Figure 3.42c, where the horizontal side of each triangle has the same length as the vertical side of the other.

EXAMPLE 2   Use trigonometric identities and the table of values of the sine and cosine in the back of the book to find an approximate value of $\sin(-2)$.

Solution.   We have $\tfrac{1}{2}\pi \approx 1.57$. By identities (4) and (6)

$$\sin(-2) = -\sin(2) = -\sin(0.43 + 1.57) \approx -\sin(0.43 + \tfrac{1}{2}\pi)$$

$$= -\cos(0.43) \approx -0.9090$$

since Table IV in the back of the book gives the approximate value 0.9090 for $\cos(0.43)$.   □

In the process of deriving differentiation formulas for $\sin x$ and $\cos x$ in the next section we will obtain the identities

(7) $\sin x - \sin y = 2 \sin[\tfrac{1}{2}(x - y)]\cos[\tfrac{1}{2}(x + y)]$

(8) $\cos x - \cos y = -2 \sin[\tfrac{1}{2}(x - y)]\sin[\tfrac{1}{2}(x + y)].$

These can then be used to derive the following 13 additional identities.

(9) $\sin(x + y) = \sin x \cos y + \cos x \sin y$

(10) $\cos(x + y) = \cos x \cos y - \sin x \sin y$

(11) $\sin(x - y) = \sin x \cos y - \cos x \sin y$

(12) $\cos(x - y) = \cos x \cos y + \sin x \sin y$

(13) $\cos(\tfrac{1}{2}\pi - x) = \sin x$

(14) $\sin(\tfrac{1}{2}\pi - x) = \cos x$

(15) $\sin(2x) = 2 \sin x \cos x$

(16) $\cos(2x) = \cos^2 x - \sin^2 x = 2 \cos^2 x - 1 = 1 - 2 \sin^2 x$

(17) $\sin^2 x = \tfrac{1}{2}[1 - \cos(2x)]$

(18) $\cos^2 x = \tfrac{1}{2}[1 + \cos(2x)]$

(19) $\sin x \sin y = \tfrac{1}{2}[\cos(x - y) - \cos(x + y)]$

(20) $\cos x \cos y = \tfrac{1}{2}[\cos(x - y) + \cos(x + y)]$

(21) $\sin x \cos y = \tfrac{1}{2}[\sin(x - y) + \sin(x + y)]$

Identities (7) through (21) need not be memorized as they are infrequently used in calculus and may be found in reference books when they are required. Of the fifteen identities, the most important for our purposes will be (17) through (21), which we will use in Chapter 8 to evaluate integrals.

We show how to derive (9) in the next example and leave the verification of (10) through (21) as exercises.

EXAMPLE 3   Derive identity (9) from (4) and (7).

Solution.   We first make the substitutions $x = u + v$ and $y = u - v$ in (7). Because $\frac{1}{2}(x - y) = v$ and $\frac{1}{2}(x + y) = u$, we obtain

(22) $\sin(u + v) - \sin(u - v) = 2 \sin v \cos u.$

Then reversing the roles of u and v in (22) and using (4) to write $\sin(v - u) = -\sin(u - v)$ gives

(23) $\sin(u + v) + \sin(u - v) = 2 \sin u \cos v.$

We obtain (9) with u and v in place of x and y by adding equations (22) and (23).   □

Amplitude, frequency, and phase shifts

Because the sine and cosine oscillate between -1 and 1, the maximum and minimum values of the functions

(24) $C \sin(kx + \psi)$ and $C \cos(kx + \psi)$ $(k \neq 0)$

are $|C|$ and $-|C|$. The number $|C|$ is called the AMPLITUDE of these functions. Each of these functions is periodic of period $2\pi/|k|$. The reciprocal $|k|/(2\pi)$ of their period is called their FREQUENCY. The constant ψ in expression (24) is called a PHASE SHIFT.

The identity $\sin(kx + \frac{1}{2}\pi) = \cos(kx)$ implies that $\cos(kx - \frac{1}{2}\pi) = \sin(kx)$, so that the sine and cosine can be obtained from each other by making phase shifts of $\frac{1}{2}\pi$ or $-\frac{1}{2}\pi$. As we illustrate in the next example, any linear combination $A \sin(kx) + B \cos(kx)$ of $\sin(kx)$ and $\cos(kx)$ can be expressed in either of the forms (24) with $C = \sqrt{A^2 + B^2}$, that is, either as a sine or a cosine with the same frequency, with amplitude $\sqrt{A^2 + B^2}$, and with a suitable phase shift.

EXAMPLE 4   Find constants C and ψ such that

(25) $\sqrt{3} \sin(6x) + \cos(6x) = C \sin(6x + \psi).$

Solution.   We will use the identity

(26) $\sin(kx + \psi) = \cos \psi \sin(kx) + \sin \psi \cos(kx)$

which follows from (9). The function on the left of (25) is of the form $A \sin(kx) + B \cos(kx)$ with $A = \sqrt{3}$ and $B = 1$. We set $C = \sqrt{A^2 + B^2} = \sqrt{(\sqrt{3})^2 + 1} = 2$ and factor this quantity from the given expression for the function:

(27) $\sqrt{3} \sin(6x) + \cos(6x) = 2[\frac{1}{2}\sqrt{3} \sin(6x) + \frac{1}{2} \cos(6x)].$

Because we divided by C, the coefficients of the sine and cosine inside the square brackets in (27) are the coordinates of a point $(\frac{1}{2}\sqrt{3}, \frac{1}{2})$ on the unit circle of a uv-plane and there are angles ψ such that $\cos\psi = \frac{1}{2}\sqrt{3}$ and $\sin\psi = \frac{1}{2}$. In fact, we may take $\psi = \frac{1}{6}\pi$ and rewrite (27) as

$$3\sin(6x) + \cos(6x) = 2[\cos(\tfrac{1}{6}\pi)\sin(6x) + \sin(\tfrac{1}{6}\pi)\cos(6x)]$$
$$= 2\sin(6x + \tfrac{1}{6}\pi).$$

This gives the expression on the right of (25) with $C = 2$ and $\psi = \frac{1}{6}\pi$. □

The laws of sines and cosines

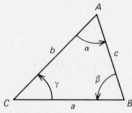

FIGURE 3.42d

Many problems in trigonometry can be conveniently solved by using either the LAW OF SINES or the LAW OF COSINES. To describe these results, we let α, β, and γ denote the positive interior angles of a triangle ABC as in Figure 3.42d, and we let a, b, and c denote the lengths of the sides opposite α, β, and γ, respectively.

Then the law of sines states that

$$(28) \qquad \frac{\sin\alpha}{a} = \frac{\sin\beta}{b} = \frac{\sin\gamma}{c}. \qquad \blacktriangleleft$$

We leave its derivation as Exercise 28. The law of sines can be used, for example, when we know two angles of a triangle and the length of the side between them (see Exercises 30 and 31).

The law of cosines

$$(29) \qquad c^2 = a^2 + b^2 - 2ab\cos\gamma \qquad \blacktriangleleft$$

enables us to find the length of the third side of a triangle if we know the lengths of the other two sides and the angle between them (see Exercises 32 and 33). We leave the derivation of (29) as Exercise 29.

Exercises

†*Answer provided.*
‡*Outline of solution provided.*

1. Sketch each of the following angles, give its radian measure, and compute its sine and cosine. **a.**‡ $-60°$ **b.**‡ $135°$ **c.**† $450°$ **d.** $-135°$ **e.** $-540°$

2. Sketch the following angles, compute their sines and cosines, and give their degree measures. **a.**† 7π radians **b.**† $-10\pi/3$ radians **c.**† $-3\pi/2$ radians **d.** $3\pi/4$ radians **e.** $25\pi/6$ radians

3.† Give the radian measures of all angles between the hands of a clock at 5:30.

4. Use the trigonometric tables to give values of $\sin x$ and $\cos x$, correct to three decimal places, for **a.**† $x = 1$ **b.**† $x = -\frac{1}{5}$ **c.**† $x = 10$ **d.** $x = 35$ **e.** $x = -4$.

5. Sketch the graphs of **a.**‡ $2\sin(x + \pi/6)$ **b.**† $-\cos(3x)$ **c.** $10\sin(\frac{1}{4}(x - \pi))$.

6.‡ What is the length of the arc subtended on a circle of radius 5 feet by an arc of $160°$?

7.† An angle of 4 radians with its vertex at the center of a circle subtends an arc 10 feet long on the circle. What is the radius of the circle?

8.† A gear of radius 20 inches and a gear of radius 30 inches are engaged. The larger gear is turning at the constant rate of 3 revolutions per minute. **a.** How fast, in inches per minute, is a point on the circumference of the larger gear moving?   **b.** How fast is a point on the circumference of the smaller gear turning?   **c.** How fast, in revolutions per minute, is the smaller gear turning?

9.† Compute the area of the sector of a circle of radius 12 inches formed by an angle of 1.865 radians with its vertex at the center of the circle.

Find all solutions x of the equations in Exercises 10 through 19.

10.‡ $\sin(5x) = -1/\sqrt{2}$ **11.**† $\cos(\tfrac{1}{7}x) = -1$

12.† $\sin(2x) = -\tfrac{1}{2}$ **13.**† $\sin^2 x = \tfrac{1}{2}$

14.† $\sin(x^2) = \tfrac{1}{2}$ **15.**† $\cos(3x + 1) = 0$

16.† $\cos^3(2x) = -1$ **17.** $\sin x - \sin^2 x = 0$

18. $\cos(x^3) = -\tfrac{1}{2}\sqrt{3}$ **19.** $\cos(1/x) = -\tfrac{1}{2}$

20. Derive identities (10) through (12) from (4) through (9).

21. **a.** Derive (13) and (14) from (11) and (12).   **b.** Derive (13) and (14) for positive acute angles x $(0 < x < \tfrac{1}{2}\pi)$ from the definitions of sines and cosines used in trigonometry.

22. Derive (15) and (16) from (9) and (10) and then (17) and (18) from (16).

23. Use (9) through (12) to verify (19) through (21).

In Exercises 24 through 27 express the given function in the form $C \sin(kx + \psi)$ and give its frequency and amplitude.

24.† $-5 \sin(-2x) + 5 \cos(-2x)$ **25.**† $7 \sin x - 7\sqrt{3} \cos x$

26. $3 \sin(16x) + 3 \cos(16x)$ **27.** $-10 \cos(\tfrac{1}{5}x)$

28. Derive the law of sines (28). (Use the angles α and β and the lengths a and b to compute in two ways the length of the perpendicular line from the vertex C to the side AB.)

29. Derive the law of cosines (29). (Use the Pythagorean theorem with uv-coordinates such that the origin is at C and the line AB is along the positive u-axis.)

In Exercises 30 through 34 give exact and approximate answers.

30.† Surveyors who want to measure the distance from a point A to a point B across a river pick a point C 100 meters away from A on the same side of the river and measure the angle at A in the triangle ABC to be 1.9 radians and the angle at C to be 0.8 radians. What is the distance $\overline{AB}$?

31.[†] One town on a lake is 5 miles directly north of a second town. A sailboat sails southeast from the first town and then turns and sails west southwest to the second town. How far does it sail?

32.[†] Two sides of a triangle are 5 and 7 feet long and the angle between them is $\frac{1}{3}\pi$ radian. How long is the other side?

33.[†] If town B is 20 kilometers north northeast of town A and town B is 30 kilometers southwest of town A, how far apart are towns A and B?

3-9   THE DERIVATIVES OF SIN x AND COS x

The derivatives of $\sin x$ and $\cos x$ at a value x_0 of x are defined to be the limits

$$(1) \qquad \left[\frac{d}{dx}\sin x\right]_{x=x_0} = \lim_{x \to x_0}\frac{\sin x - \sin x_0}{x - x_0}$$

$$(2) \qquad \left[\frac{d}{dx}\cos x\right]_{x=x_0} = \lim_{x \to x_0}\frac{\cos x - \cos x_0}{x - x_0}.$$

The quantity $x - x_0$ in the denominators of the difference quotients on the right of equations (1) and (2) represents an angle and therefore is determined by the length of an *arc of a circle*. The numerators of the difference quotients, on the other hand, are differences in the values of the sine and cosine and are determined by the *lengths of straight lines*. Because of this we can base our study of the derivatives (1) and (2) on the relationship between arclength and chord length in a circle that is described in the following lemma.

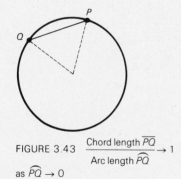

FIGURE 3.43 $\dfrac{\text{Chord length } \overline{PQ}}{\text{Arc length } \widehat{PQ}} \to 1$

as $\widehat{PQ} \to 0$

Lemma 3.1   Let P and Q be two points on a circle. Let $\widehat{PQ}$ denote the length of the arc and $\overline{PQ}$ the length of the chord joining them (Figure 3.43). Then the ratio $\overline{PQ}/\widehat{PQ}$ of the chord length to the arc length tends to 1 as the two points move together:

$$(3) \qquad \frac{\textbf{Chord length } \overline{PQ}}{\textbf{Arc length } \widehat{PQ}} \to 1 \quad \textbf{as} \quad \widehat{PQ} \to 0.$$

We can obtain an instructive partial proof of Lemma 3.1 by considering the special case where the points P and Q are adjacent vertices in an equilateral polygon of n sides inscribed in the circle (see Figure 3.43a for the case of $n = 6$, where the polygon is an equilateral hexagon). In this case the perimeter of the entire polygon is n times the length $\overline{PQ}$ of one side, and the circumference of the circle is n times the arclength $\widehat{PQ}$. Consequently,

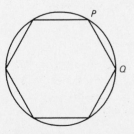

FIGURE 3.43a

$n = 6$

The circumference of the circle is $6(\widehat{PQ})$.

The perimeter of the hexagon is $6(\overline{PQ})$.

$$(4) \qquad \frac{\overline{PQ}}{\widehat{PQ}} = \frac{n(\overline{PQ})}{n(\widehat{PQ})} = \frac{\text{Perimeter of the polygon}}{\text{Circumference of the circle}}$$

and, because the perimeter of the n-sided polygon tends to the circumference of the circle as n tends to ∞, the ratio (4) tends to 1 as n tends to ∞ and $\widehat{PQ}$ tends to zero.

To establish the lemma for arbitrary chords PQ in the circle, we will use areas instead of arclength.

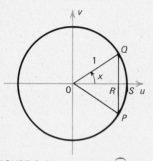

FIGURE 3.44   Arc length $\widehat{PQ} = 2x$
Chord length $\overline{PQ} = 2 \sin x$

Proof.   We may assume that the circle has radius 1 since changing the radius multiplies the chord length $\overline{PQ}$ and the arc length $\widehat{PQ}$ by the same factor. We take the circle to be the unit circle in a uv-plane, and we suppose that the points P and Q are symmetric about the u-axis as in Figure 3.44. We let x denote the positive acute angle shown in the figure.

Then the arc length $\widehat{SQ}$ is equal to x and the chord length $\overline{RQ}$ is equal to $\sin x$, so that the arc length $\widehat{PQ}$ is $2x$ and the chord length $\overline{PQ}$ is $2 \sin x$. Therefore, the ratio $\overline{PQ}/\widehat{PQ}$ is equal to the ratio $\sin x/x$, and we can establish (3) by showing that

(5) $\displaystyle \lim_{x \to 0} \frac{\sin x}{x} = 1.$

We will derive (5) by considering areas. The triangle ORQ in Figure 3.45 is contained in the sector OSQ of the circle, and the sector OSQ of the circle is contained in the triangle OST. Therefore,

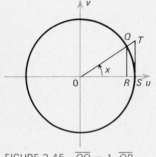

FIGURE 3.45 $\overline{OQ} = 1$, $\overline{OR}$
$= \cos x$, $\overline{RQ} = \sin x$, $\overline{OS} = 1$
$\overline{ST} = \dfrac{\sin x}{\cos x}$

(6) $\begin{array}{ccccc} \text{Area of} & & \text{Area of} & & \text{Area of} \\ \text{triangle } ORQ & \leq & \text{sector } OSQ & \leq & \text{triangle } OST. \end{array}$

The length $\overline{RQ}$ is $\sin x$ and the length $\overline{OR}$ is $\cos x$. The length $\overline{OS}$ is 1, and by similar triangles the length $\overline{ST}$ is $\sin x/\cos x$. Consequently, triangle ORQ has area $\frac{1}{2}(\sin x)(\cos x)$, whereas triangle OST has area $\frac{1}{2}(\sin x/\cos x)$. The sector OSQ is the fraction $x/2\pi$ of the unit circle, and the unit circle has area π, therefore, sector OSQ has area $x/2$.

When these formulas for the areas are put in inequalities (6) and the resulting inequalities are multiplied by 2, we obtain

(7) $(\sin x)(\cos x) \leq x \leq \dfrac{\sin x}{\cos x}.$

Dividing the first inequality in (7) by x and $\cos x$ gives

(8) $\dfrac{\sin x}{x} \leq \dfrac{1}{\cos x}.$

Multiplying the second inequality in (7) by $\cos x$ and dividing it by x gives

(9) $\cos x \leq \dfrac{\sin x}{x}.$

We have derived inequalities (8) and (9) for $0 < x < \pi/2$. They are also valid for $-\pi/2 < x < 0$ because $\cos(-x) = \cos x$ and $\big(\sin(-x)\big)/(-x) = (\sin x)/x$. Figure 3.44 gives the inequalities

$0 < 1 - \cos x = \overline{RS} < \overline{QS} < \overline{QS} = |x|$ because the length $\overline{RS}$ of the leg of the right triangle RSQ is less than the length $\overline{QS}$ of the hypotenuse, and the chord length $\overline{QS}$ is less than the arclength $\overparen{QS}$. Hence $\cos x$, and consequently $1/(\cos x)$, tend to 1 as x tends to 0. Inequalities (8) and (9) show that $(\sin x)/x$ lies between $\cos x$ and $1/(\cos x)$, so it must also tend to 1.   Q.E.D.

The derivatives of sin x and cos x

We will now use Lemma 3.1 to derive the formulas for the derivatives of $\sin x$ and $\cos x$.

Theorem 3.7   The functions sin x and cos x are continuous and have derivatives for all x given by the formulas

(10) $\dfrac{d}{dx} \sin x = \cos x$

and   ◀

(11) $\dfrac{d}{dx} \cos x = -\sin x.$

Proof.   We consider x and x_0 with $0 \le x_0 < x < \pi/2$ and the right triangle PQR with vertices $P = (\cos x_0, \sin x_0)$ and $Q = (\cos x, \sin x)$ and legs parallel to the coordinate axes as in Figure 3.46. The terminal side OT of the angle $\frac{1}{2}(x + x_0)$, which is the average of x and x_0, bisects the angle POQ and, by a result from plane geometry, is the perpendicular bisector of the chord PQ (Figure 3.47). Thus, the sides OT and OS of the angle SOT are perpendicular to the sides QR and QP of the angle at Q in the triangle PQR. By another result from plane geometry, these two angles are equal, and the angle at Q in the triangle PQR is equal to $\frac{1}{2}(x + x_0)$.

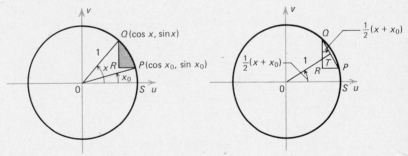

FIGURE 3.46 $\overline{RQ} = \sin x - \sin x_0.$
$\overline{PQ} = x - x_0$

FIGURE 3.47

Using the triangle PQR to compute the cosine of the angle at Q yields $\cos\left[\frac{1}{2}(x + x_0)\right] = \overline{RQ}/\overline{PQ}$, and since $\overline{RQ}$ is the difference, $\sin x - \sin x_0$, between the vertical coordinates of the points P and Q, we obtain

(12) $\sin x - \sin x_0 = \cos \left[\frac{1}{2}(x + x_0) \right] \ \overline{PQ}.$

Similarly, because the sine of $\frac{1}{2}(x + x_0)$ is $\overline{RP}/\overline{PQ}$ and $\overline{RP}$ equals $\cos x_0 - \cos x = -(\cos x - \cos x_0)$, we have

(13) $\cos x - \cos x_0 = -\sin \left[\frac{1}{2}(x + x_0) \right] \overline{PQ}.$

The proof of the theorem is based on equations (12) and (13). First, because the sine and cosine of $\frac{1}{2}(x + x_0)$ are ≤ 1, they give the inequalities

(14) $|\sin x - \sin x_0| \leq \overline{PQ}$ and $|\cos x - \cos x_0| \leq \overline{PQ}$

for $0 < x_0 < x < \pi/2$. These inequalities can be obtained for $0 < x < x_0 < \pi/2$ by interchanging the roles of x and x_0. The chord length $\overline{PQ}$ tends to zero as x tends to x_0, so by (14) $\sin x$ tends to $\sin x_0$ and $\cos x$ tends to $\cos x_0$. This shows that $\sin x$ and $\cos x$ are continuous at x_0, which is an arbitrary number with $0 < x_0 < \pi/2$.

Next we will use (12) to compute the derivative of $\sin x$ at x_0. For $0 < x_0 < x < \pi/2$ the arclength $\overset{\frown}{PQ}$ is the difference $x - x_0$ between the angles x and x_0. Dividing (12) by this amount gives

(15) $\dfrac{\sin x - \sin x_0}{x - x_0} = \cos \left[\frac{1}{2}(x + x_0) \right] \dfrac{\overline{PQ}}{\overset{\frown}{PQ}}.$

We obtain (15) for $0 < x < x_0 < \pi/2$ by switching the roles of x and x_0. As x tends to x_0, the angle $\frac{1}{2}(x + x_0)$ tends to x_0 and $\cos \left[\frac{1}{2}(x + x_0) \right]$ tends to $\cos x_0$. By Lemma 3.1, the ratio $\overline{PQ}/\overset{\frown}{PQ}$ tends to 1, so (15) yields

(16) $\left[\dfrac{d}{dx} \sin x \right]_{x=x_0} = \lim_{x \to x_0} \dfrac{\sin x - \sin x_0}{x - x_0} = \cos x_0.$

This is formula (10) at $x = x_0$, an arbitrary number with $0 < x_0 < \pi/2$. Formula (11) at $x = x_0$ follows similarly from equation (13) (see Exercise 24). We leave the proof of the theorem at $x = 0$ as Exercises 25 and 26. The theorem for x outside the interval $0 \leq x < \pi/2$ follows from identities (6) of Section 3.8 (see, for example, Exercise 27). Q.E.D.

Note.   Equations (12) and (13) yield for $0 \leq x_0 < x < \frac{1}{2}\pi$ the identities

$$\sin x - \sin x_0 = 2 \sin \left[\frac{1}{2}(x - x_0) \right] \cos \left[\frac{1}{2}(x + x_0) \right]$$

$$\cos x - \cos x_0 = -2 \sin \left[\frac{1}{2}(x - x_0) \right] \sin \left[\frac{1}{2}(x + x_0) \right]$$

which we presented as equations (7) and (8) of the last section, because the lengths $\overline{PT}$ and $\overline{TQ}$ in Figure 3.47 are both equal to $\sin \left[\frac{1}{2}(x - x_0) \right]$ and consequently the chord length $\overline{PQ}$ is $2 \sin \left[\frac{1}{2}(x - x_0) \right]$. The identities are valid if $x = x_0$ since then both sides are zero. They can be shown to hold for other ranges of x and x_0 by using identities (4) and (6) of Section 3.8.   □

EXAMPLE 1   Compute the derivative of $5 \sin x - 6 \cos x$.

Solution.   By the rule for differentiating linear combinations of functions and by formulas (10) and (11), we have

$$\frac{d}{dx}(5 \sin x - 6 \cos x) = 5\frac{d}{dx}\sin x - 6\frac{d}{dx}\cos x$$
$$= 5 \cos x - 6(-\sin x)$$
$$= 5 \cos x + 6 \sin x. \quad \square$$

Derivatives of composite functions   When we combine formulas (10) and (11) with the chain rule, we obtain the formulas

$$(17) \quad \frac{d}{dx}\sin(u(x)) = \cos(u(x))\frac{du}{dx}(x) \qquad \blacktriangleleft$$

$$(18) \quad \frac{d}{dx}\cos(u(x)) = -\sin(u(x))\frac{du}{dx}(x) \qquad \blacktriangleleft$$

for differentiating sines and cosines of functions.

EXAMPLE 2   Compute the derivative $\dfrac{d}{dx}\cos(6x^3 - 1)$.

Solution.   We apply formula (18) with $u(x) = 6x^3 - 1$ to obtain

$$\frac{d}{dx}\cos(6x^3 - 1) = -\sin(6x^3 - 1)\frac{d}{dx}(6x^3 - 1)$$
$$= -18x^2 \sin(6x^3 - 1). \quad \square$$

EXAMPLE 3   What is the derivative of $\sqrt{\sin x}$?

Solution.   By the chain rule formula for the derivative of the square root of a function we have

$$\frac{d}{dx}\sqrt{\sin x} = \frac{d}{dx}(\sin x)^{1/2} = \tfrac{1}{2}(\sin x)^{-1/2}\frac{d}{dx}\sin x$$
$$= \tfrac{1}{2}(\sin x)^{-1/2}\cos x. \quad \square$$

EXAMPLE 4   Find the derivative of $x^3 \sin^5(2x)$.

Solution.   To compute a value of $x^3 \sin^5(2x)$ from x, we first find $2x$, then the sine, then its fifth power and x^3, and finally the product. Hence, to find the derivative, we first use the product rule, then the rules for differentiating x^3 and u^5 with $u = \sin(2x)$, and finally we differentiate $\sin(2x)$ by the chain rule formula (17):

$$\frac{d}{dx}[x^3 \sin^5(2x)] = \left[\frac{d}{dx}x^3\right]\sin^5(2x) + x^3\left[\frac{d}{dx}\sin^5(2x)\right]$$

$$= 3x^2 \sin^5(2x) + x^3[5\sin^4(2x)]\frac{d}{dx}\sin(2x)$$

$$= 3x^2 \sin^5(2x) + 5x^3 \sin^4(2x) \cos(2x)\frac{d}{dx}(2x)$$

$$= 3x^2 \sin^5(2x) + 10x^3 \sin^4(2x) \cos(2x). \quad \square$$

Why radians are used in calculus

If we, for the moment, let $S(x)$ and $C(x)$ denote the sine and cosine of an angle of x *degrees*, we obtain the formulas $S(x) = \sin(\pi x/180)$ and $C(x) = \cos(\pi x/180)$. Using (17) and (18) to compute the derivatives of these functions yields $\dfrac{d}{dx}S(x) = \dfrac{\pi}{180}C(x)$ and $\dfrac{d}{dx}C(x) = -\dfrac{\pi}{180}S(x)$. The complexity of these formulas in contrast to formulas (10) and (11), shows why we use radians in calculus instead of degrees.

Exercises

In Exercises 1 through 11 compute the indicated derivatives.

†*Answer provided.*
‡*Outline of solution provided.*

1.† $\left[\dfrac{d}{dx}\cos x\right]_{x=\pi/3}$

2. $\left[\dfrac{d}{dx}\sin x\right]_{x=\pi}$

3.† $\left[\dfrac{d}{dx}(3\sin x + 10\cos x)\right]_{x=2}$

4. $\left[\dfrac{d}{dx}(5x^2 - 4\sin x)\right]_{x=1}$

5.‡ $\dfrac{d}{dx}(x^2 \cos x)$

6.† $\dfrac{d}{dx}\left(\dfrac{\sin x}{\cos x}\right)$

7. $\dfrac{d}{dx}[(\sin x)(\cos x)]$

8.‡ $\dfrac{d}{dt}[4\sin(6t) - 6\sin(4t)]$

9.† $\dfrac{d}{dw}[w\cos(w^2)]$

10. $\dfrac{d}{dx}\left[\dfrac{\sin(x^2)}{x}\right]$

11.‡ $\dfrac{d}{dx}\sin^3(5x)$

In Exercises 12 through 15 give equations of the tangent lines.

12.‡ The tangent line to the graph of $\sin x$ at $x = 3$

13.† The tangent line to the graph of $x \cos x$ at $x = \pi$

14. The tangent line to the graph of $\cos x/(\sin x + 2)$ at $x = \pi/2$

15. The tangent line to $y = \sin(x^2)$ at $x = 5$

16.† A right triangle has hypotenuse 6 feet long, and one of its acute angles is x radians. What is the rate of change of the area of the triangle with respect to x?

17. A bullet fired from the ground with a muzzle velocity of 800 feet per second with no air resistance and at an angle of α (alpha) radians with the horizontal $(0 < \alpha < \pi/2)$ will reach a height of $h = 10,000 \sin^2 \alpha$ feet above the ground. What is the rate of change of that height with respect to α at **a.** $\alpha = \pi/6$ **b.** $\alpha = \pi/4$ **c.** $\alpha = \pi/3$?

18.† The upper end of a 10 foot long ladder is sliding down a vertical wall while its base is sliding along the horizontal ground. What is the rate of change of the height of the top of the ladder with respect to the positive acute angle x between the ladder and the ground?

19.† The lengths a and b of two sides of a triangle are constant while the angle x between them is changing. What is the rate of change with respect to x of the length of the opposite side. (Use the law of cosines.)

20.† One angle in a triangle is $\frac{1}{4}\pi$ radians and the opposite side has length 5 feet. One of the other angles is x radians and the side opposite it has length y feet. **a.** Express y in terms of x (Use the law of sines.) **b.** What is the rate of change of y with respect to x?

21. What is the rate of change of one acute angle in a right triangle with respect to the other angle?

22. A triangle ABC is expanding in such a way that the angle at A is always 0.5 radians and the angle at B is always 1 radian. Let x denote the length of the side opposite A, y the length of the side opposite B, and z the length of the side opposite C. What are **a.** $\dfrac{dy}{dx}$, **b.** $\dfrac{dz}{dx}$, and **c.** $\dfrac{dz}{dy}$? Give exact and approximate answers.

23. An airplane is flying northeast from the town of Smallville, which is 50 miles east of the town of Littleton. What is the rate of change of the distance from the airplane to Littleton with respect to the distance from the airplane to Smallville?

24. Derive formula (11) for the derivative of $\cos x$ at x_0 $(0 < x_0 < \frac{1}{2}\pi)$ from equation (13).

25. According to formula (10) at $x = 0$, the derivative of $\sin x$ at 0 is 1. Verify this directly from the definition of the derivative and Lemma 3.1.

26. According to formula (11) at $x = 0$, the derivative of $\cos x$ is 0 at $x = 0$. Verify this directly from the definition of the derivative and Lemma 3.1. (Write $\cos x = \sqrt{1 - \sin^2 x}$ for x near 0 and rationalize the square root.)

27. Derive Theorem 3.7 for $\frac{1}{2}\pi \leq x < \pi$ from that result for $0 \leq x < \frac{1}{2}\pi$ and trigonometric identities (6) of Section 3.8.

Compute the derivatives of the functions in Exercises 28 through 41.

28.† $\sin^3 x \cos^5 x$

29. $\sin(x^3) \cos(x^5)$

30.† $(x^3 \cos x)^{4/3}$

31. $x \sin x + \cos x$

32.† $\cos^{1/2} x + \cos(x^{1/2})$

33.† $\sin(x^3 + x)$

34.† $\sqrt{\cos(2x)}$

35.† $\sin^2(\sqrt{x})$

36.[†] $\dfrac{\sin(3x)}{\sin(2x)}$

37.[†] $\sin(\cos x)$

38.[†] $x^5\sqrt{1 + \cos x}$

39.[†] $\sqrt{\dfrac{\sin x}{\cos x}}$

40.[†] $[x - \sin(x^3)]^3$

41. $\cos(x \sin x)$

3-10 REMARKS ON THE USE OF MATHEMATICAL MODELS (Optional)

To study an object's velocity as a derivative, we have to use a mathematical model of the motion, as when we say that a car is at $s = t^2$ (miles) on an s-axis at time t (hours). In this case we conclude that the car's velocity at time t is $\dfrac{d}{dt} t^2 = 2t$ miles per hour. This mathematical model would only approximate the car's motion since even the most accurate instruments for measuring the time and the car's position would introduce some margin of error.

Once we have chosen the mathematical model to approximate the motion of the car, we must ignore the inevitable errors that the model embodies if we are to calculate the car's velocity as a derivative. To illustrate this point, let us see what would happen if we were to allow for small but unknown errors in the measurement of the car's position when we compute the derivative.

Let us suppose that somehow we know that the car is at $s = t^2$ at time t with a possible error of up to 10^{-13} miles in the measurement. This would be an exceedingly small error because 10^{-13} miles is less than the radius of an atom. At time 3 the car would be at $s = 3^2 +$ one error $= 9 +$ one error, and at time t it would be at $s = t^2 +$ another error, where the terms labeled "one error" and "another error" are unknown quantities whose absolute values are less than 10^{-13}. Using these expressions for the car's position, we see that its average velocity in the time between 3 and t is

$$\frac{(t^2 + \text{One error}) - (9 + \text{Another error})}{t - 3}$$

$$= \frac{t^2 - 9}{t - 3} + \frac{(\text{One error}) - (\text{Another error})}{t - 3}$$

$$= (t + 3) + \frac{(\text{One error}) - (\text{Another error})}{t - 3}.$$

The two unknown error terms might be equal and cancel, but they could be of opposite signs and, in effect, add. All we can say about the difference, (one error) − (another error), is that it is an unknown quantity whose absolute value is less than $2(10^{-13})$. Therefore, we can say only that

$$(1) \quad \frac{\text{The car's average velocity}}{\text{between times 3 and } t} = (t + 3) + \frac{(\text{An error})}{t - 3}$$

where the term labeled "an error" is an unknown quantity whose absolute value is less than $2(10^{-13})$ miles.

For t close to 3, the term $t + 3$ on the right side of (1) is close to 6 and, unless t is exceedingly close to 3, the second term will be negligible. However, the mathematical procedure of differentiation calls for having t *tend* to 3, which involves values of t so close to 3 that the second term might be large. For example, if t is $3 + 10^{-13}$, the second term on the right of (1) might be as large as 2, and if t is $3 + 10^{-113}$, it might be as large as 2×10^{100}.

Therefore, if we include the unknown error terms in our calculations, we cannot say that the limit of the average velocity as t tends to 3 is 6 or even that the limit exists. In order to take the limit in the definition of the derivative, we have to deal with the mathematical model in which we suppose that the car is exactly at $s = t^2$ on the s-axis at time t.

3-11   HISTORICAL NOTES (Optional)

Tangent lines to conic sections

One of the earliest definitions of a tangent line was given by the Greek geometer Apollonius (247–222 B.C.) whose treatise *Conics* was the most important work on the conic sections (ellipses, hyperbolas, and parabolas) up to the seventeenth century. Apollonius described a tangent line to a conic as a line such that "no other straight line can fall between it and the conic section."*

Kinematic definitions of tangent lines

Another early approach to the study of tangent lines was based on the principle that an object's motion at each point as it moves along a curve is in the direction of the tangent line at that point. The Greek scientist and mathematician Archimedes (c. 287–212 B.C.) used this idea to find the lines tangent to a spiral traversed by a point that moves along a ray at a constant speed while the ray rotates about its end at a constant speed (Figure 3.48). (Such a spiral is now known as an Archimedean spiral.)

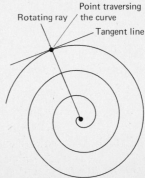

Point traversing the curve

Rotating ray

Tangent line

FIGURE 3.48   Archimedean spiral

---

*T. Heath, [1] *Apollonius of Perga*, p. 22.

This kinematic approach to tangent lines was also applied in the 1640's by an Italian physicist and student of Galileo, Evangelista Torricelli (1608–1647), and by a French mathematician Gilles Persone de Roberval (1602–1675). They used this approach to determine tangent lines to a variety of curves, including those given in modern notation by the equations $y = ax^n$ with various rational exponents n. The application of this approach to find the tangent lines to a parabola (the case of $y = ax^n$ with $n = 2$) is outlined in Exercise 1.

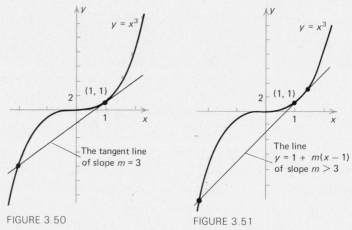

FIGURE 3.50   FIGURE 3.51

Descartes' method of equal roots

In the 1630's Descartes devised a method for finding normal lines to certain curves by studying the solutions of polynomial equations. We will describe this so-called "method of equal roots" as it would be used to find the tangent line at the point (1, 1) on the curve $y = x^3$ (Figure 3.50). (Descartes' application of the method yielded the normal lines rather than the tangent lines because he used circles where we will use lines.)

The line of slope m through the point (1, 1) has the equation $y = 1 + m(x - 1)$. The coordinates of the points of intersection of this line with the curve $y = x^3$ are the solutions of the simultaneous equations $y = x^3$ and $y = 1 + m(x - 1)$, and when we equate these two expressions for y we have the equation

$$(1) \qquad x^3 = 1 + m(x - 1)$$

for the x-coordinates of the points of intersection.

We can expect the cubic equation (1) to have three solutions. To apply the method of equal roots, we choose m so that two of these solutions are equal, that is, so that equation (1) has only two distinct solutions, and conclude that the slope of the tangent line is $m = 3$ (see Exercise 2).

The connection between this approach and the modern definition of the derivative is indicated in Figure 3.51. For m close to 3 the line of slope m through $(1, 1)$ is a secant line and intersects the curve at three distinct points. As m approaches 3, the point near $(1, 1)$ approaches $(1, 1)$ and the secant line approaches the tangent line. In the limit, for $m = 3$, the two points coincide and the line intersects the curve at only two points.

Constant and variable velocity

Aristotle believed that an object falling in a resisting medium such as air would have a constant velocity proportional to its weight. This view was commonly held through the Middle Ages, although it was challenged by various scholars who observed that the velocity is not constant but increases with time. In the fourteenth century, scholastic philosophers at Merton College, Oxford, studied motion with constant acceleration and deduced what is now known as the *Merton rule:* an object with constant acceleration travels the same distance as it would if it had constant velocity equal to the average of its initial and final velocities (see Exercise 3). In the seventeenth century Galileo and others discovered that in a void all falling objects have the same constant acceleration, so that their motion may be determined by using the Merton rule. This result, however, did not resolve the question of motion of an object falling in air. The need for mathematical descriptions of such phenomena contributed to the development of the concept of derivative.

Infinitesimals

The approaches to the study of tangent lines and velocity that led to the modern concept of the derivative were based on one notion or another of the "infinitely small." Some mathematicians considered a curve to be composed of "infinitely short" line segments and contended that the tangent line at a point is the extension of the "infinitely short" line segment at that point. Others stated that tangent lines are the lines through pairs of "infinitely close" points on the curve. Variable velocities were studied by considering the "infinitely small" distance that an object travels in an "infinitely short" period of time.

The algebraic calculations based on these ideas involved *infinitesimals*, which are quantities that are supposed to be infinitely small and yet not zero. Questions concerning the existence of the infinitely small and the infinitely large had been actively debated by philosophers in previous centuries, and the notion of infinitely small, nonzero numbers was questioned to some extent by those who used it. However, the concept of infinitesimals was accepted for its effectiveness as a mathematical tool.

Fermat used infinitesimals to find the tangent lines to curves as early as 1636, and by 1660 a number of others, including the Scottish mathematician James Gregory (1638–1675) and the English theologian Isaac Barrow (1630–1677), had done similar calculations.

FIGURE 3.52   Galileo Galilei (1564–1642)
Culver Pictures, Inc.

Newton and Leibniz

The use of infinitesimals was developed into the general operation now known as differentiation by Barrow's protégé Isaac Newton (1645–1727) in the 1660's and independently by the German philosopher Gottfried Leibniz (1646–1716) in the 1670's. Although Newton did his fundamental work earlier than Leibniz, he made little effort to make it known and did not publish his results until after 1700. Leibniz published his results promptly and took a keen interest in making his techniques comprehensible and useful to a wide audience. Consequently, he exerted a much greater influence on mathematicians of the next 150 years than did Newton. Leibniz's use of infinitesimals was perhaps related to his philosophy, which contended that the "best of all possible worlds" in which we live consists of infinitesimal, indivisible, and indestructible spiritual atoms called "Monads."

FIGURE 3.53   Isaac Newton
(1642–1727)
Culver Pictures, Inc.

FIGURE 3.54   Gottfried Leibniz
(1646–1716)
Culver Pictures, Inc.

Infinitesimals and limits

Leibniz used what he called *differentials* dx and dy to denote corresponding infinitesimal changes in quantities x and y which were related by an equation. If the equation was

$$(2) \quad y = x^2$$

he would write the same equation with x replaced by $x + dx$ and with y replaced by $y + dy$: $y + dy = (x + dx)^2$. Multiplying out the square and subtracting the original equation (2) would then give

$$(3) \quad dy = 2x(dx) + (dx)^2.$$

Leibniz would then simply drop the square $(dx)^2$ of the differential from equation (3). He would justify this by stating that the square of the differential is negligible in comparison with the differentials themselves (as a point is negligible in comparison with a line, he once said). Without the square the equation would be

$$(4) \quad dy = 2x(dx)$$

which corresponds to the modern formula $dy/dx = 2x$.

Let us compare this use of infinitesimals with the modern reasoning based on limits. We let Δx denote a nonzero (and noninfinitesimal) change in x and let Δy be the corresponding change in y. By the same algebraic calculations as above, we obtain the equation $\Delta y = 2x(\Delta x) + (\Delta x)^2$. Then instead of throwing away the $(\Delta x)^2$ term, we divide the equation by Δx and *take the limit* as Δx tends to zero to obtain

$$\frac{dy}{dx} = \lim_{\Delta x \to 0} \frac{\Delta y}{\Delta x} = \lim_{\Delta x \to 0} (2x + \Delta x) = 2x.$$

Criticism of the use of infinitesimals

The obvious criticism of the calculations with infinitesimals which led from equation (2) to equation (4) is that the differential dx is a nonzero quantity, and yet its square, which is dropped from the calculation, is somehow zero. The most famous attack on the use of infinitesimals was published in 1734 by an English bishop George Berkeley in a treatise titled *The analyst, or a discourse addressed to an infidel mathematician, wherein it is examined whether the object, principles, and inferences of the modern analyst are more distinctly conceived or more evidently deduced than religious mysteries and points of faith.* (The "infidel mathematician" was supposedly the astronomer Edmund Halley.) In this treatise Berkeley ridiculed infinitesimals as "the ghosts of departed quantities" and discussed the apparent contradictions involved in their use.

Both Newton and Leibniz did at one time or another give presentations of their "calculus" that were fairly close to the modern treatments involving limits. Newton did so, for example, in his theory of "prime and ultimate ratios" and Leibniz in a reply to a critic Nieuwentijt.

Nevertheless, Newton and Leibniz and the other prominent mathematicians of the late seventeenth and the eighteenth centuries relied generally on the use of infinitesimals. Their intuition and insight into the problems they studied enabled them to develop most of what is now known as "elementary calculus" and a great deal of more advanced mathematics, even though their arguments did not meet modern standards.

FIGURE 3.55 Augustin Louis Cauchy (1789–1857)
Culver Pictures, Inc.

The modern approach

It was not until the nineteenth century that presentations of calculus as a systematic application of the concept of limit were made popular. This change in point of view was stimulated by the work of the Swiss mathematician L. Euler (1707–1783), the French mathematicians J. L. D'Alembert (1717–1783), S. L'Huilier (1750–1840), and A. Cauchy (1789–1857), and the Czechoslovakian priest B. Bolzano (1781–1848).

A rigorous notion of infinitesimal was devised by an American mathematician Abraham Robinson in the 1960's. It involves an abstract mathematical structure that avoids consideration of the "infinitely small." Calculus based on Robinson's infinitesimals is known as *nonstandard analysis*.

FURTHER READING

J. CHILD, [1] *The Early Mathematical Manuscripts of Leibniz.*

E. BELL, [1] *Men of Mathematics* (Newton, Chapter 6; Leibniz, Chapter 7).

A. ROBINSON, [1] *Non-Standard Analysis.*

H. BOS, [1] "Differentials and the Derivative in Leibnizian Calculus."

C. BOYER, [3] *History of the Calculus.*

E. HALDANE, [1] *Descartes, His Life and Times.*

L. MORE, [1] *Isaac Newton, A Biography.*

A. DE MORGAN, [1] *Essays on the Life of Newton.*

G. GUHRAUER, [1] *Life of G. W. von Leibniz.*

Exercises

†*Answer provided.*
‡*Outline of solution provided.*

1.† The parabola $y = \frac{1}{4}x^2$ is the locus of points equidistant from $F(0, 1)$ and the line $y = -1$. Imagine the parabola to be traversed by a point that moves away from F and from the line $y = -1$ at equal speeds. The tangent line at $P(k, \frac{1}{4}k^2)$ is the line that has the direction of the motion at that point. Show that it bisects the angle between the line $x = k$ and the line PF.

2. Equation (1) has the solution $x = 1$ for any value of m.   **a.** Show that for m close to but not equal to 3, there are two other solutions.   **b.** Show that as m tends to 3, one of these solutions tends to the solution $x = 1$ and that for $m = 3$ there are just two solutions.   **c.** For what other value of m are there just two solutions? Explain.

3. Derive the Merton rule from the calculus formula $s = \frac{1}{2}at^2 + bt$ for the distance traveled in time t by an object with constant acceleration a.

MISCELLANEOUS EXERCISES

†*Answer provided.*
‡*Outline of solution provided.*

1. Show that the tangent line to the semicircle $y = \sqrt{1 - x^2}$ at $x = a\,(-1 < a < 1)$ is perpendicular to the radius.

2. Show that the tangent line is the only nonvertical line through the point (k, k^2) that intersects the parabola $y = x^2$ at only one point.

3. Show that the tangent line is the only nonvertical and nonhorizontal line through the point $(k, 1/k)$ that intersects the hyperbola $y = 1/x$ at only one point.

4.† Find the two tangent lines to the parabola $y = x^2 + 4$ that pass through the origin.

5. Show that the curves $y = |x|$ and $y = x \sin x$ are tangent where they intersect except at $x = 0$.

6. **a.** Use mathematical induction to prove that $x^n - a^n = (x^{n-1} + x^{n-2}a + x^{n-2}a^2 + \cdots + xa^{n-2} + a^{n-1})(x - a)$ for all x and a and all positive integers n.   **b.** Use the formula from part (a) to derive the differentiation formula $\dfrac{d}{dx}x^n = nx^{n-1}$ at $x = a$.

7. Derive the quotient rule from the product rule and the chain rule. (Write $f(x)/g(x) = f(x)g(x)^{-1}$.)

8.[†] Figure 3.56 shows the graph of current as a function of voltage in an ordinary light bulb. Find **a.** the approximate average rate of change of current with respect to voltage for voltage between zero and 120 volts and **b.** the approximate rate of change of current with respect to voltage at 90 volts.

9. Figure 3.57 shows the increase in atmospheric carbon dioxide since 1860 (observed and predicted). What were **a.** the approximate average rate of increase between 1900 and 1970 and **b.** the rate of increase in 1940?

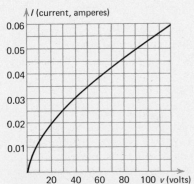

FIGURE 3.56 Current as a function of voltage in a light bulb (Adapted from A. Herrick, [1] *Mathematics for Electronics*, p. 238.)

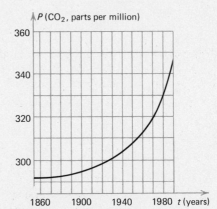

FIGURE 3.57 Increase in atmospheric CO_2 since 1860 (Adapted from A. Strahler and A. Strahler, [1] *Geography*, p. 145.)

10. Compute the following derivatives.

a.[†] $\dfrac{d}{dx}(3x-4)^{-1/3}$

b.[‡] $\dfrac{d}{dx}\left(\dfrac{x^2+1}{1-x^2}\right)$

c. $\dfrac{d}{dt}[t\sqrt{2t+3}]$

d. $\dfrac{d}{dx}\left(\dfrac{1}{1+x+x^2}\right)$

e.[†] $\dfrac{d}{dt}(t^5-3t^2+1)^{-13}$

f.[†] $\dfrac{d}{dx}\left(\dfrac{2x+3}{4x+5}\right)$

g. $\dfrac{d}{dt}\left(\dfrac{\sqrt{t}}{1-t}\right)$

h. $\dfrac{d}{dt}[\sin(t^4)-\sin^4 t]$

i.[†] $\dfrac{d}{du}\left(\dfrac{u^4+3}{u^2+1}\right)$

j.[†] $\dfrac{d}{dx}\left(\dfrac{1}{\cos x}-\dfrac{1}{\sin x}\right)$

k. $\dfrac{d}{dz}\left(\dfrac{z}{11-2z}\right)$

l. $\dfrac{d}{dt}[3\cos(\sqrt{t})+2\sqrt{\cos t}]$

m.[†] $\dfrac{d}{dx}\sin^3(5x)$

n.[†] $\dfrac{d}{dx}\sqrt{\cos(x^2)}$

o. $\dfrac{d}{dx}\sin[\cos(3x)]$

p. $\dfrac{d}{dw}\sqrt{\dfrac{\sin w}{\cos w}}$

q.[†] $G'(\pi/4)$, where $G(t)=t\sin t$

r.[†] $W'(5\pi/6)$, where $W(x)=(\sin x)/x$

11.[†] Compute   **a.** $P'(3)$ and   **b.** $P'(-3)$, where

$$P(x) = \begin{cases} x^2 + 4 & \text{for } x < 0 \\ x^3 - x & \text{for } x > 0. \end{cases}$$

12.   Compute   **a.** $\dfrac{dB}{dx}(-8)$ and   **b.** $\dfrac{dB}{dx}(8)$, where

$$B(x) = \begin{cases} x^3 & \text{for } x \le 1 \\ \sqrt[3]{x} & \text{for } x > 1. \end{cases}$$

13.   An object is at $s = t/(t^2 + 1)$ (kilometers) on an s-axis at time t (hours). What is its velocity and acceleration at time t?

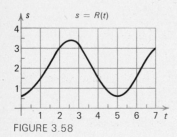

$s = R(t)$

FIGURE 3.58

14.   The density of dry air at a pressure of one atmosphere and a temperature of T degrees centigrade is

$$D = \frac{1.293}{1 + (0.00367)T} \quad \frac{\text{grams}}{\text{liter}}$$

(*Handbook of Chemistry and Physics*, p. 2150). What is the rate of change of density with respect to temperature at $T = 20°C$?

15.[†] Use the sketch in Figure 3.58 of the graph of the differentiable function R to find approximate values at $t = 2$ of   **a.** $R(t^2)$ **b.** $R(t)^2$
   c. $\dfrac{d}{dt} R(t^2)$ **d.** $\dfrac{d}{dt} R(t)^2$.

16.   The function M is defined by $M(p) = L(K(p))$, where L and K are the differentiable functions whose graphs are shown in Figures 3.59 and 3.60. Give approximate values of   **a.** $M(-1)$ **b.** $M(3)$ **c.** $M'(-1)$ **d.** $M'(3)$.

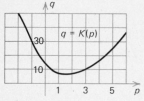

$q = K(p)$

FIGURE 3.59

17.[†] Compute $\dfrac{d}{dx}|x|$ for $x \ne 0$ and show that it does not exist at $x = 0$.

18.   Show that the derivative $\dfrac{d}{dx}|x^3|$ exists for all x and give a formula for it.

19.   If a tin rod is 10 meters long at a temperature of $0°$ C, then its length at a temperature of $T°C$ ($0 \le T \le 90$) is $L = 10 + (2.094 \times 10^{-5})\,T + (1.75 \times 10^{-8})\,T^2$ meters (*Handbook of Chemistry and Physics*, p. 2246). What is the rate of change of the length with respect to temperature at $T = 20$?

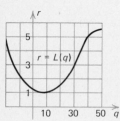

$r = L(q)$

FIGURE 3.60

20.   Use mathematical induction to derive the formula $\dfrac{d}{dx}x^n = nx^{n-1}$ for all positive integers n from the formula $\dfrac{d}{dx}x = 1$.

21.   Use the trigonometric identities $\cos(x + \tfrac{1}{2}\pi) = -\sin x$ and $\sin(x + \tfrac{1}{2}\pi) = \cos x$ to derive the formula $\dfrac{d}{dx}\cos x = -\sin x$ from the formula $\dfrac{d}{dx}\sin x = \cos x$.

22.[†] Give approximate values of   **a.** $f(-2)$ and   **b.** $f'(-2)$, where $f(x)$ $= Q(Q(x))$; the graph of Q is shown in Figure 3.61.

23.   Give approximate values of   **a.** $P(2)$ and   **b.** $P'(2)$, where $P(x)$ $= G(3x + G(x))$; the graph of G is shown in Figure 3.62.

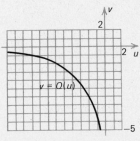

FIGURE 3.61

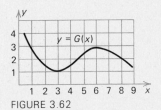

FIGURE 3.62

24.[†] Compute the derivatives at $x = 1$ of   **a.** $x^2(x^2 + 3)^{1/2}$ **b.** $[x\cos(\pi x)]^{10}$
c. $x/\sqrt{5x^3 - 1}$ **d.** $[(x^2 + x^3)(x^4 + x^5)]^{1/6}$
e. $(x + 2)(3x + 4)(5x + 6)$ **f.** $\sqrt{(x + 8)/(x + 3)}$
g. $\sin[x\sin(\pi x)]$ **h.** $\sin(\pi x)/[\sin(\pi x) + \cos(\pi x)]$ **i.** $\cos[x/(x + 1)]$

CHAPTER 4

APPLICATIONS OF DIFFERENTIATION

SUMMARY: In Chapter 3 we saw how derivatives are used to study tangent lines and rates of change. In this chapter we study a number of other applications of differentiation. Sections 4.1 through 4.4 deal with curve sketching, Section 4.5 with implicit differentiation, and Section 4.6 with inverse functions. Related rate and maximum-minimum problems are studied in Sections 4.7 through 4.9. We discuss tangent line approximations in Section 4.10 and Newton's method for finding approximate solutions of equations in Section 4.11. The Mean Value Theorem for derivatives is presented in Section 4.12, and the concept of antiderivative, which will be important in Chapter 5, is introduced in Section 4.13.

4-1   SKETCHING THE GRAPHS OF FUNCTIONS: THE FIRST DERIVATIVE TEST

In this and the following three sections we develop techniques for sketching graphs of functions. Here we use THE FIRST DERIVATIVE TEST.

The idea behind the first derivative test is indicated in Figures 4.1 and 4.2. The function f of Figure 4.1 has a positive derivative and the tangent lines to its graph have positive slopes. This implies that the

FIGURE 4.1   The graph of an increasing function (The tangent lines have positive slopes.)

FIGURE 4.2   The graph of a decreasing function (The tangent lines have negative slopes.)

function is INCREASING, that is, that $f(x)$ increases with increasing x. The function of Figure 4.2 has a negative derivative. The tangent lines to its graph have negative slopes and the function is DECREASING; its values decrease with increasing x. We will now examine these ideas more closely.

The first derivative test at a point

Definition 4.1   A function f is increasing at a point x_0 if there is an open interval I containing x_0 such that $f(x) < f(x_0)$ for x in I to the left of x_0 and $f(x_0) < f(x)$ for x in I to the right of x_0. The function f is decreasing at x_0 if $f(x) > f(x_0)$ for x in I to the left of x_0 and $f(x_0) > f(x)$ for x in I to the right of x_0.

Theorem 4.1   (The first derivative test at a point)   If the derivative $f'(x_0)$ exists and is positive, then the function f is increasing at x_0. If the derivative is negative, then f is decreasing at x_0.

Proof.   Suppose the derivative $f'(x_0)$ exists and is positive. There is an open interval I containing x_0 such that

$$(1) \quad \frac{f(x) - f(x_0)}{x - x_0} > 0 \quad \text{for } x \text{ in } I \text{ with } x \neq x_0$$

because the derivative is the limit of the difference quotient (1). For x in I and to the right of x_0, $x - x_0$ is positive, and multiplying inequality (1) by this positive number preserves its direction to give $f(x) - f(x_0) > 0$ and, therefore, $f(x_0) < f(x)$. For x in I and to the left of x_0, $x - x_0$ is negative. Multiplying (1) by this negative number reverses the direction of the inequality to show that $f(x) - f(x_0) < 0$ and hence $f(x) < f(x_0)$.

The proof for the case of $f'(x_0) < 0$ is analogous.   Q.E.D.

EXAMPLE 1   Is the function $x^3 - x$ increasing or decreasing at $x = 0$?

Solution.   We set $f(x) = x^3 - x$. Then $f'(x) = 3x^2 - 1$ and $f'(0) = -1$. The derivative is negative and the function is decreasing at $x = 0$.   □

The first derivative test in an open interval

Definition 4.2   A function f is increasing in an interval I if for any x_1 and x_2 in I with $x_1 < x_2$ we have $f(x_1) < f(x_2)$. The function is decreasing in the interval if $f(x_1) > f(x_2)$ for all such numbers x_1 and x_2.

Theorem 4.2   (The first derivative test in an open interval) If the derivative $f'(x)$ exists and is positive for all x in an open interval, then the function is increasing in that interval. If the derivative is negative in the open interval, then the function is decreasing there.

We will prove Theorem 4.2 in Section 4.12 as a special case of the first derivative test in an arbitrary interval (Theorem 4.9).

EXAMPLE 2 Determine where the function x^{-2} is increasing and where it is decreasing.

Solution.   The derivative of x^{-2} is

$$\frac{d}{dx}x^{-2} = -2x^{-3} = \frac{-2}{x^3}.$$

The derivative is negative for $x > 0$ and positive for $x < 0$. Therefore, the function x^{-2} is decreasing in the interval $x > 0$ and increasing in the interval $x < 0$ (Figure 4.3). $\square$

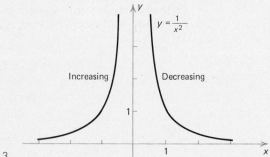

FIGURE 4.3

Local maxima and minima

Once we have determined where a function is increasing and decreasing, we can locate its LOCAL MAXIMA and MINIMA, which are defined as follows:

> **Definition 4.3**   **A function f has a local maximum at x_0 if $f(x) \le f(x_0)$ for all x in an open interval containing x_0. The function has a local minimum at x_0 if $f(x) \ge f(x_0)$ for all x in an open interval containing x_0.**

Critical points

> **Definition 4.4**   **The point x_0 is a critical point of a function f if f is defined in an open interval containing x_0 and if $f'(x_0)$ is zero or does not exist.**

If the graph of a function has a tangent line at a critical point, then the tangent line is horizontal. According to the next result, a function's local maxima and minima can only occur at its critical points.

> **Theorem 4.3**   **If f has a local maximum or minimum at x_0, then x_0 is a critical point of f.**

Proof.   If x_0 is not a critical point of f, the derivative $f'(x_0)$ exists and is not zero. By Theorem 4.1, f is either increasing or decreasing at x_0 and cannot have a local maximum or minimum there.   Q.E.D.

EXAMPLE 3   Sketch the graph of the polynomial $\frac{1}{4}x^4 - 2x^2$. Show the critical points and local maxima and minima.

Solution.   We set $f(x) = \frac{1}{4}x^4 - 2x^2$. The function's derivative is

(2) $f'(x) = \dfrac{d}{dx}\left(\dfrac{1}{4}x^4 - 2x^2\right) = x^3 - 4x = x(x^2 - 4)$

$\qquad\qquad = x(x + 2)(x - 2).$

The derivative is zero at $x = 0, 2,$ and -2, so these are the critical points of the function f.

The derivative (2) is a polynomial and is continuous for all x. By the Intermediate Value Theorem it can switch from positive to negative or vice versa only at the points $x = 0, 2,$ and -2, where it is zero.

These three values of x divide the rest of the x-axis into the four open intervals shown in Figure 4.4. We determine the sign of the derivative by computing its value at one point in each of the four intervals. We have

$$-3 < -2 \quad \text{and} \quad f'(-3) = (-3)^3 - 4(-3) = -15 < 0$$
$$-2 < -1 < 0 \quad \text{and} \quad f'(-1) = (-1)^3 - 4(-1) = 3 > 0$$
$$0 < 1 < 2 \quad \text{and} \quad f'(1) = 1^3 - 4(1) = -3 < 0$$
$$3 > 2 \quad \text{and} \quad f'(3) = 3^3 - 4(3) = 15 > 0.$$

Hence, for $x < -2, f'(x)$ is negative and $f(x)$ is decreasing; for $-2 < x < 0, f'(x)$ is positive and $f(x)$ is increasing; for $0 < x < 2,$ $f'(x)$ is negative and $f(x)$ is decreasing; for $x > 2, f'(x)$ is positive and $f(x)$ is increasing (Figure 4.5).

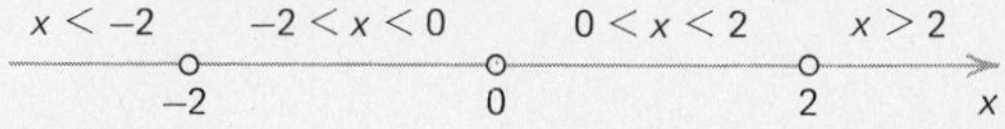

FIGURE 4.4

FIGURE 4.5

Figure 4.5 shows that the function f has local minima at the critical points $x = -2$ and $x = 2$ and a local maximum at the critical point $x = 0$. The graph is given in Figure 4.6. In making the sketch we use the information in Figure 4.5 and plot the points whose coordinates are given in the following table.

x	-3	-2	0	2	3
$y = \frac{1}{4}x^4 - 2x^2$	$\frac{9}{4}$	-4	0	-4	$\frac{9}{4}$

The points at $x = -2, 0,$ and 2 are plotted because those are the critical points of f. The points at $x = 3$ and $x = -3$ are plotted to help show the nature of the curve away from the critical points.   □

Note.   A function can have a critical point at which it does not have a local maximum or minimum. For example, the function x^3 has a critical point and a horizontal tangent line (the x-axis) at $x = 0$, but it does not have a local maximum or minimum there (Figure 4.7).   □

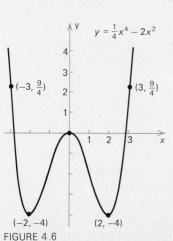

FIGURE 4.6

Exercises

†Answer provided.
‡Outline of solution provided.

In Exercises 1 through 6 sketch the graphs of the functions. Show where the functions are increasing and decreasing and label the critical points.

1.† $x^2 - 4x + 5$ 　　　　　　　　　　　**2.** $x - \frac{1}{6}x^2$

3.‡ $\frac{1}{3}x^3 + x^2$ 　　　　　　　　　　**4.** $-\frac{1}{8}x^5$

5.‡ $\frac{1}{4}x^4 - x$ 　　　　　　　　　　　**6.** $\frac{1}{2}x^4 + x^2$

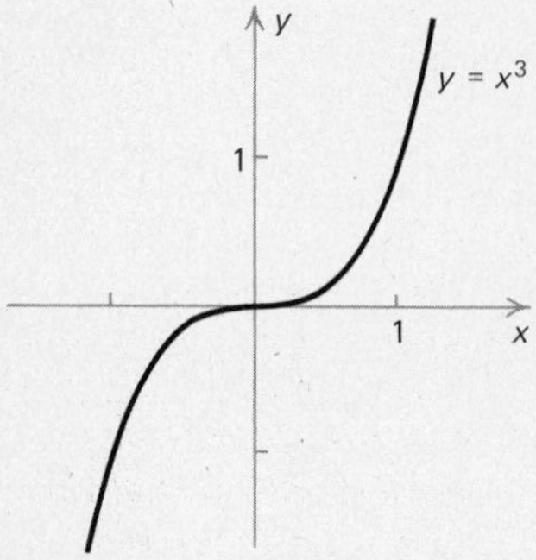

In Exercises 7 through 16 determine whether the functions are increasing or decreasing at the indicated values of x.

7.† $x^5 + x^3 + x$ at $x = -10$ 　　　**8.** $\dfrac{1}{\sqrt{x}}$ at $x = 9$

9.† $\cos x - \sin x$ at $x = \dfrac{\pi}{6}$ 　　　**10.** $\dfrac{1}{x} - \dfrac{1}{x^2}$ at $x = 1$

11.‡ $x \sin x$ at $x = \pi$ 　　　　　　**12.†** $\dfrac{x}{x^2 + 1}$ at $x = 2$

13. $\dfrac{\cos x}{x}$ at $x = \dfrac{\pi}{2}$ 　　　　**14.‡** $\sin(\sqrt[3]{x})$ at $x = 1$

15.† $\dfrac{1}{\sqrt{5 - x^2}}$ at $x = 1$ 　　　**16.** $\sin^2 x - \cos x$ at $x = -\dfrac{\pi}{4}$

Use the first derivative test to sketch the curves in Exercises 17 through 27.

17.† $y = 6x + x^2$ 　　　　　　　　**18.†** $y = \frac{1}{3}x^3 - x$

19.† $y = \frac{1}{4}x^4 - 2x^2 + 4$ 　　　　**20.†** $y = \frac{1}{3}x^3 - 2x^2 + 3x$

21.† $y = \sqrt{x^2 + 1}$ 　　　　　　　**22.†** $y = x\sqrt{x^2 + 1}$

23.‡ $y = (x - 1)^3(x - 5)^9$ 　(Factor $(x - 1)^2(x - 5)^8$ from the derivative.)

24.† $y = (x + 4)^2(x - 1)^3$ 　　　　**25.** $y = (x - 5)^4(x - 2)^2$

26.† $y = 4x^{7/3} - 7x^{4/3}$ 　　　　　**27.** $y = x^3(4 - x)^3$

FIGURE 4.7

$y = x^3$

4-2　HIGHER ORDER DERIVATIVES AND THE SECOND DERIVATIVE TEST

In this section we see how studying *second derivatives* can help in sketching the graphs of functions. The second derivative of a function $f(x)$ is the derivative of its (first) derivative $f'(x)$. The second derivative is denoted by the symbol $f''(x)$ in prime notation and by the symbol $\dfrac{d^2f}{dx^2}(x)$ in Leibniz notation. The third derivative is the derivative of the second derivative and is denoted $f^{(3)}(x)$ or $\dfrac{d^3f}{dx^3}(x)$. Higher order derivatives are defined similarly. The nth derivative where n is a positive integer, is denoted $f^{(n)}(x)$ or $\dfrac{d^nf}{dx^n}(x)$.

EXAMPLE 1   Compute the derivative $f''(x) = \dfrac{d^2f}{dx^2}(x)$ where $f(x) = x^5 - 2x$.

Solution.   The first derivative is

$$\frac{df}{dx}(x) = \frac{d}{dx}(x^5 - 2x) = 5x^4 - 2.$$

The second derivative is, therefore,

$$\frac{d^2f}{dx^2}(x) = \frac{d}{dx}\left[\frac{df}{dx}(x)\right] = \frac{d}{dx}(5x^4 - 2) = 20x^3. \quad \square$$

Concavity and the second derivative test

The first derivative $f'(x)$ of a function is the slope of the tangent line to its graph at x. If that slope increases with increasing x, then the graph curves upward and is said to be CONCAVE UP (Figure 4.8). If the slope decreases with increasing x, then the graph curves downward and is said to be CONCAVE DOWN (Figure 4.9):

Definition 4.5   The graph of a function f is concave up for x in an interval if the slope $f'(x)$ of the tangent line is an increasing function in that interval. The graph is concave down for x in the interval if $f'(x)$ is a decreasing function there.

FIGURE 4.8   Graph concave up Slope increasing. $f'' > 0$

Notice that the graph lies above its tangent lines where it is concave up (Figure 4.8) and below its tangent lines where it is concave down (Figure 4.9).

By applying Theorem 4.2 we see that the derivative $f'(x)$ is increasing in any open interval where its derivative $f''(x)$ is positive and is decreasing in any open interval where $f''(x)$ is negative. Therefore, we have the following result.

Theorem 4.4   (The second derivative test in an open interval) If the second derivative $f''(x)$ of a function f is positive in an open interval, then the graph of f is concave up for x in that interval. If the second derivative is negative in the interval, then the graph is concave down for x in the interval.

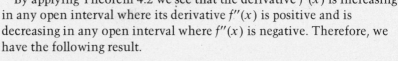

FIGURE 4.9   Graph concave down Slope decreasing. $f'' < 0$

Inflection points

To show the concavity of the graph of a function when we sketch it, we need to know where the graph switches from concave up to concave down or vice versa. Such points are called INFLECTION POINTS:

Definition 4.6   A point $(x_0, f(x_0))$ on the graph of f is an inflection point if $f''(x_0) = 0$ or the graph has a vertical tangent line at $x = x_0$ and there is an open interval I containing x_0 such that $f''(x)$ is of one sign for x in I to the left of x_0 and of the opposite sign for x in I to the right of x_0.

Vertical tangent lines are defined in Section 4.4.

EXAMPLE 2   Sketch the graph of the polynomial $x^3 - 3x^2 + 4$. Show the critical points, local maxima and minima, inflection points, where the function is increasing and decreasing, and where the graph is concave up and concave down.

Solution.   The first derivative test. We set $f(x) = x^3 - 3x^2 + 4$. The derivative

$$f'(x) = \frac{d}{dx}(x^3 - 3x^2 + 4) = 3x^2 - 6x = 3x(x - 2)$$

is zero at $x = 0$ and $x = 2$, the critical points of the function f.

The critical points divide the rest of the x-axis into the three open intervals shown in Figure 4.10. The sign of the derivative is constant in each of the three intervals, and it can be determined by computing the value of the derivative at one point in each interval. We find that $f'(-1) = 9 > 0, f'(1) = -3 < 0$, and $f'(3) = 9 > 0$. Therefore, $f'(x)$ is positive and $f(x)$ is increasing for $x < 0$; $f'(x)$ is negative and $f(x)$ is decreasing for $0 < x < 2$; and $f'(x)$ is positive and $f(x)$ is increasing for $x > 2$ (Figure 4.11). The function f has a local maximum at $x = 0$ and a local minimum at $x = 2$.

The second derivative test. The second derivative

$$f''(x) = \frac{d^2}{dx^2}(x^3 - 3x^2 + 4) = \frac{d}{dx}(3x^2 - 6x)$$
$$= 6x - 6 = 6(x - 1)$$

is zero at $x = 1$, negative for $x < 1$, and positive for $x > 1$. According to Theorem 4.4 the graph of f is concave down for $x < 1$ and is concave up for $x > 1$ (Figure 4.12). The point on the graph at $x = 1$ is an inflection point.

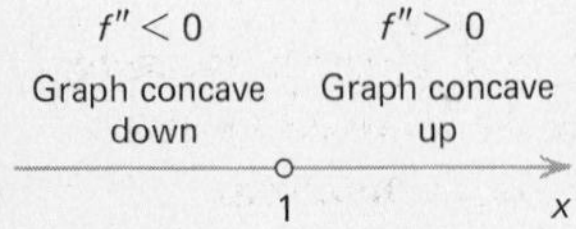

Sketching the graph. To incorporate the information from Figures 4.11 and 4.12 in a sketch of the graph of f, we plot the points at $x = 0, 1,$ and 2, where the first and second derivatives are zero, together with a few more points to improve the sketch. We plot the points given in the following table.

x	-1	0	1	2	3
$y = x^3 - 3x^2 + 4$	0	4	2	0	4

The graph is shown in Figure 4.13.   □

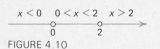

FIGURE 4.10

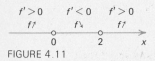

FIGURE 4.11

$f'' < 0$ $f'' > 0$

Graph concave   Graph concave
down   up

1   x

FIGURE 4.12

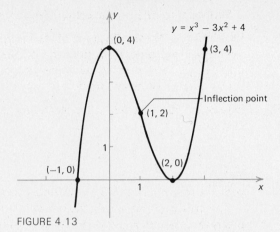

FIGURE 4.13

The second derivative test for a local maximum or minimum

We can expect a function to have a local minimum at a point where its graph has a horizontal tangent line and is concave up (Figure 4.14). We can expect a function to have a local maximum at a point where its graph has a horizontal tangent line and is concave down (Figure 4.15). This result is presented in the next theorem.

Theorem 4.5   (The second derivative test for a local maximum or minimum)   If the first derivative $f'(x_0)$ is zero and the second derivative $f''(x_0)$ is positive, then f has a local minimum at x_0. If the first derivative $f'(x_0)$ is zero and the second derivative $f''(x_0)$ is negative, then f has a local maximum at x_0.

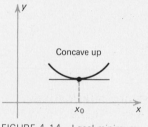

FIGURE 4.14   Local minimum
$f'(x_0) = 0, f''(x_0) > 0$

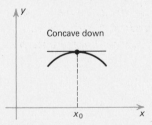

FIGURE 4.15   Local maximum
$f'(x_0) = 0, f''(x_0) < 0$

Proof.   Suppose that $f'(x_0) = 0$ and $f''(x_0) > 0$. By the first derivative test at a point (Theorem 4.2) applied to $f'(x)$, the derivative $f'(x)$ is increasing at x_0. Therefore, there are numbers a and b such that $f'(x) < f'(x_0) = 0$ for $a < x < x_0$ and $f'(x) > f'(x_0) = 0$ for $x_0 < x < b$. By the first derivative test in an arbitrary interval (Theorem 4.9, to be proved in Section 4.12), f is decreasing in the interval $a < x \leq x_0$ and increasing in the interval $x_0 \leq x < b$ and, therefore, has a local minimum at x_0. The proof for the case of $f''(x_0) < 0$ is similar.   Q.E.D.

EXAMPLE 3   The point $x = 1$ is a critical point of the polynomial $x^4 - x^3 - x$. Does the function have a local maximum or local minimum at that point?

Solution.   We set $f(x) = x^4 - x^3 - x$. Then $f'(x) = 4x^3 - 3x^2 - 1$ and $f'(1) = 4 - 3 - 1 = 0$, so that $x = 1$ is a critical point. We have $f''(x) = 12x^2 - 6x$ and $f''(1) = 12 - 6 = 6 > 0$. By Theorem 4.5 the function has a local minimum at $x = 1$.   □

Exercises

†*Answer provided.*
‡*Outline of solution provided.*

In Exercises 1 through 10 compute the derivatives.

1.† $f''(x)$, where $f(x) = 3x^4 - 6x^2$

2. $g''(0)$, where $g(x) = x^5 - x^4 + x^3 - x^2$

3.‡ $\left[\dfrac{d^2}{dx^2}(x \sin x)\right]_{x=\pi/2}$

4.† $\dfrac{d^2}{dt^2}\left[\dfrac{1}{t^2 + 1}\right]$

5. $H''(y)$, where $H(y) = \sqrt{y} - \dfrac{1}{\sqrt{y}}$

6.† $f^{(3)}(1)$, where $f(x) = 5x^7$

7. $\dfrac{d^4}{dz^4}(z^{10} - z^9)$

8.‡ $\dfrac{d^2}{dx^2}\cos(x^3)$

9.† $\dfrac{d^3}{dx^3}\sin^2 x$

10. $F''(2)$, where $F(x) = \sqrt{\cos x}$

In Exercises 11 through 20 sketch the graphs of the functions. Show where the functions are increasing and decreasing and where the graphs are concave up and concave down. Plot the critical points and inflection points.

11.† $x^3 - 12x + 7$

12.† $\frac{2}{3}x^3 - \frac{1}{5}x^5$

13. $2 - x - \frac{1}{4}x^2$

14. $-\frac{1}{4}x^3 - \frac{1}{2}x + 2$

15.† $-\frac{1}{6}x^4 + \frac{1}{3}x^3$

16. $-x^4 + 2x^2$

17. $x^3 - 3x^2 + 4$

18. $\frac{1}{4}x^4 + x^3$

19.† $x^3 + x - 5$

20. $\frac{1}{32}x^4 - x$

In Exercises 21 through 26 use the second derivative test (Theorem 4.5) to determine whether the functions have local maxima or minima at the indicated critical points.

21.‡ $\dfrac{x}{x^2 + 1}$ at $x = 1$

22.† $x^4 - 4x^3 - 2x^2 + 12x$ at $x = -1$

23. $x^5 - 6x^2$ at $x = 0$

24. $\sqrt{x^2 + 1}$ at $x = 0$

25.† $x \sin x$ at $x = 0$

26. $\sin x \cos x$ at $x = \pi/4$

4-3 SKETCHING GRAPHS OF RATIONAL FUNCTIONS

In the last two sections we concentrated on sketching the graphs of polynomials. In this section we describe techniques for sketching graphs of *rational functions*. Recall that these are functions formed from polynomials by taking sums, products, and quotients. We will consider here only rational functions that are given as *quotients of polynomials with no common factors* or those that can be put into this form.

In sketching the graph of a rational function we often devote more attention to studying the function itself than to studying its derivatives. We must study the function to determine if its graph has any vertical or horizontal asymptotes. Moreover, a careful study of the function often leads to a fairly good sketch of its graph and makes the application of the first and second derivative tests relatively easy.

We can analyze a rational function before studying its derivatives by carrying out the following procedure.

Step 1   If the rational function is given as a sum of quotients of polynomials, combine the terms into a single quotient of polynomials by taking the least common denominator.

Step 2   Determine the function's limits as x tends to ∞ and to $-\infty$. If these limits are a finite number L, then the line $y = L$ is a horizontal asymptote of the graph.

Step 3   Determine the values of x where the numerator of the function is zero. Because we assume that the numerator and denominator have no common factors, these values are where the function is zero and where its graph intersects the x-axis.

Step 4   Determine the values of x where the denominator of the function is zero. The function tends to ∞ or to $-\infty$ as x approaches any of these values from the left or from the right. Consequently, the graph has a vertical asymptote at each of these values of x.

Step 5   The values of x from Steps 3 and 4, where the numerator and denominator are zero, are the places where the function can change its sign. These values divide the rest of the x-axis into a number of open intervals. Determine whether the function is positive or negative in each of these intervals by computing its value at one point in each. This shows where the graph is above and where it is below the x-axis and determines its behavior on either side of each vertical asymptote.

With the information obtained in this procedure we can make a rough sketch of the graph, and we can improve the sketch by studying the function's first and second derivatives.

EXAMPLE 1   Sketch the graph of the function

$$\frac{6}{x^2} - \frac{6}{x}.$$

Show any horizontal or vertical asymptotes; show where the function is increasing and where it is decreasing; and, show where the graph is concave up and where it is concave down. Plot the critical points and inflection points.

Solution.   We use abbreviated notation and write y for the function

(1) $y = \dfrac{6}{x^2} - \dfrac{6}{x}$

and use y as the label on the vertical coordinate axis. We first study the function.

Step 1   We combine the two terms into one fraction by taking the common denominator x^2.

(2) $y = \dfrac{6}{x^2} - \dfrac{6}{x} = \dfrac{6 - 6x}{x^2}.$

Step 2 $\displaystyle\lim_{x \to \infty} \left(\frac{6 - 6x}{x^2} \right) = \lim_{x \to \infty} \left(\frac{-6x}{x^2} \right) = \lim_{x \to \infty} \left(\frac{-6}{x} \right) = 0$

$\displaystyle\lim_{x \to -\infty} \left(\frac{6 - 6x}{x^2} \right) = \lim_{x \to -\infty} \left(\frac{-6x}{x^2} \right) = \lim_{x \to -\infty} \left(\frac{-6}{x} \right) = 0.$

The line $y = 0$ (the x-axis) is a horizontal asymptote.

Step 3   The numerator $6 - 6x$ is zero at $x = 1$. The graph intersects the x-axis at $x = 1$.

Step 4   The denominator x^2 is zero at $x = 0$. The line $x = 0$ (the y-axis) is a vertical asymptote.

Step 5   The function is of constant sign in each of the intervals $x < 0$, $0 < x < 1$, and $x > 1$. We find that $y(-1) = 12 > 0$, $y(\frac{1}{2}) = 12 > 0$, and $y(2) = -\frac{3}{2} < 0$. The function is positive for $x < 0$, positive for $0 < x < 1$, and negative for $x > 1$. We show this information in Figure 4.16, along with the facts that the function y tends to zero as x tends to ∞ or to $-\infty$, that y is zero at $x = 1$, and that y tends either to ∞ or to $-\infty$ as x approaches zero from either side.

In making the rough sketch of the graph shown in Figure 4.17, we plot the points at $x = -1, \frac{1}{2}$, and 2 and use the information in Figure 4.16. The graph goes up on both sides of the vertical asymptote $x = 0$ (the y-axis) because the function is positive just to the right and just to the left of zero.

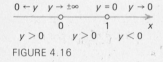

$0 \leftarrow y \quad y \to \pm\infty \qquad y = 0 \quad y \to 0$

$y > 0 \qquad y > 0 \qquad y < 0$

FIGURE 4.16

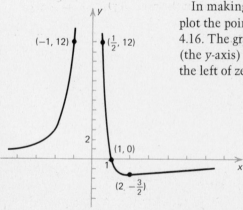

FIGURE 4.17   A rough sketch of the graph of $\dfrac{6}{x^2} - \dfrac{6}{x}$

The first derivative test. We use the original expression (1) for the function to compute its derivatives. The first derivative is

$$y' = \frac{d}{dx}\left(\frac{6}{x^2} - \frac{6}{x}\right) = \frac{d}{dx}(6x^{-2} - 6x^{-1}) = -12x^{-3} + 6x^{-2}$$

(3)

$$= \frac{-12}{x^3} + \frac{6}{x^2} = \frac{-12 + 6x}{x^3}.$$

The derivative is zero at $x = 2$, so $x = 2$ is a critical point of the original function. The derivative can change its sign at $x = 2$, where the numerator of (3) is zero, or at $x = 0$, where the denominator is zero. We find that $y'(-1) = 18 > 0$, $y'(1) = -6 < 0$, and $y'(3) = \frac{6}{27} > 0$. Therefore, y' is positive and y is increasing for $x < 0$; y' is negative and y is decreasing for $0 < x < 2$; and, y' is positive and y is increasing for $x > 2$ (Figure 4.18).

The second derivative test. The second derivative is

$$y'' = \frac{d}{dx}y' = \frac{d}{dx}(-12x^{-3} + 6x^{-2}) = 36x^{-4} - 12x^{-3}$$

$$= \frac{36}{x^4} - \frac{12}{x^3} = \frac{36 - 12x}{x^4}.$$

The second derivative can change its sign at $x = 3$, where its numerator is zero, or at $x = 0$, where its denominator is zero. We have $y''(-1) = 48 > 0$, $y''(1) = 24 > 0$, and $y''(4) = -\frac{3}{4} < 0$. The second derivative is positive and the graph is concave up for $x < 0$ and for $0 < x < 3$, and the second derivative is negative and the graph is concave down for $x > 3$ (Figure 4.19).

In making our final sketch of the graph of the function in Figure 4.20, we use the information from Figures 4.18 and 4.19 and plot the points at $x = 2$, where the first derivative is zero, and at $x = 3$, where the second derivative is zero. We have $y(2) = -\frac{3}{2}$ and $y(3) = -\frac{4}{3}$.

The function has a local minimum at the critical point $x = 2$, and the graph has an inflection point at $x = 3$. □

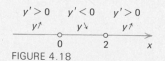

FIGURE 4.18

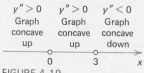

FIGURE 4.19

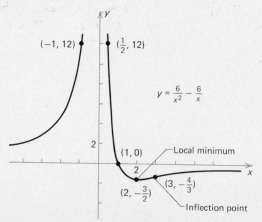

FIGURE 4.20

Exercises

In Exercises 1 through 9 sketch the graphs of the rational functions. Plot the critical points and inflection points. Show the vertical and horizontal asymptotes, where the functions are increasing and decreasing and where the graphs are concave up and concave down.

1.‡ $\dfrac{x^2}{x+2}$ **2.**† $\dfrac{5}{x^2+1}$ **3.** $\dfrac{1}{(x-2)^2}$

4. $\dfrac{x}{x+1}$ **5.** $\dfrac{x^2}{x^2+1}$ **6.**‡ $\dfrac{x^2+1}{x^2-1}$

7.† $3x+x^{-3}$ **8.** $2x^{-1}+2x^{-2}$ **9.** $x-x^{-1}$

In Exercises 10 through 21 sketch the graphs of the functions without using the second derivatives.

10.† $\dfrac{1}{x}-\dfrac{1}{x-1}$ **11.** $\dfrac{x^3}{3-x^2}$ **12.**† $\dfrac{3x}{x^2-4}$ **13.** $\dfrac{8}{x^3-3x^2}$

14.† $\dfrac{x^2}{x^2-4}$ **15.** $\dfrac{x}{x^2-2x+1}$

16.† $\sqrt{x^2+2x+5}$ **17.** $x^{-2}+(x^2+1)^{-1}$

18.† $\frac{1}{2}x-\sin x$ **19.** $\frac{1}{4}(x+\sin x)$

20.† $\cos\left(\dfrac{\pi}{x^2+1}\right)$ **21.** $(\sin x)(\cos x)$

4-4 SKETCHING GRAPHS OF OTHER FUNCTIONS

Graphs of functions can have a variety of features that do not occur in graphs of polynomials or of rational functions. We illustrate the more common of these features in this section.

Vertical tangent lines

Definition 4.7   The graph of a function f has a vertical tangent line at $x = x_0$ if f is continuous at x_0 and $f'(x)$ tends to ∞ or to $-\infty$ as x tends to x_0.

The function in the next example has a vertical tangent line at $x = 0$.

EXAMPLE 1   Sketch the graph of the function $x^{1/3}$.

Solution.   The function $x^{1/3} = \sqrt[3]{x}$ is defined and continuous for all x. Its derivative

(1) $\dfrac{d}{dx}x^{1/3} = \dfrac{1}{3}x^{-2/3} = \dfrac{1}{3x^{2/3}}$

does not exist at $x = 0$. It tends to infinity as x tends to zero, so the graph has a vertical tangent line there. The first derivative is positive and the function is increasing in the intervals $x > 0$ and $x < 0$.

The second derivative

$$\frac{d^2}{dx^2}x^{1/3} = \frac{d}{dx}\left(\frac{1}{3}x^{-2/3}\right) = \left(\frac{1}{3}\right)\left(-\frac{2}{3}\right)x^{-5/3} = \left(-\frac{2}{9}\right)\frac{1}{x^{5/3}}$$

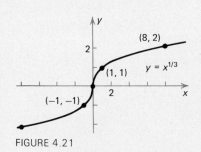

FIGURE 4.21

also does not exist at $x = 0$. For $x > 0$ it is negative and the graph of $x^{1/3}$ is concave down. For $x < 0$ it is positive and the graph is concave up. The curve has an inflection point at the origin.

The graph is shown in Figure 4.21. In making that sketch we plot the points given in the following table.

x	-8	-1	0	1	8
$y = x^{1/3}$	-2	-1	0	1	2

□

Limited extent Because the formula defining the function of the next example involves a square root, the function is defined only on a closed half line.

EXAMPLE 2 Sketch the graph of the function $\sqrt{4-x}+1$.

Solution. The function $\sqrt{4-x}+1$ is defined and continuous for $x \le 4$. Its derivative

$$\frac{d}{dx}[\sqrt{4-x}+1] = \frac{d}{dx}[(4-x)^{1/2}+1]$$

$$= \frac{1}{2}(4-x)^{-1/2}\frac{d}{dx}(4-x)$$

$$= -\frac{1}{2}(4-x)^{-1/2}$$

does not exist at $x = 4$ but is negative for $x < 4$. The function $\sqrt{4-x}+1$ is decreasing for $x < 4$.

The second derivative

$$\frac{d^2}{dx^2}[\sqrt{4-x}+1] = \frac{d}{dx}\left[-\frac{1}{2}(4-x)^{-1/2}\right]$$

$$= \left(-\frac{1}{2}\right)\left(-\frac{1}{2}\right)(4-x)^{-3/2}\frac{d}{dx}(4-x)$$

$$= -\frac{1}{4}(4-x)^{-3/2}$$

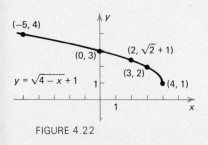

FIGURE 4.22

is negative, and the graph of the function $\sqrt{4-x}+1$ is concave down for $x < 4$.

The graph is shown in Figure 4.22. In making the sketch we plot the points from the following table.

x	-5	0	2	3	4
$y = \sqrt{4-x}+1$	4	3	$\sqrt{2}+1$	2	1

□

Periodic functions

Recall that a function f is *periodic* with period L if $f(x + L) = f(x)$ for all x. The functions $\sin x$ and $\cos x$ are periodic with period 2π. Additional periodic functions constructed from the function $\sin x$ are dealt with in the next example, where we will also employ the concept of EVEN and ODD FUNCTIONS.

Even and odd functions

Definition 4.8   A function f is even if its graph is symmetric about the y-axis. This is the case if the domain of f is symmetric about $x = 0$ and $f(-x) = f(x)$ for all x in the domain. The function is odd if its graph is symmetric about the origin. This is the case if the domain of f is symmetric about $x = 0$ and $f(-x) = -f(x)$ for all x in it.

The terminology *even* and *odd* stems from the fact that the graph of the function x^n is symmetric about the y-axis if n is an even integer and is symmetric about the origin if n is an odd integer.

EXAMPLE 3   The graph of the function $\sin x$ is shown in Figure 4.23, and the graphs of five functions constructed from $\sin x$ are shown in Figures 4.23 to 4.28. Which of these is the graph of $\sin^3 x$?

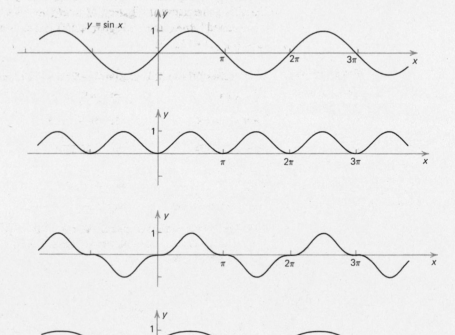

FIGURE 4.23

FIGURE 4.24

FIGURE 4.25

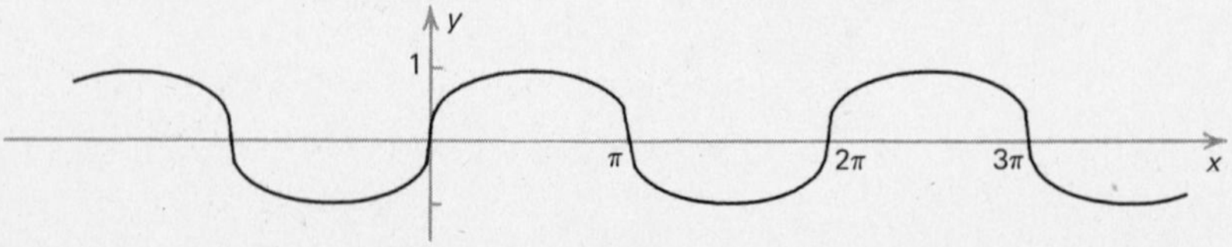

FIGURE 4.26

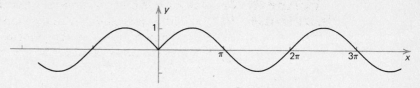

FIGURE 4.27

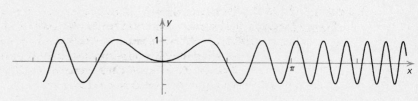

FIGURE 4.28

Solution.   The function $\sin^3 x$ is periodic with period 2π and is odd because

$$\sin^3(x + 2\pi) = [\sin(x + 2\pi)]^3 = (\sin x)^3 = \sin^3 x$$

$$\sin^3(-x) = [\sin(-x)]^3 = (-\sin x)^3 = -\sin^3 x.$$

The functions of Figures 4.27 and 4.28 are not periodic and the functions of Figures 4.24 and 4.27 are not odd. The graph of $\sin^3 x$ must be either the curve in Figure 4.25 or the curve in Figure 4.26. It is the "squashed" sine curve in Figure 4.25 because $|\sin^3 x| < |\sin x|$ when $0 < |\sin x| < 1$.   □

Exercises

†*Answer provided.*
‡*Outline of solution provided.*

In Exercises 1 through 15 sketch the graphs of the functions.

1.‡ $x^{2/3}$ **2.**† $x^{4/3}$ **3.** $1 - x^{1/3}$

4.† $x^{3/4} - 8$ **5.** $(x - 1)^{1/4}$ **6.**‡ $\sqrt{x^2 - 9}$

7.† $\sqrt{9 + x^2}$ **8.** $\sqrt{9 - x^2}$ **9.**† $\dfrac{x}{|x|}$

10. $|x|$ **11.** $x - |x|$ **12.**† $(|x| - 2)^2$

13. $|x + 4| - |x|$ **14.**‡ $(x^{1/3} - 1)^2$ **15.** $\sqrt[3]{x^2 - 1}$

16.† The graphs of the functions   **a.** $\cos x$ **b.** $\sqrt{\cos x}$ **c.** $\cos^2 x$ **d.** $\cos^3 x$ and   **e.** $\cos(\frac{1}{3}x^2)$ are shown in Figures 4.29 to 4.33. Match the functions to their graphs.

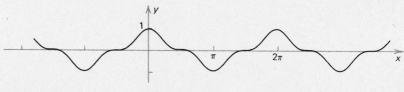

FIGURE 4.29

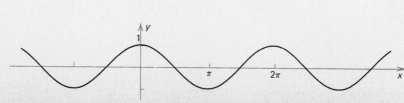

FIGURE 4.30

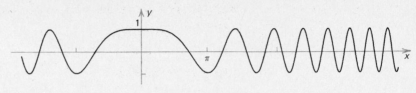

FIGURE 4.31

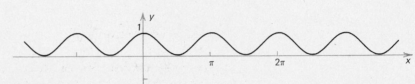

FIGURE 4.32

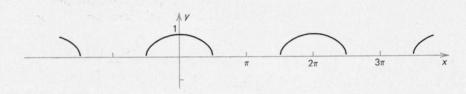

FIGURE 4.33

17. Which of the curves in Figures 4.23 to 4.28 are the graphs of **a.** $\sin^{1/3}x$ **b.** $\sin^2 x$ **c.** $\sin(\tfrac{1}{3}x^2)$ and **d.** $\sin|x|$?

18. Sketch the graph of $|\sin x|$ by tracing portions of the graph of $\sin x$ in Figure 4.23.

19. Sketch the graph of $\sqrt{\sin x}$ by tracing portions of the graph of $\sqrt{\cos x}$.

20. Sketch the graphs of $\cos|x|$ and $|\cos x|$ by tracing portions of the graph of $\cos x$.

Use the first derivative test to sketch the curves in Exercises 21 through 28.

21.† $y = 2x^{1/2} - x$ 22.† $y = \sqrt{x^2 - 4x}$

23.† $y = (4 - x)^{-1/2}$ 24. $y = x^{3/7} + 1$

25.† $y = (16 - x^4)^{1/4}$ 26. $y = 1 - x^{4/7}$

27.† $y = (x^2 - 1)^{1/3}$ 28. $y = \sqrt{x^4 - 1}$

4-5 IMPLICITLY DEFINED FUNCTIONS AND IMPLICIT DIFFERENTIATION

If the values of a function are determined by an equation that they satisfy rather than by an explicit formula, then the function is said to be DEFINED IMPLICITLY. In this section we show how to compute the derivative of an implicitly defined function at any point where we know the value of the function. This procedure is called IMPLICIT DIFFERENTIATION and is illustrated in the next example.

EXAMPLE 1   A function $y = y(x)$ is defined implicitly by the equation

(1) $x^3 + y^3 = 9xy$

and by the condition

(2) $y(4) = 2.$

Compute the derivative $\dfrac{dy}{dx}(4)$.

Solution.   We differentiate both sides of equation (1) with respect to x by using the chain rule to differentiate the cube $y^3 = y(x)^3$ of the function y and the product rule to differentiate the product $xy = xy(x)$ of the variable x and the function y. We obtain

$$\frac{d}{dx}x^3 + \frac{d}{dx}y^3 = 9x\frac{d}{dx}y + 9y\frac{d}{dx}x$$

and then

(3) $3x^2 + 3y^2\dfrac{dy}{dx} = 9x\dfrac{dy}{dx} + 9y.$

Here, as is the case with any implicit differentiation, the derivative dy/dx appears as a single factor in the terms containing it. It is, therefore, a simple matter to solve for dy/dx. We cancel a factor of 3 from each term in the equation and isolate the terms involving dy/dx on the left side: $y^2\,dy/dx - 3x\,dy/dx = 3y - x^2$. Then, factoring out the dy/dx gives $(y^2 - 3x)\,dy/dx = 3y - x^2$, or

(4) $\dfrac{dy}{dx} = \dfrac{3y - x^2}{y^2 - 3x}.$

Condition (2) states that the value of y is 2 when x is 4, and by substituting these values in (4) we obtain

$$\frac{dy}{dx}(4) = \frac{3(2) - 4^2}{2^2 - 3(4)} = \frac{5}{4}. \quad \square$$

The geometric meaning of implicit differentiation

The geometric interpretation of Example 1 is discussed in the next example.

EXAMPLE 2   The graph of the equation

(1) $x^3 + y^3 = 9xy$

is the *Folium of Descartes* shown in Figure 4.34. Let $y = y(x)$ be the function defined implicitly by equation (1) and the condition

(2) $y(4) = 2.$

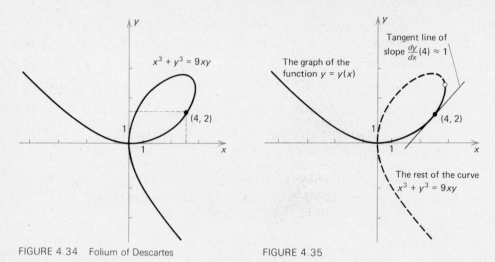

FIGURE 4.34   Folium of Descartes   FIGURE 4.35

Suppose that $y(x)$ is differentiable and that its domain is as large as possible. Trace the graph of $y(x)$ from Figure 4.34. Then find an approximate value of $\dfrac{dy}{dx}(4)$ by drawing a plausible tangent line to the graph. Compare this result with the exact value computed in Example 1.

Solution.   The graph of the function $y = y(x)$ is a portion of the curve $x^3 + y^3 = 9xy$ that includes the point $(4, 2)$. When we state that the function should be "differentiable," we mean that it should have a derivative at each x in its domain, which implies that the function must be continuous. The graph of $y(x)$ must be "unbroken" and have a nonvertical tangent line at each point on it.

The graph of the function $y(x)$ is shown in Figure 4.35. It cannot include any of the curve $x^3 + y^3 = 9xy$ beyond the circle at its upper right end because then it would contain distinct points with the same x-coordinate and would not be the graph of a function. It does not include the point indicated by the circle because the curve has a vertical tangent line there.

The tangent line to the graph at $(4, 2)$, shown in Figure 4.35, has slope of approximately 1, which is fairly close to the exact value $\dfrac{dy}{dx}(4) = \tfrac{5}{4}$ of the derivative that was computed in Example 1.   □

The dependence of $\dfrac{dy}{dx}$ on y

As is illustrated by equation (4), the formulas for dy/dx that we obtain by implicit differentiation generally involve the value y of the function. Then if there are two or more values of y that satisfy the original equation for one value of x, the formula gives the corresponding values of the derivative. For example, Figure 4.34 shows that the vertical line $x = 4$ intersects the curve $x^3 + y^3 = 9xy$ at three points. Therefore, there are three values of y that satisfy the equation $x^3 + y^3 = 9xy$ with

$x = 4$. One is the value $y = 2$ considered in Examples 1 and 2. The others are $y = -1 + \sqrt{33} \approx 4.74$ and $y = -1 - \sqrt{33} \approx -6.74$. Thus there are three functions $y(x)$ defined implicitly by $x^3 + y^3 = 9xy$ near $x = 4$, and we obtain the derivatives $\dfrac{dy}{dx}(4)$ of the three functions by setting $x = 4$ and using the corresponding values of y in (4).

Note. In the two examples of this section we have dealt with variables x and y related by an equation $x^3 + y^3 = 9xy$, and we have considered y to be a function of x. In such situations we speak of x as the INDEPENDENT VARIABLE and y as the DEPENDENT VARIABLE. If we wanted to consider x as a function of y, then y would be the independent and x the dependent variable. □

Exercises

†*Answer provided.*
‡*Outline of solution provided.*

1.‡ What is $\dfrac{df}{dx}(4)$ if $f(4) = 2$ and $x^2 f(x) = 6x + f(x)^3$ for x near 4?

2.† What is $g'(3)$ if $g(3) = 1$ and $xg(x)^3 + xg(x) = 6$ for x near 3?

3. What is $\dfrac{dh}{dt}(2)$ if $h(2) = 1$ and $th(t)^2 = 1 + \sqrt{h(t)}$ for t near 2?

In Exercises 4 through 10 differentiable functions $y = y(x)$ are defined implicitly by the given conditions. Compute the indicated derivatives.

4.‡ $\dfrac{dy}{dx}(0)$ where $xy^{3/2} + 4 = 2x + y$, $y(0) = 4$

5.† $y'(3)$ where $xy^3 = 12 + xy^2$, $y(3) = 2$

6. $\dfrac{dy}{dx}(1)$ where $x^2 y - y^3 = 6$, $y(1) = -2$

7.† $\dfrac{dy}{dx}(1)$ where $x \sin y + y^2 = \pi^2$, $y(1) = \pi$

8.† $y'(2)$ where $\sqrt{x + y} + xy = 17$, $y(2) = 7$

9. $y'(0)$ where $y = x \cos y$, $y(0) = 0$

10. $\dfrac{dy}{dx}(2)$ where $x^3 y^2 - y^5 = 7$, $y(2) = 1$

In Exercises 11 through 16 give equations for the tangent lines.

11.‡ The tangent line to $x^{2/3} + y^{2/3} = 5$ at $(8, 1)$

12.† The tangent line to $y^3 + xy^2 - x = 11$ at $(1, 2)$

13. The tangent line to $2x + y - xy^{3/2} = 4$ at $(0, 4)$

14.‡ The tangent line to $x = \cos y$ at $(\tfrac{1}{2}, \pi/3)$

15.† The tangent line to $x = \sin y / \cos y$ at $(0, 0)$

16. The tangent line to $x \sin y = 1$ at $(\sqrt{2}, \pi/4)$

17.‡ A differentiable function f with a domain as large as possible is defined implicitly by the equation $f(x)^5 - f(x) - x^2 = -1$ and by the condition $f(1) = 1$. The curve $y^5 - y - x^2 = -1$ is shown in Figure 4.36.
a. Trace the graph of f from the sketch of the curve.   **b.** Describe the domain of f (approximately) by inequalities.   **c.** Find an equation for the tangent line to the graph of f at $x = 1$ by implicit differentiation, and draw that tangent line on your sketch of the graph of f.

18.† A differentiable function h with a domain as large as possible is defined implicitly by the conditions $h(x)^5 - h(x) - x^2 = -1$ and $h(1) = 0$. Follow the instructions of Exercise 17 for this function.

19. A differentiable function k with a domain as large as possible is defined implicitly by the conditions $k(x)^5 - k(x) - x^2 = -1$ and $k(1) = -1$. Follow the instructions of Exercise 17 for this function.

20.† The *lemniscate* $3(x^2 + y^2)^2 = 25(x^2 - y^2)$ is shown in Figure 4.37. Find an equation for its tangent line at $(2, 1)$. Trace the curve and sketch its tangent line at that point.

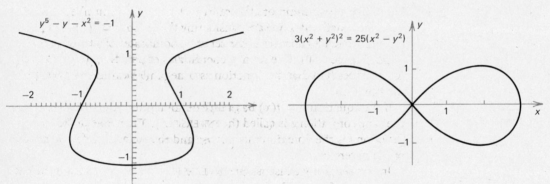

$y^5 - y - x^2 = -1$

FIGURE 4.36

$3(x^2 + y^2)^2 = 25(x^2 - y^2)$

FIGURE 4.37   A lemniscate

21. The curve in Figure 4.38 is called a *bicorn*. It has the equation $(x^2 + 8y - 16)^2 = y^2(16 - x^2)$. Verify by implicit differentiation that its tangent lines at the points $(0, 4)$ and $(0, \frac{4}{3})$ are horizontal.

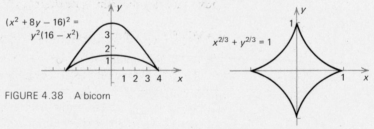

$(x^2 + 8y - 16)^2 = y^2(16 - x^2)$

FIGURE 4.38   A bicorn

$x^{2/3} + y^{2/3} = 1$

FIGURE 4.39   An astroid

22.† The curve in Figure 4.39 is known as an *astroid*. It has the equation $x^{2/3} + y^{2/3} = 1$. As the sketch suggests, the equation defines two continuous functions $y = y(x)$ in the interval $-1 \le x \le 1$.
a. Solve the equation for y to obtain formulas for the functions.

b. Give formulas for the first derivatives of the functions for $x \neq 0, \pm 1$ by differentiating the equation of the curve implicitly and then substituting in the formulas for y from part (a)   **c.** Compute the same derivatives by differentiating the formulas from part (a).

In Exercises 23 through 30 use implicit differentiation to express dy/dx in terms of x and y.

23.† $x^2 + 2x + y^4 - 4x = 10$ **24.**† $x^2 - 5xy + y^2 = 4$

25.† $\dfrac{y}{y^4 + 1} = x$ **26.**† $\dfrac{x}{y} + \dfrac{y}{x} = 1$

27.† $x^2 y^3 + y^2 x^3 + x + 1 = 0$ **28.**† $x \cos y = \sin y$

29. $\sin(xy) = y$ **30.** $2y + \cos y = x$

4-6 INVERSE FUNCTIONS

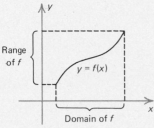

FIGURE 4.40

Recall that the *domain* of a function $f(x)$ is the set of numbers x for which it is defined. If we are considering the graph $y = f(x)$ of the function, then its domain is the set of x-coordinates of points on its graph (Figure 4.40). The set of y-coordinates of points on its graph is the set of values $f(x)$ that the function assumes and is called the RANGE of the function.

If the equation $y = f(x)$ has a unique solution $x = g(y)$ for each y in the range of f, then g is called the INVERSE of f. The range of the function f is the domain of its inverse and the domain of f is the range of the inverse.*

In this section we discuss the INVERSE FUNCTION THEOREM, which we will need in Chapter 7 where we study exponential functions, logarithms, and inverse trigonometric functions. We will present this result in two parts. The first part (Theorem 4.6a) deals with the existence of inverse functions and the second part (Theorem 4.6b) with their derivatives. Before stating and proving these theorems, we will study examples that illustrate their key features.

EXAMPLE 1   Find the inverse of the function f defined by $f(x) = x^3$.

Solution.   Solving the equation $y = x^3$ for x gives the equation $x = y^{1/3}$. The inverse of the function $f(x) = x^3$, therefore, is the function $g(y) = y^{1/3}$. □

Because x^3 takes on all values y, the domain of the inverse $y^{1/3}$ consists of all numbers y. The domain of x^3, which is the range of $y^{1/3}$, consists of all numbers x.

EXAMPLE 2   A function f is defined for $x \geq 0$ by $f(x) = 1 - x^2$. What is its inverse?

---

*Many books use the symbol f^{-1} for the inverse of f.

Solution.   To solve the equation $y = 1 - x^2$ for x, we first write $x^2 = 1 - y$ and then $x = \pm \sqrt{1 - y}$. Because the domain of the function f consists of the nonnegative numbers x, we take the plus sign to obtain the formula $g(y) = \sqrt{1 - y}$ for the inverse. The domain of g is the range of f and is the infinite interval $y \leq 1$. The range of g is the domain of f and is the infinite interval $x \geq 0$.   □

The graph of the inverse function

When we consider an inverse function $x = g(y)$, we are using y as the independent variable and x as the dependent variable. Accordingly, the graph of the inverse should be shown in a yx-plane with the y-axis horizontal and the x-axis vertical. We obtain this graph by reflecting the graph $y = f(x)$ of the original function about the line $y = x$, a process which interchanges the x- and y-axes. The reflected curve is the graph of the inverse because after it is reflected it still has the equation $y = f(x)$ which is equivalent to the equation $x = g(y)$.

The reflection process is illustrated in Figures 4.40a and 4.40b, which show the graphs of x^3 and of its inverse $y^{1/3}$, and in Figures 4.40c and 4.40d, which show the graph of the function f of Example 2 and of its inverse g.

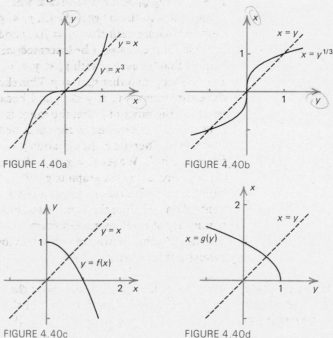

FIGURE 4.40a

FIGURE 4.40b

FIGURE 4.40c

FIGURE 4.40d

The inverse function theorem

The domains of the functions x^3 and f of Examples 1 and 2 are both intervals and each of the functions is continuous and either increasing or decreasing in its domain. (The domain of x^3 is the infinite interval consisting of all real numbers x, and it is continuous and increasing. The domain of f is the infinite interval $x \geq 0$ and it is continuous and decreasing.) The first half of the Inverse Function Theorem (Theorem 4.6a), which we will now prove, states that under these conditions the functions have continuous inverses whose domains are also intervals. Notice that this is the case with the inverse functions $y^{1/3}$ and g of Examples 1 and 2. The function $y^{1/3}$ is defined, continuous, and increasing for all y. The function g is defined, continuous, and decreasing for all $y \leq 1$.

Theorem 4.6a.   Suppose that the domain of the continuous function f is an interval I_1 and that f is either increasing in I_1 or decreasing in I_1. Then the range of f is an interval I_2 and f has a continuous inverse g with domain I_2. The inverse is increasing if f is increasing and decreasing if f is decreasing.

Proof.   We will deal with the case where f is increasing in I_1. If y_1 and y_2 are points in the range of f with $y_1 < y_2$, then there are points x_1 and x_2 in its domain such that $y_1 = f(x_1)$ and $y_2 = f(x_2)$. We have $x_1 < x_2$ because f is increasing. The Intermediate Value Theorem (Section 2.8) implies that for each y with $y_1 < y < y_2$ there is a number x with $x_1 < x < x_2$ such that $f(x) = y$. This shows that the range of f includes the entire interval $y_1 \leq y \leq y_2$ and, because y_1 and y_2 are arbitrary points in the range of f, that the range is an interval I_2.

Because f is increasing, it cannot have the same value at two different values of x. Therefore, the equation $y = f(x)$ has a unique solution x for each y in I_2. We set $x = g(y)$ to define the inverse function. We can anticipate that the graph of g will be "unbroken" and that g will be continuous because its graph is obtained by reflecting the "unbroken" graph of the continuous function f. A proof of the continuity of g follows readily from the $\epsilon\delta$-definition and is left as Exercise 24. The discussion in the previous paragraph shows that the inverse g is increasing.   Q.E.D.

The derivative of the inverse function

When we reflect the xy-plane about the line $y = x$, lines of nonzero slope m are transformed into lines of slope $1/m$. This is because the "rise" as we go from one point to another on the original line becomes the "run" on the reflected line, and vice-versa. In the reflection process the tangent lines to the graph of a function f are transformed into the tangent lines to the graph of the inverse. Therefore, if f has a nonzero derivative $f'(x_0)$ at x_0 and $y_0 = f(x_0)$ is the corresponding value of y, then the derivative $g'(y_0)$ of the inverse at y_0 is the reciprocal $1/f'(x_0)$ of the derivative of f. This is illustrated in the next two examples and is proved as a general result in the following theorem.

EXAMPLE 3   Draw the tangent lines to the graph $y = x^3$ of the function x^3 and to the graph $x = y^{1/3}$ of its inverse at $x = 1, y = 1$ and show that their slopes are reciprocals. Explain what happens at $x = 0$, $y = 0$.

Solution.   The tangent line at $x = 1$ to the curve $y = x^3$ in the xy-plane of Figure 4.41 has slope 3 because the derivative $\dfrac{d}{dx}x^3 = 3x^2$ is 3 at $x = 1$. The tangent line at $y = 1$ to the curve $x = y^{1/3}$ in the yx-plane of Figure 4.41a has the reciprocal slope $\frac{1}{3}$ because the derivative $\dfrac{d}{dy}y^{1/3} = \frac{1}{3}y^{-2/3}$ is $\frac{1}{3}$ at $y = 1$.

Because the derivative $3x^2$ of x^3 is zero at $x = 0$, the tangent line to the graph of x^3 at the origin is the horizontal x-axis (Figure 4.41). Accordingly, the vertical x-axis is the tangent line to the graph of the inverse function $y^{1/3}$ at the origin (Figure 4.41a) and $y^{1/3}$ does not have a derivative at $y = 0$.   □

EXAMPLE 4   Sketch the tangent line at $x = 1$ to the graph of the function f of Example 2. Then sketch the tangent line at the corresponding point on the graph of the inverse and show that the slopes are reciprocals.

Solution.   Near $x = 1$ the values of f are given by the formula $1 - x^2$, so

$$f'(1) = \left[\frac{d}{dx}(1 - x^2)\right]_{x=1} = [-2x]_{x=1} = -2.$$

The point at $x = 1$ on the graph of f has the y-coordinate $y = f(1) = 0$. The tangent line at that point has slope -2 and is shown in Figure 4.41b. The corresponding point on the graph of the inverse function is at $y = 0$ and the inverse function has the formula $g(y) = \sqrt{1 - y}$. We have

$$g'(0) = \left[\frac{d}{dy}(1 - y)^{1/2}\right]_{y=0} = [-\tfrac{1}{2}(1 - y)^{-1/2}]_{y=0} = -\tfrac{1}{2}$$

and this is, as expected, the reciprocal of the derivative $f'(1)$ of f. The tangent line at $y = 0, x = 1$ to the graph of g has slope $-\frac{1}{2}$ and is shown in Figure 4.41c.   □

Examples 3 and 4 suggest that the inverse function will have a derivative at all values of y that correspond to values of x where f has a *nonzero* derivative. We establish this result in the next theorem.

Theorem 4.6b.   Suppose that the domain of the continuous function $f(x)$ is an interval and that $f(x)$ is either increasing or decreasing in that interval, so it has a continuous inverse

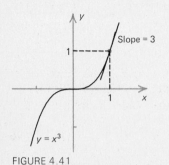

FIGURE 4.41

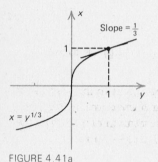

FIGURE 4.41a

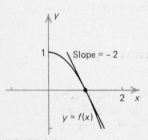

FIGURE 4.41b

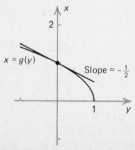

FIGURE 4.41c

$g(y)$. **If $f'(x)$ exists and is not zero at $x = g(y)$, then $g'(y)$ exists and is the reciprocal of $f'(x)$:**

(1) $g'(y) = \dfrac{1}{f'(x)}$ **where** $x = g(y)$.   ◄

Equation (1) takes a simpler form if we use abbreviated Leibniz notation. We write $y(x)$ in place of $f(x)$, dy/dx in place of f', $x(y)$ in place of $g(y)$, and dx/dy in place of g'. Then equation (1) reads

(2) $\dfrac{dx}{dy}(y) = \dfrac{1}{\dfrac{dy}{dx}(x)}$ at $x = x(y)$.   ◄

Abbreviating the notation further, we obtain the equation

(3) $\dfrac{dx}{dy} = \dfrac{1}{dy/dx}$ ◄

in which the symbolic fractions of Leibniz notation appear to obey a rule of algebra.

Proof of Theorem 4.6b.   Because $f'(x)$ is defined, x is in the interior of the domain of f and $y = f(x)$ is in the interior of the domain of g. For sufficiently small $\Delta y \neq 0$, $y + \Delta y$ is also in the domain of g and we can set $\Delta x = g(y + \Delta y) - g(y)$. Then $\Delta y = f(x + \Delta x) - f(x)$ and

(4) $\dfrac{g(y + \Delta y) - g(y)}{\Delta y} = \dfrac{\Delta x}{f(x + \Delta x) - f(x)}.$

Because g is continuous, Δx tends to zero and equation (4) becomes equation (1) as Δy tends to zero.   Q.E.D.

EXAMPLE 5   Show that $f(x) = x^3 + x$ has an inverse $g(y)$ defined for all y and that $g(2) = 1$. Compute $g'(2)$.

Solution.   The function f has an inverse because its derivative $f'(x) = 3x^2 + 1$ is positive and hence f is continuous and increasing for all x. The range of f, which is the domain of g, consists of all numbers y because $f(x)$ tends to ∞ as x tends to ∞ and tends to $-\infty$ as x tends to $-\infty$.

We have $g(2) = 1$ because $f(1) = 1^3 + 1 = 2$. Also $f'(1) = 3(1)^2 + 1 = 4$, so $g'(2) = 1/f'(1) = \frac{1}{4}$.   □

Inverse functions as functions of x

We often want to use the letter x rather than y as the variable of an inverse function. We could, for example, speak of the function $x^{1/3}$ as the inverse of the function x^3. To find a formula for the inverse g of a function f so that g has variable x, we can interchange x and y in the equation $y = f(x)$ to obtain $x = f(y)$, and then solve for y as in the next example.

EXAMPLE 6   The function $F(x) = \frac{9}{5}x + 32$ gives the temperature in degrees Fahrenheit that corresponds to x degrees Centigrade. Give the formula for the inverse function $C(x)$ which gives the temperature in degrees Centigrade corresponding to x degrees Fahrenheit. Sketch the graphs of the two functions.

Solution.   The graph of $F(x)$ is the line $y = \frac{9}{5}x + 32$ shown in Figure 4.41d. Interchanging x and y gives $x = \frac{9}{5}y + 32$ and solving for y yields the equation $y = \frac{5}{9}(x - 32)$ of the graph of the inverse function $C(x)$. This graph is the line shown in Figure 4.41e. Notice that the lines are mirror images about the line $y = x$ and that their slopes are reciprocals.   □

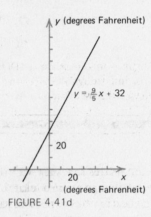

FIGURE 4.41d

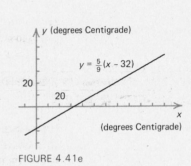

FIGURE 4.41e

Exercises

In Exercises 1 through 8 the variable y is given as a function of x. Solve the equations to express x as a function of y.

1.† $y = 2x - 3$

2. $y = 1 + \sqrt{x - 2}$

3.† $y = (2x + 3)^3 + 3$

4. $y = 3 - \sqrt[3]{x}$

5.† $y = 1 + (x + 4)^{3/4}$

6. $y = \sqrt{1 - x^3}$

7.† $y = \dfrac{1}{x - 3} + 3$ for $x > 3$

8. $y = \dfrac{x}{x + 1}$ for $x > -1$

9.† Set $f(x) = \sin x$ with domain $-\frac{1}{2}\pi \le x \le \frac{1}{2}\pi$. **a.** What is the range of f? **b.** Show that f has a continuous inverse g and that $g(\frac{1}{2}) = \frac{1}{6}\pi$. **c.** What is $g'(\frac{1}{2})$?

10.† Define $f(x) = x^{-2} + x^{-3}$ for all $x > 0$. **a.** What is the range of f? **b.** Show that f has a continuous inverse g and that $g(2) = 1$. **c.** What is $g'(2)$?

11.† Define $f(x) = 4x + \cos x$ for all (πx). **a.** Show that f has a continuous inverse g and that $g(11) = 3$. **b.** What is $g'(11)$? **c.** What is the domain of g?

12. Set $f(x) = x^{-1} - x^3$ for $x > 0$. **a.** What is the range of f? **b.** Show that f has a continuous inverse. **c.** Is g an increasing or a decreasing function? **d.** Find $g(0)$ and $g'(0)$.

In each of Exercises 13 through 17 the given function of x is increasing for x in one interval and decreasing for x in another interval, so we obtain functions with inverses by restricting the domain of the given function to each of the intervals. Give formulas for those inverse functions as functions of y.

13.‡ $(x - 3)^2 + 2$ **14.** $x^4 - 5$

15.‡ $(x^2 + 7)^{-1/2}$ **16.**† $(2x - 3)^{2/3}$

17. $\dfrac{x^2}{x^2 + 2}$

In each of Exercises 18 through 23 find the inverse of the function and express it as a function of x.

18.† $(x - 2)^{-1/2}$ **19.**† $2 + (x - 3)^5$

20.† $(2x - 1)^{1/3} - 2$ **21.** $3 + (3x + 6)^{1/5}$

22.† $(x^{1/3} + 1)^3$ **23.** $(x^3 + 1)^{1/3}$

24.† Use the $\epsilon\delta$-definition of continuity (Definition 2.12 in Section 2.9) to show that the function $g(y)$ in Theorem 4.6 is continuous for $c \le y \le d$. (Use the fact that $f(x_1) < y < f(x_2)$ implies $x_1 < g(y) < x_2$.)

4-7   RELATED RATES

If two or more quantities that vary with time are related by an equation, then their rates of change are also related. In this section we see how this principle is used in the solution of RELATED RATE PROBLEMS.

The key to solving a narrative problem in algebra is the introduction of letters to denote unknown *numbers*. The key to solving a related rate problem is the introduction of letter names for the *functions* that describe how the quantities in the problem vary. The rates of change of the quantities are the derivatives of the functions. To solve most related rate problems, we can consider all of the varying quantities to be functions of the time t and then apply the following procedure.

Step 1   Make precise definitions of the functions of time that measure the varying quantities.

Step 2   Determine one or more equations that relate the functions from Step 1 if such equations are not given explicitly in the statement of the problem. Eliminate any unnecessary functions.

Step 3   Differentiate the equation or equations from Step 2 with respect to t to obtain one or more equations that relate the derivatives. (This step usually requires the use of the chain rule.)

Step 4   Use the equations from Steps 2 and 3 to solve the problem.

EXAMPLE 1   A 13-foot ladder is leaning against a wall, and its base is slipping away from the wall at the rate of 3 feet per second when it is 5 feet from the wall. How fast is the top of the ladder dropping at that moment?

Solution.

Step 1   Since the rate at which the base of the ladder is moving is given in feet per second, we measure t in seconds. We let $x = x(t)$ be the distance (feet) from the base of the ladder to the wall and let $y = y(t)$ be the height (feet) of the top of the ladder above the ground at time t.

Step 2   The wall, the ground, and the ladder form a right triangle for any value of t (Figure 4.42). According to the Pythagorean theorem, the values $x(t)$ and $y(t)$ of the functions satisfy the equation

(1) $x(t)^2 + y(t)^2 = 13^2.$

Step 3   We differentiate equation (1) with respect to t:

$$\frac{d}{dt}[x(t)^2 + y(t)^2] = \frac{d}{dt}13^2.$$

This gives

(2) $2x(t)\dfrac{dx}{dt}(t) + 2y(t)\dfrac{dy}{dt}(t) = 0.$

Step 4   We rewrite equations (1) and (2) in abbreviated notation by dropping the symbol "(t)":

(3) $x^2 + y^2 = 13^2$

(4) $2x\dfrac{dx}{dt} + 2y\dfrac{dy}{dt} = 0.$

We are told that when the base of the ladder is 5 feet from the wall, it is moving at the rate of 3 feet per second. This means that at that moment x is 5 and dx/dt is 3, and equations (3) and (4) read

(5) $5^2 + y^2 = 13^2$

(6) $2(5)(3) + 2y\dfrac{dy}{dt} = 0.$

Solving equation (5) for y gives $y = \sqrt{13^2 - 5^2} = 12$. Then equation (6) reads $30 + 24\,(dy/dt) = 0$ and shows that $dy/dt = -\frac{5}{4}$. The top of the ladder is falling $\frac{5}{4}$ feet per second at the moment in question.   □

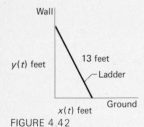

Wall

$y(t)$ feet

13 feet

Ladder

$x(t)$ feet   Ground

FIGURE 4.42

An alternate procedure   In working related rate problems it is frequently convenient to consider one of the varying quantities as a function of one of the others. If the varying quantities are u and v, we might consider the quantity u to be a function $u = u(v)$ of the quantity v. Then we would consider $v = v(t)$

as a function of the time, and u would be given as a function of the time by the composite function $u = u(v(t))$. This point of view can simplify calculations and add insight into a problem, as illustrated in the next example.

EXAMPLE 2   A toy roller coaster is in the shape of the curve $y = (1 - x/8) \cos(\pi x)$ for $0 \leq x \leq 8$ (Figure 4.43). When a roller coaster car is at $x = \frac{3}{2}$ (feet), it is moving to the right with the velocity of 3 feet per second. How rapidly is it rising at that moment?

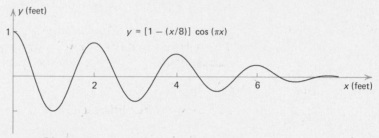

FIGURE 4.43

Solution.   We measure the time t in seconds and let $x = x(t)$ be the car's x-coordinate at time t. The equation $y = (1 - x/8) \cos(\pi x)$ of the track gives y as a function of x. By the chain rule the rate of change of $y = y(x(t))$ with respect to t is the product

(7) $$\frac{dy}{dt} = \frac{dy}{dx} \frac{dx}{dt}$$

of the slope dy/dx of the track and the car's velocity dx/dt in the x-direction.

We first compute the slope of the track. By the product and chain rules, we have

$$\frac{dy}{dx} = \left[\frac{d}{dx}\left(1 - \frac{x}{8}\right) \right] \cos(\pi x) + \left(1 - \frac{x}{8}\right)\frac{d}{dx}\cos(\pi x)$$

$$= -\frac{1}{8}\cos(\pi x) + \left(1 - \frac{x}{8}\right)\left[-\sin(\pi x)\frac{d}{dx}(\pi x) \right]$$

$$= -\frac{1}{8}\cos(\pi x) + \left(1 - \frac{x}{8}\right)[-\pi \sin(\pi x)].$$

Since $\cos(3\pi/2) = 0$ and $\sin(3\pi/2) = -1$, the slope at $x = \frac{3}{2}$ is

$$\frac{dy}{dx} = -\frac{1}{8}\cos\left(\frac{3\pi}{2}\right) + \left(1 - \frac{3/2}{8}\right)\left[-\pi \sin\left(\frac{3\pi}{2}\right) \right]$$

$$= -\frac{1}{8}(0) + \left(1 - \frac{3}{16}\right)[-\pi(-1)] = \left(1 - \frac{3}{16}\right)\pi = \frac{13}{16}\pi.$$

We are told that when the car is at $x = \frac{3}{2}$, its velocity toward the right dx/dt is 3 feet per second. By equation (7) the car's upward velocity dy/dt at that moment is

$$\frac{dy}{dt} = \frac{dy}{dx}\frac{dx}{dt} = \left(\frac{13}{16}\pi\right)(3) = \frac{39}{16}\pi \text{ feet per second.}$$

The car is rising at the rate of $\frac{39}{16}\pi$ or about 7.6 feet per second.   □

**A graphical
related rate
problem**

In the next example we deal with an empirically determined function whose values are given approximately by a sketch of its graph rather than by an exact formula.

EXAMPLE 3   The function V whose graph is sketched in Figure 4.44 gives the volume of air $V(t)$ (cubic inches) that a man has blown into a balloon after t seconds. Approximately how rapidly is the radius increasing after 6 seconds?

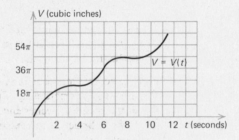

FIGURE 4.44

Solution.   We let $r = r(t)$ (inches) be the radius of the balloon and $V = V(t)$ (cubic inches) its volume at time t (seconds). The functions r and V are related by the formula for the volume of a sphere

(8) $V = \frac{4}{3}\pi r^3.$

Differentiating this equation with respect to r gives $dV/dr = 4\pi r^2$. Then the chain rule gives

(9) $\dfrac{dV}{dt} = \dfrac{dV}{dr}\dfrac{dr}{dt} = 4\pi r^2 \dfrac{dr}{dt}.$

We need to determine dr/dt at time $t = 6$ from equations (8) and (9) by finding values of V and dV/dt at $t = 6$ from the sketch of the graph in Figure 4.44. We first see that V is approximately 36π at $t = 6$, so that by equation (8), $\frac{4}{3}\pi r^3 \approx 36\pi$. Solving for r shows that it is approximately 3 at $t = 6$.

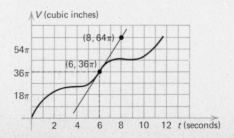

FIGURE 4.45   Tangent line of slope $\approx 14\pi$

The tangent line drawn in Figure 4.45 has an approximate slope of 14π. When this value for dV/dt and the value 3 for r are substituted into equation (9), we find that $14\pi \approx 4\pi 3^2 \, dr/dt$. We solve for dr/dt to obtain our final answer: $dr/dt \approx 14\pi/36\pi = \frac{7}{18}$ inches per second at $t = 6$. $\square$

Exercises

†*Answer provided.*
‡*Outline of solution provided.*

1.‡ A water tank is in the shape of a right circular cone with its vertex pointing downward. The radius of the top is 10 feet and the tank is 40 feet high. Water is flowing in at the rate of 7 cubic feet/minute when the water is 8 feet deep in the tank. At what rate is its depth increasing at that moment?

2.† The ends of a trough are equilateral triangles with one vertex pointing down. The trough is 9 feet long. When the water in the trough is 2 feet deep, its depth is increasing at the rate of $\frac{1}{2}$ foot/minute. At what rate is water flowing into the trough at that moment?

3. Sand is falling in a conical pile at the rate of 12 cubic feet/minute. The vertical cross section of the pile is always an isosceles right triangle. How fast is the height of the pile increasing when the height is 4 feet?

4.‡ At a certain moment the radius of the base of a right circular cone is 4 inches and is increasing at the rate of 5 inches/second, whereas the height of the cone is 3 inches and is decreasing at the rate of 6 inches/second. Is the volume of the cone increasing or decreasing at that moment? At what rate?

5.† One bicycle is 4 miles east of an intersection, and it is traveling toward the intersection at the rate of 9 miles/hour. At the same time a second bicycle is 3 miles south of the intersection, and it is traveling away from the intersection at the rate of 10 miles/hour. Is the distance between the bicycles increasing or decreasing at that moment? At what rate?

6. The area of a circle is decreasing at the rate of 5 square meters/second when its radius is 3 meters.   **a.** At what rate is the radius decreasing at that moment?   **b.** At what rate is the diameter decreasing?

7.‡ A light is at the top of a 48-foot pole. A ball is dropped from the same height from a point 10 feet away. After one second the ball has fallen 16 feet and is falling at the rate of 32 feet/second. How fast is the shadow of the ball moving on the ground at that moment?

8.† A man 7 feet tall is 20 feet from a 28-foot lamppost and is walking toward it at the rate of 4 feet/second. How fast is his shadow shrinking at that moment?

9. Suppose that the volume of a melting snowball decreases at a rate proportional to its surface area. Show that its radius decreases at a constant rate.

10.‡ If an object weighs 100 pounds on the surface of the earth, then it weighs $100 \left(1 + \frac{1}{4000} r\right)^{-2}$ pounds when it is r miles above the surface of the earth. At what rate is the weight decreasing when the object is 400 miles above the surface of the earth if its altitude is increasing at the rate of 15 miles/second?

11.† Standard barometric pressure at sea level is 760 mm. (millimeters of mercury). Water boils at 100°C at that pressure. For barometric pressures p between 660 mm. and 860 mm. the boiling point of water is given with an accuracy of one thousandth of a degree by

$$T = 100 + 28.012\left(\frac{p}{760} - 1\right) - 11.64\left(\frac{p}{760} - 1\right)^2 + 7.1\left(\frac{p}{760} - 1\right)^3.$$

How fast is the boiling point of water dropping when the barometric pressure is 836 mm. and is dropping at the rate of 5 mm./minute?

12. The thin lens equation $\dfrac{1}{s_1} + \dfrac{1}{s_2} = \dfrac{1}{f}$ relates the distances s_1 and s_2 from an object and its image to the lens. The constant f is called the *focal length* of the lens.

   An object is moving away from a lens with a focal length of 20 centimeters at the rate of 10 centimeters/minute. How fast is its image moving when the object is 60 centimeters from the lens?

13.† A hemispherical bowl has a radius of 10 inches. The volume of water in the bowl is $\frac{1}{3}\pi h^2(30 - h)$ cubic inches when it is h inches deep. When h is 5 inches, it is increasing at the rate of 6 inches/minute. At what rate is the volume increasing at that moment?

14. The force of repulsion between two positive charges of q_1 and q_2 statcoulombs is $q_1 q_2 r^{-2}$ dynes when the charges are r centimeters apart. Two charges of 3 and 5 electrostatic units are moving apart at the rate of 7 centimeters/second. At what rate is the force between them decreasing when they are 13 centimeters apart?

15.† The sides of an equilateral triangle are increasing at the rate of 5 inches/second. At what rate are   **a.** the area and   **b.** the perimeter of the triangle increasing when the sides are 10 inches long?

16. An object moves along the parabola $y = 4x^2$ in such a way that its x-coordinate increases at the constant rate of 5 units/minute. How fast is its y-coordinate decreasing when it is at the point $(-2, 16)$?

17.† Two sticks, each 3 feet long, are hinged together and are leaning against a wall so that the sticks and the floor form an isosceles triangle. The sticks slide apart, and at the moment when the triangle is equilateral, the angle at the hinge is increasing at the rate of $\frac{1}{3}$ radian/second. At what rate is the area of the triangle increasing or decreasing at that moment?

18.† As a girl 5 feet tall walks away from a 9-foot lamppost, the intensity of the light shining on her is $I = 200 \cos \theta\, r^{-2}$ candles/square foot, where r is the distance (feet) from her head to the light and θ (theta) is the angle shown in Figure 4.46. When she is 3 feet from the lamppost, she is walking away from it at the rate of 4 feet/second. At what rate is the intensity of light on her decreasing at that moment?

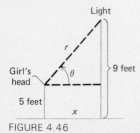

Light

r

θ

9 feet

Girl's head

5 feet

x

FIGURE 4.46

19.† According to *Bernoulli's law*, if the pressure in a water pipe is 100 pounds/square inch when the water is not flowing, then the pressure p (pounds/square inch) and the velocity of the water v (feet/second) are related by the equation $\frac{1}{2}p + v^2 = 50$ when the water is flowing. The velocity of the water in one such pipe is given by the function $v = v(t)$ of Figure 4.47. What is **a.** the approximate pressure and **b.** the approximate rate of change of the pressure at time $t = 8$?

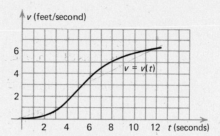

FIGURE 4.47

20.   One ship is $W(t)$ nautical miles west of a lighthouse and a second ship is $S(t)$ nautical miles south of the lighthouse at time t (hours). The graphs of W and S are shown in Figures 4.48 and 4.49. At what approximate rate is the distance between the ships increasing at $t = 1$? (Give the answer in knots = nautical miles per hour.)

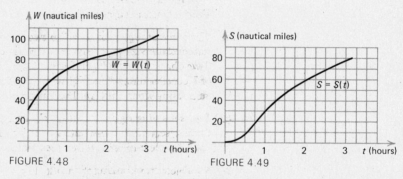

FIGURE 4.48   FIGURE 4.49

21.   At a certain moment during the compression stroke in a diesel engine one of the cylinders contains 50 cubic inches of gas vapor under a pressure of 400 pounds/square inch and the volume is decreasing at the rate of 100 cubic inches/second. The volume v and the pressure p are related by an equation of the form $pv^{1.4} = $ constant. At what rate is the pressure increasing at that moment?

22.†   What is $x'(1)$ if $x(t)$, $y(t)$, and $z(t)$ are related by the equation $x^2 + y^3 + z^4 = -3$ and $x(1) = 2, y(1) = -2, z(1) = 1, y'(1) = -3,$ and $z'(1) = 4$?

23.†   What is $p'(6)$ if $p(t)$ and $q(t)$ satisfy $\frac{1}{12}pq = \sqrt{p + q}$ and $p(6) = 12,$ $q(6) = 4$, and $q'(6) = -2$?

24.†   Suppose $x(t)$, $y(t)$, and $z(t)$ satisfy $xyz = t$ and that $x(10) = 2,$ $y(10) = -5, z(10) = -1, x'(10) = -2,$ and $z'(10) = 3$. What is $y'(10)$?

4-8   MAXIMUM-MINIMUM PROBLEMS IN FINITE CLOSED INTERVALS

In this section we use the following theorem to solve problems that call for finding the maximum or minimum value of a continuous function in a finite closed interval.

Theorem 4.7   If a function f is continuous in a finite closed interval, then its maximum and minimum values in the interval occur at the critical points of the function in the interval or at the end points of the interval.

Recall that a critical point is a value of the variable where the derivative of the function does not exist or is zero.

Proof.   The Extreme Value Theorem shows that the continuous function has a maximum and a minimum value in the finite closed interval. If the maximum or minimum does not occur at an end point of the interval, then it occurs at a point x interior to the interval where the function has a local maximum or minimum. By Theorem 4.3, x is a critical point.   Q.E.D.

To apply Theorem 4.7, we locate the critical points of the function in the interval and compare the values of the function at the critical points and end points to see which is the maximum or minimum value.

EXAMPLE 1   Find the maximum and minimum values of $x/(x^2 + 4)$ for $-1 \le x \le 4$.

Solution.   We set $f(x) = x/(x^2 + 4)$. The function f is continuous and has a derivative for all x. We have

$$f'(x) = \frac{(x^2 + 4)\dfrac{d}{dx}(x) - x\dfrac{d}{dx}(x^2 + 4)}{(x^2 + 4)^2} = \frac{x^2 + 4 - 2x^2}{(x^2 + 4)^2}$$

$$= \frac{4 - x^2}{(x^2 + 4)^2}.$$

The derivative is zero at $x = 2$ and at $x = -2$, so these are the critical points of f. The only critical point in the interval $-1 \le x \le 4$ is $x = 2$. By Theorem 4.7 the maximum value of f for $-1 \le x \le 4$ is the greatest of the numbers $f(-1)$, $f(2)$, and $f(4)$. The minimum is the least of these three numbers.

We find that $f(-1) = -\frac{1}{5}$, $f(2) = \frac{1}{4}$, and $f(4) = \frac{1}{5}$. The maximum value of the function on the interval, therefore, is $f(2) = \frac{1}{4}$ and the minimum value is $f(-1) = -\frac{1}{5}$. $\square$

EXAMPLE 2   Find the maximum and minimum values of $x - \frac{3}{2}x^{2/3}$ for $-1 \leq x \leq 1$.

Solution.   The function $x - \frac{3}{2}x^{2/3}$ is continuous for all x. Its derivative

$$\frac{d}{dx}\left(x - \frac{3}{2}x^{2/3}\right) = 1 - \left(\frac{3}{2}\right)\left(\frac{2}{3}\right)x^{-1/3} = 1 - x^{-1/3}$$

does not exist at $x = 0$ and is zero at $x = 1$. Therefore, the critical points of the function are $x = 0$ and $x = 1$. The end points of the interval are $x = -1$ and $x = 1$. We find that the function has the value $-\frac{5}{2}$ at $x = -1$, 0 at $x = 0$, and $-\frac{1}{2}$ at $x = 1$. The maximum value is zero and the minimum value is $-\frac{5}{2}$.   □

Narrative maximum-minimum problems

In a narrative maximum or minimum problem we generally want to maximize or minimize a quantity subject to a *side condition* which is given by prescribing the values of one or more other quantities. Most such problems can be solved by the following procedure.

Step 1   Choose variables to describe the object or situation being studied.

Step 2   Determine formulas that express the quantity to be maximized or minimized and those that are prescribed in terms of the variables of Step 1.

Step 3   Choose one of the variables from Step 1 to be the independent variable and use the side conditions to express the other variables (the dependent variables) in terms of the independent variable.

Step 4   Replace the dependent variables in the formula for the quantity to be maximized or minimized by their formulas from Step 3.

Step 5   Determine from the statement of the problem the range of values that the independent variable may take. Usually this is an interval.

Step 6   Use the derivative to find the maximum or minimum value of the function from Step 4 in the interval from Step 5.

EXAMPLE 3   What rectangle with a perimeter of 12 feet has the greatest area?

Solution.

Step 1   We let w be the width and L the length of the rectangle, both measured in feet (Figure 4.50).

Step 2   We are to maximize the area A, and the perimeter P is prescribed. We have $A = wL$ and $P = 2w + 2L$.

Step 3   We choose w to be the independent variable. (We could use L just as well.) The side condition is $P = 2w + 2L = 12$. Solving for L gives $2L = 12 - 2w$ or $L = 6 - w$.

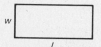

FIGURE 4.50   Area $= A = wL$ square feet. Perimeter $= P = 2w + 2L$ feet

Step 4   Replacing L by $6 - w$ in the formula for the area gives
$A = wL = w(6 - w) = 6w - w^2$.

Step 5   The width cannot be negative and cannot be greater than half
the perimeter, so w must be in the interval $0 \leq w \leq 6$.

Step 6   The function $A = 6w - w^2$ is continuous in the interval
$0 \leq w \leq 6$. Its derivative $dA/dw = 6 - 2w$ exists for all w and is zero
at $w = 3$. The only critical point of the function $A(w)$ is $w = 3$.
We find that $A(0) = 0$, $A(3) = 9$, and $A(6) = 0$. The maximum value
is 9 and occurs at $w = 3$. The rectangle of largest area is a square.   □

EXAMPLE 4   A weight on a spring moves along a vertical s-axis
according to the equation $s = \sin t + \cos t + 1$ where t is the time.
What are the highest and lowest positions of the weight?

Solution.   We need to find the maximum and minimum values of the
continuous function $s = \sin t + \cos t + 1$ for all t, but because the
function is periodic of period 2π, we need only consider t in the time
interval $0 \leq t \leq 2\pi$ of length 2π. We have

$$\frac{ds}{dt} = \frac{d}{dt}(\sin t + \cos t + 1) = \cos t - \sin t$$

so the critical points are where $\cos t = \sin t$. The critical points in the
interval $0 \leq t \leq 2\pi$ are $t = \pi/4$ and $t = 5\pi/4$. The function s has the
value $\sin(0) + \cos(0) + 1 = 2$ at the end point $t = 0$, the value
$\sin(2\pi) + \cos(2\pi) = 2$ at the end point $t = 2\pi$, the value $\sin(\pi/4) +$
$\cos(\pi/4) + 1 = 1/\sqrt{2} + 1/\sqrt{2} + 1 = 1 + \sqrt{2}$ at the critical point
$t = \pi/4$, and the value $\sin(5\pi/4) + \cos(5\pi/4) + 1 = -1/\sqrt{2} -$
$1/\sqrt{2} + 1 = 1 - \sqrt{2}$ at the critical point $t = 5\pi/4$. The greatest of
these values is $1 + \sqrt{2}$, which is the s-coordinate of the weight at its
highest point. The least of these values is $1 - \sqrt{2}$ and is the
s-coordinate at the weight's lowest point.   □

Exercises

†*Answer provided.*
‡*Outline of solution provided.*

In Exercises 1 through 8 find the maximum and minimum values of the
functions in the indicated intervals.

1.† $x^3 - 3x + 5$ for $0 \leq x \leq 2$ **2.** $6x^4 - 4x^6$ for $-2 \leq x \leq 2$

3.† $x^2/(x - 1)$ for $-1 \leq x \leq \frac{1}{2}$ **4.** $x^3 - 3x^2 + 3x$ for $0 \leq x \leq 2$

5.‡ $\frac{1}{2}x + \sin x$ for $0 \leq x \leq \pi$ **6.** $x - \cos x$ for $0 \leq x \leq 2\pi$

7.‡ $5x^{2/3} - 2x^{5/3}$ for $-1 \leq x \leq 8$ **8.** $x^{1/3} + \frac{1}{6}x^2$ for $-8 \leq x \leq 1$

9.‡ A woman wants to design a rectangular vegetable garden with an
ornamental fence around it. The fencing for three sides of the garden
costs 2 dollars/foot. The fencing for the fourth side costs 3 dollars/foot.
She has 120 dollars to spend on the fence. What dimensions should she
give the garden to maximize its area?

10.[†] For a box to be mailed by parcel post, its length plus its girth must be no greater than 84 inches. What are the dimensions of the rectangular box of largest volume that can be sent by parcel post? (Use the fact that the square has the largest area of all the rectangles with the same perimeter.)

11. A cylindrical can with no top is to be made from 12π square inches of tin. What should be the dimensions of the can to give it maximum volume?

12.[‡] Find the dimensions of the right circular cylinder of greatest volume inscribed in a right circular cone of radius 10 inches and height 24 inches.

13.[†] What is the rectangle of largest area that can be inscribed in a circle of radius r?

FIGURE 4.51

14. A farmer has 80 feet of fence to construct three adjacent rectangular pig pens as in Figure 4.51. What should the overall dimensions of the three pens be in order to maximize their total area?

15.[‡] A box without a top is to be made from a piece of cardboard 1 foot square by cutting equal squares out of the corners and then folding up the sides. What size squares should be cut out of the corners to get the box of maximum volume?

16. A rectangle has its base on the x-axis and its upper corners on the parabola $y = 12 - x^2$. Of all such rectangles, what are the dimensions of the one with largest area?

17.[†] A boy wants to make a box with square ends and no top. He has enough money to buy 600 square inches of material. He wants to paint the ends of the box red, but has only enough paint if the ends are no larger than 11 inches on each side. What dimensions should the box be given for it to have maximum volume?

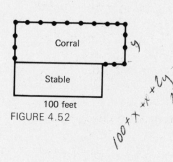

Corral

Stable

100 feet

FIGURE 4.52

18.[†] A horse breeder wants to build a corral in front of a 100-foot long stable as in Figure 4.52 with 180 feet of fence. What dimensions should he give the corral for it to have maximum area? (He does not need any fence along the front of the stable, and the corral must extend the entire length of the stable.)

19. Two light sources are located 12 feet apart. One of them is eight times as bright as the other. The intensity of the light from both sources at a point on the line between them and x feet from the brighter one is given by the formula $I = 8x^{-2} + (12 - x)^{-2}$. At what point between the light sources is the light intensity the least? (Recall $a^3 = b^3$ implies $a = b$.)

20. Rectangular beams are to be cut from logs with circular cross sections 12 inches in diameter. The strength of the beam is determined by the quantity h^2w, where h and w are the height and width of the cross section of the beam; the larger the quantity h^2w, the stronger the beam. What dimensions should the cross sections be given to yield the strongest beams?

21.[‡] A fisherman is in a row boat on a lake and 2 miles from the shore when he catches a huge fish. He wants to show the fish to his buddies in a tavern 3 miles down the (straight) shore. He can row 5 miles an hour and run 13 miles an hour. To what point on the shore should he row to get to the tavern as quickly as possible?

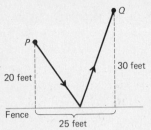

FIGURE 4.53

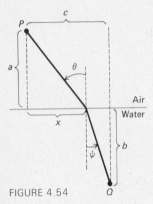

FIGURE 4.54

22.[†] At a child's party there is a contest to see which child can run the fastest from a point 20 feet from a fence, to the fence, and then to a point 30 feet from the fence as in Figure 4.53. To what point on the fence should the children run to minimize the distance?

23. According to *Fermat's principle* in optics, a ray of light that goes from a point P above water to a point Q in the water as in Figure 4.54 will take the path that requires the least time.   **a.** Let v_a denote the speed of light in air and v_w denote the speed of light in water. Express the time it would take the light to reach point Q as a function of the distance x. **b.** Show that the time is minimized if the angles θ (theta) and ψ (psi) in the figure satisfy *Snell's law:* $\sin \theta / \sin \psi = v_a / v_w$.

24.[†] The ratio v_a/v_w in Exercise 23 is about 1.33. Give an approximate value for the angle ψ when the angle θ is $\pi/4$ radians.

25. A powerhouse is on one side of a 200-foot wide straight river, and a factory is across the river and 300 feet downstream. An electric cable is to be strung from the powerhouse to the factory. It costs 5 dollars for each foot of cable in the water and 3 dollars for each foot of cable on land. No more than 100 feet of cable can be run on land by the powerhouse and none can be run on land by the factory. What path should the cable take to minimize its cost?

In each of Exercises 26 through 35 find the maximum and minimum values of the function in the given interval and the values of x where they occur.

26.[†] $x + 4x^{-1}$ for $1 \le x \le 3$ **27.**[†] $\sqrt{5 - x^2}$ for $-1 \le x \le 2$

28.[†] $\sin(x^2)$ for $-1 \le x \le 1$ **29.**[†] $x^3 + x^5 + x^7$ for $-1 \le x \le 1$

30.[†] $x^{10} - 10x$ for $-1 \le x \le 2$ **31.**[†] $x^{3/2} - 3x^{1/2}$ for $\frac{1}{4} \le x \le 4$

32.[†] $(x^2 + 1)^{-4}$ for $-3 \le x \le -1$ **33.** $x^3 + 6x^2 + 1$ for $-1 \le x \le 1$

34.[†] $x \sin x$ for $0 \le x \le \frac{1}{3}\pi$ **35.** $x/(\sin x)$ for $-\frac{1}{2}\pi \le x \le -\frac{1}{4}\pi$

4-9   OTHER MAXIMUM-MINIMUM PROBLEMS

Often we want to find the maximum or minimum value of a function in an interval that is infinite or not closed. In these situations Theorem 4.7 in the last section does not apply, and we cannot find the maximum or minimum value just by locating the critical points and comparing the values of the function at the critical points and end points of the interval. Instead, we use the derivative to find where the function is increasing and where it is decreasing as well as to find its critical points. We can also apply the second derivative test or make a rough sketch of the function's graph.

EXAMPLE 1   Find the minimum value of $x^3 - 6x^2 + 9x$ for $x \ge 0$.

Solution.   We cannot apply Theorem 4.7 because the interval $x \ge 0$ is not finite. The derivative of the function is

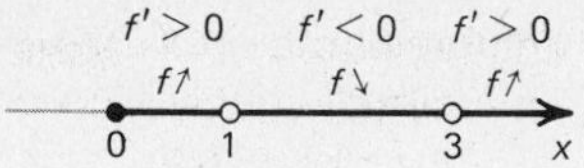

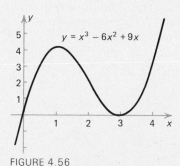

FIGURE 4.55 Behavior of $f(x) = x^3 - 6x^2 + 9x$ for $x \geq 0$

$$\frac{d}{dx}(x^3 - 6x^2 + 9x) = 3x^2 - 12x + 9 = 3(x^2 - 4x + 3)$$

$$= 3(x - 1)(x - 3)$$

and is zero at $x = 1$ and $x = 3$. For $0 \leq x < 1$ the derivative is positive, for $1 < x < 3$ it is negative, and for $x > 3$ it is positive (Figure 4.55). The minimum value of the function for $x \geq 0$ occurs either at $x = 3$, where it has a local minimum, or at the end point $x = 0$. The function $x^3 - 6x^2 + 9x$ has the value 0 at both of these points, so 0 is its minimum value. $\square$

As a check on the result of Example 1 we might compute the second derivative. It is $\dfrac{d}{dx}(3x^2 - 12x + 9) = 6x - 12$ and is negative at $x = 1$ and positive at $x = 3$, which confirms that there is a local maximum at $x = 1$ and a local minimum at $x = 3$. As a further check we could sketch the graph of the function (Figure 4.56).

FIGURE 4.56

Maximum-minimum problems where the variables are integers

In many applications the functions that arise are only defined for integer values of their variables. To apply calculus to such problems, we have to consider each function to be defined for all values of its variable in an interval. Then after our analysis based on calculus is finished, we interpret the results for the case of integer values of the variables as in the original contexts of the problems. This procedure is illustrated in the next example.

EXAMPLE 2   A company uses 600 units of a certain item each year. It costs 20 dollars each time an order is placed for the item. If the company orders x times a year, storage and other carrying costs are $3000/x$ dollars in the year. How many times a year should the company order to minimize the total ordering and carrying costs?

Solution.   If the company places an order x times a year, it costs $20x$ dollars for ordering and $3000/x$ dollars for carrying costs. The total cost is

$$(1) \qquad C(x) = 20x + \frac{3000}{x} \text{ dollars.}$$

We want the minimum value of this function for positive integers x. We consider the function (1) defined for all $x > 0$. Its derivative is

$$(2) \qquad C'(x) = 20 - \frac{3000}{x^2} = \frac{20x^2 - 3000}{x^2} = \frac{20(x^2 - 150)}{x^2}$$

and is zero at $x = \sqrt{150}$. It is negative for $0 < x < \sqrt{150}$ and positive for $x > \sqrt{150}$. The function (1), therefore, is decreasing for $0 < x < \sqrt{150}$ and increasing for $x > \sqrt{150}$. Its graph is shown in

Figure 4.57. The dots are the graph of the function considered to be defined only for positive integer values of x.

Because $12^2 = 144$ is less than 150 and $13^2 = 169$ is greater than 150, the square root of 150 lies between 12 and 13. Therefore, the minimum value of $C(x)$ for integers x occurs either at $x = 12$ or at $x = 13$. We find that $C(12) = 490$ dollars and $C(13) = 490\frac{10}{13}$ dollars. The least of these is $C(12) = 490$, so the order should be placed 12 times a year. □

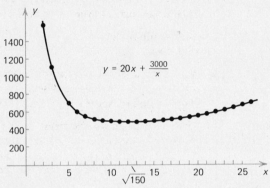

$$y = 20x + \frac{3000}{x}$$

FIGURE 4.57

Exercises

†*Answer provided.*
‡*Outline of solution provided.*

In Exercises 1 through 4 find the maximum and minimum values of the functions.

1.† $\dfrac{x-2}{x^2}$ for $x \geq 1$

2. $x^2 + \dfrac{16}{x^2}$ for $-2 < x \leq -1$

3.† $\dfrac{2\sqrt{x}}{x+1}$ for $x \geq 0$

4. $\dfrac{x}{x^4 + 48}$ for $-10 < x < 10$

5.† It costs a company $x^3 + 100x + 1500$ dollars to produce a batch of x gallons of perfume that it sells for 400 dollars a gallon.   **a.** What size batch should it produce to maximize the profit on each batch?   **b.** What is the maximum profit?

6. A bullet is fired straight up into the air with a muzzle velocity of k feet/second, and it is $6 + kt - 16t^2$ feet above the ground after t seconds. **a.** How high does it go?   **b.** How long does it take to reach that height?

7. What positive number exceeds its cube by the greatest amount?

8.‡ A rectangular box with a square bottom and no top is to have a volume of 6 cubic feet. The material for the bottom costs 3 dollars/square foot and the material for the sides costs 2 dollars/square foot. What dimensions should the box have to minimize its cost?

9.† In sixteenth and seventeenth century Austria, wine barrels were measured by how far a bung stick could be inserted through the bung hole in the middle of the side of the barrel (Figure 4.58). What cylindrical wine barrel with a volume of 8π cubic feet has the smallest bung stick measurement? (Minimize the square of the measurement.)

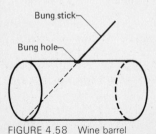

Bung stick

Bung hole

FIGURE 4.58   Wine barrel

10. A home owner wants to design a garage window with an area of 6 square feet. He has to put 2 two-by-fours on the top of the window, 1 on each side and 1 on the bottom. What size window would use the least amount of wood?

11.[†] A car driving north at 40 miles/hour reaches an intersection at the same time that a second car traveling west toward the intersection at 60 miles/hour is 100 miles directly east of the intersection.   **a.** How long will it take for the distance between the two cars to be a minimum? **b.** How far apart will they be at that time?

12. Use calculus to find the point on the line $y = 4x + 3$ that is closest to the point $(2, -6)$.

13. A toy company can manufacture x Big Bertha dolls in one day at a cost of $10 + 5x + \frac{1}{4}x^{3/2}$ dollars. It sells each doll for $8. How many dolls should be manufactured per day to maximize the profit?

14.[†] An apple orchard with 30 trees per acre yields 400 apples per tree. In planning a new orchard the grower anticipates that for each tree beyond 30 that he plants per acre the yield per tree would be reduced by 7 apples. How many trees per acre would give the largest crop? (Give an integer answer.)

15.[‡] A gutter is to be made from a long strip of tin 18 inches wide by bending up the sides, so that its cross section is an isosceles trapezoid (Figure 4.59). Through what angle should the sides be bent to maximize the volume of the gutter?

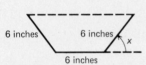

6 inches 6 inches x
6 inches

FIGURE 4.59

x

FIGURE 4.60

16.[†] A cone is made from a circular piece of paper 8 inches in diameter by cutting out a sector of x radians with its vertex at the center of the circle (Figure 4.60). What choice of x yields the cone of maximum volume?

17.[†] If you can sell $5000/(p^{3/2} + 2p^{1/2})$ items at p dollars each, what should you charge to maximize your revenue?

18.[†] A train has 1200 passengers at a fare of 2 dollars each. For every 10 cents that the fare is lowered 100 more passengers will ride the train. What fare should the railroad charge to maximize its revenue?

19. It costs $6 + \frac{1}{250}v$ dollars/mile for gasoline and maintenance to drive a truck v miles/hour, and the driver's salary is 10 dollars/hour. At what speed should the truck travel to minimize the cost per mile?

20. If the marginal revenue in producing x items is $50 - \frac{1}{80}x^2$ dollars/item and the marginal cost is $10 + \frac{1}{80}x^2$ dollars/item, how many items should be produced to maximize the profit?

21. Suppose that x units of an item cost $C(x)$ dollars to produce and that the x units sell for $R(x)$ dollars (the revenue).   **a.** Show that the marginal cost equals the marginal revenue when the profit is a maximum.   **b.** Show that the average profit on x units equals the marginal profit when the average profit is a maximum.

22. Show that the distance from the point (x_0, y_0) to the line $Ax + By + C = 0$ is $\dfrac{|Ax_0 + By_0 + C|}{\sqrt{A^2 + B^2}}$.

23. Show that if a function f has a derivative for all x, then the shortest distance to the graph of f from a point (x_0, y_0) is along a line perpendicular to the graph.

24. From theoretical considerations a researcher expects pairs (x, y) of data from his experiment to be on a line $y = mx$. He wants to determine the constant m. In four runs of the experiment he obtains the data $x_1 = 1$, $y_1 = 1.1; x_2 = 2, y_2 = 2.3; x_3 = 3, y_3 = 3.8;$ and $x_4 = 4, y_4 = 5.0$. Compute the value of m he would obtain from his data by the *method of least squares;* that is, find m that minimizes the quantity

$$E(m) = (y_1 - mx_1)^2 + (y_2 - mx_2)^2 + (y_3 - mx_3)^2 + (y_4 - mx_4)^2.$$

25. Find the point on the curve $y = \sqrt{x}$ that is closest to the point $(1, 0)$.

26.† A cylindrical can is to have a volume of 100 cubic inches. The material for the top and bottom costs $\frac{1}{10}$ cent per square inch, while the material for the side costs $\frac{1}{20}$ cent per square inch. What dimensions should the can have to minimize the cost of the material?

27. A printer is to use a page that has an area of 80 square inches with margins of one inch at the bottom and sides and one and a half inches at the top. What dimensions should the page have so that the area of the printed portion is a maximum?

28. A farmer has 40 tons of cattle that he could sell for a profit of 200 dollars/ton. For every week he waits to sell the cattle their combined weight will increase by one ton, but his profit will fall by 4 dollars/ton. He has to sell the cattle within 8 weeks. When should he sell to maximize his profit?

In each of Exercises 29 through 40 find the maximum and minimum values of the function for the indicated values of x (if they exist) and the values of x where they occur.

29.† $x^3 + 1$ for $-1 < x \leq 2$ | 30.† $(x^2 + 1)^{-1}$ for $-1 < x \leq 4$

31.† $x^2/(x^2 + 1)$ for $x \geq 0$ | 32.† $x^{-1/2}$ for $x \geq 1$

33.† $x/(x^2 + 9)$ for all x | 34.† $\sin^2 x$ for all x

35.† $x/(x^2 - 1)$ for $-3 \leq x < -1$ | 36.† $x/|x|$ for all $x \neq 0$

37.† $2x^{1/3} - x^{2/3}$ for all x | 38.† $x\sqrt{4 - x}$ for $x \leq 4$

39. $x(x^2 - 1)^{-1/2}$ for $x > 1$ | 40. $(1 - \sqrt{x})/(1 + \sqrt{x})$ for $x \geq 0$

4-10 TANGENT LINE APPROXIMATIONS AND DIFFERENTIALS

If we know the value of a function and of its derivative at a point x_0, we can give the equation of its tangent line at x_0. Then, as we will see in this section, we can use the tangent line to obtain approximate values of the function for x near x_0.

EXAMPLE 1   (a) Use a tangent line to find an approximate value for $\sqrt[3]{8.072}$. (b) Is this approximation greater than or less than the exact value?

Solution.   (a) We use the tangent line at $x = 8$, where the function $x^{1/3}$ has the value $8^{1/3} = 2$ and its derivative $\frac{1}{3}x^{-2/3}$ has the value $\frac{1}{3}(8^{-2/3}) = \frac{1}{12}$. The tangent line has the equation $y = 2 + \frac{1}{12}(x - 8)$. Because the point on the graph of $x^{1/3}$ at $x = 8.072$ is close to the point on the tangent line, we have

$$(1) \qquad \sqrt[3]{8.072} \approx [2 + \tfrac{1}{12}(x - 8)]_{x=8.072} = 2 + \tfrac{1}{12}(0.072) = 2.006.$$

(b) To determine whether the approximation is greater than or less than the exact value, we compute the second derivative:

$$\frac{d^2}{dx^2}x^{1/3} = \frac{d}{dx}\left(\frac{1}{3}x^{-2/3}\right) = -\frac{2}{9}x^{-5/3}.$$

The second derivative is negative, and the graph of $x^{1/3}$ is concave down for all $x > 0$ (Figure 4.61). The tangent line, therefore, lies above the graph, and the approximation (1) is greater than the exact value of $\sqrt[3]{8.072}$. (That value is 2.005982 . . . .)   □

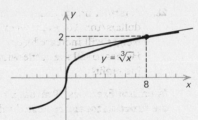

FIGURE 4.61

Differentials   We have been using the symbol dy/dx to denote the derivative of the function $y(x)$ with respect to x. The symbols dx and dy that appear in this symbolic fraction are called DIFFERENTIALS.

Mathematicians of the seventeenth and eighteenth centuries used the symbols "dx" and "dy" and the term "differentials" to designate "infinitesimally small quantities" (see the historical notes in Section 3.11), and some modern mathematicians employ this terminology. Generally, however, differentials are viewed as *related pairs of real numbers*. We use the following definition.

Definition 4.9   The differentials dx and dy at a point P on a curve are corresponding changes in x and y along the tangent line to the curve at P.

If $dx \neq 0$ and dy are differentials at a point $(x_0, f(x_0))$ on the graph of a function f, then

$$(2) \qquad \frac{dy}{dx} = \text{Slope of the tangent line} = f'(x_0)$$

(Figure 4.62). Thus, we can view the symbolic fraction dy/dx that denotes the derivative as an actual fraction, the ratio of differentials.

Multiplying equation (2) by dx leads to the following definition of differentials on the graph of a function.

Definition 4.10 If $f'(x_0)$ exists, then the differentials dx and dy at a point $(x_0, f(x_0))$ on the graph of f are a pair of numbers related by the equation

$$dy = f'(x_0)\, dx. \qquad \blacktriangleleft$$

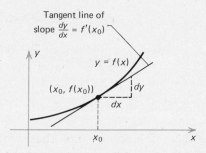

FIGURE 4.62

Differentials and error estimation

It is convenient to use differential notation when estimating errors by using tangent lines. Suppose that we measure one quantity x_0 and then use a formula to compute a related quantity $f(x_0)$. If we make an error of dx in the measurement and the exact value of x is $x_0 + dx$, then the exact value of f is $f(x_0 + dx)$ and the error in computing it is $f(x_0 + dx) - f(x_0)$. We can approximate this error by using the corresponding change $dy = f'(x_0)\, dx$ along the tangent line (Figure 4.63):

$$(3) \qquad f(x_0 + dx) - f(x_0) \approx dy = f'(x_0)\, dx. \qquad \blacktriangleleft$$

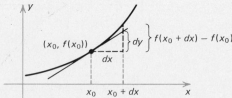

FIGURE 4.63

EXAMPLE 2   The width of a cube is measured to be 10 centimeters with a possible error of 0.03 centimeters. Use differentials to estimate the resulting error in computing the volume of the cube.

Solution.   The volume of a cube of width w is $V(w) = w^3$. We have $V'(w) = 3w^2$ and $V'(10) = 3(10)^2 = 300$. Consequently, at $w = 10$ on the graph of $V(w)$ the differentials dw and dV are related by the equation $dV = V'(10)\, dw$ or $dV = 300\, dw$. For $|dw| \le 0.03$ we have $|dV| \le 300(0.03) = 9$, so the possible error in the volume is about 9 cubic centimeters. $\square$

Differentials and tangent line approximations

To use differential notation when making a tangent line approximation of a value of a function, as in the next example, we rewrite (3) in the form

$$(4) \quad f(x_0 + dx) \approx f(x_0) + f'(x_0)\, dx. \qquad \blacktriangleleft$$

EXAMPLE 3   Use differentials to the graph of $\sin x$ at $x = \tfrac{1}{3}\pi$ to give an approximate value for $\sin(\tfrac{4}{9}\pi)$.

Solution.   We set $f(x) = \sin x$ and $x_0 = \tfrac{1}{3}\pi$. Then $f'(x) = \cos x$ and (4) reads

$$\sin(\tfrac{1}{3}\pi + dx) \approx \sin(\tfrac{1}{3}\pi) + \cos(\tfrac{1}{3}\pi)\, dx$$

$$= \tfrac{1}{2} + (\tfrac{1}{2}\sqrt{3})\, dx.$$

To obtain an approximation of $\sin(\tfrac{4}{9}\pi)$, we set $dx = \tfrac{1}{9}\pi$. Then $\tfrac{1}{3}\pi + dx = \tfrac{4}{9}\pi$ and we have

$$\sin(\tfrac{4}{9}\pi) \approx \tfrac{1}{2} + (\tfrac{1}{2}\sqrt{3})(\tfrac{1}{9}\pi) = \tfrac{1}{2} + \tfrac{1}{18}\sqrt{3}\pi. \quad \square$$

Differentials to other curves

The next example shows how to find equations relating differentials at points on curves that are not the graphs of functions.

EXAMPLE 4   The curve $9y^2 = x^3 + 3x^2$ shown in Figure 4.64 is known as *Tschirnhausen's cubic*. Find equations relating the differentials on the curve at the points $(1, \tfrac{2}{3})$, $(-2, \tfrac{2}{3})$, and $(-3, 0)$.

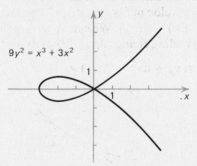

$9y^2 = x^3 + 3x^2$

FIGURE 4.64   Tschirnhausen's cubic

Solution.   We differentiate both sides of the equation $9y^2 = x^3 + 3x^2$ with respect to x, viewing y as a function of x. We obtain the equation $18y(dy/dx) = 3x^2 + 6x$. We then consider the derivative dy/dx to be the quotient of the differentials dx and dy. We divide the equation by 3 and multiply it by dx to obtain

$$(5) \quad 6y\, dy = (x^2 + 2x)\, dx.$$

This derivation is not valid where the curve has a vertical tangent line, but we could deal with that case by treating x as a function of y.

At the point $(1, \tfrac{2}{3})$ equation (5) gives the relation $4\, dy = 3\, dx$. At the point $(-2, \frac{2}{3})$ it gives $dy = 0$, which means that the tangent line at that point is horizontal. At the point $(-3, 0)$ it gives $dx = 0$, which means that the tangent line is vertical there. The tangent lines at these three points are drawn in Figure 4.65.   □

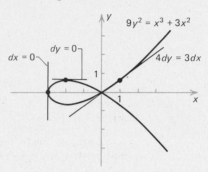

FIGURE 4.65

Exercises

†*Answer provided.*
‡*Outline of solution provided.*

In Exercises 1 through 9 give tangent line approximations of the indicated values. Determine whether each approximation is greater than or less than the exact value.

1.‡ $\sin(-1)$ (Use the tangent line to $y = \sin x$ at $x = -\pi/3$ and the approximate values $\pi \approx 3.1416$, $\sqrt{3} \approx 1.7321$.)

2.† $(15.9)^{5/4}$ (Use the tangent line to $y = x^{5/4}$ at $x = 16$.)

3. $(0.97)^{-1/10}$ (Use the tangent line to $y = x^{-1/10}$ at $x = 1$.)

4.‡ $\sqrt{1.03}$ (Use the tangent line to $y = \sqrt{x}$ at $x = 1$.)

5.† $\dfrac{1}{1.03}$ (Use the tangent line to $y = \dfrac{1}{1 + x}$ at $x = 0$.)

6. $\cos(\frac{3}{4})$ (Use the tangent line to $y = \cos x$ at $x = \pi/4$ and the approximate values $\pi \approx 3.1416, \dfrac{1}{\sqrt{2}} \approx 0.7071$.)

7.‡ $\sqrt[4]{10006}$ **8.**† $(123)^{2/3}$ **9.** $(1.02)^{4.02}$

In Exercises 10 through 13 find equations relating the differentials dx and dy at the indicated points on the curves. Then sketch the curves and the corresponding tangent lines.

10.† At the points   **a.** $(-4, 3)$ **b.** $(-3, -4)$ **c.** $(-5, 0)$ and   **d.** $(0, 5)$ on the circle $x^2 + y^2 = 25$

11. At the points   **a.** $(0, 2)$ **b.** $(1, \frac{5}{2})$ and   **c.** $(-2, -2)$ on the curve $y = \frac{1}{2}x^3 + 2$

12.† At the points   **a.** $(2, 1)$ and   **b.** $(-2, -1)$ on the curve $y = 2/x$

13. At the points   **a.** $(4, 0)$ **b.** $(0, 2)$ and   **c.** $(0, -2)$ on the cardioid $(x^2 + y^2 - 2x)^2 = 4(x^2 + y^2)$, which is shown in Figure 4.66

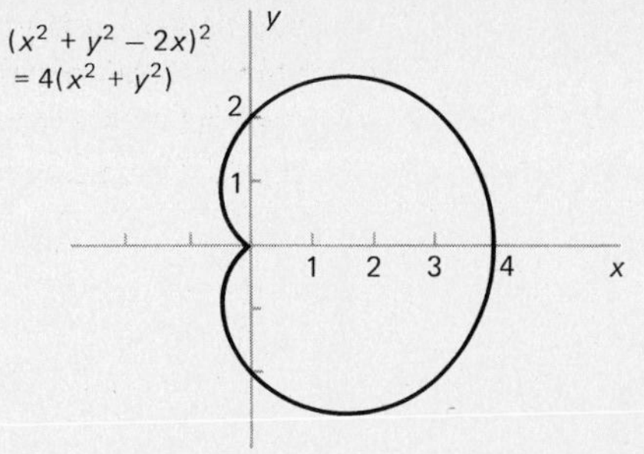

$$(x^2 + y^2 - 2x)^2 = 4(x^2 + y^2)$$

FIGURE 4.66   A cardioid

14.† Use the tangent line to $y = (\cos x)^{1/2}$ at $x = \pi/3$ to give an approximate value of $[\cos(\frac{1}{5}\pi)]^{1/2}$.

15.† Use the tangent line to the graph of $x^{19/3}$ at $x = -1$ to give an approximate value of $(-1.06)^{19/3}$.

16.† Use the tangent line to $y = x^{100}$ at $x = 1$ to give an approximate value of $(1.005)^{100}$.

17.† Use differentials to estimate the possible error in calculating $(11 + x^2)^{1/2}$ if x is measured to be 5 with a possible error of 0.005.

18.† Use differentials to estimate the possible error in calculating $\sin(\pi x)$ if x is measured to be $\frac{1}{6}$ with a possible error of 0.002.

19.† Use differentials to estimate the possible error in calculating $x \cos(\pi x)$ if x is measured to be $\frac{3}{4}$ with a possible error of $\frac{1}{64}$.

In each of Exercises 20 through 25 give an equation relating dx and dy at (x, y) on the given curve.

20.† $x^3 + y^3 = 1$

21.† $x^2y^2 - xy + 1 = 0$

22.† $\sin y = x$

23. $x \cos y = 1$

24.† $x^2 - 8xy + y^2 + x - y = 1$

25. $(x^2 + y^2)^2 = x^2 - y^2$

4-11 NEWTON'S METHOD (Optional)

In this section we discuss a method, attributed to Newton, in which tangent lines are used to find approximate solutions of equations of the form

$$(1) \qquad f(x) = 0.$$

We start by picking a value x_0 that we expect to be close to a solution of equation (1). If $f(x_0)$ is not zero, we compute the equation of the tangent line to the graph of f at $x = x_0$. We let x_1 denote the value of x where that tangent line intersects the x-axis (Figure 4.67); we let x_2 denote the point where the tangent line at $x = x_1$ intersects the x-axis (Figure 4.68); and so forth. In many cases this procedure yields successively closer approximations $x_0, x_1, x_2, x_3, \ldots$, of a solution of equation (1). To compute $x_1, x_2, x_3, \ldots$, we use the following rule.

Rule 4.1   The approximate solution x_{n+1} of the equation $f(x) = 0$ is computed in Newton's method from the previous approximate solution x_n by the formula

$$(2) \qquad x_{n+1} = x_n - \frac{f(x_n)}{f'(x_n)}. \qquad \blacktriangleleft$$

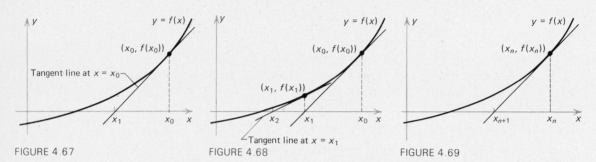

FIGURE 4.67   FIGURE 4.68   FIGURE 4.69

Derivation of Rule 4.1.   The tangent line at $x = x_n$ passes through the point $(x_n, f(x_n))$ and has slope $f'(x_n)$. Therefore, it has the equation $y - f(x_n) = f'(x_n)(x - x_n)$ (Figure 4.69). Setting $y = 0$ in this equation and solving for the x-intercept $x = x_{n+1}$ gives formula (2). $\square$

EXAMPLE 1   Use Newton's method with three applications of (2) and with an initial guess of $x_0 = 1$ to find an approximate value for $\sqrt{5}$.

Solution.   We view $\sqrt{5}$ as the solution of the equation $x^2 - 5 = 0$. Accordingly, we set $f(x) = x^2 - 5$. Then $f'(x) = 2x$ and formula (2) reads

$$(3) \qquad x_{n+1} = x_n - \frac{x_n^2 - 5}{2x_n}.$$

We take $x_0 = 1$. Then formula (3) gives, in turn, $x_1 = 3, x_2 = \frac{7}{3}$, and $x_3 = \frac{47}{21}$. The third approximation has the decimal expansion 2.23809 . . . , while $\sqrt{5}$ has the expansion 2.23606 . . . . The approximation, therefore, is accurate to two decimal places. $\square$

How Newton's method can fail

In most cases Newton's method leads to numbers that approach a solution of the equation $f(x) = 0$ if the initial "guess" x_0 is sufficiently close to a solution. Figure 4.70 shows how the method can fail if the initial guess is too far from a solution. The tangent line at x_1 in Figure 4.70 intersects the x-axis at x_0; consequently x_2 is the same number as x_0, x_3 is the same number as x_1, and so forth. The numbers given by Newton's method jump back and forth between x_0 and x_1 and never approach the solution $x = a$, of the equation, $f(x) = 0$. Other situations where Newton's method fails are described in Exercises 8, 9, and 10.

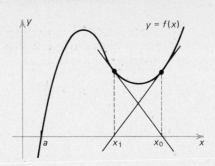

FIGURE 4.70

Exercises

†*Answer provided.*
‡*Outline of solution provided.*

In Exercises 1 through 7 find approximate values for the indicated numbers by using Newton's method with three applications of equation (2) and with the given initial guesses.

1.‡ $\sqrt[3]{2}$; $x_0 = 1$ **2.**† $\sqrt{80}$; $x_0 = 10$ **3.** $\sqrt[3]{-10}$; $x_0 = -2$

4.‡ A solution of $\sin x + \cos x = 0$; $x_0 = -1$ (Use the reference tables to obtain the values of $\sin x$ and $\cos x$ to three decimal places.)

5.† A solution of $x^5 + x - 1 = 0$; $x_0 = 0$

6. A solution of $\sin x = x \cos x$; $x_0 = 4$ (same instructions as for Exercise 4)

7.† The two solutions of $x^4 + 4x - 1 = 0$; $x_0 = -1.5$ and $x_0 = 0.5$ (Sketch the curve $y = x^4 + 4x - 1$.)

8. Newton's method applied to the function $x^{1/3}$ and with any $x_0 \neq 0$ fails to give numbers that approach the solution $x = 0$ of $x^{1/3} = 0$. Explain.

9. Explain why Newton's method with any $x_0 \neq 0$ fails to give numbers that approach the solution of $f(x) = 0$ for

$$f(x) = \begin{cases} \sqrt{x} & \text{for } x \geq 0 \\ -\sqrt{-x} & \text{for } x < 0. \end{cases}$$

10. Newton's method with $x_0 = 1$ fails to give numbers that approach a solution of the equation $2x^3 - 3x^2 + 2 = 0$. Explain.

11.† In the *midpoint method* of finding approximate solutions of an equation $f(x) = 0$ we choose an interval I_1 with f positive at one end point and negative at the other. We let x_1 be the midpoint of I_1. If $f(x_1)$ is not zero, we let I_2 be the half of I_1 with f positive at one end point and negative at the other. We let x_2 be the midpoint of I_2, and so forth. The advantage of Newton's method over the midpoint method is that, generally, it yields the desired accuracy in fewer steps. Find x_3 for $f(x) = x^2 - 5$ and I_1 the interval $2 \leq x \leq 3$, and compare your result with the result of Example 1.

12. If you have access to a computer or electronic calculator, use Newton's method with the given initial guesses and seven applications of equation (2) to find approximate solutions of the following equations.
a.‡ $\sin x = x/2$; $x_0 = 3$ **b.**† $(\sin x)/(\cos x) = 2x$; $x_0 = \frac{3}{2}$

4-12   THE MEAN VALUE THEOREM FOR DERIVATIVES

If a person averages 50 miles per hour on a car trip, then at some time during the trip he must drive 50 miles per hour. This statement, when phrased in terms of a function that gives the car's position as a function of the time, is an example of the MEAN VALUE THEOREM (or LAW OF THE MEAN) for derivatives.*

We suppose that the trip starts at time $t = a$ (hours) and ends at time $t = b$ and that the car is at $s = f(t)$ (miles) at time t. Then the average velocity is

(1) $\dfrac{[f(b) - f(a)]\ \text{miles}}{[b - a]\ \text{hours}} = \dfrac{f(b) - f(a)}{b - a}$ miles per hour.

The instantaneous velocity at time $t = c$ is

(2) $f'(c)$ miles per hour.

The statement that the instantaneous velocity must be equal to the average velocity at some time during the trip means that there must be a time c between a and b such that the derivative (2) is equal to the difference quotient (1). This is the Mean Value Theorem as applied to the function $f(t)$ on the interval $a \leq t \leq b$:

Theorem 4.8   (The Mean Value Theorem for derivatives) Suppose that f is a function that is continuous in the closed finite interval $a \leq t \leq b$ and has a derivative at each point t in the open interval $a < t < b$. Then there is at least one number c with $a < c < b$ such that

(3) $\dfrac{f(b) - f(a)}{b - a} = f'(c).$ ◄

The geometric interpretation of this result is indicated in Figure 4.71. The Mean Value Theorem asserts that there must be at least one value $t = c$ in the open interval $a < t < b$ such that the tangent line at $t = c$ is parallel to the secant line through the points at $t = a$ and $t = b$ on the graph. In Figure 4.71 there are two such points, $t = c_1$ and $t = c_2$.

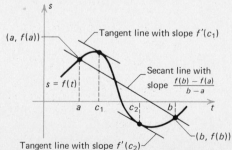

FIGURE 4.71   Tangent line with slope $f'(c_2)$

---

*We will study the Mean Value Theorem for integrals in Chapter 6.

Proof for the case where $f(a) = f(b)$. If $f(a) = f(b)$, then the left side of equation (3) is zero, and we must show that $f'(c) = 0$ for some c with $a < c < b$.

If $f(t) = f(a)$ for all t in the interval, then $f'(c) = 0$ for all c with $a < c < b$. Otherwise $f(t)$ is either greater or less than $f(a) = f(b)$ somewhere with $a < t < b$.

Suppose that $f(t)$ is greater than $f(a)$ and $f(b)$ somewhere in the interval. The Extreme Value Theorem shows that $f(t)$ has a maximum value for $a \leq t \leq b$. The maximum must occur at a point c with $a < c < b$. Then f has a local maximum at c, and by Theorem 4.3 c is a critical point. Since $f'(c)$ exists, it is zero.

Similarly, if $f(t)$ is less than $f(a)$ and $f(b)$ somewhere in the interval, then $f'(c) = 0$ where f has its minimum value. Q.E.D.

Proof for the general case. For the case where $f(a)$ is not necessarily equal to $f(b)$, we consider the auxiliary function

$$(4) \qquad g(t) = f(t) - \left[\frac{f(b) - f(a)}{b - a} \right] t.$$

The value $g(t)$ of this function is the difference between the y-coordinates of the points at t on the graph of f and on the line through the origin and parallel to the secant line through $(a, f(a))$ (Figure 4.72).

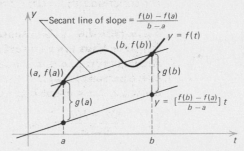

FIGURE 4.72

We have constructed the function g to have $g(a) = g(b)$. The function g is continuous for $a \leq t \leq b$, and it has a derivative at each t with $a < t < b$. Therefore, by the case of Theorem 4.8, which we have already proved, there is at least one number c with $a < c < b$ such that $g'(c) = 0$. Then, by formula (4) we have

$$0 = g'(c) = f'(c) - \left[\frac{f(b) - f(a)}{b - a} \right]$$

which gives (3) to prove the theorem. Q.E.D.

EXAMPLE 1 Verify the conclusion of the Mean Value Theorem for the function t^3 in the interval $1 \leq x \leq 3$ by finding a suitable number c.

Solution. Set $f(t) = t^3$. We need to find a number c with $1 < c < 3$ such that

$$\frac{f(3) - f(1)}{3 - 1} = f'(c).$$

Since $f(1) = 1^3 = 1$, $f(3) = 3^3 = 27$, and $f'(t) = 3t^2$, this equation reads $13 = 3c^2$. We take $c = \sqrt{13/3}$ to obtain the value between 1 and 3. □

EXAMPLE 2   Show that the conclusion of the Mean Value Theorem does not hold for the function $|t|$ in the interval $-2 \leq t \leq 2$. Which hypothesis of the theorem does this function satisfy and which does it fail to satisfy?

Solution. The graph of the function $|t|$ for $-2 \leq t \leq 2$ is shown in Figure 4.73. The slope of the secant line through the points at $t = -2$ and at $t = 2$ is 0. The conclusion of the Mean Value Theorem is not satisfied because there is no point where the function has a zero derivative. The function is continuous in the closed finite interval $-2 \leq t \leq 2$ but does not have a derivative at the point $t = 0$ in the open interval $-2 < t < 2$. □

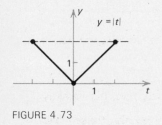

FIGURE 4.73

The last example shows that the differentiability condition cannot be dropped from the statement of the Mean Value Theorem. An example described in Exercise 13 shows that the continuity condition cannot be dropped either.

The first derivative test in arbitrary intervals

In Section 4.1 we stated the first derivative test in an open interval (Theorem 4.2). That theorem is a special case of the following result.

Theorem 4.9   (The first derivative test in an arbitrary interval)   If a function f is continuous in an interval I and if its derivative $f'(x)$ exists and is positive for all x, or for all but a finite number of points x, in the interior of I, then f is increasing in I. If the derivative exists and is negative for all x, or for all but a finite number of points x, in the interior of I, then f is decreasing in I.

Proof. We will deal with the case where $f'(x)$ exists and is positive for all but at most a finite number of points x in the interior of I. We will show that f is increasing in I by showing that $f(x_1) < f(x_2)$ for all points x_1 and x_2 in I with $x_1 < x_2$.

Consider an arbitrary pair of such points and suppose first that $f'(x)$ exists and is positive for all x with $x_1 < x < x_2$. Because f is continuous for $x_1 \leq x \leq x_2$, we can apply the Mean Value Theorem (Theorem 4.8) and conclude that there must be at least one number c with $x_1 < c < x_2$ such that

$$\frac{f(x_2) - f(x_1)}{x_2 - x_1} = f'(c).$$

Multiplying this equation by $x_2 - x_1$ gives

(5) $f(x_2) - f(x_1) = f'(c)(x_2 - x_1).$

Since $x_2 - x_1$ and $f'(c)$ are positive, $f(x_2)$ is greater than $f(x_1)$.

There may be one or more points between x_1 and x_2 where $f'(x)$ fails to exist or is not positive. Let us suppose for illustration that there are two such points, x_3 and x_4, with $x_1 < x_3 < x_4 < x_2$. We can apply the Mean Value Theorem in each of the intervals $x_1 \le x \le x_3$, $x_3 \le x \le x_4$, and $x_4 \le x \le x_2$ to conclude, as in the argument above that $f(x_1) < f(x_3)$, that $f(x_3) < f(x_4)$ and that $f(x_4) < f(x_2)$. This implies that $f(x_1) < f(x_2)$, which is what we wanted to show.   Q.E.D.

EXAMPLE 3   Use Theorem 4.9 to show that the functions x^3 and $x^{1/3}$ are increasing for all x.

Solution.   Here the interval I consists of all real numbers x. The functions x^3 and $x^{1/3}$ are continuous for all x and their derivatives

$$\frac{d}{dx}x^3 = 3x^2 \text{ and } \frac{d}{dx}x^{1/3} = \frac{1}{3}x^{-2/3} \text{ exist and are positive for all } x \text{ except}$$

$x = 0$ where the derivative of x^3 is zero and the derivative of $x^{1/3}$ does not exist. By the theorem both functions are increasing for all x.   □

Exercises

In Exercises 1 through 6 verify the conclusion of the Mean Value Theorem for derivatives by finding suitable numbers c.

1.‡ The function $\sqrt{x}$ for $1 \le x \le 9$

2.† The function $1/t$ for $-3 \le t \le -1$

3.   The function $\sin x$ for $0 \le x \le 2\pi$

4.‡ The function $\cos(2x)$ for $0 \le x \le \pi$

5.† The polynomial $x^3 + x + 7$ for $0 \le x \le 3$

6.   The polynomial $x^4 + 10$ for $-1 \le x \le 2$

7.‡ The function W of Figure 4.74 is continuous for $1 \le x \le 15$ and has a derivative for $1 < x < 15$. Give approximate values for three numbers c that verify the Mean Value Theorem for W in the interval $1 \le x \le 15$.

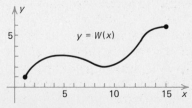

FIGURE 4.74

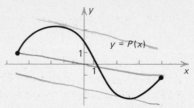

FIGURE 4.75

8. The function P of Figure 4.75 is continuous for $-6 \leq x \leq 7$ and has a derivative for $-6 < x < 7$. Give approximate values of two numbers c that verify the Mean Value Theorem for P in the interval $-6 \leq x \leq 7$.

9.[†] Use the Mean Value Theorem to prove that $(x + 1)^{1/4} - x^{1/4}$ tends to zero as x tends to infinity.

10. Suppose that f is continuous for $a \leq x \leq b$ and has a derivative for $a < x < b$ and that $f'(x) \to m$ (m finite) as $x \to a^+$. Prove that the one-sided derivative

$$D^+f(a) = \lim_{x \to a^+} \frac{f(x) - f(a)}{x - a}$$

exists and has the value m.

11. Suppose that f is continuous for $a \leq x \leq b$ and has a nonzero derivative for $a < x < b$. Prove that there is at most one solution x of the equation $f(x) = 0$ with $a \leq x \leq b$.

12. Prove that if $f'(x)$ exists for all x and $f'(x) = 0$ for only n values of x (n, a positive integer), then $f(x) = 0$ for, at most, $n + 1$ values of x.

13. Show that the function G defined by

$$G(t) = \begin{cases} 2 & \text{for } t \leq -2 \\ t & \text{for } t > -2 \end{cases}$$

does not satisfy the conclusion of the Mean Value Theorem in the interval $-2 \leq t \leq 2$. Which hypothesis of the theorem is not satisfied?

4-13 ANTIDERIVATIVES

If the function f is the derivative of the function F, then we say that F is an ANTIDERIVATIVE of f. The function x^3, for example, is an antiderivative of the function $3x^2$, because $\frac{d}{dx} x^3 = 3x^2$. If the function F is an antiderivative of the function f in an interval, then the function G defined by

(1) $G(x) = F(x) + C$

with C a constant is also an antiderivative of f in the interval, because the derivative of the constant is zero, and we have

$$G'(x) = \frac{d}{dx}[F(x) + C] = F'(x) + 0 = f(x).$$

The next theorem shows that formula (1) gives all of the antiderivatives of f in the interval.

Theorem 4.10   If two functions F and G have the same derivative in an interval, then the functions differ by a constant in that interval, that is, there is a constant C such that $G(x) = F(x) + C$ for all x in the interval.

Proof.   Set $H(x) = G(x) - F(x)$. Then $H'(x) = G'(x) - F'(x) = 0$ for all x in the interval. We consider any two numbers x_1 and x_2 in the interval with $x_1 < x_2$. According to the Mean Value Theorem for derivatives applied to H in the interval $x_1 \leq x \leq x_2$, there is a number c with $x_1 < c < x_2$ such that

$$\frac{H(x_2) - H(x_1)}{x_2 - x_1} = H'(c) = 0.$$

Therefore, $H(x_2) = H(x_1)$. Since x_1 and x_2 are arbitrary, the function H is constant on the interval: there is a constant C such that $H(x) = G(x) - F(x) = C$. This gives (1) and proves the theorem.   Q.E.D.

EXAMPLE 1   The function x^3 is an antiderivative of the function $3x^2$. Find the antiderivative G of $3x^2$ such that $G(1) = 5$.

Solution.   By Theorem 4.10 the antiderivatives of $3x^2$ are $G(x) = x^3 + C$. To have $G(1) = 5$, we must take $C = 4$. The desired antiderivative is $G(x) = x^3 + 4$.   □

The geometric meaning of Theorem 4.10

If two functions F and G have the same derivative, as in the hypotheses of Theorem 4.10, then the tangent lines to their graphs have the same slope at each value of x (Figure 4.76). Consequently, the graphs have the same shape, and the equation $G(x) = F(x) + C$ holds because each of the graphs is obtained from the other by shifting it up or down.

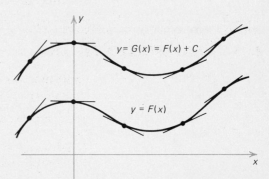

FIGURE 4.76   Corresponding tangent lines are parallel.

Exercises

†*Answer provided.*
‡*Outline of solution provided.*

1.†   Use the differentiation formula $\dfrac{d}{dx}(x^3 + 6x) = 3x^2 + 6$ to find the antiderivative g of $3x^2 + 6$ such that $g(1) = 0$.

2.   Use the differentiation formula $\dfrac{d}{dx}(x \sin x + \cos x) = x \cos x$ to find the antiderivative h of $x \cos x$ such that $h(0) = 1$.

3.‡ Use the differentiation formulas $\dfrac{d}{dx}(-\cos x) = \sin x$ and

$\dfrac{d}{dx}\left(\dfrac{1}{5}x^5\right) = x^4$ to find all the antiderivatives of $3\sin x - 2x^4$.

4.† Use the differentiation formulas $\dfrac{d}{dx}(2x^{1/2}) = x^{-1/2}$ and

$\dfrac{d}{dx}(3x^{1/3}) = x^{-2/3}$ to find all the antiderivatives of $5x^{-1/2} + 10x^{-2/3}$.

5. Use the differentiation formulas $\dfrac{d}{dx}x^{-1} = -x^{-2}$ and $\dfrac{d}{dx}x^3 = 3x^2$ to find all the antiderivatives of $3x^{-2} - 3x^2$.

6.† Suppose that $f(x)$ and $g(x)$ have continuous derivatives and that $g(x) \neq 0$ for all x. What are all the antiderivatives of   **a.** $f(x)g'(x) + f'(x)g(x)$ and   **b.** $[g(x)f'(x) - f(x)g'(x)]/[g(x)^2]$?

7.† Use the formula $\dfrac{d}{dx}\cos(x^3) = -3x^2\sin(x^3)$ to find the antiderivative of $3x^2\sin(x^3)$ that has the value 4 at $x = 0$.

8.† Use the formula $\dfrac{d}{dx}(3x^{4/3}) = 4x^{1/3}$ to find the antiderivative of $4x^{1/3}$ that has the value 10 at $x = -8$.

9.† If $F(x)$ is an antiderivative of $f(x)$ and $G(x)$ is an antiderivative of $g(x)$, then what are the antiderivatives of $Af(x) + Bg(x)$ for constants A and B?

MISCELLANEOUS EXERCISES

†*Answer provided.*
‡*Outline of solution provided.*

1. Sketch the graphs of the following functions. Show horizontal and vertical asymptotes, critical points, and inflection points:

a.† $x^2/(1 - x^2)$ **b.** $(x - 1)/x^3$
c. $x/(4 - x^2)$ **d.** $(x + 1)/(x - 2)$
e.‡ $x^4 - 6x^2 - 8x$ ($x = 2$ is a critical point.)
f. $\frac{1}{3}x^3 + 3x^2 + 9x + 1$ **g.** $x^3/(1 - x^2)$

2. Find the maximum and minimum values of the following functions in the indicated intervals.

a.† $3x^4 - 4x^3$ for $-2 \le x \le 3$
b. $x^4 + \frac{8}{3}x^3 + 2x^2$ for $-3 \le x \le 3$
c.‡ $\frac{1}{5}x^5 + \frac{1}{2}x^4 - x^3$ for $-4 \le x \le 2$
d. $x^4 - 4x^3 + 4x^2 + 1$ for $0 \le x \le 2$
e.‡ $x^4 - 4x^3 + 16x + 2$ for $-2 \le x \le 1$ ($x = -1$ is a critical point.)
f.† $\frac{1}{4}x^4 - \frac{2}{3}x^3 - \frac{1}{2}x^2 + 2x$ for $0 \le x \le 3$ ($x = 2$ is a critical point.)
g. $3x^4 + 4x^3 + 6x^2 + 12x$ for $0 \le x \le 10$ ($x = -1$ is a critical point.)
h.‡ $(x + 1)^2(x - 3)^3$ for $0 \le x \le 4$
i.† $(x - 1)^3(x + 2)^2$ for $-2 \le x \le 2$
j. $(x + 1)^3(x + 4)^3$ for $-1 \le x \le 3$
k. $(x + 3)^2(x - 5)^2$ for $-4 \le x \le 6$

3. Sketch the graphs of the following functions.

   a.[†] $\sqrt{1 + x^2}$ **b.** $|x^2 + x - 2|$ **c.**[†] $\sqrt[3]{x^2 + x - 2}$

   d. $\sqrt{x^2 + x - 2}$ **e.** $\sqrt[3]{x^3 - 4x}$

   f.[†] $\dfrac{1}{1 - 2\sin x}$ **g.** $\dfrac{1}{2 - \sin x}$ **h.**[†] $\sin\left(\dfrac{4\pi}{x}\right)$

   i. $\dfrac{\sin x}{\cos^2 x}$ **j.**[†] $\sqrt{\dfrac{x}{x + 1}}$ **k.** $\dfrac{1}{\sqrt{2 - x}}$

4. **a.**[†] Sketch the graph of the equation

$$[x^2 + y^2 - 1][x^2 + y^2 - 49][(x - 3)^2 + (y - 3)^2 - 1] \cdot$$
$$[(x + 3)^2 + (y - 3)^2 - 1][y + \sqrt{16 - x^2}] = 0.$$

(The graph of $f(x)g(x)h(x)j(x)k(x) = 0$ is formed by combining the graphs of $f(x) = 0, g(x) = 0, h(x) = 0, j(x) = 0$, and $k(x) = 0$.)
b. The equation in part (a) defines implicitly five differentiable functions with $x = 0$ in their domains and with domains as large as possible. Give formulas for these functions.

In Exercises 5 through 7 functions $y = y(x)$ are defined implicitly by the given conditions. Compute the indicated second derivatives.

5.[‡] $y^3 + 4y^2 = 11x^2, y(1) = 1; y''(1)$

6.[†] $2y^{3/2} + y = 4x, y(5) = 4; y''(5)$

7. $y^4 + y = x, y(2) = 1; y''(2)$

8.[†] A window is to be made in the shape of a rectangle with a semicircle on top of it. What dimensions should it have to maximize its area if its circumference is to be 7 feet?

9. Show that the rectangle of given area with the shortest possible diagonal is a square.

10. Show that the least possible volume of a right circular cone circumscribed about a sphere is twice that of the sphere.

11.[†] It costs 96 dollars/hour plus $2v^3$ dollars/mile to run a boat v miles/hour.
a. What is the cost per mile at v miles/hour?   **b.** At what speed is the cost per mile a minimum?

12. Toy monkeys can be manufactured for 60 cents each. At a price of 1 dollar each, 1000 can be sold. For each penny the price is lowered, fifty more monkeys can be sold. What price would maximize the profit?

13.[‡] Water is the only common liquid whose greatest density occurs at a temperature above its freezing point. (This phenomenon favors the survival of aquatic life by preventing ice from forming in the bottom of lakes.) One liter of water at 0°C occupies a volume of

$$V = 1 - 6.42(10^{-5})T + 8.51(10^{-6})T^2 - 6.79(10^{-8})T^3 \text{ liters}$$

at T°C $(0 \leq T \leq 30)$ (*Handbook of Chemistry and Physics*, p. 2248).
At what temperature is the density of water the greatest?

14.[†] The strength of the reaction to a certain drug is given by the formula $R^2 = D^2(\frac{1}{2}C - \frac{1}{3}D)$, where D is the amount of the drug administered and C is a constant. The derivative dR/dD is the *sensitivity* to the drug. For what value of D is the sensitivity the greatest?

15. A fish uses kv^3 ergs/second of energy to swim at a speed of v centimeters/second relative to the speed of a stream. Show that in order to minimize his total energy use in swimming upstream a given distance, the fish should swim at a speed of $v = \frac{3}{2}c$, where c is the speed of the stream.

16. It costs $3 + \frac{1}{100}v$ dollars/mile to drive a truck v miles/hour and the truck rents for 25 dollars/hour with driver. What speed minimizes the cost per mile?

17.[†] The demand for an item is $a/(b + cp^2)$ units if it is sold at a price of p dollars per unit (a, b, and c are positive constants). What price maximizes the revenue from the item?

18. What price gives the maximum revenue in Exercise 17 if the demand function is $\sqrt{a - bp}$? What is the maximum revenue?

19. If the marginal revenue in producing x items is $50 - \frac{1}{60}x^2$ dollars/item and the marginal cost is $10 + \frac{1}{120}x^2$ dollars/item, how many items should be produced to maximize the profit?

20. Let $D(p)$ units be the demand for an item if it is sold for p dollars/unit. Consider a change in price from p to $p + \Delta p$. The fraction by which the demand increases is $A = \big(D(p + \Delta p) - D(p)\big)/D(p)$, and the fraction by which the price increases is $B = \Delta p/p$. The limit $E(p)$ of the ratio $-A/B$ as Δp tends to zero is known as the *elasticity of demand with respect to price*. Show that it is given by the formula

$$E(p) = -\frac{p}{D(p)}\frac{dD}{dp}(p).$$

21. **a.** Show that the elasticity of demand $E(p)$ is 1 for all p under consideration if and only if the total revenue from sales of the item is independent of the price. **b.** If $E(p_0) < 1$, then the demand is called *inelastic* at price p_0. Show that raising the price from p_0 would increase the revenue in this case. **c.** If $E(p_0) > 1$, then the demand is called *elastic* at p_0. Show that raising the price from p_0 would decrease the revenue in this case. **d.** What should the elasticity of demand be to maximize the revenue?

22. Show that elasticity of demand is constant for demand functions of the form $D(p) = ap^{-k}$ with a and k being positive constants.

23. According to Newton's law of gravity, the force of attraction between an object of mass m and an object of mass M that are a distance r apart is $GmMr^{-2}$, where G is a universal constant that depends on the units used. The sun's mass is 2.7×10^7 times that of the moon, and the sun's distance from the earth is 390 times that of the moon. **a.** How many times greater is the gravitational attraction between the earth and the sun than between the earth and the moon? **b.**[†] Tides are caused by the difference in gravitational attraction of the sun or moon on the near and far sides of the earth. Use differentials to determine approximately how many times greater an effect the moon has on tides than the sun.

24. The speeds of sound c_1 and c_2 in an upper and lower layer of rock and the thickness h of the upper layer can be determined by seismic exploration if the speed of sound is greater in the lower layer (Figure 4.77). A dynamite charge is set off at point P on the surface, and the transmitted signals are recorded at a point Q located a distance L from P. Typically the first signal to arrive travels directly from P to Q along the surface of the ground and takes, say, T_1 seconds. The next signal travels from P to R in the upper layer, from R to S in the lower layer, and from S to Q in the upper layer and takes T_2 seconds. A third signal travels from P to O to Q in the upper layer.   **a.**[†] Express T_1, T_2, and T_3 in terms of L, h, c_1, c_2, and the angle θ shown in Figure 4.77.   **b.** Show, by minimizing T_2, that $\sin \theta = c_1/c_2$.   **c.**[†] Suppose that $L = 1/4$ kilometer, $T_1 = 1/8$ second, $T_2 = 1/6$ seconds, and $T_3 = 3/(8\sqrt{5})$ seconds. Find c_1 from the formula for T_1, and find h from the formula for T_3.   **d.** The equation for T_2 could now be solved for c_2. Instead, show that it is satisfied by $c_2 = 3$ kilometers/second. (Find $\tan \theta$ and $\sec \theta$ first.)

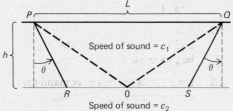

FIGURE 4.77

CHAPTER 5
THE INTEGRAL

SUMMARY: In Chapters 3 and 4 we studied the derivative and its applications; in this chapter we introduce the second basic tool of calculus, the integral. In Section 5.1 we give an informal geometric definition of the integral and in Section 5.2 the formal analytic definition. In Section 5.3 we present the Fundamental Theorem of Calculus, which we use in Section 5.4 to derive integration formulas for $x^n (n \neq -1)$, $\sin x$, and $\cos x$ from differentiation formulas. We also introduce in Section 5.4 the concept of an indefinite integral. The rules for dealing with integrals of linear combinations of functions and with integrals over adjacent intervals are described in Section 5.5, and the technique of integration by substitution is discussed in Section 5.6. An optional Section 5.7 contains historical notes on the integral, and in an optional appendix to this chapter we discuss the existence of definite integrals.

INTRODUCTION

Can you determine the volume of a barrel if you know the radii of its circular cross sections? If the cross sections all have the same radius, as in Figure 5.1, then the barrel is a cylinder and its volume is given by a formula from elementary geometry. If, however, the radii are not all the same, as in Figure 5.2, the volume is given by an integral.

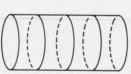

FIGURE 5.1

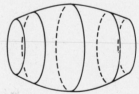

FIGURE 5.2

Can you determine how far an object travels in a given period of time if you know its velocity? If the velocity is constant, this requires only arithmetic; the distance is the velocity multiplied by the time. If the velocity varies, then the distance is given by an integral.

The study of integrals may be divided into three parts: defining integrals, developing techniques for evaluating them, and using them in applications. We concentrate on the first two parts in this chapter; Chapter 6 is devoted to applications.

We begin with a geometric definition of the integral in terms of areas, which provides an intuitive understanding of the basic concept. This definition by itself is of little practical importance, although we will see that it can be used to evaluate a few special types of integrals. We will then present some examples and exercises which illustrate how certain integrals can be found by an approximation procedure, and in Section 5.2 we use this idea to transform the geometric definition into the analytic definition of the integral as a limit.

The approximation procedure used in the analytic definition could be used to derive techniques for evaluating a wide variety of integrals, but it is easier to use the Fundamental Theorem of Calculus, to be introduced in Section 5.3. This result shows that in a certain sense integration and differentiation are inverse operations. It will enable us to use our knowledge of differentiation to study integrals and to develop the techniques of integration discussed in the last three sections of this chapter and in Chapters 7 and 8. While the Fundamental Theorem will be the key to evaluating integrals, the geometric definition will continue to provide a means of interpreting them and, as we will see in Chapter 6, the analytic definition will be used to determine which integrals are required in applications.

5-1   THE GEOMETRIC DEFINITION OF THE INTEGRAL

For a function f defined on a finite closed interval $a \leq x \leq b$, the INTEGRAL of f from a to b is a *number*, denoted by the symbol

$$(1) \qquad \int_a^b f(x)\, dx.$$

In the following informal definition the integral (1) is given in terms of the areas of regions set off by the graph of the function f.*

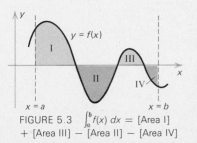

FIGURE 5.3 $\int_a^b f(x)\, dx =$ [Area I] $+$ [Area III] $-$ [Area II] $-$ [Area IV]

Definition 5.1   (The geometric definition of the integral) Consider the region that extends from the line $x = a$ on the left to the line $x = b$ on the right and that consists of the points above the x-axis and below the graph of f where $f(x)$ is positive and the points below the x-axis and above the graph of f where $f(x)$ is negative (Figure 5.3). The integral (1) of f from a to b is equal to the sum of the areas of the pieces of that region above the x-axis minus the areas of the pieces below the x-axis.

_____

*As we will see in Section 5.2 and in the appendix to this chapter, this definition may be applied to most functions f encountered in calculus.

In this definition we assume that the scales on the axes are equal.

If the function f is positive in the interval, the integral is equal to the area of the region above the x-axis and below the graph. The integral and the area are the same positive number (Figure 5.4). If the function f is negative, the integral is the negative of the area of the region below the x-axis and above the graph. The area is positive and the integral is negative (Figure 5.5).

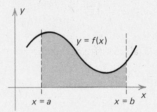

FIGURE 5.4 $\int_a^b f(x)\,dx =$ [The area of the region]

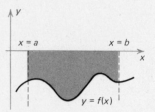

FIGURE 5.5 $\int_a^b f(x)\,dx$
= [The negative of the area of the region]

The symbol $\int$ in the integral

$$(1) \qquad \int_a^b f(x)\,dx$$

is called an INTEGRAL SIGN. The function f is called the INTEGRAND. The interval $a \leq x \leq b$ is called the INTERVAL OF INTEGRATION. The constants a and b are called the LIMITS OF INTEGRATION.* The letter "x" in the symbol (1) is called a DUMMY VARIABLE because the integral is a fixed number and not a variable function of x.

EXAMPLE 1   Sketch the graph of the function $x + 1$ and use the geometric definition to evaluate the integral

$$\int_{-3}^{3} (x + 1)\,dx.$$

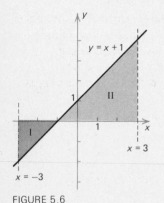

FIGURE 5.6

Solution.   The graph of the function $x + 1$ is the straight line $y = x + 1$ sketched in Figure 5.6. That graph, the x-axis, and the lines $x = -3$ and $x = 3$ set off the two triangles in the sketch. Triangle II above the x-axis is 4 units wide, 4 units high and has area 8, whereas triangle I below the x-axis is 2 units wide, 2 units high and has area 2. Therefore,

$$\int_{-3}^{3} (x + 1)\,dx = [\text{Area II}] - [\text{Area I}] = 8 - 2 = 6. \quad \square$$

---

*This use of the term "limit" has nothing to do with the limit concepts that are the basis of calculus. The "limits of integration" are where the integration begins and ends.

Integrals from a to b with $a \geq b$

Definition 5.2   **If the limits of integration in an integral are equal, then the integral is zero:**

(2) $$\int_a^a f(x)\,dx = 0.$$ ◀

An integral from a to b with $a > b$ is equal to the negative of the integral from b to a:

(3) $$\int_a^b f(x)\,dx = -\int_b^a f(x)\,dx.$$ ◀

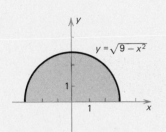

FIGURE 5.7   Region of area $\int_{-3}^{3} \sqrt{9 - x^2}\,dx = \dfrac{9\pi}{2}$

EXAMPLE 2   Evaluate the integral

(4) $$\int_3^{-3} \sqrt{9 - x^2}\,dx.$$

Solution.   The graph of the nonnegative function $\sqrt{9 - x^2}$ is the semicircle of radius 3 shown in Figure 5.7. The region below the graph and above the x-axis and from $x = -3$ on the left to $x = 3$ on the right is the half disk of area $\frac{9}{2}\pi$. Hence by Definition 5.1

(5) $$\int_{-3}^{3} \sqrt{9 - x^2}\,dx = [\text{The area of the half disk}] = \frac{9\pi}{2}.$$

By Definition 5.2 the integral (4) is equal to the negative of integral (5) or $-9\pi/2$.   □

Areas as limits

In order to evaluate an integral by applying the geometric definition, we need to know the areas of the regions set off by the graph of the function. In Examples 1 and 2 we used formulas for the areas of triangles and circles, but for most integrals we have to use a different method because there are no formulas from geometry for the areas of the corresponding regions. In the next two examples we will find the necessary areas by approximating regions with collections of rectangles.

EXAMPLE 3   In Exercise 14 of the Appendix to Chapter 1 we established the formula

(6) $$1^2 + 2^2 + 3^2 + \cdots + N^2 = \tfrac{1}{6}N(N + 1)(2N + 1)$$

for the sum of the squares of the first N positive integers. Use this formula to evaluate the integral

(7) $$\int_0^1 x^2\,dx.$$

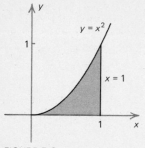

FIGURE 5.8

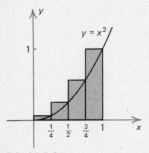

FIGURE 5.8a

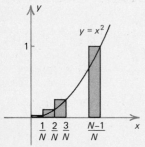

FIGURE 5.8b

Solution.   The integral (7) is equal to the area of the region bounded by the x-axis, by the line $x = 1$, and by the parabola $y = x^2$ (Figure 5.8). For each positive integer N we approximate the region by N rectangles of width $1/N$ with their bases on the x-axis and with their upper right corners on the parabola. Figure 5.8a shows the approximation in the case of $N = 4$ and Figure 5.8b is a schematic sketch of the approximation by N rectangles for an arbitrary positive integer N.

Before we deal with the general case, let us compute the area of the approximation by the four rectangles of width $\frac{1}{4}$ in Figure 5.8a. The top of the first rectangle on the left touches the parabola $y = x^2$ at $x = \frac{1}{4}$, so its height is $(\frac{1}{4})^2$. The top of the second rectangle touches the parabola at $x = \frac{1}{2}$ and its height is $(\frac{1}{2})^2$; the top of the third rectangle touches at $x = \frac{3}{4}$ and its height is $(\frac{3}{4})^2$; and the top of the last rectangle touches the parabola at $x = 1$ and its height is $1^2 = 1$.

The area of each rectangle is its width $\frac{1}{4}$ multiplied by its height, so the total area of the four rectangles is

$$\tfrac{1}{4}(\tfrac{1}{4})^2 + \tfrac{1}{4}(\tfrac{1}{2})^2 + \tfrac{1}{4}(\tfrac{3}{4})^2 + \tfrac{1}{4}(1)^2.$$

To show the pattern in this sum, we write $\frac{1}{2}$ as $\frac{2}{4}$ and 1 as $\frac{4}{4}$. Then the sum reads

$$\tfrac{1}{4}(\tfrac{1}{4})^2 + \tfrac{1}{4}(\tfrac{2}{4})^2 + \tfrac{1}{4}(\tfrac{3}{4})^2 + \tfrac{1}{4}(\tfrac{4}{4})^2 = \frac{1}{4^3}(1^2 + 2^2 + 3^2 + 4^2).$$

In the case of arbitrary N (Figure 5.8b) the top of the first rectangle touches the parabola at $x = \dfrac{1}{N}$ and its height is $\left(\dfrac{1}{N}\right)^2$; the top of the second rectangle touches the parabola at $x = \dfrac{2}{N}$ so its height is $\left(\dfrac{2}{N}\right)^2$; and in general, for $j = 1, 2, 3, \ldots, N$, the top of the jth rectangle touches the parabola at $x = \dfrac{j}{N}$ and its height is $\left(\dfrac{j}{N}\right)^2$. Each of the rectangles has width $\dfrac{1}{N}$, so the total area of the N rectangles is

$$(8) \quad \frac{1}{N}\left(\frac{1}{N}\right)^2 + \frac{1}{N}\left(\frac{2}{N}\right)^2 + \frac{1}{N}\left(\frac{3}{N}\right)^2 + \cdots + \frac{1}{N}\left(\frac{N}{N}\right)^2$$

$$= \frac{1}{N^3}(1^2 + 2^2 + 3^2 + \cdots + N^2).$$

Formula (6) enables us to express the area (8) of the approximation in the compact form

$$(9) \quad \frac{1}{N^3}[\tfrac{1}{6}N(N + 1)(2N + 1)] = \tfrac{1}{6}\left(1 + \frac{1}{N}\right)\left(2 + \frac{1}{N}\right).$$

Now imagine what happens as the number N of rectangles in the approximation tends to ∞: the width $1/N$ of the rectangles tends to zero and the approximation of the original region by rectangles becomes increasingly accurate. The area of the approximation tends to the area of the original region.

Letting N tend to ∞ in the expression on the right of (9) for the area of the approximation shows that the area of the original region, which equals the integral (7) is

$$\lim_{N \to \infty} \tfrac{1}{6}\left(1 + \frac{1}{N}\right)\left(2 + \frac{1}{N}\right) = \tfrac{2}{6} = \tfrac{1}{3}. \quad \square$$

In the next example we will use the procedure of Example 3 to evaluate the integral of x^2 over the interval $0 \le x \le a$ for an arbitrary positive number a.

EXAMPLE 4   Use formula (6) to evaluate the integral

(10) $\displaystyle\int_0^a x^2 \, dx$ for $a > 0$.

Solution.   In this case the integral is the area of the region bounded by the x-axis, the parabola $y = x^2$, and the vertical line $x = a$. The interval of integration has length a, so when we use N rectangles of equal width, each has width $\dfrac{a}{N}$ (Figure 5.8c). The height of the first rectangle on the left is $\left(\dfrac{a}{N}\right)^2$; the height of the second is $\left(\dfrac{2a}{N}\right)^2$; the height of the third is $\left(\dfrac{3a}{N}\right)^2$; and in general, for $j = 1, 2, 3, \ldots, N$, the height of the jth rectangle is $\left(\dfrac{ja}{N}\right)^2$. Therefore, the total area of the N rectangles which form the approximation is

$$\frac{a}{N}\left(\frac{a}{N}\right)^2 + \frac{a}{N}\left(\frac{2a}{N}\right)^2 + \frac{a}{N}\left(\frac{3a}{N}\right)^2 + \cdots + \frac{a}{N}\left(\frac{Na}{N}\right)^2$$

$$= \frac{a^3}{N^3}(1^2 + 2^2 + 3^2 + \cdots + N^2)$$

(11)

$$= \frac{a^3}{N^3}\left[\tfrac{1}{6}N(N+1)(2N+1)\right]$$

$$= \tfrac{1}{6}a^3\left(1 + \frac{1}{N}\right)\left(2 + \frac{1}{N}\right).$$

As N tends to ∞, the area (11) of the approximation tends to $\tfrac{1}{3}a^3$, which is the area of the region and is the integral (10). $\square$

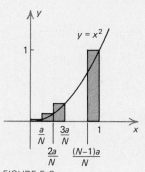

FIGURE 5.8c

Summation notation

To obtain more concise notation for sums such as in the formula

(6) $1^2 + 2^2 + 3^2 + \cdots + N^2 = \frac{1}{6}N(N+1)(2N+1)$

we replace the sum on the left with the symbol

(12) $\displaystyle\sum_{j=1}^{N} j^2.$

Here Σ is the capital Greek letter *sigma* and (12) denotes the sum of the numbers obtained by replacing j in the expression j^2 by each of the integers from 1 to N.

EXAMPLE 5   Write out the following sum. Simplify the terms but do not add them.

$$\sum_{j=1}^{5} \frac{2^j}{j+1}$$

Solution.   We have

$$\sum_{j=1}^{5} \frac{2^j}{j+1} = \frac{2^1}{1+1} + \frac{2^2}{2+1} + \frac{2^3}{3+1} + \frac{2^4}{4+1} + \frac{2^5}{5+1}$$

$$= \frac{2}{2} + \frac{4}{3} + \frac{8}{4} + \frac{16}{5} + \frac{32}{6} = 1 + \frac{4}{3} + 2 + \frac{16}{5} + \frac{16}{3}. \quad \square$$

Exercises

In Exercises 1 through 11 sketch the graphs of the integrands and use the geometric definition of the integral to evaluate the integrals.

1.† $\displaystyle\int_{-2}^{1} -2x \, dx$

2. $\displaystyle\int_{-2}^{4} \left(\frac{1}{2}x - 2\right) dx$

3.† $\displaystyle\int_{-10}^{-5} 6 \, dx$

4. $\displaystyle\int_{-3.96}^{5.72} 0 \, dx$

5.† $\displaystyle\int_{-1}^{0} \sqrt{1 - x^2} \, dx$

6. $\displaystyle\int_{0}^{5} -\sqrt{25 - x^2} \, dx$

7.‡ $\displaystyle\int_{0}^{-4} -4 \, dx$

8. $\displaystyle\int_{5}^{0} x \, dx$

9.‡ $\displaystyle\int_{-5}^{4} F(x) \, dx$ where $F(x)$ is the "step" function

$$F(x) = \begin{cases} -4 & \text{for} & x < -3 \\ 2 & \text{for } -3 \leq x \leq 0 \\ 1 & \text{for} & x > 0 \end{cases}$$

10. $\displaystyle\int_{-2}^{4} P(x) \, dx$ where $P(x) = \begin{cases} 3 & \text{for } x < -1 \\ 4 & \text{for } x = -1 \\ -2 & \text{for } x > -1 \end{cases}$

11. $\displaystyle\int_{-3}^{3} |x| \, dx$

12.‡ The graph of a function g is sketched in Figure 5.9. Which of the numbers 0, 6, −6, 12, −12, 24, and −24 is closest to the integral of g from −2 to 3?

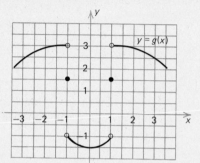

FIGURE 5.9

13.† The graph of a function R is sketched in Figure 5.10. Which of the numbers 0, 5, −5, 10, −10, 15, −15, 20, and −20 is closest to the integral of R from 2 to 12?

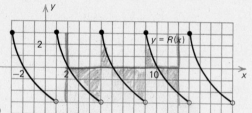

FIGURE 5.10

14. If the scales on the coordinate axes are not equal, as in Figure 5.11, integrals are given by the areas that regions would have if the scales were equal. Which of the numbers 0, 200, −200, 400, −400, 600, and −600 is closest to the integral of the function L from 5 to 20?

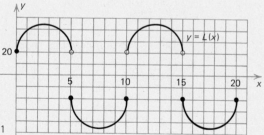

FIGURE 5.11

15.† **a.** Explain why the geometric definition of the integral (Definition 5.1) implies that

$$\int_a^b f(x)\,dx + \int_b^c f(x)\,dx = \int_a^c f(x)\,dx$$

for $a < b < c$.   **b.** Use Definition 5.2 to show that this identity is also valid for $c < a < b$.

16.[†] In Example 1 of the Appendix to Chapter 1 we established the formula

(12) $1 + 2 + 3 + \cdots + N = \frac{1}{2}N(N + 1)$

for the sum of the first N positive integers. (Another derivation is outlined in Miscellaneous Exercise 58.) Use the formula to evaluate the integral

$$\int_0^1 x \, dx$$

as the limit of the area of approximating rectangles of the type used in Example 3.

17.[†] In Exercise 19 of the appendix to Chapter 1 we established the formula

(13) $1^3 + 2^3 + 3^3 + \cdots + N^3 = \frac{1}{4}N^2(N + 1)^2$

for the sum of the cubes of the first N positive integers. (Another derivation is given in Miscellaneous Exercise 59.) Use (13) to evaluate the integral of x^3 from 0 to 1.

18. Use formula (13) to evaluate the integral of x^3 from 0 to a where a is an arbitrary positive constant.

Write out the sums in Exercises 19 through 26. Do not do the final arithmetic.

19.[†] $\displaystyle\sum_{k=1}^{5} \left(\frac{k}{12}\right)^2$

20.[†] $\displaystyle\sum_{k=1}^{5} \left(\frac{12}{k}\right)^2$

21.[†] $\displaystyle\frac{1}{5}\sum_{j=0}^{5} \frac{j}{5}$

22.[†] $\displaystyle\sum_{j=3}^{6} (1 - j)^2$

23.[†] $\displaystyle\frac{1}{10}\sum_{j=1}^{5} \left(\frac{j}{10}\right)\sin\left(\frac{j}{10}\right)$

24. $\displaystyle\sum_{n=-4}^{0} \frac{n + 1}{n + 5}$

25.[†] $\displaystyle\sum_{k=1}^{4} (-1)^{k+1}(2k + 1)$

26. $\displaystyle\sum_{m=-3}^{3} (-1)^m m^2$

27. Use the summation formulas

(14) $\displaystyle\sum_{j=1}^{N} j^4 = \frac{1}{5}N^5 + \frac{1}{2}N^4 + \frac{1}{3}N^3 - \frac{1}{30}N$

(15) $\displaystyle\sum_{j=1}^{N} j^5 = \frac{1}{6}N^6 + \frac{1}{2}N^5 + \frac{5}{12}N^4 - \frac{1}{12}N^3$

to evaluate the integrals of   **a.** x^4 and   **b.** x^5 from 0 to 1.

5-2   THE ANALYTIC DEFINITION OF THE INTEGRAL

The geometric definition of the integral which we gave in the last section would not serve as an effective basis for the theory of integrals because it did not include a precise definition of the areas of regions set off by the graph of a general function. We can obtain the required precision by defining the areas of the regions to be the limits of areas of approximations by collections of rectangles as in Examples 3 and 4 of the last section. We will now see how this approximation procedure leads us from the geometric definition of the integral to the analytic definition.

According to the geometric definition, the integral

$$(1) \qquad \int_a^b f(x)\, dx$$

for the function f of Figure 5.12 is equal to the area of region I above the x-axis minus the area of region II below the x-axis. To obtain the analytic definition, we approximate each of the regions by thin vertical rectangles as in Figure 5.13 and define the area of each of regions I and II to be the limit of the area of the rectangles that approximate it as the number of rectangles in the approximation tends to infinity and the widths of the rectangles tend to zero. We will then define the integral to be the limit of the quantity

$$(2) \qquad \left| \begin{matrix} \text{The area of the rectangles} \\ \text{above the } x\text{-axis} \end{matrix} \right| - \left| \begin{matrix} \text{The area of the rectangles} \\ \text{below the } x\text{-axis} \end{matrix} \right|$$

because this limit is the area of region I minus the area of region II.

To formulate the analytic definition, we need terminology and notation to describe the approximations by rectangles.

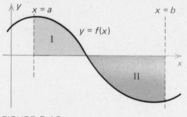

FIGURE 5.12 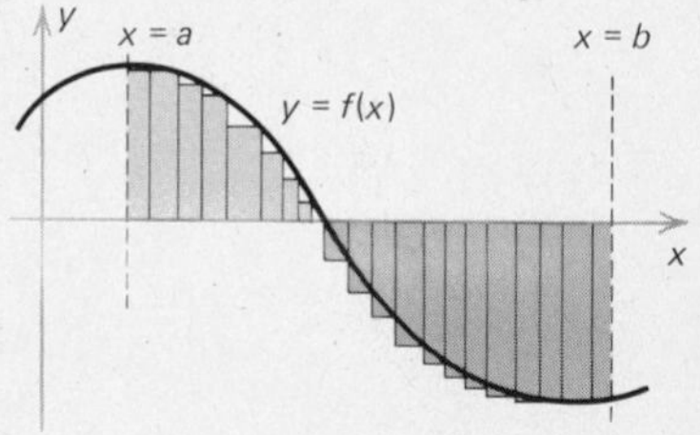   FIGURE 5.13

Partitions of intervals

A PARTITION of a finite closed interval $a \le x \le b$ is a division of that interval into subintervals. Figure 5.14 shows a partition of the interval $0 \le x \le 8$ into the four subintervals $0 \le x \le \frac{3}{2}, \frac{3}{2} \le x \le 4, 4 \le x \le 7$, and $7 \le x \le 8$. We refer to it as the partition $0 < \frac{3}{2} < 4 < 7 < 8$. The partition of the interval $0 \le x \le 8$ into four subintervals of equal length is shown in Figure 5.15. We refer to it as the partition $0 < 2 < 4 < 6 < 8$. The partition $0 < \frac{1}{10} < \frac{2}{10} < \frac{3}{10} < \frac{4}{10} < \cdots < \frac{79}{10} < 8$ is the partition of the interval $0 \le x \le 8$ into eighty equal subintervals.

Often we want to refer to a general, *unspecified* partition of an interval $a \le x \le b$. To do this we use the symbol x_0 to denote the number a, and, if there are to be N subintervals in the partition (N being a positive integer), we let the symbol x_N denote the number b. We let x_1 denote the right endpoint of the first subinterval; we let x_2 denote the right

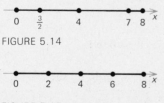

FIGURE 5.14

FIGURE 5.15

endpoint of the second subinterval; and, in general, we let x_j denote the right endpoint of the jth subinterval for $j = 1, 2, 3, \ldots, N$. Then the partition may be written

(3) $a = x_0 < x_1 < x_2 < x_3 < \cdots < x_{N-1} < x_N = b.$

A schematic sketch of this partition is shown in Figure 5.16. Notice that the jth subinterval runs from x_{j-1} to x_j for each $j = 1, 2, 3, \ldots, N$.

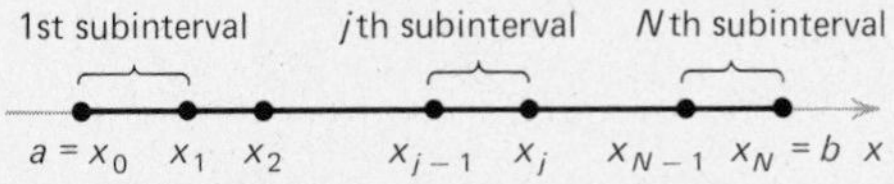

FIGURE 5.16

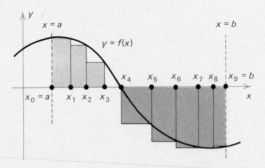

The approximating regions

We use partitions of the interval of integration to form rectangles that approximate the regions set off by the graph of the function. Figure 5.17 shows the rectangles for the function f of Figure 5.12 and a particular partition of the interval $a \leq x \leq b$ into nine subintervals. There is one rectangle for each subinterval of the partition, and each rectangle is as wide as the corresponding subinterval. The rectangles above the x-axis touch the graph of f at their upper right corners. The rectangles below the x-axis touch the graph at their lower right corners. The rectangle has zero height if the graph intersects the x-axis at the right endpoint of the subinterval.

If we start with a partition

(3) $a = x_0 < x_1 < x_2 < x_3 < \cdots < x_{N-1} < x_N = b$

into N subintervals, we obtain an approximation by N rectangles. The jth rectangle corresponds to the jth subinterval $x_{j-1} \leq x \leq x_j$. The width of the jth rectangle is the length $x_j - x_{j-1}$ of this subinterval.

FIGURE 5.17

If the value $f(x_j)$ of the function at the right endpoint of the jth subinterval is positive and the rectangle is above the x-axis, as in Figure 5.18, then the height of the rectangle is the value $f(x_j)$ of the function. In this case the area of the jth rectangle is given by the expression

(4) $f(x_j)(x_j - x_{j-1}).$

The area of the jth rectangle is also given by expression (4) if $f(x_j)$ is zero because, in this case, the rectangle has zero height and zero area and quantity (4) is zero.

If the value $f(x_j)$ of the function is negative and the rectangle is below the x-axis, as in Figure 5.19, then the height of the rectangle is the positive number $|f(x_j)| = -f(x_j)$. In this case the area of the jth rectangle is given by the expression

(5) $-f(x_j)(x_j - x_{j-1})$.

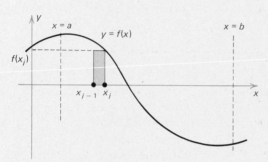

FIGURE 5.18   Rectangle of height $f(x_j)$

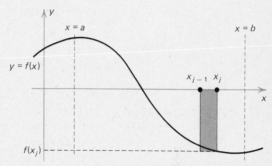

FIGURE 5.19   Rectangle of height $|f(x_j)| = -f(x_j)$

Riemann sums   The approximation of the integral that is determined by the partition is the sum of the areas of the rectangles above the x-axis minus the areas of the rectangles below the x-axis. We obtain the approximation by adding the expressions (4) for those values of j for which $f(x_j)$ is positive or zero and by subtracting the expressions (5) for those values of j for which $f(x_j)$ is negative. Subtracting the expressions (5) eliminates the minus signs in them. Therefore, the approximation of the integral is the sum

(6) $\begin{aligned} &f(x_1)(x_1 - x_0) + f(x_2)(x_2 - x_1) + f(x_3)(x_3 - x_2) + \cdots \\ &+ f(x_N)(x_N - x_{N-1}) \end{aligned}$ ◄

of the quantities $f(x_j)(x_j - x_{j-1})$ for $j = 1, 2, 3, \ldots, N$. To express this sum concisely, we use the summation notation

(7) $\displaystyle\sum_{j=1}^{N} f(x_j)(x_j - x_{j-1})$.   ◄

Expression (7) stands for the sum of the N numbers obtained by replacing j in the expression $f(x_j)(x_j - x_{j-1})$ by each of the integers $1, 2, 3, \ldots, N$.

The sum (6) or (7), which approximates the integral of f from a to b, is called a RIEMANN SUM,* according to the following definition.

Definition 5.3   Consider a function f defined in a closed finite interval $a \le x \le b$ and a partition

(3) $a = x_0 < x_1 < x_2 < x_3 < \cdots < x_{N-1} < x_N = b$

*Riemann sums are named after the German mathematician G. F. B. Riemann (1826–1866).

of that interval. For each $j = 1, 2, 3, \ldots, N$, let c_j be a point in the jth subinterval of the partition; that is, let c_j be such that $x_{j-1} \le c_j \le x_j$. Then the sum

$$(8) \qquad \sum_{j=1}^{N} f(c_j)(x_j - x_{j-1})$$

is a Riemann sum approximation of the integral of f from a to b.

The Riemann sum (8) in Definition 5.3 is more general than the Riemann sum (7) because in (8) the function is evaluated at an arbitrary point c_j in the jth subinterval rather than at its right endpoint x_j.

The analytic definition

In the analytic definition the integral is defined to be the limit of Riemann sums.

Definition 5.4   (The analytic definition of the integral) Suppose that $f(x)$ is defined for $a \le x \le b$. If the Riemann sums (8) have a finite limit L as the number of subintervals in the corresponding partitions (3) tends to infinity and the widths of the subintervals tend to zero, then that number L is the integral

$$(1) \qquad \int_{a}^{b} f(x)\,dx.$$

In this case the integral of f from b to a is the negative of the integral (1). If $a = b$, then the integral (1) is zero.

In the Appendix to this chapter we will give the $\epsilon\delta$-definition of the type of limit involved in Definition 5.4 and discuss conditions on f which guarantee that the integral (1) is defined. We will see that the integral is defined for all types of functions that are normally encountered in calculus provided f is BOUNDED in the interval of integration. This means there is a constant M such that $|f(x)| \le M$ for $a \le x \le b$. In particular, the integral (1) exists if f is continuous for $a \le x \le b$.

EXAMPLE 1   Give the Riemann sum for the integral

$$(9) \qquad \int_{-1}^{2} (x^2 - 1)\,dx$$

corresponding to the general partition $-1 = x_0 < x_1 < x_2 < \cdots < x_{N-1} < x_N = 2$ of the interval $-1 \le x \le 2$ and with the integrand evaluated at the right endpoints of the subintervals.

Solution.   The Riemann sum is the sum (7) with $f(x) = x^2 - 1$. It is

$$\sum_{j=1}^{N} (x_j^2 - 1)(x_j - x_{j-1}). \quad \square$$

EXAMPLE 2   Compute the Riemann sum of Example 1 for the particular partition $-1 < -\frac{1}{2} < 0 < \frac{1}{2} < 1 < \frac{3}{2} < 2$ of the interval $-1 \leq x \leq 2$ into six subintervals of equal length.

Solution.   The values of the function $x^2 - 1$ at the right endpoints of the six subintervals are $[x^2 - 1]_{x=-1/2} = (-\frac{1}{2})^2 - 1 = -\frac{3}{4}$, $[x^2 - 1]_{x=0} = 0^2 - 1 = -1$, $[x^2 - 1]_{x=1/2} = (\frac{1}{2})^2 - 1 = -\frac{3}{4}$, $[x^2 - 1]_{x=1} = 1^2 - 1 = 0$, $[x^2 - 1]_{x=3/2} = (\frac{3}{2})^2 - 1 = \frac{5}{4}$, and $[x^2 - 1]_{x=2} = 2^2 - 1 = 3$.

The Riemann sum is the sum of these six numbers, multiplied by the length $\frac{1}{2}$ of the subintervals. It is, therefore,

(10) $(-\frac{3}{4})(\frac{1}{2}) + (-1)(\frac{1}{2}) + (-\frac{3}{4})(\frac{1}{2}) + (0)(\frac{1}{2}) + (\frac{5}{4})(\frac{1}{2}) + 3(\frac{1}{2})$
$$= -\frac{3}{8} - \frac{1}{2} - \frac{3}{8} + 0 + \frac{5}{8} + \frac{3}{2} = \frac{7}{8}. \quad \square$$

The integral (9) is equal to the area of the region above the x-axis in Figure 5.20 minus the area of the region below the x-axis. The Riemann sum (10) is equal to the area of the rectangles above the x-axis in Figure 5.21 minus the area of the rectangles below the x-axis.

EXAMPLE 3   Give the integral for which

(11) $$\sum_{j=1}^{N} \frac{x_j}{x_j + 2}(x_j - x_{j-1})$$

is a Riemann sum corresponding to the partition

(12) $1 = x_0 < x_1 < x_2 < \cdots < x_{N-1} < x_N = 7$.

Solution.   Because the partition (12) is of the interval $1 \leq x \leq 7$, the limits of integration are 1 and 7. The jth term in the Riemann sum (11) is the value of the function $x/(x + 2)$ at the right end of the jth subinterval of the partition multiplied by the length of that subinterval. The sum is, therefore, a Riemann sum for the integral

$$\int_1^7 \frac{x}{x + 2}dx. \quad \square$$

The Δx notation in Riemann sums   We will occasionally use partitions $a = x_0 < x_1 < x_2 < x_3 < \cdots < x_{N-1} < x_N = b$ of the interval of integration into subintervals of equal length. In this case we let Δx denote the width of the subintervals: $\Delta x = (b - a)/N = x_j - x_{j-1}$ for $j = 1, 2, 3, \ldots, N$. The Riemann sum (7) then takes the form

(13) $$\sum_{j=1}^{N} f(x_j) \, \Delta x$$

and the analytic definition of the integral gives

(14) $$\int_a^b f(x) \, dx = \lim_{N \to \infty} \sum_{j=1}^{N} f(x_j) \, \Delta x. \qquad \blacktriangleleft$$

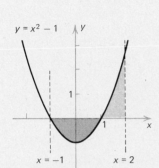

$y = x^2 - 1$

$x = -1$ $x = 2$

FIGURE 5.20

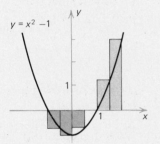

$y = x^2 - 1$

FIGURE 5.21

Estimating integrals

The results in the next proposition can be used to obtain rough estimates of the values of integrals. They are also required occasionally in theoretical discussions involving integrals.

Proposition 5.1 **(a) If the integrals of $f(x)$ and of $|f(x)|$ from a to b exist ($a < b$), then**

$$(15)\quad \left| \int_a^b f(x)\, dx \right| \le \int_a^b |f(x)|\, dx. \qquad \blacktriangleleft$$

(b) If the integrals of f and g from a to b exist and $f(x) \le g(x)$ for $a \le x \le b$, then

$$(16)\quad \int_a^b f(x)\, dx \le \int_a^b g(x)\, dx.$$

(c) If the integral of f from a to b exists and $m \le f(x) \le M$ for $a \le x \le b$, then

$$(17)\quad m(b - a) \le \int_a^b f(x)\, dx \le M(b - a).$$

Proof. For every partition $a = x_0 < x_1 < x_2 < \cdots < x_N = b$ we have

$$(18)\quad \left| \sum_{j=1}^N f(x_j)(x_j - x_{j-1}) \right| \le \sum_{j=1}^N |f(x_j)|(x_j - x_{j-1})$$

because the absolute value of a sum of numbers is no greater than the sum of their absolute values (see Miscellaneous Exercise 75) and because each $x_j - x_{j-1}$ is positive. Inequality (18) involves Riemann sums for the integrals in (15), and in the limit as N tends to ∞ and the lengths of the subintervals tend to zero, inequality (18) transforms into inequality (15). The proof of (16) is similar and (17) follows readily from (16). We leave these parts of the proof as Exercises 33 and 34.　Q.E.D.

EXAMPLE 4　Use inequalities (17) to estimate the value of

$$\int_0^\pi (3 + \cos^2 x)\, dx.$$

Solution.　As x varies between 0 and π, $\cos^2 x$ varies between 0 and 1. Hence, the minimum value of $3 + \cos^2 x$ for $0 \le x \le \pi$ is $m = 3$ and the maximum value is $M = 4$. The length of the interval of integration is π, so we obtain

$$3\pi \le \int_0^\pi (3 + \cos^2 x)\, dx \le 4\pi. \quad \square$$

Exercises

†*Answer provided.*
‡*Outline of solution provided.*

In Exercises 1 through 3 write out the sums. Simplify the terms but do not add them.

1.‡ $\displaystyle\sum_{j=1}^{6} j^2(-1)^j$

2.† $\displaystyle\sum_{j=3}^{7} \frac{2j+1}{2j-1}$

3. $\displaystyle\sum_{j=1}^{4} (j^2 + 2j + 3)$

In Exercises 4 through 12 give the Riemann sums for the integrals corresponding to the indicated partitions with the integrands evaluated at the right endpoints of the subintervals.

4.† $\displaystyle\int_1^5 (2x^3 - 3x^2)\,dx;\; 1 = x_0 < x_1 < x_2 < \cdots < x_{19} < x_{20} = 5$

5. $\displaystyle\int_2^{10} \frac{1}{x}\,dx;\; 2 = x_0 < x_1 < x_2 < \cdots < x_{N-1} < x_N = 10$

6.† $\displaystyle\int_{-\pi}^{\pi} \sin t\,dt;\; -\pi = t_0 < t_1 < t_2 < \cdots < t_{N-1} < t_N = \pi$

7. $\displaystyle\int_{-2}^{-1} (u+4)^5\,du;\; -2 = u_0 < u_1 < u_2 < \cdots < u_{999} < u_{1000} = -1$

8.‡ $\displaystyle\int_0^5 \cos x\,dx;\; 0 = \frac{0}{10} < \frac{1}{10} < \frac{2}{10} < \frac{3}{10} < \cdots < \frac{49}{10} < \frac{50}{10} = 5$

9.† $\displaystyle\int_0^1 x^3\,dx;\; 0 = \frac{0}{100} < \frac{1}{100} < \frac{2}{100} < \frac{3}{100} < \cdots < \frac{99}{100} < \frac{100}{100} = 1$

10. $\displaystyle\int_0^2 (x^2 - 1)\,dx;\; 0 = \frac{0}{8} < \frac{1}{8} < \frac{2}{8} < \frac{3}{8} < \cdots < \frac{15}{8} < \frac{16}{8} = 2$

11.† $\displaystyle\int_1^4 \sqrt{x}\,dx;\; 1 = \frac{5}{5} < \frac{6}{5} < \frac{7}{5} < \frac{8}{5} < \cdots < \frac{19}{5} < \frac{20}{5} = 4$ (Have j run from 6 to 20.)

12. $\displaystyle\int_2^3 x^{4/5}\,dx;\; 2 = \frac{200}{100} < \frac{201}{100} < \frac{202}{100} < \cdots < \frac{299}{100} < \frac{300}{100} = 3$ (Have j run from 201 to 300.)

In each of Exercises 13 through 16   **a.** sketch the regions bounded by the x-axis, the given vertical lines, and the graph of the given function.   **b.** Sketch the approximations of the regions of part (a) by rectangles whose tops or bottoms are the subintervals of the partition and which touch the graph at their upper or lower right corners.   **c.** Compute the corresponding Riemann sum for the integral. Leave the answer as a sum of fractions or give an approximate decimal value for it.

13.† $x = 1, x = 3;\; \dfrac{3}{x};\; 1 < \dfrac{4}{3} < \dfrac{5}{3} < 2 < \dfrac{7}{3} < \dfrac{8}{3} < 3;\; \displaystyle\int_1^3 \frac{3}{x}\,dx$

14. $x = -2, x = 1;\; x^2 - 4;\; -2 < -\dfrac{3}{2} < -1 < -\dfrac{1}{2} < 0 < \dfrac{1}{2} < 1;$

$\displaystyle\int_{-2}^{1} (x^2 - 4)\,dx$

15.† $x = 0, x = 2; 1 - x^2; 0 < \dfrac{1}{4} < \dfrac{1}{2} < \dfrac{3}{4} < 1 < \dfrac{5}{4} < \dfrac{3}{2} < \dfrac{7}{4} < 2;$

$$\int_0^2 (1 - x^2)\,dx$$

16. $x = -1, x = 0; 4x^3; -1 < -\dfrac{3}{4} < -\dfrac{1}{2} < -\dfrac{1}{4} < 0; \displaystyle\int_{-1}^0 4x^3\,dx$

In each of Exercises 17 through 22 find the integral for which the sum is a Riemann sum corresponding to the given partition.

17.† $\displaystyle\sum_{j=1}^N \sin(x_j)(x_j - x_{j-1}); 0 = x_0 < x_1 < x_2 < x_3 < \cdots < x_{N-1} < x_N = 3$

18. $\displaystyle\sum_{j=1}^N (3x_j^4 - 2x_j^3 + 4)(x_j - x_{j-1});$
$1 = x_0 < x_1 < x_2 < x_3 < \cdots x_{N-1} < x_N = 4$

19.† $\displaystyle\sum_{j=1}^N \dfrac{6}{1 - t_j}(t_j - t_{j-1}); -5 = t_0 < t_1 < t_2 < t_3 < \cdots < t_{N-1} < t_N = -2$

20. $\displaystyle\sum_{j=1}^N u_j^2 \cos(u_j)(u_j - u_{j-1});$
$-1 = u_0 < u_1 < u_2 < u_3 < \cdots < u_{N-1} < u_N = 1$

21.† $\displaystyle\sum_{j=1}^N [5 - (x_j^2 + 2)^2]\Delta x; 0 = x_0 < x_1 < x_2 < x_3 < \cdots < x_{N-1} < x_N = 4,$
where $x_j - x_{j-1} = \Delta x = 4/N$ for $j = 1, 2, 3, \ldots, N$

22. $\displaystyle\sum_{j=1}^N \dfrac{1 + \sin(t_j)}{2 - \cos(t_j)}\Delta t; 1 = t_0 < t_1 < t_2 < t_3 < \cdots < t_{N-1} < t_N = 10,$
where $t_j - t_{j-1} = \Delta t = 9/N$ for $j = 1, 2, 3, \ldots, N$

23. Show that the region bounded by the parabola $y = x^2$, by the x-axis, and by the line $x = 1$ is $\frac{1}{3}$ by approximating the region with N trapezoids of equal width, with their bases on the x-axis, with vertical sides, and with their upper corners on the parabola. Use summation formulas

$$1 + 2 + 3 + \cdots + N = \tfrac{1}{2}N(N + 1)$$
$$1^2 + 2^2 + 3^2 + \cdots + N^2 = \tfrac{1}{6}N(N + 1)(2N + 1).$$

24.† Use the trigonometric identity $\sin x \sin y = \frac{1}{2}[\cos(x - y) - \cos(x + y)]$ to show that for any positive integer N

$$(19) \qquad \sum_{j=1}^N \sin\left(\dfrac{j}{N}\right) = \dfrac{\cos(1/(2N)) - \cos(1 + 1/(2N))}{2\sin(1/(2N))}.$$

(Set $x = j/N$ and $y = 1/(2N)$ and sum the results.)

25.† Use formula (19) to evaluate the integral of $\sin x$ from 0 to 1.

26.† Derive and use a formula analogous to (19) to evaluate the integral of $\sin x$ from 0 to a where a is an arbitrary positive number.

27. Use the trigonometric identity $\cos x \sin y = \frac{1}{2}[\sin(x + y) - \sin(x - y)]$ to show that for any positive integer N

$$(20) \quad \sum_{j=1}^{N} \cos\left(\frac{j}{N}\right) = \frac{\sin(1 + 1/(2N)) - \sin(1/(2N))}{2 \sin(1/(2N))}.$$

28. Use formula (20) to evaluate the integral of $\cos x$ from 0 to 1.

29. Derive and use an identity analogous to (20) to evaluate the integral of $\cos x$ from 0 to a where a is an arbitrary positive number.

30. Use the summation formula

$$1^3 + 2^3 + 3^3 + \cdots + N^3 = \frac{1}{4}N^2(N + 1)^2$$

to evaluate the integral of x^3 from 0 to 1 directly as a limit of Riemann sums.

31.† **a.** Use the analytic definition of the integral to prove that if the integrals of $f(x)$ from a to b and from b to c exist $(a < b < c)$, then the integral from a to c exists and is the sum of the other two integrals.   **b.** Prove the result of part (a) for $a < c < b$.

32. Use the analytic definition of the integral to prove that if the integrals of $f(x)$ and $g(x)$ from a to b exist $(a < b)$, then for any constants A and B the integral of the linear combination $Af(x) + Bg(x)$ exists and equals A times the integral of f plus B times the integral of g.

33. Prove statement (16) in Proposition 5.1.

34. Prove statement (17) in Proposition 5.1.

35.† Use inequalities (17) to estimate the integral of $x^{1/3}$ from -8 to 27.

36. Use inequalities (17) to estimate the integral of $\sqrt{25 - x^2}$ from 0 to 3.

37.† Use inequality (16) to determine which is the greatest and which is the least of the three numbers

a. $\displaystyle\int_0^1 \sqrt{1 + x^2}\, dx,$ **b.** $\displaystyle\int_0^1 \sqrt{1 + x^3}\, dx,$ **c.** $\displaystyle\int_0^1 \sin(x^2)\, dx.$

In Exercises 38 through 41 give integrals that equal the indicated limits.

38.† $\displaystyle\lim_{N \to \infty} \frac{1}{N} \sum_{j=1}^{N} \left(\frac{j}{N}\right)\sin\left(\frac{j}{N}\right)$ 39.† $\displaystyle\lim_{N \to \infty} \frac{1}{N} \sum_{j=1}^{2N} \cos\left(\frac{j}{N}\right)$

40.† $\displaystyle\lim_{N \to \infty} \frac{1}{N} \sum_{j=1}^{N} \sqrt{1 + \left(\frac{j}{N}\right)^4}$ 41.† $\displaystyle\lim_{N \to \infty} \frac{4}{N} \sum_{j=1}^{N} \left[\frac{4j}{N} + \left(\frac{4j}{N}\right)^2\right]$

5-3   THE FUNDAMENTAL THEOREM OF CALCULUS

In this section we discuss the FUNDAMENTAL THEOREM OF CALCULUS. It shows that differentiation and integration are, in a certain sense, inverse operations, much the same as addition and subtraction or multiplication and division are inverse operations. There are two versions of this theorem. One deals with derivatives of integrals and the other with integrals of derivatives. The two results are called "versions" of the same theorem because, under suitable hypotheses, each is a consequence of the other.

The FUNDAMENTAL THEOREM FOR DERIVATIVES OF INTEGRALS shows how to construct antiderivatives of continuous functions and how to differentiate functions that are formed from integrals with variable limits of integration. It is mainly of theoretical interest in elementary calculus.

The FUNDAMENTAL THEOREM FOR INTEGRALS OF DERIVATIVES is one of the key results of calculus at all levels because it enables us to evaluate integrals by *antidifferentiation*.

The Fundamental Theorem for derivatives of integrals

Theorem 5.1 **Suppose that f is a continuous function in an open interval and that a is a point in that interval. Then for any x in the interval**

(1) $$\frac{d}{dx} \int_a^x f(t)\, dt = f(x).$$ ◀

We use the letter "t" as the dummy variable in the integral in (1) because we use "x" as its upper limit of integration.

Proof of Theorem 5.1   We will deal first with the case where f is a positive function and the x in formula (1) is greater than a. We set

(2) $$A(x) = \int_a^x f(t)\, dt.$$

Then $A(x)$ is the area of the region in a ty-plane bounded by the lines $t = a$ and $t = x$, by the t-axis, and by the graph $y = f(t)$ of the function f (Figure 5.22). To establish equation (1), we need to show that the derivative $\dfrac{dA}{dx}(x)$ of the area function $A(x)$ is equal to the height $f(x)$ of the region at its right side.

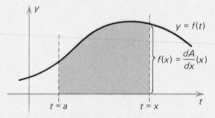

FIGURE 5.22   Region of area $A(x)$

For $\Delta x > 0$, $A(x + \Delta x)$ is the area of the region shown in Figure 5.23 and $A(x + \Delta x) - A(x)$ is the area of the region shown in Figure 5.24.

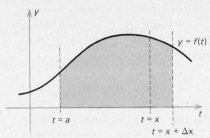

FIGURE 5.23   Region of area
= $A(x + \Delta x)$

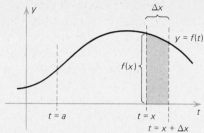

FIGURE 5.24   Region of area
= $A(x + \Delta x) - A(x)$

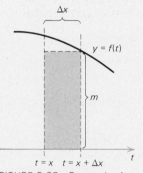

FIGURE 5.25   Rectangle of
area $m\Delta x$

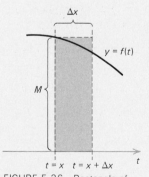

FIGURE 5.26   Rectangle of
area $M\Delta x$

By the Extreme Value Theorem the continuous function $f(t)$ has a minimum value m and a maximum value M in the finite closed interval $x \leq t \leq x + \Delta x$. The rectangle with its base on the t-axis, with its sides at $t = x$ and $t = x + \Delta x$, and with height m lies inside the region of Figure 5.24 (see Figure 5.25). The rectangle with the same base and sides but height M contains the region of Figure 5.24 (see Figure 5.26). Consequently,

(3) $\begin{bmatrix} \text{The area of the} \\ \text{rectangle in} \\ \text{Figure 5.25} \end{bmatrix} \leq \begin{bmatrix} \text{The area of} \\ \text{the region in} \\ \text{Figure 5.24} \end{bmatrix} \leq \begin{bmatrix} \text{The area of the} \\ \text{rectangle in} \\ \text{Figure 5.26} \end{bmatrix}$

The area of the smaller rectangle is $m\Delta x$; the area of the larger rectangle is $M\Delta x$; and the area of the intermediate region is $A(x + \Delta x) - A(x)$. Therefore inequalities (3) read

(4) $m\Delta x \leq A(x + \Delta x) - A(x) \leq M\Delta x.$

Dividing inequalities (4) by the positive number Δx preserves their direction and gives

(5) $m \leq \dfrac{A(x + \Delta x) - A(x)}{\Delta x} \leq M.$

Similar considerations give inequalities (5) for $\Delta x < 0$ (see Exercise 23).

To derive (5) for arbitrary continuous functions f and without direct reference to areas, we use the result of Exercise 31 in Section 5.2 to write for $a, x,$ and $x + \Delta x$ in the interval where f is continuous

$$A(x + \Delta x) = \int_a^{x+\Delta x} f(t)\, dt = \int_a^x f(t)\, dt + \int_x^{x+\Delta x} f(t)\, dt$$

$$= A(x) + \int_x^{x+\Delta x} f(t)\, dt.$$

Then for $\Delta x \neq 0$ we obtain

$$\frac{A(x + \Delta x) - A(x)}{\Delta x} = \frac{1}{\Delta x} \int_x^{x+\Delta x} f(t)\, dt.$$

Inequalities (5) then follow from an application of the third part of Proposition 5.1 in Section 5.2.

The numbers m and M in inequalities (5) are values of the function $f(t)$ in the interval between x and $x + \Delta x$. This interval shrinks to the single point $t = x$ as $\Delta x \to 0$. Hence, because $f(t)$ is a continuous function, both m and M tend to $f(x)$ as Δx tends to 0. The difference quotient in (5) is trapped between m and M, so we obtain

$$(6) \quad \frac{d}{dx} A(x) = \lim_{\Delta x \to 0} \frac{A(x + \Delta x) - A(x)}{\Delta x} = f(x).$$

Equation (1) follows from equation (6) because $A(x)$ is the integral of $f(t)$ from a to x.   Q.E.D.

EXAMPLE 1   Compute the derivative

$$\frac{d}{dx} \int_1^x (3t^4 - 6t^3)\, dt.$$

Solution.   By formula (1) with $f(t) = 3t^4 - 6t^3$ and $a = 1$, we have

$$\frac{d}{dx} \int_1^x (3t^4 - 6t^3)\, dt = [3t^4 - 6t^3]_{t=x} = 3x^4 - 6x^3. \quad \square$$

The Fundamental Theorem for integrals of derivatives

Theorem 5.2   Suppose that the function $F(t)$ has a continuous derivative $F'(t)$ for all t in an open interval. Then for a and b in that interval

$$(7) \quad \int_a^b F'(t)\, dt = F(b) - F(a). \qquad \blacktriangleleft$$

We will give two proofs of this theorem, one based on Theorem 5.1 and one based on the analytic definition of the integral and the Mean Value Theorem for derivatives.

First proof.   We define

$$(8) \quad A(x) = \int_a^x F'(t)\, dt.$$

By Theorem 5.1 the functions $A(x)$ and $F(x)$ have the same derivative $F'(x)$ for x in the interval. By Theorem 4.10 (Section 4.13) there is a constant C such that $A(x) = F(x) + C$ for all x in the interval. For $x = a$ the integral (8) has equal limits of integration and is zero, so we have $A(a) = 0$. Therefore $C = -F(a)$ and $A(x) = F(x) - F(a)$. Setting $x = b$ in the last equation gives equation (7).   Q.E.D.

Second proof.   Let $a = t_0 < t_1 < t_2 < \cdots < t_{N-1} < t_N = b$ be an arbitrary partition of the interval of integration. We will show there exist numbers $c_1, c_2, c_3, \ldots, c_N$ with $t_{j-1} \le c_j \le t_j$ for each j, such that the corresponding Riemann sum

$$(9) \quad \sum_{j=1}^N F'(c_j)(t_j - t_{j-1})$$

equals $F(b) - F(a)$. Because there is for each partition a Riemann sum that equals $F(b) - F(a)$, the integral (7), which is the limit of all the Riemann sums, must also equal $F(b) - F(a)$.

To show there exist suitable numbers c_j, we write

$$F(b) - F(a) = F(t_N) - F(t_0)$$
$$= [F(t_N) - F(t_{N-1})] + [F(t_{N-1}) - F(t_{N-2})]$$
$$+ \cdots + [F(t_1) - F(t_0)]$$
$$= \sum_{j=1}^{N} [F(t_j) - F(t_{j-1})].$$

Here we have expressed the change $F(b) - F(a)$ in $F(t)$ across the interval $a \le t \le b$ as the sum of the changes in $F(t)$ across the subintervals. By the Mean Value Theorem for derivatives, there exists for each j at least one number c_j with $t_{j-1} < c_j < t_j$ such that $F(t_j) - F(t_{j-1}) = F'(c_j)(t_j - t_{j-1})$. With these substitutions in the sum on the right of the last equation, we obtain the Riemann sum (9) to complete the proof of the theorem.   Q.E.D.

When we apply Theorem 5.2, we will often use the notation

$$[F(t)]_a^b = F(b) - F(a).$$

Conclusion (7) of the theorem then reads

$$\int_a^b F'(t)\, dt = [F(t)]_a^b$$

or, with Leibniz notation for the derivative and x replacing t,

(10) $$\int_a^b \frac{dF}{dx}(x)\, dx = [F(x)]_a^b.$$ ◄

EXAMPLE 2   Evaluate the integral

$$\int_0^1 \frac{d}{dx}(x^{10} - 6x^2)\, dx.$$

Solution.   The derivative $\dfrac{d}{dx}(x^{10} - 6x^2)$ exists and is continuous for all x. By equation (10), with $F(x) = x^{10} - 6x^2$, $a = 0$, and $b = 1$, we have

$$\int_0^1 \frac{d}{dx}(x^{10} - 6x^2)\, dx = [x^{10} - 6x^2]_0^1$$
$$= [1^{10} - 6(1)^2] - [0^{10} - 6(0)^2]$$
$$= -5. \quad \square$$

Evaluating integrals by anti-differentiation

If $F(x)$ is an antiderivative of a continuous function $f(x)$, then we can use $F(x)$ to evaluate integrals of $f(x)$. We have $dF/dx = f$ and by equation (10)

(11) $\displaystyle\int_a^b f(x)\,dx = [F(x)]_a^b.$ ◄

EXAMPLE 3   The function $x^5 - 4x$ is an anti-derivative of $5x^4 - 4$. Use this fact to evaluate the integral

$$\int_0^1 (5x^4 - 4)\,dx.$$

Solution.   Equation (11) with $F(x) = x^5 - 4x$ and $f(x) = 5x^4 - 4$ yields

$$\begin{aligned}
\int_0^1 (5x^4 - 4)\,dx &= [x^5 - 4x]_0^1 \\
&= [x^5 - 4x]_{x=1} - [x^5 - 4x]_{x=0} \\
&= [1^5 - 4(1)] - [0^5 - 4(0)] \\
&= [1 - 4] - [0] = -3. \quad \square
\end{aligned}$$

Exercises

†*Answer provided.*
‡*Outline of solution provided.*

In Exercises 1 through 4 find the derivatives.

1.† $\displaystyle\frac{d}{dx}\int_2^x \sqrt{t + 3}\,dt$

2. $\displaystyle\frac{d}{dx}\int_{-1}^x (t\sin t)\,dt$

3.† $\displaystyle\frac{d}{dy}\int_3^y \frac{x}{x^2 + 4}\,dx$

4. $\displaystyle\frac{d}{dt}\int_1^t 6\cos(u^2 + u)\,du$

5.‡ **a.** Use the formula for the area of a triangle to evaluate the integral $\int_0^x t\,dt$.   **b.** Use Theorem 4.10 to find all antiderivatives of the function x.

6. **a.**† Use the formulas for the areas of triangles and trapezoids to evaluate the integral $\int_0^x (3t + 6)\,dt$ for $x > -2$.   **b.**† Use Theorem 4.10 to find all antiderivatives of $3x + 1$ for $x > -2$.   **c.**‡ Evaluate the integral for $x < -2$.

7. **a.** Use the formula for the area of a triangle to evaluate the integral $\int_0^x -6t\,dt$.   **b.** Use Theorem 4.10 to find all antiderivatives of $-6x$.

8.† Evaluate $\displaystyle\int_0^{\pi/4} \left(\frac{d}{dx}\sin x\right)dx$.

9. Evaluate $\displaystyle\int_0^{100} \frac{d}{dx}(x - \sqrt{x})\,dx$.

In each of Exercises 10 through 20 use the given differentiation formula to evaluate the integral.

10.† $\displaystyle\frac{d}{dx}\sin(x^2) = 2x\cos(x^2); \int_0^9 2x\cos(x^2)\,dx$

11. $\displaystyle\frac{d}{dx}\left(\frac{1}{2}\sin^2 x\right) = (\sin x)(\cos x); \int_1^4 (\sin x)(\cos x)\,dx$

12.[†] $\dfrac{d}{dx}(x - x^2 + x^3) = 1 - 2x + 3x^2$; $\displaystyle\int_0^{10} (1 - 2x + 3x^2)\, dx$

13. $\dfrac{d}{dx}(3x^{4/3}) = 4x^{1/3}$; $\displaystyle\int_{-1}^{8} 4x^{1/3}\, dx$

14.[†] $\dfrac{d}{dx}\sin x = \cos x$; $\displaystyle\int_0^{\pi/3} \cos x\, dx$

15.[†] $\dfrac{d}{dx}(\tfrac{1}{10}x^{10}) = x^9$; $\displaystyle\int_0^1 x^9\, dx$

16.[†] $\dfrac{d}{dt}[-\tfrac{1}{5}\cos(5t)] = \sin(5t)$; $\displaystyle\int_0^{\pi/20} \sin(5t)\, dt$

17.[†] $\dfrac{d}{dx}(3x^{1/3} + 6x^{2/3}) = x^{-2/3} + 4x^{-1/3}$; $\displaystyle\int_1^8 (x^{-2/3} + 4x^{-1/3})\, dx$

18. $\dfrac{d}{dx}(x^2 + 9)^{1/2} = x(x^2 + 9)^{-1/2}$; $\displaystyle\int_0^4 x(x^2 + 9)^{-1/2}\, dx$

19.[†] $\dfrac{d}{dy}[\tfrac{1}{3}\sin(y^3)] = y^2\cos(y^3)$; $\displaystyle\int_0^1 y^2\cos(y^3)\, dy$

20. $\dfrac{d}{dx}[-\sin(x^{-1})] = x^{-2}\sin(x^{-1})$; $\displaystyle\int_{-2}^{-1} x^{-2}\sin(x^{-1})\, dx$

21.[†] What are   **a.** $\dfrac{d}{dx}\displaystyle\int_0^x \sqrt{1 + t^4}\, dt$ and   **b.** $\displaystyle\int_0^x \dfrac{d}{dt}\sqrt{1 + t^4}\, dt$?

22.[†] Use the formula for the area of a rectangle to evaluate the integrals

   a. $\displaystyle\int_0^x 5\, dt$ **b.** $\displaystyle\int_1^x 10\, dt$ **c.** $\displaystyle\int_0^x (-2)\, dt$

23. Derive inequalities (5) for $\Delta x < 0$ with $x + \Delta x > a$.

5-4   INDEFINITE INTEGRALS AND INTEGRALS OF x^n ($n \neq -1$), sin x and cos x

An integral such as

(1) $\displaystyle\int_a^b f(x)\, dx$

which is the type we have studied up to now, is known as a DEFINITE INTEGRAL. This terminology is used because an antiderivative is called an INDEFINITE INTEGRAL:

Definition 5.5   The indefinite integral of a function is an antiderivative of the function. The indefinite integral of $f(x)$ is denoted by the symbol

(2) $\displaystyle\int f(x)\, dx.$

We think of the antiderivative as a type of "integral" because definite integrals can be evaluated by finding antiderivatives.

Definition 5.5 means that

$$(3) \quad \frac{d}{dx} \int f(x) \, dx = f(x)$$

for all x such that the indefinite integral exists.

In this section we use differentiation formulas to derive formulas for the indefinite integrals of x^n $(n \neq -1)$, $\sin x$, and $\cos x$, and we show how indefinite integrals are used in evaluating definite integrals.

Notice that while the definite integral (1) is a single number, the indefinite integral (2) is a variable function of x for nonzero f. It is called "indefinite" because it denotes any of the infinite number of antiderivatives of f. We saw in Theorem 4.10 (Section 4.13) that two functions with the same derivative in an interval differ by a constant in that interval. *Therefore if $F(x)$ is one antiderivative of a function f in an interval, then*

$$(4) \quad \int f(x) \, dx = F(x) + C \qquad \blacktriangleleft$$

in that interval. In equation (4) the "indefiniteness" of the integral is expressed by the fact that the constant C is arbitrary and unspecified. It is called the CONSTANT OF INTEGRATION.

EXAMPLE 1   Use the differentiation formula

$$(5) \quad \frac{d}{dx} x^3 = 3x^2$$

to evaluate the indefinite integral $\int 3x^2 \, dx$.

Solution.   Equation (5) states that x^3 is one antiderivative of $3x^2$. Therefore we have $\int 3x^2 \, dx = x^3 + C$. $\square$

Integrals of x^n $(n \neq -1)$

Theorem 5.3   **For any rational number $n \neq -1$**

$$(6) \quad \int x^n \, dx = \frac{1}{n+1} x^{n+1} + C. \qquad \blacktriangleleft$$

Here x must be nonzero if n is negative and x must be positive if n is a fraction, $n = p/q$, with p an integer and q an even integer.

Proof.   We replace n by $n + 1$ in the differentiation formula $\frac{d}{dx} x^n = nx^{n-1}$ (Theorem 3.2) to obtain

$$(7) \quad \frac{d}{dx} x^{n+1} = (n+1)x^n.$$

If n is not -1, then $n + 1$ is not zero and dividing both sides of (7) by $n + 1$ gives

$$\frac{d}{dx}\left[\frac{1}{n+1}x^{n+1}\right] = x^n.$$

This gives (6) by Definition 5.5 of an indefinite integral.   Q.E.D.

Note.   If we set $n = -1$ in equation (7), we obtain the formula $\frac{d}{dx}x^0 = 0$. This just says that the function $x^0 = 1$ is an antiderivative of the function 0 and does not tell us what the antiderivatives of x^{-1} are. We will see in Chapter 7 that the antiderivatives of x^{-1} are given by logarithms.

EXAMPLE 2   Find the indefinite integral $\int x^6\,dx$.

Solution.   By formula (6) with $n = 6$ and $n + 1 = 7$, we have

(8) $\int x^6\,dx = \frac{1}{7}x^7 + C.$ $\square$

Using indefinite integrals to evaluate definite integrals

Because the indefinite integral $\int f(x)\,dx$ of a continuous function is an antiderivative of that function, we can express the Fundamental Theorem for integrals of derivatives (Theorem 5.2) in the form

(9) $\displaystyle\int_a^b f(x)\,dx = \left[\int f(x)\,dx\right]_a^b$ ◄

EXAMPLE 3   Evaluate the definite integral

$$\int_0^1 x^6\,dx.$$

Solution.   We found the corresponding indefinite integral in Example 2. Because we need only one of the antiderivatives of x^6, we take $C = 0$ in equation (8). We obtain

$$\int_0^1 x^6\,dx = \left[\int x^6\,dx\right]_0^1 = \left[\frac{1}{7}x^7\right]_0^1$$

$$= \left[\frac{1}{7}(1)^7\right] - \left[\frac{1}{7}(0)^7\right] = \frac{1}{7}.\quad \square$$

Integrals of sin x and cos x

Theorem 5.4   The indefinite integrals of cos x and sin x are given for any x by the formulas

(10) $\displaystyle\int \cos x\,dx = \sin x + C$ ◄

(11) $\displaystyle\int \sin x\,dx = -\cos x + C.$ ◄

Proof. The antidifferentiation formulas (10) and (11) are restatements of the differentiation formulas

$$\frac{d}{dx}\sin x = \cos x \quad \text{and} \quad \frac{d}{dx}\cos x = -\sin x$$

which were given in Theorem 3.7.   Q.E.D.

EXAMPLE 4   Evaluate the integral

$$\int_0^{\pi/6} \cos x \, dx.$$

Solution. With formula (10) for the indefinite integral we obtain

$$\int_0^{\pi/6} \cos x \, dx = \left[\int \cos x \, dx \right]_0^{\pi/6} = \left[\sin x \right]_0^{\pi/6}$$

$$= \sin\left(\frac{\pi}{6}\right) - \sin(0) = \frac{1}{2} - 0 = \frac{1}{2}. \quad \square$$

Exercises   In Exercises 1 through 11 evaluate the integrals.

†*Answer provided.*
‡*Outline of solution provided.*

1.† $\int x^{1/3} \, dx$

2. $\int x^{5/4} \, dx$

3.‡ $\int \sqrt{x} \, dx$

4. $\int \frac{1}{x^2} \, dx$

5.† $\int_1^4 x^2 \, dx$

6. $\int_0^1 x^{99} \, dx$

7.‡ $\int_1^9 \frac{1}{\sqrt{x}} \, dx$

8.‡ $\int_{96}^{97} 1 \, dx$

9. $\int_4^5 x \, dx$

10.† $\int_{3\pi/4}^{\pi/4} \sin x \, dx$

11. $\int_5^0 \cos x \, dx$

12.‡ Compute the area of the region bounded by the curve $y = x^2$, the x-axis, and the line $x = 2$.

13.† Compute the area of the portion of the region between the curve $y = \cos x$ and the x-axis for $-\pi/2 \le x \le \pi/2$.

14. Compute the area of the portion of the region between the curve $y = \sin x$ and the x-axis for $\pi/6 \le x \le 4\pi/3$.

15.‡ Compute the area of the region bounded by the curve $y = x^3$, by the x-axis, and by the lines $x = -2$ and $x = -1$.

16. Compute the area of the region bounded by the curve $y = -1 - x^2$, by the x-axis, and by the lines $x = -2$ and $x = -1$.

Additional drill exercises   Evaluate the integrals in Exercises 17 through 24.

17.[†] $\displaystyle\int_1^{10} x^3\,dx$

18.[†] $\displaystyle\int_0^1 x^{8/3}\,dx$

19.[†] $\displaystyle\int_{-10\pi}^{10\pi} \cos x\,dx$

20.[†] $\displaystyle\int_{\pi/4}^{\pi/3} \sin x\,dx$

21.[†] $\displaystyle\int_0^4 (\sqrt{x})^3\,dx$

22.[†] $\displaystyle\int_{-1}^1 x^{200}\,dx$

23.[†] $\displaystyle\int_{-2}^2 x^{201}\,dx$

24.[†] $\displaystyle\int_1^4 \sqrt{\frac{1}{x^5}}\,dx$

5-5   THE ALGEBRA OF INTEGRALS

In this section we discuss a rule for integrals over intervals with common endpoints and a rule for integrals of linear combinations of functions. The second rule is usually used in conjunction with integration formulas such as those derived in the last section.

Integrals over intervals with a common endpoint

Theorem 5.5   If the integrals of $f(x)$ from a to b and from b to c exist, then

(1) $\displaystyle\int_a^b f(x)\,dx + \int_b^c f(x)\,dx = \int_a^c f(x)\,dx.$ ◀

A proof of this theorem based on the analytic definition of the integral was given in the solution to Exercise 31 of Section 5.2. We will give here another proof under the assumption that f is continuous in an open interval containing the points a, b, and c.

Proof for continuous f.   Theorem 5.1 shows that f has antiderivatives in the interval where it is continuous. We let F be one of them. Then $F'(x) = f(x)$ and by Theorem 5.2

$$\int_a^b f(x)\,dx + \int_b^c f(x)\,dx = [F(b) - F(a)] + [F(c) - F(b)]$$

$$= F(c) - F(a) = \int_a^c f(x)\,dx. \quad \text{Q.E.D.}$$

EXAMPLE 1   The integrals of f from 0 to 6 and from 0 to 5 have the values

(2) $\displaystyle\int_0^6 f(x)\,dx = 3$ and $\displaystyle\int_0^5 f(x)\,dx = 4.$

What is the value of the integral of f from 5 to 6?

Solution.   We substitute the given values (2) of the integrals into the equation

$$\int_0^5 f(x)\,dx + \int_5^6 f(x)\,dx = \int_0^6 f(x)\,dx.$$

We obtain

$$4 + \int_5^6 f(x)\,dx = 3.$$

Solving for the integral shows that it has the value -1.   □

Integrals of linear combinations of functions

Integration formulas for specific functions enable us to integrate linear combinations of those functions because of the following result.

Theorem 5.6   **For any constants A and B and any functions f and g with indefinite integrals**

(3) $$\int [Af(x) + Bg(x)]\,dx = A\int f(x)\,dx + B\int g(x)\,dx.$$ ◀

If the integrals of $f(x)$ and $g(x)$ from a to b exist, then for any constants A and B

(4) $$\int_a^b [Af(x) + Bg(x)]\,dx = A\int_a^b f(x)\,dx + B\int_a^b g(x)\,dx.$$ ◀

Proof.   Differentiating the right side of equation (3) gives

(5) $$\frac{d}{dx}[A\int f(x)\,dx + B\int g(x)\,dx]$$

$$= A\frac{d}{dx}\int f(x)\,dx + B\frac{d}{dx}\int g(x)\,dx$$

$$= A f(x) + B g(x).$$

Here we have used the rule (Theorem 3.3) for differentiating linear combinations of functions. Equation (5) is equivalent to equation (3).

Equation (4) for continuous functions $f(x)$ and $g(x)$ follows from (3) and the Fundamental Theorem (Theorem 5.2). In Exercise 32 of Section 5.2 we used the analytic definition of the integral to derive (4) for arbitrary f and g whose integrals exist.   Q.E.D.

EXAMPLE 2   Evaluate the integral $\int (6x - 5\sin x)\,dx$.

Solution.   By formula (3) we have $\int (6x - 5\sin x)\,dx = 6\int x\,dx$
$- 5\int \sin x\,dx = 6(\tfrac{1}{2}x^2) - 5(-\cos x) + C = 3x^2 + 5\cos x + C.$ □

In practice, we apply equations (3) and (4) mentally, as in the solution of the next example, rather than explicitly as in the solution of Example 2.

EXAMPLE 3   Evaluate the integral

$$\int_0^8 (\cos x - 2x^{1/3})\, dx.$$

Solution.   We first find the indefinite integral:

$$\int (\cos x - 2x^{1/3})\, dx = \sin x - 2\left(\frac{1}{4/3}\right)x^{4/3} + C$$

$$= \sin x - \frac{3}{2}x^{4/3} + C.$$

Then we evaluate the definite integral:

$$\int_0^8 [\cos x - 2x^{1/3}]\, dx = \left[\sin x - \frac{3}{2}x^{4/3}\right]_0^8$$

$$= \left[\sin(8) - \frac{3}{2}(8)^{4/3}\right] - \left[\sin(0) - \frac{3}{2}(0)^{4/3}\right]$$

$$= \left[\sin(8) - \frac{3}{2}(2)^4\right] - [0]$$

$$= \sin(8) - 24.\quad \square$$

Exercises

In Exercises 1 through 23 evaluate the integrals.

†*Answer provided.*
‡*Outline of solution provided.*

1.‡ $\displaystyle\int (3x^2 - 6x^{10})\, dx$

2.† $\displaystyle\int (5x^5 - 6x^{-6})\, dx$

3. $\displaystyle\int (\sin x - 5\cos x)\, dx$

4.‡ $\displaystyle\int_0^1 (1 - 6x^7)\, dx$

5.† $\displaystyle\int_0^{\pi/4} (\cos x - 3)\, dx$

6. $\displaystyle\int_1^3 (x^{-2} - 2x)\, dx$

7.‡ $\displaystyle\int_0^1 (7x^{5/7} + 5x^{7/5})\, dx$

8.† $\displaystyle\int_8^0 (4x^{2/3} - 3)\, dx$

9. $\displaystyle\int_1^3 (x^2 + 10x)\, dx$

10.† $\displaystyle\int (3 + 4x - 5x^2)\, dx$

11.† $\displaystyle\int (x^{1/3} + x^{1/4} + x^{1/5})\, dx$

12.‡ $\displaystyle\int (x - \sqrt{x} + \sqrt[3]{x})\, dx$

13.† $\displaystyle\int_{-4}^{-1} \left(\frac{2}{y^3} - \frac{3}{y^4}\right) dy$

14. $\displaystyle\int_1^2 \left(\frac{1}{t^2} - \frac{t^2}{4}\right) dt$

15.‡ $\displaystyle\int_0^1 (1 + 3t^2)^2\, dt$

16. $\displaystyle\int_1^3 \left(1 + \frac{3}{x^2}\right)^2 dx$

17.‡ $\displaystyle\int_2^1 \frac{1 - w}{w^3}\, dw$

18. $\displaystyle\int \frac{x - x^3}{\sqrt{x}}\, dx$

19.† $\displaystyle\int_{-2}^2 (1 + x - x^2 + x^3)\, dx$

20. $\displaystyle\int (u - \sin u + 6)\, du$

21.‡ $\int_{-5}^{5} g(x)\,dx$ where $g(x) = \begin{cases} \sin x & \text{for } x \leq 0 \\ 1 - \cos x & \text{for } x > 0 \end{cases}$

22.† $\int_{0}^{3} h(x)\,dx$ where $h(x) = \begin{cases} 3x^2 & \text{for } 0 \leq x \leq 1 \\ 4x^3 & \text{for } 1 < x < 2 \\ 5x^4 & \text{for } 2 \leq x \leq 3 \end{cases}$

23. $\int_{-10}^{10} |x^3|\,dx$

In each of Exercises 24 through 34 determine whether the integral is positive, negative, or zero without evaluating it. Justify your answers.

24.† $\int_{0}^{20} \dfrac{x}{x+2}\,dx$ **25.†** $\int_{-75}^{0} x^{5/3}\,dx$

26. $\int_{-12}^{15} (5x^4 + 6x^2)\,dx$ **27.†** $\int_{-7}^{7} \dfrac{x}{1+x^2}\,dx$

28.† $\int_{-5}^{0} \dfrac{x}{1+x^2}\,dx$ **29.** $\int_{-47}^{47} \sin t\,dt$

30.† $\int_{4}^{0} \cos^2 x\,dx$ **31.†** $\int_{0}^{-1} (t^{1/3} - 5)\,dt$

32.† $\int_{100}^{-100} x \cos x\,dx$ **33.** $\int_{-3}^{-6} u^{-8/5}\,du$

34. $\int_{14}^{14} (5x^4 - 6x^{-5} + x \sin x - \cos^3 x + 457)\,dx$

35.‡ Sketch the region bounded by the curve $y = \sin x + 3$, the x-axis, and the lines $x = 0$ and $x = 5$. Compute its area.

36.† Sketch the region bounded by the curve $y = x^3 - 1$, the x-axis, and the line $x = -2$. Compute its area.

37. Compute the total area of the three regions bounded by the parabola $y = x^2 - 1$, the x-axis, and the lines $x = -2$ and $x = 2$.

5-6   INTEGRATION BY SUBSTITUTION

With the Fundamental Theorem of Calculus every differentiation formula translates into an integration formula. In this section we discuss the integration formula that is obtained from the chain rule for differentiating composite functions. It leads to a procedure called INTEGRATION BY SUBSTITUTION.

Theorem 5.7   (Integration by substitution)   Suppose that $u = u(x)$ has a continuous derivative for x in an interval I and that $g = g(u)$ is continuous at $u = u(x)$ for all x in I. Then, for x in I

(1) $$\int g(u)\frac{du}{dx}\,dx = \int g(u)\,du. \qquad \blacktriangleleft$$

Proof.   We differentiate the integral on the right of equation (1) with respect to x by using the chain rule. We obtain

$$\frac{d}{dx}\int g(u)\,du = \left[\frac{d}{du}\int g(u)\,du\right]\frac{du}{dx}.$$

Since the derivative with respect to u of the antiderivative of g is g itself, the last equation gives

$$\frac{d}{dx}\int g(u)\,du = g(u)\frac{du}{dx}.$$

This equation is a restatement of equation (1).   Q.E.D.

Theorem 5.7 shows an advantage of the notation that we use for the integral. Equation (1) is easy to remember because the symbols "dx" on the left side of the equation seem to cancel to give the right side of the equation.

In carrying out an integration by substitution, we use a symbolic shorthand and write

(2) $$du = \frac{du}{dx}\,dx. \qquad \blacktriangleleft$$

Here the "differentials" dx and du indicate the variables of integration, whereas the expression du/dx denotes the derivative of the function u with respect to x.

Theorem 5.7 applied to the integration formulas derived in Section 5.4 yields

(3) $$\int u^n\frac{du}{dx}\,dx = \int u^n\,du = \frac{1}{n+1}u^{n+1} + C(n \neq -1) \qquad \blacktriangleleft$$

(4) $$\int \cos u\,\frac{du}{dx}\,dx = \int \cos u\,du = \sin u + C \qquad \blacktriangleleft$$

(5) $$\int \sin u\,\frac{du}{dx}\,dx = \int \sin u\,du = -\cos u + C. \qquad \blacktriangleleft$$

In the simplest applications of integration formulas (3), (4) and (5) we have to choose only the appropriate formula and the appropriate substitution $u = u(x)$. This is the case in the next example.

EXAMPLE 1   Evaluate the integral $\int \sin(x^2)\,2x\,dx$.

Solution.   We use formula (5) with the substitution $u = x^2$. In carrying out the calculations we employ the symbolic equation (2) and write

$$du = \frac{du}{dx}\, dx = \left[\frac{d}{dx}(x^2)\right] dx = 2x\, dx.$$

This gives

$$\int \sin(x^2)\, 2x\, dx = \int \sin u\, du = -\cos u + C$$

$$= -\cos(x^2) + C. \quad \square$$

Adjusting a constant factor in the integrand

Once we have chosen an appropriate substitution and integration formula to perform an integration by substitution, we usually have to adjust a constant factor in the integrand in order to construct the differential $du = \dfrac{du}{dx}\, dx$. This is done through an algebraic step of the form

$$\int f(x)\, dx = \frac{1}{a} \int a f(x)\, dx$$

where we multiply the integrand by a nonzero constant a and compensate by multiplying the entire integral by $1/a$.

EXAMPLE 2   Evaluate the integral

(6) $\displaystyle\int \sqrt{2x + 1}\, dx.$

Solution.   We use formula (3) with $n = \frac{1}{2}, n + 1 = \frac{1}{2} + 1 = \frac{3}{2}$, and $u = 2x + 1$:

$$du = \frac{du}{dx}\, dx = \frac{d}{dx}(2x + 1)\, dx = 2\, dx.$$

We multiply the integrand in (6) by 2 and multiply the integral by $\frac{1}{2}$ to obtain

$$\int \sqrt{2x + 1}\, dx = \int (2x + 1)^{1/2}\, dx = \frac{1}{2} \int (2x + 1)^{1/2}\, 2\, dx$$

(7) $$= \frac{1}{2} \int u^{1/2}\, du = \frac{1}{2} \left[\frac{1}{3/2} u^{3/2} + C_1\right]$$

$$= \frac{1}{3} u^{3/2} + C = \frac{1}{3}(2x + 1)^{3/2} + C. \quad \square$$

We used C_1 as the original constant of integration to reserve the letter C for the arbitrary constant $C = \frac{1}{2} C_1$ in the final form of the answer.

The substitution $u = ax + b$

Often when an integral involves a power of x and a power of the expression $ax + b$ with constants a and b, the substitution $u = ax + b$ yields an integral of a linear combination of powers of u which can be evaluated. This is the case in the next example.

EXAMPLE 3   Evaluate the integral $\int x \sqrt{4x + 8} \, dx$.

Solution.   We set $u = 4x + 8$ for which $x = \frac{1}{4}u - 2$ and $dx = \frac{1}{4} \, du$. With these substitutions we obtain

$$\int x \sqrt{4x + 8} \, dx = \int \left(\frac{1}{4}u - 2\right) u^{1/2} \left(\frac{1}{4} \, du\right)$$

$$= \int \left(\frac{1}{16} u^{3/2} - \frac{1}{2} u^{1/2}\right) du$$

$$= \frac{1}{16}\left(\frac{1}{5/2}\right) u^{5/2} - \frac{1}{2}\left(\frac{1}{3/2}\right) u^{3/2} + C$$

$$= \frac{1}{40} u^{5/2} - \frac{1}{3} u^{3/2} + C$$

$$= \frac{1}{40}(4x + 8)^{5/2} - \frac{1}{3}(4x + 8)^{3/2} + C. \quad \square$$

In the next example the substitution procedure involves a composite function $u(x)$.

EXAMPLE 4   Evaluate $\int_0^4 \sin(5x) \cos^2(5x) \, dx$.

Solution.   We first find the indefinite integral. We make the substitution $u = \cos(5x)$ for which

$$du = \left[\frac{d}{dx} \cos(5x)\right] dx = \left[-\sin(5x) \frac{d}{dx}(5x)\right] dx = -5 \sin(5x) \, dx.$$

To construct du from $\sin(5x) \, dx$, we multiply the integrand by -5 and multiply the integral by the compensating factor $-\frac{1}{5}$:

$$\int \sin(5x) \cos^2(5x) \, dx = -\tfrac{1}{5} \int \cos^2(5x)[-5 \sin(5x) \, dx]$$

$$= -\tfrac{1}{5} \int u^2 \, du = -\tfrac{1}{5}(\tfrac{1}{3}u^3) + C$$

$$= -\tfrac{1}{15}u^3 + C = -\tfrac{1}{15}\cos^3(5x) + C.$$

We can check this much of the work by verifying that the derivative of the indefinite integral is the integrand:

$$\frac{d}{dx}[-\tfrac{1}{15}\cos^3(5x)] = -\tfrac{1}{15}\left[3 \cos^2(5x) \frac{d}{dx} \cos(5x)\right]$$

$$= -\tfrac{1}{5}\cos^2(5x)\left[-\sin(5x) \frac{d}{dx}(5x)\right]$$

$$= \sin(5x) \cos^2(5x).$$

Then, having found the indefinite integral, we see that the definite integral is

$$\int_0^4 \sin(5x)\cos^2(5x)\,dx = [-\tfrac{1}{15}\cos^3(5x)]_0^4$$

$$= [-\tfrac{1}{15}\cos^3(20)] - [-\tfrac{1}{15}\cos^3(0)]$$

$$= \tfrac{1}{15}[1 - \cos^3(20)]. \quad \square$$

Exercises

In Exercises 1 through 32 evaluate the integrals.

†Answer provided.
‡Outline of solution provided.

1.‡ $\displaystyle\int x^2\sqrt{x^3 - 1}\,dx$

2.† $\displaystyle\int \sqrt{5u + 1}\,du$

3. $\displaystyle\int \frac{1}{(3x + 4)^2}\,dx$

4.† $\displaystyle\int \sin(3t + 1)\,dt$

5.‡ $\displaystyle\int \sin^3 x \cos x\,dx$

6. $\displaystyle\int y\cos(y^2)\,dy$

7.‡ $\displaystyle\int_0^1 x\sqrt{x^2 + 9}\,dx$

8.† $\displaystyle\int_1^{10} \sqrt{2x + 1}\,dx$

9. $\displaystyle\int_0^{\pi/6} \cos(2t)\,dt$

10.‡ $\displaystyle\int \frac{t}{\sqrt{t + 10}}\,dt$

11.† $\displaystyle\int \frac{y}{(3y - 4)^3}\,dy$

12. $\displaystyle\int x\sqrt{x - 4}\,dx$

13.‡ $\displaystyle\int \frac{\sin(\sqrt{x})}{\sqrt{x}}\,dx$

14. $\displaystyle\int x^{1/3}\cos(x^{4/3})\,dx$

15.† $\displaystyle\int_0^1 (1 - t)^{12}\,dt$

16. $\displaystyle\int_0^4 (2x + 1)^{-1/2}\,dx$

17.‡ $\displaystyle\int \sin(2x)\cos(2x)\,dx$

18.† $\displaystyle\int \frac{\cos(3x)}{\sin^2(3x)}\,dx$

19. $\displaystyle\int_5^{10} \sin(5x)\sqrt{\cos(5x)}\,dx$

20.† $\displaystyle\int (4x + 1)^{1/3}\,dx$

21.† $\displaystyle\int \cos(3 - x)\,dx$

22.† $\displaystyle\int \sin(4x)\sqrt[3]{\cos(4x)}\,dx$

23.† $\displaystyle\int \frac{x}{\sqrt{1 - x}}\,dx$

24.† $\displaystyle\int (x + 1)(x - 1)^7\,dx$

25.† $\displaystyle\int (x + 1)(x^2 + 2x + 5)^{7/3}\,dx$

26.† $\displaystyle\int x^{-3}\sin(x^{-2})\,dx$

27.† $\displaystyle\int [\cos(4) - \cos(2)]\,dx$

28. $\displaystyle\int_0^2 \frac{\sin x}{(1 + \cos x)^4}\,dx$

29. $\displaystyle\int_0^1 x(x - 1)^{2/3}\,dx$

30.† $\displaystyle\int_0^1 x^{1/2}(1 - x^{3/2})^5\,dx$

31. $\displaystyle\int_1^2 \frac{x^2 + x - 1}{x^4}\,dx$

32.† $\displaystyle\int \frac{x^2 + 2x + 1}{(x + 1)^2}\,dx$

5-7   HISTORICAL NOTES (Optional)

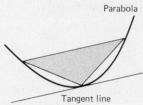

Parabola

Tangent line

FIGURE 5.27
$$\text{The area of the region bounded by the parabola} = \frac{4}{3}\left[\begin{array}{c}\text{The area of the triangle}\end{array}\right]$$

Archimedes (c. 287–212 B.C.) was the first to determine the exact area of a region with a curved boundary by an approximation procedure. He demonstrated, for example, that the area of a region bounded by a parabola and by a straight line as in Figure 5.27 has area equal to four thirds the area of the inscribed triangle whose vertex is at the point where the tangent line is parallel to the top of the region. (The Greeks did not think of areas as numbers; they merely compared areas of related regions.) In Miscellaneous Exercise 25 of Chapter 15 and Miscellaneous Exercise 37 of Chapter 18, we describe two methods that Archimedes used to obtain this result.

With modern concepts of area and integral, Archimedes' result leads to the integration formula

$$(1)\qquad \int_a^b x^2\,dx = \frac{1}{3}b^3 - \frac{1}{3}a^3$$

since the curve $y = x^2$ is a parabola (see Exercise 1). The problem of determining the areas of regions bounded by the curves $y = x^n$ for general rational constants n did not arise until the study of analytic geometry in the seventeenth century led to the consideration of these curves. Results amounting to the integration formula

$$\int_a^b x^n\,dx = \frac{1}{n+1}b^{n+1} - \frac{1}{n+1}a^{n+1}\ (0 < a < b, n \neq -1)$$

for particular positive integers n appeared in various forms during the years 1635–1660 in the work of two Italian students of Galileo, Bonaventura Cavalieri (1598–1647) and Evangelista Torricelli (1608–1647), and of two Frenchmen, Gilles Persone de Roberval (1602–1675) and the philosopher Blaise Pascal (1623–1662). The French lawyer Pierre Fermat (1601–1665) is usually given credit for establishing the integration formula (1) for arbitrary rational numbers $n \neq -1$. One of his procedures for doing this is described in Miscellaneous Exercise 38 of Chapter 18.

Pascal and others also derived results which, with modern definitions and notation, would read

$$\int_a^b \cos x\,dx = \sin(b) - \sin(a)$$

$$\int_a^b \sin x\,dx = -\cos(b) + \cos(a).$$

FIGURE 5.29   Blaise Pascal
(1623–1662)
Culver Pictures, Inc.

Indivisibles   Most of the early derivations of integration formulas employed the concept of *indivisibles*. In the context of areas, indivisibles are "infinitely narrow" rectangles of "infinitesimal" area. They are called

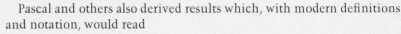

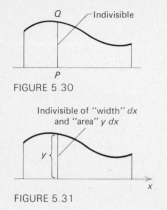

FIGURE 5.30

FIGURE 5.31

"indivisibles" because they are assumed to be so small that they cannot be divided into narrower "rectangles." A region as in Figure 5.30 was considered to consist of an infinite number of parallel indivisibles such as represented by the line PQ in the sketch. The length of each indivisible was considered to be its "area" and the area of the region was considered to be the "sum" of the "areas" of the infinite number of indivisibles in it.

Leibniz used the symbol dx to denote the "width" of an indivisible, so that the "area" of an indivisible of length y was given by the product $y \, dx$ (Figure 5.31). He then introduced the symbol $\int y \, dx$ for the "sum" or "integral" of the areas of the indivisibles which gives the area of the region.

Criticism of the use of indivisibles

The use of indivisibles in integration, like the use of infinitesimals in early treatments of the derivative, has always been subject to criticism. The French writer Voltaire, for example, once complained that calculus was "the Art of numbering and measuring exactly a Thing whose Existence cannot be conceived".* Yet even those who argued that there were logical difficulties with the use of infinitesimals and indivisibles could not deny the importance of the mathematical tools to which these concepts led. By the second half of the seventeenth century, indivisibles had been used extensively in determining the areas, volumes, and centers of gravity of a large number of geometric figures.

The mathematicians who used indivisibles to make these calculations were schooled in the classical Greek mathematics of Euclid and Archimedes. At first, they supplied rigorous proofs that were based on approximation procedures to any formulas they derived using indivisibles. But, eventually, as more results were obtained, many felt that supplying the rigorous proof to meet the standards of classical mathematics was a waste of time and they omitted this final step.

The German astronomer Johann Kepler (1571–1630) took this point of view in his popular book *Nova Stereometria Doliorum Vinariorum* (*New Solid Geometry of Wine Barrels*), in which he computed volumes of a large number of solids of revolution. Referring to the absence of classical proofs in the book, he wrote in the preface, "We could obtain absolute and in all respects perfect demonstrations from the books of Archimedes themselves, were we not repelled by the thorny reading thereof."**

The Fundamental Theorem

Special cases of the Fundamental Theorem of Calculus were discovered long before its full significance and generality were recognized. The fourteenth century French bishop Nicole Oresme (ca. 1323–1382) recognized in special cases that if a time interval were represented on a

---

*Boyer, [3] *History of the Calculus*, p. 238.
**Struik, [1] *A Source Book of Mathematics*, p. 188.

horizontal line and the corresponding variable velocity by vertical line segments along it, then the distance traveled would be equal to the resulting area. Galileo used the same idea in his study of falling bodies. Fermat used a crude version of the derivative to solve problems that we solve today with integrals, and he solved certain tangent line problems by using areas.

It was the English theologian Isaac Barrow (1630–1677), however, who was the first to clearly recognize that the finding of tangent lines and the computing of areas are inverse operations. This insight contributed to the work of his protégé Newton, and perhaps also to that of Leibniz, who purchased a copy of Barrow's *Lectures* on a trip to London in 1673. The Fundamental Theorem played a central role in the rules and procedures of calculus as developed, independently, by Newton and Leibniz.

FIGURE 5.32   Johann Kepler
(1571—1630)
Culver Pictures, Inc.

FIGURE 5.33   Isaac Barrow
(1630—1677)
Culver Pictures, Inc.

FIGURE 5.34   G. F. B. Riemann
(1826—1866)
Culver Pictures, Inc.

The analytic definition of the integral

The point of view that integrals are given either as antiderivatives or geometrically through the use of indivisibles prevailed until the early nineteenth century. The definition of the integral of a continuous function as a limit of finite sums was given by Augustin Cauchy (1789–1857) in the 1820's. Thirty years later G. F. B. Riemann (1826–1866) realized that Cauchy's definition could be extended to apply to certain discontinuous functions. He gave the analytic definition of the integral that we use today.

Exercises

1. Derive formula (1) for $a = 0$ from Archimedes' result which is stated in Figure 5.27. Consider the rectangle $-b \leq x \leq b, 0 \leq y \leq b^2$, and the triangle with vertices $(0, 0)$, $(-b, b^2)$ and (b, b^2). Then derive formula (1) for $0 < a < b$ from the result for $a = 0$.

APPENDIX TO CHAPTER 5. THE EXISTENCE OF DEFINITE INTEGRALS (Optional)

Recall that a Riemann sum for the integral

(1) $$\int_a^b f(x)\,dx$$

$(a < b)$ is a sum of the form

$$\sum_{j-1}^N f(c_j)(x_j - x_{j-1})$$

where

$$a = x_0 < x_1 < x_2 < x_3 < \cdots < x_{N-1} < x_N = b$$

is a partition of the interval of integration and c_j is in the jth subinterval ($x_{j-1} \le c_j \le x_j$) for each j. In Section 5.2 we defined the integral to be the limit of the Riemann sums. The $\epsilon\delta$-definition of this type of limit reads as follows.

Definition 5.6   The number L is the limit of the Riemann sums as the number of subintervals in the partition tend to infinity and the widths of the subintervals tend to zero if for each $\epsilon > 0$ there is a $\delta > 0$ such that

$$|[\textbf{The Riemann sum}] - L| \le \epsilon$$

for all partitions into subintervals of lengths $\le \delta$.

In this section we will see that the limit exists and the integral is defined provided f is bounded in the interval of integration and is either PIECEWISE MONOTONE or PIECEWISE CONTINUOUS. (Recall that f is bounded if there is a constant M such that $|f(x)| \le M$ for $a \le x \le b$.) The bounded functions normally encountered in calculus are both piecewise monotone and piecewise continuous.

Monotone and piecewise monotone functions   **Definition 5.7   A function f is nondecreasing in an interval if $f(x_1) \le f(x_2)$ for all x_1 and x_2 in the interval with $x_1 < x_2$ (Figure 5.35). The function is nonincreasing in the interval if $f(x_1) \ge f(x_2)$ for all such x_1 and x_2 (Figure 5.36). A function is monotone in an interval if it is either nondecreasing or nonincreasing there. A function is piecewise monotone in an interval if there is a partition of the interval such that the function is monotone in the interior of each subinterval (Figure 5.37).**

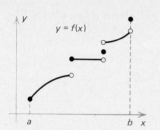

FIGURE 5.35   The graph of a nondecreasing function

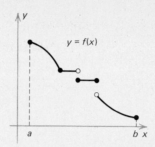

FIGURE 5.36   The graph of a nonincreasing function

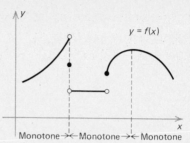

Monotone→←Monotone→←Monotone

FIGURE 5.37   The graph of a piecewise monotone function

All functions that normally arise in calculus are piecewise monotone.

Theorem 5.8   Suppose that the function f is bounded and piecewise monotone in the interval $a \leq x \leq b$. Then the integral

(1) $$\int_a^b f(x)\, dx$$

exists.

Proof.   The existence of the integral over the entire interval follows from its existence over the subintervals where the function is monotone (see Exercise 1). Accordingly, we will deal with the case of $f(x)$ bounded and nondecreasing for $a < x < b$. The case of nonincreasing $f(x)$ is similar.

Because $f(x)$ is bounded, we can redefine it at $x = a$ and $x = b$ if necessary so that it is nondecreasing for $a \leq x \leq b$. Changing its value at the two points does not affect the existence of the integral (see Exercise 2).

For each partition

(2) $a = x_0 < x_1 < x_2 < x_3 < \cdots < x_{N-1} < x_N = b$

of the interval, we obtain the greatest Riemann sum (THE UPPER SUM) by evaluating f at the right endpoint x_j where it has its greatest value in the jth subinterval. We obtain the least Riemann sum (THE LOWER SUM) by evaluating f at the left endpoint x_{j-1} of the jth subinterval:

(3)   The upper sum $= \displaystyle\sum_{j=1}^{N} f(x_j)(x_j - x_{j-1})$

(4)   The lower sum $= \displaystyle\sum_{j=1}^{N} f(x_{j-1})(x_j - x_{j-1}).$

All other Riemann sums corresponding to the partition (2) lie between the upper and lower sums.

We let $\delta > 0$ denote the greatest of the lengths $x_j - x_{j-1}$ of the subintervals of the partition. Then

$$\begin{bmatrix} \text{The upper} \\ \text{sum} \end{bmatrix} - \begin{bmatrix} \text{The lower} \\ \text{sum} \end{bmatrix} = \sum_{j=1}^{N} [f(x_j) - f(x_{j-1})](x_j - x_{j-1})$$

$$(5) \qquad\qquad \leq \sum_{j=1}^{N} [f(x_j) - f(x_{j-1})]\, \delta$$

$$= \delta \sum_{j=1}^{N} [f(x_j) - f(x_{j-1})].$$

We obtain the inequality because the numbers $f(x_j) - f(x_{j-1})$ are all ≥ 0 while the positive numbers $x_j - x_{j-1}$ are all $\leq \delta$. The sum of the changes $f(x_j) - f(x_{j-1})$ in the values of f across the subintervals is equal to the change $f(x_N) - f(x_0) = f(b) - f(a)$ across the entire interval, so (5) yields

$$(6) \qquad \begin{bmatrix} \text{The upper} \\ \text{sum} \end{bmatrix} - \begin{bmatrix} \text{The lower} \\ \text{sum} \end{bmatrix} \leq \delta\,[f(b) - f(a)].$$

The lower sum for one partition is no greater than the upper sum for any other partition (see Exercise 3). Therefore if we imagine the Riemann sums to be points on an S-axis as in Figure 5.38, then the lower sums all lie to the left of the upper sums. Inequality (6) shows that the upper sum and lower sum can be made arbitrarily close together by choosing δ sufficiently small. Consequently, there can be no "gap" between the upper and lower sums: there is a single number which is greater than or equal to all the lower sums and less than or equal to all the upper sums.* We define the integral to be that number (Figure 5.38).

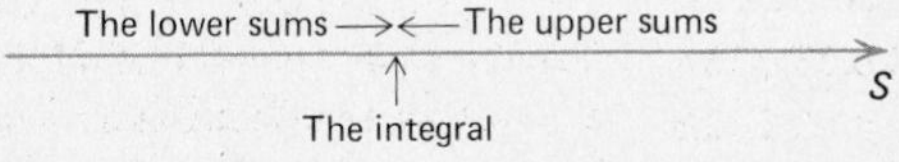

The lower sums —>←— The upper sums

FIGURE 5.38   The integral S

The distance between any Riemann sum and the integral is no greater than the distance between the upper and lower sums for the partition because both the Riemann sum and the integral lie between the upper and lower sums. Inequality (6) therefore shows that

$$(7) \qquad \left\| \begin{bmatrix} \text{The} \\ \text{integral} \end{bmatrix} - \begin{bmatrix} \text{Any Riemann} \\ \text{sum} \end{bmatrix} \right\| \leq \delta\,[f(b) - f(a)]$$

where $\delta > 0$ is the length of the longest subinterval in the partition.

Given any positive number ϵ, we set $\delta = \epsilon/[f(b) - f(a)]$. Then inequality (7) shows that Definition 5.6 is satisfied and the integral is the limit of the Riemann sums.   Q.E.D.

---

*The integral is the least upper bound of the lower sums and the greatest lower bound of the upper sums (see Axiom 18.1 in Section 18.2).

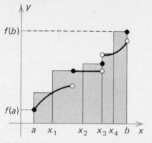

FIGURE 5.39   Area = The upper sum

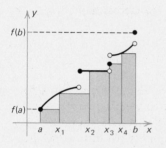

FIGURE 5.40   Area = The lower sum

A geometric interpretation of inequality (6) is indicated in Figures 5.39 through 5.41 for the positive, nondecreasing function of Figure 5.35 and for a particular partition of the interval $a \leq x \leq b$ into five subintervals. The upper sum is the area of the five rectangles in Figure 5.39 and the lower sum is the area of the five rectangles in Figure 5.40. The difference between the upper and lower sums is equal to the area of the five rectangles on the left of Figure 5.41. These rectangles can be stacked up as on the right of Figure 5.41 so that they fit inside a rectangle of height $f(b) - f(a)$ and width equal to the length δ of the largest subinterval of the partition. Inequality (6) states that the area of the five rectangles is no greater than the area of the rectangle containing them.

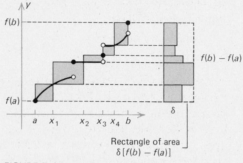

FIGURE 5.41

Area = $\begin{bmatrix} \text{The upper} \\ \text{sum} \end{bmatrix} - \begin{bmatrix} \text{The lower} \\ \text{sum} \end{bmatrix}$

Piecewise continuous functions

Definition 5.8   A function f is piecewise continuous in an interval $a \leq x \leq b$ if there is a partition $a = x_0 < x_1 < x_2 < \cdots < x_N = b$ of the interval such that f is continuous in the interior $x_{j-1} < x < x_j$ of each interval, has finite one-sided limits at x_j for $j = 1, 2, \ldots, N - 1$, and has finite limits from the right at a and from the left at b.

All bounded functions normally encountered in calculus are piecewise continuous as well as piecewise monotone. To show that piecewise continuous functions have integrals, we will use the following result from advanced calculus (see R. Courant, [1] *Differential and Integral Calculus*).

Proposition 5.2   If a function f is continuous in a finite closed interval $a \leq x \leq b$, then it is UNIFORMLY CONTINUOUS there, that is, for each $\epsilon > 0$ there is a $\delta > 0$ such that

$$|f(c) - f(d)| \leq \epsilon$$

for all c and d in the interval such that $|c - d| \leq \delta$.

Theorem 5.9   If $f(x)$ is piecewise continuous in the interval $a \leq x \leq b$, then the integral (1) of f from a to b exists.

Proof.   The proof is very similar to that of Theorem 5.8. Exercise 1 shows that if the integral exists over each of the subintervals of a partition, then it exists over the whole interval, so we only have to deal with the case where f is continuous for $a \le x \le b$ and has finite limits from the right at a and from the left at b. Then Exercise 2 shows we can redefine f at a and at b if necessary so that $f(a) = \lim\limits_{x \to a^+} f(x)$ and $f(b) = \lim\limits_{x \to b^-} f(x)$ and f is continuous in the closed interval $a \le x \le b$.

By the Extreme Value Theorem f is bounded and for each partition (2) there are numbers c_j and d_j in the jth subinterval such that $f(c_j)$ is the maximum value and $f(d_j)$ the minimum value of f in the subinterval. We use the values in $f(c_j)$ in a Riemann sum to obtain the upper sum and the values $f(d_j)$ for the lower sum, and in place of equation (5) we have

$$(8) \quad \begin{bmatrix} \text{The upper} \\ \text{sum} \end{bmatrix} - \begin{bmatrix} \text{The lower} \\ \text{sum} \end{bmatrix} = \sum_{j=1}^{N} [f(c_j) - f(d_j)](x_j - x_{j-1}).$$

Given $\epsilon > 0$, we use Proposition 5.2 to obtain a $\delta > 0$ such that

$$(9) \quad |f(c) - f(d)| \le \frac{\epsilon}{b - a}$$

for all c and d in the interval $a \le x \le b$ such that $|c - d| \le \delta$. For any partition all of whose subintervals are of length $\le \delta$, we have $|c_j - d_j| \le \delta$ and hence by (8) and (9)

$$\begin{bmatrix} \text{The upper} \\ \text{sum} \end{bmatrix} - \begin{bmatrix} \text{The lower} \\ \text{sum} \end{bmatrix} \le \frac{\epsilon}{b - a} \sum_{j=1}^{N} (x_j - x_{j-1}) = \epsilon.$$

Then, as in the proof of Theorem 5.8, we can define the integral to be the one number $\ge$ all the lower sums and $\le$ all the upper sums, and the integral is the limit of the Riemann sums.   Q.E.D.

Unbounded functions

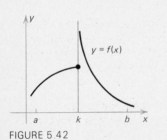

FIGURE 5.42

If a function is unbounded in the interval of integration, its integral cannot be defined directly as the limit of Riemann sums. Suppose, for example, that $f(x)$ tends to ∞ as x tends to k from the right and that $a \le k < b$ so that k is in the interval of integration (Figure 5.42). Then each partition contains a subinterval with $x_{j-1} \le k < x_j$, and the contribution $f(c_j)(x_j - x_{j-1})$ to the Riemann sum from that subinterval can be made arbitrarily large by taking c_j to the right of and sufficiently close to k (Figure 5.43). Because the Riemann sums for each partition are arbitrarily large, they cannot have a finite limit.

As we will see in Chapter 17, certain integrals of unbounded functions may be defined as *improper integrals*.

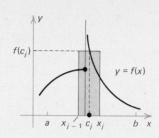

FIGURE 5.43   Rectangle of area
$f(c_j)(x_j - x_{j-1})$

Bounded functions without integrals

To define a function which is bounded and yet does not have an integral we set

$$f(x) = \begin{cases} 2 & \text{if } x \text{ is a rational number} \\ 1 & \text{if } x \text{ is an irrational number} \end{cases}$$

(Figure 5.44). Because there are rational and irrational points in each subinterval, every upper sum for the integral

$$(10) \qquad \int_0^1 f(x)\, dx$$

is equal to 2 and every lower sum is equal to 1 (Figures 5.45 and 5.46). Therefore, the Riemann sums do not have a limit and the integral (10) does not exist.

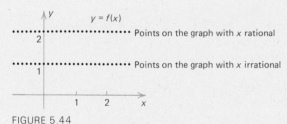

FIGURE 5.44

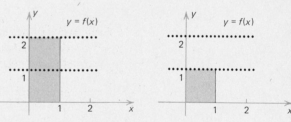

FIGURE 5.45   Upper sum = 2   FIGURE 5.46   Lower sum = 1

Exercises

†Answer provided.
‡Outline of solution provided.

1.† Prove that if the integral of $f(x)$ from a to b and from b to c exist $(a < b < c)$, then the integral from a to c exists and is equal to the sum of the other two integrals.

2. Prove that changing the value of a function at one point in the interval of integration does not affect the existence of the integral nor change its value.

3. **a.†** Prove that adding a point to a partition yields a new upper sum which is ≤ the original upper sum.   **b.** Prove that adding a point to a partition yields a new lower sum which is ≥ the original lower sum.   **c.†** Prove that the lower sum for any one partition is ≤ the upper sum for any other partition.

4. Use the Mean Value Theorem for derivatives to show that $f(x)$ is uniformly continuous in the interval $a \le x \le b$ if it is continuous there and has a bounded derivative for $a < x < b$.

MISCELLANEOUS EXERCISES

†*Answer provided.*
‡*Outline of solution provided.*

1.† Sketch the graph of the function $6 - 2x$ and evaluate its integral from -1 to 4 by using the geometric definition of the integral.

2. Use the geometric definition of the integral to evaluate the integral of the function x from -3 to 5.

3.† Use the geometric definition of the integral to show that

$$\int_a^b [-f(x)]\, dx = -\int_a^b f(x)\, dx$$

for $a < b$ and for any positive continuous function f.

4. Use the geometric definition of the integral to show that

$$\int_a^b [f(x) + 1]\, dx = \int_a^b f(x)\, dx + (b - a)$$

for $a < b$ and for any positive continuous function f.

5. **a.** Use the geometric definition of the integral to explain why

$$\left| \int_a^b f(x)\, dx \right| \le \int_a^b |f(x)|\, dx$$

for $a < b$ and any continuous function f.   **b.** What property must a continuous f have for the two sides of this inequality to be equal?

In Exercises 6 and 7 write out the sums. Do not do the final addition.

6.† $\displaystyle\sum_{j=0}^{4} \frac{2^j}{2j + 1}$ **7.** $\displaystyle\sum_{j=1}^{7} \frac{(-1)^j}{j^2}$

In Exercises 8 through 13 express the given sums in summation notation. You may need to use the symbols j, $2j$, $2j - 1$, $2j + 1$, $(-1)^j$, $(-1)^{j+1}$, or j^2.

8.‡ $1 - \frac{1}{10} + (\frac{1}{10})^2 - (\frac{1}{10})^3 + (\frac{1}{10})^4 - (\frac{1}{10})^5 + (\frac{1}{10})^6$

9.‡ $10 - \frac{1}{2}(10)^3 + \frac{1}{3}(10)^5 - \frac{1}{4}(10)^7 + \frac{1}{5}(10)^9 - \frac{1}{6}(10)^{11}$

10.† $1 - \frac{1}{4} + \frac{1}{9} - \frac{1}{16} + \frac{1}{25} - \frac{1}{36} + \frac{1}{49} - \frac{1}{64}$

11.† $\dfrac{1}{3} + \dfrac{\sqrt{3}}{5} + \dfrac{\sqrt{5}}{7} + \dfrac{\sqrt{7}}{9} + \dfrac{\sqrt{9}}{11} + \dfrac{\sqrt{11}}{13}$

12. $1 - 2^2 + 3^3 - 4^4 + 5^5 - 6^6 + 7^7 - 8^8 + 9^9$

13. $10^{-1} + 3(10)^{-4} + 5(10)^{-9} + 7(10)^{-16} + 9(10)^{-25}$

In Exercises 14 through 17 give the Riemann sums for the given integrals corresponding to the given partitions and with the integrands evaluated at the left endpoints of the subintervals.

14.† $\displaystyle\int_{-3}^{5} x^4\, dx;\ -3 = x_0 < x_1 < x_2 < \cdots < x_N = 5$

15. $\int_0^2 \cos^2 t \, dt; \; 0 = t_0 < t_1 < t_2 < \cdots < t_N = 2$

16.† $\int_0^3 \dfrac{x}{1+x} \, dx; \; 0 = \dfrac{0}{50} < \dfrac{1}{50} < \dfrac{2}{50} < \cdots < \dfrac{149}{50} < \dfrac{150}{50} = 3$

17. $\int_0^1 \cos y \, dy; \; 0 = \dfrac{0}{700} < \dfrac{1}{700} < \dfrac{2}{700} < \cdots < \dfrac{699}{700} < \dfrac{700}{700} = 1$

In Exercises 18 and 19 give the Riemann sums for the given integrals corresponding to the given partitions and with the integrands evaluated at the midpoints of the subintervals.

18.† $\int_9^{10} (x^5 - x) \, dx; \; 9 = x_0 < x_1 < x_2 < \cdots < x_N = 10$

19. $\int_{-1000}^{1000} x \sin x \, dx; \; -1000 = x_0 < x_1 < x_2 < \cdots < x_N = 1000$

In Exercises 20 and 21 give the integrals for which the sums are Riemann sums corresponding to the given partition. In each case the number c_j is an unspecified point in the jth subinterval of the partition.

20.† $\displaystyle\sum_{j=1}^N \dfrac{c_j^2}{2c_j + 4} (x_j - x_{j-1}); \; 1 = x_0 < x_1 < x_2 < \cdots < x_N = 7$

21. $\displaystyle\sum_{j=1}^N \dfrac{\sin(c_j)}{\cos(c_j)}(x_j - x_{j-1}); \; \dfrac{\pi}{6} = x_0 < x_1 < x_2 < \cdots < x_N = \dfrac{\pi}{3}$

22.† Use the analytic definition of the integral to show that

$$\int_0^a f(x) \, dx = \int_{-a}^0 f(x) \, dx$$

for any positive constant a and any even function f for which the integral exists. (Recall that f is even if $f(-x) = f(x)$ for all x.)

23. Use the analytic definition of the integral to show that

$$\int_0^a f(x) \, dx = -\int_{-a}^0 f(x) \, dx$$

for any positive constant a and any odd function f for which the integral exists. (Recall that f is odd if $f(-x) = -f(x)$ for all x.)

24. In Chapter 7 we will define the "natural logarithm" $\ln x$ $(x > 0)$ by the formula

$$\ln x = \int_1^x \dfrac{1}{t} \, dt.$$

a. Use the analytic definition of the integral to show that for $x > 0$ and $a > 0$

$$\int_a^{ax} \dfrac{1}{t} \, dt = \int_1^x \dfrac{1}{t} \, dt.$$

b. Use the result of part (a) to show that $\ln(ax) = \ln a + \ln x$.

In Exercises 25 through 32 use the chain rule and the Fundamental Theorem of Calculus to evaluate the derivatives without performing any integrations.

25.‡ $\dfrac{d}{dx} \displaystyle\int_{1}^{x^2} \sqrt{t+1}\,dt$ **26.**† $\dfrac{d}{dx} \displaystyle\int_{1}^{3x} \sin^5 t\,dt \;(x > 0)$

27. $\dfrac{d}{dx} \displaystyle\int_{1}^{\sin x} \dfrac{1}{t}\,dt \;(0 < x < \pi)$ **28.**‡ $\dfrac{d}{dx} \displaystyle\int_{x^3}^{10} \sqrt[3]{\cos t}\,dt$

29. $\dfrac{d}{dx} \displaystyle\int_{\cos x}^{100} \dfrac{1}{t^2+1}\,dt$ **30.**‡ $\dfrac{d}{dx} \displaystyle\int_{x^2}^{x^3} (1-t^2)^{1/3}\,dt$

31.† $\dfrac{d}{dx} \displaystyle\int_{4x}^{5x} t\sin t\,dt$ **32.** $\dfrac{d}{dx} \displaystyle\int_{\sin x}^{\cos x} \sqrt{t^2+1}\,dt$

33. Use Proposition 5.1c in Section 5.2 to show that

$$\tfrac{1}{3} \le \int_{0}^{2} \dfrac{1}{\sqrt{5x^2+16}}\,dx \le 1.$$

34. Use Proposition 5.1b in Section 5.2 to show that

$$0 \le \int_{0}^{1} x\sin^2 dx \le \tfrac{1}{2}.$$

In each of Exercises 35 through 47 evaluate the given integral if it is defined. If it is not defined, explain why.

35.† $\displaystyle\int_{1}^{2} \dfrac{6}{x^2}\,dx$ **36.**† $\displaystyle\int_{-2}^{-1} x^{-3}\,dx$

37.† $\displaystyle\int_{-1}^{1} \left(\dfrac{1}{4}x^4 + 4x^{1/4}\right)dx$ **38.**† $\displaystyle\int_{-1}^{1} \left(\dfrac{1}{3}x^3 + 3x^{1/3}\right)dx$

39. $\displaystyle\int_{-9}^{9} (y - \sqrt{y})\,dy$ **40.** $\displaystyle\int_{-1}^{1} (x^2 - \sqrt[7]{x})\,dx$

41. $\displaystyle\int_{0}^{8} z^{5/3}\,dz$ **42.** $\displaystyle\int_{-5}^{5} w^{-2}\,dw$

43.† $\displaystyle\int_{0}^{1} \dfrac{x}{\sqrt{x^2+1}}\,dx$ **44.** $\displaystyle\int_{0}^{1} \dfrac{x}{\sqrt{x+1}}\,dx$

45.† $\displaystyle\int_{3}^{-4} y(y+5)^{2/3}\,dy$ **46.**† $\displaystyle\int_{0}^{\pi/2} \dfrac{\sin x}{\sqrt{\cos x}}\,dx$

47. $\displaystyle\int_{0}^{\pi/4} \dfrac{\sin x}{\sqrt{\cos x}}\,dx$

In Exercises 48 through 54 evaluate the integrals.

48.‡ $\displaystyle\int (x^2 + x + 1)^2\,dx$ **49.**† $\displaystyle\int (x^2 + 1)^3\,dx$

50. $\displaystyle\int \left(x - \dfrac{1}{x}\right)^2 dx$ **51.**† $\displaystyle\int (\sqrt{x} + 1)^2\,dx$

52. $\displaystyle\int \frac{1}{x^2}\cos\left(\frac{1}{x}\right) dx$ **53.**[†] $\displaystyle\int (\cos x)\big(\cos(\sin x)\big) dx$

54. $\displaystyle\int \sin(5x) \cos^2(5x)\, dx$

55.[†] Find the area of the region bounded by the curve $y = x/\sqrt{x^2 + 4}$, by the x-axis, and by the line $x = 2$.

56.[†] Find the total area of the two regions bounded by the curve $y = x^{1/3}$, by the x-axis, and by the lines $x = -8$ and $x = 1$.

57. Find the area of the region bounded by the curve $y = 1/\sqrt{x + 16}$, by the x-axis, and by the lines $x = 0$ and $x = 20$.

58.[†] Derive the summation formula

$$(1) \qquad 1 + 2 + 3 + \cdots + N = \frac{N(N + 1)}{2} \qquad (N \text{ a positive integer}).$$

(Write out the sum twice, once forward and once backward, and add corresponding terms.)

59. The width of the square in Figure 5.47 is given by expression (1) and is divided into N pieces (N an arbitrary positive integer). The first is a square of width 1. The jth piece for $2 \le j \le N$ is an L-shaped region of width j that lies along the sides of the $(j - 1)$st piece. Show that the jth piece has area j^3. Use this fact and formula (1) to derive the summation formula*

$$(2) \qquad 1^3 + 2^3 + 3^3 + \cdots + N^3 = \left[\frac{N(N + 1)}{2}\right]^2.$$

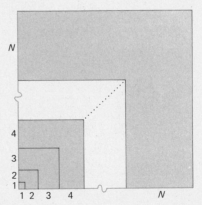

FIGURE 5.47

Exercises 60 through 64 deal with Riemann sums obtained from partitions into N equal subintervals with the functions evaluated at the right endpoints of the subintervals.   **a.** Use the reference tables to give an approximate decimal value of the Riemann sum with $N = 4$. If you have access to a computer or electronic calculator, use it to compute approximate decimal values of the Riemann sum for   **b.** $N = 10$ and   **c.** $N = 20$.   **d.** Evaluate the integral and give its approximate decimal value.

---

*This derivation was known to the Arabs in the eleventh century (see F. Woepcke, [1] *Extrait du Fakhrî par Alkarkhî*, Paris (1853), l'Imprimerie Impériale, p. 61).

60.‡ $\int_0^{10} x^2 \, dx$ **61.†** $\int_0^1 \sin x \, dx$ **62.†** $\int_0^4 \sqrt{x} \, dx$ **63.** $\int_1^2 x^{-2} \, dx$

64. $\int_0^2 x\sqrt{x^2 + 1} \, dx$

65. Compute the integral of $\sqrt{x}$ from 0 to 1 by using thin horizontal rectangles in the approximations, as shown in Figure 5.48.

66. The inequalities

(3) $$\frac{1}{n+1}N^{n+1} < 1^n + 2^n + 3^n + \cdots + N^n < \frac{1}{n+1}(N+1)^{n+1}$$

are valid for all positive integers, n and N. Use them and the analytic definition of the integral to compute the integral of x^n from 0 to 1.

67.† In Exercise 66 we used inequalities (3) to evaluate an integral of x^n. Use the formula for integrals of x^n to derive (3).

68. Use the integral $\int_0^1 \sqrt{x} \, dx$ and a Riemann sum for that integral to obtain the approximation $\frac{2}{3} N^{3/2}$ for $\sum_{j=1}^{N} \sqrt{j}$. Is the approximation greater or less than the exact value of the sum?

69.   a. Use a Riemann sum for the integral $\int_0^1 x^{1/3} \, dx$ to give an approximate value of $\sum_{j=1}^{N} j^{1/3}$ **b.** What is that approximation for $N = 1000$?

70.† Use an integral to compute the limit $\lim_{N \to \infty} N^{-3/2} (\sqrt{1} + \sqrt{2} + \sqrt{3} + \cdots + \sqrt{N})$.

71. Use an integral to compute the limit $\lim_{N \to \infty} N^{-4/3} (\sqrt[3]{1} + \sqrt[3]{2} + \sqrt[3]{3} + \cdots + \sqrt[3]{8N})$.

72.† Evaluate the integral of $f(x)$ from 0 to 3 where

$$f(x) = \begin{cases} x^2 & \text{for} & x < 1 \\ 6 & \text{for } 1 \le x \le 2 \\ 2 - 4x^3 & \text{for} & x > 2. \end{cases}$$

73. Evaluate $\int_{-3}^{-3} |x^3 - 4x| \, dx$.

74. Evaluate $\int_0^7 |\sin x| \, dx$.

75. Use mathematical induction and the "triangle inequality" (Miscellaneous Exercise 11 of Chapter 1) to prove its generalization

$$\left| \sum_{j=1}^{N} a_j \right| \le \sum_{j=1}^{N} |a_j|$$

for arbitrary positive integers $N \ge 2$ and arbitrary numbers a_j.

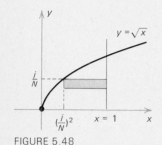

$y = \sqrt{x}$

$x = 1$

$\left(\frac{j}{N}\right)^2$ $\frac{j}{N}$

FIGURE 5.48

CHAPTER 6

APPLICATIONS OF INTEGRATION

SUMMARY: In the last chapter we used integrals to compute areas of regions bounded by graphs of functions and the x-axis. In Section 6.1 of this chapter we use integrals to compute areas of other types of regions. We use integrals to find volumes in Sections 6.2 through 6.4, lengths of curves in Section 6.5, and areas of surfaces of revolution in Section 6.6. Section 6.7 deals with integrals of rates of change, Section 6.8 with work, Section 6.9 with hydrostatic pressure, and Section 6.10 with density, moments, and centers of mass. Average values of functions and the Mean Value Theorem for integrals are discussed in Section 6.11. Other applications of integration are described in the Miscellaneous Exercises.

6-1 AREAS OF PLANE REGIONS

To compute the area of a region which is bounded on the top and bottom by graphs of functions and on the sides by vertical lines, as in Figure 6.1, we use the following rule.

Rule 6.1 (Computing areas by x-integrations) Suppose that a region in an xy-plane is bounded on the top by the curve $y = f(x)$ and on the bottom by $y = g(x)$ and extends from $x = a$ on the left to $x = b$ on the right (Figure 6.1). If the integrals of f and g from a to b exist, then the area of the region is

(1) $$\int_a^b [f(x) - g(x)] \, dx. \qquad \blacktriangleleft$$

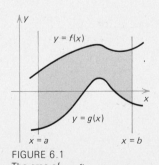

FIGURE 6.1
The area of the region $= \int_a^b [f(x) - g(x)] \, dx$

Derivation of Rule 6.1. We must assume that the functions f and g are bounded for $a \le x \le b$ so that the integrals exist. Then we can choose a constant K such that $f(x) + K$ and $g(x) + K$ are positive for

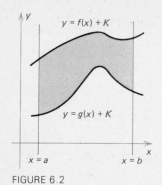

FIGURE 6.2

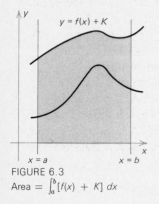

FIGURE 6.3
Area $= \int_a^b [f(x) + K]\,dx$

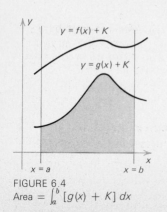

FIGURE 6.4
Area $= \int_a^b [g(x) + K]\,dx$

$a \leq x \leq b$. The region bounded by $y = f(x) + K$ and $y = g(x) + K$ in Figure 6.2 is congruent to the original region and its area is equal to the area of the region under the curve $y = f(x) + K$ in Figure 6.3 minus the area of the region under $y = g(x) + K$ in Figure 6.4. According to the geometric definition of the integral, the area of the region in Figure 6.2 is

$$\int_a^b [f(x) + K]\,dx - \int_a^b [g(x) + K]\,dx$$

$$= \int_a^b [f(x) + K - g(x) - K]\,dx$$

$$= \int_a^b [f(x) - g(x)]\,dx.$$

Since this equals the area of the region in Figure 6.1, we obtain Rule 6.1. □

EXAMPLE 1   Sketch the region bounded by the parabolas $y = x^2$ and $y = -x^2$ and by the vertical line $x = 2$. Compute its area.

Solution.   The region is sketched in Figure 6.5. Its top is formed by $y = x^2$ and its bottom by $y = -x^2$, and it extends from $x = 0$ on the left to $x = 2$ on the right. According to Rule 6.1 its area is

$$\int_0^2 [(x^2) - (-x^2)]\,dx = \int_0^2 2x^2\,dx = \left[\frac{2}{3}x^3\right]_0^2$$

$$= \left[\frac{2}{3}(2)^3\right] - \left[\frac{2}{3}(0)^3\right]$$

$$= \frac{16}{3} - 0 = \frac{16}{3}. \quad \square$$

Use of the analytic definition of the integral

It was possible for us to derive Rule 6.1 by using the geometric definition of the integral; however, in other applications we will have to use *an approximation procedure* and *the analytic definition of the integral* to determine the required integrals. In these applications the given information involves the values of one or more functions of a variable x in an interval $a \leq x \leq b$. In special cases, where the functions are constant or are first degree polynomials, the problem can be solved with arithmetic alone. In other cases we find the integral that gives the solution by first using partitions

$$(2) \qquad a = x_0 < x_1 < x_2 < \cdots < x_{N-1} < x_N = b$$

of the interval $a \leq x \leq b$ to obtain approximate solutions. Typically, the approximate solution corresponding to the partition (2) is of the form

$$(3) \qquad \sum_{j=1}^N F(x_j)(x_j - x_{j-1})$$

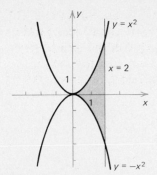

FIGURE 6.5   Region of area
$\int_0^2 [(x^2) - (-x^2)]\, dx$

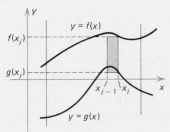

FIGURE 6.6   Rectangle of height
$f(x_j) - g(x_j)$ and width $x_j - x_{j-1}$

where F is a function constructed from the given function or functions. The sum (3) is a Riemann sum for the integral

$$(4) \qquad \int_a^b F(x)\, dx$$

and this integral is the exact solution.

Before using this procedure to study new types of problems, we illustrate its use by re-deriving Rule 6.1.

Derivation of Rule 6.1 by an approximation procedure.   We consider an arbitrary partition

$$(5) \qquad a = x_0 < x_1 < x_2 < x_3 < \cdots < x_{N-1} < x_N = b$$

of the interval $a \le x \le b$. We approximate the region by N rectangles with sides formed by the vertical lines $x = x_j$ for $j = 0, 1, 2, \ldots, N$. We put the upper right corners of the rectangles on the curve $y = f(x)$ and the lower right corners on the curve $y = g(x)$. Figure 6.6 is a schematic sketch of the jth rectangle in the approximation.

The height of the jth rectangle is the difference $f(x_j) - g(x_j)$ between the values of the functions at the right side $x = x_j$ of the rectangle. The width is the length $x_j - x_{j-1}$ of the jth subinterval of the partition. The area of the jth rectangle is $[f(x_j) - g(x_j)](x_j - x_{j-1})$ and the area of the entire approximation of the region is the sum

$$(6) \qquad \sum_{j=1}^{N} [f(x_j) - g(x_j)](x_j - x_{j-1}).$$

The sum (6) is a Riemann sum for the integral (1) and tends to that integral as the lengths of the subintervals in the corresponding partition (5) tend to zero. The integral (1) is, therefore, equal to the area of the region in Figure 6.1, as is stated in Rule 6.1.   □

In the next example we have to solve a quadratic equation to find the limits of integration in the integral that gives the area.

EXAMPLE 2   Sketch the region bounded by the curves $y = 4 - x$ and $y = x^2 - 4x + 4$ and compute its area.

Solution.   The region is shown in Figure 6.6a. Its top is formed by the line $y = 4 - x$ and its bottom by the curve $y = x^2 - 4x + 4$, so its area is

$$(6a) \qquad \int_a^b [(4 - x) - (x^2 - 4x + 4)]\, dx$$

where a and b are the x-coordinates of the intersections of the two curves. We can find the coordinates of those intersections by solving the simultaneous equations

$$y = 4 - x \quad \text{and} \quad y = x^2 - 4x + 4.$$

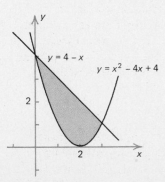

FIGURE 6.6a

Equating these two expressions for y gives the equation $4 - x = x^2 - 4x + 4$ satisfied by the x-coordinates of the intersections. This equation simplifies to $x^2 - 3x = 0$ or $x(x - 3) = 0$ and has the solutions $x = 0$ and $x = 3$. Hence we need to take $a = 0$ and $b = 3$ in (6a) and the area equals

$$\int_0^3 [(4 - x) - (x^2 - 4x + 4)]\,dx = \int_0^3 (3x - x^2)\,dx$$

$$= [\tfrac{3}{2}x^2 - \tfrac{1}{3}x^3]_0^3$$

$$= [\tfrac{3}{2}(3)^2 - \tfrac{1}{3}(3)^3]$$

$$- [\tfrac{3}{2}(0)^2 - \tfrac{1}{3}(0)^3]$$

$$= \tfrac{9}{2}. \quad \square$$

Computing areas by y-integrations

To find the area of a region such as that shown in Figure 6.7, we use an integral with respect to y.

Rule 6.2 (Computing areas by y-integrations) Suppose a region is bounded on the right by the curve $x = M(y)$ and on the left by the curve $x = N(y)$ and extends from $y = c$ on the bottom to $y = d$ on the top (Figure 6.7). If the integrals of M and N from c to d exist, then the area of the region is

(7) $$\int_c^d [M(y) - N(y)]\,dy. \qquad \blacktriangleleft$$

Derivation of Rule 6.2. We consider an arbitrary partition

(8) $$c = y_0 < y_1 < y_2 < y_3 < \cdots < y_{N-1} < y_N = d$$

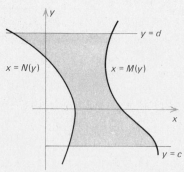

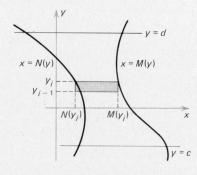

FIGURE 6.7   The area of the region
$= \int_c^d [M(y) - N(y)]\,dy$

FIGURE 6.8   Rectangle of width
$M(y_j) - N(y_j)$ and height $y_j - y_{j-1}$

of the interval $c \leq y \leq d$. We use the partition to construct an approximation of the region by thin horizontal rectangles. Figure 6.8 gives a schematic sketch of the jth rectangle in the approximation. Its width is $M(y_j) - N(y_j)$, its height is $y_j - y_{j-1}$, and its area is $[M(y_j) - N(y_j)](y_j - y_{j-1})$. The area of the entire approximation is the sum

$$(9) \qquad \sum_{j=1}^{N} [M(y_j) - N(y_j)](y_j - y_{j-1}).$$

The sum (9) is a Riemann sum for the integral (7) and tends to that integral as the lengths of the subintervals in the partition (8) tend to zero. Consequently the integral (7) is the area of the region, as stated in Rule 6.2. □

EXAMPLE 3   Sketch the region bounded by the parabola $x = y^2$ and by the lines $x = y - 1, y = -1$, and $y = 1$. Compute its area.

Solution.   The region is sketched in Figure 6.9. Its right side is formed by the parabola $x = y^2$, its left side is formed by the line $x = y - 1$, and it extends from $y = -1$ on the bottom to $y = 1$ on the top. According to Rule 6.2, its area is

$$\int_{-1}^{1} [y^2 - (y - 1)] \, dy = \int_{-1}^{1} [y^2 - y + 1] \, dy$$
$$= [\tfrac{1}{3}y^3 - \tfrac{1}{2}y^2 + y]_{-1}^{1}$$
$$= [\tfrac{1}{3}(1)^3 - \tfrac{1}{2}(1)^2 + 1] - [\tfrac{1}{3}(-1)^3 - \tfrac{1}{2}(-1)^2 - 1]$$
$$= [\tfrac{1}{3} - \tfrac{1}{2} + 1] - [-\tfrac{1}{3} - \tfrac{1}{2} - 1]$$
$$= 2\tfrac{2}{3}. \quad \square$$

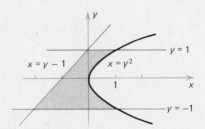

FIGURE 6.9   Region of area $\int_{-1}^{1} [y^2 - (y - 1)] \, dy$

In Exercises 1 through 26 sketch the regions bounded by the given curves in xy-planes. Then compute their areas.

1.‡ $y = 3 - x^2$ and $y = -2x$ **2.**† $y = 5 - x^2$ and $y = x^2 - 3$

3. $y = 3 - x^2$ and $y = 3 - 2x$ **4.**‡ $y = x^2$ and $y = x^3$

5.† $y = 1 - x^2$ and $y = -3$ **6.** $y = x^2 + 1$ and $y = 2x^2 - 8$

7.‡ $y = x^3 + 1$ and $y = 4x + 1$ **8.**† $y = 4 - x, y = -5x$, and $y = x$

9. $y = 9x^{-2}, y = 1$, and $y = 9$

10.‡ $y = 2 \sin x$ and $y = -3 \sin x$ for $0 \le x \le 2\pi$

11.[†] $y = \frac{1}{4}x^3$ and $y = 2x$

12. $y = \cos x, y = 0, x = 0$, and $x = 2\pi$

13.[‡] $x = y^2$ and $x = 9$ **14.**[†] $x = y^2 - 3$ and $x = 5 - y^2$

15. $x = y^3, x = y^3 + 4, y = -1$, and $y = 1$

16.[‡] $y = x^3$ and $x = y^3$ **17.**[†] $y = x^2$ and $x = y^2$

18.[‡] $y = \sqrt{9 - x}, x = 0$, and $y = 0$

19.[†] $y = \sin(3x), y = -2, x = 0, x = \dfrac{\pi}{2}$.

20. $y = \cos(4x)$ and $y = -\cos(4x)$ for $-\dfrac{\pi}{8} \le x \le \dfrac{\pi}{8}$

21.[†] $y = |\sin x|, y = 0, x = -2\pi$, and $x = 2\pi$

22. $y = \sin|x|, y = 0, x = -2\pi$, and $x = 2\pi$

23.[‡] $y = x^3 - 4x$ and $y = 0$ **24.** $y = 4x^2 - x^4$ and $y = 0$

In Exercises 25 and 26 compute the areas twice, once by using an x-integral and once by using a y-integral.

25.[‡] $y = x^3$ and $y = x^2$ **26.** $x = y^2$ and $x = 4y$

6-2   COMPUTING VOLUMES BY SLICING

In this and the following two sections we use definite integrals to compute volumes of solids. Here we use the METHOD OF SLICING, which enables us to find the volume of a solid from the *areas* of its cross sections.

We start with a solid V and an x-axis, as in Figure 6.10. The plane perpendicular to the axis and passing through the point at x on that axis intersects the solid in a plane region which we call the CROSS SECTION OF V AT x. If we know the area $A(x)$ of that cross section for each x, then we define the volume of the solid by the following rule.

Rule 6.3   (Computing volumes by slicing)   Suppose that the cross section at x of a solid V has area $A(x)$ and that the solid extends from its cross section at $x = a$ to its cross section at $x = b$ ($a < b$). If the integral of $A(x)$ from a to b exists, then the volume of the solid is that integral:

(1)   **Volume** $= \displaystyle\int_a^b A(x)\, dx.$ ◄

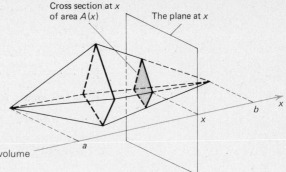

FIGURE 6.10   Solid V of volume $\int_a^b A(x)\,dx$

Motivation for Rule 6.3.   We consider an arbitrary partition

$$(2) \qquad a = x_0 < x_1 < x_2 < x_3 < \cdots < x_{N-1} < x_N = b$$

of the interval $a \le x \le b$. We use the partition to construct an approximation of the solid by N wafer-like pieces. A schematic sketch of the jth piece is shown in Figure 6.11. It extends from the plane perpendicular to the x-axis at $x = x_{j-1}$ to the plane at $x = x_j$ and its edges are perpendicular to those planes. Each cross section is congruent to the cross section of the original solid at $x = x_j$ and has area $A(x_j)$. Its width is $x_j - x_{j-1}$ and its volume is $A(x_j)(x_j - x_{j-1})$. The volume of the entire approximating solid is

$$(3) \qquad \sum_{j=1}^{N} A(x_j)(x_j - x_{j-1}).$$

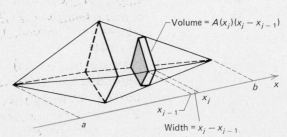

Volume $= A(x_j)(x_j - x_{j-1})$

FIGURE 6.11   Width $= x_j - x_{j-1}$

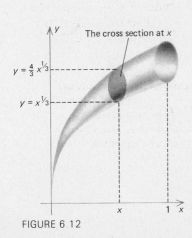

The cross section at x

$y = \frac{4}{3}x^{1/3}$

$y = x^{1/3}$

FIGURE 6.12

As the lengths of the subintervals of the partition (2) tend to zero, the approximating solid becomes a better approximation of the original solid and the sum (3) tends to the integral (1). We therefore define the volume of the solid to be the integral (1), as stated in Rule 6.3.   □

EXAMPLE 1   The cross sections perpendicular to the x-axis of the horn-shaped solid in Figure 6.12 are circles with diameters running from the curve $y = x^{1/3}$ to the curve $y = \frac{4}{3}x^{1/3}$ in the xy-plane. The solid extends from the origin to its cross section at $x = 1$. Compute its volume. (The scales are measured in feet.)

Solution.   The diameter of the cross section at x is the distance $\frac{4}{3}x^{1/3} - x^{1/3} = \frac{1}{3}x^{1/3}$ (feet) between the curves. The radius of the cross section at x is $\frac{1}{6}x^{1/3}$ feet and its area is $A(x) = \pi(\frac{1}{6}x^{1/3})^2 = (\pi/36)x^{2/3}$ square feet.

Because the solid extends from its cross section at $x = 0$ to its cross section at $x = 1$, its volume is

$$\int_0^1 A(x)\,dx = \int_0^1 \frac{\pi}{36}x^{2/3}\,dx = \left[\left(\frac{\pi}{36}\right)\frac{1}{5/3}x^{5/3}\right]_0^1$$

$$= \left[\frac{\pi}{60}x^{5/3}\right]_0^1 = \frac{\pi}{60} - 0 = \frac{\pi}{60} \text{ cubic feet.} \quad \square$$

Note.   If we assign the units of feet to the symbol "dx" in the last integral, then we obtain the correct dimensions for the integral. We obtain

$$\int_0^1 A(x)\,[\text{sq. ft.}]\,dx\,[\text{ft.}] = \text{Volume}\,[\text{cu. ft.}].$$

This is another reason for including the "dx" in the symbol for the integral. $\square$

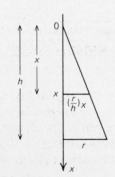

FIGURE 6.12a

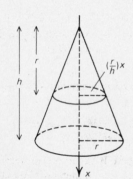

FIGURE 6.12b

EXAMPLE 2.   Use the method of slicing to compute the volume of a right circular cone of height h and with base of radius r.

Solution.   We introduce a vertical x-axis pointing downward and with its zero at the vertex of the cone (Figure 6.12a). The horizontal cross section at x ($0 \leq x \leq h$) is a circle whose center is x units from the vertex. By similar triangles (Figure 6.12b) the cross section at x has radius $\left(\frac{r}{h}\right)x$ and area $A(x) = \pi\left(\frac{rx}{h}\right)^2$. The cross sections run from $x = 0$ at the vertex to $x = h$ at the base, so the volume of the cone is

$$\int_0^h A(x)\,dx = \int_0^h \pi\left(\frac{rx}{h}\right)^2\,dx = \frac{\pi r^2}{h^2}\int_0^h x^2\,dx = \frac{\pi r^2}{h^2}[\tfrac{1}{3}x^3]_0^h$$

$$= \frac{\pi r^2}{h^2}(\tfrac{1}{3}h^3) = \tfrac{1}{3}\pi r^2 h. \quad \square$$

Exercises

In Exercises 1 through 9 compute the volumes by the method of slicing.

1.‡   The volume inside a parabolic mirror: the intersection of the mirror with the xy-plane is the portion of the parabola $y = \frac{1}{2}x^2$ for $-2 \leq x \leq 2$; the cross sections of the mirror perpendicular to the y-axis are circles with their centers on the y-axis.

2.‡   The volume of the solid whose base is the circle $x^2 + y^2 \leq 1$ and whose cross sections perpendicular to the x-axis are equilateral triangles.

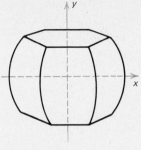

FIGURE 6.13

3. The volume of the solid whose base is the region bounded by the curves $y = x^2$ and $y = 2 - x^2$ in the xy-plane and whose cross sections perpendicular to the x-axis are semi-circles with diameters in the xy-plane.

4.‡ The volume of a general "cone" of height h and with base of area A: the vertex of the cone is a distance h above the plane in which the base lies, and the cone is formed by all the line segments which run from the vertex to the base.

5.† The volume of a stool whose intersection with the xy-plane is the portion of the circle $x^2 + y^2 \leq 2$ for $-1 \leq y \leq 1$ and whose cross sections perpendicular to the y-axis are regular hexagons with diagonals in the xy-plane (Figure 6.13).

6. The volume of a church steeple that extends from $y = -5$ to $y = 0$ and whose cross sections perpendicular to the y-axis are squares with diagonals in the xy-plane extending from the curve $x = -\frac{1}{5}y^2$ to the curve $x = \frac{1}{5}y^2$. (The scales are measured in yards.)

7.‡ The volume of the region common to two right circular cylinders of radius 4 feet whose axes intersect at right angles.

8.† The volume remaining after a two foot long cylindrical hole has been drilled through the middle of a sphere of radius r feet ($r > 1$).

9. The volume of a slice cut from a cylinder of cheese 10 inches in diameter and 10 inches high by a plane that passes through the diameter of the base and makes an angle of 45° with the base.

6-3 VOLUMES OF SOLIDS OF REVOLUTION: THE METHODS OF DISKS AND WASHERS

In this section and in Section 6.4 we use integrals to compute volumes of SOLIDS OF REVOLUTION that are formed by rotating a region in an xy-plane about a horizontal or vertical line in the plane. The line is called the AXIS of the solid. Here we use the METHOD OF DISKS and the METHOD OF WASHERS, which are special cases of computing volumes by slicing.

The method of disks

In the method of disks the axis of the solid of revolution is one side of the region being rotated and, consequently, each of the cross sections of the solid is a circular disk.

Rule 6.4   (The method of disks with an x-integration) Suppose that the top or bottom of a region R is the line $y = L$, the opposite side is the graph $y = f(x)$, and the region extends from $x = a$ to $x = b$ ($a < b$). Then the solid generated by rotating the region R about the line $y = L$ has volume

$$\text{(1)} \qquad \int_a^b A(x)\,dx = \int_a^b \pi \left[\frac{\text{Radius of the}}{\text{section at } x} \right]^2 dx$$

$$= \int_a^b \pi [f(x) - L]^2\, dx$$

◀

provided the integral exists.

Derivation of Rule 6.4 Figure 6.14 shows a region R bounded on the top by the curve $y = f(x)$, on the bottom by the line $y = L$, and extending from $x = a$ on the left to $x = b$ on the right. The solid generated by rotating that region about the line $y = L$ is shown in Figure 6.15.

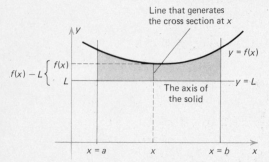

FIGURE 6.14   The region R

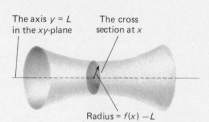

FIGURE 6.15   The solid of revolution

According to the method of slicing, the volume is the integral of the area $A(x)$ of the cross section at x (Figure 6.15). The radius of the cross section is the length $f(x) - L$ of the line which generates it (Figure 6.14). The area of the cross section is therefore $A(x) = \pi[\text{Radius}]^2 = \pi[f(x) - L]^2$ and we obtain (1).

If the graph $y = f(x)$ is below the axis of the solid, as in Figure 6.16, then the radius of the cross section is $L - f(x)$ rather than $f(x) - L$, but the square of the radius is still $[f(x) - L]^2$ as stated in equations (1).   □

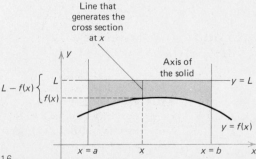

FIGURE 6.16

EXAMPLE 1   When the region bounded by the cubic $y = x^3$ and by the lines $x = 1$ and $y = -1$ is rotated about the line $y = -1$, it generates a bell-shaped solid. Compute the volume of that solid.

Solution.   The region is sketched in Figure 6.17 and the solid it generates is shown in Figure 6.18. To find the x-coordinate of the left side of the region, we solve the simultaneous equations $y = x^3$ and $y = -1$. Equating the two expressions for y gives the equation $x^3 = -1$ for the x-coordinate of the intersection of the curve $y = x^3$ and the line $y = -1$. This equation has the solution $x = -1$, thus the region extends from $x = -1$ to $x = 1$. According to Rule 6.4, the volume of the solid is

$$\int_{-1}^{1} A(x)\,dx = \int_{-1}^{1} \pi \left[\begin{array}{l} \text{Radius of the} \\ \text{section at } x \end{array} \right]^2 dx$$

$$= \int_{-1}^{1} \pi [x^3 - (-1)]^2\,dx$$

$$= \int_{-1}^{1} \pi (x^6 + 2x^3 + 1)\,dx$$

$$= \pi \left[\frac{1}{7}x^7 + 2\left(\frac{1}{4}x^4\right) + x \right]_{-1}^{1}$$

$$= \pi \left[\frac{1}{7}x^7 + \frac{1}{2}x^4 + x \right]_{-1}^{1}$$

$$= \pi \left[\frac{1}{7}(1)^7 + \frac{1}{2}(1)^4 + 1 \right] - \pi \left[\frac{1}{7}(-1)^7 + \frac{1}{2}(-1)^4 - 1 \right]$$

$$= \frac{16}{7}\pi. \quad \square$$

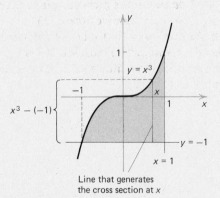

FIGURE 6.17

Line that generates
the cross section at x

$y = x^3$

$x^3 - (-1)$

$y = -1$

$x = 1$

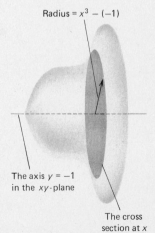

Radius $= x^3 - (-1)$

The axis $y = -1$
in the xy-plane

The cross
section at x

FIGURE 6.18   The solid of
revolution

Rule 6.5   (The method of disks with a y-integration)
**Suppose that the right or left side of a region R is the vertical
line $x = L$, the opposite side has the equation $x = M(y)$, and
the region extends from $y = c$ to $y = d(c < d)$. Then the solid
generated by rotating the region R about the line $x = L$
has volume**

$$(2) \qquad \int_c^d A(y)\, dy = \int_c^d \pi \left[\frac{\text{Radius of the}}{\text{section at } y} \right]^2 dy$$

$$= \int_c^d \pi [M(y) - L]^2\, dy \qquad \blacktriangleleft$$

provided the integral exists.

The derivation of this rule is analogous to that of Rule 6.4. It is
illustrated in the next example.

EXAMPLE 2   When the region bounded by the parabola $x = y^2$ and by
the vertical line $x = 1$ is rotated about the line $x = 1$, it generates a
top-shaped solid. Compute the volume of that solid.

Solution.   The region in the xy-plane is shown in Figure 6.19. The
intersections of the parabola $x = y^2$ and the line $x = 1$ are at $y = \pm 1$
and the region extends from $y = -1$ to $y = 1$. The solid of revolution is
shown in Figure 6.20. According to Rule 6.5, its volume is

$$\int_{-1}^1 A(y)\, dy = \int_{-1}^1 \pi \left[\frac{\text{Radius of the}}{\text{section at } y} \right]^2 dy = \int_{-1}^1 \pi [y^2 - 1]^2\, dy$$

$$= \int_{-1}^1 \pi (y^4 - 2y^2 + 1)\, dy = \pi \left[\frac{1}{5}y^5 - \frac{2}{3}y^3 + y \right]_{-1}^1$$

$$= \pi \left[\frac{1}{5}(1)^5 - \frac{2}{3}(1)^3 + 1 \right] - \pi \left[\frac{1}{5}(-1)^5 - \frac{2}{3}(-1)^3 - 1 \right]$$

$$= \frac{16\pi}{15}. \quad \square$$

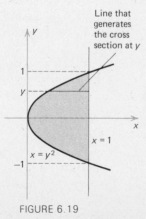

FIGURE 6.19

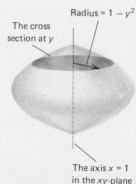

FIGURE 6.20   The solid of
revolution

The method of washers

The method of washers differs from the method of disks in only one respect. The axis of the solid is not one of the sides of the region being rotated and, consequently, at least some of the cross sections are washer-shaped regions instead of circular disks.

Rule 6.6   (The method of washers)   Suppose that the top of a region R is formed by the curve $y = f(x)$ and the bottom by the curve $y = g(x)$ and that the region extends from $x = a$ to $x = b$ $(a < b)$. Then the volume of the solid generated by rotating the region R about the horizontal line $y = L$ is

(3)
$$\int_a^b A(x)\,dx$$

$$= \int_a^b \left\{ \pi \left[\begin{matrix} \textbf{Outer radius} \\ \textbf{of the section} \\ \textbf{at } x \end{matrix} \right]^2 - \pi \left[\begin{matrix} \textbf{Inner radius} \\ \textbf{of the section} \\ \textbf{at } x \end{matrix} \right]^2 \right\} dx.$$

◀

If the region R is bounded on the left and right by curves $x = M(y)$ and $x = N(y)$ and if the axis of the solid is vertical, then the volume is given by an integral of the form

(4)
$$\int_c^d A(y)\,dy$$

$$= \int_c^d \left\{ \pi \left[\begin{matrix} \textbf{Outer radius} \\ \textbf{of the section} \\ \textbf{at } y \end{matrix} \right]^2 - \pi \left[\begin{matrix} \textbf{Inner radius} \\ \textbf{of the section} \\ \textbf{at } y \end{matrix} \right]^2 \right\} dy.$$

◀

Here, as with all rules dealing with solids of revolution, we assume that the integrals exist and that the region R lies entirely on one side or the other of the axis of the solid in the xy-plane. Rule 6.6 follows from the method of slicing because the area of a washer-shaped region between two concentric circles is equal to the area of the outer circle minus the area of the inner circle (Figure 6.21).

Expressions in terms of x or y for the inner and outer radii in formulas (3) and (4) are obtained from a sketch of the region R and the axis of the solid, as is illustrated in the next example.

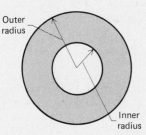

Outer radius

Inner radius

FIGURE 6.21

Area $= \pi \left[\begin{matrix} \text{Outer} \\ \text{radius} \end{matrix} \right]^2 - \pi \left[\begin{matrix} \text{Inner} \\ \text{radius} \end{matrix} \right]^2$

EXAMPLE 3   Sketch the triangle bounded by the lines $y = \frac{1}{4}x + 3$, $y = -\frac{1}{4}x + 3$, and $x = 4$. Compute the volume of the solid generated by rotating it about the line $y = 1$.

Solution.   The triangle is sketched in Figure 6.22 and the solid of revolution in Figure 6.23. The vertical line at x across the triangle in Figure 6.22 generates the washer-shaped cross section of the solid shown in Figure 6.23. Its outer radius is the distance $\frac{1}{4}x + 3 - 1 = \frac{1}{4}x + 2$ from the top of the line to the axis of the solid. Its inner radius is the

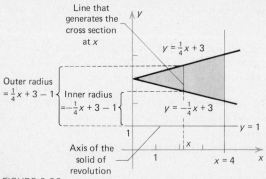

FIGURE 6.22

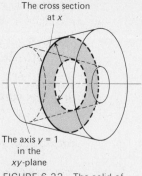

FIGURE 6.23   The solid of revolution

distance $-\frac{1}{4}x + 3 - 1 = -\frac{1}{4}x + 2$ from the bottom of the line to the axis. According to Rule 6.6, the solid has volume

$$\int_0^4 A(x)\,dx = \int_0^4 \left\{ \pi \left[\begin{matrix} \text{Outer} \\ \text{radius} \end{matrix} \right]^2 - \pi \left[\begin{matrix} \text{Inner} \\ \text{radius} \end{matrix} \right]^2 \right\} dx$$

$$= \int_0^4 \left\{ \pi \left[\frac{1}{4}x + 2 \right]^2 - \pi \left[-\frac{1}{4}x + 2 \right]^2 \right\} dx$$

$$= \int_0^4 \pi[2x]\,dx = \pi[x^2]_0^4$$

$$= \pi[4^2] - \pi[0^2] = 16\pi. \quad \square$$

Exercises

†*Answer provided.*
‡*Outline of solution provided.*

In Exercises 1 through 15 make a rough sketch of the region and compute the volume of the solid generated by rotating it about the indicated axis.

1.‡ The region bounded by $y = 1 - x^2$ and $y = -3$; axis: $y = -3$

2.‡ The region bounded by $y = \frac{1}{2}x^2$ and $y = 1$; axis: $y = 1$

3.† The region bounded by $y = 0, y = 2/x, x = 1$, and $x = 4$; axis: $y = 0$

4. The region bounded by $y = 7 - x^3, y = -1$, and $x = -1$; axis: $y = -1$

5.‡ The region bounded by $x = y^2 + 2, x = 1, y = 1$, and $y = -1$; axis: $x = 1$

6.‡ The region bounded by $x = y^{1/3}, x = 1$, and $y = -8$; axis: $x = 1$

7.† The region bounded by $x = 4/y, x = 0, y = 1$, and $y = 4$; axis: $x = 0$

8. The region bounded by $x = y, x = 3$, and $y = -2$; axis: $x = 3$

9.‡ The region bounded by $y = 3x$ and $y = x^2$; axis: $y = 0$

10.‡ The region bounded by $y = x^3, y = -1$, and $x = 1$; axis: $y = 2$

11. The region bounded by $y = 4x - x^2$ and $y = 0$; axis: $y = 4$

12.‡ The region bounded by $x = y^2$ and $x = 3y$; axis: $x = 0$

13.‡ The region bounded by $x = -y^2$ and $x = y - 2$; axis: $x = 1$

14.† The region bounded by $x = \sqrt{y}, y = 0$, and $x = 2$; axis: $x = 0$

15. The region bounded by $x = y^2 + 1$, $x = -y^2 - 1$, $y = 1$, and $y = -1$; axis: $x = 3$

In Exercises 16 and 17 give definite integrals that equal the volumes of the solids generated by rotating the regions about the indicated axes. (We will learn in later chapters how to evaluate these integrals.)

16.[†] The region bounded by $y = \sin x$, $y = -1$, $x = 0$, and $x = \pi$; axis: $y = -2$

17. The region bounded by $y = \sqrt{\dfrac{1}{x}}$, $y = x$, and $x = 2$; axis: $y = 0$

18.[†] Sketch the region bounded by the curves $y = x^3$ and $x = y^3$. Compute the volume of the solid generated by rotating it about the line $y = 2$. (Two integrals are needed.)

19. Sketch the region bounded by the parabolas, $y = 1 - x^2$ and $y = x^2 - 1$. Compute the volume of the solid generated by rotating it about the line $x = -2$. (Two integrals are needed.)

20.[†] A bullet's shape is formed by rotating about the x-axis the region bounded by the curve $x = y^4$ and $x = a$ for some positive number a. The scales are measured in inches. The volume of the bullet is 12π cubic inches. How long is it? What is the radius of its base?

21. According to tradition, Archimedes asked that the formula he derived for the volume of a sphere be inscribed on his tombstone. He based his derivation on formulas for the volumes of cones and cylinders. The method of disks is based on the formula for the area of a circle. Use it to derive the formula for the volume of a sphere.

Additional drill exercises

In Exercises 22 through 27 use the method of disks or washers to compute the volume generated by rotating the region about the x-axis.

22.[†] The region bounded by the curve $y = x - x^4$ and the x-axis

23.[†] The half ellipse described by the inequalities $0 \leq y \leq \sqrt{8 - 2x^2}$

24.[†] The triangle with vertices $(0, 1)$, $(3, 4)$, and $(3, 0)$

25.[†] The trapezoid with vertices $(0, 0)$, $(0, 4)$, $(2, 5)$, and $(2, 1)$

26.[†] The region bounded by the parabola $y = -x^2$ and the line $y = -9$

27.[†] The region bounded by $y = x^4$ and $x = y^2$

In Exercises 28 through 31 use the method of disks or washers to compute the volume generated by rotating the region about the y-axis.

28.[†] The triangle with vertices $(0, 0)$, $(0, h)$, and $(r, 0)$ with positive constants r and h (What is the solid?)

29.[†] The rectangle with corners $(0, 0)$, $(0, h)$, (r, h), and $(r, 0)$ with positive constants r and h (What is the solid?)

30.[†] The portion of a circle described by the inequalities $4 \leq x \leq \sqrt{25 - y^2}$

31.[†] The portion of a circle described by the inequalities $0 \leq x \leq \sqrt{R^2 - y^2}$ and $h \leq y \leq R$, where R and h are constants with $0 \leq h \leq R$

6-4   VOLUMES OF SOLIDS OF REVOLUTION: THE METHOD OF CYLINDERS

In the method of disks or washers we compute the volume of a solid of revolution from the areas of parallel cross sections. In the METHOD OF CYLINDERS the volume is obtained as an integral of *surface areas of concentric cylindrical sections* of the solid.

The region shown in Figure 6.24 lies between the graphs $y = f(x)$ and $y = g(x)$ of two functions. Figure 6.25 shows the solid generated by rotating that region about the line $x = L$. The vertical line that extends across the region at x is shown in Figure 6.26. When it is rotated about the axis of the solid, it generates a cylindrical section of the solid which we call the CYLINDER AT x (Figure 6.27). We define the volume of the solid to be an integral of the surface areas $A(x)$ of these cylinders.

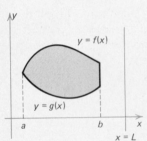

FIGURE 6.24

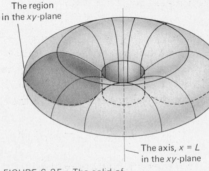

FIGURE 6.25   The solid of revolution

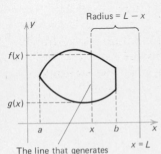

FIGURE 6.26

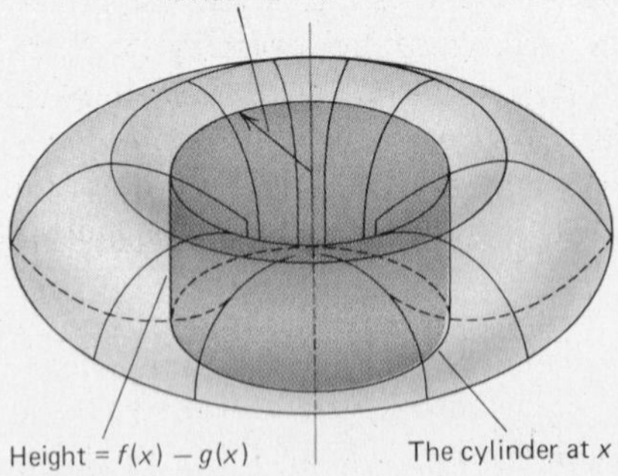

FIGURE 6.27

Rule 6.7   (The method of cylinders with an x-integration)
Suppose that the top of a region R is formed by the curve
$y = f(x)$ and the bottom by the curve $y = g(x)$ and that the
region extends from $x = a$ on the left to $x = b$ on the right.
Then the volume of the solid generated by rotating the region
R about the vertical line $x = L$ is

(1)
$$\int_a^b A(x)\,dx = \int_a^b \left|\begin{matrix}\textbf{Surface area of}\\\textbf{the cylinder at x}\end{matrix}\right|\,dx$$

$$= \int_a^b 2\pi\left|\begin{matrix}\textbf{Radius of the}\\\textbf{cylinder at x}\end{matrix}\right|\left|\begin{matrix}\textbf{Height of the}\\\textbf{cylinder at x}\end{matrix}\right|\,dx$$

◀

provided the integral exists.

The surface area of the cylinder is 2π times its radius times its height because it could be cut and flattened into a rectangle (Figure 6.28). The height of the rectangle would be the same as that of the cylinder, and the length of the rectangle would be the circumference of the cylinder or 2π times its radius.

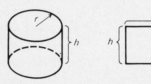

FIGURE 6.28   A cylinder and a rectangle of area $2\pi rh$

Motivation for Rule 6.7.   We consider the case illustrated in Figures 6.24 through 6.27 where the axis of the solid is on the right of the region being rotated. In this case the radius of the cylinder at x is $L - x$ and its height is $f(x) - g(x)$. Therefore, to verify formula (1) we need to show that the volume of the solid is the integral

(2) $$\int_a^b 2\pi(L - x)[f(x) - g(x)]\,dx.$$

We consider an arbitrary partition

(3) $$a = x_0 < x_1 < x_2 < x_3 < \cdots < x_{N-1} < x_N = b$$

of the interval $a \le x \le b$ and use this partition to construct an approximation of the region by a collection of N rectangles. When the rectangles are rotated about the axis of the solid, they generate cylindrical shells that fit inside each other to form an approximation of the solid. Figure 6.29 shows an approximation of the region by four rectangles, and Figure 6.30 shows the corresponding cylindrical shells.

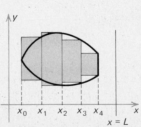

FIGURE 6.29   Approximating rectangles

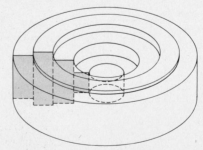

FIGURE 6.30   Approximating cylindrical shells

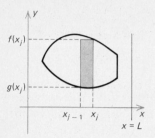

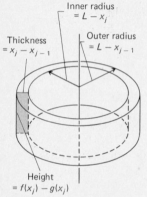

FIGURE 6.31 The jth rectangle in the approximation

Inner radius
= $L - x_j$

Thickness
= $x_j - x_{j-1}$

Outer radius
= $L - x_{j-1}$

Height
= $f(x_j) - g(x_j)$

FIGURE 6.32 The jth cylindrical shell

A schematic sketch of the jth rectangle in an approximation is shown in Figure 6.31 and the cylindrical shell it generates is shown in Figure 6.32. The inner radius of the cylindrical shell is the distance $L - x_j$ from the right side of the rectangle to the axis of the solid. The outer radius is the distance $L - x_{j-1}$ from the left side of the rectangle to the axis. The height of the shell is the height $f(x_j) - g(x_j)$ of the generating rectangle. The shell's thickness is the width $x_j - x_{j-1}$ of the rectangle.

The volume of a cylindrical shell is

$$(4) \qquad \pi \left\{ \begin{bmatrix} \text{Inner} \\ \text{radius} \end{bmatrix} + \begin{bmatrix} \text{Outer} \\ \text{radius} \end{bmatrix} \right\} [\text{Height}]\,[\text{Thickness}]$$

(see Exercise 11), so the volume of the shell generated by rotating the jth rectangle is

$$(5) \qquad \begin{aligned} \pi\{[L - x_j] &+ [L - x_{j-1}]\}\,[f(x_j) - g(x_j)]\,(x_j - x_{j-1}) \\ &= \pi(2L - x_j - x_{j-1})[f(x_j) - g(x_j)]\,(x_j - x_{j-1}). \end{aligned}$$

The volume of the entire approximation of the solid of revolution is the sum of expressions (5) for $j = 1, 2, 3, \ldots, N$:

$$(6) \qquad \sum_{j=1}^{N} \pi(2L - x_j - x_{j-1})[f(x_j) - g(x_j)]\,(x_j - x_{j-1}).$$

As the number of subintervals in the partition (3) tends to infinity and the lengths of the subintervals tend to zero, the approximation of the solid of revolution improves and the sum (6) tends to the integral

$$(7) \qquad \begin{aligned} \int_a^b \pi(2L &- x - x)[f(x) - g(x)]\,dx \\ &= \int_a^b 2\pi(L - x)[f(x) - g(x)]\,dx. \end{aligned}$$

We define the volume of the original solid to be this integral, which equals integral (2). ☐

A remark on approximating sums

The sum (6) is not exactly a Riemann sum for the integral (7). The expression

$$(8) \qquad \pi(2L - x - x)\,[f(x) - g(x)]$$

for the integrand on the left of (7) involves the variable x in four places. To obtain the approximating sum (6) for the integral (7), we replace three of the x's in (8) by x_j and the fourth by x_{j-1} rather than take the value of the integrand at a single point in the subinterval as in a Riemann sum. Nevertheless, sums of the form (6) tend to the integral (7) as the number of subintervals in the corresponding partition tends to infinity and their lengths tend to zero. The reasons for this are outlined in Exercise 12.

EXAMPLE 1   Sketch the region bounded by the parabolas $y = 1 - x^2$ and $y = x^2 - 1$, and compute the volume of the solid generated by rotating the region about the line $x = 2$.

Solution.   Figure 6.33 shows the region in the xy-plane, the axis of the solid, and the line that generates the cylinder at x for a typical value of x. The length of that line is the y-coordinate $1 - x^2$ of its top, minus the y-coordinate $x^2 - 1$ of its bottom: $(1 - x^2) - (x^2 - 1) = 2 - 2x^2$. This is the height of the cylinder at x. Its radius is the distance $x - (-3) = x + 3$ between x and -3 on the x-axis. The volume of the solid of revolution is

$$\int_{-1}^{1} A(x)\,dx = \int_{-1}^{1} 2\pi \begin{bmatrix} \text{Radius of the} \\ \text{cylinder at } x \end{bmatrix} \begin{bmatrix} \text{Height of the} \\ \text{cylinder at } x \end{bmatrix} dx$$

$$= \int_{-1}^{1} 2\pi [2 - x][2 - 2x^2]\,dx$$

$$= \int_{-1}^{1} 4\pi(x^3 - 2x^2 - x + 2)\,dx$$

$$= 4\pi[\tfrac{1}{4}x^4 - \tfrac{2}{3}x^3 - \tfrac{1}{2}x^2 + 2x]_{-1}^{1}$$

$$= 4\pi[\tfrac{1}{4}(1)^4 - \tfrac{2}{3}(1)^3 - \tfrac{1}{2}(1)^2 + 2(1)]$$

$$\quad - 4\pi[\tfrac{1}{4}(-1)^4 - \tfrac{2}{3}(-1)^3 - \tfrac{1}{2}(-1)^2 + 2(-1)]$$

$$= \tfrac{32}{3}\pi. \quad \square$$

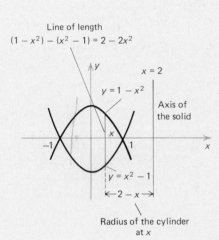

Line of length
$(1 - x^2) - (x^2 - 1) = 2 - 2x^2$

$x = 2$

$y = 1 - x^2$

Axis of the solid

$y = x^2 - 1$

$\leftarrow 2 - x \rightarrow$

Radius of the cylinder at x

FIGURE 6.33

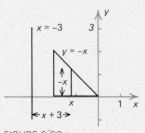

$x = -3$

$y = -x$

$-x$

x

$\leftarrow x + 3 \rightarrow$

FIGURE 6.33a

EXAMPLE 2   The triangle with vertices $(-2, 0)$, $(0, 0)$, and $(-2, 2)$ is rotated about the vertical line $x = -3$. What is the volume of the solid it generates?

Solution.   The triangle and the axis of the solid are shown in Figure 6.33a. The top of the triangle is formed by the line $y = -x$, so the length of the vertical line across the triangle at x ($-2 \le x \le 0$) is $-x$. This is the height of the cylinder at x. Its radius is the distance $x - (-3) = x + 3$ between x and -3 on the x-axis. The volume of the solid of revolution is

$$\int_{-2}^{0} 2\pi \begin{bmatrix}\text{Radius of the}\\ \text{cylinder at } x\end{bmatrix}\begin{bmatrix}\text{Height of the}\\ \text{cylinder at } x\end{bmatrix} dx$$

$$= \int_{-2}^{0} 2\pi(x+3)(-x)\,dx$$

$$= 2\pi\int_{-2}^{0}(-x^2-3x)\,dx = 2\pi[-\tfrac{1}{3}x^3 - \tfrac{3}{2}x^2]_{-2}^{0}$$

$$= 2\pi[-\tfrac{1}{3}(0)^3 - \tfrac{3}{2}(0)^2] - 2\pi[-\tfrac{1}{3}(-2)^3 - \tfrac{3}{2}(-2)^2]$$

$$= 2\pi[-\tfrac{8}{3} + 6] = \tfrac{20}{3}\pi. \quad \square$$

The method of cylinders with a y-integration

We use the method of cylinders with a y-integration if the region is bounded on the left and on the right by curves $x = M(y)$ and $x = N(y)$ and if the axis of the solid is horizontal. We let $A(y)$ denote the surface area of the cylinder generated by the horizontal line which crosses the region at y. Then the volume of the solid is given by an integral of the form

$$\int_{c}^{d} A(y)\,dy = \int_{c}^{d} 2\pi \begin{bmatrix}\text{Radius of the}\\ \text{cylinder at } y\end{bmatrix}\begin{bmatrix}\text{Height of the}\\ \text{cylinder at } y\end{bmatrix} dy. \quad ◀$$

EXAMPLE 3 Sketch the triangle bounded by the lines $x = y - 3$, $x = -y + 3$, and $y = 1$. Use the method of cylinders to compute the volume of the solid it generates when it is rotated about its base.

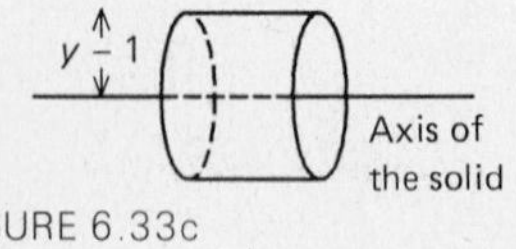

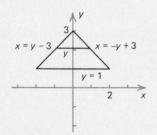

FIGURE 6.33b

FIGURE 6.33c

Solution. The triangle is shown in Figure 6.33b along with the horizontal line across it at y that generates the cylinder in Figure 6.33c. The length of the line is the x-coordinate $-y + 3$ of its right end minus the x-coordinate $y - 3$ of its left end: its length is $(-y + 3) - (y - 3) = -2y + 6$. The radius of the cylinder at y is the distance $y - 1$ between the horizontal line at y and the line $y = 1$. The triangle extends from $y = 1$ to $y = 3$, so the volume of the solid is

$$\int_{1}^{3} 2\pi \begin{bmatrix}\text{Radius of the}\\ \text{cylinder at } y\end{bmatrix}\begin{bmatrix}\text{Height of the}\\ \text{cylinder at } y\end{bmatrix} dy$$

$$= \int_{1}^{3} 2\pi(y-1)(-2y+6)\,dy = 2\pi\int_{1}^{3}(-2y^2+8y-6)\,dy$$

$$= 2\pi[-\tfrac{2}{3}y^3 + 4y^2 - 6y]_{1}^{3}$$

$$= 2\pi[-\tfrac{2}{3}(3)^3 + 4(3)^2 - 6(3)] - 2\pi[-\tfrac{2}{3}(1)^3 + 4(1)^2 - 6(1)]$$

$$= 2\pi[-18 + 36 - 18 + \tfrac{2}{3} - 4 + 6] = \tfrac{16}{3}\pi. \quad \square$$

Exercises

†Answer provided.
‡Outline of solution provided.

1. Give expressions, without absolute value signs, for the distance from the point (x, y) to **a.**† the line $x = 2$ where $x > 2$ **b.**† the line $x = -3$ where $x < -3$ **c.** the line $y = 4$ where $y < 4$ and **d.** the line $y = -6$ where $y > -6$.

In Exercises 2 through 10 sketch the regions and compute the volumes of the solids generated by rotating them about the indicated axes. (In Exercises 5 through 7 perform the calculations twice, once by an x-integration and once by a y-integration. In Exercises 8 through 10 simplify the calculations by using the symmetry of the solid.)

2.‡ The region bounded by the curves $y = x^{-1/2}$, $y = -x^{-1/2}$, $x = 1$, and $x = 2$; axes: **a.** $x = 0$ **b.** $x = 3$ **c.** $y = -2$

3.† The region bounded by $y = -\frac{1}{2}x^3$, $y = 4$ and $x = 1$; axes: **a.** $x = 2$ **b.** $y = 5$ **c.** $x = -3$

4. The region bounded by $y = x$ and $y = 3x - x^2$; axes: **a.** $y = 0$ **b.** $x = 0$ **c.** $y = 4$ **d.** $x = 4$

5.‡ The region bounded by $y = x$ and $y = \frac{1}{3}x^2$; axis: $x = 0$

6.† The region bounded by $y = x^3$, $y = 0$, and $x = 2$; axis: $x = 3$

7. The region bounded by $y = x^2$ and $y = x$; axis: $y = -1$

8.‡ The region bounded by $y = x^2$, $y = -x^2$ and $x = -2$; axis: $x = 0$

9.† The region bounded by $x = y^2$ and $x = 2 - y^2$; axis: $x = -1$

10. The region bounded by $x = y^2 + 1$, $x = -y^2 - 1$, $y = 1$, and $y = -1$; axis: $y = -2$

11.† Derive formula (4) for the volume of a cylindrical shell.

12.   **a.**† Suppose that $F(x)$ and $G'(x)$ are bounded for $a \le x \le b$ and that the integral of $F(x)G(x)$ from a to b exists. Let $a = x_0 < x_1 < x_2 < \cdots < x_N = b$ be a partition of the interval $a \le x \le b$ and let c_j and d_j be arbitrary points in the jth subinterval. Show that the sum

$$\sum_{j=1}^{N} F(c_j)G(d_j)(x_j - x_{j-1})$$

tends to the integral as the lengths of the subintervals in the partition tend to zero.   **b.** Use the result of part (a) to show that the sum (6) tends to the integral (7).

Additional drill exercises

In Exercises 13 through 17 use the method of cylinders to find the volume generated by rotating the region about the y-axis.

13.† The triangle with vertices $(0, 0)$, $(2, 2)$, and $(2, 1)$

14.† The region bounded by $y = x^2 - 3x$ and the x-axis

15.† The trapezoid with corners $(0, 3)$, $(0, -3)$, $(3, 0)$, and $(3, -3)$

16.† The region bounded by the curve $y = x^3 + 1$, the line $y = 9$, and the y-axis

17.† The region bounded by the curve $y = (\cos x)/x$, the x-axis, and the line $x = \frac{1}{6}$

In Exercises 18 through 20 use the method of cylinders to compute the volume generated by rotating the region about the x-axis.

18.[†] The region bounded by the curves $y = x^3$ and $x = y^2$

19.[†] The region bounded by the curve $xy = 3$ and the lines $x = 1, y = 1$

20.[†] The isosceles triangle with vertices $(0, 0)$, $(2, 2)$, and $(4, 0)$

6-5 LENGTHS OF CURVES

In this section we discuss a rule for computing lengths of portions of graphs of functions.

Rule 6.8   Suppose that a function f has a continuous derivative for $a \leq x \leq b$. Then the length of the portion of the graph $y = f(x)$ for $a \leq x \leq b$ is

(1) $$\int_a^b \sqrt{1 + [f'(x)]^2}\, dx.$$ ◀

Motivation for Rule 6.8.   If the graph of the function were a line segment, we could compute its length by using the Pythagorean Theorem. Therefore, we approximate the graph by a collection of line segments combined to form a polygonal line.

We consider an arbitrary partition

(2) $a = x_0 < x_1 < x_2 < x_3 < \cdots < x_{N-1} < x_N = b$

of the interval $a \leq x \leq b$. We approximate the graph of f by the line segments between successive points $(x_j, f(x_j))$ $(j = 0, 1, 2, \ldots, N)$ on the graph. Figure 6.34 shows the approximation of the graph of a function that arises from a partition into seven subintervals. The jth line segment runs from the point $(x_{j-1}, f(x_{j-1}))$ to the point $(x_j, f(x_j))$ and consequently has length

(3) $\sqrt{(x_j - x_{j-1})^2 + (f(x_j) - f(x_{j-1}))^2}.$

According to the Mean Value Theorem for derivatives there is a constant c_j with $x_{j-1} < c_j < x_j$ such that $f(x_j) - f(x_{j-1}) = f'(c_j)(x_j - x_{j-1})$. We make this substitution into formula (3) and use the fact that the number $x_j - x_{j-1}$ is positive to obtain the expression

(4) $\sqrt{(x_j - x_{j-1})^2 + [f'(c_j)]^2(x_j - x_{j-1})^2}$

$\qquad = \sqrt{1 + [f'(c_j)]^2}\,(x_j - x_{j-1})$

for the length of the jth segment of the polynomial line.

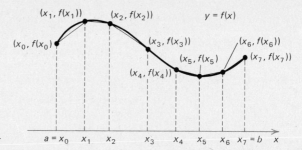

FIGURE 6.34

The length of the entire approximation of the curve is the sum of the lengths (4):

$$(5) \quad \sum_{j=1}^{N} \sqrt{1 + [f'(c_j)]^2} \, (x_j - x_{j-1}).$$

The sum (5), in turn, is a Riemann sum for the integral (1) and tends to that integral as the lengths of the subintervals in the corresponding partition (2) tend to zero. Accordingly, we define the length of the curve to be the integral (1) as stated in Rule 6.8. □

EXAMPLE 1   Compute the length of the curve $y = x^{3/2}$ for $0 \le x \le 1$.

Solution.   We set $f(x) = x^{3/2}$. Then $f'(x) = \frac{3}{2}x^{1/2}$, and formula (1) with $a = 0$ and $b = 1$ gives the length

$$(6) \quad \int_0^1 \sqrt{1 + (\tfrac{3}{2}x^{1/2})^2} \, dx = \int_0^1 \sqrt{1 + \tfrac{9}{4}x} \, dx.$$

To evaluate the indefinite integral, we set $u = 1 + \frac{9}{4}x$ and $du = \frac{9}{4} \, dx$. We obtain

$$\int \sqrt{1 + \tfrac{9}{4}x} \, dx = \tfrac{4}{9} \int (1 + \tfrac{9}{4}x)^{1/2} \, (\tfrac{9}{4}) \, dx = \tfrac{4}{9} \int u^{1/2} \, du$$

$$= \tfrac{4}{9} \tfrac{1}{3/2} u^{3/2} + C = \tfrac{8}{27} u^{3/2} + C$$

$$= \tfrac{8}{27}(1 + \tfrac{9}{4}x)^{3/2} + C.$$

Thus, the length (6) of the curve is

$$\int_0^1 \sqrt{1 + \tfrac{9}{4}x} \, dx = [\tfrac{8}{27}(1 + \tfrac{9}{4}x)^{3/2}]_0^1$$

$$= [\tfrac{8}{27}(1 + \tfrac{9}{4}(1))^{3/2}] - [\tfrac{8}{27}(1 + \tfrac{9}{4}(0))^{3/2}]$$

$$= \tfrac{1}{27}(\sqrt{2197} - 8). □$$

Example 1 deals with one of the few cases where the integral that gives the length of the curve can be evaluated by the methods at our disposal. Because of the square root, we cannot evaluate the integrals that give lengths of most curves. We can only obtain the appropriate definite integral.

EXAMPLE 2   Give a definite integral that equals the length of the curve $y = \sin(3x)$ for $0 \leq x \leq \pi$.

Solution.   We set $f(x) = \sin(3x)$. Then $f'(x) = 3\cos(3x)$, and formula (1) with $a = 0$ and $b = \pi$ gives

$$\int_0^\pi \sqrt{1 + [3\cos(3x)]^2}\, dx = \int_0^\pi \sqrt{1 + 9\cos^2(3x)}\, dx$$

as the length of the curve.   □

Exercises   In Exercises 1 and 2 find the lengths of the curves.

†*Answer provided.*
‡*Outline of solution provided.*

1.† $y = \frac{1}{3}(x^2 + 2)^{3/2}$ for $-2 \leq x \leq 3$

2. $y = \frac{1}{3}x^3 + \frac{1}{4x}$ for $1 \leq x \leq 2$

In Exercises 3 through 8 give definite integrals which equal the lengths of the curves.

3.† The portion of $y = \sin x$ for $0 \leq x \leq \pi$

4. The portion of $y = x^3 + 3x^2 - 4$ for $-10 \leq x \leq 10$

5.† The portion of the circle $x^2 + y^2 = 4$ for $-1 \leq x \leq 1$

6. The portion of the hyperbola $y = 1/x$ for $1/5 \leq x \leq 5$

7.† The portion of the ellipse $x^2 + 9y^2 = 36$ for $1 \leq y \leq 2$

8. The portion of the hyperbola $y^2 - x^2 = 3$ in the rectangle $-1 \leq x \leq 1$, $-2 \leq y \leq 2$

9.† Compute the length of the curve $y = x^{3/2}$ for $1 \leq x \leq 16$.

10. Use formula (1) of this section to compute the length of the portion of the line $y = 6x + 4$ for $0 \leq x \leq 5$.

11.† Compute the length of the graph of $\frac{2}{3}x^{3/2}$ from $x = 0$ to $x = 4$.

12.† Compute the length of the graph of $2(x^2 + \frac{1}{3})^{3/2}$ from $x = 1$ to $x = 2$.

6-6 AREAS OF SURFACES OF REVOLUTION

If a portion of the graph $y = f(x)$ is rotated about a horizontal line in the xy-plane, it generates a SURFACE OF REVOLUTION. The curve in Figure 6.35, for example, generates the surface in Figure 6.36 when it is rotated about the line $y = L$. In this section we see how integrals are used to determine the areas of such surfaces.

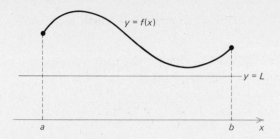

FIGURE 6.35   The portion of the
curve $y = f(x)$ for $a \leq x \leq b$

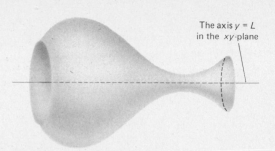

FIGURE 6.36   The surface
generated by rotating the curve of
Figure 6.35 about the line $y = L$

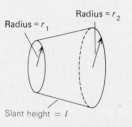

FIGURE 6.37   Area $= \pi(r_1 + r_2)\ell$

**Rule 6.9   Suppose that a function f has a continuous
derivative for $a \leq x \leq b$. The surface generated by rotating the
portion of the graph of f for $a \leq x \leq b$ about a horizontal line
has surface area**

(1) $\displaystyle\int_a^b 2\pi \left| \begin{matrix} \textbf{Radius of the} \\ \textbf{surface at x} \end{matrix} \right| \sqrt{1 + [f'(x)]^2}\, dx.$ ◄

Motivation for Rule 6.9.   If the graph of the function were a straight
line, the surface of revolution would be the frustrum of a right circular
cone (the portion of the cone between two planes perpendicular to its
axis), as in Figure 6.37. In this case we could use the formula

(2) $\pi(r_1 + r_2)\ell$

to find its area (see Exercises 1 and 2). Here r_1 and r_2 are the radii of the
ends of the frustrum of the cone and ℓ is its slant height.

To derive Rule 6.9, we use a partition

(3) $a = x_0 < x_1 < x_2 < x_3 < \cdots < x_{N-1} < x_N = b$

of the interval $a \leq x \leq b$ to construct an approximation of the graph
by a polygonal line as we did in deriving Rule 6.8 for the length of a
curve. We suppose that the curve being rotated lies above the axis of
revolution, as in Figure 6.35. In this case the radius of the surface at x
is $f(x) - L$, so we need to show that the surface area is given by the
integral

(4) $\displaystyle\int_a^b 2\pi[f(x) - L]\sqrt{1 + [f'(x)]^2}\, dx.$

If the curve is below the axis, we must replace $f(x) - L$ by $L - f(x)$
in this integral.

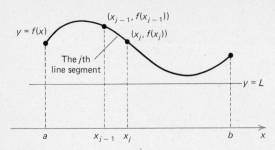

FIGURE 6.38

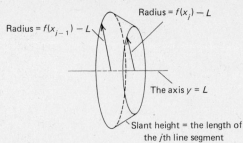

FIGURE 6.39

The jth segment of the polygonal line approximation of the graph runs between the points $(x_{j-1}, f(x_{j-1}))$ and $(x_j, f(x_j))$ (Figure 6.38). When it is rotated about the horizontal line $y = L$, it generates the frustrum of a cone shown in Figure 6.39. The radii of the ends of this cone are $r_1 = f(x_{j-1}) - L$ and $r_2 = f(x_j) - L$. The slant height is the length of the line segment in Figure 6.38. We saw in formula (4) of Section 6.5 that the length may be expressed in the form

$$\ell = \sqrt{1 + [f'(c_j)]^2}\,(x_j - x_{j-1})$$

where c_j is a point in the jth subinterval of the partition. By formula (2) the area is

$$
\begin{aligned}
(5) \quad & \pi(r_1 + r_2)\ell \\
& = \pi[(f(x_{j-1}) - L) + (f(x_j) - L)]\sqrt{1 + [f'(c_j)]^2} \\
& \quad (x_j - x_{j-1}) \\
& = \pi[f(x_{j-1}) + f(x_j) - 2L]\sqrt{1 + [f'(c_j)]^2}\,(x_j - x_{j-1}).
\end{aligned}
$$

The sum

$$(6) \quad \sum_{j=1}^{N} \pi[f(x_{j-1}) + f(x_j) - 2L]\sqrt{1 + [f'(c_j)]^2}\,(x_j - x_{j-1})$$

of the areas (5) is the area of the surface generated by rotating the entire polygonal approximation of the original curve. It is also an approximating sum for the integral

$$(7) \quad \int_a^b \pi[f(x) + f(x) - 2L]\sqrt{1 + [f'(x)]^2}\,dx$$

and tends to that integral as the lengths of the subintervals in the partition tend to zero. Since integral (7) is the same as integral (4), we are led to Rule 6.9.   □

EXAMPLE 1   When the portion of the curve $y = \cos x$ for $0 \le x \le 2\pi$ that is shown in Figure 6.40 is rotated about the line $y = -6/5$, it generates the hour-glass shaped surface in Figure 6.41. Give a definite integral which equals the area of that surface.

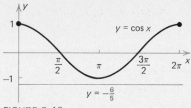

FIGURE 6.40

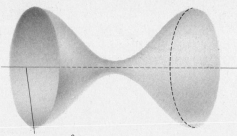

The axis $y = -\frac{6}{5}$
in the xy-plane

FIGURE 6.41   The surface of
revolution

Solution.   The radius of the surface at x is the y-coordinate $\cos x$ of the point on the curve at x minus the y-coordinate $-6/5$ of the axis of the surface:

(8)   [The radius of the surface at x] $= \cos x - \left(-\dfrac{6}{5}\right) = \cos x + \dfrac{6}{5}.$

We set $f(x) = \cos x$. Then $f'(x) = -\sin x$, and formulas (1) and (8) give the integral

$$\int_0^{2\pi} 2\pi\left[\cos x + \frac{6}{5}\right] \sqrt{1 + (-\sin x)^2}\, dx$$

$$= \int_0^{2\pi} 2\pi\left[\cos x + \frac{6}{5}\right] \sqrt{1 + \sin^2 x}\, dx$$

for the area of the surface.   □

Exercises

†*Answer provided.*
‡*Outline of solution provided.*

1.† Derive the formula for the area of the lateral surface of a right circular cone of radius r and height h from the formula for the area of a circle.

2. Derive formula (2) from the result of Exercise 1.

3.† Compute the area of the surface generated by rotating the portion of the cubic $y = x^3$ for $0 \le x \le 2$ about the x-axis.

In Exercises 4 through 11 give definite integrals that equal the areas of the surfaces generated by rotating the curves about the given axes.

4.† The portion of the parabola $y = x^2$ for $-1 \le x \le 1$; axis: $y = 0$

5. The portion of $y = x^{-2}$ for $1 \le x \le 4$; axis: $y = -2$

6.‡ The portion of the cubic $y = x^3$ for $1 \le x \le 2$; axis: $x = -1$

7.† The portion of the parabola $x = -y^2$ for $-2 \le y \le 2$; axis: $x = 2$

8. The portion of $y = \sin(\frac{1}{2}x)$ for $0 \le x \le 2\pi$; axis: $y = 0$

9.† The portion of $y = x + x^{-1}$ for $1 \le x \le 2$; axis: $y = 0$

10.† The portion of $y = (\cos x)/x$ for $\frac{1}{2}\pi \le x \le \frac{3}{2}\pi$; axis: $y = 0$

11.† The portion of the semicircle $y = \sqrt{16 - x^2}$ for $-3 \le x \le 3$; axis: $y = 0$

6-7   INTEGRALS OF RATES OF CHANGE

In this section we use integrals to obtain information about functions from their derivatives. We can work most of the examples and exercises by using either definite or indefinite integrals. We use a definite integral in our first example and an indefinite integral in the second.

EXAMPLE 1   At time $t = 0$ (hours) a car is 45 miles south of a gas station. It is driven north with the velocity of $60t - 30t^2$ miles per hour at time t until $t = 2$, when it runs out of gas. How far is it from the gas station at that time?

Solution.   We introduce an s-axis with its scale measured in miles, its positive side pointing north, and its origin at the gas station (Figure 6.42).

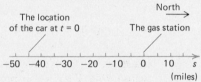

FIGURE 6.42

We let $s = s(t)$ be the car's coordinate on the s-axis at time t. Then the car's northward velocity at time t is ds/dt miles per hour, and the Fundamental Theorem gives the equation

$$\int_0^2 \frac{ds}{dt}(t)\, dt = s(2) - s(0).$$

Since $ds/dt = 60t - 30t^2$, we obtain

$$s(2) - s(0) = \int_0^2 (60t - 30t^2)\, dt = [\,60(\tfrac{1}{2}t^2) - 30(\tfrac{1}{3}t^3)\,]_0^2$$
$$= [\,30t^2 - 10t^3\,]_0^2 = [\,30(2)^2 - 10(2)^3\,]$$
$$- [\,30(0)^2 - 10(0)^3\,]$$
$$= 120 - 80 = 40.$$

Because $s(0) = -45$, we have $s(2) = -5$. The car is still 5 miles south of the gas station when it runs out of gas.   □

EXAMPLE 2   A tank of saline solution contains 50 pounds of salt at time $t = 1$ (minutes) and salt is added at the rate of $5 + 20t^{-3/2}$ pounds per minute at time t for $1 \le t \le 100$. How much salt is in the tank at $t = 100$?

Solution.   We let $f(t)$ (pounds) be the amount of salt in the tank at time t. Then we have

$$f(t) = \int \frac{df}{dt}(t)\, dt = \int (5 + 20t^{-3/2})\, dt$$

(1)

$$= 5t + \frac{20}{-1/2} t^{-1/2} + C = 5t - 40t^{-1/2} + C.$$

To determine the constant of integration, we set $t = 1$ in equation (1). Since $f(1) = 50$, we obtain $50 = f(1) = 5(1) - 40(1)^{-1/2} + C = -35 + C$. This gives $C = 85$, so we have the formula $f(t) = 5t - 40t^{-1/2} + 85$ for the function f.

At time $t = 100$ the tank contains $f(100) = 5(100) - 40(100)^{-1/2} + 85 = 581$ pounds of salt. □

Integration of acceleration Because an object's acceleration is the derivative of its velocity, we can determine its velocity by integrating its acceleration. In particular, if an object moving on an s-axis has constant acceleration a in the positive direction, then its velocity at time t is

(2) $$v(t) = \int a\, dt = at + v_0.$$ ◀

Here we use v_0 to denote the constant of integration because it is the value $v(0)$ of the velocity at $t = 0$.

Integrating formula (2) then gives the object's s-coordinate at time t

(3) $$s(t) = \int v(t)\, dt = \int (at + v_0)\, dt = \frac{1}{2}at^2 + v_0 t + s_0.$$ ◀

Here we denote the constant of integration by s_0 since it is the object's s-coordinate at $t = 0$.

EXAMPLE 3   A bullet is shot straight up into the air from the ground with a muzzle velocity of 480 feet per second. What is its height above the ground t seconds later? When does it hit the ground? (Take 32 feet per second^2 as the downward acceleration of gravity and disregard air resistance.)

Solution.   We let $s(t)$ (feet) be the bullet's height above the ground t seconds after it is shot. The bullet has the constant *upward acceleration* $s''(t) = -32$ feet/second^2 from the time it is shot until it hits the ground. Hence, by formula (3), its height above the ground is $s(t) = -16t^2 + v_0 t + s_0$ during that time period. Here v_0 is the bullet's upward velocity and s_0 is its height at $t = 0$. We have $v_0 = 480$ and $s_0 = 0$ and therefore $s(t) = -16t^2 + 480t$. This function is zero at $t = 0$ and at $t = 30$, so the bullet hits the ground 30 seconds after it is shot. □

Speed   In everyday English the terms "velocity" and "speed" have the same meaning. However, in calculus and physics these two terms are given different meanings. The SPEED of an object is defined to be the absolute value of its velocity:

$$\text{Speed} = |\text{Velocity}|. \qquad \blacktriangleleft$$

Thus, the speed is always positive or zero and only tells us how fast the object is going. The velocity can be positive or negative or zero. Its sign indicates whether the object is traveling in the positive or in the negative direction.

Total and net distance

Suppose that an object moving on an s-axis has positive velocity $v(t)$ during portions of a time interval $a \leq t \leq b$ and negative or zero velocity during the remainder of the interval. Then the *total distance* the object travels from time $t = a$ to time $t = b$ is the integral of its speed:

$$(4) \qquad \text{Total distance traveled} = \int_a^b [\text{Speed}]\, dt = \int_a^b |v(t)|\, dt. \qquad \blacktriangleleft$$

Because the integral of the velocity from a to b is the object's coordinate at $t = b$ minus its coordinate at $t = a$, the integral of the velocity is the *net distance* the object travels in the positive direction during the time interval $a \leq t \leq b$.

EXAMPLE 4   An object has velocity $10t^2 - 20t$ meters per second in the positive direction on an s-axis at time t (seconds). What is the net distance it travels in the positive direction during the time interval $0 \leq t \leq 3$? What is the total distance it travels during that time?

Solution.   The net distance is

$$\int_0^3 v(t)\, dt = \int_0^3 (10t^2 - 20t)\, dt = \left[\frac{10}{3}t^3 - 10t^2\right]_0^3$$

$$= \left[\frac{10}{3}(3)^3 - 10(3)^2\right] - \left[\frac{10}{3}(0)^3 - 10(0)^2\right]$$

$$= 90 - 90 = 0 \text{ meters.}$$

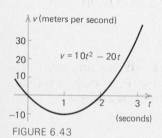

FIGURE 6.43

The object is at the same place at $t = 3$ as at $t = 0$.

The velocity $v(t) = 10t^2 - 20t = 10t(t - 2)$ is zero at $t = 0$ and at $t = 2$. It is negative for $0 < t < 2$ and positive for $2 < t \leq 3$. This is shown by the sketch of the curve $v = 10t^2 - 20t$ in Figure 6.43. Therefore the total distance the object travels during the time $0 \leq t \leq 3$ is

$$\int_0^3 |v(t)|\, dt = \int_0^2 -v(t)\, dt + \int_2^3 v(t)\, dt$$

$$= \int_0^2 (-10t^2 + 20t)\, dt + \int_2^3 (10t^2 - 20t)\, dt$$

$$= \left[-\frac{10}{3}t^3 + 10t^2 \right]_0^2 + \left[\frac{10}{3}t^3 - 10t^2 \right]_2^3$$

$$= \left[-\frac{10}{3}(2)^3 + 10(2)^2 \right] - \left[-\frac{10}{3}(0)^3 + 10(0)^2 \right]$$

$$+ \left[\frac{10}{3}(3)^3 - 10(3)^2 \right] - \left[\frac{10}{3}(2)^3 - 10(2)^2 \right]$$

$$= \frac{80}{3} \text{ meters.} \quad \square$$

Weight and mass

The WEIGHT of an object is the force of gravity on it. Weight is not an intrinsic characteristic because it depends on the object's location. On the moon, for example, an object weighs one sixth of what it weighs on the earth. To obtain an intrinsic quantity, we divide the object's weight by the acceleration of gravity at its location. This ratio is the object's MASS. It is constant because if the weight changes by a certain factor from one place to another, the acceleration of gravity changes by the same factor.

If time is measured in seconds, distances in feet, and forces in pounds, then mass is measured in slugs (a term with the same etymology as the word "sluggish"). On the surface of the earth, where the acceleration of gravity is 32 feet per second^2, an object's mass in slugs is its weight in pounds divided by 32. If distances are measured in centimeters and forces in dynes, mass is measured in grams. On the surface of the earth the acceleration of gravity is 980 centimeters per second^2, so an object's mass in grams is its weight in dynes divided by 980.*

Newton's law

The relation between weight, mass, and acceleration of gravity is a special case of NEWTON'S LAW

(5) $F = ma.$ ◄

Here we suppose that the object is moving on a coordinate line and m is its mass. Then F denotes the total force on the object in the positive direction at a certain instant and a denotes its acceleration in the positive direction at that instant.

EXAMPLE 5   An object weighing 96 pounds is subject to a force of $3t^2$ pounds in the positive direction on an x-axis at time t (seconds). The object is at $x = 3$ (feet) and has velocity 2 feet per second in the positive direction at time $t = 0$. Give the object's x-coordinate as a function of t for $t > 0$. ($g = 32$ feet per second^2.)

---

*In countries which employ the metric system, grams are also used as a unit of weight. Then one gram weight is 980 dynes or the weight of an object of one gram mass. (100 grams weight is about 3.5 ounces.)

Solution.   The object's mass is $m = 96/32 = 3$ slugs, so its acceleration and the force on it are related by Newton's law $a = (1/m)F = \frac{1}{3}F$. Since $F = 3t^2$, the object's acceleration is t^2 feet per second^2. Its velocity is of the form $\int t^2 \, dt = \frac{1}{3}t^3 + C$, and setting $t = 0$ gives $C = 2$, so the velocity is $\frac{1}{3}t^3 + 2$ feet per second. The integral of the velocity is $\frac{1}{12}t^4 + 2t + C$. Setting $t = 0$ shows that this constant of integration is 3 and the object is at $x = \frac{1}{12}t^4 + 2t + 3$ feet at time t.   □

Exercises

†*Answer provided.*
‡*Outline of solution provided.*

In Exercises 1 through 6 determine the position function $s(t)$ from the given information. Here $v(t) = \dfrac{ds}{dt}(t)$ and $a(t) = \dfrac{dv}{dt}(t)$.

1.‡ $v(t) = t^3 + 4, s(0) = 5$

2.† $v(t) = 5 - t^{-2}, s(1) = 0 \ (t \geq 1)$

3. $v(t) = \sin t, s(0) = -1$

4.‡ $a(t) = t, v(0) = 3, s(0) = -2$

5.† $a(t) = 2 - t^2, v(0) = 15, s(0) = 3$

6. $a(t) = 50 - t^{-4}, v(1) = 3, s(1) = -2 \ (t \geq 1)$

7.‡ An object moving along an s-coordinate axis with its scale measured in meters has a velocity of $6t^{2/3}$ meters per second in the positive direction at time t. At $t = 1$ it is at $s = 0$. Where is it at $t = 8$?

8.† A store takes in money at the rate of $10t - t^2$ dollars per hour t hours after it has opened. How much money does it take in during its ten hour day?

9.   At time $t = 1$ (hours) a ship is 5 nautical miles north of a lighthouse. It travels north with the velocity of $20 + \frac{1}{5}t$ knots (nautical miles per hour). How far is it from the lighthouse at $t = 20$?

10.‡ Water flows into a tank at the rate of $10t^4 + 100t + 10$ cubic feet per day at time t (days) and leaks out at the constant rate of 6 cubic feet per day. At $t = 0$ the tank contains 25 cubic feet of water. How much does it contain a day later?

11.† An object's velocity in the positive direction on an s-axis is $t^2 - t$ yards per second at time t (seconds). At $t = 0$ it is at $s = 10$ (yards).   **a.** When is its velocity zero and where is it at those times?   **b.** At what time is it again at $s = 10$?

12.† Suppose that t hours after Creation the rain was falling at the rate of $g(t)$ inches per hour. Express the total rainfall during the seventh (24 hour) day in terms of the function g.

13.   If t years from now the world's population is increasing at the rate of $120(1 + t)$ millions per year and if the world will only hold 7,200 million more people than it has now, how much time is left?

14. Air is pumped into a balloon at the rate of $3\sqrt{t}$ cubic feet per minute t minutes after its inflation has begun. The balloon breaks when its volume reaches 16 cubic feet. How long does that take?

15.[‡] A ball is thrown vertically from the ground at time $t = 0$ (seconds) with a speed of 96 feet per second.   **a.** How high will it be for $t > 0$?   **b.** How high will it go?   **c.** At what time will it hit the ground and with what speed?

16.[†] A golf ball that is dropped from a window hits the ground in 3 seconds. How high is the window?

17. An arrow is shot straight into the air from a height of 100 feet and hits the ground ten seconds later. What is its initial upward velocity?

In Exercises 18 through 21 use the given information to determine the total distance traveled by the object during the specified time interval. The time is measured in seconds.

18.[‡] $v(t) = t^3 - 4t$ ft./sec.; $-3 \leq t \leq 1$

19. $v(t) = -5t$ m./sec.; $-5 \leq t \leq 5$

20.[‡] $a(t) = 6t$ cm./sec.²; $v(0) = -48$ cm./sec.; $0 \leq t \leq 5$

21. $a(t) = \sin t$ miles/sec.²; $v(0) = 0$ miles/sec.; $0 \leq t \leq 2\pi$

22.[†] Figure 6.44 shows the downward acceleration of a spacecraft during reentry from outer space.   **a.** What was the approximate change in the spacecraft's speed between times $t = 0$ and $t = 150$?   **b.** If the spacecraft weighed 1000 pounds on the surface of the earth, what was the force on it at $t = 120$?

23. Figure 6.45 shows the rates of flow of water down the main Nile and the Blue Nile during a typical year. Approximately how much water flows down   **a.** the main Nile and   **b.** the Blue Nile each year.

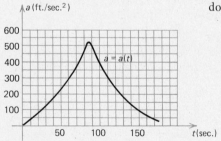

FIGURE 6.44   (Adapted from Encyclopaedia Britannica, Vol. 21, p. 1038)

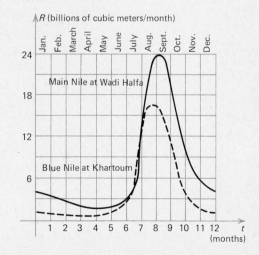

FIGURE 6.45   (Adapted from Encyclopaedia Britannica, Vol. 16, p. 521)

24.[†] A 128 pound object is subject to a force of $24t + 96$ pounds in the positive direction at time t (seconds). At $t = 0$ it is at rest. How far does it travel for $0 \leq t \leq 10$?

25. An object weighing 98 dynes is at $s = -10$ cm. and has a velocity of 20 cm./sec. in the positive direction at $t = 0$. It is subject to a force of $0.3 \sqrt{t}$ dynes in the positive direction at time t. Where is it at $t = 4$?

26. A rocket weighing 16 pounds is shot up vertically from the ground at $t = 0$ (seconds). The upward force of its engine is $47 - 21t^2$ pounds until $t = 1$ when it shuts off.   **a.** How high is the rocket and what is its velocity at $t = 1$?   **b.** What is its maximum upward velocity?   **c.** How high does it go?   **d.** When does it hit the ground? (Use the quadratic formula.)

In Exercises 27 through 30 ignore air resistance and use 32 feet/second² or 9.8 meters/second² as the acceleration due to gravity.

27.[†] A boy uses his slingshot to shoot a rock straight up in the air with an initial velocity of 14.7 meters/second. How high does it go?

28.[†] A ball is thrown down from a height of 32 feet above the ground with a speed of 16 feet per second.   **a.** How long does it take to hit the ground?   **b.** What is its speed then?

29.[†] An object that is thrown up in the air with initial velocity of 4.9 meters/second hits the ground 3 seconds later.   **a.** From what height was it thrown?   **b.** How high above the ground does it go?

30. How long does it take an object to hit the ground if it is dropped from a height of 100 meters?

31.[†] An object's acceleration on an s-axis is $12t^2$ meters per second² at time t (seconds). At $t = 0$ it is at $s = 0$ (meters) and its velocity is zero.   **a.** How long does it take to reach $s = 10$?   **b.** What is its velocity at that time?

32.[†] A man is $\frac{1}{3}t^3 - t^2 + 10$ kilometers west of an intersection on a straight road at time t (hours). What are   **a.** the net distance and   **b.** the total distance he travels from time $t = 0$ to $t = 3$?   **c.** What is his maximum speed for $0 \leq t \leq 3$ and when does it occur?

33. If an object's velocity on an s-axis is $\cos(\pi t)$ centimeters per second at time t (seconds), what is the total distance it travels between times $t = 0$ and $t = 3.5$?

34.[†] An object has constant acceleration of 2 meters/second² on an s-axis. At time $t = -1$ (seconds) its velocity is -5 meters/second and it is at $s = 8$ (meters).   **a.** At what two times is it at $s = 2$?   **b.** What is its least s-coordinate and when is it there?

6-8   WORK

Physicists introduce a concept which they call WORK to describe the transfer of energy: when a force does work on an object, it increases its kinetic or potential energy. We do not discuss energy here; we show only how work is computed. We begin with the definition of work in a case that does not require an integral.

Definition 6.1   Suppose that as an object moves a distance s in a straight line, it is subjected to a constant force F in the direction of the motion. Then the work done by the force on the object during the motion is

(1)   Work = [Force] [Distance] = Fs.   ◄

Notice that in this definition there is no reference to the nature of the object, to the nature of the force, or to the length of time it takes for the motion to occur. Also, the force being considered might be just one of a number of forces applied to the object during the motion.

If the force varies during the motion, then the work is defined by using an integral.

Rule 6.10   Suppose that an object moves on an s-axis and is subjected to a force $F(s)$ in the direction of its motion when it is at s and that the integral of $F(s)$ from a to b exists. Then the work done by the force on the object as it moves from $s = a$ to $s = b$ with a less than b is

(2) $\displaystyle\int_a^b F(s)\,ds.$ ◄

If a is greater than b, the work is

(3) $\displaystyle\int_b^a F(s)\,ds.$ ◄

Motivation for Rule 6.10.   Here we treat the case of $a < b$; the case of $a > b$ is left as Exercise 12.

We consider an arbitrary partition

(4) $a = s_0 < s_1 < s_2 < s_3 < \cdots < s_{N-1} < s_N = b$

of the interval $a \leq s \leq b$. We approximate the force on the object as it traverses the jth subinterval $s_{j-1} \leq s \leq s_j$ by the constant force $F(s_j)$. The jth subinterval has length $s_j - s_{j-1}$, so the work done by the approximating force on the object as it traverses the jth subinterval is $F(s_j)(s_j - s_{j-1})$. The work done by the approximating force on the object as it traverses the entire interval $a \leq s \leq b$ is the sum

(5) $\displaystyle\sum_{j=1}^N F(s_j)\,(s_j - s_{j-1}).$

Because the sum (5) is a Riemann sum for the integral (2), we define the work done by the original force to be that integral.   □

EXAMPLE 1   In pulling a train along a straight track, a locomotive engine exerts a force of $5 + \frac{1}{4}s$ tons when it is s miles past a station. How much work does the engine do during the first 6 miles?

Solution.   We introduce an s-axis along the track with its origin at the station, with its positive side pointing in the direction of the train's motion, and with its scale measured in miles. Then the force on the train is $F(s) = 5 + \frac{1}{4}s$ tons when the train is s miles past the station. The force is in the direction of the motion and the work done is

$$\int_0^6 F(s)\,ds = \int_0^6 \left(5 + \frac{1}{4}s\right)ds = \left[5s + \frac{1}{8}s^2\right]_0^6$$

$$= \left[5(6) + \frac{1}{8}(6)^2\right] - \left[5(0) + \frac{1}{8}(0)^2\right]$$

$$= \frac{69}{2}\text{ton-miles.}\quad\square$$

Springs   We consider a spring which is fastened to a rigid object at its left end and whose right end is free to move in a straight line (Figure 6.46). We say that the spring is at its REST POSITION and has its NATURAL LENGTH when it is neither stretched nor compressed and exerts no force in either direction. We indicate the location of the free end of the spring on an s-axis. We have the positive side of the s-axis point to the right and put its origin at the rest position of the end of the spring.

We let $F(s)$ denote the force which the spring exerts toward the right when its free end is at s. $F(s)$ is zero when s is zero and the spring is in its rest position. It is positive when s is negative because then the spring is compressed and exerts a force toward the right. $F(s)$ is negative when s is positive. Then the spring is stretched and exerts a force toward the left, which is a *negative* force toward the right.

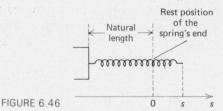

FIGURE 6.46

We suppose that the function F has the form

$$F(s) = -ks$$ ◄

with a positive constant k. This mathematical model of the forces exerted by a spring is called HOOKE'S LAW. It means that the graph of the function F is a line through the origin as in Figure 6.47. This model is used because it is easy to deal with. The graph of the function F for an actual spring might look more like the curve in Figure 6.48 in which case the approximation of this curve by a straight line would be reasonably accurate for small s but not for large s.

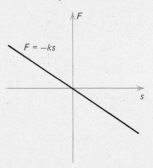

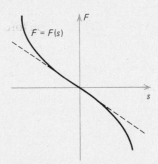

FIGURE 6.47   The function F given by Hooke's law

FIGURE 6.48   The function F for an actual spring

EXAMPLE 2   A spring which obeys Hooke's law has a natural length of 10 feet. It exerts a force of 3 pounds when it is compressed to a length of 8 feet. How much work is required to stretch it from its natural length to a length of 13 feet?

Solution.   We first need to find the constant k in Hooke's law $F(s) = -ks$. When the spring is compressed to a length of 8 feet, the coordinate of its free end is $s = -2$ and it exerts a force of 3 pounds toward the right. Therefore, we have $3 = F(-2) = -k(-2)$. This gives $k = \frac{3}{2}$ and $F(s) = -\frac{3}{2}s$.

When the end of the spring is at s the force *on* it toward the right is the negative $\frac{3}{2}s$ of the force *by* it toward the right. Therefore, stretching the spring from its natural length of 10 feet, for which $s = 0$, to a length of 13 feet, for which $s = 3$, involves

$$\int_0^3 \frac{3}{2}s\, ds = \left[\frac{3}{4}s^2\right]_0^3 = \left[\frac{3}{4}(3)^2\right] - \left[\frac{3}{4}(0)^2\right] = \frac{27}{4} \text{ foot-pounds}$$

of work.   □

Problems to which Rule 6.10 does not apply

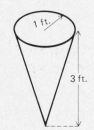

FIGURE 6.49

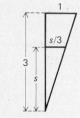

FIGURE 6.50

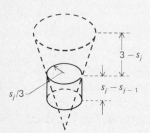

FIGURE 6.51   The jth disk

There are many problems involving the concept of work to which Rule 6.10 does not apply. In these instances the appropriate integral can be determined by using the analytic definition of the integral and an approximation procedure.

EXAMPLE 3   A water tank full of water has the shape of a right circular cone with its vertex pointing downward (Figure 6.49). It is 3 feet high and its top has a 1 foot radius. How much work is required to pump the water to the top of the tank? (Water has a density of 62.4 pounds per cubic foot.)

Solution.   We let s denote the distance (feet) from the bottom of the tank. By similar triangles (Figure 6.50), we see that the radius of the circular cross section of the tank at height s is $s/3$ feet.

We consider an arbitrary partition $0 = s_0 < s_1 < s_2 < s_3 < \cdots < s_{N-1} < s_N = 3$ of the interval $0 \leq s \leq 3$. The horizontal planes at heights $s = s_j$ ($j = 0, 1, 2, \ldots, N$) slice the tank into N pieces. We approximate the jth piece by a disk of radius equal to the radius of the tank at height s_j and of thickness equal to the thickness $s_j - s_{j-1}$ of the jth piece of the tank (Figure 6.51). The disk has volume $\pi(s_j/3)^2 (s_j - s_{j-1})$ cubic feet. It would therefore contain $[62.4 \text{ pounds/cu.ft.}][\pi(s_j/3)^2(s_j - s_{j-1}) \text{ cu.ft.}] = (62.4)\pi(s_j/3)^2 \cdot (s_j - s_{j-1})$ pounds of water. The top of this disk is $3 - s_j$ feet from the top of the tank. It would require approximately $(62.4)\pi(s_j/3)^2(3 - s_j) \cdot (s_j - s_{j-1})$ foot-pounds of work to pump the water to the top of the tank. The sum

$$\sum_{j=1}^{N} (62.4)\pi(s_j/3)^2 (3 - s_j)(s_j - s_{j-1})$$

is our approximate solution. The exact amount of work to pump the water to the top of the tank is the corresponding integral

$$\int_0^3 (62.4)\pi\left(\frac{s}{3}\right)^2 (3 - s)\,ds = (62.4)\pi \int_0^3 \left(\frac{1}{3}s^2 - \frac{1}{9}s^3\right)ds$$

$$= (62.4)\pi\left[\frac{1}{9}s^3 - \frac{1}{9(4)}s^4\right]_0^3$$

$$= (62.4)\pi\left\{\left[\frac{1}{9}(3)^3 - \frac{1}{9(4)}(3)^4\right] - \left[\frac{1}{9}(0)^3 - \frac{1}{9(4)}(0)^4\right]\right\}$$

$$= (62.4)\pi\left(\frac{3}{4}\right) = 46.8\,\pi \text{ foot-pounds.}   \square$$

Upper and lower approximate solutions (optional)

In most applications of integration an additional argument involving UPPER and LOWER APPROXIMATE SOLUTIONS can be given to make the derivation of the integral solution more convincing.

Let us consider the integral

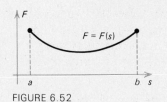

FIGURE 6.52

$$(6) \qquad \int_a^b F(s)\, ds$$

which gives the work done by a force $F(s)$ on an object that moves from $s = a$ to $s = b$ $(a < b)$ on an s-axis. We suppose that the graph of F is the curve in Figure 6.52 and that F is continuous.

A Riemann sum

$$(7) \qquad \sum_{j=1}^{N} F(c_j)\,(s_j - s_{j-1})$$

for the integral (6) is the approximate solution obtained by replacing F by the step function whose value in the jth subinterval of the corresponding partition is the value $F(c_j)$ of the function F at the point c_j in that subinterval. Figure 6.53 shows the graph of such a step function for a partition into six subintervals.

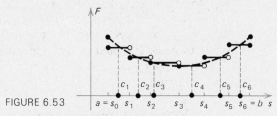

FIGURE 6.53

If we choose each c_j so that $F(c_j)$ is the *greatest value* of F in the jth subinterval (Figure 6.54), we obtain the upper approximate solution corresponding to the partition. If we choose the c_j to that $F(c_j)$ is the *least value* of F in the jth subinterval (Figure 6.55), we obtain the lower approximate solution corresponding to the partition.

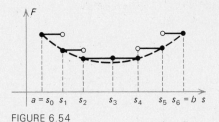

FIGURE 6.54

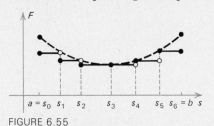

FIGURE 6.55

The force which is described by the step function of Figure 6.54 is greater than or equal to the original force for all s in the interval. The upper approximate solution is therefore greater than or equal to the exact solution, which is the work done by the original force. Similarly, the lower approximate solution is less than or equal to the exact solution:

(8) $\begin{bmatrix} \text{The lower} \\ \text{approximate} \\ \text{solution} \end{bmatrix} \leq \begin{bmatrix} \text{The exact} \\ \text{solution} \end{bmatrix} \leq \begin{bmatrix} \text{The upper} \\ \text{approximate} \\ \text{solution} \end{bmatrix}.$

These inequalities are valid for any partition of the interval. Because the upper and lower approximate solutions are Riemann sums for the integral (6), they tend to that integral as the lengths of the subintervals in the partition tend to zero. Inequalities (8) show that the exact solution is trapped between the upper and lower approximate solutions. The integral is therefore the exact solution.

Exercises

†*Answer provided.*
‡*Outline of solution provided.*

1.† The force of air resistance on an accelerating car is $\frac{1}{5}s$ pounds when the car has traveled s yards from its starting position. How much work does the air resistance do on the car in the first 100 yards?

2. A boy rolling a large boulder exerts $10 + 5 \sin s$ pounds of force on it when he has rolled it s feet. How much work does he do in rolling it 30 feet?

3.‡ A spring exerts a force of 6 ounces when it is stretched from its natural length of 15 inches to a length of 16 inches. How much work is required to compress it from a length of 14 inches to a length of 12 inches?

4.† A spring with a natural length of 10 feet exerts a force of 12 pounds when it is stretched to 12 feet. How much work is required to stretch it from its natural length to a length of 14 feet?

5. It takes 6 dyne-centimeters of work to compress a spring 2 centimeters from its natural length. What force does the spring exert when it is compressed 2 centimeters?

6.† The repulsive force between two negatively charged particles is equal to a constant divided by the square of the distance between them (Coulomb's law). Two negative charges exert a force of 2 dynes on each other when they are 10 centimeters apart. How much work is required to move one of them from a distance of 10 centimeters to a distance of 5 centimeters from the other?

7. The force of gravity on a meteorite that would weigh 1 pound on the surface of the earth is $(16 \times 10^6)s^{-2}$ pounds when it is s miles from the center of the earth. How much work does gravity do on such a meteorite as it falls from 96,000 miles above the earth to 1,000 miles above the earth? (Take 4,000 miles as the radius of the earth.)

8.‡ A bag of sand is lifted at the constant rate of 2 feet per second for 10 seconds. Initially the bag contains 100 pounds of sand but the sand leaks out at the constant rate of 3 pounds per second. How much work is done in lifting the bag?

9.† **a.** How much work is required to pump the water out of a full hemispherical fish bowl of radius 1 foot to 3 feet above the top of the bowl?   **b.** A one-tenth horsepower pump does 55 foot-pounds of work per second. How long would it take to empty the fishbowl? (The density of water is 62.4 pounds per cubic foot.)

10.   A ten foot long chain that weighs 4 pounds per foot is dangling from the top of a building. How much work is required to pull it up to the roof?

11.   A 5-pound monkey climbs up a 20 foot long dangling chain that weighs 3 pounds per foot. How much work does he do?

12.   Derive Rule 6.10 for $a > b$.

6-9   HYDROSTATIC PRESSURE

This section deals with the forces exerted by stationary bodies of water. The study of these forces is based on the concept of PRESSURE.

Pressure   If a sealed container, such as a balloon, is full of water and if external forces, such as gravity, are neglected, then the force exerted by the water on a flat plate inside the container is independent of the location and orientation of the plate. The force is directed perpendicular to the plate and is proportional to its area.

The constant of proportionality between the force exerted on the plate and its area is called the pressure of the water. It has the units of *force per area*. Thus, if the pressure is P pounds per square foot and if the plate has surface area A square feet, then the force on the plate is

(1)
$$\text{Force} = [\text{Pressure}][\text{Area}]$$
$$= \left[P\frac{\text{pounds}}{\text{sq. ft.}} \right] [A \text{ sq. ft.}] = PA \text{ pounds.} \quad \blacktriangleleft$$

Pressure in open bodies of water   In an open body of water, such as a bowl or swimming pool, pressure is caused by gravity and increases with the depth of the water. The pressure (pounds per square foot) at a depth of h feet is equal to the weight (pounds) of a column of water with a one-foot square base and a height of h feet. Since the density of water is 62.4 pounds per cubic foot, the weight of such a column of water is [62.4 pounds/cu. ft.] · [h cu. ft.] = 62.4 h pounds. Consequently,

(2)
$$\text{The pressure of water at a depth of } h \text{ feet} = 62.4\, h \frac{\text{pounds}}{\text{sq. ft.}}. \quad \blacktriangleleft$$

Because the pressure varies with the depth, the total force on a nonhorizontal plane region that is submerged in a body of water is given by an integral.

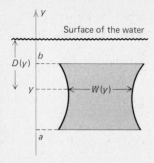

FIGURE 6.56   Submerged plate

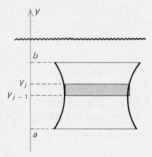

FIGURE 6.57 jth rectangle: height $= y_j - y_{j-1}$: width $= W(y_j)$

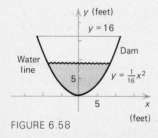

FIGURE 6.58

Rule 6.11   **Suppose that a vertical plate submerged in an open body of water extends from $y = a$ to $y = b$ $(a < b)$ on a y-axis with its scale measured in feet (Figure 6.56). Then the total force of the water on the plate is**

(3) $$\int_a^b 62.4 \left[\begin{array}{c}\textbf{Depth of the}\\ \textbf{water at } y\end{array}\right]\left[\begin{array}{c}\textbf{Width of the}\\ \textbf{plate at } y\end{array}\right] dy \textbf{ pounds} \blacktriangleleft$$

provided that the integral exists.

Motivation for Rule 6.11   We let $D(y)$ (feet) and $W(y)$ (feet) denote the depth of the water and the width of the plate at y on the coordinate axis (Figure 6.56). We want to show that the force on the plate is

(4) $$\int_a^b 62.4\, D(y)\, W(y)\, dy \text{ pounds.}$$

We consider an arbitrary partition

(5) $a = y_0 < y_1 < y_2 < y_3 < \cdots < y_{N-1} < y_N = b$

of the interval $a \leq y \leq b$. We use the partition to construct an approximation of the plate by a collection of rectangles such as we would use to find an integral with respect to y that gives the area of the region.

Figure 6.57 is a schematic sketch of the jth rectangle in the approximation. Its bottom is at $y = y_{j-1}$ and its top at $y = y_j$. Its width is taken to be the width $W(y_j)$ of the plate at $y = y_j$. Its area is $W(y_j)(y_j - y_{j-1})$. The water pressure at the top edge of the jth rectangle is $62.4\, D(y_j)$. Because the rectangle is narrow, the pressure on all of it is approximately equal to this value, and the force on the jth rectangle is approximately

(6) $[\text{Pressure}][\text{area}] = [62.4\, D(y_j)][W(y_j)(y_j - y_{j-1})]$ pounds.

The total force on the plate is approximately equal to the sum of the forces (6) or

(7) $$\sum_{j=1}^{N} 62.4\, D(y_j)\, W(y_j)\, (y_j - y_{j-1}) \text{ pounds.}$$

This is a Riemann sum for the integral (4), which gives the exact force on the plate.   □

EXAMPLE 1   A dam is in the shape of the region $\frac{1}{16}x^2 \leq y \leq 16$ with the scales on the coordinate axes measured in feet and with the y-axis pointing upward. What is the total force of the water on the dam when the water is 9 feet deep?

Solution.   The dam is sketched in Figure 6.58. Solving the equation $y = \frac{1}{16}x^2$ for x gives $x = \pm 4\sqrt{y}$. This shows that the dam is $8\sqrt{y}$ feet wide at height y. The points at height y are $9 - y$ feet below the surface of the water. By formula (4) with $D(y) = 9 - y$, $W(y) = 8\sqrt{y}$, $a = 0$, and $b = 9$, the total force of the water on the dam is

$$\int_0^9 (62.4)(9 - y)(8\sqrt{y})\, dy = (62.4)(8)\int_0^9 (9y^{1/2} - y^{3/2})\, dy$$

$$= (62.4)\,(8)\left[6y^{3/2} - \frac{2}{5}y^{5/2} \right]_0^9$$

$$= (62.4)\,(8)\left[6(9)^{3/2} - \frac{2}{5}(9)^{5/2} \right]$$

$$= (62.4)\,(518.4)$$

$$= 32{,}348.16 \text{ pounds.} \quad \square$$

Rule 6.11 applies even when the y-axis is not vertical, as in the next example, provided the width $W(y)$ is measured perpendicular to the y-axis.

EXAMPLE 2   A plate has the shape of the region bounded by the parabola $x = y^2$, the line $y = 1$, and the y-axis in an xy-plane with its scales measured in feet. The plate is submerged in water with the line $y = 1$ along the surface and with its plane making an angle of $45°$ with the vertical. What is the force of water pressure on each side of the plate?

Solution.   The plate is shown in Figure 6.58a. Its width at y on the y-axis is $W(y) = y^2$ feet. Because the point at y is $1 - y$ feet from the point at 1 on the y-axis and because the y-axis makes an angle of $45°$ with the vertical, the point at y is $D(y) = (1 - y)/\sqrt{2}$ feet beneath the surface of the water (Figure 6.58b). By Rule 6.11 the force on one side of the plate is

$$\int_0^1 62.4[(1 - y)/\sqrt{2}][y^2]\, dy = \frac{62.4}{\sqrt{2}}\int_0^1 (y^2 - y^3)\, dy$$

$$= \frac{62.4}{\sqrt{2}}[\tfrac{1}{3}y^3 - \tfrac{1}{4}y^4]_0^1$$

$$= \frac{62.4}{\sqrt{2}}(\tfrac{1}{3} - \tfrac{1}{4}) = \frac{62.4}{12\sqrt{2}} \text{ pounds} \quad \square$$

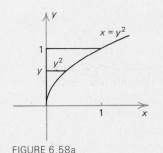

FIGURE 6.58a

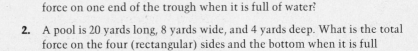

FIGURE 6.58b

Exercises

†*Answer provided.*
‡*Outline of solution provided.*

1.† The top of a trough is 10 feet long and 4 feet wide. The ends are vertical equilateral triangles with a vertex pointing downward. What is the total force on one end of the trough when it is full of water?

2. A pool is 20 yards long, 8 yards wide, and 4 yards deep. What is the total force on the four (rectangular) sides and the bottom when it is full of water?

3.† The back of a rowboat is a trapezoid which is 2 feet high, 4 feet wide at the bottom, and 6 feet wide at the top. What is the total force of the water on the back of the boat when it is 1 foot into the water?

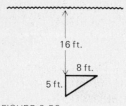

FIGURE 6.59

4.‡ An open cylindrical can of 4 inches radius and 10 inches high is full of paint weighing 110 pounds per cubic foot.   **a.** What is the force of the paint on the bottom of the can?   **b.** What is the force on the side?

5.† A rectangular can with a 2-foot square base contains water 1 foot deep with six inches of oil on top of the water. What is the total force on the sides of the can if the oil has density of 59.3 pounds per cubic foot?

6. A plate in the shape of a right triangle with legs of lengths 5 feet and 8 feet is submerged under water as in Figure 6.59. What is the force of the water on one side of the plate? (This force is balanced by an equal force on the other side of the plate.)

7.‡ A square plate, 10 feet wide, is submerged with its top horizontal and 15 feet below the surface of the water. The plate makes an angle of 30° with the vertical. What is the force of the water on one side of the plate?

8.† One side of a plate in the shape of an equilateral triangle is resting on the bottom of a pool 20 feet deep. The tip is resting on the side of the pool so that the plate makes an angle of 45° with the vertical. Each side of the plate is 1 foot long. What is the force of the water on one side of the plate?

9. The ends of a box are equilateral triangles and the three sides are squares. Each edge of the box is 4 feet long. What is the force on each of the ends when one side of the box lies on the horizontal bottom of a 10 foot deep pool? What is the force on each of the three sides?

10.† What is the force of water on one side of a vertical semi-circular disk of radius 2 feet whose flat edge is along the surface of the water?

11. A plate is in the shape of the isosceles trapezoid with vertices $(-3, 0)$, $(5, 0)$, $(2, 3)$, and $(0, 3)$ in an xy-plane with its scales measured in feet and its y-axis pointing upward. The plate is submerged in water with its surface at $y = 4$. What is the force of water pressure on each side of the plate?

12.† A plate is bounded by the curve $y = x^{2/3}$ and the line $y = 1$ in an xy-plane with its y-axis pointing upward and its scales measured in feet. The plate is submerged in oil of density 60 pounds per cubic foot with its straight edge along the surface of the oil. What is the force of the oil on each side of the plate?

13. What is the force of the water on each of the nonvertical sides of the trough in Exercise 1?

6-10   DENSITY, WEIGHT, AND CENTERS OF GRAVITY

When we speak of the LINEAR DENSITY of an object such as a straight rod, we mean its density measured in weight per unit length. If a rod of length L feet has constant linear density of ρ (rho) pounds per foot, then its weight is

(1)
$$\text{Weight} = [\text{Density}]\,[\text{Length}]$$
$$= \left[\rho\,\frac{\text{pounds}}{\text{foot}}\right][L \text{ feet}] = \rho\,L \text{ pounds.} \qquad \blacktriangleleft$$

If the rod has variable density, then its weight is defined by using an integral.

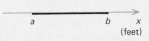

FIGURE 6.60   Rod of density $\rho(x)\dfrac{\text{pounds}}{\text{foot}}$ at x

Rule 6.12   Suppose that a rod extends from $x = a$ to $x = b$ ($a < b$) on an x-axis with its scale measured in feet (Figure 6.60), that the density of the rod is $\rho\,(x)$ pounds per foot at x, and that the integral of ρ from a to b exists. Then the weight of the rod is

(2)
$$\int_a^b \rho(x)\left[\frac{\textbf{pounds}}{\textbf{foot}}\right] dx\,[\textbf{feet}] = \int_a^b \rho(x)\,dx \textbf{ pounds.} \qquad \blacktriangleleft$$

Motivation for Rule 6.12.   We consider an arbitrary partition

(3) $a = x_0 < x_1 < x_2 < x_3 < \cdots < x_{N-1} < x_N = b$

of the interval $a \le x \le b$. If the rod had constant density equal to $\rho(x_j)$ over the jth subinterval $x_{j-1} \le x \le x_j$, then the weight of that portion of the rod would be

(4) $\left[\rho(x_j)\,\dfrac{\text{pounds}}{\text{foot}}\right][(x_j - x_{j-1}) \text{ feet}] = \rho(x_j)(x_j - x_{j-1}) \text{ pounds.}$

The sum of the weights (4) for $j = 1, 2, 3, \ldots, N$ is

(5) $\displaystyle\sum_{j=1}^{N} \rho(x_j)(x_j - x_{j-1})$

and is an approximation of the weight of the actual rod. We define the weight of the rod to be the integral (2), which is the limit of the Riemann sum (5) as the lengths of the subintervals of the partition (3) tend to zero. $\square$

EXAMPLE 1   The density of a 2 foot long rod is $2 + 6x$ pounds per foot at a distance of x feet from its left end. How much does it weigh?

Solution.   We introduce an x-axis along the rod, with its scale measured in feet and with its origin at the left end of the rod. Then the rod extends from $x = 0$ to $x = 2$ and its density at x is $\rho(x) = 2 + 6x$ pounds per foot. By Rule 6.12, its weight is

$$\int_0^2 \rho(x)\, dx = \int_0^2 (2 + 6x)\, dx = [2x + 3x^2]_0^2$$
$$= [2(2) + 3(2^2)] - [2(0) + 3(0^2)] = 16 \text{ pounds. } \square$$

Center of gravity of a rod

The CENTER OF GRAVITY (or CENTER OF MASS) of a rod is the point at which it would balance (Figure 6.61). If the rod has constant density, its center of gravity is at its midpoint. If it has variable density, its center of gravity is located by computing its MOMENT about a point on a line through the rod.

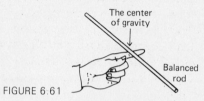

The center of gravity

Balanced rod

FIGURE 6.61

Moments of finite collections of points

Suppose that two children climb on a see-saw (Figure 6.62). It will balance if the weight of one child multiplied by his distance from the fulcrum is equal to the weight of the other child multiplied by his distance from the fulcrum.

To describe this statement algebraically, we introduce an x-axis through the see-saw with its origin at the fulcrum. We suppose that the child weighing m_1 pounds climbs on at $x = x_1$ ($x_1 > 0$), and that the child weighing m_2 pounds climbs on at $x = x_2$ ($x_2 < 0$). The see-saw balances if $m_1 x_1$ is equal to $m_2|x_2| = -m_2 x_2$, that is, if

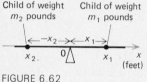

Child of weight m_2 pounds   Child of weight m_1 pounds

FIGURE 6.62

(6) $m_1 x_1 + m_2 x_2$

is zero. The positive side of the see-saw will drop if the quantity (6) is positive and the negative side will drop if (6) is negative.

The quantity (6) is the MOMENT of the children about the origin of the x-axis.

Definition 6.2   Suppose that N weights m_1, m_2, m_3, ..., m_N are on the x-axis with m_j at x_j for each j. Then the moment of the collection of weights about the origin is

(7) $$m_1 x_1 + m_2 x_2 + m_3 x_3 + \ldots + m_N x_N = \sum_{j=1}^N m_j x_j. \qquad \blacktriangleleft$$

For a rod placed along the x-axis, the moment about the origin is defined by an integral.

Rule 6.13   Suppose that a rod extends from $x = a$ to $x = b$ ($a < b$) on an x-axis and that the density of the rod is $\rho(x)$ at x. Then the moment of the rod about the origin is

(8) $\displaystyle\int_a^b x\,\rho(x)\,dx.$ ◀

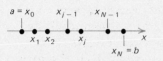

FIGURE 6.63   The jth point mass has weight $\rho(x_j)(x_j - x_{j-1})$.

Motivation for Rule 6.13.   We assume that the integral (8) exists and consider an arbitrary partition

(9) $\quad a = x_0 < x_1 < x_2 < x_3 < \cdots < x_{N-1} < x_N = b$

of the interval $a \le x \le b$. We use the partition to construct an approximation of the rod by N point masses (Figure 6.63).

We put a point mass at $x = x_j$ for $j = 1, 2, 3, \ldots, N$. We give the jth point mass the weight the portion of the rod in the jth subinterval $x_{j-1} \le x \le x_j$ would have if it had constant density equal to $\rho(x_j)$. Thus, we give the jth point mass the weight $\rho(x_j)(x_j - x_{j-1})$.

The moment of the N point masses about the origin is then given by the sum

(10) $\displaystyle\sum_{j=1}^N x_j\,\rho(x_j)\,(x_j - x_{j-1}).$

We define the moment of the rod about the origin to be the integral (8).   □

EXAMPLE 2   Compute the moment of the rod of Example 1 about the origin.

Solution.   The rod extends from $x = 0$ to $x = 2$ and its density at x is $\rho(x) = 2 + 6x$. Its moment about the origin is therefore

$$\int_0^2 x\,\rho(x)\,dx = \int_0^2 x(2 + 6x)\,dx = \int_0^2 (2x + 6x^2)\,dx$$
$$= [x^2 + 2x^3]_0^2 = [2^2 + 2(2^3)] - [0] = 20. \quad \square$$

A slight modification of the derivation of the above rule shows that the moment of the rod about a point $x = \bar{x}$ is

(11) $\displaystyle\int_a^b (x - \bar{x})\,\rho(x)\,dx = \int_a^b x\,\rho(x)\,dx - \bar{x}\int_a^b \rho(x)\,dx.$

We define the center of gravity of the rod to be the number $\bar{x}$ such that the moment (11) is zero. Setting expression (11) equal to zero and solving for $\bar{x}$ gives the following rule.

Rule 6.14   **If a rod extends from $x = a$ to $x = b$ ($a < b$) and if its density at x is $\rho(x)$, then its center of gravity is at $x = \bar{x}$ where**

$$(12) \quad \bar{x} = \frac{\displaystyle\int_a^b x\rho(x)\,dx}{\displaystyle\int_a^b \rho(x)\,dx}. \qquad \blacktriangleleft$$

We assume that the integrals in (12) exist and that the integral of the nonnegative function $\rho(x)$ is not zero. Notice that the center of gravity $\bar{x}$ is equal to the rod's moment about the origin divided by its weight.

EXAMPLE 3   Locate the center of gravity of the rod of Examples 1 and 2.

Solution.   By Example 1, the rod's weight is 16 and by Example 2, its moment about the origin is 20. Its center of gravity is therefore at $\bar{x} = \frac{20}{16} = \frac{5}{4}$. $\square$

Weight, moments, and centers of gravity of flat plates

The CENTER OF GRAVITY of a flat plate is the point at which it would balance on a pin (Figure 6.64). If the plate is lying in an xy-plane (Figure 6.65), we can locate its center of gravity by computing its *weight* and its *moments* about the x-axis and about the y-axis.

We restrict our attention to plates such as that shown in Figure 6.65, which occupy regions between the graphs of functions of x. We also suppose that the PLANAR DENSITY (the weight per unit area) of each plate is a function of x alone. (We will deal with the case where the density depends on y as well in Chapter 15, where we study two-dimensional integrals). We have the following rule.

Rule 6.15   **Suppose that a plate occupies the region beneath the graph $y = f(x)$ and above the graph $y = g(x)$ and extending from $x = a$ on the left to $x = b$ on the right. Suppose that the planar density of the plate is $\rho(x)$ at the point (x, y). Then**

$$(13) \quad \begin{array}{l}\textbf{The weight of}\\\textbf{the plate}\end{array} = \int_a^b \rho(x)\,[f(x) - g(x)]\,dx \qquad \blacktriangleleft$$

$$(14) \quad \begin{array}{l}\textbf{The moment of}\\\textbf{the plate about}\\\textbf{the y-axis}\end{array} = \int_a^b x\rho(x)[f(x) - g(x)]\,dx \qquad \blacktriangleleft$$

$$(15) \quad \begin{array}{l}\textbf{The moment of}\\\textbf{the plate about}\\\textbf{the x-axis}\end{array} = \frac{1}{2}\int_a^b \rho(x)\,[f(x)^2 - g(x)^2]\,dx \qquad \blacktriangleleft$$

and the coordinates $(\bar{x}, \bar{y})$ of the center of gravity are

$$(16) \quad \bar{x} = \frac{\textbf{Moment of the plate about the y-axis}}{\textbf{Weight of the plate}} \qquad \blacktriangleleft$$

$$(17) \quad \bar{y} = \frac{\textbf{Moment of the plate about the x-axis}}{\textbf{Weight of the plate}}. \qquad \blacktriangleleft$$

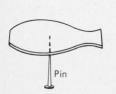

FIGURE 6.64   Balanced plate

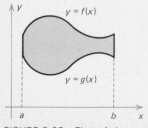

FIGURE 6.65   Plate of planar density $\rho(x)$ at (x, y)

Motivation for Rule 6.15.   We assume that the integrals exist and that integral (13) is not zero. Starting with an arbitrary partition $a = x_0 < x_1 < x_2 < x_3 < \cdots < x_N < x_{N-1} = b$ of the interval $a \le x \le b$, we use the partition to construct an approximation of the region by N rectangles such as we would use to study the region's area (Figure 6.66). We then approximate the jth rectangle by a uniform rod, located along the right side of the rectangle at $x = x_j$, and with weight equal to the weight

$$(18) \quad \rho(x_j) \left[f(x_j) - g(x_j) \right] (x_j - x_{j-1})$$

that the rectangle would have if it had constant density $\rho(x_j)$ (Figure 6.67).

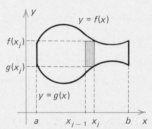

FIGURE 6.66   Rectangle of area $[f(x_j) - g(x_j)](x_j - x_{j-1})$

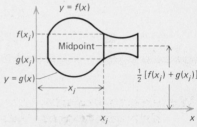

FIGURE 6.67   Rod of weight $\rho(x_j)[f(x_j) - g(x_j)](x_j - x_{j-1})$

This gives us integral (13) for the weight of the plate because the sum of the quantities (18) is the approximate weight and is a Riemann sum for that integral.

The moment of the jth rod about the y-axis is the product

$$x_j \rho(x_j) \left[f(x_j) - g(x_j) \right] (x_j - x_{j-1})$$

of its weight (18) with its x-coordinate. This gives integral (14) for the moment of the original plate about the y-axis.

The moment of the jth rod about the x-axis is the product

$$\frac{1}{2}[f(x_j) + g(x_j)] \, \rho(x_j) [f(x_j) - g(x_j)](x_j - x_{j-1})$$

$$= \frac{1}{2}\rho(x_j) [f(x_j)^2 - g(x_j)^2] (x_j - x_{j-1})$$

of its weight and the y-coordinate of its midpoint. This gives integral (15) for the moment of the original plate about the x-axis.

The center of gravity is the point $(\bar{x}, \bar{y})$ such that the moments of the plate about the lines $x = \bar{x}$ and $y = \bar{y}$ are zero. This leads to formulas (16) and (17). $\square$

EXAMPLE 4   A flat plate occupies the quarter circle $x^2 + y^2 \le 1$, $x \ge 0, y \ge 0$ in an xy-plane with its scales measured in feet. The density of the plate is 7 pounds per square foot. Compute the moment of the plate about the x-axis and the coordinates of its center of gravity.

Solution. The top of the plate is the graph of the function $f(x) = \sqrt{1 - x^2}$ and its bottom is the graph of the function $g(x) = 0$. It extends from $x = 0$ on the left to $x = 1$ on the right. Hence, by formula (15), its moment about the x-axis is

$$\frac{1}{2} \int_0^1 \left(7 \frac{\text{pounds}}{\text{sq. ft.}} \right) (\sqrt{1 - x^2}\,\text{ft.})^2 \, dx \, [\text{feet}]$$

$$= \frac{7}{2} \int_0^1 (1 - x^2) \, dx = \frac{7}{2} \left[x - \frac{1}{3}x^3 \right]_0^1 = \frac{7}{2} \left(1 - \frac{1}{3} \right) = \frac{7}{3}\text{foot-pounds.}$$

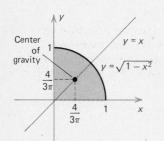

Center of gravity

$y = x$

$y = \sqrt{1 - x^2}$

$\dfrac{4}{3\pi}$

FIGURE 6.68

The area of the quarter circle is $\frac{1}{4}\pi$ sq. ft., so the plate weighs (7 pounds/sq. ft.) ($\frac{1}{4}\pi$ sq. ft.) $= 7\,\pi/4$ pounds. By formula (17), the y-coordinate of the center of gravity is

$$\overline{y} = \frac{\text{Moment about the } x\text{-axis}}{\text{Weight}} = \frac{7/3 \text{ foot-pounds}}{7\pi/4 \text{ pounds}} = \frac{4}{3\pi} \text{ feet.}$$

The symmetry of the plate about the line $y = x$ (Figure 6.68) implies that $\overline{x} = \overline{y}$. Hence the coordinates of the center of gravity are $(4/3\pi, 4/3\pi)$. ☐

Centroids

The center of gravity of an object with *constant density* is also known as the CENTROID of the object. Its location does not depend on the value of the constant density (see Exercise 9).

Centroids of graphs of functions

We can find the centroid of a curve which is the graph of a function by using the following rule. Its derivation is left as Exercise 14.

Rule 6.16 Suppose that *f(x)* has a continuous derivative for $a \le x \le b$ and that the curve

(19) $y = f(x)$ $a \le x \le b$

has density of 1 unit of weight per unit of length. Then its moment about the y-axis is

(20)   Moment about the y-axis $= \displaystyle\int_a^b x\sqrt{1 + [f'(x)]^2} \, dx$ ◄

and its moment about the x-axis is

(21)   Moment about the x-axis $= \displaystyle\int_a^b f(x)\sqrt{1 + [f'(x)]^2} \, dx.$ ◄

The weight of curve (19) is equal to its length, so its centroid has coordinates

(22) $\overline{x} = \dfrac{\text{Moment about the y-axis}}{\text{Length}}$ ◄

(23) $\overline{y} = \dfrac{\text{Moment about the x-axis}}{\text{Length}}.$ ◄

The theorems of Pappus

The Greek geometer Pappus (ca. 300 A.D.) derived the following results dealing with solids and surfaces of revolution.

Pappus' first theorem   **If a region R lies on one side of a line L in its plane, then the volume of the solid generated by rotating R about L is equal to the area of R multiplied by the distance traveled by its centroid.**

Pappus' second theorem   **If a curve C lies on one side of a line L in its plane, then the surface area of the surface generated by rotating C about L is equal to the length of the curve multiplied by the distance traveled by its centroid.**

These results are derived in Exercises 10 and 11 and are applied in Exercises 12 and 13.

Exercises

†*Answer provided*
‡*Outline of solution provided*

1. Find the centers of gravity of the following collections of point masses.
   a.† 2 pounds at $(-3, 0)$, 1 pound at $(0, 0)$, and 5 pounds at $(2, 0)$ **b.** 5 pounds at each of the points $(-10, 0)$, $(2, 0)$, $(3, 0)$, and $(4, 0)$ **c.**‡ 6 pounds at $(2, -2)$, 3 pounds at $(1, 1)$, and 1 pound at $(3, 4)$ **d.**† 1 pound at $(5, 5)$, 3 pounds at $(-1, 0)$, and 3 pounds at $(0, -1)$ **e.** 1 pound at $(-3, -4)$, 2 pounds at $(-3, 2)$, and 3 pounds at $(7, 0)$

2. Find the centroids of   **a.**‡ the region in Figure 6.69   **b.**† the region in Figure 6.70   **c.** the region in Figure 6.71.

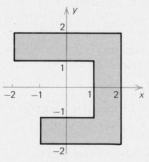

FIGURE 6.69

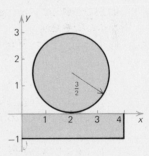

FIGURE 6.70

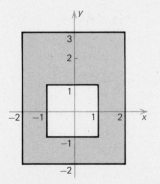

FIGURE 6.71

3. Find the weight and center of gravity of   **a.**† a rod extending from $x = -3$ (feet) to $x = 3$ (feet) and of density $3 + x^2$ pounds per foot at x **b.** a rod of length 3π (meters) and of density $5 + \sin(3x)$ kilograms per meter at a distance of x meters from its left end.

4. Derive a formula for the center of gravity of a rod lying along an x-axis that uses the moment about an arbitrary point $x = x_0$ rather than about the origin.

5. Show that formula (12) for the center of gravity of a rod gives the midpoint when the density is constant.

6. Find the weights and centers of gravity of the following plates. The scales on the coordinate axes are measured in feet.   **a.**[†] The plate bounded by $y = x^2$ and $y = x^3$ with density $\sqrt{x}$ pounds per square foot   **b.** The plate bounded by $y = 4 - x^2$ and $y = x^2 - 4$ with density $x^2 + 2$ ounces per square foot   **c.**[†] The plate bounded by $y = 1 + x^3$, $y = -1 - x^3$, and $x = 1$ with density $3 - x^2$ pounds per square foot   **d.** The plate bounded by $x = y^2$ and $x = 9$ with density $2x$ pounds per square foot

7. Locate the centroids of the regions bounded by the following collections of curves. Use symmetry whenever possible.   **a.**[†] $y = \sqrt{x}$, $y = 0$, and $x = 4$ **b.** $y = x^3$, $y = -1$, and $x = 1$ **c.**[†] $y = x^{1/3}$, $y = 0$, $x = -8$, and $x = 8$ **d.** $y = x^2$ and $x = y^2$

8. Work Example 4 by using formula (14).

9. Show that the center of gravity of a plate is not changed when its density is multiplied by a constant factor.

10. Prove the first theorem of Pappus for regions R bounded by the graphs of functions and horizontal axes of revolution.

11. Prove the second theorem of Pappus for graphs of functions and horizontal axes of revolution.

12. Use Pappus' theorems, the formulas $V = \frac{4}{3}\pi r^3$ and $A = 4\pi r^2$ for the volume and surface area of a sphere of radius r, and the formulas $A = \pi r^2$ and $C = 2\pi r$ for the area and circumference of a circle of radius r to locate the centroids of   **a.**[†] the region $0 \le y \le \sqrt{r^2 - x^2}$ **b.** the curve $y = \sqrt{r^2 - x^2}$.

13. Use the theorems of Pappus to compute the volume and surface area of the torus generated by rotating the circle $(x - b)^2 + y^2 \le a^2$ $(0 < a < b)$ about the y-axis.

14. Derive formulas (20) and (21) for the moments of a portion of the graph of a function.

15. Find the centroid of the circumference $y = \sqrt{1 - x^2}$ of a semicircle of radius 1 with its center at the origin.

6-11 AVERAGE VALUES OF FUNCTIONS AND THE MEAN VALUE THEOREM FOR INTEGRALS

If a student makes one A, two B's, and two D's in one term, his grade point average for that term is $\frac{1}{5}(4 + 3 + 3 + 1 + 1) = \frac{12}{5} = 2.4$. If a football team scores 10, 17, 21, and 3 points in four games, then it scores an average of $\frac{1}{4}(10 + 17 + 21 + 3) = \frac{51}{4} = 12.75$ points per game in the four games. These uses of the term "average" deal with what mathematicians call ARITHMETIC MEANS. The arithmetic mean of N numbers $a_1, a_2, a_3, \ldots, a_N$ is defined to be the number

$$\frac{1}{N}(a_1 + a_2 + a_3 + \cdots + a_N). \qquad \blacktriangleleft$$

We are dealing with a different sort of "average" when we say that a person who drives two hundred miles in the five hours between 3:00 and 8:00 has an "average velocity" of 200 miles/5 hours = 40 miles per hour. Unless the driver goes at the constant velocity of 40 miles per hour, he has an infinite number of different velocities during the trip, and we cannot take an arithmetic mean of an infinite number of numbers.

In what sense is the average velocity of the car an "average" of its velocities during the trip? We saw in Section 6.7 that if $v(t)$ is the car's velocity at t o'clock, the distance the car travels between 3:00 and 8:00 is the integral

$$\int_3^8 v(t) \left[\frac{\text{miles}}{\text{hour}}\right] dt\,[\text{hours}] = \int_3^8 v(t)\, dt\,[\text{miles}]$$

and the average velocity is

$$\left[\begin{array}{l}\text{The average} \\ \text{velocity}\end{array}\right] = \frac{1}{5\,[\text{hours}]} \int_3^8 v(t)\, dt\,[\text{miles}]$$

$$= \frac{1}{5} \int_3^8 v(t)\, dt\, \frac{\text{miles}}{\text{hour}}.$$

Thus, the average velocity is the integral of the velocity divided by the length of the interval of integration. We can apply this concept of "average" to any function that has an integral:

Definition 6.3   **If the integral of f over the interval $a \le x \le b$ $(a < b)$ exists, then the average value of $f(x)$ in the interval is**

(1) $$\frac{1}{b-a} \int_a^b f(x)\, dx.$$ ◄

EXAMPLE 1   Compute the average value of x^2 for $-2 \le x \le 3$.

Solution.   By definition (1) with $a = -2$, $b = 3$, and $f(x) = x^2$ we have

$$\left[\begin{array}{l}\text{The average value of} \\ x^2 \text{ for } -2 \le x \le 3\end{array}\right] = \frac{1}{3 - (-2)} \int_{-2}^3 x^2\, dx$$

$$= \frac{1}{5} \int_{-2}^3 x^2\, dx = \frac{1}{5}\left[\frac{1}{3} x^3\right]_{-2}^3$$

$$= \frac{1}{5}\left[\frac{1}{3}(3)^3\right] - \frac{1}{5}\left[\frac{1}{3}(-2)^3\right] = \frac{7}{3}. \quad \square$$

A geometric interpretation of average values

The graph of the function x^2 of Example 1 is sketched in Figure 6.72. That graph, the vertical lines $x = -2$ and $x = 3$, and the horizontal line $y = 7/3$ bound the three shaded regions in the sketch. The fact that $7/3$ is the average value of x^2 for $-2 \le x \le 3$ means that the sum of the areas of regions I and III above the line $y = 7/3$ is equal to the area of region II below that line. Roughly speaking, this means that the graph of x^2 dips below the line $y = 7/3$ as much as it rises above it for $-2 \le x \le 3$. This geometric interpretation of average value can be applied to any function.

Rule 6.17   The region between the graph of a function f and the horizontal line $y = L$ and between the vertical lines $x = a$ and $x = b$ is in one or more pieces. The number L is the average value of f for $a \le x \le b$ if the total area of those pieces above the line $y = L$ is equal to the total area of the pieces below that line.

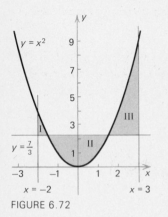

FIGURE 6.72

We leave the justification of this interpretation of average value as Exercise 3.

Notice that if L is the average value of $f(x)$ for $a \le x \le b$, then multiplying definition (1) by $b - a$ gives

$$(2) \qquad \int_a^b f(x)\, dx = L(b - a) \qquad \blacktriangleleft$$

$$= \begin{bmatrix} \text{The average value} \\ \text{of } f \text{ for} \\ a \le x \le b \end{bmatrix} \begin{bmatrix} \text{The length of} \\ \text{the interval} \\ a \le x \le b \end{bmatrix}.$$

This formula with the geometric interpretation of average value provides a valuable intuitive interpretation of integrals.

The Mean Value Theorem for integrals

The concept of average value of a function leads us to the following result, which is useful in theoretical discussions of the integral.

Theorem 6.1   (The Mean Value Theorem for integrals) If the function f is continuous for $a \le x \le b$, then there is at least one number c with $a < c < b$ such that

$$(3) \qquad \int_a^b f(x)\, dx = f(c)(b - a). \qquad \blacktriangleleft$$

Proof.   Equation (3) is immediate if f is constant in the interval. If f is not constant, we let L denote its average value for $a \le x \le b$. The geometric interpretation of the average value shows that a portion of the graph of f lies above the horizontal line $y = L$ and a portion below it. The function f therefore has values greater than L and less than L in the interval $a \le x \le b$. By the Intermediate Value Theorem, the function f cannot "jump" the value L and there must be at least one point c with $a < c < b$ such that

$$f(c) = L = \frac{1}{b - a} \int_a^b f(x)\, dx.$$

Multiplying this equation by $b - a$ gives equation (3) and proves the theorem.   Q.E.D.

A geometric interpretation of the Mean Value Theorem for Integrals

If the continuous function f has positive values for $a \le x \le b$, then equation (3) states that the area under the graph of f is equal to the area of the rectangle with width $b - a$ and height $f(c)$ (Figure 6.73).

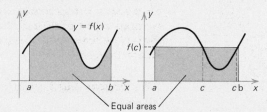

FIGURE 6.73

EXAMPLE 2   Verify the conclusion of the Mean Value Theorem for the integral of the function x^3 from 0 to 2 by finding a suitable number c.

Solution.   Because the interval of integration has length 2, we want to find a number c with $0 < c < 2$ such that

(4) $\int_0^2 x^3\, dx = 2c^3.$

Because

$$\int_0^2 x^3\, dx = \left[\frac{1}{4}x^4\right]_0^2 = \frac{1}{4}(2^4) - \frac{1}{4}(0^4) = 4$$

we can satisfy (4) by taking $c = \sqrt[3]{2}$.   □

Exercises

†*Answer provided.*
‡*Outline of solution provided.*

1. Compute the average values of   **a.**† $\sqrt{x}$ for $0 \le x \le 9$ **b.** $x^4 - x^2$ for $-2 \le x \le 2$ **c.**† $\sin x$ for $0 \le x \le \pi$ **d.** $x \cos(x^2)$ for $-1 \le x \le 1$ **e.** $(2x + 3)^7$ for $-1 \le x \le 0$.

2. Verify the conclusion of the Mean Value Theorem for integrals by finding suitable numbers c for   **a.**† x^{-2} in $-3 \le x \le -1$ **b.** $\sin(4x)$ in $0 \le x \le \pi/2$.

3. Derive Rule 16.17 from the geometric definition of the integral.

4. The statement "To average 40 miles per hour on a trip you have to drive 40 miles an hour at some time during the trip" can be considered an example of the Mean Value Theorem for integrals or of the Mean Value Theorem for derivatives. Explain.

5. The following functions do not satisfy the hypotheses of the Mean Value Theorem for integrals in the interval $0 \leq x \leq 5$. Why? Do they satisfy the conclusion of the theorem?

a.[†] $f(x) = \begin{cases} x & \text{for } 0 \leq x \leq 4 \\ -1 & \text{for } 4 < x \leq 5 \end{cases}$

b. $g(x) = \begin{cases} x^2 & \text{for } 0 \leq x < 3 \\ -2 & \text{for } 3 \leq x \leq 5 \end{cases}$

6.[†] What is the average value of x^{-5} for $\frac{1}{10} \leq x \leq 1$?

7. What is the average value of x^2 for $-2 \leq x \leq 2$?

8.[†] If a falling object has velocity $32t + 10$ feet per second at time t, what is its average velocity for $0 \leq t \leq 3$?

9. If the temperature is $15 + 5 \sin(\frac{1}{12}\pi t)$ degrees Centigrade at time t, what is the average temperature for $0 \leq t \leq 12$?

10.[†] Show that the function $x^{2/3}$ satisfies the conclusion of the Mean Value Theorem for integrals in the interval $-8 \leq x \leq 1$ by finding a suitable number c.

11. Show that the function

$$f(x) = \begin{cases} x^2 + 1 & \text{for } x \leq 0 \\ -x^2 - 1 & \text{for } x > 0 \end{cases}$$

does not satisfy the conclusion of the Mean Value Theorem for integrals in the interval $-2 \leq x \leq 2$.

12. Does the function f of Example 11 satisfy the conclusion of the Mean Value Theorem for integrals in the interval $-1 \leq x \leq 2$?

MISCELLANEOUS EXERCISES

†*Answer provided.*
‡*Outline of solution provided.*

In Exercises 1 through 19 make rough sketches of the regions bounded by the given curves in xy-planes. Then compute their areas.

1.[‡] $y = 2x^2$ and $y = x^2 + 3x + 4$

2. $y = 2x^2 - x^4$ and $y = -2x^2$

3.[†] $y = x^2 - 2x + 2$ and $y = 2$

4. $y = x^2$ and $y = \frac{1}{4}x^4$

5.[‡] $\sqrt{x} + \sqrt{y} = 1, x = 0$, and $y = 0$

6. $y = \sqrt{x}, y = 1/x^2$, and $x = 4$

7.[‡] $y = x^4 - 2x^2$ and $y = -1$

8. $y = x^{2/3}$ and $y = |x|$

9.[‡] $y = 9x^{-2}, y = 1$, and $y = 9$

10. $y = x^{4/5}$ and $y = 1$

11. $y = \sin x$ and $y = x^2 - \pi x$

12.‡ $y = 1/2$ and one arch of $y = \sin x$

13.† $y = 1/\sqrt{2}$ and one arch of $y = \cos x$

14. One of the regions bounded by $y = \sin x$ and $y = \cos x$

15.† $y = x, y = 2 - x$, and $y = -3x$ (A sum of integrals is required.)

16. $y = x^2, y = x$, and $x = 2$ (A sum of integrals is required.)

In Exercises 17 through 19 compute the areas once by an x-integration and once by a y-integration. In some instances a sum of integrals is required.

17.‡ $x = y^2$ and $x = y + 2$

18.† $y = x, y = x^{-2}$, and $x = 2$

19. $y = 3 - x^2$ and $y = -2x$

In each of Exercises 20 through 27 a region is bounded by the given curves. Compute the volumes of the solids generated by rotating the region about the indicated axes.

20.† $y = x + x^2, y = x^2 - 1$, and $x = 0$; axes: **a.** $y = 1$ **b.** $y = -1$
  c. $x = 0$ **d.** $x = -2$

21. $y = x^{2/3}$, and $y = 4$; axes: **a.** $x = -9$ **b.** $y = 0$ **c.**‡ $x = 0$

22.† $2x = 4 - y^2$ and $2x = y^2 - 4$; axes: **a.** $x = -3$ **b.** $x = 0$ **c.** $y = 0$
  d. $y = -3$

23. $y = \sqrt{x^2 + 1}, x = 1, x = 0$, and $y = 0$; axes: **a.** $x = 0$ **b.** $y = 0$

24.† $y = x^{2/3}$ and $y = 1$; axis: $x = 2$ (Compute the volume once by an x-integration and once by a y-integration.)

25. $y = x^2, y = x^2 + 2, x = 0$, and $x = 1$; axis: $x = 2$ (Compute the volume two ways.)

26.† $y = x^3, y = 0, x = -1$ and $x = 1$; axis: $x = 1$ (Two integrals are required.)

27. $y = x^{1/3}, y = 0, x = 8$, and $x = -8$; axis: $y = 3$ (Two integrals are required.)

28.‡ A hemispheric bowl of 10-inch radius contains water to a depth of 6 inches. What is the volume of the water?

29.† Water is being added to the bowl of Exercise 28 at the rate of 25 cubic inches per minute. How fast is the depth increasing when it is 6 inches?

30. Use the formula for the area of a circle to evaluate the integral

$$\int_{-a}^{a} \sqrt{a^2 - x^2}\,dx.$$

31.† When the circle $x^2 + y^2 \le a^2$ is rotated about the line $x = b\,(0 < a < b)$, it generates a *torus*. Use the result of Exercise 30 to compute its volume.

32.† How much does a sphere of radius 1 foot weigh if it sinks 4 inches below the surface when it is floating in water? (By Archimedes' principle the sphere displaces its weight in water. Water weighs 62.4 pounds per cubic foot.)

33.† When the region bounded by the graph $y = f(x)$ of a positive continuous function f, by the x- and y-axis, and by the line $x = a$ $(a > 0)$ is rotated about the x-axis, it generates a solid of volume $\pi(\frac{1}{5}a^5 + \frac{2}{3}a^3 + a)$. What is the function?

34. **a.** Give a formula for the function whose graph is the upper half of the astroid $x^{2/3} + y^{2/3} = 1$.   **b.** Why can we not apply formula (1) of Section 6.5 directly to find the length of the graph of this function?   **c.†** Find the length of the graph by computing the length for $a \le x \le b (0 < a < b < 1)$ and letting $a \to 0^+$ and $b \to 1^-$.   **d.** Use the same method to find the centroid of the upper half of the astroid.

35. The portion of a rod from 0 to x (feet) on a coordinate axis $(x > 0)$ weighs $6x^2 + 3x$ pounds. What is its density?

36. When the ellipse $x^2/a^2 + y^2/b^2 = 1$ $(a > 0, b > 0)$ is rotated about the x-axis it generates an *ellipsoid*. Compute its volume.

37.† A stone which is thrown straight into the air hits the ground with the speed of 64 feet per second. How high does it go?

38.† An arrow that is shot straight up into the air strikes the ground 4 seconds later with a speed of 100 feet per second. From what height was it shot and with what speed?

39. An object is at the origin on an s-axis at time $t = 0$ and has acceleration $a(t) = -3t^2$ in the positive direction for $t > 0$. Its greatest s-coordinate is $s = 3/4$. What is its velocity at $t = 0$?

40.† The net distance an object travels from time $t = a$ to $t = b$ $(a < b)$ is no greater than its total distance traveled. Express this as an inequality involving the integrals of its velocity $v(t)$ and of $|v(t)|$.

41. A small airplane is flying north with the constant air speed of 125 miles per hour. The wind is blowing south at the speed of t^3 miles per hour at time t (hours). Where is the plane at $t = 6$ relative to its position at $t = 0$?

42.† The daily overhead in the manufacture of a certain product is $200 and the marginal cost to produce x units of the product is $50 - \frac{1}{50}x^2$ dollars per unit. What is the total cost of producing 20 units in one day?

43. Use the differentiation formula $\dfrac{d}{dx}\left[\dfrac{1}{2}(\sin x)(\cos x) + \dfrac{1}{2}x\right] = \cos^2 x$ to compute the volume inside the hour glass of Example 1 in Section 6.6.

44.† We might try to find the integral which equals the length of the curve $y = f(x)$, $a \le x \le b$ by approximating the portion of the curve between points $(x_{j-1}, f(x_{j-1}))$ and $(x_j, f(x_j))$ by the horizontal line from $(x_{j-1}, f(x_{j-1}))$ to $(x_j, f(x_{j-1}))$ and the vertical line from $(x_j, f(x_{j-1}))$ to $(x_j, f(x_j))$. What (incorrect) integral would we obtain for the length?

45. A child drinks a half pound of milk through a 10-inch straw that he keeps vertical and with its end on the bottom of his cylindrical glass. The depth of the milk in the glass changes from 6 inches to 2 inches as he drinks. How much work does he do?

46.‡ The weight of an object on the surface of the earth is the force of gravity exerted on it. As the object is lifted from the surface of the earth, the force is equal to a constant times r^{-2} where r is the distance to the center of the earth (Newton's law). How much work does it take to lift a 100 pound satellite 300 miles above the surface of the earth? (Use 4000 miles as the radius of the earth.)

47. Sand leaks out of a bag at a constant rate. As the bag is lifted at the constant rate of 3 feet per minute for 6 minutes, the amount of sand in the bag decreases from 500 pounds to 250 pounds. How much work is done lifting the bag?

48.† It takes twice as much work to compress a spring from a length of 5 inches to a length of 4 inches than it takes to compress it from a length of 6 inches to a length of 5 inches. What is its natural length?

49. A cylinder with a piston in one end contains air at a pressure of 100 pounds per square foot when the piston is 1 foot from the closed end of the cylinder. The area of the end of the cylinder is one square foot. The pressure p and the volume V satisfy an equation of the form $pV^{1.4} = $ constant. How much work is done by the air on the piston as the piston moves to a distance of two feet from the end of the cylinder?

50.† A boat 16 feet from a pier is moved to a distance of 9 feet by a sailor pulling on a rope with the constant force of 20 pounds (Figure 6.74). The rope passes through a pulley at the end of the pier and is fastened to the bow of the boat. The pulley is 12 feet above the level of the boat. How much work does the sailor do?

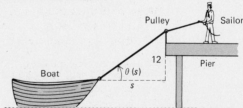

FIGURE 6.74

51. Prove that the total force on one side of a submerged vertical plate is equal to its area multiplied by the pressure at its centroid.

52.† A rectangular plate, submerged in water, makes an angle α with the vertical. Its top is horizontal, w feet long, and H feet beneath the surface of the water. The adjacent sides of the plate are L feet long. Give an expression for the force of the water on one side of the plate in terms of α, w, H, and L. What are the maximum and minimum forces for

$$0 \le \alpha \le \frac{\pi}{2} \text{ and fixed } w, H, \text{ and } L?$$

53.† Let $P(y)$ kilograms per square meter be the air pressure and $\rho(y)$ kilograms per cubic meter be the density of the air y meters above the surface of the earth.   **a.** Use an integral to express $P(y)$ in terms of $\rho(t)$ for $y \leq t \leq Y$, where Y is so large that we can assume that the pressure and density are zero at that altitude.   **b.** A reasonable model for studying atmospheric pressure is obtained by assuming that the density is proportional to the pressure. Using data at the surface of the earth to determine the constant of proportionality, we obtain the equation $\rho(y) = 0.00012\,P(y)$. Use this and the Fundamental Theorem of Calculus to express $\dfrac{dP}{dy}(y)$ in terms of $P(y)$.

54.† Describe the upper and lower approximations in the derivation of Rule 6.11 of Section 6.9. (Upper and lower approximations are discussed at the end of Section 6.8.)

55. Describe the upper and lower approximations in the derivation of Rule 6.12 in Section 6.10.

56.† The density of a hemispherical solid of radius 5 meters is $6x$ kilograms per cubic meter at the points x meters from its flat base.   **a.** How much does it weigh?   **b.** Where is its center of gravity?

57. Find the x-coordinate of the centroid of the curve $y = \frac{1}{8}x^4 + \frac{1}{4}x^{-2}$, $1 \leq x \leq 3$.

58. Show that the centroid of a triangle is at the point where its three medians intersect.

In Exercises 59 and 60, $f = f(x)$ denotes a positive function with a continuous derivative for $a \leq x \leq b$.

59.† Let R be the region bounded by the curve $y = f(x)$, by the x-axis, and by the lines $x = a$ and $x = b$. Derive formulas for the coordinates of the centroid of the solid generated by rotating R about the x-axis.

60. Derive formulas for the coordinates of the centroid of the surface generated by rotating the curve $y = f(x)$, $a \leq x \leq b$ about the x-axis.

61.† The *moment of inertia* about the origin of a point of weight M at x on a coordinate axis is defined to be the number Mx^2. Derive an integral for the moment of inertia about the origin of a rod extending from $x = a$ to $x = b\,(a < b)$ and with density $\rho(x)$ at x.

62.† A boy drops a rock in a river from a bridge. He hears the rock hit the water $4\frac{3}{8}$ seconds later. How high is the bridge above the water? (The speed of sound is 1089 feet per second.)

63.‡ The cross sections of a 6 foot long boat have the shape of the region bounded by $y = x^2$ and $y = 1$ in an xy-plane with the coordinates measured in feet. How deep does it settle into the water if it weighs 200 pounds? (Water weighs 62.4 pounds per cubic foot.)

64. When the region bounded by $y = x^{1/3}$, $x = 0$, and $y = 1$ (feet) is rotated about the y-axis, it generates a top-shaped region. A top with this shape and a weighted tip to keep it upright is placed in a swimming pool. The top weighs 1 pound. How far does it settle into the water?

65.‡ Suppose that the demand for a product is $D(p)$ units at a price of p dollars per unit where $D(p) = 0$ for $p \geq p_0$. If the product is sold for p_1 dollars per unit with $p_1 < p_0$, then some customers would have paid a higher price. The amount they save is called the *consumer surplus*. Express it as an integral.

66. If a product with demand function $D(p)$ is sold for p_0 dollars per unit, the revenue is the area of a rectangle determined by the graph of D. What is that rectangle?

67. A company has a machine whose productivity decreases at the rate of $f(t)$ dollars per month t months after it is overhauled. It costs A dollars to overhaul it. **a.** Why would the company want to overhaul it every T months with T such that $g(T) = \dfrac{1}{T}\left(A + \displaystyle\int_0^T f(t)\, dt\right)$ is a minimum? **b.** Show that the minimum occurs for a T such that $g(T) = f(T)$. (Assume that f is continuous.)

68. Suppose that of a population of P animals, $s(t)\,P$ are alive t years later $(0 \leq s(t) \leq 1)$. (The function $s(t)$ is called a *survival function*.) Show that if $P(t)$ is the population and $r(t)$ the birth rate at time t, then
$$P(t) = P(0)s(t) + \int_0^t s(t - u)r(u)\, du.$$

69. An object's *momentum* is its mass times its velocity. Suppose that the total force on an object moving along an x-axis is $F(t)$ pounds at time t (sec.). Show that the integral of $F(t)$ from a to b is the increase in the object's momentum.

70. The *root mean square value* of $f(x)$ for $a \leq x \leq b$ is
$$\left[\frac{1}{b - a}\int_a^b f(x)^2\, dx\right]^{1/2}.$$
Compute **a.** the average value and **b.** the root mean square value of $\sin x$ for $0 \leq x \leq \pi$. (Use the trigonometric identity $\sin^2 x = [1 - \cos(2x)]/2$.)

71. Figure 6.75 shows the rate of oxygen intake by a man who bicycles for 20 minutes starting at $t = 0$ (minutes). Approximately how much oxygen does he take in for $0 \leq t \leq 35$? (Adapted from *Handbook of Physiology*, Vol II, p. 945.)

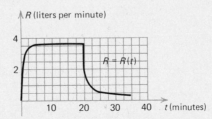

FIGURE 6.75

CHAPTER 7

TRANSCENDENTAL FUNCTIONS

SUMMARY: A function $y = y(x)$ is called *algebraic* if it can be defined implicitly by a polynomial equation in x and y. Other functions are called *transcendental*. We already have studied two transcendental functions, the sine and cosine. In this chapter we study the other trigonometric functions as well as exponential functions, logarithms, and hyperbolic functions. In Section 7.1 we discuss the algebraic properties of exponential functions and logarithms and show how the calculus of logarithms leads to the introduction of the number e and of the natural logarithm. Section 7.2 contains the basic theory of logarithms. We use the natural logarithm to evaluate integrals in Section 7.3 and to perform logarithmic differentiation in Section 7.4. The exponential function e^x is studied in Section 7.5. Two related topics, the differential equation $y' = ry$ and compound interest, are discussed in Sections 7.6 and 7.7. Sections 7.8 and 7.9 deal with trigonometric functions and inverse trigonometric functions, and Section 7.10 with hyperbolic functions. Section 7.11 is an optional section with notes on the history of the transcendental functions.

7-1 THE ALGEBRA OF EXPONENTIAL FUNCTIONS AND LOGARITHMS

An EXPONENTIAL FUNCTION is a function of the form b^x with b a positive constant. Exponential functions and their inverses, the LOGARITHM functions, arise in the study of radioactive decay, population growth, falling bodies, chemical reaction rates, electrical circuits, and other phenomena in physics, economics, chemistry, and biology. They also occur in many branches of pure mathematics. This section provides a brief discussion of the basic algebraic properties of these functions.

Irrational exponents

Up to this point we have considered the expression b^x $(b > 0)$ only for rational exponents x. We define b^x for irrational x to be the limit

$$(1) \qquad b^x = \lim_{n \to \infty} b^{r_n} \qquad \blacktriangleleft$$

where $r_1, r_2, r_3, \ldots$ are rational numbers tending to x:

$$(2) \qquad x = \lim_{n \to \infty} r_n \qquad \blacktriangleleft$$

(In the next section we will show that the limit (1) exists and is independent of the choice of the rational numbers r_n in (2).)

One way to choose suitable rational numbers r_n in (2) is to have r_n be the first n digits in a decimal expansion of the irrational number x. To define $3^{\sqrt{2}}$, for example, we can take r_n to be the first n digits in the decimal expansion $\sqrt{2} = 1.4142135 \ldots$ of the irrational number $\sqrt{2}$. The resulting values of 3^{r_n} for $n = 1, 2, 3, \ldots, 8$ are listed in Table 7.1. The limit $3^{\sqrt{2}}$ of the numbers 3^{r_n} has the decimal expansion $3^{\sqrt{2}} = 4.7288043 \ldots$.

The basic algebraic properties of b^x are stated in the following proposition.

TABLE 7.1

$3^{r_1} = 3^1 = 3.0000000 \ldots$
$3^{r_2} = 3^{1.4} = 4.6555367 \ldots$
$3^{r_3} = 3^{1.41} = 4.706950 \ldots$
$3^{r_4} = 3^{1.414} = 4.7276950 \ldots$
$3^{r_5} = 3^{1.4142} = 4.7287339 \ldots$
$3^{r_6} = 3^{1.41421} = 4.7287858 \ldots$
$3^{r_7} = 3^{1.414213} = 4.7288014 \ldots$
$3^{r_8} = 3^{1.4142135} = 4.7288040 \ldots$

Proposition 7.1   **For arbitrary positive b and c and arbitrary numbers x, x_1, and x_2**

$$(3) \qquad b^{x_1} b^{x_2} = b^{x_1 + x_2} \qquad \blacktriangleleft$$

$$(4) \qquad (b^{x_1})^{x_2} = b^{x_1 x_2} \qquad \blacktriangleleft$$

$$(5) \qquad (bc)^x = b^x c^x \qquad \blacktriangleleft$$

$$(6) \qquad b^{-x} = 1/(b^x) = (1/b)^x \qquad \blacktriangleleft$$

For $b > 1$, the function b^x is positive, continuous and increasing and its value tends to ∞ as $x \to \infty$ and to 0 as $x \to -\infty$.

We know that these properties hold for rational exponents. We will prove Proposition 7.1 for arbitrary exponents in the next section. For $b = 1$, $b^x = 1^x$ is the constant function 1. For $0 < b < 1$, $1/b$ is greater than 1, and by Proposition 7.1, $b^x = (1/b)^{-x}$ is positive, continuous, and decreasing and tends to 0 as $x \to \infty$ and to ∞ as $x \to -\infty$ (Figure 7.1).

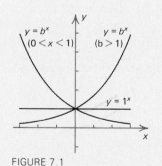

$y = b^x$
$(0 < x < 1)$

$y = b^x$
$(b > 1)$

$y = 1^x$

FIGURE 7.1

EXAMPLE 1   Solve the equation $2^{x^2} = 32$ for x.

Solution.   Because 2^5 is 32 and the increasing function 2^x cannot have the same value at two different values of x, the equation $2^{x^2} = 32$ is equivalent to the equation $x^2 = 5$ and has the solutions $x = \pm\sqrt{5}$.   □

Logarithms

Proposition 7.1 shows that for any constant $b > 1$, b^x is continuous and increasing and takes on all positive values. By the Inverse Function Theorem (Theorem 4.6a), it has a continuous inverse defined for positive x. The inverse is the LOGARITHM TO THE BASE b and is denoted $\log_b x$. It is defined for $x > 0$ by the condition

(7) $y = \log_b x$ if and only if $x = b^y$.

The relation between the functions b^x and $\log_b x$ can also be expressed by the equations

(8) $\log_b(b^x) = x$ for all x

(9) $b^{\log_b x} = x$ for all $x > 0$.

EXAMPLE 2   What is $\log_3(\frac{1}{9})$?

Solution.   The equation $\log_3(\frac{1}{9}) = x$ is equivalent to $3^x = \frac{1}{9}$, which has the solution $x = -2$. Consequently, $\log_3(\frac{1}{9}) = -2$.   □

The basic algebraic properties of the logarithm are stated in the next proposition, to be proved in Section 7.2.

Proposition 7.2   For $b > 1$, positive a and c, and arbitrary n

(10) $\log_b(1) = 0$

(11) $\log_b(b) = 1$

(12) $\log_b(ac) = \log_b(a) + \log_b(c)$

(13) $\log_b(a/c) = \log_b(a) - \log_b(c)$

(14) $\log_b(a^n) = n \log^b(a)$.

The number $\log_b x$ tends to ∞ as x tends to ∞ and tends to $-\infty$ as x tends to 0^+.

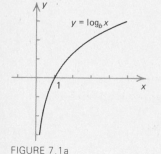

FIGURE 7.1a

The graph of $\log_b x$ for one value of the constant b is shown in Figure 7.1a.

Common logarithms   The most convenient logarithm for arithmetic calculations is the logarithm to the base 10, which is known as the COMMON (or BRIGGSIAN) LOGARITHM. The base 10 is used because multiplying a positive number by a power of 10, as when we shift a decimal point, adds the exponent to the common logarithm:

(15) $\log_{10}[x(10)^n] = \log_{10} x + n$ ◀

(see Exercise 14). This makes it easy to determine the common logarithm of any positive number from a table of common logarithms of numbers between 1 and 10 (see the reference tables at the back of this book).

EXAMPLE 3   Solve $4 \log_{10}(x^2) - \log_{10}(\sqrt{x}) = 3$ for x.

Solution.   We rewrite the equation in the equivalent forms

$$\log_{10}[(x^2)^4] - \log_{10}(x^{1/2}) = \log_{10}(10^3)$$

and then

$$\log_{10}(x^8/x^{1/2}) = \log_{10}(1000).$$

Because $\log_{10}x$ is an increasing function, the last equation implies that $x^{8-1/2} = 1000$, which gives $x^{15/2} = 1000$ and then $x = (1000)^{2/15}$. □

EXAMPLE 4   Use the common logarithm to find the solutions x of the equation $3^{x^2} = 7^x$. Use the table of values of the common logarithm in the back of the book to give approximate values for the solutions.

Solution.   We take logarithms of both sides of the equation to obtain $\log_{10}(3^{x^2}) = \log_{10}(7^x)$ and then $x^2 \log_{10}(3) = x \log_{10}(7)$. This has one solution $x = 0$. We can cancel x from the equation if it is not zero to obtain $x \log_{10}(3) = \log_{10}(7)$ or $x = [\log_{10}(7)]/[\log_{10}(3)]$. The table gives $\log_{10}(7) \approx 0.8451$ and $\log_{10}(3) \approx 0.4771$, so the solutions are $x = 0$ and $x = [\log_{10}(7)]/[\log_{10}(3)] \approx (0.8451)/(0.4771) \approx 1.7713$. □

The number e

The calculus of exponential functions and logarithms involves a special number called "e." This number, which is as important to calculus as the number π is to geometry and trigonometry, is defined to be the limit

$$(16) \quad e = \lim_{n \to \infty} \left(1 + \frac{1}{n}\right)^n.$$

It is irrational and has the decimal expansion $e = 2.71828182 \ldots$ We will show in the next section that the limit (16) exists.

To see how the number e arises, let us use Propositions 7.1 and 7.2 and the existence of limit (16) to compute the derivative of $\log_b x$ $(b > 1)$.

For $x > 0$ the derivative of $\log_b x$ is defined by

$$(17) \quad \frac{d}{dx}\log_b x = \lim_{\Delta x \to 0} \frac{\log_b(x + \Delta x) - \log_b x}{\Delta x}.$$

By property (13) in Proposition 7.2, the difference of logarithms is the logarithm of the quotient and we have

$$(18) \quad \frac{\log_b(x + \Delta x) - \log_b x}{\Delta x} = \frac{1}{\Delta x}\log_b\left(\frac{x + \Delta x}{x}\right) = \frac{1}{\Delta x}\log\left(1 + \frac{\Delta x}{x}\right).$$

We set $\Delta x = x/n$ with n positive and use property (14) to express the difference quotient (18) in the form

$$\frac{\log_b(x + \Delta x) - \log_b x}{\Delta x} = \frac{n}{x}\log_b\left(1 + \frac{1}{n}\right) = \frac{1}{x}\log_b\left[\left(1 + \frac{1}{n}\right)^n\right].$$

Since x is a fixed positive number, Δx tends to 0 from the right as n tends to ∞, so by the continuity of the logarithm

$$\lim_{\Delta x \to 0+} \frac{\log_b(x + \Delta x) - \log_b x}{\Delta x} = \lim_{n \to \infty} \frac{1}{x}\log_b\left[\left(1 + \frac{1}{n}\right)^n\right]$$

$$= \frac{1}{x}\log_b\left[\lim_{n \to \infty}\left(1 + \frac{1}{n}\right)^n\right].$$

We could show that the same limit is obtained as Δx tends to 0 from the left. Then definition (17) of the derivative yields

$$\frac{d}{dx}\log_b x = \frac{1}{x}\log_b\left[\lim_{n\to\infty}\left(1 + \frac{1}{n}\right)^n\right].$$

In order to obtain a concise expression for the derivative, we use definition (16) of e and obtain

$$(19) \quad \frac{d}{dx}\log_b x = \frac{1}{x}\log_b e \quad \text{for} \quad x > 0$$

which shows that the derivative of the logarithm to any base involves the number e.

The natural logarithm

We like to deal whenever possible with functions whose derivatives are given by simple formulas. This preference led us to use radians instead of degrees when we defined the trigonometric functions $\sin x$ and $\cos x$. It will now lead us to use the number e as a base for logarithms. Because $\log_e e$ is 1, formula (19) with $b = e$ simplifies to the form

$$(20) \quad \frac{d}{dx}\log_e x = \frac{1}{x} \quad \text{for} \quad x > 0.$$

The logarithm to the base e is called the NATURAL (or NAPERIAN) LOGARITHM. It is "natural" in the sense that it is the most convenient logarithm for use in calculus and its applications. It is usually given the special notation $\ln x$.* With this notation formula (20) reads

$$(21) \quad \frac{d}{dx}\ln x = \frac{1}{x} \quad \text{for} \quad x > 0.$$

An easier approach

The above derivation does not prove differentiation formula (21) because we have not yet shown that the definition (1) of b^x for irrational x is justified, nor have we proved Propositions 7.1 and 7.2 nor shown that the limit (16) defining e exists. Establishing these facts by algebra alone requires a relatively difficult and tedious string of agruments. Fortunately, there is an easier approach using tools of calculus which we will follow in the next section to supply the formal proofs.

In this alternate approach, we will not define $\ln x$ algebraically as the inverse of an exponential function but rather as an integral, chosen so we can use the Fundamental Theorem of Calculus to show immediately that it has the expected derivative $1/x$. Then we will use this function to derive the results listed at the beginning of the preceding paragraph and, as a last step, we will show that $\ln x$ is, in fact, the logarithm to the base e.

---

*Some books use $\log x$ for the natural logarithm, while others use $\log x$ for $\log_{10} x$.

Exercises

1. Give exact values of   **a.**† $\log_4(2)$ **b.**† $\log_2(4)$ **c.**† $\log_5(5)$
   d. $\log_3(27)$ **e.** $\log_{27}(3)$ **f.** $\log_7(\frac{1}{7})$ **g.** $\log_{10}(1000)$

2. Solve the following equations for x.
   a.† $\log_{10}x = 5$ **b.** $\log_5 x = 10$
   c.† $2\log_{10}(x/3) = 1$ **d.** $\log_2(1/x) = 5$
   e.† $\log_{10}x - \log_{10}(1/x) + \log_{10}(\sqrt{x}) = -10$
   f. $\log_2(2x^2) = 4$ **g.** $\log_5(25/x) = -1$
   h.† $[\log_{10}x]^2 = 16$ **i.** $\sqrt{\log_{10}(\sqrt{x})} = 3$
   j.† $\log_6(4x) = 1/5$ **k.** $\log_4(1 - x) = -1.2$

3. Use the reference tables at the back of this book to give numerical values, correct to four decimal places, for the logarithms below. Leave the logarithm of numbers between 0 and 1 expressed as a decimal between 0 and 1 (the *mantissa*) minus an integer (the *characteristic*).
   a.† $\log_{10}(4.56)$ **b.** $\log_{10}(6.12)$ **c.**† $\log_{10}(375)$
   d. $\log_{10}(132,000)$ **e.**† $\log_{10}(0.0034)$ **f.** $\log_{10}(0.965)$

4. The common logarithm $\log_{10}x$ is increasing for all $x > 0$, tends to $-\infty$ as $x \to 0^+$, and tends to ∞ as $x \to \infty$. Sketch its graph in an xy-plane by plotting the points at $x = \frac{1}{10}, \frac{1}{4}, \frac{1}{2}, 1, 2, 4,$ and 10.

5. Use the table of common logarithms at the back of this book to give decimal expressions for the following quantities to whatever accuracy is possible.
   a.‡ $\dfrac{2.16}{57.3}$ **b.**† $\dfrac{981}{0.0321}$ **c.** $\dfrac{(987)(654)}{456,000}$
   d.† $\sqrt[5]{8.62}$ **e.** $(4.9)^{0.9}$ **f.**† $(1560)^{11}$
   g. $(0.0045)^{12}$ **h.**‡ $\sqrt[4]{0.0123}$ **i.** $(11.1)^{1/11}$
   j.† $(1 + \frac{1}{10})^{10}$ **k.** $(1 + \frac{1}{100})^{100}$ **l.**‡ $5^{\sqrt{10}}$
   m. $10^{\sqrt{5}}$

6. Use the table of common logarithms to give approximate values for the solutions x of the equations
   a.‡ $3^{\sqrt{x}} = 7$ **b.**† $4^x = \frac{1}{3}$ **c.** $12^{3x} = 5$ **d.** $100^x = 0.07$.

7.† Given that $\dfrac{1}{x}\log_{10}x \to 0$ as $x \to \infty$, show that $\dfrac{1}{\sqrt{x}}\log_{10}x \to 0$ as $x \to \infty$.

8. Derive properties (10) through (14) of $\log_b x$ from properties of b^x.

9. In chemistry the pH of a solution is defined by the equation $pH = -\log_{10}[H^+]$ where $[H^+]$ is the concentration of hydrogen ions, measured in moles per liter. (One mole $= 6 \times 10^{23}$ molecules.) Distilled water has a pH of 7 and is called *neutral*.   **a.**   What is $[H^+]$ for water?   **b.**   Solutions with $[H^+]$ greater than that for water are called *acidic*. Those with $[H^+]$ less than that of water are called *alkaline* or *basic*. What are the pH's of acidic and basic solutions?

10.† The magnitude M of an earthquake on the Richter scale may be defined by the formula $M = 0.67\log_{10}(0.37E) + 1.46$ where E is the energy of the earthquake in kilowatt hours. What is the energy of   **a.**   an earthquake of magnitude 5 and   **b.**   an earthquake of magnitude 6?

11. The loudness of sound is measured in *decibels*. One decibel is supposed to be the smallest change in loudness that a human can detect. (One *bel*, named in honor of Alexander Graham Bell, equals 10 decibels.) A sound reaches the eardrum as an oscillation in the air pressure. If the variation in the air pressure is P pounds per square inch, then the loudness of the sound is $D = 20 \log_{10}[3.45(10^8)P]$ decibels. What is the variation in the air pressure caused by   **a.**   a whisper at 10 decibels and   **b.**   a rock band at 120 decibels?

12.† The magnitude of a star is defined by the formula $M = -2.5 \log_{10}(kI)$ where k is a constant and I is the intensity of the light from the star. What is the ratio of the intensities of light from the brightest star Sirius, which has a magnitude of -1.6, and from the star Betelgeuse, which has a magnitude of 0.9?

13. The A above middle C on the piano is denoted $A^{(1)}$; the A below middle C is denoted $A^{(0)}$, and so forth. The frequencies of various A's are listed in Table 7.2.   **a.**   Give a formula for the frequency of $A^{(n)}$ in terms of the integer n.   **b.**   In the "well-tempered" scale the ratio between the frequencies of successive notes $A, A\sharp, B, C, C\sharp, D, D\sharp, E, F, F\sharp, G, G\sharp,$ and A is constant. What is that ratio?   **c.**   What is the frequency of middle C in the well-tempered scale?

14. Derive formula (15).

15.† **a.**   Often, in fields such as engineering, the graphs of functions $f(x)$ are plotted on logarithm paper so that, in effect, it is the graph of $\log_{10}[f(x)]$ that is shown. This is done because the graph is then a straight line. What functions $f(x)$ have this property?   **b.**   What functions have straight line graphs when logarithmic scales are used on both axes?

TABLE 7.2

NOTE	FREQUENCY (CYCLES PER SECOND)
$A^{(-2)}$	55
$A^{(-1)}$	110
$A^{(0)}$	220
$A^{(1)}$	440
$A^{(2)}$	880
$A^{(3)}$	1760
$A^{(4)}$	3520

Find all solutions x of the equations in Exercises 16 through 25.

16.† $2^{3x} = \frac{1}{8}$ **17.**† $3^x = 4^x$

18.† $(10^x)^x = 100$ **19.**† $2^x 4^x 8^x = \frac{1}{16}$

20.† $\sqrt{16^x} = 2$ **21.**† $\dfrac{100^x}{1000^x} = 10{,}000$

22.† $7^{x^2} = (7^x)^2$ **23.**† $x^{x-2} = 1$

24.† $3^{x^2} = 9(3^x)$ **25.**† $(2^x)^2 = 4^x$

Use the logarithm to the base 2 to give all solutions x of the equations in Exercises 26 through 31.

26.† $2^{3x} = 7$ **27.**† $4^{\sqrt{x}} = 5$

28.† $\sqrt{2^x} = 9$ **29.**† $4^x = 5(8^x)$

30.† $2^{-x} 4^x 8^{-x} = 10$ **31.**† $3^{x^2} = 7$

Use the common logarithm to give all solutions x of the equations in Exercises 32 through 37.

32.† $(0.001)^x = 2$ **33.**† $5^{x^3} = 6$

34.† $10^{x^2} = \dfrac{1000^x}{100}$ **35.**† $(0.01)^{1/x} = (0.1)^x$

36.† $\dfrac{1000^x}{10^x} = \dfrac{75}{100^x}$ 　　　　　　　**37.**† $10^{x^2} = 7^x$

Solve the equations in Exercises 38 through 43 for x.

38.† $\log_4(x^5) = -6$ 　　　　　　　**39.**† $[\log_2(5x)]^2 = 9$

40.† $\log_7(x^2) = 3\log_7(2x)$ 　　　　　**41.**† $\log_{10}x = \log_5 x$

42.† $\log_{10}(x^3) + \log_{10}(\sqrt{x}) = 5 + 2\log_{10}x$

43.† $\log_2(x + 1) - \log_2(x^2 + 2x + 1) = 4$

7-2   THE NATURAL LOGARITHM

The discussion at the end of the last section showed we can expect the natural logarithm $\ln x$ to have the derivative

(1)　　$\dfrac{d}{dx}\ln x = \dfrac{1}{x}$ 　for 　$x > 0$.

Because it is a logarithm, it should also have the value 0 at $x = 1$:

(2)　　$\ln(1) = 0$.

We will begin our formal discussion of the natural logarithm by defining it as an integral which has properties (1) and (2).

For any continuous function f, the integral $\int_a^x f(t)\,dt$ is zero at $x = a$ and, by the Fundamental Theorem of Calculus for derivatives of integrals, its derivative is $f(x)$. Accordingly, to obtain a function $\ln x$ with properties (1) and (2), we can take $a = 1$ and $f(t) = 1/t$ in the integral:

Definition 7.1　For $x > 0$, the natural logarithm of x is

(3)　　$\ln x = \displaystyle\int_1^x \dfrac{1}{t}\,dt.$ 　　　　　　◀

In Theorems 7.1 and 7.2 we will use this definition to derive some of the basic properties of $\ln x$. Then we will use this function to justify the definition of b^x for irrational x that was given in the last section and to prove Propositions 7.1 and 7.2. In Theorem 7.3 we will show that the limit defining the number e exists and that $\ln x$ is the logarithm to the base e. Finally, we will obtain the differentiation formulas for logarithms to other bases.

Theorem 7.1　The natural logarithm, defined by (3), has the value 0 at $x = 1$ and for $x > 0$ is continuous, increasing, and has the derivative

(4)　　$\dfrac{d}{dx}\ln x = \dfrac{1}{x}.$ 　　　　　　◀

Proof. The natural logarithm has the value $\ln(1) = 0$ at 1 because for $x = 1$ the limits of integration in (3) are equal. For $x > 0$, the Fundamental Theorem (Theorem 5.1) gives

$$\frac{d}{dx} \ln x = \frac{d}{dx} \int_1^x \frac{1}{t} \, dt = \left[\frac{1}{t}\right]_{t=x} = \frac{1}{x}.$$

The function $\ln x$ is continuous and increasing because it has a positive derivative.   Q.E.D.

EXAMPLE 1   Compute the derivative of $\ln x$ at $x = 5$.

Solution.   By Theorem 7.1

$$\left[\frac{d}{dx} \ln x\right]_{x=5} = \left[\frac{1}{x}\right]_{x=5} = \frac{1}{5}. \quad \square$$

Derivatives of logarithms of functions   If $u(x)$ is a positive differentiable function, then by formula (4) and the chain rule

$$(5) \qquad \frac{d}{dx} \ln[u(x)] = \frac{1}{u(x)} \frac{du}{dx}(x). \qquad\qquad \blacktriangleleft$$

EXAMPLE 2   Compute the derivative of $\ln(x^2 + 1)$.

Solution.   By (5) with $u(x) = x^2 + 1$, we have

$$\frac{d}{dx} \ln(x^2 + 1) = \frac{1}{x^2 + 1} \frac{d}{dx}(x^2 + 1) = \frac{2x}{x^2 + 1}. \quad \square$$

In the next example the natural logarithm is the "inside" function of the composite function.

EXAMPLE 3   Find the derivative of $\sin(\ln x)$.

Solution.   By the chain rule formula for differentiating sines of functions we have for $x > 0$

$$\frac{d}{dx} \sin(\ln x) = \cos(\ln x) \frac{d}{dx} \ln x = \cos(\ln x)\left(\frac{1}{x}\right). \quad \square$$

The graph of ln x   If we describe the integral (3) defining $\ln x$ in terms of areas, we see that for $x > 1$, $\ln x$ is the area of the region bounded by the hyperbola $y = 1/t$, the t-axis, and the vertical lines $t = 1$ and $t = x$ (Figure 7.2). For $0 < x < 1$, $\ln x$ is the negative of the area of the region bounded by the hyperbola, the t-axis, and the lines $t = x$ and $t = 1$ (Figure 7.2a). This shows geometrically why $\ln x$ is an increasing function which is negative for $0 < x < 1$, zero at $x = 1$, and positive for $x > 1$.

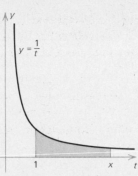

FIGURE 7.2   For $x > 1$, $\ln x$ is this area.

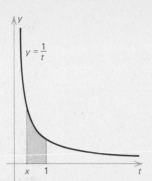

FIGURE 7.2a   For $0 < x < 1$, $\ln x$ is the negative of this area.

The graph of $\ln x$ is shown in Figure 7.3. It is concave down because the second derivative $-1/x^2$ of $\ln x$ is negative.

Algebraic properties of $\ln x$ We will now use Theorem 7.1 to derive algebraic properties of $\ln x$ and to justify the definition of b^x for irrational x that was given in Section 7.1.

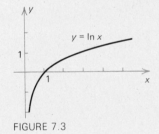

FIGURE 7.3

Theorem 7.2   (a) For any positive numbers a and b and any rational number r

(6) $\ln(ab) = \ln(b) + \ln(a)$ ◄

and

(7) $\ln(b^r) = r \ln b.$ ◄

(b) $\ln x \to \infty$ as $x \to \infty$ and $\ln x \to -\infty$ as $x \to 0^+$.

(c) If x is irrational and $r_1, r_2, r_3, \ldots, r_n, \ldots$ are rational numbers tending to x as n tends to ∞, then $b^{r_1}, b^{r_2}, \ldots, b^{r_n}, \ldots$ have a limit b^x, which is the unique number such that

(8) $\ln(b^x) = x \ln(b).$ ◄

Proof.   (a) Differentiation formula (5) with $u(x) = ax$ gives

$$\frac{d}{dx}\ln(ax) = \frac{1}{ax}\frac{d}{dx}(ax) = \frac{a}{ax} = \frac{1}{x}.$$

This shows that $\ln(ax)$ and $\ln x$ have the same derivative for $x > 0$ so that $\ln(ax) = \ln x + C$. Setting $x = 1$ and using (2) shows that $C = \ln(a)$ and $\ln(ax) = \ln(x) + \ln(a)$. Setting $x = b$ yields (6).

Similarly, formula (5) with $u(x) = x^r$ and r a rational number yields

$$\frac{d}{dx}\ln(x^r) = \frac{1}{x^r}\frac{d}{dx}(x^r) = \frac{rx^{r-1}}{x^r} = \frac{r}{x}$$

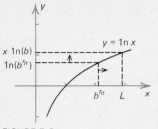

FIGURE 7.3a

and shows that $\ln(x^r)$ and $r \ln x$ have the same derivative. This implies that $\ln(x^r) = r \ln x + C$, and setting $x = 1$ shows that $C = 0$ to give (7).
(b) Because $\ln x$ is an increasing function, we can show that its values tend to ∞ as x tends to ∞ by showing that it has arbitrarily large values. This is the case because $\ln(2)$ is positive and $\ln(2^n) = n \ln(2)$ tends to ∞ as the integer n tends to ∞. The equation $\ln x = -\ln(x^{-1})$ then shows that $\ln x$ tends to $-\infty$ as x tends to 0^+.
(c) By property (7) of $\ln x$ we have $\ln(b^{r_n}) = r_n \ln(b)$ for each n and

$$(9) \qquad \lim_{n \to \infty} \ln(b^{r_n}) = \lim_{n \to \infty} [r_n \ln(b)] = x \ln(b).$$

The continuous, increasing function $\ln x$ has a continuous inverse (by the Inverse Function Theorem, Theorem 4.6a), and consequently the numbers b^{r_n} have a finite limit L as n tends to ∞ (see Figure 7.3a). Because $\ln x$ is continuous, (9) implies that $\ln(L) = x \ln(b)$. L is the only number that satisfies this equation since $\ln x$ is increasing. We define b^x to be the number L to complete the proof of statement (c).   Q.E.D.

Calculations with the natural logarithm

A table of values of the natural logarithm $\ln x$ for numbers x between 1 and 10 is given at the back of this book. To determine $\ln x$ for positive x out of this range, we use properties of the logarithm and the value $\ln(10) = 2.30258 \ldots$ of the natural logarithm of 10.

EXAMPLE 4   Give the numerical value of $\ln(241)$, correct to four decimal places.

Solution.   The table gives the value $\ln(2.41) \approx 0.8796$. Therefore we have

$$\ln(241) = \ln[(10)^2(2.41)] = 2 \ln(10) + \ln(2.41)$$
$$\approx 2(2.3026) + 0.8796 = 5.4848. \quad \square$$

Proofs of Propositions 7.1 and 7.2

We will now use the properties of $\ln x$ that we have established to prove Propositions 7.1 and 7.2 of the last section.

Proof of Proposition 7.1.   By Theorem 7.2 we have for positive b and arbitrary x_1 and x_2

$$\ln(b^{x_1}b^{x_2}) = \ln(b^{x_1}) + \ln(b^{x_2}) = x_1\ln(b) + x_2\ln(b)$$
$$= (x_1 + x_2) \ln(b) = \ln(b^{x_1+x_2}).$$

Because $\ln x$ is increasing, it cannot have the same value for two different values of x and the last equation implies that $b^{x_1}b^{x_2} = b^{x_1+x_2}$, which is equation (3) of Proposition 7.1. The equation

$$\ln[(b^{x_1})^{x_2}] = x_2\ln(b^{x_1}) = x_1x_2\ln(b) = \ln(b^{x_1x_2})$$

follows from Theorem 7.2 and implies that $(b^{x_1})^{x_2} = b^{x_1x_2}$, which is equation (4) of Proposition 7.1. Similar arguments yield equations (5) and (6) of that proposition (see Exercise 25).

Now we rewrite equation (8) of Theorem 7.2 for $b > 1$ in the form

(10) $\dfrac{1}{\ln b} \ln(b^x) = x.$

Because $\ln b$ is positive for $b > 1$, the function $\ln(x)/\ln(b)$ is continuous and increasing and tends to ∞ as $x \to \infty$ and tends to $-\infty$ as $x \to 0^+$. Equation (10) shows that b^x is the inverse of $\ln(x)/\ln(b)$ and, consequently, b^x is defined for all x, is positive, continuous, and increasing and its values tend to ∞ as x tends to ∞ and tend to 0 as x tends to $-\infty$. This completes the proof of Proposition 7.1.   Q.E.D.

Proof of Proposition 7.2.   Because $\ln(x)/\ln(b)$ is the inverse of the exponential function b^x, it is the logarithm to the base b:

(11) $\log_b x = \dfrac{\ln x}{\ln b}$ for $b > 1$ and $x > 0$.   ◀

Because $\ln b$ is positive, the properties of $\log_b x$ that are stated in Proposition 7.2 follow from (11) and the corresponding properties of $\ln x$ that were established in Theorems 7.1 and 7.2.   Q.E.D.

The function ln x is the logarithm to the base *e*

We can now complete our derivation of the basic properties of logarithms by proving the following result.

Theorem 7.3   The limit

(12) $e = \lim\limits_{n \to \infty} \left(1 + \dfrac{1}{n}\right)^n$

defining the number *e* exists, ln(*e*) = 1, and ln x is the logarithm to the base *e*.

We may consider n to be a real variable that passes through all positive values as it tends to infinity in the limit (12).

Proof.   Formula (8) of Theorem 7.2 yields for positive n

(13) $\ln\left[\left(1 + \dfrac{1}{n}\right)^n\right] = n \ln\left(1 + \dfrac{1}{n}\right).$

If we set $\Delta x = 1/n$ and use the fact that $\ln(1)$ is zero, we obtain

(14)
$$\ln\left[\left(1 + \dfrac{1}{n}\right)^n\right] = n\left[\ln\left(1 + \dfrac{1}{n}\right) - \ln(1)\right]$$
$$= \frac{\ln(1 + \Delta x) - \ln(1)}{\Delta x}.$$

As n tends to infinity, Δx tends to zero from the right, so the quotient on the right of equation (14) tends to the derivative of $\ln x$ *at* $x = 1$:

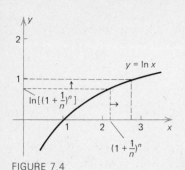

FIGURE 7.4

$$(15) \quad \lim_{n \to \infty} \ln\left[\left(1 + \frac{1}{n}\right)^n\right] = \left[\frac{d}{dx}\ln x\right]_{x=1} = \left[\frac{1}{x}\right]_{x=1} = 1.$$

Because $\ln x$ has a continuous inverse, (15) implies that $(1 + 1/n)^n$ has a limit as n tends to ∞ and that the limit is the number x such that $\ln x = 1$ (Figure 7.4). Thus, the limit (12) defining e exists and $\ln e = 1$. The fact that $\ln x$ is $\log_e x$ then follows from (11) with $b = e$.   Q.E.D.

Since $\ln x$ is the logarithm to the base e, the functions $\ln x$ and e^x are inverse functions and are related by the equations

$$(16) \quad \ln(e^x) = x \quad \text{for all } x \qquad \blacktriangleleft$$

$$(17) \quad e^{\ln x} = x \quad \text{for } x > 0. \qquad \blacktriangleleft$$

EXAMPLE 5   Solve the equation $e^{5x} = 4e^{2x}$ for x.

Solution.   When we divide both sides of the equation $e^{5x} = 4e^{2x}$ by e^{2x}, we obtain $e^{3x} = 4$. Taking the natural logarithm of both sides then yields $\ln(e^{3x}) = \ln(4)$ or $3x = \ln(4)$. The solution is $x = \frac{1}{3}\ln(4)$.   □

EXAMPLE 6   Solve the equation $\ln(x^3 + 1) = 2$ for x.

Solution.   Because $\ln x$ and e^x are inverses, the equation $\ln(x^3 + 1) = 2$ is equivalent to the equation $x^3 + 1 = e^2$, which gives $x^3 = e^2 - 1$ and then $x = (e^2 - 1)^{1/3}$.   □

Logarithms to other bases   Equation (11) implies that logarithms to bases b and c ($b > 1, c > 1$) are related by the equations

$$(18) \quad \log_b x = (\log_b c)(\log_c x) = \frac{\log_c x}{\log_c b}. \qquad \blacktriangleleft$$

To remember these formulas, think of the letter "c" as cancelling in the second two expressions:

$$\log_b x = (\log_b \cancel{c}) \log_{\cancel{c}} x = \frac{\log_{\cancel{c}} x}{\log_{\cancel{c}} b}.$$

In particular, with $c = e$ in (18) we obtain two expressions

$$(19) \quad \log_b x = (\log_b e)(\ln x) = \frac{\ln x}{\ln b} \qquad \blacktriangleleft$$

for $\log_b x$ in terms of $\ln x$. Because the derivative of $\ln x$ is $1/x$, equations (19) yield the formulas

$$(20) \quad \frac{d}{dx}\log_b x = \frac{1}{x}\log_b e = \frac{1}{x \ln(b)} \qquad \blacktriangleleft$$

for the derivative of $\log_b x$.

EXAMPLE 7   What is the slope of the tangent line to the graph of $\log_{10} x$ at $x = 1$?

Solution.   By formula (20) the derivative is

$$\left[\frac{d}{dx}\log_{10}x\right]_{x=1} = \left[\frac{1}{x\,\ln(10)}\right]_{x=1} = \frac{1}{\ln(10)} \approx 0.434. \quad \square$$

The graph of $\log_b x$ Identity (19) shows that for any $b > 1$, $\log_b x$ is $\ln x$ multiplied by the positive number $1/\ln b$. Therefore, the graph of $\log_b x$ is obtained from the graph of $\ln x$ by either contracting it or stretching it in the y-direction (Figure 7.5). (We can think of $\ln x$ as the logarithm whose graph has slope 1 at $x = 1$ where it crosses the x-axis.)

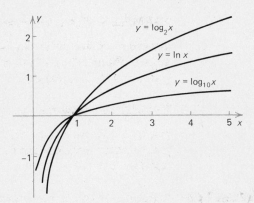

FIGURE 7.5

EXAMPLE 8   Sketch the graph of the function $x \ln x$. (Its values tend to 0 as $x \to 0^+$.)

Solution.   The function $x \ln x$ is defined for $x > 0$. It is negative for $0 < x < 1$, has the value 0 at $x = 1$, and is positive for $x > 1$. By the product rule its derivative is

$$\frac{d}{dx}(x \ln x) = \left(\frac{d}{dx}x\right)\ln x + x\left(\frac{d}{dx}\ln x\right) = \ln x + 1.$$

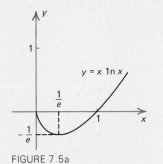

FIGURE 7.5a

The derivative is zero at $x = 1/e$ since $\ln(1/e) = -\ln e = -1$. The derivative is negative and $x \ln x$ is decreasing for $0 < x < 1/e$; the derivative is positive and $x \ln x$ is increasing for $x > 1/e$. The function has a minimum at $x = 1/e$ where its value is $(1/e)\ln(1/e) = -1/e$. Its graph is shown in Figure 7.5a. It is concave up for all $x > 0$ because the second derivative $1/x$ is positive. $\square$

Exercises

†*Answer provided.*
‡*Outline of solution provided.*

1.† Compute exact values of $\left(1 + \dfrac{1}{n}\right)^n$ for $n = 2, 3,$ and 4.

2. Use the table of common logarithms to compute approximate values of $\left(1 + \dfrac{1}{n}\right)^n$ for $n = 20$ and $n = 50$.

3. Use the table of natural logarithms to give approximate values of
   a.[†] $\ln(5)$ **b.**[†] $\ln(1/5)$ **c.**[†] $\ln(765,000)$ **d.** $\ln(2.71)$
   e. $\ln(0.0045)$.

4. Sketch the graph of the function $\ln x$. Plot the points and draw the tangent lines at $x = 1/4, x = 1, x = 4,$ and $x = 8$.

5. Use the table of natural logarithms to give approximate values of
   a.[‡] $\log_5(1.89)$ **b.**[†] $\log_{1.9}(9.1)$ **c.** $\log_2(1/9)$.

In Exercises 6 through 18 compute the derivatives.

6.[†] $L'(300)$ for $L(x) = \ln x$ 7.[†] $M'(100)$ for $M(x) = \log_{10}x$

8. $\left[\dfrac{d}{dx}\ln x\right]_{x=0.004}$ 9.[‡] $\dfrac{d}{dx}\ln(\sin x)$

10.[‡] $\dfrac{d}{dx}\sin(\ln x)$ 11.[†] $\dfrac{d}{dx}[\ln\sqrt{x} + \sqrt{\ln x}]$

12.[†] $F'(3)$ for $F(x) = x\ln x$ 13. $\dfrac{d}{du}\left[\dfrac{\ln u}{u}\right]$

14. $\dfrac{d}{dt}\ln(6t^5 - 5t^6)$ 15.[†] $\dfrac{d}{dy}\sqrt{\ln\sqrt{y}}$

16. $\dfrac{d}{dx}\ln(\ln x)$ 17.[†] $\dfrac{d}{dx}[\log_{10}x]^{10}$

18. $\left[\dfrac{d}{dx}\log_4 x\right]_{x=4}$

19.[†] Sketch the graph of the function $(\ln x)/x$. Show where it is increasing and decreasing. Classify its critical points. (The function $(\ln x)/x$ tends to zero as x tends to ∞.)

20. Sketch the graph of the function $(\ln x)^2$. Show where it is increasing and decreasing and show where its graph is concave up and concave down. Classify its critical points and find its inflection points.

21. Sketch the region in a ty-plane bounded by the hyperbola $y = 1/t$, by the t-axis, and by the lines $t = 1$ and $t = 5$. Sketch an approximation of this region by four rectangles of width 1, with their bases on the t-axis, and touching the hyperbola at their upper left corners. Use these rectangles to give an approximate value for $\ln(5)$.

22.[†] Show that for any fixed number $k > 0$, $\lim\limits_{n\to\infty}\left(1 + \dfrac{1}{n}\right)^k = 1$ and
   $$\lim_{n\to\infty}\left(1 + \dfrac{1}{k}\right)^n = \infty.$$

23. Use definition (7) of $\ln x$ and the symmetry of the hyperbola about the line $y = t$ to show that $\ln(1/x) = -\ln x$ for $x > 0$.

24.[†] Translate the equation $\ln(x_1) = \ln(x_1x_2) - \ln(x_2)$ for x_1 and $x_2 > 1$ into a statement about the areas of regions bounded by the hyperbola $y = 1/t$ in a ty-plane. Prove this result directly by approximating each of the regions by N rectangles of equal width which have their bases on the t-axis and touch the hyperbola at their upper right corners.

25. Use Theorem 7.2 to derive formulas (5) and (6) of Proposition 7.2.

26.[†] For what value of b does the graph of $\log_b x$ have slope 2 at $x = 1$?

27. A positive integer greater than 1 is called *prime* if it is not divisible by any other integers greater than 1. According to the *Prime Number Theorem*, the number of primes less than N for large N is approximately $N/\ln N$. Approximately how many primes are there less than 1,000,000?

28. For large integers n, *Stirling's formula* $n! \approx \sqrt{2\pi n}\,(n/e)^n$ gives an approximate value of $n! = n(n-1)(n-2)\cdots(2)(1)$ (*n factorial*). Use it to find an approximate value of 1000!.

Give all solutions x of the equations in Exercises 29 through 38.

29.[†] $\ln(x+4) = \ln(x^2 + x)$ **30.**[†] $(e^x)^4 = e^{x^3}$

31.[†] $2\ln x - \ln(\sqrt{x}) = 17$ **32.**[†] $\sqrt{\ln(x^2)} = 5$

33.[†] $e^{5x^3} = 6$ **34.**[†] $\ln(5^{1/x}) = 7$

35.[†] $e^{4\ln x} = 3$ **36.** $\ln(1 + x^2) = -1$

37.[†] $e^{\sqrt{x}} = 4e^{-\sqrt{x}}$ **38.** $(4 - \ln x)(1 + e^x) = 0$

Describe the domains and find the derivatives of the functions in Exercises 39 through 46.

39.[†] $[\ln(x^3)]^{1/3}$ **40.**[†] $[\ln(x^2)]^{1/2}$

41.[†] $\ln(2 + \cos x)$ **42.**[†] $\log_{10}(\ln x)$

43.[†] $\ln(\log_{10} x)$ **44.** $[\ln(x^2 - 1)]^{-1/2}$

45.[†] $\sin(x \ln x)$ **46.** $\dfrac{x}{\ln x}$

7-3 THE INTEGRAL OF x^{-1}

The integration formula

$$(1) \qquad \int x^n \, dx = \frac{1}{n+1} x^{n+1} + C$$

which we derived in Chapter 5, is not valid for $n = -1$. The integral of x^{-1} is given by the natural logarithm:

Theorem 7.4   For any $x \neq 0$

$$(2) \qquad \frac{d}{dx} \ln|x| = \frac{1}{x}. \qquad \blacktriangleleft$$

Consequently,

$$(3) \qquad \int \frac{1}{x} \, dx = \ln|x| + C \quad \text{for } x \neq 0. \qquad \blacktriangleleft$$

Proof. For $x > 0$, we have $|x| = x$ and equation (2) is the equation $\dfrac{d}{dx} \ln x = 1/x$, which we proved in Theorem 7.1. For $x < 0$, we have $|x| = -x$ and, by the chain rule,

$$\frac{d}{dx} \ln|x| = \frac{d}{dx} \ln(-x) = \frac{1}{-x} \frac{d}{dx}(-x) = \frac{1}{-x}(-1) = \frac{1}{x}. \quad \text{Q.E.D.}$$

The geometric interpretation of formula (2) is indicated in Figure 7.6, where we show the graph of the function $\ln|x|$. It consists of the graph of $\ln x$ to the right of the y-axis and its mirror image about the y-axis. Equation (2) states that for any $x \neq 0$, the tangent line to $y = \ln|x|$ at x has slope $1/x$. This is illustrated in Figure 7.6 by the tangent lines at $x = -2, -\frac{1}{2}, \frac{1}{2}$, and 2.

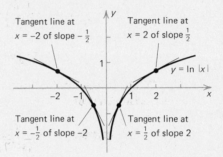

FIGURE 7.6

EXAMPLE 1   Compute the derivative of $\ln|x|$ at $x = -5$.

Solution.   By Theorem 7.2 we have

$$\left[\frac{d}{dx} \ln|x| \right]_{x=-5} = \left[\frac{1}{x} \right]_{x=-5} = -\frac{1}{5}. \quad \square$$

EXAMPLE 2   Evaluate the definite integral

$$\int_5^{10} \frac{1}{x} \, dx.$$

Solution.   The indefinite integral is given by formula (3), and we have

$$\int_5^{10} \frac{1}{x} \, dx = \left[\int \frac{1}{x} \, dx \right]_5^{10} = [\ln|x|]_5^{10} = \ln(10) - \ln(5) = \ln(2). \quad \square$$

EXAMPLE 3   Evaluate

$$\int_{-7}^{-2} \frac{1}{x} \, dx.$$

Solution.   We have

$$\int_{-7}^{-2} \frac{1}{x} \, dx = [\ln|x|]_{-7}^{-2} = \ln|-2| - \ln|-7| = \ln(2) - \ln(7). \quad \square$$

In the next example we combine integration formula (3) with the method of substitution.

EXAMPLE 4   Evaluate

$$\int \frac{1}{1-4x} \, dx.$$

Solution.   We set $u = 1 - 4x$. Then $du = -4dx$ and

$$\int \frac{1}{1-4x} \, dx = -\frac{1}{4} \int \frac{1}{1-4x} (-4dx) = -\frac{1}{4} \int \frac{1}{u} \, du$$

$$= -\frac{1}{4} \ln|u| + C = -\frac{1}{4} \ln|1-4x| + C. \quad \square$$

When differentiation formula (2) is combined with the chain rule, we obtain the rule

(4) $$\frac{d}{dx} \ln|u(x)| = \frac{1}{u(x)} \frac{du}{dx}(x)$$ ◄

valid for differentiable nonzero functions $u(x)$.

EXAMPLE 5   Describe the domain of $\ln|x^2 - 4|$ and compute its derivative.

Solution.   The function $\ln|x^2 - 4|$ is defined for all x such that $|x^2 - 4|$ is positive. This is for all $x \neq \pm 2$. By rule (4) with $u(x) = x^2 - 4$, we have

$$\frac{d}{dx} \ln|x^2 - 4| = \frac{1}{x^2 - 4} \frac{d}{dx}(x^2 - 4) = \frac{2x}{x^2 - 4}. \quad \square$$

EXAMPLE 6   Sketch the region bounded by the curve $y = \dfrac{2x}{x^2 + 1}$, the x-axis, and the line $x = 2$. Compute its area.

Solution.   The region is shown in Figure 7.6a. Its area is

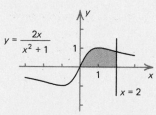

$y = \dfrac{2x}{x^2 + 1}$

$x = 2$

FIGURE 7.6a

(5) $$\int_0^2 \frac{2x}{x^2 + 1} \, dx.$$

To evaluate the indefinite integral, we make the substitution $u = x^2 + 1, du = 2x\,dx$:

$$\int \frac{2x}{x^2 + 1}\,dx = \int \frac{1}{u}\,du = \ln|u| + C = \ln(x^2 + 1) + C.$$

We did not put absolute value signs around $x^2 + 1$ in the expression on the right of this equation because $x^2 + 1$ is positive and $|x^2 + 1| = x^2 + 1$ for all x. Having found the indefinite integral, we see that the area (5) is

$$\int_0^2 \frac{2x}{x^2 + 1}\,dx = [\ln(x^2 + 1)]_0^2 = \ln(2^2 + 1) - \ln(0^2 + 1)$$

$$= \ln(5). \quad \square$$

Exercises

In Exercises 1 through 10 evaluate the integrals.

1.† $\displaystyle\int_1^6 \frac{1}{x}\,dx$ **2.** $\displaystyle\int_{-6}^{-1} \frac{1}{x}\,dx$ **3.**‡ $\displaystyle\int_0^5 \frac{x}{x^2 + 4}\,dx$

4.† $\displaystyle\int_0^6 \frac{1}{x + 2}\,dx$ **5.** $\displaystyle\int_0^1 \frac{x^2}{4 - x^3}\,dx$ **6.**‡ $\displaystyle\int \frac{1}{x \ln x}\,dx$

7.† $\displaystyle\int \frac{\sqrt{\ln x}}{x}\,dx$ **8.** $\displaystyle\int \frac{\sin(6x)}{\cos(6x)}\,dx$ **9.**† $\displaystyle\int_1^8 \frac{x^{1/3}}{1 + x^{4/3}}\,dx$

10. $\displaystyle\int_8^9 \frac{7x}{10 - x^2}\,dx$

11.† Sketch the region bounded by the line $x + y = 3$ and by the hyperbola $xy = 2$. Compute its area.

12. Sketch the region bounded by the curves $y = 1/x$ and $y = -1/x$ and by the lines $y = 1/e$ and $y = e$. Compute its area.

13.‡ Compute the area of the region bounded by the curve $y = 1/(1 - x)$, by the y-axis, and by the line $y = 3$.

14.† Sketch the region bounded by the y-axis, by the lines $y = 1$ and $y = 4$, and by the graph of the function $1/x$. Compute the volume of the solid generated by rotating that region about the line $y = -1$.

15. Sketch the curves $y = \ln(x - 4)$ and $y = \ln|x - 4|$.

16. Sketch the graph of the function $1/(2x + 6)$ and of one of its antiderivatives.

17. Compute the derivatives   **a.**† $\displaystyle\frac{d}{dx}\ln|x^2 - 9|$ **b.** $\displaystyle\frac{d}{dx}\ln|\sin x|$.

18. Sketch the graph of $\frac{1}{2}\ln(x^2)$. Show its tangent lines at $x = -2, -\frac{1}{2}, \frac{1}{2}$, and 2.

19. Sketch the graphs of   **a.** $\ln(x^2 + 1)$ and   **b.** $\ln(x^2 - 1)$.

20.† A region R is bounded by the curve $y = (x^2 + 4)^{-1}$, by the y-axis and by the line $y = \frac{1}{8}$. What volume does it generate when it is rotated about the y-axis?

21.† A rod extends from $x = 1$ to $x = 4$ on an x-axis with its scale measured in feet. Its density at x is $2/x$ pounds per foot. Where is its center of gravity?

22.† What is the average value of $2x - 3 + 4x^{-1}$ for $-3 \le x \le -1$?

23. What is the area of the region bounded by the x-axis, the curve $y = x^2/(x^3 + 2)$, and the line $x = 2$?

24. What is the average value of $(\sin x)/(1 - 3\cos x)$ for $0 \le x \le \frac{1}{4}\pi$?

7-4 LOGARITHMIC DIFFERENTIATION (Optional)

If a function is expressed as a product or quotient of powers of simpler functions, we can compute its derivative *by computing the derivative of the logarithm of its absolute value.* This procedure is called LOGARITHMIC DIFFERENTIATION.

EXAMPLE 1   Use logarithmic differentiation to compute the derivative of the function

(1) $f(x) = \dfrac{\sin x \, \sqrt{x + 3}}{x^3 + 4}.$

Solution.   We have

$$\ln|f(x)| = \ln|\sin x| + \frac{1}{2}\ln|x + 3| - \ln|x^3 + 4|.$$

Differentiating both sides yields

$$\frac{\dfrac{df}{dx}(x)}{f(x)} = \frac{\dfrac{d}{dx}\sin x}{\sin x} + \frac{1}{2}\frac{\dfrac{d}{dx}(x + 3)}{x + 3} - \frac{\dfrac{d}{dx}(x^3 + 4)}{x^3 + 4}$$

or

$$\frac{\dfrac{df}{dx}(x)}{f(x)} = \frac{\cos x}{\sin x} + \frac{1}{2}\left[\frac{1}{x + 3}\right] - \frac{3x^2}{x^3 + 4}.$$

Multiplying both sides by $f(x)$ and using expression (1) for that function gives

$$\frac{df}{dx}(x) = f(x)\left[\frac{\cos x}{\sin x} + \frac{1}{2x + 6} - \frac{3x^2}{x^3 + 4}\right]$$

$$= \frac{\sin x \, \sqrt{x + 3}}{x^3 + 4}\left[\frac{\cos x}{\sin x} + \frac{1}{2x + 6} - \frac{3x^2}{x^3 + 4}\right]. \quad \square$$

Exercises

†Answer provided.
‡Outline of solution provided.

In Exercises 1 through 8 compute the derivatives by logarithmic differentiation.

1.† $f'(x)$ where $f(x) = \dfrac{(2x + 3)(3x + 4)}{(5x + 6)}$

2. $\dfrac{dg}{dy}(y)$ where $g(y) = (y^3 + 1)^{1/3}(y^4 + 1)^{1/4}(y^5 + 1)^{1/5}$

3.† $\dfrac{d}{dx}\sqrt{\dfrac{x^2 + 1}{x^2 - 1}}$

4. $\dfrac{d}{dt}[\cos(t)\cos(2t)\cos(3t)]$

5.† $\dfrac{d}{dx}x^x$

6. $\dfrac{d}{dx}x^{\sin x}$

7.‡ $F'(1)$ where $F(x) = (3 - 2x)^{10}(6 - 5x)^{20}(9 - 8x)^{30}$

8. $G'(0)$ where $G(t) = \sqrt{2t + 1}\,\sqrt[3]{3t + 1}\,\sqrt[4]{4t + 1}\,\sqrt[5]{5t + 1}$

Find the derivatives of the functions in Exercises 9 through 14.

9.† $x^{1/7}(2x + 1)^{1/6}(x + 1)^{10}$

10.† $\sqrt{(x + 1)(x - 3)(x + 2)(x - 4)}$

11.† $x(\sin x)(\cos x)(\ln x)$

12. $(\sin x)(1 - \cos x)^4(\sin x + \cos x)^{1/3}$

13.† $\dfrac{x \sin x}{(x^2 + 1)^5 \cos x}$

14. $\dfrac{(2x + 1)(3x + 1)}{(4x + 1)(5x + 1)}$

7-5 THE EXPONENTIAL FUNCTION

Of all exponential functions b^x $(b > 0)$ it is the exponential function e^x with base $e = 2.71828 \ldots$ that is most frequently used in calculus. This is because it has the simplest formulas for its derivative and integral. The function e^x is therefore called *the* exponential function.

Theorem 7.5   The derivative and integral of the exponential function e^x are given for all x by the formulas

(1) $\dfrac{d}{dx}e^x = e^x$ ◀

(2) $\displaystyle\int e^x\,dx = e^x + C.$ ◀

Proof.   Because $\ln x$ has a positive derivative for all $x > 0$, tends to ∞ as $x \to \infty$, and tends to $-\infty$ as $x \to 0^+$, the Inverse Function Theorem implies that e^x has a derivative for all x. To compute the derivative, we differentiate the equation $\ln(e^x) = x$ with respect to x. This gives

$$\frac{\dfrac{d}{dx}e^x}{e^x} = 1.$$

When we multiply both sides of this equation by e^x, we obtain equation (1). Formula (2) is the same result phrased in terms of antiderivatives.   Q.E.D.

The graph of e^x

The graph of e^x and of its inverse function $\ln x$ are shown in Figure 7.7. The two graphs are mirror images about the line $y = x$.

The first and second derivatives of e^x are e^x itself, which is positive for all x. The function e^x is therefore positive and increasing and its graph is concave up for all x. The y-intercept of the graph is 1 because $e^0 = 1$. The x-axis is an asymptote because $e^x \to 0$ as $x \to -\infty$. Also $e^x \to \infty$ as $x \to \infty$.

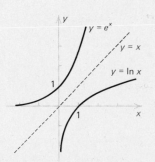

FIGURE 7.7

EXAMPLE 1   Give an equation for the tangent line to the curve $y = e^x$ at $x = 0$.

Solution.   We have $\dfrac{d}{dx} e^x = e^x$ and $e^0 = 1$. The tangent line, therefore, passes through the point $(0, 1)$ and has slope 1. It has the equation $y - 1 = 1(x - 0)$ or $y = x + 1$.   □

EXAMPLE 2   Evaluate the integral

$$\int_{-2}^{2} e^x \, dx.$$

Solution.   With formula (2) we have

$$\int_{-2}^{2} e^x \, dx = \left[\int e^x \, dx \right]_{-2}^{2} = [e^x]_{-2}^{2} = e^2 - e^{-2}.   □$$

Derivatives of composite functions

With the chain rule formula (1) gives

(3) $\dfrac{d}{dx} e^{u(x)} = e^{u(x)} \dfrac{du}{dx}(x)$ ◀

for any differentiable function $u(x)$.

EXAMPLE 3   Compute the derivatives of (a) $e^{\sin x}$ and (b) $\sin(e^x)$.

Solution.   (a) We use rule (3) with $u(x) = \sin x$ to obtain $\dfrac{d}{dx} e^{\sin x} = e^{\sin x} \dfrac{d}{dx} \sin x = \cos x \, e^{\sin x}$. (b) We use the rule for differentiating the sine of a function: $\dfrac{d}{dx} \sin(e^x) = \cos(e^x) \dfrac{d}{dx} e^x$

$= e^x \cos(e^x)$.   □

EXAMPLE 4   Sketch the graph of the function $x^2 e^{-x}$.

Solution.   The function's derivative

$$\frac{d}{dx}(x^2 e^{-x}) = \left(\frac{d}{dx}x^2\right)e^{-x} + x^2\left(\frac{d}{dx}e^{-x}\right) = 2xe^{-x} - x^2 e^{-x}$$

$$= (2x - x^2)e^{-x} = x(2 - x)e^{-x}$$

is zero at $x = 0$ and at $x = 2$. The derivative is negative and the function $x^2 e^{-x}$ is decreasing for $x < 0$ and for $x > 2$. The derivative is positive and the function is increasing for $0 < x < 2$. The function has a minimum at $x = 0$ where its value is 0 and a local maximum at $x = 2$ where its value is $2^2 e^{-2} \approx 0.54$. The graph is shown in Figure 7.8. (In Chapter 17 we will see how to show that $x^2 e^{-x^2}$ tends to zero as x tends to infinity.)   □

In the next example we use integration formula (2) with the method of substitution.

EXAMPLE 5   Evaluate

$$\int xe^{-x^2}\, dx.$$

Solution.   We set $u = -x^2$. Then $du = \dfrac{du}{dx}dx = \dfrac{d}{dx}(-x^2)\, dx$

$= -2x\, dx$. Therefore,

$$\int xe^{-x^2}\, dx = -\frac{1}{2}\int e^{-x^2}(-2x\, dx) = -\frac{1}{2}\int e^u\, du$$

$$= -\frac{1}{2}e^u + C = -\frac{1}{2}e^{-x^2} + C.\quad □$$

We use the exponential function to prove the following two theorems.

Theorem 7.6   The differentiation formula

(4) $\dfrac{d}{dx}x^n = nx^{n-1}$ **(x > 0)**   ◀

is valid for irrational exponents n as well as for rational exponents.*

---

*If n is irrational, then x^n is not defined for negative x, so we do not consider its derivative for $x \le 0$.

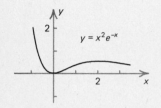

$y = x^2 e^{-x}$

FIGURE 7.8

Proof.   We have for $x > 0$, $x^n = e^{\ln(x^n)} = e^{n \ln x}$ and

$$\frac{d}{dx} x^n = \frac{d}{dx} e^{n \ln x} = e^{n \ln x} \frac{d}{dx} (n \ln x)$$

$$= x^n \left(\frac{n}{x} \right) = nx^{n-1}. \quad \text{Q.E.D.}$$

Theorem 7.7   For any positive constant b

(5) $\dfrac{d}{dx} b^x = (\ln b)\, b^x$ ◄

and

(6) $\displaystyle\int b^x\, dx = \frac{1}{\ln b} b^x + C.$ ◄

Proof.   To derive (6) we write $b^x = e^{\ln(b^x)} = e^{x \ln b}$. Then we obtain

$$\frac{d}{dx} b^x = \frac{d}{dx} e^{x \ln b} = e^{x \ln b} \frac{d}{dx} [x \ln b]$$

$$= b^x \frac{d}{dx} [x \ln b] = (\ln b)\, b^x.$$

This is formula (5). Formula (6) is an immediate consequence.   Q.E.D.

EXAMPLE 6   Compute the derivative of 10^x.

Solution.   By formula (5) with $b = 10$, $\dfrac{d}{dx} (10^x) = [\ln 10] 10^x.$ □

EXAMPLE 7   Evaluate the integral

$$\int_3^4 2^x\, dx.$$

Solution.   Formula (6) gives

$$\int_3^4 2^x\, dx = \left[\int 2^x\, dx \right]_3^4 = \left[\frac{1}{\ln(2)} 2^x \right]_3^4$$

$$= \frac{1}{\ln(2)} (2^4 - 2^3) = \frac{8}{\ln(2)}. \quad \square$$

Note.   Rather than referring to Theorem 7.7 in the solution of Example 6, we could find the derivative by expressing the power of 10 as a power of e. Because $y = e^{\ln y}$ for any positive y, we have for all x

$$10^x = e^{\ln(10^x)} = e^{x \ln(10)}$$

and

$$\frac{d}{dx} 10^x = \frac{d}{dx} e^{x \ln(10)} = e^{x \ln(10)} \frac{d}{dx} [x \ln(10)]$$

$$= \ln(10) \, e^{x \ln(10)} = [\ln(10)] \, 10^x.$$

We use this technique to find another derivative in the next example.   □

EXAMPLE 8   What is the derivative of x^x?

Solution.   For positive x

$$x^x = e^{\ln(x^x)} = e^{x \ln x}$$

so that

$$\frac{d}{dx} x^x = \frac{d}{dx} e^{x \ln x} = e^{x \ln x} \frac{d}{dx} (x \ln x)$$

$$= x^x \left[\left(\frac{d}{dx} x \right) \ln x + x \left(\frac{d}{dx} \ln x \right) \right] = x^x (\ln x + 1).   □$$

In the next two examples we evaluate integrals by the method of substitution.

EXAMPLE 9   Find the indefinite integral $\int x^{1/4} e^{x^{5/4}} \, dx$.

Solution.   We made the substitution $u = x^{5/4}, du = \frac{5}{4} x^{1/4} \, dx$:

$$\int x^{1/4} e^{x^{5/4}} \, dx = \frac{4}{5} \int e^{x^{5/4}} (\tfrac{5}{4} x^{1/4} \, dx) = \frac{4}{5} \int e^u \, du$$

$$= \frac{4}{5} e^u + C = \frac{4}{5} e^{x^{5/4}} + C.   □$$

EXAMPLE 10   Evaluate $\int_0^5 e^{-x} \sin(e^{-x}) \, dx$.

Solution.   We use $u = e^{-x}$, for which $du = -e^{-x} \, dx$, to obtain

$$\int e^{-x} \sin(e^{-x}) \, dx = - \int \sin(e^{-x})(-e^{-x} \, dx) = - \int \sin u \, du$$

$$= \cos u + C = \cos(e^{-x}) + C.$$

Hence

$$\int_0^5 e^{-x} \sin(e^{-x}) \, dx = [\cos(e^{-x})]_0^5$$

$$= \cos(e^5) - \cos(e^0) = \cos(e^5) - \cos(1).   □$$

Exercises

†*Answer provided.*
‡*Outline of solution provided.*

1. Simplify the following expressions.   **a.**‡ $e^{3 \ln x}$ **b.**† $\ln(e^{-x^2})$
   c. $e^{(\ln 3 - \ln x)}$ **d.** $5 \ln(e^{1/x})$

2. Solve the following equations for x.   **a.**‡ $e^{x/2} = 4$
   b.† $\ln(1 + x) = 5$ **c.** $e^{x^2} = 100$ **d.** $\ln(x^3 + 4) = 1$
   e. $x^e = 1$

Compute the derivatives of the functions in Exercises 3 through 20.

3.† e^{6x} **4.** $e^{\cos x}$ **5.**† xe^{-x}

6. $\sqrt{1 + e^x}$ **7.**† $\ln(1 - e^x)$ **8.** $e^{-1/x}$

9.† $\dfrac{e^x}{x + 3}$ **10.** e^4 **11.**‡ 5^{-x}

12. $2^{\sin x}$ **13.**† $e^x + x^e$ **14.** $[\sin x]^{\pi}$

15.† $[\sin x]^x$ **16.** x^{-x} **17.**† $[3 + x]^{3-x}$

18. $\log_{10}(e^x)$ **19.**† $e^{\sin(3x)}$ **20.** $\sqrt[3]{x - e^{5x}}$

21. A function $y = y(x)$ is defined implicitly by the equation $e^y - e^{-y} = x$ and by the condition $y(0) = 0$. What is $y'(0)$?

22.† The atmospheric pressure at a height of h (meters) above sea level is $P = 10^4 \, e^{-0.00012\,h}$ kilograms per square meter. What is the rate of change of the pressure with respect to height at $h = 1000$?

23.† Give an equation of the tangent line to the curve $y = x^2 e^{2x}$ at $x = 1$.

24. In one diagram sketch the graphs of the functions 2^x, e^x, and 3^x. Plot the points at $x = 0, \pm 1$, and ± 2 on each graph.

25. Sketch the graphs of the following functions. Show any critical points, inflection points, and horizontal or vertical asymptotes. **a.**† xe^x **b.** e^{-x^2} **c.**† $1/(1 - e^x)$ **d.** $3 - e^{-x^2/2}$

Evaluate the integrals in Exercises 26 through 33.

26.† $\displaystyle\int_0^2 e^{-3t} \, dt$ **27.** $\displaystyle\int x^2 \, e^{x^3} \, dx$

28.† $\displaystyle\int e^t \cos(e^t) \, dt$ **29.** $\displaystyle\int \frac{e^y}{1 + e^y} \, dy$

30.† $\displaystyle\int_1^{\pi} (x^{\pi} - \pi^x) \, dx$ **31.** $\displaystyle\int [\sin x]^{\sqrt{2}} \cos x \, dx$

32.† $\displaystyle\int_0^1 \sin(7x) \, e^{\cos(7x)} \, dx$ **33.** $\displaystyle\int_1^9 \frac{1}{\sqrt{x}} e^{\sqrt{x}} \, dx$

34.† Sketch the region bounded by the x-axis, by the lines $x = -1$ and $x = 1$, and by the curve $y = e^{2x}$. Compute its area.

35. Compute the volume of the solid generated by rotating region of Exercise 34 about the x-axis.

36.† Make a rough sketch of the region bounded by the curves $y = e^x$, $y = e^{-x}$, and $y = 4$. Compute its area.

37. Make a rough sketch of the region bounded by the graphs of the functions 6^x and $4(3^x)$ and by the y-axis. Compute its area.

38.† Find a constant a such that the region bounded by the x- and y-axes, by the graph of the function e^{-x}, and by the line $x = a$ has area 0.99.

39. Find a constant r such that $y = e^{rx}$ satisfies the differential equation $y'' + 2y' + y = 0$.

40. Find the maximum value of the function $x - e^x$.

41.[†] If, on the average, one car crosses a bridge every two minutes, then the probability that n cars cross in a time interval of t minutes is $(t/2)^n(1/n!)e^{-t/2}$. Here $0! = 1$ and $n! = n(n-1) \ldots (2)(1)$ for positive integers n. **a.** What is the probability that no cars cross the bridge in a four minute period? **b.** What is the probability that eight cars will cross the bridge during one minute? (In this mathematical model we disregard the effect of traffic congestion.)

42. The probability of a radio transistor failing after a months and before b months is

$$k \int_a^b e^{-kt}\, dt$$

for some constant k. There is a 20% chance it will fail within one year. **a.** What is k? **b.** What is the probability that the transistor will fail during the second year? (In this mathematical model we assume that the likelihood of the transistor failing does not increase with age as is generally the case.)

43. A vertical beam of light, which has intensity I_0 when it hits the ocean surface, has intensity $I = I_0\, e^{-1.4y}$ at a depth of y meters. At what depth is the light intensity equal to 1% of its value at the surface?

Describe the domains and find the derivatives of the functions in Exercises 44 through 51.

44.[†] $\dfrac{x}{1 - e^x}$

45.[†] $\sin(x - e^x)$

46.[†] $\sqrt{1 - e^{3x}}$

47.[†] $(1 + x^2)^{\pi}$

48.[†] $10^{\ln x} + (\ln x)^{10}$

49. $\dfrac{2 + e^{-x}}{2 - e^x}$

50.[†] $x^{\sqrt{2}}$

51. $x \ln(e^{-x})$

Evaluate the integrals in Exercises 52 through 57.

52.[†] $\displaystyle\int_0^2 x e^{x^2}\, dx$

53.[†] $\displaystyle\int (3^x + x^3)\, dx$

54.[†] $\displaystyle\int \dfrac{e^x + 1}{e^x}\, dx$

55. $\displaystyle\int \dfrac{1}{x} e^{\ln x}\, dx$

56.[†] $\displaystyle\int (e^{3x} + 1)^2\, dx$

57. $\displaystyle\int_0^{\pi} (\sin x) e^{\cos x}\, dx$

7-6   THE DIFFERENTIAL EQUATION $dy/dt = ry$ AND APPLICATIONS

Radioactive materials *decompose* at a rate that is proportional to the amount of the material present at each moment. For example, if 100 grams of a radioactive substance decays at the rate of 3 grams per year, then 200 grams of the same substance will decay at the rate of 6 grams per year. Consequently, if $y = y(t)$ is the weight of the radioactive material in a sample at time t, then the function y satisfies a DIFFERENTIAL EQUATION of the form

$$(1) \qquad \frac{dy}{dt} = ry$$

where r is a negative constant. Equation (1) is called a "differential equation" because it involves the unknown function $y = y(t)$ and its derivative.

Under certain circumstances populations of animals or of bacteria may be considered to *grow* at a rate which is always proportional to the number present. In this case, the size $y = y(t)$ of the population at time t satisfies a differential equation of the form (1) with a positive constant r.

These are but two of the variety of phenomena that are governed by differential equations of the form (1) with constants r. We study this differential equation and several of its applications in this section.

Theorem 7.8   **For any constant r the solutions of the differential equation**

$$(1) \qquad \frac{dy}{dt} = ry \qquad\qquad\qquad \blacktriangleleft$$

are of the form

$$(2) \qquad y = Ce^{rt} \qquad\qquad\qquad \blacktriangleleft$$

where C is an arbitrary constant.

Proof.   If $y = Ce^{rt}$, then

$$\frac{dy}{dt} = \frac{d}{dt}[Ce^{rt}] = Ce^{rt}\frac{d}{dt}(rt) = rCe^{rt} = ry.$$

Conversely, if y is a solution of (1), then

$$\frac{d}{dt}[e^{-rt}y] = e^{-rt}\frac{dy}{dt} + y\frac{d}{dt}e^{-rt} = e^{-rt}\left(\frac{dy}{dt} - ry\right) = 0.$$

This shows that $e^{-rt}y$ is equal to a constant C and gives (2).   Q.E.D.

Often a differential equation such as (1) is given together with the requirement that the solution have a prescribed value at one particular point. Such a requirement is called an INITIAL CONDITION, and the differential equation with the initial condition is called an INITIAL VALUE PROBLEM.

EXAMPLE 1   Solve the initial value problem

(3) $\dfrac{dy}{dt} = 2y \qquad y(0) = 4.$

Solution.   By Theorem 7.8 the solutions of the differential equation (3) are $y(t) = Ce^{2t}$. The value of this function at $t = 0$ is $y(0) = Ce^0 = C$. To have $y(0) = 4$, we set $C = 4$ and obtain the solution $y(t) = 4e^{2t}$.   □

Radioactive decay

The rate of decay of a radioactive substance is usually described by giving the HALF-LIFE of the substance. This is the time it takes for half of a sample to decompose. Table 7.3 gives half-lives of a few radioactive substances. Carbon-14 (also known as radio-carbon) is used to date archaeological finds. Uranium, Potassium, and Rubidium are used in geological dating.

TABLE 7.3

*Handbook of Chemistry and Physics, pp. 451–501.

ELEMENT	HALF-LIFE
Krypton-91	10 seconds
Radium-226	1620 years
Carbon-14	5700 years
Uranium-238	4.5×10^9 years
Potassium-40	14×10^9 years
Rubidium-87	6×10^{10} years

EXAMPLE 2   A radioactive substance has a half-life of T years, and the amount $y(t)$ in a sample satisfies the differential equation

(1) $\dfrac{dy}{dt} = ry.$

How are the constants T and r related?

Solution.   The solutions of equation (1) are of the form

(2) $y(t) = Ce^{rt}.$

Since T is the half-life of the substance, we have $y(t + T) = \frac{1}{2}y(t)$ for all t. With equation (2) this gives $Ce^{r(t+T)} = \frac{1}{2}Ce^{rt}$. We cancel Ce^{rt} from both sides of the last equation to obtain the equation

(4) $e^{rT} = \frac{1}{2}$ ◀

relating r and T.

To put relation (4) in a more convenient form, we take the natural logarithm of both sides to obtain $rT = \ln\left(\frac{1}{2}\right)$ or

(5) $r = \dfrac{1}{T}\ln\left(\frac{1}{2}\right).$ ◀

Notice that, since $\frac{1}{2}$ is less than 1, this gives a negative value of r, as expected.   □

We can solve equation (4) for $e^r = (\frac{1}{2})^{1/T}$ and substitute the result into equation (2) to express the solution in the form $y = C[(\frac{1}{2})^{1/T}]^t$ or

(6) $y(t) = C(\frac{1}{2})^{t/T}$ ◀

which shows clearly that one half of the sample decomposes every T years.

EXAMPLE 3   The half-life of radium is 1620 years. If a sample contains 10 grams of radium initially, how much will it contain 162 years later?

Solution.   The solution may be put in the form (6) with $T = 1620$: $y(t) = C[\frac{1}{2}]^{t/1620}$. If we let $t = 0$ denote the time when there are 10 grams of radium in the sample, then we must take $C = 10$ and we have $y(t) = 10[\frac{1}{2}]^{t/1620}$. At $t = 162$ years there will be $y(162)$ $= 10[\frac{1}{2}]^{162/1620} = 10[\frac{1}{2}]^{1/10} \approx 9.330$ grams in the sample.   □

In the next example we are given the differential equation for the radioactive substance rather than its half life.

EXAMPLE 4   Radioactive Carbon-14 decomposes at a rate equal to 0.0122% of the amount present. How much of a 3 gram sample decomposes in 1000 years?

Solution.   If $y(t)$ is the number of grams in a sample at time t years, then

$$\frac{dy}{dt} = -0.000122y$$

and therefore $y = Ce^{-0.000122t}$. Since $y = 3$ at $t = 0$, we have $C = 3$ and $y = 3e^{-0.000122t}$. At $t = 1000$ the sample contains $y(1000)$ $= 3e^{-0.122}$ grams of Carbon-14 and $3 - 3e^{-0.122} = 3(1 - e^{-0.122})$ ≈ 0.345 grams have decomposed.

Biological growth   Biological growth is generally a more complicated phenomenon than radioactive decay. The rate at which a population of animals or bacteria multiplies usually depends on factors that vary both with time and the size of the population relative to its surroundings. Nevertheless, in many situations a good mathematical model for biological growth can be obtained by assuming that the rate of growth is proportional to the size, and we can study such a model by using Theorem 7.8.

EXAMPLE 5   The population of a certain country in 1950 was 1 million, and the instantaneous rate of growth at any moment is equal to 3% of the current population. What will the country's population be in the year 2000?

Solution.   We let t (years) be the time with t equal to the year (A.D.) at the beginning of each year. We let $P(t)$ be the population (millions) of the country at time t. Then we have the initial value problem

$dP/dt = 0.03\,P$, $P(1950) = 1$. The solutions of the differential equation are $P = Ce^{0.03t}$, and the initial condition gives $C = e^{-0.03(1950)}$. Therefore, $P(t) = e^{0.03(t-1950)}$. In the year $t = 2000$, the population will be $P(2000) = e^{0.03(2000-1950)} = e^{0.03(50)} = e^{1.5}$ million. This is a population of about 4.4817 million.   □

EXAMPLE 6   A culture contains initially 1000 bacteria and the number of bacteria increases at a rate proportional to the number present. After 5 hours there are 1250 bacteria in the culture. How many will it contain 3 hours later?

Solution.   We let $y(t)$ denote the number of bacteria in the culture at time t (hours) with $t = 0$ the time when $y = 1000$. The differential equation (1) gives the solution $y = Ce^{rt}$ and the initial condition $y(0) = 1000$ shows that $C = 1000$ and $y = 1000e^{rt}$. Because y is 1250 when t is 5, we have $1250 = 1000e^{5r}$ or $e^{5r} = 1.25$. Solving for e^r gives $e^r = (1.25)^{1/5}$ and substituting this expression in the formula for y gives

$$y = 1000e^{rt} = 1000[(1.25)^{1/5}]^t = 1000(1.25)^{t/5}.$$

Notice that this formula gives the correct values for y at $t = 0$ and $t = 5$. Three hours after there are 1250 bacteria in the culture, t is 8 and $y = 1000(1.25)^{8/5} = 1000(1.25)^{1.6} \approx 1429$ bacteria are present.   □

Note.   In Examples 1, 4, and 5, where we were given explicit values for the constant r in the differential equation (1), we left the solutions in the form (2) involving the exponential function e^x. Because we were not given r in Examples 2 and 6, but rather the half life or the value of y at two different times, the formula for r involves the natural logarithm. Equation (5) shows that in Example 2, $r = \frac{1}{1620}\ln(\frac{1}{2})$, and in Example 6 we have $e^{5r} = 1.25$ so that $r = \frac{1}{5}\ln(1.25)$. In these cases we found it more convenient to solve for e^r instead of for r and to express the solutions in terms of exponential functions with bases other than e.   □

Exercises

In Exercises 1 through 4 solve the initial value problems.

†Answer provided.
‡Outline of solution provided.

1.† $\dfrac{dy}{dt} = -4y$, $y(0) = -3$

2.† $\dfrac{dz}{dx} = (\ln 4)z$, $z(2) = 15$

3. $\dfrac{dy}{dt} = (0.006)y$, $y(50) = 3$

4. $\dfrac{dy}{dx} = -(\ln 6)y$, $y(-10) = -6$

In Exercises 5 through 17 give exact answers and numerical approximations.

5.‡ The instantaneous rate of decrease of the value of a currency is 5 percent per year. What is the ratio of the value of the currency at time $t + 10$ to its value at time t?

6. If the rate of increase of the population of a country is always equal to 3% of the population, by what factor does it increase every 100 years?

7.† The population of the United States was 4 million in 1790 and 180 million in 1960. If the rate of growth of the population was at all times proportional to the population at that time, what was the constant of proportionality?

8.[†] Carbon-14 (C^{14}) is formed in the upper atmosphere from Nitrogen-14 (N^{14}) by the bombardment of neutrons in cosmic rays. The C^{14} then disintegrates back into N^{14} with the emission of beta particles and with a half-life of 5700 years. The result is an equilibrium of about one C^{14} atom per 10^{12} atoms of carbon in the atmosphere and in living things. When a plant or animal dies, it is no longer supplied with C^{14} at the equilibrium ratio, so the ratio of C^{14} in it diminishes with time. (The amount of C^{14} in a sample is determined by counting the beta particles with a Geiger counter.) Tests show that 20% of the C^{14} in a piece of charcoal from an archaeological site has decomposed since the charcoal was a living tree. How old is the charcoal?

9. A 10-inch deep bucket is filled with water. It leaks so that the depth of the water (inches) decreases at a rate (inches per second) equal to $\frac{1}{3}$ times the depth. How long does it take for the depth to drop to 1 inch?

10.[‡] The rate of growth of a population of bacteria is at all times proportional to the number present. There are 5000 bacteria present initially and the number doubles every ten days. How many are there after 14 days?

11.[†] The number of bacteria in a test tube triples every 10 hours. At the end of the first 10 hours there are 30,000 bacteria present. How many were present initially?

12. The population of the earth was about one billion in 1850 and about two billion in 1925. If the population doubles every 75 years, when will there be one person for each of the $1.5(10^{14})$ square meters of land on the earth?

13.[†] Nitrogen pentoxide (N_2O_5) is a solid that decomposes into the gases nitrogen dioxide (NO_2) and oxygen (O_2). If $N(t)$ denotes the amount of N_2O_5 present in a sample at time t (seconds), then $dN/dt = -0.0005\,N$.* What fraction of a sample of N_2O_5 decomposes in 1000 seconds?

14. If at time $t = 0$ (seconds) the source of current is removed from a circuit of resistance R ohms and inductance L henries, then for $t > 0$ the current satisfies the differential equation $L\,di/dt + Ri = 0$. If the resistance R is 2 ohms and the inductance L is 3 henries, and if the current is 5 amps at $t = 0$, what is the current at $t = 6$?

15. The air pressure $P(y)$ (kilograms per square meter) at an altitude of y (meters) above the surface of the earth satisfies the differential equation $dP/dy = -0.00012\,P$ (see Miscellaneous Exercise 53 of Chapter 6.) The air pressure at the surface of the earth is 10^4 kilograms per square meter. What is the air pressure at an altitude of 10 kilometers (10,000 meters)?

16. If the buying power $B(t)$ of a currency decreases at a rate $0.05\,B(t)$ at time t (years), how long does it take for $B(t)$ to be cut in half?

17. Uranium-238 (U^{238}) decomposes into lead-206 (Pb^{206}) with a half life of 4.5 billion years. In a volcanic eruption the lead is removed from the lava. If a sample of lava has one part Pb^{206} to 99 parts U^{238}, when was the volcanic eruption that formed it?

---

*Adapted from W. J. Moore, [1] *Physical Chemistry*, p. 261.

18. According to *Newton's law of cooling* if an object is placed in an environment at a constant temperature T_0, the temperature $T(t)$ of the object changes at a rate proportional to the difference $T_0 - T(t)$ between the temperatures of the environment and the object:

$$\frac{dT}{dt} = k(T_0 - T) \quad \text{with} \quad k > 0.$$

Show that this differential equation has the solutions $T = T_0 + Ce^{-kt}$. (Solve a differential equation for $y = T - T_0$.)

In Exercises 19 through 21 use Newton's law of cooling.

19.† A box at a temperature of $50°F$ is placed in a room at a constant temperature of $70°F$, and 2 minutes later its temperature is $60°F$. What is its temperature 5 minutes after it is placed in the room?

20.† A cake at $300°F$ is removed from an oven and placed in a room at a constant temperature of $65°F$. One minute later the temperature of the cake is $200°F$. What is the rate of change of the temperature of the cake at that time?

21. A box at a temperature of $10°C$ is placed in ice water at $0°C$. The temperature of the box is falling at the rate of 7 degrees per minute when it is $4°C$. At what rate does the temperature fall when it is $1°C$?

7-7 COMPOUND INTEREST (Optional)

Suppose that you have $100,000 to put in a savings account and, whereas all the banks in town pay 5% annual interest and most of them compound the interest quarterly, one bank compounds the interest every day. How much more interest would you earn at the end of the year from the bank which compounds interest daily? How much more would you earn if the bank compounded interest every hour?

The answers to these questions are related to the definition of the number e and are given by the following theorem.

Theorem 7.9   An initial deposit of C dollars in a savings account paying 100r% annual interest compounded n times a year yields a balance

(1) $$C\left[1 + \frac{r}{n}\right]^{nN} \textbf{ dollars}$$ ◄

after N years.

Proof.   After $1/n$ of the year the bank pays $C[r/n]$ dollars on the initial deposit of C dollars. This leaves a balance

$$C + C\left[\frac{r}{n}\right] = C\left[1 + \frac{r}{n}\right] \text{ dollars.}$$

After another $1/n$ of the year the bank pays $C[1 + r/n][r/n]$ dollars interest, leaving a balance

$$C\left[1 + \frac{r}{n}\right] + C\left[1 + \frac{r}{n}\right]\left[\frac{r}{n}\right] = C\left[1 + \frac{r}{n}\right]^2 \text{ dollars.}$$

Similarly, every time the bank pays interest it increases the balance in the account by a factor of $1 + r/n$. After one year the account has a balance

$$C\left[1 + \frac{r}{n}\right]^n \text{ dollars}$$

and after N years it has the balance given by expression (1).   Q.E.D.

EXAMPLE 1   If the \$24 purportedly paid for Manhattan by Peter Minuit in 1626 had earned interest at the annual rate of 6%, compounded quarterly, how much would it have become by 1976? Give the exact answer and a numerical approximation.

Solution.   It was 350 years from 1626 to 1976, so the balance B in 1976 is given by expression (1) with $C = 24, r = 0.06, n = 4$, and $N = 350$:

$$B = 24\left(1 + \frac{0.06}{4}\right)^{4(350)} = 24(1.015)^{1400} \text{ dollars.}$$

To obtain an approximate value for B, we write

$$\log_{10} B = \log_{10}(24) + 1400 \log_{10}(1.015)$$
$$\approx 1.3802 + 1400(0.0065) = 10.4802.$$

Since $\log_{10}(3.02)$ is approximately 0.4802, we obtain the approximate value $\$3.02(10^{10}) = \$30,200,000,000$ for B.   □

Interest compounded continuously

According to Theorem 7.9, the expression

(2) $C\left[1 + \dfrac{r}{n}\right]^{nt}$

is the balance after t years on an initial deposit of C dollars when $100r\%$ annual interest is compounded n times a year. The limit of expression (2) as n tends to ∞ is referred to as the balance after t years if the interest is COMPOUNDED CONTINUOUSLY. In this case the balance is given in terms of the exponential function, as is stated in the next theorem.

Theorem 7.10   A deposit of C dollars which earns $100r\%$ annual interest, compounded continuously, leaves a balance

(3) Ce^{rt} **dollars**   ◄

after t years.

Proof.   We replace n by kr in (2) to obtain

$$\lim_{n \to \infty} C\left[1 + \frac{r}{n}\right]^{nt} = \lim_{k \to \infty} C\left[1 + \frac{1}{k}\right]^{krt}$$

$$= C\left\{\lim_{k \to \infty}\left[1 + \frac{1}{k}\right]^{k}\right\}^{rt} = Ce^{rt}. \quad \text{Q.E.D.}$$

We have listed in Table 7.4 the balances after one year from a deposit of \$100,000 at 5% annual interest for various frequencies of compounding. Notice that the bank which compounds interest daily would pay \$32.22 more interest than a bank which compounds quarterly; a bank which compounds hourly would pay only 33¢ more than a bank which compounds daily; and a bank which compounds continuously would pay only 3¢ more than a bank which compounds hourly.

TABLE 7.4   Balance after one year from an initial deposit of \$100,000 at 5% annual interest

FREQUENCY OF COMPOUNDING	BALANCE
Annually (once a year)	\$105,000.00
Semi-annually (twice a year)	105,062.50
Quarterly (four times a year)	105,094.53
Monthly (twelve times a year)	105,116.19
Weekly (fifty-two times a year)	105,124.58
Daily (365 times a year)	105,126.75
Hourly (8760 times a year)	105,127.08
Continuous	105,127.11

EXAMPLE 2   How much would the \$24 paid in 1626 for Manhattan have become by 1976 if it earned 6% annual interest compounded continuously?

Solution.   In the 350 years the \$24 would have become $\$24e^{(0.06)350} = \$24e^{21} \approx \$31{,}650{,}000{,}000.$ $\square$

Present value   An income of I dollars t years from now could be achieved with a deposit of Ie^{-rt} dollars now in a savings account which pays $100r\%$ annual interest compounded continuously. Therefore, the quantity Ie^{-rt} is called the PRESENT VALUE of I dollars t years from now at $100r\%$ interest compounded continuously.

If you will receive income at the rate of $R(t)$ dollars per year t years from now for $0 \le t \le T$, then the present value of the income is given by the integral

(4)   Present value $= \displaystyle\int_{0}^{T} e^{-rt} R(t)\,dt$ dollars   ◄

(see Exercise 8). Exercises 7 through 12 deal with present value.

Exercises

†*Answer provided.*
‡*Outline of solution provided.*

In each of Exercises 1 through 6 give exact answers and numerical approximations.

1.† How much would a $100 deposit become after 10 years at 6% interest compounded annually?

2. How much would a $100 deposit become after 10 years at 6% interest compounded continuously?

3.‡ What annual rate of interest compounded continuously would give the same balance at the end of each year as 5% interest compounded annually?

4. What rate of interest compounded annually would give the same balance at the end of each year as 5% compounded continuously?

5.† How long does it take for a deposit to double in a savings account which pays 10% annual interest compounded semi-annually?

6. With what annual rate of interest, compounded quarterly, will a deposit in a savings account double every ten years? How long does it take for such a deposit to triple?

7.† A stand of trees will yield $5,000 \, e^{\sqrt{t}}$ dollars profit if cut in t years. When should the trees be cut to maximize the present value of the profit at 10% interest compounded continuously?

8.‡ Derive formula (4) for present value.

9. What is the present value at 5% interest compounded continuously of a constant income of $5,000 per year for 30 years? (Give the exact answer and a numerical approximation.)

10. If a gift includes future income, such as rents or annuities, the government computes the gift tax on the present value at 3.5% interest. Suppose that a gift consists of a constant income of $10,000 per year for ten years plus $25,000 to be paid at the end of 10 years.   **a.**‡ What is the present value of the gift at 3.5% interest compounded continuously? **b.** What would the present value be if the interest rate were 5%?

11.† What is the present value at 3.5% interest compounded annually of a gift of $15,000 to be made four years from now?

12. What is the total present value at 8% interest compounded quarterly of gifts of $1000 each to be made one, two, and three years from now?

7-8   THE TRIGONOMETRIC FUNCTIONS

In Chapters 3 through 6 we studied the sine and cosine. There are many situations where it is convenient to use the other trigonometric functions, the *tangent, cotangent, secant,* and *cosecant*. They may be defined as follows.

Definition 7.2   For any x such that the denominator is not zero:

(1) $\quad \tan x = \dfrac{\sin x}{\cos x}$ ◄

(2) $\quad \cot x = \dfrac{1}{\tan x} = \dfrac{\cos x}{\sin x}$ ◄

(3) $\quad \sec x = \dfrac{1}{\cos x}$ ◄

(4) $\quad \csc x = \dfrac{1}{\sin x}.$ ◄

The functions $\tan x$ and $\sec x$ are not defined at $x = (n + \frac{1}{2})\pi$ $(n = 0, \pm 1, \pm 2, \ldots)$ where $\cos x$ is zero, and the functions $\cot x$ and $\csc x$ are not defined at $x = n\pi$ $(n = 0, \pm 1, \pm 2, \ldots)$ where $\sin x$ is zero.

EXAMPLE 1 Find all solutions x of the equation $\sec x = 2$.

Solution. The equation $\sec x = 2$ is equivalent to the equation $\cos x = \frac{1}{2}$, which has the solutions $x = \pm \frac{1}{3}\pi + 2n\pi$ with integers n. □

The derivatives of tan x, cot x, sec x, and csc x

In Chapter 3 we derived the differentiation formulas:

(5) $\quad \dfrac{d}{dx} \sin x = \cos x$ ◄

(6) $\quad \dfrac{d}{dx} \cos x = -\sin x.$ ◄

Formulas (5) and (6) are valid for all x. With definitions (1) through (4) and the quotient rule, they yield the formulas for differentiating the other trigonometric functions.

Theorem 7.11 For any x such that the functions are defined:

(7) $\quad \dfrac{d}{dx} \tan x = \sec^2 x$ ◄

(8) $\quad \dfrac{d}{dx} \cot x = -\csc^2 x$ ◄

(9) $\quad \dfrac{d}{dx} \sec x = \sec x \tan x$ ◄

(10) $\quad \dfrac{d}{dx} \csc x = -\csc x \cot x.$ ◄

We leave the proof of Theorem 7.11 as Exercises 5 and 6.

Notice that formulas (6), (8), and (10) for the derivatives of the "cofunctions" $\cos x$, $\cot x$, and $\csc x$ can be obtained from formulas (5), (7), and (9) for the derivatives of $\sin x$, $\tan x$, and $\sec x$ by replacing sin x by cos x, tan x by cot x, sec x by csc x, and vice versa, and by adding minus signs. This makes the differentiation formulas easier to remember.

Formulas (7) through (10) should be memorized. They may, however, be recalled whenever necessary by working out their derivations from formulas (5) and (6) for the derivatives of sin x and cos x.

EXAMPLE 2 Compute $\dfrac{d}{dx} \tan x$ at $x = \dfrac{\pi}{4}$.

Solution. Formulas (7) and (5) give

$$\left[\frac{d}{dx} \tan x \right]_{x=\pi/4} = [\sec^2 x]_{x=\pi/4} = \left[\frac{1}{\cos^2 x} \right]_{x=\pi/4} = (\sqrt{2})^2 = 2$$

since $\cos(\pi/4) = 1/\sqrt{2}$. □

Derivatives of composite functions

The chain rule with formulas (7) through (10) gives the rules for differentiating tangents, cotangents, secants, and cosecants of functions:

(11) $\dfrac{d}{dx} \tan[f(x)] = \sec^2[f(x)] \dfrac{df}{dx}(x)$ ◄

(12) $\dfrac{d}{dx} \cot[f(x)] = -\csc^2[f(x)] \dfrac{df}{dx}(x)$ ◄

(13) $\dfrac{d}{dx} \sec[f(x)] = \sec[f(x)] \tan[f(x)] \dfrac{df}{dx}(x)$ ◄

(14) $\dfrac{d}{dx} \csc[f(x)] = -\csc[f(x)] \cot[f(x)] \dfrac{df}{dx}(x)$.   ◄

EXAMPLE 3 Compute (a) $\dfrac{d}{dx} \sec(x^3)$ and (b) $\dfrac{d}{dx} \sec^3 x$.

Solution.   (a) We apply formula (13) with $f(x) = x^3$ to obtain

$$\frac{d}{dx} \sec(x^3) = \sec(x^3) \tan(x^3) \frac{d}{dx} x^3 = 3x^2 \sec(x^3) \tan(x^3).$$

(b) We apply the rule for differentiating a power of a function and formula (9):

$$\frac{d}{dx} \sec^3 x = \frac{d}{dx} [\sec x]^3 = 3 \sec^2 x \frac{d}{dx} \sec x$$
$$= 3 \sec^2 x [\sec x \tan x] = 3 \sec^3 x \tan x. □$$

Integration formulas

Differentiation formulas (7) through (10) translate into the following integration (antidifferentiation) formulas.

(15) $\displaystyle\int \sec^2 x \, dx = \tan x + C$ ◄

(16) $\displaystyle\int \csc^2 x \, dx = -\cot x + C$ ◀

(17) $\displaystyle\int \sec x \tan x \, dx = \sec x + C$ ◀

(18) $\displaystyle\int \csc x \cot x \, dx = -\csc x + C$ ◀

EXAMPLE 4   Evaluate the definite integral

$$\int_0^{\pi/6} \sec x \tan x \, dx.$$

Solution.   The formula (17) for the indefinite integral gives

$$\int_0^{\pi/6} \sec x \tan x \, dx = [\sec x]_0^{\pi/6} = \sec\left(\frac{\pi}{6}\right) - \sec(0)$$

$$= \frac{1}{\cos(\pi/6)} - \frac{1}{\cos(0)} = \frac{1}{\sqrt{3}/2} - \frac{1}{1}$$

$$= \frac{2}{\sqrt{3}} - 1$$

since $\cos(\pi/6) = \sqrt{3}/2$ and $\cos(0) = 1$.   □

EXAMPLE 5   Evaluate the indefinite integral $\int \csc^2(6x) \, dx$.

Solution.   We make the substitution $u = 6x$. Then

$$du = \frac{du}{dx} dx = \frac{d}{dx}(6x) \, dx = 6 \, dx \text{ and}$$

$$\int \csc^2(6x) \, dx = \tfrac{1}{6}\int \csc^2(6x)(6 \, dx) = \tfrac{1}{6}\int \csc^2 u \, du$$

$$= \tfrac{1}{6}(-\cot u) + C = -\tfrac{1}{6}\cot(6x) + C.$$

Here we have used the integration formula (16) with x replaced by u.   □

The graphs of tan x, cot x, sec x, and csc x

The graphs of $\tan x$, $\cot x$, $\sec x$, and $\csc x$ are shown in Figures 7.9 through 7.12. The graphs of $\cos x$ and $\sin x$ are drawn with dashed lines in the sketches of the graphs of their reciprocals $\sec x$ and $\csc x$. The graphs of $\tan x$ and $\sec x$ have vertical asymptotes at $x = (n + \tfrac{1}{2})\pi$ ($n = 0, \pm 1, \pm 2, \ldots$) where $\cos x$ is zero, whereas the graphs of $\cot x$ and $\csc x$ have vertical asymptotes at $x = n\pi$ ($n = 0, \pm 1, \pm 2, \ldots$) where $\sin x$ is zero.

All six trigonometric functions are periodic of period 2π. The functions $\tan x$ and $\cot x$, however, are also periodic of period π:

(19) $\tan(x + \pi) = \tan x$ and $\cot(x + \pi) = \cot x$.   ◀

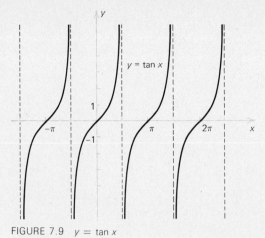

FIGURE 7.9 $y = \tan x$

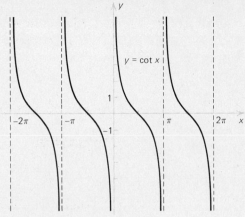

FIGURE 7.10 $y = \cot x$

This is because $\sin(x + \pi) = -\sin x$ and $\cos(x + \pi) = -\cos x$.

The graphs of $\sec x$ and $\csc x$ do not extend between the horizontal lines $y = 1$ and $y = -1$ because $|\cos x|$ and $|\sin x|$ are never greater than 1 and, consequently, $|\sec x|$ and $|\csc x|$ are never less than 1.

The function $\tan x$ is increasing for all x where it is defined because its derivative $\sec^2 x$ is positive. Cot x is decreasing for all x where it is defined, because its derivative $-\csc^2 x$ is negative.

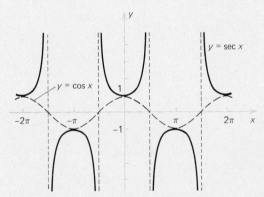

FIGURE 7.11 $y = \sec x$

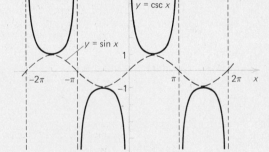

FIGURE 7.12 $y = \csc x$

A direct definition of tan x

We defined $\sin x$ and $\cos x$ in Chapter 3 by using a unit circle in an auxiliary uv-plane (Figure 7.13). We can use the same circle to give a geometric definition of $\tan x$. This direct definition is useful for finding values of $\tan x$ and for determining properties of this function. We draw the vertical line $u = 1$ which is tangent to the unit circle where it crosses the positive u-axis (Figure 7.14). Then we extend the terminal side of the angle x (forward or backwards) to where it intersects the line $u = 1$. The v-coordinate of the intersection is the tangent of x (Figures 7.14 and 7.15). The proof that this alternate definition of $\tan x$ is consistent with definition (1) involves similar triangles and is left as Exercise 42.

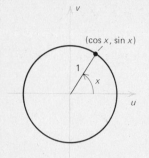

FIGURE 7.13

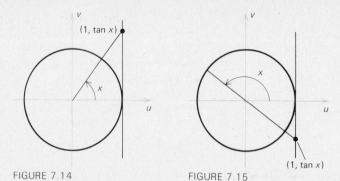

FIGURE 7.14

FIGURE 7.15

EXAMPLE 6   Find all solutions x of the equation $\tan x = \sqrt{3}$.

Solution.   The line through the origin and through the point at $v = \sqrt{3}$ on the vertical line $u = 1$ in a uv-plane forms the terminal sides of the angles whose tangents are $\sqrt{3}$ (Figure 7.15a). The leg OP of the right triangle OPQ has length 1 and the leg PQ has length $\sqrt{3}$. Therefore, the hypotenuse OQ has length $\sqrt{1^2 + (\sqrt{3})^2} = 2$ and the triangle is a 30°-60°-right triangle. The angle POQ is 60° or $\frac{1}{3}\pi$ radians, so $x = \frac{1}{3}\pi$ is one solution of the equation $\tan x = \sqrt{3}$. The general solution is $x = \frac{1}{3}\pi + n\pi$ with integers n.   □

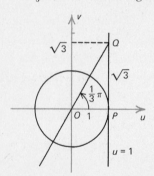

FIGURE 7.15a

Trigonometric identities   When we divide the Pythagorean identity $\sin^2 x + \cos^2 x = 1$ by $\cos^2 x$ and by $\sin^2 x$, in turn, we obtain the additional Pythagorean identities

(20) $\tan^2 x + 1 = \sec^2 x$ ◄

(21) $1 + \cot^2 x = \csc^2 x$.   ◄

When we combine the two identities

$$\sin(x - y) = \sin x \cos y - \cos x \sin y$$

$$\cos(x - y) = \cos x \cos y + \sin x \sin y$$

we obtain

$$\tan(x - y) = \frac{\sin(x - y)}{\cos(x - y)} = \frac{\sin x \cos y - \cos x \sin y}{\cos x \cos y + \sin x \sin y}.$$

Then dividing the numerator and denominator of the last expression by $\cos x \cos y$ yields the identity

$$(22) \quad \tan(x - y) = \frac{\tan x - \tan y}{1 + \tan x \tan y}$$

for the tangent function.

Angles between curves

When we speak of the *angles between two curves* at a point of intersection, we mean the angles between their tangent lines at that point. Angles between curves can be found by using identity (22).

EXAMPLE 7   What is the positive acute angle between the curves $y = x^2$ and $y = x^{1/3}$ at $(1, 1)$?

Solution.   The tangent line to $y = x^2$ at $x = 1$ has slope $[2x]_{x=1} = 2$ and the tangent line to $y = x^{1/3}$ has slope $[\frac{1}{3}x^{-2/3}]_{x=1} = \frac{1}{3}$ at that point. Hence if α is the angle from the positive x-axis to the tangent line to $y = x^2$ and β is the angle from the positive x-axis to the tangent line to $y = x^{1/3}$, then $\tan \alpha = 2$, $\tan \beta = \frac{1}{3}$ and the angle $\alpha - \beta$ between the tangent lines is given by

$$\tan(\alpha - \beta) = \frac{\tan \alpha - \tan \beta}{1 + \tan \alpha \tan \beta} = \frac{2 - \frac{1}{3}}{1 + 2(\frac{1}{3})} = 1.$$

The positive acute angle with tangent equal to 1 is $\frac{1}{4}\pi$, so this is the angle $\alpha - \beta$ between the curves.   □

Exercises

1. Give exact values of:
    a.† $\sin(7\pi/4)$ b. $\cos(2\pi/3)$ c.† $\tan(-3\pi)$ d.† $\sec(4\pi/3)$
    e. $\cot(\pi/6)$ f.† $\csc(-3\pi/4)$ g. $\tan(-5\pi/6)$
    h. $\sec(-3\pi/2)$

2. Use the reference tables to give approximate numerical values for:
    a.‡ $\tan(5\pi/7)$ b.† $\sec(3\pi/10)$ c. $\cot(\pi/8)$ d.‡ $\csc(5)$
    e.† $\tan(-1.2)$ f. $\sec(3)$ g.† $\cot(15\pi/7)$
    h. $\csc(-31\pi/10)$

3. Give exact values for all solutions x of the following equations.
    a.‡ $\sec(3x) = \sqrt{2}$ b.† $\csc(-x) = 2$ c. $\sec(x - 1) = 1$
    d.‡ $\tan(x/5) = 1/\sqrt{3}$ e.† $\cot(1 - x) = -1$
    f. $\sec(4x) = -2$ g.† $\tan(10x) = \sqrt{3}$ h. $\cot(5x) = 0$
    i. $\csc(6x - 1) = 2/\sqrt{3}$.

4. Use the reference tables to give approximate numerical values for all solutions x of the following equations.
    a.‡ $\tan x = 2.3$ b.† $\sec x = -4$ c. $\cot x = -10$
    d.† $\tan(2x) = -\frac{1}{4}$ e. $\sec(10x) = \frac{3}{2}$

5.[†] Derive differentiation formulas (7) and (9) from (5) and (6).

6. Derive differentiation formulas (8) and (10) from (5) and (6).

Compute the derivatives in Exercises 7 through 17.

7.[†] $\left[\dfrac{d}{dx}\cot x\right]_{x=\pi/4}$ **8.** $\left[\dfrac{d}{dx}\csc x\right]_{x=\pi/3}$ **9.** $\left[\dfrac{d}{dx}\tan x\right]_{x=0}$

10.[†] $\dfrac{d}{dt}\tan(4t)$ **11.** $\dfrac{d}{dx}\sec(3x+1)$ **12.**[†] $\dfrac{d}{dy}\cot^2 y$

13. $\dfrac{d}{dx}\sqrt{\cot x}$ **14.**[†] $\dfrac{d}{dx}[4\tan^2 x - 2\tan^4 x]$

15. $\dfrac{d}{dz}[\csc^5 z + \csc(z^5)]$ **16.**[‡] $\dfrac{d}{dx}\tan^3(6x)$

17. $\dfrac{d}{dx}\sqrt{\sec(\sqrt{x})}$

18.[†] Give an equation for the tangent line to the curve $y = \sec(3x)$ at $x = \pi/12$.

19. Give equations for the tangent lines to the graph of $\tan x$ at $x = -\pi/4$, $x = 0$, and $x = \pi/4$. Sketch the tangent lines and the portion of the graph for $-\pi/2 < x < \pi/2$.

Evaluate the integrals in Exercises 20 through 29.

20.[†] $\displaystyle\int \sec(2x)\tan(2x)\,dx$ **21.** $\displaystyle\int x\sec^2(x^2)\,dx$

22.[†] $\displaystyle\int \sqrt{\tan x}\,\sec^2 x\,dx$ **23.** $\displaystyle\int (3 - 4\cot x)^{10}\csc^2 x\,dx$

24.[‡] $\displaystyle\int \sec^5 x\,\tan x\,dx$ **25.** $\displaystyle\int \csc^7 x\,\cot x\,dx$

26.[†] $\displaystyle\int_0^{\pi/4} \sec^2 t\,dt$ **27.** $\displaystyle\int_{-1}^1 \sec y\,\tan y\,dy$

28.[‡] $\displaystyle\int_{5\pi/8}^{7\pi/8} \dfrac{1}{\sin^2(2x)}\,dx$ **29.** $\displaystyle\int_{-1/2}^{1/2} x^2\sec(x^3)\,\tan(x^3)\,dx$

30.[‡] Sketch one of the regions bounded by the curves $y = 8\sin x$ and $y = \csc^2 x$. Compute its area.

31. Sketch the region bounded by the curves $y = \sec^2 x$, $x = -\pi/3$, $x = \pi/3$, and $y = 0$. Compute its area.

32.[‡] A searchlight 15 miles off a straight shore is revolving at the rate of 2 revolutions per minute. At what speed does the beam of light pass a point 20 miles down the shore from the lighthouse?

33. An airplane is flying 100 miles per hour in a straight line at an altitude of $\frac{1}{2}$ mile. It passes over a searchlight which is shining on it. At what rate is the searchlight turning when the airplane is a horizontal distance of 3 miles past it?

34.† Find the maximum and minimum values of $\sqrt{3}\cos(2x) - \sin(2x)$.

35. Find the maximum and minimum values of $x + \sin x$ for $\pi/2 \le x \le 3\pi$.

36.† An object traveling around the ellipse $9x^2 + 4y^2 = 36$ has coordinates $(2\cos t, 3\sin t)$ at time t. At what rate is its distance to the origin increasing at $t = \pi/4$?

37.‡ A function $y = y(x)$ is defined implicitly by the equation $\cot(x + y) = \sqrt{3}$ and the condition $y(0) = \pi/6$. What is $y'(0)$?

38. A function $y = y(x)$ is defined implicitly by the equation $\sec(\pi y) = 2x$ and the condition $y(1) = \frac{1}{3}$. What is $y'(1)$?

39. Compute the second derivatives of   **a.**† $\sin^2(4x)$ **b.** $\sec(x^2)$
c.† $x^2\sin x$ **d.** $\tan x \cot x$.

40. Use the indicated tangent lines to give approximate values for the following quantities.   **a.**‡ $\sin(5\pi/16)$; tangent line at $x = \pi/4$
b.† $\sec(4\pi/9)$; tangent line at $x = \pi/3$ **c.** $\cot(5\pi/16)$; tangent line at $x = \pi/4$

41.† In a circuit containing a source of alternating current and a capacitor (Figure 7.16) the counterclockwise current $i(t)$ (amperes) is the derivative with respect to time of the charge $q(t)$ (coulombs) on the capacitor. What is the maximum charge on the capacitor if $q(0) = 0$ and $i(t) = 10\sin(\pi t/30)$?

42. Show that the alternate definition of $\tan x$ which was described at the end of the section is consistent with Definition 7.2.

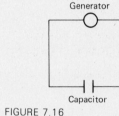

Generator

Capacitor

FIGURE 7.16

Find all solutions x of the equations in Exercises 43 through 52.

43.† $\tan^3 x = 3\tan x$

44.† $\sec^3 x = 2\sec x$

45.† $\cot x = \tan x$

46.† $\sec^2 x + \tan^2 x = 3$

47.† $\csc^2(5x) = \frac{4}{3}$

48.† $\sec^3 x = -8$

49.† $\tan x + \sec x = 0$

50. $e^{2\cos x} = 1$

51.† $\sec^2 x - 3\sec x + 2 = 0$

52. $\ln(\sec x) = 0$

Find the derivatives of the functions in Exercises 53 through 58.

53.† $\tan(\ln x)$

54.† $\ln(\tan x)$

55.† $\sec^3(\sin^2 x)$

56. $xe^{\csc x}$

57.† $\sqrt{\sec^2 x - \tan^2 x}$

58. $\dfrac{\cot x}{x}$

Evaluate the integrals in Exercises 59 through 64.

59.† $\displaystyle\int \tan^3 x \sec^2 x \, dx$

60.† $\displaystyle\int \sec^3 x \tan x \, dx$

61.† $\displaystyle\int \frac{\sec^2 x}{1 + \tan x} \, dx$

62. $\displaystyle\int \csc^2 x \sqrt{\cot^3 x} \, dx$

63.† $\displaystyle\int \frac{\sin^2 x + x^2}{x^2 \sin^2 x}\, dx$ **64.** $\displaystyle\int_0^1 x \tan(x^2) \sec(x^2)\, dx$

65.† What is the angle between the curves $y = 16x^{-2}$ and $y = x^2$ at $(-2, 4)$?

66.† Give equations of the two lines that make an angle of $\frac{1}{6}\pi$ with the curve $y = x^3$ at $(1, 1)$.

67. What are all of the angles between the line $y = \sqrt{3}\, x$ and the curve $y = \tan x$ at $x = \frac{1}{3}\pi$?

68.† Use the table of values of the tangent function in the back of the book to give an approximate value for the positive acute angle between the curves $y = x^{-1}$ and $y = x^{-2}$ at $x = 1$.

7-9 THE INVERSE TRIGONOMETRIC FUNCTIONS

The inverse trigonometric functions are the ARCSINE, ARCCOSINE, ARCTANGENT, ARCCOTANGENT, ARCSECANT, and ARCCOSECANT. They arise from the study of the trigonometric functions and also are required for the evaluation of certain integrals.

The arcsine function

To define the function $\arcsin x$ we start with the curve $x = \sin y$ which is obtained from the curve $y = \sin x$ by reflecting it about the line $y = x$ (Figure 7.17). The curve $x = \sin y$ is not the graph of a function of x because there are an infinite number of points on it for each x with $-1 \le x \le 1$. The graph of the function $\arcsin x$ is taken to be the portion of the curve $x = \sin y$ for $-\pi/2 \le y \le \pi/2$ (Figure 7.18) as in the following definition.

Definition 7.3   **For $-1 \le x \le 1$, arcsin x is the angle y such that sin y = x and $-\pi/2 \le y \le \pi/2$.**

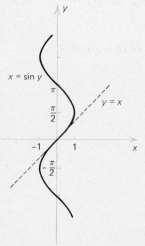

FIGURE 7.17

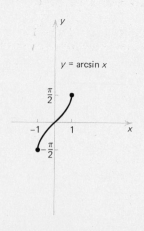

FIGURE 7.18

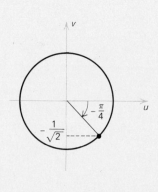

FIGURE 7.19

EXAMPLE 1   Find $\arcsin(-1/\sqrt{2})$.

Solution.   The angle y such that $\sin y = -1/\sqrt{2}$ and $-\pi/2 \le y \le \pi/2$ is $y = -\pi/4$ (Figure 7.19). Therefore, $\arcsin(-1/\sqrt{2}) = -\pi/4$. $\square$

The derivative of arcsin x

Because $\sin x$ is increasing for $-\pi/2 \le x \le \pi/2$ and has a positive derivative for $-\pi/2 < x < \pi/2$, the Inverse Function Theorem shows that $\arcsin x$ is continuous for $-1 \le x \le 1$ and has a derivative for $-1 < x < 1$. To find the derivative, we differentiate the identity $\sin(\arcsin x) = x$ with respect to x. This gives

(1) $\cos(\arcsin x) \dfrac{d}{dx} \arcsin x = 1.$

Since $\arcsin x$ is defined to lie between $-\pi/2$ and $\pi/2$, its cosine is positive, and the Pythagorean identity $\sin^2 y + \cos^2 y = 1$ with $y = \arcsin x$ shows that

(2) $\cos(\arcsin x) = \sqrt{1 - \sin^2(\arcsin x)} = \sqrt{1 - x^2}.$

When we substitute equation (2) into (1) and divide by $\sqrt{1 - x^2}$, we obtain the differentiation formula

(3) $\dfrac{d}{dx} \arcsin x = \dfrac{1}{\sqrt{1 - x^2}}$ for $-1 < x < 1$.   ◀

EXAMPLE 2   Compute the derivative of $\arcsin x$ at $x = \tfrac{3}{5}$.

Solution.   We have

$$\left[\dfrac{d}{dx} \arcsin x \right]_{x=3/5} = \left[\dfrac{1}{\sqrt{1 - x^2}} \right]_{x=3/5} = \dfrac{1}{\sqrt{1 - \frac{9}{25}}} = \dfrac{5}{4}. \quad \square$$

The arccosine function

The curve $x = \cos y$ is shown in Figure 7.20. The portion of that curve for $0 \le y \le \pi$ is the graph of the function $\arccos x$ (Figure 7.21):

Definition 7.4   **For $-1 \le x \le 1$, arccos x is the angle y such that cos y = x and $0 \le y \le \pi$.**

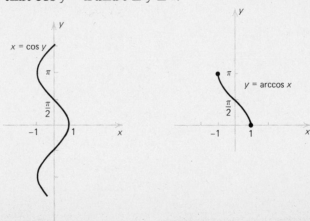

FIGURE 7.20   FIGURE 7.21

A calculation similar to that used to derive formula (3) yields the differentiation formula

$$(4) \quad \frac{d}{dx}\arccos x = \frac{-1}{\sqrt{1-x^2}} \quad \text{for } -1 < x < 1 \qquad \blacktriangleleft$$

(see Exercise 33).

The functions $\arcsin x$ and $\arccos x$ are related by the identity

$$(5) \quad \arccos x = \frac{\pi}{2} - \arcsin x \quad \text{for } -1 \le x \le 1 \qquad \blacktriangleleft$$

(see Exercise 34).

The arctangent function

The curve $x = \tan y$ has an infinite number of pieces, one above the other (Figure 7.22). The graph of the function $\arctan x$ is the piece of that graph which lies in the region $-\pi/2 < y < \pi/2$ (Figure 7.23):

Definition 7.5　**For any number x, arctan x is the angle y such that tan y = x and $-\pi/2 < y < \pi/2$.**

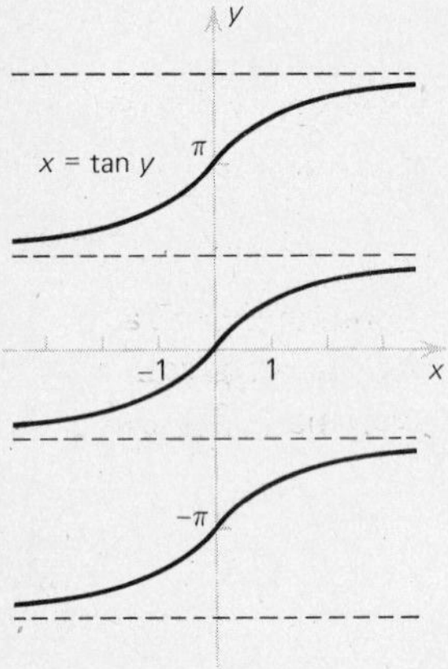

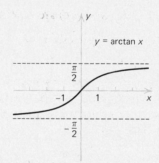

FIGURE 7.23

FIGURE 7.22

Arctan x has a derivative for all x because $\tan x$ tends to ∞ as $x \to (\pi/2)^-$, tends to $-\infty$ as $x \to (-\pi/2)^+$, and has a positive derivative for $-\pi/2 < x < \pi/2$. To obtain a formula for the derivative, we differentiate the identity $\tan(\arctan x) = x$ with respect to x. We obtain

$$\sec^2(\arctan x)\frac{d}{dx}\arctan x = 1.$$

By a Pythagorean identity, we have

$$\sec^2(\arctan x) = 1 + \tan^2(\arctan x) = 1 + x^2$$

and with this we obtain the differentiation formula

$$(6) \quad \frac{d}{dx}\arctan x = \frac{1}{1+x^2} \quad \text{for all } x. \qquad \blacktriangleleft$$

The arcsecant function

The curve $x = \sec y$ is shown in Figure 7.24. The graph of the function arcsec x is the portion of that curve for $0 \leq y \leq \pi$. It is in two pieces (Figure 7.25). Arcsec x is defined for $x \leq -1$ and for $x \geq 1$:

Definition 7.6   **For $|x| \geq 1$, arcsec x is the angle y such that sec $y = x$ and $0 \leq y \leq \pi$.**

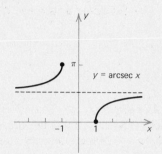

FIGURE 7.25

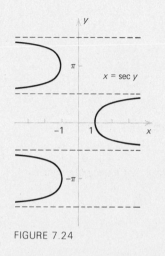

FIGURE 7.24

The derivative of arcsec x is given by the formula (Exercise 35)

$$(6a) \qquad \frac{d}{dx} \operatorname{arcsec} x = \frac{1}{|x| \sqrt{x^2 - 1}}.$$

The arccotangent and arccosecant functions

The remaining inverse trigonometric functions, the arccotangent and the arccosecant, may be given definitions analogous to those given for the other inverse trigonometric functions, or they may be defined by the formulas

$$(7) \qquad \operatorname{arccot} x = \frac{\pi}{2} - \arctan x \quad \text{for all } x \qquad \blacktriangleleft$$

and

$$(8) \qquad \operatorname{arccsc} x = \frac{\pi}{2} - \operatorname{arcsec} x \quad \text{for } |x| \geq 1. \qquad \blacktriangleleft$$

The formulas for the derivatives of these functions follow from these definitions.

Derivatives of composite functions

With the chain rule, formulas (3), (4), and (6) give

$$(9) \qquad \frac{d}{dx} \arcsin[u(x)] = \frac{1}{\sqrt{1 - u(x)^2}} \frac{du}{dx}(x) \qquad \blacktriangleleft$$

$$(10) \qquad \frac{d}{dx} \arccos[u(x)] = \frac{-1}{\sqrt{1 - u(x)^2}} \frac{du}{dx}(x) \qquad \blacktriangleleft$$

$$(11) \qquad \frac{d}{dx} \arctan[u(x)] = \frac{1}{1 + u(x)^2} \frac{du}{dx}(x). \qquad \blacktriangleleft$$

EXAMPLE 3   Compute $\dfrac{d}{dx}$ arctan (x^3).

Solution.   Formula (11) with $u(x) = x^3$ gives

$$\frac{d}{dx} \arctan(x^3) = \frac{1}{1 + (x^3)^2} \frac{d}{dx} x^3 = \frac{3x^2}{1 + x^6}. \quad \square$$

Integrals requiring inverse trigonometric functions   One of the most important uses of inverse trigonometric functions is for evaluating integrals. The differentiation formulas (3) and (6) for the arcsine and arctangent translate into the integration formulas

(12) $\displaystyle\int \frac{1}{\sqrt{1 - x^2}} \, dx = \arcsin x + C$ ◀

(13) $\displaystyle\int \frac{1}{1 + x^2} \, dx = \arctan x + C.$ ◀

   The following generalizations of formulas (12) and (13) are used frequently and are worth remembering. Each of them is valid for any choice of the positive constant a.

(14) $\displaystyle\int \frac{1}{\sqrt{a^2 - x^2}} \, dx = \arcsin\left(\frac{x}{a}\right) + C$ ◀

(15) $\displaystyle\int \frac{1}{a^2 + x^2} \, dx = \frac{1}{a} \arctan\left(\frac{x}{a}\right) + C.$ ◀

These may be derived from formulas (12) and (13) by making the substitution $u = x/a$ (see Exercise 36). Differentiation formula (6a) yields

(16) $\displaystyle\int \frac{1}{|x| \sqrt{x^2 - a^2}} \, dx = \frac{1}{a} \operatorname{arcsec}\left(\frac{x}{a}\right) + C.$

EXAMPLE 4   Evaluate

$$\int_1^{10} \frac{1}{25 + x^2} \, dx.$$

Solution.   By formula (15) with $a = 5$, we have

$$\int \frac{1}{25 + x^2} \, dx = \frac{1}{5} \arctan\left(\frac{x}{5}\right) + C.$$

Therefore,

$$\int_1^{10} \frac{1}{25 + x^2} \, dx = \left[\frac{1}{5} \arctan\left(\frac{x}{5}\right) \right]_1^{10}$$

$$= \left[\frac{1}{5} \arctan\left(\frac{10}{5}\right) \right] - \left[\frac{1}{5} \arctan\left(\frac{1}{5}\right) \right]$$

$$= \frac{1}{5}\left[\arctan(2) - \arctan\left(\frac{1}{5}\right) \right]. \quad \square$$

Exercises

†*Answer provided.*
‡*Outline of solution provided.*

1. Use the definitions of the inverse trigonometric functions to give the exact values of the following expressions.

a.‡ $\arctan[\tan(4\pi/5)]$ **b.**† $\arccos[\cos(4\pi/5)]$

c. $\arcsin[\sin(4\pi/5)]$ **d.**‡ $\sin[\arctan(5)]$

e.† $\tan[\arcsin(-\frac{1}{4})]$ **f.** $\sin[\text{arcsec}(-3)]$

g.† $\tan[\arctan(-5)]$ **h.** $\sin[\arcsin(0.03)]$

i.† $\arcsin(\sqrt{3}/2)$ **j.** $\arctan(\sqrt{3})$

k.† $\text{arcsec}(-2)$ **l.** $\arctan(-1/\sqrt{3})$

2. Use the reference tables to give approximate numerical values of the following expressions.

a.‡ $\arcsin(0.32)$ **b.**† $\arctan(5)$

c. $\text{arcsec}(5/4)$ **d.**‡ $\arctan(-2.3)$

e. $\arcsin(-0.87)$ **f.** $\text{arcsec}(-4.5)$

Compute derivatives 3 through 12.

3.† $\dfrac{d}{dx}\arcsin(x^2)$ **4.** $\dfrac{d}{dx}(\arcsin x)^2$

5.† $\dfrac{d}{dt}\arctan(3t-4)$ **6.** $\dfrac{d}{dy}\text{arcsec}(\sqrt{y})$

7.‡ $\dfrac{d}{dx}[\arctan(6x^2)]^{1/3}$ **8.** $\dfrac{d}{dx}\sin[\arctan(2x)]$

9.† $\dfrac{d}{dz}\arctan(7\sec z)$ **10.** $\dfrac{d}{dw}[\arcsin(w^{-3})]^3$

11.† $\dfrac{d}{dx}[x\arcsin x]$ **12.** $\dfrac{d}{dx}\left[\dfrac{\arctan x}{x}\right]$

Evaluate integrals 13 through 22.

13.‡ $\displaystyle\int\dfrac{3x}{\sqrt{1-x^4}}\,dx$ **14.**† $\displaystyle\int\dfrac{x^2}{1+x^6}\,dx$

15. $\displaystyle\int(9-x^2)^{-1/2}\,dx$ **16.** $\displaystyle\int\dfrac{\sin x}{25+\cos^2 x}\,dx$

17.† $\displaystyle\int_0^1\dfrac{1}{\sqrt{4-x^2}}\,dx$ **18.** $\displaystyle\int_1^{100}\dfrac{1}{100+y^2}\,dy$

19.† $\displaystyle\int\dfrac{\arctan x}{1+x^2}\,dx$ **20.** $\displaystyle\int\sqrt{\dfrac{\arcsin x}{1-x^2}}\,dx$

21.‡ $\displaystyle\int_{-1}^2\dfrac{1}{\sqrt{25-4x^2}}\,dx$ **22.** $\displaystyle\int_0^1\dfrac{1}{9+16x^2}\,dx$

23. What is wrong with the formula

$$\int_0^4\dfrac{1}{\sqrt{1-x^2}}\,dx=\arcsin 4?$$

24.[†] Give an equation for the tangent line to the graph of arcsin x at $x = \frac{1}{2}$.

25. Give equations for the tangent lines to the graph of arctan x at $x = -2$, $x = 0$, and $x = 2$. Sketch these tangent lines and the graph of the function. Label the inflection points.

In Exercises 26 through 31 give exact answers and numerical approximations.

26.[†] Sketch the region bounded by the x-axis, by the lines $x = -3$ and $x = 3$, and by the graph of the function $(x^2 + 1)^{-1}$. Then compute the area of the region.

27. Sketch the curve $y = 1/\sqrt{1 - x^2}$. Compute the area of the region bounded by this curve and the lines $y = 0$, $x = \frac{1}{2}$ and $x = -\frac{1}{2}$.

28.[‡] A boy and a girl of equal height are standing 30 feet apart at a county fair watching a balloon the boy has released rise over his head.   **a.** What is the angle of the girl's line of sight with the horizontal when the balloon is 40 feet above the level of their eyes?   **b.** How fast is the angle increasing at that moment if the balloon is rising 10 feet per second?

29.[‡] A picture 3 feet high is placed on a wall with its bottom 10 feet from the floor. A critic's eye is 5 feet from the floor. How far should he stand from the wall so that the vertical angle subtended by the picture at his eye be the greatest?

30. The top of a 13 foot ladder is sliding down a wall. When the base of the ladder is 5 feet from the wall, it is sliding away at the rate of 2 feet per second.   **a.** What is the angle between the ladder and the wall at that time?   **b.** How rapidly is the angle increasing at that time?

31. A billboard is perpendicular to a road and its nearest edge is 18 feet from the road. The billboard is 10 feet wide. At what points on the road does the billboard subtend the greatest angle from a driver's eyes?

32.[†] The least force acting at an angle θ with the horizontal which will pull an object of weight W along a floor is given by the formula $F = \mu W/(\cos \theta + \mu \sin \theta)$. Here μ is a positive constant called the *coefficient of static friction*. What angle requires the least force?

33. Derive formula (4) for the derivative of arccos x.

34. Derive identity (5) relating arcsin x and arccos x.

35. Derive the formula for the derivative of arcsec x.

36.[†] Derive integration formulas (14) and (15) from (12) and (13).

37.[†] What is the derivative of arcsec x **a.** at $x = 3$,   **b.** at $x = -3$?

38.[†] Evaluate the integral $\displaystyle\int_2^4 \frac{1}{x\sqrt{x^2 - 1}}\,dx$

39.[†] Evaluate the integral $\displaystyle\int_{-100}^{-10} \frac{1}{x\sqrt{x^2 - 1}}\,dx$.

7-10   THE HYPERBOLIC FUNCTIONS

The HYPERBOLIC FUNCTIONS are constructed from the functions e^x and e^{-x}. They are of interest because they have many properties analogous to those of trigonometric functions and because they arise in the study of falling bodies, hanging cables, ocean waves, and other topics in science and engineering.

The hyperbolic sine and cosine

The HYPERBOLIC SINE and COSINE are denoted $\sinh x$ and $\cosh x$ (usually pronounced "cinch" x and "cosh" x). They have the following definitions:

Definition 7.7   For any number x

(1) $\sinh x = \dfrac{e^x - e^{-x}}{2}$ and $\cosh x = \dfrac{e^x + e^{-x}}{2}.$ ◀

Notice that $\sinh x$, like $\sin x$, has the value 0 at $x = 0$ and that $\cosh x$, like $\cos x$, has the value 1 at $x = 0$.

Definition (1) and the formulas $\dfrac{d}{dx} e^x = e^x$ and $\dfrac{d}{dx} e^{-x} = -e^{-x}$ yield the differentiation formulas

(2) $\dfrac{d}{dx} \sinh x = \cosh x$ and $\dfrac{d}{dx} \cosh x = \sinh x$ ◀

(see Exercise 1). From these we obtain the integration formulas

(3) $\displaystyle\int \cosh x \, dx = \sinh x + C$ and $\displaystyle\int \sinh x \, dx = \cosh x + C.$ ◀

Equations (2) and the chain rule give the rules

(4)
$$\frac{d}{dx} \sinh[u(x)] = \cosh[u(x)] \frac{du}{dx}(x)$$
$$\frac{d}{dx} \cosh[u(x)] = \sinh[u(x)] \frac{du}{dx}(x)$$
◀

for computing derivatives of hyperbolic sines or hyperbolic cosines of functions.

EXAMPLE 1   Compute the derivative of $\sinh(6x^7 - 5x)$.

Solution.   We have

$$\frac{d}{dx} \sinh(6x^7 - 5x) = \cosh(6x^7 - 5x) \frac{d}{dx}(6x^7 - 5x)$$
$$= \cosh(6x^7 - 5x)(42x^6 - 5). \quad \square$$

EXAMPLE 2   Evaluate

$$\int_0^5 x^2 \sinh(x^3)\, dx.$$

Solution.   To evaluate the indefinite integral, we make the substitution $u = x^3$ for which $du = 3x^2\, dx$. We obtain:

$$\int x^2 \sinh(x^3)\, dx = \tfrac{1}{3}\int \sinh(x^3)(3x^2\, dx) = \tfrac{1}{3}\int \sinh u\, du$$

$$= \tfrac{1}{3}\cosh u + C = \tfrac{1}{3}\cosh(x^3) + C.$$

With this we can evaluate the definite integral:

$$\int_0^5 x^2 \sinh(x^3)\, dx = \left[\int x^2 \sinh(x^3)\, dx\right]_0^5 = \left[\tfrac{1}{3}\cosh(x^3)\right]_0^5$$

$$= \tfrac{1}{3}\cosh(5^3) - \tfrac{1}{3}\cosh(0^3) = \tfrac{1}{3}\cosh(125) - \tfrac{1}{3}. \quad \square$$

The graphs of sinh x and cosh x

The graphs of $\sinh x$ and $\cosh x$ are shown in Figures 7.26 and 7.27. Their key features can be easily obtained from definition (1) and the differentiation formulas (2) by recalling that e^x and e^{-x} are positive for all x, that e^x tends to ∞ as x tends to ∞ and tends to 0 as x tends to $-\infty$, and that e^{-x} tends to 0 as x tends to ∞ and tends to ∞ as x tends to $-\infty$.

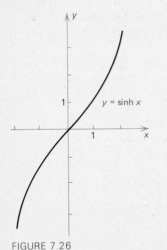

FIGURE 7.26

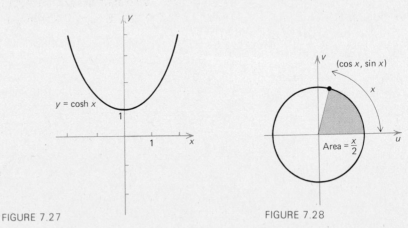

FIGURE 7.27

FIGURE 7.28

The hyperbolic functions and hyperbolas

We defined $\cos x$ and $\sin x$ as the coordinates of the intersection of the terminal side of the angle x with the unit circle in a uv-plane (Figure 7.28). The area swept out in a counterclockwise direction by the terminal side of the angle is $x/2$, and we can consider $\cos x$ and $\sin x$ as functions of this area. We can then make an analogous definition of the hyperbolic sine and cosine by using the hyperbola $u^2 - v^2 = 1$ in the uv-plane: For positive x, the point $(\cosh x, \sinh x)$ is the point on the hyperbola $u^2 - v^2 = 1$ such that the region bounded by the line from that point to the origin, by the u-axis, and by the hyperbola has area $x/2$ (Figure 7.29). (See Exercise 27.) The point $\big(\cosh(-x), \sinh(-x)\big)$ is the symmetric point below the u-axis.

The other hyperbolic functions

The definitions of the HYPERBOLIC TANGENT, COTANGENT, SECANT, and COSECANT are analogous to the definitions of the corresponding trigonometric functions.

Definition 7.8 The hyperbolic tangent and secant are defined for all x and the hyperbolic cotangent and cosecant for all x ≠ 0 by

(5)　　$\tanh x = \dfrac{\sinh x}{\cosh x} = \dfrac{e^x - e^{-x}}{e^x + e^{-x}}$　　◀

(6)　　$\coth x = \dfrac{1}{\tanh x} = \dfrac{\cosh x}{\sinh x} = \dfrac{e^x + e^{-x}}{e^x - e^{-x}}$　　◀

(7)　　$\operatorname{sech} x = \dfrac{1}{\cosh x} = \dfrac{2}{e^x + e^{-x}}$　　◀

and

(8)　　$\operatorname{csch} x = \dfrac{1}{\sinh x} = \dfrac{2}{e^x - e^{-x}}.$　　◀

The formulas for the derivatives of these four functions are similar to those for the corresponding trigonometric functions but are not very important and are not worth memorizing. Their derivations are left as Exercise 25. The graphs of the functions $\tanh x$, $\coth x$, $\operatorname{sech} x$, and $\operatorname{csch} x$ are shown in Figures 7.30 through 7.33.

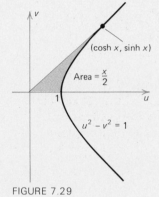

FIGURE 7.29

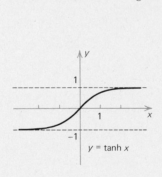

FIGURE 7.30

FIGURE 7.31

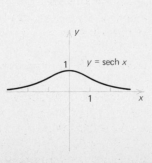

FIGURE 7.32

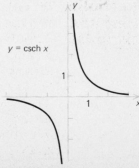

FIGURE 7.33

Identities

The fact that the point $(\cosh x, \sinh x)$ lies on the hyperbola $u^2 - v^2 = 1$ in a uv-plane is expressed in the identity

$$(9) \quad \cosh^2 x - \sinh^2 x = 1. \qquad \blacktriangleleft$$

This, and other identities for the hyperbolic functions which are analogous to identities for the trigonometric functions, are derived in Exercise 23.

Hanging cables

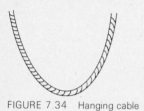

FIGURE 7.34   Hanging cable

The hyperbolic cosine has an interesting special property. A cable of uniform density which is suspended at its ends takes the form of the graph of a hyperbolic cosine (Figure 7.34). If the two ends of the cable lie in an xy-plane with its y-axis vertical and with its scale measured in feet, and if the lowest point on the cable is where it crosses the y-axis, then the cable takes the shape of the curve $y = a \cosh(x/a) + b$ for certain constants a and b. This curve is called a *catenary* (Figure 7.35).

The verification that the function $a \cosh(x/a) + b$ satisfies the differential equation of a hanging cable is given as Exercise 29. (The differential equation is derived in Miscellaneous Exercise 28 of Chapter 11. The derivation of the formula for the catenary from the differential equation is outlined in Miscellaneous Exercise 17 of Chapter 9.)

Inverse hyperbolic functions

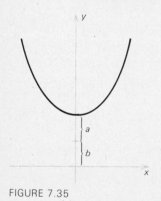

FIGURE 7.35
$y = a \cosh\left(\dfrac{x}{a}\right) + b$

Just as the hyperbolic functions are defined in terms of exponential functions, the inverse hyperbolic functions ARCSINH x, ARCCOSH x, and so forth, may be expressed in terms of the natural logarithm.

EXAMPLE 3   Give an expression for $\operatorname{arcsinh} x$ in terms of the natural logarithm.

Solution.   We set $x = \sinh y = \frac{1}{2}(e^y - e^{-y})$ and solve for $y = \operatorname{arcsinh} x$. Multiplying by $2e^y$ gives the equation $2xe^y = e^{2y} - 1$, which we rewrite in the form $(e^y)^2 - 2x(e^y) - 1 = 0$. Then, by the quadratic formula,

$$e^y = \frac{2x \pm \sqrt{4x^2 + 4}}{2} = x \pm \sqrt{x^2 + 1}.$$

We must take the plus sign since e^y is positive. Therefore,

$$(10) \quad y = \operatorname{arcsinh} x = \ln(x + \sqrt{x^2 + 1}). \quad \square$$

Exercises

†*Answer provided.*
‡*Outline of solution provided.*

1.[†] Derive differentiation formulas (2).

Compute the derivatives of the functions in Exercises 2 through 11.

2.[†] $\sinh(6x)$

3. $\cosh^6 x$

4.[†] $\cosh^3(x^{1/3})$

5. $\dfrac{\sinh x}{x}$

6.[†] $\ln(\cosh x)$

7. $\sinh(e^{-x})$

8.[†] $\sqrt{\sinh^2 x + 1}$

9. $3 \sinh(4x) - 4 \cosh(3x)$

10.[†] $\sinh(\ln x)$

11. $\cos(\cosh x)$

Evaluate integrals 12 through 21.

12.† $\displaystyle\int_0^{100} \cosh(37x)\,dx$ **13.** $\displaystyle\int \sinh x \cosh x\,dx$

14.† $\displaystyle\int \frac{\sinh(\sqrt{x})}{\sqrt{x}}\,dx$ **15.** $\displaystyle\int x \sinh(5x^2)\,dx$

16.† $\displaystyle\int \frac{\cosh x}{\sinh x + 1}\,dx$ **17.‡** $\displaystyle\int \tanh x\,dx$

18. $\displaystyle\int \sinh(\sin x)\cos x\,dx$ **19.** $\displaystyle\int_{-1}^{1} e^{3x}\cosh(e^{3x})\,dx$

20.† $\displaystyle\int_1^{10} \frac{1}{x}\cosh(\ln x)\,dx$ **21.** $\displaystyle\int_{-5}^{5} x^2 \sinh(x^3)\sqrt{\cosh(x^3)}\,dx$

22. Use the table of values of the exponential function to give approximate numerical values of   **a.†** $\sinh(1)$,   **b.** $\cosh(2)$,   **c.†** $\tanh(-\tfrac{1}{2})$,   **d.** $\operatorname{sech}(3)$.

23. Derive the identities: **a.†** $\cosh^2 x - \sinh^2 x = 1$
   b.† $1 - \tanh^2 x = \operatorname{sech}^2 x$ **c.** $\coth^2 x - 1 = \operatorname{csch}^2 x$
   d.† $\cosh(x \pm y) = \cosh x \cosh y \pm \sinh x \sinh y$
   e. $\sinh(x \pm y) = \sinh x \cosh y \pm \cosh x \sinh y$
   f. $\cosh(2x) = 2\cosh^2 x - 1$ **g.** $\sinh(2x) = 2\sinh x \cosh x$
   h. $\cosh^2 x = \tfrac{1}{2}\cosh(2x) + \tfrac{1}{2}$ **i.** $\sinh^2 x = \tfrac{1}{2}\cosh(2x) - \tfrac{1}{2}$
   j. $(\cosh x + \sinh x)^n = \cosh(nx) + \sinh(nx)$ for any constant n

24. Use identities involving hyperbolic functions to give exact values of
   a.‡ $\cosh x$ where $\sinh x = 4$,   **b.†** $\sinh x$ where $\cosh x = 3$,
   c. $\tanh x$ where $\sinh x = -100$,   **d.** $\coth x$ where $\cosh x = 4$.

25.† Derive formulas for the derivatives of   **a.** $\tanh x$,   **b.** $\coth x$,
   c. $\operatorname{sech} x$,   **d.** $\operatorname{csch} x$.

26.   **a.** Show that $y = \sin x$ satisfies the differential equation and initial conditions $y'' + y = 0$, $y(0) = 0$, $y'(0) = 1$.   **b.** Show that $y = \sinh x$ satisfies $y'' - y = 0$, $y(0) = 0$, $y'(0) = 1$.   **c.** Show that $y = \cos x$ satisfies $y'' + y = 0$, $y(0) = 1$, $y'(0) = 0$.   **d.** Show that $y = \cosh x$ satisfies $y'' - y = 0$, $y(0) = 1$, $y'(0) = 0$.

27. Show, by a v-integration, that the region bounded by the line through the origin and the point $(\cosh k, \sinh k)$ $(k > 0)$, by the u-axis, and by the right half of the hyperbola $u^2 - v^2 = 1$ in a uv-plane has area $k/2$. (Use the substitution $v = \sinh t$, $t = \operatorname{arcsinh} v$ and the identities $\cosh^2 t = \tfrac{1}{2}(\cosh(2t) + 1)$ and $\sinh(2t) = 2\sinh t \cosh t$ to evaluate the integral.)

28.† Show that $\tanh x$ tends to 1 as x tends to ∞ and tends to -1 as x tends to $-\infty$.

29. A hanging cable of constant density ρ pounds per foot and horizontal tension H pounds takes the shape of the graph of a solution of the differential equation

$$\frac{d^2y}{dx^2} = \frac{\rho}{H} \sqrt{1 + \left(\frac{dy}{dx}\right)^2}.$$

We use an xy-plane with its y-axis vertical and its scales measured in feet. Show that the catenary $y = a \cosh(x/a) + b$ with $a = H/\rho$ satisfies this differential equation.

30.† Find an equation for the cable of density ρ and horizontal tension H which is fastened at $(-1, 0)$ and $(1, 0)$.

31. A hanging cable has the shape of the catenary $y = 100 \cosh(x/100) + 50$, $-200 \leq x \leq 200$ in an xy-plane with its x-axis along the ground, its y-axis vertical, and its scales measured in feet.   **a.** How high are the ends of the cable above the ground?   **b.** How long is it?   **c.** The density of the cable is 3 pounds per foot. How much does it weigh? (Give exact and approximate numerical answers.)

32. For $x \geq 1$, $y = \operatorname{arccosh} x$ is the nonnegative solution of the equation $x = \cosh y$. For $-1 < x < 1$, $y = \operatorname{arctanh} x$ is the solution of the equation $x = \tanh y$. Derive formulas for   **a.**† $\operatorname{arccosh} x$ and   **b.** $\operatorname{arctanh} x$ in terms of the natural logarithm.

33. Derive formulas for the derivatives of   **a.**‡ $\operatorname{arcsinh} x$,   **b.**† $\operatorname{arccosh} x$,   and   **c.** $\operatorname{arctanh} x$, first by using implicit differentiation and then by using the formulas of Example 3 and Exercise 32.

34. Use the results of Exercise 33 to derive the integration formulas

$$\int \frac{1}{\sqrt{x^2 + a^2}}\, dx = \ln(x + \sqrt{x^2 + a^2}) + C \quad \text{for all } x$$

$$\int \frac{1}{\sqrt{x^2 - a^2}}\, dx = \ln(x + \sqrt{x^2 - a^2}) + C \quad \text{for } x > a.$$

35.‡ Find the minimum value of $\sinh x + 2 \cosh x$.

36. Find the maximum value of $2 \sinh x - 3 \cosh x$.

37.† Locate the centroid of the portion of the catenary $y = \cosh x$ for $-1 \leq x \leq 1$.

38. Sketch the region R bounded by the curve $y = \sinh x$, by the x-axis, and by the line $x = \ln(10)$.   **a.** Compute the area of R.   **b.** Compute the volume of the solid formed by rotating R about the x-axis.

39.† When an arc $y = \cosh x$ ($a \leq x \leq b$) of a catenary is rotated about the x-axis, it generates a surface called a *catenoid*, which has the least area of all surfaces generated by rotating curves having the same endpoints. What is its area?

40. The *wavelength L* (meters) of a string of ocean waves is the distance between crests. The *frequency f* (seconds^{-1}) is the number of crests which pass a fixed point each second. In shallow water of depth h meters, the frequency and wave length are related by the equation

$$f^2 = \frac{g}{2\pi L} \tanh\left(\frac{2\pi h}{L}\right)$$

where g is the acceleration of gravity (9.80 meters per second^2).*

a. Show that f is a decreasing function of L for fixed h.   **b.** Suppose your boat is anchored where the ocean is 10 meters deep and a wave crest passes every 12 seconds. Use the graph of f^2 as a function of L for $h = 10$, which is shown in Figure 7.36, to determine the approximate wave length and speed of the waves.   **c.**[†]   Use the tangent line to $y = \tanh x$ at $x = 0$ to give an approximate formula for f^2 for L much larger than h.

d.[†]   Give an approximate formula for f^2 for L much smaller than h.

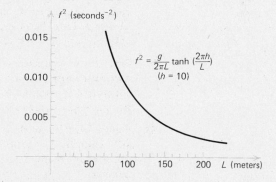

FIGURE 7.36

7-11 HISTORICAL NOTES (Optional)

Trigonometry   The beginnings of trigonometry are recorded in Egyptian tables from the thirteenth century B.C., which relate the time of day to the length of shadows. Similar tables were used in Mesopotamia. The tables give a version of the cotangent because the length of the shadow is the cotangent of the angle of the sun multiplied by the height of the object making the shadow.

Later, trigonometry developed as a tool in astronomy. Around 150 B.C. the Greek astronomer Hipparchus (ca. 180–125 B.C.) formulated techniques for finding angles and lengths of sides of plane and spherical triangles. He used tables relating lengths of chords to lengths of arcs in circles. These techniques were transmitted by Ptolemy (second century A.D.) in his *Almagest*, which was the standard work on astronomy until the introduction of the heliocentric theory of the solar system in the sixteenth century. A "sine" function was introduced in India in the fifth century, and from the ninth century through the fifteenth, the sine

*B. Kinsman, [1] *Wind Waves*, p. 125.

function and the "shadow" function (the cotangent) were tabulated as sexagesimals by Arab astronomers ranging from Spain to Iran.

Trigonometry began to be viewed as a subject in its own right during the thirteenth century. In the sixteenth century, Viète and others introduced symbolic notation to replace the clumsy verbal statements of rules and procedures that had been used until then. Modern trigonometric notation was established in the eighteenth century.

Logarithms

Logarithms were invented at the beginning of the seventeenth century by a Scottish mathematician, John Napier (1550–1617), and independently by a Swiss clockmaker, Jobst Bürgi (1552–1632). Napier was the first to make his ideas known and the first to publish a table of logarithms. In 1614, after twenty years of work on the project, he published his work under the title *Mirifici Logarithmorum Canonis Descriptio (A Description of the Marvelous Rule of Logarithms).*

Napier and Bürgi devised their logarithms to simplify the calculation of products and quotients. Tables of trigonometric functions and "product to sum" formulas such as the trigonometric identity $2 \cos x \cos y = \cos(x + y) + \cos(x - y)$ had been used extensively by astronomers and mathematicians for this purpose in the sixteenth century, but logarithms are simpler to deal with and soon were widely accepted.

Napier's and Bürgi's logarithms were constructed arithmetically and without calculus or reference to exponents. Except for the location of decimal points, Napier used the logarithm to the base $(1 - 10^{-7})^{10^7}$, which is close to $1/e$. Because of this, natural logarithms are now referred to as *Naperian logarithms*. Bürgi's logarithm was essentially the logarithm to the base $(1 + 10^{-4})^{10^4}$, which is close to e.

Tables of logarithms to the base 10 were first constructed by the English mathematician Henry Briggs (1561–1639). He was led to the idea of using 10 as a base from conversations with Napier in 1615. In

FIGURE 7.37   John Napier
(1550–1617)
Culver Pictures, Inc.

FIGURE 7.39   John Wallis
(1616–1703)
Culver Pictures, Inc.

honor of Briggs, logarithms to the base 10 are now called *Briggsian logarithms.*

Exponential notation

Exponential notation was developed after the invention of logarithms. René Descartes popularized the use of positive integer exponents with his treatise *La Géométrie*, which appeared in 1637. Negative and fractional exponents were first used systematically in the second half of the seventeenth century by the English mathematician John Wallis (1616–1703) and by Newton (1642–1727). The modern definition of logarithms in terms of exponential functions was first given by the eighteenth century Swiss mathematician Leonard Euler (1707–1783).

The Bernoullis and Euler

FIGURE 7.40   Leonard Euler (1701–1783)
The Granger Collection

Euler was one of the most productive mathematicians of all time. His collected works, consisting of 856 titles, are still in the process of publication. They will fill 72 volumes. Even in the last twelve years of his life, when he was totally blind, his productivity hardly slackened.

Euler studied under the Swiss mathematician Jean Bernoulli (1667–1748). Jean and his elder brother Jacques (1654–1705) had extended the work of Leibniz (1646–1716), with whom they had maintained an active correspondence. Euler worked with Jean's son Daniel (1700–1782), who did basic work in fluid mechanics and other branches of physics.

Euler put plane analytic geometry, trigonometry, and the calculus of functions of one variable into essentially the forms that are studied today. He did fundamental work in algebra, number theory, differential equations, the geometry of curves and surfaces, and many branches of mathematical physics, including the mechanics of fluids and of rigid and elastic bodies, optics, and celestial mechanics. Euler's contributions were of significant practical as well as theoretical importance. His study of the motion of the moon under the gravitational attraction of both the

FIGURE 7.41   Jean Bernoulli (1667–1748)
The Granger Collection

FIGURE 7.42   Jacques Bernoulli (1654–1705)
The Granger Collection

FIGURE 7.43   Daniel Bernoulli (1700–1782)
Culver Pictures, Inc.

earth and the sun, for example, enabled astronomers to compile lunar tables that improved the accuracy of navigation.

Euler made the concept of function the basis of calculus, and he introduced much of the notation we use today, including the letter "*e*" for the base of the natural logarithm, the letter "*i*" for $\sqrt{-1}$, the symbol "*f(x)*" for the value of a function, the summation sign Σ, and the symbols for the trigonometric functions. He also popularized the use of the symbol π.

FURTHER READING.

E. KENNEDY, [1] "The History of Trigonometry."

"LEONHARD EULER," *A Dictionary of Scientific Biography.*

MISCELLANEOUS EXERCISES

[†]*Answer provided.*
[‡]*Outline of solution provided.*

Describe the domains of the functions in Exercises 1 through 17 and compute their derivatives.

1.‡ $\sqrt{\sin(3x)}$ **2.**† $\ln[\cos(x^2)]$ **3.** $\ln|\sin(x^3)|$

4. $\cos(1/x) + 1/\cos x$ **5.**† $[\cos x]^\pi$

6. 3^{x^3} **7.**† $x^{\ln x}$ **8.**‡ $\ln(x^2 - x - 2)$

9. $(x^2 + 3x - 4)^e$ **10.**† $\tan(6x)\cot(6x)$ **11.** $\cos(\arcsin x)$

12.† $(1 + x^2)^{\tan x}$ **13.** $\ln[\arctan(x^2)]$ **14.**† $\arcsin(\sec x)$

15. $\arcsin(\cos x)$ **16.**† $\arctan(\tan x)$ **17.** $\tan(\arctan x)$

Evaluate integrals 18 through 23.

18.† $\displaystyle\int_e^\pi (x^e + e^x + x^\pi + \pi^x)\,dx$ **19.** $\displaystyle\int_e^{e^2} \frac{1}{x\ln x}\,dx$

20.† $\displaystyle\int_0^{\pi/2} \frac{\cos x}{\sin x - 3}\,dx$ **21.** $\displaystyle\int_0^{1/2} \frac{4^x}{4^x - 4}\,dx$

22.† $\displaystyle\int_0^2 \left|\sin x - \frac{1}{\sqrt{2}}\right|\,dx$ **23.** $\displaystyle\int_{-1}^1 |e^x - 1|\,dx$

24.† Match the following functions to their graphs shown in Figures 7.44 through 7.48.   **a.** $\ln(1/x)$ **b.** $\ln(1 + x^2)$ **c.** $1/(\ln x)$ **d.** $|\ln x|$ **e.** $\ln(x^2 - 1)$

25. Match the following functions to their graphs shown in Figures 7.49 through 7.53.   **a.** $(2x)^{-1}e^{-x/2}$ **b.** xe^{-x} **c.** $e^{1/x}$ **d.** e^{-x^2} **e.** xe^{-x^2}

26.† Match the following functions to their graphs shown in Figures 7.54 through 7.58.   **a.** $e^{\sin x}$ **b.** $e^{(\sec x)/4}$ **c.** $\ln(\sec x)$ **d.** $(\tan x)(\sec x)$ **e.** $\ln(\sin x)$

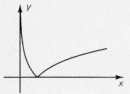

FIGURE 7.44

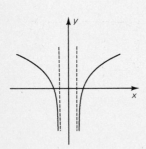

FIGURE 7.45

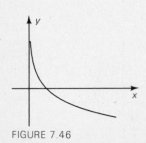

FIGURE 7.46

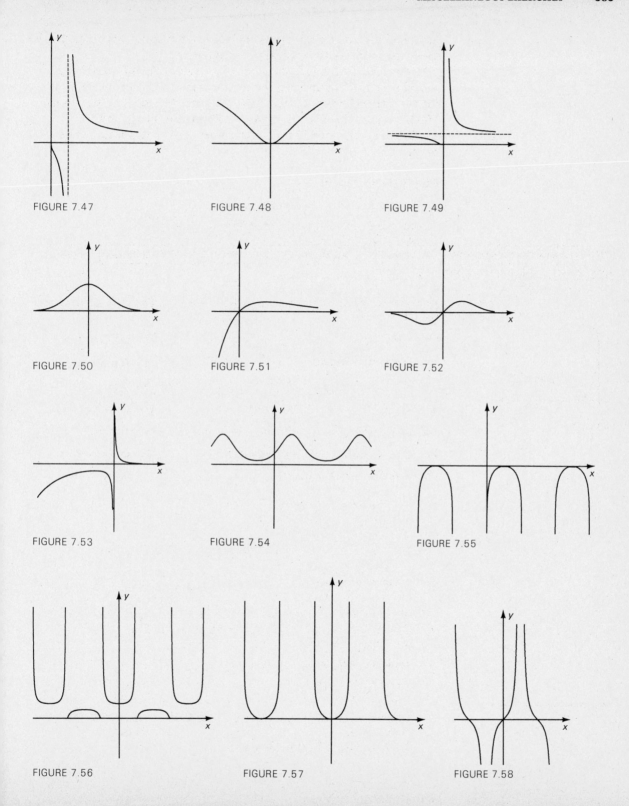

FIGURE 7.47

FIGURE 7.48

FIGURE 7.49

FIGURE 7.50

FIGURE 7.51

FIGURE 7.52

FIGURE 7.53

FIGURE 7.54

FIGURE 7.55

FIGURE 7.56

FIGURE 7.57

FIGURE 7.58

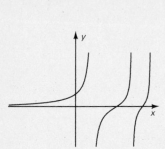

FIGURE 7.59

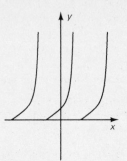

FIGURE 7.60

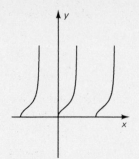

FIGURE 7.61

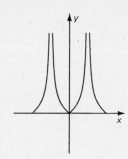

FIGURE 7.62

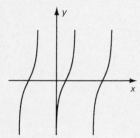

FIGURE 7.63

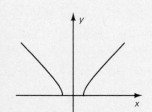

FIGURE 7.64

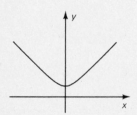

FIGURE 7.65

27. Match the following functions to their graphs shown in Figures 7.59 through 7.63.   **a.** $|\tan x|$ **b.** $\ln(\tan x)$ **c.** $e^{\tan x}$
   d. $\sqrt{\tan x}$ **e.** $\tan(e^{x/4})$

28. Match the following functions to their graphs shown in Figures 7.64 through 7.68.   **a.** $|x^2 - 1|$ **b.** $\sqrt[4]{(x-1)(x+1)(x-2)(x+2)}$
   c. $\sqrt{x^2 - 1}$ **d.** $\sqrt{x^2 + 1}$ **e.** $\sqrt[3]{x^2 - 1}$

29. Sketch the curves $y = e^{-x}$ and $y = e^{-x}\sin(5x)$, and show that they are tangent (have the same tangent line) where they touch.

30.† Find a number k such that the curves $y = k \sin x$ and $y = k \cos x$ intersect at right angles.

31. Show that the curves $y = \sec^2 x$ and $y = 2 \tan x$ intersect at only one point for $-\pi/2 < x < \pi/2$ and that they are tangent there.

32.† Use the trigonometric identity

$$\tan(\phi - \psi) = (\tan \phi - \tan \psi)/(1 + \tan \phi \tan \psi)$$

to find an angle between the tangent lines to the parabolas, $y = x^2$ and $y = 2 - x^2$ at the point $(1, 1)$. Give the exact answer and a numerical approximation.

33.† Let R be the region bounded by the x-axis, by the lines $x = 1$ and $x = 3$, and by the curve $y = x^{-2}$. Compute the volume of the solid generated by rotating R about the y-axis.

34. Sketch the region bounded by the curve $y = \csc x$ and by the lines $y = 0, x = \pi/4$, and $x = (3\pi)/4$. Compute the volume of the solid generated by rotating this region about the x-axis.

35.† Let R be the region bounded by the x- and y-axes, by the line $x = \pi/3$, and by the graph of the function $\sec^2(x^2)$. Compute the volume of the region generated by rotating R about the y-axis.

36. Sketch the region R bounded by the x- and y-axes, by the line $x = 1$, and by the curve $y = (x^2 + 1)^{-1/2}$. Compute the volumes of the solids generated by rotating R **a.** about the x-axis   **b.** about the y-axis.

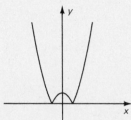

FIGURE 7.66

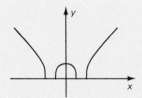

FIGURE 7.67

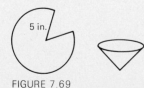

FIGURE 7.68

37.[†] Evaluate the limit

$$\lim_{N \to \infty} \int_{-N}^{N} \frac{1}{x^2 + 1} \, dx.$$

Give a geometric interpretation of your result.

38. A rod extends from $x = 0$ to $x = 10$ on an x-axis with its scale measured in inches. The density of the rod is $1/(30 - 2x)$ ounces per inch at x. How much does the rod weigh?

39.[‡] An aviarist is watching one of his birds that has escaped and is flying directly south from the aviary at a height of 100 feet above the aviarist's eye level. When the aviarist's line of vision makes an angle of $\pi/3$ radians with the horizontal, the angle is decreasing at the rate of $\frac{1}{4}$ radians per second. How fast is the bird flying at that moment?

40. An airplane is flying at an altitude of 400 feet at the speed of 200 feet per second directly away from a searchlight on the ground. The searchlight is shining on the plane. At what rate is the angle between the ray of light and the ground changing when the angle is 30°?

41.[†] The light from the rising sun is shining horizontally on a falling ball. The shadow of the ball is on a hemispherical building of radius 100 feet. The ball is $100 - 16t^2$ feet above the ground at time t (seconds). What is the speed of the shadow at $t = 2$?

42.[†] A surveyor standing 100 feet along the horizontal ground from the base of a vertical cliff measures the height of the cliff by measuring the angle between the base and the top. His measurement is 0.94 radians with a possible error of 0.01 radians. How high does he determine the cliff to be and approximately how accurate is his answer? (Use the reference tables at the back of this book.)

43.[‡] Give the dimensions of the right circular cone of greatest volume that can be inscribed in a sphere of radius r.

44.[†] A cone is to be formed from a circular piece of paper of radius 5 inches by cutting out a segment of the circle and joining its edges (Figure 7.69). What is the maximum volume of a cone that can be formed in this way?

45.[†] A conical glass is $\sqrt{3}$ inches high and the radius of its top is 1 inch. It is full of cola and someone puts a spherical cherry in it. What size cherry would make the biggest mess?

46. What is the height of the right circular cone of maximum lateral surface area that can be inscribed in a sphere of radius r?

47.[†] Of all pie slices of perimeter 10 inches, which has the greatest area? (You cannot have a whole pie.)

48.[†] A rectangular piece of tin 10 inches wide and 5 feet long is to be bent into a gutter with the arc of a circle as its cross section. Give an equation in r which would have to be solved to find the radius of the circle to maximize the volume of the gutter.

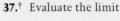

FIGURE 7.69

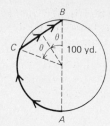

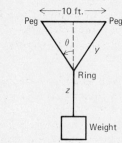

FIGURE 7.70

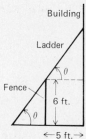

FIGURE 7.71

49. Sketch the graphs of   **a.**[†] $\sin(\arcsin x)$,   **b.**[†] $\arcsin(\sin x)$,
c.[†] $\cos(\arcsin x)$,   **d.** $\arcsin(\cos x)$,   **e.** $\tan(\arcsin x)$,
f. $\tan(\arctan x)$,   **g.** $\arctan(\tan x)$.

50.[†] A girl wants to get from point A, beside a circular lake of radius 100 yards, to the diametrically opposite point B as fast as possible. She can run 5 yards per second and swim 1 yard per second.   **a.** Express the time it would take her to run around the lake to the point C in Figure 7.70 and swim from C to B as a function of the angle θ.   **b.** For what angle θ will this time be a minimum?

51.[†] A projectile is fired from the origin in an xy-plane with its y-axis vertical and with the scales measured in feet. Its initial speed is v_0 feet per second and it is fired from an angle θ with the x-axis. If there is no air resistance, the projectile will travel along the parabola $y = x \tan \theta - 16x^2 v_0^{-2} \sec^2\theta$. What angle θ gives the projectile the maximum range?

52.[‡] A 60 foot rope with a ring at one end is placed over two pegs that are at the same height and 10 feet apart. Then the end without the ring is threaded through the ring and fastened to a weight that pulls the rope taut (Figure 7.71). What angle will the rope form at the ring? (Disregard the weight of the rope and the thickness of the pegs and ring.)

53.[†] What is the length of the shortest ladder that will reach from the ground over a 6 foot fence to a building which is 5 feet behind the fence (Figure 7.72)?

54. What rectangle with its base on the x-axis and its upper corners on the curve $y = e^{-x^2}$ has the maximum area? What is the area?

55.[†] The minute hand on a school clock is 11 inches long and the hour hand is 10 inches long.   **a.** How fast do the hands turn in radians per hour?   **b.** How rapidly is the distance between the tips of the hands increasing at 8:00?

56. Which is larger: e^π or π^e?

57.[†] A capacitor of capacitance C farads is in the open circuit in Figure 7.73 and has a charge of q_0 coulombs on it. The circuit is closed at time $t = 0$ (seconds) (Figure 7.74) and subsequently the charge $q(t)$ on the capacitor satisfies the differential equation $R\, dq/dt + 1/C\, q = 0$ where R (ohms) is the resistance of the circuit. Find $q(t)$.

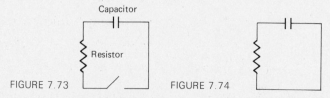

FIGURE 7.73   FIGURE 7.74

58. A battery is supplying a direct current of i_0 amperes in a circuit with a resistance of R ohms and an inductor of inductance L henries (Figure 7.75). At time $t = 0$ (seconds) the battery is switched out of the circuit (Figure 7.76). Subsequently the current $i(t)$ in the circuit satisfies the differential equation $L\, di/dt + Ri = 0$. Find $i(t)$.

Battery

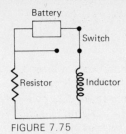

Switch

Resistor   Inductor

FIGURE 7.75

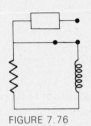

FIGURE 7.76

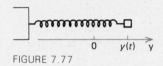

$$0 \quad y(t) \quad y$$

FIGURE 7.77

59.[†] Before Galileo discovered that the speed of a falling body is proportional to the time from when it is dropped, he conjectured that the speed was proportional to the distance it had fallen. What would this imply?

60.[†] Lambert's law of absorption states that the fraction of light absorbed by a layer of liquid is proportional to the thickness of the layer. Use this to derive the formula $I = I_0 e^{-ky}$ for the intensity of light a distance y beneath the surface of the ocean.

61. Consider an object of mass M that is attached to a spring and moves parallel to a y-axis (Figure 7.77). The object's y-coordinate $y = y(t)$ at time t satisfies the differential equation $M \, d^2y/dt^2 + ky = 0$ where k is the "spring constant."   **a.** For what values of the constants A, B, and ω (omega) is $y = A \cos(\omega t) + B \sin(\omega t)$ a solution of the differential equation?   **b.** For what values of these constants does it also satisfy the initial conditions $y(0) = y_0$ and $dy/dt(0) = v_0$?

62.[†] According to *Boyle's law*, given an ideal gas in a cylinder with a piston in one end, the pressure p and the volume V satisfy a relation of the form $pV = $ constant. A cylinder has a radius of 2 inches and when the piston is 4 inches from the closed end of the cylinder, the pressure is 100 pounds per square inch. How much work is required to compress the piston from a distance of 4 inches to a distance of 2 inches from the closed end of the cylinder?

63. Find two constants r such that $y = e^{rx}$ satisfies the differential equation $y'' + 2y' - 3y = 0$.

64. Show that $y = e^x$ and $y = xe^x$ are solutions of the differential equation $y'' - 2y' + y = 0$.

65. Show that $y = e^x \cos(2x)$ and $y = e^x \sin(2x)$ are solutions of the differential equation $y'' - 2y' + 5y = 0$.

66.[†] An object is at $s = 0$ on an s-axis and has velocity 0 at $t = 1$. Its acceleration in the positive direction is t^{-2} at time t for $t \geq 1$. Give its position as a function of t for $t \geq 1$.

67.[†] Show that the equation $x \ln 5 = 5 \ln x$ has two solutions.

68. Solve the equation $\ln(2 + x) = \ln(1 - x)$ for x.

69. Show that the equation $e^x = x^3$ has one solution x between 1 and 2 and one solution greater than 2.

70.[†] Find a curve such that its length from $(0,1)$ to (x,y) is $y - e^x$.

71.[†] Find $\lim\limits_{N \to \infty} [(3^{1/N} + 3^{2/N} + \ldots + 3^{N/N})/N]$.

72.[†] Find   **a.** $\lim\limits_{b \to \infty} \log_b e$ **b.** $\lim\limits_{b \to 1^+} \log_b e$

73.[†] **a.** Find the maximum value of $|x \ln x|$ for $0 < x \leq 1$.   **b.** Show that $x^2 \ln x \to 0$ as $x \to 0^+$.   **c.** Show that for any positive constant a, $x^a \ln x \to 0$ as $x \to 0^+$. (Replace x by $y = x^{a/2}$.)

74. Show that for any $x > 1$ and $a > 0$,

$$0 < \ln x < \int_1^x t^{a-1} dt = \frac{1}{a} x^a - \frac{1}{a}.$$

Then use these inequalities to show that $x^{-b} \ln x$ tends to zero as x tends to ∞ for any $b > 0$. (Take $0 < a < b$.)

75. Use the result of Exercise 74 to show that $x^b \ln x \to 0$ as $x \to 0^+$ for any $b > 0$.

76. Solve the equation $\int_1^T t^{a-1} dt = 1$ $(a \neq 0)$ for T as a function of a. Find the limit of T as $a \to 0^+$

77. Find the limit of $\ln(1 + x)/x$ as $x \to 0^+$ **a.** by the definition of e. **b.** by the definition of the derivative of $\ln(1 + x)$ at $x = 0$.

78. Define $\log_b x$ for $b = \frac{1}{2}$ and sketch its graph.

79. Suppose that $L(x)$ is continuous and has the "logarithm property" that $L(x_1 x_2) = L(x_1) + L(x_2)$ for all x_1 and x_2 in its domain.   **a.**   Show that if 0 is in the domain of L, then $L(x) = 0$ for all x in its domain. **b.**   Show that if the domain of L consists of all $x \neq 0$ and $L(b) = 1$ with $b > 0$, then $L(x) = \log_b |x|$.

80. Use the results in Proposition 7.1 in Section 7.1 to prove Proposition 7.2.

81. Show that $\left(1 + \dfrac{1}{n}\right)^n$ tends to e as n tends to $-\infty$. (Set $m = -n - 1$ and let m tend to ∞.) Use this fact and the argument at the end of Section 7.1 to show that for $x > 0$

$$\lim_{\Delta x \to 0^-} \left\{ \frac{1}{\Delta x} [\log_b(x + \Delta x) - \log_b x] \right\} = \frac{1}{x} \log_b e.$$

82. A country has enough resources to last 700 years at the present rate of comsumption. If the population increases at the continuous rate of 5% per year and if the rate of consumption per capita does not change, how long will the resources last?

83.† Suppose that t years from now a machine will provide income at the rate of $f(t)$ dollars per year and that it could be sold for $S(t)$ dollars. **a.**   Show that the present value of the machine if it is sold T years from now is

$$v(T) = \int_0^T f(t) e^{-rt} dt + S(T) e^{-rT} \text{ dollars}$$

where $100r\%$ is the interest rate.   **b.**   What condition must f and S satisfy at a maximum of v?

84. What constant rate of deposit in a savings account paying 6% interest compounded continuously is required for a parent to save \$10,000 in ten years to send his child to college?

85. Find the equation of the tangent line to $y = \log_b x$ which passes through the origin. How does it vary with b?

86. Use the fact that $x^n e^{-x} \to 0$ as $x \to \infty$ for all integers n to prove that e^x is not an algebraic function. [Show that $y = e^x$ cannot satisfy an equation of the form $a_n(x)y^n + a_{n+1}(x)y^{n-1} + \ldots + a_0(x) = 0$ with polynomials $a_n(x), a_{n-1}(x), \ldots, a_0(x)$.]

87. In one method of measuring the rate of flow of blood from the heart, a few milligrams of dye are injected in a vein near the heart and the dye concentration is measured as a function of time in an artery in the arm (the solid curve in Figure 7.78). The dye appears after about 10 seconds, increases to a maximum, decays exponentially, and then increases again due to recirculation of the dye in the blood stream. The logarithm of the concentration is plotted as in Figure 7.79, where the exponential decay appears as a straight line. To determine how the concentration would decay without recirculation, the line is extended (the broken line in Figure 7.79) and the corresponding curve is plotted with nonlogarithmic scales in Figure 7.78.   **a.**[†] Express the total mass of injected dye as an integral involving the rate of flow from the heart and the dye concentration $C(t)$ without recirculation. (Assume $C(t)$ is zero for $t > T$.)   **b.**   Use the result of part (a) to determine the approximate rate of flow from the heart in the case of Figures 7.78 and 7.79, where 24 milligrams of dye were injected, under the assumption that the rate of flow is constant.

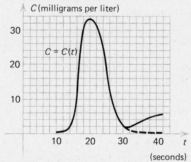

FIGURE 7.78

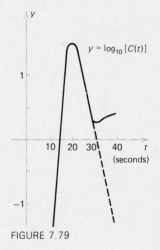

FIGURE 7.79

CHAPTER 8

TECHNIQUES OF INTEGRATION

SUMMARY: In this chapter we develop techniques which will greatly increase the variety of integrals that we can evaluate. In Section 8.1 we review the method of substitution and the basic integration formulas that were derived in previous chapters. (These formulas are listed in Section 8.1 and inside the back cover for reference.) The related rule for changing variables in definite integrals is given in an optional Section 8.2. The new integration techniques are discussed in Sections 8.3 through 8.6. Each of these is a means of transforming integrals into forms that can be evaluated by the methods reviewed in Section 8.1. Section 8.3 deals with integration by parts, Section 8.4 with integration of powers of trigonometric functions, Section 8.5 with integration by inverse trigonometric substitutions, and Section 8.6 with integration by partial fractions. The last three sections are optional: Section 8.7 deals with integrals of rational functions of $\sin x$ and $\cos x$; in Section 8.8 we illustrate the use of integral tables,* and in Section 8.9 we discuss approximate integration.

8.1 REVIEW OF INTEGRATION BY SUBSTITUTION

In this section we review the use of the following integration formulas, which were derived in Chapters 5 and 7.

$$(1) \qquad \int u^n \frac{du}{dx}\, dx = \int u^n\, du = \begin{cases} \dfrac{1}{n+1} u^{n+1} + C & \text{for } n \neq -1 \\ \ln|u| + C & \text{for } n = -1 \end{cases}$$

---

*In courses where time is at a premium, instructors might choose to emphasize Section 8.8 instead of Sections 8.2 through 8.7.

$$\cdot(2) \quad \int \cos u \, \frac{du}{dx} \, dx = \int \cos u \, du = \sin u + C$$

$$\cdot(3) \quad \int \sin u \, \frac{du}{dx} \, dx = \int \sin u \, du = -\cos u + C$$

$$\cdot(4) \quad \int \sec^2 u \, \frac{du}{dx} \, dx = \int \sec^2 u \, du = \tan u + C$$

$$\cdot(5) \quad \int \csc^2 u \, \frac{du}{dx} \, dx = \int \csc^2 u \, du = -\cot u + C$$

$$\cdot(6) \quad \int \sec u \tan u \, \frac{du}{dx} \, dx = \int \sec u \tan u \, du = \sec u + C$$

$$\cdot(7) \quad \int \csc u \cot u \, \frac{du}{dx} \, dx = \int \csc u \cot u \, du = -\csc u + C$$

$$\cdot(8) \quad \int \frac{1}{\sqrt{a^2 - u^2}} \, \frac{du}{dx} \, dx = \int \frac{1}{\sqrt{a^2 - u^2}} \, du = \arcsin\left(\frac{u}{a}\right) + C$$

$$\cdot(9) \quad \int \frac{1}{u^2 + a^2} \, \frac{du}{dx} \, dx = \int \frac{1}{u^2 + a^2} \, du = \frac{1}{a} \arctan\left(\frac{u}{a}\right) + C$$

$$(10) \quad \int e^u \, \frac{du}{dx} \, du = \int e^u \, du = e^u + C$$

$$(11) \quad \int b^u \, \frac{du}{dx} \, dx = \int b^u \, du = \frac{1}{\ln b} b^u + C$$

$$(12) \quad \int \sinh u \, \frac{du}{dx} \, dx = \int \sinh u \, du = \cosh u + C$$

$$(13) \quad \int \cosh u \, \frac{du}{dx} \, dx = \int \cosh u \, du = \sinh u + C$$

Integrals of tan x, cot x, sec x, and csc x
The integrals of $\sin x$ and $\cos x$ are given in formulas (2) and (3). The integrals of the other four trigonometric functions may be obtained from formula (1) with $n = -1$ and the substitutions which are indicated in the next two rules.

Rule 8.1   To integrate tan x, make the substitution $u = \cos x$, and to integrate cot x, make the substitution $u = \sin x$.

With these substitutions we obtain

$$(14) \quad \int \tan x \, dx = \int \frac{\sin x}{\cos x} \, dx = -\int \frac{1}{\cos x} [-\sin x \, dx]$$

$$= -\int \frac{1}{u} \, du = -\ln|u| + C = -\ln|\cos x| + C$$

and by a similar calculation

*(15) $\displaystyle\int \cot x \, dx = \int \frac{\cos x}{\sin x} \, dx = \ln|\sin x| + C.$

Rule 8.2   To integrate sec x, multiply and divide it by sec x + tan x and make the substitution u = sec x + tan x. To integrate csc x, multiply and divide it by csc x + cot x and make the substitution u = csc x + cot x.

The use of Rule 8.2 is illustrated in the next example.

EXAMPLE 1   Evaluate $\displaystyle\int \sec(4x) \, dx.$

Solution.   We multiply and divide by $\sec(4x) + \tan(4x)$ and make the substitution $u = \sec(4x) + \tan(4x)$ for which

$$du = \frac{d}{dx}[\sec(4x) + \tan(4x)] \, dx$$
$$= 4[\sec(4x)\tan(4x) + \sec^2(4x)] \, dx.$$

We obtain

$$\int \sec(4x) \, dx = \int \frac{\sec^2(4x) + \sec(4x)\tan(4x)}{\sec(4x) + \tan(4x)} \, dx$$
$$= \frac{1}{4} \int \frac{1}{u} \, du = \frac{1}{4} \ln|u| + C$$
$$= \frac{1}{4} \ln|\sec(4x) + \tan(4x)| + C. \quad \square$$

Exercises

†*Answer provided.*
‡*Outline of solution provided.*

In Exercises 1 through 56   **a.**   give substitutions $u = u(x)$ and the numbers of appropriate integration formulas (1) through (13) which could be used to evaluate the integrals.   **b.**   Evaluate the integrals.

1.† $\displaystyle\int x\sqrt{x^2 + 6} \, dx$

2. $\displaystyle\int_0^\pi \sin\left(\frac{x}{2}\right) dx$

3.† $\displaystyle\int \cos x \, \sqrt{\sin x} \, dx$

4. $\displaystyle\int \csc^2(3x + 5) \, dx$

5.† $\displaystyle\int \tan^2 x \sec^2 x \, dx$

6. $\displaystyle\int_0^1 \frac{x}{x^2 - 4} \, dx$

7.† $\displaystyle\int \frac{1 + \cos x}{x + \sin x} \, dx$

8. $\displaystyle\int_{-\pi}^\pi x \sin(x^2) \, dx$

9.† $\displaystyle\int \cot(3x) \, dx$

10. $\displaystyle\int_{-1}^0 \frac{2}{6x - 1} \, dx$

11.† $\displaystyle\int \frac{1}{x\sqrt{\ln x}} \, dx$

12. $\displaystyle\int_1^5 \frac{\sqrt{\ln x}}{x} \, dx$

13.† $\displaystyle\int \frac{x}{\sqrt{9-x^4}}\,dx$

14. $\displaystyle\int_0^7 \frac{3}{4+x^2}\,dx$

15.† $\displaystyle\int \frac{\arcsin x}{\sqrt{1-x^2}}\,dx$

16. $\displaystyle\int_0^4 \frac{1}{\sqrt{25-x^2}}\,dx$

17.† $\displaystyle\int e^{-x}\sin(e^{-x})\,dx$

18. $\displaystyle\int_0^1 xe^{-x^2}\,dx$

19.† $\displaystyle\int \csc(6x)\,dx$

20. $\displaystyle\int_0^1 (7-10x)^{29}\,dx$

21.† $\displaystyle\int_0^1 \frac{x^3}{\sqrt{4-x^8}}\,dx$

22. $\displaystyle\int e^x\sqrt{2+e^x}\,dx$

23.† $\displaystyle\int_0^1 \frac{x}{\sqrt{4-x^2}}\,dx$

24. $\displaystyle\int \cos(3x)\,e^{\sin(3x)}\,dx$

25.† $\displaystyle\int \frac{\sin x}{1+\cos^2 x}\,dx$

26. $\displaystyle\int \frac{\sin(3x)}{1+\cos(3x)}\,dx$

27.† $\displaystyle\int_2^3 \frac{2x^2}{1-x^3}\,dx$

28. $\displaystyle\int \frac{\sec^2 x}{\sqrt{\tan x}}\,dx$

29.† $\displaystyle\int \frac{1}{\sqrt{x}}\sec\sqrt{x}\,\tan\sqrt{x}\,dx$

30. $\displaystyle\int \frac{1}{\sqrt{1-9x^2}}\,dx$

31.† $\displaystyle\int_0^2 \sin x\cos x\,dx$

32. $\displaystyle\int x\csc(3x^2)\cot(3x^2)\,dx$

33.† $\displaystyle\int \frac{x^2}{9+x^6}\,dx$

34. $\displaystyle\int_0^{\pi/2} \cos(3x)\,dx$

35.† $\displaystyle\int_0^1 \frac{\sin x}{\sqrt{\cos x}}\,dx$

36. $\displaystyle\int x\sinh(x^2+1)\,dx$

37.† $\displaystyle\int e^{3x}(6+e^{3x})^7\,dx$

38. $\displaystyle\int \frac{1-e^{-x}}{x+e^{-x}}\,dx$

39.† $\displaystyle\int \frac{\arctan x}{1+x^2}\,dx$

40. $\displaystyle\int_0^{10} e^x\cosh(3-e^x)\,dx$

41.† $\displaystyle\int_0^1 \frac{x^2}{6+x^3}\,dx$

42. $\displaystyle\int_{-1}^1 \frac{1}{\sqrt{9-4x^2}}\,dx$

43.† $\displaystyle\int \frac{\ln x}{x}\,dx$

44. $\displaystyle\int_1^2 \frac{4x^3+1}{1-x-x^4}\,dx$

45.† $\displaystyle\int \frac{\sec^2 x}{\sqrt{1-\tan^2 x}}\,dx$

46. $\displaystyle\int_0^2 (2^x+x^2)\,dx$

47.† $\displaystyle\int \cos x\,[\pi^{\sin x}+(\sin x)^\pi]\,dx$

48. $\displaystyle\int_{0.01}^{200} \frac{1}{x}\log_{10}x\,dx$

49.‡ $\displaystyle\int \frac{x}{\sqrt{1-x}}\,dx$

50. $\displaystyle\int \frac{x}{(x+3)^2}\,dx$

51.† $\displaystyle\int_0^1 x(1-2x)^{30}\,dx$ **52.** $\displaystyle\int \frac{x}{(x^2+3)^2}\,dx$

53.† $\displaystyle\int_0^{10} \frac{x}{\sqrt{16+2x}}\,dx$ **54.** $\displaystyle\int_0^1 x\sqrt{1-x}\,dx$

55. $\displaystyle\int \frac{x^2}{\sqrt{5-x}}\,dx$ **56.** $\displaystyle\int_{-2}^1 \frac{x}{\sqrt{x+3}}\,dx$

Evaluate the integrals in Exercises 57 through 63 by first making a substitution of the form $u = x^r$ or $u = f(x)^r$ with r a rational number.

57.‡ $\displaystyle\int \frac{1}{3\sqrt{x}+x}\,dx$ **58.**† $\displaystyle\int \frac{5}{2x^{3/4}-3x}\,dx$

59. $\displaystyle\int \frac{1}{\sqrt{x^{8/5}-x^2}}\,dx$ **60.**‡ $\displaystyle\int \frac{\cos x}{3\sin x+4\sqrt{\sin x}}\,dx$

61.† $\displaystyle\int \frac{e^{2x}}{\sqrt{e^x-2}}\,dx$ **62.** $\displaystyle\int \frac{3}{x-x^{4/7}}\,dx$

63.† $\displaystyle\int_0^8 \frac{1}{x^{4/3}+x^{2/3}}\,dx$

8-2 CHANGING VARIABLES IN DEFINITE INTEGRALS (Optional)

It is sometimes convenient to change the variable in a definite integral before evaluating it, and in theoretical discussions it can be necessary to change variables in a definite integral which is not to be evaluated. To make such a change of variables, we perform a substitution as we would in an indefinite integral and change the limits of integration as described in the following theorem.

Theorem 8.1   Suppose that $u(x)$ is either increasing or decreasing and has a continuous derivative for $a \le x \le b$ and $F(u)$ is continuous in an interval containing the values $u(x)$ for $a \le x \le b$. Then

(1) $$\int_a^b F\big(u(x)\big)\frac{du}{dx}(x)\,dx = \int_{u(a)}^{u(b)} F(u)\,du. \qquad \blacktriangleleft$$

Proof.   We let $G(u)$ be an antiderivative of $F(u)$ with respect to u: $\dfrac{dG}{du}(u) = F(u)$. Then, by the chain rule and the Fundamental Theorem of Calculus, we have

$$\int_a^b F(u(x)) \frac{du}{dx}(x) \, dx = \int_a^b \frac{dG}{du}(u(x)) \frac{du}{dx}(x) \, dx$$

$$= \int_a^b \frac{d}{dx} G(u(x)) \, dx$$

$$= G(u(b)) - G(u(a))$$

$$= \int_{u(a)}^{u(b)} \frac{dG}{du}(u) \, du = \int_{u(a)}^{u(b)} F(u) \, du.$$

This gives formula (1).   Q.E.D.

EXAMPLE 1   Make the change of variables $u = x^{1/3}$ in the definite integral

(2) $$\int_1^8 \frac{x^{2/3}}{4 + x^{1/3}} \, dx.$$

Solution.   To indicate that the integration with respect to x in the integral (2) goes from 1 to 8, we write $1 \xrightarrow{x} 8$. As x goes from 1 to 8, $u = x^{1/3}$ goes from $1^{1/3} = 1$ to $8^{1/3} = 2$, so we write

(3) $$1 \xrightarrow{u} 2.$$

Solving the equation $u = x^{1/3}$ for x gives $x = u^3$. From this we obtain the symbolic equation $dx = 3u^2 \, du$. We make these substitutions and use the limits of integration indicated in (3). We obtain

$$\int_1^8 \frac{x^{2/3}}{4 + x^{1/3}} \, dx = \int_1^2 \frac{u^2}{4 + u}(3u^2 \, du) = \int_1^2 \frac{3u^4}{4 + u} \, du. \quad \square$$

EXAMPLE 2   Make the substitution $u = 9 - \sqrt{x}$ in the definite integral

$$\int_4^{25} \sin^2(9 - \sqrt{x}) \, dx.$$

Solution.   Because x runs from 4 to 25 in the integral, we write $4 \xrightarrow{x} 25$. As x runs from 4 to 25, $u = 9 - \sqrt{x}$ runs from $9 - \sqrt{4} = 7$ to $9 - \sqrt{25} = 4$, so we write $7 \xrightarrow{u} 4$.

The equation $u = 9 - \sqrt{x}$ gives $\sqrt{x} = 9 - u$ and then $x = (9 - u)^2$. Therefore, we have $dx = 2(9 - u)(-du) = -2(9 - u) \, du$ and

$$\int_4^{25} \sin^2(9 - \sqrt{x}) \, dx = \int_7^4 -2(9 - u)\sin^2 u \, du$$

$$= 2 \int_4^7 (9 - u) \sin^2 u \, du. \quad \square$$

Exercises   Make the indicated changes in Exercises (1) through (8).

†*Answer provided.*
‡*Outline of solution provided.*

1.† $\displaystyle\int_0^{16} e^{\sqrt{x}}\, dx;\ u = \sqrt{x}$

2. $\displaystyle\int_1^8 \frac{x}{2 + x^{1/3}}\, dx;\ u = x^{1/3}$

3.† $\displaystyle\int_1^9 (1 + \sqrt{x})^{10}\, dx;\ u = 1 + \sqrt{x}$

4. $\displaystyle\int_1^{16} \frac{x^2}{x^{3/4} + x^{1/4}}\, dx;\ u = x^{1/4}$ **5.**† $\displaystyle\int_0^2 \ln(9 - 3x)\, dx;\ u = 9 - 3x$

6. $\displaystyle\int_1^9 \sqrt{10 - \sqrt{x}}\, dx;\ u = -\sqrt{x}$

7.† $\displaystyle\int_0^{\ln 5} \sqrt{e^x + 3}\, dx;\ u = e^x + 3$ **8.** $\displaystyle\int_e^{e^3} \frac{x}{\ln x}\, dx;\ u = \ln x$

9.† The function $1/(\sigma\sqrt{2\pi})\, e^{-(x-k)^2/2\sigma^2}$ is the *normal probability density* with *mean* k and *standard deviation* $\sigma > 0$. Show that the distance between the inflection points of the graph of this function is 2σ, so that σ is a measure of how "spread out" the graph is. Sketch the graph for $k = 0$ and $\sigma = 1$ and $\sigma = 2$.

10.† Use the fact that

$$\lim_{X \to \infty} \int_{-X}^X e^{-x^2/2}\, dx = \sqrt{2\pi}$$

to show that for any $\sigma > 0$

$$\lim_{X \to \infty} \frac{1}{\sigma\sqrt{2\pi}} \int_{-X}^X e^{-x^2/2\sigma^2}\, dx = 1.$$

What is the geometric interpretation of this result?

11.† A gun is clamped in a fixed position and fired repeatedly at a wall with a horizontal x-axis on it. The scale on the x-axis is measured in feet. The bullet holes are equally distributed on both sides of $x = 0$, and the probability that a bullet will hit the wall in the interval $a \le x \le b$ is

$$\frac{1}{\sigma\sqrt{2\pi}} \int_a^b e^{-x^2/2\sigma^2}\, dx.$$

a. What does the result of Exercise 10 say about this formula?
b. The bullet patterns from two guns have mean 0 and standard deviations σ_1 and σ_2. The probability that a bullet from the first gun will hit within one foot of 0 is equal to the probability that a bullet from the second gun will hit within two feet. What is the ratio σ_1/σ_2? (Use $x = y\sigma_1/\sigma_2$.)

12. **a.** Express the area of a circle of radius 1 in terms of an integral of the function $(1 - x^2)^{1/2}$.   **b.** Compute the area of the region bounded by the ellipse $x^2/a^2 + y^2/b^2 = 1$ by expressing it as an integral and then changing variables in the integral and using the result of part (a).

8-3   INTEGRATION BY PARTS

In this section we study the integration rule which is obtained from the product rule for derivatives. The application of this rule is known as INTEGRATION BY PARTS.

For functions $u = u(x)$ and $v = v(x)$ the product rule reads

$$\frac{d}{dx}(uv) = u\frac{dv}{dx} + v\frac{du}{dx}.$$

It translates into the integration formula

$$(1) \qquad \int u\frac{dv}{dx}\,dx + \int v\frac{du}{dx}\,dx = uv + C.$$

The integration by parts formula

$$(2) \qquad \int u\frac{dv}{dx}\,dx = uv - \int v\frac{du}{dx}\,dx \qquad\qquad \blacktriangleleft$$

is obtained from equation (1) by subtracting the integral $\int u\dfrac{dv}{dx}\,dx$ from both sides. There is no "$+ C$" term on either side of (2) because there is an indefinite integral on each side and both indefinite integrals are determined only up to arbitrary constants.

To apply the integration by parts formula (2) in evaluating an integral, we view the integrand as the product of two functions. We call one of them u and the other dv/dx. Formula (2) then expresses the given integral, on the left, in terms of a new integral on the right. In the new integral, u has been differentiated and dv/dx has been integrated. This is useful if the new integral is easier to integrate than the original one.

We illustrate integration by parts by showing how it is used to evaluate three types of integrals.

Type 1. Integrals of $x^n e^{ax}$, $x^n \sin(ax)$, and $x^n \cos(ax)$

Rule 8.3   To evaluate an integral of $x^n e^{ax}$, $x^n \sin(ax)$, or $x^n \cos(ax)$ with n a positive integer and constant a, integrate by parts n times, differentiating the power of x and integrating the e^{ax}, $\sin(ax)$, or $\cos(ax)$ each time.

Each integration by parts lowers the power of x in the integral.

EXAMPLE 1   Evaluate

$$(3) \qquad \int x \sin(6x)\,dx.$$

Solution. Because we want to differentiate x and integrate $\sin(6x)$, we set

$$(4) \qquad u = x \quad \text{and} \quad \frac{dv}{dx} = \sin(6x).$$

From this we obtain

$$(5) \qquad \frac{du}{dx} = 1 \quad \text{and} \quad v = -\frac{1}{6}\cos(6x).$$

We find the formula for v by making the substitution $w = 6x$, $dw = 6dx$ in the integral of $dv/dx = \sin(6x)$: $v = \int \sin(6x)\,dx = \frac{1}{6}\int \sin(6x)[6\,dx] = \frac{1}{6}\int \sin w\,dw = -\frac{1}{6}\cos w + C = -\frac{1}{6}\cos(6x) + C$. We put $C = 0$ in (5) because we need only one of the possible functions v to perform the integration by parts.

Making substitutions (4) and (5) in formula (2) yields

$$\int x \sin(6x)\,dx = \int u\,\frac{dv}{dx}\,dx = uv - \int v\,\frac{du}{dx}\,dx$$

$$(6) \qquad\qquad = x\left[-\tfrac{1}{6}\cos(6x)\right] - \int \left[-\tfrac{1}{6}\cos(6x)\right][1]\,dx$$

$$= -\tfrac{1}{6}x\cos(6x) + \tfrac{1}{6}\int \cos(6x)\,dx.$$

Equation (6) expresses the integral of $x\sin(6x)$, which we cannot evaluate by substitution, in terms of the integral of $\cos(6x)$, which we can evaluate. We set $w = 6x$ again and obtain

$$\int x\sin(6x)\,dx = -\tfrac{1}{6}x\cos(6x) + \tfrac{1}{6}\left(\tfrac{1}{6}\right)\int \cos(6x)[6\,dx]$$

$$= -\tfrac{1}{6}x\cos(6x) + \tfrac{1}{36}\int \cos w\,dw$$

$$= -\tfrac{1}{6}x\cos(6x) + \tfrac{1}{36}\sin w + C$$

$$= -\tfrac{1}{6}x\cos(6x) + \tfrac{1}{36}\sin(6x) + C. \quad \square$$

Type 2. Integrals involving ln x, arcsin x, and arctan x

Rule 8.4 **Certain integrals involving ln x, arcsin x, or arctan x can be evaluated by using the substitution $u = \ln x$, $u = \arcsin x$, or $u = \arctan x$, respectively. Other such integrals (but not all) can be evaluated by integration by parts in which the function ln x, arcsin x, or arctan x is differentiated.**

The integration by parts procedure is illustrated in the next example.

EXAMPLE 2 Evaluate $\int x^2 \ln x\,dx$.

Solution. Because we want to differentiate $\ln x$, we have to integrate x^2. We set $u = \ln x$ and $dv/dx = x^2$. Then we have $du/dx = 1/x$ and $v = \frac{1}{3}x^3$. With these substitutions, formula (2) gives

$$\int x^2 \ln x \, dx = \int u \frac{dv}{dx} dx = uv - \int v \frac{du}{dx} dx$$

$$= [\tfrac{1}{3}x^3][\ln x] - \int [\tfrac{1}{3}x^3]\left[\frac{1}{x}\right] dx$$

$$= \tfrac{1}{3}x^3 \ln x - \tfrac{1}{3} \int x^2 \, dx$$

$$= \tfrac{1}{3}x^3 \ln x - \tfrac{1}{3}[\tfrac{1}{3}x^3] + C$$

$$= \tfrac{1}{3}x^3 \ln x - \tfrac{1}{9}x^3 + C. \quad \square$$

Type 3. Products of e^{ax}, $\sin(ax)$, $\cos(ax)$

Rule 8.5   **To evaluate an integral of $\sin(ax)\sin(bx)$ or $\cos(ax)\cos(bx)$ with $a^2 \neq b^2$ or of $e^{ax} \sin(bx)$, $e^{ax} \cos(bx)$, or $\sin(ax) \cos(bx)$ with arbitrary constants a and b, integrate by parts twice. This yields an equation which can be solved for the original integral.**

EXAMPLE 3   Evaluate $\int e^{2x} \cos x \, dx$.

Solution.   In our initial integration by parts it does not matter which of the functions e^{2x} and $\cos x$ we differentiate and which we integrate. To differentiate e^{2x} and integrate $\cos x$, we set $u = e^{2x}$ and $dv/dx = \cos x$. Then we have $du/dx = 2e^{2x}$ and $v = \sin x$. With these equations, the integration by parts formula yields

$$\int e^{2x} \cos x \, dx = \int u \frac{dv}{dx} dx = uv - \int v \frac{du}{dx} dx$$

(7)
$$= e^{2x} \sin x - \int \sin x \, [2e^{2x}] \, dx$$

$$= e^{2x} \sin x - 2 \int e^{2x} \sin x \, dx.$$

We now integrate by parts, differentiating e^{2x} and integrating the trigonometric function as in the initial integration by parts. (If we were to integrate e^{2x} and differentiate the trigonometric function, we would be reversing the steps in equations (7) and the procedure would lead us nowhere.) We set $u = e^{2x}$ and $dv/dx = \sin x$. Then $du/dx = 2e^{2x}$ and $v = -\cos x$, and hence

$$\int e^{2x} \sin x \, dx = \int u \frac{dv}{dx} dx = uv - \int v \frac{du}{dx} dx$$

(8)
$$= e^{2x}[-\cos x] - \int [-\cos x][2e^{2x}] \, dx$$

$$= -e^{2x}\cos x + 2 \int e^{2x} \cos x \, dx.$$

Next we substitute equation (8) into equation (7) (while being careful with the unusually difficult algebra). This gives

$$\int e^{2x} \cos x \, dx = e^{2x} \sin x - 2 \left[-e^{2x} \cos x + 2 \int e^{2x} \cos x \, dx \right]$$

$$= e^{2x} \sin x + 2e^{2x} \cos x - 4 \int e^{2x} \cos x \, dx.$$

Solving for the integral $\int e^{2x} \cos x \, dx$ yields

$$\int e^{2x} \cos x \, dx = \tfrac{1}{5}[e^{2x} \sin x + 2e^{2x} \cos x] + C. \quad \Box$$

Integrals of this type also can be evaluated by the *method of undetermined coefficients*, which is described in Exercises 41 through 44. Integrals of $\sin(ax)\sin(bx)$, $\cos(ax)\cos(bx)$, and $\sin(ax)\cos(bx)$ can be integrated in a third way: by the use of trigonometric identities. This technique will be discussed in the next section.

Exercises   Evaluate the integrals in Exercises 1 through 34.

†*Answer provided.*
‡*Outline of solution provided.*

1.† $\displaystyle \int x \cos(5x) \, dx$

2. $\displaystyle \int x e^{4x} \, dx$

3.‡ $\displaystyle \int x \sin(x^2) \, dx$

4. $\displaystyle \int x \sin(2x) \, dx$

5.† $\displaystyle \int x \ln x \, dx$

6. $\displaystyle \int_1^2 x^3 \ln x \, dx$

7.† $\displaystyle \int \frac{1}{x \ln x} \, dx$

8. $\displaystyle \int \arcsin(2x) \, dx$

9.† $\displaystyle \int \frac{\arcsin(2x)}{\sqrt{1 - 4x^2}} \, dx$

10. $\displaystyle \int_0^1 \arctan x \, dx$

11.† $\displaystyle \int e^x \sin x \, dx$

12. $\displaystyle \int x \sin(3x) \, dx$

13.‡ $\displaystyle \int \sin(2x) \sin x \, dx$

14. $\displaystyle \int \sin(5x) \cos x \, dx$

15.† $\displaystyle \int x^4 \ln x \, dx$

16. $\displaystyle \int_0^{10} x e^{5x} \, dx$

17.† $\displaystyle \int x e^{x^2} \, dx$

18. $\displaystyle \int \frac{(\ln x)^4}{x} \, dx$

19.† $\displaystyle \int \cos(4x) \sin(3x) \, dx$

20. $\displaystyle \int \ln(5x) \, dx$

21.† $\displaystyle \int_0^{\pi/2} x^2 \sin x \, dx$

22. $\displaystyle \int e^x \cos x \, dx$

23.† $\displaystyle\int \sin(3x)\cos(3x)\,dx$ **24.** $\displaystyle\int \frac{\ln x}{x}\,dx$

25.‡ $\displaystyle\int_0^\pi \sin^2 x\,dx$ **26.** $\displaystyle\int \cos^2(3x)\,dx$

27.† $\displaystyle\int x2^x\,dx$ **28.** $\displaystyle\int x\arctan(x^2)\,dx$

29.† $\displaystyle\int x\sinh x\,dx$ (Differentiate x.)

30. $\displaystyle\int x\cosh(3x^2)\,dx$

31.† $\displaystyle\int \cos(\ln x)\,dx$ (Integrate by parts twice.)

32. $\displaystyle\int \sin(\ln x)\,dx$ **33.**† $\displaystyle\int \frac{x^3}{\sqrt{1-x^2}}\,dx$ (Differentiate x^2.)

34. $\displaystyle\int x\sec^2(2x)\,dx$

35.† Find the area of the region bounded by the x-axis, by the line $x = 11$, and by the curve $y = \ln x$.

36. Sketch the region bounded by the curves $y = xe^x$, $y = e$, and $x = -2$ and compute its area.

37.† A rod extends from $x = 0$ to $x = 3$ (centimeters) and its density at x is e^{-x} grams (weight) per centimeter.   **a.**   How much does it weigh?   **b.**   Where is its center of gravity?

38. A rod extends from $x = -\pi$ to $x = \pi$ (inches) and its density at x is $5 + \sin x$ ounces per inch.   **a.**   How much does it weigh?   **b.**   Where is its center of gravity?

39.† The region R is bounded by the curves $x = 1$, $y = e^3$, and $y = e^x$. Compute the volume of the solid generated by rotating R about the y-axis.

40. Compute the volumes of the solids generated by rotating the region of Exercise 35   **a.**   about the x-axis, and   **b.**   about the y-axis.

The method of undetermined coefficients is described in Exercises 41 through 44.

41.‡ Without carrying through the calculations, we can see that the integration by parts procedure for evaluating the integral of $e^{ax}\sin(bx)$ will give a result of the form

$$\int e^{ax}\sin(bx)\,dx = A\,e^{ax}\sin(bx) + B\,e^{ax}\cos(bx) + C.$$

Find the constants A and B (in terms of a and b) by differentiating this equation.

42. Use the method of Exercise 41 to evaluate the integral of $e^{ax} \cos(bx)$.

43.[†] The integration by parts procedure for evaluating the integral of $\sin(ax) \sin(bx)$ $(a^2 \neq b^2)$ yields an integration formula of the form

$$\int \sin(ax) \sin(bx) \, dx = A \sin(ax) \cos(bx) + B \cos(ax) \sin(bx) + C.$$

Find the coefficients A and B by differentiating this equation.

44. Use the method of Exercise 43 to find the integrals of $\cos(ax) \cos(bx)$ and $\sin(ax) \cos(bx)$ $(a^2 \neq b^2)$.

8-4 INTEGRALS OF POWERS AND PRODUCTS OF TRIGONOMETRIC FUNCTIONS

In this section we see how integration by parts and trigonometric identities can be used to evaluate certain powers and products of trigonometric functions. We illustrate the techniques by applying them to several types of such integrals.

Integrals of $\sin^n x \cos^m x$ with either n or m a positive odd integer

Rule 8.6 **To evaluate an integral of $\sin^n x \cos^m x$ with n an odd positive integer, use the Pythagorean identity**

(1) $\quad \sin^2 x + \cos^2 x = 1$

to change all but one of the sines to cosines. Then use the substitution $u = \cos x$. If m is an odd positive integer, use (1) to change all but one of the cosines to sines and use the substitution $u = \sin x$.

EXAMPLE 1   Evaluate $\int \sin^2 x \cos^3 x \, dx$.

Solution.   We use the identity $\cos^2 x = 1 - \sin^2 x$ to express two of the factors of $\cos x$ in terms of $\sin x$. We have

$$\int \sin^2 x \cos^3 x \, dx = \int \sin^2 x \, [1 - \sin^2 x] \cos x \, dx$$

$$= \int [\sin^2 x - \sin^4 x] \cos x \, dx.$$

The substitution $u = \sin x$, $du = \cos x \, dx$ then gives the result

$$\int \sin^2 x \cos^3 x \, dx = \int [\sin^2 x - \sin^4 x] \cos x \, dx$$

$$= \int [u^2 - u^4] \, du = \tfrac{1}{3} u^3 - \tfrac{1}{5} u^5 + C$$

$$= \tfrac{1}{3} \sin^3 x - \tfrac{1}{5} \sin^5 x + C. \quad \square$$

Integrals of $\sin^n x \cos^n x$ with n and m positive even integers

Rule 8.7   To evaluate an integral of $\sin^n x \cos^m x$ with n and m positive even integers, use the Pythagorean identity (1) to switch all of the sines to cosines or all of the cosines to sines. Then express the resulting integrals in terms of integrals of lower powers of sines or cosines. This may be done either by integration by parts and application of identity (1) or by the use of double angle formulas.

The integration by parts technique and the use of double angle formulas are illustrated in Examples 2 and 3, respectively.

EXAMPLE 2   Evaluate $\int \sin^2 x \, dx$ by one integration by parts.

Solution.   We have

$$\int \sin^2 x \, dx = \int \sin x \, \frac{d}{dx}(-\cos x) \, dx$$

$$(2) \qquad\qquad = \sin x \, (-\cos x) - \int (-\cos x) \frac{d}{dx} \sin x \, dx$$

$$= -\sin x \cos x + \int \cos^2 x \, dx.$$

Here we applied the integration by parts formula without writing out the choices of the functions u and v. With the Pythagorean identity (1), equation (2) gives

$$\int \sin^2 x \, dx = -\sin x \cos x + \int (1 - \sin^2 x) \, dx$$

$$= -\sin x \cos x + x - \int \sin^2 x \, dx.$$

Then, solving the last equation for the integral of $\sin^2 x$, we obtain

$$\int \sin^2 x \, dx = \tfrac{1}{2}[x - \sin x \cos x] + C. \quad \square$$

The double-angle formulas we use in the next example are

$$(3) \qquad \sin^2 \theta = \frac{1 - \cos(2\theta)}{2} \qquad\qquad\qquad \blacktriangleleft$$

$$(4) \qquad \cos^2 \theta = \frac{1 + \cos(2\theta)}{2}. \qquad\qquad\qquad \blacktriangleleft$$

EXAMPLE 3   Use formulas (3) and (4) to evaluate $\int \sin^4(3x) \, dx$.

Solution. We think of $\sin^4(3x)$ as the square of $\sin^2(3x)$. Identity (3) with $\theta = 3x$ gives

$$\int \sin^4(3x)\,dx = \int \left[\frac{1 - \cos(6x)}{2}\right]^2 dx$$

$$= \tfrac{1}{4} \int [1 - 2\cos(6x) + \cos^2(6x)]\,dx.$$

Then, with identity (4) and $\theta = 6x$, we obtain

$$\int \sin^4(3x)\,dx = \tfrac{1}{4} \int [1 - 2\cos(6x) + \tfrac{1}{2} + \tfrac{1}{2}\cos(12x)]\,dx$$

$$= \tfrac{1}{4} \int [\tfrac{3}{2} - 2\cos(6x) + \tfrac{1}{2}\cos(12x)]\,dx$$

$$= \tfrac{1}{4}[\tfrac{3}{2}x - 2(\tfrac{1}{6})\sin(6x) + \tfrac{1}{2}(\tfrac{1}{12})\sin(12x)] + C$$

$$= \tfrac{3}{8}x - \tfrac{1}{12}\sin(6x) + \tfrac{1}{96}\sin(12x) + C. \quad \square$$

If we had used the technique of Example 2 in Example 3, we would have obtained an expression in terms of $\sin(3x)$ and $\cos(3x)$. We could put the result of Example 3 in this form by using the trigonometric identities

(5) $\sin(2\theta) = 2\sin\theta\cos\theta$ ◀

(6) $\cos(2\theta) = 1 - 2\sin^2\theta.$ ◀

These identities can be derived easily from identities (1), (3), and (4) when needed.

Integrals of $\sin(ax)\sin(bx)$, $\cos(ax)\cos(bx)$, and $\sin(ax)\cos(bx)$ We saw in the last section that integrals of $\sin(ax)\sin(bx)$, $\cos(ax)\cos(bx)$, and $\sin(ax)\cos(bx)$ can be evaluated by a double integration by parts or by the method of undetermined coefficients. We can also use the trigonometric identities

(7) $\sin\theta\sin\psi = \tfrac{1}{2}[\cos(\theta - \psi) - \cos(\theta + \psi)]$ ◀

(8) $\cos\theta\cos\psi = \tfrac{1}{2}[\cos(\theta - \psi) + \cos(\theta + \psi)]$ ◀

(9) $\sin\theta\cos\psi = \tfrac{1}{2}[\sin(\theta - \psi) + \sin(\theta + \psi)].$ ◀

This technique is illustrated in the exercises.

Integrals of $\sec^n x$ or $\csc^n x$ with n a positive even integer   **Rule 8.8   To evaluate an integral of a positive even power of sec x or csc x, use the appropriate Pythagorean identity to express all but two of the powers in terms of tan x or cot x. Then use the substitution $u = \tan x$ or $u = \cot x$.**

This technique is illustrated in the exercises.

Reduction formulas

An integral of a positive integer power of $\tan x$ or $\cot x$ or of an odd positive power of $\sec x$ or $\csc x$ can be evaluated by using a REDUCTION FORMULA. This is a formula that expresses the integral of a power of a function in terms of an integral of a lower power of that function or a related function. We derive one such reduction formula in the next example. Others are derived in the exercises.

EXAMPLE 4   Derive the reduction formula

(10) $\displaystyle \int \sec^n x \, dx = \frac{1}{n-1} \sec^{n-2} x \tan x + \frac{n-2}{n-1} \int \sec^{n-2} x \, dx.$

Here n is any integer greater than 1.

Solution.   We first integrate by parts:

$$\int \sec^n x \, dx = \int \sec^{n-2} x \, \frac{d}{dx} \tan x \, dx$$

$$= \sec^{n-2} x \tan x - \int \tan x \, \frac{d}{dx} \sec^{n-2} x \, dx$$

$$= \sec^{n-2} x \tan x$$

$$- (n-2) \int \tan x \sec^{n-3} x \sec x \tan x \, dx$$

$$= \sec^{n-2} x \tan x - (n-2) \int \tan^2 x \sec^{n-2} x \, dx.$$

Then we employ the identity $\tan^2 x = \sec^2 x - 1$ to obtain

$$\int \sec^n x \, dx = \sec^{n-2} x \tan x - (n-2) \int (\sec^2 x - 1) \sec^{n-2} x \, dx$$

$$= \sec^{n-2} x \tan x - (n-2) \int \sec^n x \, dx$$

$$+ (n-2) \int \sec^{n-2} x \, dx.$$

Solving this equation for the integral of $\sec^n x$ yields formula (10). $\square$

Exercises

†*Answer provided.*
‡*Outline of solution provided.*

Evaluate the integrals in Exercises 1 through 18.

1.† $\displaystyle \int \sin^3 (4x) \, dx$

2. $\displaystyle \int \cos^3 (5x) \, dx$

3.‡ $\displaystyle \int (\sin x)^{\sqrt{2}} \cos^3 x \, dx$

4. $\displaystyle \int_0^\pi \sin^{5/2} x \cos^3 x \, dx$

5.† $\displaystyle \int \sin^4 x \, dx$

6. $\displaystyle \int \cos^4 (3x) \, dx$

7.‡ $\displaystyle\int \sin(3x)\cos x\, dx$ **8.** $\displaystyle\int \sin(3x)\sin(5x)\, dx$

9.† $\displaystyle\int \cos(7x)\cos(8x)\, dx$ **10.** $\displaystyle\int_0^1 \sin(5x)\cos(3x)\, dx$

11.‡ $\displaystyle\int \tan^2(6x)\, dx$ **12.** $\displaystyle\int \sin^3(2x)\cos^2(2x)\, dx$

13.† $\displaystyle\int x\cot^2(x^2)\, dx$ **14.** $\displaystyle\int \sin^2(7x)\, dx$

15.† $\displaystyle\int \sec^4 x\, dx$ **16.** $\displaystyle\int e^x\cot^3(e^x)\, dx$

17.† $\displaystyle\int_{\pi/4}^{\pi/2} \csc^3 x\, dx$ **18.** $\displaystyle\int \tan^3(2x)\, dx$

Derive the reduction formulas in Exercises 19 through 25.

19. $\displaystyle\int \csc^n x\, dx = -\frac{1}{n-1}\csc^{n-2}x\cot x + \frac{n-2}{n-1}\int \csc^{n-2}x\, dx$

20. $\displaystyle\int \sin^n x\, dx = -\frac{1}{n}\sin^{n-1}x\cos x + \frac{n-1}{n}\int \sin^{n-2}x\, dx$

21. $\displaystyle\int \cos^n x\, dx = \frac{1}{n}\cos^{n-1}x\sin x + \frac{n-1}{n}\int \cos^{n-2}x\, dx$

22.† $\displaystyle\int \sin^m x\cos^n x\, dx = -\frac{1}{m+n}\sin^{m-1}x\cos^{n+1}x + \frac{m-1}{m+n}\int \sin^{m-2}x\cos^n x\, dx$

23. $\displaystyle\int \tan^n x\, dx = \frac{1}{n-1}\tan^{n-1}x - \int \tan^{n-2}x\, dx$

24. $\displaystyle\int x^n\sin x\, dx = -x^n\cos x + n\int x^{n-1}\cos x\, dx$

25. $\displaystyle\int x^n\cos x\, dx = x^n\sin x - n\int x^{n-1}\sin x\, dx$

Use the reduction formulas in Example 4 and Exercises 19 through 25 to evaluate the integrals in Exercises 26 through 29.

26.‡ $\displaystyle\int \sec^6 x\, dx$ **27.** $\displaystyle\int \sin^4 x\, dx$

28.† $\displaystyle\int \tan^4 x\, dx$ **29.** $\displaystyle\int x^3\sin x\, dx$

8-5 INVERSE TRIGONOMETRIC SUBSTITUTIONS

An integral which involves integer powers of x and an integer power of one of the square roots $\sqrt{a^2 - x^2}$, $\sqrt{x^2 + a^2}$, or $\sqrt{x^2 - a^2}$ can be transformed into an integral involving integer powers of trigonometric functions and no square roots. This is done by using an ARCSINE, ARCTANGENT, or ARCSECANT SUBSTITUTION. The resulting integral then can be evaluated by one of the other techniques of integration.

Arcsine substitutions

Rule 8.9   Integrals involving integer powers of x and of $\sqrt{a^2 - x^2}$ $(a > 0)$ are transformed into integrals involving integer powers of sines and cosines by the substitution

(1) $\theta = \arcsin\left(\dfrac{x}{a}\right)$ ◀

for which

(2) $x = a \sin \theta$ ◀

(3) $dx = \dfrac{d}{d\theta}[a \sin \theta]\, d\theta = a \cos \theta\, d\theta$ ◀

(4) $\sqrt{a^2 - x^2} = \sqrt{a^2(1 - \sin^2 \theta)} = a \cos \theta.$ ◀

In deriving (4) from (1) we use the fact that the arcsine is defined to be between $-\pi/2$ and $\pi/2$, so that $\cos \theta$ is nonnegative.

Formulas (2) and (4) may be recalled by examining the case of $0 < \theta < \pi/2$ for which a is the length of the hypotenuse and x and $\sqrt{a^2 - x^2}$ are the lengths of the legs of a right triangle (Figure 8.1).

When making an arcsine substitution, we use formulas (2), (3), and (4) to express x, dx, and $\sqrt{a^2 - x^2}$ in terms of $\sin \theta$ and $\cos \theta$. Then when the integration has been performed, we use formulas (1), (2), and (4) to put the result in terms of the original variable x.

EXAMPLE 1   Use the integration formula

(5) $\displaystyle\int \cot^2 \theta\, d\theta = -\cot \theta - \theta + C$

to evaluate the integral

(6) $\displaystyle\int \dfrac{\sqrt{1 - x^2}}{x^2}\, dx.$

Solution.   We make the substitution

(7) $\theta = \arcsin x.$

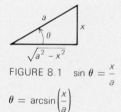

FIGURE 8.1 $\sin \theta = \dfrac{x}{a}$

$\theta = \arcsin\left(\dfrac{x}{a}\right)$

Then we have

(8) $x = \sin\theta,$ $dx = \cos\theta\,d\theta,$ $\sqrt{1 - x^2} = \cos\theta.$

Making substitutions (8) in integral (6) gives

$$\int \frac{\sqrt{1 - x^2}}{x^2}\,dx = \int \frac{\cos\theta}{\sin^2\theta}\cos\theta\,d\theta = \int \cot^2\theta\,d\theta.$$

Then we apply the integration formula (5) to obtain

$$\int \frac{\sqrt{1 - x^2}}{x^2}\,dx = \int \cot^2\theta\,d\theta = -\cot\theta - \theta + C.$$

To express the result in terms of x, we write $\cot\theta = \cos\theta/\sin\theta$ and use formulas (7) and (8). This yields

$$\int \frac{\sqrt{1 - x^2}}{x^2}\,dx = -\frac{\cos\theta}{\sin\theta} - \theta + C$$

$$= -\frac{\sqrt{1 - x^2}}{x} - \arcsin x + C. \quad \square$$

Arctangent substitutions

Rule 8.10   Integrals involving integer powers of x and of $\sqrt{x^2 + a^2}$ ($a > 0$) are transformed into integrals involving integer powers of tangents and secants by the substitution

(9) $\theta = \arctan\left(\dfrac{x}{a}\right)$ ◄

for which

(10) $x = a\tan\theta$ ◄

(11) $dx = \dfrac{d}{d\theta}[a\tan\theta]\,d\theta = a\sec^2\theta\,d\theta$ ◄

(12) $\sqrt{x^2 + a^2} = \sqrt{a^2(\tan^2\theta + 1)} = a\sec\theta.$ ◄

FIGURE 8.2 $\tan\theta = \dfrac{x}{a}$

$\theta = \arctan\left(\dfrac{x}{a}\right)$

Calculation (12) uses the fact that the arctangent is defined to lie between $-\pi/2$ and $\pi/2$, so that $\sec\theta$ is positive.

Formulas (10) and (11) may be recalled by examining the case of $\theta > 0$ for which x and a are the lengths of the legs of a right triangle and $\sqrt{x^2 + a^2}$ is the length of the hypotenuse (Figure 8.2).

To perform an arctangent substitution, use formulas (10), (11), and (12) to express the integral in terms of θ, evaluate that integral, and then use formulas (9), (10), and (12) to put the result in terms of x.

Arcsecant substitutions

Rule 8.11   Integrals involving integer powers of x and of $\sqrt{x^2 - a^2}\,(a > 0)$ are transformed into integrals involving integer powers of secants and tangents by the substitution

(13) $\theta = \text{arcsec}\left(\dfrac{x}{a}\right)$ ◄

for which

(14) $x = a \sec \theta$ ◄

(15) $dx = \dfrac{d}{d\theta}\left[a \sec \theta\right]d\theta = a \sec \theta \tan \theta\, d\theta$ ◄

(16) $\sqrt{x^2 - a^2} = \sqrt{a^2(\sec^2 \theta - 1)} = a\,|\tan \theta|.$ ◄

FIGURE 8.3 $\sec \theta = \dfrac{x}{a}$

$\theta = \text{arcsec}\left(\dfrac{x}{a}\right)$

We need the absolute value signs in (16) because, for $x < -a$, θ is between $\pi/2$ and π and has a negative tangent.

The geometric interpretation of equations (14) and (16) for the case of $x > a$ is given in Figure 8.3.

Completing the square

An integral involving integer powers of x and of $\sqrt{ax^2 + bx + c}$ can be evaluated by the process called COMPLETING THE SQUARE, followed by the application of one of the basic integration formulas or by an arcsine, arctangent, or arcsecant substitution.

Rule 8.12   To complete the square in the quadratic polynomial $ax^2 + bx + c$, factor out the coefficient of x^2. Then add and subtract the square of half the resulting coefficient of x. This leaves the polynomial expressed in the form $ax^2 + bx + c = a[x^2 + 2rx + r^2 + s] = a[(x + r)^2 + s]$ for some constants r and s.

EXAMPLE 2   Evaluate

$$\int \frac{1}{x^2 + 2x + 5}\, dx.$$

Solution.   The coefficient of x^2 in $x^2 + 2x + 5$ is 1, so we do not have to factor it out. Half of the coefficient of x is $\frac{1}{2}(2) = 1$ and its square is $1^2 = 1$, so we add and subtract 1 in the polynomial. We obtain $x^2 + 2x + 5 = x^2 + 2x + 1 - 1 + 5 = (x^2 + 2x + 1) + 4 = (x + 1)^2 + 4$. When we use this equation and the substitution $u = x + 1$, for which $du = dx$, we obtain

$$\int \frac{1}{x^2 + 2x + 5}\, dx = \int \frac{1}{(x + 1)^2 + 4}\, dx = \int \frac{1}{u^2 + 4}\, du$$

$$= \tfrac{1}{2}\arctan\left(\tfrac{1}{2}u\right) + C$$

$$= \tfrac{1}{2}\arctan\left[\tfrac{1}{2}(x + 1)\right] + C. \quad \square$$

To concentrate attention on the techniques of inverse trigonometric substitution, rather than on the techniques for evaluating the resulting integrals of trigonometric functions, we supply the following list of integrals to be used in working the exercises of this section.

(17) $\displaystyle\int \cos^2 \theta \, d\theta = \frac{1}{2} [\theta + \sin \theta \cos \theta] + C$

(18) $\displaystyle\int \sin^2 \theta \, d\theta = \frac{1}{2} [\theta - \sin \theta \cos \theta] + C$

•(19) $\displaystyle\int \sin^2 \theta \cos^2 \theta \, d\theta = \frac{1}{8} [\theta - \sin \theta \cos^3 \theta + \sin^3 \theta \cos \theta] + C$

(20) $\displaystyle\int \sin^3 \theta \cos^2 \theta \, d\theta = -\frac{1}{3} \cos^3 \theta + \frac{1}{5} \cos^5 \theta + C$

(21) $\displaystyle\int \sin^3 \theta \, d\theta = -\cos \theta + \frac{1}{3} \cos^3 \theta + C$

(22) $\displaystyle\int \sin^3 \theta \cos^4 \theta \, d\theta = -\frac{1}{5} \cos^5 \theta + \frac{1}{7} \cos^7 \theta + C$

(23) $\displaystyle\int \sec \theta \, d\theta = \ln|\sec \theta + \tan \theta| + C$

(24) $\displaystyle\int \csc \theta \, d\theta = -\ln|\csc \theta + \cot \theta| + C$

•(25) $\displaystyle\int \cos^2 \theta \csc \theta \, d\theta = -\ln|\csc \theta + \cot \theta| + \cos \theta + C$

(26) $\displaystyle\int \sec^2 \theta \, d\theta = \tan \theta + C$

•(27) $\displaystyle\int \sec^3 \theta \, d\theta = \frac{1}{2} \tan \theta \sec \theta + \frac{1}{2} \ln|\sec \theta + \tan \theta| + C$

(28) $\displaystyle\int \tan^3 \theta \sec \theta \, d\theta = \frac{1}{3} \sec^3 \theta - \sec \theta + C$

(29) $\displaystyle\int \cos \theta \, d\theta = \sin \theta + C$

(30) $\displaystyle\int \cot \theta \, d\theta = \ln|\sin \theta| + C$

(31) $\displaystyle\int \tan^2 \theta \, d\theta = \tan \theta - \theta + C$

(32) $\displaystyle\int \cot \theta \csc \theta \, d\theta = -\csc \theta + C$

(33) $\displaystyle\int \csc^2 \theta \, d\theta = -\cot \theta + C$

(34) $\displaystyle\int \cot^2 \theta \, d\theta = -\cot \theta - \theta + C$

(35) $\displaystyle\int \tan \theta \, d\theta = -\ln|\cos \theta| + C$

Exercises   Evaluate the following integrals. Give your answers in terms of x.

†Answer provided.
‡Outline of solution provided.

1.‡ $\displaystyle\int \frac{1}{x(x^2 + 16)} \, dx$ **2.** $\displaystyle\int \sqrt{x^2 + 4} \, dx$ **3.**‡ $\displaystyle\int \frac{1}{x\sqrt{9 - x^2}} \, dx$

4. $\displaystyle\int \frac{\sqrt{16 - x^2}}{x^2} \, dx$ **5.**† $\displaystyle\int x^2 \sqrt{25 - x^2} \, dx$ **6.** $\displaystyle\int \frac{x^2}{x^2 + 1} \, dx$

7.† $\displaystyle\int x^3(16 - x^2)^{3/2} \, dx$ **8.** $\displaystyle\int \frac{1}{\sqrt{x^2 + 9}} \, dx$

9.† $\displaystyle\int \frac{x^3}{\sqrt{x^2 + 1}} \, dx$ **10.** $\displaystyle\int \frac{1}{x\sqrt{x^2 + 9}} \, dx$ **11.**† $\displaystyle\int \sqrt{4 - x^2} \, dx$

12. $\displaystyle\int \frac{1}{x^2\sqrt{1 - x^2}} \, dx$ **13.**† $\displaystyle\int \frac{1}{1 - x^2} \, dx$ **14.** $\displaystyle\int \frac{1}{(4 - x^2)^{3/2}} \, dx$

15.† $\displaystyle\int \frac{x^2}{(1 - x^2)^{3/2}} \, dx$ **16.** $\displaystyle\int \frac{x^2}{\sqrt{4 - x^2}} \, dx$

17.‡ $\displaystyle\int \frac{1}{x^2 - 6x + 25} \, dx$

18. $\displaystyle\int \frac{1}{9x^2 + 6x + 2} \, dx$

19.‡ $\displaystyle\int \frac{1}{\sqrt{84 - 8x - x^2}} \, dx$ **20.** $\displaystyle\int \frac{1}{(x^2 + 1)^2} \, dx$

21.† $\displaystyle\int \frac{1}{(2x - x^2)^{3/2}} \, dx$ **22.** $\displaystyle\int \frac{1}{\sqrt{x^2 - 8x + 25}} \, dx$

8-6   INTEGRATION BY PARTIAL FRACTIONS

The method of PARTIAL FRACTIONS is an algebraic procedure for expressing a rational function as a sum of simpler rational functions that can be integrated by using the techniques discussed in Sections 8-1 through 8-5. The resulting expression is called the partial fraction decomposition of the original rational function.

We begin this section by explaining what we mean by a *proper rational function*. Next, we discuss briefly the *factorization of polynomials*, and then we show how to find partial fraction decompositions.

Proper rational functions

A rational function is said to be PROPER if it is expressed as the sum of a polynomial and of a quotient of polynomials, where the degree of the polynomial in the numerator is *less* than the degree of the polynomial in the denominator. Thus, the rational functions

$$x^2 + \frac{1}{x^2 - 4} \quad \text{and} \quad \frac{1 + x + x^2}{(x - 3)(x - 2)(x - 1)}$$

are proper, whereas the rational functions

$$\frac{x}{2x + 3} \quad \text{and} \quad x + \frac{x^3}{x^2 + x + 1}$$

are not proper.

Factorization of quadratic polynomials

The basic tool in studying the factorization of polynomials is the QUADRATIC FORMULA

$$(1) \quad x = \frac{-b \pm \sqrt{b^2 - 4ac}}{2a} \qquad \blacktriangleleft$$

which gives the solutions (or "roots") x of the quadratic equation

$$(2) \quad ax^2 + bx + c = 0. \qquad \blacktriangleleft$$

The form of the factorization of the polynomial $ax^2 + bx + c$ depends on whether the number $b^2 - 4ac$ inside the square root sign in the quadratic formula is positive, zero, or negative. The number $b^2 - 4ac$ is called the DISCRIMINANT of the equation or of the polynomial.

The case of $b^2 - 4ac > 0$. If the discriminant $b^2 - 4ac$ is positive, then the quadratic formula (1) yields two distinct real solutions, $x = r_1$ and $x = r_2$, of equation (2). In this case the polynomial $ax^2 + bx + c$ has the factorization

$$(3) \quad ax^2 + bx + c = a(x - r_1)(x - r_2). \qquad \blacktriangleleft$$

To remember this formula notice that the expression on the right is zero at $x = r_1$ and at $x = r_2$, as is the polynomial on the left; moreover, notice that if the right side were multiplied out, it would involve the term ax^2 as in the polynomial on the left.

The case of $b^2 - 4ac = 0$. If the discriminant is zero, then the quadratic formula (1) gives one real solution $x = r = -b/(2a)$ of (2). In this case the polynomial has the factorization

$$(4) \quad ax^2 + bx + c = a(x - r)^2. \qquad \blacktriangleleft$$

The case of $b^2 - 4ac < 0$. If the discriminant is negative, the numbers given by the quadratic formula are not real and (2) has *no real solutions*. Because we consider only real-valued functions, we cannot factor the polynomial $ax^2 + bx + c$. We say that it is IRREDUCIBLE.

Factorization of polynomials of degree > 2

The FUNDAMENTAL THEOREM OF ALGEBRA states that any real-valued polynomial of a real variable can, in principle, be factored into linear (first-degree) and irreducible quadratic factors. Finding the factors can be a very difficult matter, but in some cases they can be found by using the quadratic formula. This is illustrated in the exercises.

The partial fraction decomposition

We use the following procedure to find the partial fraction decompositions of rational functions.

Step 1   Express the rational function as a proper rational function if it is not already in that form. This is accomplished by long division (or by the equivalent shorthand process called "synthetic division"). The proper rational function is of the form

(5) $p(x) + \dfrac{r(x)}{q(x)}$

where $p(x)$, $r(x)$, and $q(x)$ are polynomials and the degree of $r(x)$ is less than the degree of $q(x)$. The rest of the procedure deals only with the quotient $r(x)/q(x)$ of polynomials in (5).

Step 2   Factor the denominator $q(x)$ of the quotient $r(x)/q(x)$ into linear and irreducible quadratic factors.

Step 3   Write out the partial fraction decomposition of the quotient $r(x)/q(x)$ with letters for the coefficients that are to be determined: for each linear factor $ax + b$ of $q(x)$ which appears to the kth power add the k terms

(6) $\dfrac{A}{ax + b} + \dfrac{B}{(ax + b)^2} + \cdots + \dfrac{C}{(ax + b)^k}.$ ◀

For each irreducible quadratic factor $ax^2 + bx + c$ which appears to the kth power, add the k terms

(7) $\dfrac{Ax + B}{ax^2 + bx + c} + \dfrac{Cx + D}{(ax^2 + bx + c)^2} + \cdots$

$+ \dfrac{Ex + F}{(ax^2 + bx + c)^k}.$ ◀

Step 4   Combine the sum of terms from Step 3 into a single quotient of polynomials by taking their least common denominator, which is the denominator $q(x)$ of the original rational function. This gives an equation of the form

(8) $\dfrac{r(x)}{q(x)} = \dfrac{s(x)}{q(x)}$

where the polynomial $s(x)$ involves the unknown coefficients from Step 3.

Step 5   For equation (8) to be valid for all x in the domain of the rational function, the numerators must be equal for all x. Solve the equation

(9) $r(x) = s(x)$

for the coefficients in the partial fraction decomposition. We do this by taking special values for x or by comparing coefficients of powers of x, as we will see in the examples and exercises.

Step 6   Substitute the values of the coefficients into the partial fraction decomposition of $r(x)/q(x)$ from Step 3 and add the polynomial $p(x)$ to obtain the decomposition of the original rational function $p(x) + r(x)/q(x)$.

Step 1 of this procedure is illustrated in the next example.

EXAMPLE 1   Express the rational function $(x^3 + 2)/(x^2 + 3x + 2)$ as a proper rational function.

Solution.   We divide the denominator $x^2 + 3x + 2$ into the numerator. We carry out the calculation until the remainder is a polynomial of degree less than that of the denominator:

$$
\begin{array}{r}
x - 3 \\
x^2 + 3x + 2 \overline{)\, x^3 + 0x^2 + 0x + 2} \\
\underline{x^3 + 3x^2 + 2x} \\
-3x^2 - 2x + 2 \\
\underline{-3x^2 - 9x - 6} \\
7x + 8.
\end{array}
$$

(We write $x^3 + 2$ as $x^3 + 0x^2 + 0x + 2$ to leave room for the x and x^2 terms in the calculations.) Thus, dividing $x^2 + 3x + 2$ into $x^3 + 2$ yields $x - 3$ with a remainder of $7x + 8$:

$$\frac{x^3 + 2}{x^2 + 3x + 2} = x - 3 + \frac{7x + 8}{x^2 + 3x + 2}.$$

The expression on the right of the last equation is a proper rational function because the degree of the numerator is 1 and the degree of the denominator is 2.   □

Our second example is an illustration of Step 3 of the procedure.

EXAMPLE 2   Give the form of the partial fraction decomposition of the proper rational function

(10) $$\frac{x^4 - 5}{x(x + 1)^2(x^2 + 4)^2}.$$

Solution.   The rational function (10) is proper because the degree of its numerator is 4 and the degree of its denominator is 7. The denominator is factored because the quadratic polynomial $x^2 + 4$ is irreducible. The partial fraction decomposition is of the form

$$\frac{x^4 - 5}{x(x + 1)^2(x^2 + 4)^2} = \frac{A}{x} + \frac{B}{x + 1} + \frac{C}{(x + 1)^2}$$
$$+ \frac{Dx + E}{x^2 + 4} + \frac{Fx + G}{(x^2 + 4)^2}. \quad \square$$

Determining the coefficients in the decomposition

Finding the coefficients in the partial fraction decomposition is simplified if the denominator of the rational function has linear factors. As the next example shows, we can easily determine one coefficient for each distinct linear factor. We then compare the coefficients of powers of x to obtain simultaneous equations for any remaining coefficients.

EXAMPLE 3   Give the partial fraction decomposition of

$$\frac{3x^2 - 4x + 1}{(x - 2)(x^2 + 1)}.$$

Solution.
Step 1   Unnecessary. The rational function is proper.

Step 2   Unnecessary. The denominator is already factored.

Step 3   The decomposition has the form

(11) $$\frac{3x^2 - 4x + 1}{(x - 2)(x^2 + 1)} = \frac{A}{x - 2} + \frac{Bx + C}{x^2 + 1}.$$

Step 4   When we take the common denominator on the right of (11), we obtain

(12) $$\frac{3x^2 - 4x + 1}{(x - 2)(x^2 + 1)} = \frac{A(x^2 + 1) + (Bx + C)(x - 2)}{(x - 2)(x^2 + 1)}.$$

Step 5   Equating the numerators of (12) gives for $x \neq 2$

(13) $$3x^2 - 4x + 1 = A(x^2 + 1) + (Bx + C)(x - 2).$$

This equation must also hold at $x = 2$ because both sides are continuous functions. We find one of the coefficients by setting $x = 2$ so that the linear factor $x - 2$ is zero. We obtain $3(2)^2 - 4(2) + 1 = A(2^2 + 1)$ or $12 - 8 + 1 = 5A$, which gives the value $A = 1$ for A. Next, we put $A = 1$ in (13), multiply out the products, and collect the coefficients of powers of x on the right side. This gives

(14)
$$\begin{aligned} 3x^2 - 4x + 1 &= (x^2 + 1) + (Bx + C)(x - 2) \\ &= x^2 + 1 + Bx^2 + Cx - 2Bx - 2C \\ &= (1 + B)x^2 + (C - 2B)x + (1 - 2C). \end{aligned}$$

In order for equation (14) to hold, the coefficients of x^2 and of x and the constant terms on the two sides of the equation must be equal. This gives the three equations

(15) $1 + B = 3,$ $C - 2B = -4,$ $1 - 2C = 1$

for the two coefficients B and C. (There are three equations in two unknowns because we have already found the coefficient A.) The first of equations (15) implies that $B = 2$. The third shows that $C = 0$. The second equation serves as a check on the calculations.

Step 6   Substituting the values $A = 1$, $B = 2$, and $C = 0$ in equation (11) gives the result

(16) $\dfrac{3x^2 - 4x + 1}{(x - 2)(x^2 + 1)} = \dfrac{1}{x - 2} + \dfrac{2x}{x^2 + 1}.$ $\square$

Integrating the rational function   Once the partial fraction decomposition of the rational function has been determined, the terms in it can be integrated by the techniques of previous sections.

EXAMPLE 4   Evaluate

$$\int \frac{3x^2 - 4x + 1}{(x - 2)(x^2 + 1)}\, dx.$$

Solution.   We found the partial fraction decomposition of the integrand in Example 3. We have

$$\int \frac{3x^2 - 4x + 1}{(x - 2)(x^2 + 1)}\, dx = \int \frac{1}{x - 2}\, dx + \int \frac{2x}{x^2 + 1}\, dx.$$

We make the substitutions $u = x - 2$, $du = dx$, $v = x^2 + 1$, $dv = 2x\, dx$ to obtain

$$\int \frac{3x^2 - 4x + 1}{(x - 2)(x^2 + 1)}\, dx = \int \frac{1}{u}\, du + \int \frac{1}{v}\, dv = \ln|u| + \ln|v| + C$$

$$= \ln|x - 2| + \ln(x^2 + 1) + C.\quad \square$$

A special procedure   **Rule 8.13   To evaluate an integral**

$$\int \frac{Ax + B}{(ax^2 + bx + c)^k}\, dx$$

where k is a positive integer, $ax^2 + bx + c$ is irreducible, and A is not zero, add and subtract a constant in the numerator if necessary to express the integral in the form

$$C \int \frac{1}{u^k}\, du + D \int \frac{1}{(ax^2 + bx + c)^k}\, dx$$

with $u = ax^2 + bx + c$.

The second integral can then be evaluated by completing the square in the quadratic polynomial and by using an arctangent (if $k = 1$) or an arctangent substitution (if $k > 1$).

EXAMPLE 5   Evaluate

$$\int \frac{x}{x^2 + 6x + 10}\, dx.$$

Solution.   We want to make the substitution $u = x^2 + 6x + 10$, $du = (2x + 6)\, dx$, so we construct $2x + 6$ from x by multiplying and dividing by 2 and then adding and subtracting 6:

$$\int \frac{x}{x^2 + 6x + 10}\, dx = \tfrac{1}{2} \int \frac{2x}{x^2 + 6x + 10}\, dx$$

$$= \tfrac{1}{2} \int \frac{2x + 6 - 6}{x^2 + 6x + 10}\, dx$$

$$= \tfrac{1}{2} \int \frac{2x + 6}{x^2 + 6x + 10}\, dx - \tfrac{1}{2} \int \frac{6}{x^2 + 6x + 10}\, dx$$

$$= \tfrac{1}{2} \int \frac{1}{u}\, du - 3 \int \frac{1}{x^2 + 6x + 10}\, dx.$$

To evaluate the second integral, we complete the square and make the substitution $v = x + 3$, $dv = dx$. We obtain

$$\int \frac{x}{x^2 + 6x + 10}\, dx = \tfrac{1}{2} \int \frac{1}{u}\, du - 3 \int \frac{1}{(x + 3)^2 + 1}\, dx$$

$$= \tfrac{1}{2} \int \frac{1}{u}\, du - 3 \int \frac{1}{v^2 + 1}\, dv$$

$$= \tfrac{1}{2} \ln|u| - 3 \arctan v + C$$

$$= \tfrac{1}{2} \ln(x^2 + 6x + 10)$$
$$\qquad - 3 \arctan(x + 3) + C. \quad \square$$

Exercises

Give the partial fraction decompositions of rational functions in Exercises 1 through 22.

1.‡ $\dfrac{13x + 6}{x(x + 2)(x + 3)}$

2.† $\dfrac{2x^2 + x - 22}{x^2 - 4}$

3.‡ $\dfrac{x - 2}{(x + 1)^2}$

4. $\dfrac{x^2 + 3x + 1}{x^3 + 2x^2 + x}$

5.† $\dfrac{x^5}{x^2 + 16}$

6. $\dfrac{3x^3 + 5x^2 - x - 1}{x^3 + x^2}$

7.† $\dfrac{x}{(x + 1)^3}$

8. $\dfrac{2x^2 + 1}{x^3 + x}$

9.‡ $\dfrac{x^2 + x + 1}{2x^3 - x^2 - x}$

10. $\dfrac{4x}{x^2 - 4x - 12}$

11.‡ $\dfrac{x^3}{(x^2 + 1)^2}$

12. $\dfrac{x^2 - 2x}{(x - 1)(x^2 + 2)}$

13.† $\dfrac{x^4 + 3x^3 + 1}{x^3 + 2x^2 + 5x}$

14. $\dfrac{x^3}{x^2 + 2x - 3}$

15.† $\dfrac{3x}{x^2 + 4x + 4}$

16. $\dfrac{2x^2}{x^2 - 3x + 5}$

17.† $\dfrac{2x + 1}{x^4 + 2x^3 + x^2}$

18. $\dfrac{x^3}{x^3 + x^2 + x}$

19.† $\dfrac{3x^2 + 12x + 2}{(x^2 + 4)^2}$

20. $\dfrac{x^2 + 2x + 2}{(x + 1)(x + 2)(x + 3)}$

21.† $\dfrac{2x^3 - 4x^2 + 6}{x(x - 1)(x - 2)(x - 3)}$

22. $\dfrac{3x^3 - 3x - 4}{x^4 + 3x^3 + 4x^2}$

Evaluate the integrals in Exercises 23 through 46.

23.† $\displaystyle\int \dfrac{5x - 1}{x^2 + x}\,dx$

24. $\displaystyle\int \dfrac{2x + 3}{x^2 - 1}\,dx$

25.† $\displaystyle\int \dfrac{2x + 3}{x^2 + 3x + 2}\,dx$

26. $\displaystyle\int \dfrac{x - 14}{x^2 - 4}\,dx$

27.‡ $\displaystyle\int \dfrac{x + 3}{x^2 + 4}\,dx$

28.† $\displaystyle\int \dfrac{x^2 + 13x}{x^2 + 7}\,dx$

29.† $\displaystyle\int \dfrac{x^2 + 3x + 1}{x^3 + 2x^2 + x}\,dx$

30. $\displaystyle\int \dfrac{x + 4}{x^2 - 2x}\,dx$

31.‡ $\displaystyle\int \dfrac{6x}{(x^2 + 1)^3}\,dx$

32. $\displaystyle\int \dfrac{2x^2 - x - 1}{x^3 + x^2}\,dx$

33.† $\displaystyle\int_1^5 \dfrac{x^2 - 3}{x^3 + x}\,dx$

34. $\displaystyle\int \dfrac{x^3 + 4x + 4}{x^2 + 4}\,dx$

35.† $\displaystyle\int \dfrac{x + 1}{x^2 + 2x + 10}\,dx$

36. $\displaystyle\int_0^3 \dfrac{x^4 + 3x^2 + 3}{x^2 + 1}\,dx$

37.‡ $\displaystyle\int \dfrac{x}{x^2 + 2x + 5}\,dx$

38. $\displaystyle\int \dfrac{x - 2}{(x + 1)^2}\,dx$

39.† $\displaystyle\int_{-3}^{-2} \dfrac{3x^2 - 5x - 2}{x^3 - x}\,dx$

40. $\displaystyle\int_0^1 \dfrac{x^3}{x^2 - 9}\,dx$

41.† $\displaystyle\int \dfrac{x^2 - 2x + 1}{(x^2 + 1)^2}\,dx$

42. $\displaystyle\int \dfrac{x}{x^2 - 2x + 10}\,dx$

43.† $\displaystyle\int_0^{10} \dfrac{x^2 - 1}{x^2 + 1}\,dx$

44. $\displaystyle\int_0^1 \dfrac{x^5 - 35x}{x^2 + 6}\,dx$

45.[†] $\displaystyle\int_{-10}^{-1} \frac{2x^2 - 9x + 4}{x^3 - 4x^2 + 4x} dx$ **46.** $\displaystyle\int_0^5 \frac{7x + 23}{x^2 + 7x + 10} dx$

47. Sketch the parabola $y = ax^2 + bx + c$ for values of a, b, and c with $b^2 - 4ac$ positive, negative, and zero to show the geometric significance of the quadratic formula.

48.[†] Derive the quadratic formula by completing the square in the equation $ax^2 + bx + c = 0$.

8-7   INTEGRALS OF RATIONAL FUNCTIONS OF SINES AND COSINES (Optional)

Rule 8.14   An integral of a rational function of sin x and cos x is transformed into an integral of a rational function of u by the substitution

(1) $u = \tan\left(\dfrac{x}{2}\right)$ ◄

for which

(2) $\sin x = \dfrac{2u}{1 + u^2}$ ◄

(3) $\cos x = \dfrac{1 - u^2}{1 + u^2}$ ◄

(4) $dx = \dfrac{2}{1 + u^2} du.$ ◄

The integral of the rational function of u may then be evaluated by the method of partial fractions.

Derivation of formulas (2) through (4)   To obtain formula (2), we use the trigonometric identities

(5) $\sin x = 2 \sin\left(\dfrac{x}{2}\right)\cos\left(\dfrac{x}{2}\right), \qquad 1 + \tan^2\left(\dfrac{x}{2}\right) = \sec^2\left(\dfrac{x}{2}\right)$

and write

$$\sin x = 2 \sin\left(\frac{x}{2}\right)\cos\left(\frac{x}{2}\right) = \frac{2 \tan\left(\dfrac{x}{2}\right)}{\sec^2\left(\dfrac{x}{2}\right)} = \frac{2u}{1 + u^2}.$$

Using (5) and the identity $\cos x = 2 \cos^2(x/2) - 1$, we obtain

$$\cos x = 2 \cos^2\left(\frac{x}{2}\right) - 1 = \frac{2}{\sec^2\left(\dfrac{x}{2}\right)} - 1 = \frac{2 - \sec^2\left(\dfrac{x}{2}\right)}{\sec^2\left(\dfrac{x}{2}\right)}$$

$$= \frac{1 - \tan^2\left(\dfrac{x}{2}\right)}{1 + \tan^2\left(\dfrac{x}{2}\right)} = \frac{1 - u^2}{1 + u^2}$$

which is (3).

To derive (4), we solve (1) for x. This gives $x = 2 \arctan u$ and then

$$dx = \frac{d}{du}[2 \arctan u]\, du = \frac{2}{1 + u^2}\, du.$$

EXAMPLE 1 Use substitution (1) to evaluate $\int \sec x \, dx$.

Solution. Formulas (3) and (4) give

$$\int \sec x \, dx = \int \frac{1}{\cos x}\, dx = \int \left[\frac{1 + u^2}{1 - u^2}\right]\left[\frac{2}{1 + u^2}\, du\right]$$

$$= \int \frac{2}{1 - u^2}\, du = \int \left[\frac{1}{1 + u} + \frac{1}{1 - u}\right] du$$

$$= \ln|1 + u| - \ln|1 - u| + C = \ln\left|\frac{1 + u}{1 - u}\right| + C.$$

To put the result in the form of Example 1 of Section 8.1, we multiply the numerator and denominator inside the logarithm by $1 + u$. This yields

$$\int \sec x \, dx = \ln\left|\frac{1 + u}{1 - u}\right| + C = \ln\left|\frac{1 + 2u + u^2}{1 - u^2}\right| + C$$

$$= \ln\left|\frac{1 + u^2}{1 - u^2} + \frac{2u}{1 - u^2}\right| + C = \ln\left|\frac{1}{\cos x} + \frac{\sin x}{\cos x}\right| + C$$

$$= \ln|\sec x + \tan x| + C. \quad \square$$

Exercises

†*Answer provided.*
‡*Outline of solution provided.*

Evaluate the integrals in Exercises 1 through 8.

1.‡ $\displaystyle\int \frac{1}{\tan x - 1}\, dx$

2. $\displaystyle\int \frac{1}{1 + \cos x}\, dx$

3.‡ $\displaystyle\int \frac{1}{5 + 3 \cos x}\, dx$

4. $\displaystyle\int \frac{1}{\tan x + \sin x}\, dx$

5.† $\displaystyle\int \frac{1}{\tan x - \sin x}\, dx$

6. $\displaystyle\int \frac{1 - \cos x}{1 + \cos x}\, dx$

7.† $\displaystyle\int_0^{\pi/3} \frac{1}{5 \sec x - 3}\, dx$

8. $\displaystyle\int_0^{\pi/4} \frac{1}{1 - \sin x}\, dx$

8-8   THE USE OF INTEGRAL TABLES   (Optional)

The use of a table of integrals, with the method of substitution or other integration techniques as required, can improve efficiency and accuracy in the evaluation of integrals. This section provides drill in the use of the integral table at the back of this book.

EXAMPLE 1   Use the table of integrals to evaluate

$$\int_1^3 \frac{1}{x(3x+6)}\,dx.$$

Solution.   Formula (19) in the table reads

$$\int \frac{1}{x(ax+b)}\,dx = \frac{1}{b}\ln\left|\frac{x}{ax+b}\right| + C.$$

With $a = 3$ and $b = 6$ it gives

$$\int_1^3 \frac{1}{x(3x+6)}\,dx = \left[\int \frac{1}{x(3x+6)}\,dx\right]_1^3 = \left[\tfrac{1}{6}\ln\left|\frac{x}{3x+6}\right|\right]_1^3$$

$$= \left[\tfrac{1}{6}\ln\left|\frac{3}{3(3)+6}\right|\right] - \left[\tfrac{1}{6}\ln\left|\frac{1}{3(1)+6}\right|\right]$$

$$= \tfrac{1}{6}[\ln(\tfrac{3}{15}) - \ln(\tfrac{1}{9})] = \tfrac{1}{6}\ln(\tfrac{9}{5}). \quad \square$$

EXAMPLE 2   Use the table of integrals to evaluate

$$\int x^5\, e^{x^2}\, dx.$$

Solution.   There are no integrals involving e^{x^2} in the table, so we try the substitution $u = x^2$, $du = 2x\,dx$. We obtain

$$\int x^5\, e^{x^2}\, dx = \tfrac{1}{2}\int (x^2)^2\, e^{x^2}\,(2x\,dx) = \tfrac{1}{2}\int u^2\, e^u\, du.$$

Repeated application of formula (45) in the table gives

$$\int u^2\, e^{au}\, du = \frac{1}{a^3}\, e^{au}(a^2u^2 - 2au + 2) + C.$$

Setting $a = 1$, we obtain

$$\int x^5\, e^{x^2}\, dx = \tfrac{1}{2}\int u^2\, e^u\, du = \tfrac{1}{2}e^u(u^2 - 2u + 2) + C$$

$$= \tfrac{1}{2}e^{x^2}(x^4 - 2x^2 + 2) + C. \quad \square$$

Exercises

†*Answer provided.*
‡*Outline of solution provided.*

Use Table VIII in the back of the text to evaluate the integrals in Exercises 1 through 44 and to solve Exercises 45 through 51.

1.‡ $\displaystyle\int_0^{12} \sqrt{169 - x^2}\, dx$

2.† $\displaystyle\int \frac{1}{4 - x^2}\, dx$

3. $\displaystyle\int_{-1}^{1} \frac{1}{9 - x^2}\, dx$

4.‡ $\displaystyle\int x \sqrt{1 - x^4}\, dx$

5.‡ $\displaystyle\int \frac{e^x}{1 - e^{2x}}\, dx$

6.† $\displaystyle\int_0^{10} \sqrt{x^2 + 144}\, dx$

7. $\displaystyle\int_0^4 \frac{1}{(25 - x^2)^{3/2}}\, dx$

8. $\displaystyle\int \frac{1}{(2x + 3)(4x + 5)}\, dx$

9.† $\displaystyle\int x \sin^3(x^2)\, dx$

10. $\displaystyle\int_{-1}^{1} \tan^2 x\, dx$

11.† $\displaystyle\int \sin x \cos(5x)\, dx$

12.‡ $\displaystyle\int_0^{10} \frac{1}{9 + (2x + 5)^2}\, dx$

13. $\displaystyle\int_{-1/4}^{1/4} \sec(\pi x)\, dx$

14.† $\displaystyle\int_0^{10} \frac{x}{x^4 + 100}\, dx$

15.† $\displaystyle\int_0^{10} \frac{x^3}{x^4 + 100}\, dx$

16.† $\displaystyle\int \frac{\sin x}{\cos^2 x + \cos x}\, dx$

17. $\displaystyle\int \arctan(3x)\, dx$

18. $\displaystyle\int_0^5 \frac{1}{\sqrt{169 - x^2}}\, dx$

19. $\displaystyle\int_0^5 \frac{x}{\sqrt{169 - x^2}}\, dx$

20.† $\displaystyle\int \frac{1 + e^x}{x + e^x}\, dx$

21. $\displaystyle\int \frac{1}{x^2(x + 1)}\, dx$

22. $\displaystyle\int \frac{x}{\sqrt{x + 1}}\, dx$

23.‡ $\displaystyle\int \frac{1}{(x^2 + 1)^2}\, dx$

24.† $\displaystyle\int x\, 3^{x^2}\, dx$

25.† $\displaystyle\int e^x \arcsin(e^x)\, dx$

26.† $\displaystyle\int_0^{\pi} \cos x \cos(2x)\, dx$

27. $\displaystyle\int_0^{2\pi} \cos(2x) \sin(2x)\, dx$

28.† $\displaystyle\int x^2 \sin x\, dx$

29.† $\displaystyle\int \frac{\csc^2(\sqrt{x})}{\sqrt{x}}\, dx$

30. $\displaystyle\int \sec(6x + 7)\, dx$

31. $\displaystyle\int \sin^3\left(\frac{x}{3}\right) dx$

32. $\displaystyle\int x \sin(2x)\, dx$

33. $\displaystyle\int x \sin(x^2)\, dx$

34.† $\displaystyle\int \ln(\sin x) \cos x\, dx$

35. $\displaystyle\int_0^1 \sin(2x + 3)\, dx$

36. $\displaystyle\int_{-4}^{1} \frac{1}{\sqrt{5 - x}}\, dx$

37.‡ $\displaystyle\int_1^3 \frac{1}{5 - 10x}\, dx$ **38.**† $\displaystyle\int \frac{1}{x^2 + x - 12}\, dx$

39.† $\displaystyle\int \frac{1}{x^2 - 4x + 13}\, dx$ **40.** $\displaystyle\int \frac{1}{x^2 - 4x + 4}\, dx$

41. $\displaystyle\int_1^{16} \frac{1}{\sqrt{x}}\, e^{\sqrt{x}}\, dx$ **42.**† $\displaystyle\int_0^2 \cos^2 x\, dx$

43.† $\displaystyle\int_0^5 \frac{1}{\sqrt{x^2 + 144}}\, dx$ **44.** $\displaystyle\int_2^{10} \frac{x}{1 - x^2}\, dx$

45.† Find the length of the parabola $y = x^2$ from $x = 0$ to $x = 3$.

46. Sketch the region bounded by $y = xe^x$, the x-axis, and the line $x = 1$ and compute its area.

47. A rod extends from $x = 1$ to $x = 10$ (inches) on an x-axis and its density at x is $\ln x$ pounds per inch. How much does it weigh and where is its center of gravity?

48.† Sketch the region bounded by $y = 1/(x^2 - 4)$, the x-axis, and the lines $x = \pm 1$. Compute its area.

49. Sketch the region R bounded by $y = 1/(x^2 + 1)$, the x-axis, and the lines $x = \pm 1$. Compute the volumes of the solids generated by rotating R
a. about the y-axis and   **b.** about the x-axis.

50.† Water is flowing out of a pipe at the rate of arctan t gallons per minute at time t for $0 \le t \le 60$. How much water flows out of the pipe in that hour?

51. Derive the formula for the area of the circle $x^2 + y^2 \le r^2$ by evaluating an integral.

8-9 APPROXIMATE INTEGRATION (Optional)

The integral

(1) $\displaystyle\int_0^1 e^{-x^2}\, dx$

which arises in probability theory, is one of many integrals that cannot be evaluated by the techniques of integration we have been studying. We have to rely on techniques of APPROXIMATE INTEGRATION to find approximate numerical values for such integrals. We describe in this section three of these techniques known as THE MIDPOINT RULE, THE TRAPEZOID RULE, and SIMPSON'S RULE. In each of these rules we obtain an approximation of the integral

(2) $\displaystyle\int_a^b f(x)\, dx$

from a partition

(3) $a = x_0 < x_1 < x_2 < x_3 < \cdots < x_{N-1} < x_N = b$

of the interval of integration into N equal subintervals of length Δx.

The midpoint rule

With the midpoint rule, we approximate the integral (2) by the Riemann sum

(4) $\displaystyle\sum_{j=1}^{N} f(\tfrac{1}{2}[x_j + x_{j-1}]) \, \Delta x$ ◄

which is obtained by evaluating the integrand $f(x)$ at the midpoint $\frac{1}{2}[x_j + x_{j-1}]$ of each subinterval of the partition. This amounts to approximating the regions between the graph of f and the x-axis by rectangles of the type shown in Figure 8.4.

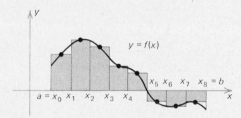

FIGURE 8.4

The trapezoid rule

With the trapezoid rule, we approximate the regions whose areas give the integral by trapezoids with corners at the points $(x_j, f(x_j))$ (Figure 8.5). This leads to the approximation

(5) $[\tfrac{1}{2} f(x_0) + f(x_1) + f(x_2) + f(x_3) + \ldots + f(x_{N-1})$
         $+ \tfrac{1}{2} f(x_N)] \, \Delta x$ ◄

of the integral (2) (see Exercise 11).

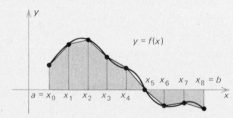

FIGURE 8.5

Simpson's rule

With Simpson's rule, we take a partition into an even number $N = 2M$ of subintervals (M a positive integer). We then approximate the graph of f by M segments of parabolas, as shown in Figure 8.6. The jth parabola is the parabola through the points $(x_{2j-2}, f(x_{2j-2}))$, $(x_{2j-1}, f(x_{2j-1}))$, and $(x_{2j}, f(x_{2j}))$ on the graph of f. This leads to the approximation

(6) $\tfrac{1}{3}[f(x_0) + 4f(x_1) + 2f(x_2) + 4f(x_3) + 2f(x_4)$
         $+ \ldots + 4f(x_{2M-3}) + 2f(x_{2M-2}) + 4f(x_{2M-1})$
         $+ f(x_{2M})] \, \Delta x$ ◄

(see Exercise 12). In this formula the value $f(x_j)$ is multiplied by 4 for every odd integer j and is multiplied by 2 for every even integer j other than the first and the last.

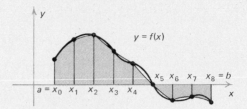

FIGURE 8.6

EXAMPLE 1   Find the approximate values of the integral

$$\int_0^1 \sqrt{x}\, dx$$

which are given (a) by the midpoint rule, (b) by the trapezoid rule, and (c) by Simpson's rule with $N = 4$.

Solution.   The partition of the interval of integration into four equal subintervals is $0 < \frac{1}{4} < \frac{1}{2} < \frac{3}{4} < 1$. Each subinterval is of length $\Delta x = \frac{1}{4}$.

(a) The midpoints of the subintervals are $\frac{1}{8}, \frac{3}{8}, \frac{5}{8},$ and $\frac{7}{8}$. The midpoint rule gives

$$\int_0^1 \sqrt{x}\, dx \approx [\sqrt{\tfrac{1}{8}} + \sqrt{\tfrac{3}{8}} + \sqrt{\tfrac{5}{8}} + \sqrt{\tfrac{7}{8}}]\tfrac{1}{4}$$

$$= \frac{1}{4\sqrt{8}}[\sqrt{1} + \sqrt{3} + \sqrt{5} + \sqrt{7}] \approx 0.6730.$$

(b) The trapezoid rule gives

$$\int_0^1 \sqrt{x}\, dx \approx [\tfrac{1}{2}\sqrt{0} + \sqrt{\tfrac{1}{4}} + \sqrt{\tfrac{1}{2}} + \sqrt{\tfrac{3}{4}} + \tfrac{1}{2}\sqrt{1}]\tfrac{1}{4}$$

$$= \tfrac{1}{8}[0 + 1 + \sqrt{2} + \sqrt{3} + 1] \approx 0.6434.$$

(c) Simpson's rule gives

$$\int_0^1 \sqrt{x}\, dx \approx \tfrac{1}{3}[\sqrt{0} + 4\sqrt{\tfrac{1}{4}} + 2\sqrt{\tfrac{1}{2}} + 4\sqrt{\tfrac{3}{4}} + \sqrt{1}]\tfrac{1}{4}$$

$$= \tfrac{1}{12}[0 + 2 + \sqrt{2} + \sqrt{12} + 1] \approx 0.6565. \quad \square$$

Notice that each approximation in Example 1 is reasonably close to the exact value

$$\int_0^1 \sqrt{x}\, dx = [\tfrac{2}{3}x^{3/2}]_0^1 = \tfrac{2}{3} = 0.6666 \ldots$$

of the integral.

Exercises

†*Answer provided.*
‡*Outline of solution provided.*

Use   **a.** the midpoint rule,   **b.** the trapezoid rule, and   **c.** Simpson's rule to give approximate values of the integrals in Exercises 1 through 10 with the indicated number of subintervals in the partitions.

1.‡ $\int_0^1 e^x \, dx$; $N = 4$ **2.** $\int_0^1 x^2 \, dx$; $N = 4$

3.† $\int_0^2 \sin x \, dx$; $N = 4$ **4.** $\int_0^1 e^{-x^2} \, dx$; $N = 4$

5.† $\int_0^4 \sinh x \, dx$; $N = 4$ **6.** $\int_1^4 \ln x \, dx$; $N = 6$

7.† $\int_0^1 \tanh x \, dx$; $N = 8$ **8.** $\int_1^4 \log_{10} x \, dx$; $N = 6$

9.† $\int_{-4}^2 e^{-x} \, dx$; $N = 6$ **10.** $\int_0^3 \cos \sqrt{x} \, dx$; $N = 6$

11.† Derive the trapezoid rule (5).

12.† Derive Simpson's rule (6).

13. Use the fact that integral (1) cannot be evaluated exactly by techniques of this chapter to show that $\int_0^1 x^n e^{-x^2} \, dx$ cannot be evaluated exactly by those techniques for n an even positive integer but can be evaluated for n an odd positive integer.

If you have access to a computer or electronic calculator, use Simpson's rule to give approximate values of the integrals in Exercises 14 through 19 with   **a.** 10 and   **b.** 20 subintervals in the partitions.

14.† $\int_1^2 \sin(\ln x) \, dx$ **15.** $\int_0^1 \frac{1}{\sqrt{\cos x}} \, dx$

16.† $\int_0^\pi \sqrt[3]{\sin x} \, dx$ **17.** $\int_0^{10} (\arctan x)^3 \, dx$

18.† $\int_{-2}^0 \cos(\sinh x) \, dx$ **19.** $\int_0^1 [\sin(e^x) - e^{\sin x}] \, dx$

MISCELLANEOUS EXERCISES

†*Answer provided.*
‡*Outline of solution provided.*

Part I   Describe integration procedures which could be used to evaluate the integrals in Exercises 1 through 49.
Part II   Evaluate the integrals in Exercises 1 through 49.

1.‡ $\int \frac{1}{x^3 - 8} \, dx$ (Note, $2^3 = 8$.)   **2.** $\int x^2 \cos(x^3) \, dx$

3.† $\int \frac{(\ln x)^3}{x} \, dx$ **4.** $\int x^3 \ln x \, dx$

5.† $\int x^2 \sin(3x) \, dx$ **6.** $\int e^{37x} \, dx$

7.† $\displaystyle\int \arcsin(3x)\,dx$

8. $\displaystyle\int \frac{\arctan x}{x^2+1}\,dx$

9.† $\displaystyle\int (\sin x)\,e^{4\cos x}\,dx$

10. $\displaystyle\int x\,e^{-x}\,dx$

11.‡ $\displaystyle\int \frac{x^5}{x^2+1}\,dx$

12. $\displaystyle\int \frac{x+1}{(x^2+2x+7)^3}\,dx$

13.† $\displaystyle\int \cos^2(4x)\,dx$

14. $\displaystyle\int x\,\tan^2(x^2)\,dx$

15.† $\displaystyle\int \sin^5 x\,\cos^3 x\,dx$

16. $\displaystyle\int (2^x+x^2)\,dx$

17.‡ $\displaystyle\int \arctan(x-1)\,dx$

18. $\displaystyle\int \sec(5x)\,dx$

19.† $\displaystyle\int x\,\tan(x^2)\,dx$

20. $\displaystyle\int \frac{1}{x^2-x}\,dx$

21.† $\displaystyle\int \frac{\cos x}{1+\sin^2 x}\,dx$

22. $\displaystyle\int \sin(3x)[5-\cos(3x)]^{4/9}\,dx$

23.† $\displaystyle\int \cos^3(5x)\,dx$

24. $\displaystyle\int \frac{1}{x\ln(3x)}\,dx$

25.† $\displaystyle\int x\,\sin(3x)\,dx$

26. $\displaystyle\int x\sqrt{9-x^2}\,dx$

27.† $\displaystyle\int \frac{x}{(1+x^2)^2}\,dx$

28. $\displaystyle\int \frac{1}{\sqrt{4+x^2}}\,dx$

29.‡ $\displaystyle\int x^3(1-x^2)^{5/2}\,dx$

30. $\displaystyle\int \frac{x^2+3}{x(x^2+x-2)}\,dx$

31.† $\displaystyle\int \frac{x^4+3x^2+1}{x(x^2+1)^2}\,dx$

32. $\displaystyle\int \frac{x^2+1}{(x+1)(x^2-4)}\,dx$

33.† $\displaystyle\int x^2\,\csc^2(x^3)\,dx$

34. $\displaystyle\int (\sinh(3x))^5\,\cosh(3x)\,dx$

35.† $\displaystyle\int x\,5^{x^2}\,dx$

36. $\displaystyle\int \frac{1}{x^2-2x+17}\,dx$

37.‡ $\displaystyle\int \frac{x}{x^2-2x+17}\,dx$

38. $\displaystyle\int \sqrt{9-x^2}\,dx$

39.† $\displaystyle\int x(9-x^2)^{2/5}\,dx$

40. $\displaystyle\int \frac{1}{\sqrt{16+x^2}}\,dx$

41.† $\displaystyle\int \frac{x^2}{\sqrt{1-x^2}}\,dx$

42. $\displaystyle\int \frac{\sin x}{1+\cos x}\,dx$

43.† $\displaystyle\int \frac{e^{\sqrt{x}}}{\sqrt{x}}\,dx$

44. $\displaystyle\int \cos(4x-3)\,dx$

45.[†] $\int x^2 \sqrt{x+1} \, dx$ **46.** $\int \dfrac{x^3+1}{x^3-x^2} \, dx$

47.[†] $\int \dfrac{\sin^3 x}{1+\cos^2 x} \, dx$ **48.** $\int \dfrac{x^4+2x^2+x+1}{(x^2+1)^2} \, dx$

49.[†] $\int \cosh(1-3x) \, dx$

50. Factor each of the following polynomials into real linear and irreducible quadratic factors.   **a.**[†] $4x^2+6x-2$
   b.[‡] x^3+3x^2+6x+8 (This is zero at $x=-2$.)
   c.[†] $x^3-6x^2-11x-4$ (This is zero at $x=-1$.)
   d.[‡] x^4+5x-6 (This is zero at $x=1$ and $x=-2$.)
   e.[†] $7x^6-4x^5-4x^4$ **f.**[†] $2x^2+4x+1$
   g. x^3-2x^2+1 (This is zero at $x=1$.)
   h. $2x^3+x^2+x-4$ (This is zero at $x=1$.)
   i.[†] x^4-4x^2+4 **j.**[†] x^4-16 **k.** $5x^2+8x+3$
   l. x^3+4x^2+4x+3 (This is zero at $x=-3$.)

Evaluate the integrals in Exercises 51 through 64

51.[†] $\int \dfrac{6x^2+9x+4}{x(x+1)^2} \, dx$ **52.** $\int \dfrac{x^4}{x^2-9} \, dx$

53.[†] $\int \dfrac{2x^2-9x+4}{x(x-2)^2} \, dx$ **54.** $\int \dfrac{x^2-1}{x^2+1} \, dx$

55.[‡] $\int \dfrac{2x^2+x+2}{(x^2+1)^2} \, dx$ **56.** $\int \dfrac{6x^2+22x+18}{(x+1)(x+2)(x+3)} \, dx$

57.[†] $\int \dfrac{x}{x^2-2x+10} \, dx$ **58.** $\int \dfrac{x+2}{(x^2+4x+7)^2} \, dx$

59.[†] $\int \dfrac{7x+23}{x^2+7x+10} \, dx$ **60.** $\int \dfrac{x^3+4x^2+7x}{x^2+4x+5} \, dx$

61.[†] $\int \dfrac{x^4-2x^3+5x^2+26x+4}{x^2-4x+13} \, dx$

62. $\int \dfrac{x^5-2x^4+17x^3+x}{x^2-2x+17} \, dx$ **63.**[†] $\int \dfrac{x^4-2x^3-2x^2+x-4}{(x+1)(x-3)} \, dx$

64. $\int \dfrac{x^5-35x}{x^2+6} \, dx$

Evaluate the integrals in Exercises 65 through 74 by using appropriate special substitutions.

65.[‡] $\int \dfrac{1}{3-\sqrt{x}} \, dx$ **66.**[‡] $\int \dfrac{1}{x(1+x^{1/3})} \, dx$

67.[‡] $\int \sqrt{1-e^x} \, dx$ **68.** $\int \dfrac{\sqrt{3+x}+7}{\sqrt{3+x}-7} \, dx$

69.‡ $\displaystyle\int \frac{e^x + 2}{2 - e^x} \, dx$ **70.** $\displaystyle\int \sqrt{1 + \sqrt{x}} \, dx$

71.† $\displaystyle\int \frac{x^{1/2}}{1 - x^{1/4}} \, dx$ **72.** $\displaystyle\int \frac{x}{1 + \sqrt{x}} \, dx$

73.† $\displaystyle\int \frac{1}{x^{1/4} + x^{1/2}} \, dx$ **74.** $\displaystyle\int \sin x \, \ln(\cos x) \, dx$

75.† A rod extends from $x = 0$ to $x = 5$ (inches) on the x-axis and its density at x is $3/(5 + x)$ pounds per inch. Find the x-coordinate of the center of mass of the rod.

76.† A cylindrical can of radius 2 feet and of length 5 feet is full of oil that weighs 100 pounds per cubic foot. What is the force of the oil on one end of the can when it is lying on its side?

77.† When the circle $x^2 + y^2 = 1$ is rotated about the line $y = -3$ in an xy-plane, it generates a torus. Find its surface area.

78.† Find the length of the parabola $y = \frac{1}{2}x^2$ from $x = 0$ to $x = 10$.

79.† Find the length of the curve $y = \frac{4}{5}x^{5/4}$ for $0 \le x \le 9$.

80. Explain why any integral of the form $\int R(e^x) \, dx$ with $R(u)$ a rational function can be evaluated by using the substitution $u = e^x$.

81.† The voltage in a circuit with alternating current is $V(t) = V_{max} \cos(\omega t)$ volts and the current is $i(t) = i_{max} \cos(\omega t)$ amperes at time t (seconds). The *period* of the current is $T = (2\pi/\omega)$ seconds. Compute the *average power* of the circuit, which is the average value of $V(t)i(t)$ (watts) in an interval of length T.

82. The voltage and current in Exercise 81 are *in phase* because they peak at the same times. Compute the average power for $V(t) = V_{max} \cos(\omega t + \theta)$ and $i(t) = i_{max} \cos(\omega t + \psi)$ with $\theta \ne \psi$, where the current and voltage are out of phase.

CHAPTER 9

SEPARABLE DIFFERENTIAL EQUATIONS

SUMMARY: In this chapter we study separable differential equations. The technique for solving them is described in Section 9.1. Applications to mathematics, physics, biology, chemistry, economics, and psychology are given in Section 9.2 and in the Miscellaneous Exercises.

9-1 SEPARABLE DIFFERENTIAL EQUATIONS

A DIFFERENTIAL EQUATION is an equation involving an unknown function and one or more of its derivatives. The order of the highest derivative is the *order* of the equation. A first order differential equation is called SEPARABLE if it is of the form

(1) $\qquad \dfrac{dy}{dx} = f(x)g(y)$ ◀

where the right side is the product of a function of x and a function of y. To solve (1) we divide by $g(y)$ to put the unknown function $y = y(x)$ and its derivative on the left. Then we integrate with respect to x to obtain

$$\int \frac{1}{g(y)} \frac{dy}{dx} \, dx = \int f(x) \, dx.$$

Making the substitution $y = y(x)$, $dy = dy/dx \, dx$ gives an equation

(2) $\qquad \displaystyle\int \dfrac{1}{g(y)} \, dy = \int f(x) \, dx$ ◀

involving one integral with respect to x and another with respect to y. We solve the differential equation by evaluating the integrals.

An abbreviated procedure

In practice we take advantage of the Leibniz notation to go directly from an equation of the form (1) to the corresponding equation (2): We treat the derivative dy/dx as a fraction and move the expressions involving x to one side of the equation and the expressions involving y to the other side. This gives

$$(3) \qquad \frac{1}{g(y)} \, dy = f(x) \, dx. \qquad \blacktriangleleft$$

With integral signs equation (3) becomes equation (2).

EXAMPLE 1   Solve the differential equation

$$(4) \qquad \frac{dy}{dx} = -2xy^2.$$

Solution.   To separate the variables, we divide equation (4) by y^2 and multiply it by the symbol dx. We obtain $y^{-2} \, dy = -2x \, dx$. With integration signs this becomes $\int y^{-2} \, dy = \int -2x \, dx$.

Next, evaluating the integrals gives

$$(5) \qquad -y^{-1} = -x^2 + C.$$

We multiply by -1 and take reciprocals to obtain the result

$$(6) \qquad y = \frac{1}{x^2 - C}. \qquad \square$$

Note.   To check the result of Example 1, we write $y = (x^2 - C)^{-1}$. Then we have $\dfrac{dy}{dx} = \dfrac{d}{dx} (x^2 - C)^{-1}$

$$= -(x^2 - C)^{-2} \frac{d}{dx}(x^2 - C) = -2x(x^2 - C)^{-2} = -2xy^2. \qquad \square$$

Initial-value problems

The constant of integration which appears in the solution of the differential equation is determined by prescribing, as an INITIAL CONDITION, the value of the unknown function $y = y(x)$ at one value of x. The differential equation and the initial condition constitute an INITIAL-VALUE PROBLEM.

EXAMPLE 2   Solve the initial-value problem

$$(7) \qquad \frac{dy}{dx} = -2xy^2, \qquad y(0) = 1.$$

Solution.   We found the solutions of the differential equation in Example 1. We now only have to determine the constant C of integration. We could use the final form (6) of the solution, but it is easier to find C from equation (5), which resulted from evaluating the

integrals. Setting $x = 0$ and $y = 1$ in (5) gives $C = -1$. Then equation
(6) yields

$$y = \frac{1}{x^2 - C} = \frac{1}{x^2 - (-1)} = \frac{1}{x^2 + 1}. \quad \square$$

Direction fields

The differential equation

$$(1) \quad \frac{dy}{dx} = f(x)g(y)$$

requires that the slope of the tangent line to the graph of a solution at
(x, y) be $f(x)g(y)$. Properties of the differential equation are shown
geometrically by a sketch of its DIRECTION FIELD. The direction field
consists of short segments of tangent lines to the graphs of solutions.
The line with midpoint (x, y) in the direction field has slope $f(x)g(y)$,
given by the right side of the differential equation (1).

The direction field for the differential equation

$$(4) \quad \frac{dy}{dx} = -2xy^2$$

of Examples 1 and 2 is sketched in Figure 9.1. Only some of the lines are
shown; conceptually, there is one line through each point in the
xy-plane. The slope of the line with midpoint (x, y) is $-2xy^2$.

We can visualize the graphs of solutions of the differential equation
by examining the direction field. At each point on the graph of a
solution, the graph is tangent to the line in the direction field, and, in
this sense, the graph "follows" the directions indicated by the direction
field.

Figure 9.2 shows the graph of the solution $y = 1/(x^2 + 1)$ of the
$(-2 < x < 2)$ of the differential equation that satisfies the initial
condition $y(0) = -\frac{1}{4}$. We do not consider the second solution of the
initial value problem to extend outside the interval $-2 < x < 2$ even
though the function $1/(x^2 - 4)$ is defined for $x < -2$ and for $x > 2$
because the function is discontinuous and does not satisfy the
differential equation at $x = 2$ or $x = -2$.

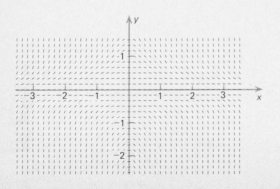

FIGURE 9.1

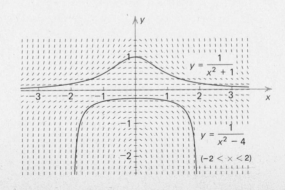

FIGURE 9.2

Curves as solutions of differential equations

Separable differential equations are often given in the form

(8) $p(x)q(y) \, dx + r(x)s(y) \, dy = 0.$

When we divide this equation by $q(y)$ and by $r(x)$ and add integral signs, we obtain the equation

(9) $\int \dfrac{p(x)}{r(x)} \, dx + \int \dfrac{s(y)}{q(y)} \, dy = C.$

The "C" on the right side of this equation is the constant of integration which we would obtain from evaluating the integrals.

After we evaluate the integrals in (9), we obtain an equation involving the variables x and y. Solving this equation for y as a function of x would yield the solutions $y = y(x)$ of the differential equation (8).

We can also adopt a different point of view. We can consider the solutions of the differential equation (8) to be the *curves* whose equations are obtained by evaluating the integrals in equation (9). The curves may or may not be the graphs of functions $y = y(x)$.

EXAMPLE 3   (a) Find an equation for the curve which is the solution of the differential equation

(10) $4 \cos x \, dx - y \, dy = 0$

and which passes through the point $(0, 3)$. (b) Find the solution which passes through $(0, 2)$.

Solution.   The variables are already separated in (10) and with integral signs it becomes $4 \int \cos x \, dx - \int y \, dy = C.$ When we evaluate the integrals, we obtain the equation

(11) $4 \sin x - \tfrac{1}{2} y^2 = C$

for the solutions of (10).

(a) To find the solution through $(0, 3)$, we set $x = 0$ and $y = 3$ in (11). This gives $C = -\tfrac{9}{2}$ and the equation $4 \sin x - \tfrac{1}{2} y^2 = -\tfrac{9}{2}$ or $y^2 = 8 \sin x + 9$. We consider the solution to be a continuous curve. Because $8 \sin x + 9$ is always positive, the curve cannot cross the x-axis. It is above the x-axis at $x = 0$, so it remains there and it has the equation

(12) $y = \sqrt{8 \sin x + 9}.$

This curve and the direction field for the differential equation are shown in Figure 9.3.

(b) To find the solution through $(0, 2)$, we set $x = 0$ and $y = 2$ in (11). We obtain $C = -2$ and the equation $4 \sin x - \tfrac{1}{2} y^2 = -2$, or

(13) $y^2 = 8 \sin x + 4.$

The curve (13) has the egg-like shape shown in Figure 9.3. Its two halves are the curves $y = \sqrt{8 \sin x + 4}$ and $y = -\sqrt{8 \sin x + 4}$. They meet at $x = -\pi/6$ and $x = 7\pi/6$ on the x-axis because $8 \sin x + 4$ is zero there. The curve (13) does not extend beyond the interval $-\pi/6 \leq x \leq 7\pi/6$ because $8 \sin x + 4$ is negative and (13) has no real solutions y for x outside that interval. □

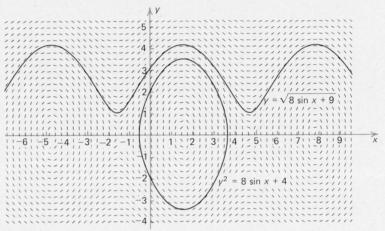

FIGURE 9.3

Differential equations as equations between differentials

Recall from Section 4.10 that we can consider the differentials dx and dy as corresponding to changes in x and y along a tangent line to a curve (Figure 9.4).

With this interpretation, we can think of

(8) $p(x)q(y)\,dx + r(x)s(y)\,dy = 0$

as a differential equation for a curve expressed in terms of its differentials: A curve is a solution of equation (8) if at each point (x, y) on it the differentials dx and dy satisfy that equation.

Exercises

Solve the differential equations and initial-value problems in Exercises 1 through 16.

1.† $\dfrac{dy}{dt} = y \sin t$ $y(0) = 1$

2. $\dfrac{dv}{dt} = 5 - v$ $v(0) = 10$

3.‡ $f'(x) = -3f(x)^4$

4. $\dfrac{dQ}{dx} = -3Q^{1/4}$

5.† $y^2\,dx + (x^2 + 1)\,dy = 0$

6.† $\sqrt{1 - y^2}\,dx - dy = 0$

7.† $\dfrac{dy}{dx} = \cos^2 y$

8. $\dfrac{dy}{dx} = (-\cos x)\,y$

9.† $K'(x) = \sqrt{xK(x)}$ $K(1) = 1$

10. $Q'(t) = tQ(t)^{1/3}$ $Q(0) = 1$

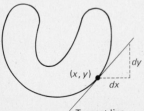

FIGURE 9.4   Differentials dx and dy at a point (x,y) on a curve

11.† $\dfrac{dF}{dx} = 4 + F^2$ $F(0) = 2$ **12.** $dy + \sqrt{y}\, dx = 0$ $y(0) = 4$

13.† $x\,dy + y^2\, dx = 0$ $y(1) = 2$

14. $\dfrac{dy}{dx} = -2xy$

15.‡ $P'(t) = P(t)[10 - P(t)]$ $P(0) = 2$

16.† $\dfrac{df}{dt} = 5(f - 1)(3 - f)$ $f(0) = 2$

Find equations for the curves which are solutions of differential equations in Exercises 17 through 20.

17. $x(4 - y^2)\, dx + y\, dy = 0$ **18.†** $(x + 1)y\, dx + x(y - 1)\, dy = 0$

19.† $(y^2 + 1)\, dx - y(x^2 + 1)\, dy = 0$

20. $\dfrac{dp}{dq} = \dfrac{2 + \cos q}{2 - \sin p}$

21.† Figure 9.5 shows the direction field of $dy/dx = \frac{1}{4} y^2 \sin x$. Find the solutions of this differential equation which satisfy the initial conditions
    a. $y(0) = 2$ **b.** $y(0) = -2$ **c.** Use the direction field to sketch the graphs of these two solutions.

22.† Figure 9.6 shows the direction field of $(x^3 - 6x)\, dx + 2y\, dy = 0$. Find the solution through the point $(0, 3)$. Use the direction field to sketch it.

23. Figure 9.7 shows the direction field of $7 \cos x\, dx - y^2\, dy = 0$. Find the solutions through the points   **a.** $(0, 3)$ **b.** $(0, -1)$
    c. Use the direction field to sketch the curves.

24.† Figure 9.8 shows the direction field of $x^2 y^2\, dx = 9\, dy$. Find the solution through the point $(-2, 1)$ and use the direction field to sketch it.

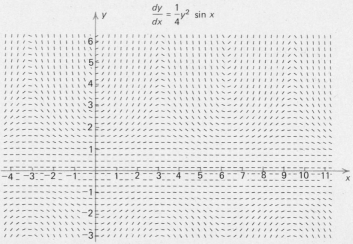

$$\frac{dy}{dx} = \frac{1}{4} y^2 \sin x$$

FIGURE 9.5

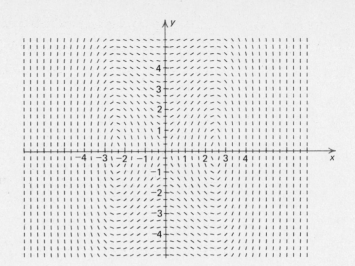

FIGURE 9.6
$(x^3 - 6x)\, dx + 2y\, dy = 0$

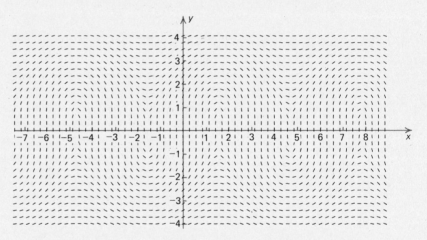

FIGURE 9.7
$7\cos x\, dx - y^2\, dy = 0$

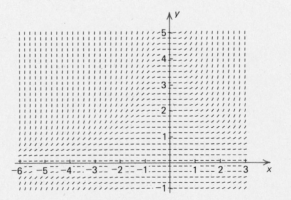

FIGURE 9.8
$x^2 y^2\, dx = 9\, dy$

9-2   APPLICATIONS OF SEPARABLE DIFFERENTIAL EQUATIONS

In this section we discuss problems from various fields that are solved through the use of separable differential equations.*

Air resistance

EXAMPLE 1   An 800 lb. balloon gondola has zero velocity at time $t = 0$ (seconds) when it tears loose from the balloon. As it falls, it is subject to the force of gravity and to the force of air resistance, which is $\frac{1}{50} v^2$ (pounds) when the gondola has velocity v (feet per second). Find the downward velocity as a function of t. What is the terminal velocity (the velocity the gondola would approach if it fell forever)?

Solution.   We let $v(t)$ (feet per second) be the downward velocity of the gondola at time t. Then $\frac{dv}{dt}(t)$ is the downward acceleration at time t and by Newton's law

$$(1) \quad m\frac{dv}{dt}(t) = F(t)$$

where m is the mass of the gondola and $F(t)$ is the total downward force on it at time t.

We take 32 feet per second^2 as the acceleration due to gravity. Then the mass of the gondola is $\frac{800}{32} = 25$ slugs. The downward force of gravity is equal to the gondola's weight, 800 pounds. Air resistance exerts an upward force of $\frac{1}{50} v(t)^2$ pounds or a downward force of $-\frac{1}{50} v(t)^2$ pounds at time t. Therefore equation (1) yields

$$25\frac{dv}{dt}(t) = 800 - \tfrac{1}{50} v(t)^2.$$

We rewrite this equation in the form

$$(2) \quad 1250\frac{dv}{dt} = 200^2 - v^2.$$

Separating the variables then gives

$$\int \frac{dv}{200^2 - v^2} = \frac{1}{1250} \int dt.$$

We evaluate the integral on the left by the method of partial fractions or by using a table of integrals. We obtain the equation

$$(3) \quad \frac{1}{400} \ln\left|\frac{200 + v}{200 - v}\right| = \frac{1}{1250} t + C.$$

---

*Instructors: It is not necessary to cover all the applications in this section and in the Miscellaneous Exercises. Topics may be chosen according to the interests of the class.

Setting $t = 0$ and $v = 0$ gives the relation $\frac{1}{400} \ln(1) = C$, which implies that $C = 0$. Setting $C = 0$ in (3) and multiplying the equation by 400 yields

(4) $\ln \left| \dfrac{200 + v}{200 - v} \right| = 0.32t.$

The velocity is initially less than 200, and equation (4) shows that the only way it can approach 200 is for t to tend to infinity. We conclude that v must remain less than 200 and we can drop the absolute value signs in equation (4). That equation then gives

$$\frac{200 + v}{200 - v} = e^{0.32t}.$$

Solving for v gives the desired formula

(5) $v = 200 \left(\dfrac{e^{0.32t} - 1}{e^{0.32t} + 1} \right).$

To find the terminal velocity, we multiply the numerator and denominator of (5) by $e^{-0.32t}$ and obtain

$$v = 200 \left(\frac{1 - e^{-0.32t}}{1 + e^{-0.32t}} \right).$$

This equation shows that $v(t)$ tends to 200 as t tends to ∞. The terminal velocity is 200 feet per second. At this velocity the force of air resistance $\frac{1}{50} v^2 = \frac{1}{50}(200)^2 = 800$ pounds would just balance the force of gravity on the gondola. $\square$

The direction field for the differential equation (2) and the graph of the solution (5) are shown in Figure 9.9.

Note.   If we multiply the numerator and denominator of (5) by $e^{-0.16t}$, we can see that the solution $v(t)$ is a hyperbolic tangent:

$$v = 200 \left(\frac{e^{0.16t} - e^{-0.16t}}{e^{0.16t} + e^{-0.16t}} \right) = 200 \tanh(0.16t). \quad \square$$

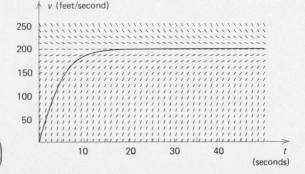

FIGURE 9.9

$v = 200 \left(\dfrac{e^{0.32t} - 1}{e^{0.32t} + 1} \right)$

Population growth in a crowded environment

EXAMPLE 2 Suppose that rabbits are introduced to a South Pacific Island where they have no natural enemies. The island can only support 1000 rabbits; because of this, we suppose that the rate of growth $\frac{dP}{dt}(t)$ of the population at time t (months) is proportional to the product of the population at that time $P(t)$ and of $1000 - P(t)$, which represents how much room there is for more rabbits. At time $t = 0$ there are 20 rabbits and the population is increasing at the rate of 9.8 rabbits per month. How many rabbits are there at time t?

Solution. We are told that $dP/dt = aP(1000 - P)$ for some constant a, and that at $t = 0$, $P = 20$ and $dP/dt = 9.8$. This gives the equation $9.8 = a(20)(1000 - 20) = (20)(980)a$, which shows that $a = 1/2000$.

To solve the resulting differential equation

$$(6) \qquad \frac{dP}{dt} = \frac{1}{2000} P(1000 - P)$$

we separate the variables and obtain

$$\int \frac{dP}{P(1000 - P)} = \frac{1}{2000} \int dt.$$

The integral on the left can be evaluated by partial fractions or by using a table of integrals. This yields the equation

$$\frac{1}{1000} \ln \left| \frac{P}{1000 - P} \right| = \frac{1}{2000} t + C_1$$

or

$$(7) \qquad \ln \left| \frac{P}{1000 - P} \right| = \tfrac{1}{2} t + C.$$

Setting $t = 0$ and $P = 20$ in equation (7) gives the value

$$(8) \qquad C = \ln \left| \frac{20}{1000 - 20} \right| = \ln \left(\frac{1}{49} \right)$$

for the constant C. By the reasoning employed in Example 1, we can drop the absolute value signs in (7); then, with the value (8) for C, we obtain

$$\frac{P}{1000 - P} = e^{t/2} \, e^{\ln(1/49)} = \frac{1}{49} e^{t/2}.$$

When we solve for P, we obtain the formula

$$(9) \qquad P = 1000 \left[\frac{e^{t/2}}{49 + e^{t/2}} \right] = 1000 \left[\frac{1}{49e^{-t/2} + 1} \right]$$

for the population of rabbits at time t. We obtained the final expression for P by multiplying and dividing the previous expression by $e^{-t/2}$. The final expression shows that P tends to 1000 as t tends to ∞. $\square$

The direction field for the differential equation (6) and the graph of the solution (9) are sketched in Figure 9.10. The differential equation is called a *logistic differential equation*, and the graph is called a *logistic curve*.

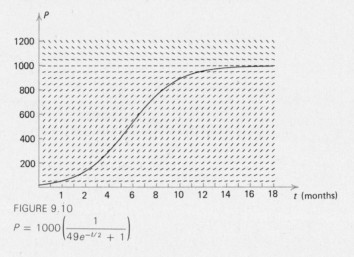

FIGURE 9.10

$$P = 1000\left(\frac{1}{49e^{-t/2} + 1}\right)$$

Stimulus and sensation

Weights of objects, loudness of sounds, and brightness of lights can be measured by instruments. In what way do these measurements relate to how heavy an object *feels* when someone picks it up, or to how loud a sound or how bright a light *seems* to an individual? Can we set up scales to measure these sensations? How would such scales compare with the scales used to measure the stimuli?

One approach to this problem is to have a change of one unit correspond to the smallest change in the stimulus that most individuals can distinguish. This is the idea behind the decibel scale of loudness of sounds: one decibel is intended to be the smallest change in loudness that the human ear can discern.

This approach often leads to the use of *Weber's law* to relate the scale for measuring the sensation to the scale for measuring the stimulus.* Weber's law states that a small change ds in the stimulus from s to $s + ds$ produces a change dS in the sensation proportional to the fraction ds/s by which the stimulus is increased:

(10) $dS = K\dfrac{ds}{s}.$ ◀

In stating this law, we have followed a common practice of referring to the differentials ds and dS as corresponding small changes in the related quantities s and S. We can interpret this in two ways: we can think of ds and dS as corresponding changes of s and S along a tangent line to the graph of the equation relating s and S. Then, when ds and dS

---

*Formulated in 1846 by the German physiologist E. H. Weber.

are small, they will be approximately equal to actual changes in s and S. Or, we can interpret (10) as an alternate way of writing the differential equation

(10) $$\frac{dS}{ds} = K\frac{1}{s}$$

which involves the limit

$$\frac{dS}{ds} = \lim_{\Delta s \to 0} = \frac{\Delta S}{\Delta s}$$

of the ratio of actual changes Δs and ΔS of s and S.

EXAMPLE 3   In most cases a person can distinguish between two weights only if the heavier one weighs at least 5% more than the lighter one. Under this assumption use Weber's law to define a scale S of apparent weight such that a difference of one unit in S corresponds to the smallest noticeable change in actual weight s. Have $S = 1$ correspond to a weight of $s = 1$ pound.

Solution.   Equation (10) yields $\int dS = K \int ds/s$ or $S = S(s)$ $= K \ln s + C$. The condition $S(1) = 1$ implies that $C = 1$. The requirement that $S(s)$ increase by one unit for a 5% increase in s gives the equation

$$1 = S(1.05s) - S(s) = [K \ln(1.05s) + 1] - [K \ln s + 1]$$
$$= K \ln\left[\frac{1.05s}{s}\right] = K \ln(1.05).$$

Therefore $K = 1/\ln(1.05)$ and we have

$$S(s) = \frac{1}{\ln(1.05)} \ln s + 1$$
$$\approx 20.5 \ln s + 1. \quad \square$$

Chemical reaction rates   The rate at which a chemical reaction proceeds depends in part on how frequently the component molecules collide. In the simplest mathematical model of a chemical reaction the rate of the reaction is considered to be proportional to the product of the concentrations of the component molecules. This model serves for the chemical reaction described in the next example.

Numbers of molecules are given in MOLES. One mole is $6(10^{23})$ molecules. The concentration of a type of molecule A in a solution is measured in *moles per liter* and is denoted by the symbol [A].

EXAMPLE 4   The chemical reaction, $CH_3COCH_3 + I_2 \to$ $CH_3COCH_2I + HI$, between acetone (CH_3COCH_3) and iodine (I_2) is governed by the differential equation

(11) $\dfrac{d}{dt}[I_2] = -0.0004[CH_3COCH_3][I_2]$

where t is the time, measured in seconds.* At time $t = 0$ the concentrations of acetone and iodine are both a moles per liter. Find the concentration of iodine as a function of t.

Solution.   We let $y(t) = [I_2]$ be the concentration of iodine at time t. Because equal amounts of acetone and iodine are involved in the reaction, the concentration $[CH_3COCH_3^*]$ of acetone at time t is also $y(t)$. The differential equation (11) therefore reads $\dfrac{dy}{dt}(t)$ $= -0.0004\, y(t)^2$. Separating the variables, we obtain

$$\int y^{-2}\, dy = -0.0004 \int dt$$

or $-1/y = -0.0004t + C$. Setting $t = 0$ and $y = a$ gives the value $C = -1/a$ for the constant of integration. Giving C this value and solving for y yields the formula

$$y = \frac{1}{0.0004t + (1/a)}.$$

The concentration of iodine tends to zero as the reaction proceeds and t tends to infinity.   □

Differential equations of predator-prey

The Austrian mathematician A. J. Lotka (1880–1949) and the Italian mathematician Vito Volterra (1860–1940) proposed the following simple model for how the fluctuations of populations of a predator and its prey affect each other. We let $x(t)$ denote the population of predators and $y(t)$ the population of prey at time t. The predator might be a type of insect and its prey a type of scale which is its food, or the predator might be a type of bird and the prey a type of insect which is its diet.

We assume that if there were no predators present, the population of prey would increase at a rate $py(t)$ proportional to their number but that the predators eat the prey at a rate $qx(t)y(t)$ proportional to the product of the number of prey and the number of predators. This gives the differential equation

(12) $\dfrac{dy}{dt} = py - qxy$

with positive constants p and q.

We assume that if there were no prey, the predators would starve and their population would decrease at a rate $rx(t)$ proportional to their

---

*Adapted from G. W. Castellan, [1] *Physical Chemistry*, p. 769.

number, but that in the presence of prey the rate of growth of the population of predators is increased by an amount $sx(t)y(t)$. We obtain the second differential equation

$$(13) \quad \frac{dx}{dt} = -rx + sxy$$

with additional positive constants r and s.

Equations (12) and (13) form a system of differential equations known as the *Lotka-Volterra equations*. We cannot solve them for x and y as functions of t, but we can derive an equation relating the variables x and y. Dividing equation (12) by equation (13) gives the separable differential equation

$$(14) \quad \frac{dy}{dx} = \frac{dy/dt}{dx/dt} = \frac{py - qxy}{-rx + sxy} = \frac{y(p - qx)}{x(sy - r)}.$$

We solve (14) for a particular choice of the constants p, q, r, and s in the next example.

EXAMPLE 5   Suppose that $x = x(t)$ (hundreds of predators) and $y = y(t)$ (thousands of prey) satisfy the Lotka-Volterra equations (12) and (13) with $p = 3, q = 1, r = 4$, and $s = 2$. Suppose that at a time when there are 100 predators ($x = 1$), there are 1000 prey ($y = 1$). Find an equation relating x and y.

Solution.   With the given values of the constants, p, q, r, and s, equation (14) reads

$$(15) \quad \frac{dy}{dx} = \frac{y(3 - x)}{x(2y - 4)}.$$

Separating the variables yields $[(2y - 4)/y]\,dy = [(3 - x)/x]\,dx$ and then

$$\int \left(2 - \frac{4}{y}\right) dy = \int \left(\frac{3}{x} - 1\right) dx.$$

Evaluating the integrals gives $2y - 4 \ln y = 3 \ln x - x + C$. We set $x = 1$ and $y = 1$ to see that $C = 3$ and to obtain the desired equation

$$(16) \quad x + 2y - 3 \ln x - 4 \ln y = 3. \quad \square$$

The curve (16) and the direction field of the differential equation (15) are shown in Figure 9.11. Notice that the variations in the populations of the predator and prey are *cyclic*. The population of prey tends to rise when the population of predators is low. Then, with the resulting abundance of food, the population of predators increases. As a result, the population of prey decreases again, and this forces the population of predators to decline due to the lack of food. The pattern then repeats.

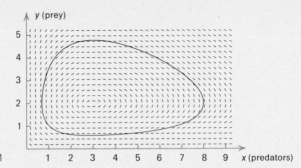

FIGURE 9.11

Orthogonal trajectories

An equation involving the two variables x and y and a parameter c is an equation of a FAMILY OF CURVES. We obtain one curve in the family for each value of c. Figure 9.12 shows some of the curves in the family $y = ce^{-x/4}$.

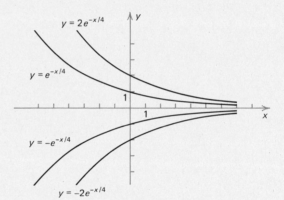

FIGURE 9.12

The ORTHOGONAL TRAJECTORIES of a family of curves is a second family of curves such that at each point where curves from the two families intersect they are orthogonal (have perpendicular tangent lines). In the next example we solve a separable differential equation to find the orthogonal trajectories of the family of curves shown in Figure 9.12.

EXAMPLE 6   Find the orthogonal trajectories of the family of curves

(17) $y = ce^{-x/4}$.

Solution.   We first find a differential equation for the given family of curves. Differentiating equation (17) with respect to x gives

(18) $\dfrac{dy}{dx} = -\tfrac{1}{4} ce^{-x/4}$.

This is not a single differential equation for the entire family of curves because it involves the parameter c. Solving equation (17) for c gives $c = ye^{x/4}$. With this substitution equation (18) yields

(19) $\dfrac{dy}{dx} = -\tfrac{1}{4}y$

which is a differential equation for the family (17).

Equation (19) shows that the slope of the tangent line to the curve in the family (17) passing through (x, y) is $-\tfrac{1}{4}y$. The slope of the tangent line to the orthogonal trajectory through that point is therefore $4/y$, and the orthogonal trajectories satisfy the differential equation

(20) $\dfrac{dy}{dx} = \dfrac{4}{y}.$

Separating variables gives $\int y\,dy = \int 4\,dx$ and then evaluating the integrals yields

(21) $y^2 = 8x + C.$ □

A few of the curves in the family (21) of orthogonal trajectories are shown in Figure 9.13 along with curves of the original family.

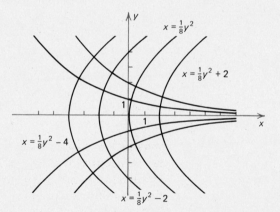

FIGURE 9.13

Exercises

1.‡ (Water resistance). The force of water resistance on a boat with speed v (feet per second) is given by $F = -kv^2$ (pounds). The constant k is called the *drag coefficient* of the boat.

As the yacht "Gimcrack" is coasting with its sails down and with no forces on it other than water resistance, it slows down from a speed of 10 feet per second to a speed of 5 feet per second in 20 seconds time. The Gimcrack weighs 6,560 pounds. What is its drag coefficient? (Adapted from C. A. Marchaj, [1] *Sailing Theory and Practice*, p. 244.)

2. (Ice formation). The thickness $x(t)$ (inches) of ice forming on a lake satisfies the differential equation $x'(t) = 3/x(t)$. At $t = 0$ (days) the ice is one inch thick. When is it two inches thick?

3.† (Population growth). A newly created lake is stocked with 400 fish at time $t = 0$ (months). Thereafter the population increases at the rate of $\sqrt{P}$ fish per month when there are P fish in the lake. What is the fish population at time t?

4. (Population growth). The amount of a type of mountain moss increases at the rate of $M/12$ pounds per month when there are M pounds of it present on April 1. It decreases at the same rate when there are M pounds of it present on October 1. As a mathematical model of the moss growth and decline, we take the differential equation $dM/dt = \frac{1}{12}M \sin(\pi t/6)$, where t is the time in months and $t = 0$ is January 1. There are 600 pounds of the moss on a hillside on January 1. How much is there on October 1?

5.† (Population growth). If a population satisfies the logistic differential equation $dP/dt = kP(L - P)$, what is the population when its rate of growth is a maximum?

6.† (Population growth). Figure 9.14 shows the graph of the population of a culture of yeast in a limited environment, considered as a function of the time. Find a logistic differential equation which would serve as a mathematical model of this yeast culture.

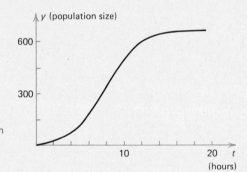

FIGURE 9.14   The logistic growth of a population of yeast (Adapted from *Encyclopaedia Britannica*, Vol. 18, p. 238)

7.† (Stimulus and sensation). The loudness D (decibels) of a sound and the variation in air pressure P (pounds per square inch) caused by the sound wave are related by Weber's law $dD = 8.7\,(dp/p)$. A sound is said to be at zero decibels if the pressure variation is 0.0083 pounds per square inch. Give the formula for D in terms of p.

8. (Stimulus and sensation). Certain scales for measuring stimuli s are related to the scales for measuring the corresponding sensations S by differential equations of the form $dS/S = k\,(ds/s)$. What is the form of the resulting equations relating the stimulus and the sensation?

9.‡ (Chemical reaction rates). The chemical reaction $C_2H_4Br_2 + 3KI \rightarrow C_2H_4 + 2KBr + KI_3$ between ethylene bromide ($C_2H_4Br_2$) and potassium iodide (KI) in 99% methanol is governed by the differential equation

$$\frac{d}{dt}[KI_3] = 0.3[C_2H_4Br_2][KI].$$

At time $t = 0$ the solution contains a moles of $C_2H_4Br_2$, b moles of KI ($b \neq 3a$), and no KI_3.   **a.**   Find an equation relating $y = y(t) = [KI_3]$ and t. (Do not solve it for y.) What happens to $[KI_3]$ as $t \to \infty$?   **b.**   if $b > 3a$ **c.**   if $b < 3a$? Explain. (Adapted from W. J. Moore, [1] *Physical Chemistry*, p. 262.)

10.[†]   (Chemical reaction rates). The oxidation of nitric oxide (NO) into nitrous oxide (NO_2), $2NO + O_2 \to 2NO_2$, is governed by the differential equation $\dfrac{d}{dt}[NO_2] = 10^{10}[NO]^2[O_2]$. At time $t = 0$ a solution contains a moles per liter of O_2, $2a$ moles per liter of NO, and no NO_2. Give $y = [NO_2]$ as a function of t for $t > 0$. (Adapted from F. T. Gucker and R. L. Seifert, [1] *Physical Chemistry*, p. 696.)

11.   (Chemical reaction rates). In an *autocatalytic reaction*, $A \to B + \ldots$, the product B hastens the reaction. The reaction is governed by a differential equation of the form $(d/dt)[A] = -k[A][B]$. Show that this leads to the logistic differential equation of population growth.

12.[†]   (Chemical reaction rates). The decomposition of acetaldehyde, $CH_3CHO \to CH_4 + CO$, is governed by the differential equation $\dfrac{d}{dt}[CH_4] = k[CH_3CHO]^{3/2}$. If at $t = 0$ $[CH_3CHO] = 0.01$, what is $[CH_3CHO]$ at time $t > 0$?

13.[†]   (Chemical reaction rates). At high temperatures the formation of hydrogen iodide, $H_2 + I_2 \to 2HI$, and its decomposition back into hydrogen and iodine, $2HI \to H_2 + I_2$, are *competing reactions* which occur simultaneously. At 293° centigrade the reactions are governed by the differential equation

(22) $\dfrac{d}{dt}[HI] = 0.160[H_2][I_2] - 0.00253[HI]^2$

with the time measured in hours.   **a.**   What is the equilibrium concentration if the solution contains $2a$ moles per liter of H atoms and of I atoms?   **b.**   figure 9.15 shows the direction field for equation (22) expressed in terms of $y(t) = [HI]$ under the conditions of part (a). Use the direction field to sketch the graphs of $y(t)$ with $y(0) = 0$ and with

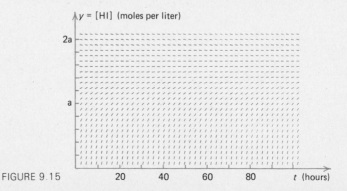

FIGURE 9.15

$y(0) = 2a$ without attempting to solve the differential equation. Explain the nature of the graphs. (Adapted from W. J. Moore, [1] *Physical Chemistry*, p. 266.)

14. (Predator-prey). Examine the differential equations (12) and (13) with the values of the constants from Example 5 to determine whether the point (x, y) travels around the curve in Figure 9.11 in a clockwise or in a counterclockwise direction as time increases.

15. (Predator-prey). Suppose that the predator in Example 5 is a type of bird and the prey a type of insect. What would happen if at a time when (x, y) is on the lower part of the curve in Figure 9.11, insecticides were used which decreased the number of insects?

16.[†] (Population decline). Due to a curse imposed by a neighboring tribe, the members of a village are gradually driven to murder and suicide. The rate of change of the population is $-2\sqrt{P}$ people per month when the population is P. When the curse is made, the population is 1600. When will they all be dead?

17. (Orthogonal trajectories). Find the orthogonal trajectories of the following families of curves:

   a.[†] $x^2 - y^2 = c$ **b.** $x^2 + 2y^2 = c$
   c. $x^2 + 3y^2 = c$ **d.** $x^2 + y^2 = c$
   e.[†] $x = cy^4$ **f.** $x = cy^2$
   g.[†] $x^4 + y^4 = c$ **h.** $y = cx + 3$
   i. $x^2 + 8x + y^2 - 2y = c$ **j.**[†] $y = e^{cx}$

MISCELLANEOUS EXERCISES

1. Solve the following initial-value problems.   **a.**[†] $dy/dx = (1 + \ln x)y$ $y(1) = 1$ **b.** $\cos^2 y \, dx + \csc x \, dy = 0$ $y(0) = \pi/4$ **c.**[†] $(y^2 + 1) \, dx + (2x^2y + 2y) \, dy = 0$ $y(0) = 0$ **d.** $dy/dx = (4x^3 + 1)(y^2 + 1)$ $y(0) = -1$

2. (Population growth). Explain why the differential equation $dP/dt = rP + a$ provides a model for population growth in a country that has steady immigration. Then solve the differential equation.

3.[†] (Newton's law of cooling). According to Newton's law of cooling, if an object is placed in an environment at a different temperature, the rate at which the object's temperature changes is proportional to the difference between its temperature and that of the environment. Suppose that a box at a temperature of 100°F is put in a room at 70°F and that one minute later the temperature of the box is 90°F. What is the temperature of the box t seconds after it is put into the room?

4.[‡] (A mixing problem). A 120 gallon tank is full of water in which 20 pounds of salt is dissolved. Starting at time $t = 0$ (minutes) a salt solution containing $\frac{1}{2}$ pound of salt per gallon is added to the tank at the constant rate of 6 gallons per minute. The two solutions mix instantly and the excess drains off.   **a.**   How much salt is in the tank at time $t (t > 0)$?

b. What is the limit of the concentration of salt in the tank as t tends to ∞?

5. (A mixing problem). Children put 5 pounds of soap in a water fountain containing 300 gallons of water. When the soap is discovered, clean water is run into the tank at the rate of 10 gallons per minute. The solutions mix instantly and the excess drains off. When will there be only one pound of soap in the water?

6.[†] (Bernoulli's equation of physical and moral wealth). In the 1738 *Commentari* of the St. Petersberg Academy of Sciences, Daniel Bernoulli proposed a mathematical model to relate moral fortune, which we might think of as happiness, to physical fortune, or wealth. He suggested that an increase in a person's wealth from M to $M + dM$ causes an increase dH in his happiness which is proportional, not to the amount dM by which his wealth is increased, but to the fraction dM/M by which it is increased. Use this principle to express a person's happiness H as a function of his wealth M measured in dollars.

7.[†] (A melting snowball). Suppose that a snowball melts at a rate proportional to its surface area. Express this as a differential equation for the volume V and solve it under the condition that the volume is V_0 at $t = 0$. What happens to the snowball?

8. (Rumor spreading). Let $P = P(t)$ be the number of people in a population of P_1 who know a rumor at time t (days). Suppose that the rate P increases is proportional to the product of the number of people who know the rumor and of the number who do not yet know it. Give a differential equation for $P(t)$ and describe the nature of its solutions.

9. (Diffusion). Suppose that a cell in an organism has (fixed) volume V and surface area A. Suppose that c_1 is the (constant) concentration of a substance outside the cell and that $c = c(t)$ is the concentration inside the cell at time t. *Fick's law* states that the rate at which the substance crosses the cell wall is proportional to the difference in concentrations, inside and outside. The constant k of proportionality depends on the nature of the cell. Give a differential equation for $c(t)$ and solve it to express $c(t)$ in terms of V, A, k, c_1, t, and $c_0 = c(0)$.

10.[‡] (Genetics). Suppose that the fraction of type A of a hereditary gene in a population is $a(t)$, while the fraction of type B is $b(t)$ $(a(t) + b(t) = 1)$. Suppose that the A genes mutate into B genes at a rate (in genes per year) equal to 0.1% of the number of A genes present and the B genes mutate into A genes at a rate equal to 0.5% of the number of B genes present. Assuming that mutation is the only cause for change in $a(t)$ and $b(t)$, **a.** give a differential equation for $a(t)$; **b.** solve it for $a(t)$ in terms of $a(0)$; and **c.** find the equilibrium proportion of genes A and B.

11. (Genetics). Sickle cell anemia is caused by abnormal hemoglobin in the red blood cells. The abnormal hemoglobin, in turn, is caused by type B of a hereditary gene, whereas normal hemoglobin is caused by type A. An AA gene combination is normal. Persons with a BB combination have the abnormal hemoglobin exclusively by adulthood and develop severe anemia. Those with an AB combination are susceptible to mild anemia in certain circumstances only. They are also immune to a type of malaria, and in West Africa where the malaria is common, they have a better chance of survival than those with an AA combination.

If we assume that the probabilities that persons with AB, AA, and BB gene combinations reproduce are $1, 1 - s$, and $1 - r$, respectively, we have the differential equation

(1) $$\frac{dp}{dt} = \frac{rpq^2 - sp^2q}{1 - sp^2 - rq^2}$$

where $p = p(t)$ is the fraction of the genes which are of type A in a population at time t and $q = q(t)$ the fraction of type B $(p(t) + q(t) = 1)$.   **a.**   What is the equilibrium value of p in West Africa, where $s = 0.15$ and $r = 0.80$?   **b.**   In the United States, where there is almost no malaria and better medical care, we can take $s = 0$ and $r = 0.50$. Express (1) as a differential equation for p with these constants and solve it to give an equation that defines $p(t)$ implicitly.   **c.**   According to this mathematical model, what will happen to sickle cell anemia in the United States? (Adapted from Cavalli-Sforza and Bodmer, [1] *The Genetics of Human Populations.*)

12.   (Dissolution). When a powder is dissolved in a liquid, the rate at which the amount $x = x(t)$ of dissolved substance increases is given by the equation $dx/dt = kx(L - x)$ with k a positive constant.   **a.**   Sketch the direction field for this differential equation and discuss the nature of its solutions.   **b.**[†]   What interpretation does the constant L have? What is the explanation of the case where $x(0) > L$?

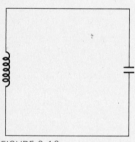

FIGURE 9.16

13.   (A time-varying resistor). Consider an electric circuit containing a capacitor of capacitance C (farads) and a resistor of time-varying resistance $R(t)$ (ohms) (Figure 9.16). The charge $q = q(t)$ (coulombs) on the capacitor at time t (seconds) satisfies the differential equation $R(t)\frac{dq}{dt}(t) + \frac{1}{C}q(t) = 0$. Find $q(t)$ in terms of $q(0)$ and C under the assumption that $R(t) = 1/(2 - \cos t)$.

14.   (A nonlinear, time invariant resistor). Suppose that the voltage drop across the resistor in Figure 9.16 is ki^3 volts when the current is $i = dq/dt$ (amperes). Then the charge $q = q(t)$ satisfies the differential equation $k(dq/dt)^3 + (1/C)q = 0$. Solve this differential equation under the condition that $q(0) = q_0$.

15.   (Economics). Find the demand function $D(p)$ such that the elasticity of demand is the positive constant k (see Miscellaneous Exercise 20 of Chapter 4).

16.   (Chemical reaction rates). The formation of Hydrogen Bromide, $H_2 + Br_2 \rightarrow 2HBr$, is governed by the differential equation

$$\frac{d}{dt}[H_2] = \frac{k[H_2][Br_2]^{1/2}}{m + [HBr]/[Br_2]}.$$

Find an important equation for $[H_2]$ under the condition that at $t = 0$, $[H_2] = [Br_2] = a$ and $[2HBr] = 0$. (W. J. Moore, [1] *Physical Chemistry*, p. 284.)

17.[‡] (Hanging cable). In Miscellaneous Exercise 28 of Chapter 11, we will derive the differential equation $f''(x) = (\rho/k) \sqrt{1 + [f'(x)]^2}$ for the function whose graph is the shape of a hanging cable. Solve this differential equation under the conditions that $f(0) = b$ and $f'(0) = 0$. (Solve for $m(x) = f'(x)$ first.)

18.[†] (Probability). Suppose that pieces of equipment fail at the rate of kN pieces per month when there are N pieces of equipment. A large quantity of new equipment is purchased at time $t = 0$ (months), and $F(t)$ is the fraction that has failed by time t. Give a differential equation for $F(t)$ and solve it.

19.[†] (Air resistance) An object of mass 100 slugs is subject to air resistance of $2v^2$ pounds when its velocity is v feet per second. At $t = 0$ (seconds) it has upward velocity of 40 feet per second. Find its upward velocity for $t > 0$.

20. (Air resistance). An object weighing 64 pounds is dropped from a hovering helicopter at time $t = 0$ (seconds). When its downward velocity is v, it is subject to $\frac{1}{4} v$ pounds of air resistance. What is its downward velocity for $t > 0$ (until it hits something)? What is its terminal velocity?

21. (Population growth). Due to periodic changes in the environment, the rate of change of the population P of a bacteria culture at time t is equal to $(2 + \sin(kt)) P$ if the population is P at time t. At $t = 0$, $P = 1000$. What is P for $t > 0$?

22. (Air and rolling resistance). The force of air and rolling resistance on a certain car is $\frac{2}{81} v^2$ pounds when the car's speed is v feet per second. The car's engine produces a maximum force of 200 pounds and the car weighs 1600 pounds.   **a.**   What is the speed that the car can approach but never achieve on a level road under only its own power?   **b.**   What is its speed after t seconds if it starts from rest and exerts its maximum force?

23. (Orthogonal trajectories). Sketch the family of curves $y = \sqrt[3]{x + c}$ and the family of their orthogonal trajectories.

CHAPTER 10

CONIC SECTIONS AND POLAR COORDINATES

SUMMARY: In the first four sections of this chapter we present the basic
analytic geometry of conic sections (circles, ellipses, parabolas, and
hyperbolas). We prove in Section 10.5 that these curves can be obtained as the
intersections of planes and cones. In Sections 10.6 and 10.7 we study polar
coordinates and their use in describing conic sections and other curves. In
Section 10.8 we use polar coordinates to compute areas.

INTRODUCTION

Circles, ellipses, parabolas, and hyperbolas are called CONIC SECTIONS
because each can be obtained as the intersection of a plane and the cone
formed by all lines through a point and the circumference of a circle
(Figure 10.1). The point is the VERTEX of the cone and lies on the AXIS,
which is the line through the center of the circle and perpendicular to
it. The lines that form the cone are its ELEMENTS.

When we intersect the cone with a plane that is perpendicular to the
axis and does not contain the vertex, we obtain a *circle* (Figure 10.2). If
the plane is tilted and cuts across one half of the cone, the intersection
is an *ellipse* (Figure 10.3). If the plane is parallel to an element of the
cone, the intersection is a *parabola* (Figure 10.4). If the plane cuts both
halves of the cone, the intersection is a *hyperbola* (Figure 10.5).

Conic sections were studied as early as the fourth century B.C., when
the Greek geometer Menaechmus used them to solve, geometrically, the
problem of constructing a cube of volume twice that of a given cube.
Later Greek mathematicians, notably Euclid (ca. 300 B.C.), Archimedes
(ca. 287–212 B.C.), Apollonius (ca. 225 B.C.), and Pappus (ca. A.D. 300),
developed the theory of conic sections with a style of proof similar to

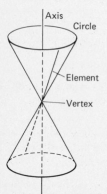

FIGURE 10.1   Cone

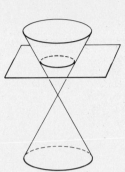

FIGURE 10.2   Circle

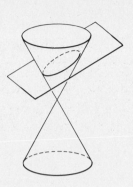

FIGURE 10.3   Ellipse

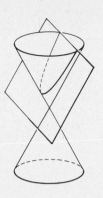

FIGURE 10.4   Parabola

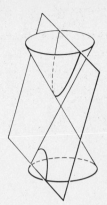

FIGURE 10.5   Hyperbola

that used in modern plane geometry. In calculus, conic sections are studied by using analytic geometry. This approach was initiated by Descartes and Fermat in the seventeenth century.

During the 2,000 years from Menaechmus to Descartes and Fermat, conic sections were studied primarily for their own sake, as a topic in geometry. In the seventeenth century, Galileo and others showed that if there were no air resistance, cannon balls and other projectiles would have parabolic trajectories, and Kepler discovered that the planets move in elliptical orbits. Newton showed that objects under the force of gravity can also have parabolic or hyperbolic paths, and he invented the reflecting telescope, which uses a mirror with parabolic cross sections.

Many large modern telescopes use mirrors with hyperbolic cross sections in addition to those with parabolic cross sections, and navigational systems such as LORAN (Long Range Navigation) are based on properties of hyperbolas. Conic sections are also important in pure mathematics. We showed in Chapter 7 that the graphs of derivatives of logarithms are hyperbolas, and in Chapters 12 and 14 we will see that conic sections arise in the study of surfaces and of maxima and minima of functions of two variables.

10-1   PARABOLAS

Definition 10.1   A parabola is the locus of points P in a plane that are equidistant from a line and a point in the plane.

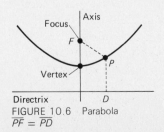

FIGURE 10.6   Parabola
$\overline{PF} = \overline{PD}$

The line is the DIRECTRIX and the point is the FOCUS (Figure 10.6). The line through the focus and perpendicular to the directrix is the *axis* of the parabola. The point where the axis intersects the parabola is the *vertex*.

A parabola in a coordinate plane has the simplest equation if, like one of the parabolas in Figures 10.7 through 10.10, its focus is on a

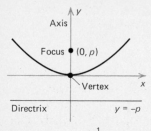

FIGURE 10.7 $y = \dfrac{1}{4p}x^2$

$(p > 0)$

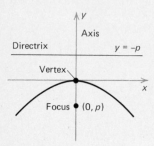

FIGURE 10.8 $y = \dfrac{1}{4p}x^2$

$(p < 0)$

coordinate axis and its vertex is at the origin. The equations of such parabolas are given in the following theorem.

Theorem 10.1   The parabola with its vertex at the origin in an xy-plane, with its focus at $(0, p)$ on the y-axis, and with the line $y = -p$ as directrix (Figures 10.7 and 10.8) has the equation

$$\text{(1)} \qquad y = \frac{1}{4p}x^2. \qquad\qquad\qquad\qquad \blacktriangleleft$$

If the vertex is at the origin, if the focus is at $(p, 0)$ on the x-axis, and if the line $x = -p$ is the directrix (Figures 10.9 and 10.10), then the parabola has the equation

$$\text{(2)} \qquad x = \frac{1}{4p}y^2. \qquad\qquad\qquad\qquad \blacktriangleleft$$

Proof.   The distance from a point (x, y) to the focus $(0, p)$ of the parabola in Figure 10.7 or 10.8 is

$$\text{(3)} \qquad \sqrt{(x - 0)^2 + (y - p)^2}.$$

The distance from (x, y) to the directrix $y = -p$ is

$$\text{(4)} \qquad |y - (-p)| = |y + p|.$$

The point (x, y) is on the parabola if and only if distance (3) is equal to distance (4). Equating the squares of the distances gives $(x - 0)^2 + (y - p)^2 = (y + p)^2$. When we compute the squares and simplify the result, we obtain equation (1). We obtain (2) by switching the roles of x and y.   Q.E.D.

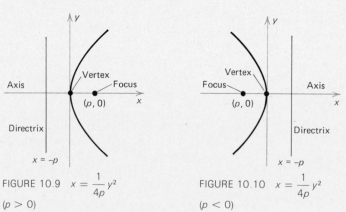

FIGURE 10.9 $x = \dfrac{1}{4p}y^2$

$(p > 0)$

FIGURE 10.10 $x = \dfrac{1}{4p}y^2$

$(p < 0)$

The following example shows how we can use (1) or (2) to find the focus and directrix of a parabola from its equation.

EXAMPLE 1   Find the focus and directrix of the parabola $x = -y^2$.

Solution.   The equation $x = -y^2$ is of the form (2) with $p = -\frac{1}{4}$.

Therefore, the focus of the parabola is at $(p, 0) = (-\frac{1}{4}, 0)$ and the directrix is the line $x = -p = \frac{1}{4}$. The parabola, its focus, and its directrix are shown in Figure 10.11   □

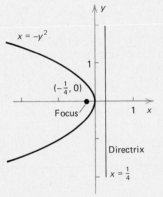

$x = -y^2$

1

$(-\frac{1}{4}, 0)$

Focus

1   x

Directrix

$x = \frac{1}{4}$

FIGURE 10.11

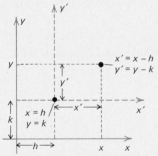

y'

y

y

$x' = x - h$
$y' = y - k$

y'

$x = h$
$y = k$

k

$\longleftarrow x' \longrightarrow$

x'

$\longleftarrow h \longrightarrow$

x x

FIGURE 10.12   Translation of axes

Translation of axes

If we are given the equation of a parabola with a horizontal or vertical axis but whose vertex is not at the origin, we locate its focus and directrix by introducing auxiliary $x'y'$-axes that are parallel to the original xy-axes and have their origin at the vertex of the parabola. This procedure is called a TRANSLATION OF AXES.

If the origin of the $x'y'$-axes is at the point with original coordinates $x = h$ and $y = k$ (Figure 10.12), then the new and old coordinates of a point are related by the equations

(5) $x' = x - h, \quad y' = y - k.$ ◀

The appropriate formulas (5) are found by *completing the square* in the equation of the parabola, as is illustrated in the next example.

EXAMPLE 2   Sketch the parabola $x^2 + 6x - 8y + 17 = 0$. Show its vertex, focus, and directrix.

Solution.   To put the y and the constant term on the left, we add $8y - 17$ to both sides of the equation. This gives $x^2 + 6x = 8y - 17$. We complete the square on the left by putting it in the form $(x + k)^2 = x^2 + 2kx + k^2$. We accomplish this by adding the square of half the coefficient of x to both sides of the equation. The coefficient of x is 6, so we add $3^2 = 9$ to both sides to obtain $x^2 + 6x + 9 = 8y - 17 + 9$ or $(x + 3)^2 = 8(y - 1)$. We write this as $y - 1 = \frac{1}{8}(x + 3)^2$.

To put the last equation in the form

(6) $y' = \frac{1}{8}(x')^2$

we introduce the auxiliary coordinates

$x' = x + 3, \quad y' = y - 1.$

The new variables x' and y' are zero at the point where $x = -3$ and $y = 1$. We draw the $x'y'$-axes with their origin at that point (Figure 10.13).

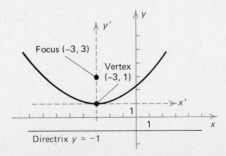

y' y

Focus $(-3, 3)$

Vertex $(-3, 1)$

1   x'

1   x

FIGURE 10.13
$x^2 + 6x - 8y + 17 = 0$

Directrix $y = -1$

Equation (6) is of the form $y' = \dfrac{1}{4p}(x')^2$ with $p = 2$. Therefore, the focus of the parabola is 2 units above and the directrix is 2 units below the origin of the $x'y'$-plane. The parabola, its focus, and its directrix are sketched in Figure 10.13. Relative to the original coordinates the focus has coordinates $(-3, 3)$ and the directrix has the equation $y = -1$.   □

EXAMPLE 3   Give an equation for the locus of points P equidistant from the point $(0, -1)$ and the line $x = 4$.

Solution.   The locus is the parabola with focus $(0, -1)$ and directrix $x = 4$ that is shown in Figure 10.13a. Its axis is the line $y = -1$ through the focus and perpendicular to the directrix. Its vertex is the point $(2, -1)$ on the axis midway between the focus and directrix. Because the parabola opens up to the left, its equation is of the form $x' = \dfrac{1}{4p}(y')^2$ with x' and y' translated coordinates and p negative. The distance from the vertex to the focus and directrix is 2, so $p = -2$. Because the origin of the translated coordinates is at the vertex, which has xy-coordinates $x = 2, y = -1$, we have $x' = x - 2, y' = y + 1$ and the parabola has the equation $x - 2 = -\frac{1}{8}(y + 1)^2$ or $\frac{1}{8}y^2 + x + \frac{1}{4}y = \frac{15}{8}$.   □

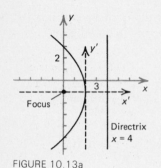

FIGURE 10.13a

An optical property of parabolas

The simplest reflecting telescopes have one mirror with parabolic cross sections. The parallel rays of light from the stars are reflected by the mirror to an eyepiece or camera at the focus of the parabolas (Figure 10.14). This occurs because the tangent line at a point P on the parabola makes equal angles with the line parallel to the axis and with the FOCAL RADIUS, which is the line from P to the focus (Figure 10.15). (See Exercise 29.)

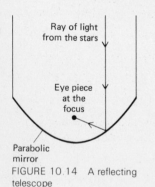

FIGURE 10.14   A reflecting telescope

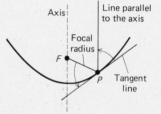

FIGURE 10.15   The tangent line makes equal angles with the focal radius and the line parallel to the axis.

Parabolic trajectories of projectiles

We saw in Section 6.7 that the coordinate at time t of an object moving on a vertical y-axis under the force of gravity is

$$(7) \quad y = -16t^2 + v_0t + y_0$$

where y_0 is the y-coordinate and v_0 is the upward velocity of the object at $t = 0$. Here, time is measured in seconds and the scale on the y-axis in feet.

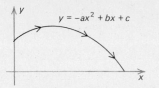

FIGURE 10.16   Parabolic trajectory of a projectile

If the object does not move on a vertical line, it moves in an xy-plane with its y-axis vertical (Figure 10.16). The object's y-coordinate is given by equation (7) and, because there are no horizontal forces, its horizontal velocity is constant and its x-coordinate at time t is

$$(8) \quad x = v_1 t + x_0.$$

Here, x_0 is the object's x-coordinate at $t = 0$ and v_1 is its horizontal velocity.

To obtain an equation for the object's path, we solve (8) for t and substitute the result in equation (7). We obtain

$$y = -16\left(\frac{x - x_0}{v_1}\right)^2 + v_0\left(\frac{x - x_0}{v_1}\right) + y_0 = -ax^2 + bx + c$$

where a, b, and c are constants with $a > 0$. This is an equation for the parabolic trajectory of the object (Figure 10.16). We will return to this topic in the next chapter, where we will be able to employ *vector notation* in studying it.

Exercises

Give equations of the parabolas in Exercises 1 through 10 and sketch them. Show their foci, vertices, and directrices.

1.‡ The parabola with focus $(0, 4)$ and directrix $y = 0$

2.† The parabola with focus $(3, 3)$ and directrix $x = 1$

3. The parabola with focus $(4, -5)$ and directrix $y = 1$

4.† The parabola with vertex $(-2, 0)$ and focus $(4, 0)$

5. The parabola with vertex $(-1, 0)$ and directrix $y = 4$

6.‡ The parabolas with axis $y = 0$, focus $(0, 0)$, and passing through $(3, 4)$

7.† The parabola with vertex $(0, 0)$, axis $x = 0$, and passing through $(3, 1)$

8. The parabolas with axis $x = 0$, focus $(0, -2)$, and passing through $(2, 2)$

9.† The parabola with axis $x = 0$ and passing through $(1, 4)$ and $(2, 7)$

10. The parabola with axis $x = 0$ and passing through $(2, -1)$ and $(-4, 5)$

Sketch the parabolas in Exercises 11 through 21. Show their foci, vertices, and directrices.

11.† $8x^2 + y = 0$ **12.** $3y^2 - x = 0$

13.† $-x^2 + 2x + y = 0$ **14.** $y^2 - 4y - x + 7 = 0$

15.† $x^2 + 6x + 6y = 0$ **16.** $y^2 - 6y + 2x + 13 = 0$

17.† $x^2 + 3x + y + 4 = 0$ **18.** $16y^2 + x - 16y = 1$

19.† $x^2 = 6x + 9y$ **20.** $y^2 + 25 = 10y + 25x$

21. $y = 4x^2 - 4x$

We saw in Exercise 22 of Section 4.9 that the distance from the point (x_0, y_0) to the line $Ax + By + C = 0$ is

(9) $$\frac{|Ax_0 + By_0 + C|}{\sqrt{A^2 + B^2}}.$$

Use formula (9) to find equations for the parabolas in Exercises 22 through 26. Then sketch the curves.

22.‡ The parabola with focus $(0, 3)$ and directrix $y = x$

23.† The parabola with focus $(0, 0)$ and directrix $x + 2y = 4$

24. The parabola with focus $(0, -4)$ and directrix $y = 2x$

25.† The parabola with focus $(-2, 0)$ and vertex $(0, -4)$

26. The parabola with vertex $(0, 0)$ and focus $(-3, 4)$

27.† A headlight reflector has parabolic cross sections and a circular face. The face has a radius of 3 inches and the reflector is 3 inches deep. What is its volume?

28. The shape of a parabolic telescope mirror is obtained by rotating the portion of the curve $y = 10\sqrt{x}$ for $0 \le x \le 9$ about the x-axis. Give an integral which equals the surface area of the mirror.

29. Show that the tangent line to a parabola makes equal angles with the focal radius and a line parallel to the axis. (Choose coordinates so the parabola has the equation $y = \dfrac{1}{4p}x^2$. Show that the tangent line at $P = (x_0, y_0)$ on the parabola intersects the y-axis at $Q = (0, -y_0)$ and that the triangle PQF is isosceles. Here $F = (0, p)$ is the focus.)

10-2   ELLIPSES

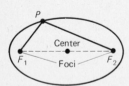

FIGURE 10.17   Ellipse
$\overline{PF_1} + \overline{PF_2} = \text{constant}$

Definition 10.2   **An ellipse is the locus of points P in a plane such that the sum of the distances from P to two points F_1 and F_2 in the plane is a constant (Figure 10.17).**

The points F_1 and F_2 are the *foci* of the ellipse. The point half way between the foci is the CENTER of the ellipse. A circle is the special case of an ellipse where the two foci coincide.

An ellipse in a coordinate plane has the simplest equation if its center is at the origin and its foci are on a coordinate axis (Figures 10.18 and 10.19). The equations of such ellipses are given in the next theorem.

Theorem 10.2   **Suppose that the foci of an ellipse are the points $(c, 0)$ and $(-c, 0)$ on the x-axis (Figure 10.18) and that the sum of the distances from an arbitrary point on the ellipse to the two foci is the constant $2a$ $(a > c \ge 0)$. Then the ellipse has the equation**

(1) $$\frac{x^2}{a^2} + \frac{y^2}{b^2} = 1$$ ◄

where

(2) $b = \sqrt{a^2 - c^2}.$ ◀

If the foci are the points $(0, c)$ and $(0, -c)$ on the y-axis (Figure 10.19), the ellipse has the equation

(3) $\dfrac{x^2}{b^2} + \dfrac{y^2}{a^2} = 1.$ ◀

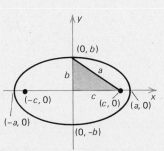

FIGURE 10.18
$\dfrac{x^2}{a^2} + \dfrac{y^2}{b^2} = 1 \ (a > b)$

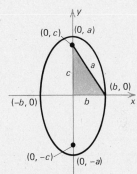

FIGURE 10.19
$\dfrac{x^2}{b^2} + \dfrac{y^2}{a^2} = 1 \ (a > b)$

Equations (1) and (2) are of the same form. The distinction is that in both cases a and b are related by equation (2) and, therefore, $a \geq b > 0$.

Proof.   The equation for the ellipse is $\overline{PF_1} + \overline{PF_2} = 2a$ and expressed in coordinates reads

$$\sqrt{(x - c)^2 + (y - 0)^2} + \sqrt{(x + c)^2 + (y - 0)^2} = 2a.$$

Moving the second square root to the right of the equation and squaring both sides gives

$$(x - c)^2 + y^2 = (x + c)^2 + y^2 - 4a\sqrt{(x + c)^2 + y^2} + 4a^2.$$

This equation simplifies to $cx + a^2 = a\sqrt{(x + c)^2 + y^2}$. Squaring both sides again and simplifying yields

$$\frac{x^2}{a^2} + \frac{y^2}{a^2 - c^2} = 1$$

which becomes equation (1) with substitution (2). Switching the roles of x and y gives equation (3).   Q.E.D.

Notice that the constants a, b, and c are the lengths of the sides of right triangles in Figures 10.18 and 10.19.

Vertices and axes of ellipses

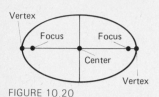

Vertex

Focus   Focus

Center

Vertex

FIGURE 10.20

The line segment extending through the foci and across the ellipse is called the MAJOR AXIS of the ellipse (Figure 10.20). The ends of the major axis are the *vertices* of the ellipse. The MINOR AXIS is perpendicular to the major axis and passes through the center. If the ellipse has an equation of the form (1) or (3), then the length of its major axis is $2a$ and the length of its minor axis is $2b$.

EXAMPLE 1 Sketch the ellipse $9x^2 + 4y^2 = 36$. Show its foci and vertices.

Solution. We put the equation in form (1) or (3) by dividing it by 36 and obtain

$$(4) \qquad \frac{x^2}{2^2} + \frac{y^2}{3^2} = 1.$$

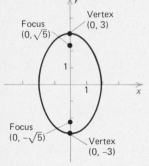

Focus
$(0, \sqrt{5})$

Vertex
$(0, 3)$

Focus
$(0, -\sqrt{5})$

Vertex
$(0, -3)$

FIGURE 10.21 $9x^2 + 4y^2 = 36$

Because 3 is larger than 2, we have $a = 3$ and $b = 2$. The equation of the ellipse is in form (3) and the ellipse has a vertical major axis of length 6 and a horizontal minor axis of length 4. The major axis runs from the vertex at $(0, 3)$ to the vertex at $(0, -3)$, and the minor axis runs from $(2, 0)$ to $(-2, 0)$ (Figure 10.21). (We can check this by finding the x- and y-intercepts from equation (4): setting $y = 0$ gives $x = \pm 2$ as the x-intercepts, and setting $x = 0$ gives $y = \pm 3$ as the y-intercepts.)

Solving equation (2) for c (or examining the shaded right triangle in Figure 10.19) shows that $c = \sqrt{a^2 - b^2} = \sqrt{9 - 4} = \sqrt{5}$ $= 2.236 \ldots$. The foci are $(0, \sqrt{5})$ and $(0, -\sqrt{5})$, as shown in Figure 10.21. $\square$

Translation of axes

If the axes of an ellipse are parallel to the coordinate axes but its center is not at the origin, we complete the squares to determine auxiliary $x'y'$-axes such that the new origin is at the center of the ellipse.

EXAMPLE 2 Sketch the ellipse $x^2 + 4y^2 - 4x + 8y + 4 = 0$. Show its foci and vertices.

Solution. We first group together the terms in x and the terms in y and move the constant to the right side of the equation to obtain $[x^2 - 4x] + [4y^2 + 8y] = -4$. Factoring the coefficient of y^2 from the terms involving y gives $[x^2 - 4x] + 4[y^2 + 2y] = -4$. We complete the square inside the first pair of brackets by adding and subtracting the square $(-2)^2 = 4$ of half the coefficient of x. We complete the square inside the second pair of brackets by adding and subtracting the square $1^2 = 1$ of half the coefficient of y. We obtain first $[x^2 - 4x + 4 - 4]$ $+ 4[y^2 + 2y + 1 - 1] = -4$ and then $[(x - 2)^2 - 4]$ $+ 4[(y + 1)^2 - 1] = -4$. Next, we move the constant terms to the right side, remembering to multiply the -1 by the 4. This gives $(x - 2)^2 + 4(y + 1)^2 = 4$ and then

$$\frac{(x - 2)^2}{2^2} + \frac{(y + 1)^2}{1^2} = 1.$$

To put this equation in the form

(5) $\dfrac{(x')^2}{2^2} + \dfrac{(y')^2}{1^2} = 1$

we introduce the auxiliary coordinates

(6) $x' = x - 2, \quad y' = y + 1.$

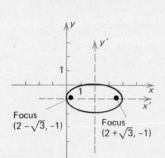

FIGURE 10.22
$x^2 + 4y^2 - 4x + 8y + 4 = 0$

The center of the ellipse is at the origin of the new coordinates. This is the point with original coordinates $x = 2$ and $y = -1$. Equation (6) shows that the x'- and y'-intercepts are $x' = \pm 2$ and $y' = \pm 1$. The ellipse has a horizontal major axis of length 4 and a vertical minor axis of length 2. The vertices are at $x = 0, y = -1$ and at $x = 4, y = -1$ (Figure 10.22). We have $a = 2$ and $b = 1$ in (5). Therefore $c = \sqrt{a^2 - b^2} = \sqrt{3}$ and the foci are $\sqrt{3} = 1.732 \ldots$ to the left and right of the center at $x = 2 - \sqrt{3}, y = -1$ and $x = 2 + \sqrt{3}$, $y = -1$. □

EXAMPLE 3   Give an equation of the locus of points P such that the sum of the distances from P to the points $(-1, -2)$ and $(-1, 4)$ is 8.

Solution.   The locus is the ellipse with foci $(-1, -2)$ and $(-1, 4)$ that is shown in Figure 10.22a. Because its major axis is vertical, its equation is of the form

$$\dfrac{(x')^2}{b^2} + \dfrac{(y')^2}{a^2} = 1$$

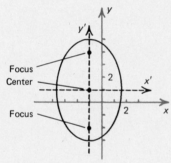

FIGURE 10.22a

with x' and y' auxiliary translated coordinates and $0 < b < a$. The origin of the auxiliary coordinates is at the center $(-1, 1)$ of the ellipse midway between the foci, so we have $x' = x + 1, y' = y - 1$. The constant a is 4 because the sum of the distances from points on the ellipse to the foci is 8. The constant $c = 3$ is the distance from the center to the foci. We have $b = \sqrt{a^2 - c^2} = \sqrt{4^2 - 3^2} = \sqrt{7}$, so the ellipse has the equation

$$\dfrac{(x + 1)^2}{7} + \dfrac{(y - 1)^2}{16} = 1$$

or $16x^2 + 7y^2 + 32x - 14y = 89$. □

An optical property of ellipses

If a ray of light passes through one focus of an ellipse and reflects off the ellipse, then it passes through the other focus (Figure 10.23). This is because the tangent line at a point P on an ellipse makes equal angles with the focal radii at that point (Figure 10.24 and Exercise 27).

This property of ellipses is used in "whispering rooms". In these rooms the ceilings have cross sections which are arcs of ellipses. A person standing at one focus of the ellipses at one end can hear another person whispering at the other focus because all of the sound from one focus that hits the ceiling is reflected to the other focus.

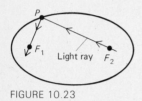

FIGURE 10.23

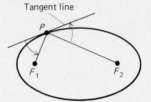

FIGURE 10.24   The tangent line makes equal angles with the focal radii.

Exercises

†*Answer provided.*
‡*Outline of solution provided.*

Sketch the ellipses in Exercises 1 through 8. Show their major and minor axes and foci.

1.† $x^2 + 4y^2 = 16$ **2.** $9x^2 + 16y^2 = 144$

3.† $9x^2 + y^2 - 18x + 2y + 1 = 0$

4. $x^2 + 8x + 2y^2 + 12 = 0$ **5.**† $4x^2 + y^2 - 16x - 4y = 4$

6. $10x^2 + 3y^2 + 18y = 3$ **7.**† $9x^2 + 25y^2 + 18x + 50y = 191$

8. $49x^2 + y^2 - 14y = 0$

Give equations for the ellipses in Exercises 9 through 19.

9.† The ellipse with foci $(-3, 0)$ and $(3, 0)$ and major axis of length 10

10. The ellipse with foci $(0, 4)$ and $(0, -4)$ and minor axis of length 6

11.‡ The ellipse with foci $(0, 0)$ and $(24, 0)$ and major axis of length 26

12.† The ellipse with foci $(1, 1)$ and $(1, 3)$ and minor axis of length 2

13. The ellipse with foci $(0, -1)$ and $(0, -5)$ and major axis of length 6

14.† The ellipse with its center at the point $(1, 10)$, a horizontal axis of length 8, and a vertical axis of length 2

15. The ellipse with its center at $(2, 2)$, a horizontal axis of length 6, and a vertical axis of length 8

16.† The ellipse with vertices $(-8, 3)$ and $(0, 3)$ and minor axis of length 6

17. The ellipse with vertices $(2, 1)$ and $(2, 11)$ and minor axis of length 2

18.‡ The locus of points P such that the sum of the distances from P to $(2, -1)$ and to $(2, 3)$ is 12

19. The locus of points P such that the sum of the distances from P to $(0, 0)$ and to $(0, -4)$ is 10

Give equations in the form $Ax^2 + Bxy + Cy^2 + Dx + Ey + F = 0$ for the ellipses in Exercises 20 through 23.

20.‡ The locus of points P such that the sum of the distances from P to $(0, 0)$ and to $(1, 1)$ is 2

21.† The locus of points P such that the sum of the distances from P to $(-1, 0)$ and $(0, 3)$ is 10

22. The locus of points P such that the sum of the distances from P to $(1, 0)$ and to $(0, 1)$ is 6

23.† Find the area of the ellipse (1).

24. Give an integral that equals the circumference of the ellipse (1). (For $a \neq b$, this is an *elliptic integral*, whose value may be found in tables of these integrals.)

25.† When the ellipse (1) is rotated about the y-axis, it generates an *oblate spheroid*, and when it is rotated about the x-axis, it generates a *prolate spheroid* (two types of *ellipsoids*). Compute their volumes.

26. Show that the point $(a \cos t, b \sin t)$ lies on the ellipse (1) for any value of t.

27.‡ Show that the tangent line at the point (x_0, y_0) on the ellipse (1) makes equal angles with the focal radii at that point.

28. Explain why, when you look at a circle from an angle, you see an ellipse. (Show that if your line of sight is perpendicular to the x-axis and makes an angle α with the y-axis, a circle of radius r in the xy-plane looks like the ellipse $x^2 + y^2/\sin^2\alpha = r^2$ viewed from a point perpendicular to the plane.)

10-3   HYPERBOLAS

Definition 10.3   A hyperbola is the locus of points P in a plane such that the difference between the distances from P to two points F_1 and F_2 in the plane is constant (Figure 10.25).

Here when we say the "difference" of the distances, we mean the larger minus the smaller one. The points F_1 and F_2 are the *foci* of the hyperbola; the line through the foci is the *axis*; the point half way between the foci is the *center*; and the points where the axis intersects the hyperbola are the *vertices*.

The hyperbolas with the simplest equations are those with centers at the origin and foci on a coordinate axis. The equations of such hyperbolas are given in the following theorem.

Theorem 10.3   Suppose that the foci of a hyperbola are the points $(c, 0)$ and $(-c, 0)$ on the x-axis (Figure 10.26) and that the difference between the distances from an arbitrary point on the hyperbola to the two foci is $2a$ ($0 < a < c$). Then the hyperbola has the equation

(1) $\dfrac{x^2}{a^2} - \dfrac{y^2}{b^2} = 1$ ◀

with

(2) $b = \sqrt{c^2 - a^2}.$ ◀

If the foci are $(0, c)$ and $(0, -c)$ on the y-axis (Figure 10.27), then the hyperbola has the equation

(3) $\dfrac{y^2}{a^2} - \dfrac{x^2}{b^2} = 1.$ ◀

The distinction between equations (1) and (3) is that in both cases b^2 goes with the minus sign. Either a or b may be the larger number, or they may be equal.

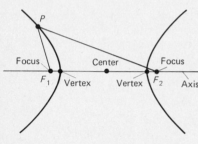

FIGURE 10.25   Hyperbola $|\overline{PF_1} - \overline{PF_2}| = $ constant

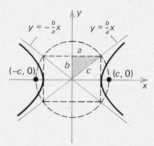

FIGURE 10.26 $\dfrac{x^2}{a^2} - \dfrac{y^2}{b^2} = 1$

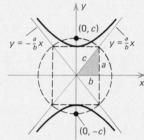

FIGURE 10.27 $\dfrac{y^2}{a^2} - \dfrac{x^2}{b^2} = 1$

Proof.   The hyperbola has the equation $|\overline{PF_1} - \overline{PF_2}| = 2a$, which in coordinates reads

(4) $|\sqrt{(x - c)^2 + y^2} - \sqrt{(x + c)^2 + y^2}| = 2a.$

This equation can be simplified by a calculation similar to that used for the ellipse (see Exercise 26). It becomes

$$\frac{x^2}{a^2} - \frac{y^2}{c^2 - a^2} = 1$$

which gives (1) with substitution (2). Interchanging the roles of x and y gives equation (3).   Q.E.D.

Asymptotes   Hyperbolas have one feature not possessed by parabolas or ellipses; they have *asymptotes*. If we solve equation (1) for y and then factor out $(b/a)\, x$, we obtain

$$y = \pm \sqrt{\left(\frac{b}{a} x\right)^2 - b^2} = \pm \left(\frac{b}{a} x\right) \sqrt{1 - \left(\frac{a}{x}\right)^2}.$$

From this equation it follows that $y \mp (b/a)\, x$ tends to 0 as x tends to ∞ or to $-\infty$ (see Exercise 30). Therefore, the lines $y = \pm (b/a)\, x$ through the corners of the rectangle in Figure 10.26 are asymptotes of the hyperbola. Similarly, the hyperbola of Figure 10.27 has the asymptotes $y = \pm(a/b)x$.

Equation (2) shows that c is the length of the hypotenuse of the shaded right triangles in Figures 10.26 and 10.27. Consequently, the circles through the corners of the rectangles pass through the foci of the hyperbolas.

EXAMPLE 1   Sketch the hyperbola $9x^2 - 16y^2 = 144$. Show its foci, vertices, and asymptotes.

Solution.   We divide the equation by 144 to obtain

$$\frac{x^2}{4^2} - \frac{y^2}{3^2} = 1.$$

This is in the form of equation (1) with $a = 4$ and $b = 3$. The center of the hyperbola is at the origin and its vertices are at $(4, 0)$ and $(-4, 0)$. (We can also see this by setting $y = 0$ in the equation to find the x-intercepts $x = \pm 4$.) The asymptotes are the lines $y = \pm\frac{3}{4}x$. Equation (2) gives $c = \sqrt{4^2 + 3^2} = 5$, so the foci are at $(5, 0)$ and $(-5, 0)$ (Figure 10.28).   □

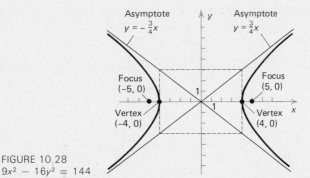

FIGURE 10.28
$9x^2 - 16y^2 = 144$

Translation of axes

If the axis of the hyperbola is horizontal or vertical but its center is not at the origin, we find its center, foci, vertices, and asymptotes by translating axes. We complete the squares, just as we did in the case of an ellipse in Section 10.2, to find $x'y'$-coordinates with their origin at the center of the hyperbola.

EXAMPLE 2   Sketch the hyperbola $y^2 - x^2 - 10y + 6x = 0$. Show its center, foci, vertices, and asymptotes.

Solution.   We complete the two squares by writing $[y^2 - 10y]$ $- [x^2 - 6x] = 0$, $[y^2 - 10y + 25] - [x^2 - 6x + 9] = 25 - 9$, and then $(y - 5)^2 - (x - 3)^2 = 16$. Dividing by 16 gives

$$\frac{(y - 5)^2}{4^2} - \frac{(x - 3)^2}{4^2} = 1.$$

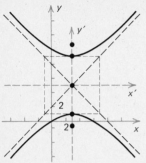

FIGURE 10.29   The hyperbola
$y^2 - x^2 - 10y + 6x = 0$
Foci: $(3, 5 + \sqrt{32})$, $(3, 5 - \sqrt{32})$
Vertices: $(3, 9)$, $(3, 1)$
Center: $(3, 5)$
Asymptotes: $y = x + 2$,
$y = -x + 8$

We introduce the auxiliary coordinates

$$(5) \quad x' = x - 3, \quad y' = y - 5$$

to put the equation in the form

$$(6) \quad \frac{(y')^2}{4^2} - \frac{(x')^2}{4^2} = 1.$$

The center of the hyperbola is at $x = 3$, $y = 5$ where x' and y' are zero (Figure 10.29). Setting $x' = 0$ in (6) shows that the vertices are at $x' = 0$, $y' = \pm 4$ on the y'-axis. By equations (5) their xy-coordinates are $x = 3$, $y = 9$ and $x = 3$, $y = 1$. Equation (6) gives $a = b = 4$. Therefore $c = \sqrt{a^2 + b^2} = \sqrt{4^2 + 4^2}$ is $\sqrt{32} = 5.656 \ldots$, and the foci are $\sqrt{32}$ units above and below the center. They have xy-coordinates $x = 3$, $y = 5 + \sqrt{32}$ and $x = 3$, $y = 5 - \sqrt{32}$. The asymptotes have equations $y' = \pm x'$ or, by equations (5), $y = x + 2$ and $y = -x + 8$. $\square$

A hyperbola, such as in Figure 10.29, whose asymptotes are perpendicular to each other is known as a *rectangular hyperbola*.

EXAMPLE 3   (a) What is the locus of points P such that the difference between the distances from P to $(1, 0)$ and $(-5, 0)$ is 4? (b) What is the locus of points that are 4 units closer to $(1, 0)$ than to $(-5, 0)$?

Solution.   (a) The locus is the hyperbola with foci $(1, 0)$ and $(-5, 0)$ that is shown in Figure 10.29a. Its equation is of the form

$$\frac{(x')^2}{a^2} - \frac{(y')^2}{b^2} = 1$$

with translated coordinates x' and y'. The origin of the auxiliary coordinates is at the center $(-2, 0)$ of the hyperbola midway between the foci. Therefore, $x' = x + 2$ and $y' = y$. The constant a is 2 because the difference between the distances from points on the hyperbola to the foci is 4. The constant c is the distance 3 from the center to the foci. Since $b = \sqrt{c^2 - a^2} = \sqrt{3^2 - 2^2} = \sqrt{5}$, the hyperbola has the equation

$$(7) \quad \frac{(x + 2)^2}{4} - \frac{y^2}{5} = 1$$

or $5x^2 + 20x - 4y^2 = 0$.

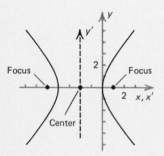

FIGURE 10.29a

(b) The locus of points which are 4 units closer to $(1, 0)$ than to $(-5, 0)$ is the right half of the hyperbola in Figure 10.29a. Solving equation (7) for x gives $x = -2 \pm 2\sqrt{1 + \frac{1}{3}y^2}$, so the right half of the hyperbola has the equation $x = -2 + 2\sqrt{1 + \frac{1}{3}y^2}$. □

An optical property of hyperbolas

A ray of light that approaches a hyperbola from the side opposite a focus and in a line pointing toward that focus reflects off the hyperbola in a line pointing toward the other focus (Figure 10.30). This is because the tangent line at a point P on the hyperbola bisects the focal radii at P (Figure 10.31). (See Exercise 27).

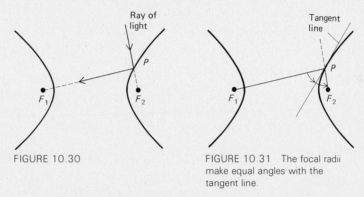

FIGURE 10.30

FIGURE 10.31   The focal radii make equal angles with the tangent line.

This property of hyperbolas is employed in the Cassegrain telescope, which uses a mirror with hyperbolic cross sections and a mirror with parabolic cross sections (Figure 10.32). One focus of the hyperbolic mirror is at the focus of the parabolic mirror and the other is at the vertex of the parabolic mirror where there is an eyepiece or camera. The parallel rays of light from the stars reflect off the parabolic mirror toward its focus and then off the hyperbolic mirror toward the eyepiece or camera.

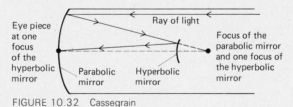

Eye piece at one focus of the hyperbolic mirror

Ray of light

Focus of the parabolic mirror and one focus of the hyperbolic mirror

Parabolic mirror   Hyperbolic mirror

FIGURE 10.32   Cassegrain telescope

Hyperbolas in navigation

The LORAN (Long Range Navigation) system enables ships to determine their location by radio. Typically, a "master" station transmits pulses at 0.05 second intervals and a "slave" station transmits pulses 0.001 seconds later. A ship has equipment that measures the delay between reception of signals from the two stations. By using two pairs of stations, or one master and two slaves, as in Figure 10.33, the ship finds its location as an intersection of hyperbolas. A ship at F in Figure 10.33, for example, receives the signals from the master station 3000 microseconds (0.003 seconds) before the signals from slave 1 and 2500 microseconds before the signals from slave 2. Its position is the intersection of the hyperbolas along which these time delays occur. (Delay times for the master and slave 1 are constant along the half hyperbolas drawn with broken lines. Delay times for the master and slave 2 are constant along the half hyperbolas drawn with solid lines. Times are measured in microseconds.)

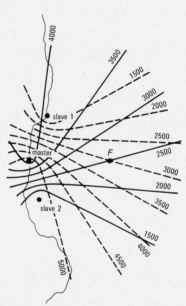

FIGURE 10.33   (Adapted from *McGraw Hill Encyclopedia of Science*, Vol. 7, p. 662.)

Exercises

†*Answer provided.*
‡*Outline of solution provided.*

Sketch the hyperbolas in Exercises 1 through 10. Show their foci, vertices, and asymptotes.

1.† $16x^2 - y^2 = 16$

2. $36y^2 - 9x^2 = 36$

3.† $x^2 - y^2 + 2x + 2y = 1$

4. $9y^2 - 16x^2 - 64x = 208$

5.† $y^2 - 25x^2 - 10y = 0$

6. $9x^2 - y^2 + 18x - 4y = 4$

7.† $x^2 - y^2 + 2x + 6y = 12$

8. $4y^2 - 9x^2 - 36x + 24y - 36 = 0$

9.† $x^2 - 100y^2 + 20x - 200y = 100$

10. $81y^2 - x^2 + 18x + 162y - 81 = 0$

Derive equations for the hyperbolas in Exercises 11 through 21.

11.[†] The hyperbola with foci $(-4, 0)$ and $(4, 0)$ and vertices $(-3, 0)$ and $(3, 0)$

12. The hyperbola with foci $(0, 2)$ and $(0, -2)$ and vertices two units apart

13.[†] The hyperbola with vertices $(1, 3)$ and $(9, 3)$ and foci 10 units apart

14. The hyperbola with foci $(-4, 5)$ and $(-4, -15)$ and vertices $(-4, -3)$ and $(-4, -7)$

15.[‡] The hyperbola with asymptotes $y = \pm 2x$ and vertices $(1, 0)$ and $(-1, 0)$

16.[†] The hyperbola with asymptotes $y = \pm x$ and foci $(3, 0)$ and $(-3, 0)$

17. The hyperbola with asymptotes $y = \pm \frac{3}{4}x$ and foci $(0, 5)$ and $(0, -5)$

18.[†] The locus of points P such that the difference between the distances from P to $(4, 5)$ and $(-2, 5)$ is 2

19. The locus of points P such that the difference between the distances from P to $(3, 6)$ and $(3, 0)$ is 4

20.[†] The locus of points P such that the difference between the distances from P to $(-4, 1)$ and $(6, 1)$ is 6

21. The locus of points P such that the difference between the distances from P to $(2, -3)$ and $(2, 7)$ is 4

Give equations in the form $Ax^2 + Bxy + Cy^2 + Dx + Ey + F = 0$ for the hyperbolas in Exercises 22 through 24.

22.[‡] The hyperbola with vertices $(3, 4)$ and $(6, 8)$ and with foci $(0, 0)$ and $(9, 12)$

23.[†] The locus of points P such that the difference between the distances from P to $(4, 4)$ and $(-4, -4)$ is 8

24. The locus of points P such that the difference between the distances from P to $(3, -4)$ and $(-3, 4)$ is 6

25. Find the asymptotes of the hyperbolas   **a.**[‡] $4x^2 + 5xy + y^2 = 10$, **b.**[†] $5x^2 - 6xy + y^2 = 9$,   **c.** $4x^2 + 5xy - y^2 = 1$.

26. Derive equation (1) from equation (4).

27.[‡] Prove that the tangent line at the point (x_0, y_0) on the hyperbola (1) bisects the focal radii at that point.

28. What condition must the constants a and b satisfy for the hyperbolas (1) and (3) to be rectangular?

29. Let R denote the region bounded by the hyperbola $y^2 - x^2 = 1$ and by the lines $x = \pm 1$. Compute the volumes of the solids generated by rotating R **a.** about the x-axis and   **b.** about the y-axis.

30.[†] Show that $y = \pm (b/a)\, x$ are asymptotes of hyperbola (1).

10-4   ROTATION OF AXES

Every conic section has an equation of the form

(1) $Ax^2 + Bxy + Cy^2 + Dx + Ey + F = 0$ ◀

with at least one of the constants A, B, and C not zero. If the coefficient B of xy is zero, the axis of the conic section is parallel to one of the coordinate axes and, as we saw in the previous sections, we can determine its properties by translating axes. If the coefficient B is not zero, the axis of the conic section is not parallel to either coordinate axis and we initiate our study of the curve by introducing auxiliary $x'y'$-axes such that one of the new axes is parallel to the axis of the conic. The $x'y'$-axes are obtained from the original xy-axes by rotating them about the origin. We use the following result.

The formulas for rotation of axes

Theorem 10.4 **(a) If the auxiliary x′y′-axes are obtained by rotating the original xy-axes through an angle α (alpha) (Figure 10.34), then the old and new coordinates of a point are related by the equations**

(2) $\begin{aligned}x &= x'\cos\alpha - y'\sin\alpha \\ y &= x'\sin\alpha + y'\cos\alpha.\end{aligned}$ ◀

(b) Making substitution (2) in equation (1) for a conic section yields an equation of the form

(3) $A'x'^2 + B'x'y' + C'y'^2 + D'x' + E'y' + F' = 0$

with coefficients such that

(4) $B'^2 - 4A'C' = B^2 - 4AC.$ ◀

(c) If B is not zero in equation (1) and α is such that

(5) $\tan\alpha = \dfrac{C-A}{B} + \sqrt{\left(\dfrac{C-A}{B}\right)^2 + 1}$ ◀

then B′ is zero in equation (3) and the axis of the conic section is parallel to the x′- or y′-axis.

Note. An angle α which makes B' zero may also be determined from the equation

(6) $\cot(2\alpha) = \dfrac{A-C}{B}$

(see Exercise 19). The advantage of formula (5) is that it enables us to determine $\sin\alpha$ and $\cos\alpha$ directly from $\tan\alpha$ without finding the angle α, as we will see in Examples 1 and 2.   □

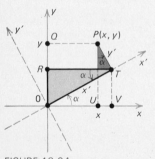

FIGURE 10.34
$x = \overline{OU} = \overline{OV} - \overline{UV}$
$\quad = x'\cos\alpha - y'\sin\alpha$
$y = \overline{OQ} = \overline{OR} + \overline{RQ}$
$\quad = x'\sin\alpha + y'\cos\alpha$

Proof. (a) We will derive formulas (2) for a point P with positive coordinates as in Figure 10.34. Similar arguments would give the formulas for any location of P. (They can also be derived by using trigonometric identities, as in Exercise 21.)

With the notation of Figure 10.34 we have $x = \overline{OU}$, $y = \overline{OQ}$, $x' = \overline{OT}$, and $y' = \overline{TP}$. Using the triangles which contain the angle α, we then find that $\overline{OV} = x'\cos\alpha$, $\overline{UV} = y'\sin\alpha$, $\overline{OR} = \overline{TV} = x'\sin\alpha$, and $\overline{RQ} = y'\cos\alpha$. Therefore,

$$x = \overline{OU} = \overline{OV} - \overline{UV} = x'\cos\alpha - y'\sin\alpha$$

$$y = \overline{OQ} = \overline{OR} + \overline{RQ} = x'\sin\alpha + y'\cos\alpha.$$

These are equations (2).

(b) When we substitute formulas (2) into equation (1) and use the Pythagorean identity $\sin^2\alpha + \cos^2\alpha = 1$ to simplify the resulting equation, we obtain (3) with

(7) $A' = A\cos^2\alpha + B\cos\alpha\sin\alpha + C\sin^2\alpha$

(8) $B' = B(\cos^2\alpha - \sin^2\alpha) + 2(C - A)\sin\alpha\cos\alpha$

(9) $C' = A\sin^2\alpha - B\cos\alpha\sin\alpha + C\cos^2\alpha.$

Using (7), (8), and (9) to express $(B')^2 - 4A'C'$ in terms of A, B, and C then yields equation (4). (We leave these calculations as Exercises 17 and 18.)

(c) If B is not zero, we set the right side of equation (8) equal to zero and divide by $\cos^2\alpha$ to obtain the equation

$$B(1 - \tan^2\alpha) + 2(C - A)\tan\alpha = 0$$

and ther

$$B\tan^2\alpha - 2(C - A)\tan\alpha + B = 0$$

for the choices of $\tan\alpha$ that make B' zero. By the quadratic formula, the last equation has the solutions

(10) $\tan\alpha = \dfrac{C - A}{B} \pm \sqrt{\left(\dfrac{C - A}{B}\right)^2 + 1}.$

Taking the plus sign in (10) gives formula (5).   Q.E.D.

Note. Because the quantity (5) is positive, we can use a positive acute angle α and obtain this tangent. If we had taken the minus sign in (10), we would have a negative $\tan\alpha$, which would correspond to a negative acute angle α of rotation.   □

To sketch a conic section whose equation involves an xy-term (an equation of the form (1) with $B \neq 0$), we carry out the following steps.

Step 1 Compute $\tan\alpha$ from equation (5).

Step 2 Find $\sin\alpha$ and $\cos\alpha$ for a corresponding positive acute angle α by using a right triangle as in Figure 10.34a.

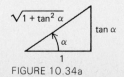

FIGURE 10.34a

Step 3   Substitute formulas (2) with cos α and sin α from Step 2 into the equation for the conic section and combine coefficients of $(x')^2$, $x'y'$, $(y')^2$, x', and y'. The $x'y'$ terms should drop out leaving an equation of the form

(11) $A'(x')^2 + B'(y')^2 + D'x' + E'y' + F' = 0.$

Step 4   Use equation (11) and the techniques of Sections 10.1 through 10.3 to study the curve relative to the $x'y'$-coordinates.

Step 5   If necessary, use the formulas

(12)
$$x' = x \cos \alpha + y \sin \alpha$$
$$y' = -x \sin \alpha + y \cos \alpha$$

to express equations for directrices or asymptotes in terms of x and y.

(Because we can obtain the xy-axes from the $x'y'$-axes by a rotation of $-\alpha$ and because $\cos(-\alpha) = \cos \alpha$ and $\sin(-\alpha) = -\sin \alpha$, equations (12) are obtained from equations (2) by interchanging x with x' and y with y' and by replacing $\sin \alpha$ with $-\sin \alpha$.)

Unless the coefficients D' and E' of x' and y' in (11) are both zero, Step 4 involves a translation of axes after the rotation. This is illustrated in the exercises. The following examples involve only rotation of axes.

EXAMPLE 1   Sketch the hyperbola $xy = 1$. Show its foci, vertices, and asymptotes.

Solution.   In the equation $xy = 1$ we have $A = 0$, $B = 1$, and $C = 0$. Formula (5) gives $\tan \alpha = 1$, which we can satisfy by taking $\alpha = \pi/4$. With this α we have $\cos \alpha = \cos(\pi/4) = 1/\sqrt{2}$ and $\sin \alpha = \sin(\pi/4) = 1/\sqrt{2}$, and equations (2) read

(13) $x = \dfrac{1}{\sqrt{2}} x' - \dfrac{1}{\sqrt{2}} y', \quad y = \dfrac{1}{\sqrt{2}} x' + \dfrac{1}{\sqrt{2}} y'.$

Making substitutions (13) in the equation $xy = 1$ gives the equation for the hyperbola in terms of x' and y':

$$\left(\frac{1}{\sqrt{2}} x' - \frac{1}{\sqrt{2}} y' \right)\left(\frac{1}{\sqrt{2}} x' + \frac{1}{\sqrt{2}} y' \right) = 1.$$

When we compute the product, we obtain

(14) $\frac{1}{2}(x')^2 - \frac{1}{2}(y')^2 = 1.$

We do not have to perform a translation of axes because there are no x' or y' terms in equation (14).

To sketch the curve, we draw the auxiliary $x'y'$-axes as in Figure 10.35. Setting $y' = 0$ in (14) gives $x' = \pm\sqrt{2}$, so the vertices of the hyperbola are at $x' = \pm\sqrt{2}$ on the x'-axis. From equations (13) we find that the vertices are at $x = 1, y = 1$ and at $x = -1, y = -1$.

Equation (14) is of the form of equation (1) in Section 10.3 with x and y replaced by x' and y' and with $a = b = \sqrt{2}$. Equation (2) of Section 10.3 gives $c = \sqrt{a^2 + b^2} = \sqrt{2 + 2} = 2$, so the foci of the hyperbola are at $x' = \pm 2$ on the x'-axis. Equations (13) show that the foci are at $x = \sqrt{2}, y = \sqrt{2}$ and at $x = -\sqrt{2}, y = -\sqrt{2}$.

Because $a = b$, the asymptotes have the equations $y' = \pm x'$ in the auxiliary coordinates and are at an angle of $\pi/4$ with the auxiliary axes. Thus the asymptotes are the x- and y-axes. The curve is the rectangular hyperbola shown in Figure 10.35. □

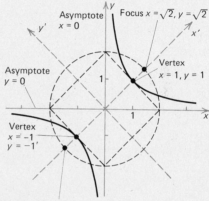

FIGURE 10.35   The hyperbola
$xy = 1$ or $\frac{1}{2}(x')^2 - \frac{1}{2}(y')^2 = 1$

EXAMPLE 2   Sketch the conic section

(15) $17x^2 - 12xy + 8y^2 = 20.$

Solution.   Equation (15) is of the form (1) with $A = 17, B = -12$, and $C = 8$, so formula (5) gives

$$\tan \alpha = \frac{C - A}{B} + \sqrt{\left(\frac{C - A}{B}\right)^2 + 1} = \tfrac{3}{4} + \sqrt{(\tfrac{3}{4})^2 + 1}$$

$$= \tfrac{3}{4} + \sqrt{\tfrac{25}{16}} = \tfrac{3}{4} + \tfrac{5}{4} = 2.$$

The triangle in Figure 10.35a shows that for the positive acute angle α with $\tan \alpha = 2$, we have $\sin \alpha = 2/\sqrt{5}$ and $\cos \alpha = 1/\sqrt{5}$. Then equations (2) read

FIGURE 10.35a

(16) $x = \dfrac{1}{\sqrt{5}}(x' - 2y'), \qquad y = \dfrac{1}{\sqrt{5}}(2x' + y').$

Substituting formulas (16) into (15) gives

$$\tfrac{17}{5}(x' - 2y')^2 - \tfrac{12}{5}(x' - 2y')(2x' + y') + \tfrac{8}{5}(2x' + y')^2 = 20$$

which simplifies to

$$(17 - 24 + 32)(x')^2 + (-68 + 36 + 32)x'y'$$

$$+ (68 + 24 + 8)(y')^2 = 100$$

and then

(17) $\frac{1}{4}(x')^2 + (y')^2 = 1.$

Equation (17) does not involve any x' or y' terms, so we do not have to introduce additional translated coordinates. It is of the form of equation (1) of section 10.2 with $a = 2$ and $b = 1$, so the curve is an ellipse whose major axis runs from $x' = 2$ to $x' = -2$ on the x'-axis and whose minor axis runs from $y' = 1$ to $y' = -1$ on the y'-axis. The constant c is $\sqrt{a^2 - b^2} = \sqrt{2^2 - 1^2} = \sqrt{3}$, so the foci are at $x' = \pm\sqrt{3}$ on the x'-axis. We use the fact that $\tan\alpha$ is 2 to draw the x'- and y'-axes and then sketch the ellipse as in Figure 10.35b. We can use formulas (16) with $x' = \pm\sqrt{3}$ and $y' = 0$ to find the xy-coordinates $(\sqrt{\frac{3}{5}}, 2\sqrt{\frac{3}{5}})$ and $(-\sqrt{\frac{3}{5}}, -2\sqrt{\frac{3}{5}})$ of the foci. $\square$

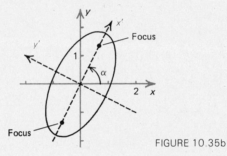

FIGURE 10.35b

Classification of conic sections by their discriminants

We have not yet made use of equation (4)

(4) $B'^2 - 4A'C' = B^2 - 4AC$

of Theorem 10.4. The number $B^2 - 4AC$ is the DISCRIMINANT of the conic section. Equation (4) shows that the discriminant is not changed when we rotate coordinate axes. Because it is also not changed by translations of axes (see Exercise 20), we can determine whether a conic section is a parabola, an ellipse, or a hyperbola without any rotations or translations of axes by computing its discriminant. We employ the following result.

Theorem 10.5   The conic section

(1) $Ax^2 + Bxy + Cy^2 + Dx + Ey + F = 0$

is a parabola if its discriminant $B^2 - 4AC$ is zero, an ellipse if its discriminant is negative, and a hyperbola if its discriminant is positive.

Proof. We rotate the axes to obtain an equation

(18) $A'(x')^2 + C'(y')^2 + D'x' + E'y' + F' = 0$

with no $x'y'$ term. We then have, by equation (4),

(19) $B^2 - 4AC = (B')^2 - 4A'C' = -4A'C'.$

If the discriminant (19) is zero then either A' or C' is zero and the other is not, so completing the square in equation (18) gives the equation of a parabola. If the discriminant is negative, then A' and C' have the same sign and completing the squares gives the equation of an ellipse. If the discriminant is positive, then A' and C' are of opposite signs, and completing the squares gives the equation of a hyperbola.   Q.E.D.

EXAMPLE 3   Is the conic section

(20) $x^2 - 2xy + 5y^2 - 6x + y - 4 = 0$

a parabola, an ellipse, or a hyperbola?

Solution.   We have $A = 1, B = -2,$ and $C = 5$ in equation (20). The discriminant $B^2 - 4AC = (-2)^2 - 4(1)(5) = -16$ is negative, so the conic section is an ellipse.   □

Degenerate conic sections

For certain choices of the constants the graph of the equation

(1) $Ax^2 + Bxy + Cy^2 + Dx + Ey + F = 0$

is not what we would normally call a parabola, an ellipse, or a hyperbola but is a line, a pair of lines, a point, or the empty set. We refer to such graphs as DEGENERATE CONIC SECTIONS, and we classify them by their discriminants.

The equations $(x - y)^2 = 0, x^2 - 4 = 0,$ and $y^2 + 1 = 0$ have $B^2 - 4AC = 0$ and their graphs are *degenerate parabolas.* The graph of the first is the line $y = x$; the graph of the second is the pair of parallel lines $x = 2$ and $x = -2$; and the graph of the third is the empty set. The equations $x^2 + y^2 = 0$ and $x^2 + y^2 + 1 = 0$ have $B^2 - 4AC < 0$ and their graphs are *degenerate ellipses.* The graph of the first is the single point $(0, 0)$ and the graph of the second is the empty set. The equation $xy = 0$ has $B^2 - 4AC > 0$ and its graph is a *degenerate hyperbola.* It consists of the intersecting lines $x = 0$ and $y = 0$.

Exercises

†*Answer provided.*
‡*Outline of solution provided.*

Use the discriminant to classify the conic sections in Exercises 1 through 6.

1.† $x^2 + 3xy - y^2 + 6x - 6 = 0$ **2.** $3x^2 - xy + 4y^2 - 16x - 3y = 0$

3.† $-x^2 + 4xy - 4y^2 + x = 1$ **4.** $x^2 + xy + y^2 = x + y$

5.† $x^2 + 2xy + y^2 = x - y$ **6.** $x^2 + 3xy + y^2 = -x + y + 2$

In Exercises 7 through 15 sketch the conic sections. Use either formula (5) or formula (6) to find a suitable angle α of rotation.

7.‡ $x^2 - 2\sqrt{3}xy - y^2 = 8$ **8.**† $2x^2 + \sqrt{3}xy + y^2 = 2$

9. $x^2 + xy + y^2 = 8$ **10.** $x^2 + 4xy + y^2 = 1$

11.‡ $2xy + 2\sqrt{2}x = 1$ **12.†** $x^2 + 2xy + y^2 = 4\sqrt{2}y$

13. $x^2 + 2xy + y^2 + 8x - 8y = 0$

14.† $x^2 + \sqrt{3}xy + 2y^2 + \sqrt{3}x - y = \frac{1}{2}$

15. $5x^2 + 6xy + 5y^2 + 8\sqrt{2}x + 8\sqrt{2}y = 0$

16. For what values of the constant B is the conic section $x^2 + Bxy + y^2 = 1$ an ellipse? A degenerate parabola? A hyperbola? A circle? Sketch the parabolas and the circle. Sketch the curve for $B = 0, 1, 2,$ and 4.

17. Derive formulas (7), (8), and (9) for the coefficients in equation (3).

18. Use equations (7), (8), and (9) to show that under a rotation of axes $B^2 - 4AC = (B')^2 - 4A'C'$.

19. Use the trigonometric identities $\cos(2\alpha) = \cos^2\alpha - \sin^2\alpha$ and $\sin(2\alpha) = 2\sin\alpha\cos\alpha$ to show that B' given by equation (8) can be made zero by choosing α with $\cot(2\alpha) = (A - C)/B$.

20. Show that the discriminant of a conic section is not changed by a translation of axes.

21.† Derive equations (2) by using the trigonometric identities
$$\cos(\Theta + \alpha) = \cos\Theta\cos\alpha - \sin\Theta\sin\alpha$$
$$\sin(\Theta + \alpha) = \cos\Theta\sin\alpha + \sin\Theta\cos\alpha.$$

In Exercises 22 through 34 sketch the conic sections. Use formula (5) to find suitable $\sin\alpha$ and $\cos\alpha$.

22.‡ $4x^2 - 6xy - 4y^2 = 5$

23. $x^2 + 10xy + 25y^2 + 5\sqrt{26}x - \sqrt{26}y = 0$

24.† $4x^2 - 4xy + 7y^2 = 72$ **25.** $41x^2 - 24xy + 9y^2 = 45$

26.† $4x^2 - 4xy + y^2 - \sqrt{5}x - 2\sqrt{5}y = 0$

27.† $x^2 - 2xy + y^2 - 2x - 2y = 0$

28.† $7x^2 - 6\sqrt{3}xy + 13y^2 = 16$

29.† $32x^2 - 60xy + 7y^2 = 52$

30.† $4xy - 3x^2 = 20$

31.† $3x^2 + 2\sqrt{2}xy + 2y^2 = 12$

32. $-7x^2 + 48xy + 7y^2 = 100$ (Note that $24^2 + 7^2 = 25^2$.)

33. $9x^2 + 6xy + y^2 + \sqrt{10}x - 3\sqrt{10}y = 0$

34. $-7x^2 + 4\sqrt{8}xy + 7y^2 = 81$

Sketch the degenerate conic sections in Exercises 35 through 38.

35.† $x^2 - 2xy + y^2 = 0$ **36.†** $\sqrt{3}x^2 - 2xy - \sqrt{3}y^2 = 0$

37.† $x^2 + \frac{1}{4}y^2 - 2x - y + 2 = 0$ **38.†** $x^2 + 2xy + y^2 = 2$

10-5   SECTIONS OF CONES; DIRECTRICES AND ECCENTRICITY

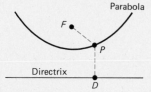

FIGURE 10.36

We defined a parabola to be the locus of points P equidistant from its focus F and its directrix (Figure 10.36). If we let D denote the foot of the perpendicular line from the point P to the directrix, the definition gives the equation

$$(1)\qquad \overline{PF} = \overline{PD}$$

for the parabola.

In the next theorem we show that every hyperbola and every ellipse that is not a circle has a pair of directrices, one for each focus (Figures 10.37 and 10.38), and that each of these curves can be defined by an equation analogous to (1) by using one focus and the corresponding directrix. At the same time we show that parabolas, ellipses and hyperbolas are formed by the intersection of double cones and planes. We use arguments discovered in 1822 by the Belgian mathematician G. P. Dandelin (1794–1847).

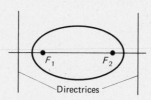

FIGURE 10.37   An ellipse

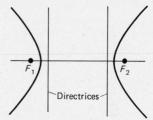

FIGURE 10.38   A hyperbola

Theorem 10.6   (a) Consider a double right circular cone whose elements make an angle β with its axis ($0 < \beta < \pi/2$). A plane that does not pass through the vertex of the cone and that makes an angle α with the axis intersects the cone in a circle if $\alpha = \pi/2$, in an ellipse if $\beta < \alpha < \pi/2$, in a parabola if $\alpha = \beta$, and in a hyperbola if $0 \le \alpha < \beta$.

(b) Except in the case of a circle the conic section may be characterized as the locus of points P in the plane such that

$$(2)\qquad \overline{PF} = e\,\overline{PD}. \qquad\blacktriangleleft$$

Here $\overline{PF}$ is the distance from P to a focus, $\overline{PD}$ is the distance from P to the corresponding directrix, and the constant e is given by the formula

$$(3)\qquad e = \frac{\cos \alpha}{\cos \beta}. \qquad\blacktriangleleft$$

The constant e is called the ECCENTRICITY of the conic section. For ellipses other than circles we have $0 < e < 1$, for parabolas we have

$e = 1$, and for hyperbolas we have $e > 1$. To have (3) satisfied for circles, we define their eccentricity to be zero.

Proof.   (a) We deal first with the case of $\beta < \alpha < \pi/2$, in which the plane cuts across half of the cone. Imagine two spheres inside the cone, one above the plane and one below it as in Figure 10.39, and just the right size so they are tangent to the plane and to the cone. We let F_1 and F_2 denote the points where the spheres touch the planes and let P be an arbitrary point on the conic section. In Figure 10.40 we show the cone turned so that the element through P is the edge of the cone as we see it. We let Q and R denote the points of intersection of the element of the cone and the circles where the spheres touch the cone. The line segments PF_1 and PR have the same length because they are both tangent to the lower sphere and they emanate from the same point. Similarly, PF_2 and PQ are of the same length because they are tangent to the upper sphere. Hence,

$$\overline{PF_1} + \overline{PF_2} = \overline{PR} + \overline{PQ} = \overline{QR}.$$

This is the constant distance between the circles measured along elements of the cone and is independent of the point P. The conic section is therefore an ellipse with foci F_1 and F_2.

In the case of $0 \leq \alpha < \beta$, we again use two spheres, tangent to the plane and to the cone, but with one in the upper half of the cone and one in the lower half (Figure 10.41). We let F_1 and F_2 denote the points where the spheres touch the plane. Then an argument similar to that used for the case of an ellipse shows that the conic section is a hyperbola with foci F_1 and F_2 (see Exercise 5).

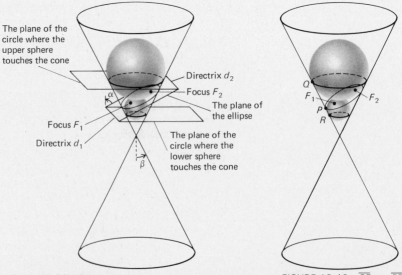

FIGURE 10.39   The case of an ellipse ($\beta < \alpha < \pi/2$)

FIGURE 10.40 $\overline{PF_1} = \overline{PR}$
$\overline{PF_2} = \overline{PQ}$

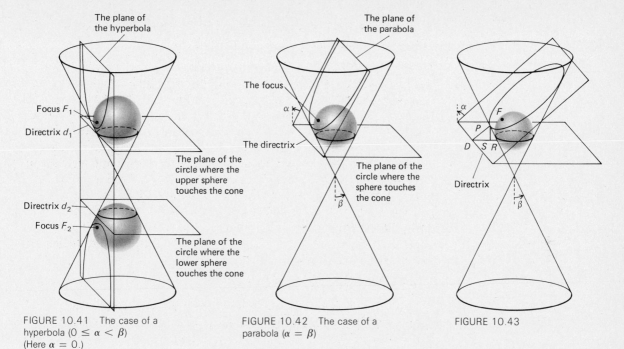

FIGURE 10.41   The case of a hyperbola $(0 \le \alpha < \beta)$ (Here $\alpha = 0$.)

FIGURE 10.42   The case of a parabola $(\alpha = \beta)$

FIGURE 10.43

In the case of $\alpha = \beta$, we use only one sphere and let F denote the point where it touches the plane of the conic section (Figure 10.42). We consider the plane through the circle where the sphere touches the cone. The proof given below of part (b) of the theorem shows that the conic section is the parabola with focus F and with the line of intersection of the two planes as directrix.

(b) As in the case of the parabola, the directrices of ellipses and hyperbolas are the lines of intersection of the planes of the conic sections and the planes through the circles where the spheres touch the cone (Figures 10.39 and 10.41). We will derive the formula

$$(2) \quad \overline{PF} = e\overline{PD}$$

with reference to the ellipse and lower sphere shown in Figure 10.43. The same argument applies to the upper sphere for the ellipse and to the hyperbola and parabola.

As before, we let D be the foot of the perpendicular line from the point P on the conic section to the directrix, and we let R be the point where the element of the cone containing P intersects the circle of tangency of the sphere. We let S be the foot of the perpendicular line from P to the plane of the circle. The points P, S, and D form a right triangle, with α at the upper vertex; the points P, S, and R form a second right triangle, with β at the upper vertex (Figure 10.44). Therefore, we have $\overline{PS} = \overline{PR} \cos \beta = \overline{PF} \cos \beta$ and $\overline{PS} = \overline{PD} \cos \alpha$. Equating these two expressions for the length $\overline{PS}$ yields the desired equation $\overline{PF} = (\cos \alpha / \cos \beta)\,\overline{PD} = e\,\overline{PD}$ with e given by (3).   Q.E.D.

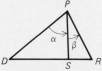

FIGURE 10.44 $\overline{PS} = \overline{PD} \cos \alpha$, $\overline{PS} = \overline{PR} \cos \beta = \overline{PF} \cos \beta$

Finding eccentricity and directrices

Recall that the foci of the ellipse

$$(4) \quad \frac{x^2}{a^2} + \frac{y^2}{b^2} = 1 \quad (0 < b < a)$$

are at $(c, 0)$ and $(-c, 0)$ with

$$(5) \quad c = \sqrt{a^2 - b^2}$$

whereas the foci of the hyperbola

$$(6) \quad \frac{x^2}{a^2} - \frac{y^2}{b^2} = 1$$

are at $(c, 0)$ and $(-c, 0)$ with

$$(7) \quad c = \sqrt{a^2 + b^2}.$$

In either case the eccentricity of the conic section is

$$(8) \quad e = \frac{c}{a} \qquad \blacktriangleleft$$

and its directrices are the lines

$$(9) \quad x = \pm \frac{a}{e} = \pm \frac{a^2}{c} \qquad \blacktriangleleft$$

(see Exercise 7).

Exercises

†Answer provided.
‡Outline of solution provided.

1. Give equations for and sketch the conic sections with foci at $(-1, 0)$ and $(1, 0)$ and with eccentricities **a.†** $e = \frac{1}{3}$, **b.†** $e = 3$, **c.** $e = \frac{12}{13}$, **d.** $e = \frac{13}{12}$.

2. Compute the eccentricity of the following conic sections.
   a.† $x^2 + y^2 = 1$ **b.†** $x^2 + 9y^2 = 1$ **c.†** $x^2 - 10y^2 = 100$
   d. $4x^2 + y^2 = 36$ **e.** $y = 97x^2$ **f.** $x^2 - 25y^2 = 100$

3. Show that the conic sections with eccentricity $\sqrt{2}$ are the rectangular hyperbolas.

4.† What happens to the conic section with focus $(0, 0)$ and corresponding directrix $x = 1$ as its eccentricity increases through all positive values? (Describe what happens to its other focus and directrix.)

5. Prove part (a) of Theorem 10.6 for $0 \le \alpha < \beta$.

6. What is the greatest eccentricity of conic sections of a right circular cone whose elements make an angle β with its axis?

7. Derive formulas (8) and (9) from formula (2) with P the vertices of an ellipse or hyperbola.

8. Sketch the following ellipses and hyperbolas, showing their foci and directrices.
   a.† $\dfrac{x^2}{9} + \dfrac{y^2}{4} = 1$ **b.†** $x^2 - y^2 = 16$

   c. $\dfrac{x^2}{25} + \dfrac{y^2}{169} = 1$ **d.** $\dfrac{y^2}{16} - \dfrac{x^2}{9} = 1$

Find the eccentricity, foci, and directrices of the conic sections in Exercises 9 through 16.

9.[†] $x^2 - 8y^2 = 8$ **10.**[†] $y^2 - 3x^2 = 3$

11.[†] $5x^2 + 9y^2 = 45$ **12.** $x^2 - y^2 + 2x - 2y = 1$

13.[†] $x^2 + 2y^2 - 8y + 6 = 0$ **14.** $x^2 - 2x + y = 0$

15. $4x^2 + 3y^2 = 12$ **16.**[†] $4x^2 + 4y^2 = 7$

10-6 POLAR COORDINATES

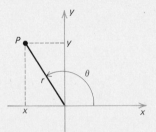

FIGURE 10.45   Point with rectangular coordinates (x, y) and polar coordinates $[r, \Theta]$

Many problems in analytic geometry and calculus are simplified by the use of POLAR COORDINATES in place of rectangular coordinates. The polar coordinates of a point in a coordinate plane give its distance and direction from the origin.

Definition 10.4   The polar coordinates of a point P other than the origin in an xy-plane are $[r, \Theta]$ where r is the distance from P to the origin and Θ is an angle from the positive x-axis to the line between the origin and P (Figure 10.45). The origin has polar coordinates $[0, \Theta]$ where Θ is an arbitrary angle.

The definitions of sine and cosine give the formulas

(1) $x = r \cos \Theta, \qquad y = r \sin \Theta$ ◀

which express the rectangular coordinates (x, y) of a point in terms of its polar coordinates $[r, \Theta]$. The equation expressing r in terms of x and y

(2) $r = \sqrt{x^2 + y^2}$ ◀

comes from the Pythagorean theorem.

The best way to find suitable angles Θ is to plot the point (x, y), as in Example 2 below. We can also use the formulas

(3) $\Theta = \arctan\left(\dfrac{y}{x}\right) + 2n\pi \quad \text{for} \quad x > 0$

(4) $\Theta = \arctan\left(\dfrac{y}{x}\right) + (2n + 1)\pi \quad \text{for} \quad x < 0$

where n denotes an arbitrary integer, or similar formulas involving the arccotangent for $y \neq 0$.

EXAMPLE 1   Plot the point with polar coordinates $[5, 2\pi/3]$ and compute its rectangular coordinates.

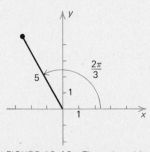

FIGURE 10.46   The point with polar coordinates $[5, \frac{2}{3}\pi]$ and rectangular coordinates $(-\frac{5}{2}, \frac{5}{2}\sqrt{3})$

Solution.   The point is plotted in Figure 10.46. From equations (1) we obtain the rectangular coordinates $x = 5 \cos(2\pi/3) = -\frac{5}{2}$, $y = 5 \sin(2\pi/3) = \frac{5}{2}\sqrt{3}$. □

Negative values
of r

In the next section we will consider the graphs of equations given in polar coordinates. In this context it is convenient to allow *negative* values of r, and we use the following definition in place of Definition 10.4.

Definition 10.5   To locate a point in an xy-plane with polar coordinates [r, Θ] where r may be negative, introduce an auxiliary r-axis whose positive side makes an angle Θ with the positive x-axis (Figure 10.47). The point [r, Θ] has coordinate r on that axis.

Figure 10.48 shows, for example, the points with polar coordinates $[3, \pi/4]$, $[0, \pi/4]$, and $[-3, \pi/4]$. For negative r, equations (1) are valid, but equations (2), (3), and (4) must be replaced by

$$(5) \quad r = -\sqrt{x^2 + y^2}, \qquad \Theta = -\arctan\left(\frac{y}{x}\right) = -\operatorname{arccot}\left(\frac{x}{y}\right).$$

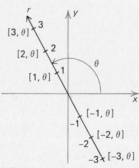

FIGURE 10.47   The auxiliary
r-axis

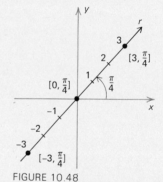

FIGURE 10.48

Possible choices
of Θ

Because there are an infinite number of angles with the same terminal side, every point has an infinite number of possible polar coordinates. If a point has polar coordinates $[r, \Theta]$, then it also has polar coordinates $[r, \Theta + 2n\pi]$ for any integer n because the angles Θ and $\Theta + 2n\pi$ have the same terminal sides. If we allow negative values of r, then the point also has polar coordinates $[-r, \Theta + (2n + 1)\pi]$ because the terminal sides of the angles $\Theta + (2n + 1)\pi$ and Θ are opposite each other.

EXAMPLE 2   Give all possible polar coordinates $[r, \Theta]$ of the point with rectangular coordinates $(-4, 4)$.

Solution.   The point $(-4, 4)$ is plotted in Figure 10.49. Its distance to the origin is $[4^2 + (-4)^2]^{1/2} = 4\sqrt{2}$. One angle from the positive x-axis to the line between the origin and the point is $3\pi/4$ (Figure 10.49), so the polar coordinates of the point with r positive are $[4\sqrt{2}, 3\pi/4 + 2n\pi]$ with n an integer. The polar coordinates of the point with r negative are $[-4\sqrt{2}, 3\pi/4 + (2n + 1)\pi]$ or $[-4\sqrt{2}, 7\pi/4 + 2n\pi]$ (Figure 10.50).   □

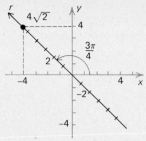

FIGURE 10.49

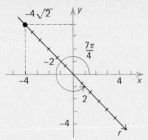

FIGURE 10.50

Exercises

1. Plot the points with the following polar coordinates and give their rectangular coordinates. In parts (g) through (j) give exact and approximate numerical values.

   a.† $\left[3, \dfrac{3\pi}{4}\right]$ **b.** $[2, \pi]$ **c.**† $\left[-\tfrac{1}{2}, \dfrac{\pi}{6}\right]$

   d.† $[0, \pi]$ **e.** $\left[-2, \dfrac{15\pi}{6}\right]$ **f.** $[\pi, 0]$

   g.‡ $[2, 3]$ **h.** $[3, 2]$ **i.**† $[-4, -10]$

   j. $[\tfrac{1}{4}, \tfrac{1}{10}]$

2. Give all possible polar coordinates of the points with the following rectangular coordinates. In parts (g) through (j) give exact and approximate numerical values.

   a.† $(-3, -3)$ **b.** $(0, -4)$ **c.**† $(7, 0)$
   d. $(-2, -2)$ **e.**† $(-1, \sqrt{3})$ **f.** $(3, 3\sqrt{3})$
   g.‡ $(3, 4)$ **h.** $(5, -12)$ **i.**† $(-3, -6)$
   j. $(-1, 30)$

Find the rectangular coordinates of the points whose polar coordinates are given in Exercises 3 through 6.

   3.† $[1, 7\pi/4]$ **4.**† $[\pi, \pi]$

   5.† $[3, 3]$ **6.**† $[-4, 2\pi/3]$

Find all polar coordinates of the points whose rectangular coordinates are given in Exercises 7 through 14.

   7.† $(0, 0)$ **8.**† $(11, 5)$

   9.† $(0, 2)$ **10.**† $(-3, -\sqrt{3})$

   11.† $(-4, -4)$ **12.**† $(-1, 10)$

   13.† $(-1, -10)$ **14.**† $(10, -1)$

10-7   GRAPHS OF EQUATIONS IN POLAR COORDINATES

Many curves can be conveniently described with polar coordinates. We use the following definition.

Definition 10.6   The graph of an equation in polar coordinates is the set of points for which at least one choice of polar coordinates [r, Θ] satisfy the equation.

We generally allow negative values of r in Definition 10.6.

An equation in rectangular coordinates can be transformed to one in polar coordinates by using the equations

$$(1) \quad x = r \cos \Theta, \qquad y = r \sin \Theta.$$

EXAMPLE 1   Give an equation in polar coordinates for the circle of radius 1 with its center at the point $(0, 1)$ in an xy-plane.

Solution.   The circle is shown in Figure 10.51. It has the equation

$$(2) \quad x^2 + (y - 1)^2 = 1$$

in rectangular coordinates. Making substitutions (1) gives $(r \cos \Theta)^2 + (r \sin \Theta - 1)^2 = 1$. When this is multiplied out and the relation $\sin^2 \Theta + \cos^2 \Theta = 1$ is used, we obtain $r^2 = 2r \sin \Theta$. Cancelling one factor of r gives the equation

$$(3) \quad r = 2 \sin \Theta$$

for the circle.   □

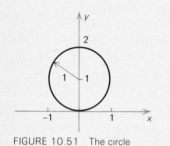

FIGURE 10.51   The circle
$x^2 + (y - 1)^2 = 1$ or $r = 2 \sin \Theta$

Note.   Equation (3) for the circle in Example 1 shows why we do not require that all possible polar coordinates of a point satisfy the equation. The origin is on the circle and has polar coordinates $[0, \Theta]$ for any Θ but the polar coordinates $[0, \Theta]$ satisfy the equation only for those Θ such that $\sin \Theta = 0$.   □

Symmetry in polar coordinates

The points with polar coordinates $[r, \Theta]$ and $[-r, \Theta]$ are symmetric about the origin (Figure 10.52). Therefore, the graph of an equation in polar coordinates is symmetric about the origin if replacing r by $-r$ gives an equivalent equation. The points with polar coordinates $[r, \Theta]$ and $[r, -\Theta]$ are symmetric about the x-axis, (Figure 10.53), so the graph of an equation is symmetric about the x-axis if replacing Θ by $-\Theta$ gives an equivalent equation. Finally, the points with polar coordinates $[r, \Theta]$ and $[r, \pi - \Theta]$ are symmetric about the y-axis (Figure 10.54), so the graph of an equation is symmetric about the y-axis if replacing Θ by $\pi - \Theta$ gives an equivalent equation.

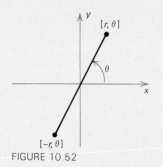

FIGURE 10.52

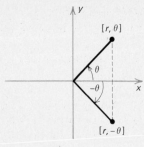

FIGURE 10.53

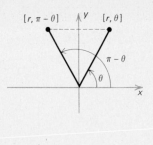

FIGURE 10.54

Sketching graphs in polar coordinates

When sketching the graph of an equation in polar coordinates, it is often useful to first sketch the graph in rectangular coordinates in a Θr-plane. We use this procedure in the next example.

EXAMPLE 2   Sketch the graph of $r = 3 \cos(2\Theta)$ in an xy-plane.

Solution.   We make the table below of values of r and Θ for one period of the function $3 \cos(2\Theta)$. We list only the points where $3 \cos(2\Theta)$ has its maximum and minimum value and where it is zero because we know the graph of $r = 3 \cos(2\Theta)$ in a Θr-plane is a modified cosine curve. It is drawn in Figure 10.55.

2Θ	0	$\pi/2$	π	$3\pi/2$	2π
Θ	0	$\pi/4$	$\pi/2$	$3\pi/4$	π
$r = 3\cos(2\Theta)$	3	0	-3	0	3

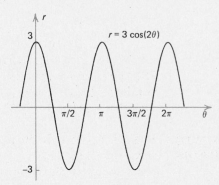

FIGURE 10.55

To sketch the graph of $r = 3 \cos(2\Theta)$ in an xy-plane from the sketch in Figure 10.55, we imagine an auxiliary r-axis with a marker that is at $r = 3 \cos(2\Theta)$ when the angle from the positive x-axis to the positive r-axis is Θ. As Θ increases and the r-axis rotates, the marker slides up and down and traces the graph of the equation.

This process is indicated in Figures 10.56 through 10.60. From Figure 10.55 we see that as Θ increases from 0 to $\pi/4$, r decreases from 3 to 0. Consequently, the marker tracing the curve swings around from

$x = 3$ on the x-axis to the origin (Figure 10.56). As Θ increases from $\pi/4$ to $\pi/2$, r decreases from 0 to -3 and the marker swings around from the origin to $y = -3$ on the y-axis (Figure 10.57). As Θ increases to 2π, r increases to 3, decreases to -3, and increases to 3, while the marker passes through the origin three more times (Figures 10.58, 10.59, and 10.60) and traces the rest of the curve shown in Figure 10.61. That curve is the graph of the equation $r = 3\cos(2\Theta)$ and is known as a *four-petaled rose.* □

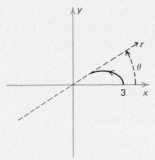

FIGURE 10.56 $0 < \Theta < \frac{1}{4}\pi$
$r > 0$

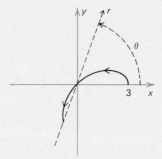

FIGURE 10.57 $\frac{1}{4}\pi < \Theta < \frac{1}{2}\pi$
$r < 0$

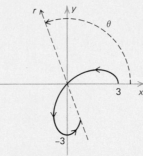

FIGURE 10.58 $\frac{1}{2}\pi < \Theta < \frac{3}{4}\pi$
$r < 0$

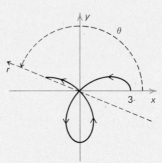

FIGURE 10.59 $\frac{3}{4}\pi < \Theta < \pi$
$r > 0$

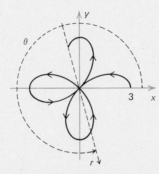

FIGURE 10.60 $\frac{3}{2}\pi < \Theta < \frac{7}{4}\pi$
$r < 0$

FIGURE 10.61 $r = 3\cos(2\Theta)$

A brief catalog of curves

Any curve with an equation in rectangular coordinates can be given an equation in polar coordinates by making the substitutions $x = r \cos \Theta$, $y = r \sin \Theta$, but in many cases the equations in polar coordinates are too complicated to be very useful. We will now list a few types of curves whose equations in polar coordinates are relatively simple and are frequently used.

As we saw in Example 2, the graph of an equation of the form $r = a \cos(2\Theta)$ with nonzero a is a four-petaled rose whose "petals" lie along the x- and y-axes. Curves with equations of the form $r = a \sin(2\Theta)$ are four-petaled roses whose petals lie along the lines $y = \pm x$. In general, the graphs of $r = a \cos(n\Theta)$ and $r = a \sin(n\Theta)$ with n an even positive integer are "roses" of $2n$ petals and those with n an odd positive integer are roses with n petals. If n is even, the curve is described once as Θ goes from 0 to 2π. If n is odd, the entire curve is described as Θ goes from 0 to π and is repeated as Θ goes from π to 2π.

The "one-petaled roses" $r = a \cos \Theta$ and $r = a \sin \Theta$ are circles of radius $\frac{1}{2}|a|$ tangent to the x- or y-axis at the origin (see Figures 10.62 and 10.63 and Exercise 28).

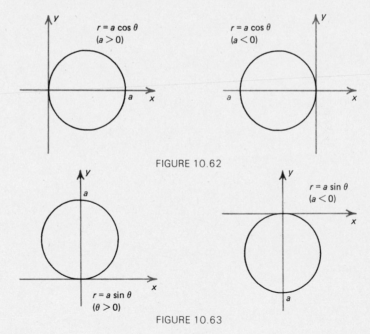

FIGURE 10.62

FIGURE 10.63

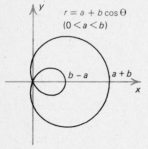

FIGURE 10.64   A limaçon

Curves with equations of the form $r = a + b \cos \Theta$ and $r = a + b \sin \Theta$ with nonzero a and b are called *limaçons*. If $|b|$ is greater than $|a|$, as in Figure 10.64, the curve has two loops; if $|b|$ is less than $|a|$, it has a "dent" as in Figure 10.65. If $|a| = |b|$, the limaçon has a heart-like shape and is called a *cardioid* (Figure 10.66).

When we replace x by $r \cos \Theta$ in the equation $x = a$ $(a \neq 0)$ for a vertical line other than the y-axis, we obtain $r \cos \Theta = a$ and then $r = a \sec \Theta$, which is an equation for the line in polar coordinates (Figure 10.67). Similarly, the substitution $y = r \sin \Theta$ transforms the equation $y = a$ $(a \neq 0)$ for a horizontal line other than the x-axis into its polar form $r = a \csc \Theta$ (Figure 10.68).

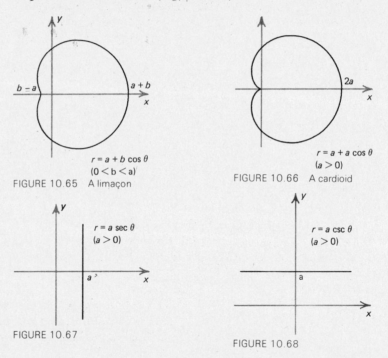

$r = a + b \cos \theta$
$(0 < b < a)$
FIGURE 10.65   A limaçon

$r = a + a \cos \theta$
$(a > 0)$
FIGURE 10.66   A cardioid

$r = a \sec \theta$
$(a > 0)$
FIGURE 10.67

$r = a \csc \theta$
$(a > 0)$
FIGURE 10.68

The line through the origin at an angle Θ_0 with the positive x-axis has the equation $\Theta = \Theta_0$ (Figure 10.69). The circle of radius a ($a > 0$) with its center at the origin has the equation $r = a$ (Figure 10.70).

Tangent lines through the origin

We can improve a sketch of a curve $r = r(\Theta)$ which passes through the origin one or more times by drawing its tangent line(s) at that point. The curve passes through the origin when $\Theta = \Theta_0$ if $r(\Theta_0) = 0$ and $r(\Theta)$ is continuous and increasing or decreasing for $\alpha \le \Theta \le \beta$ ($\alpha < \Theta_0 < \rho$). In this case the line $\Theta = \Theta_0$ is tangent to the curve at the origin. This occurs because the line $\Theta = \Theta_1$ is a secant line to the curve for Θ_1 near Θ_0 and approaches the line $\Theta = \Theta_0$ as Θ_1 tends to Θ_0 (Figure 10.71). For example, the lines $\Theta = \pi/4$ and $\Theta = 3\pi/4$ (which are also the lines $\Theta = 5\pi/4$ and $\Theta = 7\pi/4$) are tangent to the four-petaled rose $r = 3\cos(2\Theta)$ in Figure 10.61 because the function $3\cos(2\Theta)$ is zero at $\Theta = \pi/4, 3\pi/4, 5\pi/4$, and $7\pi/4$.

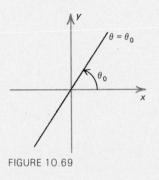

FIGURE 10.69

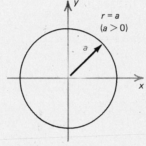

FIGURE 10.70

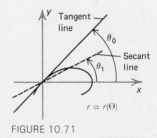

FIGURE 10.71

Polar equations for conic sections

In Section 10.5 we showed that any conic section other than a circle can be described by the equation

(4) $\overline{PF} = e\overline{PD}$

where P is a point on the conic section, e is the eccentricity, F is a focus, and $\overline{PD}$ is the distance from P to the corresponding directrix (Figure 10.72). Equation (4) transforms easily into an equation for the conic section in polar coordinates.

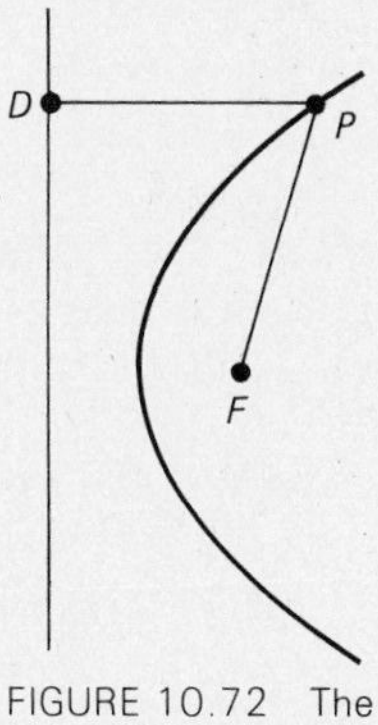

We suppose that the focus is at the origin and that the directrix is the vertical line $x = -k$ (Figure 10.73). We consider a point P with polar coordinates $[r, \Theta]$ ($r > 0$) and rectangular coordinates (x, y). The distance $\overline{PF}$ from P to the origin is r. The distance $\overline{PD}$ from P to the directrix is $x + k = r\cos\Theta + k$. Therefore, equation (4) reads $r = e(r\cos\Theta + k)$.

Solving for r gives the equation

(5) $$r = \frac{ke}{1 - e\cos\Theta}$$

for the conic section.

FIGURE 10.72   The conic section
$\overline{PF} = e\overline{PD}$

Exercises

†Answer provided.
‡Outline of solution provided.

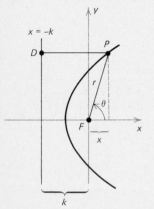

FIGURE 10.73 $r = \dfrac{ke}{1 - e \cos \Theta}$

In Exercises 1 through 10 put the equations in polar coordinates.

1.† $x = 2$ **2.** $y = 3$

3.† $x = 3y$ **4.** $2x^2 + y^2 = 1$

5.† $x^2 - y^2 = 1$ **6.** $y = x^2$

7.† $x^2 + 2x + y^2 = 0$ **8.** $x^2 = y^2$

9.† $(x^2 + y^2 - 2x)^2 = 4(x^2 + y^2)$

10. $(x^2 + y^2)^2 + 8(x^2 + y^2) = 4x^2$

In Exercises 11 through 26 sketch the graphs of the equations in xy-planes.

11.† $r = 1 - 2 \sin \Theta$ **12.** $r = 4 \sin(3\Theta)$

13.† $r = 4 - 4 \cos \Theta$ **14.** $r = \sin(2\Theta)$

15.† $r = \cos\left(\dfrac{\Theta}{2}\right)$ **16.** $r = 5 + 3 \cos \Theta$

17.† $r = \sec \Theta$ **18.** $r = 5$

19.† $r^2 - 5r + 4 = 0$ **20.** $r = \dfrac{\Theta}{2}$

21. $\Theta = 1$ **22.** $r = -5 \csc \Theta$

23.† $r = \dfrac{3}{1 - 2 \cos \Theta}$ **24.**† $r = \dfrac{3}{1 - \cos \Theta}$

25. $r = \dfrac{6}{2 - \cos \Theta}$ **26.** $r = e^{\Theta/6}$

27. Show that (5) is the equation of an ellipse for $0 < e < 1$, a parabola for $e = 1$, and a hyperbola for $e > 1$ by transforming the equation to rectangular coordinates and computing the discriminant of the resulting equation.

28. Use the fact that any triangle inscribed in a semi-circle with one side along a diameter is a right triangle to show that the graph of the equation $r = a \cos \Theta$ (a "one-petaled rose") is a circle.

Additional drill exercises

In Exercises 29 through 36 put the equations in polar coordinates.

29.† $(x^2 + y^2)^2 = 8xy$ **30.**† $x^2 + y^2 = 7$

31.† $y^2 = 6x + 9$ **32.**† $x^2 - 4x + y^2 = 0$

33.† $y = 2x$ **34.** $x = -4$

35.† $y = -5$ **36.** $x^2 y^2 + y^4 = x^2$

Sketch in xy-planes the curves whose equations are given in Exercises 37 through 44.

37.† $r = -3 \csc \Theta$ **38.**† $r = -\sec \Theta$

39.† $\Theta = 4\pi/3$ **40.** $r = 2 + 2 \cos \Theta$

41.[†] $r = \cos(2\Theta)$

42. $r = -4 \sin \Theta$

43.[†] $r = 2|\sin \Theta|$

44.[†] $r^2 = \cos \Theta \sin \Theta$ (a lemniscate)

10-8   AREA IN POLAR COORDINATES

We use the following rule to compute the area of a region bounded by a curve $r = r(\Theta)$ given in polar coordinates.

Rule 10.1   The region bounded by the lines $\Theta = a$ and $\Theta = b$ $(0 < b - a \le 2\pi)$ and by the curve $r = r(\Theta)$ (Figure 10.74) has area

(1) $$\int_a^b \tfrac{1}{2} r(\Theta)^2 \, d\Theta.$$ ◄

Here we assume that $r(\Theta)$ is continuous and either nonnegative or nonpositive for $a \le \Theta \le b$.

Derivation of Rule 10.1   We consider an arbitrary partition

$$a = \Theta_0 < \Theta_1 < \Theta_2 < \Theta_3 < \cdots < \Theta_{N-1} < \Theta_N = b$$

of the interval $a \le \Theta \le b$. The lines $\Theta = \Theta_j$ divide the region into N pieces (Figure 10.75). We approximate the jth piece by the sector of the circle of radius $r(\Theta_j)$ that lies between the lines $\Theta = \Theta_{j-1}$ and $\Theta = \Theta_j$ (Figure 10.76). The angle $\Theta_j - \Theta_{j-1}$ is the fraction $(\Theta_j - \Theta_{j-1})/2\pi$ of an entire revolution, so the sector is that fraction of a circle and has area $\pi r(\Theta_j)^2 (\Theta_j - \Theta_{j-1})/2\pi = \tfrac{1}{2} r(\Theta_j)^2 (\Theta_j - \Theta_{j-1})$. The area of the entire approximation of the original region is

(2) $$\sum_{j=1}^{N} \tfrac{1}{2} r(\Theta_j)^2 (\Theta_j - \Theta_{j-1}).$$

The integral (1) gives the area of the original region because the sum (2) is a Riemann sum for that integral.   □

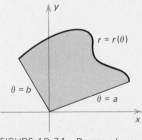

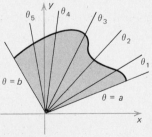

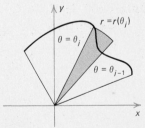

FIGURE 10.74   Region of area $\int_a^b \tfrac{1}{2} r(\Theta)^2 \, d\Theta$

FIGURE 10.75

FIGURE 10.76

EXAMPLE 1   Compute the area of the region bounded by the cardioid $r = 1 + \cos \Theta$.

Solution. The region is shown in Figure 10.77. We obtain the entire cardioid by taking $0 \leq \Theta \leq 2\pi$. By Rule 10.1 the area of the region is

$$\int_0^{2\pi} \tfrac{1}{2}(1 + \cos \Theta)^2 \, d\Theta = \int_0^{2\pi} (\tfrac{1}{2} + \cos \Theta + \tfrac{1}{2}\cos^2 \Theta) \, d\Theta.$$

We employ the trigonometric identity $\cos^2 \Theta = \tfrac{1}{2}(1 + \cos(2\Theta))$ to put the integral in the form

$$\int_0^{2\pi} (\tfrac{3}{4} + \cos \Theta + \tfrac{1}{4}\cos(2\Theta)) \, d\Theta = [\tfrac{3}{4}\Theta + \sin \Theta + \tfrac{1}{8}\sin(2\Theta)]_0^{2\pi}$$

$$= \frac{3\pi}{2}. \quad \square$$

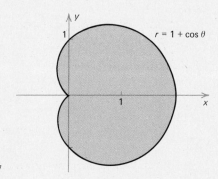

FIGURE 10.77

EXAMPLE 2   Compute the area inside one loop of the three-petaled rose $r = 4 \sin(3\Theta)$.

Solution. The curve, shown in Figure 10.78, is traversed as Θ goes from 0 to π. Because $4 \sin(3\Theta)$ is zero at $\Theta = 0, \tfrac{1}{3}\pi, \tfrac{2}{3}\pi$, and π, the curve passes through the origin at these values of Θ. The upper right petal is traversed for $0 \leq \Theta \leq \tfrac{1}{3}\pi$, so the area inside it is

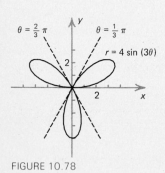

FIGURE 10.78

$$\int_0^{\pi/3} \tfrac{1}{2}[4 \sin(3\Theta)]^2 \, d\Theta = 8 \int_0^{\pi/3} \sin^2(3\Theta) \, d\Theta$$

$$= 4 \int_0^{\pi/3} [1 - \cos(6\Theta)] \, d\Theta$$

$$= 4[\Theta - \tfrac{1}{6}\sin(6\Theta)]_0^{\pi/3}$$

$$= 4[\tfrac{1}{3}\pi - \tfrac{1}{6}\sin(2\pi)] - 4[0 - \tfrac{1}{6}\sin(0)]$$

$$= \tfrac{4}{3}\pi.$$

In the second step of this calculation we used the trigonometric identity $\sin^2 \alpha = \tfrac{1}{2}[1 - \cos(2\alpha)]$ with $\alpha = 3\Theta$. $\quad \square$

Exercises

Compute the areas of the regions described in Exercises 1 through 7.

1.† The circle bounded by $r = 6 \sin \Theta$

2. The region bounded by the limaçon $r = 5 + 4 \sin \Theta$

3.† One leaf of the three-leaved rose $r = 2 \cos(3\Theta)$

4. All eight leaves of the eight-leaved rose $r = \cos(4\Theta)$

5.† The region inside the outer loop and outside the inner loop of the limaçon $r = 1 + 2 \sin \Theta$

6. The region bounded by the lines $\Theta = 0$ and $\Theta = \pi/2$ and by the portion of the hyperbolic spiral e^{Θ} for $0 \le \Theta \le \pi/2$

7.† The region inside the circle $r = 2 \cos \Theta$ and to the right of the line $x = \frac{3}{2}$

8. According to *Poiseuille's law*, the velocity of a fluid at a distance r (feet) from the center of a pipe of radius R is $k(R^2 - r^2)$(feet/minute) where kR^2 is the velocity at the center. What is the total rate of flow through the pipe?

9.† Compute the area of the region lying inside the circle $r = 2 \cos \Theta$ and outside the circle $r = \sqrt{2}$.

10. Compute the area inside one petal of the five-petaled rose $r = -2 \cos(5\Theta)$.

11.† What is the area of the region inside the limaçon $r = 2 - \sin \Theta$?

12.† Compute the area inside one loop of the lemniscate $r^2 = 4 \cos(2\Theta)$.

13.† What is the area of the region lying inside both circles $x^2 - 6x + y^2 = 0$ and $x^2 + y^2 - 6y = 0$?

14. Compute the area of the region between the circles $r = 3 \sin \Theta$ and $r = 7 \sin \Theta$.

15. Use polar coordinates to find the area of the triangle with vertices $(0, 0)$, $(-1, 1)$, and $(1, 1)$.

MISCELLANEOUS EXERCISES

1.† Find an equation of the parabola with axis parallel to the y-axis and which passes through $(0, 0)$, $(1, 2)$, and $(3, -6)$.

2. Find an equation of the parabola with axis parallel to the y-axis and which passes through $(-2, 1)$, $(2, 1)$, and $(4, 4)$.

3. **a.** Show that for any constant $k > -1$ the curve

$$\frac{x^2}{10 + k} + \frac{y^2}{1 + k} = 1$$

is an ellipse with foci $(3, 0)$ and $(-3, 0)$. **b.** Does this equation give all such ellipses? **c.** What is the graph of this equation for other values of k?

4. Show that the ellipse $\frac{1}{169}x^2 + \frac{1}{144}y^2 = 1$ and the hyperbola $\frac{1}{9}x^2 - \frac{1}{16}y^2 = 1$ have the same foci and intersect at right angles.

5. In the equation of a conic section

$$y^2 = Lx + kx^2 \qquad (k \geq -1)$$

the number L is the "latus rectum" of the conic. Under what conditions on k is the conic a parabola, an ellipse, or a hyperbola? Show that in any case L is the "width" of the conic at a focus.

6.† The orbit of the earth is an ellipse of eccentricity 0.0167, with the sun at one focus, and with the major axis 185.8 million miles long. What is the earth's   **a.**   aphelion (farthest distance from the sun) and **b.**   perihelion (closest distance to the sun)?

7. Of all the planets, Pluto has the elliptical orbit of the largest eccentricity. That eccentricity is 0.2486. What is the ratio of the lengths of its major and minor axes? Sketch such an ellipse.

8. The earth satellite Lunik 3 had an apogee (farthest distance from the surface of the earth) of 250,000 miles and a perigee (closest distance to the surface of the earth) of 25,000 miles. The radius of the earth is 4000 miles. What was the eccentricity of the elliptical orbit of Lunik 3?

9. The orbit of the satellite Explorer 6 had eccentricity 0.75 and apogee 26,400 miles. What was its perigee?

10. The astroid Icarus has a perihelion of 17 million miles (inside the orbit of Mercury) and aphelion of 183 million miles (outside the orbit of Mars). (See Figure 10.79.) What is the eccentricity of its elliptic orbit?

11. Sketch   **a.†**   the *hyperbolic spiral* $r = 2\pi/\Theta$ **b.**   the *Archimedean spiral* $r = \Theta/2\pi$ and   **c.**   the *logarithmic spiral* $r = e^{\Theta/4\pi}$.

12. Sketch   **a.**   the *lemniscate* $r^2 = 4\cos(2\Theta)$ **b.**   the *cissoid* $r = 2\sin\Theta\tan\Theta$, and   **c.**   the *conchoid of Nicomedes* $r = 2\sec\Theta - 1$.

13. Sketch the curves   **a.** $r = \sin(\Theta/3)$ and   **b.** $r = |\cos\Theta|$.

14.† Sketch *Freeth's nephroid* $r = 1 + 2\sin(\Theta/2)$.

15. Sketch the graph of the equation $(x^2 + y^2)^2 - 3x^2 - 3y^2 + 2 = 0$, by first switching the equation to polar coordinates.

16. The curve $x^2 + y^2/(1 - k) = 1$ is an ellipse for $k < 1$ and a hyperbola for $k > 1$. What happens to its foci and directrices as k crosses 0? What happens to them as k crosses 1?

17. Find the asymptotes of the following hyperbolas without rotating or translating axes.   **a.‡** $x^2 + 3xy + y^2 = 5$ **b.†** $xy - y^2 = 4$ **c.** $2x^2 - 5xy + y^2 = -2$

18. **a.** Show that the curve $y = 1/(1 - x)$ is a hyperbola and sketch it. What are its asymptotes?   **b.** Archimedes solved the cubic equation which in modern notation reads $x^3 - x^2 + 1 = 0$ by finding the intersections of the hyperbola of part (a) and the parabola $y = x^2$. Show that this gives the solutions. How many are there? Sketch the two curves.

19. Descartes' first work in analytic geometry was in the solution of problems such as the following posed by Pappus: let PA, PB, and PC be the line segments from a point P to three given lines which meet the lines at given angles. Find the locus of points P such that $\overline{PA}\,\overline{PB} = \overline{PC}^2$. In every case the locus consists of conic sections.* Find the locus where the three lines are the x-axis, the y-axis and the line $x = 1$ and the three angles are right angles.

20. In Fermat's first treatise on analytic geometry, *Ad Locos Planos et Solidos Isagoge*, he solved the following problem: given any number of fixed lines, find the locus of points P such that the sum of the squares of the lengths of the line segments from P to the given lines at given angles is a constant. Explain why the locus is a conic section.

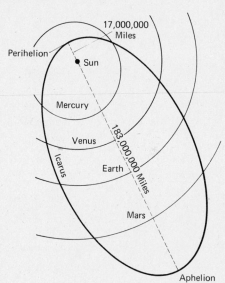

FIGURE 10.79   The orbit of Icarus
(Adapted from O. Gingerich, [1]
Frontiers in Astronomy, p. 61.)

---

*Unknown to Pappas and Descartes, this problem was probably solved by Apollonius in the third century B.C. (see Boyer, [2] *History of Analytic Geometry*, p. 27).

CHAPTER 11

CURVES AND VECTORS IN THE PLANE

SUMMARY: In this chapter we discuss vectors, curves, and the motion of objects in planes. Section 11.1 deals with the algebra of vectors and Section 11.2 with the dot product. We study parametric representations of curves, arclength, and speed in Section 11.3, and vector representations of curves, velocity vectors, and curvature in Sections 11.4 and 11.5. Section 11.6 deals with the vector form of Newton's law and the tangential and normal components of acceleration vectors. In an optional Section 11.7 we discuss the polar decomposition of vectors and Kepler's and Newton's laws of planetary motion.

11-1 VECTORS

Often a direction and a number appear together in the description of a geometric or physical situation, as when we say that one town is a certain distance northeast of another or when we describe the wind by giving its direction and velocity. In such situations we can treat the direction and number together as a VECTOR according to the following informal definition.

Definition 11.1   A nonzero vector consists of a direction and a positive number. The zero vector corresponds to the number zero and has no direction.

The number is called the LENGTH or MAGNITUDE of the vector. We use bold-faced letters such as A and B and letters with arrows over them such as the symbol $\overrightarrow{PQ}$ to denote vectors, and we designate lengths of vectors by using absolute value signs. Thus, the length of the vector A is $|A|$. In diagrams we represent a vector with an arrow pointing in the direction of the vector (Figure 11.1). The length of the vector is the

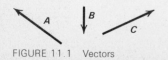

FIGURE 11.1   Vectors

FIGURE 11.2

length of the arrow (Figure 11.2) measured by a designated scale which is usually taken to be the scale on the coordinate axes. The zero vector is denoted O.

We think of vectors that arise in different contexts as different types of vectors. We begin by describing four common types.

Displacement vectors

The relative positions of points are described by DISPLACEMENT VECTORS.

FIGURE 11.3   The displacement vector $\vec{PQ}$

Definition 11.2   The displacement vector $\vec{PQ}$ is the vector with its base at the point P and tip at the point Q (Figure 11.3); its length is the distance between the points and its direction is that from P toward Q.

Position vectors

The POSITION VECTOR of a point in a coordinate plane gives its position relative to the origin.

Definition 11.3   The position vector of a point P in a coordinate plane is the displacement vector $\vec{OP}$ from the origin $O = (0, 0)$ to the point P (Figure 11.4).

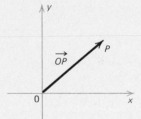

FIGURE 11.4   The position vector $\vec{OP}$ of the point P

Force vectors

We use a FORCE VECTOR to describe a force applied to an object.

Definition 11.4   The length of a force vector is the strength of the force (measured in pounds or other units), and its direction is that in which the force is applied.

Velocity vectors

An object's *speed* is a nonnegative number that indicates how fast the object is moving. Its VELOCITY VECTOR gives both its speed and direction of motion.

Definition 11.5   The length of an object's velocity vector is its speed, and the direction of the velocity vector is the direction of the object's motion.

We will study speed and velocity vectors more closely in Sections 11.3 and 11.4.

Translation of vectors

If two vectors, such as the vectors A and B in Figure 11.5, have the same length and direction, we consider them to be *the same vector in different locations* and we write $A = B$. We could move A to coincide with B by moving it parallel to its original position without changing its length. This is called TRANSLATION of the vector A.

FIGURE 11.5   The same vector in different locations

Arithmetic of vectors

The following definitions show how a vector can be multiplied by a number or two vectors can be added to yield other vectors.

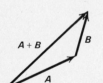

FIGURE 11.6   Multiplication of A by a positive number k

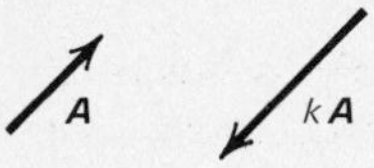

Definition 11.6   The product kA of a vector A and a number k is the vector of length $|k|\,|A|$ and with the direction of A if k is positive and the opposite direction if k is negative (Figures 11.6 and 11.7).

The vector $A + B$ is determined by placing the base of B at the tip of A (Figure 11.8). The sum $A + B$ then runs from the base of A to the tip of B.

We can also obtain the sum $A + B$ by placing A and B with a common base, as in Figure 11.9, and by drawing the parallelogram with sides formed by the vectors. Then $A + B$ runs from the base of A and B to the opposite vertex of the parallelogram.

The difference of two vectors is defined by the equation $A - B = A + (-1)B$. We can view $A - B$ as the vector we add to B to obtain the vector A. Thus, if we place A and B with a common base, then $A - B$ runs from the tip of B to the tip of A (Figure 11.10).

FIGURE 11.7   Multiplication of A by a negative number k

FIGURE 11.8   Addition of vectors

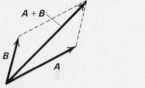

FIGURE 11.9   Addition of vectors

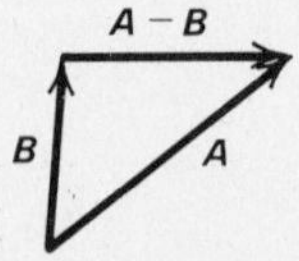

FIGURE 11.10   Subtraction of vectors

The unit vectors i and j

A vector of length 1 is called a UNIT VECTOR. To use rectangular coordinates in dealing with vectors in an xy-plane, we introduce a unit vector $\mathbf{i}$ that points in the direction of the positive x-axis and a unit vector $\mathbf{j}$ that points in the direction of the positive y-axis (Figure 11.11). Any vector A in the xy-plane can be placed with its base at the origin, and it is then the position vector $\overrightarrow{OP}$ of the point P at its tip (Figure

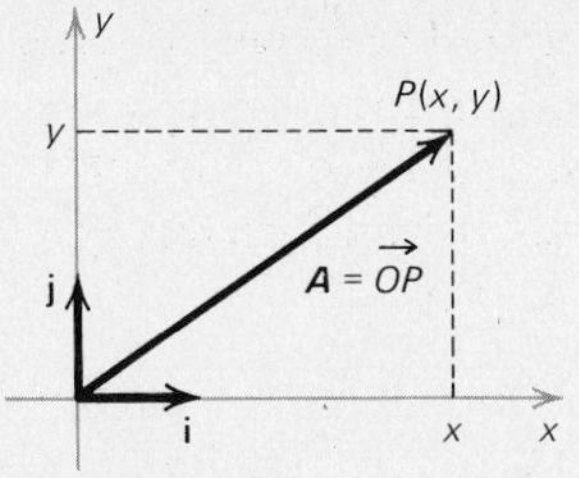

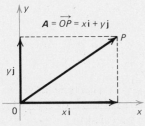

FIGURE 11.11

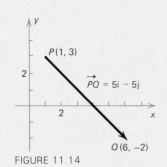

FIGURE 11.12

11.11). The following rule shows how the vector can then be expressed in terms of **i** and **j** and the coordinates of the point P.

Rule 11.1   If the base of a vector A is at the origin in an xy-plane and its tip is at the point $P(x, y)$, then

(1) $A = \overrightarrow{OP} = x\mathbf{i} + y\mathbf{j}.$ ◀

We obtain (1) because the vectors $x\mathbf{i}$ and $y\mathbf{j}$ run from the origin to the points at x and y on the coordinate axes, whereas $A = \overrightarrow{OP}$ runs along the diagonal of the rectangle they form (Figure 11.12).

The number x in (1) is called the x-COMPONENT or **i**-COMPONENT of the vector A. The number y is the y-COMPONENT or **j**-COMPONENT. We will often use the notation $\langle x, y \rangle$ for the vector

(2) $\langle x, y \rangle = x\mathbf{i} + y\mathbf{j}$ ◀

whose components are x and y. In particular, we have $\mathbf{i} = \langle 1, 0 \rangle$ and $\mathbf{j} = \langle 0, 1 \rangle$.

If a vector does not have its base at the origin, its components are the coordinates of its tip minus the coordinates of its base, as is stated in the next rule.

Rule 11.2   If the base of a vector is at $P(x_0, y_0)$ and its tip is at $Q(x_1, y_1)$ (Figure 11.13), then it is the displacement vector

(3) $\overrightarrow{PQ} = (x_1 - x_0)\mathbf{i} + (y_1 - y_0)\mathbf{j} = \langle x_1 - x_0, y_1 - y_0 \rangle.$ ◀

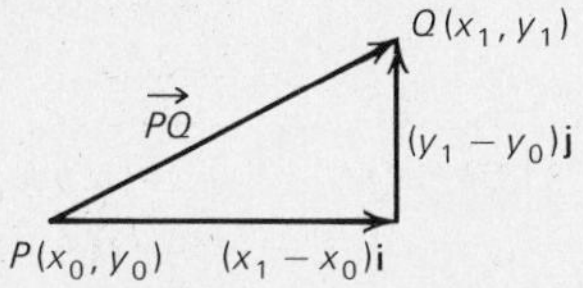

FIGURE 11.13   The displacement vector
$\overrightarrow{PQ} = (x_1 - x_0)\mathbf{i} + (y_1 - y_0)\mathbf{j}$
$\quad = \langle x_1 - x_0, y_1 - y_0 \rangle$

EXAMPLE 1   Express the displacement vector $\overrightarrow{PQ}$ in terms of its x- and y-components, where P is the point $(1, 3)$ and Q is the point $(6, -2)$.

Solution. $\overrightarrow{PQ} = (6 - 1)\mathbf{i} + (-2 - 3)\mathbf{j} = 5\mathbf{i} - 5\mathbf{j}$. The vector is shown in Figure 11.14.   □

FIGURE 11.14

Doing vector arithmetic by components

According to the next rule, we can add or subtract vectors and we can multiply vectors by numbers by performing these operations on each of the components.

Rule 11.3   For any vectors $\langle x_1, y_1 \rangle, \langle x_2, y_2 \rangle$, and $\langle x, y \rangle$ in an xy-plane and any number k

(4) $\langle x_1, y_1 \rangle + \langle x_2, y_2 \rangle = \langle x_1 + x_2, y_1 + y_2 \rangle$ ◀

(5) $k\langle x, y \rangle = \langle kx, ky \rangle.$ ◀

The derivation of (4) for positive $x_1, y_1, x_2,$ and y_2 is indicated in Figure 11.15, and the derivation of (5) for positive x and y and $k > 1$ is indicated in Figure 11.16. Similar diagrams would yield these identities for other values of the numbers.

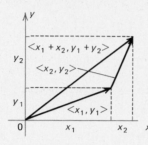

FIGURE 11.15   FIGURE 11.16

EXAMPLE 2   Compute $2A - 3B$ where $A = \langle 4, -1 \rangle$ and $B = \langle 2, 5 \rangle$.

Solution.   We have $2A - 3B = 2\langle 4, -1 \rangle - 3\langle 2, 5 \rangle$
$= \langle 2(4) - 3(2), 2(-1) - 3(5) \rangle = \langle 2, -17 \rangle$. ☐

The formal definition of vectors

The representation of vectors by their components leads us to the following formal definition in which vectors in a coordinate plane are defined without reference to the context in which they appear.

Definition 11.7   A vector A in an xy-plane is an ordered pair of numbers denoted either $\langle x, y \rangle$ or $x\mathbf{i} + y\mathbf{j}$.

The laws of vector arithmetic

Because the addition of two vectors and the multiplication of a vector by a number can be accomplished by performing the operations on their components, the laws of arithmetic of numbers give the following analogous laws for dealing with vectors.

$$A + B = B + A, \qquad A + (B + C) = (A + B) + C$$
$$A + O = A, \qquad A - A = O, \qquad (1)A = A, \qquad (0)A = O \qquad \blacktriangleleft$$
$$aO = O, \qquad (ab)A = a(bA), \qquad a(A + B) = aA + aB$$
$$(a + b)A = aA + bA.$$

The length of a vector

The Pythagorean theorem gives the formula

$$|\langle x, y \rangle| = \sqrt{x^2 + y^2} \qquad \blacktriangleleft$$

which expresses the length of a vector in terms of its components. Given a nonzero vector A, we can find the unit vector with the same direction as A by dividing A by its length.

EXAMPLE 3   Find the unit vector with the same direction as $3\mathbf{i} - 4\mathbf{j}$.

Solution.   The length of $3\mathbf{i} - 4\mathbf{j}$ is $|3\mathbf{i} - 4\mathbf{j}| = \sqrt{(3)^2 + (-4)^2}$
$= \sqrt{9 + 16} = 5$. Therefore, the unit vector with the same direction is
$\frac{1}{5}(3\mathbf{i} - 4\mathbf{j}) = \frac{3}{5}\mathbf{i} - \frac{4}{5}\mathbf{j}$. ☐

The polar form of a vector

The next rule shows how to compute the components of a vector from its length and the angle between it and the vector **i**.

Rule 11.4   If Θ is the angle from the direction of the vector i to the direction of the nonzero vector A, then

(6) $A = |A| (\cos \Theta\, \mathbf{i} + \sin \Theta\, \mathbf{j})$.   ◄

Derivation of Rule 11.4   We place the vector **A** with its base at the origin as in Figure 11.17. Then the polar coordinates of its tip are $[|A|, \Theta]$, and the rectangular coordinates are $(|A|\cos \Theta, |A|\sin \Theta)$. With Rule 11.1 this gives (6).   □

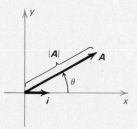

FIGURE 11.17
$A = |A|(\cos \Theta\, \mathbf{i} + \sin \Theta\, \mathbf{j})$

EXAMPLE 4   Give the x- and y-components of the vector **A** of length 7 such that the angle from the direction of **i** to the direction of **A** is $\pi/4$ radians.

Solution.   We have $A = 7[\cos(\pi/4)\mathbf{i} + \sin(\pi/4)\mathbf{j}]$
$= (7/\sqrt{2})\mathbf{i} + (7/\sqrt{2})\mathbf{j}$.   □

The next example illustrates how sums of vectors can arise in the study of velocities.

EXAMPLE 5   A passenger on a train moving 50 kilometers per hour in the direction 60° north of east rolls a ball with velocity relative to the train of 2 kilometers per hour toward the west. What is the velocity of the ball relative to the ground? (Use an xy-plane with the positive x-axis pointing east and the positive y-axis pointing north.)

Solution.   The velocity **A** of the train relative to the ground has length 50 kilometers per hour and makes an angle of $\frac{1}{3}\pi$ with the positive x-axis. Therefore,

$$A = 50[\cos(\tfrac{1}{3}\pi)\mathbf{i} + \sin(\tfrac{1}{3}\pi)\mathbf{j}]$$

$$= 50(\tfrac{1}{2}\mathbf{i} + \tfrac{1}{2}\sqrt{3}\mathbf{j}) \text{ kilometers per hour.}$$

The velocity **B** of the ball relative to the train has length 2 kilometers per hour and points in the direction of the negative x-axis. Hence $B = -2\mathbf{i}$. The velocity of the ball relative to the ground is the sum

$$A + B = 50(\tfrac{1}{2}\mathbf{i} + \tfrac{1}{2}\sqrt{3}\mathbf{j}) - 2\mathbf{i}$$

$$= 23\mathbf{i} + 25\sqrt{3}\mathbf{j} \text{ kilometers per hour}$$

of the velocities **A** and **B**.   □

Exercises

†Answer provided.
‡Outline of solution provided.

1. Express the displacement vector $\overrightarrow{PQ}$ in the form $\langle x, y \rangle$ for   **a.**† $P(2, 5)$ and $Q(4, -3)$ **b.**† $P(0, -1)$ and $Q(-7, -6)$ **c.** $P(-1, 1)$ and $Q(1, 1)$.

2. Compute $A + 2B$, $|A + 2B|$, and the unit vector in the direction of $A + 2B$ for each of the following pairs of vectors.   **a.**† $A = \langle 1, 6 \rangle$ and $B = \langle 2, 1 \rangle$ **b.**† $A = 2\mathbf{i} - 4\mathbf{j}$ and $B = O$ **c.** $A = 3\mathbf{i} - 4\mathbf{j}$ and $B = \langle 1, -1 \rangle$.

3.‡ Give the two unit vectors which are parallel to the line $y = 2x + 1$.

4. Give the two unit vectors which are parallel to the tangent line to $y = x^3$ at $x = 1$.

5. Use vectors to find the vertex opposite Q of the parallelogram $PQRS$ for   **a.**‡ $P(3, 4)$, $Q(1, 1)$, and $R(5, 2)$ **b.** $P(3, 1)$, $Q(5, 3)$, and $R(7, -4)$.

6. Give the x- and y-components of   **a.**† the vector of length 5 whose direction is at an angle of $\frac{2}{3}\pi$ radians from the direction of $\mathbf{i}$, and   **b.** the vector of length 17 whose direction is at an angle of $\frac{5}{4}\pi$ from the direction of $\mathbf{i}$.

7. Find numbers s and t such that   **a.**‡ $s\langle 2, 5 \rangle + t\langle 3, 2 \rangle = \langle 6, -7 \rangle$ **b.** $s\langle 5, 10 \rangle - t\langle -1, 3 \rangle = \langle 7, 4 \rangle$.

8.‡ The point P lies one fourth of the way from the point Q toward the point R. Use vectors to express the coordinates of P in terms of those of Q and R.

9. Find the coordinates of   **a.**† the point one third of the way from $(2, 4)$ toward $(6, 3)$ and   **b.** the point one fifth of the way from $(1, -3)$ toward $(7, 0)$. (See Exercise 8.)

10.‡ Use vectors to show that the midpoints of the sides of any quadrilateral are the vertices of a parallelogram.

11.† Use vectors to show that the line joining the midpoints of two sides of a triangle is parallel to the third side and is half as long.

12. Use vectors to show that the medians of any triangle intersect at a point which is $\frac{2}{3}$ of the distance from each vertex to the midpoint of the opposite side.

13.† A car is traveling 60 miles per hour down a straight road in the direction of the vector $\mathbf{i}$ when a boy throws a ball out the window with velocity, relative to the car, in the direction of $\mathbf{j}$ and of magnitude 10 miles per hour. What are   **a.** the velocity and   **b.** the speed of the ball relative to the ground when he lets it go?

14. An airplane's air velocity is 200 miles per hour toward the northwest, and the wind is blowing 50 miles per hour toward the northeast. The vector $\mathbf{i}$ points east. What are the airplane's   **a.** velocity and   **b.** speed relative to the ground?

15. A man can swim 3 miles per hour, and he wants to swim in the direction of the vector $\mathbf{j}$ straight across a river that is $\frac{1}{4}$ mile wide and flowing at the rate of 2 miles per hour in the direction of $\mathbf{i}$.   **a.** In what direction should he swim?   **b.** How long will it take him?

16.[†] Two forces of $F_1 = 3i - 2j$ pounds and $F_2 = 5i + j$ pounds are applied at the same point on an object.   **a.**   What single force F_3 applied at the same point would have the same effect?   **b.**   What single force F_4 applied at the same point would cancel the effect of forces F_1 and F_2 combined?

17.   **a.**   Show that three forces applied at the same point are in equilibrium (cancel each other out) if and only if the arrows representing the forces can be translated to form the sides of a triangle with the tip of each vector touching the base point of the next.   **b.**   What is the corresponding statement for four forces?

18.[†] Compute   **a.** $2C - 3A + B$,   **b.** $4C - \langle 3, 2 \rangle$ and   **c.** $2(A + B + \langle 1, 3 \rangle)$ for $A = \langle 1, -1 \rangle$, $B = \langle -2, 4 \rangle$, $C = \langle 2, 0 \rangle$.

19.[†] Compute   **a.** $(2A - B + 4C)/|C|$,   **b.** $|A|^2 + |B|^2$, and   **c.** $|A - 3B + C|$ for $A = i - j$, $B = 2i + j$, $C = 3i - 4j$.

20.[†] Find the fourth vertex of each of the three parallelograms with the vertices $(1, 2)$, $(2, 0)$, and $(4, 4)$.

21.[†] Solve the following equations for the vector A.

   a. $2A + 3\langle 4, 6 \rangle = \langle 2, -1 \rangle$ **b.** $A - \langle 1, 3 \rangle = 3A + \langle 4, 6 \rangle$
   c. $A - i + j = 4(3i - 2j - 3A)$

22.[†] Find the angle $\Theta (0 \le \Theta \le 2\pi)$ from the position vector i to the position vector A for   **a.** $A = i + j$,   **b.** $A = -i + j$,   **c.** $A = i - j$,   **d.** $A = \langle 9, 3\sqrt{3} \rangle$,   **e.** $A = \langle 2, -4 \rangle$.

23.[†] A train is travelling 50 miles per hour toward the northeast when a ball is rolled from right to left across the aisle with the speed of one mile per hour relative to the train. What is the ball's velocity relative to the ground? (Have j point north.)

24.[†] As you are sailing your boat directly south at 10 knots (nautical miles per hour), the wind gauge shows that the wind's velocity relative to the boat is 2 knots toward the west. What is the wind's velocity relative to the stationary water? (Have j point north.)

25.[†] A car is travelling counterclockwise around the circular track $x^2 + y^2 = 2$. The force of friction on it has one component of 50 pounds directed toward the center of the track and another component of 10 pounds tangent to the track in the clockwise direction. The force of the car's engine is 100 pounds tangent to the track in the counterclockwise direction. What are the total forces on the car when it reaches the point $(1, 1)$? (Distances are measured in miles.)

26.   The force of the wind on a rubber raft is 25 pounds toward the southwest. What force must the raft's engine exert for the combined force of the wind and engine to be 20 pounds toward the south? (Have j point north.)

11-2   THE DOT PRODUCT

In this section we discuss the DOT PRODUCT (or SCALAR PRODUCT), which is used to study angles between vectors.

Definition 11.8   The dot product $A \cdot B$ of the vectors $A = \langle x_1, y_1 \rangle$ and $B = \langle x_2, y_2 \rangle$ is the number

(1) $A \cdot B = x_1 x_2 + y_1 y_2.$ ◀

The dot product can be used to relate the angle between vectors and their components because it can be expressed either in terms of the components, as in Definition 11.8, or in terms of the lengths of the vectors and an angle between them, as in the next theorem.

Theorem 11.1   For any nonzero vectors A and B

(2) $A \cdot B = |A|\,|B| \cos \Theta$ ◀

where Θ is an angle between A and B.

Proof.   We will derive expression (2) for the dot product by computing the length of the vector $A - B$ in two ways. First we place the vectors $A = \langle x_1, y_1 \rangle$ and $B = \langle x_2, y_2 \rangle$ with their bases at the origin as in Figure 11.18. We have

$$
\begin{aligned}
|A - B|^2 &= |\langle x_1 - x_2, y_1 - y_2 \rangle|^2 \\
&= (x_1 - x_2)^2 + (y_1 - y_2)^2 \\
&= (x_1^2 - 2x_1 x_2 + x_2^2) + (y_1^2 - 2y_1 y_2 + y_2^2) \\
&= (x_1^2 + y_1^2) - 2(x_1 x_2 + y_1 y_2) + (x_2^2 + y_2^2).
\end{aligned}
$$

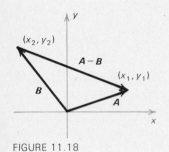

FIGURE 11.18

With definition (1) this yields the equation

(3) $|A - B|^2 = |A|^2 - 2A \cdot B + |B|^2.$

To compute $|A - B|^2$ in a second way, we introduce uv-coordinates with A along the positive u-axis as in Figure 11.19. In the new coordinates the tip of A is the point $(|A|, 0)$ and the tip of B is the point $(|B|\cos \Theta, |B|\sin \Theta)$. The length of $A - B$ is the distance between those points:

$$
\begin{aligned}
|A - B|^2 &= (|B|\cos \Theta - |A|)^2 + (|B|\sin \Theta - 0)^2 \\
&= |B|^2 \cos^2\Theta - 2|A|\,|B|\cos \Theta + |A|^2 + |B|^2 \sin^2\Theta.
\end{aligned}
$$

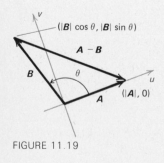

FIGURE 11.19

Since $\cos^2\Theta + \sin^2\Theta = 1$, this may be written

(4) $|A - B|^2 = |A|^2 - 2|A|\,|B|\cos \Theta + |B|^2.$

A comparison of formulas (3) and (4) gives expression (2) for the dot product.

We can take any angle Θ from A to B or from B to A because all such angles have the same cosine.   Q.E.D.

Note. Formula (4) is called the *law of cosines* when it is presented in the context of trigonometry. □

EXAMPLE 1   (a) Find the cosine of the angles between the vectors $A = \langle 4, 3 \rangle$ and $B = \langle 3, -1 \rangle$. (b) Use the table of cosines to give an approximate value for one such angle.

Solution.   (a) We have $A \cdot B = \langle 4, 3 \rangle \cdot \langle 3, -1 \rangle = 4(3) + 3(-1) = 9$, $|A| = |\langle 4, 3 \rangle| = \sqrt{25} = 5$, and $|B| = |\langle 3, -1 \rangle| = \sqrt{10}$. Therefore, with formula (2) we have

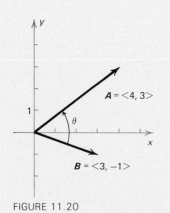

$$(5) \qquad \cos \Theta = \frac{A \cdot B}{|A| \, |B|} = \frac{9}{5\sqrt{10}}$$

where Θ is an angle between A and B.

(b) The approximate value of the quantity (5) is 0.57. The table of cosines gives 0.96 radians as an approximate value of an angle with this cosine. This positive acute angle from B to A is shown in Figure 11.20. □

FIGURE 11.20

Algebra of the dot product

Definition (1) of the dot product shows that it has the properties

$$(6) \qquad A \cdot B = B \cdot A \qquad \blacktriangleleft$$

$$(7) \qquad (aA) \cdot (bB) = (ab)A \cdot B \qquad \blacktriangleleft$$

$$(8) \qquad A \cdot (B + C) = A \cdot B + A \cdot C \qquad \blacktriangleleft$$

for any vectors A, B, and C and any numbers a and b (Exercise 11). Formula (2) and the fact that $|\cos \Theta| \leq 1$ yield the inequality

$$(9) \qquad |A \cdot B| \leq |A| \, |B|. \qquad \blacktriangleleft$$

If we set $A = B$ in either definition (1) or formula (2), we obtain the equation

$$(10) \qquad A \cdot A = |A|^2. \qquad \blacktriangleleft$$

Perpendicular vectors

Because the cosine of a right angle is zero, formula (2) shows that two nonzero vectors are perpendicular if their dot product is zero. We consider the zero vector to be perpendicular to all vectors. We then have the following rule.

Rule 11.5   Two vectors *A* and *B* are perpendicular if and only if their dot product *A* · *B* is zero.

EXAMPLE 2   Find the two unit vectors perpendicular to $\langle 1, 4 \rangle$.

Solution.   If we interchange the components of $\langle a, b \rangle$ and multiply one of the components by -1, we obtain vectors $\langle -b, a \rangle$ and $\langle b, -a \rangle$ with opposite directions, which are perpendicular to $\langle a, b \rangle$ because their dot products with it are zero. Thus, the vectors $\langle -4, 1 \rangle$ and

$\langle 4, -1 \rangle$ are perpendicular to $\langle 1, 4 \rangle$. To obtain unit vectors, divide these vectors by their lengths. We obtain

$$\frac{\langle -4, 1 \rangle}{\sqrt{17}} \quad \text{and} \quad \frac{\langle 4, -1 \rangle}{\sqrt{17}}. \quad \square$$

The component of one vector in the direction of another

Definition 11.9   For nonzero vectors A and B

(11) $\left[\begin{array}{l} \textbf{The component of } A \\ \textbf{in the direction of } B \end{array} \right] = |A| \cos \Theta$

where Θ is an angle between A and B. For $A = 0$, the component of A in the direction of B is 0.

The geometric interpretation of Definition 11.9 is indicated in Figures 11.21 and 11.22. We place the vectors A and B with a common base point and draw an auxiliary t-axis with its origin at the bases of the vectors and with B lying along its positive side. We drop a perpendicular line from the tip of the vector A to the t-axis. The component of A in the direction of B is then the t-coordinate of the foot of the perpendicular line. If the angle Θ between the vectors is acute, as in Figure 11.21, then the quantity (11) is positive. If the angle Θ is obtuse, as in Figure 11.22, then (11) is negative. It is zero if Θ is a right angle and the vectors are perpendicular.

Theorem 11.1 enables us to compute the components of vectors by using the dot product, according to the following rule.

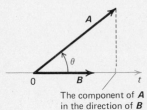

The component of A in the direction of B

FIGURE 11.21

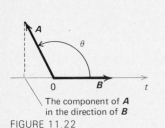

The component of A in the direction of B

FIGURE 11.22

Rule 11.6   For any vector A and any nonzero vector B

(12) $\left[\begin{array}{l} \textbf{The component of } A \\ \textbf{in the direction of } B \end{array} \right] = \dfrac{A \cdot B}{|B|}.$ ◀

Notice that the quantity on the right of (12) is the dot product of A with the unit vector $B/|B|$ in the direction of B. Also notice that the component of a vector $A = x\mathbf{i} + y\mathbf{j}$ in the direction of the vector $\mathbf{i}$ is its x-component and its component in the direction of $\mathbf{j}$ is its y-component.

EXAMPLE 3   Find the component of $A = 6\mathbf{i} - 7\mathbf{j}$ in the direction of $B = -3\mathbf{i} + 2\mathbf{j}$.

Solution.   By formula (12) the component of A in the direction of B is

$$\frac{A \cdot B}{|B|} = \frac{\langle 6, -7 \rangle \cdot \langle -3, 2 \rangle}{|\langle -3, 2 \rangle|} = \frac{(6)(-3) + (-7)(2)}{\sqrt{(-3)^2 + 2^2}}$$

$$= \frac{-18 - 14}{\sqrt{9 + 4}} = \frac{-32}{\sqrt{13}}. \quad \square$$

The projection of one vector along another

When we multiply the component of A in the direction of B by the unit vector $B/|B|$ in the direction of B, we obtain a vector that is called the PROJECTION OF A ALONG B (Figure 11.23).

Definition 11.10   For any vector A and any nonzero vector B

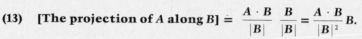

(13)   [The projection of A along B] $= \dfrac{A \cdot B}{|B|} \dfrac{B}{|B|} = \dfrac{A \cdot B}{|B|^2} B.$ ◄

FIGURE 11.23   The projection C of A along B

EXAMPLE 4   Find the projection of A along B for the A and B of Example 3.

Solution.   The desired vector is

$$\frac{A \cdot B}{|B|^2} B = -\frac{32}{13}\langle -3, 2 \rangle = \left\langle \frac{96}{13}, -\frac{64}{13} \right\rangle. \quad \square$$

Exercises

†*Answer provided.*
‡*Outline of solution provided.*

1. Compute   **a.**† $\langle 3, 1 \rangle \cdot \langle -6, 2 \rangle$ **b.** $(2\mathbf{i} - 3\mathbf{j}) \cdot (-7\mathbf{j})$
    c.† $(\mathbf{i} + \mathbf{j}) \cdot (5\mathbf{i} - 6\mathbf{j})$ and   **d.** $\langle -1, 6 \rangle \cdot \langle 4, -2 \rangle$.

2. Give approximate values for angles between the following pairs of vectors.   **a.**† $\langle 3, 4 \rangle$ and $\langle 4, 3 \rangle$ **b.** $\langle -1, 2 \rangle$ and $\langle -4, 2 \rangle$
    c.† $6\mathbf{i} + \mathbf{j}$ and $-6\mathbf{i} + \mathbf{j}$ **d.** $5\mathbf{i} + 12\mathbf{j}$ and $-4\mathbf{j}$.
    (Use the reference tables.)

3. Give the two unit vectors perpendicular to   **a.**† $6\mathbf{i} + 3\mathbf{j}$
    b. $-2\mathbf{i} - 5\mathbf{j}$ **c.**† $\langle 3, -7 \rangle$ **d.** $\langle \pi, e \rangle$.

4. For each of the following pairs of vectors compute the component of A in the direction of B and the projection of A along B. Then sketch the three vectors with a common base point.
    a.† $A = \langle 1, 4 \rangle, B = \langle 2, 1 \rangle$ **b.** $A = 6\mathbf{i} + 8\mathbf{j}, B = 4\mathbf{i} + 3\mathbf{j}$
    c.† $A = \langle -6, -3 \rangle, B = \langle 1, 1 \rangle$ **d.** $A = -\mathbf{i} + 6\mathbf{j}, B = \mathbf{i} - 6\mathbf{j}$.

5. Vectors A, B, C, and D and a scale for measuring their lengths are shown in Figure 11.24. Give approximate values for the components of
    a.‡ A,   **b.**† B, and   **c.** C in the direction of D.

6.† The *work* done by a constant force F on an object as it moves on the line segment from point P to point Q is $F \cdot \overrightarrow{PQ}$. A man walks 100 feet directly north as the wind exerts a constant force of 10 pounds on him toward the northwest. How much work does the wind do?

7.‡ Find two unit vectors which make an angle of 45° with $3\mathbf{i} + 2\mathbf{j}$. Sketch the three vectors.

8. Find two vectors of length 6 which make an angle of 30° with $\langle 3, -4 \rangle$. Sketch the three vectors.

9.† Use vectors to show that a parallelogram is a rhombus if and only if its diagonals are perpendicular.

10. Use vectors to prove that the altitudes of a triangle meet in a point.

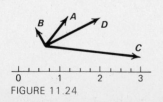

FIGURE 11.24

11. Derive formulas (6), (7), and (8).

12.† Use vectors to show that the distance from the point (x_0, y_0) to the line $Ax + By + C = 0$ is

(14) $\dfrac{|Ax_0 + By_0 + C|}{\sqrt{A^2 + B^2}}.$

13. Use formula (14) to compute   **a.**† the distance from the point $(2, -1)$, to the line $3x - 4y + 6 = 0$ **b.**† the distance from the point $(1, 5)$ to the line $y = -2x$ **c.** the distance from the origin to the line $x + y = 6$ **d.** the distance from the line $5x = 12y$ to the circle $(x + 3)^2 + (y - 3)^2 = 4$.

14. Find a constant k such that $\langle -2, 4 \rangle$ and $\langle 3, k \rangle$ are perpendicular.

15.† Compute $A \cdot B$ for   **a.** $A = \langle 2, 3 \rangle$, $B = \langle -1, 6 \rangle$,   **b.** $A = 2\mathbf{i} - 3\mathbf{j}$, $B = \mathbf{i} - 2\mathbf{j}$ and   **c.** $A = 2(3\mathbf{i} - 4\mathbf{j})$, $B = -(\mathbf{i} - \mathbf{j})$.

16.† Find all vectors perpendicular to   **a.** $\langle 2, -5 \rangle$,   **b.** $\mathbf{j}$, and   **c.** $\langle 4, 0 \rangle$.

17.† Find a vector A perpendicular to $B = \langle 3, 2 \rangle$ such that $A + B$ has a zero $\mathbf{j}$-component.

18.† Find numbers x such that $\langle 3, 4 \rangle$ and $\langle x, 2 \rangle$ are   **a.** parallel and   **b.** perpendicular.

19.† Find a number x such that the angle between $\langle 1, 1 \rangle$ and $\langle x, 1 \rangle$ is $45°$.

20.† Find the two unit vectors tangent to the curve $y = \sin x$ at $x = \pi/3$.

21.† Find the component of A in the direction of B for   **a.** $A = \mathbf{i} - \mathbf{j}$, $B = \mathbf{i} + 2\mathbf{j}$,   **b.** $A = \langle 1, 3 \rangle$, $B = \langle -6, 2 \rangle$, and   **c.** $A = \langle 1, 101 \rangle$, $B = \langle 100, -1 \rangle$.

22.† Find a vector A parallel to $B = \langle 1, 5 \rangle$ such that $A + B$ has component $12\sqrt{2}$ in the direction of $\langle 1, 1 \rangle$.

23.† Find the projection of A along the line through B for   **a.** $A = \mathbf{i} - 3\mathbf{j}$, $B = 3\mathbf{i} + 4\mathbf{j}$,   **b.** $A = \langle 2, 4 \rangle$, $B = \langle -2, 1 \rangle$, and   **c.** $A = \mathbf{i}$, $B = -4\mathbf{i} + 2\mathbf{j}$.

24. Find two numbers y such that the triangle with vertices $(1, 1)$, $(2, 2)$, and $(0, y)$ is a right triangle.

25.† Use the formula in Exercise 6 to determine the work done by the constant force F (pounds) on an object that traverses the line segment from P to Q for   **a.** $F = \mathbf{i} + \mathbf{j}$, $P = (1, 1)$, $Q = (4, 0)$,   **b.** $F = -3\mathbf{j}$, $P = (0, 0)$, $Q = (-3, 4)$ and   **c.** $F = -\mathbf{i} - 2\mathbf{j}$, $P = (4, 1)$, $Q = (8, -1)$. (The scales on the axes are measured in feet.)

26.† If the wind is blowing 10 knots toward $30°$ south of east, what is the component of the wind's velocity toward the northeast?

27. What is the velocity of the wind if its component toward east is 10 knots (nautical miles per hour) and its component toward the northeast is $7\sqrt{2}$ knots? (Have $\mathbf{j}$ point north.)

11-3 PARAMETRIC EQUATIONS OF CURVES, ARCLENGTH, SPEED

The most effective means for applying calculus to the study of a curve which is not the graph of a function is to use PARAMETRIC EQUATIONS for the curve. The coordinates of points on the curve are given as functions $x = x(t)$ and $y = y(t)$ of an auxiliary variable t, which is called the PARAMETER. The curve is said to be ORIENTED in the direction of increasing t.

Definition 11.11 An oriented curve

(1) $C: x = x(t), y = y(t), a \leq t \leq b$

consists of the points $(x(t), y(t))$ in an xy-plane for $a \leq t \leq b$. The positive direction along C is that given by having t increase.

EXAMPLE 1   Sketch the curve $C: x = \frac{1}{8}t^3 - \frac{3}{2}t, y = \frac{3}{8}t^2, -4 \leq t \leq 4$.

Solution.   We first use our techniques for sketching graphs of functions to sketch the curve $x = \frac{1}{8}t^3 - \frac{3}{2}t$ in a tx-plane (Figure 11.25) and the curve $y = \frac{3}{8}t^2$ in a ty-plane (Figure 11.26). We plot the points on these curves at integer values of t. We then plot the points at integer values of t on the curve C and use the graphs in Figures 11.25 and 11.26 to see where $x(t)$ and $y(t)$ are increasing and decreasing as we draw the curve through the plotted points (Figure 11.27). The arrows indicate the positive direction along C.   □

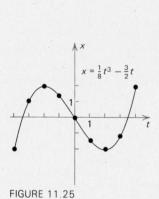

FIGURE 11.25

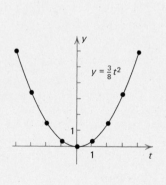

FIGURE 11.26

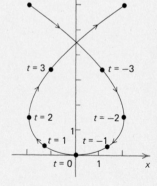

FIGURE 11.27 $C: x = \frac{1}{8}t^3 - \frac{3}{2}t,$ $y = \frac{3}{8}t^2, -4 \leq t \leq 4$

Parametric equations in polar coordinates

If we substitute an equation $r = r(\Theta)$ of a curve in polar coordinates into the formulas $x = r \cos \Theta$, $y = r \sin \Theta$, we obtain parametric equations

$$x = r(\Theta) \cos \Theta, \quad y = r(\Theta) \sin \Theta \qquad \blacktriangleleft$$

of the curve with Θ as parameter.

EXAMPLE 2 The cardioid in Figure 11.28 has the equation $r = 2 + 2 \cos \Theta$ in polar coordinates. Give it parametric equations with Θ as parameter.

Solution. The cardioid is traversed once for $0 \leq \Theta \leq 2\pi$, so it has the equations $C: x = (2 + 2 \cos \Theta)\cos \Theta, y = (2 + 2 \cos \Theta)\sin \Theta$, $0 \leq \Theta \leq 2\pi$. $\square$

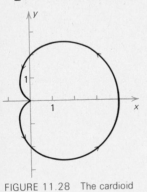

FIGURE 11.28   The cardioid
$r = 2 + 2 \cos \Theta$

Deriving parametric equations

In the next example we derive parametric equations for the curve described by a point on the rim of a rolling wheel.

EXAMPLE 3 The point P on a circle of radius k is at the origin in an xy-plane when the center of the circle is at $(0, k)$ on the y-axis (Figure 11.29). Give parametric equations for the curve traversed by the point p as the circle rolls on the x-axis. (The curve is called a CYCLOID.)

Solution. Figure 11.30 shows the cycloid and the circle after it has rolled a short distance to the right. When the circle has turned an angle t, its center has moved the length kt of the arc subtended by the angle t. Therefore, with the notation of Figure 11.30 we have $\overline{OT} = kt$, $\overline{CT} = k$, $\overline{CQ} = k \cos t$, and $\overline{PQ} = k \sin t$. Hence, the coordinates of the point P are $x = \overline{OT} - \overline{PQ} = kt - k \sin t$ and $y = \overline{CT} - \overline{CQ} = k - k \cos t$. These equations are valid for any t, so the cycloid has the parametric equations

$$(2) \quad x = k(t - \sin t), \quad y = k(1 - \cos t).$$

We do not prescribe any interval of values of t in (2) because t can take any value. The cycloid extends infinitely far to the left and to the right. $\square$

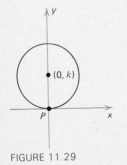

FIGURE 11.29

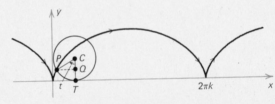

FIGURE 11.30   The cycloid
$x = k(t - \sin t)$
$y = k(1 - \cos t)$

Historical Note (optional)

The cycloid is of particular historical interest because it provides the solution of two famous problems, the *tautochrone* (Greek: equal times) and the *brachistochrone* (Greek: shortest time) problems. The tautochrone problem is to find the shape of a wire from a point *P* to a lower point *Q* (Figure 11.31) such that a frictionless bead released at a point *R* between *P* and *Q* on the wire will take the same time to reach *Q* regardless of the location of *R*. The brachistochrone problem is to find among all wires from *P* to *Q* the one along which a bead will slide from *P* to *Q* in the shortest time. Both problems are solved by taking the wire to have the shape of half an arc of a cycloid, as was discovered in the seventeenth century by Jacques and Jean Bernoulli.

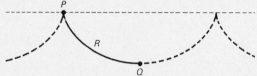

FIGURE 11.31

FIGURE 11.32 Christiaan Huygens (1629–1695)
Culver Pictures, Inc.

The Dutch mathematician, physicist, and astronomer Christiaan Huygens (1629–1695), who invented the pendulum clock, used the tautochrone property of the cycloid to construct a clock in which the period of the pendulum is independent of its amplitude. (See Boyer, [3] *A History of Mathematics*, pp. 410–414.) □

Length of a curve

In Section 6.5 we derived the formula

$$(3) \quad \text{Length} = \int_a^b \sqrt{1 + \left[\frac{df}{dx}(x)\right]^2} \, dx$$

for the length of the portion of the graph $y = f(x)$ of a function f for $a \le x \le b$. The next theorem, which we will prove at the end of this section, is the corresponding result for curves given by parametric equations.

Theorem 11.2 **If $x(t)$ and $y(t)$ have continuous derivatives for $a \le t \le b$, then the curve $C\colon x = x(t), y = y(t), a \le t \le b$ has length**

$$(4) \quad \int_a^b \sqrt{\left[\frac{dx}{dt}(t)\right]^2 + \left[\frac{dy}{dt}(t)\right]^2} \, dt. \quad \blacktriangleleft$$

Formula (3) is the special case of formula (4) where $x = x(t) = t$ and $y(t) = f(t)$.

EXAMPLE 4   Give a definite integral which equals the length of the curve $x = \frac{1}{8}t^3 - \frac{3}{2}t, y = \frac{3}{8}t^2, -4 \le t \le 4$ of Example 1.

Solution. We have $dx/dt = \frac{3}{8}(t^2 - 4)$ and $dy/dt = \frac{3}{4}t$. Therefore, by formula (3) with $a = -4$ and $b = 4$ the length of the curve is

$$\int_{-4}^{4} \sqrt{[\tfrac{3}{8}(t^2 - 4)]^2 + [\tfrac{3}{4}t]^2}\, dt = \int_{-4}^{4} \tfrac{3}{8}\sqrt{t^4 - 4t^2 + 16}\, dt. \quad \square$$

Speed   Suppose that an object has coordinates $x = x(t)$, $y = y(t)$ at time t for $a \le t \le b$. Then, by formula (4) the length of its path from time a to time t is

$$(5) \quad s(t) = \int_{a}^{t} \sqrt{\left[\frac{dx}{dt}(u)\right]^2 + \left[\frac{dy}{dt}(u)\right]^2}\, du.$$

This is the total distance the object travels up to time t. (We use u in place of t as the dummy variable in the integral because t is the upper limit of integration.) The derivative of the distance function $s(t)$ is the object's *speed*, which we can compute by using the following result.

Theorem 11.3   Suppose that $x(t)$ and $y(t)$ have continuous derivatives and that an object is at $x = x(t)$, $y = y(t)$ at time t. Then its speed at time t is

$$(6) \quad \frac{ds}{dt}(t) = \sqrt{\left[\frac{dx}{dt}(t)\right]^2 + \left[\frac{dy}{dt}(t)\right]^2}. \quad \blacktriangleleft$$

Proof. By the Fundamental Theorem for derivatives of integrals, we have

$$\frac{ds}{dt}(t) = \frac{d}{dt}\int_{a}^{t} \sqrt{\left[\frac{dx}{dt}(u)\right]^2 + \left[\frac{dy}{dt}(u)\right]^2}\, du = \sqrt{\left[\frac{dx}{dt}(t)\right]^2 + \left[\frac{dy}{dt}(t)\right]^2}. \quad \text{Q.E.D.}$$

EXAMPLE 5   An object's coordinates at time t (seconds) are $x = e^{2t}$ (feet) and $y = \sin(3t)$ (feet). What is its speed at $t = 0$?

Solution. We have $\dfrac{dx}{dt}(t) = 2e^{2t}$, $\dfrac{dy}{dt}(t) = 3\cos(3t)$, $\dfrac{dx}{dt}(0) = 2$,

and $\dfrac{dy}{dt}(0) = 3$. Therefore, the speed at $t = 0$ is

$$\frac{ds}{dt}(0) = \sqrt{\left[\frac{dx}{dt}(0)\right]^2 + \left[\frac{dy}{dt}(0)\right]^2} = \sqrt{2^2 + 3^2} = \sqrt{13}\,\frac{\text{feet}}{\text{second}}. \quad \square$$

Notice that if the curve C is the path of an object that is at $x = x(t)$, $y = y(t)$ at time t, then the length (4) of the curve for $a \le t \le b$ is the integral of the object's speed (6).

EXAMPLE 6   An object is at $x = \frac{1}{2}t^2$ (meters), $y = \frac{1}{3}t^3$ (meters) at time t (seconds). (a) Compute its speed at time t from formula (6). (b) What is the total distance it travels from time 0 to a positive time t? (c) Compute its speed for positive t from the result of part (b).

Solution.   (a) We have

$$\frac{dx}{dt} = \frac{d}{dt}(\tfrac{1}{2}t^2) = t \quad \text{and} \quad \frac{dy}{dt} = \frac{d}{dt}(\tfrac{1}{3}t^3) = t^2$$

so the object's speed at time t is

(7) $$\sqrt{\left(\frac{dx}{dt}\right)^2 + \left(\frac{dy}{dt}\right)^2} = \sqrt{t^2 + (t^2)^2}$$

$$= |t|\sqrt{1 + t^2} \text{ meters per second.}$$

(b) Because the object's speed at time $u \geq 0$ is $u\sqrt{1 + u^2}$, the distance it travels from time 0 to a positive time t is

(8) $$\int_0^t \left[\text{Speed at time } u\left(\frac{\text{meters}}{\text{second}}\right)\right] du(\text{seconds})$$

$$= \int_0^t u\sqrt{1 + u^2} \, du \text{ meters.}$$

With the substitution $v = 1 + u^2$, $dv = 2u \, du$, we obtain

$$\int u\sqrt{1 + u^2} \, du = \tfrac{1}{2}\int v^{1/2} \, dv = (\tfrac{1}{2})(\tfrac{2}{3})v^{3/2} + C$$

$$= \tfrac{1}{3}(1 + u^2)^{3/2} + C.$$

Therefore the distance (8) is

(9) $$[\tfrac{1}{3}(1 + u^2)^{3/2}]_0^t = \tfrac{1}{3}(1 + t^2)^{3/2} - \tfrac{1}{3} \text{ meters.}$$

(c) The speed is the derivative of the total distance traveled, which for positive t is given by (9). Hence for $t > 0$ the speed is

$$\frac{d}{dt}[\tfrac{1}{3}(1 + t^2)^{3/2} - \tfrac{1}{3}] = \tfrac{1}{3}(\tfrac{3}{2})(1 + t^2)^{1/2}\frac{d}{dt}(1 + t^2)$$

$$= t\sqrt{1 + t^2} \text{ meters per second}$$

and this is, of course, the same as the result of part (a) for $t > 0$.   □

Arclength as parameter   Occasionally it is convenient to study a curve with the arclength s as parameter. If the derivatives $\dfrac{dx}{dt}(t)$ and $\dfrac{dy}{dt}(t)$ are never both zero for the same value of t, then the derivative (6) of the arclength with respect to t is always positive. In this case the function $s(t)$ has a differentiable inverse which we denote $t(s)$. The curve can then be described with s as parameter by writing $x = x(t(s))$, $y = y(t(s))$.

EXAMPLE 7   Give parametric equations of the circle $x = 5 \cos t$, $y = 5 \sin t, 0 \leq t \leq 2\pi$ with s as parameter.

Solution.   We choose s to be 0 at $t = 0$. We then have

$$s(t) = \int_0^t \sqrt{\left[\frac{dx}{dt}(u)\right]^2 + \left[\frac{dy}{dt}(u)\right]^2}\, du$$

$$= \int_0^t \sqrt{(-5 \sin u)^2 + (5 \cos u)^2}\, du = \int_0^t 5\, du = 5t.$$

We solve the equation $s = 5t$ for t to obtain the inverse function $t = s/5$. Making this substitution gives the equations $x = 5 \cos(s/5)$, $y = 5 \sin(s/5), 0 \leq s \leq 10\pi$ for the circle with s as parameter.   □

Proof of Theorem 11.2   We determine the length of the curve as the limit of the lengths of approximations by polygonal lines, as we did in the case of graphs of functions. We consider an arbitrary partition

(10) $a = t_0 < t_1 < t_2 < t_3 < \cdots < t_{N-1} < t_N = b$

of the interval $a \leq t \leq b$. We approximate the curve C by the polygonal line joining the successive points $(x(t_j), y(t_j))$ $(j = 0, 1, 2, \ldots, N)$ along the curve. The jth segment of the approximation then goes from the point $(x(t_{j-1}), y(t_{j-1}))$ to $(x(t_j), y(t_j))$ and, by the Pythagorean theorem, the length of the entire approximation is

(11) $\displaystyle\sum_{j=1}^{N} \sqrt{[x(t_j) - x(t_{j-1})]^2 + [y(t_j) - y(t_{j-1})]^2}.$

According to the Mean Value Theorem for derivatives, there exist for each j numbers c_j and d_j in the interval $t_{j-1} < t < t_j$ such that

(12) $x(t_j) - x(t_{j-1}) = \dfrac{dx}{dt}(c_j)(t_j - t_{j-1})$

and

(13) $y(t_j) - y(t_{j-1}) = \dfrac{dy}{dt}(d_j)(t_j - t_{j-1}).$

We use equations (12) and (13) and the fact that each $t_j - t_{j-1}$ is positive to express the sum (11) in the form

(14)
$$\sum_{j=1}^{N} \sqrt{\left[\frac{dx}{dt}(c_j)(t_j - t_{j-1})\right]^2 + \left[\frac{dy}{dt}(d_j)(t_j - t_{j-1})\right]^2}$$

$$= \sum_{j=1}^{N} \sqrt{\left[\frac{dx}{dt}(c_j)\right]^2 + \left[\frac{dy}{dt}(d_j)\right]^2}\, (t_j - t_{j-1}).$$

The integral (4) gives the exact length of the curve because the approximate length (14) is an approximating sum for that integral and tends to it as the number of subintervals in the partition (10) tends to infinity and their lengths tend to zero.   Q.E.D.

Exercises

†Answer provided.
‡Outline of solution provided.

1.‡ Sketch the curve $x = 6t - \frac{1}{2}t^3$, $y = \frac{1}{4}t^4 - \frac{13}{4}t^2 + 9$, $-3 \le t \le 3$.

2. Sketch *Tschirnhausen's cubic* $x = t^2 - 3$, $y = \frac{1}{3}t^3 - t$.

3.‡ Show that the curve $x = 3 \cosh t$, $y = 5 \sinh t$ is half a hyperbola and sketch it. Give an equation for the hyperbola in rectangular coordinates.

4.† Show that the curve $x = 3 \cos t$, $y = 5 \sin t$, $0 \le t \le 2\pi$ is an ellipse by giving an equation for it in rectangular coordinates.

5. Sketch the curve $x = 4 + 4 \cos \Theta$, $y = -3 - 3 \sin \Theta$, $0 \le \Theta \le 2\pi$.

6.† Sketch the curve $x = \cos t + (t/\pi)$, $y = \sin t$, $0 \le t \le 4\pi$.

7. Choose numbers a and b such that the curve $x = \sec t$, $y = \tan t$, $a < t < b$ is   **a.**   the right half of a hyperbola and   **b.**   the left half of that hyperbola.

8.† Compute the length of the curve $x = \frac{1}{3}t^3$, $y = \frac{1}{2}t^2$, $0 \le t \le 2$.

9.   a.   Give a definite integral that equals the length of the ellipse $x = a \cos t$, $y = b \sin t$, $0 \le t \le 2\pi$. (This is known as an *elliptic integral* if $a \ne b$.)   **b.**   Compute the length for the case of a circle ($a = b$).

10. Show that the length of a curve given in polar coordinates by $r = r(\Theta)$, $a \le \Theta \le b$ is

$$\int_a^b \sqrt{\left[\frac{dr}{d\Theta}(\Theta) \right]^2 + r(\Theta)^2} \, d\Theta.$$

11.† Sketch the portion of the *logarithmic spiral* $r = e^{\Theta/6}$ for $0 \le \Theta \le 6\pi$ and use the formula from Exercise 10 to compute its length.

12. An object's coordinates are $x = x(t)$, $y = y(t)$ (meters) at time t (minutes). Compute its speed at time t_0 for   **a.**† $x(t) = t^3 - 3t^2 + 1$, $y(t) = t^5 - t^7$, $t_0 = 1$ **b.** $x(t) = 6 \cos t$, $y(t) = 5 \sin t$, $t_0 = \pi/3$ **c.**† $x(t) = t \sin t$, $y(t) = t \cos t$, $t_0 = \pi/4$ **d.** $x(t) = \ln t$, $y(t) = e^t$, $t_0 = 3$.

13.† The *cross curve* $4x^{-2} + 4y^{-2} = 1$ is shown in Figure 11.33. An object's coordinates at time t are $x = 2 \sec t$ and $y = 2 \csc t$ for $-\pi/2 < t < 0$.   **a.**   Show that it travels on the cross curve.   **b.**   Where is it at $t = -\pi/4$?   **c.**   What is its speed then?

14. An object's coordinates at time t are $x = 6 \cos t + 5 \cos(3t)$ and $y = 6 \sin t - 5 \sin(3t)$ and its path is the *hypotrochoid* in Figure 11.34. What are its speeds when it crosses the x-axis? (Use the trigonometric identities $\cos(3t) = 4 \cos^3 t - 3 \cos t$, $\sin(3t) = 3 \sin t - 4 \sin^3 t$.)

15. The line $x = ky$ in Figure 11.35 intersects the circle $x^2 + (y - \frac{1}{2})^2 = \frac{1}{4}$ at the point R and intersects the line $y = 1$ at the point S. As k varies, the vertex P of the right triangle PRS traverses a curve known as the *witch of Agnesi*.   **a.**†   Give parametric equations of the curve with k as parameter.   **b.**†   Give an equation for the curve in rectangular coordinates.   **c.**   An object is at $x = \tan t$, $y = \cos^2 t$ at time t for $-\pi/2 < t < \pi/2$. Show that its path is the witch of Agnesi.

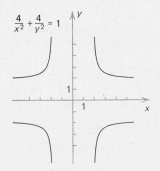

$$\frac{4}{x^2} + \frac{4}{y^2} = 1$$

FIGURE 11.33

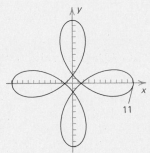

FIGURE 11.34   A hypotrochoid

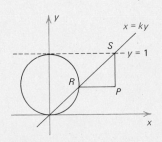

FIGURE 11.35

d. Where is it and what is its speed at $t = \pi/4$? (Maria Gaetana Agnesi (1718–1799) was the first prominent woman mathematician in modern Europe and wrote one of the most popular calculus books of her time. She gave the curve of this example the name "versiera" after the Latin word "vertere", to turn. The curve came to be known as the "witch" of Agnesi because "versiera" is also an Italian word for "she-devil". See D. Struik, [1] *A Source Book in Mathematics*, pp. 178–180.)

16. A disk of radius 5 feet rolls to the right on the x-axis with an angular velocity of one radian per second. At $t = 0$ its center is at $(0, 5)$ and the point P on the disk is at $(0, 2)$. The curve described by the point P is called a *curtate cycloid.*   **a.** Sketch the curve and derive parametric equations for it.   **b.** Where is the point P relative to the disk when its speed is a maximum and what is that maximum speed?

17.[†] Give parametric equations of the quarter circle $x = 7 \cos(t/4)$, $y = 7 \sin(t/4)$, $0 \leq t \leq 2\pi$ with arclength as parameter.

18. Sketch the curve $x = 3t^3$, $y = 4t^3$, $0 \leq t \leq 2$ and give it parametric equations with arclength as parameter.

19. If an object is at $x = e^{2t} \cos t$ (centimeters), $y = e^{2t} \sin t$ (centimeters) at time t (hours), what is the total distance it travels from time $t = -3$ to $t = 3$?

11-4   VECTOR REPRESENTATIONS OF CURVES; VELOCITY VECTORS

In this section we show how VECTOR-VALUED FUNCTIONS and their derivatives can be used to study curves and velocity vectors.

Definition 11.12   A vector-valued function R of the real variable t is a rule that assigns a unique vector $R(t)$ to each t in its domain.

The domain of $\boldsymbol{R}(t)$ is an interval or other collection of real numbers t.
    To study a curve $x = x(t)$, $y = y(t)$, $a \leq t \leq b$ using vectors, we introduce the vector-valued function

$$\boldsymbol{R}(t) = x(t)\mathbf{i} + y(t)\mathbf{j} = \langle x(t), y(t) \rangle.$$

For each t, $\boldsymbol{R}(t)$ is the position vector of the point $(x(t), y(t))$ on the curve (Figure 11.37). If we keep the base of the vector $\boldsymbol{R}(t)$ at the origin, then, as t runs from a to b, the tip of $\boldsymbol{R}(t)$ traces out the curve C.

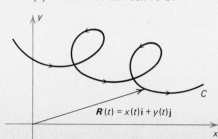

FIGURE 11.37   The position vector of the point $(x(t), y(t))$ on the curve C

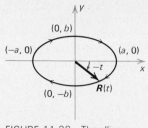

FIGURE 11.38   The ellipse
$R(t) = a \cos(-t)\mathbf{i} + b \sin(-t)\mathbf{j}$

EXAMPLE 1   Show that the curve C: $\mathbf{R}(t) = a \cos(-t)\mathbf{i} + b \sin(-t)\mathbf{j}, 0 \le t \le 2\pi$ $(a > 0, b > 0)$ is an ellipse oriented clockwise.

Solution.   The curve is an ellipse because we have $x = a \cos(-t)$, $y = b \sin(-t)$, and

$$\frac{x^2}{a^2} + \frac{y^2}{b^2} = \cos^2(-t) + \sin^2(-t) = 1.$$

The angle from the positive x-axis to the position vector $\mathbf{R}(t)$ is $-t$ (Figure 11.38). Therefore, the curve is traversed clockwise as t increases from 0 to 2π. $\square$

Limits of vector-valued functions

Definition 11.13   **If $x(t)$ tends to a and $y(t)$ tends to b as t tends to t_0, then the limit of the vector-valued function $\langle x(t), y(t) \rangle$ as t tends to t_0 is the vector $\langle a, b \rangle$.**

If the limit of $\langle x(t), y(t) \rangle$ is $\langle a, b \rangle$, and if we hold the base of the vector $\langle x(t), y(t) \rangle$ at a fixed point, then the arrow representing $\langle x(t), y(t) \rangle$ approaches the arrow representing $\langle a, b \rangle$ as t tends to t_0.

EXAMPLE 2   Find the limit $\lim\limits_{t \to 2} [(t^2 + 3t)\mathbf{i} + (4/t)\mathbf{j}]$.

Solution.   As t tends to 2, $t^2 + 3t$ tends to $2^2 + 3(2) = 10$ and $4/t$ tends to $\frac{4}{2} = 2$. Therefore, $\lim\limits_{t \to 2} [(t^2 + 3t)\mathbf{i} + (4/t)\mathbf{j}] = 10\mathbf{i} + 2\mathbf{j}$. $\square$

Derivatives of vector-valued functions

Because we can subtract vectors and divide them by nonzero real numbers, we can give a definition of derivative of a vector-valued function that is analogous to that for real-valued functions:

Definition 11.14   **The derivative of the vector-valued function $R(t)$ is the limit**

(1) $$\frac{dR}{dt}(t) = \lim_{\Delta t \to 0} \frac{R(t + \Delta t) - R(t)}{\Delta t}.$$ ◄

The next result shows that we can differentiate a vector-valued function by differentiating its components.

Theorem 11.4   **If $x(t)$ and $y(t)$ have derivatives, then the derivative of the vector-valued function $R(t) = x(t)\mathbf{i} + y(t)\mathbf{j}$ is**

(2) $$\frac{dR}{dt}(t) = \frac{d}{dt}[x(t)\mathbf{i} + y(t)\mathbf{j}] = \frac{dx}{dt}(t)\mathbf{i} + \frac{dy}{dt}(t)\mathbf{j}.$$ ◄

Proof. By Definition 11.14 we have

$$\frac{d\mathbf{R}}{dt}(t) = \lim_{\Delta t \to 0}\left[\frac{x(t+\Delta t)-x(t)}{\Delta t}\right]\mathbf{i} + \left[\frac{y(t+\Delta t)-y(t)}{\Delta t}\right]\mathbf{j}$$

$$= \left[\lim_{\Delta t \to 0}\frac{x(t+\Delta t)-x(t)}{\Delta t}\right]\mathbf{i} + \left[\lim_{\Delta t \to 0}\frac{y(t+\Delta t)-y(t)}{\Delta t}\right]\mathbf{j}$$

$$= \frac{dx}{dt}(t)\mathbf{i} + \frac{dy}{dt}(t)\mathbf{j}. \quad \text{Q.E.D.}$$

EXAMPLE 3   Compute the derivative $\dfrac{d\mathbf{R}}{dt}(-1)$ where
$\mathbf{R}(t) = t^4\mathbf{i} + t^3\mathbf{j}$.

Solution.   We have

$$\frac{d\mathbf{R}}{dt}(t) = \left[\frac{d}{dt}t^4\right]\mathbf{i} + \left[\frac{d}{dt}t^3\right]\mathbf{j} = (4t^3)\mathbf{i} + (3t^2)\mathbf{j}.$$

Therefore,

$$\frac{d\mathbf{R}}{dt}(-1) = [(4t^3)\mathbf{i} + (3t^2)\mathbf{j}]_{t=-1} = -4\mathbf{i} + 3\mathbf{j}. \quad \square$$

Velocity vectors

In Section 11.1 we defined an object's velocity vector as the vector with length equal to the object's speed and with the direction of the object's motion, which is the direction tangent to the object's path and pointing in the direction of increasing t (Figure 11.39). The next theorem shows that the velocity vector may be computed as a derivative.

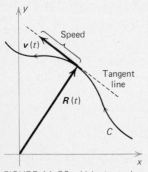

FIGURE 11.39   Velocity and position vectors

Theorem 11.5   If $x(t)$ and $y(t)$ have continuous derivatives and an object's position vector at time t is $R(t) = x(t)\mathbf{i} + y(t)\mathbf{j}$, then its velocity vector is the derivative

$$(3) \qquad \mathbf{v}(t) = \frac{d\mathbf{R}}{dt}(t). \qquad \blacktriangleleft$$

Proof.   The vector $\mathbf{R}(t+\Delta t) - \mathbf{R}(t)$ runs from the point at t to the point at $t+\Delta t$ on the object's path. For $\Delta t > 0$ this vector points in the direction of increasing t (Figure 11.40), and therefore the vector

$$(4) \qquad \frac{\mathbf{R}(t+\Delta t)-\mathbf{R}(t)}{\Delta t} = \frac{1}{\Delta t}[\mathbf{R}(t+\Delta t)-\mathbf{R}(t)]$$

points along the secant line in the direction of increasing t (Figure 11.41). The vector (4) also points along a secant line in the direction of increasing t for $\Delta t < 0$ because, whereas $\mathbf{R}(t+\Delta t) - \mathbf{R}(t)$ points in the direction of decreasing t, we obtain (4) from it by dividing by the *negative* number Δt.

The derivative

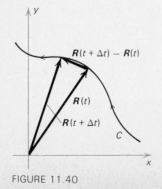

FIGURE 11.40

$$(5) \qquad \frac{d\mathbf{R}}{dt}(t) = \frac{dx}{dt}(t)\mathbf{i} + \frac{dy}{dt}(t)\mathbf{j}$$

is the limit of the vector (4) as Δt tends to zero. Because the secant line approaches the tangent line, the derivative (5) points along the tangent line to the curve in the direction of increasing t (Figure 11.42).

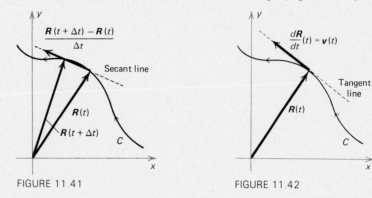

FIGURE 11.41   FIGURE 11.42

The length of the derivative (5) is

$$\left| \frac{d\mathbf{R}}{dt}(t) \right| = \sqrt{\left[\frac{dx}{dt}(t) \right]^2 + \left[\frac{dy}{dt}(t) \right]^2}$$

which, by Theorem 11.3 of Section 11.3, is the object's speed. Therefore, the derivative (5) is the velocity vector.   Q.E.D.

EXAMPLE 4   An object's position vector at time t (seconds) is $\mathbf{R}(t) = t^3\mathbf{i} + t^2\mathbf{j}$ (feet). Give an equation in rectangular coordinates for its path. Sketch the path and the velocity vector $\mathbf{v}(\frac{1}{2})$.

Solution.   The object's coordinates at time t are $x = t^3, y = t^2$. Solving the first equation for t gives $t = x^{1/3}$, and making this substitution in the second equation shows that the path is the curve $y = x^{2/3}$ shown in Figure 11.43. The velocity vector at time t is

$$\mathbf{v}(t) = \frac{d\mathbf{R}}{dt}(t) = \frac{d}{dt}(t^3\mathbf{i} + t^2\mathbf{j}) = 3t^2\mathbf{i} + 2t\mathbf{j}.$$

At $t = \frac{1}{2}$ it is $\mathbf{v}(\frac{1}{2}) = \frac{3}{4}\mathbf{i} + \mathbf{j}$. This vector is shown in Figure 11.43 with its base at $x = \frac{1}{8}, y = \frac{1}{4}$, the object's position at $t = \frac{1}{2}$. The length of the velocity vector is $|\mathbf{v}(\frac{1}{2})| = \sqrt{(\frac{3}{4})^2 + 1^2} = \frac{5}{4}$.   □

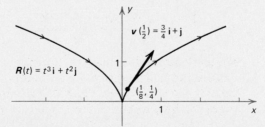

FIGURE 11.43   A velocity vector

The unit tangent vector

The UNIT TANGENT VECTOR T at a point P on a curve is the vector of length 1 that is tangent to the curve at P and points in the direction of its orientation. If for a curve $C: R = R(t)$ the vector $\dfrac{dR}{dt}(t)$ is not zero, then $T = T(t)$ is the unit vector with the same direction:

$$(6) \quad T(t) = \frac{\dfrac{dR}{dt}(t)}{\left|\dfrac{dR}{dt}(t)\right|}. \qquad \blacktriangleleft$$

EXAMPLE 5   Find the unit tangent vector $T(\tfrac{1}{2})$ at $t = \tfrac{1}{2}$ on the curve of Example 4.

Solution.   We saw in Example 4 that $\dfrac{dR}{dt}(\tfrac{1}{2}) = \tfrac{3}{4}\mathbf{i} + \mathbf{j}$. Hence

$$T(\tfrac{1}{2}) = \frac{\tfrac{3}{4}\mathbf{i} + \mathbf{j}}{|\tfrac{3}{4}\mathbf{i} + \mathbf{j}|} = \frac{\tfrac{3}{4}\mathbf{i} + \mathbf{j}}{\tfrac{5}{4}} = \tfrac{3}{5}\mathbf{i} + \tfrac{4}{5}\mathbf{j}.$$

This tangent vector is shown in Figure 11.44.   □

Note.   A curve may fail to have a tangent line at a point where $\dfrac{dR}{dt}(t)$ is zero even though the functions $x(t)$ and $y(t)$ have derivatives at that value of t. Such is the case for the curve of Examples 4 and 5 at $t = 0$.   □

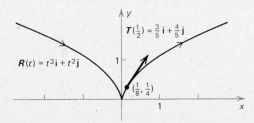

FIGURE 11.44   A unit tangent vector

Differentiation rules

The basic rules for differentiating vector-valued functions follow easily from the rules for differentiating real-valued functions. We have the rule

$$(7) \quad \frac{d}{dt}[aF(t) + bG(t)] = a\frac{dF}{dt}(t) + b\frac{dG}{dt}(t) \qquad \blacktriangleleft$$

for differentiating linear combinations of vector-valued functions (see Exercise 10), the product rules

$$(8) \quad \frac{d}{dt}[f(t)F(t)] = \frac{df}{dt}(t)F(t) + f(t)\frac{dF}{dt}(t) \qquad \blacktriangleleft$$

$$(9) \quad \frac{d}{dt}[F(t) \cdot G(t)] = \frac{dF}{dt}(t) \cdot G(t) + F(t) \cdot \frac{dG}{dt}(t) \qquad \blacktriangleleft$$

(see Exercise 11), and the chain rule

(10) $\dfrac{d}{du} F(t(u)) = \dfrac{dF}{dt}(t(u)) \dfrac{dt}{du}(u)$ ◀

(see Exercise 12). Here we assume that the real-valued functions $f(t)$ and $t(u)$ and the vector-valued functions $F(t)$ and $G(t)$ have derivatives.

EXAMPLE 6   Use rule (9) to show that $F(t)$ and $\dfrac{dF}{dt}(t)$ are perpendicular if $F(t)$ has constant length.

Solution.   The equation $F(t) \cdot F(t) = |F(t)|^2$ and rule (9) yield

$$\frac{d}{dt}|F(t)|^2 = \frac{d}{dt}[F(t) \cdot F(t)] = \frac{dF}{dt}(t) \cdot F(t) + F(t) \cdot \frac{dF}{dt}(t)$$

$$= 2\,F(t) \cdot \frac{dF}{dt}(t).$$

If $|F(t)|^2$ is constant, then the dot product is zero and the vectors $F(t)$ and $\dfrac{dF}{dt}(t)$ are perpendicular.   □

Antiderivatives of vector-valued functions

We use the indefinite integral notation $\int F(t)\,dt$ for antiderivatives of the vector-valued function $F(t)$. For $F(t) = \langle x(t), y(t) \rangle$ we have $\int F(t)\,dt = \langle \int x(t)\,dt, \int y(t), dt \rangle$.

EXAMPLE 7   Find the antiderivative $G(t)$ of the vector $6t\,\mathbf{i} + 5\cos t\,\mathbf{j}$ such that $G(0) = 3\mathbf{i} - 7\mathbf{j}$.

Solution.   The antiderivatives $G(t)$ of $6t\,\mathbf{i} + 5\cos t\,\mathbf{j}$ are $G(t) = \int (6t\,\mathbf{i} + 5\cos t\,\mathbf{j})\,dt = 3t^2\,\mathbf{i} + 5\sin t\,\mathbf{j} + C$ where C is a constant vector. To satisfy the condition $G(0) = 3\mathbf{i} - 7\mathbf{j}$, we must take $C = 3\mathbf{i} - 7\mathbf{j}$. Therefore, $G(t) = (3t^2\,\mathbf{i} + 5\sin t\,\mathbf{j}) + 3\mathbf{i} - 7\mathbf{j}$ $= (3t^2 + 3)\mathbf{i} + (5\sin t - 7)\mathbf{j}$.   □

We could have dealt separately with the components of the vectors in Example 7 rather than use vector notation since the solution amounted to finding the antiderivative $3t^2 + 3$ of $6t$ that has the value 3 at $t = 0$ and the antiderivative $5\sin t - 7$ of $5\cos t$ that has the value -7 at $t = 0$.

In the next example we find a position vector as an antiderivative of a velocity vector.

EXAMPLE 8   A car in an xy-plane has the velocity vector $\langle 30 + 7t^{1/3},$ $20 + 4t^{4/3} \rangle$ kilometers per hour at time t (hours). The car is at $x = 50$, $y = 10$ (kilometers) at $t = 1$. What is its position vector as a function of t?

Solution.   The position vector $R(t)$ of the car is an antiderivative of its velocity vector:

$$R(t) = \int \langle 30 + 4t^{1/3}, 20 + 7t^{4/3} \rangle \, dt$$

$$= \left\langle \int (30 + 4t^{1/3}) \, dt, \int (20 + 7t^{4/3}) \, dt \right\rangle$$

$$= \langle 30t + 3t^{4/3}, 20t + 3t^{7/3} \rangle + C.$$

Because the car is at $(50, 10)$ at $t = 1$, $R(1)$ must equal $\langle 50, 10 \rangle$. The formula above for $R(t)$ gives $R(1) = \langle 30(1) + 3(1)^{4/3}, 20(1) + 3(1)^{7/3} \rangle + C = \langle 33, 23 \rangle + C$. Hence, we have $C = \langle 17, -13 \rangle$ and $R(t) = \langle 30t + 3t^{4/3} + 17, 20t + 3t^{7/3} - 13 \rangle$.   □

Exercises

†*Answer provided.*
‡*Outline of solution provided.*

1. Sketch the following curves and their velocity vectors at the indicated values of t.
    a.† $R(t) = t^2 \mathbf{i} - (1/t^2) \mathbf{j}, \frac{1}{2} \le t \le 2; t = 1$
    b. $R(t) = (3 \cos t)\mathbf{i} + (2 \sin t)\mathbf{j}, 0 \le t \le \pi; t = \pi/3$
    c.† $R(t) = (e^t)\mathbf{i} + (e^{2t})\mathbf{j}, -1 \le t \le 1; t = 0$
    d. $R(t) = [2 + \cos(6t)]\mathbf{i} + [2 + \sin(6t)]\mathbf{j}, 0 \le t \le \pi/3; t = \pi/9$
    e.† $R(t) = [2 + \cos(-t)]\mathbf{i} + [2 + \sin(-t)]\mathbf{j}, 0 \le t \le 2\pi; t = \frac{3}{4}\pi$
    f. $R(t) = (e^t)\mathbf{i} + [\sin(e^t)]\mathbf{j}, \ln(\pi/2) \le t \le \ln(4\pi); t = \ln(\pi)$

2. Give the unit tangent vectors and equations of the tangent lines to the following curves at the indicated values of t.

    a.‡ $R(t) = (t \sin t)\mathbf{i} + \left(\dfrac{\cos t}{t}\right)\mathbf{j}; t = \pi/2$

    b. $R(t) = (t^3 - t)\mathbf{i} + (t^2 - 4t^4)\mathbf{j}; t = 1$
    c.† $R(t) = (e^{2t})\mathbf{i} + (t^2)\mathbf{j}; t = 0$
    d. $R(t) = [\sin(3t)]\mathbf{i} - [2 \cos t]\mathbf{j}; t = \pi/4$

3. Find dR/dt and d^2R/dt^2 where
    a.† $R(t) = t^5 \mathbf{i} - t^7 \mathbf{j}$ **b.** $R(t) = (te^t)\mathbf{i} - (\sin t)\mathbf{j}$
    c.† $R(t) = [\ln(3t)]\mathbf{i} + [\arctan(3t)]\mathbf{j}$
    d. $R(t) = [t \sin t]\mathbf{i} + [t \cos t]\mathbf{j}$.

4. Find the position vectors $R(t)$ from the velocity vectors $v(t)$ and particular values of $R(t)$:   **a.**† $v(t) = t\mathbf{i} - t^2\mathbf{j}; R(1) = \mathbf{i}$
    b. $v(t) = [\sin(3t)]\mathbf{i} + [\cos(4t)]\mathbf{j}; R(0) = 4\mathbf{i} + 5\mathbf{j}$
    c.† $v(t) = [e^{3t}]\mathbf{i} - [e^{-4t}]\mathbf{j}; R(0) = 5\mathbf{i} + 5\mathbf{j}$

    d. $v(t) = \left[\dfrac{3}{t}\right]\mathbf{j} + \left[\dfrac{5}{t}\right]\mathbf{j}; R(1) = \mathbf{i} + \mathbf{j}$

5.† An object's coordinates at time t are $x = 8 \cos(\frac{1}{4}t) - 7 \cos t$ and $y = 8 \sin(\frac{1}{4}t) - 7 \sin t \, (0 \le t \le 8\pi)$, and its path is the *epitrochoid* shown in Figure 11.45. Compute its velocity vectors at   **a.** $t = 2\pi$
    and   **b.** $t = 4\pi$.   **c.** Trace the curve and draw these velocity vectors on the sketch.

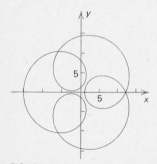

FIGURE 11.45   An epitrochoid

6. An object's coordinates at time t are $x = 1 + \sin t$, $y = \cos t(1 + \sin t)$
    $(-\pi/2 \le t \le 3\pi/2)$, and its path is the *piriform* shown in Figure 11.46.
    Compute its velocity vector at   **a.** $t = 0$,   **b.** $\pi/4$,   **c.** $\pi/2$.
    d. Trace the curve and sketch these velocity vectors.

7.† An object has position vector $\boldsymbol{R}(t) = [8 \sin(\tfrac{3}{4}t)]\mathbf{i} + [7 \sin t]\mathbf{j}$ at time t.
    Its path (Figure 11.47) is the *Bowditch curve* (or *curve of Lissajous*).
    a. How long does it take the object to traverse the entire curve?
    b. What is its velocity vector as it passes through the origin from the
    right?   **c.** From the left?

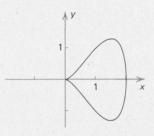

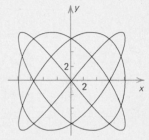

FIGURE 11.46   A piriform   FIGURE 11.47   A Bowditch curve

8.‡ The graphs of differentiable functions $f(t)$ and $g(t)$ are shown in
    Figures 11.48 and 11.49. Sketch the curve $C: x = f(t)$, $y = g(t)$,
    $1 \le t \le 7$ and its velocity vectors at $t = \tfrac{7}{2}$ and $t = 5$.

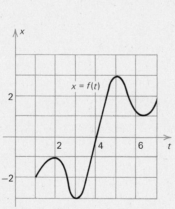

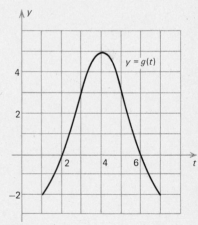

FIGURE 11.48   FIGURE 11.49

9. The graphs of differentiable functions $x(t)$ and $y(t)$ are shown in Figures 11.50 and 11.51. Sketch the curve C: $\mathbf{R}(t) = x(t)\mathbf{i} + y(t)\mathbf{j}$, $-2 \le t \le 6$ and its velocity vectors at $t = 3$ and $t = 5$.

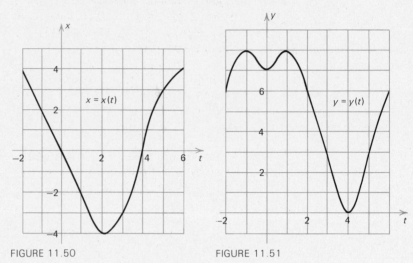

FIGURE 11.50   FIGURE 11.51

10. Derive differentiation formula (7).

11. Derive the product rule formulas (8) and (9).

12.† Derive the chain rule formula (10).

11-5 CURVATURE

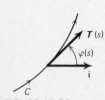

FIGURE 11.52

The CURVATURE κ (*kappa*) of a plane curve at a point P is a number that indicates how rapidly the curve is bending at P and whether it is bending toward the left or right as viewed by an object traversing the curve. We suppose that the curve has been given parametric equations with arclength s as parameter:

(1) $C: x = x(s), \qquad y = y(s).$

We let $\phi(s)$ denote the angle from the direction of the vector $\mathbf{i}$ to the direction of the unit tangent vector $\mathbf{T}(s)$ at $(x(s), y(s))$ on the curve (Figure 11.52). Then we have the following definition.

Definition 11.15   **The curvature of a curve $x = x(s)$, $y = y(s)$ at the point $\big(x(s), y(s)\big)$ is the derivative of the angle $\phi(s)$ with respect to arclength s:**

(2) $\kappa(s) = \dfrac{d\phi}{ds}(s).$ ◀

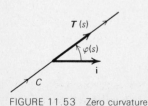

FIGURE 11.53   Zero curvature

The curvature is zero if the curve is a straight line and $\phi(s)$ is constant (Figure 11.53). If the curvature is positive, then $\phi(s)$ increases

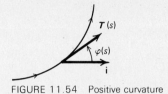

FIGURE 11.54   Positive curvature

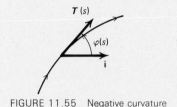

FIGURE 11.55   Negative curvature

with increasing s and the curve bends toward the left as viewed in the positive direction along the curve (Figure 11.54). If the curvature is negative, then $\phi(s)$ decreases as s increases and the curve bends toward the right (Figure 11.55). The curve bends sharply if $|\kappa(s)|$ is large and bends gradually if $|\kappa(s)|$ is small. We use the following result to compute curvature of graphs of functions.

Theorem 11.6   The curvature of the graph $y = f(x)$ at the point $(x, f(x))$ is

(3) $$\kappa(x) = \frac{f''(x)}{[1 + [f'(x)]^2]^{3/2}}.$$ ◄

Here we consider $y = f(x)$ to be oriented in the direction of increasing x, and we assume that f has continuous first- and second-order derivatives.

Proof.   Since $f'(x)$ is the slope of the tangent line to the graph, we obtain a suitable angle ϕ by setting $\phi(x) = \arctan[f'(x)]$. Differentiating this equation with respect to x then gives

(4) $$\frac{d\phi}{dx}(x) = \frac{f''(x)}{1 + [f'(x)]^2}.$$

The curvature is the derivative of ϕ, not with respect to x, but with respect to arclength s, and we have

$$\frac{dx}{ds} = \frac{1}{ds/dx} = \frac{1}{\sqrt{1 + [f'(x)]^2}}.$$

Therefore, with equation (2) and the chain rule we obtain

$$\kappa(x) = \frac{d\phi}{ds}(x) = \frac{d\phi}{dx}(x)\frac{dx}{ds} = \left[\frac{f''(x)}{1 + [f'(x)]^2}\right]\frac{1}{\sqrt{1 + [f'(x)]^2}}$$

which gives equation (3).   Q.E.D.

EXAMPLE 1   Compute the curvature of the parabola $y = -x^2$ at $x = 0$.

Solution.   We set $f(x) = -x^2$. Then $f'(x) = -2x$ and $f''(x) = -2$, so that $f'(0) = 0$ and $f''(0) = -2$. By formula (3)

$$\kappa(0) = \frac{f''(0)}{[1 + [f'(0)]^2]^{3/2}} = \frac{-2}{[1 + 0]^{3/2}} = -2. \quad \square$$

A calculation similar to that used in the proof of Theorem 11.6 yields the following result for curves given by parametric equations (see Exercise 13).

Theorem 11.7   If $x(t)$ and $y(t)$ have continuous first- and second-order derivatives and if not both $x'(t)$ and $y'(t)$ are zero,

then the curvature of $C: x = x(t), y = y(t)$ at $(x(t), y(t))$ is

(5) $\kappa(t) = \dfrac{x'(t)y''(t) - y'(t)x''(t)}{[[x'(t)]^2 + [y'(t)]^2]^{3/2}}.$

EXAMPLE 2   Compute the curvature of the ellipse $x = 3 \cos t$, $y = 5 \sin t$ at $t = \pi/2$.

Solution.   We have $x'(t) = -3 \sin t, x''(t) = -3 \cos t, y'(t) = 5 \cos t$, and $y''(t) = -5 \sin t$. Therefore, $x'(\pi/2) = -3, x''(\pi/2) = 0$, $y'(\pi/2) = 0$, and $y''(\pi/2) = -5$. By formula (5)

$$\kappa\left(\frac{\pi}{2}\right) = \frac{(-3)(-5) - (0)(0)}{[3^2 + 0^2]^{3/2}} = \frac{15}{27} = \frac{5}{9}. \quad \square$$

EXAMPLE 3   Show that the curvature of a circle of radius ρ (rho) is $1/\rho$ if the circle is oriented counterclockwise and $-1/\rho$ if the circle is oriented clockwise.

Solution.   If the center of the circle is at (x_0, y_0) and it is oriented counterclockwise, then it has the parametric equations $x = x_0 + \rho \cos t, y = y_0 + \rho \sin t$, for which $x' = -\rho \sin t$, $y' = \rho \cos t, x'' = -\rho \cos t$, and $y'' = -\rho \sin t$. Therefore, by (5) and the Pythagorean identity $\sin^2 t + \cos^2 t = 1$ we have

$$\kappa(t) = \frac{(-\rho \sin t)(-\rho \sin t) - (\rho \cos t)(-\rho \cos t)}{[(-\rho \sin t)^2 + (\rho \cos t)^2]^{3/2}} = \frac{1}{\rho}.$$

Similarly, we obtain $\kappa(t) = -1/\rho$ for the circle oriented clockwise, which has the parametric equations $x = x_0 + \rho \cos(-t)$, $y = y_0 + \rho \sin(-t)$. $\square$

The unit normal vector

The UNIT NORMAL VECTOR $N = N(t)$ at a point on a curve is the vector of length 1 that is perpendicular to, and points to the left of, the unit tangent vector $T(t)$ (Figure 11.56). Because the tangent vector is at an angle ϕ from the direction of the vector $\mathbf{i}$, it is given by

(6) $T = \cos \phi\, \mathbf{i} + \sin \phi\, \mathbf{j}.$ ◀

The unit vectors perpendicular to T are $-\sin \phi\, \mathbf{i} + \cos \phi\, \mathbf{j}$ and $\sin \phi\, \mathbf{i} - \cos \phi\, \mathbf{j}$. Comparing the signs of the components of the vectors T and N in a sketch such as Figure 11.56 shows that

(7) $N = -\sin \phi\, \mathbf{i} + \cos \phi\, \mathbf{j}.$ ◀

Equations (6) and (7) yield the differentiation formula

(8) $\dfrac{dT}{d\phi} = N.$ ◀

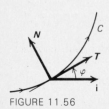

FIGURE 11.56

Circles of curvature

Recall that the tangent line to a curve at a point P is the line that best approximates the curve near that point and that it is the limiting position of the secant line through the point P and a nearby point Q on the curve as Q tends to P. If the curvature at P is not zero, we can obtain a better approximation of the curve near P by using the CIRCLE OF CURVATURE, which is the circle that best approximates the curve near P.

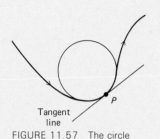

Tangent line

FIGURE 11.57   The circle of curvature

Definition 11.16   The circle of curvature at a point P on a curve where the curvature κ is not zero is the circle of radius $1/|\kappa|$ which has the same tangent line as the curve at P and which lies on the side toward which the curve is bending (Figure 11.57).

The calculations in Example 3 show that the circle of curvature has the same curvature as the curve at the point P. The radius of the circle of curvature is called the RADIUS OF CURVATURE and is denoted by the letter ρ (rho). The center is called the CENTER OF CURVATURE at P.

It can be shown that the circle of curvature is the limiting position of a circle through P and two nearby points Q and R on the curve as Q and R approach P (Figure 11.58). (See Miscellaneous Exercise 16 of Chapter 17.)

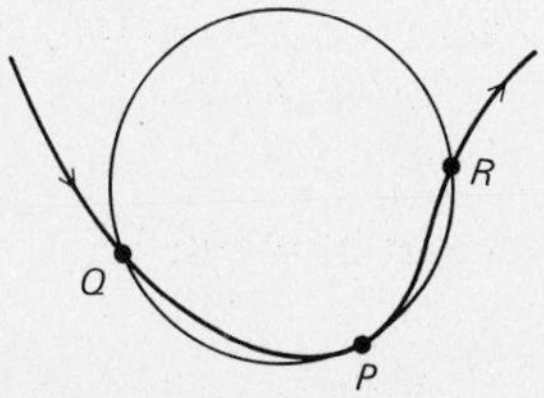

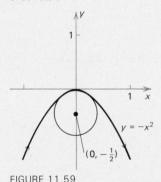

FIGURE 11.58   Circle approximating the circle of curvature

EXAMPLE 4   Sketch the parabola $y = -x^2$ of Example 1 and its circle of curvature at $x = 0$.

Solution.   In Example 1 we saw that the curvature of the parabola at $x = 0$ is -2. Its radius of curvature is therefore $\frac{1}{2}$. The tangent line at $x = 0$ is horizontal, so the center of curvature is at the point $(0, -\frac{1}{2})$. The parabola and the circle of curvature are sketched in Figure 11.59. The curvature is negative because the curve bends to the right as we move along it in the direction of its orientation.   □

FIGURE 11.59

The case of zero curvature

If the curvature of a curve is zero at a point P, then there is no circle that approximates the curve better near P than does the tangent line. In this case we think of the tangent line itself as the "circle" of curvature at P, and we say that the radius of curvature at that point is "infinite".

Exercises

†*Answer provided.*
‡*Outline of solution provided.*

In Exercises 1 through 4 compute the curvature and radius of curvature of the curves at the indicated points.

1.† $y = \frac{1}{4}x^4 - \frac{1}{2}x^2$ at   **a.** $x = 0$ and   **b.** $x = 1$

2. $y = x - (3/x)$ at $x = 1$

3.† $x = \frac{1}{2}t^2, y = -\frac{1}{3}t^3$, at $t = 2$

4. $x = e^{-t}, y = e^{3t}$, at $t = 0$

5. Find the center of curvature of the hyperbola $xy = 1$ at   **a.‡** (1, 1) and   **b.†** $(-1, -1)$.   **c.†** Sketch the hyperbola and its circles of curvature at the two points.

6.† Sketch the ellipse $x = 3 \cos t, y = 5 \sin t, 0 \le t \le 2\pi$ and its circle of curvature at $t = \pi/2$.

7. Find the curvature of $y = \sin x$ at   **a.** $x = 0$,   **b.** $x = \pi/2$, and   **c.** $x = 3\pi/2$.   **d.** Sketch the curve and its circles of curvature at those points.

8.† Find the maximum curvature of $y = \ln x$ oriented from right to left. Sketch the curve and the corresponding circle of curvature.

9. Find the minimum radius of curvature of $y = \frac{1}{3}x^3$. Sketch the curve and the circle(s) of curvature of minimum radius.

10.‡ A curve C is sketched in Figure 11.60. Give approximate values for its curvature at the points   **a.** P **b.** Q **c.** R.   **d.** Trace the curve and sketch the vectors T and N at those points. (The dots are the centers of curvature at $P, Q,$ and R.)

11. A curve C and the centers of curvature at points $P, Q,$ and R are shown in Figure 11.61. Give approximate values for the curvature at   **a.** P **b.** Q and   **c.** R.   **d.** Trace the curve and draw the vectors T and N at the three points.

12. Give the unit tangent vector $T(t)$ and the unit normal vector $N(t)$ as functions of the parameter t for   **a.†** the circle $x = 5 \cos t$, $y = 5 \sin t$ and   **b.** the circle $x = 5 \cos(-t), y = 5 \sin(-t)$.   **c.** Sketch the circles of parts (a) and (b) and the vectors $T(t)$ and $N(t)$ at $t = \pi/4, t = \pi,$ and $t = 7\pi/4$.

13. Prove Theorem 11.7.

14. Work Example 3 by expressing the angle ϕ as a function of arclength and using Definition 11.15 of the curvature.

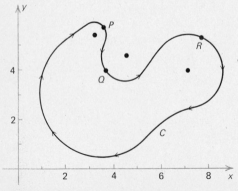

FIGURE 11.60

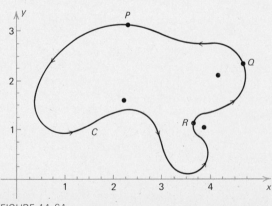

FIGURE 11.61

11-6 ACCELERATION VECTORS AND NEWTON'S LAW

In Chapter 4 we defined the *acceleration* of an object moving along a coordinate line to be the derivative of its velocity. We have an analogous definition of the ACCELERATION VECTOR of an object moving along a curved path.

Definition 11.17*   If an object's velocity vector at time t is $v(t)$, then its acceleration vector is the derivative

$$(1) \qquad a(t) = \frac{d}{dt}\, v(t). \qquad \blacktriangleleft$$

EXAMPLE 1   An object's position vector in an xy-plane is $R(t) = t^3\mathbf{i} + t^5\mathbf{j}$ (feet) at time t (seconds). Give its acceleration vector as a function of t.

Solution.   The velocity vector is

$$\mathbf{v}(t) = \frac{d\mathbf{R}}{dt}(t) = \frac{d}{dt}(t^3\mathbf{i} + t^5\mathbf{j}) = 3t^2\mathbf{i} + 5t^4\mathbf{j} \text{ feet/second.}$$

Therefore, the acceleration vector is

$$a(t) = \frac{d\mathbf{v}}{dt}(t) = \frac{d}{dt}(3t^2\mathbf{i} + 5t^4\mathbf{j}) = 6t\mathbf{i} + 20t^3\mathbf{j} \text{ feet/second}^2. \quad \square$$

Newton's law   According to NEWTON'S LAW in vector form, the total force $F(t)$ on an object at time t is related to the object's acceleration vector $a(t)$ by the equation

$$(2) \qquad F(t) = m\, a(t) \qquad \blacktriangleleft$$

where m is the object's *mass*. (Recall that the mass is the object's weight divided by the acceleration g of gravity.)

EXAMPLE 2   An object weighs 160 pounds and the total force on it at time t (seconds) is $15t^2\mathbf{i} - \sin t\,\mathbf{j}$ (pounds). What is its acceleration vector at time t?

Solution.   The mass of the object is $160/32 = 5$ slugs, and by Newton's law (2) its acceleration vector is

$$\mathbf{a}(t) = \tfrac{1}{5} F(t) = \tfrac{1}{5}(15t^2\mathbf{i} - \sin t\,\mathbf{j})$$
$$= 3t^2\mathbf{i} - \tfrac{1}{5}(\sin t)\mathbf{j} \text{ feet/second}^2. \quad \square$$

---

*In this and the next section we assume that all vector-valued functions considered have continuous derivatives.

Motion of projectiles

Suppose that a projectile has velocity vector $\mathbf{v}_0$ at time $t = 0$, and that subsequently the only force on it is due to gravity. We may assume that at $t = 0$ the projectile is in an xy-plane that is parallel to $\mathbf{v}_0$ and that has a vertical y-axis (Figure 11.62). The projectile's acceleration vector is then the constant vector

(3) $\mathbf{a}(t) = -32\mathbf{j}$ feet/second².   ◀

Integrating with respect to t shows that the velocity vector is $-32t\mathbf{j} + \mathbf{C}$ with a constant vector $\mathbf{C}$. We set $t = 0$ to see that $\mathbf{C} = \mathbf{v}_0$ and hence the velocity vector is

(4) $\mathbf{v}(t) = -32t\mathbf{j} + \mathbf{v}_0$ feet/second.   ◀

Integrating again shows that the projectile's position vector is

(5) $\mathbf{R}(t) = -16t^2\mathbf{j} + \mathbf{v}_0 t + \mathbf{s}_0$ feet   ◀

where $\mathbf{s}_0$ is its position vector at $t = 0$. Writing (5) in components and expressing y in terms of x shows that the path of the projectile is a parabola (see Exercise 17).

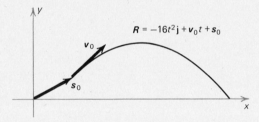

FIGURE 11.62   Path of a projectile

EXAMPLE 3   An arrow is shot from a bow with initial speed of 150 feet per second from a height of 7 feet above the ground and at an angle of 45° with the horizontal. Describe its path. (Ignore air resistance.)

Solution.   We take t to be 0 when the arrow is shot and introduce an xy-plane with a vertical y-axis so that the arrow is shot from the point $(0, 7)$ at an angle of 45° with the horizontal x-axis. Then $\mathbf{v}_0 = (150/\sqrt{2})\mathbf{i} + (150/\sqrt{2})\mathbf{j}$ and $\mathbf{s}_0 = 7\mathbf{j}$ (Figure 11.63), so that the arrow's position vector at time t is

$$\begin{aligned}\mathbf{R}(t) &= -16t^2\mathbf{j} + (150/\sqrt{2})t\mathbf{i} + (150/\sqrt{2})t\mathbf{j} + 7\mathbf{j} \\ &= (150/\sqrt{2})t\mathbf{i} + [-16t^2 + (150/\sqrt{2})t + 7]\mathbf{j}\end{aligned}$$

The arrow has traveled a horizontal distance of $(150/\sqrt{2})t$ feet and is $-16t^2 + (150/\sqrt{2})t + 7$ feet above the ground at time t (up to the time it hits something). □

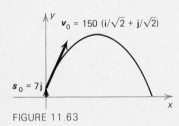

FIGURE 11.63

Tangential and normal components of acceleration

The next theorem enables us to study the motion of an object by studying the components of its acceleration vector in the directions tangent and normal to its path.

Theorem 11.8   The acceleration vector $a = a(t)$ may be expressed in the form

$$(6) \qquad a = \frac{d^2s}{dt^2} T + \kappa \left(\frac{ds}{dt}\right)^2 N \qquad\qquad\qquad \blacktriangleleft$$

where $s = s(t)$ is the length of the object's path up to time t, and $T = T(t), N = N(t)$, and $\kappa = \kappa(t)$ are the unit tangent vector, unit normal vector and curvature of the path at time t.

The quantity $\dfrac{ds}{dt}$ is the object's speed, so formula (6) states that the normal component of the acceleration vector is the curvature multiplied by the square of the speed and the tangential component is the rate of change $\dfrac{d}{dt}\left(\dfrac{ds}{dt}\right)$ of the speed (the rate at which the object is speeding up or slowing down).

Proof.   The velocity vector $\mathbf{v} = \dfrac{ds}{dt} T$ is the speed multiplied by the unit tangent vector. We use the product rule to differentiate v with respect to t and obtain

$$(7) \qquad \boldsymbol{a} = \frac{d^2s}{dt^2} \boldsymbol{T} + \frac{ds}{dt}\frac{d\boldsymbol{T}}{dt}.$$

We consider $\boldsymbol{T} = \boldsymbol{T}(\phi)$ as a function of the angle ϕ, $\phi = \phi(s)$ as a function of the arclength s, and $s = s(t)$ as a function of the time t. By the chain rule and equations (2) and (8) of Section 11.5 we have $\dfrac{d\boldsymbol{T}}{ds} = \dfrac{d\boldsymbol{T}}{d\phi}\dfrac{d\phi}{ds} = \kappa \, \boldsymbol{N}$. To compute the derivative with respect to t, we apply the chain rule again to obtain

$$(8) \qquad \frac{d\boldsymbol{T}}{dt} = \frac{d\boldsymbol{T}}{ds}\frac{ds}{dt} = \kappa\left(\frac{ds}{dt}\right)\boldsymbol{N}.$$

Equations (7) and (8) combine to give (6).   Q.E.D.

Before we apply Theorem 11.8, we will rederive it in two special cases in the next examples.

EXAMPLE 4   Derive (6) for an object whose path is a straight line.

Solution.   If the object's path is a straight line, then its tangent vector $\boldsymbol{T}$ is constant. Differentiating the equation $\mathbf{v} = \dfrac{ds}{dt} \boldsymbol{T}$ with respect to t gives $\boldsymbol{a} = \dfrac{d^2s}{dt^2} \boldsymbol{T}$, which is (6) in this case since straight lines have

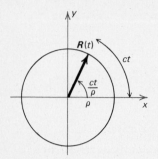

FIGURE 11.64

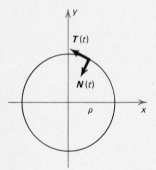

FIGURE 11.65

zero curvature. In this case the acceleration vector is parallel to the path. □

EXAMPLE 5   Derive (6) for an object which is traveling counterclockwise around the circle $x^2 + y^2 = \rho^2$ at the constant speed c.

Solution.   We suppose that the object is at $(\rho, 0)$ at $t = 0$. At time t it has traveled a distance ct in the counterclockwise direction around the circle, so its position vector has traveled through an angle of ct/ρ radians (Figure 11.64). Therefore, its position vector at time t is $\mathbf{R}(t) = \rho \cos(ct/\rho)\mathbf{i} + \rho \sin(ct/\rho)\mathbf{j}$. Differentiating with respect to t gives $\mathbf{v}(t) = -c \sin(ct/\rho)\mathbf{i} + c \cos(ct/\rho)\mathbf{j}$ and then

$$(9) \quad \boldsymbol{a}(t) = \frac{c^2}{\rho}\left[-\cos\!\left(\frac{ct}{\rho}\right)\mathbf{i} - \sin\!\left(\frac{ct}{\rho}\right)\mathbf{j}\right] = \frac{c^2}{\rho}\,\boldsymbol{N}(t)$$

where $\boldsymbol{N}(t) = -\cos(ct/\rho)\mathbf{i} - \sin(ct/\rho)\mathbf{j}$ is the unit normal vector to the circle (Figure 11.65). Equation (9) is (6) in this case because the curvature of the circle is $1/\rho$, the speed is c, and the rate of change of the speed is zero. □

Examples 4 and 5 give us a means of interpreting formula (6): the acceleration vector is equal to that of an object moving along the tangent line with its speed increasing at the rate d^2s/dt^2 plus that of an object moving around the circle of curvature at the constant speed ds/dt.

Note.   When we refer to an object's *acceleration* in everyday speech, we generally mean the rate of change of its speed, which is only the tangential component of its *acceleration vector*, and we do not have in mind the normal component of the acceleration vector, which is due to the object's path being curved. □

We apply Theorem 11.8 in the next example.

EXAMPLE 6   An object weighing 8 pounds is moving along the curve $y = -x^2$ from left to right. The total force on it is $\mathbf{i} - 2\mathbf{j}$ pounds when it is at the origin. What is its speed and at what rate is it speeding up or slowing down at that point?

Solution.   The mass of the object is $\frac{8}{32} = \frac{1}{4}$ slug. By Newton's law its acceleration vector at the origin is

$$(10) \quad \boldsymbol{a} = \frac{1}{m}\,\boldsymbol{F} = \frac{1}{1/4}\,(\mathbf{i} - 2\mathbf{j}) = 4\mathbf{i} - 8\mathbf{j}.$$

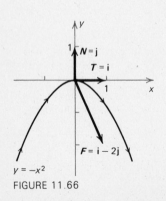

$y = -x^2$

FIGURE 11.66

On the other hand, the unit tangent vector at the origin is $\mathbf{i}$ and the unit normal vector is $\mathbf{j}$ (Figure 11.66), and we saw in Example 1 of Section 11.5 that the curvature of the parabola at that point is -2. Therefore, formula (6) gives

$$(11) \quad a = \frac{d^2s}{dt^2}\mathbf{i} - 2\left(\frac{ds}{dt}\right)^2\mathbf{j}.$$

Comparing equations (10) and (11) shows that $d^2s/dt^2 = 4$ and $-2(ds/dt)^2 = -8$. Solving the second equation gives $ds/dt = 2$. The object has speed 2 feet per second and is speeding up at the rate of 4 feet per second^2 when it is at the origin. □

EXAMPLE 7   The force of friction exerted toward the center of a circular track by the tires of a car can be no greater than 450 pounds. The radius of the track is 200 feet and the car weighs 3200 pounds. What is the fastest constant speed the car can travel without skidding?

Solution.   The acceleration of gravity is 32 feet/second^2, so the mass m of the car is $3200/32 = 100$ slugs. We assume that the car is moving counterclockwise around the track. Then its path has curvature $\frac{1}{200}$ and because the car's speed is constant, d^2s/dt^2 is zero. By formula (6) the car's acceleration is $\frac{1}{200}\left(\frac{ds}{dt}\right)^2 \mathbf{N}$ feet/second^2 and by Newton's law the total force on it is $\mathbf{F} = m\mathbf{a} = \frac{1}{2}\left(\frac{ds}{dt}\right)^2 \mathbf{N}$ pounds, where $\mathbf{N}$ is the inward pointing unit normal to the track. This force must be exerted by the friction on the car's tires, so $\frac{1}{2}\left(\frac{ds}{dt}\right)^2$ can be no greater than 450 pounds.

Solving the equation $\frac{1}{2}\left(\frac{ds}{dt}\right)^2 = 450$ gives $\frac{ds}{dt} = 30$ feet per second as the maximum constant speed the car can have. (The tangential component of friction is balanced by the force of the engine that is maintaining the car's constant speed.) □

Exercises

†*Answer provided.*
‡*Outline of solution provided.*

1.   Compute the acceleration vectors of objects whose position vectors are
   a.‡ $\sin(2t)\mathbf{i} + 3\cos(4t)\mathbf{j}$ millimeters at time t (minutes),
   b.† $(te^{2t})\mathbf{i} - (e^{-3t})\mathbf{j}$ yards at time t (years), and
   c. $\cos(t^2)\mathbf{i} - \sin(t^2)\mathbf{j}$ leagues at time t (lifetimes).
In the following exercises the scales on coordinate axes are measured in feet and the time is measured in seconds. Ignore air resistance.

2.   A projectile moves in an xy-plane with its y-axis vertical. At $t = 0$ it is at the origin, has speed M, and is moving at an angle ψ with the vector $\mathbf{i}$. The only force on it is gravity.   **a.**†   Give its coordinates as functions of t.   **b.**   What is its range? (How far from the origin does its path cross the x-axis again?)   **c.**   What angle ψ gives the maximum range for fixed M?

3.†   Find two angles of fire such that the shell from a cannon with muzzle velocity of 400 feet per second will hit the ground 2500 feet away. [Use the trigonometric identity $2 \sin \psi \cos \psi = \sin(2\psi)$.]

4.†   A boy throws a ball at an angle of 45° with the horizontal and with an initial speed of 30 feet per second toward a building 35 feet away. The ball leaves the boy's hand from a height of three feet above the ground. Will it reach the building?

5. An arrow is shot from a height of 5 feet with an initial speed of 100 feet per second and at an angle of 30° with the horizontal. How high does it go?

6. An object weighing 16 pounds is at rest at the origin in an xy-plane at time $t = 0$ (seconds). Then a force $F(t)$ pounds is applied to the object so that for $t > 0$ its coordinates are $x = 1 - \cos t, y = t - \sin t$. What is $F(t)$? The path of the object is a cycloid. Sketch the cycloid and the force vectors at   **a.** $t = 0$ **b.** $t = \pi/2$ **c.** $t = \pi$ **d.** $t = 3\pi/2$. (Neglect gravity.)

7.† A bead slides down a ramp that makes an angle of 5° with the horizontal. There is no friction, so the only force exerted by the ramp is the force which cancels the component of the force of gravity perpendicular to the ramp. How far does the bead slide in 6 seconds if its initial speed is zero?

8.‡ Find the force of friction required to keep a car on a circular race track of radius 300 feet if the car weighs 3200 pounds and travels at the constant speed of 200 feet per second.

9.† A satellite is traveling at a constant speed in a circular orbit 400 miles above the surface of the earth where the acceleration of gravity is 30 feet per second². What is its speed? (The radius of the earth is 4000 miles and one mile is 5280 feet.)

10. A dime weighing k pounds will not slide off a turning phonograph record provided the force on it is no greater than $\frac{1}{9}k$ pounds. How far from the center of a record turning at 45 revolutions per minute can a dime be without sliding off?

11. Show that a car is just as likely to skid traveling at a constant speed $2v_0$ around a circular track of radius ρ feet as at a constant speed v_0 around a circular track of radius $\frac{1}{4}\rho$ feet.

12.† What is the maximum magnitude of the force required to cause an object weighing 2 pounds to move at the constant speed of 3 feet per second along the parabola $y = x^2$?

13. **a.** Show that the speed of an object is constant if and only if at all times the force applied to it is perpendicular to its path.   **b.** Show that the path of an object is a straight line if and only if the force applied to it is always tangent to its path.

14.‡ An object is traveling in the counterclockwise direction around the ellipse $4x^2 + 9y^2 = 36$. When it reaches the point $(0, -2)$ its acceleration vector is $a = 3\mathbf{i} + 5\mathbf{j}$. What is its speed and at what rate is it speeding up or slowing down at that time?

15.† An object weighing 16 pounds has the velocity vector $4\mathbf{j}$ feet per second and the total force on it is $\mathbf{i} - 3\mathbf{j}$ pounds when it reaches the origin in an xy-plane. Find the curvature of its path at that moment.

16. A man weighing 160 pounds is running in the clockwise direction around the circle $x^2 + y^2 = 25$. When he gets to the point $(3, 4)$, he is running 5 feet per second and he is speeding up at the rate of 10 feet per second². What is his acceleration vector at that moment?

17. Show that the path of a projectile with position vector (5) is a parabola.

Exercises 18 through 22 deal with the curve shown in Figure 11.67 and with an object weighing 96 pounds that travels around the curve in the direction of its orientation. Trace the curve and use your sketch to find approximate answers.

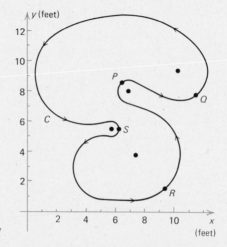

FIGURE 11.67

18.† Draw the unit tangent and normal vectors T and N to the curve at the points P, Q, R, and S. Estimate the curvature of the curve at those points. (The dots off the curve are the corresponding centers of curvature.)

19.‡ The acceleration vector of the object at the point P is $(5/\sqrt{2})\mathbf{i} - (3/\sqrt{2})\mathbf{j}$ feet per second². How fast is it traveling and at what rate is it speeding up or slowing down at that point?

20. The force on the object when it is at the point Q is $-\frac{23}{4}\mathbf{i} + \frac{13}{4}\mathbf{j}$ pounds. How fast is it moving and at what rate is it speeding up or slowing down at that point?

21.† When the object is at the point R, it is neither speeding up nor slowing down and its speed is 3 feet per second. Give the x- and y-components of its acceleration vector at that point.

22. When the object is at the point S, its speed is zero and it is speeding up at the rate of 2 feet per second². Give the x- and y-components of the force on it at that point.

In each of Exercises 23 through 30 an object is moving in an xy-plane with its y-axis pointing upward. Find the position vector of the object as a function of the time t, measured in seconds. Assume that there are no forces on the object other than gravity and that the acceleration due to gravity is 32 feet/second² or 9.8 meters/second².

23.† A ball that is at $x = 3, y = 4$ (feet) and has velocity $2\mathbf{i} - 3\mathbf{j}$ (feet per second) at $t = 0$

24.† A rock that is at $x = 10, y = 3$ (feet) at time $t = 0$ and at $x = 5, y = 2$ at $t = 1$

25. An object that is at $x = 3, y = 2$ (feet) at $t = 0$ and has velocity $5\mathbf{i} + 5\mathbf{j}$ feet per second at $t = 2$

26.[†] A ball that passes through the point $x = 40$, $y = 50$ (feet) at times $t = 0$ and $t = 1$

27.[†] An object that is at $x = 10$, $y = 0$ (meters) and has velocity $3\mathbf{i} + 2\mathbf{j}$ meters per second at $t = 0$

28. A bullet that is at the origin at $t = 0$ and whose highest point is $x = 250$, $y = 490$ (meters)

29.[†] An arrow whose path intersects the x-axis at $x = 0$ (meters) and at $x = 100$ five seconds later

30.[†] A bullet that is shot from the origin at $t = 0$ with a speed of 20 meters per second at an angle of 30° with the positive x-axis

31.[†] At time $t = 1$ (hours) a boat at $(10, -10)$ in a horizontal xy-plane with distances measured in nautical miles has velocity $\langle 5, -1 \rangle$ knots. Its acceleration for $t \geq 1$ is $\langle 3\sqrt{t}, t^{-2} \rangle$ knots per hour. Where is it for $t \geq 1$?

32.[†] An object is at $(1, 0)$ in an xy-plane and has velocity $\langle -1, 1 \rangle$ at $t = 0$. Its acceleration for $t \geq 0$ is $\langle 2(t + 1)^{-3}, -2t(t^2 + 1)^{-2} \rangle$. What happens to it as $t \to \infty$?

33. An object's acceleration vector at time t is $20t^3\mathbf{i} + 2\mathbf{j}$. At $t = 0$ it is at $(2, 3)$ and at $t = 1$ it is at $(5, 0)$. Give its velocity as a function of t.

11-7   POLAR FORMS OF VECTORS; PLANETARY MOTION   (Optional)

In this section we use the POLAR FORMS of velocity and acceleration vectors to study Newton's law of gravitation and Kepler's laws of planetary motion.

Polar forms of vectors

In the last section we studied acceleration vectors by expressing them in terms of the unit tangent and normal vectors along a curve (Figure 11.68). Here we study velocity and acceleration vectors by expressing them in terms of unit vectors $\mathbf{u}_r$ and $\mathbf{u}_\Theta$ which arise from polar coordinates.

FIGURE 11.68 The unit tangent and normal vectors

Definition 11.18   At the point with polar coordinates $[r, \Theta]$, the vector $u_r = u_r(\Theta)$ is the unit vector which points away from the origin (Figure 11.69). The unit vector $u_\Theta = u_\Theta(\Theta)$ is perpendicular to u_r and points to its left.

The definitions of the functions $\sin\Theta$ and $\cos\Theta$ show that the vector $\mathbf{u}_r$ may be expressed in the form

(1) $\mathbf{u}_r = \cos\Theta\,\mathbf{i} + \sin\Theta\,\mathbf{j}.$ ◄

Because $\mathbf{u}_\Theta$ is a unit vector perpendicular to $\mathbf{u}_r$ and pointing to its left, we then have

(2) $\mathbf{u}_\Theta = -\sin\Theta\,\mathbf{i} + \cos\Theta\,\mathbf{j}.$ ◄

Equations (1) and (2) yield the differentiation formulas

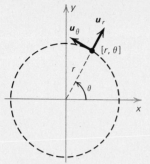

FIGURE 11.69   The unit vectors u_r and u_θ

(3) $\dfrac{d}{d\Theta} \boldsymbol{u}_r = \boldsymbol{u}_\Theta$ and $\dfrac{d}{d\Theta} \boldsymbol{u}_\Theta = -\boldsymbol{u}_r.$ ◀

The formulas for velocity and acceleration vectors in terms of $\boldsymbol{u}_r$ and $\boldsymbol{u}_\Theta$ are given in the next theorem.

Theorem 11.9   Let $[r(t), \Theta(t)]$ be the polar coordinates of an object at time t. Then its position, velocity, and acceleration vectors at time t are

(4) $\boldsymbol{R} = r\boldsymbol{u}_r$ ◀

(5) $v = \dfrac{dr}{dt} \boldsymbol{u}_r + r \dfrac{d\Theta}{dt} \boldsymbol{u}_\Theta$ ◀

(6) $a = \left[\dfrac{d^2 r}{dt^2} - r\left(\dfrac{d\Theta}{dt}\right)^2 \right] \boldsymbol{u}_r + \left[r\dfrac{d^2\Theta}{dt^2} + 2\dfrac{dr}{dt}\dfrac{d\Theta}{dt} \right] \boldsymbol{u}_\Theta.$ ◀

In formulas (4) through (6), $r = r(t)$ and $\Theta = \Theta(t)$ are functions of t whereas $\boldsymbol{u}_r = \boldsymbol{u}_r(\Theta(t))$ and $\boldsymbol{u}_\Theta = \boldsymbol{u}_\Theta(\Theta(t))$ are composite functions.

Proof of (4) and (5).   Formula (4) follows from the definitions of polar coordinates and of the vectors $\boldsymbol{R}$ and $\boldsymbol{u}_r$. To obtain (5) we differentiate equation (4) with respect to t. This requires first the product rule and then the chain rule and gives

$$\boldsymbol{v}(t) = \dfrac{d\boldsymbol{R}}{dt}(t) = \dfrac{d}{dt}[r(t)\boldsymbol{u}_r(\Theta)]$$

$$= \dfrac{dr}{dt}(t)\,\boldsymbol{u}_r(\Theta) + r(t)\dfrac{d}{dt}\boldsymbol{u}_r(\Theta)$$

$$= \dfrac{dr}{dt}(t)\,\boldsymbol{u}_r(\Theta) + r(t)\left[\dfrac{d}{d\Theta}\boldsymbol{u}_r(\Theta)\right]\dfrac{d\Theta}{dt}.$$

With the first of differentiation formulas (3), we obtain (5).   Q.E.D.

Differentiating equation (5) with respect to t leads to equation (6). We leave this calculation as Exercise 5.

Speed in polar coordinates

Because $\boldsymbol{u}_r$ and $\boldsymbol{u}_\Theta$ are perpendicular unit vectors, we have

(7) $|a\boldsymbol{u}_r + b\boldsymbol{u}_\Theta| = \sqrt{a^2 + b^2}$ ◀

for any numbers a and b (see Exercise 6). Therefore, if an object has the velocity vector (5), its speed is

(8) $\dfrac{ds}{dt} = |\mathbf{v}| = \sqrt{\left(\dfrac{dr}{dt}\right)^2 + r^2 \left(\dfrac{d\Theta}{dt}\right)^2}.$ ◀

EXAMPLE 1   The curve with polar equation $r = \sin(2\Theta)$ is a "four-leaved rose". What are the maximum and minimum speeds of a point $[r, \Theta]$ traversing the curve if Θ increases at the constant rate of π radians ($\frac{1}{2}$ revolution) per second.

Solution.   We have $d\Theta/dt = \pi$, and we compute dr/dt from the equation of the curve and the chain rule:

$$\frac{dr}{dt} = \left[\frac{d}{d\Theta}\sin(2\Theta)\right]\frac{d\Theta}{dt} = 2\cos(2\Theta)\frac{d\Theta}{dt} = 2\pi\cos(2\Theta).$$

By formula (8) the speed is

$$\frac{ds}{dt} = \sqrt{4\pi^2\cos^2(2\Theta) + \pi^2\sin^2(2\Theta)} = \pi\sqrt{4\cos^2(2\Theta) + \sin^2(2\Theta)}$$

$$= \pi\sqrt{1 + 3\cos^2(2\Theta)}.$$

Here we have used the identity $\cos^2(2\Theta) + \sin^2(2\Theta) = 1$. Because the minimum value of $\cos^2(2\Theta)$ is 0 and its maximum value is 1, the minimum speed is π and the maximum speed is 2π.   □

Kepler's laws

In 1609 and 1619 the German astronomer Johann Kepler (1571–1630) published his famous three laws of planetary motion:

I. The planets have elliptical orbits with the sun at one focus (discovered in 1605).

II. The line segment joining a planet and the sun sweeps out equal areas in equal times (discovered in 1602). (See Figure 11.70.)

III. The ratio a^3/T^2 is the same for all planets, where a is the average of the planet's closest and farthest distances from the sun and T is the time it takes for the planet to make one revolution (discovered in 1618).

Kepler's laws were the beginning of modern astronomy. Most earlier astronomers thought that the orbits of the planets were circles or circles formed by circles turning on circles (Figure 11.71). Kepler formulated his laws after studying voluminous data collected (without a telescope) by the Danish astronomer Tycho Brahe (1546–1601). (The telescope was invented in 1608.)

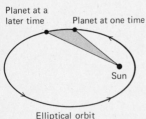

Planet at a later time   Planet at one time

Sun

Elliptical orbit

FIGURE 11.70   Area swept out by the line from the sun to a planet

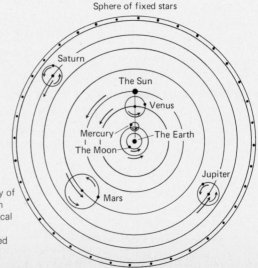

Sphere of fixed stars

Saturn

The Sun

Venus

Mercury   The Earth

The Moon

Jupiter

Mars

FIGURE 11.71   Ptolemy's theory of the solar system, in which the sun goes around the earth in an elliptical path and the planets in epicycles (circles turning on circles) (Adapted from *The Flammarion Book of Astronomy*, p. 253.)

**Newton's inverse
square law**

Probably as early as 1667 (at age 23) Newton knew how to derive the inverse square law of gravitation from Kepler's laws and how to derive Kepler's laws from the inverse square law. The inverse square law states that two points of masses m and M attract each other with a force directed along the line between them and of magnitude

(9) $\dfrac{1}{r^2} GmM$

where G is a universal constant. (If m and M are measured in grams, r in centimeters, and the force in dynes, then $G = 6.670 \times 10^{-8}$ dyne-centimeter^2/gram^2.)

Earlier scientists had proposed versions of inverse-square laws to explain the motion of the planets, but Newton was the first to state the law in its definitive form and to carry out the mathematics to discuss it thoroughly. He published his theory of gravitation in 1687, some twenty years after he had developed it, in his treatise *Principia Mathematica*. Several explanations of the delay have been proposed. One is that his theory was not consistent with early, inaccurate astronomical data. Another is that he could not see at first how to show that the planets and the sun behave like point masses. (We will obtain this result as Miscellaneous Exercise 20 in Chapter 15, where we study three-dimensional integrals.) A third explanation is that it was only upon the urging of his friend, the astronomer Edmund Halley, that he published his results. (The first edition of the *Principia* was financed by Halley.)

We will show here how Kepler's first two laws can be derived from the inverse square law.

Theorem 11.10   Kepler's second law is a consequence of the inverse square law.

Proof.   We will use only the fact that the force of attraction on the planet is directed toward the sun and not the fact that the force is given by an inverse square law. We assume that the sun and the planet act like point masses and that the only force on the planet is its attraction by the sun. Then it can be shown by using vectors in three-dimensional space that the planet moves in a plane containing the sun (see Exercise 11 of Section 12.4). We take the plane to be an xy-plane with the sun at the origin and let $[r(t), \Theta(t)]$ be the polar coordinates of the planet at time t.

According to Rule 10.1 the area swept out by the line from the sun to the planet from time t_0 to time t is

$$A(t) = \int_{\Theta(t_0)}^{\Theta(t)} \tfrac{1}{2} r(\Theta)^2 \, d\Theta.$$

By the Fundamental Theorem of Calculus and the chain rule, the derivative of this integral with respect to t is

$$(10) \quad \frac{dA}{dt}(t) = \tfrac{1}{2} r(\Theta(t))^2 \frac{d\Theta}{dt}(t) = \tfrac{1}{2} r^2 \frac{d\Theta}{dt}.$$

Differentiating again, we obtain

$$(11) \quad \frac{d^2 A}{dt^2}(t) = \tfrac{1}{2} r^2 \frac{d^2\Theta}{dt^2} + r \frac{dr}{dt}\frac{d\Theta}{dt} = \frac{r}{2}\left[r \frac{d^2\Theta}{dt^2} + 2\frac{dr}{dt}\frac{d\Theta}{dt} \right].$$

Because the force and acceleration vectors are directed toward the origin, the component $r\dfrac{d^2\Theta}{dt^2} + 2\dfrac{dr}{dt}\dfrac{d\Theta}{dt}$ of the acceleration vector (6) in the direction of $\mathbf{u}_\theta$ is zero. Therefore, the second derivative (11) is zero and the rate of change (10) of the area is constant.   Q.E.D.

For future reference we let k denote the constant such that

$$(12) \quad r^2 \frac{d\Theta}{dt} = k.$$

By equation (10), k is twice the rate of change of the area $A(t)$.

Theorem 11.11   Kepler's first law is a consequence of the inverse square law.

Proof. *   We choose coordinate axes and the time scale so that at $t = 0$ the planet is at its closest point to the sun and is on the positive x-axis as in Figure 11.72. Because $r(t)$ is a minimum at $t = 0$, we have $\dfrac{dr}{dt}(0) = 0$, and since $\mathbf{u}_\theta(0) = \mathbf{j}$, formula (5) for the velocity vector gives $\mathbf{v}(0) = r(0)\dfrac{d\Theta}{dt}(0)\,\mathbf{u}_\theta(0) = (\text{constant})\mathbf{j}$.

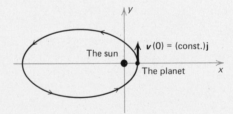

FIGURE 11.72

Newton's law may be expressed in the form

$$(13) \quad \frac{d\mathbf{v}}{dt} = -\frac{\mu}{r^2}\mathbf{u}_r$$

where μ (*mu*) is a positive constant. We will integrate this equation to

---

*This proof is adapted from D. E. Richmond, [1] "Inverse square laws: a simple treatment".

find a formula for $\mathbf{v}$. Equation (3) and the chain rule give the identity

$$\mathbf{u}_r = -\frac{1}{d\Theta/dt}\frac{d}{dt}\mathbf{u}_\theta.$$

With this and (12), Newton's law (13) gives

$$\frac{d\mathbf{v}}{dt} = \frac{\mu}{r^2\,d\Theta/dt}\frac{d}{dt}\mathbf{u}_\theta = \frac{\mu}{k}\frac{d}{dt}\mathbf{u}_\theta \quad \text{or} \quad \frac{k}{\mu}\frac{d}{dt}\mathbf{v} = \frac{d}{dt}\mathbf{u}_\theta.$$

Integrating with respect to t yields the equation $(k/\mu)\,\mathbf{v} = \mathbf{u}_\theta + \mathbf{C}$, where $\mathbf{C}$ is a constant vector. Because $\mathbf{u}_\theta = \mathbf{j}$ and $\mathbf{v} = (\text{constant})\mathbf{j}$ at $t = 0$, we have $\mathbf{C} = e\mathbf{j}$ with some constant e, and hence

$$(14) \quad \mathbf{v} = \frac{\mu}{k}\mathbf{u}_\theta + \frac{\mu e}{k}\mathbf{j}.$$

Rather than integrating equation (14), we will equate the components of $\mathbf{v}$ in the direction of $\mathbf{u}_\theta$ as given by (5) and (14) and solve for r. Because $\mathbf{u}_\theta \cdot \mathbf{u}_\theta = 1$ and $\mathbf{u}_\theta \cdot \mathbf{j} = \cos\Theta$, equation (14) yields

$$(15) \quad \mathbf{u}_\theta \cdot \mathbf{v} = \frac{\mu}{k}(1 + e\cos\Theta)$$

and, since $\mathbf{u}_\theta \cdot \mathbf{u}_r = 0$, equation (5) gives

$$(16) \quad \mathbf{u}_\theta \cdot \mathbf{v} = r\frac{d\Theta}{dt}.$$

Since by (12), $d\Theta/dt = kr^{-2}$, equations (15) and (16) give

$$\frac{\mu}{k}(1 + e\cos\Theta) = \frac{k}{r}.$$

We solve for r and set $p = k^2/\mu$ to obtain

$$r = \frac{p}{1 + e\cos\Theta}$$

which is the equation of a conic section (see Section 10.7). Since the planet's orbit is closed, it must be an ellipse, as stated in Kepler's first law.   Q.E.D.

Exercises

†*Answer provided.*
‡*Outline of solution provided.*

In Exercises 1 and 2 give the unit tangent vectors to the curves at the indicated points. Express the tangent vectors first in terms of $\mathbf{u}_r$ and $\mathbf{u}_\theta$ and then in terms of $\mathbf{i}$ and $\mathbf{j}$.

1.† The Archimidean spiral $r = \Theta$ at $\Theta = \frac{3}{4}\pi$

2. The logarithmic spiral $r = e^\Theta$ at $\Theta = \pi$

3.† An object traverses the limaçon $r = 3 - 2\sin\Theta$ in such a way that $d\Theta/dt = 5$. What is its speed at $(0, 1)$?

4. An object traverses the logarithmic spiral $r = e^\theta$ in such a way that $dr/dt = 2$. What is its speed at the point $(e^{6\pi}, 0)$?

5. Derive the polar decomposition (6) of the acceleration vector.

6.† Show that for any perpendicular unit vectors $\boldsymbol{A}$ and $\boldsymbol{B}$ and any numbers a and b the generalized Pythagorean Theorem, $|a\boldsymbol{A} + b\boldsymbol{B}|^2 = a^2 + b^2$, is valid.

7. Show that Kepler's second law implies that the force of gravity is in line with the sun.

8. Show that Kepler's second and third laws imply an inverse square law valid for all circular orbits.

9. Show that Kepler's first and second laws imply an inverse square law for each orbit.

10. Show that Newton's inverse square law implies Kepler's third law for circular orbits.

11.† Show that if $r(\Theta_0) = 0$ and $\dfrac{dr}{d\Theta}(\Theta_0) \neq 0$, then the line $\Theta = \Theta_0$ is tangent to the curve $r = r(\Theta)$ at the origin.

MISCELLANEOUS EXERCISES

†Answer provided.
‡Outline of solution provided.

1.† The forces applied to the book in Figure 11.73 are the negatives of each other, but the book is not in equilibrium. Explain.

Use vectors to prove the results stated in Exercises 2 through 6.

2.† The sum of the distances from any point inside an equilateral triangle to the three sides is constant.

3. If M is the midpoint of side AC and T is one-third of the distance from the vertex B to the vertex C in the triangle ABC of Figure 11.74, then P is three-fourths of the distance from A to T.

4. The diagonals of a parallelogram bisect each other.

5. The sum of the squares of the lengths of the sides of a parallelogram is equal to the sum of the squares of the lengths of the diagonals.

6. The line joining the midpoints of the nonparallel sides of a trapezoid is parallel to the other two sides and its length is the average of their lengths.

7.‡ a. Derive parametric equations of the *hypocycloid* traced out by a point on the circumference of a circle of radius $\frac{1}{4}a$ which rolls inside a circle of radius a. [Put the center of the larger circle at the origin and have the point on the smaller circle touch the larger circle at $(a, 0)$.]   **b.** Use the identities $\cos^3\alpha = \frac{3}{4}\cos\alpha + \frac{1}{4}\cos(3\alpha)$ and $\sin^3\alpha = \frac{3}{4}\sin\alpha - \frac{1}{4}\sin(3\alpha)$ to show that the hypocycloid is the astroid $x^{2/3} + y^{2/3} = a^{2/3}$.

8. Use the trigonometric identity $\cos^2(\frac{1}{2}\Theta) = \frac{1}{2}(1 + \cos\Theta)$ to compute the length of the cardioid $r = a(1 + \cos\Theta)$.

9.† Give the catenary $y = \cosh x$ a parametric representation for $x \geq 0$ with arclength as parameter. (Use the inverse hyperbolic sine function arcsinh x.)

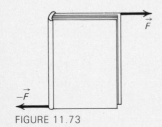

FIGURE 11.73

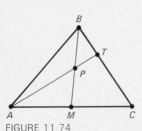

FIGURE 11.74

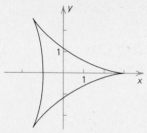

FIGURE 11.75   A deltoid

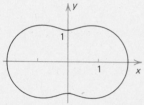

FIGURE 11.76   A hippopede

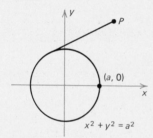

FIGURE 11.77

10. Give the curve $y = \frac{2}{3}x^{3/2}$ a parametric representation with arclength as parameter.

11. Each of the following curves is the graph of a function, $y = y(x)$. Compute dy/dx and d^2y/dx^2 in terms of the parameter t.
    a.‡ $x = e^{-t} - t^3, y = \sin t$ **b.**† $x = t(\ln t), y = t^5$
    c. $x = \arctan t + t, y = e^t - t$

12. When a circle of radius 1 rolls inside a circle of radius 3, a point on the circumference of the smaller circle traces out the hypocycloid shown in Figure 11.75. This curve is also known as a *deltoid* and has the parametric equations $\mathbf{R}(t) = (2\cos^2 t + 2\cos t - 1)\mathbf{i} + 2(\sin t - \sin t \cos t)\mathbf{j}$. Show that $\dfrac{d\mathbf{R}}{dt}(t) = \mathbf{O}$ where the curve does not have a tangent line.

13.† The *hippopede* in Figure 11.76 has the equation $r^2 = 4 - 3\sin^2\Theta$.
    a. Use implicit differentiation to compute $dr/d\Theta$ in terms of Θ.
    b. Compute $dx/d\Theta$ and $dy/d\Theta$ and use the result to find a unit tangent vector to the curve at $\Theta = \pi/4$. Trace the curve and sketch the tangent vector.

14. Use the parametric equations $x = a\cos^3 t, y = a\sin^3 t$ to compute the length of the astroid $x^{2/3} + y^{2/3} = a^{2/3}$.

15.‡ Find the intersection of the curves $C_1: x = t^3 + 6, y = 3t + 4$ and $C_2: x = t^2 - 3, y = 3t - 5$.

16. The position vectors of two bugs at time t are $3(t-1)\mathbf{i} + 9(t-1)^2\mathbf{j}$ and $t^6\mathbf{i} + 2t^3\mathbf{j}$. Show that they never collide but their paths cross twice.

17.† The end P of the thread that is unwound from the fixed spool in Figure 11.77 traces out the *involute* of the circle $x^2 + y^2 = a^2$. Give parametric equations of the involute. [Have the end of the thread be on the spool at the point $(a, 0)$.]

18. The locus of the centers of curvature of a curve is called the *evolute* of the curve. Show that the evolute of the curve $x = f(t), y = g(t)$ of positive curvature has the equations

$$x = f - \rho \frac{dg}{dt} \Bigg/ \sqrt{\left(\frac{df}{dt}\right)^2 + \left(\frac{dg}{dt}\right)^2}, y = g + \rho \frac{df}{dt} \Bigg/ \sqrt{\left(\frac{df}{dt}\right)^2 + \left(\frac{dg}{dt}\right)^2}$$

where $\rho = \rho(t)$ is the radius of curvature of the curve at t.

19.† Find the evolute of the cycloid $x = t - \sin t, y = \cos t - 1$. Show that it is another cycloid of the same shape.

20.† A man who can swim 3 miles/hour and run 4 miles/hour must get to a point directly across a one-half mile wide river as quickly as possible. The river is flowing 2 miles per hour. In what direction (relative to the water) should he swim to minimize his time?

21. Show that on the logarithmic spiral $r = e^\Theta$ the angle between the position vector $\mathbf{R}(t)$ and the acceleration vector $\mathbf{a}(t)$ is constant.

22.† If the maximum range of a cannon is 3 miles, what is its muzzle velocity? (Ignore air resistance.)

23. At what angle should a cannon be fired to obtain the maximum range up a hill that makes an angle of α radians with the horizontal? (Ignore air resistance.)

24. The best English archers with longbows could shoot arrows 1200 feet. What is the least speed with which an arrow can leave a bow and travel that far? (Ignore air resistance.)

25.[‡] If $\mathbf{v} = \mathbf{v}(t)$ is the velocity of a projectile of mass m that is subject to air resistance proportional to its velocity, then $\mathbf{v}$ satisfies the vector differential equation $m \, d\mathbf{v}/dt = -mg\mathbf{j} - k\mathbf{v}$.   **a.**   Solve for $\mathbf{v}(t)$ by putting both terms involving $\mathbf{v}$ on the left, multiplying the equation by $1/m \, e^{kt/m}$, and noting that then the left side is the derivative of a product.   **b.**   What happens to $\mathbf{v}(t)$ as t tends to ∞?

FIGURE 11.78

26.[†] Let $\Theta = \Theta(t)$ be the angle made by a pendulum with the vertical (Figure 11.78).   **a.**   Show that if the pendulum is of length 1, then the function Θ satisfies the differential equation $\dfrac{d^2\Theta}{dt^2}(t) = -g \sin \Theta$.

Multiplying both sides of this equation by $d\Theta/dt$ turns them into t-derivatives.   **b.**   Use this fact to solve for $d\Theta/dt$ in terms of Θ and the maximum value Θ_0 of Θ.   **c.**   What is the maximum speed of the end of the pendulum?   **d.**   Find the derivative $dt/d\Theta$ of the inverse function $t = t(\Theta)$ and use it to express the period of the pendulum as an integral.

27.[†] A straight frictionless wire is to be fastened from a point on a ceiling to a wall ten feet away. What angle should the wire make with the ceiling so a bead will slide down the wire from the ceiling to the wall in the least time?

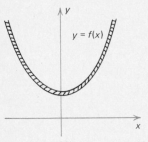

FIGURE 11.79   A hanging cable

28. Suppose that a hanging cable of constant density ρ pounds per foot has the shape of the curve $y = f(x)$ and that the lowest point on the cable is at $x = 0$ (Figure 11.79).   **a.**   Show that the weight of the portion of the cable for $0 \le x \le u$ is

$$w(u) = \rho \int_0^u \sqrt{1 + [f'(x)]^2} \, dx.$$

b.   If the horizontal tension of the cable is k pounds, then the force at $x = u$ on the portion of the cable for $0 \le x \le u$ is the vector $k\mathbf{i} + w(u)\mathbf{j}$. Show that the condition that the force vector be tangent to the cable implies $f'(u) = w(u)/k$.   **c.**   Use the results of parts (a) and (b) to demonstrate that $f(x)$ satisfies the differential equation $f''(x) = (\rho/k)\sqrt{1 + [f'(x)]^2}$.

29. A bucket is swung at a constant speed in a vertical circle of radius 4 feet. At what rate must it be swung (in revolutions per second) so a coin will not fall out of it?

30.   **a.**[†]   Give vector equations of the motions of the earth and of the planet Mercury under the simplifying assumptions that the orbits are circles in the same plane with the sun at their centers; that the radius of the earth's orbit is 93 million miles and the radius of Mercury's orbit is 36 million

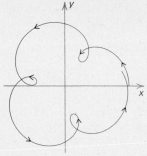

FIGURE 11.80

miles; that the earth's year is 365 days long; and that Mercury's year is 88 days long. Use an xy-plane with scales measured in millions of miles and the origin at the sun. Have both planets move counterclockwise with Mercury on the positive x-axis and the earth on the negative x-axis at $t = 0$ (days).   **b.**   Give the equation of the motion of Mercury as viewed from the earth (with the earth at the origin). (This curve, shown in Figure 11.80 for $0 \le t \le 365$, is an *epicycle*. The actual path of Mercury, relative to the earth, during one year is shown in Figure 11.81, where the sun's path is drawn with a broken line and Mercury's with a solid line and where the numbers give the month-by-month locations.)

31. Alternating current is generated by rotating an armature in a uniform magnetic field. If, for example, the loop of wire in Figure 11.82 is rotated with constant velocity, the current is proportional to the vertical velocity of the segment PQ and flows from P to Q when PQ is on the right moving up and from Q to P when PQ is on the left moving down. Show that the current is given as a function of time by a cosine.

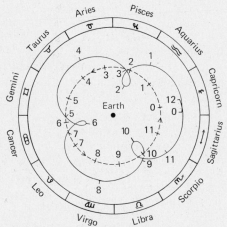

FIGURE 11.81   The paths of the Sun and Mercury relative to the earth in 1945 (Adapted from *The Flammarion Book of Astronomy*, p. 255.)

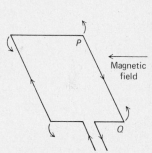

FIGURE 11.82   Current in a loop of wire that is rotating in a magnetic field

CHAPTER 12

VECTORS IN SPACE AND SOLID ANALYTIC GEOMETRY

SUMMARY: In Chapters 13, 14, and 15 we will use lines, curves, planes, and surfaces in three-dimensional space to study functions of two and three variables. This chapter contains the necessary background material on solid analytic geometry. Section 12.1 deals with coordinates, vectors, and the dot product of vectors and Section 12.2 with the cross product. We use vectors to study lines and planes in Section 12.3 and to study curves in Section 12.4. In Section 12.5 we discuss several types of surfaces which will be used in the following chapters. Cramer's rule for solving systems of three linear equations in three unknowns is derived in an optional appendix to the chapter.

12-1   RECTANGULAR COORDINATES AND VECTORS IN SPACE

In this section we show how rectangular coordinates are used to study geometric objects and vectors in three-dimensional space. The formulas we present here are very similar to those derived in previous chapters for coordinate planes, but they involve one more variable.

Rectangular coordinates in space

Rectangular coordinates in three-dimensional space are obtained by introducing three mutually perpendicular coordinate axes that intersect at their zero points (Figure 12.1). The z-axis is usually placed vertical with its positive side on the top. Then the xy-plane is horizontal. We choose the x- and y-axes so that they have their usual orientation when viewed from above. The point where the axes intersect is called the *origin*.

The coordinates (x, y, z) of a point in space are determined by planes that pass through the point and are perpendicular to the coordinate

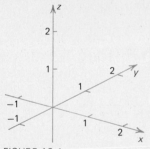

FIGURE 12.1

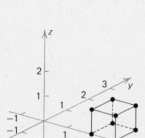

FIGURE 12.3

axes. The plane perpendicular to the x-axis intersects it at the x-coordinate of the point; the plane perpendicular to the y-axis intersects it at the y-coordinate; and the plane perpendicular to the z-axis intersects it at the z-coordinate. If none of the coordinates are zero, the planes through the point and the coordinate planes form a rectangular box. The origin is at one corner of the box and the point (x, y, z) is at the diagonally opposite corner (Figure 12.2). Thus, we can think of xyz-space as the set of ordered triples (x, y, z).

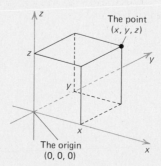

FIGURE 12.2

EXAMPLE 1   Sketch the cube consisting of the points (x, y, z) with $2 \le x \le 3, 0 \le y \le 1$, and $0 \le z \le 1$. What are the coordinates of the eight corners of the cube?

Solution.   The cube is sketched in Figure 12.3. The corners of its base are $(2, 0, 0)$, $(3, 0, 0)$, $(3, 1, 0)$, and $(2, 1, 0)$. The corners of its top are $(2, 0, 1)$, $(3, 0, 1)$, $(3, 1, 1)$ and $(2, 1, 1)$.   □

The Pythagorean theorem in space

Suppose that a rectangular box has length a, width b, and height c as in Figure 12.4. By the Pythagorean theorem the length of the diagonal of the base is $\sqrt{a^2 + b^2}$. The diagonal of the box is the diagonal of the vertical rectangle of width $\sqrt{a^2 + b^2}$ and height c which is shown in Figure 12.5. By the Pythagorean theorem again, the length of the

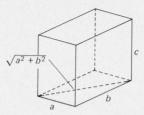

FIGURE 12.4

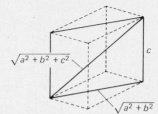

FIGURE 12.5

diagonal of the box is the square root of $[\sqrt{a^2 + b^2}]^2 + c^2$ $= a^2 + b^2 + c^2$. This gives the three-dimensional version of the Pythagorean theorem: for a rectangular box of sides a, b, and c

(1)   The length of the diagonal $= \sqrt{a^2 + b^2 + c^2}$.   ◀

The distance formula

Two points $P(x_1, y_1, z_1)$ and $Q(x_2, y_2, z_2)$ are at diagonally opposite corners of a rectangular box with sides of lengths $|x_2 - x_1|$, $|y_2 - y_1|$, and $|z_2 - z_1|$ (Figure 12.6). Hence, by the Pythagorean theorem (1)

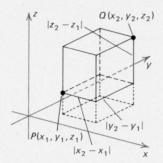

FIGURE 12.6

(2)   The distance $\overline{PQ}$ between $P(x_1, y_1, z_1)$ and $Q(x_1, y_1, z_1)$
$$= \sqrt{(x_2 - x_1)^2 + (y_2 - y_1)^2 + (z_2 - z_1)^2}.$$
◀

EXAMPLE 2   Compute the distance between the points $(4, 5, 0)$ and $(-1, 3, 2)$.

Solution.   By formula (2) the distance is
$$\sqrt{(-1 - 4)^2 + (3 - 5)^2 + (2 - 0)^2} = \sqrt{(-5)^2 + (-2)^2 + 2^2}$$
$$= \sqrt{33}. \ \square$$

Vectors in space

To employ the coordinate system to study vectors in space, we use three unit vectors $\mathbf{i}, \mathbf{j},$ and $\mathbf{k}$. The vector $\mathbf{i}$ points in the direction of the positive x-axis; the vector $\mathbf{j}$ points in the direction of the positive y-axis; and the vector $\mathbf{k}$ points in the direction of the positive z-axis (Figure 12.7).

The POSITION VECTOR $\overrightarrow{OP}$ of a point (x, y, z) in space is the vector from the origin to the point (Figure 12.8). It can be expressed in the

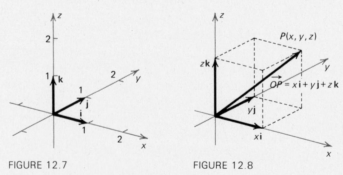

FIGURE 12.7   FIGURE 12.8

form $\overrightarrow{OP} = x\mathbf{i} + y\mathbf{j} + z\mathbf{k}$. We use the symbol $\langle x, y, z \rangle$ for this linear combination of the vectors $\mathbf{i}, \mathbf{j},$ and $\mathbf{k}$:

(3) $\vec{OP} = x\mathbf{i} + y\mathbf{j} + z\mathbf{k} = \langle x, y, z \rangle$ for $P = (x, y, z)$.   ◄

Thus, we have $\mathbf{i} = \langle 1, 0, 0 \rangle$, $\mathbf{j} = \langle 0, 1, 0 \rangle$, and $\mathbf{k} = \langle 0, 0, 1 \rangle$.

The displacement vector $\vec{PQ}$ from the point $P = (x_1, y_1, z_1)$ to the point $Q = (x_2, y_2, z_2)$ (Figure 12.9) can be computed from the coordinates of the points by the formula

(4) $\vec{PQ} = (x_2 - x_1)\mathbf{i} + (y_2 - y_1)\mathbf{j} + (z_2 - z_1)\mathbf{k}$
        $= \langle x_2 - x_1, y_2 - y_1, z_2 - z_1 \rangle$.   ◄

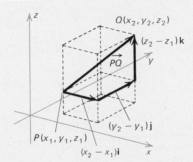

FIGURE 12.9

EXAMPLE 3   Find the displacement vector $\vec{PQ}$ for $P = (6, 5, 8)$ and $Q = (7, 3, 9)$.

Solution.   By formula (4) the displacement vector is
$\vec{PQ} = \langle 7 - 6, 3 - 5, 9 - 8 \rangle = \langle 1, -2, 1 \rangle$.   □

Velocity, force, and acceleration vectors in three dimensions are defined as in two dimensions.

The distance formula (2) shows that the length of the vector $\mathbf{A} = x\mathbf{i} + y\mathbf{j} + z\mathbf{k}$ is

(5) $|\mathbf{A}| = |x\mathbf{i} + y\mathbf{j} + z\mathbf{k}| = \sqrt{x^2 + y^2 + z^2}$.   ◄

The sum $\mathbf{A} + \mathbf{B}$ of two vectors in space and the product $a\mathbf{A}$ of a number and a vector are defined geometrically as in two dimensions. The sum $\mathbf{A} + \mathbf{B}$ lies along the diagonal of a parallelogram with sides formed by placing $\mathbf{A}$ and $\mathbf{B}$ with a common base point (Figure 12.10). If instead we place the base of $\mathbf{B}$ at the tip of $\mathbf{A}$, the sum runs from the base of $\mathbf{A}$ to the tip of $\mathbf{B}$ (Figure 12.11). The product $a\mathbf{A}$ has length $|a|\,|\mathbf{A}|$ and the same direction as $\mathbf{A}$ if a is positive and the opposite direction if a is negative (Figure 12.12).

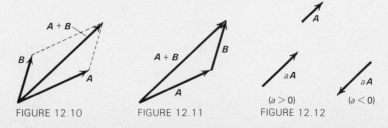

FIGURE 12.10   FIGURE 12.11   FIGURE 12.12

The formal definition

The representation of the vectors in terms of their components leads us to the following formal definition.

Definition 12.1   A vector in xyz-space is an ordered triple of numbers expressed either in the form xi + yj + zk or ⟨x, y, z⟩.

As in an xy-plane, we can add vectors in xyz-space and multiply them by numbers by performing the operations on the components. We have

(6) $\langle x_1, y_1, z_1 \rangle + \langle x_2, y_2, z_2 \rangle = \langle x_1 + x_2, y_1 + y_2, z_1 + z_2 \rangle$ ◄

(7) $a\langle x, y, z \rangle = \langle ax, ay, az \rangle$.   ◄

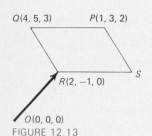

$Q(4, 5, 3)$ $P(1, 3, 2)$

S

$R(2, -1, 0)$

$O(0, 0, 0)$

FIGURE 12.13

EXAMPLE 4   Three vertices of a parallelogram $PQRS$ are $P = (1, 3, 2)$, $Q = (4, 5, 3)$, and $R = (2, -1, 0)$. What is the point S opposite Q?

Solution.   The schematic sketch of the parallelogram in Figure 12.13 serves to guide our calculations. Because $PQRS$ is a parallelogram, the displacement vectors $\overrightarrow{QP}$ and $\overrightarrow{RS}$ are equal. Thus, $\overrightarrow{RS} = \overrightarrow{QP}$ $= \langle 1 - 4, 3 - 5, 2 - 3 \rangle = \langle -3, -2, -1 \rangle$. On the other hand, the position vector $\overrightarrow{OS}$ is equal to the sum of the position vector $\overrightarrow{OR}$ and the displacement vector $\overrightarrow{RS}$. Hence, $\overrightarrow{OS} = \overrightarrow{OR} + \overrightarrow{RS}$ $= \langle 2, -1, 0 \rangle + \langle -3, -2, -1 \rangle = \langle -1, -3, -1 \rangle$ and the point S has coordinates $S = (-1, -3, -1)$.   □

The dot product in space

Definition 12.2   The dot product between two vectors $A = \langle x_1, y_1, z_1 \rangle$ and $B = \langle x_2, y_2, z_2 \rangle$ is

(8) $A \cdot B = \langle x_1, y_1, z_1 \rangle \cdot \langle x_2, y_2, z_2 \rangle = x_1 x_2 + y_1 y_2 + z_1 z_2$.   ◄

The proof of Theorem 11.1, when applied to the case of three dimensions, gives the formula

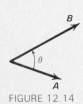

B

θ

A

FIGURE 12.14

(9) $A \cdot B = |A||B| \cos \Theta$ ◄

which expresses the dot product in terms of the lengths of the vectors and the cosine of an angle Θ between them (Figure 12.14).

EXAMPLE 5   Show that the triangle with vertices $(4, 1, -1)$, $(3, 3, 5)$, and $(1, 0, 2)$ is a right triangle.

Solution.   We set $P = (4, 1, -1)$, $Q = (3, 3, 5)$, and $R = (1, 0, 2)$ and consider the schematic sketch of the triangle in Figure 12.15. The sides of the triangle are formed by the vectors

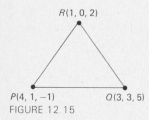

$R(1, 0, 2)$

$P(4, 1, -1)$ $Q(3, 3, 5)$

FIGURE 12.15

$$\overrightarrow{PQ} = \langle 3 - 4, 3 - 1, 5 - (-1) \rangle = \langle -1, 2, 6 \rangle$$
$$\overrightarrow{PR} = \langle 1 - 4, 0 - 1, 2 - (-1) \rangle = \langle -3, -1, 3 \rangle$$
$$\overrightarrow{QR} = \langle 1 - 3, 0 - 3, 2 - 5 \rangle = \langle -2, -3, -3 \rangle.$$

Consequently,

$$\vec{PQ} \cdot \vec{PR} = \langle -1, 2, 6 \rangle \cdot \langle -3, -1, 3 \rangle$$
$$= (-1)(-3) + (2)(-1) + (6)(3) = 19$$

$$\vec{PQ} \cdot \vec{QR} = \langle -1, 2, 6 \rangle \cdot \langle -2, -3, -3 \rangle$$
$$= (-1)(-2) + (2)(-3) + 6(-3) = -22$$

$$\vec{QR} \cdot \vec{PR} = \langle -2, -3, -3 \rangle \cdot \langle -3, -1, 3 \rangle$$
$$= (-2)(-3) + (-3)(-1) + (-3)(3) = 0.$$

The last of these is zero, so the vectors $\vec{QR}$ and $\vec{PR}$ are perpendicular, and the triangle is a right triangle with its right angle at the point R.   □

Components and projections of vectors

The formulas for the component of one vector in the direction of another and for the projection of one vector along another involve the dot product and are the same for vectors in xyz-space as for vectors in the plane: for any vector A and any nonzero vector B

(10) $\left[\begin{array}{l} \text{The component of } A \\ \text{in the direction of } B \end{array} \right] = \dfrac{A \cdot B}{|B|}$ ◄

(11) $\left[\begin{array}{l} \text{The projection of} \\ A \text{ along } B \end{array} \right] = \dfrac{A \cdot B}{|B|^2} B.$ ◄

Direction angles and direction cosines of a nonzero vector

Whereas we can describe the direction of a nonzero vector in a plane with one angle, we need at least two angles to describe the direction of a nonzero vector A in space. We let α (*alpha*), β (*beta*), and γ (*gamma*) denote the angles between A and the unit vectors $\mathbf{i}$, $\mathbf{j}$, and $\mathbf{k}$ (Figure 12.16). We choose angles with $0 \le \alpha \le \pi, 0 \le \beta \le \pi$, and $0 \le \gamma \le \pi$. If $A = a\mathbf{i} + b\mathbf{j} + c\mathbf{k}$, then $a = A \cdot \mathbf{i} = |A| \cos \alpha, b = A \cdot \mathbf{j} = |A| \cos \beta$, and $c = A \cdot \mathbf{k} = |A| \cos \gamma$, so that

$$A = |A| [\cos \alpha \, \mathbf{i} + \cos \beta \, \mathbf{j} + \cos \gamma \, \mathbf{k}].$$ ◄

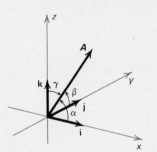

FIGURE 12.16   Direction angles of a vector A

The numbers

(12) $\cos \alpha = \dfrac{A \cdot \mathbf{i}}{|A|}, \quad \cos \beta = \dfrac{A \cdot \mathbf{j}}{|A|}, \quad \cos \gamma = \dfrac{A \cdot \mathbf{k}}{|A|}$ ◄

are called the DIRECTION COSINES of the vector A. The arccosines α, β, and γ of the numbers (12) are called the DIRECTION ANGLES of A.

EXAMPLE 6   Give the direction cosines of $A = 2\mathbf{i} - 2\mathbf{j} + \mathbf{k}$ and approximate values of its direction angles.

Solution.   The length of A is $\sqrt{2^2 + 2^2 + 1^2} = 3$ and its components are $A \cdot \mathbf{i} = 2, A \cdot \mathbf{j} = -2$, and $A \cdot \mathbf{k} = 1$. Consequently, $\cos \alpha = \frac{2}{3}$, $\cos \beta = -\frac{2}{3}$, and $\cos \gamma = \frac{1}{3}$.

The closest value to $\frac{2}{3}$ in the table of cosines in the back of the book is $\cos(0.84) \approx 0.6675$, so $\alpha \approx 0.84$ radians. The same value in the table yields $-\frac{2}{3} \approx -\cos(0.84) = \cos(\pi - 0.84)$ and, therefore, $\beta \approx \pi - 0.84 \approx 3.14 - 0.84 = 2.30$ radians. Because the closest value to $\frac{1}{3}$ in the table is $\cos(1.23) \approx 0.3342$, we have $\gamma \approx 1.23$ radians. $\square$

Formula (5) for the length of a vector and formula (12) yield the identity

(13) $\cos^2\alpha + \cos^2\beta + \cos^2\gamma = 1$ ◄

relating the direction cosines. Because of this identity we cannot choose the angles $\alpha, \beta,$ and γ independently. If we are given α and β, for example, with $\cos^2\alpha + \cos^2\beta = 1$, then $\cos\gamma = 0$ and all vectors with the same α and β have the same direction. If $\cos^2\alpha + \cos^2\beta$ is less than 1, identity (13) gives two choices of $\cos\gamma$, and the vectors have two possible directions. If $\cos^2\alpha + \cos^2\beta$ is greater than 1, there are no such vectors. (See Exercise 32 for a geometric interpretation.)

EXAMPLE 7   Give the two unit vectors that make an angle $\frac{1}{3}\pi$ with **i** and an angle $\frac{3}{4}\pi$ with **j**. What angles do they make with **k**?

Solution.   We have $\cos\alpha = \cos(\frac{1}{3}\pi) = \frac{1}{2}$ and $\cos\beta = \cos(\frac{3}{4}\pi) = -1/\sqrt{2}$. Therefore identity (13) gives $\cos^2\gamma = 1 - \cos^2\alpha - \cos^2\beta = 1 - (\frac{1}{2})^2 - (-1/\sqrt{2})^2 = \frac{1}{4}$. From this we see that $\cos\gamma = \pm\frac{1}{2}$. If we take $\cos\gamma = \frac{1}{2}$, then γ is $\frac{1}{3}\pi$ and the vector is $\frac{1}{2}\mathbf{i} - (1/\sqrt{2})\mathbf{j} + \frac{1}{2}\mathbf{k}$. If we take $\cos\gamma = -\frac{1}{2}$, then γ is $\frac{2}{3}\pi$ and the vector is $\frac{1}{2}\mathbf{i} - (1/\sqrt{2})\mathbf{j} - \frac{1}{2}\mathbf{k}$. $\square$

Exercises

† *Answer provided.*
‡ *Outline of solution provided.*

In Exercises 1 through 12 trace the coordinate axes in Figure 12.1 and use them to sketch the indicated geometrical object. (A lined transparent ruler might help you in drawing parallel lines.)

1.† The tetrahedron with vertices, $(0, 0, 0)$, $(2, 0, 0)$, $(0, 3, 0)$, and $(0, 0, 4)$

2. The triangle with vertices $(0, 0, 0)$, $(0, 4, 0)$, and $(0, 2, 2)$

3.† The parallelogram with vertices $(2, 0, 0)$, $(2, 1, 0)$, $(0, 1, 2)$, and $(0, 2, 2)$

4. The box with vertices $(1, 0, 0)$, $(2, 1, 0)$, $(1, 2, 0)$, $(0, 1, 0)$, $(1, 0, 3)$, $(2, 1, 3)$, $(1, 2, 3)$, and $(0, 1, 3)$

5.† The circular disk consisting of all points (x, y, z) such that $x^2 + y^2 \le 1$ and $z = 0$

6. The circular disk consisting of all (x, y, z) such that $y^2 + z^2 \le 1$ and $x = 0$

7.† The right cylinder whose base is the disk of Exercise 5 and whose height is 2 units

8. The right cone whose base is the disk of Exercise 5 and whose height is 3 units

9.† The sphere of radius 3 and center at the origin

10. The wedge-shaped region with vertices $(0, -1, 0)$, $(0, 0, 0)$, $(3, 0, 0)$, $(3, -1, 0)$, $(3, 0, -3)$, and $(0, 0, -3)$

11. The rectangle consisting of all (x, y, z) such that $x = 0, 0 \le y \le 4$, and $0 \le z \le 3$

12.[†] The hemisphere consisting of all (x, y, z) such that $x^2 + y^2 + z^2 = 1$ and $z \ge 0$

13. a. Explain why $\sqrt{x^2 + y^2}$ is the distance from the point (x, y, z) to the z-axis. Give formulas for the distance from (x, y, z) to   **b.** the y-axis and   **c.** the x-axis.

14.[†] Compute the distances from the point $(3, -4, -1)$ to   **a.** the z-axis,   **b.** the x-axis,   **c.** the y-axis,   **d.** the xy-plane,   **e.** the xz-plane,   **f.** the yz-plane, and   **g.** the origin.

15.[†] Sketch the cone consisting of all points equidistant to the z-axis and to the xy-plane. What is the angle between the axis and elements of this cone?

16. Sketch the cone consisting of all points whose distance to the x-axis is equal to half the distance to the origin. What is the angle between the axis and elements of this cone?

17.[†] Compute   **a.** $A + 3B$,   **b.** $A - (A \cdot B)B$,   **c.** $|A|$,   **d.** $|B|$, and   **e.** $(A + B) \cdot (A - B)$ for $A = \langle 3, 1, -2 \rangle$ and $B = \langle -1, 6, 4 \rangle$.   **f.** Why is the answer to part (e) the difference of the squares of the answers to (c) and (d)?

18. Which of the angles of the triangle with vertices $(1, -2, 0)$, $(2, 1, -2)$, and $(6, -1, -3)$ is a right angle?

19.[†] Find a number k so that the quadrilateral $PQRS$ with $P = (0, 1, 0)$, $Q = (1, 3, k)$, $R = (3, k, k)$ and $S = (2, 0, 0)$ is a parallelogram. Is it a rectangle? Is it a square?

20.[†] Use the table of cosines to give the approximate radian measure of the interior angles of the triangle with vertices $(1, 2, 3)$, $(4, 0, 2)$, and $(2, 1, 3)$.

21.[†] Three vertices of the parallelogram $PQRS$ are $P = (6, 8, -2)$, $Q = (3, 4, 0)$, and $R = (4, -7, -10)$. What is the vertex S opposite Q?

22.[†] Explain why the equation $A \cdot B = A \cdot C$ with $A \ne O$ does not imply $B = C$ for vectors in space. What does the equation imply?

23.[†] Which of the vectors $A = \langle 4, -6, 2 \rangle$, $B = \langle 1, 2, 4 \rangle$, and $C = \langle -6, 9, -3 \rangle$ are parallel? Which are perpendicular?

24.[†] Find   **a.**[‡] the acute angle between the diagonal of a cube and an edge, and   **b.** the acute angle between the diagonal and a diagonal of one of the faces.

25. A grocer sells a gallons of milk, b pounds of potatoes, and c bags of flour in one day. He makes p dollars profit per gallon of milk, r dollars profit per pound of potatoes, and s dollars profit per bag of flour. Use the vectors $A = \langle a, b, c \rangle$ and $B = \langle p, r, s \rangle$ to express his total profit on the three items as a dot product.

26. Are the points $(3, 0, 1)$, $(2, 4, 5)$, and $(-1, 4, 4)$ on the same line?

27.[†] Set $A = \langle 4, 3, 1 \rangle$ and $B = \langle -1, 0, 5 \rangle$. Find   **a.** the component of A in the direction of B and   **b.** the projection of A along B.

28. Find the direction cosines and approximate values for the direction angles of the vectors   **a.**[†] $\langle 3, -2, -3 \rangle$,   **b.**[†] $\langle 1, -2, 4 \rangle$, and   **c.** $\langle -4, 3, 2 \rangle$.

29.[†] Find a unit vector whose direction angles are equal.

30. A rectangular room is 10 feet high, 12 feet wide, and 20 feet long. Two strings from the corners of the ceiling at one end are stretched to the diagonally opposite corners on the floor. What is the acute angle between the strings where they cross? (Give an approximate value.)

31.[‡] A constant force of $F = 3\mathbf{i} - 4\mathbf{j} + \mathbf{k}$ pounds is applied to an object as it moves along the line segment from $P = (5, 7, 0)$ to $Q = (6, 6, 6)$. How much work is done by the force? (Coordinates are given in feet.)

32. **a.**[†] Suppose that you want to find a unit vector $\mathbf{u}$ that makes given acute angles α and β with $\mathbf{i}$ and $\mathbf{j}$ $(0 < \alpha < \pi/2, 0 < \beta < \pi/2)$. Explain, by considering intersections of cones, why there are two such vectors $\mathbf{u}$ if $\alpha + \beta > \pi/2$, one such vector if $\alpha + \beta = \pi/2$, and no such vector if $\alpha + \beta < \pi/2$.   **b.** Give the corresponding results for $0 < \alpha < \pi/2$ and $\pi/2 < \beta < \pi$.   **c.** Do the same for $\pi/2 < \alpha < \pi$ and $\pi/2 < \beta < \pi$.

Additional drill exercises

33.[†] Compute the area of the right triangle with vertices $(3, 5, 1)$, $(2, 4, 3)$, and $(5, 7, 3)$.

34.[†] Find the vertex S opposite Q in the parallelogram $PQRS$ where $P = (1, 2, 1)$, $Q = (3, 0, 4)$, and $R = (4, -3, 2)$.

35.[†] Find the component of A in the direction of B for   **a.** $A = \langle 1, 2, 4 \rangle$ and $B = \langle -1, 2, 5 \rangle$,   **b.** $A = \mathbf{i} - 2\mathbf{j} + 10\mathbf{k}$ and $B = -\mathbf{i} - \mathbf{k}$,   **c.** $A = \langle 4, 3, 2 \rangle$ and $B = \langle -3, 1, -1 \rangle$.

36.[†] Find the projection of A along the line through B for   **a.** $A = \langle 5, 4, 3 \rangle$ and $B = \langle 1, 1, 2 \rangle$,   **b.** $A = 5\mathbf{i} - 4\mathbf{k}$ and $B = 6\mathbf{j} + \mathbf{k}$,   **c.** $A = \langle 2, 3, 1 \rangle$ and $B = \langle 2, -1, -1 \rangle$.

37.[†] Find a vector whose component in the direction of $\mathbf{i}$ is 1, whose component in the direction of $\mathbf{j} + \mathbf{k}$ is $2\sqrt{2}$, and whose component in the direction of $\mathbf{i} - 3\mathbf{j}$ is $\sqrt{10}$.

38.[†] Compute the work done by the constant force $\mathbf{i} - 3\mathbf{j} + \mathbf{k}$ pounds on an object that traverses the line segment from $(2, 3, 6)$ to $(7, -4, 0)$ in xyz-space with distances measured in feet.

39.[†] **a.** What is the component of the force $\langle 3, -2, -5 \rangle$ dynes in the direction of the vector $\langle 1, 2, 2 \rangle$?   **b.** How much work is done by the constant force $\langle 3, -2, -5 \rangle$ dynes on an object that traverses the line segment from $(4, 4, 4)$ to $(5, 6, 6)$ in xy-space with distances measured in centimeters?

40.[†] What is an object's velocity if its speed is 10 kilometers per hour, the components of its velocity in the directions of $\mathbf{i}$ and $\mathbf{k}$ are 2 and -1, respectively, and the component of its velocity in the $\mathbf{j}$ direction is negative?

12-2   THE CROSS PRODUCT OF TWO VECTORS IN SPACE

FIGURE 12.17   Directions perpendicular to $A = \langle a, b \rangle$ in a plane

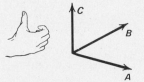

FIGURE 12.18   Directions perpendicular to A and B in space

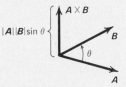

FIGURE 12.19   A vector C with direction determined by the right-hand rule from A toward B

FIGURE 12.20

There are two directions perpendicular to a nonzero vector $A = \langle a, b \rangle$ in an xy-plane. They are the directions of the vectors $\langle b, -a \rangle$ and $\langle -b, a \rangle$ obtained by interchanging the components of $\langle a, b \rangle$ and multiplying one of them by -1 (Figure 12.17). Similarly, there are two directions perpendicular to two nonzero and nonparallel vectors A and B in space (Figure 12.18). In this section we study the CROSS PRODUCT, which can be used to determine these directions.

We distinguish between the two directions perpendicular to A and B by using the RIGHT-HAND RULE: a vector C perpendicular to A and B has the direction given by the right-hand rule from A toward B if, when the fingers of a right hand curl from A toward B, the thumb points in the direction of C (Figure 12.19).

The cross product $A \times B$ is a vector perpendicular to A and B whose direction is determined by the right-hand rule and whose length is determined by the lengths of A and B and the angle between them:

Definition 12.3   If A and B are nonzero, nonparallel vectors in space and Θ is the angle between them ($0 < \Theta < \pi$), then the cross product $A \times B$ is the vector perpendicular to A and B, of length

(1) $|A \times B| = |A|\,|B|\,\sin \Theta$ ◄

and with direction determined by the right-hand rule from A toward B (Figure 12.20). If A and B are parallel or one of them is zero, then $A \times B$ is the zero vector.

EXAMPLE 1   Find $A \times B$ for $A = 5\mathbf{i}$ and $B = \mathbf{i} + \mathbf{j}$.

Solution.   The vectors perpendicular to $5\mathbf{i}$ and to $\mathbf{i} + \mathbf{j}$ are perpendicular to the xy-plane and have either the direction of $\mathbf{j}$ or the direction of $-\mathbf{j}$. The right-hand rule from $5\mathbf{i}$ toward $\mathbf{i} + \mathbf{j}$ picks out the direction of $\mathbf{j}$.

The length of $5\mathbf{i}$ is 5 and the length of $\mathbf{i} + \mathbf{j}$ is $\sqrt{1^2 + 1^2} = \sqrt{2}$. The angle Θ between $5\mathbf{i}$ and $\mathbf{i} + \mathbf{j}$ with $0 < \Theta < \pi$ is $\frac{1}{4}\pi$ and its sine is $1/\sqrt{2}$. Therefore,

$$A \times B = (5\mathbf{i}) \times (\mathbf{i} + \mathbf{j}) = |5\mathbf{i}|\,|\mathbf{i} + \mathbf{j}|\,\sin \Theta\,\mathbf{j}$$

$$= 5(\sqrt{2})(1/\sqrt{2})\mathbf{j} = 5\mathbf{j}. \quad \square$$

We could use Definition 12.3 to find the cross product in Example 1 because we could easily visualize the vectors and find the directions perpendicular to them. We will see later in this section how to compute $A \times B$ algebraically and without geometric considerations for arbitrary A and B. Then we will be able to use the cross product to determine the directions perpendicular to vectors.

The length of the cross product as an area

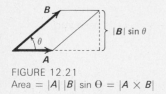

FIGURE 12.21
Area $= |A|\,|B|\sin\Theta = |A \times B|$

The next theorem gives the geometric meaning of the length of $A \times B$.

Theorem 12.1   If A and B are nonzero, nonparallel vectors, then the length (1) of the cross product $A \times B$ is the area of a parallelogram with sides formed by the vectors.

Proof.   If the base of the parallelogram is formed by the vector A, then the length of the base is $|A|$ and the height is $|B|\sin\Theta$ (Figure 12.21), so the area is $|A|\,|B|\sin\Theta = |A \times B|$.   Q.E.D.

Algebra of the cross product

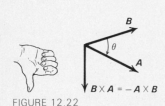

FIGURE 12.22

The procedure for computing $A \times B$ from the components of A and B is based on properties of the cross product given in the next theorem.

Theorem 12.2   The cross product obeys the laws

(2) $B \times A = -A \times B$ ◄

(3) $(aA) \times B = A \times (aB) = a(A \times B)$ ◄

(4) $A \times (B + C) = A \times B + A \times C.$ ◄

Because of the minus sign in (2) we must be careful to preserve the order of the factors when dealing with cross products.

Proof.   Formula (2) follows from the right-hand rule: when the fingers of the right hand are directed from B toward A instead of from A toward B, the thumb points in the opposite direction (Figure 12.22). To verify (3) we have to show that the three vectors $(aA) \times B$, $A \times (aB)$, and $a(A \times B)$ have the same length and direction. Formula (1) for the length of the cross product shows that each of them has length $|a|\,|A|\,|B|\sin\Theta$. We leave the verification that they have the same direction as Exercise 29.

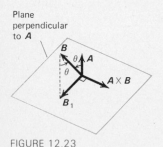

Plane perpendicular to A

FIGURE 12.23

Once (4) has been derived for unit vectors A, we can obtain it for vectors of arbitrary length by using (3). To derive (4) for a unit vector A we will show that $A \times B$ is obtained from B by a projection and a rotation and then use the fact that the projection and rotation procedure transforms a parallelogram into another parallelogram.

We consider the plane perpendicular to A and through the base of A and B (Figure 12.23). The PROJECTION B_1 of B onto the plane is the vector whose base is the base of B and whose tip is at the foot of the perpendicular line from the tip of B to the plane. Its length is $|B|\sin\Theta$, and because A is a unit vector, this is the length of the cross product $A \times B$. We then obtain the vector $A \times B$ by rotating B_1 by $\pi/2$ radians in the plane, as shown in Figure 12.23.

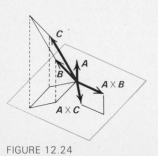

FIGURE 12.24

When we apply this projection and rotation procedure to the parallelogram with sides formed by B and C and diagonal formed by $B + C$ (Figure 12.24), we obtain the parallelogram in the plane with sides formed by $A \times B$ and $A \times C$ and hence with diagonal formed by $A \times B + A \times C$. The diagonal is also formed by $A \times (B + C)$ because it is obtained from the diagonal of the original parallelogram by projection and rotation. Therefore, $A \times (B + C) = A \times B + A \times C$, as stated in (4).   Q.E.D.

Computing cross products

Because the cross product of any vector with itself is the zero vector, we have the formulas

(5) $\mathbf{i} \times \mathbf{i} = O, \quad \mathbf{j} \times \mathbf{j} = O, \quad \mathbf{k} \times \mathbf{k} = O.$

Because the vectors $\mathbf{i}, \mathbf{j}$, and $\mathbf{k}$ are mutually perpendicular unit vectors, oriented as in Figure 12.25, we also have

(6) $\mathbf{i} \times \mathbf{j} = \mathbf{k}, \quad \mathbf{j} \times \mathbf{k} = \mathbf{i}, \quad \mathbf{k} \times \mathbf{i} = \mathbf{j}.$

Formulas (6) and (2) then yield

(7) $\mathbf{j} \times \mathbf{i} = -\mathbf{k}, \quad \mathbf{k} \times \mathbf{j} = -\mathbf{i}, \quad \mathbf{i} \times \mathbf{k} = -\mathbf{j}.$

Formulas (2) through (7) can be used to compute cross products as in the next example.

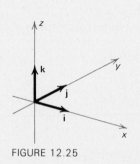

FIGURE 12.25

EXAMPLE 2   Compute $(2\mathbf{i} - \mathbf{j}) \times (3\mathbf{j} + \mathbf{k})$.

Solution.   We have

$$\begin{aligned}
(2\mathbf{i} - \mathbf{j}) \times (3\mathbf{j} + \mathbf{k}) &= (2\mathbf{i}) \times (3\mathbf{j}) + (2\mathbf{i}) \times \mathbf{k} \\
&\quad - \mathbf{j} \times (3\mathbf{j}) - \mathbf{j} \times \mathbf{k} \\
&= 6\mathbf{i} \times \mathbf{j} + 2\mathbf{i} \times \mathbf{k} - 3\mathbf{j} \times \mathbf{j} - \mathbf{j} \times \mathbf{k} \\
&= 6\mathbf{k} - 2\mathbf{j} - \mathbf{i} = -\mathbf{i} - 2\mathbf{j} + 6\mathbf{k}. \quad \square
\end{aligned}$$

(8)

The procedure used in Example 2 is awkward and difficult to carry out with accuracy. It is easier to compute cross products by using the notation of DETERMINANTS, which we now describe.

2 × 2 determinants

A 2×2 (two-by-two) determinant is an expression of the form

(9) $\begin{vmatrix} x_1 \ y_1 \\ x_2 \ y_2 \end{vmatrix}.$

It denotes a number computed from x_1, y_1, x_2, and y_2 according to the following definition.

Definition 12.4   The determinant (9) is the number

(10) $\begin{vmatrix} \mathbf{x_1} \ \mathbf{y_1} \\ \mathbf{x_2} \ \mathbf{y_2} \end{vmatrix} = \mathbf{x_1 y_2} - \mathbf{x_2 y_1}.$ ◀

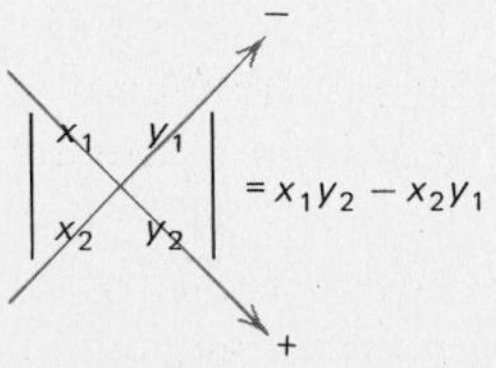

Notice that the determinant (10) is equal to the product of its upper left and lower right entries minus the product of its lower left and upper right entries (Figure 12.26).

FIGURE 12.26

EXAMPLE 3   Evaluate

$$\begin{vmatrix} 5 \ -7 \\ -4 \ 2 \end{vmatrix}.$$

Solution.

$$\begin{vmatrix} 5 & -7 \\ -4 & 2 \end{vmatrix} = 5(2) - (-4)(-7) = 10 - 28 = -18. \quad \square$$

3 × 3 determinants

A 3 × 3 determinant is an expression of the form

$$(11) \qquad \begin{vmatrix} x_1 & y_1 & z_1 \\ x_2 & y_2 & z_2 \\ x_3 & y_3 & z_3 \end{vmatrix}$$

It is defined as follows.

Definition 12.5   The determinant (11) is a number computed from 2 × 2 determinants by the formula

$$(12) \qquad \begin{vmatrix} x_1 & y_1 & z_1 \\ x_2 & y_2 & z_2 \\ x_3 & y_3 & z_3 \end{vmatrix} = x_1 \begin{vmatrix} y_2 & z_2 \\ y_3 & z_3 \end{vmatrix} - y_1 \begin{vmatrix} x_2 & z_2 \\ x_3 & z_3 \end{vmatrix} + z_1 \begin{vmatrix} x_2 & y_2 \\ x_3 & y_3 \end{vmatrix}. \qquad \blacktriangleleft$$

Each of the determinants on the right side of equation (12) is obtained by crossing out the row and column of one of the numbers in the first row of the 3 × 3 determinant. The right side consists of the first number in the first row multiplied by the corresponding 2 × 2 determinant, minus the second number in the first row multiplied by the corresponding 2 × 2 determinant, plus the third number in the first row multiplied by the corresponding 2 × 2 determinant (Figure 12.27).

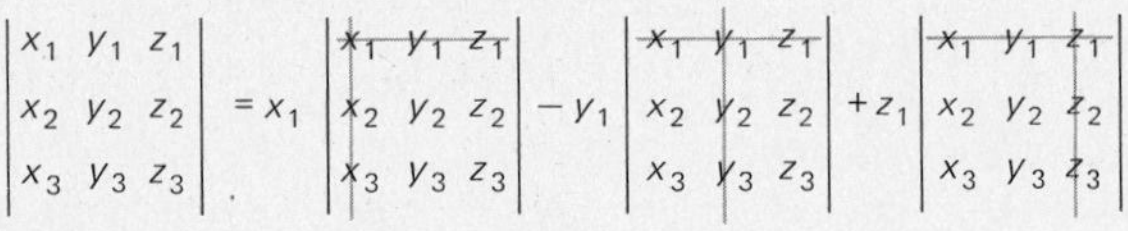

FIGURE 12.27

EXAMPLE 4   Evaluate

$$\begin{vmatrix} 3 & 2 & 4 \\ -1 & 0 & 6 \\ 5 & 1 & -2 \end{vmatrix}.$$

Solution.   By rule (12) the determinant equals

$$3\begin{vmatrix} 0 & 6 \\ 1 & -2 \end{vmatrix} - 2\begin{vmatrix} -1 & 6 \\ 5 & -2 \end{vmatrix} + 4\begin{vmatrix} -1 & 0 \\ 5 & 1 \end{vmatrix}$$
$$= 3[0(-2) - 6(1)] - 2[(-1)(-2) - 6(5)]$$
$$\qquad + 4[(-1)(1) - 0(5)]$$
$$= 3(-6) - 2(-28) + 4(-1) = -18 + 56 - 4 = 34. \quad \square$$

Computing cross products by determinants

Theorem 12.3   The cross product of vectors
$A = x_1\mathbf{i} + y_1\mathbf{j} + z_1\mathbf{k}$ and $B = x_2\mathbf{i} + y_2\mathbf{j} + z_2\mathbf{k}$ is

$$(13) \qquad A \times B = \begin{vmatrix} \mathbf{i} & \mathbf{j} & \mathbf{k} \\ x_1 & y_1 & z_1 \\ x_2 & y_2 & z_2 \end{vmatrix}$$

**where the top row of the determinant is formed by the
vectors i, j, and k, the second row by the components of A,
and the third row by the components of B.**

Proof.   We apply Definition 12.5 as if $\mathbf{i}, \mathbf{j}$, and $\mathbf{k}$ were numbers. We
must show that the cross product is equal to

$$(14) \quad \mathbf{i}\begin{vmatrix} y_1 & z_1 \\ y_2 & z_2 \end{vmatrix} - \mathbf{j}\begin{vmatrix} x_1 & z_1 \\ x_2 & z_2 \end{vmatrix} + \mathbf{k}\begin{vmatrix} x_1 & y_1 \\ x_2 & y_2 \end{vmatrix}$$
$$= (y_1 z_2 - z_1 y_2)\mathbf{i} - (x_1 z_2 - z_1 x_2)\mathbf{j} + (x_1 y_2 - y_1 x_2)\mathbf{k}.$$

This follows from formulas (2) through (7) and Definition 12.4 of 2×2
determinants (see Exercise 27).   Q.E.D.

EXAMPLE 5   Compute the cross product $\langle 3, 1, -2 \rangle \times \langle 0, 4, 2 \rangle$.

Solution.   Formula (13) gives

$$\langle 3, 1, -2 \rangle \times \langle 0, 4, 2 \rangle = \begin{vmatrix} \mathbf{i} & \mathbf{j} & \mathbf{k} \\ 3 & 1 & -2 \\ 0 & 4 & 2 \end{vmatrix}$$
$$= \mathbf{i}\begin{vmatrix} 1 & -2 \\ 4 & 2 \end{vmatrix} - \mathbf{j}\begin{vmatrix} 3 & -2 \\ 0 & 2 \end{vmatrix} + \mathbf{k}\begin{vmatrix} 3 & 1 \\ 0 & 4 \end{vmatrix}$$
$$= [1(2) - (-2)(4)]\mathbf{i} - [3(2) - (-2)(0)]\mathbf{j}$$
$$\quad + [3(4) - 1(0)]\mathbf{k}$$
$$= 10\mathbf{i} - 6\mathbf{j} + 12\mathbf{k} = \langle 10, -6, 12 \rangle. \quad \square$$

Note.   As a partial check on the calculations of Example 5, we can
compute the dot products $\langle 3, 1, -2 \rangle \cdot \langle 10, -6, 12 \rangle$ and
$\langle 0, 4, 2 \rangle \cdot \langle 10, -6, 12 \rangle$ between the given vectors and their computed
cross product. These dot products are zero, as they should be, since the
cross product is perpendicular to each of the original vectors. $\square$

EXAMPLE 6   Compute the area of the parallelogram whose sides are
formed by the vectors of Example 5.

Solution.   The area of the parallelogram is the length of the cross
product or $|\langle 10, -6, 12 \rangle| = \sqrt{(10)^2 + (-6)^2 + (12)^2} = \sqrt{280}$
$= 2\sqrt{70}$. $\square$

EXAMPLE 7   Give the two unit vectors that are perpendicular to the
vectors of Example 5.

Solution.   The two vectors are the cross product of Example 5 divided
by its length and the negative of that unit vector:

$$\pm \frac{\langle 10, -6, 12 \rangle}{|\langle 10, -6, 12 \rangle|} = \frac{\pm 1}{2\sqrt{70}}\langle 10, -6, 12 \rangle = \pm \frac{1}{\sqrt{70}}\langle 5, -3, 6 \rangle. \quad \square$$

In Example 7 we used the direction of a cross product. In the next
example, as in Example 6, we use its magnitude.

EXAMPLE 8   What is the area of the triangle with vertices $P(1, 0, 3)$, $Q(2, 1, 2)$, and $R(0, 3, 1)$?

Solution.   The area of the triangle is half the area $|\overrightarrow{PQ} \times \overrightarrow{PR}|$ of a parallelogram with P, Q, and R as three of its corners. The displacement vector $\overrightarrow{PQ}$ is $\langle 2 - 1, 1 - 0, 2 - 3 \rangle = \langle 1, 1, -1 \rangle$ and the vector $\overrightarrow{PR}$ is $\langle 0 - 1, 3 - 0, 1 - 3 \rangle = \langle -1, 3, -2 \rangle$. The cross product is

$$\overrightarrow{PQ} \times \overrightarrow{PR} = \begin{vmatrix} \mathbf{i} & \mathbf{j} & \mathbf{k} \\ 1 & 1 & -1 \\ -1 & 3 & -2 \end{vmatrix} = \mathbf{i} \begin{vmatrix} 1 & -1 \\ 3 & -2 \end{vmatrix} - \mathbf{j} \begin{vmatrix} 1 & -1 \\ -1 & -2 \end{vmatrix} + \mathbf{k} \begin{vmatrix} 1 & 1 \\ -1 & 3 \end{vmatrix}$$

$$= (-2 + 3)\mathbf{i} - (-2 - 1)\mathbf{j} + (3 + 1)\mathbf{k}$$
$$= \mathbf{i} + 3\mathbf{j} + 4\mathbf{k}.$$

Therefore, the area of the triangle is $\frac{1}{2}|\overrightarrow{PQ} \times \overrightarrow{PR}| = \frac{1}{2}\sqrt{1^2 + 3^2 + 4^2} = \frac{1}{2}\sqrt{26}$.   □

The triple product $A \cdot (B \times C)$

The number $A \cdot (B \times C)$ is called a SCALAR TRIPLE PRODUCT of the vectors A, B, and C. If the vectors have the components $A = \langle x_1, y_1, z_1 \rangle$, $B = \langle x_2, y_2, z_2 \rangle$, and $C = \langle x_3, y_3, z_3 \rangle$, then

$$B \times C = \begin{vmatrix} \mathbf{i} & \mathbf{j} & \mathbf{k} \\ x_2 & y_2 & z_2 \\ x_3 & y_3 & z_3 \end{vmatrix}$$

and, because we can obtain $A \cdot (B \times C)$ by replacing $\mathbf{i}$, $\mathbf{j}$, and $\mathbf{k}$ by x_1, x_2, and x_3 in an expression for $B \times C$, the triple product is given by the determinant

$$(15) \quad A \cdot (B \times C) = \begin{vmatrix} x_1 & y_1 & z_1 \\ x_2 & y_2 & z_2 \\ x_3 & y_3 & z_3 \end{vmatrix} \qquad \blacktriangleleft$$

whose rows are the components of the three vectors in the order they appear on the left.

In the next theorem we express the triple product in terms of the volume of the parallelopiped with sides formed by the vectors A, B, and C (Figure 12.28).

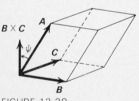

FIGURE 12.28
Volume $= |A \cdot (B \times C)|$

Theorem 12.4   **If the angle ψ between A and $B \times C$ is acute $(0 \le \psi \le \pi/2)$, then $A \cdot (B \times C)$ is the volume of the parallelopiped with sides formed by A, B, and C. If ψ is obtuse $(\pi/2 \le \psi \le \pi)$, then $A \cdot (B \times C)$ is the negative of that volume.**

Theorem 12.4 shows that in any case the parallelopiped has volume $|A \cdot (B \times C)|$.

Proof.   If ψ is acute as in Figure 12.28, then the height of the parallelopiped is $|A| \cos \psi$ and the area of its base is $|B \times C|$, so its volume is $|A||B \times C| \cos \psi = A \cdot (B \times C)$ by equation (9) of Section 12.1. If ψ is obtuse, the volume is $-A \cdot (B \times C)$ because the height is $-|A| \cos \psi$.   Q.E.D.

Exercises

In Exercises 1 through 3 evaluate the determinants.

1.† $\begin{vmatrix} 1 & 3 & 0 \\ 2 & -1 & 1 \\ 0 & 4 & 5 \end{vmatrix}$ **2.**† $\begin{vmatrix} 5 & 0 & 1 \\ 0 & 4 & -2 \\ 1 & 2 & 3 \end{vmatrix}$ **3.** $\begin{vmatrix} 5 & 0 & 0 \\ 0 & 6 & 4 \\ -3 & 2 & 8 \end{vmatrix}$

In Exercises 4 through 7 compute the cross products.

4.† $\langle 1, 4, -6 \rangle \times \langle 3, 2, -4 \rangle$ **5.** $\langle 3, 1, -2 \rangle \times \langle 1, 4, 0 \rangle$

6.† $\langle 2, -1, 6 \rangle \times \langle 1, 3, -2 \rangle$ **7.** $\langle -1, 0, 7 \rangle \times \langle 6, 4, 1 \rangle$

8.† Compute the area of the parallelogram $PQRS$ with vertices $P(1, 4, 0)$, $Q(3, 2, 1)$, and $R(5, 2, 1)$.

9. Compute the area of the triangle with vertices $P(5, 3, 0)$, $Q(6, 0, 2)$, and $R(4, 1, 1)$.

10.† Show that the area of a triangle in an xy-plane with vertices (x_1, y_1), (x_2, y_2), and (x_3, y_3) is $\frac{1}{2}|D|$ where D is the determinant

$$D = \begin{vmatrix} x_1 & y_1 & 1 \\ x_2 & y_2 & 1 \\ x_3 & y_3 & 1 \end{vmatrix}.$$

11.† Use the identity $\sin^2\Theta + \cos^2\Theta = 1$ to derive the identity $|A \times B|^2 = |A|^2|B|^2 - (A \cdot B)^2$.

In Exercises 12 through 15 find the two unit vectors that are perpendicular to the given pair of vectors.

12.† $\langle 4, 3, 1 \rangle$ and $\langle 5, 7, 2 \rangle$ **13.** $\langle 5, 1, -2 \rangle$ and $\langle 0, 4, -1 \rangle$

14.† $\langle 1, 0, 1 \rangle$ and $\langle 2, 3, 3 \rangle$ **15.** $\langle 0, 5, 6 \rangle$ and $\langle 0, 6, 5 \rangle$

In each of Exercises 16 through 19 one vertex of a parallelopiped is P and the three adjacent vertices are Q, R, and S. Find the coordinates of the vertex opposite P.

16.‡ $P(1, 2, 3), Q(2, 2, 5), R(5, 3, 3), S(4, 3, 3)$

17.† $P(1, 1, 1), Q(0, 0, -6), R(8, 1, 2), S(-4, 3, 1)$

18. $P(-1, 2, -3), Q(-1, -1, -3), R(5, 3, -3), S(4, 2, 0)$

19. $P(10, 8, -5), Q(11, 9, -4), R(11, 10, -4), S(12, 10, -6)$

20. Compute the volumes of the parallelopipeds **a.**‡ of Exercise 16, **b.**† of Exercise 17, **c.** of Exercise 18, and **d.** of Exercise 19.

21.† Show that the volume of the tetrahedron with sides formed by the vectors A, B, and C is $\frac{1}{6}|(A \times B) \cdot C|$.

22. Compute the volume of the tetrahedrons with vertices **a.**† $P(1, 2, 3)$, $Q(3, 3, 6), R(4, 2, -1), S(6, 3, 5)$ and **b.** $P(0, 0, 0), Q(1, 3, 4)$, $R(2, 1, 5), S(3, 4, -2)$.

23.† Show that the points P, Q, R, and S lie in the same plane if and only if the triple product $\overrightarrow{PQ} \cdot (\overrightarrow{PR} \times \overrightarrow{PS})$ is zero.

24.† Are $(1, 2, -1)$, $(3, 3, -4)$, $(2, 2, 1)$, and $(5, 3, 0)$ in the same plane?

25. Are $(3, 2, 1)$, $(2, 4, 6)$, $(4, 0, 1)$ and $(5, -2, 7)$ in the same plane?

26.[†] Use Theorem 12.4 to show that $(A \times B) \cdot C = A \cdot (B \times C)$.

27.[†] Use formulas (2) through (7) to show that
$(x_1\mathbf{i} + y_1\mathbf{j} + z_1\mathbf{k}) \times (x_2\mathbf{i} + y_2\mathbf{j} + z_2\mathbf{k})$ is equal to
$(y_1z_2 - z_1y_2)\mathbf{i} - (x_1z_2 - z_1x_2)\mathbf{j} + (x_1y_2 - y_1x_2)\mathbf{k}$.

28. Use formula (12) to show that interchanging the rows and columns of a
3×3 determinant does not change its value: show that

$$\begin{vmatrix} x_1 & x_2 & x_3 \\ y_1 & y_2 & y_3 \\ z_1 & z_2 & z_3 \end{vmatrix} = \begin{vmatrix} x_1 & y_1 & z_1 \\ x_2 & y_2 & z_2 \\ x_3 & y_3 & z_3 \end{vmatrix}.$$

29. Show that $(aA) \times B$, $A \times (aB)$, and $a(A \times B)$ have the same direction
for any a, A, and B.

Additional drill exercises

Evaluate the determinants in Exercises 30 through 32.

30.[†] $\begin{vmatrix} 1 & 1 & 2 \\ 2 & 1 & 0 \\ 1 & 3 & 1 \end{vmatrix}$ **31.**[†] $\begin{vmatrix} -1 & 1 & 0 \\ 0 & 2 & 1 \\ 3 & 3 & 3 \end{vmatrix}$ **32.**[†] $\begin{vmatrix} 1 & 2 & 3 \\ 4 & 5 & 6 \\ 7 & 8 & 9 \end{vmatrix}$

Compute the cross products in Exercises 33 through 36.

33.[†] $\langle 1, 1, 1 \rangle \times \langle 2, 1, 2 \rangle$ **34.**[†] $\langle 1, 0, 5 \rangle \times \langle 5, 2, 3 \rangle$

35.[†] $(\mathbf{i} + \mathbf{j} - \mathbf{k}) \times (6\mathbf{i} - 7\mathbf{k})$ **36.**[†] $(-\mathbf{i} + \mathbf{j} - 3\mathbf{k})(2\mathbf{i} + 2\mathbf{j} - 5\mathbf{k})$

In each of Exercises 37 through 40 compute the area of the triangle PQR.

37.[†] $P(1, 2, 4)$, $Q(5, 2, 1)$, $R(0, 4, 4)$ **38.**[†] $P(1, 2, 3)$, $Q(3, 4, 6)$, $R(0, 2, 1)$

39.[†] $P(0, 0, 0)$, $Q(1, 3, -1)$, $R(1, 0, 7)$ **40.**[†] $P(1, 0, 0)$, $Q(0, 1, 0)$, $R(0, 0, 1)$.

41.[†] Compute the volume of the tetrahedron with vertices $(1, 2, 3)$, $(2, 1, 4)$,
$(6, 0, 2)$, and $(4, 4, 4)$.

42.[†] Compute the volume of the tetrahedron with vertices $(0, 1, 0)$, $(1, 0, 1)$,
$(0, 0, 1)$, and $(1, 1, 1)$.

12-3 EQUATIONS OF LINES AND PLANES

In this section we use vectors to study lines and planes in
three-dimensional space.

Parametric equations of lines

We can describe a line in a coordinate plane by giving a point on it and
its slope. The most convenient way to describe a line in space is to give a
point on it and a nonzero vector parallel to it. From this information we
obtain PARAMETRIC EQUATIONS of the line, as described in the next
theorem.

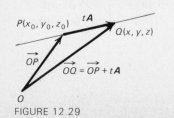

FIGURE 12.29

**Theorem 12.5   The line through the point $P(x_0, y_0, z_0)$ and
parallel to the nonzero vector $A = \langle a, b, c \rangle$ (Figure 12.29) has the
parametric equations**

(1) $x = x_0 + at,\quad y = y_0 + bt,\quad z = z_0 + ct.$ ◀

Proof.   The position vector of P is $\overrightarrow{OP} = \langle x_0, y_0, z_0 \rangle$. If $Q(x, y, z)$ is another point on the line, then the displacement vector $\overrightarrow{PQ}$ is parallel to A and is of the form tA for some number t (Figure 12.29). Therefore the position vector of Q is

$$\langle x, y, z \rangle = \overrightarrow{OQ} = \overrightarrow{OP} + \overrightarrow{PQ} = \overrightarrow{OP} + tA$$
$$= \langle x_0, y_0, z_0 \rangle + t\langle a, b, c \rangle$$
$$= \langle x_0 + at, y_0 + bt, z_0 + ct \rangle.$$

This gives equations (1) for the coordinates of Q.   Q.E.D.

Note.   If we let $\mathbf{R}_0 = \langle x_0, y_0, z_0 \rangle$ denote the position vector of $P(x_0, y_0, z_0)$ and $\mathbf{R} = \langle x, y, z \rangle$ the position vector of the arbitrary point $Q(x, y, z)$ on the line, then we can write equations (1) in the vector form

$$\mathbf{R} = \mathbf{R}_0 + t\mathbf{A}. \quad \square$$

EXAMPLE 1   Give parametric equations for the line through the point $(6, 4, 3)$ and parallel to the vector $\langle 2, 0, -7 \rangle$.

Solution.   By formula (1) with $(x_0, y_0, z_0) = (6, 4, 3)$ and $\langle a, b, c \rangle = \langle 2, 0, -7 \rangle$, we obtain the equations

$$x = 6 + 2t,\quad y = 4 + 0t = 4,\quad z = 3 - 7t. \quad \square$$

EXAMPLE 2   Give parametric equations for the line through the points $P = (5, 3, 1)$ and $Q = (7, -2, 0)$.

Solution.   The vector $\overrightarrow{PQ} = \langle 7 - 5, -2 - 3, 0 - 1 \rangle = \langle 2, -5, -1 \rangle$ is parallel to the line. Using this vector and the point P in formula (1) gives the equations

$$x = 5 + 2t,\quad y = 3 - 5t,\quad z = 1 - t$$

for the line. (As a check, we can see that the equations give the point P at $t = 0$ and the point Q at $t = 1$.) $\square$

Equations of planes   We can describe a plane in space by giving a point on it and a nonzero vector perpendicular to it (Figure 12.30). The vector is called a NORMAL VECTOR to the plane. The next theorem shows how to obtain an equation for the plane from this information.

Theorem 12.6   The plane through the point $P(x_0, y_0, z_0)$ and with the nonzero normal vector $n = \langle a, b, c \rangle$ has the equation

FIGURE 12.30   The plane through P and with normal vector n

(2) $a(x - x_0) + b(y - y_0) + c(z - z_0) = 0.$ ◀

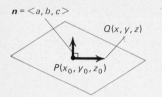

FIGURE 12.31 Q is in the plane if $\vec{PQ}$ is perpendicular to $\boldsymbol{n}$.

Proof.   A point $Q(x, y, z)$ is on the plane if and only if the displacement vector $\vec{PQ} = \langle x - x_0, y - y_0, z - z_0 \rangle$ is perpendicular to $\boldsymbol{n}$ (Figure 12.31). This is the case if and only if the dot product

$$\boldsymbol{n} \cdot \vec{PQ} = \langle a, b, c \rangle \cdot \langle x - x_0, y - y_0, z - z_0 \rangle$$
$$= a(x - x_0) + b(y - y_0) + c(z - z_0)$$

is zero, so we obtain equation (2).   Q.E.D.

Note.   If we let $\boldsymbol{R}_0 = \langle x_0, y_0, z_0 \rangle$ and $\boldsymbol{R} = \langle x, y, z \rangle$ denote the position vectors of the point $P(x_0, y_0, z_0)$ and of the arbitrary point $Q(x, y, z)$ on the plane, then we can write equation (2) in the vector form

$$\boldsymbol{n} \cdot (\boldsymbol{R} - \boldsymbol{R}_0) = 0. \quad \square$$

EXAMPLE 3   Give an equation for the plane through the point $(2, 3, 4)$ and perpendicular to the vector $\langle -6, 5, -4 \rangle$.

Solution.   By formula (2) with $(x_0, y_0, z_0) = (2, 3, 4)$ and $\langle a, b, c \rangle = \langle -6, 5, -4 \rangle$, we obtain the equation
$-6(x - 2) + 5(y - 3) - 4(z - 4) = 0. \quad \square$

We cannot describe a plane by giving a point on it and one vector parallel to it because there are an infinite number of such planes (Figure 12.32). We have to give the point and *two nonparallel* vectors that are parallel to the plane (Figure 12.33). We illustrate in the next example how an equation of the plane can be obtained from this information by using a cross product.

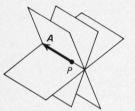

FIGURE 12.32   Planes through P and parallel to A

EXAMPLE 4   Give an equation for the plane through the point $(4, 5, 1)$ and parallel to the vectors $\boldsymbol{A} = \langle -2, 0, 1 \rangle$ and $\boldsymbol{B} = \langle 0, 1, -4 \rangle$.

Solution.   The cross product $\boldsymbol{A} \times \boldsymbol{B}$ is perpendicular to the plane, and we have

FIGURE 12.33   The plane through P and parallel to A and B

$$\boldsymbol{A} \times \boldsymbol{B} = \begin{vmatrix} \mathbf{i} & \mathbf{j} & \mathbf{k} \\ -2 & 0 & 1 \\ 0 & 1 & -4 \end{vmatrix}$$

$$= \mathbf{i} \begin{vmatrix} 0 & 1 \\ 1 & -4 \end{vmatrix} - \mathbf{j} \begin{vmatrix} -2 & 1 \\ 0 & -4 \end{vmatrix} + \mathbf{k} \begin{vmatrix} -2 & 0 \\ 0 & 1 \end{vmatrix}$$

$$= [0(-4) - 1(1)]\mathbf{i} - [(-2)(-4) - 1(0)]\mathbf{j}$$
$$+ [(-2)(1) - 0(0)]\mathbf{k}$$
$$= -\mathbf{i} - 8\mathbf{j} - 2\mathbf{k}.$$

With this normal vector and the point $(4, 5, 1)$ on the plane, formula (2) yields the equation $-(x - 4) - 8(y - 5) - 2(z - 1) = 0$ or $(x - 4) + 8(y - 5) + 2(z - 1) = 0$ for the plane. $\square$

Equation (2) for a plane with normal vector $\langle a, b, c \rangle$ can be simplified to the form $ax + by + cz + d = 0$. Therefore, if we are given an equation for a plane in this form, we can obtain a normal vector by taking as its components $\langle a, b, c \rangle$ the coefficients of x, y, and z in the equation. This fact is used in the next example.

EXAMPLE 5   Give parametric equations for the line through the point $(5, -3, 2)$ and perpendicular to the plane $6x + 2y - 7z = 5$.

Solution.   The vector $\langle 6, 2, -7 \rangle$ is perpendicular to the plane and parallel to the line. By Theorem 12.5, the line has the equations $x = 5 + 6t, y = -3 + 2t, z = 2 - 7t.$ $\square$

The distance between parallel planes

If P is a point on one plane and Q a point on a parallel plane, then the vector $\overrightarrow{PQ}$ runs from one plane to the other. Therefore, if $\boldsymbol{n}$ is a nonzero normal vector to the planes, the absolute value of the component of $\overrightarrow{PQ}$ in the direction of $\boldsymbol{n}$

(3) $\dfrac{|\overrightarrow{PQ} \cdot \boldsymbol{n}|}{|\boldsymbol{n}|}$ ◀

is the distance between the planes (Figure 12.34).

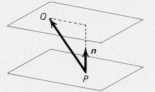

FIGURE 12.34   Parallel planes a distance $\dfrac{|\overrightarrow{PQ} \cdot n|}{|n|}$ apart

EXAMPLE 6   Compute the distance between the parallel planes $-x - 2y + 3z = 3$ and $-x - 2y + 3z = 10$.

Solution.   The vector $\boldsymbol{n} = \langle -1, -2, 3 \rangle$ is perpendicular to the planes. We can set $y = 0$ and $z = 0$ in the equations and solve for x to find points $P(-3, 0, 0)$ on the plane $-x - 2y + 3z = 3$ and $Q(-10, 0, 0)$ on the plane $-x - 2y + 3z = 10$. We then have $\overrightarrow{PQ} = \langle -7, 0, 0 \rangle$ and formula (3) gives

$$\frac{|\overrightarrow{PQ} \cdot \boldsymbol{n}|}{|\boldsymbol{n}|} = \frac{|\langle -7, 0, 0 \rangle \cdot \langle -1, 2, 3 \rangle|}{|\langle -1, 2, 3 \rangle|} = \frac{7}{\sqrt{14}}$$

as the distance between the planes. $\square$

Skew lines

Two lines in space which are not parallel and do not intersect are called SKEW LINES (Figure 12.35). Any two skew lines lie in parallel planes (Figure 12.36). (The plane containing line L_1 in Figure 12.36, for example, is determined by L_1 and a line intersecting L_1 and parallel to L_2.) The distance between the lines is the distance between the planes.

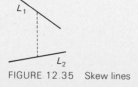

FIGURE 12.35   Skew lines

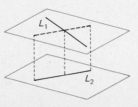

FIGURE 12.36   Skew lines lie in parallel planes.

EXAMPLE 7 Compute the distance between the skew lines $x = 7t$, $y = 2 + t, z = 4 - 3t$ and $x = 3 - t, y = 5, z = 6 + 2t$.

Solution. The vector $\mathbf{A} = \langle 7, 1, -3 \rangle$ is parallel to the first line and the vector $\mathbf{B} = \langle -1, 0, 2 \rangle$ to the second. Each of these vectors is parallel to the planes containing the lines, so the cross product

$$\mathbf{n} = \mathbf{A} \times \mathbf{B} = \begin{vmatrix} \mathbf{i} & \mathbf{j} & \mathbf{k} \\ 7 & 1 & -3 \\ -1 & 0 & 2 \end{vmatrix} = \mathbf{i} \begin{vmatrix} 1 & -3 \\ 0 & 2 \end{vmatrix} - \mathbf{j} \begin{vmatrix} 7 & -3 \\ -1 & 2 \end{vmatrix} + \mathbf{k} \begin{vmatrix} 7 & 1 \\ -1 & 0 \end{vmatrix}$$

$$= [1(2) - (-3)(0)]\mathbf{i} - [7(2) - (-3)(-1)]\mathbf{j} + [7(0) - 1(-1)]\mathbf{k} = 2\mathbf{i} - 11\mathbf{j} + \mathbf{k}$$

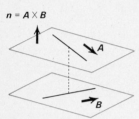

$\mathbf{n} = \mathbf{A} \times \mathbf{B}$

FIGURE 12.37

is perpendicular to the two planes (Figure 12.37).

Setting $t = 0$ in the equations for the lines shows that $P(0, 2, 4)$ is on the first line and $Q(3, 5, 6)$ on the second line. By formula (3), the distance between the planes, and hence the lines, is

$$\frac{|\overrightarrow{PQ} \cdot \mathbf{n}|}{|\mathbf{n}|} = \frac{|\langle 3, 3, 2 \rangle \cdot \langle 2, -11, 1 \rangle|}{|\langle 2, -11, 1 \rangle|} = \frac{|-25|}{\sqrt{126}} = \frac{25}{\sqrt{126}}. \quad \square$$

Intercept equations of planes

The x-, y-, and z-*intercepts* of a plane are the coordinates where it intersects the x-, y-, and z-axes. If any of the intercepts are zero, the plane passes through the origin and has an equation of the form $Ax + By + Cz = 0$. If the plane does not go through the origin and intersects all three axes, then it has the equation

$$(4) \qquad \frac{x}{a} + \frac{y}{b} + \frac{z}{c} = 1 \qquad \blacktriangleleft$$

where a is its x-intercept, b its y-intercept, and c its z-intercept. One way to sketch this plane is to draw the triangle with vertices at the intercepts (Figure 12.38).

If the plane intersects two of the axes but is parallel to the third, then it has an equation in one of the forms

$$(5) \qquad \frac{x}{a} + \frac{y}{b} = 1, \quad \frac{x}{a} + \frac{z}{c} = 1, \quad \frac{y}{b} + \frac{z}{c} = 1. \qquad \blacktriangleleft$$

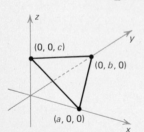

FIGURE 12.38   The plane $\dfrac{x}{a} + \dfrac{y}{b} + \dfrac{z}{c} = 1$

If the plane intersects only one of the axes, then it is perpendicular to that axis and has an equation in one of the forms

$$(6) \qquad x = a, \quad y = b, \quad z = c. \qquad \blacktriangleleft$$

The plane $x = a$ is parallel to the yz-plane; the plane $y = b$ is parallel to the xz-plane; and the plane $z = c$ is parallel to the xy-plane. In particular, the plane $x = 0$ is the yz-plane; the plane $y = 0$ is the xz-plane; and the plane $z = 0$ is the xy-plane (Figure 12.39).

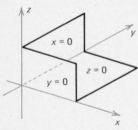

FIGURE 12.39

Exercises

†Answer provided.
‡Outline of solution provided.

1.† Which of the following statements are true and which are false for lines and planes in space?   **a.**   Two lines parallel to a third line are parallel.   **b.**   Two lines perpendicular to a third line are parallel.   **c.**   Two planes parallel to a third plane are parallel.   **d.**   Two planes perpendicular to a third plane are parallel.   **e.**   Two lines parallel to a plane are parallel.   **f.**   Two lines perpendicular to a plane are parallel.   **g.**   Two planes parallel to a line are parallel.   **h.**   Two planes perpendicular to a line are parallel.   **i.**   Two lines are either parallel or intersect.   **j.**   Two planes are either parallel or intersect.   **k.**   A line and a plane are either parallel or intersect.

In each of Exercises 2 through 14 give parametric equations of the lines.

2.† The line through $(2, 4, 3)$ and parallel to the vector $\langle 4, 0, -7 \rangle$

3. The line through $(4, 5, 0)$ and parallel to the y-axis

4.† The line through $(-3, -2, 0)$ and parallel to the line $x = 2, y = 3t - 4, z = 5t + 10$

5. The line through the origin and parallel to the line $x = t + 5, y = 2t - 3, z = 3t + 1$

6.‡ The line through $(3, -4, 5)$ and perpendicular to the plane $6x - 4y = 10$

7. The line through the origin and perpendicular to the plane $3x - 4y + 5z - 18 = 0$

8.† The line through $(6, 2, 3)$ and $(7, 0, -10)$

9. The line through $(0, 5, -4)$ and $(2, 0, 3)$

10.‡ The line through $(6, 0, -3)$ and parallel to the planes $2x - 4y = 7$ and $3y + 5z = 0$

11. The line through $(3, 2, 0)$ and parallel to the planes $x - 2y + 3z = 0$ and $4x - 5y - z = 0$

12.‡ The line of intersection of the planes $x + y + z = 1$ and $x - 2y - z = 1$

13.† The line of intersection of the planes $2x - y = 3$ and $3y - 4z = 10$

14. The line of intersection of the planes $x + y - z = 0$ and $x - y + z = 0$

In Exercises 15 through 28 give equations of the planes.

15.† The plane through $(3, 0, 8)$ and perpendicular to the vector $\langle 4, 0, 5 \rangle$

16. The plane through $(0, 2, -5)$ and perpendicular to the vector $\langle -1, 2, -3 \rangle$

17.† The plane through the origin and parallel to the plane $3x - y + z = 1000$

18.‡ The plane through the line $x = 2t - 3, y = 4t - 2, z = 6$ and parallel to the plane $2x - y + z = 0$

19. The plane containing the line $x = 3t - 5, y = 2t - 4, z = t - 3$ and parallel to the plane $y - 2z = 15$

20.‡ The plane through the point $(0, 0, 0)$ and containing the intersection of the planes $x - y = 4$ and $3x - 2y + z = 5$

21.   The plane containing the line $x = 2t$, $y = 3t$, $z = 4t$ and the intersection of the planes $x + y + z = 0$ and $2y - z = 0$

22.[†]   The plane through $(5, -1, -2)$ and perpendicular to the planes $y - z = 4$ and $x + z = 3$

23.   The plane through $(2, 0, 0)$ and perpendicular to the planes $z = 4$ and $x + y + z = 0$

24.[‡]   The plane through the points $(1, -1, 2)$, $(4, -1, 5)$, and $(1, 2, -1)$

25.   The plane through the points $(2, 1, 3)$, $(3, 2, 4)$, and $(2, 6, 3)$

26.[†]   The plane with x-intercept $= 2$, y-intercept $= -3$, and z-intercept $= 5$

27.[†]   The plane parallel to the x-axis, with y-intercept $= 5$, and with z-intercept $= -2$

28.   The plane parallel to the yz-plane and with x-intercept $= -10$

In each of Exercises 29 through 34 sketch the plane by plotting its intercepts and drawing a portion of it.

29.[†] $2x - 3y + z = 6$ **30.** $6x - 5y + 3z = 30$

31.[†] $2x - 5z = 10$ **32.** $x - 5y = 15$

33.[†] $x = 3$ **34.** $z = -2$

35.[†]   Give equations for the four planes that form the sides of the tetrahedron with vertices $(0, 0, 0)$, $(3, 0, 0)$, $(0, -4, 0)$, and $(0, 0, 5)$.

36.[†]   Find the distance between the planes $x - 3y = 10$ and $x - 3y = -6$.

37.   Find the distance from the point $(3, 6, 2)$ to the plane $2x - 3y + z = 4$.

38.[†]   Find the distance between the lines $x = 2 - t$, $y = 3 + 4t$, $z = 2t$ and $x = -1 + t$, $y = 2$, $z = -1 + 2t$.

39.   Find the distance between the lines $x = 5$, $y = 3t$, $z = 4t$ and $x = t - 2$, $y = 0$, $z = 3t + 2$.

40.   If the line $x = t$, $y = 2t + 1$, $z = 3t - 3$, and the plane $3x + 2y - z = 4$ are parallel, find the distance between them. If they intersect, find their point of intersection.

41.   Show with sketches how three distinct planes can divide xyz-space into 4, 6, 7, or 8 regions.

42.[†]   One airplane is at $x = 300t$, $y = 1670t + 10t^2$, $z = 500 + 60t$ (kilometers) at time t (hours) when another airplane is at $x = 100 + 100t$, $y = -80t + 1840$, $z = 280t$. Do the planes collide? At what point do their paths intersect?

43.[†]   Give an equation for the line through the origin and the intersection of lines $x = 2t + 3$, $y = -4t$, $z = t - 3$ and $x = 10t + 3$, $y = 5t - 25$, and $z = 4 - 2t$.

12-4   CURVES IN SPACE

The discussion of arclength, speed, and velocity and acceleration vectors that was given in Chapter 11 generalizes with little change to curves in xyz-space. We summarize these results in this section.

A curve in space can be given by parametric equations curves in xyz-space. We summarize these results in this section. We assume that all functions considered have continuous derivatives.

A curve in space can be given by parametric equations

(1) $C: x = x(t), y = y(t), z = z(t), \qquad a \leq t \leq b.$

These equations can also be given as a single vector equation

(2) $C: \mathbf{R} = \mathbf{R}(t), \qquad a \leq t \leq b$

with $\mathbf{R}(t) = x(t)\mathbf{i} + y(t)\mathbf{j} + z(t)\mathbf{k}$ the position vector of the point $(x(t), y(t), z(t))$ on the curve C.

The length of the curve C is given by the integral

(3) $\text{Length} = \displaystyle\int_a^b \sqrt{\left[\frac{dx}{dt}(t)\right]^2 + \left[\frac{dy}{dt}(t)\right]^2 + \left[\frac{dz}{dt}(t)\right]^2}\, dt = \int_a^b \left|\frac{d\mathbf{R}}{dt}(t)\right| dt$ ◄

(see Exercise 1). If we consider the parameter t to be the time and the parametric equations to give an object's position on the curve at time t, then the object's speed at time t is

(4) $\text{Speed} = \sqrt{\left[\frac{dx}{dt}(t)\right]^2 + \left[\frac{dy}{dt}(t)\right]^2 + \left[\frac{dz}{dt}(t)\right]^2} = \left|\frac{d\mathbf{R}}{dt}(t)\right|$ ◄

(see Exercise 2). The object's velocity vector at time t is

(5) $\mathbf{v}(t) = \dfrac{d\mathbf{R}}{dt}(t)$ ◄

and the unit tangent vector to the curve is

(6) $\mathbf{T}(t) = \dfrac{1}{\left|\dfrac{d\mathbf{R}}{dt}(t)\right|} \dfrac{d\mathbf{R}}{dt}(t).$ ◄

EXAMPLE 1   The curve $C: \mathbf{R} = \cos t\, \mathbf{i} + \sin t\, \mathbf{j} + (t/3)\, \mathbf{k}$, $0 \leq t \leq 6\pi$ is a portion of a *helix*. Give its velocity vector and speed as functions of the time t. Compute its length. Then sketch the curve and its velocity vector at $t = \frac{7}{4}\pi$.

Solution.   The velocity vector at time t is

$$\mathbf{v}(t) = \frac{d\mathbf{R}}{dt}(t) = \frac{d}{dt}[\cos t\, \mathbf{i} + \sin t\, \mathbf{j} + (t/3)\mathbf{k}]$$

$$= -\sin t\, \mathbf{i} + \cos t\, \mathbf{j} + \tfrac{1}{3}\mathbf{k}.$$

The speed is

$$|\mathbf{v}(t)| = \sqrt{(-\sin t)^2 + (\cos t)^2 + (\tfrac{1}{3})^2} = \sqrt{1 + \tfrac{1}{9}} = \frac{\sqrt{10}}{3}.$$

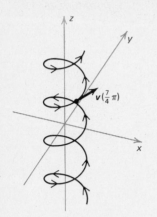

FIGURE 12.40   A helix

The length of the curve is

$$\int_0^{6\pi} |\mathbf{v}(t)|\, dt = \int_0^{6\pi} \frac{\sqrt{10}}{3}\, dt = 6\pi\left(\frac{\sqrt{10}}{3}\right) = 2\sqrt{10}\,\pi.$$

The velocity vector at $t = \frac{7}{4}\pi$ is

$$\mathbf{v}(\tfrac{7}{4}\pi) = -\sin(\tfrac{7}{4}\pi)\,\mathbf{i} + \cos(\tfrac{7}{4}\pi)\,\mathbf{j} + \tfrac{1}{3}\mathbf{k}$$

$$= \frac{1}{\sqrt{2}}\mathbf{i} + \frac{1}{\sqrt{2}}\mathbf{j} + \frac{1}{3}\mathbf{k}.$$

The curve and this vector are shown in Figure 12.40.   □

Curvature and normal vectors of space curves (optional)

In Section 11.5 we defined the normal $\mathbf{N}$ to a plane curve to be the unit vector perpendicular to the tangent vector $\mathbf{T}$ and pointing to the left from the direction of $\mathbf{T}$. We then showed in Section 11.6 that

$$(7) \qquad \frac{d\mathbf{T}}{ds} = \kappa\,\mathbf{N}$$

where s is arclength and κ is the curvature of the curve. Here κ is $d\phi/ds$, with ϕ the angle of inclination of $\mathbf{T}$, and can be positive, negative, or zero. From equation (7) we derived the formula

$$(8) \qquad \mathbf{a} = \frac{d^2s}{dt^2}\,\mathbf{T} + \kappa\left[\frac{ds}{dt}\right]^2 \mathbf{N}$$

which gives the decomposition of an acceleration vector into tangential and normal components.

We have to modify this development for a curve in space because in this case there are an infinite number of unit vectors perpendicular to $\mathbf{T}$. We define κ to be the (nonnegative) length of the vector $d\mathbf{T}/ds$ and, for $\kappa \neq 0$, define $\mathbf{N}$ to be the unit vector in the direction of $d\mathbf{T}/ds$. The vector $\mathbf{N}$ is perpendicular to $\mathbf{T}$ because $\mathbf{T} = \mathbf{T}(s)$ has constant length (see Exercise 9).

For $\kappa \neq 0$ formula (7) follows from the definition of $\mathbf{N}$, and formula (8) has the same derivation as given in Section 11.6 for plane curves. If κ is zero, then $\mathbf{N}$ is not defined, but equation (8) remains valid if we interpret the term involving κ to be the zero vector.

EXAMPLE 2   Use (8) to derive the formula

$$(9) \qquad \kappa = \frac{\left|\dfrac{d\mathbf{R}}{dt} \times \dfrac{d^2\mathbf{R}}{dt^2}\right|}{\left|\dfrac{d\mathbf{R}}{dt}\right|^3} \qquad\qquad ◀$$

for the curvature of a curve in space.

Solution. We have $d\mathbf{R}/dt = \mathbf{v} = (ds/dt)\,\mathbf{T}$ and $d^2\mathbf{R}/dt^2 = \mathbf{a}$. Therefore, by (8)

(10) $\dfrac{d\mathbf{R}}{dt} \times \dfrac{d^2\mathbf{R}}{dt^2} = \left[\dfrac{ds}{dt}\dfrac{d^2s}{dt^2}\right]\mathbf{T} \times \mathbf{T} + \left[\kappa\left(\dfrac{ds}{dt}\right)^3\right]\mathbf{T} \times \mathbf{N}.$

The vector $\mathbf{T} \times \mathbf{T}$ is zero and, because $\mathbf{T}$ and $\mathbf{N}$ are perpendicular unit vectors, $\mathbf{T} \times \mathbf{N}$ is also a unit vector. Therefore (10) yields

$$\left|\dfrac{d\mathbf{R}}{dt} \times \dfrac{d^2\mathbf{R}}{dt^2}\right| = \kappa\left(\dfrac{ds}{dt}\right)^3.$$

Solving for κ gives (9) because $ds/dt = |d\mathbf{R}/dt|$.   □

EXAMPLE 3   What is the (constant) curvature of the helix of Example 1?

Solution.   We have $d\mathbf{R}/dt = -\sin t\,\mathbf{i} + \cos t\,\mathbf{j} + \tfrac{1}{3}\mathbf{k}$. Therefore $d^2\mathbf{R}/dt^2 = -\cos t\,\mathbf{i} - \sin t\,\mathbf{j}$ and

$$\dfrac{d\mathbf{R}}{dt} \times \dfrac{d^2\mathbf{R}}{dt^2} = \begin{vmatrix} \mathbf{i} & \mathbf{j} & \mathbf{k} \\ -\sin t & \cos t & \tfrac{1}{3} \\ -\cos t & -\sin t & 0 \end{vmatrix} = \tfrac{1}{3}\sin t\,\mathbf{i} - \tfrac{1}{3}\cos t\,\mathbf{j} + \mathbf{k}.$$

This cross product has length $\tfrac{1}{3}\sqrt{10}$. We saw in Example 1 that this is also the length of $d\mathbf{R}/dt$. Consequently formula (9) gives $\tfrac{9}{10}$ as the curvature of the helix.   □

Exercises

1.† Verify formula (3) for the length of a curve in space for $x(t)$, $y(t)$, and $z(t)$ with continuous first-order derivatives.

2.† Verify formula (4) for speed under the hypothesis stated in Exercise 1.

In each of Exercises 3 through 5 find the velocity vector of the curve, the speed, and the acceleration vector as functions of the time t.

3.† $x = 3t^2 + 1$, $y = \sin t$, $z = \cos^2 t$

4.† $x = \sin(2t)$, $y = \sin(3t)$, $z = \sin(3t)$

5. $x = e^t$, $y = t^3 - t^2$, $z = \ln(t^2 + 1)$

In Exercises 6 and 7 give definite integrals which equal the lengths of the curves. Do not evaluate the integrals.

6.† $x = \sin t$, $y = te^t$, $z = t^3 - 3t$, $0 \le t \le 10$

7. $x = t^2$, $y = t^3$, $z = t - 2\cos t$, $-5\pi \le t \le 5\pi$

8. Use the definitions of derivatives and algebraic properties of the dot and cross products to derive the product rules

a.† $\dfrac{d}{dt}[\mathbf{A}(t) \cdot \mathbf{B}(t)] = \dfrac{d\mathbf{A}}{dt}(t) \cdot \mathbf{B}(t) + \mathbf{A}(t) \cdot \dfrac{d\mathbf{B}}{dt}(t)$

b. $\dfrac{d}{dt}[\mathbf{A}(t) \times \mathbf{B}(t)] = \dfrac{d\mathbf{A}}{dt}(t) \times \mathbf{B}(t) + \mathbf{A}(t) \times \dfrac{d\mathbf{B}}{dt}(t)$

for differentiable vector-valued functions $\mathbf{A}(t)$ and $\mathbf{B}(t)$ in space.

9.† Show that if the length of the vector-valued function $A(t)$ is constant then $\dfrac{dA}{dt}(t)$ is perpendicular to $A(t)$ for each t.

10. Show that if the acceleration vector of an object is the zero vector for all t, then the object moves along a straight line.

11.† Suppose that the force vector $F(t)$ on an object is at all times directed toward the origin in space. Show that the object moves in a plane containing the origin. (Show that $v(t) \times R(t)$ is a constant vector.)

12.† Suppose that the velocity and position vectors of an object in space are related by the equation $v(t) = kR(t)$ with k a constant. Express $R(t)$ in terms of k and $R(0)$ and show that the motion is on a line through the origin. (Consider the derivative of $e^{-kt}R(t)$.)

13. Suppose that the position vector $R(t)$ of an object satisfies the differential equation $\dfrac{dR}{dt}(t) = A \times R(t)$, where A is a constant vector. Show that the object's acceleration vector is perpendicular to A and that the object's speed is a constant.

14.† Find the curvature of the curve $x = \sin t, y = \cos t, z = \frac{1}{3}t^3$ at $t = 1$.

15. The vector N for a space curve is called the *principal normal* to the curve. The *binormal* to the curve is the vector B defined by $B = T \times N$. Find T, N, and B for the helix $x = \cos t, y = \sin t, z = t$ at $t = \pi/3$.

In each of Exercises 16 through 19 determine the position vector as a function of t from the acceleration vector and the specified values of the position vector and velocity vector.

16.† $a(t) = \langle t^4, t^3, t^2 \rangle$; $v(0) = \langle 1, 2, 3 \rangle$; $R(0) = \langle 7, 6, 5 \rangle$

17.† $a(t) = \sin(2t)(i - k)$; $v(0) = 5j$; $R(0) = 5i$

18.† $a(t) = \langle 0, 0, 1 \rangle$; $v(10) = \langle 0, 1, 0 \rangle$; $R(10) = \langle 1, 0, 0 \rangle$

19. $a(t) = \langle e^t, e^{2t}, e^{3t} \rangle$; $v(1) = \langle e, \frac{1}{2}e^2, \frac{1}{3}e^3 \rangle$; $R(1) = \langle e, \frac{1}{4}e^2, \frac{1}{9}e^3 \rangle$

In each of Exercises 20 through 27 find the position vector of the object as a function of t (seconds) in xyz-space with the z-axis pointing upward. Assume that the only force on the object is the force of gravity and use the value 32 feet/second² or 9.8 meters/second² as the acceleration due to gravity.

20.† A ball that is at $(4, 3, 2)$ and has velocity $5i + 6j + 10k$ feet per second at $t = 0$ (The coordinates are given in feet.)

21.† An arrow whose highest point is $(50, 40, 200)$ at $t = 3$ when its velocity is $10i$ feet per second (The coordinates are given in feet.)

22.† A bullet whose velocity is $\langle 50, 60, 120 \rangle$ feet per second at $t = 0$ and whose highest point is $(300, 400, 600)$ with coordinates measured in feet.

23.[†] A rock that is at the origin at $t = 0$ and at $(12, -16, 10)$ at $t = 2$ (The coordinates are measured in feet.)

24. An object that is at $(10, 15, 20)$ and has velocity $-5\mathbf{i} - 4\mathbf{j} - 3\mathbf{k}$ at $t = 0$ (The coordinates are given in meters.)

25.[†] An object is at the origin at $t = 0$ and at $(50, -100, 30)$ at $t = 5$ (The coordinates are measured in meters.)

26.[†] An arrow that is shot horizontally at $t = 0$ and passes through the point $(-10, -20, 4.2)$ with distances measured in meters.

27. A rock that is at the origin at $t = 0$ reaches its highest point at $t = 1$ and hits the ground at the point $(31, 16, 0)$ (The coordinates are given in meters.)

12-5   A BRIEF CATALOG OF SURFACES

The graphs $y = f(x)$ of functions of one variable are curves in xy-planes. As we will see in the next chapter, the graphs of functions $f(x,y)$ of the two variables x and y are *surfaces* in xyz-space. In this section we describe several types of surfaces that we will frequently encounter and we show how to determine the general shape of each surface by examining its equation.

Cylinders

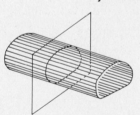

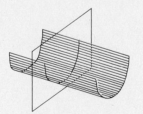

FIGURE 12.41   Cylinders

We use the term CYLINDER for any surface which can be formed from parallel lines. Typically, a cylinder is formed by taking all parallel lines through a curve in a plane (Figure 12.41).

Any surface whose equation involves only two of the three variables x, y, and z is a cylinder. If the equation does not involve z, the surface consists of lines parallel to the z-axis since, if (x_0, y_0, z_0) satisfies the equation, then (x_0, y_0, z) also satisfies it for any z. Similarly, the surface is a cylinder formed from lines parallel to the x-axis if the equation does not involve x and is a cylinder formed from lines parallel to the y-axis if the equation does not involve y.

EXAMPLE 1   Sketch the surface $x^2 + y^2 = 16$ in xyz-space.

Solution. The graph of $x^2 + y^2 = 16$ in the xy-plane is the circle of radius 4 with its center at the origin. The graph in xyz-space is formed from all lines through the circle and parallel to the z-axis. It is the right circular cylinder shown in Figure 12.42. □

EXAMPLE 2 Sketch the surface $z = -\frac{1}{8}x^3$ in xyz-space.

Solution. The surface is the cylinder formed by the lines parallel to the y-axis and passing through the curve $z = -\frac{1}{8}x^3$ in the xz-plane (Figure 12.43). □

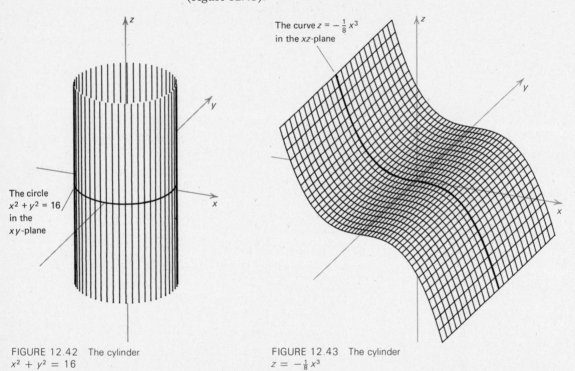

FIGURE 12.42 The cylinder $x^2 + y^2 = 16$

FIGURE 12.43 The cylinder $z = -\frac{1}{8}x^3$

Quadric surfaces

QUADRIC SURFACES are those with equations of the form

(1) $Ax^2 + By^2 + Cz^2 + Dxy + Exz + Fyz + Gx + Hy + Iz + J = 0.$

We do not need to go into the general theory of these surfaces, so we will restrict our attention to a few basic types. We will examine surfaces with equations of the form

(2) $z = Ax^2 + Bxy + Cy^2 + Dx + Ey + F$

and of the form

(3) $z^2 = Ax^2 + By^2 + C.$

Surfaces of type (2) are PARABOLOIDS and those of type (3) are ELLIPSOIDS, HYPERBOLOIDS, and CONES.

Paraboloids   One way to visualize a surface is to picture its CROSS SECTIONS, which are the intersections of the surface with planes. This is an especially suitable approach to studying quadric surfaces because their cross sections are parabolas, hyperbolas, and ellipses. For example, the vertical cross sections of the surfaces (2) are parabolas, and it is for this reason that the surfaces are called paraboloids.

The paraboloid $z = \frac{1}{2}x^2 + \frac{1}{2}y^2$ in Figure 12.44 has been drawn by showing some of its cross sections in vertical planes parallel to the x- and y-axes. We will show that these cross sections are parabolas by studying simultaneous equations for them. A vertical plane parallel to the x-axis has an equation $y = c$ with c a constant. Consequently, its intersection with the surface is given by the simultaneous equations

$$z = \tfrac{1}{2}x^2 + \tfrac{1}{2}y^2, \quad y = c.$$

Replacing y by c in the first of these gives the equivalent equations

$$z = \tfrac{1}{2}x^2 + \tfrac{1}{2}c^2, \quad y = c.$$

Since the new first equation is of the form $z = \frac{1}{2}x^2 +$ constant, the cross section is a parabola in the plane $y = c$ with the shape of the parabola $z = \frac{1}{2}x^2$ in an xz-plane. Similarly, a cross section in a vertical plane parallel to the y-axis is given by the equations $z = \frac{1}{2}x^2 + \frac{1}{2}y^2$, $x = c$ or $z = \frac{1}{2}c^2 + \frac{1}{2}y^2$, $x = c$ and hence is a parabola in the plane $x = c$ with the shape of the parabola $z = \frac{1}{2}y^2$ in a yz-plane. In both cases the cross sections are parabolas that open upward.

The surface $z = \frac{1}{2}x^2 - \frac{1}{2}y^2$ in Figure 12.45 is a second type of paraboloid. Its cross sections perpendicular to the y-axis are given by

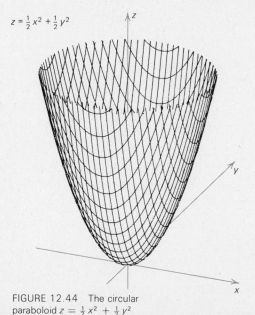

$z = \frac{1}{2}x^2 + \frac{1}{2}y^2$

FIGURE 12.44   The circular paraboloid $z = \frac{1}{2}x^2 + \frac{1}{2}y^2$

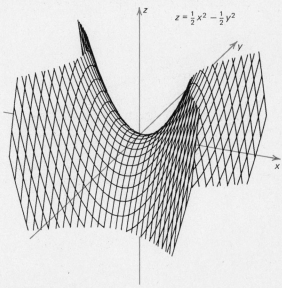

$z = \frac{1}{2}x^2 - \frac{1}{2}y^2$

FIGURE 12.45   The hyperbolic paraboloid $z = \frac{1}{2}x^2 - \frac{1}{2}y^2$

simultaneous equations $z = \frac{1}{2}x^2 - \frac{1}{2}y^2, y = c$ or $z = \frac{1}{2}x^2 - \frac{1}{2}c^2, y = c$ and are parabolas that open upward. The cross sections perpendicular to the x-axis are given by simultaneous equations $z = \frac{1}{2}x^2 - \frac{1}{2}y^2, x = c$ or $z = \frac{1}{2}c^2 - \frac{1}{2}y^2, x = c$ and are parabolas that open downward.

Figure 12.46 shows the surface $z = -\frac{1}{4}x - y^2$, which is a third type of paraboloid. Its cross sections in vertical planes parallel to the y-axis have equations $z = -\frac{1}{4}x - y^2, x = c$ or $z = -\frac{1}{4}c - y^2, x = c$ and are parabolas that open downward. Its cross sections in vertical planes parallel to the x-axis have equations $z = -\frac{1}{4}x - y^2, y = c$ or $z = -\frac{1}{4}x - c^2, y = c$ and are parallel lines (degenerate parabolas). Because the surface is formed by parallel lines, it is called a PARABOLIC CYLINDER.

The three paraboloids in Figures 12.44, 12.45, and 12.46 have different types of horizontal cross sections. Horizontal planes have equations $z = c$ so the horizontal cross sections of $z = \frac{1}{2}x^2 + \frac{1}{2}y^2$ have the equations $z = \frac{1}{2}x^2 + \frac{1}{2}y^2, z = c$ or $c = \frac{1}{2}x^2 + \frac{1}{2}y^2, z = c$ and are circles (Figure 12.47). Because of this the surface is called a CIRCULAR PARABOLOID (a special case of an ELLIPTIC PARABOLOID, which has ellipses for horizontal cross sections). The horizontal cross sections of $z = \frac{1}{2}x^2 - \frac{1}{2}y^2$ have equations $z = \frac{1}{2}x^2 - \frac{1}{2}y^2, z = c$ or $c = \frac{1}{2}x^2 - \frac{1}{2}y^2, z = c$ and are hyperbolas (Figure 12.48), so this surface

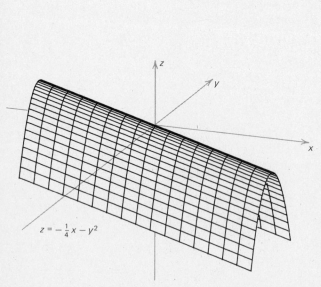

FIGURE 12.46   The parabolic cylinder $z = -\frac{1}{4}x - y^2$

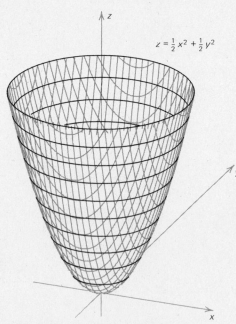

$z = \frac{1}{2}x^2 + \frac{1}{2}y^2$

FIGURE 12.47   Circular horizontal cross sections of a circular paraboloid

is called a HYPERBOLIC PARABOLOID. The horizontal cross sections of the parabolic cylinder $z = -\frac{1}{4}x - y^2$ have equations $z = -\frac{1}{4}x^2 - y^2, z = c$ or $c = -\frac{1}{4}x^2 - y^2, z = c$ and are parabolas (Figure 12.49).

As is shown in the next theorem, we can determine the type of a paraboloid by computing the *discriminant* $B^2 - 4AC$ in its equation

(4) $z = Ax^2 + Bxy + Cy^2 + Dx + Ey + F.$

Theorem 12.7   The surface (4) is an elliptic paraboloid if $B^2 - 4AC$ is negative, a hyperbolic paraboloid if $B^2 - 4AC$ is positive, and a parabolic cylinder if $B^2 - 4AC$ is zero.

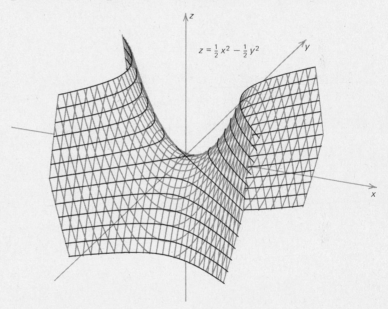

$z = \frac{1}{2}x^2 - \frac{1}{2}y^2$

FIGURE 12.48   Hyperbolic horizontal cross sections of a hyperbolic paraboloid

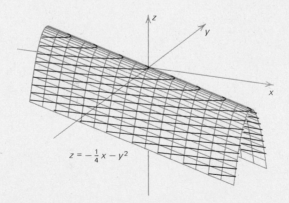

$z = -\frac{1}{4}x - y^2$

FIGURE 12.49   Parabolic horizontal cross sections of a parabolic cylinder

Proof.   The planes perpendicular to the x-axis have equations $x = c$ and the other vertical planes have equations $y = ax + b$. Making these substitutions in equation (2) shows that the vertical cross sections of the surface are parabolas.

The horizontal cross sections have the equations $c = Ax^2 + Bxy + Cy^2 + Dx + Ey + F, z = c$. Theorem 10.5 shows that they are ellipses if $B^2 - 4AC$ is negative, hyperbolas if $B^2 - 4AC$ is positive, and parabolas if $B^2 - 4AC$ is zero. In the first case the surface is an elliptic paraboloid and in the second case a hyperbolic paraboloid. In the third case we have $B^2 - 4AC = 0$ and hence $B = \pm 2\sqrt{AC}$. Suppose, for example, that A, B, and C are all positive. Then $B = 2\sqrt{AC}$ and making this substitution in (2) yields $z = [(\sqrt{A})x + (\sqrt{C})y]^2 + Dx + Ey + F$. Consequently, the cross sections in the parallel planes $(\sqrt{A})x + (\sqrt{C})y = c$ are straight lines, and the surface is a parabolic cylinder.   Q.E.D.

Ellipsoids, cones, and hyperboloids

The quadric surfaces

$$(3) \qquad z^2 = Ax^2 + By^2 + C$$

are less important in calculus than paraboloids. They are named according to their types of cross sections. If A and B are negative, the cross sections are all ellipses and the surface is an ELLIPSOID (Figure 12.50). In the other cases the cross sections are ellipses and hyperbolas, and the surface is either an ELLIPTIC CONE (Figure 12.51), a HYPERBOLOID OF ONE SHEET (Figure 12.52), or a HYPERBOLOID OF TWO SHEETS (Figure 12.53).

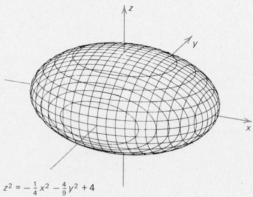

$$z^2 = -\tfrac{1}{4}x^2 - \tfrac{4}{9}y^2 + 4$$

FIGURE 12.50   An ellipsoid

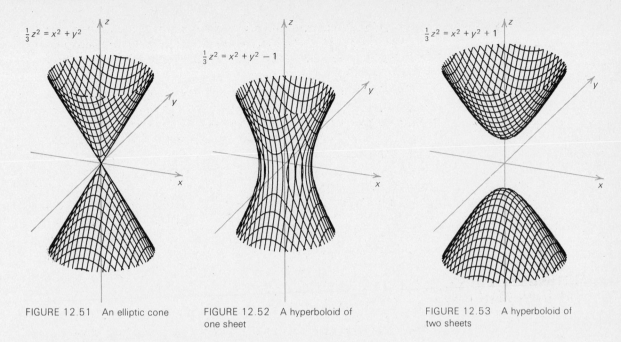

FIGURE 12.51   An elliptic cone

FIGURE 12.52   A hyperboloid of one sheet

FIGURE 12.53   A hyperboloid of two sheets

Traces of a surface

The TRACES of a surface are its intersections with the coordinate planes. The trace in the xy-plane can be found by setting z equal to zero in an equation for the surface; the trace in the yz-plane can be found by setting x equal to zero; and the trace in the xz-plane can be found by setting y equal to zero. Knowing the traces makes it easier to visualize or sketch a surface.

EXAMPLE 3   What are the traces of the cone and hyperboloids in Figures 12.51, 12.52, and 12.53?

Solution.   Setting $z = 0$ in the equation $\frac{1}{3}z^2 = x^2 + y^2$ gives $x^2 + y^2 = 0$ which has only the solution $x = 0, y = 0$. The trace of the cone in the xy-plane consists of a single point, the origin. Setting x equal to zero yields $\frac{1}{3}z^2 = y^2$ which has the solutions $z = \pm \sqrt{3}y$ and setting y equal to zero gives $\frac{1}{3}z^2 = x^2$ which has the solutions $z = \pm \sqrt{3}x$. The traces in the yz- and xz-planes are pairs of intersecting lines.

When we set $z = 0$ in the equation $\frac{1}{3}z^2 = x^2 + y^2 - 1$ for the hyperboloid of one sheet in Figure 12.52, we obtain $x^2 + y^2 = 1$. The trace in the xy-plane is a circle. Setting $x = 0$ yields $y^2 - \frac{1}{3}z^2 = 1$ and setting $y = 0$ gives $x^2 - \frac{1}{3}z^2 = 1$. The traces in the yz- and xz-planes are hyperbolas.

The hyperboloid of two sheets in Figure 12.53 has the equation $\frac{1}{3}z^2 = x^2 + y^2 + 1$. Putting $z = 0$ gives the equation $x^2 + y^2 + 1 = 0$ which has no solutions. The surface does not intersect the xy-plane and has no trace there. When we set $x = 0$, we obtain $\frac{1}{3}z^2 - y^2 = 1$ and when we set $y = 0$, we obtain $\frac{1}{3}z^2 - x^2 = 1$, so the traces in the yz- and xz-planes are hyperbolas.   □

Exercises

†Answer provided.
‡Outline of solution provided.

1.[†] What condition must an equation in x, y, and z satisfy for the surface it represents to be   **a.** symmetric about the xy-plane,   **b.** symmetric about the xz-plane,   **c.** symmetric about the yz-plane,   **d.** symmetric about the z-axis,   **e.** symmetric about the y-axis,   **f.** symmetric about the x-axis, and   **g.** symmetric about the origin?

In each of Exercises 2 through 20 give equations for the traces of the surface and state what types of curves they are. Then classify the surface and sketch it. Note any symmetry.

2.[†] $x^2 + z^2 = 1$

3.[†] $x = y^2$

4.[†] $4x^2 + y^2 = 4$

5. $y = x^3$

6.[†] $z = x^4 - 2x^2$

7. $x = z^2$

8.[†] $z = 1 - x^2 - y^2$

9. $x^2 + y^2 + z^2 = 9$

10.[†] $z = xy$

11. $z = x - y^2$

12.[†] $z^2 = x^2 + y^2 + 1$

13.[†] $z^2 = x^2 + y^2 - 1$

14. $z^2 = y^2 - x^2 - 1$

15.[†] $z^2 = x^2 + y^2$

16. $z^2 = x^2 - y^2$

17.[†] $z = \sqrt{16 - x^2 - y^2}$

18. $z = -\sqrt{16 - x^2 - y^2}$

19.[†] $z = -\sqrt{x^2 + y^2}$

20. $z = \sqrt{x^2 + y^2 + 1}$

APPENDIX TO CHAPTER 12:   CRAMER'S RULE

In this appendix we derive CRAMER'S RULE for solving systems of three linear equations

$$(1) \quad \begin{cases} a_1x + b_1y + c_1z = d_1 \\ a_2x + b_2y + c_2z = d_2 \\ a_3x + b_3y + c_3z = d_3 \end{cases}$$

in three unknowns. We let A, B, C, and D denote the vectors $A = \langle a_1, a_2, a_3 \rangle$, $B = \langle b_1, b_2, b_3 \rangle$, $C = \langle c_1, c_2, c_3 \rangle$, and $D = \langle d_1, d_2, d_3 \rangle$, whose components are the coefficients of x, y, and z and the constant terms in the equations. We express the vectors as columns by writing

$$(2) \quad A = \begin{bmatrix} a_1 \\ a_2 \\ a_3 \end{bmatrix}, B = \begin{bmatrix} b_1 \\ b_2 \\ b_3 \end{bmatrix}, C = \begin{bmatrix} c_1 \\ c_2 \\ c_3 \end{bmatrix}, \text{and } D = \begin{bmatrix} d_1 \\ d_2 \\ d_3 \end{bmatrix}.$$

Then equations (1) may be written as the vector equation

$$(3) \quad xA + yB + zC = D.$$

We let $\det[A, B, C]$ denote the determinant

$$(4) \quad \det[A, B, C] = \begin{vmatrix} a_1 & b_1 & c_1 \\ a_2 & b_2 & c_2 \\ a_3 & b_3 & c_3 \end{vmatrix}$$

whose columns are the vectors A, B, and C. In cases where this determinant is not zero, the system of equations (1) has unique solutions x, y, and z, and Cramer's rule gives the solutions in terms of determinants.

Theorem 12.8   (Cramer's rule) Suppose that the determinant $\det(A, B, C)$ is not zero. Then the system of equations (1) has unique solutions given by the formulas

$$(5) \quad x = \frac{\det[D, B, C]}{\det[A, B, C]} = \frac{\begin{vmatrix} d_1 & b_1 & c_1 \\ d_2 & b_2 & c_2 \\ d_3 & b_3 & c_3 \end{vmatrix}}{\begin{vmatrix} a_1 & b_1 & c_1 \\ a_2 & b_2 & c_2 \\ a_3 & b_3 & c_3 \end{vmatrix}} \quad \blacktriangleleft$$

$$(6) \quad y = \frac{\det[A, D, C]}{\det[A, B, C]} = \frac{\begin{vmatrix} a_1 & d_1 & c_1 \\ a_2 & d_2 & c_2 \\ a_3 & d_3 & c_3 \end{vmatrix}}{\begin{vmatrix} a_1 & b_1 & c_1 \\ a_2 & b_2 & c_2 \\ a_3 & b_3 & c_3 \end{vmatrix}} \quad \blacktriangleleft$$

$$(7) \quad z = \frac{\det[A, B, D]}{\det[A, B, C]} = \frac{\begin{vmatrix} a_1 & b_1 & d_1 \\ a_2 & b_2 & d_2 \\ a_3 & b_3 & d_3 \end{vmatrix}}{\begin{vmatrix} a_1 & b_1 & c_1 \\ a_2 & b_2 & c_2 \\ a_3 & b_3 & c_3 \end{vmatrix}} \cdot \quad \blacktriangleleft$$

The determinant $\det[A, B, C]$ that appears in the denominators of formulas (5), (6), and (7) is called the DETERMINANT OF THE SYSTEM OF EQUATIONS (1). Notice that the determinants in the numerators are obtained by putting the column D in place of the column of $\det[A, B, C]$ that corresponds to the unknown being calculated.

Proof.   We will derive (7) in the case where all four determinants and hence the numbers x, y, and z in formulas (5) through (7) are positive. The other cases and formulas (5) and (6) can be obtained by similar arguments.

When we place the vectors A, B, and C with a common base point as in Figure 12.54, they form the edges of a parallelepiped, which we call Solid 1. It follows from Theorem 12.4 and Exercise 28 of Section 12.2 that

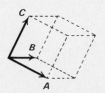

FIGURE 12.54   Solid 1

$(8) \quad$ The volume of Solid 1 $= \det[A, B, C]$.

The vectors $x\mathbf{A}$, $y\mathbf{B}$, and $z\mathbf{C}$ form a parallelepiped which we call Solid 2 (Figure 12.55). Equation (3) states that the diagonal of that parallelepiped is formed by the vector $\mathbf{D}$.

We denote by Solid 3 the parallelepiped whose base is the same as that of Solid 1 and whose nonhorizontal edge is formed by the vector $\mathbf{D}$ (Figure 12.56). We then have

(9)   The volume of Solid 3 $= \det[\mathbf{A}, \mathbf{B}, \mathbf{D}]$.

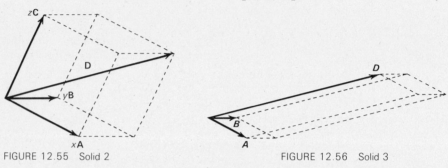

FIGURE 12.55   Solid 2   FIGURE 12.56   Solid 3

The number z is the ratio $|z\mathbf{C}|/|\mathbf{C}|$ of the lengths of the vectors $z\mathbf{C}$ and $\mathbf{C}$, which form the lateral sides of Solids 1 and 2. By similar triangles, that ratio is equal to the ratio of the heights of the solids:

$$z = \frac{\text{Height of Solid 2}}{\text{Height of Solid 1}}.$$

Then, because Solid 3 has the same height as Solid 2 and the same base as Solid 1, we have

$$z = \frac{\text{Height of Solid 3}}{\text{Height of Solid 1}} = \frac{[\text{Height of Solid 3}]\,[\text{Area of base}]}{[\text{Height of Solid 1}]\,[\text{Area of base}]}$$

$$= \frac{\text{Volume of Solid 3}}{\text{Volume of Solid 1}}.$$

With formulas (8) and (9) for the volumes, we obtain formula (7).   Q.E.D.

EXAMPLE 1   The determinant of the system of equations

$$\begin{cases} x + 2y + 3z = 7 \\ 4x - 3y - 2z = 8 \\ 5x - \ y + 2z = 9 \end{cases}$$

is not zero. Use Cramer's rule to express the solutions in terms of determinants.

Solution.

$$x = \frac{\begin{vmatrix} 7 & 2 & 3 \\ 8 & -3 & -2 \\ 9 & -1 & 2 \end{vmatrix}}{\begin{vmatrix} 1 & 2 & 3 \\ 4 & -3 & -2 \\ 5 & -1 & 2 \end{vmatrix}}, y = \frac{\begin{vmatrix} 1 & 7 & 3 \\ 4 & 8 & -2 \\ 5 & 9 & 2 \end{vmatrix}}{\begin{vmatrix} 1 & 2 & 3 \\ 4 & -3 & -2 \\ 5 & -1 & 2 \end{vmatrix}}, z = \frac{\begin{vmatrix} 1 & 2 & 7 \\ 4 & -3 & 8 \\ 5 & -1 & 9 \end{vmatrix}}{\begin{vmatrix} 1 & 2 & 3 \\ 4 & -3 & -2 \\ 5 & -1 & 2 \end{vmatrix}}. \quad \square$$

Exercises

†*Answer provided.*
‡*Outline of solution provided.*

Evaluate the determinants in Exercises 1 through 6.

1.‡ $\begin{vmatrix} 2 & 4 & 2 \\ 0 & 1 & -4 \\ -3 & 0 & 4 \end{vmatrix}$

2. $\begin{vmatrix} 2 & 4 & -4 \\ 1 & 7 & -2 \\ -1 & 3 & -5 \end{vmatrix}$

3.† $\begin{vmatrix} \frac{1}{2} & 2 & 3 \\ 1 & 5 & 0 \\ 0 & 3 & \frac{1}{4} \end{vmatrix}$

4. $\begin{vmatrix} 4 & 3 & 1 \\ 1 & 2 & 3 \\ 0 & 5 & -2 \end{vmatrix}$

5.† $\begin{vmatrix} 0 & 3 & 2 \\ 1 & 1 & 6 \\ 4 & 4 & 3 \end{vmatrix}$

6. $\begin{vmatrix} 1 & 4 & 6 \\ 2 & -3 & 1 \\ -1 & 7 & 5 \end{vmatrix}$

Use Cramer's rule to solve the systems of equations in Exercises 7 through 14.

7.‡ $\begin{cases} x - y + z = 4 \\ x \quad\ + 2z = 0 \\ 2x - y + z = 6 \end{cases}$

8. $\begin{cases} x + 2y - 3z = 0 \\ x - y + 4z = 1 \\ x + y + 5z = 3 \end{cases}$

9.† $\begin{cases} x + 2y - z = 0 \\ 2x - 3y - z = 1 \\ 3x - y - z = 2 \end{cases}$

10. $\begin{cases} -y + 3z = 1 \\ x + y - 4z = 0 \\ x + y - z = -1 \end{cases}$

11.† $\begin{cases} x + y + z = 2 \\ x - y + z = 4 \\ 2x + y - 2z = -3 \end{cases}$

12.† $\begin{cases} x - 5y + z = 7 \\ 2x + y - z = 5 \\ x + y - z = 1 \end{cases}$

13.† $\begin{cases} x + y + z = 1 \\ x - y - z = 0 \\ 2x + 2y - z = 5 \end{cases}$

14.† $\begin{cases} x - 2y + z = 0 \\ 5x + 5y \quad\ = 1 \\ 2x + y + z = 1 \end{cases}$

MISCELLANEOUS EXERCISES

†*Answer provided.*
‡*Outline of solution provided.*

1. Find the center and radius of the intersection of the sphere $x^2 + y^2 + z^2 = 16$ with the plane $z = -3$.

2. Show that nonzero vectors A and B are perpendicular if and only if $|A + tB| \geq |A|$ for all numbers t.

3. Find the unit vectors that make an angle of 60° with $A = \langle -2, -2, 1 \rangle$ and an angle of 45° with $B = \langle 0, -1, 1 \rangle$.

4. Show that the position vector $C = aA + bB$ bisects the angle between the position vectors A and B in space if $a = |B|/(|A| + |B|)$ and $b = |A|/(|A| + |B|)$.

5. Use vectors to prove that the distance from a point (x_0, y_0, z_0) to a plane $ax + by + cz + d = 0$ is

$$\frac{|ax_0 + by_0 + cz_0 + d|}{\sqrt{a^2 + b^2 + c^2}}.$$

6. Use the result of Exercise 5 to find the distances   **a.**[†] from $(1, 2, 3)$ to the plane $2x - y - z + 1 = 0$,   **b.**[†] from the origin to the plane $z = 2x + 3y + 4$,   **c.** from $(0, 5, 2)$ to the plane $x - y = 10$, and   **d.** from $(-1, 2, 1)$ to the plane $6x + 2y + 4z = 6$.

7. Let $P(x_0, y_0, z_0)$ and $Q(x_1, y_1, z_1)$ be distinct points and L the distance between them. Show that for each number t with $0 \le t \le 1$, the point on the line segment PQ a distance tL from Q and $(1 - t)L$ from P is $(tx_0 + (1 - t)x_1, ty_0 + (1 - t)y_1, tz_0 + (1 - t)z_1)$.

8. Use the result of Exercise 7 to find the point that is one tenth of the distance along the line segment PQ from $P(1, 3, 4)$ to $Q(5, -1, 0)$.

9. Show that the curve $C: x = 2 + \tan t, y = 3 - \tan t, z = 4 + 2\tan t$, $-\pi/2 < t < \pi/2$ is a line.

10. Show that $L_1: x = 3 + t, y = 4 - 2t, z = 5 - t$ and $L_2: x = 2 - 2t$, $y = 6 + 4t, z = 6 + 2t$ are the same line.

11. Show that the curve $C: x = \sin t \cos t, y = \sin^2 t, z = \cos t$ lies on a sphere with its center at the origin. What is its radius?

12. The curve $x = t \cos t, y = t \sin t, z = 4t$ lies on a circular cone. Give an equation for the cone.

13. Give an equation for the locus of points equidistant from $(3, 2, 4)$ and $(-1, 8, -6)$.

14.[‡] Find the point where the lines $x = 2t, y = -t + 3, z = 4t - 2$ and $x = t + 4, y = 2t + 1, z = 6$ intersect.

15.[†] Find the distance from the point $(1, 2, 3)$ to the line $x = 2t - 1$, $y = t + 1, z = 3t + 2$.

16. Suppose that π_1 and π_2 are intersecting planes and L their line of intersection, and that π_3 is a plane perpendicular to L. The angle $\psi (0 < \psi \le \pi/2)$ between π_1 and π_2 is the angle between the intersections of π_1 and π_3 and of π_2 and π_3.   **a.** Show ψ is the angle between the normal vectors to π_1 and π_2.   **b.**[†] Give an approximate value for the angle between the planes $2x - y + z = 0$ and $3x + z = 0$.   **c.** Give an approximate value for the angle between the planes $y = 5$ and $2x + 3y - 4z = 10$.

17. When we speak of the angle between a line and a plane, we mean the least angle between that line and lines in the plane.   **a.**[†] What is the angle if A is a vector parallel to the line and n is a normal vector to the plane?   **b.**[†] Give an approximate value for the angle between the plane $2x - y + 2z = 4$ and the line $x = t, y = 2t + 1, z = 3t + 2$.   **c.** Give an approximate value for the angle between the plane $x + y = 0$ and the line $x = -t, y = 4, z = 5t + 5$.

18. Show that $(A \times B) \times A = B$ if A and B are perpendicular unit vectors.

19. Show that $A \cdot B = A \cdot C$ and $A \times B = A \times C$ with $A \neq 0$ imply $B = C$.

20.[†] Prove the identity $A \times (B \times C) = (C \cdot A)B - (B \cdot A)C$.

21.[†] Prove the identity $(A \times B) \cdot (C \times D)$
$= (A \cdot C)(B \cdot D) - (A \cdot D)(B \cdot C)$.

22.[†] Prove the identity $(A \times B) \times (C \times D)$
$= [(A \times B) \cdot D]C - [(A \times B) \cdot C]D$.

23. Use the identities $|A \times B|^2 = |A|^2 |B|^2 - (A \cdot B)^2$ and $|A - B|^2 = |A|^2 - 2A \cdot B + |B|^2$ to derive Heron's formula $\sqrt{s(s - a)(s - b)(s - c)}$ for the area of a triangle with sides of length a, b, and c $(s = \frac{1}{2}(a + b + c))$.

24. Use the cross product $\langle \cos \alpha, \sin \alpha, 0 \rangle \times \langle \cos \beta, \sin \beta, 0 \rangle$ to derive the trigonometric identity $\sin(\alpha - \beta) = \sin \alpha \cos \beta - \cos \alpha \sin \beta$.

25.[†] Find the center and radius of the sphere $x^2 + y^2 + z^2 - 6x + 4y - 2z = 0$ by completing the squares in its equation.

26. Determine the types of the following quadric surfaces by completing the squares in their equations. Then sketch them.
   a.[‡] $z = x^2 + y^2 - 2x - 2y + 2$
   b. $z = 4x - 2y - x^2 - y^2$
   c.[†] $z = 2x - x^2 - 1$
   d. $z^2 = x^2 + y^2 - 4x - 4y + 8$

27.[†] Give parametric equations for the locus of points equidistant from $(1, 6, 6)$, $(4, 2, 7)$, and $(5, 5, 0)$.

28. Explain why the line, $x = at + x_0, y = bt + y_0, z = ct + z_0$, with a, b, and c nonzero numbers, can be described by the equations

$$\frac{x - x_0}{a} = \frac{y - y_0}{b} = \frac{z - z_0}{c}.$$

These are known as the *symmetric equations* of the line.

29. Show that the planes $a_1 x + b_1 y + c_1 z = 0, a_2 x + b_2 y + c_2 z = 0$, and $a_3 x + b_3 y + c_3 z = 0$ with normal vectors $n_1 = \langle a_1, b_1, c_1 \rangle$, $n_2 = \langle a_2, b_2, c_2 \rangle$, and $n_3 = \langle a_3, b_3, c_3 \rangle$ **a.** coincide if the normal vectors are parallel,   **b.** intersect in a line if the vectors are not parallel but $(n_1 \times n_2) \cdot n_3 = 0$, and   **c.** intersect in a single point if $(n_1 \times n_2) \cdot n_3 \neq 0$.

30. Show that if $A = \overrightarrow{OP}, B = \overrightarrow{OQ}$, and $C = \overrightarrow{OR}$ with points P, Q, and R in space, then $A \times B + B \times C + C \times A$ is perpendicular to the plane containing P, Q, and R.

CHAPTER 13

PARTIAL DERIVATIVES

SUMMARY: In the first five sections of this chapter we study functions of two variables. We discuss graphs and level curves in Section 13.1, partial derivatives in Section 13.2, the chain rule in Section 13.3, directional derivatives and gradient vectors in Section 13.4, and tangent planes in Section 13.5. We study level surfaces, directional derivatives, and gradient vectors of functions of three variables in Section 13.6. Examples of functions of more than three variables are dealt with in the Miscellaneous Exercises.

13-1 FUNCTIONS OF TWO VARIABLES

Although a great variety of situations can be described by using functions of one variable, many others call for the use of two or more independent variables. For example, the height of a mountain or the depth of a lake can be given as a function of two horizontal coordinates x and y; the density of the earth is a function of three coordinates x, y, and z; the pressure in a gas-filled balloon is a function of its temperature and volume; and if a factory manufactures several items, its income is a function of the amounts of those items it sells.

In this section we begin our study of functions of two independent variables. The rules for calculating derivatives and integrals of such functions are based on the corresponding rules for functions of one variable and are not much more difficult to apply. The significant new element in calculus of two variables is that the graphs of functions are surfaces in three-dimensional space, and we often have to be able to visualize these surfaces to determine the differentiation or integration procedures that are required. In this section we discuss examples of functions of two variables and show how to determine the shapes of their graphs by studying their vertical and horizontal cross sections. We begin with the following formal definition of a function of two variables.

Definition 13.1 A function f of the variables x and y is a rule that assigns a unique number $f(x, y)$ to each point (x, y) in its domain. The domain is a portion or all of the xy-plane.

EXAMPLE 1   Compute $f(-2, 4)$ where $f(x, y) = \frac{1}{2}x^2 + \frac{1}{2}y^2$.

Solution.   We have $f(-2, 4) = [\frac{1}{2}x^2 + \frac{1}{2}y^2]_{x=-2,\,y=4}$
$= \frac{1}{2}(-2)^2 + \frac{1}{2}(4)^2 = 10.$ □

When the values of a function are given by a formula and we do not explicitly describe the domain of the function, we assume the domain consists of all points (x, y) for which the formula is defined.

EXAMPLE 2   Sketch the domain of the function $\sqrt{x - y^2}$.

Solution.   The expression $\sqrt{x - y^2}$ is defined for all (x, y) such that $x - y^2 \geq 0$. The equivalent inequality $x \geq y^2$ shows that the domain of the function consists of the points on and to the right of the parabola $x = y^2$ in the xy-plane (Figure 13.1).   □

The graph of a function of two variables

Just as the graph of a function $f(x)$ of one variable is the curve $y = f(x)$, the graph of a function $f(x, y)$ of two variables is the surface $z = f(x, y)$.

Definition 13.2   **The graph of the function $f(x, y)$ is the set of points (x, y, z) in xyz-space with (x, y) in the domain of f and $z = f(x, y)$.**

To find the value of f at the point (x, y), we move vertically (parallel to the z-axis) from the point (x, y) in the xy-plane to the graph and then horizontally (parallel to the xy-plane) to the z-axis, where we read off the value of the functions (Figure 13.2).

EXAMPLE 3   Sketch the graph of the function $\frac{1}{2}x^2 + \frac{1}{2}y^2$.

Solution.   The graph is the circular paraboloid $z = \frac{1}{2}x^2 + \frac{1}{2}y^2$ studied in Section 12.5 (Figure 13.3).   □

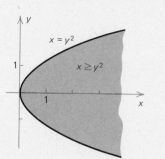

FIGURE 13.1   The domain of the function $\sqrt{x - y^2}$

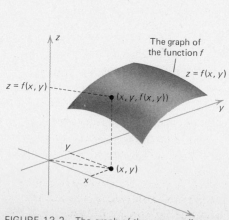

FIGURE 13.2   The graph of the function f

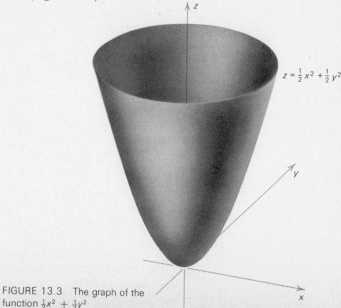

FIGURE 13.3   The graph of the function $\frac{1}{2}x^2 + \frac{1}{2}y^2$

EXAMPLE 4   Sketch the graph of the function $\frac{1}{2}x^2 - \frac{1}{2}y^2$.

Solution.   The graph is the hyperbolic paraboloid $z = \frac{1}{2}x^2 - \frac{1}{2}y^2$ studied in Section 12.5 (Figure 13.4).   □

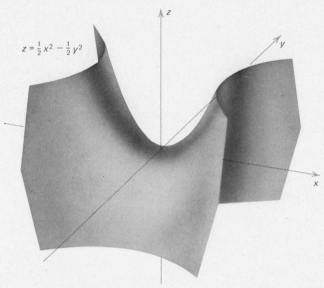

FIGURE 13.4   The graph of the function $\frac{1}{2}x^2 - \frac{1}{2}y^2$

Vertical cross sections of graphs

In Section 12.5, we determined the shapes of the paraboloids in Figures 13.3 and 13.4 by studying their vertical cross sections. We saw that cross sections of $z = \frac{1}{2}x^2 + \frac{1}{2}y^2$ in vertical planes parallel to the x- and y-axes are parabolas that open upward (Figure 13.5), whereas cross sections of $z = \frac{1}{2}x^2 - \frac{1}{2}y^2$ in vertical planes parallel to the x-axis are parabolas that open upward and those in vertical planes parallel to the y-axis are parabolas that open downward (Figure 13.6). We study vertical cross sections of another type of surface in the next example.

EXAMPLE 5   The graph of the function $x - \frac{1}{12}x^3 - \frac{1}{4}y^2 + \frac{1}{2}$ is the surface $z = x - \frac{1}{12}x^3 - \frac{1}{4}y^2 + \frac{1}{2}$, which has been drawn in Figure 13.7 by showing some of its cross sections in vertical planes parallel to the x- and y-axes. Sketch a curve in an xz-plane that has the shape of the cross sections in vertical planes parallel to the x-axis and a curve in a yz-plane that has the shape of the cross sections in vertical planes parallel to the y-axis.

Solution.   A vertical plane parallel to the x-axis has the equation $y = c$, so its intersection with the surface is given by the simultaneous equations $z = x - \frac{1}{12}x^3 - \frac{1}{4}y^2 + \frac{1}{2}$, $y = c$. Replacing y by c in the first equation gives the equivalent equations

(1) $z = x - \frac{1}{12}x^3 - \frac{1}{4}c^2 + \frac{1}{2}$, $y = c$.

$z = \frac{1}{2}x^2 + \frac{1}{2}y^2$

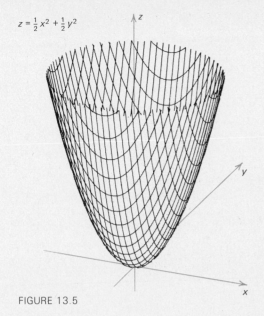

FIGURE 13.5

$z = \frac{1}{2}x^2 - \frac{1}{2}y^2$

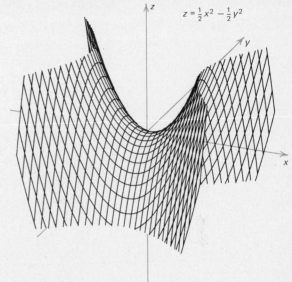

FIGURE 13.6

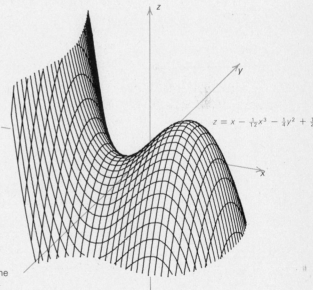

FIGURE 13.7   The graph of the function $x - \frac{1}{12}x^3 - \frac{1}{4}y^2 + \frac{1}{2}$

$z = x - \frac{1}{12}x^3 - \frac{1}{4}y^2 + \frac{1}{2}$

The first of equations (1) is of the form $z = x - \frac{1}{12}x^3 + $ constant, so these cross sections have the shape of the cubic $z = x - \frac{1}{12}x^3$ in an xz-plane (Figure 13.8). The cubic passes through the origin and has a local minimum at $x = -2$ and a local maximum at $x = 2$.

A cross section in a vertical plane parallel to the y-axis has simultaneous equations $z = x - \frac{1}{12}x^3 - \frac{1}{4}y^2 + \frac{1}{2}$, $x = c$ or $z = c - \frac{1}{12}c^3 - \frac{1}{4}y^2 + \frac{1}{2}$, $x = c$. The first of the last pair of equations is of the form $z = -\frac{1}{4}y^2 + $ constant, so the cross sections have the shape of the parabola $z = -\frac{1}{4}y^2$ in a yz-plane (Figure 13.9).   □

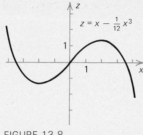

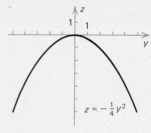

FIGURE 13.8   FIGURE 13.9

Level curves

In Section 12.5 we studied quadric surfaces by examining their horizontal as well as their vertical cross sections. To study horizontal cross sections of the graph of a function $f(x, y)$ of two variables, we draw LEVEL CURVES of the function. These are obtained by projecting horizontal cross sections of the graph vertically from the surface to the xy-plane. Figure 13.10 shows, for example, the graph of a function $f(x, y)$, its horizontal cross section at $z = c$, and the corresponding level curve in the xy-plane. The level curve consists of the points (x, y) in the xy-plane where the function has the value c, so it has the equation $f(x, y) = c$. This definition of a level curve can be made for any function.

Definition 13.3   A level curve (or contour curve) of a function $f(x, y)$ is the curve $f(x, y) = c$ in the xy-plane. It consists of the points (x, y) where the function has the value c.

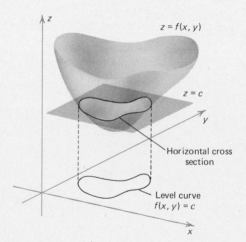

FIGURE 13.10

EXAMPLE 6   Sketch the level curves of $\frac{1}{2}x^2 + \frac{1}{2}y^2$.

Solution.   The equation $\frac{1}{2}x^2 + \frac{1}{2}y^2 = c$ for a level curve may be rewritten as $x^2 + y^2 = 2c$. For $c > 0$ the level curve is the circle of radius $\sqrt{2c}$ with its radius at the origin, and for $c = 0$ it consists of the single point $(0, 0)$. For $c < 0$ there are no level curves because the equation $x^2 + y^2 = 2c$ has no real solutions. Figure 13.11 shows the level curves for $c = 0, 1, 2, \ldots, 10$.   □

Figure 13.12 shows the cross sections of the graph of $\frac{1}{2}x^2 + \frac{1}{2}y^2$ that correspond to the level curves in Figure 13.11. There are no level curves for $c < 0$ because, for negative c, the plane $z = c$ does not intersect the graph.

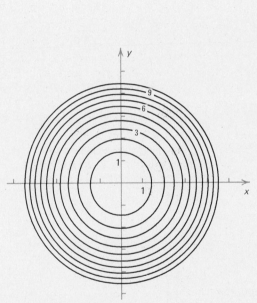

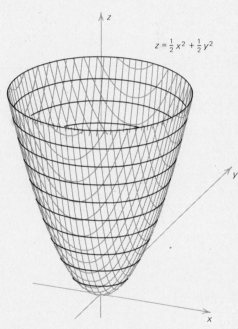

$$z = \frac{1}{2}x^2 + \frac{1}{2}y^2$$

FIGURE 13.11   Level curves of $\frac{1}{2}x^2 + \frac{1}{2}y^2$

FIGURE 13.12   Horizontal cross sections of $z = \frac{1}{2}x^2 + \frac{1}{2}y^2$

EXAMPLE 7   Sketch the level curves of $\frac{1}{2}x^2 - \frac{1}{2}y^2$.

Solution.   The level curves have the equations $\frac{1}{2}x^2 - \frac{1}{2}y^2 = c$ or $x^2 - y^2 = 2c$. For $c = 0$ we have $y^2 = x^2$ and the level curve is the degenerate hyperbola consisting of the intersecting lines $y = x$ and $y = -x$. For $c \neq 0$ the level curve is a nondegenerate hyperbola. Figure 13.13 shows the level curves for $c = 0, \pm\frac{1}{2}, \pm 1, \pm\frac{3}{2}, \ldots, \pm 3$.   □

Figure 13.14 shows the horizontal cross sections of the graph of $\frac{1}{2}x^2 - \frac{1}{2}y^2$ that correspond to the level curves of Figure 13.13.

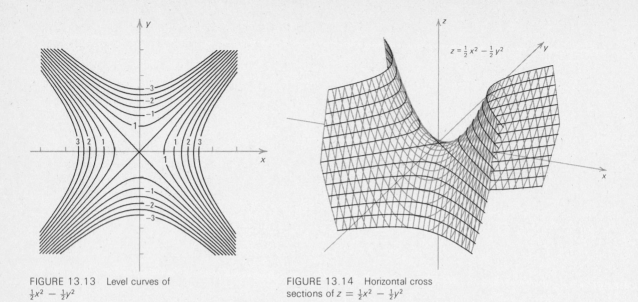

FIGURE 13.13   Level curves of $\frac{1}{2}x^2 - \frac{1}{2}y^2$

FIGURE 13.14   Horizontal cross sections of $z = \frac{1}{2}x^2 - \frac{1}{2}y^2$

EXAMPLE 8   Figure 13.15 shows the level curves $x - \frac{1}{12}x^3 - \frac{1}{4}y^2 + \frac{1}{2} = c$ of $x - \frac{1}{12}x^3 - \frac{1}{4}y^2 + \frac{1}{2}$ for $c = 0, \pm 1, \pm 2, \pm 3, \pm 4$, and ± 5. Trace the curves and label those for $c = 0, \pm 2, \pm 4$.

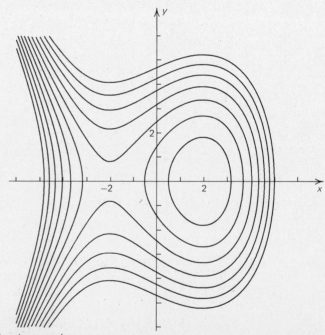

FIGURE 13.15   Level curves of $x - \frac{1}{12}x^3 - \frac{1}{4}y^2 + \frac{1}{2}$

Solution.   Along the x-axis, where $y = 0$, the values of the function are given by $x - \frac{1}{12}x^3 + \frac{1}{2}$. We compute the following table of these values.

x	-5	-4	-3	-2	-1	0	1	2	3	4	5
$x - \frac{1}{12}x^3 + \frac{1}{2}$	$5\frac{11}{12}$	$1\frac{5}{6}$	$-\frac{1}{4}$	$-\frac{5}{6}$	$-\frac{5}{12}$	$\frac{1}{2}$	$1\frac{5}{12}$	$1\frac{5}{6}$	$1\frac{1}{4}$	$-\frac{5}{6}$	$-4\frac{11}{12}$

In order for the function to have these values at the corresponding points on the x-axis, the level curves must be as labeled in Figure 13.16. □

The horizontal cross sections of the graph of $x - \frac{1}{12}x^3 - \frac{1}{4}y^2 + \frac{1}{2}$ that correspond to the level curves of Figures 13.15 and 13.16 are shown in Figure 13.17.

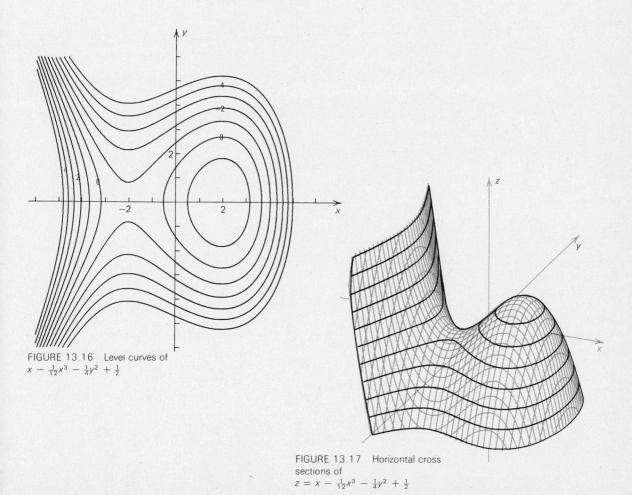

FIGURE 13.16   Level curves of
$x - \frac{1}{12}x^3 - \frac{1}{4}y^2 + \frac{1}{2}$

FIGURE 13.17   Horizontal cross
sections of
$z = x - \frac{1}{12}x^3 - \frac{1}{4}y^2 + \frac{1}{2}$

Topographical maps

Figure 13.18 is a topographical map of the Mount Shasta area in California and Figure 13.19 is a photograph of the mountain. We can consider the map as a sketch of level curves. The points on each curve of the map correspond to points at constant elevation on the mountain, and the curves are level curves of the function whose value at each point is the elevation of the corresponding point on the mountain.

EXAMPLE 9   What is the approximate elevation of the highest point on Mount Shasta?

Solution.   An examination of Figure 13.18 shows that the level curves drawn with heavy lines represent elevations 2,500 feet apart, whereas the other curves represent elevations 500 feet apart. The summit is, therefore, approximately 14,000 feet above sea level.   □

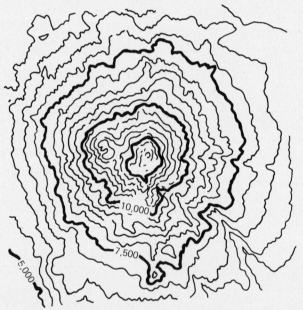

FIGURE 13.18   Topographical map of Mount Shasta

FIGURE 13.19   Photograph of
Mount Shasta (Courtesy of J. and
W. Goff)

Exercises

†Answer provided.
‡Outline of solution provided.

1.† Compute   **a.** $f(2, 3)$ and   **b.** $f(-1, 10)$ for $f(x, y) = x^2y^3$.

2.† Compute $g(4, \pi)$ for $g(x, y) = e^x \sin(y/3)$.

3. Compute $h(-1, 0)$ for $h(x, y) = x^2y^3 - y^2x^3 + x^7 - y + 10$.

4. When a small spherical pebble falls under the force of gravity in a deep body of still water, it quickly approaches its terminal speed, which by *Stokes' law* is $(2.17 \times 10^4)r^2(\rho - 1)$ centimeters per second if r (centimeters) is the radius of the pebble and ρ (grams per cubic centimeter) is its density. What is the terminal speed of a quartz pebble for which $r = 0.01$ and $\rho = 2.60$?

Sketch the domains of the functions in Exercises 5 through 11.

5.‡ $\dfrac{1}{\sqrt{x - y}}$ **6.** $\sqrt{1 - x^2 - y^2}$ **7.**† $\ln(xy - 1)$

8.† $\arcsin(x - y)$ **9.** $\dfrac{1}{y(x + 1)}$ **10.**† $\sqrt{xy}$

11. $\ln[\ln(x/y)]$

In Exercises 12 through 20 sketch the graphs of the functions.

12. $3x + 2y - 6$ **13.**† $\sqrt{1 - x^2 - y^2}$ **14.**† x^2

15.† xy **16.** $4 - x^2 - y^2$ **17.**† $\frac{1}{2}\sqrt{x^2 + y^2}$

18. $-\frac{1}{2}\sqrt{x^2 + y^2}$ **19.** $\sqrt{1 - x^2}$

20. $f(x, y) = \begin{cases} 2 & \text{for} \quad x^2 + y^2 \leq 1 \\ 1 & \text{for } 1 < x^2 + y^2 \leq 4 \end{cases}$

21. The graphs of   **a.**† $\sin x + \frac{1}{9} y^3 + \frac{1}{8}$ **b.**† $\sin x$ **c.** $\sin x \sin y$ **d.**† $\sin^2 x + \frac{1}{2} y^2$ **e.** $\frac{1}{9} y^3 \sin x$ and   **f.** $3 \sin(x) e^{y/5}$ are shown in Figures 13.20 through 13.25. Match the functions to their graphs.

22. Level curves of the six functions of Exercise 21 are shown in Figures 13.26 through 13.31. Match the functions to their level curves.

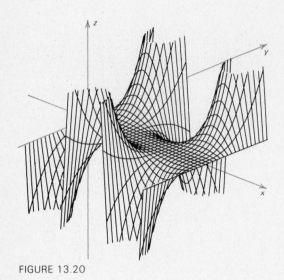

FIGURE 13.20

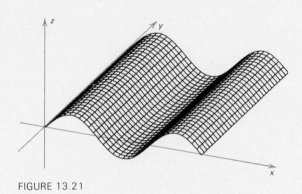

FIGURE 13.21

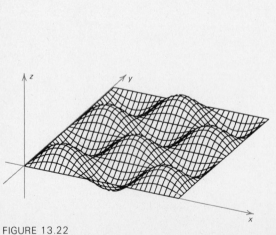

FIGURE 13.22

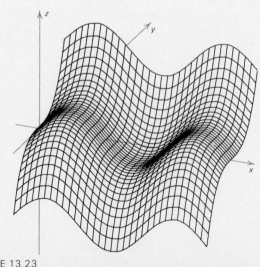

FIGURE 13.23

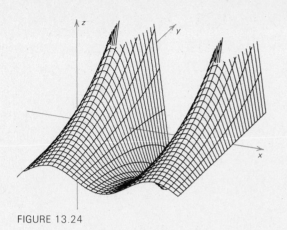

FIGURE 13.24

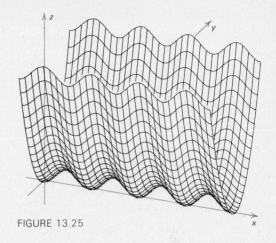

FIGURE 13.25

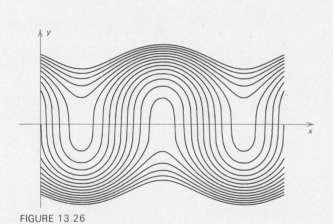

FIGURE 13.26

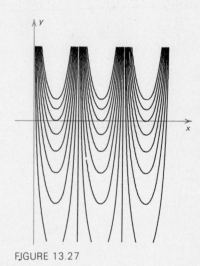

FIGURE 13.27

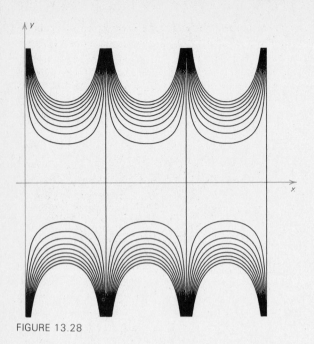

FIGURE 13.28

FIGURE 13.29

FIGURE 13.30

FIGURE 13.31

23. In some cases it helps in visualizing the graph of a function to study its vertical cross sections in various directions through a fixed point. The graphs of the functions   **a.** $5(x^2 + y^2)^{-1/2}$,   **b.** $\ln(x^2 + y^2)$, and **c.** $\sin^2(\sqrt{x^2 + y^2})$ are shown in Figures 13.32 through 13.34. Match the functions to their graphs by studying vertical cross sections through the origin.

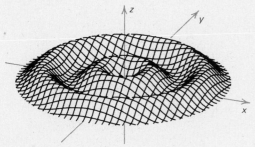

FIGURE 13.32

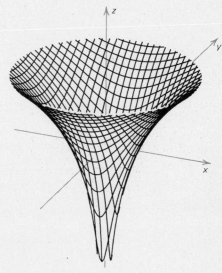

FIGURE 13.33

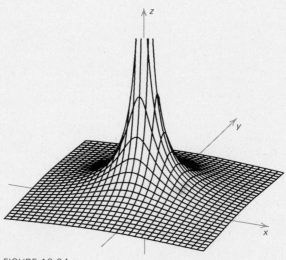

FIGURE 13.34

24. Figure 13.35 shows the graph of the function $-5x/(x^2 + y^2 + 1)$ and Figure 13.36 shows the level curves where it has the values 0, $\pm\frac{1}{4}$, $\pm\frac{1}{2}$, $\pm\frac{3}{4}$, ± 1, $\pm\frac{5}{4}$, $\pm\frac{3}{2}$, $\pm\frac{7}{4}$, ± 2, $\pm\frac{9}{4}$.   **a.**†   Trace and label the level curves with values 0, $\pm\frac{1}{2}$, ± 1, $\pm\frac{3}{2}$, and ± 2.   **b.**   Show that one of the level curves is a line and the others are circles.

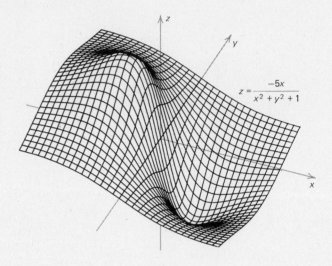

$$z = \frac{-5x}{x^2 + y^2 + 1}$$

FIGURE 13.35

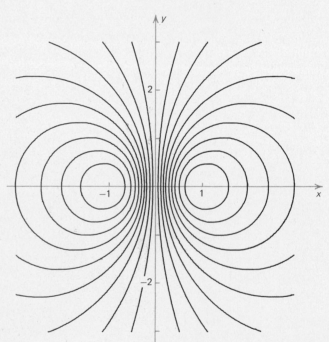

FIGURE 13.36

25. The surface $z = \frac{1}{3}x^3 - xy^2$ shown in Figure 13.37 is called a *monkey saddle* because a monkey sitting on it would have a place for his tail as well as for his legs. The level curves $\frac{1}{3}x^3 - xy^2 = c$ of $\frac{1}{3}x^3 - xy^2$ for $c = 0, \pm\frac{1}{4}, \pm\frac{1}{2}, \pm\frac{3}{4}, \pm1, \pm\frac{5}{4}, \pm\frac{3}{2}, \pm\frac{7}{4}, \pm2$ are shown in Figure 13.38. Trace and label those for $c = 0, \pm1, \pm2$.

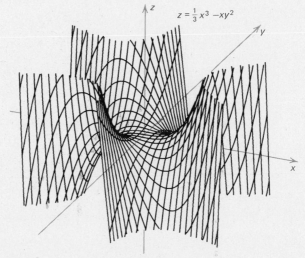

$$z = \frac{1}{3}x^3 - xy^2$$

FIGURE 13.37   A monkey saddle

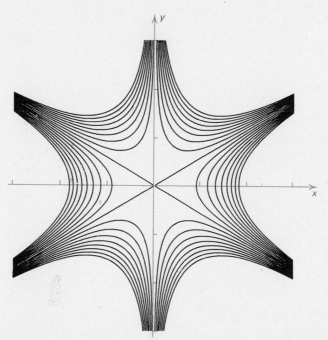

FIGURE 13.38   Level curves of
$\frac{1}{3}x^3 - xy^2$

Sketch the level curves of the functions in Exercises 26 through 36.

26.† xy **27.** $x + 3y$ **28.** $4 - x^2 - y^2$

29.† $x^2 + y$ **30.** $y - \sin x$ **31.** x/y

32.† $\sqrt[3]{x - y}$ **33.** 2^{x-y} **34.** $x^2 y$

35.‡ $y \sin x$ **36.** $x \csc y$

37. The map of North America in Figure 13.39 shows level curves of the number of species of mammals native to each region. What is the approximate maximum number for regions in the United States?

38.† Large earthquakes with centers near coastlines can generate "tsunami" waves which travel at speeds of up to 600 miles per hour in very deep water and are up to 90 feet high near shores. The curves in Figure 13.40 show the successive hourly positions of the tsunami wave front generated by the Alaskan earthquake of 1964. What function has those level curves?

39. The *specific gravity* of an object is its weight divided by the weight of an equal volume of water. In water its weight is reduced by the weight of water it displaces. Express the specific gravity of an object as a function of its weight x in air and its weight y submerged in water. Sketch the level curves of this function.

FIGURE 13.39   Number of species of mammals   (Adapted from C. J. Krebs, [1] *Ecology*, p. 511.)

FIGURE 13.40   Tsunami wave front   (The dot shows the center of the earthquake.) (Adapted from A. Strahler and A. Strahler, [1] *Environmental Geoscience*, p. 221.)

13-2   PARTIAL DERIVATIVES

We can apply calculus of one variable to a function $f(x, y)$ of two variables by studying functions of one variable obtained from $f(x, y)$. We can, for example, hold either x or y constant and consider $f(x, y)$ as a function of the other variable. The derivatives of the resulting functions are called the PARTIAL DERIVATIVES of $f(x, y)$.

Definition 13.4   The partial derivative $\dfrac{\partial f}{\partial x}(x, y)$ of $f(x, y)$ with respect to x is obtained by holding y constant and differentiating with respect to x. The partial derivative $\dfrac{\partial f}{\partial y}(x, y)$ with respect to y is obtained by holding x constant and differentiating with respect to y.

The partial derivative $\dfrac{\partial f}{\partial x}(x, y)$ is the rate of change of $f(x, y)$ with respect to x and $\dfrac{\partial f}{\partial y}(x, y)$ is the rate of change with respect to y.

When we combine Definition 13.4 with the definition of the derivative of a function of one variable, we obtain the direct definitions of the partial derivatives as limits of difference quotients:

$$(1) \qquad \frac{\partial f}{\partial x}(x, y) = \lim_{\Delta x \to 0} \frac{f(x + \Delta x, y) - f(x, y)}{\Delta x} \qquad \blacktriangleleft$$

$$(2) \qquad \frac{\partial f}{\partial y}(x, y) = \lim_{\Delta y \to 0} \frac{f(x, y + \Delta y) - f(x, y)}{\Delta y}. \qquad \blacktriangleleft$$

In most cases we do not have to calculate the limits (1) and (2) to find a function's partial derivatives. Instead, we use rules for differentiating functions of one variable, as is illustrated in the next example.

EXAMPLE 1   Compute the partial derivatives $\dfrac{\partial f}{\partial x}(x, y)$ and $\dfrac{\partial f}{\partial y}(x, y)$ of $f(x, y) = x^3y - y^2x^2 + x$.

Solution.   Holding y constant and differentiating with respect to x, we obtain

$$\frac{\partial f}{\partial x}(x, y) = \frac{\partial}{\partial x}(x^3y - y^2x^2 + x) = 3x^2y - 2y^2x + 1.$$

Holding x constant and differentiating with respect to y gives

$$\frac{\partial f}{\partial y}(x, y) = \frac{\partial}{\partial y}(x^3y - y^2x^2 + x) = x^3 - 2yx^2. \quad \square$$

The graph of the function $f(x, y)$ of x with y held constant has the shape of the intersection of the graph of f with the corresponding vertical plane $y = $ constant (parallel to the x-axis), and the partial derivative $\dfrac{\partial f}{\partial x}(x, y)$ is the slope of the tangent line to that curve (Figure 13.41). Similarly, the graph of $f(x, y)$ considered a function of y with x held constant has the shape of the intersection of the graph of f with the corresponding vertical plane $x = $ constant (parallel to the y-axis), and the partial derivative $\dfrac{\partial f}{\partial y}(x, y)$ is the slope of the tangent line to that curve (Figure 13.42).

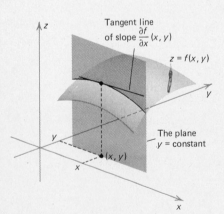

FIGURE 13.41   Vertical cross section of the graph of f with $y = $ constant

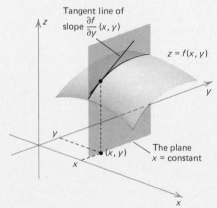

FIGURE 13.42   Vertical cross section of the graph of f with $x = $ constant

Alternate notation

The x-derivative $\dfrac{\partial f}{\partial x}(x, y)$ is also denoted by the symbols $f_x(x, y)$ or $f_1(x, y)$. The y-derivative $\dfrac{\partial f}{\partial y}(x, y)$ is also denoted $f_y(x, y)$ or $f_2(x, y)$.

The subscript "1" on f_1 means the derivative is taken with respect to the first variable x. The subscript "2" on f_2 means the derivative is taken with respect to the second variable y.

EXAMPLE 2   Compute $g_x(3, 5)$ and $g_y(3, 5)$ for $g(x, y) = \sin(x^2 - y)$.

Solution.   We use the chain rule for functions of one variable and obtain

$$g_x(x, y) = \frac{\partial g}{\partial x}(x, y) = \frac{\partial}{\partial x} \sin(x^2 - y)$$

$$= \cos(x^2 - y) \frac{\partial}{\partial x}(x^2 - y) = 2x \cos(x^2 - y)$$

$$g_y(x, y) = \frac{\partial g}{\partial y}(x, y) = \frac{\partial}{\partial y} \sin(x^2 - y)$$

$$= \cos(x^2 - y) \frac{\partial}{\partial y}(x^2 - y) = -\cos(x^2 - y).$$

Setting $x = 3$ and $y = 5$ yields $g_x(3, 5) = 2(3)\cos(3^2 - 5) = 6 \cos(4)$ and $g_y(3, 5) = -\cos(3^2 - 5) = -\cos(4)$.   □

EXAMPLE 3   What are $h_1(x, y)$ and $h_2(x, y)$ for $h(x, y) = (2x - 3y)^{10}$?

Solution.

$$h_1(x, y) = \frac{\partial}{\partial x}(2x - 3y)^{10} = 10(2x - 3y)^9 \frac{\partial}{\partial x}(2x - 3y)$$

$$= 10(2x - 3y)^9(2) = 20(2x - 3y)^9$$

$$h_2(x, y) = \frac{\partial}{\partial y}(2x - 3y)^{10} = 10(2x - 3y)^9 \frac{\partial}{\partial y}(2x - 3y)$$

$$= 10(2x - 3y)^9(-3) = -30(2x - 3y)^9.   □$$

Estimating partial derivatives

The *average rate of change* of $f(x, y)$ with respect to x between the points (x_0, y_0) and (x_1, y_0) with the same y-coordinate is the ratio

(3) $$\frac{f(x_1, y_0) - f(x_0, y_0)}{x_1 - x_0}.$$

The average rate of change with respect to y between the points (x_0, y_0) and (x_0, y_1) with the same x-coordinate is the ratio

(4) $$\frac{f(x_0, y_1) - f(x_0, y_0)}{y_1 - y_0}.$$

According to definitions (1) and (2) of the partial derivatives with (x, y) replaced by (x_0, y_0), $x + \Delta x$ replaced by x_1, and $y + \Delta y$ replaced by y_1, the (instantaneous) rate of change $\dfrac{\partial f}{\partial x}(x_0, y_0)$ is the limit of the average rate of change (3) as x_1 tends to x_0 and the (instantaneous) rate of change $\dfrac{\partial f}{\partial y}(x_0, y_0)$ is the limit of the average rate of change (4) as y_1 tends to y_0. In the next example we use average rates of change to find approximate values of partial derivatives.

EXAMPLE 4   Figure 13.43 shows level curves of the temperature $T(t, h)$ (degrees Centigrade) as a function of time t (hours) and the depth h (centimeters) beneath the surface of the ground at O'Neil, Nebraska, from midnight one day ($t = 0$) to midnight the next ($t = 24$). (a) Describe the partial derivatives $\dfrac{\partial T}{\partial t}(14, 10)$ and $\dfrac{\partial T}{\partial h}(14, 10)$ as rates of change. (b) Use the sketch of level curves to give approximate values of these derivatives.

Solution.   (a) The partial derivative $\dfrac{\partial T}{\partial t}(14, 10)$ is the rate of change of the temperature with respect to time at time $t = 14$ (2:00 *P.M.*) and at a depth of $h = 10$ centimeters. The partial derivative $\dfrac{\partial T}{\partial h}(14, 10)$ is the rate of change of the temperature with respect to depth at that time and depth.

(b) We draw the horizontal and vertical lines through the point $(14, 10)$ as in Figure 13.44. The point $(14, 10)$ lies between the level curves $T = 28$ and $T = 29$ of the temperature. Along the horizontal line $h = 10$ the distance between these level curves is approximately one unit (one hour) on the t-axis, and the level curve $T = 29$ is to the right of the level curve $T = 28$. T increases approximately one degree as t

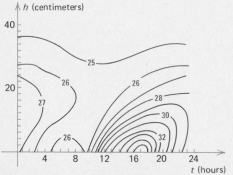

FIGURE 13.43   Soil temperatures near ground level   (Adapted from S. Williamson, [1] *Fundamentals of Air Pollution*.)

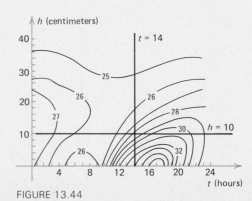

FIGURE 13.44

increases one hour, so the average rate of change and the derivative $\frac{\partial T}{\partial t}(14, 10)$ are approximately one degree per hour.

Along the vertical line $t = 14$ the distance between the two level curves is approximately 2 centimeters on the h-axis, and the level curve $T = 29$ is below the level curve $T = 28$. The temperature decreases by approximately one degree as h increases 2 centimeters, so the average rate of change and the derivative $\frac{\partial T}{\partial h}(14, 10)$ are approximately $(-1$ degree$)/(2$ centimeters$) = -\frac{1}{2}$ degree/centimeter. The temperature decreases with increasing depth at the rate of approximately $\frac{1}{2}$ degree per centimeter. $\square$

Higher-order partial derivatives

The first-order partial derivatives $\frac{\partial f}{\partial x}$ and $\frac{\partial f}{\partial y}$ of a function of the variables x and y can be differentiated with respect to x or y to obtain the second-order partial derivatives $\frac{\partial^2 f}{\partial x^2} = \frac{\partial}{\partial x}\left(\frac{\partial f}{\partial x}\right)$, $\frac{\partial^2 f}{\partial y \partial x} = \frac{\partial}{\partial y}\left(\frac{\partial f}{\partial x}\right)$, $\frac{\partial^2 f}{\partial x \partial y} = \frac{\partial}{\partial x}\left(\frac{\partial f}{\partial y}\right)$, $\frac{\partial^2 f}{\partial y^2} = \frac{\partial}{\partial y}\left(\frac{\partial f}{\partial y}\right)$. We also use the symbols f_{xx} and f_{11} for $\frac{\partial^2 f}{\partial x^2}$, the symbols f_{xy} and f_{12} for $\frac{\partial^2 f}{\partial y \partial x}$, the symbols f_{yx} and f_{21} for $\frac{\partial^2 f}{\partial x \partial y}$, and the symbols f_{yy} and f_{22} for $\frac{\partial^2 f}{\partial y^2}$.

The derivatives f_{xy} and f_{yx} are called the MIXED SECOND-ORDER DERIVATIVES of f. As is illustrated in the next example, they are equal for any functions f encountered in routine calculus. (A proof of this fact is outlined in Miscellaneous Exercise 16.)

EXAMPLE 5   Compute the second-order partial derivatives of $f(x, y) = xy^2 + x^3 y^5$.

Solution.   We have

$$f_x = \frac{\partial}{\partial x}(xy^2 + x^3 y^5) = y^2 + 3x^2 y^5$$

$$f_y = \frac{\partial}{\partial y}(xy^2 + x^3 y^5) = 2xy + 5x^3 y^4.$$

Therefore,

$$f_{xx} = \frac{\partial}{\partial x} f_x = \frac{\partial}{\partial x}(y^2 + 3x^2 y^5) = 6xy^5$$

$$f_{yy} = \frac{\partial}{\partial y} f_y = \frac{\partial}{\partial y}(2xy + 5x^3 y^4) = 2x + 20x^3 y^3$$

$$f_{xy} = \frac{\partial}{\partial y} f_x = \frac{\partial}{\partial y}(y^2 + 3x^2 y^5) = 2y + 15x^2 y^4.$$

As a check on the last calculation, we compute the mixed derivative by first differentiating with respect to y. We obtain

$$f_{yx} = \frac{\partial}{\partial x} f_y = \frac{\partial}{\partial x}(2xy + 5x^3y^4) = 2y + 15x^2y^4.$$

This is, as expected, equal to f_{xy}. □

Third- and higher-order derivatives have analogous definitions and notation. We have, for example,

$$f_{xxx} = \frac{\partial^3 f}{\partial x^3} = \frac{\partial}{\partial x} f_{xx} \quad \text{and} \quad f_{xxy} = \frac{\partial^3 f}{\partial y \partial x^2} = \frac{\partial}{\partial y} f_{xx}.$$

We can also compute f_{xxy} by differentiating first with respect to y and then twice with respect to x, or by differentiating first with respect to x, then with respect to y, and finally with respect to x again. This is because the mixed third-order derivatives, $f_{xxy}, f_{yxx},$ and $f_{xyx},$ are equal.

Exercises

†*Answer provided.*
‡*Outline of solution provided.*

In Exercises 1 through 13 compute the partial derivatives.

1.† $\dfrac{\partial}{\partial x}(xy^5 - 4y^2 + 6x^4y^7)$

2.† $\dfrac{\partial}{\partial y}(x^2\sin(xy) + y - x)$

3. $\dfrac{\partial}{\partial x}(x^3y^2 - x + y)$

4. $\dfrac{\partial}{\partial y}(x^2e^{3y} + y^2e^{3x})$

5.† $\dfrac{\partial f}{\partial x}(x, y)$ and $\dfrac{\partial f}{\partial y}(x, y)$ for $f(x, y) = \sin(x^2y^4)$

6.† $G_x(x, y)$ and $G_y(x, y)$ for $G(x, y) = \ln(1 - xy)$

7. $f_1(x, y)$ and $f_2(x, y)$ for $f(x, y) = x \arctan y$

8.† $p_u(1, 2)$ and $p_v(1, 2)$ for $p(u, v) = e^{u^2} \cos(v^2)$

9. $q_1(10, -5)$ and $q_2(10, -5)$ for $q(u, v) = u^2 \ln(uv^2)$

10.† $\dfrac{\partial}{\partial x}[x \ln(xy)]$

11. $\dfrac{\partial}{\partial y} \sqrt{x^2 + y^2}$

12.† $\dfrac{\partial}{\partial r}\left[\dfrac{r + s}{r - s}\right]$

13. $\dfrac{\partial}{\partial s}[\sin(r + s) + \sin(r - s)]$

In Exercises 14 through 21 compute the second-order partial derivatives of the functions.

14.† $f(x, y) = x^4y^5$

15.† $f(x, y) = \ln(2x - 3y)$

16.† $f(x, y) = (1 + xy)^{10}$

17. $f(x, y) = \arctan(xy)$

18. $g(u, v) = \sinh(3u - 4v)$

19. $g(u, v) = e^{u^2v} + u^2 + v$

20.† $P(r, s) = \dfrac{\sin(rs)}{rs}$

21. $P(r, s) = \sec\left(\dfrac{r}{s}\right)$

22.[†] Compute   **a.** $\dfrac{\partial f}{\partial x}(2, 10),$ **b.** $\dfrac{\partial^2 f}{\partial x^2}(2, 10),$ and   **c.** $\dfrac{\partial^3 f}{\partial y \partial x^2}(2, 10)$

for $f(x, y) = x^3 y^4$.

23. Compute $g_{xyy}(0, 2)$ for $g(x, y) = y^3 e^{-4x}$.

24. Compute $f_{222}(2, 2)$ for $f(x, y) = x^3 y^5$.

25.[†] The volume of a right circular cone of height h and with base of radius r is $V(r, h) = \frac{1}{3} \pi r^2 h$. What is the rate of change of the volume with respect to the radius?

26. If a constant current of I amperes flows through a circuit of resistance R ohms for t seconds, it will produce $H(I, t) = RI^2 t/(4.18)$ calories of heat. What is the rate of heat production with respect to time?

27.[†] If a gas has density ρ_0 grams per cubic centimeter at $0°$ C and 760 millimeters of mercury (mm), then its density at $T°$ C and pressure P mm is $\rho(T, P) = \rho_0\left(1 + \dfrac{T}{273}\right)\dfrac{760}{P}\dfrac{g.}{cm.^3}$. What are the rates of change of the density with respect to temperature and pressure?

28.[†] Figure 13.45 shows level curves of the function $R(L, t)$ that gives the amount of solar radiation (measured in gram-calories) received by a horizontal plate one centimeter square during a cloudless day as a function of the latitude L and the time of year t. The latitude is measured in degrees with negative L for the southern hemisphere. The time is given in months with $t = 0$ corresponding to January 1. The curve labeled "Sun's declination" gives the latitude of the sun as a function of t. (The amount of radiation varies with L and t because it depends on the angle of the sun and the length of the day.) What are the approximate values at a latitude of $40°$ N on May 1 (at $t = 4$) of   **a.**   the amount of radiation,   **b.**   the rate of change of the radiation with respect to L, and   **c.**   the rate of change of the radiation with respect to t?

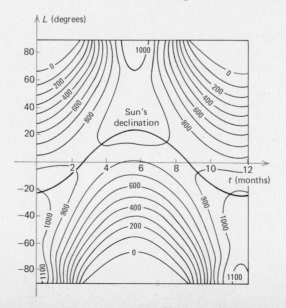

FIGURE 13.45 Solar radiation as a function of latitude and time of year (Adapted from A. Strahler, [1] *Physical Geography*, p. 189.)

29. Figure 13.46 shows level curves of the function $Y(x, t)$ that gives the yield (cubic feet per acre) from a pine plantation with x trees per acre that are harvested t years after planting. **a.**[†] Describe $\dfrac{\partial Y}{\partial t}(1000, 20)$ as a rate of change. **b.** Give an approximate value for this rate of change.

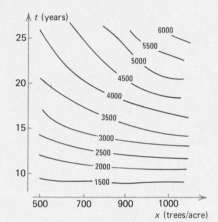

FIGURE 13.46 Yield from a loblolly pine plantation (Adapted from D. Lenhart, [1] *Yields for Loblolly Pine Plantations in the Interior West Gulf Coastal Plain*, p. 17.)

30. In a triode vacuum tube (Figure 13.47) the plate voltage V can be considered a function of the grid voltage x and the plate current y. Figure 13.48 shows level curves of $V(x, y)$. What are the approximate rates of change of the plate voltage **a.** with respect to the grid voltage and **b.** with respect to plate current when the grid voltage is -5 volts and the plate current is 4 amperes?

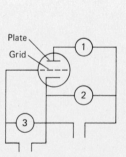

FIGURE 13.47 A triode vacuum tube (1) Ammeter for measuring plate current (2) Voltmeter for measuring plate voltage (3) Voltmeter for measuring grid voltage

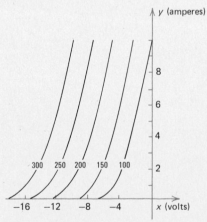

FIGURE 13.48 Level curves of plate voltage (volts) (Adapted from Herrick, [1] *Mathematics for Electronics*, p. 800.)

31. *Scherk's minimal surface*, shown in Figure 13.49, is called "minimal" because it has the least surface area of all surfaces with the same boundary. If we were to dip a piece of wire with the shape of its boundary (Figure 13.50) into soapy water, the film of soap that would form on the wire would have the shape of the minimal surface. Because the surface is the graph of $f(x, y) = \ln(\cos x) - \ln(\cos y)$, the function f satisfies the *minimal surface differential equation* $(1 + f_y^2)f_{xx} - 2f_x f_y f_{xy} + (1 + f_x^2)f_{yy} = 0$. Show that this is the case.

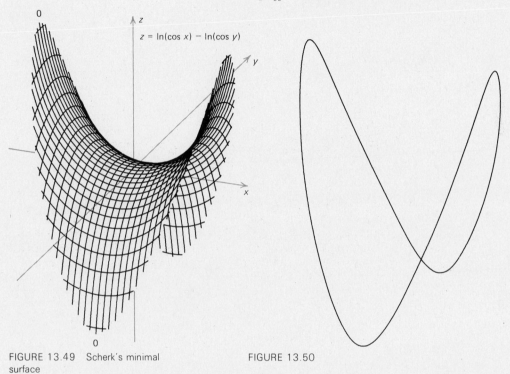

FIGURE 13.49   Scherk's minimal surface

FIGURE 13.50

32. **a.** Show that the temperature $T(x, y) = \dfrac{\sin x \sinh y}{\sinh 3}$ of a metal plate that occupies the region $0 \le x \le \pi, 0 \le y \le 3$ satisfies Laplace's differential equation $T_{xx} + T_{yy} = 0$.   **b.**   What is the temperature of the plate on its sides, bottom, and top? (Level curves of $T(x, y)$ are shown in Figure 13.51.)

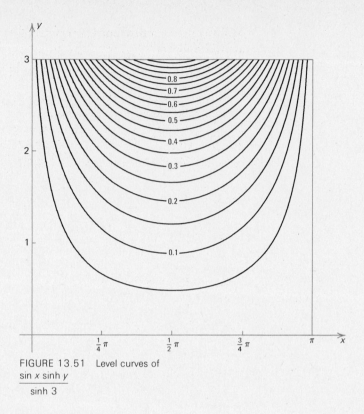

FIGURE 13.51   Level curves of
$$\frac{\sin x \sinh y}{\sinh 3}$$

Additional drill exercises

Find the first-order partial derivatives of the functions in Exercises 33 through 40.

33.[†] $f = x^{1/2}y^{1/4} + x^2y^4$ **34.**[†] $f = x\sqrt{x^2 + y^2}$

35.[†] $f = \sin(x/y) + \cos(xy)$ **36.**[†] $f = (x + y)/(x - y)$

37.[†] $g = \arctan(y/x)$ **38.**[†] $g = xye^{x-y}$

39.[†] $g = x^2y^3/(x^4 - y^5)$ **40.**[†] $g = \ln(\sec x + \tan y)$

Find the second-order partial derivatives of the functions in Exercises 41 through 46.

41.[†] $f = x^2y^3 - y^2x^3$ **42.**[†] $f = xy/(x + y)$

43.[†] $f = \sin(xy)$ **44.**[†] $g = x^{1/2}(2 - y)^{1/3}$

45.[†] $g = \ln(xy + y^2)$ **46.**[†] $g = \sqrt{4 - x^2 - y^2}$

13-3   LIMITS, CONTINUITY, AND THE CHAIN RULE

Whereas a variable point x on a coordinate line can approach a fixed point x_0 in only two directions, namely either from the left or from the right, a variable point (x, y) in a coordinate plane can approach a fixed point (x_0, y_0) along an infinite number of paths (Figure 13.52). In defining limits of functions of two variables we will say that (x, y) is approaching (x_0, y_0) if the distance between the points is tending to zero, regardless of the particular path (x, y) follows on its way toward (x_0, y_0).

Recall that we could speak of the limit of $f(x)$ as x tends to x_0 whether or not f was defined at x_0. We required only that $f(x)$ be defined in open intervals $a < x < x_0, x_0 < x < b$ on both sides of x_0. Also, if $f(x)$ was defined on only one side of x_0, we could consider its one-sided limit as x tended to x_0 from the left or the right.

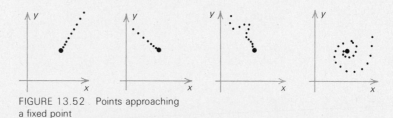

FIGURE 13.52   Points approaching a fixed point

Similarly, in defining the limit of a function $f(x, y)$ of two variables as (x, y) approaches a fixed point (x_0, y_0), it is not necessary that $f(x, y)$ be defined at (x_0, y_0). We require only that there be points $(x, y) \neq (x_0, y_0)$ in the domain of f that approach (x_0, y_0). Then we say that the limit of $f(x, y)$ is the number L and we write

$$\lim_{(x,y)\to(x_0,y_0)} f(x, y) = L$$

or

$$f(x, y) \to L \quad \text{as} \quad (x, y) \to (x_0, y_0)$$

provided the value $f(x, y)$ of the function at (x, y) tends to L as (x, y) approaches (x_0, y_0) along all paths that lie in the domain of f.*

Many questions about such limits can be answered by using the following two theorems. The first deals with functions that depend on only one of the two variables. The second deals with sums, products, and quotients of functions.

Theorem 13.1   **If $F(x)$ tends to a number L as x tends to x_0, then $F(x)$ considered as a function of x and of y tends to L as (x, y) approaches (x_0, y_0) for arbitrary y_0.**

If $G(y)$ tends to a number M as y tends to y_0, then $G(y)$ considered as a function of x and of y tends to M as (x, y) approaches (x_0, y_0) for arbitrary x_0.

---

*In the $\epsilon\delta$-definition of this limit we require that for each $\epsilon > 0$ there be a $\delta > 0$ such that $|f(x, y) - L| < \epsilon$ for all (x, y) in the domain of f with $0 < (x - x_0)^2 + (y - y_0)^2 < \delta^2$.

The domain of $F(x)$ viewed as a function of x and y consists of all (x, y) with x in the domain of $F(x)$ and arbitrary y. The domain of $G(y)$ viewed as a function of the two variables consists of all (x, y) with y in the domain of $G(y)$ and arbitrary x.

Theorem 13.2   **If there are points common to the domains of $f(x, y)$ and $g(x, y)$ that approach (x_0, y_0) and if $f(x, y)$ tends to a number L and $g(x, y)$ tends to a number M as (x, y) approaches (x_0, y_0), then**

> $f(x, y) + g(x, y) \to L + M$

> $f(x, y)g(x, y) \to LM$

as (x, y) approaches (x_0, y_0). If M is not zero, then

$$\frac{f(x, y)}{g(x, y)} \to \frac{L}{M}.$$

The domains of the sum and product of the two functions are formed by all points that are in the domains of both functions. The domain of the quotient consists of all points in the domains of both functions for which the denominator is not zero.

The proofs of these theorems are based on the $\epsilon\delta$-definitions of limits and are left as Exercises 28 and 29.

EXAMPLE 1   Use Theorems 13.1 and 13.2 to show that

$$\lim_{(x,y)\to(2,3)} \frac{x^2 - y}{xy} = \frac{1}{6}.$$

Solution.   Theorem 13.1 shows that $x \to 2$, $y \to 3$, and $x^2 \to 2^2 = 4$ as $(x, y) \to (2, 3)$. Then Theorem 13.2 implies that $x^2 - y \to 4 - 3 = 1$, $xy \to (2)(3) = 6$, and, because the limit of the denominator is not zero, $(x^2 - y)/(xy) \to \frac{1}{6}$.   □

In the next example, the limit of the denominator is zero, so we cannot apply Theorem 13.2. We find the limit by estimating the values of the function.

EXAMPLE 2   Show that

$$\lim_{(x,y)\to(0,0)} \frac{x^3}{x^2 + y^2} = 0.$$

Solution.   Because $|x| = (x^2)^{1/2}$ is less than or equal to $(x^2 + y^2)^{1/2}$, we have for $(x, y) \neq (0, 0)$

$$\left| \frac{x^3}{x^2 + y^2} \right| \leq \frac{(x^2 + y^2)^{3/2}}{x^2 + y^2} = (x^2 + y^2)^{1/2}.$$

Since $(x^2 + y^2)^{1/2}$ tends to zero as (x, y) approaches the origin, $x^3/(x^2 + y^2)$ must tend to zero also.   □

Recall that if the one-sided limits $\lim_{x \to x_0^-} f(x)$ and $\lim_{x \to x_0^+} f(x)$ of a function of one variable exist but are different numbers, then the two-sided limit $\lim_{x \to x_0} f(x)$ does not exist. The next example deals with a similar case involving a function of two variables: the function $f(x, y)$ does not have a limit as (x, y) approaches the origin because $f(x, y)$ tends to different numbers as (x, y) approaches the origin from different directions.

EXAMPLE 3   Show that the function

$$f(x, y) = \frac{x^2}{x^2 + y^2}$$

does not have a limit as (x, y) tends to $(0, 0)$.

Solution.   The function $f(x, y)$ is not defined at the origin. At all other points on the y-axis, x is zero and the function has the value 0: $f(0, y) = 0$. The nonvertical line through the origin with slope m has the equation $y = mx$. At all points on this line other than the origin f has the constant value

$$f(x, mx) = \frac{x^2}{x^2 + (mx)^2} = \frac{1}{1 + m^2}$$

so $f(x, y)$ has this limit as (x, y) approaches the origin along this line. Thus, $f(x, y)$ approaches different numbers depending on the path (x, y) takes toward the origin, and the function does not have a limit.   □

Continuity   As with functions of one variable, continuity of functions of two variables is defined in terms of limits.

Definition 13.5   A function $f(x, y)$ is continuous at a point (x_0, y_0) in its domain if $f(x, y) \to f(x_0, y_0)$ as $(x, y) \to (x_0, y_0)$.

Theorem 13.1 implies that a continuous function of one variable x or y is continuous when considered a function of the two variables. Theorem 13.2 shows that the sum, product, and quotient of continuous functions are continuous wherever they are defined.

Composite functions   Composing two functions of one or two variables leads to composite functions of the four types

(1) $F\big(x(t)\big)$

(2) $F\big(x(u, v)\big)$

(3) $f\big(x(t), y(t)\big)$

(4) $f\big(x(u, v), y(u, v)\big).$

Function (1) is the composition of functions $F(x)$ and $x(t)$ of one variable and is the type we studied in Chapter 3. In (2) we have obtained a function of the two variables u and v by composing $F(x)$ with a function $x(u, v)$ of u and v. We obtain the function (3) of the one variable t by composing a function $f(x, y)$ of two variables with two functions $x(t)$ and $y(t)$ of t. Finally, composing $f(x, y)$ with two functions $x(u, v)$ and $y(u, v)$ of u and v yields the function of two variables in (4).

EXAMPLE 4   Give a formula for $F(x(u, v))$ in terms of u and v where $F(x) = x^3 + 2x$ and $x(u, v) = u^3 \sin v$.

Solution. $F(x(u, v)) = x(u, v)^3 + 2x(u, v) = (u^3 \sin v)^3 + 2(u^3 \sin v)$
$= u^9 \sin^3 v + 2u^3 \sin v.$ $\square$

EXAMPLE 5   Express $f(x(t), y(t))$ directly in terms of t where $f(x, y) = x^2 y - y^3, x(t) = e^t$ and $y(t) = \ln t$.

Solution. $f(x(t), y(t)) = x(t)^2 y(t) - y(t)^3 = (e^t)^2 (\ln t) - (\ln t)^3$
$= e^{2t} \ln t - (\ln t)^3.$ $\square$

EXAMPLE 6   What is the value of $f(x(u, v), y(u, v))$ at $u = 1, v = 2$ if $x(1, 2) = 3, y(1, 2) = 5$ and $f(3, 5) = 10$?

Solution. $f(x(1, 2), y(1, 2)) = f(3, 5) = 10.$ $\square$

The domain for each composite function consists of all values of its variable for which the defining expression makes sense. For example, the domain of $f(x(u, v), y(u, v))$ consists of all (u, v) that are in the domains of both $x(u, v)$ and $y(u, v)$ and are such that $(x(u, v), y(u, v))$ is in the domain of f.

Theorem 13.3   The composition of continuous functions is continuous.

Proof.   We will deal with the case of functions of the form $f(x(u, v), y(u, v))$. The proof for the other cases is similar. Suppose that (u_0, v_0) is in the domain of the composite function and that the points (u, v) in the domain approach (u_0, v_0). Because $x(u, v)$ and $y(u, v)$ are continuous functions, the points $(x(u, v), y(u, v))$ approach the point $(x(u_0, v_0), y(u_0, v_0))$ and then, because $f(x, y)$ is continuous, $f(x(u, v), y(u, v))$ tends to $f(x(u_0, v_0), y(u_0, v_0))$. This shows that the composite function is continuous at the arbitrary point (u_0, v_0) in its domain.   Q.E.D.

The chain rule for $F(x(u,v))$

The chain rule for functions of one variable (Theorem 3.6) states that if $x(t)$ has a derivative at t and $F(x)$ has a derivative at $x(t)$, then the composite function $F(x(t))$ has a derivative at t given by

$$(5) \quad \frac{d}{dt} F(x(t)) = \frac{dF}{dx}(x(t)) \frac{dx}{dt}(t).$$

This result also tells how to differentiate the composite function $F(x(u, v))$ because in computing its partial derivatives we hold either u or v constant and the "inside" function $x(u, v)$ becomes a function of only one variable. The notation is different because we use the symbols $\partial x / \partial u$ and $\partial x / \partial v$ for the derivatives of $x(u, v)$:

Theorem 13.4   If $x(u, v)$ has first order partial derivatives at (u, v) and $F(x)$ has a derivative at $x(u, v)$, then the composite function $F(x(u, v))$ has first order partial derivatives at (u, v) given by

$$(6) \quad \frac{\partial}{\partial u} F(x(u, v)) = \frac{dF}{dx}(x(u, v)) \frac{\partial x}{\partial u}(u, v) \qquad \blacktriangleleft$$

$$(7) \quad \frac{\partial}{\partial v} F(x(u, v)) = \frac{dF}{dx}(x(u, v)) \frac{\partial x}{\partial v}(u, v). \qquad \blacktriangleleft$$

EXAMPLE 7   What are the first order derivatives of $g(u, v) = x(u, v)^{4/3}$ at $u = 10, v = 20$ if $x(10, 20) = 8, \dfrac{\partial x}{\partial u}(10, 20) = 5,$ and $\dfrac{\partial x}{\partial v}(10, 20) = -7$?

Solution.   By (6) and (7) with $F(x) = x^{4/3}$ and $\dfrac{dF}{dx}(x) = \frac{4}{3} x^{1/3}$, we have

$$\frac{\partial g}{\partial u}(u, v) = \frac{\partial}{\partial u} x(u, v)^{4/3} = \tfrac{4}{3} x(u, v)^{1/3} \frac{\partial x}{\partial u}(u, v)$$

$$\frac{\partial g}{\partial v}(u, v) = \frac{\partial}{\partial v} x(u, v)^{4/3} = \tfrac{4}{3} x(u, v)^{1/3} \frac{\partial x}{\partial v}(u, v).$$

Then setting $u = 10$ and $v = 20$ yields

$$\frac{\partial g}{\partial u}(10, 20) = \tfrac{4}{3} x(10, 20)^{1/3} \frac{\partial x}{\partial u}(10, 20) = \tfrac{4}{3}(8)^{1/3}(5) = \tfrac{4}{3}(2)(5)$$

$$= \tfrac{40}{3}$$

$$\frac{\partial g}{\partial v}(10, 20) = \tfrac{4}{3} x(10, 20)^{1/3} \frac{\partial x}{\partial v}(10, 20) = \tfrac{4}{3}(8)^{1/3}(-7)$$

$$= \tfrac{4}{3}(2)(-7) = -\tfrac{56}{3}. \quad \square$$

The chain rule for $f(x(t), y(t))$

We cannot use the chain rule (5) for functions of one variable to find the derivative of a composite function of the form $f(x(t), y(t))$ because both variables, $x = x(t)$ and $y = y(t)$, of f vary when we vary t. We use the version of the chain rule proved in the next theorem.

Theorem 13.5   Suppose that $x(t)$ and $y(t)$ have first-order derivatives at t and that $f(x, y)$ has continuous first-order partial derivatives in a circle centered at $(x(t), y(t))$. Then

$$\frac{d}{dt} f(x(t), y(t)) = \frac{\partial f}{\partial x}(x(t), y(t))\frac{dx}{dt}(t)$$

(8)   ◀

$$+ \frac{\partial f}{\partial y}(x(t), y(t))\frac{dy}{dt}(t).$$

If we abbreviate the notation by dropping the references to where the derivatives are evaluated, the chain rule (8) reads

(9) $\dfrac{d}{dt}f = \dfrac{\partial f}{\partial x}\dfrac{dx}{dt} + \dfrac{\partial f}{\partial y}\dfrac{dy}{dt}.$ ◀

It states that the rate of change of f with respect to t is equal to the rate of change of f with respect to x multiplied by the rate of change of x with respect to t, plus the rate of change of f with respect to y multiplied by the rate of change of y with respect to t.

Proof of Theorem 13.5   Because $x(t)$ and $y(t)$ have derivatives at t, they are continuous there and

(10) $\lim\limits_{\Delta t \to 0} x(t + \triangle t) = x(t), \qquad \lim\limits_{\Delta t \to 0} y(t + \triangle t) = y(t).$

Consequently, for $|\triangle t|$ small enough the point $(x(t + \triangle t), y(t + \triangle t))$ is contained in the circle where $f(x, y)$ has continuous derivatives. We set $(x_0, y_0) = (x(t), y(t))$ and $(x_1, y_1) = (x(t + \triangle t), y(t + \triangle t))$ and write

(11)
$$\begin{aligned} f(x(t + \Delta t), y(t + \Delta t)) - f(x(t), y(t)) &= f(x_1, y_1) - f(x_0, y_0) \\ &= [f(x_1, y_0) - f(x_0, y_0)] + [f(x_1, y_1) - f(x_1, y_0)]. \end{aligned}$$

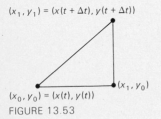

$(x_1, y_1) = (x(t + \Delta t), y(t + \Delta t))$

(x_1, y_0)

$(x_0, y_0) = (x(t), y(t))$

FIGURE 13.53

Equation (11) states that the change in $f(x, y)$ from (x_0, y_0) to (x_1, y_1) is equal to the sum of the changes as (x, y) moves parallel to the x-axis from (x_0, y_0) to (x_1, y_0) and then parallel to the y-axis from (x_1, y_0) to (x_1, y_1) (Figure 13.53). [Algebraically, we obtain equation (11) by adding $f(x_1, y_0) - f(x_1, y_0)$, which equals zero, to $f(x_1, y_1) - f(x_0, y_0)$.]

According to the Mean Value Theorem for derivatives applied to the function $f(x, y_0)$ of x and to the function $f(x_1, y)$ of y, there are numbers c_1 between x_0 and x_1 and c_2 between y_0 and y_1 such that

(12) $f(x_1, y_0) - f(x_0, y_0) = \dfrac{\partial f}{\partial x}(c_1, y_0)(x_1 - x_0)$

(13) $f(x_1, y_1) - f(x_1, y_0) = \dfrac{\partial f}{\partial y}(x_1, c_2)(y_1 - y_0).$

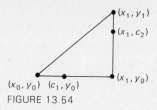

FIGURE 13.54

The points (c_1, y_0) and (x_1, c_2) are on the sides of the right triangle shown in Figure 13.54 and, therefore, approach (x_0, y_0) as (x_1, y_1) approaches (x_0, y_0).

When we substitute equations (12) and (13) into (11) and divide by Δt, we obtain

(14)
$$\frac{f(x(t + \Delta t), y(t + \Delta t)) - f(x(t), y(t))}{\Delta t}$$

$$= \frac{\partial f}{\partial x}(x_0, c_1) \left[\frac{x(t + \Delta t) - x(t)}{\Delta t} \right]$$

$$+ \frac{\partial f}{\partial y}(x_1, c_2) \left[\frac{y(t + \Delta t) - y(t)}{\Delta t} \right].$$

Here we have also replaced $x_1 - x_0$ by $x(t + \Delta t) - x(t)$ and $y - y_0$ by $y(t + \Delta t) - y(t)$. As Δt tends to zero, the left side of (14) approaches the derivative of $f(x(t), y(t))$ with respect to t; the points (x_0, c_1) and (x_1, c_2) approach $(x_0, y_0) = (x(t), y(t))$, so the partial derivatives of $f(x, y)$ on the right side of (14) tend to their values at $(x(t), y(t))$; and the expressions in square brackets tend to the derivatives of $x(t)$ and $y(t)$ with respect to t. Therefore, in the limit equation (14) becomes equation (8).   Q.E.D.

EXAMPLE 8   If $x(1) = 2$, $y(1) = 3$, $\dfrac{dx}{dt}(1) = -4$, and $\dfrac{dy}{dt}(1) = 5$, and if $f(x, y)$ has continuous first-order partial derivatives with the values $\dfrac{\partial f}{\partial x}(2, 3) = -6$ and $\dfrac{\partial f}{\partial y}(2, 3) = 7$ at $x = 2$, $y = 3$, then what is the derivative $\dfrac{d}{dt} f(x(t), y(t))$ at $t = 1$?

Solution.   By the chain rule (8) we have

$$\left[\frac{d}{dt} f(x(t), y(t)) \right]_{t=1} = \frac{\partial f}{\partial x}(x(1), y(1)) \frac{dx}{dt}(1)$$

$$+ \frac{\partial f}{\partial y}(x(1), y(1)) \frac{dy}{dt}(1)$$

$$= \frac{\partial f}{\partial x}(2, 3) \frac{dx}{dt}(1) + \frac{\partial f}{\partial y}(2, 3) \frac{dy}{dt}(1)$$

$$= (-6)(-4) + (7)(5) = 59. \quad \square$$

EXAMPLE 9   What is the derivative of $G(t) = H(t^3, 5t)$ at $t = 1$ if $H(x, y)$ has continuous first order derivatives and $H_x(1, 5) = 4$, $H_y(1, 5) = -2$?

Solution.   By formula (8) with f replaced by H and $x(t) = t^3$, $y(t) = 5t$

$$\frac{dG}{dt}(t) = \frac{d}{dt} H(t^3, 5t) = H_x(t^3, 5t) \frac{d}{dt}(t^3) + H_y(t^3, 5t) \frac{d}{dt}(5t)$$

$$= 3t^2 H_x(t^3, 5t) + 5H_y(t^3, 5t).$$

When we set $t = 1$ and use the given values of the partial derivatives, we obtain

$$\frac{dG}{dt}(1) = 3H_x(1, 5) + 5H_y(1, 5) = 3(4) + 5(-2) = 2. \quad \square$$

EXAMPLE 10   Let $P(b, g)$ dollars denote the profit a store makes on potatoes and milk if it sells b bushels of potatoes and g gallons of milk in one day. Let $b(t)$ bushels and $g(t)$ gallons be the amounts of potatoes and milk it has sold t hours after opening. Express the profit after t hours in terms of the functions $P(b, g)$, $b(t)$, and $g(t)$ and express the rate of increase of the profit in terms of their derivatives.

Solution.   After t hours the profit is $P(b(t), g(t))$ dollars and by the chain rule the rate of increase of the profit is

$$(15) \quad \frac{d}{dt} P(b(t), g(t)) = \frac{\partial P}{\partial b}(b(t), g(t)) \frac{db}{dt}(t) + \frac{\partial P}{\partial g}(b(t), g(t)) \frac{dg}{dt}(t)$$

dollars per hour. $\square$

Dimension analysis

The partial derivative $\dfrac{\partial P}{\partial b}$ in Example 10 has the units of dollars per bushel; the derivative $\dfrac{\partial P}{\partial g}$ has the units of dollars per gallon; the derivative $\dfrac{db}{dt}$ has the units of bushels per hour; and the derivative $\dfrac{dg}{dt}$ has the units of gallons per hour. When we express equation (15) in abbreviated notation and include the units, it reads

$$\frac{d}{dt} P\left(\frac{\text{dollars}}{\text{hour}}\right) = \frac{\partial P}{\partial g}\left(\frac{\text{dollars}}{\text{gallon}}\right) \frac{dg}{dt}\left(\frac{\text{gallons}}{\text{hour}}\right)$$
$$+ \frac{\partial P}{\partial b}\left(\frac{\text{dollars}}{\text{bushel}}\right) \frac{db}{dt}\left(\frac{\text{bushels}}{\text{hour}}\right)$$

where the units of "gallons" and "bushels" on the right cancel to give the units of dollars per hour on the left:

$$\frac{\text{dollars}}{\text{hour}} = \left(\frac{\text{dollars}}{\text{gallon}}\right)\left(\frac{\text{gallons}}{\text{hour}}\right) + \left(\frac{\text{dollars}}{\text{bushel}}\right)\left(\frac{\text{bushels}}{\text{hour}}\right).$$

The chain rule for $f(x(u, v), y(u, v))$

If we start with a function $f(x, y)$ of two variables and consider x and y as functions $x = x(u, v), y = y(u, v)$ of two other variables, we obtain a composite function $f(x(u, v), y(u, v))$, whose partial derivatives can be found by using the next theorem.

Theorem 13.6   Suppose that $x(u, v)$ and $y(u, v)$ have first-order partial derivatives at (u, v) and that $f(x, y)$ has continuous first-order partial derivatives in a circle centered at $(x(u, v), y(u, v))$. Then

(16) $\dfrac{\partial}{\partial u} f(x(u, v), y(u, v)) = \dfrac{\partial f}{\partial x}(x(u, v), y(u, v)) \dfrac{\partial x}{\partial u}(u, v)$

$$+ \dfrac{\partial f}{\partial y}(x(u, v), y(u, v)) \dfrac{\partial y}{\partial u}(u, v)$$

(17) $\dfrac{\partial}{\partial v} f(x(u, v), y(u, v)) = \dfrac{\partial f}{\partial x}(x(u, v), y(u, v)) \dfrac{\partial x}{\partial v}(u, v)$

$$+ \dfrac{\partial f}{\partial y}(x(u, v), y(u, v)) \dfrac{\partial y}{\partial v}(u, v).$$

Formulas (16) and (17) are more easily remembered in abbreviated notation

(18) $\dfrac{\partial}{\partial u} f = \dfrac{\partial f}{\partial x} \dfrac{\partial x}{\partial u} + \dfrac{\partial f}{\partial y} \dfrac{\partial y}{\partial u}$

(19) $\dfrac{\partial}{\partial v} f = \dfrac{\partial f}{\partial x} \dfrac{\partial x}{\partial v} + \dfrac{\partial f}{\partial y} \dfrac{\partial y}{\partial v}$

where the symbols "∂x" and "∂y" on the right sides of the equations seem to cancel.

Theorem 13.6 is an immediate consequence of Theorem 13.5 because, when we hold u or v constant to compute a partial derivative with respect to the other variable, we are considering functions $x(u, v)$ and $y(u, v)$ of one variable.

EXAMPLE 11   The function $f(x, y)$ has continuous first-order partial derivatives with the values $\dfrac{\partial f}{\partial x}(2, -4) = 10$ and $\dfrac{\partial f}{\partial y}(2, -4) = -5$ at $x = 2, y = -4$. Compute $\dfrac{\partial}{\partial u} f(ue^v, v - u^2)$ and $\dfrac{\partial}{\partial v} f(ue^v, v - u^2)$ at $u = 2, v = 0$.

Solution.   By (16) and (17) we have

$$\dfrac{\partial}{\partial u} f(ue^v, v - u^2) = \dfrac{\partial f}{\partial x}(ue^v, v - u^2) \dfrac{\partial}{\partial u}(ue^v)$$

$$+ \dfrac{\partial f}{\partial y}(ue^v, v - u^2) \dfrac{\partial}{\partial u}(v - u^2)$$

$$= \dfrac{\partial f}{\partial x}(ue^v, v - u^2)(e^v) + \dfrac{\partial f}{\partial y}(ue^v, v - u^2)(-2u)$$

$$\dfrac{\partial}{\partial v} f(ue^v, v - u^2) = \dfrac{\partial f}{\partial x}(ue^v, v - u^2) \dfrac{\partial}{\partial v}(ue^v)$$

$$+ \dfrac{\partial f}{\partial y}(ue^v, v - u^2) \dfrac{\partial}{\partial v}(v - u^2)$$

$$= \dfrac{\partial f}{\partial x}(ue^v, v - u^2)(ue^v) + \dfrac{\partial f}{\partial y}(ue^v, v - u^2)(1).$$

At $u = 2, v = 0$ these derivatives are

$$\left[\frac{\partial}{\partial u} f(ue^v, v - u^2) \right]_{u=2,\, v=0} = \frac{\partial f}{\partial x}(2e^0, 0 - 2^2)(e^0)$$

$$+ \frac{\partial f}{\partial y}(2e^0, 0 - 2^2)(-2(2))$$

$$= \frac{\partial f}{\partial x}(2, -4)(1) + \frac{\partial f}{\partial y}(2, -4)(-4)$$

$$= (10)(1) + (-5)(-4) = 30$$

$$\left[\frac{\partial}{\partial v} f(ue^v, v - u^2) \right]_{u=2,\, v=0} = \frac{\partial f}{\partial x}(2e^0, 0 - 2^2)(2e^0)$$

$$+ \frac{\partial f}{\partial y}(2e^0, 0 - 2^2)(1)$$

$$= \frac{\partial f}{\partial x}(2, -4)(2) + \frac{\partial f}{\partial y}(2, -4)(1)$$

$$= (10)(2) + (-5)(1) = 15. \quad \square$$

Exercises

In Exercises 1 through 6 find the limits.

1.† $\displaystyle \lim_{(x,\, y) \to (-5,\, 3)} \frac{x^3 y}{x^2 - y}$

2. $\displaystyle \lim_{(x,\, y) \to (-2,\, 7)} x \sin(xy)$

3.† $\displaystyle \lim_{(x,\, y) \to (0,\, 0)} e^{3x - 2y}$

4. $\displaystyle \lim_{(x,\, y) \to (1,\, 3)} \ln(x^2 + 3y)$

5.† $\displaystyle \lim_{(x,y) \to (10,0)} \arcsin(xy)$

6. $\displaystyle \lim_{(x,y) \to (3,\pi)} \tan(\tfrac{1}{4} xy)$

7.† Express $F(t) = f(x(t), y(t))$ in terms of t for $f(x, y) = (x/y)\, e^{xy}$, $x(t) = \sin t$, and $y(t) = t^3$.

8. Express $F(t) = f(x(t), y(t))$ in terms of t for $f(x, y) = x \sin(xy)$, $x(t) = t^5$, and $y(t) = t^{-3}$.

9.† Express $P(u, v) = f(x(u, v), y(u, v))$ in terms of u and v for $f(x, y) = x \ln(xy + 1)$, $x(u, v) = ue^v$, and $y(u, v) = u^2 v^3$.

10. Express $P(u, v) = f(x(u, v), y(u, v))$ in terms of u and v for $f(x, y) = \sqrt{x^2 + y^2}$, $x(u, v) = u^2 v^2$, and $y(u, v) = 2u - 3v$.

In the remaining exercises all functions are assumed to have continuous derivatives.

11.‡ Compute **a.** $g(2)$ and **b.** $\dfrac{dg}{dt}(2)$ for $g(t) = f(t^3, t^4)$ where $f(8, 16) = 3, f_x(8, 16) = 5,$ and $f_y(8, 16) = -7$.

12.† Compute **a.** $h(3)$ and **b.** $\dfrac{dh}{dt}(3)$ for $h(t) = g(t^3 - 5t, 11t - 1)$ where $g(12, 32) = 0, g_x(12, 32) = -3,$ and $g_y(12, 32) = 2$.

13. Compute **a.** $K(0)$ and **b.** $K'(0)$ for $K(t) = L(\sin t, \cos t)$ where $L(0, 1) = 50$, $L_x(0, 1) = 10$, and $L_y(0, 1) = -7$.

14. Compute **a.** $Z(0)$ and **b.** $Z'(0)$ for $Z(t) = W(\ln(1 + t), e^{5t})$ where $W(0, 1) = -3$, $W_x(0, 1) = 2$, and $W_y(0, 1) = 3$.

15.‡ Compute **a.** $F(1, 2)$, **b.** $F_u(1, 2)$, and **c.** $F_v(1, 2)$ for $F(u, v) = f(u^2 v^3, 3u + 7v)$ with $f(8, 17) = 10$, $f_x(8, 17) = 5$, and $f_y(8, 17) = -1$.

16.† Compute **a.** $F(0, 2)$, **b.** $F_u(0, 2)$, and **c.** $F_v(0, 2)$ for $F(u, v) = f(v \sin u, u \sin v)$ with $f(0, 0) = 4$, $f_x(0, 0) = 10$, and $f_y(0, 0) = 2$.

17. Compute **a.** $F(3, 0)$, **b.** $F_u(3, 0)$, and **c.** $F_v(3, 0)$ for $F(u, v) = f(2u + e^{3v}, u + e^{4v})$ with $f(7, 4) = -3$, $f_x(7, 4) = 4$, and $f_y(7, 4) = -5$.

18.‡ Compute **a.** $P(3, 2)$, **b.** $P_u(3, 2)$, and **c.** $P_v(3, 2)$ for $P(u, v) = R(x(u, v), y(u, v))$ where $x(3, 2) = 1$, $y(3, 2) = 0$, $x_u(3, 2) = 5$, $x_v(3, 2) = 6$, $y_u(3, 2) = 7$, $y_v(3, 2) = 4$, $R(1, 0) = 8$, $R_x(1, 0) = 9$, and $R_y(1, 0) = 10$.

19. Compute **a.** $W(11, 10)$, **b.** $W_u(11, 10)$, and **c.** $W_v(11, 10)$ for $W(u, v) = Z(x(u, v), y(u, v))$ with $x(11, 10) = 9$, $y(11, 10) = 8$, $x_u(11, 10) = 7$, $x_v(11, 10) = 6$, $y_u(11, 10) = 5$, $y_v(11, 10) = 4$, $Z(9, 8) = 3$, $Z_x(9, 8) = 2$, and $Z_y(9, 8) = 1$.

20. The temperature $T(x, y)$ degrees Farenheit at (x, y) on a metal plate does not vary with time. An ant crossing the plate is at $x = t^2$ (feet), $y = 4t + 1$ (feet) at time t (minutes). The temperature has the properties $T(9, 13) = 50$, $T_x(9, 13) = 5$, and $T_y(9, 13) = -1$. **a.** What is the temperature at the ant's position at $t = 3$? **b.** What is the rate of change of that temperature with respect to time at $t = 3$?

21.† Suppose that a person derives $E(x, y)$ calories of energy from eating x pounds of chocolate candy and drinking y pints of cola and that he eats $x(t)$ pounds of candy and drinks $y(t)$ pints of cola from time 0 to time t (hours). **a.** How much energy does he derive from the candy and cola between time $t = 0$ and time $t = 90$? **b.** At what rate does he obtain energy from the candy and cola at $t = 90$? (Express your answer in terms of E_x, E_y, x' and y'.)

22.† Suppose that $y = y(x)$ is defined implicitly by the equation $f(x, y) = c$ with c a constant. **a.** Express $y'(x)$ in terms of $f_x(x, y(x))$ and $f_y(x, y(x))$. **b.** Express $y''(x)$ in terms of $f_x(x, y(x))$, $f_y(x, y(x))$, $f_{xx}(x, y(x))$, $f_{xy}(x, y(x))$, and $f_{yy}(x, y(x))$.

23.† Show that $f(x, y) = F(x + y) + F(x - y)$ satisfies the differential equation $f_{xx} = f_{yy}$ for any function $F(t)$.

24. Show that $f(x, y) = F(x^2 - y^2)$ satisfies the differential equation $xf_y + yf_x = 0$ for any function $F(t)$.

25. Show that $f(x, y) = yF(y/x)$ satisfies the differential equation $xf_x + yf_y = f$ for any function $F(t)$.

26.† Apply the Mean Value Theorem for derivatives of functions of one variable to the function $F(t) = f(x_0 + t(x_1 - x_0), y_0 + t(y_1 - y_0))$ for $0 \le t \le 1$ to derive the *Mean Value Theorem for derivatives of functions of two variables*: if $f(x, y)$ has continuous first-order derivatives in a circle containing (x_0, y_0) and (x_1, y_1), then there is a point (X, Y) on the line segment between (x_0, y_0) and (x_1, y_1) such that

$$f(x_1, y_1) - f(x_0, y_0) = \frac{\partial f}{\partial x}(X, Y)(x_1 - x_0) + \frac{\partial f}{\partial y}(X, Y)(y_1 - y_0).$$

27. Use the result of Exercise 26 to show that if $f_x(x, y)$ and $f_y(x, y)$ are zero throughout a disk, then $f(x, y)$ is constant in the disk.

28. Use the $\epsilon\delta$-definitions of the limits given in Section 2.9 and in a footnote at the beginning of this section to prove Theorem 13.1.

29. Use the $\epsilon\delta$-definition of limit given in a footnote at the beginning of this section to prove Theorem 13.2. (Use the proof of Theorem 2.1 in Section 2.9 as a model.)

In Exercises 30 through 34 use properties of functions of one variable and Theorems 13.1, 13.2, and 13.3 to verify the limits.

30.† $\displaystyle \lim_{(x,y) \to (0,3)} \frac{x^3 y - y}{2x + y + 1} = -\frac{3}{4}$ **31.**† $\displaystyle \lim_{(x,y) \to (1,2)} \frac{(x + y)(x - 2y)}{6x + 7y} = -\frac{9}{20}$

32. $\displaystyle \lim_{(x,y) \to (4,-1)} [e^{xy}\sin(x + y)] = e^{-4}\sin(3)$

33. $\displaystyle \lim_{(x,y) \to (2,3)} [\tan(xy) \ln(y - x)] = 0$

34. $\displaystyle \lim_{(x,y) \to (4,3)} x^y = 64$ (Write $x^y = e^{y \ln x}$.)

In each of Exercises 35 through 41 describe the domain of the function and either find its indicated limit or show that it does not exist.

35.† $\displaystyle \lim_{(x,y) \to (0,0)} \frac{x + e^y}{y - e^x}$ **36.**† $\displaystyle \lim_{(x,y) \to (0,0)} \sqrt{y - x}$

37.† $\displaystyle \lim_{(x,y) \to (0,0)} \frac{y}{x}$ **38.**† $\displaystyle \lim_{(x,y) \to (0,0)} \frac{\sqrt{xy}}{\sqrt{x^2 + y^2}}$

39. $\displaystyle \lim_{(x,y) \to (0,0)} \frac{x^3 y + y^3 x}{x^3 + y^3}$ **40.**† $\displaystyle \lim_{(x,y) \to (2,3)} \arcsin(y - x)$

41.† $\displaystyle \lim_{(x,y) \to (0,0)} f(x, y)$ where $f(x, y) = \frac{y}{x}$ for $|y| < x^2$ and $f(x, y) = 0$ for $|y| \ge x^2$. (Compare with Exercise 37.)

42. Show that $f(x, y) = \dfrac{x^3}{x^4 + y^2}$ tends to zero as (x, y) approaches the origin along any line through the origin and yet $f(x, y)$ does not have a limit as $(x, y) \to (0, 0)$. (Consider points that approach the origin along parabolas.)

13-4   DIRECTIONAL DERIVATIVES AND GRADIENTS

The x-derivative $\dfrac{\partial f}{\partial x}(x, y)$ and y-derivative $\dfrac{\partial f}{\partial y}(x, y)$ only tell us the rates of change of $f(x, y)$ as (x, y) moves parallel to the x- or y-axis. To have a complete understanding of the function, we need to know its rates of change as (x, y) moves in other directions. Such rates of change are called DIRECTIONAL DERIVATIVES.

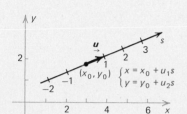

FIGURE 13.55

To define the directional derivative of $f(x, y)$ at a point (x_0, y_0) in the direction of the unit vector $\boldsymbol{u} = <u_1, u_2>$, we introduce an auxiliary s-coordinate axis with the same scale as on the x- and y-axes, with its zero at (x_0, y_0), and with its positive side pointing in the direction of $\boldsymbol{u}$ (Figure 13.55). The point at s on the s-axis then has x- and y-coordinates

$$(1) \qquad x = x_0 + u_1 s, \qquad y = y_0 + u_2 s$$

so the composite function

$$(2) \qquad F(s) = f(x_0 + u_1 s, y_0 + u_2 s)$$

gives the value of $f(x, y)$ as a function of s along the s-axis.

We call the function (2) the CROSS SECTION of $f(x, y)$ at (x_0, y_0) in the direction of $\boldsymbol{u}$. The directional derivative of $f(x, y)$ is defined as the derivative of the cross section:

Definition 13.6   The derivative $D_{\boldsymbol{u}}f(x_0, y_0)$ of $f(x, y)$ at (x_0, y_0) in the direction of the unit vector $u = \langle u_1, u_2 \rangle$ is the derivative

$$(3) \qquad D_{\boldsymbol{u}}f(x_0, y_0) = \frac{dF}{ds}(0) = \left[\frac{d}{ds} f(x_0 + u_1 s, y_0 + u_2 s) \right]_{s=0}$$

at $s = 0$ of its cross section (2).

The value $s = 0$ is used in equation (3) because the point (x_0, y_0) is at $s = 0$ on the s-axis. The derivative (3) is the rate of change of $f(x, y)$ with respect to distance in the direction $\boldsymbol{u}$.

EXAMPLE 1   What is the derivative of $f(x, y) = x^5 + \sin y$ at $(1, 2)$ in the direction of the unit vector $\boldsymbol{u} = \tfrac{3}{5}\boldsymbol{i} - \tfrac{4}{5}\boldsymbol{j}$?

Solution.   We introduce an s-axis with its origin at $(1, 2)$, with the same scale as on the x- and y-axes, and pointing in the direction of the unit vector $\boldsymbol{u} = \tfrac{3}{5}\boldsymbol{i} - \tfrac{4}{5}\boldsymbol{j}$. Then the point at s on the s-axis has xy-coordinates $x = 1 + \tfrac{3}{5}s$, $y = 2 - \tfrac{4}{5}s$, so that $F(s) = (1 + \tfrac{3}{5}s)^5 + \sin(2 - \tfrac{4}{5}s)$ is the value of $f(x, y)$ at s on the s-axis. We have

$$\frac{dF}{ds}(s) = \frac{d}{ds}[(1 + \tfrac{3}{5}s)^5 + \sin(2 - \tfrac{4}{5}s)]$$

$$= 5(1 + \tfrac{3}{5}s)^4 \frac{d}{ds}(1 + \tfrac{3}{5}s) + \cos(2 - \tfrac{4}{5}s)\frac{d}{ds}(2 - \tfrac{4}{5}s)$$

$$= 3(1 + \tfrac{3}{5}s)^4 - \tfrac{4}{5}\cos(2 - \tfrac{4}{5}s)$$

and by Definition 13.6

$$D_u f(1, 2) = \frac{dF}{ds}(0) = [3(1 + \tfrac{3}{4}s)^4 - \tfrac{4}{5}\cos(2 - \tfrac{4}{5}s)]_{s=0}$$

$$= 3 - \tfrac{4}{5}\cos(2). \quad \square$$

The next theorem will enable us to compute directional derivatives from a function's partial derivatives without introducing the auxiliary s-axis and the function $F(s)$.

Theorem 13.7   If $f(x, y)$ has continuous first-order partial derivatives in a circle centered at (x_0, y_0), then for any unit vector $u = \langle u_1, u_2 \rangle$, $D_u f(x_0, y_0)$ exists and is given by the formula

(4) $\quad D_u f(x_0, y_0) = \dfrac{\partial f}{\partial x}(x_0, y_0)u_1 + \dfrac{\partial f}{\partial y}(x_0, y_0)u_2.$ ◄

Proof.   By the chain rule (Theorem 13.5) we have

$$\frac{dF}{ds}(s) = \frac{d}{ds} f(x_0 + u_1 s, y_0 + u_2 s)$$

$$= \frac{\partial f}{\partial x}(x_0 + u_1 s, y_0 + u_2 s)\frac{d}{ds}(x_0 + u_1 s)$$

$$+ \frac{\partial f}{\partial y}(x_0 + u_1 s, y_0 + u_2 s)\frac{d}{ds}(y_0 + u_2 s)$$

$$= \frac{\partial f}{\partial x}(x_0 + u_1 s, y_0 + u_2 s)\, u_1 + \frac{\partial f}{\partial y}(x_0 + u_1 s, y_0 + u_2 s)\, u_2.$$

We set $s = 0$ in this equation and use Definition 13.6 to obtain (4).   Q.E.D.

Notice that the derivative in the direction of $\mathbf{i}$ is the x-derivative; the derivative in the direction of $\mathbf{j}$ is the y-derivative; the derivative in the direction of $-\mathbf{i}$ is the negative of the x-derivative; and the derivative in the direction of $-\mathbf{j}$ is the negative of the y-derivative.

EXAMPLE 2   Find the directional derivative of Example 1 by using Theorem 13.7.

Solution.   Because $u = \tfrac{3}{5}\mathbf{i} - \tfrac{4}{5}\mathbf{j}$, we have

$$D_u f(1, 2) = \tfrac{3}{5}f_x(1, 2) - \tfrac{4}{5}f_y(1, 2).$$

The function $f = x^5 + \sin y$ has partial derivatives $f_x(x, y) = 5x^4$ and $f_y(x, y) = \cos y$ with the values $f_x(1, 2) = 5$, $f_y(1, 2) = \cos(2)$ at $(1, 2)$. Therefore,

$$D_u f(1, 2) = \tfrac{3}{5}(5) - \tfrac{4}{5}\cos(2) = 3 - \tfrac{4}{5}\cos(2). \quad \square$$

EXAMPLE 3   What is the derivative of x^2y^5 at the point $(3, 1)$ in the direction toward the point $(4, -3)$?

Solution.   The displacement vector from $(3, 1)$ to $(4, -3)$ is $\langle 4 - 3, -3 - 1 \rangle = \langle 1, -4 \rangle$, so the unit vector in that direction is

$$\boldsymbol{u} = \frac{\langle 1, -4 \rangle}{|\langle 1, -4 \rangle|} = \left\langle \frac{1}{\sqrt{17}}, \frac{-4}{\sqrt{17}} \right\rangle$$

or $\langle u_1, u_2 \rangle$ with $u_1 = 1/\sqrt{17}$ and $u_2 = -4/\sqrt{17}$.
  We set $f(x, y) = x^2y^5$. Then $f_x(x, y) = 2xy^5$ and $f_y(x, y) = 5x^2y^4$, so that $f_x(3, 1) = 2(3)(1)^5 = 6$ and $f_y(3, 1) = 5(3)^2(1)^4 = 45$. By formula (4) the directional derivative is

$$D_{\boldsymbol{u}}f(3, 1) = f_x(3, 1)u_1 + f_y(3, 1)u_2$$
$$= 6\left(\frac{1}{\sqrt{17}}\right) + 45\left(\frac{-4}{\sqrt{17}}\right) = -\frac{174}{\sqrt{17}}. \quad \square$$

Note.   Because the unit vector at an angle Θ from the vector $\boldsymbol{i}$ is $\boldsymbol{u} = \cos\Theta\,\boldsymbol{i} + \sin\Theta\,\boldsymbol{j}$, the derivative of f at (x_0, y_0) in that direction may be written

$$D_{\boldsymbol{u}}f(x_0, y_0) = \frac{\partial f}{\partial x}(x_0, y_0)\cos\Theta + \frac{\partial f}{\partial y}(x_0, y_0)\sin\Theta. \quad \square$$

  In the next example we approximate a directional derivative by an average rate of change.

EXAMPLE 4   Figure 13.56 shows level curves of the temperature $T(x, y)$ of the surface of the ocean off the west coast of the United States one summer day. (a) Express the rate of change of the temperature at the point P toward the northeast as a directional derivative. (b) Give an approximate value for this rate of change.

Solution.   (a) The point P has coordinates $(1240, 1000)$, and the unit vector pointing toward the northeast is $\boldsymbol{u} = (1/\sqrt{2})\boldsymbol{i} + (1/\sqrt{2})\boldsymbol{j}$. With this notation the rate of change of the temperature is $D_{\boldsymbol{u}}T\,(1240, 1000)$ degrees per mile.
  (b) We introduce an s-axis through the point P and directed toward the northeast as in Figure 13.57. We approximate the directional derivative at P by the average rate of change of the temperature between the points on either side of P where the level curves cross the s-axis. The level curves $T = 17°$ and $T = 18°$ cross the s-axis at points approximately 200 miles apart, and the temperature at the point northeast of P is $17°$ whereas the temperature at the point southwest of P is $18°$. Therefore, the average rate of change of the temperature toward the northeast between the points and the directional derivative at P is approximately

$$\frac{17° - 18°}{200 \text{ miles}} = -\frac{1}{200}\frac{\text{degrees}}{\text{mile}}.$$

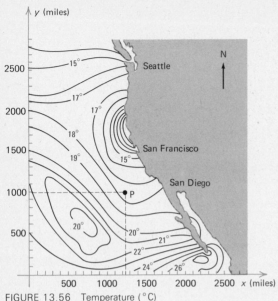

FIGURE 13.56   Temperature (°C)
on the surface of the ocean
(Adapted from S. Ekman, [1]
Zoogeography of the Sea, p. 144.)

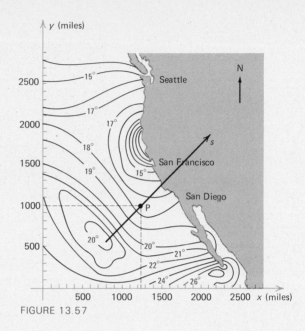

FIGURE 13.57

The temperature is decreasing at approximately 1/200 degrees per mile toward the northeast.   □

Graphs of cross sections

The graph of the cross section

(2) $F(s) = f(x_0 + u_1 s, y_0 + u_2 s)$

of $f(x, y)$ has the shape of the intersection of the surface $z = f(x, y)$ with the vertical plane through the s-axis in the xy-plane (Figure 13.58). We make that plane into an sw-coordinate plane by introducing a vertical w-axis with its zero at (x_0, y_0) in the xy-plane. Then the graph of the cross section has the equation $w = F(s)$ and the directional derivative $D_{\mathbf{u}}f(x_0, y_0) = \dfrac{dF}{ds}(0)$ is the slope of its tangent line at $s = 0$.

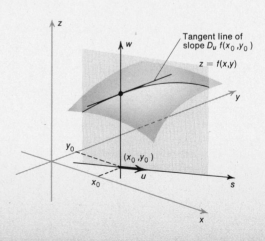

FIGURE 13.58   The graph of the cross section of $f(x, y)$

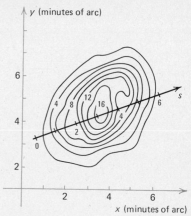

FIGURE 13.59   The Crab Nebula in the Constellation Taurus (Hale Observatory)

FIGURE 13.60   Intensity of radio signals (thousands of degrees Kelvin) with wavelength 21.3 cm. (Adapted from J. S. Hey, [1] *The Radio Universe*, p. 157.)

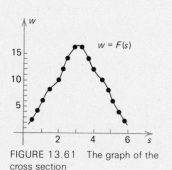

FIGURE 13.61   The graph of the cross section

EXAMPLE 5   Figure 13.59 is a photograph of the Crab Nebula, which is believed to be the remnant of a "supernova" explosion recorded in Chinese annals from A.D. 1054. The Crab Nebula emits radio waves as well as visible light, and Figure 13.60 shows level curves of the intensity $f(x, y)$ of radio signals from the same portion of the sky as shown in Figure 13.59. Sketch the graph of the cross section $F(s)$ of $f(x, y)$ determined by the s-axis shown in Figure 13.60.

Solution.   We plot approximate values of $F(s)$ at the points where the level curves cross the s-axis and then draw a curve through the plotted points (Figure 13.61).   □

Gradient vectors

The formula $D_{\boldsymbol{u}}f = f_x u_1 + f_y u_2$ for a directional derivative has the form of the dot product of the vector $\boldsymbol{u} = \langle u_1, u_2 \rangle$ with the vector $\langle f_x, f_y \rangle$. This leads us to study the latter vector, which is called the GRADIENT of f and is denoted $\overrightarrow{\text{grad}} f$.

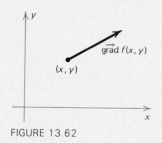

FIGURE 13.62

Definition 13.7   **The gradient of $f(x, y)$ at the point (x, y) is the vector**

(5) $$\overrightarrow{\text{grad}}\, f(x, y) = \frac{\partial f}{\partial x}(x, y)\, \mathbf{i} + \frac{\partial f}{\partial y}(x, y)\, \mathbf{j}.$$ ◄

We represent the gradient $\overrightarrow{\text{grad}}\, f(x, y)$ by an arrow with its base at the point (x, y) (Figure 13.62). (Often the symbol $\overrightarrow{\nabla} f$ is used for $\overrightarrow{\text{grad}}\, f$.)

EXAMPLE 6   Sketch $\overrightarrow{\text{grad}}\, f(1, 1)$, $\overrightarrow{\text{grad}}\, f(-1, 2)$, and $\overrightarrow{\text{grad}}\, f(-2, -1)$ for $f(x, y) = x^2 y$.

Solution.   We have

$$\overrightarrow{\text{grad}}\, f(x, y) = \left\langle \frac{\partial f}{\partial x}(x, y), \frac{\partial f}{\partial y}(x, y) \right\rangle = \left\langle \frac{\partial}{\partial x}(x^2 y), \frac{\partial}{\partial y}(x^2 y) \right\rangle$$
$$= \langle 2xy, x^2 \rangle.$$

Therefore, $\overrightarrow{\text{grad}}\, f(1, 1) = \langle 2, 1 \rangle$, $\overrightarrow{\text{grad}}\, f(-1, 2) = \langle -4, 1 \rangle$, and $\overrightarrow{\text{grad}}\, f(-2, -1) = \langle 4, 4 \rangle$. These vectors are shown in Figure 13.63.   □

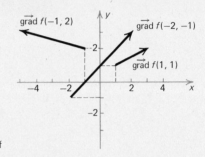

FIGURE 13.63   Gradient vectors of $f(x, y) = x^2 y$

With Definition 13.7 of the gradient, formula (4) for a directional derivative may be written

(6) $D_{\mathbf{u}} f(x, y) = \overrightarrow{\text{grad}}\, f(x, y) \cdot \mathbf{u}$.   ◄

The next theorem, which gives the meaning of the length and direction of the gradient vector, is based on formula (6).

Theorem 13.8   Suppose that $f(x, y)$ has continuous first-order derivatives in a circle centered at (x, y). If $\overrightarrow{\text{grad}}\, f(x, y)$ is the zero vector, then all directional derivatives of f at (x, y) are zero. If $\overrightarrow{\text{grad}}\, f(x, y)$ is not the zero vector, then the maximum directional derivative of f at (x, y) is the length

(7) $\left| \overrightarrow{\text{grad}}\, f(x, y) \right| = \sqrt{ \left[\frac{\partial f}{\partial x}(x, y) \right]^2 + \left[\frac{\partial f}{\partial y}(x, y) \right]^2 }$

of $\overrightarrow{\text{grad}}\, f(x, y)$ and occurs for u with the direction of $\overrightarrow{\text{grad}}\, f(x, y)$.

Proof.   If $\overrightarrow{\text{grad}}\, f(x, y) = \mathbf{O}$, then $f_x(x, y)$, $f_y(x, y)$ and hence all directional derivatives of f at (x, y) are zero. If $\overrightarrow{\text{grad}}\, f(x, y) \neq \mathbf{O}$, then formula (2) of Section 11.2 shows that the dot product (6) is the length of $\overrightarrow{\text{grad}}\, f(x, y)$ multiplied by the length of $\mathbf{u}$ and by the cosine of an angle ψ between $\overrightarrow{\text{grad}}\, f(x, y)$ and $\mathbf{u}$. Because $\mathbf{u}$ has length 1, this yields

(8) $D_{\mathbf{u}} f(x, y) = \left| \overrightarrow{\text{grad}}\, f(x, y) \right| \cos \psi$.

Since the maximum value of $\cos \psi$ is 1 and occurs for $\psi = 0$, the maximum directional derivative is the length of the gradient and occurs when $\boldsymbol{u}$ has the same direction as the gradient.   Q.E.D.

Because the minimum value of $\cos \psi$ is -1 and occurs for $\psi = \pi$, formula (8) shows that the minimum directional derivative of f at (x, y) is $-|\overrightarrow{\text{grad}}\, f(x, y)|$ and occurs when $\boldsymbol{u}$ has the direction opposite that of $\overrightarrow{\text{grad}}\, f(x, y)$. Formula (8) also shows that the directional derivative is zero for the vectors $\boldsymbol{u}$ that are perpendicular to $\overrightarrow{\text{grad}}\, f(x, y)$.

EXAMPLE 7   Find the maximum directional derivative of $y^3 e^{2x}$ at $(2, -1)$ and the unit vector in the direction of that derivative.

Solution.   We have

$$\overrightarrow{\text{grad}}\,(y^3 e^{2x}) = \left\langle \frac{\partial}{\partial x}(y^3 e^{2x}), \frac{\partial}{\partial y}(y^3 e^{2x}) \right\rangle = \langle 2y^3 e^{2x}, 3y^2 e^{2x} \rangle.$$

At $(2, -1)$ this gradient is

(9) $\langle 2(-1)^3 e^{2(2)}, 3(-1)^2 e^{2(2)} \rangle = \langle -2e^4, 3e^4 \rangle$

and by Theorem 13.8 its length

$$|\langle -2e^4, 3e^4 \rangle| = \sqrt{(-2e^4)^2 + (3e^4)^2} = \sqrt{13e^8} = \sqrt{13}\, e^4$$

is the maximum directional derivative.

The maximum directional derivative is in the direction of the gradient (9). The unit vector in that direction is the gradient divided by its length:

$$\boldsymbol{u} = \frac{\langle -2e^4, 3e^4 \rangle}{|\langle -2e^4, 3e^4 \rangle|} = \frac{\langle -2e^4, 3e^4 \rangle}{\sqrt{13}\, e^4} = \left\langle -\frac{2}{\sqrt{13}}, \frac{3}{\sqrt{13}} \right\rangle. \quad \square$$

Gradient vectors and level curves

The next theorem shows that the gradient vectors of a function are perpendicular to its level curves.

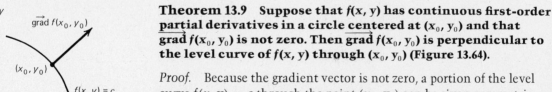

FIGURE 13.64   The gradient of f at (x_0, y_0) is perpendicular to the level curve through (x_0, y_0).

Theorem 13.9   Suppose that $f(x, y)$ has continuous first-order partial derivatives in a circle centered at (x_0, y_0) and that $\overrightarrow{\text{grad}}\, f(x_0, y_0)$ is not zero. Then $\overrightarrow{\text{grad}}\, f(x_0, y_0)$ is perpendicular to the level curve of $f(x, y)$ through (x_0, y_0) (Figure 13.64).

Proof.   Because the gradient vector is not zero, a portion of the level curve $f(x, y) = c$ through the point (x_0, y_0) can be given parametric equations $x = x(t), y = y(t)$ with $(x(0), y(0)) = (x_0, y_0)$ and with a nonzero tangent vector $\left\langle \dfrac{dx}{dt}(0), \dfrac{dy}{dt}(0) \right\rangle$.* Then, for t near 0, the point

_____

*This is a consequence of the Implicit Function Theorem, which is derived in courses on advanced calculus.

$(x(t), y(t))$ is on the level curve $f(x, y) = c$ and consequently $f(x(t), y(t)) = c$. Differentiating this equation with respect to t gives

$$\frac{\partial f}{\partial x}(x(t), y(t))\frac{dx}{dt}(t) + \frac{\partial f}{\partial y}(x(t), y(t))\frac{dy}{dt}(t) = \frac{d}{dt}c = 0$$

which at $t = 0$ reads

$$\frac{\partial f}{\partial x}(x_0, y_0)\frac{dx}{dt}(0) + \frac{\partial f}{\partial y}(x_0, y_0)\frac{dy}{dt}(0) = 0$$

and shows that the dot product of the gradient vector $\overrightarrow{\text{grad}}\, f(x_0, y_0)$ with the tangent vector $\left\langle \dfrac{dx}{dt}(0), \dfrac{dy}{dt}(0) \right\rangle$ is zero. Because the tangent vector is not zero, it follows that the gradient vector is perpendicular to the level curve.   Q.E.D.

EXAMPLE 8   Sketch the gradient of xy at the points $(1, 2)$ and $(-3, 1)$ and the level curves of xy through those points.

Solution.   We have $\overrightarrow{\text{grad}}\,(xy) = \left\langle \dfrac{\partial}{\partial x}(xy), \dfrac{\partial}{\partial y}(xy) \right\rangle = \langle y, x \rangle$. At $(1, 2)$

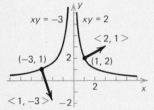

FIGURE 13.65   Level curves and gradient vectors of xy

the gradient vector is $\langle 2, 1 \rangle$, and at $(-3, 1)$ it is $\langle 1, -3 \rangle$. Because xy is 2 at $(1, 2)$, the level curve through $(1, 2)$ is the hyperbola $xy = 2$. Because xy is -3 at $(-3, 1)$, the level curve through $(-3, 1)$ is the hyperbola $xy = -3$. The gradient vectors and level curves are drawn in Figure 13.65. We have drawn only the half of each hyperbola which goes through the corresponding point.   □

Estimating gradient vectors

In the next example we use Theorems 13.8 and 13.9 to estimate a gradient vector from a sketch of level curves of a function.

EXAMPLE 9   Level curves of a function $f(x, y)$ are shown in Figure 13.66. Sketch $\overrightarrow{\text{grad}}\, f(3, 2)$.

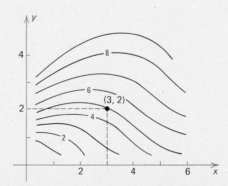

FIGURE 13.66   Level curves of $f(x, y)$

Solution.   We draw a plausible tangent line at $(3, 2)$ to the level curve of $f(x, y)$ through that point and introduce an s-axis perpendicular to the tangent line and with its positive side pointing in the direction of increasing f (Figure 13.67). By Theorems 13.8 and 13.9 the gradient vector at $(3, 2)$ is in the direction of the positive s-axis and its length is the corresponding directional derivative of f.

The distance between the level curves $f = 5$ and $f = 6$ along the s-axis is approximately $\frac{1}{2}$ unit, so the directional derivative in that direction is approximately $\dfrac{6 - 5}{1/2} = \dfrac{1}{1/2} = 2$, and $\overrightarrow{\text{grad}}\, f(3, 2)$, sketched in Figure 13.68, has length approximately 2.   □

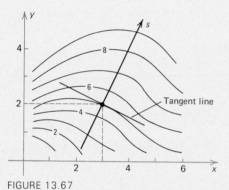

FIGURE 13.67   FIGURE 13.68

Exercises

†*Answer provided.*
‡*Outline of solution provided.*

In Exercises 1 through 7 compute   **a.** the gradient of the function at the given point and   **b.** the derivative at the point in the indicated direction.

1.† $x^3 y^2$ at $(2, -3)$; the direction of $\frac{3}{5}\mathbf{i} - \frac{4}{5}\mathbf{j}$

2. $\sin(xy)$ at $(\frac{3}{4}, \pi)$; the direction of $(-\mathbf{i} + 2\mathbf{j})/\sqrt{5}$

3.‡ $x^2 e^{2y}$ at $(4, 3)$; the direction of $2\mathbf{i} - 3\mathbf{j}$

4.† $\arctan(y/x)$ at $(-1, -1)$; the direction of $-\mathbf{i} - \mathbf{j}$

5. $\ln(x^2 - y^2)$ at $(4, 1)$; the direction of $-6\mathbf{j}$

6.‡ $x\cos(xy)$ at $(2, 3)$; the direction toward $(0, 0)$

7. $x^2 + xy^2$ at $(10, -10)$; the direction toward $(11, -11)$

8.† Draw the level curves of $\frac{1}{4}x^2 + \frac{1}{4}y^2$ through the points $(1, 1)$, $(1, -2)$, and $(-3, -1)$ and $\overrightarrow{\text{grad}}(\frac{1}{4}x^2 + \frac{1}{4}y^2)$ at those points.

9. Draw the level curves of $y - \frac{1}{4}x^2$ through the points $(0, 0)$, $(-1, -2)$, and $(-1, 3)$ and $\overrightarrow{\text{grad}}(y - \frac{1}{4}x^2)$ at those points.

In Exercises 10 through 13 find   **a.** the maximum directional derivative of the function at the given point and the unit vector in that direction and   **b.** the minimum directional derivative and the unit vector in that direction.

10.‡ $x^5 y^{20}$ at $(-1, 1)$ **11.** $\sqrt{x^2 + y^2}$ at $(3, 4)$

12.[†] $2x - \ln y$ at $(0, 3)$ **13.** e^{xy} at $(5, 6)$

In Exercises 14 through 16 find the two unit vectors **u** for which the directional derivative is zero.

14.[‡] $D_{\mathbf{u}}f(10, 10)$ for $f(x, y) = x^3 y^3 - xy$

15.[†] $D_{\mathbf{u}}g(3, 2)$ for $g(x, y) = \dfrac{x}{x + y}$

16. $D_{\mathbf{u}}h(4, 0)$ for $h(x, y) = e^{4x - 3y}$

17.[‡] Give the two unit normal vectors to the curve $xy^3 + 6x^2 y = -7$ at the point $(1, -1)$ in an xy-plane.

18. **a.** Give the two unit normal vectors to the curve $x - y^2 = 0$ at the point $(4, 2)$.   **b.** Do the same for the curve $e^{x-y^2} = 1$.

19.[‡] If $D_{\mathbf{u}}f(1, 2) = 7/\sqrt{2}$ for $\mathbf{u} = (\mathbf{i} - \mathbf{j})/\sqrt{2}$ and $D_{\mathbf{v}}f(1, 2) = 0$ for $\mathbf{v} = \frac{4}{5}\mathbf{i} + \frac{3}{5}\mathbf{j}$, then what are $f_x(1, 2)$ and $f_y(1, 2)$?

20.[†] If $D_{\mathbf{u}}P(0, 0) = 2$ for $\mathbf{u} = (\sqrt{3}/2)\mathbf{i} + \frac{1}{2}\mathbf{j}$ and $D_{\mathbf{v}}P(0, 0) = 8$ for $\mathbf{v} = (-\sqrt{3}/2)\mathbf{i} + \frac{1}{2}\mathbf{j}$, then what are $P_x(0, 0)$ and $P_y(0, 0)$?

21. If $D_{\mathbf{u}}W(5, 10) = -17$ for $\mathbf{u} = \frac{12}{13}\mathbf{i} - \frac{5}{13}\mathbf{j}$ and $D_{\mathbf{v}}W(5, 10) = 13\sqrt{2}$ for $\mathbf{v} = (-\mathbf{i} + \mathbf{j})/\sqrt{2}$, what are $W_x(5, 10)$ and $W_y(5, 10)$?

22.[‡] Level curves of a function $f(x, y)$ are sketched in Figure 13.69. Find an approximate value of $D_{\mathbf{u}}f(3, 1)$ with $\mathbf{u} = (-2\mathbf{i} + \mathbf{j})/\sqrt{5}$.

23.[†] Level curves of a function $g(x, y)$ are sketched in Figure 13.70. Find an approximate value of $D_{\mathbf{u}}g(-1, 1)$ with $\mathbf{u} = (\mathbf{i} - 3\mathbf{j})/\sqrt{10}$.

24. Level curves of a function $h(x, y)$ are sketched in Figure 13.71. Find an approximate value of $D_{\mathbf{u}}h(50, 30)$ with $\mathbf{u} = (-4\mathbf{i} - 3\mathbf{j})/5$.

25.[‡] Give approximate components of $\overrightarrow{\text{grad}}\, f(2, 1)$ for the function $f(x, y)$ of Figure 13.69.

26.[†] Give approximate components of $\overrightarrow{\text{grad}}\, g(2, 2)$ for the function $g(x, y)$ of Figure 13.70.

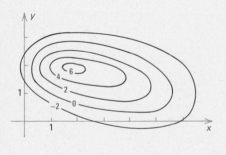

FIGURE 13.69 Level curves of $f(x, y)$

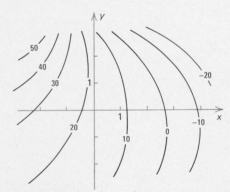

FIGURE 13.70 Level curves of $g(x, y)$

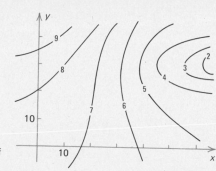

FIGURE 13.71   Level curves of $h(x, y)$

27.   Give approximate components of $\overrightarrow{\text{grad}}\, h(30, 30)$ for the function $h(x, y)$ of Figure 13.71.

28.   Figure 13.72 shows the gradient of $F(x, y)$ at $(2, 1)$, $(3, 3)$, and $(5, -1)$. Give approximate values of   **a.**‡ $F_x(5, -1)$,   **b.**† $F_y(3, 3)$, **c.**† $F_x(2, 1)$,   **d.** $F_y(2, 1)$,   **e.**† $D_{\boldsymbol{u}} F(2, 1)$ with $\boldsymbol{u} = (\mathbf{i} + \mathbf{j})/\sqrt{2}$, and   **f.** $D_{\boldsymbol{u}} F(3, 3)$ with $\boldsymbol{u} = (\mathbf{i} - 3\mathbf{j})\sqrt{10}$.

29.   Set $g(x, y) = F(f(x, y))$ where $f(x, y)$ has continuous first-order derivatives and a nonzero gradient and where $F(t)$ has a continuous nonzero first derivative. Show that $\overrightarrow{\text{grad}}\, g(x, y)$ and $\overrightarrow{\text{grad}}\, f(x, y)$ are parallel. How are the level curves of $f(x, y)$ and $g(x, y)$ related?

30.   Figure 13.73 shows level curves of the depth $f(x, y)$ (meters) of the ocean in the Monterey, Carmel, and Soquel canyons off the coast of California. Sketch the graph of the cross section $F(s)$ of $f(x, y)$ determined by the s-axis in the sketch.

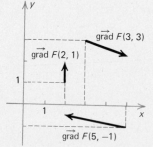

FIGURE 13.72

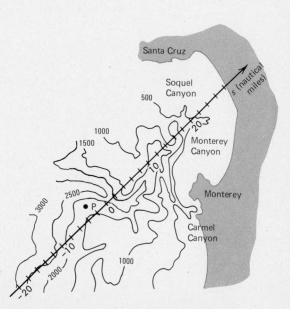

FIGURE 13.73   (Adapted from F. P. Shepard and R. F. Dill, [1] *Submarine Canyons*, p. 82.)

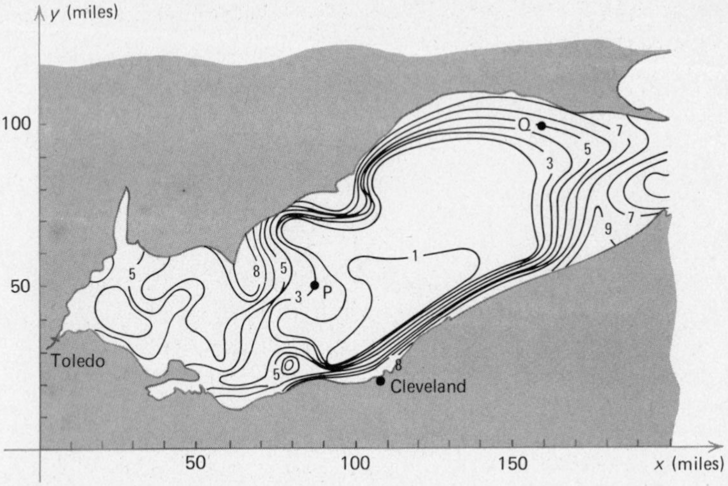

FIGURE 13.74   Dissolved oxygen
in the bottom of Lake Erie (1960)
(Adapted from A. M. Beeton, [1]
*Eutrophication: Causes,
Consequences, Correctives,* p. 172.)

31. Figure 13.74 shows level curves of the concentration $C(x, y)$ (parts per
million) of oxygen in the bottom waters of Lake Erie.   **a.**   What is the
approximate rate of change of $C(x, y)$ with respect to distance at point P
in the direction of Cleveland?   **b.**   Give approximate components of
$\overrightarrow{\text{grad}}\, C(x, y)$ at the point Q.

32. The path of steepest ascent follows the directions of the gradient vectors,
perpendicular to the level curves. Trace the level curves and draw a
plausible path of steepest ascent from point P at the bottom of the ocean
in Figure 13.73.

33. If the temperature $T(x, y)$ at a point (x, y) in a plate does not vary with
time, then the *heat current density* at (x, y) is the vector
$q(x, y) = -K \overrightarrow{\text{grad}}\, T(x, y)$, where K is the *thermal conductivity* of the
plate.   **a.**   Compute $q(\frac{3}{4}\pi, 2)$ for $T(x, y) = (\sin x)(\sinh y)/\sinh 3$ and
$K = 3$.   **b.**   Trace the level curves of $T(x, y)$ from Figure 13.51 of
Section 13.2 and draw $q(\frac{3}{4}\pi, 2)$ on your sketch.

34. An ant is on a metal plate whose temperature at (x, y) is $3x^2 y - y^3$
degrees Centigrade. When he is at the point $(5, 1)$, he is anxious to move
in the direction in which the temperature drops the most rapidly. Give
the unit vector in that direction.

13-5   TANGENT PLANES

We have seen that the x-derivative $\dfrac{\partial f}{\partial x}(x_0, y_0)$ is the slope of the tangent line to the cross section of the surface $z = f(x, y)$ in the vertical plane parallel to the x-axis, that the y-derivative $\dfrac{\partial f}{\partial y}(x_0, y_0)$ is the slope of the tangent line to the cross section in the vertical plane parallel to the y-axis, and that the directional derivative $D_{\boldsymbol{u}}f(x_0, y_0)$ is the slope of the tangent line to the cross section in the vertical plane parallel to the vector $\boldsymbol{u}$ (Figure 13.75).

We will define the TANGENT PLANE to the surface to be the plane containing the tangent lines to the cross sections in the planes parallel to the x- and y-axes. Then we will prove that the tangent lines to all the cross sections lie in the tangent plane (Figure 13.76). Because of this, the tangent plane is the plane that best approximates the surface near the point of tangency.

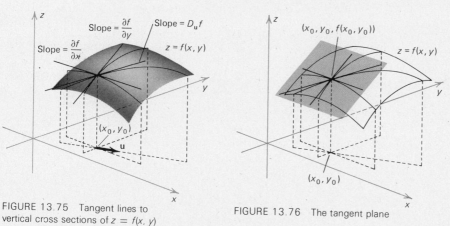

FIGURE 13.75   Tangent lines to vertical cross sections of $z = f(x, y)$

FIGURE 13.76   The tangent plane

Definition 13.8   Suppose that $f(x, y)$ has continuous first-order partial derivatives in a circle centered at (x_0, y_0). The tangent plane to $z = f(x, y)$ at $\big(x_0, y_0, f(x_0, y_0)\big)$ is the plane containing the tangent lines at that point to the cross sections of the surface in vertical planes parallel to the x- and y-axes.

Theorem 13.10   Under the conditions of Definition 13.8 the tangent plane at $\big(x_0, y_0, f(x_0, y_0)\big)$ has the equation

(1) $$z = f(x_0, y_0) + \frac{\partial f}{\partial x}(x_0, y_0)(x - x_0) + \frac{\partial f}{\partial y}(x_0, y_0)(y - y_0) \qquad \blacktriangleleft$$

and contains the tangent lines to all vertical cross sections of the graph through $\big(x_0, y_0, f(x_0, y_0)\big)$.

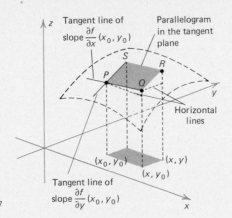

FIGURE 13.77

Proof. We consider an arbitrary point (x, y) in the xy-plane and the rectangle with sides parallel to the x- and y-axes and with (x_0, y_0) and (x, y) at opposite corners (Figure 13.77). We let $P, Q, R,$ and S denote the vertices of the parallelogram in the tangent plane above the rectangle, as shown in Figure 13.77. The side PQ is on the tangent line to the cross section of the surface in the vertical plane parallel to the x-axis, so it has slope $\dfrac{\partial f}{\partial x}(x_0, y_0)$. The side QR has slope $\dfrac{\partial f}{\partial y}(x_0, y_0)$ because it is parallel to the side PS on the tangent line to the cross section in the vertical plane parallel to the y-axis. In going from P to Q the x-coordinate is increased by $x - x_0$ and hence the z-coordinate by $\dfrac{\partial f}{\partial x}(x_0, y_0)(x - x_0)$. In going from Q to R the y-coordinate is increased by $y - y_0$ and hence the z-coordinate by $\dfrac{\partial f}{\partial y}(x_0, y_0)(y - y_0)$. Because the z-coordinate of P is $f(x_0, y_0)$, it follows that

$$(2) \qquad \left[\begin{matrix} \text{The } z\text{-coordinate} \\ \text{of } R \end{matrix} \right]$$
$$= f(x_0, y_0) + \frac{\partial f}{\partial x}(x_0, y_0)(x - x_0) + \frac{\partial f}{\partial y}(x_0, y_0)(y - y_0).$$

Equation (2) gives equation (1) because R has coordinates (x, y) and R is an arbitrary point on the tangent plane.

Now suppose that the point (x, y) is a distance 1 from (x_0, y_0) in the direction of the unit vector $\mathbf{u} = \langle u_1, u_2 \rangle$. Then $x - x_0 = u_1$, $y - y_0 = u_2$, and equation (1) shows that

$$(3) \qquad \left[\begin{matrix} \text{The } z\text{-coordinate} \\ \text{of } R \end{matrix} \right] - \left[\begin{matrix} \text{The } z \text{ coordinate} \\ \text{of } P \end{matrix} \right]$$
$$= \frac{\partial f}{\partial x}(x_0, y_0)u_1 + \frac{\partial f}{\partial f}(x_0, y_0)u_2.$$

Because (x, y) is a distance 1 from (x_0, y_0), the quantity (3) is the slope of the line PR. It is also equal to the directional derivative $D_{\mathbf{u}} f(x_0, y_0)$, so PR is on the tangent line to the cross section of the surface in the direction $\mathbf{u}$ and the tangent line lies in the tangent plane. Since $\mathbf{u}$ is an arbitrary unit vector, the tangent lines to all vertical cross sections through P lie in the tangent plane.   Q.E.D.

Normal vectors   We can rewrite equation (1) for the tangent plane in the form

$$(4) \quad \frac{\partial f}{\partial x}(x_0, y_0)(x - x_0) + \frac{\partial f}{\partial y}(x_0, y_0)(y - y_0) - (z - f(x_0, y_0)) = 0.$$

Therefore, the vector

$$(5) \quad \mathbf{n} = \left\langle \frac{\partial f}{\partial x}(x_0, y_0), \frac{\partial f}{\partial y}(x_0, y_0), -1 \right\rangle \quad \blacktriangleleft$$

whose components are the coefficients of x, y, and z in (4), is perpendicular (normal) to the tangent plane and to the graph of f at $x = x_0, y = y_0$.

EXAMPLE 1   Give an equation for the tangent plane and a normal vector to the graph of the function $x^3 y^4$ at $x = -1, y = 2$.

Solution.   The tangent plane to the graph of $f(x, y)$ at $x = -1, y = 2$ has the equation

$$z = f(-1, 2) + f_x(-1, 2)(x - (-1)) + f_y(-1, 2)(y - 2).$$

We set $f(x, y) = x^3 y^4$. Then $f_x(x, y) = 3x^2 y^4$, $f_y(x, y) = 4x^3 y^3$, $f(-1, 2) = (-1)^3 (2)^4 = -16$, $f_x(-1, 2) = 3(-1)^2 (2)^4 = 48$, and $f_y(-1, 2) = 4(-1)^3 (2)^3 = -32$. The tangent plane therefore has the equation $z = -16 + 48(x + 1) - 32(y - 2)$. By formula (5) the vector $\mathbf{n} = \langle f_x(-1, 2), f_y(-1, 2), -1 \rangle = \langle 48, -32, -1 \rangle$ is normal to the graph at $x = -1, y = 2$.   □

In the next example we use a tangent plane to give an approximate value of a function.

EXAMPLE 2   Use the tangent plane at $x = 5, y = 12$ to the graph of $\sqrt{x^2 + y^2}$ to obtain an approximate value of $\sqrt{(5.01)^2 + (11.97)^2}$.

Solution.   We set $f(x, y) = \sqrt{x^2 + y^2} = (x^2 + y^2)^{1/2}$. Then

$$f_x(x, y) = \frac{\partial}{\partial x}(x^2 + y^2)^{1/2} = \tfrac{1}{2}(x^2 + y^2)^{-1/2} \frac{\partial}{\partial x}(x^2 + y^2)$$

$= x(x^2 + y^2)^{-1/2}$, and, similarly, $f_y(x, y) = y(x^2 + y^2)^{-1/2}$. Therefore $f(5, 12) = (5^2 + 12^2)^{1/2} = 13$, $f_x(5, 12) = 5(5^2 + 12^2)^{-1/2} = \frac{5}{13}$, and $f_y(5, 12) = 12(5^2 + 12^2)^{-1/2} = \frac{12}{13}$.

These calculations show that the tangent plane has the equation $z = 13 + \frac{5}{13}(x - 5) + \frac{12}{13}(y - 12)$. For x close to 5 and y close to 12 the tangent plane is close to the graph of the function. In particular, the value

$$z = 13 + \tfrac{5}{13}(5.01 - 5) + \tfrac{12}{13}(11.97 - 12)$$

$$= 13 + \tfrac{5}{13}(0.01) + \tfrac{12}{13}(-0.03) = 13 - \frac{0.31}{13} \approx 12.976$$

of z at $x = 5.01, y = 11.97$ on the tangent plane is approximately equal to the value $z = \sqrt{(5.01)^2 + (11.97)^2}$ on the surface. ☐

Differentials   We say that the numbers dx, dy, and dz are *differentials* at a point P on a surface if they are corresponding changes in x, y, and z along the tangent plane to the surface at P. In the case where the surface is the graph of a function $f(x, y)$, we write df for dz and obtain the following definition.

Definition 13.9   If $f(x, y)$ has continuous first-order partial derivatives in a circle centered at (x_0, y_0), then the differentials dx, dy, and df at $x = x_0$, $y = y_0$ are numbers related by the equation

(6) $$df = \frac{\partial f}{\partial x}(x_0, y_0)\, dx + \frac{\partial f}{\partial y}(x_0, y_0)\, dy. \qquad \blacktriangleleft$$

The differential df in equation (6) is often called the TOTAL DIFFERENTIAL of f. In the next example we use a total differential in carrying out an error analysis.

EXAMPLE 3   The radius of a right circular cylinder is measured to be 10 centimeters with a possible error of 0.01 centimeters, and the height is measured to be 15 centimeters with a possible error of 0.005 centimeters. Use differentials to give an approximate value for the possible error in the computed volume of the cylinder.

Solution.   The volume of the cylinder of radius r and height h is $V = \pi r^2 h$, for which $V_r = 2\pi rh$ and $V_h = \pi r^2$. At $r = 10, h = 15$ these derivatives are $V_r = 300\pi$ and $V_h = 100\pi$, so the differentials dr, dh, and dV are related by the equation $dV = 300\pi\, dr + 100\pi\, dh$. An error of $dr = \pm 0.01$ in measuring r and an error of $dh = \pm 0.005$ in measuring h would yield an error of approximately

(7) $$dV = 300\pi(\pm 0.01) + 100\pi(\pm 0.005)$$

in the computed volume. We obtain the largest $|dV|$ by taking dr and dh to have the same sign, so the approximate maximum error in the computed volume is $|dV| = 300\pi(0.01) + 100\pi(0.005) = 3.5\pi$. ☐

A graph without a tangent plane (optional) A function $f(x)$ of one variable has a derivative at a point x_0 if and only if its graph has a nonvertical tangent line at that point, but in the case of two variables the relationship between derivatives and tangent planes is more complicated. As the next example shows, it is possible that the graph of a function $f(x, y)$ has no tangent plane at $x = x_0, y = y_0$ even though the partial derivatives f_x and f_y and all directional derivatives $D_{\mathbf{u}}f$ exist at (x_0, y_0). To know that the tangent plane exists we have to show that the function satisfies a stronger condition, such as the hypothesis in Theorem 13.10 that its derivatives be continuous in a disk centered at (x_0, y_0).

EXAMPLE 4   Consider the function f given by

$$f(x, y) = \frac{x^2 y}{x^2 + y^2} \quad \text{for} \quad (x, y) \neq (0, 0)$$

and defined to be zero at $(0, 0)$. Show that its graph consists of straight lines through the origin, so that the directional derivatives $D_{\mathbf{u}}f(0,0)$ exist for all $\mathbf{u}$, but that these lines do not lie in a plane and the function does not have a tangent plane at the origin.

Solution.   If x or y is zero, $f(x, y)$ is zero, so the x- and y-axes are on the graph and the partial derivatives $f_x(0, 0)$ and $f_y(0, 0)$ exist and are zero. If the graph were to have a tangent plane, it would have to contain the x- and y-axes and hence be the xy-plane.

On an s-axis through the origin in the direction of an arbitrary unit vector $\mathbf{u} = \langle u_1, u_2 \rangle$ we have $x = u_1 s, y = u_2 s$ and

$$f(x, y) = \frac{(u_1 s)^2 (u_2 s)}{(u_1 s)^2 + (u_2 s)^2} = \left\{ \frac{u_1^2 u_2}{u_1^2 + u_2^2} \right\} s$$

so the cross section of the graph in that direction is a line of slope

$$m = \frac{u_1^2 u_2}{u_1^2 + u_2^2}.$$

This line must lie in the xy-plane if it is to be the tangent plane, but does not unless m is zero, and that occurs only if u_1 or u_2 is zero. Therefore, the tangent plane does not exist.   □

Example 4 also shows that the formula $D_{\mathbf{u}}f(x_0, y_0) = u_1 f_x(x_0, y_0) + u_2 f_y(x_0, y_0)$ for the directional derivative may not be valid if the function does not have continuous derivatives in a disk centered at (x_0, y_0). The partial derivatives of the function in Example 4 are both zero at $(0, 0)$ but all the directional derivatives are not. Other formulas based on the chain rule might also fail for this function at the origin.

Exercises

†*Answer provided.*
‡*Outline of solution provided.*

In Exercises 1 through 8 give normal vectors and equations of tangent planes to the graphs of the functions at the indicated values of x and y.

1.‡ x^2y^{-3}; $x = 2, y = 1$

2.† xe^{-y}; $x = 3, y = 0$

3. $\arctan(2x + 4y)$; $x = 4, y = -2$

4. $y \ln(\cos x)$; $x = 0, y = 5$

5.† $\sin(x)\cos(y)$; $x = \pi/4, y = \frac{2}{3}\pi$

6. $(x^3 + y^2)^{1/3}$; $x = \frac{1}{7}, y = -\frac{3}{7}$

7.† $\ln\sqrt{x^2 + y^2}$; $x = 3, y = 4$

8. $y/(5 - 7x)$; $x = 2, y = 4$

In Exercises 9 through 13 use the indicated tangent planes to find approximate values for the given expressions.

9.‡ $\sqrt[3]{(3.02)^2 + (1.97)^3 + 10}$; tangent plane to $z = \sqrt[3]{x^2 + y^3 + 10}$ at $x = 3$, $y = 2$

10.† $\sqrt{\dfrac{9.003}{3.998}}$; tangent plane to $z = \sqrt{\dfrac{x}{y}}$ at $x = 9, y = 4$

11. $[(5.003)^2 - (4.002)^2]^{3/2}$; tangent plane to $z = (x^2 - y^2)^{3/2}$ at $x = 5$, $y = 4$

12.† $\ln[(0.0003)^{75} + (0.9995)^{100}]$; tangent plane to $z = \ln(x^{75} + y^{100})$ at $x = 0, y = 1$

13. $[(1.99)^3 + (1.02)^5 + 23]^{1/5}$; tangent plane to $z = (x^3 + y^5 + 23)^{1/5}$ at $x = 2, y = 1$

14.‡ The side AB of a triangle is known to have length 10 inches. The side AC is measured to be 8 inches with a possible error of 0.001 inches, and the angle BAC is measured to be $\pi/6$ radians with a possible error of 0.02 radians. Use differentials to estimate the error made in the calculated area of the triangle.

15.† The side opposite the angle Θ in a right triangle is measured to be 4 inches and the hypotenuse is measured to be 8 inches with possible errors of 0.01 inches in each measurement. Use differentials to estimate the possible error in the calculated value of Θ as an arcsine.

16. The radius of the base of a right circular cone is measured to be 2 feet with a possible error of 0.002 feet and the height is measured to be 7 feet with a possible error of 0.01 feet. Use differentials to estimate the possible error in the calculated volume of the cone.

Give the total differentials of the functions in Exercises 17 through 20.

17.† $f = x^2y^5$ **18.**† $G = \ln(u - 3v)$

19.† $H = r^2 - s^3 + 4rs$ **20.** $W = pq/(p + q)$

In each of Exercises 21 through 26 use differentials to estimate the largest error in the value of the function that could be caused by the possible errors in the values of the variables.

21.[†] The area $A = wh$ of a rectangle of width $w = 3 \pm 0.02$ meters and height $h = 5 \pm 0.03$ meters

22.[†] The lateral surface area $A = \pi r \sqrt{r^2 + h^2}$ of a right circular cone of radius $r = 3 \pm 0.01$ inches and height $h = 4 \pm 0.005$ inches

23.[†] The area $A = \frac{1}{2}x^2 \tan \Theta$ of a right triangle containing an angle $\Theta = \frac{1}{4}\pi \pm 0.1$ radians and adjacent side of length $x = 10 \pm 0.03$ centimeters

24.[†] The *geometric mean* $M = \sqrt{xy}$ of the numbers $x = 9 \pm 0.003$ and $y = 4 \pm 0.002$

25. $z = x^2 y^{-3}$ where $x = 4 \pm \frac{1}{16}$ and $y = -2 \pm \frac{1}{8}$

26. $x^2 \ln(10 - 3y)$ where $x = 2 \pm 0.02$ and $y = 3 \pm 0.015$

13-6   FUNCTIONS OF THREE VARIABLES

A function f of three variables is a rule that assigns a number $f(x, y, z)$ to each point (x, y, z) in its domain. The domain is a portion or all of xyz-space. We cannot visualize the graph of a function of three variables because the graph would be in four-dimensional space. Instead, we consider its LEVEL SURFACES.

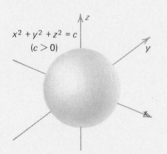

$x^2 + y^2 + z^2 = c$
$(c > 0)$

FIGURE 13.78   Level surface of $x^2 + y^2 + z^2$

Definition 13.10   A level surface of $f(x, y, z)$ is a surface $f(x, y, z) = c$ in xyz-space where the function has a constant value.

EXAMPLE 1   Describe the level surfaces of $x^2 + y^2 + z^2$.

Solution.   For $c > 0$ the level surface $x^2 + y^2 + z^2 = c$ is the sphere of radius $\sqrt{c}$ with its center at the origin (Figure 13.78). The level surface $x^2 + y^2 + z^2 = 0$ with $c = 0$ is the single point $(0, 0, 0)$, and there is no level surface for $c < 0$.   □

EXAMPLE 2   Describe the level surfaces of $\frac{1}{3}z^2 - x^2 - y^2$.

Solution.   For $c = 0$ the level surface $\frac{1}{3}z^2 - x^2 - y^2 = c$ is the double infinite cone $\frac{1}{3}z^2 = x^2 + y^2$. For $c = 1$ it is the hyperboloid of two sheets $\frac{1}{3}z^2 = x^2 + y^2 + 1$ and for $c = -1$ it is the hyperboloid of one sheet $\frac{1}{3}z^2 = x^2 + y^2 - 1$ (Figure 13.79). Similarly, for any positive c the level surface is a hyperboloid of two sheets inside the cone and for any negative c a hyperboloid of one sheet surrounding the cone.   □

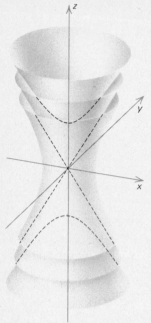

FIGURE 13.79   Level surfaces of
$\frac{1}{3}z^2 - x^2 - y^2$

EXAMPLE 3   Figure 13.80 shows a cross sectional view of level surfaces of cosmic radiation (Van Allen belts) around the earth. (To visualize the entire surfaces imagine the curves in Figure 13.80 to be rotated around the axis of the earth.) The radiation is measured in counts per second and the scale shows distance in earth radii (4000 miles). At approximately what distances from the equator is the radiation the greatest?

Solution.   The radiation is greatest at approximately $\frac{1}{2}$ and $2\frac{1}{2}$ earth radii or 2000 miles and 10,000 miles from the equator.   □

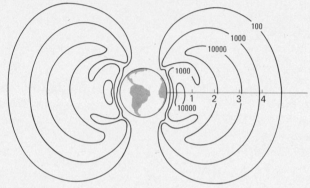

FIGURE 13.80   Van Allen radiation belts (Adapted from J. A. Van Allen, [1] "Radiation Belts around the Earth," pp. 40–41.)

Limits, continuity, partial derivatives

The concepts of limit, continuity, and partial derivatives for functions of three variables are essentially the same as for functions of two variables. We say that the limit of $f(x, y, z)$ is L as (x, y, z) approaches (x_0, y_0, z_0) if $f(x, y, z)$ tends to L as the distance between the points (x, y, z) and (x_0, y_0, z_0) tends to zero. If f is defined in a sphere centered at (x_0, y_0, z_0), we say that f is continuous at (x_0, y_0, z_0) if $f(x, y, z)$ tends to $f(x_0, y_0, z_0)$ as (x, y, z) approaches (x_0, y_0, z_0). The partial derivative

$$\frac{\partial f}{\partial x}(x, y, z) = f_x(x, y, z) = f_1(x, y, z)$$

is obtained by holding y and z constant and differentiating with respect to x. Partial derivatives with respect to y and z and higher-order derivatives are defined similarly.

EXAMPLE 4   Compute $f_{xyz}(1, 2, 3)$ for $f(x, y, z) = x^4y^3z^2$.

Solution.   We have

$$f_x(x, y, z) = \frac{\partial}{\partial x}(x^4y^3z^2) = 4x^3y^3z^2$$

$$f_{xy}(x, y, z) = \frac{\partial}{\partial y}f_x(x, y, z) = \frac{\partial}{\partial y}(4x^3y^3z^2) = 12x^3y^2z^2$$

$$f_{xyz}(x, y, z) = \frac{\partial}{\partial z}f_{xy}(x, y, z) = \frac{\partial}{\partial z}(12x^3y^2z^2) = 24x^3y^2z.$$

Therefore,

$$f_{xyz}(1, 2, 3) = [24x^3y^2z]_{x=1,\, y=2,\, z=3} = 24(1)^3(2)^2(3) = 288. \quad \square$$

The chain rule   The proof of Theorem 13.5 can be modified to yield the chain rule for functions of three variables: if $f(x, y, z)$ has continuous first-order derivatives in a sphere centered at $(x(t), y(t), z(t))$ and if $\dfrac{dx}{dt}(t)$, $\dfrac{dy}{dt}(t)$, and $\dfrac{dz}{dt}(t)$ exist, then

(1) $$\frac{d}{dt}f(x(t), y(t), z(t)) = \frac{\partial f}{\partial x}\frac{dx}{dt} + \frac{\partial f}{\partial y}\frac{dy}{dt} + \frac{\partial f}{\partial z}\frac{dz}{dt} \qquad \blacktriangleleft$$

where the partial derivatives of f are evaluated at $(x(t), y(t), z(t))$, and the derivatives of x, y and z are evaluated at t.

EXAMPLE 5   Suppose that the partial derivatives of $f(x, y, z)$ are continuous and that $f_x(1, 1, 1) = 4$, $f_y(1, 1, 1) = 5$, and $f_z(1, 1, 1) = 6$. What is the derivative $\dfrac{d}{dt}f(t^7, t^8, t^9)$ at $t = 1$?

Solution.   By the chain rule (1), we have

$$\frac{d}{dt}f(t^7, t^8, t^9)$$

$$= f_x(t^7, t^8, t^9)\frac{d}{dt}t^7 + f_y(t^7, t^8, t^9)\frac{d}{dt}t^8 + f_z(t^7, t^8, t^9)\frac{d}{dt}t^9$$

$$= f_x(t^7, t^8, t^9)\,7t^6 + f_y(t^7, t^8, t^9)\,8t^7 + f_z(t^7, t^8, t^9)\,9t^8.$$

At $t = 1$ this equals

$$f_x(1, 1, 1)(7) + f_y(1, 1, 1)(8) + f_z(1, 1, 1)(9)$$
$$= 4(7) + 5(8) + 6(9) = 122. \quad \square$$

Formula (1) yields versions of the chain rule where x, y, and z are functions of two or three variables. Suppose, for example, that $x = x(u, v, w)$, $y(u, v, w)$, and $z(u, v, w)$ have first-order derivatives at (u, v, w), while $f(x, y, z)$ has continuous first-order derivatives in a sphere centered at $(x(u, v, w), y(u, v, w), z(u, v, w))$. Set $G(u, v, w) = f(x(u, v, w), y(u, v, w), z(u, v, w))$. Then

$$(2) \quad \frac{\partial G}{\partial u} = \frac{\partial f}{\partial x}\frac{\partial x}{\partial u} + \frac{\partial f}{\partial y}\frac{\partial y}{\partial u} + \frac{\partial f}{\partial z}\frac{\partial z}{\partial u}$$

$$(3) \quad \frac{\partial G}{\partial v} = \frac{\partial f}{\partial x}\frac{\partial x}{\partial v} + \frac{\partial f}{\partial y}\frac{\partial y}{\partial v} + \frac{\partial f}{\partial z}\frac{\partial z}{\partial v} \qquad \blacktriangleleft$$

$$(4) \quad \frac{\partial G}{\partial w} = \frac{\partial f}{\partial x}\frac{\partial x}{\partial w} + \frac{\partial f}{\partial y}\frac{\partial y}{\partial w} + \frac{\partial f}{\partial z}\frac{\partial z}{\partial w}.$$

Directional derivatives and gradients

If we introduce an s-axis with the same scale as on the coordinate axes, with its zero at (x_0, y_0, z_0), and pointing in the direction of the unit vector $\boldsymbol{u} = \langle u_1, u_2, u_3 \rangle$, then the point at s on the s-axis has coordinates

$$x = x_0 + u_1 s, \quad y = y_0 + u_2 s, \quad z = z_0 + u_3 s.$$

The function

$$(5) \quad F(s) = f(x_0 + u_1 s, y_0 + u_2 s, z_0 + u_3 s)$$

which we call the *cross section* of $f(x, y, z)$ at (x_0, y_0, z_0) in the direction of $\boldsymbol{u}$, gives the value of f at s on the s-axis. We define the derivative of f at (x_0, y_0, z_0) in the direction $\boldsymbol{u}$ to be the derivative

$$(6) \quad D_{\boldsymbol{u}}f(x_0, y_0, z_0) = \frac{dF}{ds}(0)$$

of the cross section (5). Then the chain rule (1) shows that

$$(7) \quad D_{\boldsymbol{u}}f(x_0, y_0, z_0) = \frac{\partial f}{\partial x}u_1 + \frac{\partial f}{\partial y}u_2 + \frac{\partial f}{\partial z}u_3 \qquad \blacktriangleleft$$

with the partial derivatives of f evaluated at (x_0, y_0, z_0).

We define the *gradient* of $f(x, y, z)$ to be the vector

$$(8) \quad \overrightarrow{\text{grad}} f(x, y, z) = \frac{\partial f}{\partial x}(x, y, z)\mathbf{i} + \frac{\partial f}{\partial y}(x, y, z)\mathbf{j} + \frac{\partial f}{\partial z}(x, y, z)\mathbf{k} \qquad \blacktriangleleft$$

in xyz-space. With this definition, formula (7) yields the formula

$$(9) \quad D_{\boldsymbol{u}}f(x, y, z) = \overrightarrow{\text{grad}} f(x, y, z) \cdot \boldsymbol{u} \qquad \blacktriangleleft$$

for the derivative of f at (x, y, z) in the direction of the unit vector $\boldsymbol{u}$. As in the case of two variables (Theorem 13.8), it then follows that the maximum directional derivative of f at (x, y, z) is $|\overrightarrow{\text{grad}} f(x, y, z)|$ and occurs in the direction of $\overrightarrow{\text{grad}} f(x, y, z)$.

EXAMPLE 6   Set $f(x, y, z) = x^3 y^4 z^5$. Compute $\overrightarrow{\text{grad}} f(1, 1, 1)$, $D_{\boldsymbol{u}}f(1, 1, 1)$ for $\boldsymbol{u} = (4\mathbf{i} - 2\mathbf{j} - \mathbf{k})/\sqrt{21}$, the maximum directional derivative of f at $(1, 1, 1)$, and the unit vector in the direction of that maximum directional derivative.

Solution.   We have

$$\overrightarrow{\text{grad}} f = f_x \mathbf{i} + f_y \mathbf{j} + f_z \mathbf{k} = (3x^2 y^4 z^5) \mathbf{i} + (4x^3 y^3 z^5) \mathbf{j} + (5x^3 y^4 z^4) \mathbf{k}.$$

Therefore, $\overrightarrow{\text{grad}} f(1, 1, 1) = 3\mathbf{i} + 4\mathbf{j} + 5\mathbf{k}$. By formula (7), $D_{\mathbf{u}} f(1, 1, 1) = (3\mathbf{i} + 4\mathbf{j} + 5\mathbf{k}) \cdot (4\mathbf{i} - 2\mathbf{j} - \mathbf{k})/\sqrt{21} = -1/\sqrt{21}$. The maximum directional derivative is $|\overrightarrow{\text{grad}} f(1, 1, 1)| = |3\mathbf{i} + 4\mathbf{j} + 5\mathbf{k}| = \sqrt{50}$ and occurs for $\mathbf{u} = \overrightarrow{\text{grad}} f(1, 1, 1)/|\overrightarrow{\text{grad}} f(1, 1, 1)| = (3\mathbf{i} + 4\mathbf{j} + 5\mathbf{k})/\sqrt{50}$.   □

Tangent planes to level surfaces

If $f(x, y, z)$ has continuous derivatives in a sphere centered at (x_0, y_0, z_0) and if $\text{grad} f(x_0, y_0, z_0)$ is not the zero vector, then the Implicit Function Theorem, which is studied in advanced calculus, shows that the level surface $f(x, y, z) = c$ through (x_0, y_0, z_0) has a tangent plane at that point and the directional derivatives $D_{\mathbf{u}} f(x_0, y_0, z_0)$ are zero for $\mathbf{u}$ parallel to that plane. It then follows from formula (9) that $\text{grad} f(x_0, y_0, z_0)$ is perpendicular to all such $\mathbf{u}$ and hence is a normal vector to the tangent plane. This yields the following result.

Theorem 13.11   If $f(x, y, z)$ has continuous first-order partial derivatives in a sphere centered at (x_0, y_0, z_0) and if $\overrightarrow{\text{grad}} f(x_0, y_0, z_0)$ is not the zero vector, then it is a normal vector to the level surface $f(x, y, z) = c$ at (x_0, y_0, z_0) and the tangent plane to that level surface has the equation

(10) $$\frac{\partial f}{\partial x}(P_0)(x - x_0) + \frac{\partial f}{\partial y}(P_0)(y - y_0) + \frac{\partial f}{\partial z}(P_0)(z - z_0) = 0 \qquad \blacktriangleleft$$

where P_0 denotes the point (x_0, y_0, z_0).

Equation (10) is a consequence of formula (2) of Section 12.3 for the equation of the plane through a given point with a given normal vector.

EXAMPLE 7   Give an equation for the tangent plane to the hyperboloid of two sheets $z^2 - x^2 - y^2 = 4$ at $(1, 2, 3)$.

Solution.   The hyperboloid is a level surface of $f(x, y, z) = z^2 - x^2 - y^2$, for which $f_x = -2x, f_y = -2y, f_z = 2z$ and $f_x(1, 2, 3) = -2, f_y(1, 2, 3) = -4, f_z(1, 2, 3) = 6$. By (10) the tangent plane has the equation $-2(x - 1) - 4(y - 2) + 6(z - 3) = 0$ or $(x - 1) + 2(y - 2) - 3(z - 3) = 0$.   □

Differentials

In situations where we would make tangent plane approximations of functions of two variables, we use *differentials* to give approximate values of functions of three variables. The differential (or *total differential*) of $f(x, y, z)$ at (x_0, y_0, z_0) is the expression

(11) $$df = \frac{\partial f}{\partial x}(x_0, y_0, z_0) \, dx + \frac{\partial f}{\partial y}(x_0, y_0, z_0) \, dy + \frac{\partial f}{\partial z}(x_0, y_0, z_0) \, dz \qquad \blacktriangleleft$$

and is the approximate change in $f(x, y, z)$ as (x, y, z) changes from (x_0, y_0, z_0) to $(x_0 + dx, y_0 + dy, z_0 + dz)$.

EXAMPLE 8   Use the differential of the function $(x^2 + y^2 + z^2)^{1/2}$ $(3, 4, 12)$ to give an approximate value of $[(3.01)^2 + (4.02)^2 + (12.03)^2]^{1/2}$.

Solution.   We set $f(x, y, z) = (x^2 + y^2 + z^2)^{1/2}$. Then $f_x = x(x^2 + y + z^2)^{-1/2}, f_y = y(x^2 + y^2 + z^2)^{-1/2}$, and $f_z = z(x^2 + y^2 + z^2)^{-1/2}$. At $(3, 4, 12)$, we have $f = (3^2 + 4^2 + 12^2)^{1/2} = 169^{1/2} = 13, f_x = \frac{3}{13}, f_y = \frac{4}{13}$, and $f_z = \frac{12}{13}$ and hence $df = \frac{3}{13}dx + \frac{4}{13}dy + \frac{12}{13}dz$. For $dx = 0.01, dy = 0.02$, and $dz = 0.03$, we have $df = 0.47/13$, and

$$[(3.01)^2 + (4.02)^2 + (12.03)^2]^{1/2} = f(3.01, 4.02, 12.03)$$
$$\approx f(3, 4, 12) + df = 13 + \frac{0.47}{13}. \quad \square$$

Exercises

1.   For a yacht to be in the R–5.5 class at the Olympic games its rating

$$R(L, V, A) = 0.9\left(\frac{L\sqrt{A}}{12\sqrt[3]{V}} + \frac{L + \sqrt{A}}{4}\right)$$

must be less than 5.5. Here L (meters) is the length of the yacht, A (square meters) the surface area of its sails, and V (cubic meters) the volume of water it displaces. (C. A. Marchaj, [1] *Sailing Theory and Practice*, p. 75). If your yacht is 10 meters long, has 36 square meters of sail, and displaces 8 cubic meters of water, would it qualify for the R–5.5 class?

In Exercises 2 through 6 compute the partial derivatives.

2.†   **a.** $f_{xy}(x, y, z)$,   **b.** $f_{yz}(x, y, z)$, and   **c.** $f_{xz}(x, y, z)$ for $f(x, y, z) = x^2 e^{-3y} \sin(4z)$

3.   **a.** $g_{11}(2, 2, 2)$,   **b.** $g_{22}(1, 1, 1)$, and   **c.** $g_{33}(3, 3, 3)$ for $g(x, y, z) = x^2 y^3 z^4$

4.† $H_{xxyyzz}(x, y, z)$ for $H(x, y, z) = \sin(x)\cos(y)e^z$

5. $f_{xx} + f_{yy} - 2f_{zz}$ for $f(x, y, z) = \sin(x)\sin(y)\sin(z)$

6. $\dfrac{\partial^6}{\partial x^2 \partial y^2 \partial z^2}(x^{10}y^{20}z^{30})$

7.†   If Θ is the angle between sides of length x and y in a triangle, then by the law of cosines the length of the third side is $z = \sqrt{x^2 + y^2 - 2xy\cos\Theta}$. What are the rates of change of z with respect to x, y, and Θ?

8.   What are the rates of change of the area of the triangle of Exercise 7 with respect to x, y, and Θ?

In Exercises 9 through 14 compute   **a.**   the gradient of the function at the given point and   **b.**   the derivative at that point in the indicated direction.

9.‡ $x^4 y^{-2} z^3$ at $(1, 2, 3)$; the direction of $3\mathbf{i} - 4\mathbf{j} - 12\mathbf{k}$

10. $x\sin y + y\sin z + z\sin x$ at $(\pi/6, \pi/3, \pi/2)$; the direction of $\mathbf{i} - \mathbf{k}$

11.† $\ln(xyz)$ at $(5, 2, 6)$; the direction toward $(4, 3, 7)$

12. $e^{3x-2y+z}$ at $(0, 0, 0)$; the direction toward $(100, -100, 100)$

13.† $(x + y)\sinh(y - z)$ at $(5, 4, 4)$; the direction toward $(-5, -4, -4)$

14. $[x^7 - y^6 + z^5]^{96}$ at $(1, 1, 1)$; the direction toward $(0, 0, 0)$

15.† Sketch the level surfaces $x^2 + y^2 - z = c$ of $x^2 + y^2 - z$ for $c = -2, 0, 2$.

16. Sketch the level surfaces $4x^2 + y^2 + z^2 = c$ of $4x^2 + y^2 + z^2$ for $c = 4$ and 16.

In Exercises 17 through 21 give   **a.**   a normal vector and   **b.**   an equation of the tangent plane to the surface at the given point.

17.‡ $x^3y^2 + y^3z^2 = 76$ at $(1, 2, 3)$

18. $\sqrt{x + 2y + 3z} = 3$ at $(5, -1, 2)$

19.† $x \ln(xz) = 0$ at $(5, 10, \frac{1}{5})$

20.† $\arctan(xyz) = \arctan(10)$ at $(1, 2, 5)$

21. $3xy^2 + 2yz^3 + xz = 93$ at $(-1, 2, 3)$

22. Prove that the tangent plane to the quadric surface $Ax^2 + By^2 + Cz^2 + D = 0$ at (x_0, y_0, z_0) has the equation $Ax_0x + By_0y + Cz_0z + D = 0$.

23.† Give parametric equations of the normal line to the surface $x^2y^3z^4 + xyz = 2$ at $(2, 1, -1)$.

24. Give parametric equations of the normal line to the hyperboloid of two sheets $z^2 - 4x^2 - 9y^2 = 11$ at $(2, 1, 6)$.

25.‡ Use the differential of $x^2y^4/(1 + z^2)$ at $(5, 1, 2)$ to give an approximate value of $(4.996)^2(1.003)^4/[1 + (1.995)^2]$.

26. Use the differential of $(\sin x + \cos y)/\sin z$ at $(\pi/6, \pi/3, \pi/2)$ to give an approximate value of $[\sin(\frac{3}{24}\pi) + \cos(\frac{19}{27}\pi)]/\sin(\frac{9}{20}\pi)$.

27.† The height h of the frustrum of a right circular cone is measured to be 3 inches with a possible error of 0.01 inches. The radius R of its base and the radius r of its top are measured to be 10 inches and 5 inches, respectively, with a possible error of 0.03 inches in each measurement. Use differentials to estimate the possible resulting error in the computed volume $V = \frac{1}{3}\pi h(R^2 + Rr + r^2)$ of the solid.

28. The work done by a constant force of magnitude F on an object that moves a distance S along a straight line at an angle Θ with the direction of the force is $W = FS \cos \Theta$. The force F is measured to be 10 pounds with a possible error of 0.1 pounds; the distance S is measured to be 100 feet with a possible error of one inch; and the angle Θ is measured to be 30° with a possible error of 1 degree. Use differentials to estimate the possible resulting error in the calculated work.

Additional drill exercises

Compute the first-order derivatives of the functions in Exercises 29 through 34.

29.† $f = x^{1/2}y^{1/4} + y^{1/5}z^{1/6}$

30.† $f = \ln(x + 3y - 4z)$

31.† $f = x \sin y - y \cos z$

32.† $g = xyz/(x + y + z)$

33.† $g = xy \arctan(yz)$

34.† $g = xy\, e^{x^2 z}$

Compute the second-order derivatives of the functions in Exercises 35 through 38.

35.† $f = x^2 y - y^3 z^2 + xz$

36.† $f = (x^2 - y^3 + z^4)^{10}$

37.† $g = \sin(xyz)$

38.† $g = x^2 y^3 \tan z$

39.† Compute the derivative of $x^{1/2}y^{1/3}z^{1/4}$ at $(9, 8, 1)$ in the direction toward $(10, 7, 3)$.

40.† Compute the derivative of $(x + y)/z$ at $(2, 3, 4)$ in the direction of the vector $\langle 1, -1, 2 \rangle$.

41.† Find the greatest directional derivative of $xy \sin z$ at $(5, 10, \frac{1}{6}\pi)$ and the unit vector in that direction.

42.† Find the greatest directional derivative of $x^4 y^5 z^2$ at $(2, -1, -3)$ and the unit vector in that direction.

In Exercises 43 through 48 assume that all functions have continuous first-order derivatives.

43.† Compute $P(2)$ and $P'(2)$ where $P(t) = f(t, t^2, t^3)$ and $f(x, y, z)$ satisfies $f(2, 4, 8) = 3, f_x(2, 4, 8) = 4, f_y(2, 4, 8) = 5, f_z(2, 4, 8) = -6$.

44.† Compute $g(1, 3), g_x(1, 3)$, and $g_y(1, 3)$ where $g(x, y) = H\big(u(x, y), v(x, y), w(x, y)\big)$ and $u(1, 3) = 4, v(1, 3) = 5, w(1, 3) = 6, u_x(1, 3) = 1,$ $u_y(1, 3) = 2, v_x(1, 3) = 3, v_y(1, 3) = 4, w_x(1, 3) = 5, w_y(1, 3) = 6,$ $H(4, 5, 6) = 7, H_u(4, 5, 6) = -7, H_v(4, 5, 6) = 10,$ and $H_w(4, 5, 6) = -4$.

45.† Compute $H(0, 0, 0), H_x(0, 0, 0), H_y(0, 0, 0)$, and $H_z(0, 0, 0)$ where $H(x, y, z) = \sqrt{L(x, y, z)}$ and $L(0, 0, 0) = 9, L_x(0, 0, 0) = 5,$ $L_y(0, 0, 0) = 4, L_z(0, 0, 0) = -3$.

46.† Compute $G(1, 2, 3), G_x(1, 2, 3), G_y(1, 2, 3)$, and $G_z(1, 2, 3)$ where $G(x, y, z) = f(x) f(y)^2 + f(z)^3$ and $f(1) = 4, f(2) = 3, f(3) = -1,$ $f'(1) = 2, f'(2) = -5,$ and $f'(3) = 10$.

47.† Compute $A(1, 10, 100), A_u(1, 10, 100), A_v(1, 10, 100), A_w(1, 10, 100)$ where $A(u, v, w) = u^2 v^2 B(w)^{-1}, B(100) = 1000,$ and $B'(100) = 30$.

48.† Express the first order derivatives of $K(x, y, z) = A(x^2) + A(y)^3 + A(z^4)$ in terms of A and A'.

Find the total differentials of the functions in Exercises 49 through 52.

49.† $f = xyz \sin(x + y + z)$

50.† $g = x^2 e^{3y - 4z}$

51.† $h = \dfrac{\sinh x + \cosh y}{\sinh z}$

52.† $K = (x^2 + y^3 - z^4)^{1/5}$

53.† Suppose that the angles α and β (radians) describe the positions of the volume and tuning knobs on a radio and that the radio produces $D(\alpha, \beta, t)$ decibels of sound if the knobs are at α and β at time t (minutes). A child is playing with the knobs so they are at positions $\alpha = \alpha(t)$, $\beta = \beta(t)$ and the radio is producing $N(t) = D\big(\alpha(t), \beta(t), t\big)$ decibels of sound at time t. Express the derivative of N in terms of derivatives of D, α, and β and describe each derivative as a rate of change. (Because t is the variable of the "inside" functions α and β as well as a variable of the "outside" function D, use the notation $D_1, D_2,$ and D_3 for the derivatives of D.)

MISCELLANEOUS EXERCISES

†*Answer provided.*
‡*Outline of solution provided.*

1. Partial derivatives of functions of more than three variables are obtained by holding all but one variable fixed and differentiating with respect to that variable. Compute all of the first-order partial derivatives of the following functions.   **a.**‡ $\sin(xyzw)$ **b.**† $x^2y^2 + y^2z^2 + z^2u^2 + u^2v^2$ **c.** $(x + y + z + w)/(x - y - z - w)$
d. $(\sin x)(\sin y)(\sin z)(\sin w)$

2. Four-dimensional vectors are expressions of the form $\langle x, y, z, w \rangle$, and the gradient of a function $f(x, y, z, w)$ of four variables is the vector $\langle f_x(x, y, z, w), f_y(x, y, z, w), f_z(x, y, z, w), f_w(x, y, z, w) \rangle$. Compute the gradients of the following functions.   **a.**† $x^7y^6z^5w^4$
b. $xe^y - y \sin z + z \tan w$ **c.** $6x - 5y + 4z - 3w$

3.† Compute $\dfrac{\partial}{\partial x}(xy)^v$ and $\dfrac{\partial}{\partial y}(xy)^v$ for $x > 0, y > 0$.

4. What are the unit vectors $\boldsymbol{u}$ such that $D_{\boldsymbol{u}}f(x_0, y_0) = 1$ if $\overrightarrow{\mathrm{grad}}\, f(x_0, y_0) = \boldsymbol{i} + \sqrt{3}\boldsymbol{j}$?

5. Use the geometric interpretation of the gradient to show that for $(x, y, z) \neq (0, 0, 0)$, $\overrightarrow{\mathrm{grad}}\,(x^2 + y^2 + z^2)^{1/2}$ is a unit vector pointing away from the origin. Then obtain this result by differentiating.

6. Prove that if $\boldsymbol{u}$ and $\boldsymbol{v}$ are perpendicular unit vectors in the xy-plane, then $|\overrightarrow{\mathrm{grad}}\, f(x, y)| = \big[[D_{\boldsymbol{u}}f(x, y)]^2 + [D_{\boldsymbol{v}}(x, y)]^2\big]^{1/2}$ for any function $f(x, y)$ with continuous first derivatives.

7. If a sound is emitted at frequency f from a source with velocity v on an x-axis and if the wind has velocity w, an observer with velocity V would hear sound at frequency $F = \left[\dfrac{c + w - V}{c + w - v}\right] f$. (This is called the *Doppler effect.*) Here the constant c is the speed of sound. Compute the derivatives of F with respect to $v, w, V,$ and f.

8. **a.**† Give a formula for the volume of a right circular cone as a function of its height h and the angle Θ between its elements and its axis.
b. The height is measured to be 15 meters with a possible error of 1 centimeter and the angle is measured to be 30° with a possible error of 1°. Use the differentials to estimate the possible resulting error in the computed volume.

9.   a.† Suppose that $z = z(x, y)$ is defined implicitly by the equation $f(x, y, z) = c$ with c a constant. Express z_x and z_y in terms of derivatives of f and x, y, and z.   **b.**   Suppose $z(1, 2) = 3$, $f_x(1, 2, 3) = 6$, $f_y(1, 2, 3) = 7$, and $f_z(1, 2, 3) = 8$. What are $z_x(1, 2)$ and $z_y(1, 2)$?

10.† A function $z = z(x, y)$ is defined implicitly by the equation $xyz + 5x^2y^2z^5 = 6$ and by the condition $z(-1, -1) = 1$. Compute $z_x(-1, -1)$ and $z_y(-1, -1)$.

11.   A function $z = z(x, y)$ is defined implicitly by the equation $xe^{yz} + ye^{xz} = 5$ and by the condition $z(2, 3) = 0$. Compute $z_x(2, 3)$ and $z_y(2, 3)$.

12.   a.   Show that if $\mathbf{A} = [\overrightarrow{\operatorname{grad}} f(x_0, y_0, z_0)] \times [\overrightarrow{\operatorname{grad}} g(x_0, y_0, z_0)]$ is not the zero vector, then the tangent planes to the level surfaces of f and g at (x_0, y_0, z_0) intersect in a line that is parallel to $\mathbf{A}$. This is a tangent line to the curve of intersection of the level surfaces.   **b.**   Give parametric equations of the tangent line to the intersection of the cone $x^2 + y^2 - z^2 = 0$ and the sphere $x^2 + y^2 + z^2 = 338$ at $(5, -12, 13)$.   **c.**   Sketch the curve of intersection and the tangent line.

13.   When the parabola $z = ax^2$ in the xz-plane is rotated about the z-axis it generates the circular paraboloid $z = ax^2 + ay^2$.   **a.**   Show that the normal lines to this paraboloid intersect the z-axis.   **b.**   Explain how this result relates to telescope mirrors (see Section 10.1).

14.   A surface can be given parametrically by equations $x = x(u, v)$, $y = y(u, v)$, $z = z(u, v)$ with (u, v) varying over a region in the uv-plane, or it can be given by the vector function $\mathbf{R}(u, v) = x(u, v)\mathbf{i} + y(u, v)\mathbf{j} + z(u, v)\mathbf{k}$. Show that the vector $\dfrac{\partial \mathbf{R}}{\partial u}(u, v) \times \dfrac{\partial \mathbf{R}}{\partial v}(u, v)$ is normal to the surface at $(x(u, v), y(u, v), z(u, v))$. (Consider tangent lines to the curves $u = $ constant and $v = $ constant on the surface.)

15.   A function $f(x, y)$ is called *homogeneous* of degree n with n a real number if $f(tx, ty) = t^n f(x, y)$ for all $t \neq 0$ and all x and y in the domain of the function. What are the degrees of homogeneity of **a.**† $x^2y - y^2x$,   **b.** $\ln x - \ln y$, and   **c.** $x^{1/4}y^{-2} + (xy)^{-7/8}$?   **d.**   Show that any function $f(x, y)$ that has continuous first-order derivatives and is homogeneous of degree n satisfies *Euler's differential equation* $xf_x + yf_y = nf$. (Differentiate the equation $f(tx, ty) = t^n f(x, y)$ with respect to t and set $t = 1$.)   **e.**   Show that the functions of parts (a), (b), and (c) satisfy the corresponding Euler's differential equation by computing their derivatives.

16.† (Equality of mixed partial derivatives.) Show that $f_{xy}(x_0, y_0) = f_{yx}(x_0, y_0)$ for any function $f(x, y)$ with continuous first- and second-order derivatives in a circle centered at (x_0, y_0). (Apply the Mean Value Theorem for derivatives of functions of one variable four times to show that the quantity $f(x, y) + f(x_0, y_0) - f(x, y_0) - f(x_0, y)$ can be expressed either in the form $f_{xy}(c_1, c_2)(x - x_0)(y - y_0)$ or in the form $f_{yx}(c_3, c_4)(x - x_0)(y - y_0)$ with (c_1, c_2) and (c_3, c_4) in the rectangle with (x_0, y_0) and (x, y) as opposite corners and with sides parallel to the x- and y-axes.)

17. Figure 13.81 shows level curves of the temperature (isotherms) at time $t = 0$ (minutes) before an afternoon thunderstorm. Figures 13.82 and 13.83 show the temperature at times $t = 20$ and $t = 40$ during the storm. (The heavier curves in Figures 13.82 and 13.83 outline the region of heavy rainfall.) Figure 13.84 gives the temperature at $t = 80$, thirty minutes after the last rain had fallen. What was the approximate minimum temperature at each of the four times?

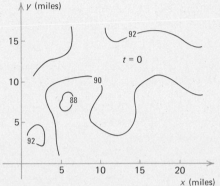

FIGURE 13.81   Isotherms (°F)
(Adapted from *The Thunderstorm*, p. 64.)

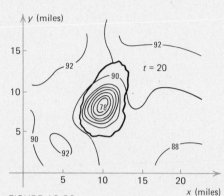

FIGURE 13.82

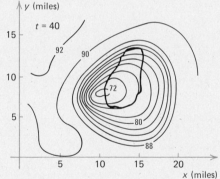

FIGURE 13.83

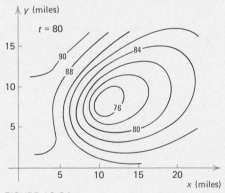

FIGURE 13.84

18. If for each fixed time t the curve $y = f(x, t)$ is the shape of a vibrating string, then $f(x, t)$ satisfies the *wave equation* $f_{xx} = k^{-2}f_{tt}$.   **a.**   Show that $f(x, t) = (\sin x)[\cos(kt)]$ satisfies this differential equation.   **b.**   Sketch the shape of such a string with $k = 1$ at times $t = 0, \frac{1}{4}\pi, \frac{1}{2}\pi, \frac{3}{4}\pi, \pi, \frac{5}{4}\pi, \frac{3}{2}\pi, \frac{7}{4}\pi$, and 2π.

19. If $T(x, t)$ is the temperature of a rod at point x and time t, then $T(x, t)$ satisfies the *heat-flow equation* $T_t = kT_{xx}$. Show that the functions **a.** $T(x, t) = (\sin x)e^{-kt}$ and   **b.** $T(x, t) = t^{-1/2}e^{-x^2/4kt}$ satisfy this differential equation.

20. If for each fixed t the surface $z = f(x, y, t)$ is the shape of a vibrating drum head, then $f(x, y, t)$ satisfies the *wave equation* $f_{xx} + f_{yy} = k^{-2}f_{tt}$ with a constant k. Show that $f(x, y, t) = (\sin x)(\sin y)[\cos(kt\sqrt{2})]$ satisfies this equation. (The shape of such a drum head at $t = 0$ is shown in Figure 13.22 of Section 13.1.)

21.[†]   **a.** What is the domain of $\arctan\left(\dfrac{x}{y^2 - 2y + 1}\right)$? What are its limits
     b. as $(x, y) \to (3, 1)$ and   **c.** as $(x, y) \to (-3, 1)$?

22.   Do **a.** $|xy|$ and   **b.** $|x + y|$ have continuous first order partial derivatives for all (x, y)?

23.   Define $f(x, y)$ to be $(x^3y - y^3x)/(x^2 + y^2)$ for $(x, y) \neq (0, 0)$ and to be zero at $(0, 0)$.   **a.** Show that $f_x(0, y) = -y$ for all y and $f_y(x, 0) = x$ for all x.   **b.** Show that $f_{xy}(0, 0) \neq f_{yx}(0, 0)$.   **c.** Prove that $f_{xy}(x, y) = f_{yx}(x, y)$ for $(x, y) \neq (0, 0)$ by using the result stated in Exercise 16.   **d.** Show that $f_{xy}(x, y)$ is discontinuous at $(0, 0)$ by using parts (a) and (c).

24.[†]   It is shown in advanced calculus that

$$\frac{d}{dx} \int_a^b f(x, t)\, dt = \int_a^b \frac{\partial f}{\partial x}(x, t)\, dt$$

for constants a and b and functions $f(x, t)$ with continuous first-order derivatives. Use this result and the chain rule to show that

$$\frac{d}{dx} \int_{a(x)}^{b(x)} f(x, t)\, dt = f(x, b(x)) \frac{db}{dx}(x) - f(x, a(x)) \frac{da}{dx}(x)$$

$$+ \int_{a(x)}^{b(x)} \frac{\partial f}{\partial x}(x, t)\, dt$$

for functions $a(x)$, $b(x)$, and $f(x, t)$ with continuous first-order derivatives. (Consider the function $G(x, u, v) = \int_u^v f(x, t)\, dt$.)

25.   Use the formula from Exercise 24 to compute the following derivatives. (The answers involve integrals.)

    a.[‡] $\dfrac{d}{dx} \displaystyle\int_{x^2}^{x^3} e^{-t^2}\, dt$ **b.** $\dfrac{d}{dx} \displaystyle\int_{\sin x}^{\cos x} \sqrt{t^3 + 1}\, dt$

    c. $\dfrac{d}{dx} \displaystyle\int_{2x}^{3x} \sin^{10}t\, dt$ **d.**[‡] $\dfrac{d}{dx} \displaystyle\int_{-x}^{5x} e^{xt^2}\, dt$

    e.[†] $\dfrac{d}{dx} \displaystyle\int_x^{10} \sqrt[3]{\sin(xt)}\, dt$ **f.** $\dfrac{d}{dx} \displaystyle\int_0^{\ln x} \sqrt{x^4 + t^4}\, dt$

26.   Use the formula from Exercise 24 to show that

$$\frac{d}{dx} \int_a^x f(x, t)\, dt = f(x, x) + \int_a^x \frac{\partial f}{\partial x}(x, t)\, dt$$

for a function $f(x, t)$ with continuous first-order derivatives.

27.   Figure 13.85 shows the formation and decay of the type of "spike" in the electric potential of the brain that appears every few seconds in electroencephalograms of persons subject to epilepsy. The 12 drawings show the level curves at 0.005-second intervals and in each drawing the curves show the potential at 3.8 microvolt intervals. If $V(x, y)$ denotes the voltage as a function of the coordinates x and y (centimeters) in one of the drawings, then the surface current density at that time is approximately $\sigma \overrightarrow{\text{grad}}\, V(x, y)$ microamperes per centimeter, where σ is the conductivity of the scalp and is approximately $\frac{1}{350}$ ohms^{-1}. What is the approximate maximum magnitude of the current density in the fifth drawing (when the spike is the greatest)?

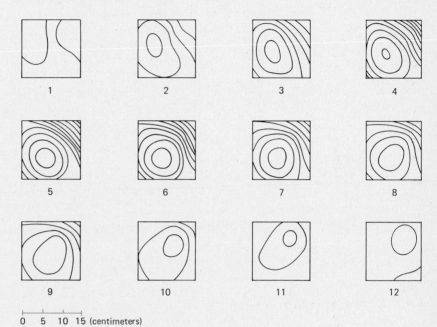

FIGURE 13.85   Electric potential on the surface of the brain (Adapted from material supplied by Dr. R. Bickford, U.C.S.D. Medical School.)

CHAPTER 14

MAXIMA AND MINIMA WITH TWO OR THREE VARIABLES

SUMMARY: In this chapter we study techniques for finding maximum and minimum values of functions of two or three variables. We discuss the first-derivative test in Section 14.1, the second-derivative test in Section 14.2, and maximum-minimum problems with constraints in Section 14.3.

14-1 THE FIRST-DERIVATIVE TEST

As we saw in Chapter 4, the techniques for finding the maximum or minimum value of a function $f(x)$ in an interval depend on whether the interval is *closed* (includes its endpoints) or *open* (does not include its endpoints) and whether it is *bounded* (finite) or *unbounded* (infinite). For functions of one variable the Extreme Value Theorem (Theorem 2.5) applies to closed, bounded intervals, and the first derivative test is used to find maxima and minima in open intervals or in the *interiors* of intervals (the open intervals obtained by removing the endpoints).

The Extreme Value Theorem and the first derivative test for functions of two or three variables employ analogous concepts of "closed" sets, "open" sets, and "interiors" of sets in the xy-plane and in xyz-space. In these cases the BOUNDARIES of the sets correspond to the endpoints in the case of intervals.

The boundary of a set

Definition 14.1   **A point (x_0, y_0) is in the boundary of a set D in an xy-plane if every disk**

$$(x - x_0)^2 + (y - y_0)^2 < \delta^2 \qquad (\delta > 0)$$

centered at (x_0, y_0), no matter how small it is, contains at least one point in D and one point not in D.

A point (x_0, y_0, z_0) is in the boundary of a set D in xyz-space if every sphere

$$(x - x_0)^2 + (y - y_0)^2 + (z - z_0)^2 < \delta^2 \qquad (\delta > 0)$$

centered at (x_0, y_0, z_0), no matter how small, contains at least one point in D and one point not in D.

We say that a set in an xy-plane or in xyz-space is *closed* if it includes all its boundary points and that it is *open* if it includes none of them. We define the *interior* of a set to be the open set obtained by removing its boundary points from it.

EXAMPLE 1   What are the boundary and the interior of the closed disk $x^2 + y^2 \leq 1$?

Solution.   The boundary of the disk (Figure 14.1) is the circle $x^2 + y^2 = 1$. The disk $x^2 + y^2 \leq 1$ is closed because it includes the circle. The interior is the open disk $x^2 + y^2 < 1$ obtained by removing the circle.   □

EXAMPLE 2   What are the boundary and the interior of the rectangle $1 \leq x < 3, 1 \leq y < 2$? Is the rectangle closed? Is it open?

Solution.   The boundary of the rectangle (Figure 14.1a) consists of its four sides. The interior is the open rectangle $1 < x < 3, 1 < y < 2$. The rectangle $1 \leq x < 3, 1 \leq y < 2$ is neither open nor closed because it contains some of its boundary points but not all of them.   □

EXAMPLE 3   What are the boundary and interior of the open half plane $x > 1$?

Solution.   The boundary (Figure 14.1b) is the line $x = 1$. The half plane is open and is its own interior because it does not include any of the boundary.   □

EXAMPLE 4   What are the boundary and interior of the closed sphere $x^2 + y^2 + z^2 \leq 4$?

Solution.   The boundary is the surface of the sphere $x^2 + y^2 + z^2 = 4$. The interior is the open sphere $x^2 + y^2 + z^2 < 4$.   □

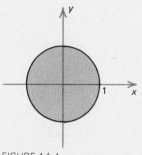

FIGURE 14.1
The closed disk $x^2 + y^2 \leq 1$

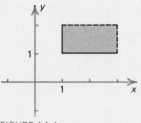

FIGURE 14.1a
The rectangle $1 \leq x < 3$,
$1 \leq y < 2$

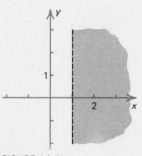

FIGURE 14.1b
The open half plane $x > 1$

The Extreme Value Theorem

A set in the xy-plane is said to be *bounded* if it can be enclosed in a circle. A set in xyz-space is bounded if it can be enclosed in a sphere. The disk and rectangle in Figures 14.1 and 14.1a are bounded and the half plane in Figure 14.1b is not. The Extreme Value Theorem for functions of two or three variables applies to continuous functions in closed bounded sets:

Theorem 14.1 (The Extreme Value Theorem for functions of two or three variables) If a function f is continuous in a closed and bounded set in an xy-plane or in xyz-space, then it has a maximum and a minimum value in that set.

This result is proved in books on advanced calculus (see, for example, R. Courant [1]).

EXAMPLE 5 What are the maximum and minimum of $\sqrt{x^2 + y^2}$ in the closed rectangle $-1 \le x \le 3, -1 \le y \le 4$?

Solution. The maximum and minimum exist because the function is continuous for all (x, y) and the rectangle (Figure 14.1c) is closed and bounded. The function $\sqrt{x^2 + y^2}$ gives the distance from (x, y) to the origin, and the farthest point from the origin is the upper right corner $(3, 4)$ of the rectangle. Therefore the maximum of $\sqrt{x^2 + y^2}$ in the rectangle is its value $\sqrt{3^2 + 4^2} = 5$ at $(3,4)$. The minimum is 0 and occurs at the origin. □

FIGURE 14.1c
The closed rectangle
$-1 \le x \le 3, -1 \le y \le 4$

Local maxima and minima

As the last example illustrates, the maximum or minimum of a function can occur on the boundary of a region or in its interior. In this section and in Section 14.2 we will deal primarily with finding maxima and minima in the interiors of regions. Techniques for finding maxima and minima on boundaries of sets and other curves and surfaces will be discussed in Section 14.3. Recall that a function f of one variable has a *local maximum* at a point x_0 in the interior of its domain if $f(x_0)$ is the greatest value of f in an open interval containing x_0, and that f has a *local minimum* at x_0 if $f(x_0)$ is the least value of f in such an interval. Local maxima and minima for functions of two or three variables are defined similarly:

We say that $f(x, y)$ has a local maximum at a point (x_0, y_0) in the interior of its domain if there is a circle centered at (x_0, y_0) such that $f(x_0, y_0)$ is the greatest value of f in the circle. The function has a local minimum at (x_0, y_0) if $f(x_0, y_0)$ is the least value of f in such a circle. A function $f(x, y, z)$ has a local maximum at (x_0, y_0, z_0) in the interior of its domain if $f(x_0, y_0, z_0)$ is the greatest value of f in a sphere centered at (x_0, y_0, z_0). The function has a local minimum at (x_0, y_0, z_0) if $f(x_0, y_0, z_0)$ is the least value of f in such a sphere.

A point (x_0, y_0) in the interior of the domain of $f(x, y)$ is a *critical point* of f if the partial derivatives $f_x(x_0, y_0)$ and $f_y(x_0, y_0)$ do not exist or are zero. A point (x_0, y_0, z_0) in the interior of the domain of $f(x, y, z)$ is a critical point of f if the partial derivatives $f_x(x_0, y_0, z_0)$, $f_y(x_0, y_0, z_0)$, and $f_z(x_0, y_0, z_0)$ do not exist or are zero. As is the case with one variable, local maxima and minima of functions of two or three variables occur at critical points.

Theorem 14.2   (The first derivative test for functions of two or three variables)   If a function f of two or three variables has a local maximum or minimum at a point P, then P is a critical point of f.

Proof.   If $f(x, y)$ has a local maximum or minimum at (x_0, y_0), then the function $f(x, y_0)$ of x has a local maximum or minimum at x_0 and the function $f(x_0, y)$ of y has a local maximum or minimum at y_0. By the first-derivative test for functions of one variable (Theorem 4.3), the derivatives $f_x(x_0, y_0)$ and $f_y(x_0, y_0)$ either do not exist or are zero and, therefore, (x_0, y_0) is a critical point of $f(x, y)$.

Similarly, if $f(x, y, z)$ has a local maximum or minimum at (x_0, y_0, z_0), then the three functions $f(x, y_0, z_0)$, $f(x_0, y, z_0)$, and $f(x_0, y_0, z)$ have local maxima or minima at $x = x_0$, $y = y_0$, and $z = z_0$, respectively, and hence the derivatives $f_x(x_0, y_0, z_0)$, $f_y(x_0, y_0, z_0)$, and $f_z(x_0, y_0, z_0)$ do not exist or are zero.   Q.E.D.

EXAMPLE 6   Find the critical points of $x^2 + y^2 - 2x - 4y + 6$.

Solution.   We have

$$\frac{\partial}{\partial x}(x^2 + y^2 - 2x - 4y + 6) = 2x - 2$$

$$\frac{\partial}{\partial y}(x^2 + y^2 - 2x - 4y + 6) = 2y - 4.$$

At a critical point both partial derivatives must be zero, so x and y must satisfy the simultaneous equations $2x - 2 = 0$ and $2y - 4 = 0$. The solutions are $x = 1$ and $y = 2$, and the one critical point is $(1, 2)$.   □

EXAMPLE 7   Show by completing the squares that the function $x^2 + y^2 - 2x - 4y + 6$ of Example 6 has an absolute minimum at its critical point.

Solution.   We complete the squares by writing

$$\begin{aligned}
x^2 + y^2 - 2x - 4y + 6 &= (x^2 - 2x) + (y^2 - 4y) + 6 \\
&= (x^2 - 2x + 1 - 1) \\
&\quad + (y^2 - 4y + 4 - 4) + 6 \\
&= (x - 1)^2 + (y - 2)^2 + 1.
\end{aligned}$$

The sum $(x - 1)^2 + (y - 2)^2$ is positive except at $x = 1$, $y = 2$, where it is zero. Therefore, the absolute minimum of the original function is 1 and occurs at the point $(1, 2)$.   □

A geometric interpretation

We can think of the critical points of a function $f(x, y)$ of two variables as the points where either its graph has no tangent plane or the tangent plane is horizontal. If (x_0, y_0) is a critical point because either or both of the partial derivatives $f_x(x_0, y_0)$ or $f_y(x_0, y_0)$ do not exist, then the graph has no tangent plane at $x = x_0, y = y_0$. If the graph has a tangent plane $z = f(x_0, y_0) + f_x(x_0, y_0)(x - x_0) + f_y(x_0, y_0)(y - y_0)$ and (x_0, y_0) is a critical point, then both derivatives exist and are zero, the tangent plane has the equation $z = f(x_0, y_0)$ and it is horizontal. For example, the graph $z = x^2 + y^2 - 2x - 4y + 6$ of the function of Examples 6 and 7 is the circular paraboloid shown in Figure 14.1d and has a horizontal tangent plane at $x = 1, y = 2$, where the function has a critical point.

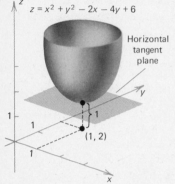

$z = x^2 + y^2 - 2x - 4y + 6$

Horizontal tangent plane

$(1, 2)$

FIGURE 14.1d

EXAMPLE 8   Show that the minimum of $f(x, y) = \sqrt{x^2 + y^2}$ occurs at a critical point.

Solution.   Because $\sqrt{x^2 + y^2}$ is 0 at $(0, 0)$ and is positive for $(x, y) \neq (0, 0)$ the minimum of f is 0 and occurs at the origin.

When we set $y = 0$ in the formula for $f(x, y)$, we obtain $f(x, 0) = \sqrt{x^2} = |x|$. Since $|x|$ does not have a derivative at $x = 0$, $f_x(0, 0)$ does not exist. Similarly, setting $x = 0$ gives $f(0, y) = |y|$, so $f_y(0, 0)$ does not exist either and $(0, 0)$ is a critical point.

The graph of f does not have a tangent plane at the origin because that point is the vertex of the cone $z = \sqrt{x^2 + y^2}$ that is the graph (Figure 14.1e).   □

In the next example we find the critical points by solving a pair of simultaneous equations.

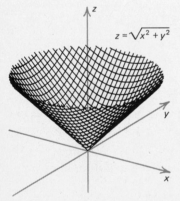

$z = \sqrt{x^2 + y^2}$

FIGURE 14.1e   The cone $z = \sqrt{x^2 + y^2}$ has no tangent plane at the origin.

EXAMPLE 9   Find the maximum and minimum of

$$f(x, y) = \frac{x}{x^2 + y^2 + 4}$$

and the points where they occur.

Solution.   To find the x-derivative, we use the quotient rule:

(1)
$$\frac{\partial f}{\partial x} = \frac{(x^2 + y^2 + 4)\dfrac{\partial}{\partial x}(x) - x\dfrac{\partial}{\partial x}(x^2 + y^2 + 4)}{(x^2 + y^2 + 4)^2}$$

$$= \frac{x^2 + y^2 + 4 - x(2x)}{(x^2 + y^2 + 4)^2} = \frac{y^2 + 4 - x^2}{(x^2 + y^2 + 4)^2}.$$

Because the numerator of f does not depend on y, we write $f = x(x^2 + y^2 + 4)^{-2}$ to compute the y-derivative:

(2) $$\frac{\partial f}{\partial y} = -x(x^2 + y^2 + 4)^{-2}\frac{\partial}{\partial y}(x^2 + y^2 + 4) = \frac{-2xy}{(x^2 + y^2 + 4)^2}.$$

Since the derivatives exist for all x, the critical points are where both derivatives are zero, and equations (1) and (2) show these are where x and y satisfy the simultaneous equations

(3) $y^2 + 4 - x^2 = 0$ and $2xy = 0$.

In solving such a system of equations it is often convenient to begin by dealing with the simpler equation, which in this case is the second one. The equation $2xy = 0$ is satisfied if and only if either x or y is zero. We study the first equation in each of these cases separately. If x is zero, the first equation reads $y^2 + 4 = 0$ and has no solutions. If y is zero, the first equation reads $4 - x^2 = 0$ and has the solutions $x = \pm 2$. Hence the solutions of equations (3) are $x = 2, y = 0$ and $x = -2$, $y = 0$. The critical points are $(2, 0)$ and $(-2, 0)$.

Because we are considering the function in the whole xy-plane, which has no boundary points, a maximum or minimum must be a local maximum or minimum and occur at one of the two critical points. The function has the value $f(2, 0) = 2/(2^2 + 4) = \frac{1}{4}$ at $(2, 0)$ and the value $f(-2, 0) = -2/(2^2 + 4) = -\frac{1}{4}$ at $(-2, 0)$. If the function has a maximum, it must be the value $\frac{1}{4}$ at $(2, 0)$, and if it has a minimum, it must be $-\frac{1}{4}$ at $(-2, 0)$.

We cannot apply the Extreme Value Theorem directly to show that the maximum and minimum exist because the xy-plane is not a bounded set. Instead, we note that $|x| = \sqrt{x^2}$ is less than $\sqrt{x^2 + y^2 + 4}$ and consequently

(4) $|f(x, y)| = \dfrac{|x|}{x^2 + y^2 + 4} < \dfrac{1}{\sqrt{x^2 + y^2 + 4}}.$

This shows that $|f(x, y)|$ is small when $x^2 + y^2$ is large. In particular, (4) shows that $|f(x, y)|$ is less than $\frac{1}{8}$ for $x^2 + y^2 \geq 60$ and hence

(5) $-\frac{1}{8} < f(x, y) < \frac{1}{8}$ for $x^2 + y^2 \geq 60$.

Applying the Extreme Value Theorem to the continuous function f in the closed bounded disk $x^2 + y^2 \leq 60$ shows that f has a maximum and a minimum in that disk. Inequalities (5) imply that the maximum must be $\frac{1}{4}$ and the minimum $-\frac{1}{4}$ because $f(x, y)$ lies between these values for (x, y) on the boundary $x^2 + y^2 = 60$ of the disk. Finally, inequalities (5) also show that $\frac{1}{4}$ and $-\frac{1}{4}$ are the maximum and minimum for all (x, y) because $f(x, y)$ lies between these values for (x, y) outside the disk as well.   □

Narrative problems   In the next example we are to find the minimum of a quantity that depends on three variables, subject to a side condition. We use the side condition to express one of the two variables in terms of the other two and then use the first derivative test to find the minimum of the resulting function of two variables.

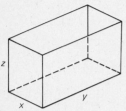

FIGURE 14.2
Area of bottom $= xy$
Area of front $= xz$
Area of side $= yz$

EXAMPLE 10   A rectangular box with no top and a volume of 6 cubic feet is to be constructed from material that costs $6 per square foot for the bottom, $2 per square foot for the front and back, and $1 per square foot for the sides. What dimensions would minimize the cost of the box?

Solution.   We denote the width of the front and back by x (feet), the length of the sides by y (feet), and the height by z(feet). The area of the bottom is xy square feet, the combined area of the front and back is $2xz$ square feet, and the combined area of the sides is $2yz$ square feet (Figure 14.2). The box with these dimensions would cost

$$
(6) \quad \left[6 \, \frac{\text{dollars}}{\text{sq. ft.}} \right] [xy \text{ sq. ft.}] + \left[2 \, \frac{\text{dollars}}{\text{sq. ft.}} \right] [2xz \text{ sq. ft.}]
$$

$$
+ \left[1 \, \frac{\text{dollar}}{\text{sq. ft.}} \right] [2yz \text{ sq. ft.}] = 6xy + 4xz + 2yz \text{ dollars.}
$$

In order for the volume xyz of the box to be 6 cubic feet, we must have $xyz = 6$ or $z = 6/(xy)$. We make this substitution in formula (6) to obtain a formula for the cost

$$
(7) \quad C(x, y) = 6xy + 4x\left(\frac{6}{xy}\right) + 2y\left(\frac{6}{xy}\right) = 6xy + \frac{24}{y} + \frac{12}{x}
$$

of a box of volume 6, width x, and length y. We are to find the minimum of $C(x, y)$ in the open quarter plane $x > 0, y > 0$.

Because the region is open, the minimum must occur at a critical point, which we find by computing the derivatives of C:

$$
C_x(x, y) = \frac{\partial}{\partial x}(6xy + 24y^{-1} + 12x^{-1}) = 6y - 12x^{-2}
$$

$$
C_y(x, y) = \frac{\partial}{\partial y}(6xy + 24y^{-1} + 12x^{-1}) = 6x - 24y^{-2}.
$$

The critical points are the solutions (x, y) of the simultaneous equations $6y - 12x^{-2} = 0$ and $6x - 24y^{-2} = 0$. Solving the first of these for y and the second for x yields

$$
(8) \quad y = \frac{2}{x^2} \text{ and } x = \frac{4}{y^2}.
$$

Substituting the first of these equations into the second gives the equation

$$
x = \frac{4}{(2/x^2)^2} = \frac{4x^4}{4} = x^4
$$

which has the solutions $x = 0$ and $x = 1$. The function is not defined for $x = 0$, so the only relevant solution is $x = 1$. The first of equations (8) then shows that $y = 2$ and that the one critical point is $(1, 2)$.

The corresponding value of the height $z = 3$ can be found from the equation $xyz = 6$ for the volume. Since we expect the problem to have a solution and have found only one critical point, we can assume it gives the solution and that the cost is minimized by a box of width 1, length 2, and height 3.

If we want to supply a more formal demonstration that $C(x, y)$ has a minimum at $(1, 2)$, we can use the type of argument given in the solution of Example 9: We can show that $C(x, y)$ is greater than $C(1, 2)$ for all (x, y) on the boundary and outside the closed square $a \leq x \leq b, a \leq y \leq b$ for suitable positive numbers a and b (see Exercise 45). This implies that the minimum in the square occurs at the critical point and then that the value there is the minimum for all $x > 0, y > 0$.   □

EXAMPLE 11   The function $f(x, y, z) = x^4 + y^4 + z^4 - 4xyz$ has a minimum value for $x \geq 0, y \geq 0, z \geq 0$. What is it?

Solution.   We have $f_x = 4x^3 - 4yz, f_y = 4y^3 - 4xz, f_z = 4z^3 - 4xy$, so the critical points in the interior of the region are given by the positive solutions of the simultaneous equations

$$4x^3 - 4yz = 0, \quad 4y^3 - 4xz = 0, \quad 4z^3 - 4xy = 0.$$

When we multiply the first of these equations by x, the second by y, and the third by z and rewrite the results, we obtain

(9) $4x^4 = 4xyz, \quad 4y^4 = 4xyz, \quad 4z^4 = 4xyz.$

Since the right sides of these equations are the same, we have $x^4 = y^4 = z^4$ and then, because x, y, and z are positive, $x = y = z$. Finally the first of equations (9) with y and z replaced by x yields $4x^4 = 4x^3$ to show that $x = 1$ and hence $y = 1$ and $z = 1$.

The function has one critical point $(1, 1, 1)$, where it has the value $f(1, 1, 1) = 1^4 + 1^4 + 1^4 - 4(1)(1)(1) = -1$. At every point of the boundary of the region $x \geq 0, y \geq 0, z \geq 0$, either x, y, or z is zero and hence f equals $x^4 + y^4 + z^4$ and is nonnegative. Therefore, the minimum is the value -1 at the critical point.   □

Exercises

Find the critical points of the functions in Exercises 1 through 15.

1.† $x^2 - xy + y$

2. e^{xy}

3.† $x \sin y$

4. $x^4 + 32x - y^3$

5.‡ $4xy - x^4 - y^4$

6.† $3xy^2 + x^3 - 3x$

7.‡ $2y^3 - 3x^4 - 6x^2y$

8. $3xy - x^3 + y^3$

9.† $x^3 + y^4 - 36y^2 - 12x$

10. $2y^2 + x^4 - 8xy$

11. $e^{x^2 - 2y^2 + y - 6x}$

12.† $xye^{2x + 3y}$

13.† xy^3

14.† $\cos(x - y)$

15. $x^2 + 6xy + 9y^2$

16.[‡] Find the maximum of xye^{-x-y} for $x \geq 0$, $y \geq 0$.

17. Use the first-derivative test to find the point on the plane $z = 2x + 3y$ closest to the point $(4, 2, 0)$.

18.[†] Find the points on the surface $z = 1/xy$ that are closest to the origin.

19. Suppose that it costs $C(x)$ dollars to produce x units of a product, that x_1 units sold in one market bring in $R_1(x_1)$ dollars revenue, and that x_2 units sold in a second market bring in $R_2(x_2)$ dollars.   **a.**   What is the profit $P(x_1, x_2)$ on x_1 units sold in the first market and x_2 units sold in the second market?   **b.**   Show that $R_1'(x_1) = R_2'(x_2)$ for sales that maximize the profit.

20.[‡] Use the first-derivative test to show that the rectangular box of surface area A and maximum volume is a cube.

21. A rectangular box of volume 24 cubic feet is to be constructed. Material for the sides costs 50 cents per square foot, for the front and back 75 cents per square foot, and for the top and bottom a dollar per square foot. What dimensions should the box have to minimize its cost?

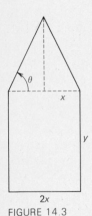

FIGURE 14.3

22.[‡] A window of total area 1 square yard is to be constructed in the shape of an isosceles triangle resting on a rectangle as in Figure 14.3. What choices of x, y, and Θ would give it the smallest perimeter?

23. A rectangular box with corners at $(\pm x, \pm y, \pm z)$ is inscribed in the ellipsoid $z^2 = 1 - x^2 - 4y^2$. What choices of x, y, and z would maximize the total length of its twelve edges?

24. Find the maximum and minimum values of $(x + 2y)e^{-x^2-y^2}$.

25. (The method of least squares) In N runs of an experiment a researcher obtains N pairs of data $(x_1, y_1), (x_2, y_2), \ldots, (x_N, y_N)$. From theoretical considerations he expects the pairs of data to lie on a line $y = mx + b$, but they do not because of experimental error. He determines a line to fit the data by the *method of least squares*: he chooses m and b to minimize the sum

$$E(m, b) = \sum_{j=1}^{N} (mx_j + b - y_j)^2$$

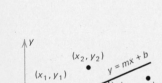

FIGURE 14.4

of the squares of the vertical distances between the points (x_j, y_j) and the line (Figure 14.4). Show that m and b are given by the equations

$$(9) \quad m \sum_{j=1}^{N} x_j^2 + b \sum_{j=1}^{N} x_j = \sum_{j=1}^{N} x_j y_j$$

$$(10) \quad m \sum_{j=1}^{N} x_j + Nb = \sum_{j=1}^{N} y_j.$$

26.[†] Use the method of least squares [equations (9) and (10)] to find a line $y = mx + b$ to fit the data $(1.1, 2.4)$, $(1.9, 3.1)$, $(4.1, 3.8)$, $(4.9, 4.4)$.

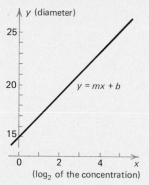

FIGURE 14.5

TABLE 14.1

$x = $ LOG$_2$ OF THE CONCENTRATION	$y = $ DIAMETER OF THE CIRCLE
0	15.87
1	17.78
2	19.52
3	21.35
4	23.13
5	24.77

27. The concentration of a sample of penicillin can be determined by placing a drop of specified volume on a plate containing a culture of bacteria. The penicillin inhibits the growth of the bacteria in a circular region, whose diameter y is measured. The number x determined from the line in Figure 14.5 is then the logarithm to the base 2 of the concentration. The line is determined by the method of least squares with the data in Table 14.1, which was obtained by testing samples of known concentration. What is the equation of the line? (Adapted from O. L. Davies and P. L. Goldsmith, [1] *Statistical Methods in Research and Production*, p. 207.)

28. Derive equations analogous to equations (9) and (10) for the plane $z = Ax + By + C$ that makes the least-squares fit to the data (x_j, y_j, z_j) ($j = 1, 2, 3, \ldots, N$); that is, find equations for $A, B,$ and C that minimize

$$\sum_{j=1}^{N} (Ax_j + By_j + C - z_j)^2.$$

29. Find the point (x, y) such that the sum of the squares of its distances to $(x_1, y_1), (x_2, y_2),$ and (x_3, y_3) is a minimum.

30.[†] Find the minimum of $2x^2 + 3y^2 + z^2$ for $x + 2y + z = 17$.

31. Find the maximum or minimum values of the following functions by completing the squares.   **a.**[†] $x^2 + y^2 - 4x + 6y + 8$
b. $-x^2 - y^2 + 2x - 8y - 7$ **c.** $3x^2 + 3y^2 - 12x - 6y$
d. $6x - 3y - 5x^2 - 5y^2$

32.[†] **a.** Show that $x^2 + xy + y^2 + 3x - 9y$ has a minimum value by showing that its graph is an elliptic paraboloid that opens upward.
b. Find the minimum value.

33. Find the vertices of the tetrahedron(s) of least volume formed by the coordinate planes and by a plane with positive intercepts and passing through the point $(1, 2, 3)$.

34. The level curves in Figure 14.6 give the average number of days per year with high pollution potential because of weather conditions. What are the approximate values of this function at its local maxima in the United States?

FIGURE 14.6

Find the required maxima and minima in Exercises 35 through 43 and the points where they occur.

35.[†] The maximum of $6x - x^2 - 8y - 2y^2$

36. The minimum of $x^2 + y^2 + xy - 2x$

37.[†] The maximum of $\dfrac{\sin x}{1 + y^2}$ **38.**[†] The minimum of $\dfrac{y^2}{\cos x + 2}$

39. The maximum and minimum of $\dfrac{\sin y - 3}{\cos x + 2}$

40.[†] The maximum and minimum of $xe^{-x^2-y^2}$

41.[†] The minimum of $x^4 + y^4 - (x + y)^2$

42. The minimum of $\ln(x^2 + y^2 + 1)$

43.[†] The maximum and minimum of $(\cos x)e^{\sin y}$

44.[†] Find the dimensions of the rectangular box of maximum volume whose base is the rectangle with corners $(\pm x, \pm y)$ in the xy-plane and whose upper corners are on the elliptic paraboloid $z = 4 - 9x^2 - y^2$.

45.[†] Find positive numbers a and b such that the function $C(x, y)$ of Example 10 has values greater than its value at its critical point for $0 < x \le a$ and all y, for $0 < y \le a$ and all x, for $x \ge a$ and $y \ge b$, and for $y \ge a$ and $x \ge b$.

46.[†] The function $\dfrac{4}{x} + \dfrac{1}{y} + \dfrac{1}{z} + xyz$ has a minimum for $x > 0, y > 0,$ $z > 0$. What is it?

47. The function $4x^4 + y^3 + z^2 - x^2 - 3y - 4z$ has a minimum for $x \ge 0, y \ge 0, z \ge 0$. What is it?

14-2   THE SECOND-DERIVATIVE TEST

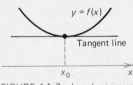

FIGURE 14.7   Local minimum
$f'(x_0) = 0, f''(x_0) > 0$

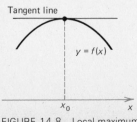

FIGURE 14.8   Local maximum
$f'(x_0) = 0, f''(x_0) < 0$

To determine whether a function $f(x)$ of one variable has a local maximum or minimum at a critical point x_0, we can compute the second derivative $f''(x_0)$. If $f''(x_0)$ is positive, the graph is concave up and lies above the horizontal tangent line for x near x_0 and the function has a local minimum at x_0 (Figure 14.7). If $f''(x_0)$ is negative, the graph is concave down and lies below the tangent line and the function has a local maximum (Figure 14.8). If $f''(x_0)$ is zero, the second-derivative test fails; the function may or may not have a local maximum or minimum at x_0.

The idea behind the second-derivative test for a function $f(x, y)$ of two variables is to use the second derivatives $f_{xx}(x_0, y_0), f_{xy}(x_0, y_0),$ $f_{yy}(x_0, y_0)$ to determine the type of paraboloid that best approximates the graph of the function near a critical point $x = x_0, y = y_0$. If the approximating surface is an elliptic paraboloid that opens upward as in

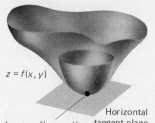

$z = f(x, y)$

$(x_0, y_0, f(x_0, y_0))$

Horizontal tangent plane

FIGURE 14.9   Graph approximated by an elliptic paraboloid near a local minimum

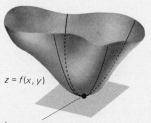

$z = f(x, y)$

$(x_0, y_0, f(x_0, y_0))$

FIGURE 14.10   Vertical cross sections that are concave up at a local minimum

Figure 14.9, then the function has a local minimum at the critical point. Near the critical point the vertical cross sections of the graph are concave up and the graph lies above its horizontal tangent plane (Figure 14.10). If the approximating surface is an elliptic paraboloid that opens downward as in Figure 14.11, the function has a local maximum at the critical point. Near the critical point the vertical cross sections of the graph are concave down and the graph lies below its tangent plane (Figure 14.12). If the approximating surface is a hyperbolic paraboloid as in Figure 14.13, the function has neither a local maximum or minimum at the critical point. Some of the vertical cross sections of the graph are concave and some are concave down, and the graph lies on both sides of its tangent plane (Figure 14.14). In this case the function is said to have a SADDLE POINT at the critical point (x_0, y_0).

In the one remaining case, where the paraboloid which best approximates the graph is a parabolic cylinder, the second-derivative test fails. The function may or may not have a local maximum or minimum at its critical point.

The criterion for determining the type of paraboloid that best approximates the graph is stated in the next theorem.

Horizontal tangent plane

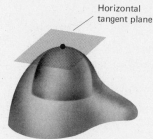

FIGURE 14.11   Graph approximated by an elliptic paraboloid near a local maximum

FIGURE 14.12   Vertical cross sections that are concave down near a local maximum

Horizontal tangent plane

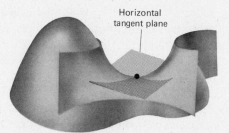

FIGURE 14.13   Graph approximated by a hyperbolic paraboloid near a saddle point

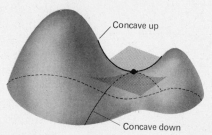

Concave up

Concave down

FIGURE 14.14   Vertical cross sections near a saddle point

Theorem 14.3   Suppose that $f(x, y)$ has continuous second-order derivatives in a circle centered at a critical point (x_0, y_0). Set

(1) $A = f_{xx}(x_0, y_0), \qquad B = f_{xy}(x_0, y_0), \qquad C = f_{yy}(x_0, y_0)$.

The function f has a local minimum at (x_0, y_0) if $B^2 - AC$ is negative and A and C are positive; f has a local maximum if $B^2 - AC$, A, and C are negative; and f has a saddle point if $B^2 - AC$ is positive. If $B^2 - AC$ is zero, the second-derivative test fails.

Notice that Theorem 14.3 involves the expression $B^2 - AC$ and *not* the similar expression $B^2 - 4AC$, which we discussed in Section 10.4 as the discriminant of a conic section. All possible combinations of signs of $A, B, C,$ and $B^2 - AC$ are dealt with in the theorem because if $B^2 - AC$ is negative, then A and C must be both positive or both negative.

We do not include a full proof of Theorem 14.3, as it is more appropriately a topic for advanced calculus. We give only the portion of the proof that explains the role of the expression $B^2 - AC$ in the criterion.

Partial proof of Theorem 14.3.   The paraboloid that best approximates the graph of $f(x, y)$ near the critical point (x_0, y_0) is the graph of the second-degree polynomial

(2) $g(x, y) = \frac{1}{2} Ax^2 + Bxy + \frac{1}{2} Cy^2 + Dx + Fy + G$

that has the same value and the same first- and second-order derivatives as $f(x, y)$ at (x_0, y_0). This is the paraboloid that has the same tangent plane as the graph of f and whose vertical cross sections through $(x_0, y_0, f(x_0, y_0))$ have the same curvature as the corresponding cross sections of the graph of f.

A short calculation shows that the second derivatives of $g(x, y)$ are the constants $g_{xx} = A, g_{xy} = B,$ and $g_{yy} = C$. Therefore, to have the second derivatives of g be equal to those of f, the numbers $A, B,$ and C must be those given in formulas (1). The level curves of $g(x, y)$ have the equations

(3) $\frac{1}{2} Ax^2 + Bxy + \frac{1}{2} Cy^2 + Dx + Ey + F = \text{Constant}$

and by Theorem 10.5 (with A replaced by $\frac{1}{2} A$ and C replaced by $\frac{1}{2} C$), are hyperbolas, ellipses, or parabolas according to whether the discriminant

(4) $B^2 - 4(\frac{1}{2} A)(\frac{1}{2} C) = B^2 - AC$

is positive, negative, or zero.

If $B^2 - AC$ is positive, the level curves are hyperbolas and the graph of g is a hyperbolic paraboloid, so the function f has a saddle point at (x_0, y_0), as stated in the theorem. If $B^2 - AC$ is negative and A and C are positive, the level curves are ellipses and the graph of g is an elliptic paraboloid that opens upward, so f has a local minimum at (x_0, y_0). If $B^2 - AC, A$, and C are negative, the graph of g is an elliptic paraboloid that opens downward, so f has a local maximum. (In the case of $B^2 - AC = 0$, where the second-derivative test fails, the level curves of g are parabolas and its graph is a parabolic cylinder.) □

EXAMPLE 1   Use the second-derivative test to classify the critical points of $4xy - x^4 - y^4 + \frac{1}{16}$.

Solution.   We set $f(x, y) = 4xy - x^4 - y^4 + \frac{1}{16}$. Then

(5)
$$f_x = \frac{\partial}{\partial x}(4xy - x^4 - y^4 + \frac{1}{16}) = 4y - 4x^3 = 4(y - x^3)$$

$$f_y = \frac{\partial}{\partial y}(4xy - x^4 - y^4 + \frac{1}{16}) = 4x - 4y^3 = 4(x - y^3).$$

Setting $f_x = 0$ and $f_y = 0$ gives the equations $y = x^3, x = y^3$. We substitute the first of these equations into the second to obtain the equation $x = (x^3)^3$ or $x = x^9$ for x. This has solutions $x = 0, 1, -1$, so using the equation $y = x^3$, we see that the critical points are $(0, 0)$, $(1, 1)$, and $(-1, -1)$.

From equations (5) we compute $f_{xx} = \frac{\partial}{\partial x}(4y - 4x^3) = -12x^2$,

$f_{xy} = \frac{\partial}{\partial y}(4y - 4x^3) = 4$, and $f_{yy} = \frac{\partial}{\partial y}(4x - 4y^3) = -12y^2$. We then

make the following table of values of $A = f_{xx}, B = f_{xy}, C = f_{yy}$ and $B^2 - AC$ at the critical points.

Critical point	$A = f_{xx}$	$B = f_{xy}$	$C = f_{yy}$	$B^2 - AC$
$(0, 0)$	0	4	0	$4^2 - 0(0) = 16$
$(1, 1)$	-12	4	-12	$4^2 - (-12)(-12) = -128$
$(-1, -1)$	-12	4	-12	$4^2 - (-12)(-12) = -128$

At $(0, 0)$, $B^2 - AC$ is positive and there is a saddle point. At $(1, 1)$ and $(-1, -1)$ there are local maxima because $B^2 - AC, A$, and C are negative. These properties of the function can be seen in Figure 14.15, where we show its graph, and in Figure 14.16 where we show a sketch of its level curves. □

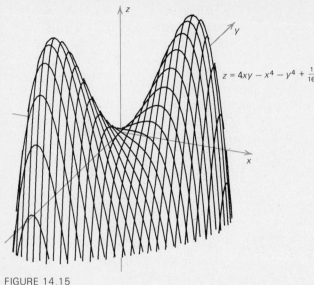

$$z = 4xy - x^4 - y^4 + \tfrac{1}{16}$$

FIGURE 14.15

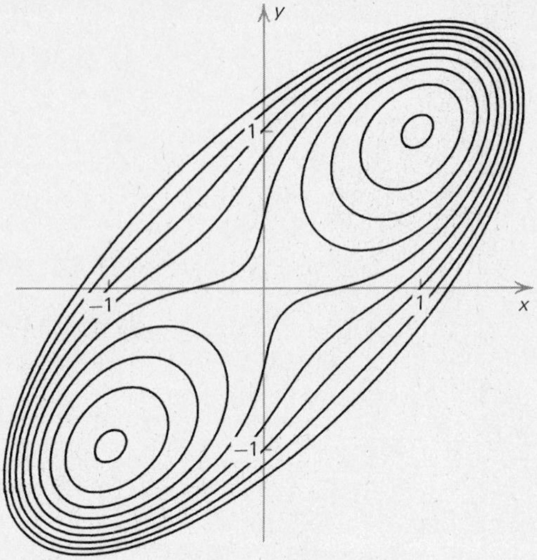

FIGURE 14.16   Level curves of $4xy - x^4 - y^4 + \tfrac{1}{16}$

Exercises

†*Answer provided.*
‡*Outline of solution provided.*

1. The graphs of the functions   **a.**‡ $3xy - x^3 - y^3 + \tfrac{1}{8}$,
b.† $\tfrac{3}{2}x - \tfrac{1}{2}x^3 - xy^2 + \tfrac{1}{16}$,   **c.** $x^4 - 2x^2 + y^2 - \tfrac{17}{16}$,
d.‡ $2y^3 - 3x^4 - 6x^2y + \tfrac{1}{16}$,   **e.** $x^4 - y^4 - 2x^2 + 2y^2 + \tfrac{1}{16}$, and
f. $-x^4 + 4xy - 2y^2 + \tfrac{1}{16}$ are shown in Figures 14.17 through 14.22.
Classify the critical points and match the function to their graphs.

2. Level curves of the six functions of Exercise 1 are shown in Figures 14.23 through 14.28. Match the functions to their level curves.

Find and classify the critical points of the functions in Exercises 3 through 11.

3.† $x \sin y$

4. e^{xy}

5.† $x^2 - e^{-y^2}$

6.‡ $x^2y - x^2 - 2y^2$

7. $xy\, e^{-x-y}$

8. $x^2 - 2x + xy$

9.† $4x^3 + y^2 - 12x^2 - 36x$

10.† $4x^2y + 3xy^2 - 12xy$

11.† $(x + 1)(y + 1)(x + y + 1)$ (Use the product rule to facilitate factoring the derivatives.)

12. Show that $\sqrt[3]{xyz} \le \tfrac{1}{3}(x + y + z)$ for $x \ge 0, y \ge 0, z \ge 0$ by finding the minimum of $x + y + z$ for $xyz = k$ with $x \ge 0, y \ge 0, z \ge 0$. (Set $z = k/xy$ and find the minimum for $x > 0, y > 0$.)

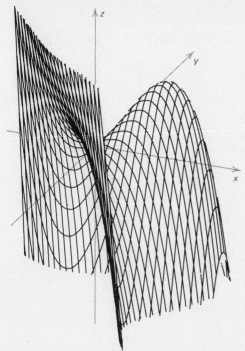

FIGURE 14.17

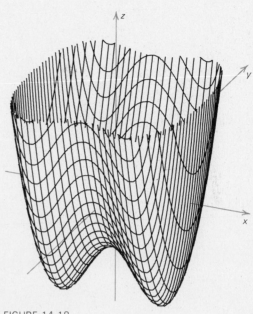

FIGURE 14.18

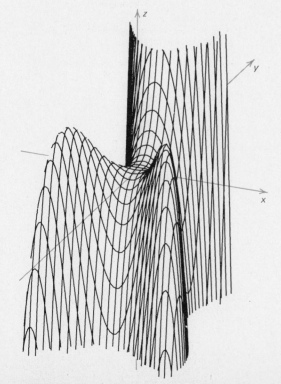

FIGURE 14.19

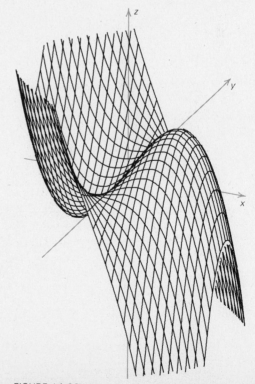

FIGURE 14.20

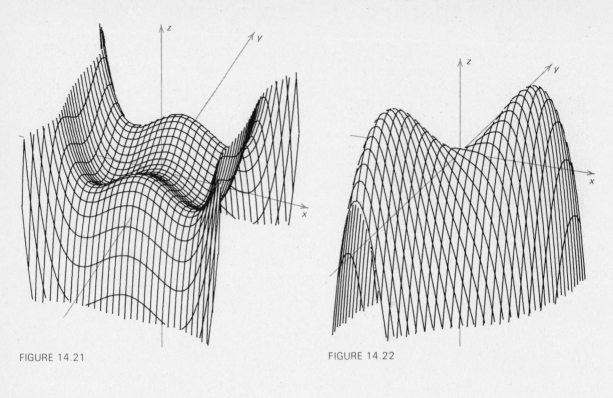

FIGURE 14.21

FIGURE 14.22

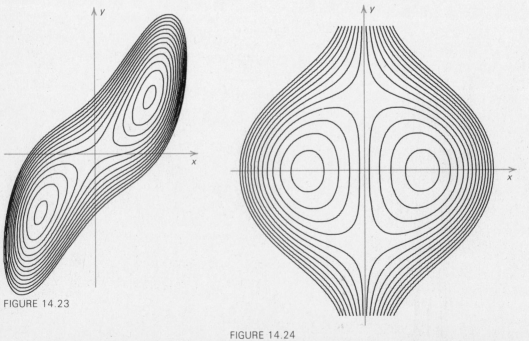

FIGURE 14.23

FIGURE 14.24

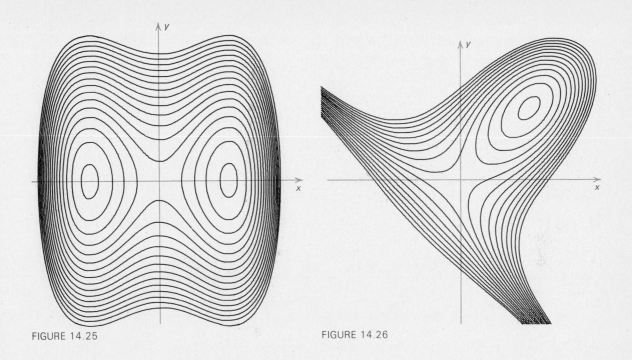

FIGURE 14.25

FIGURE 14.26

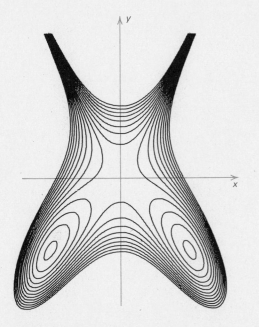

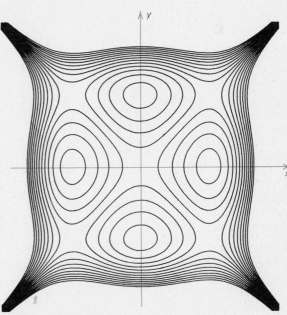

FIGURE 14.27

FIGURE 14.28

13. Show that the second-derivative test fails at the critical point of $(x - y)^4 + (x + y + 2)^2$ and that the function has a local minimum there.

14. Show that the second-derivative test fails at the critical point of $1 - x^4 - y^4$ and that the function has a local maximum there.

15. Show that the second-derivative test fails at the critical point of $x^3 + (x - y)^2$ and that the function has neither a local maximum or minimum there.

16. **a.** Use the second-derivative test to classify the critical points of $-\frac{1}{2}x^2y - x^2 - y^2 - 2y + \frac{1}{16}$. **b.** The level curves of the function of part (a) that are shown in Figure 14.29 are at levels $z = k/4$ with integers k. Use the information from part (a) to label the level curves.

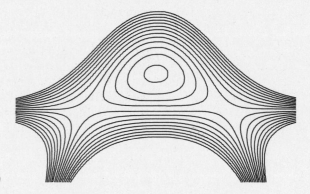

FIGURE 14.29

Find and classify the critical points of the functions in Exercises 17 through 26.

17.[†] $\cos x \sinh(y^2)$ **18.**[†] $y^2 + xe^{-x^2}$

19. $\cos x \cosh y$ **20.** xe^{x+y^2}

21.[†] $(x^3 - 3x)\cosh y$ **22.**[†] $2x^2 - x^4 - \sinh(y^2)$

23. $e^{-x^2-y^2}$ **24.**[†] $\sinh(xy)$

25.[†] $e^{-x}\ln(y^2 + 1)$ **26.**[†] $(x - 2)^3 + (5 - y)^3$

14-3 LAGRANGE MULTIPLIERS

In this section we discuss the technique of LAGRANGE MULTIPLIERS* for finding the maximum or minimum value of a function of two variables on a curve or of a function of three variables on a surface. The idea behind this technique can be described by analyzing the solution of the following graphical example.

---

*Named after the French mathematician J. L. Lagrange (1736–1813).

EXAMPLE 1   (The milkmaid problem) A milkmaid sitting on her porch at point F_1 in Figure 14.30 has to get a bucket of water in the nearby stream and take it to her cow at point F_2 in the barn. How could she determine graphically where on the stream to fill her bucket so as to minimize the total distance she has to walk?

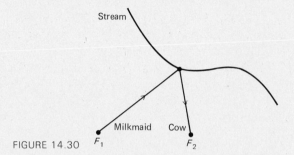

FIGURE 14.30

Solution.   For each number k greater than the distance $\overline{F_1F_2}$, the locus of points P such that she can walk from her porch at F_1 to P and then to the cow at F_2 in a total distance k is the ellipse $\overline{PF_1} + \overline{PF_2} = k$ with foci F_1 and F_2. She should draw ellipses with these foci on a map of the farm, as in Figure 14.31, and determine the smallest such ellipse that intersects the stream. She should fill her bucket at that point of intersection.   □

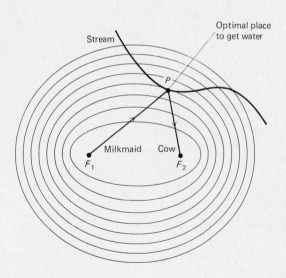

FIGURE 14.31

Analysis of Example 1

We introduce coordinate axes and define $f(x, y)$ to be the sum of the distances from the point (x, y) to the points F_1 and F_2. Then the ellipses of Figure 14.31 are the level curves of $f(x, y)$. We let C denote the curve which represents the stream. (We refer to C as the CONSTRAINT CURVE.) The milkmaid is to find the minimum of $f(x, y)$ for (x, y) on C. We suppose that C is the level curve $g(x, y) = c$ of another function $g(x, y)$ with nonzero gradient.

The idea behind the method of Lagrange multipliers in this case is that the smallest ellipse that intersects the stream C is *tangent* to it. A slightly larger ellipse would cross C without being tangent to it and a smaller ellipse would not touch C at all. Because $\overrightarrow{\text{grad}}\, f(x, y)$ is perpendicular to its level curves (the ellipses) and $\overrightarrow{\text{grad}}\, g(x, y)$ is perpendicular to its level curve C (the stream), we can express the condition that the ellipse be tangent to C as the condition that $\overrightarrow{\text{grad}}\, f(x, y)$ be *parallel* to $\overrightarrow{\text{grad}}\, g(x, y)$ (Figure 14.32). We express this condition algebraically by requiring that $\overrightarrow{\text{grad}}\, f(x, y) = \lambda\,\text{grad}\, g(x, y)$ for some number λ (lambda), which is called the LAGRANGE MULTIPLIER. Analogous geometric considerations could be made for the general problem of determining the maximum or minimum of any function $f(x, y)$ on a level curve of another function $g(x, y)$ and lead to the following result.

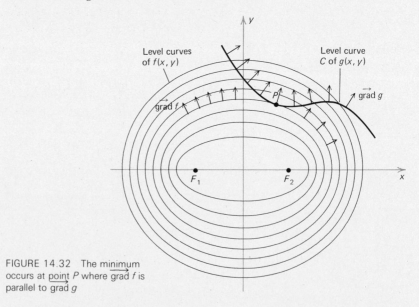

FIGURE 14.32 The minimum occurs at point P where $\overrightarrow{\text{grad}}\, f$ is parallel to $\overrightarrow{\text{grad}}\, g$

Theorem 14.4 **Suppose that C is a level curve of a function $g(x, y)$ with nonzero gradient and that $f(x, y)$ and $g(x, y)$ have continuous first order derivatives in an open set containing C. If $f(x, y)$ has a maximum or minimum value for (x, y) on C, then that extreme value occurs at a point where**

(1) $\quad \overrightarrow{\text{grad}}\, f(x, y) = \lambda\, \overrightarrow{\text{grad}}\, g(x, y)$ ◄

for some number λ.

Proof. Suppose that a maximum or minimum value of $f(x, y)$ for (x, y) on C occurs at (x_0, y_0). By an Implicit Function Theorem discussed in advanced calculus courses a portion of C including (x_0, y_0) may be described by parametric equations $x = x(t), y = y(t)$ with $x(0) = x_0, y(0) = y_0$ and with the tangent vector $\langle x'(0), y'(0) \rangle$ not zero. Set $F(t) = f(x(t), y(t))$. Then $F(t)$ has a maximum or minimum at $t = 0$, so $F'(0) = 0$. By the chain rule this condition yields the equation

$$0 = F'(0) = \left[\frac{d}{dt} f(x(t), y(t)) \right]_{t=0}$$
$$= f_x(x(0), y(0)) x'(0) + f_y(x(0), y(0)) y'(0)$$
$$= \overrightarrow{\text{grad}} f(x_0, y_0) \cdot \langle x'(0), y'(0) \rangle.$$

Thus, $\overrightarrow{\text{grad}} f(x_0, y_0)$ is perpendicular to the tangent vector and hence to the curve C at (x_0, y_0). Since $\overrightarrow{\text{grad}} g(x_0, y_0)$ is also perpendicular to the level curve C of $g(x, y)$, the two gradients are parallel and we obtain condition (1) at (x_0, y_0).   Q.E.D.

EXAMPLE 2   Use Lagrange multipliers to find the rectangle of perimeter 12 and maximum area.

Solution.   We let x denote the length and y the width of the rectangle (Figure 14.33). Then its area is xy and its perimeter is $2x + 2y$, so we want the maximum of xy subject to the constraint

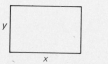

y

x

FIGURE 14.33   Area $= xy$
Perimeter $= 2x + 2y$

(2) $2x + 2y = 12$.

By Theorem 14.4 with $f = xy, g = 2x + 2y, \overrightarrow{\text{grad}} f = \langle y, x \rangle$, and $\overrightarrow{\text{grad}} g = \langle 2, 2 \rangle$, the maximum occurs at (x, y) where

(3) $\langle y, x \rangle = \lambda \langle 2, 2 \rangle$

for some real number λ. We can express the vector equation (3) as the two numerical equations

(4) $y = 2\lambda$, $x = 2\lambda$.

We then need to solve equations (2) and (4) for x and y.

As in any Lagrange multiplier problem with two variables, there are two basic approaches to carrying out the necessary algebra. We can use equations (4) to express x and y in terms of λ and substitute in equation (2) to find λ. This gives $2(2\lambda) + 2(2\lambda) = 12$ or $8\lambda = 12$ and then $\lambda = \frac{3}{2}$. Then equations (4) give $x = 2(\frac{3}{2}) = 3, y = 2(\frac{3}{2}) = 3$ as the solution. Alternately, we can eliminate λ from equations (4) to express y in terms of x and substitute into (2) to find x: eliminating λ shows that $x = y$, and then equation (2) gives $2x + 2x = 12$. We obtain again the solution $x = y = 3$. Either procedure shows that the rectangle of maximum area is a square.   □

The constraint curve $2x + 2y = 12$ in Example 2 is the line shown in Figure 14.34, and the level curves of xy are the hyperbolas shown there. The maximum occurs at a point $(3, 3)$ where the hyperbola is tangent to the line.

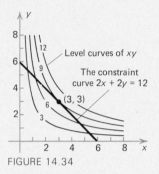

FIGURE 14.34

Reversing the roles of *f(x, y)* and *g(x, y)*

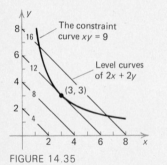

FIGURE 14.35

In many cases the maximum value of $f(x, y)$ for (x, y) on a level curve $g(x, y) = c$ of g occurs at the same point as the *minimum* value of $g(x, y)$ for (x, y) on a level curve $f(x, y) = k$ of f. This is because, in either case, the maximum or minimum occurs at a point where $\overrightarrow{\mathrm{grad}}\, f(x, y)$ and $\overrightarrow{\mathrm{grad}}\, g(x, y)$ are parallel.

Suppose, for example, we want to find the rectangle of minimum perimeter and area 9; that is, suppose we want to find the minimum value of $2x + 2y$ subject to the constraint $xy = 9$. We find the minimum occurs for $x = y = 3$ just as in Example 1 where we found the rectangle of maximum area and perimeter 12. In the case of minimizing the perimeter, the constraint curve is the hyperbola $xy = 9$ and the level curves are the lines $2x + 2y = c$ (Figure 14.35), and, as in the case of maximizing the area, the extreme value occurs where a hyperbola $xy = k$ and a line $2x + 2y = c$ are tangent.

Lagrange multipliers for three variables

In the case of three variables we find that the maximum or minimum values of one function $f(x, y, z)$ on a level surface $g(x, y, z) = c$ of another function occur at points where the level surface of f is tangent to the level surface of g. Because $\overrightarrow{\mathrm{grad}}\, f$ and $\overrightarrow{\mathrm{grad}}\, g$ are perpendicular to their respective level surfaces, the extreme value occurs where $\overrightarrow{\mathrm{grad}}\, f$ and $\overrightarrow{\mathrm{grad}}\, g$ are parallel and hence $\overrightarrow{\mathrm{grad}}\, f = \lambda\, \overrightarrow{\mathrm{grad}}\, g$.

Theorem 14.5   Suppose that S is a level surface of a function $g(x, y, z)$ with nonzero gradient and that $f(x, y, z)$ and $g(x, y, z)$ have continuous derivatives in an open set containing S. If $f(x, y, z)$ has a maximum or minimum value for (x, y, z) on S, then that extreme value occurs at a point where

(5) $\overrightarrow{\mathrm{grad}}\, f(x, y, z) = \lambda\, \overrightarrow{\mathrm{grad}}\, g(x, y, z)$ ◄

for some number λ.

Proof.   Suppose that the maximum or minimum of $f(x, y, z)$ occurs at (x_0, y_0, z_0) on S and let T be an arbitrary nonzero vector tangent to S at (x_0, y_0, z_0). By an Implicit Function Theorem studied in advanced calculus, there is a curve $x = x(t), y = y(t), z = z(t)$ on S with $(x(0), y(0), z(0)) = (x_0, y_0, z_0)$ and $\langle x'(0), y'(0), z'(0) \rangle = T$. Because $f(x, y, z)$ has an extreme value at (x_0, y_0, z_0), the function $F(t) = f(x(t), y(t), z(t))$ has an extreme value at $t = 0$. Therefore,

$$0 = \frac{dF}{dt}(0) = f_x(x(0), y(0), z(0))x'(0) + f_y(x(0), y(0), z(0))y'(0)$$
$$+ f_z(x(0), y(0), z(0))z'(0)$$
$$= \overrightarrow{\text{grad}}\, f(x_0, y_0, z_0) \cdot T.$$

Hence $\overrightarrow{\text{grad}}\, f(x_0, y_0, z_0)$ is perpendicular to T and, since T is an arbitrary vector tangent to S at (x_0, y_0, z_0), $\overrightarrow{\text{grad}}\, f(x_0, y_0, z_0)$ is perpendicular to S at that point. Because S is a level surface of g, $\overrightarrow{\text{grad}}\, g(x_0, y_0, z_0)$ is also perpendicular to S at (x_0, y_0, z_0) and is parallel to $\overrightarrow{\text{grad}}\, f(x_0, y_0, z_0)$. This gives equation (5).   Q.E.D.

EXAMPLE 3   Find the maximum and minimum values of $6x + 3y + 2z - 5$ on the surface $4x^2 + 2y^2 + z^2 = 70$.

Solution.   We have $\dfrac{\partial}{\partial x}(6x + 3y + 2z - 5) = 6$,

$\dfrac{\partial}{\partial y}(6x + 3y + 2z - 5) = 3$, and $\dfrac{\partial}{\partial z}(6x + 3y + 2z - 5) = 2$, so

$\overrightarrow{\text{grad}}(6x + 3y + 2z - 5) = \langle 6, 3, 2 \rangle$. Also $\dfrac{\partial}{\partial x}(4x^2 + 2y^2 + z^2) = 8x$,

$\dfrac{\partial}{\partial y}(4x^2 + 2y^2 + z^2) = 4y$, and $\dfrac{\partial}{\partial z}(4x^2 + 2y^2 + z^2) = 2z$, so
$\overrightarrow{\text{grad}}(4x^2 + 2y^2 + z^2) = \langle 8x, 4y, 2z \rangle$. By Theorem 14.5 with $f = 6x + 3y - 2z - 5$ and $g = 4x^2 + 2y^2 + z^2$, the maximum and minimum occur at points where

(6) $\langle 6, 3, 2 \rangle = \lambda \langle 8x, 4y, 2z \rangle.$

This vector equation yields the three numerical equations

(7) $6 = 8\lambda x, \qquad 3 = 4\lambda y, \qquad 2 = 2\lambda z.$

We are to solve these for (x, y, z) with the constraint

(8) $4x^2 + 2y^2 + z^2 = 70.$

We will use equations (7) to express x, y, and z in terms of λ, use (8) to determine the possible values of λ, and then use (7) again to find the points (x, y, z).

Equations (7) yield

(9) $x = \dfrac{3}{4\lambda}, \qquad y = \dfrac{3}{4\lambda}, \qquad z = \dfrac{1}{\lambda}$

and then equation (8) gives

$$4\left(\frac{3}{4\lambda}\right)^2 + 2\left(\frac{3}{4\lambda}\right)^2 + \left(\frac{1}{\lambda}\right)^2 = 70$$

which simplifies to $\frac{35}{8}\lambda^{-2} = 70$ and then $\lambda^2 = \frac{1}{16}$. The choices of λ are $\lambda = \frac{1}{4}$ and $\lambda = -\frac{1}{4}$. Taking $\lambda = \frac{1}{4}$ in equations (9) yields $x = 3$, $y = 3$, $z = 4$ and taking $\lambda = -\frac{1}{4}$ yields $x = -3$, $y = -3$, $z = -4$. The extreme values of $6x + 3y + 2z - 5$ on the surface $4x^2 + 2y^2 + z^2 = 70$ occur at $(3, 3, 4)$ and $(-3, -3, -4)$. To find the maximum and minimum, we compare the values at these points. At $(3, 3, 4)$ the function $6x + 3y + 2z - 5$ has the value $6(3) + 3(3) + 2(4) - 5 = 30$, and at $(-3, -3, -4)$ it has the value $6(-3) + 3(-3) + 2(-4) -5 = -40$. The maximum is 30 and the minimum is -40. □

Graphical problems   EXAMPLE 4   Figure 14.36 shows level curves of the yield of corn per acre (measured in thousands of pounds) that a farmer obtains if he applies x acre-feet of irrigation water and y pounds of fertilizer per acre during the growing season. Suppose that the water costs $20 per acre-foot, the fertilizer costs $3 per pound, and the farmer has $60 to invest per acre for water and fertilizer. How much water and how much fertilizer should he buy to maximize the yield?

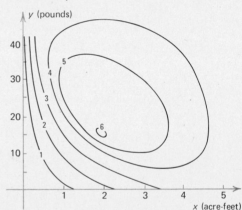

FIGURE 14.36   Yield (thousands of pounds) of corn per acre

Solution.   It would cost him $C(x, y) = 20x + 3y$ dollars for x acre-feet of water and y pounds of fertilizer, so he must choose x and y to satisfy the condition $20x + 3y \le 60$, represented by the shaded triangle in Figure 14.37. As the figure shows, the maximum occurs on the boundary line $20x + 3y = 60$. To estimate where the maximum occurs, we draw a plausible level curve tangent to the line $20x + 3y = 60$ and use the coordinates of the point of tangency. The farmer should buy about $1\frac{3}{4}$ acre-feet of water and $8\frac{1}{3}$ pounds of fertilizer per acre of corn. □

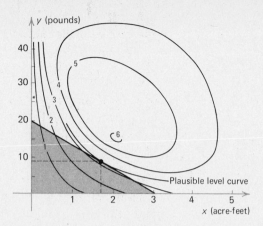

FIGURE 14.37

Exercises

†*Answer provided.*
‡*Outline of solution provided.*

In Exercises 1 through 10 use Lagrange multipliers to find the maximum and minimum values of the functions and the points where they occur.

1.‡ $2x + y$ for $x^2 + 2y^2 = 18$

2.† $3x - y + 1$ for $3x^2 + y^2 = 16$

3. $4x + y - 2$ for $x^2 + 2y^2 = 66$

4.‡ $5 - 6x - 3y - 2z$ for $4x^2 + 2y^2 + z^2 = 70$

5.† $x + 5z$ for $2x^2 + 3y^2 + 5z^2 = 198$

6.‡ xy for $2x^2 + 3y^2 = 12$

7.† xyz for $6x^2 + 5y^2 + 4z^2 = 90$

8.‡ $x^2 + 2x + y^2$ for $3x^2 + 2y^2 = 48$

9. x^2y for $x^2 + 8y^2 = 24$

10.† $x^2 + y^2$ for $x^2 + 2x + y^2 - 4y = 0$

11. What is the minimum value of $2x^2 + 3y^2$ for $3x + 4y = 59$ and where does it occur?

In Exercises 12 through 14 use the first-derivative test and Lagrange multipliers to find the maximum and minimum values of the functions in the given regions.

12.‡ x^2y^2 for $x^2 + 4y^2 \le 24$

13.† $x^2 + 2x + y^2$ for $x^2 + y^2 \le 4$

14. xy for $4x^2 + y^2 \le 32$

15. Find the maximum volume of a rectangular box with sides parallel to the coordinate planes, with one corner at $(0, 0, 0)$ and with the diagonally opposite corner on the plane $\dfrac{x}{a} + \dfrac{y}{b} + \dfrac{z}{c} = 1$ $(a > 0, b > 0, c > 0)$.

16. Find the maximum of $(\sin x)(\sin y)(\sin z)$ for $x, y,$ and z the interior angles of a triangle.

17. Show that the rectangular box of maximum volume that can be inscribed in a sphere of radius R is a cube.

18. Show by sketching the level curves of $x - y$ and the constraint curve $y - x - x^5 = 2$ that the method of Lagrange multipliers yields a point where $x - y$ has neither a maximum nor a minimum.

19. Show by sketching the level curves of e^y and the constraint curve $y = 3x - x^3$ that the method of Lagrange multipliers yields points where e^y has a local maximum and a local minimum but not an absolute maximum or absolute minimum.

20.[†] How would the farmer of Example 4 maximize the yield from his corn fields if he had $100 to invest per acre for water and fertilizer?

21.[†] Explain why in the case of Example 4, where the farmer maximizes his yield by spending all his available capital, he would probably also maximize his profit, whereas in the case of Exercise 20, where the farmer maximizes his yield without spending all of his capital, he would probably not maximize his profit. (Assume the price he pays for corn is constant.)

22.[†] Figure 14.38 shows level curves of wind chill in North America one January day. What was the approximate maximum and minimum wind chill within 1000 miles of Chicago? Draw plausible level curves through the points where the maximum and minimum occur.

23. Figure 14.39 shows level curves of the intensity of the earthquake at Agadir, Morocco, February 29, 1960. (The levels of intensity were determined by the amount of damage caused, with the worst damage inside the curve labeled 10.) At approximately what point within a circle of radius 5 miles with center at point P was the intensity the greatest? The least? Draw the circle and plausible level curves at those points.

FIGURE 14.38   Wind chill in January (kilogram-calories per square meter per hour)(Adapted from Critchfield, [1] p. 352.)

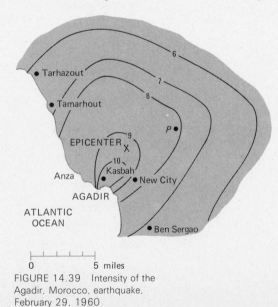

FIGURE 14.39   Intensity of the Agadir, Morocco, earthquake, February 29, 1960

24. Figure 14.40 shows level curves of the declination of a compass (the number of degrees its reading differs from the direction of true north) with positive values where a compass points to the west of true north and negative values where a compass points east of true north. What are the approximate maximum and minimum declinations in South America? Sketch the plausible level curves of the declinations at the corresponding points.

25.† What is the significance of the case of $\lambda = 0$ in Theorem 14.4 or 14.5?

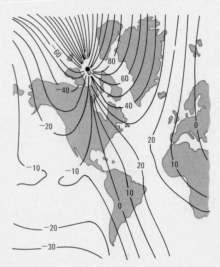

FIGURE 14.40   Declination of a compass (The magnetic north pole is at N.) (Adapted from A. Strahler, [1] *Physical Geography*, p. 50.)

MISCELLANEOUS EXERCISES

†*Answer provided.*
‡*Outline of solution provided.*

Use the second-derivative test to classify the critical points of the functions in Exercises 1 through 4.

1.† $x^2 + \sin y$

2. $x^3y - 12xy$

3. $(\cos x)(\cos y)$ for $-\pi/4 < x, y < 5\pi/4$

4. $(x^2 + y^2 + 1)^{-1}$

5.† Find the maximum and minimum values of $x^2 - xy + y^2$ for $x^2 + y^2 \le 8$.

6. Find the maximum and minimum values of x^2y^4 for $16x^2 + 2y^2 = 12$.

7. Find the maximum and minimum values of $xy + \sqrt{9 - x^2 - y^2}$ for $x^2 + y^2 \le 9$.

8.† Use a sketch of the level curves of $x^2 + 2x + y^2$ to determine its maximum value on the square $|x| + |y| = 4$ and the method of Lagrange multipliers to find its minimum value on that square. Why can Lagrange multipliers not be used to find the maximum?

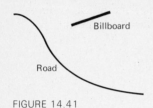

FIGURE 14.41

FIGURE 14.42

9. Show that in the method of least squares for finding a line to approximate data (x_j, y_j) $(j = 1, 2, 3, \ldots, N)$ the line passes through the center of gravity of the data points. (See Exercise 25 of Section 14.1.)

10. The map in Figure 14.41 shows a road and a billboard. Let $f(x, y)$ be the angle subtended by the billboard at the point (x, y).   **a.** Show that the level curves of f are the arcs of circles with their end points at the ends of the billboard.   **b.** Use the level curves to find the approximate point on the road where the billboard subtends the largest angle.

11. A boy at a fair will eat x ounces of cotton candy and ride y minutes on the ferris wheel. The *indifference curves* in Figure 14.42 show his preference for various combinations of ounces of cotton candy and minutes of ferris wheel riding: the points $(2, 15)$ and $(10, 5)$ are on the same curve because he would just as soon eat 2 ounces of cotton candy and ride the ferris wheel 15 minutes as eat 10 ounces of cotton candy and ride 5 minutes. The curves are level curves of the corresponding *utility function* $U(x, y)$; he would prefer combinations (x, y) for which $U(x, y)$ is larger. Suppose that cotton candy costs 20¢ per ounce, the ferris wheel costs 10¢ per minute, and he has $2.00 to spend. What combination (x, y) of ounces of cotton candy and minutes of ferris wheel riding would he prefer of those he can afford?

12.[†] A consumer's utility function for x units of item A and y units of item B is $6x^{1/3}y^{1/2}$. Each unit of item A costs $80 and each unit of item B costs $30. What choice of (x, y) would maximize his utility function if he spends $200?

13. A company buys x items of new equipment, utilizes y man-hours of labor, and distributes z advertising brochures in the manufacture and promotion of their product at a cost of $C(x, y, z)$ dollars. The resulting revenue is $R(x, y, z)$ dollars. The partial derivatives C_x and R_x are called the *marginal cost* and *marginal revenue* with respect to x, the partial derivatives C_y and R_y are the marginal cost and revenue with respect to y, and C_z and R_z are the marginal cost and revenue with respect to z. Show that if Lagrange multipliers are used to find the maximum of $R(x, y, z)$ subject to the constraint $C(x, y, z) = $ constant, then the Lagrange multiplier is the ratio of the marginal revenue and the marginal cost with respect to each variable at the maximum.

14.[†] (Lagrange multipliers with two constraints) Show that at a maximum or minimum of $f(x, y, z)$ subject to the two constraints $g(x, y, z) = $ constant and $h(x, y, z) = $ constant, the numbers $x, y,$ and z satisfy

$$\overrightarrow{\text{grad}}\, f(x, y, z) = \lambda\, \overrightarrow{\text{grad}}\, g(x, y, z) + \mu\, \overrightarrow{\text{grad}}\, h(x, y, z)$$

with real numbers λ and μ(mu) provided $\overrightarrow{\text{grad}}\, g(x, y, z)$ and $\overrightarrow{\text{grad}}\, h(x, y, z)$ are not parallel.

15.[‡] The intersection of the paraboloid $x^2 + y^2 - z = 0$ and the plane $x + 2y + 2z = 5$ is an ellipse. Use the equation of Exercise 14 to find the points on the ellipse that are closest to and farthest from the origin.

16.[†] Use the equation of Exercise 14 to find the maximum and minimum of $3x + 2y + z$ on the ellipse that is the intersection of the circular paraboloids $x^2 - 2x + y^2 - z = -1$ and $x^2 + 2x + y^2 + z = 41$.

17.[†] Show that the solutions λ of the determinant equation

$$\begin{vmatrix} A - \lambda & B \\ B & C - \lambda \end{vmatrix} = 0$$

are the maximum and minimum of $Ax^2 + 2Bxy + Cy^2$ for $x^2 + y^2 = 1$.

18.[†] (The fly problem) A fly in a circus tent wants to go from a camel to the ceiling of the tent and then to an elephant in the least possible total distance. To do so it will go to a point where the ceiling is tangent to a level surface. What is the level surface?

CHAPTER 15

MULTIPLE INTEGRALS

SUMMARY: In this chapter we study double and triple integrals. The definition of double integrals and the procedures for evaluating them as iterated single integrals are given in Section 15.1. Section 15.2 deals with applications and Section 15.3 with double integrals in polar coordinates. We study triple integrals in Section 15.4 and cylindrical and spherical coordinates in Section 15.5. We omit many of the details of the theory of multiple integrals which are more appropriately a part of advanced calculus.

15-1 DOUBLE INTEGRALS

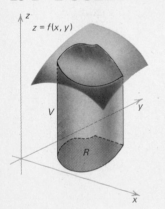

FIGURE 15.1   Solid of volume
$\iint_R f(x, y) \, dA \; (f(x, y) > 0)$

The DOUBLE INTEGRAL

$$(1) \qquad \int \int_R f(x, y) \, dA$$

of a function $f(x, y)$ over a region R in the xy-plane may be described in terms of volumes. If $f(x, y)$ is positive for (x, y) in R, then the integral is the volume of the solid lying above the region R in the xy-plane and below the graph of f (Figure 15.1). If $f(x, y)$ is negative for (x, y) in R, the integral is the negative of the volume of the solid lying under the region R and above the graph. In general, the integral is equal to the volume of the solid below the graph of f and above the portion of R where $f(x, y)$ is positive, minus the volume of the solid above the graph of f and below the portion of R where $f(x, y)$ is negative.

Piecewise smooth curves

We will say that a curve in an xy-plane is PIECEWISE SMOOTH if it consists of a finite number of horizontal or vertical line segments or graphs $y = f(x)$ of continuous functions defined in finite closed intervals such that the function f is either increasing or decreasing and either it or its inverse has a continuous derivative. The curve in Figure 15.1a is piecewise smooth.

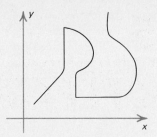

FIGURE 15.1a   A piecewise
smooth curve

Although double integrals exist for more general regions, we will
consider them only over bounded regions with piecewise smooth
boundaries. (Recall that R is bounded if it can be enclosed in a circle.)

A double integral, like a definite integral of a function of one variable,
is defined formally as a limit of approximating sums. To form an
approximating sum, we use a number of piecewise smooth curves to
divide the region R into subregions

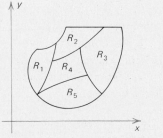

FIGURE 15.2   A partition of R into
five subregions

(2) $R_1, R_2, R_3, \ldots, R_N$

(Figure 15.2). We call the collection of subregions a *partition* of R. The
approximating sums corresponding to this partition are of the form

(3) $$\sum_{j=1}^{N} f(x_j, y_j) [\text{Area of } R_j]$$

where (x_j, y_j) is a point in R_j for $j = 1, 2, 3, \ldots, N$ (Figure 15.3). The
integral is the limit of such sums:

Definition 15.1   The integral

(1) $$\int \int_R f(x, y) \, dA$$

**of a function $f(x, y)$ over a region R in the xy-plane is the limit
of the approximating sums (3) as the number of subregions
in the corresponding partitions (2) tend to infinity and their
diameters tend to zero.**

Here by the "diameter" of a subregion we mean the diameter of the
smallest circle containing it.

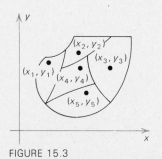

FIGURE 15.3

The interpretation of Definition 15.1 in terms of volumes is indicated in Figures 15.4 and 15.5 for the region R, the positive function $f(x, y)$ of Figure 15.1 and the partition of Figure 15.3. The base of the solid in Figure 15.4 is the subregion R_4 of the partition shown in Figure 15.2. Its height is the value $f(x_4, y_4)$ of the function at the point from R_4 that is shown in Figure 15.3. Its sides are vertical, its top is horizontal and congruent to R_4, and its volume is $f(x_4, y_4)$ [Area of R_4]. The solids corresponding to the five subregions of the partition in Figure 15.2 combine to form the solid of volume

$$\sum_{j=1}^{5} f(x_j, y_j)[\text{Area of } R_j]$$

which is shown in Figure 15.5. This solid approximates the solid in Figure 15.1, and we would obtain better approximations by taking large numbers of subregions with small diameters. The volume of each such approximating solid is given by an approximating sum (3), and, consequently, the integral (1), which is the limit of the approximating sums, is the volume of the original region in Figure 15.1.

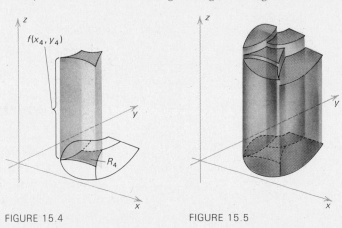

FIGURE 15.4   FIGURE 15.5

In the next example we evaluate a double integral by computing a volume.

EXAMPLE 1   Evaluate the double integral

$$(4) \qquad \int\int_R \sqrt{9 - x^2 - y^2} \, dA$$

where R is the circle $x^2 + y^2 \leq 9$ in the xy-plane.

Solution.   The graph $z = \sqrt{9 - x^2 - y^2}$ is the upper surface and the region R is the base of the hemisphere of radius 3 that is shown in Figure 15.6, so the integral is its volume. Since the volume of a hemisphere of radius r is $\frac{2}{3}\pi r^3$, the integral equals $\frac{2}{3}\pi \, 3^3 = 18\pi$.   □

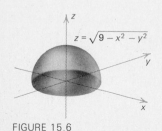

FIGURE 15.6

The existence of double integrals

We saw in the Appendix to Chapter 5 that the integral $\int_a^b f(x)\,dx$ of a function of one variable exists as the limit of Riemann sums if f is *piecewise continuous* in the closed interval $a \leq x \leq b$. Similarly, the double integral (1) exists as the limit of approximating sums if the region R is bounded and has a piecewise smooth boundary and if the integrand $f(x, y)$ is bounded and piecewise continuous in R. In this case f is bounded if there is a constant M such that $|f(x, y)| \leq M$ for all (x, y) in R, and f is piecewise continuous if there is a partition R_1, $R_2, \ldots, R_N$ of R such that the interior of each R_j, f is equal to a function g_j that is continuous in R_j including its boundary. (Figure 15.3a shows the graph of a piecewise continuous function $f(x, y)$.)

We will omit the proof that the double integral is defined under these conditions, and we will assume in all definitions, rules, and theorems dealing with double integrals that these conditions are satisfied.

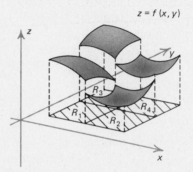

$z = f(x, y)$

FIGURE 15.6a   The graph of a piecewise continuous function

Double integrals as iterated integrals

Whereas a few double integrals, such as in Example 1, can be evaluated by using formulas for volumes from geometry, most double integrals are evaluated by using the techniques of integration of functions of one variable. We do this by expressing a double integral as an ITERATED INTEGRAL involving one integration with respect to x and another with respect to y. The next rule describes the procedure that is used when we integrate first with respect to y.

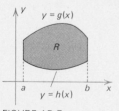

FIGURE 15.7

Rule 15.1. **Let R denote the region in the xy-plane that is bounded on the top by the graph $y = g(x)$ and on the bottom by the graph $y = h(x)$ and that extends from $x = a$ on the left to $x = b$ on the right (Figure 15.7). Double integrals over R may be expressed as iterated integrals by the formula**

(5) $$\int \int_R f(x, y)\,dA = \int_{x=a}^{x=b} \left\{ \int_{y=h(x)}^{y=g(x)} f(x, y)\,dy \right\} dx. \qquad \blacktriangleleft$$

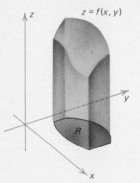

FIGURE 15.8   Solid of volume
$= \int\int_R f(x, y)\, dA$

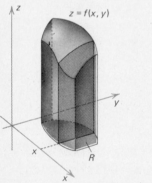

FIGURE 15.9   Cross section of area
$A(x)$

Derivation of Rule 15.1.   We will consider the case of the region R of Figure 15.7 and the positive function $f(x, y)$ of Figure 15.8, for which the integral is the volume of the solid in Figure 15.8. We let $A(x)$ denote the area of the intersection of the solid with the plane perpendicular to the x-axis at x (Figure 15.9). By the method of slicing (Section 6.2),

$$(6)\quad \int\int_R f(x, y)\, dA = \begin{bmatrix} \text{The volume} \\ \text{of the solid} \end{bmatrix} = \int_{x=a}^{x=b} A(x)\, dx.$$

We have written the limits of integration as "$x = a$" and "$x = b$" in the integral on the right to emphasize the fact that the integration is with respect to x.

If we look straight down the x-axis, we obtain the view of the cross section that is shown in Figure 15.10. As that view shows, the bottom of the cross section is at $z = 0$, the top is formed by the curve $z = f(x, y)$, and the cross section extends from $y = h(x)$ on the left to $y = g(x)$ on the right. Here $z = f(x, y)$ is the equation of a curve and $h(x)$ and $g(x)$ are numbers because we are considering a fixed value of x. The area $A(x)$ is, therefore, equal to the integral

$$A(x) = \int_{y=h(x)}^{y=g(x)} f(x, y)\, dy$$

and making this substitution in equation (6) yields formula (5).   □

In practice we do not make a three-dimensional sketch to determine how to express a double integral as an iterated integral. We make only a sketch of the region of integration as in Figure 15.11, where the vertical line crossing the region at x indicates that in the iterated integral (5) we first integrate with respect to y for fixed x from the bottom $y = h(x)$ to the top $y = g(x)$ of the region. The horizontal arrow in the figure indicates that we then integrate with respect to x from $x = a$ at the left of the region to $x = b$ at the right. Notice that the vertical line sweeps out the region as x runs from a to b.

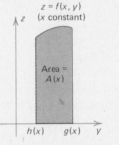

FIGURE 15.10   Area $= A(x)$
$$= \int_{y=h(x)}^{y=g(x)} f(x, y)\, dy$$

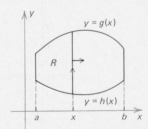

FIGURE 15.11   The integration procedure in the iterated integral (5)

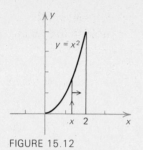

FIGURE 15.12

EXAMPLE 2   Evaluate the double integral

$$(7) \qquad \int\int_R 2xy \, dA$$

where R is the region bounded by the curve $y = x^2$ and the lines $y = 0$ and $x = 2$.

Solution.   Figure 15.12 shows the region R and the arrows that indicate the integration procedure. We first integrate with respect to y from $y = 0$ to $y = x^2$ with x fixed. Then we integrate with respect to x from $x = 0$ to $x = 2$. We obtain

$$\int\int_R 2xy \, dA = \int_{x=0}^{x=2} \left\{ \int_{y=0}^{y=x^2} 2xy \, dy \right\} dx.$$

In the inner integral x is constant, so we have

$$\int_{y=0}^{y=x^2} 2xy \, dy = [xy^2]_{y=0}^{y=x^2} = [x(x^2)^2] - [x(0)^2] = x^5.$$

Therefore,

$$\int_{x=0}^{x=2} \left\{ \int_{y=0}^{y=x^2} 2xy \, dy \right\} dx = \int_{x=0}^{x=2} x^5 \, dx = [\tfrac{1}{6} x^6]_{x=0}^{x=2} = \tfrac{32}{3}. \qquad \square$$

An argument similar to the derivation of Rule 15.1 gives the following result for evaluating double integrals by first integrating with respect to x.

Rule 15.2   Consider a region R that is bounded on the left by a curve $x = M(y)$ and on the right by a curve $x = N(y)$ and that extends from $y = c$ on the bottom to $y = d$ on the top (Figure 15.13). Double integrals over R may be expressed as iterated integrals by the formula

$$(8) \qquad \int\int_R f(x, y) \, dA = \int_{y=c}^{y=d} \left\{ \int_{x=M(y)}^{x=N(y)} f(x, y) \, dx \right\} dy. \qquad \blacktriangleleft$$

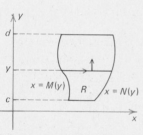

FIGURE 15.13   The integration procedure in formula (8)

The arrows in Figure 15.13 indicate the integration procedure in formula (8). We first integrate with respect to x for fixed y from the left $x = M(y)$ to the right $x = N(y)$ of the region. Then we integrate with respect to y from the bottom $y = c$ to the top $y = d$. As y runs from c to d the horizontal line at y sweeps out the region.

EXAMPLE 3   Evaluate the double integral of Example 2 by performing an x-integration first.

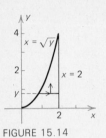

FIGURE 15.14

Solution.   We can think of the left side of the region as the curve $x = \sqrt{y}$ (Figure 15.14). By Rule 15.2 we integrate with respect to x from the left side $x = \sqrt{y}$ to the right side $x = 2$. Then we integrate with respect to y from the bottom of the region at $y = 0$ to the top at $y = 4$. The integral of $2xy$ with respect to x is $x^2 y$, so the double integral equals

$$\int\int_R f(x, y)\, dA = \int_{y=0}^{y=4} \left\{ \int_{x=\sqrt{y}}^{x=2} 2xy\, dx \right\} dy = \int_{y=0}^{y=4} [x^2 y]_{x=\sqrt{y}}^{x=2}\, dy$$

$$= \int_{y=0}^{y=4} [(2)^2 y - (\sqrt{y})^2 y]\, dy$$

$$= \int_{y=0}^{y=4} [4y - y^2]\, dy = [2y^2 - \tfrac{1}{3}y^3]_{y=0}^{y=4}$$

$$= 2(4)^2 - \tfrac{1}{3}(4)^3 = 32 - \tfrac{64}{3} = \tfrac{32}{3}.$$

This is, of course, the answer we obtained in Example 2.   □

Reversing the order of integration

In the next example we cannot perform the inner integration in the given iterated integral but we can evaluate the integral by reversing the order of integration so that we integrate first with respect to x rather than with respect to y.

EXAMPLE 4   Evaluate the integral

(9) $$\int_{x=0}^{x=8} \int_{y=x^{1/3}}^{y=2} \sin(y^4)\, dy\, dx.$$

FIGURE 15.14a

Solution.   In (9) we integrate first with respect to y from $y = x^{1/3}$ to $y = 2$ for each x between 0 and 8 and then integrate with respect to x from 0 to 8. The integral is over the region bounded by the curve $y = x^{1/3}$, the y-axis and the line $y = 2$ as shown in Figure 15.14a. The integration procedure starting with an x-integral is indicated in Figure 15.14b. Solving $y = x^{1/3}$ for x gives $x = y^3$, so we integrate with respect to x from $x = 0$ to $x = y^3$ for each y between 0 and 2 and then integrate with respect to y from 0 to 2. The integral (9) equals

(10) $$\int_{y=0}^{y=2} \int_{x=0}^{x=y^3} \sin(y^4)\, dx\, dy.$$

Because y is constant in the inner integral, we have

(11) $$\int_{x=0}^{x=y^3} \sin(y^4)\, dx = [x \sin(y^4)]_{x=0}^{x=y^3} = y^3 \sin(y^4).$$

Then to evaluate the integral of $y^3 \sin(y^4)$ with respect to y, we make the substitution $u = y^4$, $du = 4y^3\, dy$:

$$\int y^3 \sin(y^4)\, dy = \tfrac{1}{4} \int \sin(y^4)(4y^3\, dy) = \tfrac{1}{4} \int \sin u\, du$$

$$= -\tfrac{1}{4} \cos u + C = -\tfrac{1}{4} \cos(y^4) + C.$$

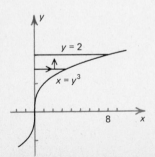

FIGURE 15.14b

With equation (11) we see that the double integral (10) equals

$$\int_{y=0}^{y=2} \int_{x=0}^{x=y^3} \sin(y^4) \, dx \, dy = \int_{y=0}^{y=2} y^3 \sin(y^4) \, dy$$

$$= \left[-\tfrac{1}{4} \cos(y^4) \right]_{y=0}^{y=2}$$

$$= -\tfrac{1}{4} \cos(16) + \tfrac{1}{4} \cos(0)$$

$$= \tfrac{1}{4}[1 - \cos(16)]. \quad \square$$

$\Delta x \, \Delta y$-notation in approximating sums

Often it is convenient to view a double integral

$$(1) \qquad \int \int_R f(x, y) \, dA$$

as a limit of approximating sums formed by partitions into rectangles. We choose a rectangle $S: a \leq x \leq b, c \leq y \leq d$ that contains the region of integration R (Figure 15.15) and define (or redefine) $f(x, y)$ to be zero in the portion of S outside of R. Then

$$\int \int_R f(x, y) \, dA = \int \int_S f(x, y) \, dA.$$

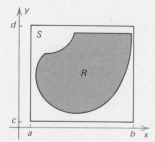

FIGURE 15.15

For positive integers N and M, we consider the partitions

$$a = x_0 < x_1 < x_2 < \cdots < x_N = b$$
$$c = y_0 < y_1 < y_2 < \cdots < y_M = d$$

of the intervals $a \leq x \leq b$ and $c \leq y \leq d$ into subintervals of lengths $\Delta x = (b - a)/N$ and $\Delta y = (d - c)/M$, respectively. The lines $x = x_j \, (j = 0, 1, 2, \dots, N)$ and $y = y_k \, (k = 0, 1, 2, \dots, M)$ partition S into NM subrectangles (Figure 15.16), for which

$$(12) \qquad \sum_{j=1}^{N} \sum_{k=1}^{M} f(x_j, y_k) \Delta x \Delta y$$

is an approximating sum. The integral of $f(x, y)$ over R is the limit of the approximating sums (12) as N and M tend to infinity.

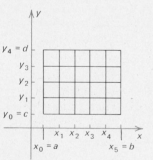

FIGURE 15.16   A partition into subrectangles ($N = 5$, $M = 4$)

Alternate notation for double integrals

Because the double integral is the limit approximating sums of the form (12), the notation $dx\,dy$ is usually used in place of dA in the symbol for the integral. Thus, we have

$$\int \int_R f(x, y)\,dx\,dy = \int \int_R f(x, y)\,dA$$

$$= \lim_{N,M \to \infty} \sum_{j=1}^{N} \sum_{k=1}^{M} f(x_j, y_k)\Delta x\,\Delta y.$$

We will use this alternate notation in the exercises for this section and in the remainder of the book.

Note. Often the notation $\int_a^b \int_{h(x)}^{g(x)} f(x, y)\,dy\,dx$ or $\int_a^b dx \int_{h(x)}^{g(x)} f(x, y)\,dy$ is used for the iterated integral that we denote by $\int_{x=a}^{x=b} \int_{y=h(x)}^{y=g(x)} f(x, y)\,dy\,dx.$ $\square$

Exercises

†*Answer provided.*
‡*Outline of solution provided.*

Evaluate double integrals (1) through (19) for the indicated regions R.

1.‡ $\int \int_R 4x^3 y\,dx\,dy$; R bounded by $y = x^2$, $y = 2x$

2.† $\int \int_R 3x^2 y^2\,dx\,dy$; R bounded by $y = x$, $y = 2x$, $x = 1$

3. $\int \int_R 2y\,dx\,dy$; R bounded by $y = x^2$, $y = 2x^2 - 1$

4.‡ $\int \int_R e^x \sin y\,dx\,dy$; R bounded by $x = -3$, $x = 4$, $y = 0$, $y = 5$

5.† $\int \int_R 3y^2 \sqrt{x}\,dx\,dy$; R bounded by $y = x^2$, $y = -x^2$, $x = 4$

6. $\int \int_R 10x^3 y^4\,dx\,dy$; R bounded by $y = x^3$, $y = 0$, $x = 1$

7. $\int \int_R y^2 \sin(x^2)\,dx\,dy$; R bounded by $y = x^{1/3}$, $y = -x^{1/3}$, $x = 8$

8.‡ $\int \int_R y^2\,dx\,dy$; R bounded by $x = y^2$, $x = 2 - y^2$

9.† $\int \int_R e^{y^2}\,dx\,dy$; R bounded by $x = -y$, $x = y$, $y = 2$

10.‡ $\int \int_R \sqrt{xy}\,dx\,dy$; R the triangle with vertices $(0, 0)$, $(1, 1)$, $(4, 1)$

11.† $\int \int_R \cos\sqrt{y}\,dx\,dy$; R bounded by $y = x^{-2}$, $y = 1$, $y = 4$. (Use $u = \sqrt{y}$ in the final integration.)

12. $\int \int_R 12x^2 y\,dx\,dy$; R bounded by $x = -1 - y^2$, $x = 1 + y^2$, $y = 0$, $y = -1$

13. $\int \int_R (x + 2y)\,dx\,dy$; R the portion of the circle $x^2 + y^2 \le 1$ in the first quadrant

14. $\int \int_R (3y^2 + 4)\,dx\,dy$; R bounded by $x = 1$, $x = 4$, $y = 1/x$, $y = -1/x$

15.† $\int \int_R x\,dy\,dx$; R bounded by $y = 2/x$, $x + y = 3$

16.† $\int \int_R y^{-2} e^{x/\sqrt{y}}\,dx\,dy$; R the rectangle $0 \le x \le 1$, $1 \le y \le 2$ (Only one order of integration will work.)

17. $\int \int_R \dfrac{x}{y}\,dx\,dy$; R bounded $x = 1$, $x = 3$, $y = x^3$, $y = 4x^3$ (Use a special property of logarithms.)

18.‡ $\int \int_R x^2 \, dx \, dy$; R the triangle with vertices $(0, 0)$, $(1, 3)$, $(2, 2)$ (Two integrals are required.)

19. $\int \int_R 2y \, dx \, dy$; R bounded by $y = 2 - x^2$, $y = |x|$

20.† Express $\int \int_R f(x, y) \, dx \, dy$ as a sum of iterated integrals for R bounded by $y = 3/x$, $y = x/3$, $y = 3x$.

Switch the order of integration in Exercises 21 through 26.

21.‡ $\int_{x=-2}^{x=2} \int_{y=x^2}^{y=4} f(x, y) \, dy \, dx$ **22.**† $\int_{x=-1}^{x=1} \int_{y=-1}^{y=x^3} f(x, y) \, dy \, dx$

23. $\int_{x=0}^{x=1} \int_{y=0}^{y=2x} f(x, y) \, dy \, dx$ **24.** $\int_{x=0}^{x=4} \int_{y=-\sqrt{x}}^{y=\sqrt{x}} f(x, y) \, dy \, dx$

25.† $\int_{y=-1}^{y=3} \int_{x=-1}^{x=y} f(x, y) \, dx \, dy$ **26.** $\int_{y=-2}^{y=1} \int_{x=y^3}^{x=1} f(x, y) \, dx \, dy$

27.† Evaluate the integral $\int_{y=0}^{y=4} \int_{x=y}^{x=4} e^{-x^2} \, dx \, dy$ by first reversing the order of integration.

28. Evaluate $\int \int_R x \sin(y^3) \, dx \, dy$ with R the triangle with vertices $(0, 0)$, $(0, 2)$, $(1, 2)$ by choosing the appropriate order of integration.

Reverse the order of integration in Exercises 29 through 32. (Two integrals may be required.)

29.‡ $\int_{x=-1}^{x=2} \int_{y=x^2}^{y=x+2} f(x, y) \, dy \, dx$ **30.**† $\int_{y=0}^{y=1} \int_{x=y}^{x=2y} f(x, y) \, dx \, dy$

31. $\int_{x=0}^{x=2} \int_{y=x}^{y=4-x} f(x, y) \, dy \, dx$ **32.** $\int_{x=-\pi/2}^{x=\pi/2} \int_{y=-1}^{y=\sin x} f(x, y) \, dy \, dx$

33.† Show that for $f(x, y) = F(x) G(y)$ and R the rectangle $a \leq x \leq b$, $c \leq y \leq d$, $\int \int_R f(x, y) \, dx \, dy = \int_a^b F(x) \, dx \int_c^d G(y) \, dy$.

In Exercises 34 through 36 use the result of Exercise 33 to evaluate the integrals.

34.† $\int \int_R (\sin x)(\sin y) \, dx \, dy$; R the rectangle $0 \leq x \leq \pi$, $0 \leq y \leq \pi$

35.† $\int \int_R \frac{y}{x} \, dx \, dy$; R the rectangle $1 \leq x \leq 10$, $0 \leq y \leq 7$

36. $\int \int_R e^{2x+3y} \, dx \, dy$; R the rectangle $0 \leq x \leq 5$, $0 \leq y \leq 6$

15-2   APPLICATIONS OF DOUBLE INTEGRALS

In this section we show how double integrals are used to compute the volumes of solids, the weight and centers of gravity of flat plates of variable density, the average values of functions of two variables, and the surface areas of portions of graphs of functions of two variables.

Volumes

Rule 15.3 **Suppose that $g(x, y)$ and $h(x, y)$ are piecewise continuous with $g(x, y) \leq h(x, y)$ for all (x, y) in a region R in the xy-plane and that V is the solid consisting of all points (x, y, z) with (x, y) in R and $g(x, y) \leq z \leq h(x, y)$ (Figure 15.17). Then the volume of V is**

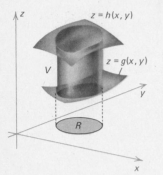

(1) $$\int \int_R [h(x, y) - g(x, y)]\, dx\, dy. \qquad \blacktriangleleft$$

FIGURE 15.17   Solid of volume $\int \int_R [h(x, y) - g(x, y)]\, dx\, dy$

To obtain Rule 15.3 we consider a partition of R into subregions R_1, $R_2, R_3, \ldots, R_N$ and pick a point (x_j, y_j) from R_j for each j. The solid consisting of the points (x, y, z) with (x, y) in R_j and $h(x_j, y_j) \leq z \leq h(x_j, y_j)$ has volume $[h(x_j, y_j) - g(x_j, y_j)][\text{Area of } R_j]$. The N solids for $j = 1, 2, 3, \ldots, N$ approximate the solid of Figure 15.17, and the sum of their volumes is an approximating sum for the integral (1). Therefore, the integral is the volume of the solid of Figure 15.17.

The region R in Rule 15.3 is called the PROJECTION of the solid V on the xy-plane. To apply this rule, we determine the projection R and the functions $g(x, y)$ and $h(x, y)$ whose graphs form the top and bottom of the solid. We generally make a rough sketch of the solid to determine the functions $g(x, y)$ and $h(x, y)$ and a more accurate sketch of the projection R in order to compute the double integral (1) as an iterated integral.

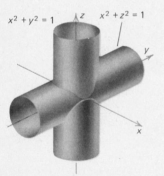

$x^2 + y^2 = 1$ $x^2 + z^2 = 1$

EXAMPLE 1   Compute the volume of the solid V bounded by the circular cylinders $x^2 + y^2 = 1$ and $x^2 + z^2 = 1$.

Solution.   The cylinders are shown in Figure 15.18. The projection of V on the xy-plane is the disk $x^2 + y^2 \leq 1$ (Figure 15.19). The top of V is formed by the half cylinder $z = \sqrt{1 - x^2}$ (Figure 15.20), and its bottom by the half cylinder $z = -\sqrt{1 - x^2}$ (Figure 15.21). By Rule 15.3 the volume of V is

FIGURE 15.18 V is bounded by these cylinders.

(2) $$\int \int_R [\sqrt{1 - x^2} - (-\sqrt{1 - x^2})]\, dx\, dy = \int \int_R 2\sqrt{1 - x^2}\, dx\, dy$$

where R is the disk $x^2 + y^2 \leq 1$.

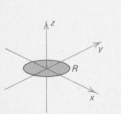

FIGURE 15.19   The projection of V is the disk $x^2 + y^2 \leq 1$ in the xy-plane.

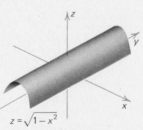

$z = \sqrt{1 - x^2}$

FIGURE 15.20

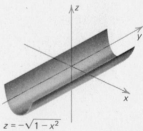

$z = -\sqrt{1 - x^2}$

FIGURE 15.21

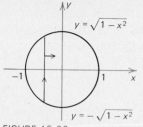

$y = \sqrt{1 - x^2}$

$y = -\sqrt{1 - x^2}$

FIGURE 15.22

The top of the disk R is the semi-circle $y = \sqrt{1 - x^2}$ in the xy-plane and the bottom is the semi-circle $y = -\sqrt{1 - x^2}$ (Figure 15.22). By the integration procedure indicated in Figure 15.22, the double integral (2) is equal to the iterated integral

$$\int_{x=-1}^{x=1} \left\{ \int_{y=-\sqrt{1-x^2}}^{y=\sqrt{1-x^2}} 2\sqrt{1 - x^2} \, dy \right\} dx$$

$$= \int_{x=-1}^{x=1} \left[2y\sqrt{1 - x^2} \right]_{y=-\sqrt{1-x^2}}^{y=\sqrt{1-x^2}} dx$$

$$= \int_{x=-1}^{x=1} \left[2\sqrt{1 - x^2}\,\sqrt{1 - x^2} - 2(-\sqrt{1 - x^2})\,\sqrt{1 - x^2} \right] dx$$

$$= \int_{x=-1}^{x=1} 4(1 - x^2) \, dx$$

$$= \left[4x - \tfrac{4}{3}x^3 \right]_{x=-1}^{x=1} = [4(1) - \tfrac{4}{3}(1)^3]$$
$$- [4(-1) - \tfrac{4}{3}(-1)^3] = \tfrac{16}{3}. \quad \square$$

We can also compute volumes of solids by integrating over regions in the xz- or yz-planes. If the solid V consists of the points (x, y, z) with (x, z) in a region R of the xz-plane and with $g(x, z) \le y \le h(x, z)$ as in Figure 15.23, then its volume is

$$(3) \qquad \int \int_R [h(x, z) - g(x, z)] \, dx \, dz. \qquad \blacktriangleleft$$

If the solid V consists of the points (x, y, z) with (y, z) in a region R of the yz-plane and with $g(y, z) \le x \le h(y, z)$ as in Figure 15.24, then its volume is

$$(4) \qquad \int \int_R [h(y, z) - g(y, z)] \, dy \, dz. \qquad \blacktriangleleft$$

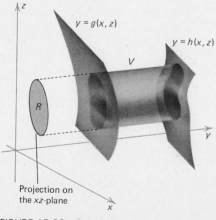

FIGURE 15.23   Solid of volume
$\int\int_R [h(x, z) - g(x, z)] \, dx \, dz$

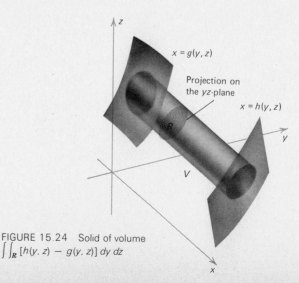

FIGURE 15.24   Solid of volume
$\int\int_R [h(y, z) - g(y, z)] \, dy \, dz$

Weight, moments, and centers of gravity

Rule 15.4 **Suppose that a flat plate occupies the bounded region R in the xy-plane and that the density of the plate is $\rho(x, y)$ at (x, y). Then the weight of the plate is**

$$(5) \qquad \int\int_R \rho(x, y)\, dx\, dy;$$

its moment about the y-axis is

$$(6) \qquad \int\int_R x\rho(x, y)\, dx\, dy;$$

its moment about the x-axis is

$$(7) \qquad \int\int_R y\rho(x, y)\, dx\, dy;$$

and its center of gravity is $(\overline{x}, \overline{y})$ where

$$(8) \qquad \overline{x} = \frac{\displaystyle\int\int_R x\rho(x, y)\, dx\, dy}{\displaystyle\int\int_R \rho(x, y)\, dx\, dy} \text{ and } \overline{y} = \frac{\displaystyle\int\int_R y\rho(x, y)\, dx\, dy}{\displaystyle\int\int_R \rho(x, y)\, dx\, dy}.$$

The concepts involved in Rule 15.4 were discussed in Section 6.10, where we considered weight, moments, and centers of gravity of plates whose densities are functions of one of the two variables. We leave its derivation as Exercise 26.

Average value of $f(x, y)$

If a function $f(x, y)$ is constant in each of the subregions $R_1, R_2, R_3,$ . . . , R_N in a partition of a region R, and if (x_j, y_j) is a point in R_j for each j, then we define the *average value* of f in R to be the sum of the value $f(x_j, y_j)$ in each R_j, multiplied by [area of R_j]/[area of R], which is the fraction of the area of R that is in R_j:

$$(9) \qquad \text{The average value} = \sum_{j=1}^{n} f(x_j, y_j)\, \frac{[\text{Area of } R_j]}{[\text{Area of } R]}$$

$$= \frac{1}{[\text{Area of } R]} \sum_{j=1}^{n} f(x_j, y_j)[\text{Area of } R_j].$$

For a general function, we define its average value to be the limit of the sums (9) as the number of subregions in the partition tends to infinity and their diameters tend to zero. In other words, we define the average value to be the integral of $f(x, y)$ over R divided by the area of R.

Definition 15.2 **The average value of $f(x, y)$ in a region R of the xy-plane is**

$$(10) \qquad \frac{1}{[\text{Area of } R]} \int\int_R f(x, y)\, dx\, dy. \qquad \blacktriangleleft$$

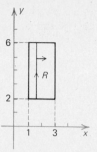

FIGURE 15.25

EXAMPLE 2   What is the average value of $e^x \sin y$ in the rectangle $R: 1 \le x \le 3, 2 \le y \le 6$?

Solution.   The rectangle is shown in Figure 15.25. Its area is 8, so the average value of $e^x \sin y$ in it is

$$\tfrac{1}{8} \int \int_R e^x \sin y \, dx \, dy = \tfrac{1}{8} \int_{x=1}^{x=3} \left\{ \int_{y=2}^{y=6} e^x \sin y \, dy \right\} dx$$

$$= \tfrac{1}{8} \left[\int_1^3 e^x \, dx \right] \left[\int_2^6 \sin y \, dy \right]$$

$$= \tfrac{1}{8} [e^3 - e][\cos(2) - \cos(6)].$$

We have used here the result of Exercise 33 of Section 15.1, which states that for R the rectangle $a \le x \le b, c \le y \le d$,

$$\int \int_R F(x) G(y) \, dx \, dy = \left[\int_a^b F(x) \, dx \right] \left[\int_c^d G(y) \, dy \right]. \quad \square$$

The mean value theorem

It is shown in courses in advanced calculus that if $f(x, y)$ is a continuous function in a region R, then there must be points (X, Y) in R such that $f(X, Y)$ is the average value of $f(x, y)$ in R and hence

(11) $\quad \int \int_R f(x, y) \, dx \, dy = f(X, Y)[\text{Area of } R].$ ◀

This result is known as the *Mean Value Theorem for double integrals.*

EXAMPLE 3   Show by evaluating the double integral that the conclusion of the Mean Value Theorem for double integrals is valid for the integral of $x - y$ over the rectangle $0 \le x \le 1, 0 \le y \le 2$.

Solution.   The Mean Value Theorem applies to this integral because $x - y$ is continuous for all (x, y).
   The double integral is

$$\int_{x=0}^{x=1} \int_{y=0}^{y=2} (x - y) \, dy \, dx = \int_{x=0}^{x=1} [xy - \tfrac{1}{2} y^2]_{y=0}^{y=2} \, dx$$

$$= \int_{x=0}^{x=1} \{ [(x)(2) - \tfrac{1}{2}(2)^2] - [(x)(0) - \tfrac{1}{2}(0)^2] \} \, dx$$

$$= \int_{x=0}^{x=1} (2x - 2) \, dx = [x^2 - 2x]_{x=0}^{x=1}$$

$$= [1^2 - 2(1)] - [0^2 - 2(0)] = -1.$$

The Mean Value Theorem states that there are points (x, y) in the rectangle such that the double integral equals the value $x - y$ of the function multiplied by the area of the rectangle. Because the integral is -1 and the area is 2, we are looking for points (x, y) such that $-1 = (x - y)(2)$ or $y = x + \tfrac{1}{2}$. The suitable points lie along the line $y = x + \tfrac{1}{2}$ that crosses the rectangle (Figure 15.25a). $\quad \square$

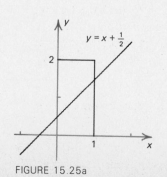

FIGURE 15.25a

Surface area

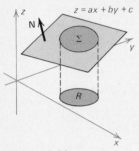

FIGURE 15.26

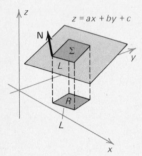

FIGURE 15.27

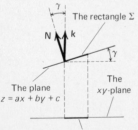

The rectangle Σ

The plane
$z = ax + by + c$

The
xy-plane

The rectangle R

FIGURE 15.28   [Width of Σ]
= sec γ[Width of R]

The next rule enables us to express the surface areas of portions of graphs $z = f(x, y)$ of functions of two variables as double integrals.

Rule 15.5   **If Σ is the portion of the surface $z = f(x, y)$ for (x, y) in a region R of the xy-plane and if f has continuous first order derivatives, then**

(12)   **Surface area of $\Sigma = \displaystyle\int\int_R \sqrt{[f_x(x, y)]^2 + [f_y(x, y)]^2 + 1}\, dx\, dy$.** ◄

To obtain Rule 15.5 we consider first the case where the graph of $f(x, y)$ is a plane $z = ax + by + c$ (Figure 15.26). In this case $\sqrt{f_x^2 + f_y^2 + 1}$ is the constant $\sqrt{a^2 + b^2 + 1}$ and (12) may be written

(13)   Area of $\Sigma = \sqrt{a^2 + b^2 + 1}$ [Area of R].

The normal vector $\mathbf{N} = \langle -a, -b, 1 \rangle$ to the plane makes an angle γ with the unit vector $\mathbf{k}$ where

(14) $\cos \gamma = \dfrac{\mathbf{N} \cdot \mathbf{k}}{|\mathbf{N}|} = \dfrac{\langle -a, -b, 1 \rangle \cdot \langle 0, 0, 1 \rangle}{\sqrt{a^2 + b^2 + 1}} = \dfrac{1}{\sqrt{a^2 + b^2 + 1}}.$

If R is a rectangle of length L and if the sides of length L are perpendicular to $\mathbf{N}$, as in Figure 15.27, then Σ is also a rectangle of length L. When we look at R and Σ in the direction parallel to the sides of length L, we obtain the edge view shown in Figure 15.28. As that sketch shows, the width of Σ is equal to the width of R multiplied by sec γ, so the area of Σ is the area of R multiplied by sec γ and equation (13) follows from (14). Equation (13) for the plane $z = ax + by + c$ and an arbitrary region R is obtained by approximating R by collections of rectangles with sides perpendicular to $\mathbf{N}$.

To derive (12) for a function $f(x, y)$ whose graph is not a plane (Figure 15.29), we take a partition $R_1, R_2, R_3, \ldots, R_N$ of R and for each j let Σ_j denote the portion of Σ over R_j (Figure 15.30). We pick a point (x_j, y_j) from R_j and consider the tangent plane

(15) $z = f(x_j, y_j) + f_x(x_j, y_j)(x - x_j) + f_y(x_j, y_j)(y - y_j)$

to $z = f(x, y)$ at $x = x_j, y = y_j$. We approximate Σ_j by the portion Π_j of this plane over R_j (Figure 15.31). The plane regions $\Pi_1, \Pi_2, \Pi_3, \ldots, \Pi_N$ form an approximation of the original surface (Figure 15.31). Equation (15) is of the form $z = ax + by + c$ with $a = f_x(x_j, y_j)$ and $b = f_y(x_j, y_j)$, so by formula (11) with Σ replaced by Π_j and R replaced by R_j, the area of the approximating surface is

(16) $\displaystyle\sum_{j=1}^{N} [\text{Area of } \Pi_j]$

$= \displaystyle\sum_{j=1}^{N} \sqrt{[f_x(x_j, y_j)]^2 + [f_y(x_j, y_j)]^2 + 1}\ [\text{Area of } R_j].$

The integral (12) is the surface area of Σ because it is the limit of the approximating sums (16).   □

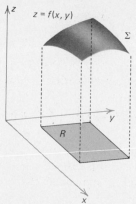

FIGURE 15.29

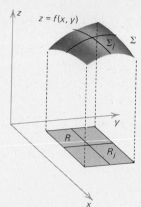

FIGURE 15.30

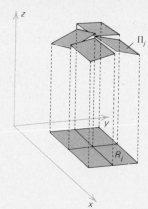

FIGURE 15.31

EXAMPLE 4   Show that the portion of the circular paraboloid $z = \frac{1}{2}x^2 + \frac{1}{2}y^2$ and the portion of the hyperbolic paraboloid $z = xy$ for (x, y) in any bounded region R of the xy-plane have the same surface area.

Solution.   We set $f(x, y) = \frac{1}{2}x^2 + \frac{1}{2}y^2$ and $g(x, y) = xy$. The surface area of the portion of the circular paraboloid is

$$\int \int_R \sqrt{[f_x(x, y)]^2 + [f_y(x, y)]^2 + 1} \, dx \, dy$$

$$= \int \int_R \sqrt{x^2 + y^2 + 1} \, dx \, dy$$

and the area of the portion of the hyperbolic paraboloid is

$$\int \int_R \sqrt{[g_x(x, y)]^2 + [g_y(x, y)]^2 + 1} \, dx \, dy$$

$$= \int \int_R \sqrt{y^2 + x^2 + 1} \, dx \, dy.$$

These integrals are equal, so the surface areas are equal. $\square$

Exercises

†*Answer provided.*
‡*Outline of solution provided.*

In Exercises 1 through 10 use double integrals to compute the volumes of the solids bounded by the given surfaces.

1.‡ $z = -1, z = x^2 + y^2, y = x^2, y = 1$

2.† $z = (\sin x)(\sin y), z = -2, x = 0, x = \pi, y = 0, y = \pi$

3. $z = xe^y, z = 0, y = 0, x = 1, y = x^2$

4.‡ $y = x^2 + z^2, y = -3, x = 0, x = 4, z = 0, z = 5$

5. $x = -z^2, x = z^2 + 2, y = 0, y = 4, z = 0, z = 1$

6.‡ $3x + 2y + z = 6, x = 0, y = 0, z = 0$

7.† $z = 2x - y - 4, x = 0, y = 0, z = 0$

8.‡ $x = 0, y = 0, z = -2$, the plane with x-intercept 2, y-intercept -2, and z-intercept 4

9. $x = 0, y = 0, 2x + 2y + z = 2, 4x + 4y - z = 4$

10. $z = 1 - x, z = 1 + 3x, y = 0, y = 5, x = 2$

In each of Exercises 11 through 13 compute the weight of a flat plate that occupies the region R in an xy-plane and has density $\rho(x, y)$ at (x, y).

11.‡ R the quarter circle $x^2 + y^2 \le 1, x \ge 0, y \ge 0$; $\rho(x, y) = xy$

12.† R the rectangle $0 \le x \le 2, 0 \le y \le 1$; $\rho(x, y) = x + y^2$

13. R bounded by $y = x^2$ and $y = x^3$; $\rho(x, y) = \sqrt{xy}$

14.‡ Find the center of gravity of the plate of Exercise 11.

15.† Find the center of gravity of the plate of Exercise 12.

16. Find the center of gravity of the plate of Exercise 13.

In Exercises 17 through 19 express the surface areas of the indicated surfaces as iterated integrals.

17.† The portion of $z = (\sin x)(\sin y)$ for $0 \le x \le 3\pi, 0 \le y \le 3\pi$

18.† The portion of $z = 1 - x^4 - y^2$ above the xy-plane

19. The portion of the monkey saddle $z = \frac{1}{3}x^3 - xy^2$ for $0 \le x \le 100$, $0 \le y \le \sqrt{x}$

20. **a.**† Use Rule 15.5 to compute the surface area of the portion of the half-cylinder $z = \sqrt{R^2 - x^2}$ for $0 \le y \le h, -k \le x \le k$ $(0 < k < R)$.
    b. Compute the area of the half cylinder $z = \sqrt{R^2 - x^2}$, for $0 \le y \le h$ by taking the limit as $k \to R^-$ in part (a).

21.† (Moments of inertia) The *moment of inertia* about a point P of a point mass of weight m at point Q is $I = m\,\overline{PQ}^2$. Derive the integral $I = \int \int_R \rho(x, y)[(x - x_0)^2 + (y - y_0)^2]\,dx\,dy$ for the moment of inertia about the point (x_0, y_0) of a flat plate of density $\rho(x, y)$ occupying the region R in an xy-plane.

In Exercises 22 through 24 use the result of Exercise 21 to compute the moment of inertia about the point (x_0, y_0) of a flat plate occupying the region R and having density $\rho(x, y)$ at (x, y).

22.‡ $(x_0, y_0) = (0, 0)$; R bounded by $y = x^2$ and $y = x$; $\rho(x, y) = 15xy^2$

23.† $(x_0, y_0) = (1, 1)$; R the triangle $0 \le x \le 1, 0 \le y \le x$; $\rho(x, y) = 12y$

24. $(x_0, y_0) = (0, 0)$; R the triangle with vertices $(0, 0), (1, 1), (-1, 1)$; $\rho(x, y) = 6$

25. **a.**† Compute the moment of inertia of a rectangle of length a and width b about its corners.   **b.** Of all rectangles with the same area, does a square have the maximum or minimum moment of inertia about its corners? Justify your answer.

26. Derive Rule 15.4.

15-3   DOUBLE INTEGRALS IN POLAR COORDINATES

In this section we show how to express double integrals in polar coordinates, a technique that is often employed when the integrand or equations for the boundary of the region of integration involve the distance $\sqrt{x^2 + y^2}$ to the origin or when the region of integration is given in polar coordinates.

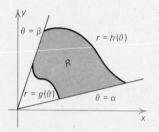

FIGURE 15.32

We will consider regions of the type shown in Figure 15.32, which is bounded by the rays $\theta = \alpha$ and $\theta = \beta$ through the origin with $0 < \beta - \alpha \leq 2\pi$ and by the curves given in polar coordinates by $r = g(\theta)$ and $r = h(\theta)$ with continuous g and h such that $0 \leq g(\theta) \leq h(\theta)$ for $\alpha \leq \theta \leq \beta$. We assume that $g(\theta)$ and $h(\theta)$ are nonnegative because in dealing with double integrals we use only polar coordinates with $r \geq 0$.

To express an integral over the region of Figure 15.32 in polar coordinates we make the substitutions

(1) $x = r \cos \theta, \quad y = r \sin \theta, \quad dx\,dy = r\,dr\,d\theta$ ◄

and assign limits of integration as described in the next rule.

Rule 15.6   Double integrals over regions of the type shown in Figure 15.32 are expressed in polar coordinates by using the formula

(2) $\displaystyle \int \int_R f(x, y)\,dx\,dy = \int_{\theta=\alpha}^{\theta=\beta} \left\{ \int_{r=g(\theta)}^{r=h(\theta)} f(r \cos \theta, r \sin \theta) r\,dr \right\} d\theta.$ ◄

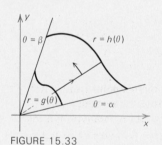

FIGURE 15.33

The integration procedure in equation (2) is indicated in Figure 15.33: we integrate with respect to r for fixed θ from $r = g(\theta)$ to $r = h(\theta)$ and then integrate with respect to θ from $\theta = \alpha$ to $\theta = \beta$.

Derivation of Rule 15.6.   We pick constants a and b so that $0 \leq a \leq g(\theta)$ and $h(\theta) \leq b$ for $\alpha \leq \theta \leq \beta$. Then the region S bounded by the circles $r = a$ and $r = b$ and by the rays $\theta = \alpha$ and $\theta = \beta$ contains the region R (Figure 15.34). We define (or redefine) $f(x, y)$ to be zero in the portion of S which is not in R, so that

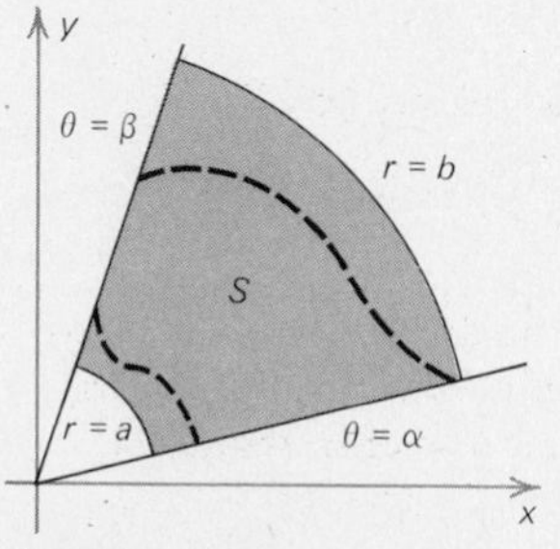

(3) $\displaystyle \int \int_R f(x, y)\,dx\,dy = \int \int_S f(x, y)\,dx\,dy.$

FIGURE 15.34

We let N and M denote arbitrary positive integers and set $\Delta r = (b - a)/N$, $\Delta \theta = (\beta - \alpha)/M$. We let

(4) $a = r_0 < r_1 < r_2 < r_3 < \cdots < r_{N-1} < r_N = b$

denote the partition of $a \leq r \leq b$ into N subintervals of length Δr, and let

(5) $\alpha = \theta_0 < \theta_1 < \theta_2 < \theta_3 < \cdots < \theta_{M-1} < \theta_M = \beta$

denote the partition of $\alpha \le \theta \le \beta$ into M subintervals of length $\Delta\theta$.

The circles $r = r_j$ ($j = 1, 2, 3, \ldots, N$) and the rays $\theta = \theta_k$ ($k = 1, 2, 3, \ldots, M$) through the origin divide S into subregions such as in Figure 15.35. The point $(r_j \cos \theta_k, r_j \sin \theta_k)$ is in the subregion bounded by the circles $r = r_{j-1}$ and $r = r_j$ and by the rays $\theta = \theta_{k-1}$ and $\theta = \theta_k$ (Figure 15.36), and the area of that subregion is

(6) $\frac{1}{2}(r_{j-1} + r_j)\, \Delta r \, \Delta\theta$

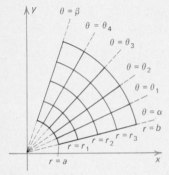

FIGURE 15.35

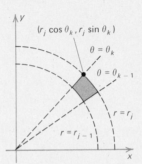

FIGURE 15.36

(see Exercise 21), and consequently the sum

(7) $\displaystyle\sum_{j=1}^{N} \sum_{k=1}^{M} f(r_j \cos \theta_k, r_j \sin \theta_k)\, \tfrac{1}{2}(r_{j-1} + r_j)\, \Delta r \, \Delta\theta$

is an approximating sum for the integral of $f(x, y)$ over S.

On the other hand, the region S in the xy-plane corresponds to a rectangle $W\colon a \le r \le b, \alpha \le \theta \le \beta$ in an $r\theta$-plane (Figure 15.37), and the lines $r = r_j$, $\theta = \theta_k$ divide it into subrectangles (Figure 15.38). The area of the rectangle bounded by $r = r_{j-1}, r = r_j, \theta = \theta_{k-1}$, and $\theta = \theta_k$ is $\Delta r \Delta\theta$, so the sum (7) is also an approximating sum for

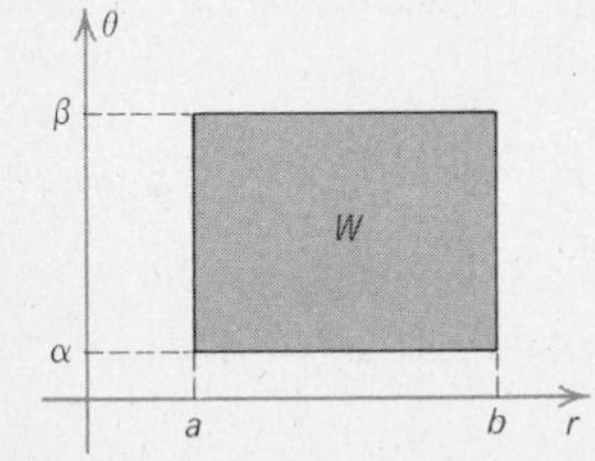

FIGURE 15.37

(8) $\displaystyle\int\!\!\!\int_{W} f(r \cos \theta, r \sin \theta)\, \tfrac{1}{2}(r + r)\, dr\, d\theta$

$$= \int\!\!\!\int_{W} f(r \cos \theta, r \sin \theta)\, r\, dr\, d\theta$$

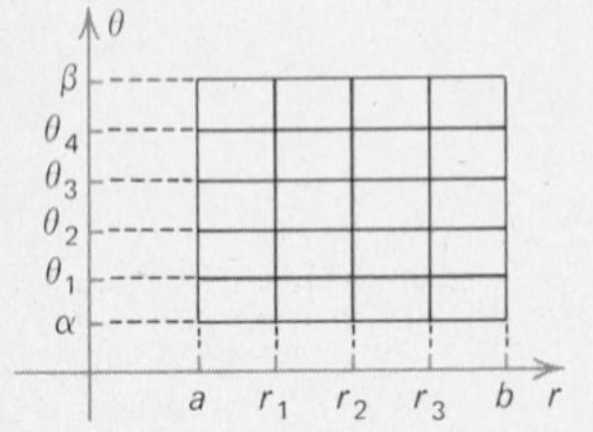

and the integral (8) is equal to the integral of $f(x, y)$ over R. When we express integral (8) as an iterated integral, we obtain

FIGURE 15.38

$$\int\!\!\!\int_{R} f(x, y)\, dx\, dy = \int_{\theta=\alpha}^{\theta=\beta} \left\{ \int_{r=a}^{r=b} f(r \cos \theta, r \sin \theta)\, r\, dr \right\} d\theta$$

which gives equation (2) because $f(r \cos \theta, r \sin \theta)$ is zero for $a \le r < g(\theta)$ and $h(\theta) < r \le b$. $\square$

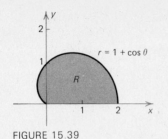

FIGURE 15.39

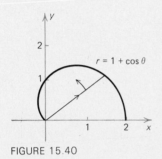

FIGURE 15.40

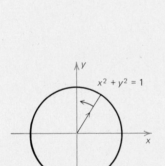

FIGURE 15.41

EXAMPLE 1   Use polar coordinates to evaluate the integral

$$\int\int_R y \, dx \, dy$$

where R is the upper half of the region bounded by the cardioid $r = 1 + \cos\theta$ (Figure 15.39).

Solution.   The integration procedure is indicated in Figure 15.40. We make the substitutions $y = r\sin\theta$ and $dx \, dy = r \, dr \, d\theta$ and integrate with respect to r from $r = 0$ to $r = 1 + \cos\theta$ for fixed θ and then with respect to θ from $\theta = 0$ to $\theta = \pi$:

$$\int\int_R y \, dx \, dy = \int_{\theta=0}^{\theta=\pi} \left\{ \int_{r=0}^{r=1+\cos\theta} (r\sin\theta) r \, dr \right\} d\theta$$

$$= \int_{\theta=0}^{\theta=\pi} \left\{ \int_{r=0}^{r=1+\cos\theta} r^2 \sin\theta \, dr \right\} d\theta$$

$$= \int_{\theta=0}^{\theta=\pi} \left[\tfrac{1}{3} r^3 \sin\theta \right]_{r=0}^{r=1+\cos\theta} d\theta$$

$$= \int_{\theta=0}^{\theta=\pi} \tfrac{1}{3}(1 + \cos\theta)^3 \sin\theta \, d\theta.$$

To evaluate the last integral, we make the substitutions $u = 1 + \cos\theta$, $du = -\sin\theta \, d\theta$ and obtain

$$[-\tfrac{1}{12}(1 + \cos\theta)^4]_{\theta=0}^{\theta=\pi} = [-\tfrac{1}{12} 0^4] - [-\tfrac{1}{12} 2^4] = \tfrac{4}{3}. \quad \square$$

In the next example we use polar coordinates to evaluate an integral that we could not evaluate with rectangular coordinates.

EXAMPLE 2   Compute the surface area of the portion of the hyperbolic paraboloid $z = xy$ for (x, y) in the circle $x^2 + y^2 \le 1$.

Solution.   By Example 3 of Section 15.2, the surface area is $\int\int_R \sqrt{x^2 + y^2 + 1} \, dx \, dy$, where R is the circle $x^2 + y^2 \le 1$. We make the substitutions $x^2 + y^2 = r^2$ and $dx \, dy = r \, dr \, d\theta$ and use the integration procedure indicated in Figure 15.41 to see that the surface area is

$$(9) \qquad \int_{\theta=0}^{\theta=2\pi} \left\{ \int_{r=0}^{r=1} \sqrt{r^2 + 1} \, r \, dr \right\} d\theta.$$

To evaluate the inner integral, we make the substitutions $u = r^2 + 1$, $du = 2r \, dr$ and obtain

$$\int \sqrt{r^2 + 1} \, r \, dr = \tfrac{1}{2} \int \sqrt{r^2 + 1} \, 2r \, dr = \tfrac{1}{2} \int u^{1/2} \, du$$

$$= \tfrac{1}{2}(\tfrac{2}{3}) u^{3/2} + C = \tfrac{1}{3}(r^2 + 1)^{3/2} + C.$$

This calculation shows that the surface area (9) is

$$\int_{\theta=0}^{\theta=2\pi} \left[\tfrac{1}{3}(r^2+1)^{3/2}\right]_{r=0}^{r=1} d\theta = \int_{\theta=0}^{\theta=2\pi} \left(\tfrac{1}{3}\, 2^{3/2} - \tfrac{1}{3}\, 1^{3/2}\right) d\theta$$

$$= \tfrac{2}{3}\pi\, (2^{3/2}-1). \quad \square$$

Exercises

†Answer provided.
‡Outline of solution provided.

Use polar coordinates to evaluate the integrals in Exercises 1 through 4.

1.‡ $\int \int_R (x^2+y^2)^{-2}\, dx\, dy;\ R: 2 \le x^2+y^2 \le 4$

2.† $\int_{x=-4}^{x=4} \int_{y=-\sqrt{16-x^2}}^{y=\sqrt{16-x^2}} e^{-x^2-y^2}\, dy\, dx$

3. $\int_{x=0}^{x=3} \int_{y=x}^{y=\sqrt{18-x^2}} \sin(x^2+y^2+1)\, dy\, dx$

4.‡ $\int \int_R (x^2+y^2)\, dx\, dy;\ R$ inside $x^2-4x+y^2=0$ and outside $x^2-2x+y^2=0$

In Exercises 5 through 12 use polar coordinates to compute the indicated areas and volumes.

5.‡ The area of the three-leaved rose bounded by $r = \cos(3\theta)$

6. The area of the outer loop of the limaçon $r = 1 - 2\cos\theta$

7.† The volume of the ellipsoid $x^2+y^2+4z^2 \le 4$

8. The volume of a sphere of radius R

9. The surface area of the portion of the hemisphere $z = \sqrt{R^2-x^2-y^2}$ inside the cylinder $x^2+y^2 = k^2\ (0 < k < R)$

10.‡ The volume of the solid inside the sphere $x^2+y^2+z^2 = 4$ and the cylinder $(x-1)^2+y^2 = 1$

11. The volume of the solid bounded by $x^2+y^2 = 25,\ z = 0,\ x+y+z = 8$

12.‡ The volume of the solid volume bounded by the circular paraboloid $z = x^2+y^2$ and the plane $z = 2y$

Express the integrals in Exercises 13 and 14 as iterated integrals in rectangular coordinates.

13.† $\int_{\theta=0}^{\theta=\pi/4} \int_{r=\sec\theta}^{r=2\cos\theta} \dfrac{r^2}{1+r\sin\theta}\, dr\, d\theta$

14. $\int_{\theta=\pi/4}^{\theta=3\pi/4} \int_{r=0}^{r=4\csc\theta} r^5 \sin^2\theta\, dr\, d\theta$

15. Switch the following integral to rectangular coordinates and then evaluate it.

$$\int_{\theta=3\pi/4}^{\theta=4\pi/3} \int_{r=0}^{r=-5\sec\theta} r^3 \sin^2\theta\, dr\, d\theta$$

16. The solid V is defined by the inequalities $0 \le z \le xy,\ x \ge 0,\ y \ge 0,$ $x^2+y^2 \le 9$. Express its volume as iterated integrals   **a.** in rectangular coordinates and   **b.** in polar coordinates.   **c.** Evaluate both integrals.

17.† Use the result of Exercise 9 to compute the surface area of the hemisphere $z = \sqrt{R^2 - x^2 - y^2}$.

18. Compute the surface area of the cone $z = \sqrt{x^2 + y^2}, z \leq h$ by finding the limit of the surface area of the frustrum of the cone for $k \leq z \leq h \ (0 < k < h)$.

19. A flat plate occupies the disk $x^2 - 2x + y^2 \leq 0$ and its density at (x, y) is $5\sqrt{x^2 + y^2}$. Where is its center of gravity?

20. Find the moment of inertia about the origin of the four-leaved rose bounded by $r = \sin(2\theta)$. (Consider $0 \leq \theta \leq \pi/2$.)

21.† Derive formula (6) for the area of the region given in polar coordinates by the conditions $r_{j-1} \leq r \leq r_j, \theta_{k-1} \leq \theta \leq \theta_k \ (\Delta r = r_j - r_{j-1}, \Delta\theta = \theta_k - \theta_{k-1})$.

15-4   TRIPLE INTEGRALS

This section deals with TRIPLE INTEGRALS

$$(1) \quad \int \int \int_V f(x, y, z) \, dx \, dy \, dz$$

of functions $f(x, y, z)$ over bounded solids V in xyz-space. We will consider solids V of the type shown in Figure 15.42 where the top is formed by the surface $z = h(x, y)$ and the bottom by the surface $z = g(x, y)$ and the solid consists of the points (x, y, z) between the two surfaces such that (x, y) is in a region R of the xy-plane. We assume that the region R is bounded and has a piecewise smooth boundary and that the functions $g(x, y)$ and $h(x, y)$ are piecewise continuous. We will also consider regions of the same type with the variables $x, y,$ and z interchanged. For example, the solid might be defined by the requirement that (y, z) be in a region R of the yz-plane and that the point (x, y, z) lie between surfaces $x = g(y, z)$ and $x = h(y, z)$.

To define the triple integral (1), we divide V into subregions

$$(2) \quad V_1, V_2, V_3, \ldots, V_N$$

of the same type as V. We pick a point (x_j, y_j, z_j) from V_j for each j and form the approximating sum

$$(3) \quad \sum_{j=1}^{N} f(x_j, y_j, z_j)[\text{Volume of } V_j].$$

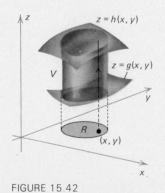

FIGURE 15.42

The integral is the limit of such sums.

Definition 15.3   The triple integral (1) is the limit of the approximating sums (3) as the number of subregions in the partition (2) tends to infinity and their diameters tend to zero.

Here the diameter of a subregion is the diameter of the smallest sphere containing it. The limit exists and the integral is defined if $f(x, y, z)$ is bounded and piecewise continuous in V. In this case f is bounded if there is a constant M such that $|f(x, y, z)| \leq M$ for all (x, y, z) in V and f is piecewise continuous if there is a partition V_1, $V_2, \ldots, V_N$ of V such that in the interior of each subregion f is equal to a function that is continuous in the subregion and its boundary. We will not give the proof of the existence of the integral under these conditions and we will assume that these conditions are satisfied in all definitions, rules, and theorems that deal with triple integrals.

Iterated integrals   The next rule shows how triple integrals over solids V of the type shown in Figure 15.42 can be evaluated by first performing an integration with respect to z and then evaluating a double integral with respect to x and y. If the double integral can be expressed as an iterated integral, then the triple integral can be expressed as a succession of three one-dimensional integrals.

Rule 15.7   Suppose that $g(x, y) \leq h(x, y)$ for (x, y) in a region R of the xy-plane and that the solid V consists of the points (x, y, z) with (x, y) in R and $g(x, y) \leq z \leq h(x, y)$ (Figure 15.42). Then triple integrals over V can be expressed as iterated integrals by the formula

$$(4) \quad \int \int \int_V f(x, y, z) \, dx \, dy \, dz = \int \int_R \left\{ \int_{z=g(x, y)}^{z=h(x, y)} f(x, y, z) \, dz \right\} dx \, dy. \quad \blacktriangleleft$$

The integration procedure in the iterated integral (4) is indicated by the vertical line with an arrow on it in Figure 15.42. The line is above the point (x, y) in the xy-plane and extends from the bottom of the solid to the top. As suggested by the line, we integrate with respect to z from $z = g(x, y)$ at the bottom to $z = h(x, y)$ at the top of the solid and then integrate with respect to x and y over the region R.

Derivation of Rule 15.7.   We choose constants so that $a \leq g(x, y)$ and $h(x, y) \leq b$ for all (x, y) in R and let W denote the solid consisting of all (x, y, z) with (x, y) in R and $a \leq z \leq b$ (Figure 15.43). W contains V, and we define (or redefine) $f(x, y, z)$ to be zero in the part of W which is not in V, so that

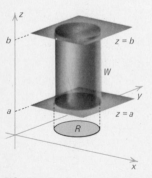

FIGURE 15.43   The solid W that contains the solid V

$$(5) \quad \int \int \int_V f(x, y, z) \, dx \, dy \, dz = \int \int \int_W f(x, y, z) \, dx \, dy \, dz.$$

We consider a partition

(6) $R_1, R_2, R_3, \ldots, R_N$

of the region R and a partition

(7) $a = z_0 < z_1 < z_2 < z_3 < \cdots < z_{M-1} < z_M = b$

of the interval $a \leq z \leq b$. For each j and k we let W_{jk} denote the solid formed by the points (x, y, z) with (x, y) in R_j and $z_{k-1} \leq z \leq z_k$ (Figure 15.44). The solids W_{jk} for $j = 1, 2, \ldots, N$ and $k = 1, 2, \ldots, M$ form a partition of W. We choose a point (x_j, y_j) in R_j for each j, so that (x_j, y_j, z_k) is in W_{jk}. The volume of W_{jk} is $(z_k - z_{k-1})[\text{area of } R_j]$, so we obtain the approximating sum

$$
\begin{aligned}
\text{(8)} \quad & \sum_{j=1}^{N} \sum_{k=1}^{M} f(x_j, y_j, z_k)(z_k - z_{k-1})[\text{Area of } R_j] \\
& = \sum_{j=1}^{N} \left\{ \sum_{k=1}^{M} f(x_j, y_j, z_k)(z_k - z_{k-1}) \right\} [\text{Area of } R_j]
\end{aligned}
$$

for the integral of $f(x, y, z)$ over W and, hence, over V.

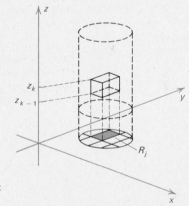

FIGURE 15.44   The subsolid W_{jk}:
(x, y) in R_j, $z_{k-1} \leq z \leq z_k$

If we hold the partition (6) of R fixed and take partitions (7) of $a \leq z \leq b$ with the lengths of their subintervals tending to zero, the inner sum on the right of (8) tends to the integral of $f(x_j, y_j, z)$ with respect to z from a to b and the entire sum tends to

(9) $\displaystyle \sum_{j=1}^{n} \left\{ \int_{z=a}^{z=b} f(x_j, y_j, z)\, dz \right\} [\text{Area of } R_j].$

This, in turn, is an approximating sum for the integral

(10) $\displaystyle \int \int_R \left\{ \int_{z=a}^{z=b} f(x, y, z)\, dz \right\} dx\, dy$

and tends to it as we take partitions of R with the diameters of their subregions tending to zero. Hence the integral (10) is equal to the triple integral of $f(x, y, z)$ on the left of (4). It also is equal to the iterated integral on the right of (4) because $f(x, y, z)$ is zero for $a \leq z < g(x, y)$ and for $h(x, y) < z \leq b$. □

EXAMPLE 1 Evaluate the triple integral

$$(11) \quad \int \int \int_V \tfrac{1}{9} x \, dx \, dy \, dz$$

where V is the tetrahedron bounded by the plane $3x + y + z = 3$ and the coordinate planes.

Solution. The plane $3x + y + z = 3$ has x-intercept 1, y-intercept 3, and z-intercept 3, and it forms the top of the tetrahedron V that is sketched in Figure 15.45. The base is a triangle R in the xy-plane. To apply Rule 15.7, we write the equation for the top of V in the form $z = 3 - 3x - y$ and note that the xy-plane has the equation $z = 0$. The triple integral (11) is equal to the iterated integral

$$(12) \quad \int \int_R \left\{ \int_{z=0}^{z=3-3x-y} \tfrac{1}{9} x \, dz \right\} dx \, dy.$$

Setting $z = 0$ in the equation of the plane $3x + y + z = 3$ shows that its intersection with the xy-plane is the line $y = 3 - 3x$. This line forms the top of the triangle R (Figure 15.46), and its other sides are formed by the x- and y-axes. The integrals (11) and (12) equal the triple iterated integral

$$(13) \quad \int_{x=0}^{x=1} \left\{ \int_{y=0}^{y=3-3x} \left\{ \int_{z=0}^{z=3-3x-y} \tfrac{1}{9} x \, dz \right\} dy \right\} dx.$$

In practice we will often omit the brackets in expressions such as (13). Notice that the limits of integration on the innermost integral in (13) depend on x and y, the limits on the intermediate integral depend only on x, and the limits on the outer integral are constants.

We first evaluate the innermost integral in (13), which is an integral with respect to z with x and y constant:

$$\int_{z=0}^{z=3-3x-y} \tfrac{1}{9} x \, dz = [\tfrac{1}{9} xz]_{z=0}^{z=3-3x-y} = \tfrac{1}{9} x (3 - 3x - y)$$

$$= \tfrac{1}{3} x - \tfrac{1}{3} x^2 - \tfrac{1}{9} xy.$$

Next, we use this result and evaluate the y-integral with x constant:

$$\int_{y=0}^{y=3-3x} \int_{z=0}^{z=3-3x-y} \tfrac{1}{9} x \, dz \, dy = \int_{y=0}^{y=3-3x} (\tfrac{1}{3} x - \tfrac{1}{3} x^2 - \tfrac{1}{9} xy) \, dy$$

$$= [\tfrac{1}{3} xy - \tfrac{1}{3} x^2 y - \tfrac{1}{18} xy^2]_{y=0}^{y=3-3x}$$

$$= \tfrac{1}{3} x (3 - 3x) - \tfrac{1}{3} x^2 (3 - 3x) - \tfrac{1}{18} x (3 - 3x)^2$$

$$= \tfrac{1}{2} x^3 - x^2 + \tfrac{1}{2} x.$$

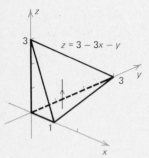

FIGURE 15.45 The solid V

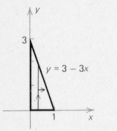

FIGURE 15.46 The region R

With this calculation we see that the integral (11) equals

$$\int_{x=0}^{x=1} (\tfrac{1}{2}x^3 - x^2 + \tfrac{1}{2}x)\,dx = [\tfrac{1}{8}x^4 - \tfrac{1}{3}x^3 + \tfrac{1}{4}x^2]_{x=0}^{x=1}$$

$$= \tfrac{1}{8} - \tfrac{1}{3} + \tfrac{1}{4} = \tfrac{1}{24}. \quad \square$$

Other orders of integration

Arguments similar to that used to derive Rule 15.7 show that integrals over solids of the type shown in Figure 15.47 may be expressed as iterated integrals by the formula

$$(14) \quad \int\int\int_V f(x,y,z)\,dx\,dy\,dz$$

$$= \int\int_R \left\{ \int_{y=g(x,z)}^{y=h(x,z)} f(x,y,z)\,dy \right\} dx\,dz \quad \blacktriangleleft$$

and that for solids of the type in Figure 15.48 we may use

$$(15) \quad \int\int\int_V f(x,y,z)\,dx\,dy\,dz$$

$$= \int\int_R \left\{ \int_{x=g(y,z)}^{x=h(y,z)} f(x,y,z)\,dx \right\} dy\,dz. \quad \blacktriangleleft$$

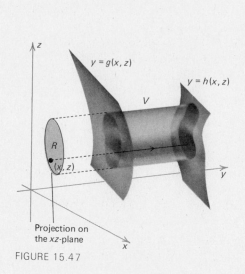

Projection on the xz-plane

FIGURE 15.47

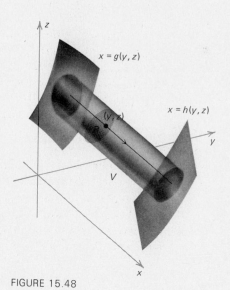

FIGURE 15.48

Volumes and weights

Definition 15.3 with $f(x,y,z)$ the constant function 1 shows that

$$(16) \quad \text{The volume of } V = \int\int\int_V dx\,dy\,dz. \quad \blacktriangleleft$$

If V has density $\rho(x,y,z)$ at (x,y,z), then

$$(17) \quad \text{The weight of } V = \int\int\int_V \rho(x,y,z)\,dx\,dy\,dz. \quad \blacktriangleleft$$

We can see this by considering partitions $V_1, V_2, V_3, \ldots, V_N$ of V and noting that if we take the density to be $\rho(x_j, y_j, z_j)$ in V_j with (x_j, y_j, z_j) in V_j, the weight is an approximating sum for the integral (17).

Formula (17) gives us a means for interpreting triple integrals of nonnegative functions: The triple integral of $f(x, y, z)$ over V is the weight of the solid V with density $f(x, y, z)$ at (x, y, z). If $f(x, y, z)$ has negative values, we can think of it as an electric charge density with positive charge where $f(x, y, z)$ is positive and negative charge where it is negative. The triple integral is then the total charge on V.

Moments and centers of gravity

We define the moment about the xy-plane of a point mass of weight m at (x, y, z) to be the product zm of its z-coordinate and weight, the moment about the xz-plane to be ym, and the moment about the yz-plane to be xm. This leads to the formulas

(18) The moment about the xy-plane $= \displaystyle\int\int\int_V z\rho(x, y, z)\, dx\, dy\, dz$ ◄

(19) The moment about the xz-plane $= \displaystyle\int\int\int_V y\rho(x, y, z)\, dx\, dy\, dz$ ◄

(20) The moment about the yz-plane $= \displaystyle\int\int\int_V x\rho(x, y, z)\, dx\, dy\, dz$ ◄

for the moments of a solid V with density $\rho(x, y, z)$ (see Exercise 24). Then we can find the center of gravity $(\bar{x}, \bar{y}, \bar{z})$ of V by using the equations

(21) $\quad \bar{z} = \dfrac{\text{The moment about the } xy\text{-plane}}{\text{The weight}}$

(22) $\quad \bar{y} = \dfrac{\text{The moment about the } xz\text{-plane}}{\text{The weight}}$

(23) $\quad \bar{x} = \dfrac{\text{The moment about the } yz\text{-plane}}{\text{The weight}}.$

Exercises

Evaluate the integrals in Exercises 1 through 7.

1.‡ $\int\int\int_V z\, dx\, dy\, dz$; V bounded by $z = \sqrt{x^2 + y^2}, z = 0, x = \pm 1, y = \pm 1$

2.† $\int\int\int_V 15\, x^2 z^2\, dx\, dy\, dz$; V bounded by $x^2 + y^2 = 1, x^2 + z^2 = 1$

3. $\int\int\int_V (\sin x)(\sin y)(\sin z)\, dx\, dy\, dz$; V bounded by $z = y, z = 0, x = 0,$
$x = \pi/2, y = \pi$

4.‡ $\int\int\int_V 15\sqrt{yz}\, dx\, dy\, dz$; V bounded by $z = 0, z = y, y = x^2, y = 1$

5.† $\int\int\int_V (\sin y)e^{3z}\, dx\, dy\, dz$; V bounded by $y = x, y = 0, x = 4, z = 0, z = 6$

6. $\int\int\int_V y\cos z\, dx\, dy\, dz$; V bounded by $x = 0, x = 5, z = y^2, z = 4$

7.† $\int\int\int_V x^2 y^2 z\, dx\, dy\, dz$; V defined by $0 \le z \le x^2 - y^2, 0 \le x \le 1$

8.[†] Show that if V is the rectangular box $a_1 \le x \le b_1$, $a_2 \le y \le b_2$, $a_3 \le z \le b_3$, then

$$\int \int \int_V f(x)g(y)h(z)\, dx\, dy\, dz$$

$$= \left[\int_{a_1}^{b_1} f(x)\, dx \right] \left[\int_{a_2}^{b_2} g(y)\, dy \right] \left[\int_{a_3}^{b_3} h(z)\, dz \right].$$

Use the result of Exercise 8 to evaluate integrals (9) through (11).

9.[‡] $\int \int \int_V xz^2 \sin y\, dx\, dy\, dz$; $V\colon 0 \le x \le 3, 1 \le y \le 5, -3 \le z \le 0$

10.[†] $\int \int \int_V e^{2x+3y-4z}\, dx\, dy\, dz$; $V\colon -4 \le x \le 4, -3 \le y \le 3, -2 \le z \le 2$

11. $\int \int \int_V (x^2yz + yz)\, dx\, dy\, dz$; $V\colon 0 \le x \le 5, 0 \le y \le 10, 0 \le z \le 20$

In Exercises 12 through 17 describe the regions of integration in the iterated integrals.

12.[‡] $\int_{x=-1}^{x=1} \int_{y=-\sqrt{1-x^2}}^{y=\sqrt{1-x^2}} \int_{z=-5}^{z=x^2-y^2} f(x, y, z)\, dz\, dy\, dx$

13.[†] $\int_{x=0}^{x=2} \int_{y=0}^{y=\sqrt{4-x^2}} \int_{z=0}^{z=\sqrt{4-x^2-y^2}} f(x, y, z)\, dz\, dy\, dx$

14. $\int_{x=-1}^{x=1} \int_{y=x^2}^{y=1} \int_{z=0}^{z=1-y} f(x, y, z)\, dz\, dy\, dx$

15.[‡] $\int_{y=0}^{y=1} \int_{z=0}^{z=2-2y} \int_{x=0}^{x=4-4y-2z} f(x, y, z)\, dx\, dz\, dy$

16.[†] $\int_{z=-2}^{z=0} \int_{x=0}^{x=z+2} \int_{y=0}^{y=2-x+z} f(x, y, z)\, dy\, dx\, dz$

17. $\int_{x=-1}^{x=1} \int_{z=-\sqrt{1-x^2}}^{z=\sqrt{1-x^2}} \int_{y=0}^{y=4} f(x, y, z)\, dy\, dz\, dx$

18.[‡] A solid occupies the region bounded by the parabolic cylinder $z = y^2$ and by the planes $x = 0, x = 2$, and $z = 1$. Its density at (x, y, z) is $z + 2$. What are the coordinates $(\bar{x}, \bar{y}, \bar{z})$ of its center of gravity?

19.[†] A solid is bounded by the planes $x = 0, x = 1, y = 0, y = 1$, and $z = 0$ and by the paraboloid $z = x^2 + y^2$. Its density at (x, y, z) is xy. What are the coordinates $(\bar{x}, \bar{y}, \bar{z})$ of its center of gravity?

20. The tetrahedron with vertices $(0, 0, 0), (1, 0, 0), (0, 1, 0), (0, 0, 1)$ has density x^2 at (x, y, z). What are the coordinates $(\bar{x}, \bar{y}, \bar{z})$ of its center of gravity?

In Exercises 21 through 23 switch the orders of integration to orders in which the z-integration is performed last.

21.[‡] $\int_{x=0}^{x=4} \int_{y=0}^{y=4-x} \int_{z=0}^{z=4-x-y} f(x, y, z)\, dz\, dy\, dx$

22.[†] $\int_{x=-2}^{x=2} \int_{y=x^3}^{y=8} \int_{z=0}^{z=3} f(x, y, z)\, dz\, dy\, dx$

23. $\int_{x=0}^{x=5} \int_{y=x}^{y=5} \int_{z=0}^{z=5-y} f(x, y, z)\, dz\, dy\, dx$

24. Derive formulas (18), (19), and (20).

Additional drill exercises

Evaluate the integrals in Exercises 25 through 31.

25.[†] $\int \int \int_V x\, dx\, dy\, dz$ where V is bounded by $z = -x^2 - y^2, z = x^2 + y^2$, $y = -x, y = x^3$, and $x = 1$

26.[†] $\int \int \int_V xy^2\, e^{-z}\, dx\, dy\, dz$ with $V\colon 0 \le x \le 2, 0 \le y \le 3, 0 \le z \le 4$

27.[†] $\int\int\int_V x^2z\,dx\,dy\,dz$ with V bounded by $z = 0, z = x, x = 0, y = 0,$ and $x + y = 1$

28.[†] $\int\int\int_V (x^2 + y^2 + z^2)\,dx\,dy\,dz$ with $V: 1 \le x \le 2, 0 \le y \le 1, 0 \le z \le 2$

29.[†] $\int\int\int_V x^2y\,e^{xyz}\,dx\,dy\,dz$ with $V: 0 \le x \le 1, 0 \le y \le 4, 0 \le z \le 3$

30.[†] $\int\int\int_V \dfrac{1}{x^3y^2z}\,dx\,dy\,dz$ with $V: 1 \le x \le 2, 2 \le y \le 3, 3 \le z \le 4$

31.[†] $\int\int\int_V e^{x^4}\,dx\,dy\,dz$ with V bounded by $z = 0, z = y^2, y = 0, y = x, x = 1$

15-5   CYLINDRICAL AND SPHERICAL COORDINATES

In this section we study triple integrals in CYLINDRICAL and SPHERICAL COORDINATES. Cylindrical coordinates are often used when there is symmetry about the z-axis or when the distance $\sqrt{x^2 + y^2}$ to the z-axis plays a special role. Spherical coordinates are often used when there is symmetry about the origin or the distance $\sqrt{x^2 + y^2 + z^2}$ to the origin plays a special role.

Cylindrical coordinates

We obtain cylindrical coordinates by replacing x and y by polar coordinates r and θ. The coordinate z remains the same as in rectangular coordinates:

Definition 15.4   A point with rectangular coordinates (x, y, z) has cylindrical coordinates $[r, \theta, z]$ where $[r, \theta]$ are polar coordinates of (x, y) with $r \ge 0$.

The formulas relating polar and rectangular coordinates in an xy-plane yield the formulas

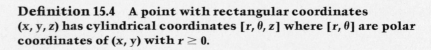

(1) $x = r\cos\theta,\qquad y = r\sin\theta,\qquad z = z$ ◀

(2) $dx\,dy\,dz = r\,dr\,d\theta\,dz$ ◀

(3) $r = \sqrt{x^2 + y^2}$ ◀

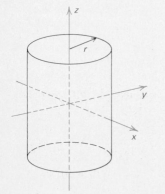

FIGURE 15.49   The coordinate surface r = Constant

relating cylindrical and rectangular coordinates in xyz-space.

Coordinate surfaces

To use cylindrical coordinates in triple integrals we need to be familiar with their COORDINATE SURFACES, which are the surfaces where one of the variables $r, \theta,$ and z is constant. The coordinate surface r = constant is an infinite circular cylinder of radius r and with the z-axis as its axis (Figure 15.49). The coordinate surface θ = constant is the vertical half plane with its edge along the z-axis and which intersects the xy-plane in a ray which makes an angle θ with the positive x-axis (Figure 15.50). The coordinate surface z = constant is a horizontal plane, as it is in rectangular coordinates (Figure 15.51).

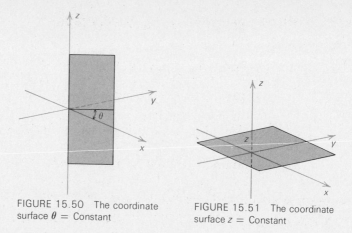

FIGURE 15.50   The coordinate
surface $\theta = $ Constant

FIGURE 15.51   The coordinate
surface $z = $ Constant

**Integrals in
cylindrical
coordinates**

To express a triple integral as an iterated integral in cylindrical
coordinates we make substitutions (1) and (2) and determine the
appropriate limits of integration from the geometry of the region. The
next result shows how the limits are assigned in the special case where
the region is bounded by a cylinder $r = a$ and by horizontal planes.

**Rule 15.8   If V is the solid bounded by the cylinder
$r = a$ $(a > 0)$ and by the planes $z = b$ and $z = c$ $(b < c)$
(Figure 15.52), then**

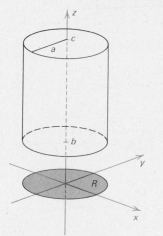

FIGURE 15.52   The cylinder
$0 \le r \le a, 0 \le \theta \le 2\pi,$
$b \le z \le c$

$$\int \int \int_V f(x, y, z)\, dx\, dy\, dz$$

(4)

$$= \int_{r=0}^{r=a} \left\{ \int_{\theta=0}^{\theta=2\pi} \left\{ \int_{z=b}^{z=c} f(r\cos\theta, r\sin\theta, z)\, r\, dz \right\} d\theta \right\} dr.$$

◀

**The iterated integral on the right of (4) may also be given with
any of the five other possible orders of integration.**

Notice that the cylindrical region in Rule 15.8 consists of the points
whose cylindrical coordinates satisfy $0 \le r \le a, 0 \le \theta \le 2\pi, b \le z \le c$
and that the integral on the right of (4) has the corresponding limits of
integration.

Derivation of Rule 15.8.   We express the triple integral as

$$\int \int_R \left\{ \int_{z=b}^{z=c} f(x, y, z)\, dz \right\} dx\, dy$$

where R is the disk $x^2 + y^2 \le a^2$, which is the projection of the solid
on the xy-plane. We obtain (4) by expressing the integral over R as an
iterated integral in polar coordinates. The iterated integral (4) may be
computed with the integration performed in any order because the
limits of integration are constant.   □

The six orders of integration

The possible orders of integration in the iterated integral of Rule 15.8 are indicated in Figures 15.53 through 15.58. In each case we first hold two of the variables fixed and integrate with respect to the third. In this process the point $[r, \theta, z]$ describes a line segment or a circle. When we integrate with respect to a second variable, the line or circle sweeps out a portion of the coordinate surface of the remaining variable. Finally, as we integrate with respect to the third variable, the portion of its coordinate surface sweeps out the cylindrical solid.

If the region of integration is not a cylinder, we determine the limits of integration from a sketch of the region. Unless the limits of integration are constant, they will depend on the order in which the integration is performed, and in many cases one order of integration will be easier to carry out than another.

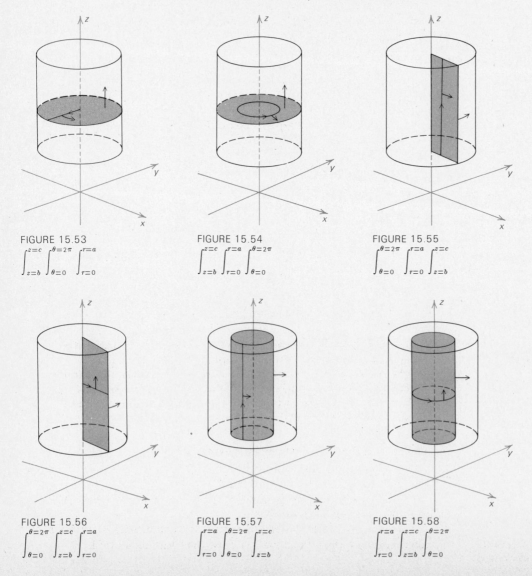

FIGURE 15.53
$$\int_{z=b}^{z=c} \int_{\theta=0}^{\theta=2\pi} \int_{r=0}^{r=a}$$

FIGURE 15.54
$$\int_{z=b}^{z=c} \int_{r=0}^{r=a} \int_{\theta=0}^{\theta=2\pi}$$

FIGURE 15.55
$$\int_{\theta=0}^{\theta=2\pi} \int_{r=0}^{r=a} \int_{z=b}^{z=c}$$

FIGURE 15.56
$$\int_{\theta=0}^{\theta=2\pi} \int_{z=b}^{z=c} \int_{r=0}^{r=a}$$

FIGURE 15.57
$$\int_{r=0}^{r=a} \int_{\theta=0}^{\theta=2\pi} \int_{z=b}^{z=c}$$

FIGURE 15.58
$$\int_{r=0}^{r=a} \int_{z=b}^{z=c} \int_{\theta=0}^{\theta=2\pi}$$

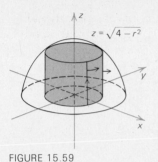

FIGURE 15.59

$z = \sqrt{4 - r^2}$

EXAMPLE 1 Use cylindrical coordinates to evaluate the integral

$$(5) \qquad \int \int \int_V z\sqrt{x^2 + y^2}\, dx\, dy\, dz$$

where V is the hemisphere $x^2 + y^2 + z^2 \leq 4, \quad z \geq 0$.

Solution. Figure 15.59 shows the region of integration and the integration procedure where we first integrate with respect to z, then with respect to θ, and last with respect to r. The base of the cylinder in Figure 15.59 is on the xy-plane and its top edge is on the hemisphere $z = \sqrt{4 - x^2 - y^2} = \sqrt{4 - r^2}$. Therefore, we integrate first with respect to z from $z = 0$ to $z = \sqrt{4 - r^2}$, then with respect to θ from $\theta = 0$ to $\theta = 2\pi$, and finally with respect to r from $r = 0$ to $r = 2$ (the radius of the base of the hemisphere). We make the substitutions $\sqrt{x^2 + y^2} = r$ and $dx\, dy\, dz = r\, dr\, d\theta\, dz$ to express the triple integral (5) as the iterated integral

$$\int_{r=0}^{r=2} \left\{ \int_{\theta=0}^{\theta=2\pi} \left\{ \int_{z=0}^{z=\sqrt{4-r^2}} (zr)\, r\, dz \right\} d\theta \right\} dr.$$

The integral with respect to z is

$$\int_{z=0}^{z=\sqrt{4-r^2}} zr^2\, dz = \left[\tfrac{1}{2} z^2 r^2 \right]_{z=0}^{z=\sqrt{4-r^2}} = \tfrac{1}{2}(4 - r^2)r^2 = 2r^2 - \tfrac{1}{2} r^4.$$

Next, the θ-integration gives

$$\int_{\theta=0}^{\theta=2\pi} (2r^2 - \tfrac{1}{2} r^4)\, d\theta = \left[(2r^2 - \tfrac{1}{2} r^4)\theta \right]_{\theta=0}^{\theta=2\pi}$$
$$= 2\pi(2r^2 - \tfrac{1}{2} r^4) = 4\pi r^2 - \pi r^4.$$

Finally, we perform the r-integration to see that the triple integral equals

$$\int_{r=0}^{r=2} (4\pi r^2 - \pi r^4)\, dr = \left[\tfrac{4}{3}\pi r^3 - (\pi/5)r^5 \right]_{r=0}^{r=2}$$
$$= \tfrac{32}{3}\pi - \tfrac{32}{5}\pi = \tfrac{64}{15}\pi. \quad \square$$

Spherical coordinates

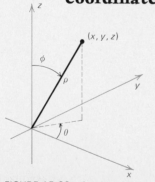

FIGURE 15.60 A point with rectangular coordinates (x, y, z) and spherical coordinates $[\rho, \theta, \phi]$

Definition 15.5 **The point with rectangular coordinates (x, y, z) has spherical coordinates $[\rho, \theta, \phi]$ where ρ is the distance from the point to the origin, where θ is the same angle as in cylindrical coordinates, and where ϕ is the angle from the positive z-axis to the line segment between the point and the origin (Figure 15.60).**

By the distance formula, the coordinate ρ is

$$(6) \qquad \rho = \sqrt{x^2 + y^2 + z^2}. \qquad \blacktriangleleft$$

We do not give formulas for θ and ϕ in terms of x, y, and z since such formulas are not often needed.

To derive the formulas for the rectangular coordinates in terms of the spherical coordinates of a point, we use the coordinate r from

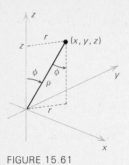

FIGURE 15.61
$r = \rho \sin \phi, z = \rho \cos \phi$

cylindrical coordinates. The angle ϕ lies in the interval $0 \le \phi \le \pi$, so $\sin \phi$ is nonnegative and $r = \rho \sin \phi$ (Figure 15.61). Since $x = r \cos \theta$ and $y = r \sin \theta$, we have $x = \rho \sin \phi \cos \theta$ and $y = \rho \sin \phi \sin \theta$. Also $z = \rho \cos \phi$ (Figure 15.61). Thus, the rectangular coordinates of a point may be computed from it spherical coordinates by the formulas

$$(7) \quad x = \rho \sin \phi \cos \theta, \quad y = \rho \sin \phi \sin \theta, \quad z = \rho \cos \phi. \quad \blacktriangleleft$$

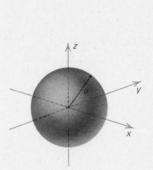

FIGURE 15.62   The coordinate surface $\rho = $ Constant

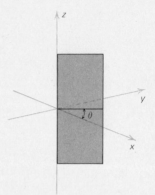

FIGURE 15.63   The coordinate surface $\theta = $ Constant

Coordinate surfaces

The coordinate surface $\rho = $ constant is the sphere of radius ρ with its center at the origin (Figure 15.62). As in cylindrical coordinates, the coordinate surface $\theta = $ constant is a half plane with its edge along the z-axis (Figure 15.63). The coordinate surface $\phi = $ constant is the positive z-axis for $\phi = 0$, is a cone with its vertex at the origin and opening upward for $0 < \phi < \pi/2$ (Figure 15.64), is the xy-plane for $\phi = \pi/2$, a cone that opens downward for $\pi/2 < \phi < \pi$ (Figure 15.65), and the negative z-axis for $\phi = \pi$.

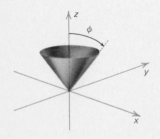

FIGURE 15.64

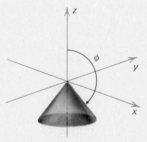

FIGURE 15.65

Integrals in spherical coordinates

To express a triple integral in spherical coordinates we have to make substitutions (7) along with

$$(8) \quad dx \, dy \, dz = \rho^2 \sin\phi \, d\rho \, d\theta \, d\phi \quad \blacktriangleleft$$

and determine the limits of integration from the geometry of the region. The next rule deals with the special case where the region of integration is a sphere.

Rule 15.9   **If V is the sphere of radius ρ with its center at the origin, then**

(9) $$\iiint_V f(x, y, z)\, dx\, dy\, dz \qquad\blacktriangleleft$$

$$= \int_{\rho=0}^{\rho=a} \int_{\theta=0}^{\theta=2\pi} \int_{\phi=0}^{\phi=\pi} f(\rho\sin\phi\cos\theta, \rho\sin\phi\sin\theta, \rho\cos\phi)\, \rho^2\sin\phi\, d\phi\, d\theta\, d\rho.$$

The iterated integral may also be given any of the five other possible orders of integration.

Notice that the sphere consists of the points whose spherical coordinates satisfy the conditions $0 \le \rho \le a$, $0 \le \theta \le 2\pi$, and $0 \le \phi \le \pi$, and that the iterated integral (9) has the corresponding limits of integration.

Derivation of Rule 15.9.   We consider partitions $0 = \rho_0 < \rho_1 < \rho_2 < \cdots < \rho_J = a$, $0 = \theta_0 < \theta_1 < \theta_2 < \cdots < \theta_K = 2\pi$, and $0 = \phi_0 < \phi_1 < \phi_2 < \cdots < \phi_N = \pi$ of the intervals $0 \le \rho \le a$, $0 \le \theta \le 2\pi$, and $0 \le \phi \le \pi$ into subintervals of lengths $\Delta\rho$, $\Delta\theta$, and $\Delta\phi$, respectively. The coordinate surfaces $\rho = \rho_j$, $\theta = \theta_k$, $\phi = \phi_n$ divide the sphere into subregions of the type shown in Figure 15.66, which is bounded by the spheres $\rho = \rho_{j-1}$ and $\rho = \rho_j$, the half planes $\theta = \theta_{k-1}$ and $\theta = \theta_k$, and the cones $\phi = \phi_{n-1}$ and $\phi = \phi_n$.

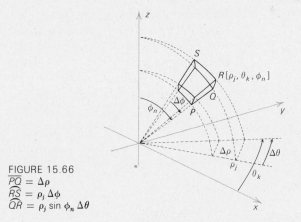

FIGURE 15.66
$\overline{PQ} = \Delta\rho$
$\widehat{RS} = \rho_j\,\Delta\phi$
$\widehat{QR} = \rho_j \sin\phi_n\,\Delta\theta$

The volume of the subregion in Figure 15.66 is given approximately by the product $\overline{PQ}\,\widehat{RS}\,\widehat{QR}$ of the lengths of three of its edges. Because P is on the sphere of radius ρ_{j-1} and Q is on the sphere of radius $\rho_j = \rho_{j-1} + \Delta\rho$, we have $\overline{PQ} = \Delta\rho$. The points R and S are on a circle of radius ρ_j with its center at the origin, and the arc RS subtends an angle $\Delta\phi$ at the center of that circle. Hence $RS = \rho_j\,\Delta\phi$. The points Q and R are on a horizontal circle of radius $\rho_j\sin\phi_n$ with its center on the z-axis (Figure 15.67), and the arc $\widehat{QR}$ subtends an angle $\Delta\theta$ at the center of that circle. Therefore, $QR = \rho_j\sin\phi_n\,\Delta\theta$ and the volume of the subregion is approximately

$$\overline{PQ}\,\widehat{RS}\,\widehat{QR} = (\Delta\rho)(\rho_j\,\Delta\phi)(\rho_j\sin\phi_n\,\Delta\theta) = \rho_j^2\sin\phi_n\,\Delta\rho\,\Delta\theta\,\Delta\phi.$$

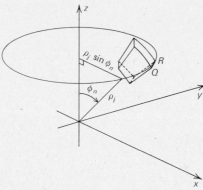

FIGURE 15.67 Q and R are on a
circle of radius $\rho_j\sin\phi_n$.

The point R is in the subregion and, by formulas (7), has rectangular coordinates

$$(\rho_j\sin\phi_n\cos\theta_k,\ \rho_j\sin\phi_n\sin\theta_k,\ \rho_j\cos\phi_n)$$

so the sum

$$\sum_{j=1}^{J}\sum_{k=1}^{K}\sum_{n=1}^{N} f(\rho_j\sin\phi_n\cos\theta_k,\ \rho_j\sin\phi_n\sin\theta_k,\ \rho_j\cos\phi_n)\,\rho_j^2\sin\phi_n\,\Delta\rho\,\Delta\theta\,\Delta\phi$$

is an approximating sum for the integral of $f(x, y, z)$ over V. This is also an approximating sum for the iterated integral on the right of (9), so equation (9) must hold.* Any order of integration may be used because the limits of integration are constant.   □

The six possible orders of integration in Rule 15.9 are indicated in Figures 15.68 through 15.73. The next example illustrates the use of spherical coordinates for a region other than a sphere.

-----

*An alternate derivation of formula (9) based on an exact expression for the volume of the subregions is outlined in Exercise 30.

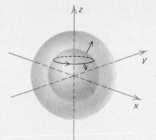

FIGURE 15.68

$$\int_{\rho=0}^{\rho=a} \int_{\phi=0}^{\phi=\pi} \int_{\theta=0}^{\theta=2\pi}$$

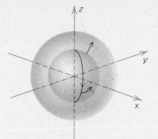

FIGURE 15.69

$$\int_{\rho=0}^{\rho=a} \int_{\theta=0}^{\theta=2\pi} \int_{\phi=0}^{\phi=\pi}$$

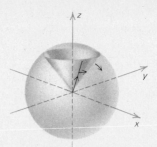

FIGURE 15.70

$$\int_{\phi=0}^{\phi=\pi} \int_{\theta=0}^{\theta=2\pi} \int_{\rho=0}^{\rho=a}$$

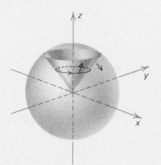

FIGURE 15.71

$$\int_{\phi=0}^{\phi=\pi} \int_{\rho=0}^{\rho=a} \int_{\theta=0}^{\theta=2\pi}$$

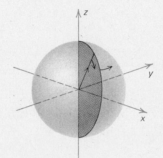

FIGURE 15.72

$$\int_{\theta=0}^{\theta=2\pi} \int_{\rho=0}^{\rho=a} \int_{\phi=0}^{\phi=\pi}$$

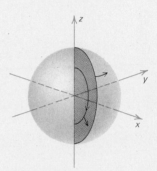

FIGURE 15.73

$$\int_{\theta=0}^{\theta=2\pi} \int_{\phi=0}^{\phi=\pi} \int_{\rho=0}^{\rho=a}$$

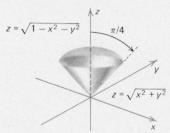

FIGURE 15.74

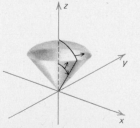

FIGURE 15.75

EXAMPLE 2   Use spherical coordinates to evaluate the integral

$$(10) \quad \int \int \int_V z \sqrt{x^2 + y^2 + z^2}\, dx\, dy\, dz$$

where V is defined by the inequalities $\sqrt{x^2 + y^2} \leq z \leq \sqrt{1 - x^2 - y^2}$.

Solution.   The solid V is bounded on the top by the hemisphere $z = \sqrt{1 - x^2 - y^2}$ and on the bottom by the cone $z = \sqrt{x^2 + y^2}$ (Figure 15.74). The hemisphere is the upper half of the sphere with equation $\rho = 1$ in spherical coordinates and the cone has the equation $\phi = \pi/4$. The solid consists of the points whose spherical coordinates $[\rho, \theta, \phi]$ satisfy the conditions $0 \leq \rho \leq 1, 0 \leq \theta \leq 2\pi, 0 \leq \phi \leq \pi/4$. These conditions determine the limits of integration in the iterated integral (Figure 15.75). We make substitutions (6), (7), and (8) to obtain

$$\int_{\theta=0}^{\theta=2\pi} \int_{\rho=0}^{\rho=1} \int_{\phi=0}^{\phi=\pi/4} (\rho \cos\phi)\rho(\rho^2 \sin\phi)\, d\phi\, d\rho\, d\theta$$

$$= \left(\int_0^{2\pi} d\theta \right) \left(\int_0^1 \rho^4\, d\rho \right) \left(\int_0^{\pi/4} \cos\phi \sin\phi\, d\phi \right)$$

$$= (2\pi)(\tfrac{1}{5})(\tfrac{1}{4}) = \tfrac{1}{10}\pi. \quad \square$$

Spherical coordinates in geography

Spherical coordinates are closely related to the coordinates of latitude and longitude used in geography. We introduce xyz-coordinates with the origin at the center of the earth, with the positive z-axis passing through the North Pole, and with the positive x-axis passing through the prime meridian (Figure 15.76). The meridians are the great circles through the North and South Poles and are given by θ = constant. The parallels are the horizontal circles given by ϕ = constant. Longitudes are given in degrees east or west of the prime meridian, so we can obtain suitable angles θ by switching from degrees to radians and by taking the longitudes west of the prime meridian as negative angles. Latitudes are given in degrees north or south of the equator; we can obtain the angles ϕ by using radians instead of degrees and by measuring the angles from the North Pole.

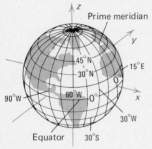

FIGURE 15.76

EXAMPLE 3   Suppose that the scales on the coordinate axes in Figure 15.76 are given in miles. What are the rectangular coordinates of the town of Esquina, Argentina, which has a longitude of 60° W and a latitude of 30° S? (Use 4000 miles as the radius of the earth.)

Solution.   The town has spherical coordinates $[\rho, \theta, \phi]$ with $\rho = 4000$. Because 60° W is $\frac{1}{3}\pi$ radians west of the prime meridian, θ is $-\frac{1}{3}\pi$. Because 30° S is $\frac{1}{6}\pi$ radians south of the equator and the equator is at a polar angle of $\frac{1}{2}\pi$ radians, the angle ϕ is $\frac{1}{2}\pi + \frac{1}{6}\pi = \frac{2}{3}\pi$.

Formulas (7) give the rectangular coordinates

$$x = 4000 \sin\left(\frac{2\pi}{3}\right) \cos\left(-\frac{\pi}{3}\right) = 4000 \left(\frac{\sqrt{3}}{2}\right)\left(\frac{1}{2}\right) = 1000\sqrt{3} \text{ miles}$$

$$y = 4000 \sin\left(\frac{2\pi}{3}\right) \sin\left(-\frac{\pi}{3}\right) = 4000 \left(\frac{\sqrt{3}}{2}\right)\left(-\frac{\sqrt{3}}{2}\right) = -3000 \text{ miles}$$

$$z = 4000 \cos\left(\frac{2\pi}{3}\right) = 4000 \left(-\frac{1}{2}\right) = -2000 \text{ miles.} \quad \square$$

Great circle distances

The shortest path along the surface of the earth between two points P and Q is an arc of the GREAT CIRCLE formed by the intersection of the surface of the earth with the plane through P, Q and the center of the earth. In the next example we show how to use spherical coordinates and the dot product of vectors to find the great circle distance between two points on the earth if we know their latitudes and longitudes.

EXAMPLE 4   What is the great circle distance between Leningrad, Russia (30°E, 60°N) and Wassau, Wisconsin (90°E, 45°N)?

Solution.   We introduce xyz-coordinates as in Figure 15.76 and let A and B denote the unit vectors directed from the origin (the center of the earth) toward the two cities. If ψ is the angle between A and B such that $0 \le \psi \le \pi$, then the great circle path between the two cities is subtended by the angle ψ at the center of the great circle, and because the circle has radius 4000 miles, the great circle distance is 4000 ψ. Thus we can find the distance by finding ψ.

Because Leningrad has longitude 30°E, we can take $\theta = \frac{1}{6}\pi$ in its spherical coordinates, and because its latitude is 60°N, its polar angle is $\phi = \frac{1}{6}\pi$. The components of the unit vector A directed from the center of the earth toward Leningrad are the rectangular coordinates of the point with spherical coordinates $\rho = 1, \theta = \frac{1}{6}\pi, \phi = \frac{1}{6}\pi$ a distance 1 from the origin. Therefore

(11)
$$A = \langle \sin(\tfrac{1}{6}\pi)\cos(\tfrac{1}{6}\pi), \sin(\tfrac{1}{6}\pi)\sin(\tfrac{1}{6}\pi), \cos(\tfrac{1}{6}\pi)\rangle$$
$$= \langle (\tfrac{1}{2})(\tfrac{1}{2}\sqrt{3}), (\tfrac{1}{2})(\tfrac{1}{2}), \tfrac{1}{2}\sqrt{3}\rangle = \langle \tfrac{1}{4}\sqrt{3}, \tfrac{1}{4}, \tfrac{1}{2}\sqrt{3}\rangle.$$

Similarly, because Wassau, Wisconsin has longitude 90°W, we can take $\theta = -\frac{1}{2}\pi$ in its spherical coordinates, and because its latitude is 45°N, it has polar angle $\phi = \frac{1}{4}\pi$. The unit vector B from the origin toward Wassau is

(12)
$$B = \langle \sin(\tfrac{1}{4}\pi)\cos(-\tfrac{1}{2}\pi), \sin(\tfrac{1}{4}\pi)\sin(-\tfrac{1}{2}\pi), \cos(\tfrac{1}{4}\pi)\rangle$$
$$= \langle 0, -\tfrac{1}{2}\sqrt{2}, \tfrac{1}{2}\sqrt{2}\rangle.$$

Since A and B are unit vectors, their dot product $A \cdot B = |A||B|\cos\psi = \cos\psi$ is the cosine of the angles between them. Formulas (11) and (12) give

$$\cos\psi = A \cdot B = \langle \tfrac{1}{4}\sqrt{3}, \tfrac{1}{4}, \tfrac{1}{2}\sqrt{3}\rangle \cdot \langle 0, -\tfrac{1}{2}\sqrt{2}, \tfrac{1}{2}\sqrt{2}\rangle$$
$$= (\tfrac{1}{4}\sqrt{3})(0) + \tfrac{1}{4}(-\tfrac{1}{2}\sqrt{2}) + (\tfrac{1}{2}\sqrt{3})(\tfrac{1}{2}\sqrt{2})$$
$$= -\tfrac{1}{8}\sqrt{2} + \tfrac{1}{4}\sqrt{6}.$$

The exact angle ψ is the arccosine of this amount. To find an approximate value for ψ, we compute $-\frac{1}{8}\sqrt{2} + \frac{1}{4}\sqrt{6} \approx 0.436$ and from the table of cosines use the value $\cos(1.12) \approx 0.4357$ to conclude that $\psi \approx 1.12$ radians and that the great circle distance is

$$4000 \arccos(-\tfrac{1}{8}\sqrt{2} + \tfrac{1}{4}\sqrt{6}) \approx 4000(1.12) = 4480 \text{ miles.} \quad \square$$

Exercises

†Answer provided.
‡Outline of solution provided.

1. Give cylindrical coordinates of the points with rectangular coordinates
    a.‡ $(2, -2, 3)$ **b.**† $(-\sqrt{3}, -1, -5)$ **c.** $(0, -7, 8)$
    d. $(-3, -3, -3)$

2. Give rectangular coordinates of the points with cylindrical coordinates
    a.‡ $[3, 5\pi/6, 4]$ **b.**† $[1, 3\pi/2, -2]$ **c.** $[2, 5\pi/4, 0]$
    d. $[4, \pi, e]$

3. Give rectangular coordinates of the points with spherical coordinates
    a.‡ $[4, 3\pi/4, \pi/3]$ **b.**† $[0, 5, \pi]$ **c.** $[4, \pi, 5\pi/6]$
    d.† $[5, 1, \pi]$ **e.** $[4, 7\pi/4, \pi/2]$ **f.**† $[\sqrt{2}, \pi, 0]$
    g. $[10, 0, \pi/2]$

4. Give spherical coordinates of the points with rectangular coordinates
   a.‡ $(-3, -3, 0)$ **b.**‡ $(\sqrt{2}, \sqrt{2}, -2)$ **c.**† $(-2, -2\sqrt{3}, 0)$
   d.† $(0, 0, -100)$ **e.** $(-5, 0, 5)$ **f.**† $(-\sqrt{3}, 1, -2\sqrt{3})$
   g. $(-1, -1, -\sqrt{2})$ **h.**† $(\sqrt{\tfrac{3}{2}}, \sqrt{\tfrac{3}{2}}, 1)$ **i.** $(0, 0, 4)$
   j. $(0, -5, 0)$

5. Describe the following coordinate surfaces in rectangular coordinates.
   a.† $r = 2$ **b.**† $\theta = \pi/4$ **c.**† $\theta = 2\pi/3$ **d.** $\theta = 3\pi/2$
   e.† $\rho = 2$ **f.**† $\phi = \pi/4$ **g.**† $\phi = 2\pi/3$ **h.** $\phi = \pi/2$
   i. $\phi = \pi$

6. Describe the following lines and surfaces with cylindrical coordinates.
   a.† $z = -\sqrt{x^2 + y^2}$ **b.**† $y = x, x \geq 0$
   c.† $z = -\sqrt{4 - x^2 - y^2}$ **d.** $z = 5$
   e.† $x^2 - 2x + y^2 = 0$ **f.** $x^2 + y^2 = 9$
   g.† $y = 3$ **h.** $x = 0, y = 0, z \leq 0$
   i.† $z = \sqrt{3x^2 + 3y^2}$

7. Describe the lines and surfaces of Exercise 6 with spherical coordinates.

Use cylindrical coordinates to evaluate the integrals in Exercises 8 through 10.

8.‡ $\iiint_V \sqrt{x^2 + y^2}\, dx\, dy\, dz$; V bounded by $z = \sqrt{x^2 + y^2}, z = 0$,
   $x^2 + y^2 = 1$

9.† $\iiint_V z^2\, dx\, dy\, dz$; V the hemisphere bounded by $z = \sqrt{1 - x^2 - y^2}$,
   $z = 0$

10. $\iiint_V z\sqrt{x^2 + y^2}\, dx\, dy\, dz$; V bounded by $z = 4, z = x^2 + y^2$

In Exercises 11 through 13 use cylindrical coordinates to evaluate the volumes of the solids V.

11.‡ V bounded below by $z = x^2 + y^2$ and above by $x^2 + y^2 + z^2 = 2$

12.† V bounded by $z^2 = 1 + x^2 + y^2$ and $x^2 + y^2 = 1$

13. V bounded by $z = e^{x^2 + y^2}, z = 0, x^2 + y^2 = 16$

14.† A solid occupies the region $x^2 + y^2 \leq 25, x \geq 0, 0 \leq z \leq 1$ and its density at (x, y, z) is z^2. Use cylindrical coordinates to compute its weight.

The integrals in Exercises 15 and 16 are given in cylindrical coordinates. Express them as iterated integrals in rectangular coordinates. Do not evaluate them.   •

15.† $\int_{\theta=0}^{\theta=\pi/2} \int_{r=0}^{r=2} \int_{z=-r^2}^{z=4} \dfrac{zr^3}{r\sin\theta + 4}\, dz\, dr\, d\theta$

16. $\int_{\theta=0}^{\theta=2\pi} \int_{r=0}^{r=3} \int_{z=-r}^{z=\sqrt{9-r^2}} e^z\, r^6 \cos^4\theta\, dz\, dr\, d\theta$

Use spherical coordinates to evaluate the integrals in Exercises 17 through 19.

17.‡ $\iiint_V z^2\, dx\, dy\, dz$; V the quarter sphere $x^2 + y^2 + z^2 \leq 1, y \geq 0, z \geq 0$

18.‡ $\iiint_V \sqrt{x^2 + y^2 + z^2}\, dx\, dy\, dz$; V the region above the cone
   $z = -\sqrt{3x^2 + 3y^2}$ and inside the sphere $x^2 + y^2 + z^2 = 4$

19. $\int\int\int_V \sin([x^2 + y^2 + z^2]^{3/2})\, dx\, dy\, dz$; V the spherical shell between the spheres $x^2 + y^2 + z^2 = 1$ and $x^2 + y^2 + z^2 = 9$

20.† Use spherical coordinates to compute the moment of inertia of a sphere of radius R and constant density k about a diameter.

The integrals in Exercises 21 and 22 are given in spherical coordinates. Express them as iterated integrals in rectangular coordinates. Do not evaluate them.

21.† $\int_{\theta=0}^{\theta=\pi} \int_{\phi=3\pi/4}^{\phi=\pi} \int_{\rho=0}^{\rho=1} \rho^5 \cos\theta \sin^2\phi \, d\rho\, d\phi\, d\theta$

22. $\int_{\theta=\pi/2}^{\theta=3\pi/2} \int_{\phi=\pi/2}^{\phi=\pi} \int_{\rho=0}^{\rho=5} \dfrac{\rho^4 \sin^2\phi}{\rho^2 + 1} \, d\rho\, d\phi\, d\theta$

23.† The integral

$$\int_{\theta=0}^{\theta=\pi} \int_{z=0}^{z=5} \int_{r=0}^{r=\sqrt{25-z^2}} r^2 \sin\theta \, dr\, dz\, d\theta$$

is given in cylindrical coordinates. Express it as an iterated integral in spherical coordinates. Do not evaluate it.

24. A solid occupies the region $\sqrt{x^2 + y^2} \le z \le 1$ and has density $z\sqrt{x^2 + y^2 + z^2}$ at (x, y, z). How much does it weigh?

25.† A solid of density xyz at (x, y, z) occupies the one-eighth sphere $x^2 + y^2 + z^2 \le 4, x \ge 0, y \le 0, z \le 0$. Where is its center of gravity?

26. Compute the volume of a sphere of radius R by using   **a.**   cylindrical coordinates and   **b.**   spherical coordinates.

27. A sphere of radius 4 with its center at the origin has density $2x^2 + 2y^2$ at (x, y, z). How much does it weigh?

28.† Evaluate the integral of z over the region bounded by the half cone $z = \sqrt{2x^2 + 2y^2}$ and the half hyperboloid $z = \sqrt{x^2 + y^2 + 1}$.

29. Evaluate the integral of z over the region V: $3 \le z \le \sqrt{25 - x^2 - y^2}$.

30.† Consider $0 \le \rho_{j-1} < \rho_j, 0 \le \theta_{k-1} < \theta_k \le 2\pi, 0 < \phi_{n-1} < \phi_n < \pi$ and set $\Delta\rho = \rho_j - \rho_{j-1}, \Delta\theta = \theta_k - \theta_{k-1}, \Delta\phi = \phi_n - \phi_{n-1}$.   **a.**   Show that the region in the xz-plane bounded by $\rho = \rho_{j-1}, \rho = \rho_j, \phi = \phi_{n-1}$, and $\phi = \phi_n$ has area $\frac{1}{2}(\rho_j + \rho_{j-1})\Delta\rho\Delta\phi$ (Figure 15.77).   **b.**   Show that the center of gravity of the region of part (a) is at a distance $\rho_j^* \sin(\phi_n^*)$ from the z-axis for some ρ_j^* and ϕ_n^* with $\rho_{j-1} < \rho_j^* < \rho_j$ and $\phi_{n-1} \le \phi_n^* \le \phi_n$.   **c.**   Use a theorem of Pappus (Section 6.10) to show that the subregion shown in Figure 15.66 has volume $\frac{1}{2}\rho_j^* \sin(\phi_n^*)(\rho_j + \rho_{j-1})\Delta\rho\Delta\theta\Delta\phi$.   **d.**   Use the result of part (c) to derive Rule 15.9.

In Exercises 31 through 33 use the coordinate axes of Example 3 and the longitudes and latitudes in Table 15.1 to find the rectangular coordinates of the cities. Use 4000 miles as the radius of the earth.

31.‡ Leningrad, Russia   **32.†** Wassau, Wisconsin

33. Tambu, Indonesia

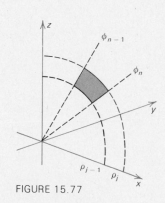

FIGURE 15.77

TABLE 15.1

Bennett, British Columbia	135°W, 60°N
Calais, France	0°E, 45°N
Chiangchunmiao, China	90°E, 45°N
Davyhurst, Australia	120°E, 30°S
Esquina, Argentina	60°W, 30°S
Leningrad, Russia	30°E, 60°N
Tambu, Indonesia	120°E, 0°S
Timimoun, Algeria	0°E, 30°N
Wassau, Wisconsin	90°W, 45°N
Yartsevo, Russia	90°E, 60°N

In Exercises 34 through 37 use the longitudes and latitudes in Table 15.1, vector dot products, and the table of cosines to give the approximate great circle distances between the cities.

34.‡ Timimoun, Algeria, and Bennett, British Columbia

35.† Chiangchunmiao, China, and Esquina, Argentina

36.† Davyhurst, Australia, and Calais, France

37. Yartsevo, Russia, and Tambu, Indonesia

Additional drill exercises

Use cylindrical or spherical coordinates to evaluate the integrals in Exercises 38 through 44.

38.† $\int \int \int_V z\sqrt{x^2 + y^2}\, dx\, dy\, dz$ with $V: (x - 1)^2 + y^2 \le 1, 0 \le z \le 4$

39.† $\int \int \int_V z\sqrt{x^2 + y^2 + z^2}\, dx\, dy\, dz$ with $V: x^2 + y^2 \le 16, 0 \le z \le 3$

40.† $\int_{x=-3}^{x=0} \int_{y=0}^{y=\sqrt{9-x^2}} \int_{z=0}^{z=\sqrt{x^2+y^2}} xy\, dz\, dy\, dx$

41.† $\int_{x=1}^{x=2} \int_{y=0}^{y=x} \int_{z=0}^{z=x^2+y^2} \frac{z}{(x^2 + y^2)^2}\, dz\, dy\, dx$

42.† $\int \int \int_V \cos[(x^2 + y^2 + z^2)^{3/2}]\, dx\, dy\, dz$ with $V: 1 \le x^2 + y^2 + z^2 \le 36$

43.† $\int \int \int_V \sin z\, dx\, dy\, dz$ with $V: 0 \le z \le 4x^2 + 4y^2,$
$x \le y \le \sqrt{1 - x^2}, 0 \le x \le 2^{-1/2}$

44.† $\int_{x=-2}^{x=2} \int_{y=-\sqrt{4-x^2}}^{y=\sqrt{4-x^2}} \int_{z=0}^{z=\sqrt{4-x^2-y^2}} z(x^2 + y^2 + z^2)^{3/2}\, dz\, dy\, dx$

MISCELLANEOUS EXERCISES

†*Answer provided.*
‡*Outline of solution provided.*

Reverse the order of integration in Exercises 1 through 4.

1.† $\int_{x=-1}^{x=1} \int_{y=\arctan x}^{y=\pi/4} f(x, y)\, dy\, dx$

2. $\int_{x=0}^{x=5} \int_{y=1}^{y=e^x} f(x, y)\, dy\, dx$

3. $\int_{y=-1}^{y=1} \int_{x=\arcsin y}^{x=\pi/2} f(x, y)\, dx\, dy$

4.† $\int_{x=0}^{x=4} \int_{y=0}^{y=4x-x^2} f(x, y)\, dy\, dx$

5.† Evaluate the integral of $\sqrt{x^2 + y^2}$ over the square $0 \le x \le 1, 0 \le y \le 1$ by using polar coordinates.

6.† Compute the volume of the solid lying under the plane $z = y + 4$ and over the region R in the xy-plane with R bounded by the limaçon $r = 3 + \cos \theta$.

7. A plate occupies the region inside the circle $r = 2$ and outside the cardioid $r = 1 + \sin \theta$ and its density at (x, y) is $x^2 + y^2$. How much does it weigh?

8.† Use polar coordinates to evaluate the integral of $3\sqrt{x^2 + y^2}$ over the region bounded by $y = x^2$ and $y = x$.

Evaluate the integrals in Exercises 9 through 12.

9.† $\int\int\int_V xz \, dx \, dy \, dz$; V bounded by $z = -\sqrt{x^2 + y^2}, x = 0, x = 1,$
$y = -1, y = 1,$ and $z = 4$

10. $\int\int\int_V xy \, dx \, dy \, dz$; V bounded by $z = xy, z = 2, x = -4, x = -2,$
$y = 0,$ and $y = 3$

11.‡ $\int\int\int_V z \, dx \, dy \, dz$; V bounded by $z = xy, z = 2, x = 1, y = 1$

12. $\int\int\int_V x \, dx \, dy \, dz$; V bounded by $z = x - y^2, z = 0,$ and $x = 1$

13.† Compute the volume of the solid bounded by $z = x^2 + y^2$ and $z = 2x$.

14. Evaluate the integral of $x^2 y^2 \cos z$ over the solid V bounded by $z = y^3$,
$z = -1, x = 0, x = 1, y = 1$.

15.† Use spherical coordinates to evaluate the integral of e^{-z} over the sphere
$x^2 + y^2 + z^2 \leq 36$. (Do the ϕ-integration first.)

16. Compute the volume of the solid bounded by $z = x^2 + y^2, z = 0,$
$x^2 + (y - 1)^2 = 1$.

17. Evaluate the integral of $(x^2 + y^2)^{-1/2}$ over the solid bounded by
$z = \sqrt{x^2 + y^2}, z = 1,$ and $x^2 + y^2 = 4$.

18.† Compute the volume of the portion $k \leq z \leq \sqrt{R^2 - x^2 - y^2}$
$(0 < k < R)$ of a sphere.

19. Use Rule 15.5 to show that if $f(x)$ has a continuous derivative and
$f(x) > L$ for $a \leq x \leq b$, then the area of the surface generated by rotating
that portion of the graph $y = f(x)$ about the line $y = L$ in the xy-plane is

$$2\pi \int_a^b [f(x) - L] \sqrt{[f'(x)]^2 + 1} \, dx.$$

20. By Newton's law, the force of gravitational attraction between points of
mass m and M a distance R apart is $GmMR^{-2}$ with G a universal
constant.   **a.**   Show that the vertical component of the gravitational
attraction between a point of mass m at $(0, 0, b)$ $(b > 0)$ and a sphere
$x^2 + y^2 + z^2 \leq a^2$ $(0 < a < b)$ of constant density μ is the integral over
the sphere of $Gm\mu(\cos\alpha) R^{-2}$ with $R^2 = \rho^2 + b^2 - 2\rho b \cos\phi$ and $\cos\alpha$
$= (b - \rho \cos\phi)/R$.   **b.†**   Show that with the notation of part (a)

$$\int_0^\pi \frac{\cos\alpha}{R^2} \sin\phi \, d\phi = \frac{2}{b^2}$$

by introducing R as the variable of integration.   **c.**   Use the results of
parts (a) and (b) to show that the gravitational attraction between the
sphere and the point mass is the same as that between the point mass and
a point at the center of the sphere with mass equal to that of the sphere.

21. Use the techniques of Exercise 20 to show that a homogeneous spherical
shell $(c^2 < x^2 + y^2 + z^2 < a^2)$ exerts no gravitational force inside it (for
$x^2 + y^2 + z^2 < c^2$).

22. Show that the results of Exercises 20 and 21 are valid when the density of
the sphere is a function of the distance ρ to its center.

23.[†] Find the volume of a solid whose projection on the xy-plane in a circle of diameter 1, whose projection on the xz-plane is a square of side 1, and whose projection on the yz-plane is an isosceles triangle.

24. Show that the center of gravity of a triangle is two-thirds of the way along a median from one vertex to the midpoint of the opposite side.

Note. Mathematicians of the seventeenth century felt they were breaking with tradition when they used indivisibles to compute areas and volumes and abandoned the style of proof based on approximation procedures that had come down from ancient Greece (see the Historical Notes in Section 5.7). It was not known until 1908 that Archimedes himself had used indivisibles. In that year a treatise (now known as *The Method*) was found in which Archimedes described how he used indivisibles and properties of levers to derive a number of results to which he later supplied formal proofs (see S. H. Gould, [1] "The Method of Archimedes" and T. L. Heath, [2] *A History of Greek Mathematics*). In the next exercise we outline the technique used in *The Method* to determine the area of a region bounded by a parabola and a straight line. □

25. Figure 15.78 shows the region bounded by a parabola and a line AB and the triangle ABC whose side AC is parallel to the axis of the parabola and whose side BC is tangent to the parabola. Imagine that the line BJ in Figure 15.79 through the midpoint F of AC is a horizontal lever or balance beam with fulcrum at F and with arms BF and FJ of equal length. Consider the indivisible DE of the curved region, the corresponding indivisible DG of the triangle, and the indivisible $D'E'$ of length equal to that of DE and with its midpoint at J on the lever. Under the assumption that the weights of the indivisibles are proportional to their lengths, the indivisibles $D'E'$ and DG would balance on the lever.   **a.** Show that this is the case for the region bounded by the parabola $y = x^2$ and the line $y = 1$.

We suppose that the weights of the curved region and triangle are proportional to their areas. Because the corresponding indivisibles would balance, we conclude that the entire curved region, if suspended from the end of lever at J, would balance the triangle lying along the other half of the lever.   **b.** Use the result of Exercise 24 to show that the area of the triangle is three times that of the curved region.

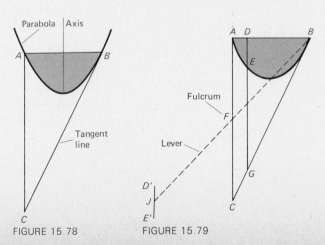

FIGURE 15.78   FIGURE 15.79

CHAPTER 16

VECTOR ANALYSIS

SUMMARY: Vector analysis is the branch of calculus dealing with vector fields and line and surface integrals and is a basic tool in the study of phenomena such as gravitational attraction, electricity and magnetism, and fluid flow. We discuss line integrals in Section 16.1 and vector fields and criteria for recognizing gradient vector fields and path-independent line integrals in Section 16.2. Green's Theorem, the Divergence Theorem, and Stokes' Theorem in the plane are derived in Section 16.3. The basic results concerning surface integrals and the three-dimensional versions of these theorems are described in the Miscellaneous Exercises.

16-1 LINE INTEGRALS

When we want to compute the rate of flow of a fluid across an oriented curve in an xy-plane or the work done on an object as it traverses an oriented curve in an xy-plane or in xyz-space, we use LINE INTEGRALS. There are three types of line integrals, denoted $\int_C f\,dx$, $\int_C f\,dy$, and $\int_C f\,ds$, for curves C in the xy-plane and functions $f(x, y)$ defined on them, and there are four types, denoted $\int_C f\,dx$, $\int_C f\,dy$, $\int_C f\,dz$, and $\int_C f\,ds$ for curves C in space and functions $f(x, y, z)$ defined on them.

Each of these line integrals is defined as a limit of approximating sums. We pick points $P_0, P_1, P_2, \ldots, P_N$ on the curve with P_0 the beginning of the curve, P_N the end, and the other points numbered in order following the orientation of the curve. These points partition the curve into N subcurves (Figure 16.1). We denote the coordinates of P_j by (x_j, y_j) in the case of two variables and by (x_j, y_j, z_j) in the case of three variables. We let Q_j denote an arbitrary point on the jth subcurve (the portion of C lying between P_{j-1} and P_j) (Figure 16.2), and we write $f(Q_j)$ for the value of $f(x, y)$ or $f(x, y, z)$ at Q_j. Then we have the following definition.

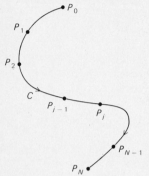

FIGURE 16.1   A partition of the curve C

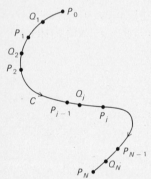

FIGURE 16.2   A point Q_j in the jth subcurve

Definition 16.1   For a curve C in the xy-plane and a function $f(x, y)$ of two variables or for a curve C in xyz-space and a function $f(x, y, z)$ of three variables

(1) $\quad \displaystyle\int_C f \, dx = \lim \sum_{j=1}^{N} f(Q_j)(x_j - x_{j-1})$ ◀

(2) $\quad \displaystyle\int_C f \, dy = \lim \sum_{j=1}^{N} f(Q_j)(y_j - y_{j-1})$ ◀

(3) $\quad \displaystyle\int_C f \, ds = \lim \sum_{j=1}^{N} f(Q_j) \, \overline{P_j P_{j-1}}.$ ◀

For a curve in space and a function of three variables, we have also

(4) $\quad \displaystyle\int_C f \, dz = \lim \sum_{j=1}^{N} f(Q_j)(z_j - z_{j-1}).$ ◀

The limits are taken as the number of points in the partition $P_1, P_2, \ldots, P_N$ tends to infinity and the distances between the points tend to zero.

We will see in Theorem 16.1 that the limits exist and the integrals are defined if the curve C is piecewise smooth and the function f is piecewise continuous. Notice that the approximating sum in (1) is formed from the values of f at the point on the jth subcurve multiplied by the change $x_j - x_{j-1}$ in x from the beginning to the end of the subcurve, whereas the approximating sums (2) and (4) involve the changes in y and z and the approximating sum (3) involves the length of the line segment between the endpoints of the jth subcurve. We refer to integrals (1) through (4) as the integrals with respect to dx, dy, ds, and dz. Integral (3) is also called the integral with respect to arclength.

Interpretations of line integrals

In the case of two variables we can consider the line integral $\int_C f \, ds$ with respect to arclength as a difference of areas, just as a definite integral $\int_a^b f(x) \, dx$ of a function of one variable can be interpreted as a difference of areas. If the function $f(x, y)$ is positive on the curve C, then the integral is equal to the area of the vertical curtain-like surface that extends from the curve in the xy-plane up to the graph of f (Figure 16.3). To see why this is the case, we observe that an approximating sum on the right of the definition (3) of the integral is equal to the area of an approximation of the "curtain" surface by a collection of vertical rectangles (Figure 16.4). For a general function $f(x, y)$, the integral $\int_C f \, ds$ is equal to the area of the "curtain" surface above the curve and below the graph of f where $f(x, y)$ is positive, minus the area of the "curtain" surface below the curve and above the graph of f where $f(x, y)$ is negative.

If the function f has only positive values on the curve C, we can also think of the integral $\int_C f \, ds$ as the weight the curve would have if its

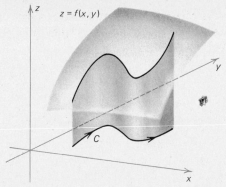

FIGURE 16.3 "Curtain" surface
of area $\int_C f(x, y)\, ds$

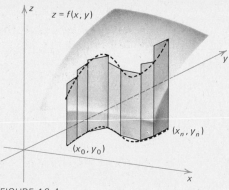

FIGURE 16.4

density were given by the function f. This interpretation applies to space curves as well as to curves in the xy-plane. In particular, if the function f is the constant function 1, then the integral is equal to the length of the curve:

(5)   Length of $C = \displaystyle\int_C ds.$ ◄

If f has negative or both positive and negative values on C, we can think of it as an electrical charge density. The integral $\int_C f\, ds$ is then the net charge (positive, negative, or zero) on the entire curve.

Another interpretation of integrals with respect to ds that can be applied to space as well as to plane curves involves the *average value* of the functions on the curves.

Rule 16.1 **For a function $f(x, y)$ on a curve C in an xy-plane or a function $f(x, y, z)$ on a curve C in xyz-space**

(6)   **The average value of f on C = $\dfrac{1}{\textbf{The length of } C}\displaystyle\int_C f\, ds.$**   ◄

Because of formula (6) we can view the integral $\int_C f\, ds$ as the average value of f on the curve multiplied by the length of the curve.

If f is the constant function 1, then the approximating sum in (1) for the integral $\int_C f\, dx = \int_C dx$ is the sum of the changes $x_j - x_{j-1}$ in x as we go from P_{j-1} to P_j on the curve. This sum is equal to the change in x from P_0 to P_N:

(7) $\displaystyle\sum_{j=1}^{N} (x_j - x_{j-1}) = x_N - x_0.$

The approximating sum (7) is the same for all partitions, so the integral has the same value and we have

(8) $\displaystyle\int_C dx = \begin{bmatrix} \text{The value of } x \\ \text{at the end of } C \end{bmatrix} - \begin{bmatrix} \text{The value of } x \text{ at} \\ \text{the beginning of } C \end{bmatrix}.$ ◄

Similarly,

(9) $\displaystyle\int_C dy = \left[\begin{array}{l}\text{The value of } y \text{ at}\\ \text{the end of } C\end{array}\right] - \left[\begin{array}{l}\text{The value of } y \text{ at}\\ \text{the beginning of } C\end{array}\right]$ ◀

and, for a curve in space

(10) $\displaystyle\int_C dz = \left[\begin{array}{l}\text{The value of } z \text{ at}\\ \text{the end of } C\end{array}\right] - \left[\begin{array}{l}\text{The value of } z \text{ at}\\ \text{the beginning of } C\end{array}\right].$ ◀

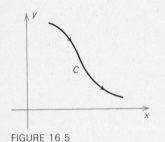

FIGURE 16.5

Later in this section we will see how any integral with respect to dx, dy, or dz can be expressed as an integral with respect to ds, and this will provide us with a means of interpreting such integrals.

We can also study the integrals by examining their approximating sums, as illustrated in the next example.

EXAMPLE 1   A function $f(x, y)$ has negative values on the curve C shown in Figure 16.5. One of the three integrals (a) $\int_C f\, dx$, (b) $\int_C f\, dy$, and (c) $\int_C f\, ds$ is positive, and the other two are negative. Which one is positive?

Solution.   The changes $x_j - x_{j-1}$ in x as we go from one point to the next in a partition of the curve are positive because the curve is oriented toward the right. The changes $y_j - y_{j-1}$ in y are negative because y decreases along the curve. The distances $\overline{P_j P_{j-1}}$ between successive points are positive. Since the function f has negative values, the approximating sums for the integrals with respect to dx and ds are negative and those for the integral with respect to dy are positive. Hence, it is the integral with respect to dy that is positive.   □

The negative
of a curve

Definition 16.2   The negative $-C$ of a curve C is obtained by reversing the orientation of the original curve (Figure 16.6).

An examination of Definition 16.1 shows that reversing the orientation of a curve multiplies line integrals with respect to dx, dy, and dz by -1 and leaves integrals with respect to ds unchanged:

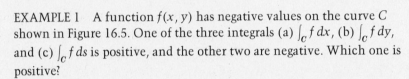

(11) $\displaystyle\int_{-C} f\, dx = -\int_C f\, dx, \quad \int_{-C} f\, dy = -\int_C f\, dy,$

$\displaystyle\int_{-C} f\, dz = -\int_C f\, dz$ ◀

(12) $\displaystyle\int_{-C} f\, ds = \int_C f\, ds.$ ◀

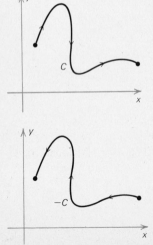

FIGURE 16.6   A curve C and its negative $-C$

Orientation by decreasing parameters

Up to now we have always given curves the orientations obtained by having their parameters increase. To make it possible to have orientations where the parameters decrease, we can describe a curve in an xy-plane by writing $C: x = x(t), y = y(t), a \underset{t}{\rightarrow} b$. This curve is oriented by decreasing t if a is greater than b. We will use similar notation for curves in space.

With this notation we can obtain parametric equations of the negative of a curve from parametric equations of the curve by switching the limits of its parameter: if C has parametric equations $x = x(t)$, $y = y(t), a \underset{t}{\rightarrow} b$, then $-C$ has the parametric equations $x = x(t)$, $y = y(t), b \underset{t}{\rightarrow} a$.

Occasionally we will use one of the variables x or y as the parameter of a curve, as illustrated in the next example.

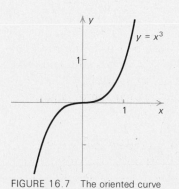

EXAMPLE 2   Use the variable x as the parameter to describe the curve $y = x^3$ for $-1 \leq x \leq 1$, oriented from right to left.

Solution.   The oriented curve is shown in Figure 16.7 and may be described by writing $y = x^3, 1 \underset{x}{\rightarrow} -1$. $\square$

FIGURE 16.7   The oriented curve $y = x^3, 1 \underset{x}{\rightarrow} -1$

The sum of two curves

Definition 16.3   If the end point of a curve C_1 is the beginning of a curve C_2, then the sum $C_1 + C_2$ of the curves is the curve obtained by following C_1 by C_2 (Figure 16.8).

Definitions 16.1 and 16.3 show that any line integral over $C_1 + C_2$ is equal to the sum of the corresponding line integrals over C_1 and C_2.

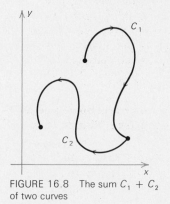

FIGURE 16.8   The sum $C_1 + C_2$ of two curves

Evaluating line integrals

The next result enables us to use parametric equations for curves to evaluate line integrals.

Theorem 16.1   If $f(x, y)$ is piecewise continuous on the piecewise smooth oriented curve $C: x = x(t), y = y(t), a \underset{t}{\rightarrow} b$, then the integrals of f with respect to dx, dy, and ds exist and

$$(13) \quad \int_C f \, dx = \int_a^b f(x(t), y(t)) \frac{dx}{dt}(t) \, dt \qquad \blacktriangleleft$$

$$(14) \quad \int_C f \, dy = \int_a^b f(x(t), y(t)) \frac{dy}{dt}(t) \, dt \qquad \blacktriangleleft$$

$$(15) \quad \int_C f\,ds = \begin{cases} \displaystyle\int_a^b f(x(t), y(t)) \sqrt{\left[\frac{dx}{dt}(t)\right]^2 + \left[\frac{dy}{dt}(t)\right]^2}\; dt \text{ if } a < b \\[2em] \displaystyle\int_b^a f(x(t), y(t)) \sqrt{\left[\frac{dx}{dt}(t)\right]^2 + \left[\frac{dy}{dt}(t)\right]^2}\; dt \text{ if } b < a. \end{cases}$$ ◄

In the case of a piecewise continuous function $f(x, y, z)$ on a piecewise smooth curve C: $x = x(t)$, $y = y(t)$, $z = z(t)$, $a \underset{t}{\to} b$ in xyz-space, the integrals (13), (14), and (15) exist with $f(x(t), y(t))$ replaced by $f(x(t), y(t), z(t))$ and

$$\sqrt{\left[\frac{dx}{dt}(t)\right]^2 + \left[\frac{dy}{dt}(t)\right]^2} \text{ replaced by } \sqrt{\left[\frac{dx}{dt}(t)\right]^2 + \left[\frac{dy}{dt}(t)\right]^2 + \left[\frac{dz}{dt}(t)\right]^2}$$

in the formulas. The integral with respect to dz also exists and is given by

$$(16) \quad \int_C f\,dz = \int_a^b f(x(t), y(t), z(t)) \frac{dz}{dt}(t)\,dt.$$ ◄

Proof of (13) for $a < b$. We will deal with the case of a curve C with parametric equations $x = x(t)$, $y = y(t)$, $a \le t \le b$, where $x(t)$ and $y(t)$ have continuous first derivatives for $a \le t \le b$ and where f is continuous on C. In the general case of piecewise smooth C and piecewise continuous f, we can express C as a sum of such curves.*

We consider a partition

$$a = t_0 < t_1 < t_2 < t_3 < \cdots < t_{N-1} < t_N = b$$

of the interval $a \le t \le b$. The points $(x_j, y_j) = (x(t_j), y(t_j))$ $(j = 0, 1, 2, 3, \ldots, N)$ form a partition of the curve and the sum

$$(17) \quad \sum_{j=1}^N f(x_j, y_j)(x_j - x_{j-1}) = \sum_{j=1}^N f(x(t_j), y(t_j))[x(t_j) - x(t_{j-1})]$$

is a corresponding approximating sum for the line integral on the left of equation (13). According to the Mean Value Theorem for derivatives, there is, for each j, a point c_j with $t_{j-1} < c_j < t_j$ such that

$$(18) \quad x(t_j) - x(t_{j-1}) = \frac{dx}{dt}(c_j)(t_j - t_{j-1}).$$

When we substitute (18) in the right side of (17), we obtain an approximating sum which tends to the definite integral on the right of (13). Therefore, both integrals in (13) exist and are equal.   Q.E.D.

We leave the derivation of (14) and (15) for $a < b$ as Exercises 18 and 19. Equations (13), (14), and (15) for $b < a$ are consequences of formulas (11) and (12). Analogous arguments yield the results for line integrals in space.

We will assume that the curves are piecewise smooth and the functions piecewise continuous in all definitions, rules, and theorems that deal with line integrals.

---

*After changing the values of f at the endpoints of the subcurves if necessary.

The results of Theorem 16.1 may be summarized by the symbolic equations

$$(19) \quad dx = \frac{dx}{dt}\, dt, \quad dy = \frac{dy}{dt}\, dt, \quad dz = \frac{dz}{dt}\, dt \qquad \blacktriangleleft$$

for either plane or space curves,

$$(20) \quad ds = \pm\sqrt{\left(\frac{dx}{dt}\right)^2 + \left(\frac{dy}{dt}\right)^2}\, dt \qquad \blacktriangleleft$$

for plane curves, and

$$(21) \quad ds = \pm\sqrt{\left(\frac{dx}{dt}\right)^2 + \left(\frac{dy}{dt}\right)^2 + \left(\frac{dz}{dt}\right)^2}\, dt \qquad \blacktriangleleft$$

for space curves. The plus signs are used in (20) and (21) if the curve is oriented in the direction of increasing t and the minus signs if the orientation is in the direction of decreasing t.

EXAMPLE 3   Evaluate the line integral $\int_C x^2 y\, dx$, where C is the curve $x = t^3, y = t^2 - 2t, 0 \underset{t}{\to} 1$.

Solution.   We have $dx = \dfrac{dx}{dt}\, dt = \left(\dfrac{d}{dt}\, t^3\right) dt = 3t^2\, dt$. Therefore

$$\int_C x^2 y\, dx = \int_0^1 (t^3)^2 (t^2 - 2t)(3t^2\, dt)$$

$$= \int_0^1 (3t^{10} - 6t^9)\, dt = -\frac{18}{55}. \quad \square$$

Expressing dx, dy, and dz in terms of ds

We can use Theorem 16.1 to show how to express integrals with respect to dx, dy, and dz as integrals with respect to arclength:

Rule 16.2   Line integrals with respect to dx and dy over curves in the xy-plane may be expressed as integrals with respect to ds by using the symbolic equations

$$(22) \quad dx = u_1\, ds, \quad dy = u_2\, ds \qquad \blacktriangleleft$$

where $T = \langle u_1, u_2 \rangle$ is the unit tangent vector to the curve that points in the direction of the curve's orientation. For curves in space, we have

$$(23) \quad dx = u_1\, ds, \quad dy = u_2\, ds, \quad dz = u_3\, dz \qquad \blacktriangleleft$$

where $T = \langle u_1, u_2, u_3 \rangle$ is the unit tangent vector to the curve that points in the direction of the curve's orientation.

Derivation of Rule 16.2.   Consider a curve $C: x = x(t), y = y(t)$, $a \underset{t}{\to} b$ in the plane with $a < b$. We suppose that the parametric equations have been chosen so that the tangent vector $\left\langle \dfrac{dx}{dt}(t), \dfrac{dy}{dt}(t) \right\rangle$ is not zero and hence

$$(24) \quad \mathbf{T} = \langle u_1, u_2 \rangle = \frac{\left\langle \dfrac{dx}{dt}, \dfrac{dy}{dt} \right\rangle}{\sqrt{\left(\dfrac{dx}{dt}\right)^2 + \left(\dfrac{dy}{dt}\right)^2}}.$$

We obtain $dx = u_1\, ds$ because $dx = \dfrac{dx}{dt}\, dt$, $ds = \sqrt{\left(\dfrac{dx}{dt}\right)^2 + \left(\dfrac{dy}{dt}\right)^2}\, dt$,

and by (24), $u_1 = \dfrac{dx}{dt} \Big/ \sqrt{\left(\dfrac{dx}{dt}\right)^2 + \left(\dfrac{dy}{dt}\right)^2}$. The other statements in Rule 16.2 follow similarly. $\square$

We can use Rule 16.2 to interpret integrals with respect to dx, dy, and dz. According to that rule, the integral $\int_C f\, dx$ of f with respect to dx is equal to $\int_C f u_1\, ds$, which is the average value of $f u_1$ multiplied by the length of C. The quantity, $f u_1$, in turn, is obtained by multiplying the value of f by $u_1 = \mathbf{T} \cdot \mathbf{i}$, which is the cosine of the angle between $\mathbf{T}$ and $\mathbf{i}$ and is a number between -1 and 1: u_1 is 1 where the tangent line to the curve is parallel to the x-axis and the curve is oriented in the positive x-direction; u_1 is 0 where the tangent line is perpendicular to the x-axis; and u_1 is -1 where the tangent line is parallel to the x-axis and the curve is oriented in the negative x-direction. Similarly, the integrals of f with respect to dy and dz can be viewed as the length of the curve multiplied by the average values of $f u_2$ and $f u_3$, where u_2 and u_3 are the cosines of the angles between $\mathbf{T}$ and $\mathbf{j}$ and $\mathbf{k}$, respectively.

Line integrals of linear combinations

It follows from Definition 16.1 that a line integral of a linear combination $Af + Bg$ of functions is equal to the same linear combination of the line integrals of the functions. In the case of integrals with respect to dx, for example, we have

$$(25) \quad \int_C (Af + Bg)\, dx = A \int_C f\, dx + B \int_C g\, dx \qquad \blacktriangleleft$$

and there are analogous identities for integrals with respect to dy, dz, and ds.

Exercises

Evaluate the line integrals in Exercises 1 through 10.

1.‡ $\int_C (x + y)\, dx$ for $C: x = t^3, y = t^2, 0 \underset{t}{\to} 3$

2.† $\int_C e^{x+y-z}\, dy$ for $C: x = 2 + t, y = 3 - 2t, z = 4t, 0 \underset{t}{\to} 2$

3. $\int_C xy\, dy$ for $C: x = t^3 + t, y = -t^5, 0 \underset{t}{\to} 1$

4.‡ **a.** $\int_C x\, dy$ and   **b.** $\int_C xy\, ds$ with C the line segment from $(1, 1)$ to $(4, 5)$

5.† **a.** $\int_C e^{x+2y}\, ds$ and   **b.** $\int_C \sin x\, dy$ with C the line segment from $(0, 0)$ to $(7, -6)$

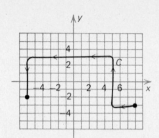

FIGURE 16.9

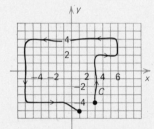

FIGURE 16.10

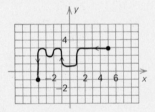

FIGURE 16.11

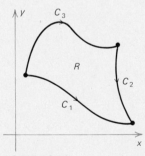

FIGURE 16.12

6. **a.** $\int_C \sqrt{x+y}\, dz$ and   **b.** $\int_C \sqrt{x+z}\, ds$ with C the line segment from $(0,0,0)$ to $(4,5,6)$

7.‡ The integrals of xy^2 **a.** with respect to dx,   **b.** with respect to dy, and   **c.** with respect to ds for $C: x = t^3, y = t^2, 1 \underset{t}{\to} -1$

8. The integrals of xy/z **a.** with respect to dz and   **b.** with respect to ds for $C: x = t, y = t^2, z = \tfrac{2}{3}t^3, 2 \underset{t}{\to} 1$

9.‡ The integrals of x **a.** with respect to dx,   **b.** with respect to dy,   **c.** with respect to ds for C the semicircle $y = \sqrt{1-x^2}$ oriented from left to right

10. The integrals of y **a.** with respect to dx,   **b.** with respect to dy, and   **c.** with respect to ds for C the portion of the cubic $y = x^3$ for $-1 \le x \le 1$, oriented from right to left

Give definite integrals that equal the line integrals in Exercises 11 and 12. Do not evaluate them.

11.† **a.** $\int_C \sin(xy)\, dz$ and   **b.** $\int_C e^z\, ds$ for $C: x = \sin t, y = \sin(2t), z = \sin(3t), \pi \underset{t}{\to} 0$

12. **a.** $\int_C \ln(x^2 + y^2 + z^2)\, ds$ and   **b.** $\int_C e^{xyz}\, dx$ for $C: x = t\cos t, y = t^2, z = t, 10 \underset{t}{\to} 1$

13.‡ A function f satisfies $7.9 \le f(x,y) \le 8.0$ for all (x,y) on the curve C in Figure 16.9. Which of the numbers $0, \pm 10, \pm 50, \pm 100, \pm 200$ are closest to the integrals   **a.** $\int_C f\, dx$,   **b.** $\int_C f\, dy$,   **c.** $\int_C f\, ds$?

14.† A function g satisfies $-11.01 \le g(x,y) \le -10.97$ for all (x,y) on the curve C in Figure 16.10. Which of the numbers $0, \pm 10, \pm 20, \pm 80, \pm 250, \pm 400$ are closest to   **a.** $\int_C g\, dx$,   **b.** $\int_C g\, dy$,   **c.** $\int_C g\, ds$?

15. A function h satisfies $-4.03 \le h(x,y) \le -3.97$ for all (x,y) on the curve C in Figure 16.11. Which of the numbers $0, \pm 15, \pm 30, \pm 80, \pm 125$ are closest to   **a.** $\int_C h\, dx$,   **b.** $\int_C h\, dy$,   **c.** $\int_C h\, ds$?

16.‡ The curves $C_1, C_2,$ and C_3 in Figure 16.12 form the boundary of a region R. C is the entire boundary oriented counterclockwise.   **a.** Express C in terms of C_1, C_2, C_3.   **b.** If $\int_{C_1} f\, dx = 5, \int_{C_2} f\, dx = -10$, and $\int_C f\, dx = 12$, then what is $\int_{C_3} f\, dx$?   **c.** If $\int_{C_1} f\, ds = -2, \int_{C_2} f\, ds = 6$, and $\int_C f\, ds = 14$, then what is $\int_{C_3} f\, ds$?

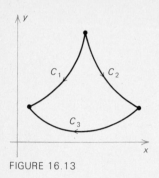

FIGURE 16.13

17. The curves C_1, C_2, and C_3 in Figure 16.13 form the boundary of a region R. C is the entire boundary oriented clockwise.   **a.**   Express C in terms of C_1, C_2, and C_3.   **b.**   If $\int_{C_1} f\,dy = -1$, $\int_{C_2} f\,dy = 4$, and $\int_C f\,dy = 3$, then what is $\int_{C_3} f\,dy$?   **c.**   If $\int_{C_1} f\,ds = 10$, $\int_{C_2} f\,ds = -4$, and $\int_C f\,ds = 0$, then what is $\int_{C_3} f\,ds$?

18. Derive (14) for curves C in the xy-plane with $a < b$.

19.[†] Derive (15) for curves C in the xy-plane with $a < b$.

20.[†] Compute the average value of xy^2 on the line segment between $(1,0)$ and $(-3,-6)$.

21. Compute the average value of x^2 on the circle $x^2 + y^2 = 1$.

22.[†] Compute the average value of $(1 + \frac{2}{3}y)^2$ on the curve $y = \frac{3}{2}x^{2/3}$, $1 \le x \le 8$.

Additional drill exercises

Evaluate the line integrals in Exercises 23 through 30.

23.[†] $\int_C (xy\,dx - x^2\,dy)$ with $C: x = t^3, y = t^4 - 2t^2, -1 \underset{t}{\to} 1$

24.[†] $\int_C xe^{xy}\,dy$ with $C: y = x^2, 0 \underset{x}{\to} -1$

25.[†] $\int_C (\sin x\,dy + \cos y\,dx)$ with C the line segment from $(2,4)$ to $(7,-5)$

26.[†] $\int_C (x\,dx + y^2\,dy + z^3\,dz)$ with $C: x = \sin t, y = e^{3t}, z = t^4, 1 \underset{t}{\to} 0$

27.[†] $\int_C (xy\,dx - xz\,dy + y\,dz)$ with C the line segment from $(1,2,3)$ to $(2,4,6)$

28.[†] $\int_C (xy + z)\,ds$ with C the line segment from $(0,0,0)$ to $(3,-7,4)$

29.[†] $\int_C (3x)^{-1/3}\,ds$ with $C: x = \frac{1}{3}t^3, y = \frac{1}{4}t^4, 1 \to 5$

30.[†] $\int_C xy\,ds$ with $C: x = 2\sin t, y = 3\cos t, z = 4t, 10 \underset{t}{\to} 0$

16-2   VECTOR FIELDS AND PATH-INDEPENDENT LINE INTEGRALS

A VECTOR FIELD is a function whose values are vectors.

Definition 16.4   A two-dimensional vector field is a rule that assigns a two-dimensional vector $A(x, y) = p(x, y)i + q(x, y)j$ to each point (x, y) in a region in the xy-plane. A three-dimensional vector field assigns a three-dimensional vector $A(x, y, z) = p(x, y, z)i + q(x, y, z)j + r(x, y, z)k$ to each point (x, y, z) in a region in xyz-space.

Vector fields are used to describe a variety of phenomena including gravitational attraction and fluid flow. To describe the gravitational attraction of the sun, for example, we define a FORCE VECTOR FIELD by assigning to each point in space the vector that equals the gravitational force the sun would exert on an object of unit mass at that point.

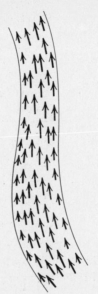

FIGURE 16.14   The gravitational field of the sun

FIGURE 16.15   The velocity field of a river

FIGURE 16.16   The vector field $\frac{1}{2} y\mathbf{i} - \frac{1}{2} x\mathbf{j}$

Figure 16.14 is a sketch of the portion of that vector field corresponding to the points in a plane through the center of the sun. Only a few of the vectors are shown; we should imagine that there is a vector at each point. All of the vectors are directed toward the center of the sun because that is the direction of the force. The vectors are longer near the sun because the force is greater there. Figure 16.15 is a sketch of the VELOCITY VECTOR FIELD of a river. The vectors are the velocity vectors of the particles of water at various points in the river.

EXAMPLE 1   Sketch the vector field $\mathbf{A}(x, y) = -\frac{1}{2} y\mathbf{i} + \frac{1}{2} x\mathbf{j}$.

Solution.   Figure 16.16 shows the vectors in the vector field at the points $(\pm 2, 0)$, $(0, \pm 2)$, $(\pm 2, \pm 2)$, $(\pm 4, 0)$, $(0, \pm 4)$, and $(\pm 4, \pm 4)$.   □

Gradient fields

One way to obtain a vector field is to take the gradient of a function of two or three variables:

Definition 16.5   The gradient field of a function $U(x, y)$ of two variables is the vector field $\overrightarrow{\mathbf{grad}}\, U(x, y) = U_x(x, y)\mathbf{i} + U_y(x, y)\mathbf{j}$. The gradient field of a function $U(x, y, z)$ of three variables is the vector field $\overrightarrow{\mathbf{grad}}\, U(x, y, z) = U_x(x, y, z)\mathbf{i} + U_y(x, y, z)\mathbf{j} + U_z(x, y, z)\mathbf{k}$.

EXAMPLE 2   Sketch the vector field $\overrightarrow{\mathrm{grad}}(\frac{1}{2} xy)$.

Solution.   We have

$$\overrightarrow{\mathrm{grad}}\,(\tfrac{1}{2}xy) = \frac{\partial}{\partial x}(\tfrac{1}{2}xy)\mathbf{i} + \frac{\partial}{\partial y}(\tfrac{1}{2}xy)\mathbf{j} = \tfrac{1}{2}y\mathbf{i} + \tfrac{1}{2}x\mathbf{j}.$$

This vector field is sketched in Figure 16.17.   □

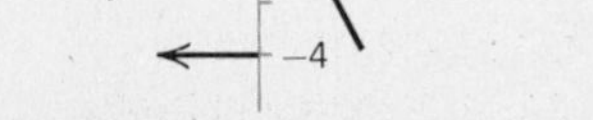

FIGURE 16.17   The vector field $\overrightarrow{\mathrm{grad}}(\frac{1}{2} xy)$

Our next results provide criteria for determining whether a vector field is a gradient field. Theorem 16.2 deals with the case of two variables and Theorem 16.3 with the case of three variables.

Theorem 16.2   Suppose that $p(x, y)$ and $q(x, y)$ have continuous first-order derivatives in a rectangle with sides parallel to the coordinate axes. The vector field $p\mathbf{i} + q\mathbf{j}$ is a gradient field if and only if

$$(1) \qquad p_y = q_x \qquad \blacktriangleleft$$

in the rectangle.

Theorem 16.3   Suppose that $p(x, y, z), q(x, y, z)$, and $r(x, y, z)$ have continuous first-order derivatives in a box with sides parallel to the coordinate planes. The vector field $p\mathbf{i} + q\mathbf{j} + r\mathbf{k}$ is a gradient field if and only if

$$(2) \qquad p_y = q_x, \quad p_z = r_x, \quad q_z = r_y \qquad \blacktriangleleft$$

in the box.*

Proof of Theorem 16.2.   If $p\mathbf{i} + q\mathbf{j}$ is the gradient field of a function $U(x, y)$, then we have $U_x = p$ and $U_y = q$ and by the equality of mixed partial derivatives

$$(3) \qquad p_y = (U_x)_y = (U_y)_x = q_x$$

so we obtain (1).

Conversely, suppose that p and q satisfy (1) and let (x_0, y_0) and (x, y) be arbitrary points in the rectangle. We can express the value of any function at (x, y) as its value at (x_0, y_0), plus the change in its value as we go in the x-direction from (x_0, y_0) to (x, y_0), plus the change in its value as we go in the y-direction from (x, y_0) to (x, y) (Figure 16.18):

$$U(x, y) = U(x_0, y_0) + [U(x, y_0) - U(x_0, y_0)]$$
$$+ [U(x, y) - U(x, y_0)].$$

If the function has continuous first derivatives, we can then express its changes in the x- and y-directions as integrals of its x- and y-derivatives:

$$(4) \qquad U(x, y) = U(x_0, y_0) + \int_{x_0}^{x} U_x(t, y_0) \, dt + \int_{y_0}^{y} U_y(x, t) \, dt.$$

Therefore, to define a function $U(x, y)$ with $U_x = p$ and $U_y = q$, we set

$$(5) \qquad U(x, y) = K + \int_{x_0}^{x} p(t, y_0) \, dt + \int_{y_0}^{y} q(x, t) \, dt.$$

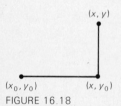

(x, y)

(x_0, y_0) (x, y_0)

FIGURE 16.18

---

*Theorems 16.2 and 16.3 are valid for more general regions, provided the regions have no "holes" in them. These cases are discussed in advanced calculus courses.

We have written "K" in (5) where there is the term "$U(x_0, y_0)$" in (4) because we are considering the point (x_0, y_0) to be fixed and the point (x, y) to be variable. We will complete the proof by showing that the function U, defined by (5), has p and q as its x- and y-derivatives.

For constant x, the first integral in (5) is constant, so by the Fundamental Theorem for derivatives of integrals

(6) $U_y(x, y) = \dfrac{\partial}{\partial y} \displaystyle\int_{y_0}^{y} q(x, t)\, dt = q(x, y).$

By a result from advanced calculus, we may differentiate the second integral on the right of (4) with respect to x by differentiating under the integral sign. We obtain

$$U_x(x, y) = \frac{\partial}{\partial x} \int_{x_0}^{x} p(t, y_0)\, dt + \frac{\partial}{\partial x} \int_{y_0}^{y} q(x, t)\, dt$$

$$= p(x, y_0) + \int_{y_0}^{y} q_x(x, t)\, dt.$$

By equation (1), we can replace q_x by p_y to obtain

(7) $U_x(x, y) = p(x, y_0) + \displaystyle\int_{y_0}^{y} p_y(x, t)\, dt$

$$= p(x, y_0) + [p(x, y) - p(x, y_0)] = p(x, y).$$

Equations (6) and (7) show that $p\mathbf{i} + q\mathbf{j}$ is the gradient field of U.   Q.E.D.

The gradient field $p\mathbf{i} + q\mathbf{j} + r\mathbf{k} = U_x\mathbf{i} + U_y\mathbf{j} + U_z\mathbf{k}$ for a function of three variables must satisfy conditions (2) because pairs of the mixed second-order derivatives of $U(x, y, z)$ are equal: we have $p_y = (U_x)_y = (U_y)_x = q_x$, $p_z = (U_x)_z = (U_z)_x = r_x$, and $q_z = (U_y)_z = (U_z)_y = r_y$. We leave the rest of the proof of Theorem 16.3 as Exercise 33.

EXAMPLE 3   One of the vector fields (a) $\langle xy^2, 2x \rangle$ and (b) $\langle y + \cos x, x - 1 \rangle$ is the gradient field of a function $U(x, y)$. Find such a function.

Solution.   For the vector field (a) we have $p = xy^2$ and $q = 2x$ and the derivatives $p_y = 2xy$ and $q_x = 2$ are not equal. By Theorem 16.2 this is not a gradient vector field.

In the case of vector field (b) we have $p = y + \cos x$ and $q = x - 1$ for which $p_y = \dfrac{\partial}{\partial y}(y + \cos x) = 1$ and $q_x = \dfrac{\partial}{\partial x}(x - 1) = 1$ are equal. This is a gradient field.

The function $U(x, y)$ must satisfy the conditions

(8) $U_x = y + \cos x, \quad U_y = x - 1.$

If we hold y fixed and integrate with respect to x in the first of equations (8) we obtain $U = \int (y + \cos x)\, dx$ or

(9) $\quad U(x, y) = xy + \sin x + \phi(y).$

The function $\phi(y)$ is the constant of integration which may be different for different values of y and hence is an arbitrary function of y.

Differentiating (9) with respect to y yields the equation

$$U_y = \frac{\partial}{\partial y}[xy + \sin x + \phi(y)] = x + \phi'(y).$$

Comparison with the second of equations (8) shows that we must take $\phi'(y) = -1$ or $\phi(y) = -y + C$. We make this substitution in (9) to see that the suitable functions $U(x, y)$ are $U(x, y) = xy + \sin x - y + C$. □

Path-independent line integrals

Theorems 16.2 and 16.3 can be used to determine whether a line integral

(10) $\quad \displaystyle\int_C (p\, dx + q\, dy)$

in the xy-plane or a line integral

(11) $\quad \displaystyle\int_C (p\, dx + q\, dy + r\, dz)$

in space is PATH-INDEPENDENT, according to the following definition.

Definition 16.6 **A line integral (10) or (11) is path-independent in a region if its value for any curve C in the region depends only on the location of the beginning and end points of C and not on the rest of its path.**

The basic results concerning path-independence are stated in the next theorems.

Theorem 16.4 **Suppose that $p(x, y)$ and $q(x, y)$ have continuous first-order derivatives in a rectangle with sides parallel to the coordinate axes. The integral (10) is path-independent in the rectangle if and only if**

(1) $\quad p_y = q_x$ ◄

there. In this case

(12) $\quad \displaystyle\int_C (p\, dx + q\, dy) = U(x_1, y_1) - U(x_0, y_0)$ ◄

where (x_0, y_0) is the beginning point and (x_1, y_1) the end point of C and $U(x, y)$ is a function whose gradient is $p\mathbf{i} + q\mathbf{j}$.

**Theorem 16.5   Suppose that $p(x, y, z)$, $q(x, y, z)$, and $r(x, y, z)$
have continuous first-order derivatives in a box with sides
parallel to the coordinate planes. The integral (11) is
path-independent in the box if and only if**

(2) $p_y = q_x$, $p_z = r_x$, $q_z = r_y$ ◀

there. In this case

(13) $\displaystyle\int_C (p\,dx + q\,dy + r\,dz) = U(x_1, y_1, z_1) - U(x_0, y_0, z_0)$ ◀

**where (x_0, y_0, z_0) is the beginning point and (x_1, y_1, z_1) the end
point of C and $U(x, y, z)$ is a function whose gradient is
$p\mathbf{i} + q\mathbf{j} + r\mathbf{k}$.**

Theorems 16.2 and 16.3 show that under conditions (1) or (2) there
exist suitable functions U.

Proof of Theorem 16.4.   Suppose first that condition (1) is satisfied. We
let $U(x, y)$ be a function such that $\overrightarrow{\operatorname{grad}} U = p\mathbf{i} + q\mathbf{j}$ so that $U_x = p$
and $U_y = q$ and let $C: x = x(t), y = y(t), a \underset{t}{\longrightarrow} b$ be any curve in the
rectangle such that $x(t)$ and $y(t)$ have continuous first derivatives.
We let $(x_0, y_0) = (x(a), y(a))$ denote the beginning point and
$(x_1, y_1) = (x(b), y(b))$ the end point of C. The function $U(x(t), y(t))$
of t has a continuous derivative, so by the Fundamental Theorem of
Calculus and the chain rule, we have

$$U(x_1, y_1) - U(x_0, y_0) = U(x(b), y(b)) - U(x(a), y(a))$$
$$= \int_a^b \frac{d}{dt} U(x(t), y(t))\,dt$$
$$= \int_a^b \left[U_x(x(t), y(t)) \frac{dx}{dt}(t) + U_y(x(t), y(t)) \frac{dy}{dt}(t) \right] dt$$
$$= \int_C [U_x(x, y)\,dx + U_y(x, y)\,dy] = \int_C (p\,dx + q\,dy).$$

This gives equation (12), which shows that the integral is path
independent.

Conversely, suppose that the integral $\int_C (p\,dx + q\,dy)$ is
path-independent. We let (x_0, y_0) be a fixed point in the rectangle and
define $U(x, y)$ to be $\int_C (p\,dx + q\,dy)$ where C is any curve from
(x_0, y_0) to (x, y). If we choose C as in Figure 16.19 and use y as the
parameter on the vertical line segment and x as parameter on the
horizontal line segment, we obtain

$$U(x, y) = \int_C (p\,dx + q\,dy)$$

(14)

$$= \int_{y_0}^y q(x_0, t)\,dt + \int_{x_0}^x p(t, y)\,dt.$$

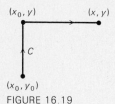

(x_0, y) → (x, y)

C

(x_0, y_0)

FIGURE 16.19

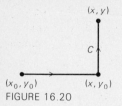

FIGURE 16.20

Because the first integral in the last sum does not depend on x, equation (14) gives

(15) $U_x(x, y) = \dfrac{\partial}{\partial x} \displaystyle\int_{x_0}^{x} p(t, y)\, dt = p(x, y).$

Similarly, if we use a curve C as in Figure 16.20, we obtain

$$U(x, y) = \int_C (p\, dx + q\, dy)$$

$$= \int_{x_0}^{x} p(t, y_0)\, dt + \int_{y_0}^{y} q(x, t)\, dt$$

which yields

(16) $U_y(x, y) = \dfrac{\partial}{\partial y} \displaystyle\int_{y_0}^{y} q(x, t)\, dt = q(x, y).$

Equations (15) and (16) show that $p\mathbf{i} + q\mathbf{j}$ is a gradient field. Condition (1) is satisfied because of Theorem 16.2.   Q.E.D.

The proof of Theorem 16.5 is similar. We leave it as Exercise 34.

EXAMPLE 4   The curves C in the line integrals below run from $(1, 2)$ to $(3, 4)$, and one of the integrals is path-independent. Evaluate that integral:

(a) $\displaystyle\int_C \left(y \sin(xy)\, dx + x \sin(xy)\, dy \right)$ (b) $\displaystyle\int_C (2ye^x\, dx + y^2 e^x\, dy)$

Solution.   Line integral (a) is path independent because

$p_y = \dfrac{\partial}{\partial y}\left(y \sin(xy) \right) = \sin(xy) + xy \cos(xy)$ and $q_x = \dfrac{\partial}{\partial x}\left(x \sin(xy) \right)$

$= \sin(xy) + xy \cos(xy)$ are equal. A suitable function $U(x, y)$ must satisfy

(17) $U_x = y \sin(xy), \quad U_y = x \sin(xy).$

To satisfy the first of equations (17) we must have

(18) $U(x, y) = \displaystyle\int y \sin(xy)\, dx = -\cos(xy) + \phi(y)$

where $\phi(y)$ is an arbitrary function of y. Differentiating (18) with respect to y then yields

$$U_y = \frac{\partial}{\partial y}[-\cos(xy) + \phi(y)] = x \sin(xy) + \phi'(y).$$

Comparing this equation with the second of equations (17) shows that we must take $\phi'(y) = 0$, which implies $\phi(y) = C$. We need only one function U, so we take $C = 0$ and obtain $U(x, y) = -\cos(xy)$. Because the curve runs from $(1, 2)$ to $(3, 4)$, formula (12) gives

$$\int_C (y \sin(xy)\, dx + x \sin(xy)\, dy) = U(3, 4) - U(1, 2)$$

$$= -\cos(12) + \cos(2).$$

Integral (b) is not path independent because in this case

$$p_y = \frac{\partial}{\partial y}(2ye^x) = 2e^x \text{ and } q_x = \frac{\partial}{\partial x}(y^2e^x) = y^2e^x \text{ are not equal.} \quad \square$$

Work

Recall that the *work* done by a constant force F on an object as it traverses a line segment is defined to be $|F| \cos \theta\, \Delta s$, where θ is an angle between the force vector F and the direction of motion and Δs is the length of the line segment (Figure 16.21). We can use a dot product to express the work as

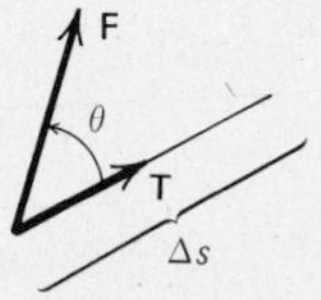

(19)   Work $= F \cdot T \Delta s$

where T is the unit vector in the direction of the motion.

    If the force is not constant or the path is curved, we compute the work by using the following rule.

FIGURE 16.21
Work $= |F| \cos \theta\, \Delta s$
$= F \cdot T \Delta s$

Rule 16.3   Suppose that an object is subject to a force $F(x, y) = p(x, y)\mathbf{i} + q(x, y)\mathbf{j}$ pounds when it is at (x, y) on a curve C in the xy-plane and that the scales on the coordinate axes are measured in feet. Then the work done by the force field $F(x, y)$ on the object as it traverses C in the direction of its orientation is

(20)   **Work $= \displaystyle\int_C F \cdot T\, ds = \int_C (p\, dx + q\, dy)$ foot-pounds.**   ◄

    In the case of a curve C in space and a three-dimensional force field $F(x, y, z) = p(x, y, z)\mathbf{i} + q(x, y, z)\mathbf{j} + r(x, y, z)\mathbf{k}$, the work is given by

(21)   **Work $= \displaystyle\int_C F \cdot T\, ds = \int_C (p\, dx + q\, dy + r\, dz)$ foot-pounds.**   ◄

    The formula $\int_C F \cdot T\, ds$ for the work in the case of either a plane or space curve follows from formula (19) and Definition 16.1 of the line integral (see Exercise 35). To obtain the second integrals in (20) and (21), we use Rule 16.2 to express dx, dy, and dz in terms of ds. In the case of a plane curve, we have $T = \langle u_1, u_2 \rangle$, $dx = u_1\, ds$, and $dy = u_2\, ds$, so that

$$F \cdot T\, ds = \langle p, q \rangle \cdot \langle u_1, u_2 \rangle\, ds = (pu_1 + qu_2)\, ds = p\, dx + q\, dy.$$

In the case of a curve in space, we have $T = \langle u_1, u_2, u_3 \rangle$, $dx = u_1\,ds$, $dy = u_2\,ds$, and $dz = u_3\,ds$. Consequently,

$$F \cdot T\,ds = \langle p, q, r \rangle \cdot \langle u_1, u_2, u_3 \rangle\,ds$$
$$= (pu_1 + qu_2 + ru_3)\,ds = p\,dx + q\,dy + r\,dz.$$

EXAMPLE 5   Figure 16.22 is a sketch of a force field $F(x, y)$ (pounds) along a curve C. The lengths of the arrows are measured by the scales used on the coordinate axes. Which of the numbers $0, \pm 3, \pm 9, \pm 12, \pm 21, \pm 27$ (foot-pounds) is closest to the work done by the force field on an object traversing the curve?

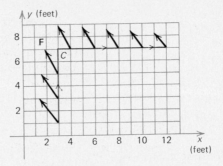

FIGURE 16.22   A force field F along a curve C

Solution.   Along the vertical portion C_1 of the curve C, $F \cdot T$ is approximately 2, whereas along the horizontal portion C_2, $F \cdot T$ is approximately -1. Therefore, the work is

$$\int_C F \cdot T\,ds = \int_{C_1} F \cdot T\,ds + \int_{C_2} F \cdot T\,ds$$

$$\approx \int_{C_1} 2\,ds + \int_{C_2} (-1)\,ds$$
$$= 2[\text{Length of } C_1] - [\text{Length of } C_2].$$

The length of C_1 is 6 and the length of C_2 is 9, so the work is approximately $2(6) - 9 = 3$ foot-pounds. $\square$

Conservative force fields   A curve C is called CLOSED if its beginning and end points coincide (Figure 16.23). This terminology is used in the definition of a CONSERVATIVE FORCE FIELD.

Definition 16.7   **A force field F is conservative in a region if the work $\int_C F \cdot T\,ds$ is zero for every closed curve C in the region.**

The force field is conservative if and only if the line integrals $\int_C F \cdot T\,ds$ are path-independent for all curves, closed or not, in the region (see Exercise 36). Therefore, Theorems 16.4 and 16.5 provide criteria for determining whether force fields are conservative.

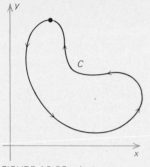

FIGURE 16.23   A closed curve

Corollary 16.1   **A two-dimensional force field $F = p\mathbf{i} + q\mathbf{j}$ with continuous first-order derivatives is conservative in a rectangle with sides parallel to the coordinate axes if and only if**

(1) $p_y = q_x$ ◀

there. A three-dimensional force field $F = p\mathbf{i} + q\mathbf{j} + r\mathbf{k}$ with continuous first-order derivatives is conservative in a box with sides parallel to the coordinate planes if and only if in the box

(2) $p_y = q_x, \quad p_z = r_x, \quad q_z = r_y.$ ◀

If a force field $\mathbf{F}$ is conservative, then a function U such that $\mathbf{F} = \overrightarrow{\text{grad}}\, U$ is called a POTENTIAL FUNCTION of the force field. Theorems 16.4 and 16.5 show that the work done by a conservative force field on an object that traverses a curve C is equal to the value of the potential function at the end of the curve minus its value at the beginning.

Exercises

†*Answer provided.*
‡*Outline of solution provided.*

In Exercises 1 through 9 sketch the vector fields. Use the scales on the axes to measure the lengths of the arrows.

1.† $\mathbf{i} + y\mathbf{j}$ **2.** $y\mathbf{i} + y\mathbf{j}$

3.† $\langle -\tfrac{1}{2}x, \tfrac{1}{2}y \rangle$ **4.** $\langle y, 1 \rangle$

5.† $\langle \tfrac{1}{2}(x + y), \tfrac{1}{2}(y - x) \rangle$ **6.** $\langle \tfrac{1}{2}y, -\tfrac{1}{2}x \rangle$

7.† $\overrightarrow{\text{grad}}(\tfrac{1}{2}x^2 + \tfrac{1}{2}y^2)$ **8.** $\overrightarrow{\text{grad}}(\tfrac{1}{2}xy)$

9. $\overrightarrow{\text{grad}}(x - 2y)$

In each of Exercises 10 through 19 determine whether the given vector field F is a gradient field. If so, find the functions U such that $\mathbf{F} = \overrightarrow{\text{grad}}\, U$.

10.† $\langle \sin(3y) + x, 3x\cos(3y) - y \rangle$

11. $\langle 3x\cos(3y) - 1, \sin(3y) + 1 \rangle$

12.† $\left\langle \dfrac{y}{1 + xy}, \dfrac{x}{1 + xy} \right\rangle$ **13.** $\left\langle \ln y + e^x, \dfrac{x}{y} - y^2 \right\rangle$

14.† $\langle ye^{xy} - 1, xe^{xy} \rangle$ **15.** $\langle 3x^2y^4 + 2y, 4x^3y^3 - x \rangle$

16.‡ $\langle yz + y + 1, xz + x + z, xy + y - 2 \rangle$

17.† $\langle \cos(x + z) + 1, ze^{yz}, \cos(x + z) + ye^{yz} \rangle$

18.† $\langle yz + 2xy^2, xz + 2x^2y, 2x^2y^2 \rangle$

19. $\langle e^y, xe^y + e^z, ye^z - 2e^{-2z} \rangle$

In Exercises 20 through 25 determine which integrals are path-independent and evaluate those integrals.

20.† $\int_C \left((2xy^3 - 3x^2)\,dx + (3xy^2 + 2y)\,dy\right)$ with C running from $(0, 0)$ to $(1, 2)$

21. $\int_C \left((x - xe^y)\,dx + (y - ye^x)\,dy\right)$ with C running from $(2, 3)$ to $(-3, -2)$

22.† $\int_C \left(2x \sin z\,dx + (z^3 - e^y)\,dy + (x^2 \cos z + 3yz^2)\,dz\right)$ with C running from $(0, 0, 0)$ to $(1, 2, 3)$

23. $\int_C \left((e^y - y^3 \cos x + 1)\,dx + (xe^y - 3y^2 \sin x)\,dy\right)$ with C running from $(0, 0)$ to $(10, 0)$

24.† $\int_C \left((y + \cos z)\,dx + (y \cos x + \cos z)\,dy - (x \sin z)\,dz\right)$ with C running from $(1, 2, 3)$ to $(4, 5, 6)$

25. $\int_C \left((xy^2z^2 + x^2y^3)\,dx + (x^2yz^2 + x^3y^2)\,dy + (x^2y^2z)\,dz\right)$ with C running from $(1, 1, 1)$ to $(0, 0, 0)$

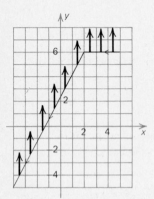

FIGURE 16.24   A vector field **F** along a curve C

26.† Figure 16.24 shows a curve C and a vector field **F** along it. Which of the numbers $0, \pm 5, \pm 10, \pm 25, \pm 50, \pm 100$ is closest to the integral $\int_C \mathbf{F} \cdot \mathbf{T}\,ds$?

27. Figure 16.25 shows a curve C and a force field **G** along it. Which of the numbers $0, \pm 5, \pm 10, \pm 15, \pm 25, \pm 50$ is closest to the work done by this force field on an object traversing the curve?

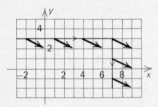

FIGURE 16.25   A force field **G** along a curve C

28.† Is the force field $yz \cos(xyz)\mathbf{i} + xz \cos(xyz)\mathbf{j} + xy \cos(xyz)\mathbf{k}$ conservative? If so, find a potential function for it and compute the work it does on an object that moves from $(0, 0, 0)$ to $(2, 3, 4)$.

29.† Is the force field $\langle 3x^2y^2z, 2x^2y^2z, x^3y^2 + x \rangle$ conservative? If so, find a potential function for it and compute the work it does on an object that moves from $(1, 4, 3)$ to $(-3, 2, 8)$.

30. Is the force field $\langle e^x + ye^{xy}, e^{-y} + xe^{xy} \rangle$ conservative? If so, find a potential function for it and compute the work it does on an object that moves from $(0, 1)$ to $(1, 0)$.

31.† Compute the work done by the nonconservative force field $\langle x^2y, xy^2 \rangle$ on an object that moves from $(0, 0)$ to $(1, 1)$ **a.**   along the parabola $y = x^2$,   **b.**   along the line $y = x$, and   **c.**   along the line segment from $(0, 0)$ to $(1, 0)$ followed by the line segment from $(1, 0)$ to $(1, 1)$.

32.† Two of the force fields sketched in Figures 16.26, 16.27, and 16.28 are not conservative. Which are they and why?

33. Prove the part of Theorem 16.3 that was not proved in the text: if $p(x, y, z)\mathbf{i} + q(x, y, z)\mathbf{j} + r(x, y, z)\mathbf{k}$ has continuous first derivatives and satisfies conditions (2) in a box with sides parallel to the coordinate planes, then $p\mathbf{i} + q\mathbf{j} + r\mathbf{k}$ is a gradient field there.

34. Prove Theorem 16.5.

FIGURE 16.26

FIGURE 16.27

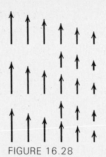

FIGURE 16.28

35. Derive Rule 16.3.

36.[†] Show that the line integral $\int_C (p\,dx + q\,dy)$ is path-independent in a region if and only if the line integral is zero for all closed curves in the region.

16-3   GREEN'S THEOREM, THE DIVERGENCE THEOREM, AND STOKES' THEOREM IN THE PLANE

In this section we study GREEN'S THEOREM, which enables us to express the double integral of a derivative $f_x(x, y)$ or $f_y(x, y)$ over a bounded region in the xy-plane as a line integral of f around the boundary of the region. Then we introduce the concepts of FLUX, CIRCULATION, DIVERGENCE, and SCALAR CURL of a two-dimensional vector field and put Green's Theorem in vector form as the DIVERGENCE THEOREM and STOKES' THEOREM, which are two of the most important tools of vector analysis.*

Green's Theorem in the plane

We say that the boundary C of a bounded region R in the xy-plane is ORIENTED POSITIVELY if the region is on the left as the boundary curve is traversed. For example, the boundary of the region in Figure 16.29 is oriented positively. The boundary is in two pieces with the outer curve oriented counterclockwise and the inner curve oriented clockwise.

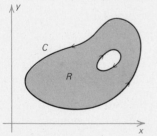

FIGURE 16.29   A region R with its boundary C oriented positively

Theorem 16.6   (Green's Theorem)   Suppose that $f(x, y)$ has continuous first-order derivatives in a bounded region R with a piecewise smooth boundary C which is oriented positively. Then

(1) $\displaystyle \int\int_R \frac{\partial f}{\partial x}(x, y)\,dx\,dy = \int_C f(x, y)\,dy$ ◀

(2) $\displaystyle \int\int_R \frac{\partial f}{\partial y}(x, y)\,dx\,dy = -\int_C f(x, y)\,dx.$ ◀

---

*Green's Theorem, the Divergence Theorem, and Stokes' Theorem are also known by a variety of other names.

Theorem 16.6 is valid for the bounded regions that are normally encountered in calculus. Complete statements of the theorem, including conditions that the region and its boundary should satisfy, may be found in books on advanced calculus. We will not go into these details and will only outline the proof.

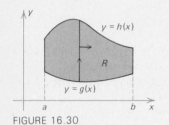

FIGURE 16.30

Partial proof of (2).   Green's Theorem is a consequence of the Fundamental Theorem of Calculus combined with the procedures for expressing double integrals as iterated integrals and the definitions of line integrals. We will first derive (2) for a region R that is bounded on the top by a curve $y = h(x)$, on the bottom by a curve $y = g(x)$, and that extends from $x = a$ on the left to $x = b$ on the right (Figure 16.30). We can express the double integral of $f_y(x, y)$ over R as an iterated integral by writing

$$(3) \qquad \int\int_R \frac{\partial f}{\partial y}(x, y) \, dx \, dy = \int_{x=a}^{x=b} \left\{ \int_{y=g(x)}^{y=h(x)} \frac{\partial f}{\partial y}(x, y) \, dy \right\} dx.$$

The inner integral on the right of (3) is the integral with respect to y of the derivative of $f(x, y)$ with respect to y, so with the Fundamental Theorem that equation gives

$$(4) \qquad \int\int_R \frac{\partial f}{\partial y}(x, y) \, dx \, dy = \int_a^b [f(x, h(x)) - f(x, g(x))] \, dx.$$

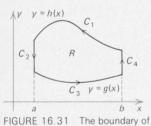

FIGURE 16.31   The boundary of R
is $C = C_1 + C_2 + C_3 + C_4$.

On the other hand, we can also express the integral $\int_C f \, dx$ around the boundary of R as a definite integral with respect to x. We let $C_1, C_2, C_3,$ and C_4 be the portions of the boundary shown in Figure 16.31. The integrals with respect to dx over C_2 and C_4 are zero because those curves are vertical line segments. We use x as parameter on C_1 and C_3 by writing $C_1: y = h(x), b \underset{x}{\to} a; C_3: y = g(x), a \underset{x}{\to} b$ to show that

$$\int_C f \, dx = \left[\int_{C_1} + \int_{C_2} + \int_{C_3} + \int_{C_4} \right] f \, dx$$

$$(5) \qquad = \int_b^a f(x, h(x)) \, dx + 0 + \int_a^b f(x, g(x)) \, dx + 0$$

$$= -\int_a^b [f(x, h(x)) - f(x, g(x))] \, dx.$$

We obtained the last equation from the fact that the integral from b to a of $f(x, h(x))$ is the negative of the integral from a to b.

Comparing equations (4) and (5) gives equation (2) for this type of region R. To derive (2) for a more general region, we divide it into a finite number of pieces of the type for which we have derived the equation. In Figure 16.32, for example, the region of Figure 16.29 has been divided into four regions R_1, R_2, R_3, R_4 of that type. We let $C_1, C_2, C_3,$ and C_4 be their boundaries oriented positively and apply (2) to the four regions to obtain

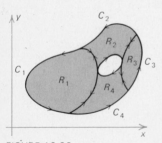

FIGURE 16.32

$$\int \int_R \frac{\partial f}{\partial y} \, dx \, dy = \int \int_{R_1} + \int \int_{R_2} + \int \int_{R_3} + \int \int_{R_4} \frac{\partial f}{\partial y} \, dx \, dy$$

$$= -\left[\int_{C_1} + \int_{C_2} + \int_{C_3} + \int_{C_4} \right] f \, dx$$

$$= -\int_C f \, dx$$

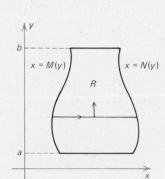

FIGURE 16.33

which is equation (2) for the region of Figure 16.32. The sum of the line integrals over C_1, C_2, C_3, and C_4 is equal to the line integral over C because in traversing the boundaries of the subregions the vertical line segments that are not part of C are traversed twice, once in each direction, and those portions of the line integrals cancel. □

Formula (1) can be derived by a similar procedure, starting with regions of the type shown in Figures 16.33 and 16.34 (see Exercise 1).

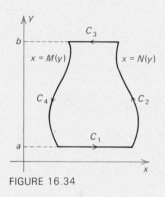

FIGURE 16.34

EXAMPLE 1   Express $\int \int_R \left[\frac{\partial p}{\partial x}(x, y) + \frac{\partial q}{\partial y}(x, y) \right] dx \, dy$ in terms of line integrals of p and q over C, where R is a bounded region and C is its boundary oriented positively.

Solution.   By equations (1) with f replaced by p and (2) with f replaced by q, we have

$$\int \int_R \left[\frac{\partial p}{\partial x}(x, y) + \frac{\partial q}{\partial y}(x, y) \right] dx \, dy$$

$$= \int_C [p(x, y) \, dy - q(x, y) \, dx]. \quad □$$

EXAMPLE 2   Express the line integrals (a) $\int_C x \, dy$ and (b) $\int_C y \, dx$ in terms of the area of R, where R is a bounded region and C is its boundary oriented positively.

Solution.   Formula (1) applied to integral (a) yields

(6) $\int_C x \, dy = -\int \int_R \frac{\partial}{\partial x}(x) \, dx \, dy = \int \int_R dx \, dy = [\text{Area of } R].$ ◄

Formula (2) applied to integral (b) shows that

(7) $\int_C y \, dx = -\int \int_R \frac{\partial}{\partial y}(y) \, dx \, dy = -\int \int_R dx \, dy$

$$= -[\text{Area of } R]. \quad □$$

Flux across a curve

The concept of the *flux* of a vector field across an oriented curve arises from the case of a velocity vector field of a fluid, where the flux is the rate of flow across the curve. Consider first an oriented line segment C in a constant velocity field $\mathbf{v}$ of a fluid (Figure 16.35). Let us suppose that distances are measured in feet and the velocity in feet per second. Every second enough fluid crosses the line segment to fill the shaded parallelogram in Figure 16.36. The area of the parallelogram is, consequently, equal to the rate of flow of fluid across the line.

We let $\mathbf{n}$ denote the unit normal that is directed toward the *right* of the tangent vector $\mathbf{T}$. (Recall that in Chapter 11 we used $\mathbf{N}$ for the unit normal vector that points to the left of $\mathbf{T}$.) Then in the case of Figure 16.36, where $\mathbf{n}$ and $\mathbf{v}$ make an acute angle θ, the area of the parallelogram is $|\mathbf{v}| \cos\theta \, \Delta s = \mathbf{v} \cdot \mathbf{n} \, \Delta s$ with Δs the length of the line segment. Thus,

$$(8) \quad \begin{bmatrix} \text{The rate of flow across } C \\ \text{in the direction of } \mathbf{n} \end{bmatrix} = \mathbf{v} \cdot \mathbf{n} \, \Delta s.$$

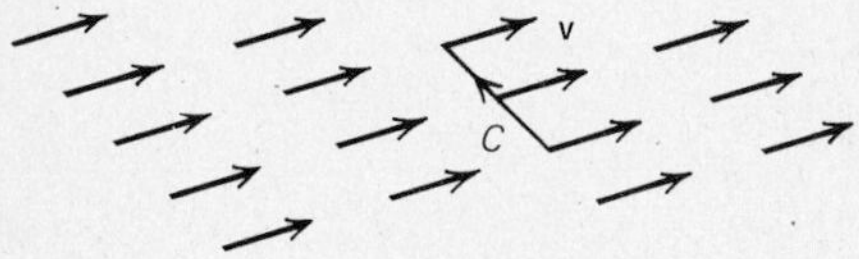

FIGURE 16.35   A constant velocity vector field $\mathbf{v}$

FIGURE 16.36   Parallelogram of area $\mathbf{v} \cdot \mathbf{n} \, \Delta s$

Formula (8) is also valid if θ is negative, since in this case $\mathbf{v} \cdot \mathbf{n}$ and the rate of flow are both negative.

If we approximate a general curve by collections of line segments and a general velocity field by fields that are constant on the line segments, we are led to the formula

$$(9) \quad \begin{bmatrix} \text{The rate of flow across } C \\ \text{in the direction of } \mathbf{n} \end{bmatrix} = \int_C \mathbf{v} \cdot \mathbf{n} \, ds \quad \blacktriangleleft$$

(see Exercise 26). In the case of an arbitrary vector field we refer to the line integral (9) as the *flux* of $\mathbf{v}$ across C and obtain the following rule.

Rule 16.4   The flux of a vector field $v = p\mathbf{i} + q\mathbf{j}$ across an oriented curve C is

$$(10) \quad \text{Flux} = \int_C \mathbf{v} \cdot \mathbf{n} \, ds = \int_C (p \, dy - q \, dx) \quad \blacktriangleleft$$

where n is the unit normal vector to C that points to the right of the unit tangent vector.

We obtain the second integral in (10) by using Rule 16.2 to express dx and dy in terms of ds. We have, on the one hand, $\mathbf{T} = \langle u_1, u_2 \rangle$ and

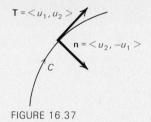

FIGURE 16.37

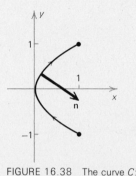

FIGURE 16.38   The curve C:
$x = y^2$, $-1 \underset{y}{\to} 1$

$n = \langle u_2, -u_1, \rangle$ (Figure 16.37) and, on the other hand, $dx = u_1\, ds$ and $dy = u_2\, ds$. Therefore,

$$\mathbf{v} \cdot \mathbf{n}\, ds = \langle p, q \rangle \cdot \langle u_2, -u_1 \rangle\, ds = (pu_2 - qu_1)\, ds$$
$$= p\, dy - q\, dx$$

and this gives the second integral in (10).

EXAMPLE 3   What is the rate of flow of the velocity vector field $x^3\mathbf{i} + \mathbf{j}$ (feet/second) from left to right across the portion of the parabola $x = y^2$ for $-1 \le y \le 1$? (Here x and y are measured in feet.)

Solution.   We use y as parameter and describe the curve as $C\colon x = y^2$, $-1 \underset{y}{\to} 1$. The unit normal $\mathbf{n}$ points to the right (Figure 16.38), so by the second integral in (10) with $p = x^3$ and $q = 1$, the rate of flow is

$$\int_C (x^3\, dy - dx) = \int_{-1}^{1} ((y^2)^3 - 2y)\, dy = \int_{-1}^{1} (y^6 - 2y)\, dy$$
$$= [\tfrac{1}{7} y^7 - y^2]_{-1}^{1} = \tfrac{2}{7} \text{ square feet/second.}$$

In this calculation we have used the equation $dx = 2y\, dy$.   □

Circulation around a closed curve

In the case of a velocity field $\mathbf{v}$ of a fluid, the integral $\int_C \mathbf{v} \cdot \mathbf{T}\, ds$ of the tangential component of $\mathbf{v}$ around a closed curve C is known as the *circulation* of the field around the curve. We use this terminology for an arbitrary vector field and obtain the following rule.

Rule 16.5   The circulation of a vector field $v = p\mathbf{i} + q\mathbf{j}$ around a closed oriented curve C is

(11)   Circulation $= \displaystyle\int_C \mathbf{v} \cdot \mathbf{T}\, ds = \int_C (p\, dx + q\, dy)$.   ◄

Notice that in the case of a force field $\mathbf{F}$ the circulation is the work done by the field on an object that goes around the curve.

The Divergence Theorem

The *divergence* of a vector field $\mathbf{v}$ is a numerical-valued function (not a vector-valued function), that is denoted div $\mathbf{v}$ and is defined as follows.

Definition 16.8   The divergence of the vector field $v(x, y) = p(x, y)\mathbf{i} + q(x, y)\mathbf{j}$ is

(12)   div $v = \dfrac{\partial p}{\partial x}(x, y) + \dfrac{\partial q}{\partial y}(x, y)$.   ◄

EXAMPLE 4   Compute the divergence of the vector field $\langle x^3y, x^4y^7 \rangle$.

Solution.   Formula (12) with $p = x^3y$ and $q = x^4y^7$ yields

$$\operatorname{div}\langle x^3y, x^4y^7 \rangle = \frac{\partial}{\partial x}(x^3y) + \frac{\partial}{\partial y}(x^4y^7) = 3x^2y + 7x^4y^6. \quad \square$$

The Divergence Theorem expresses the flux of a vector field across the boundary of a region as the double integral of the divergence.

Theorem 16.7   (The Divergence Theorem)   For a bounded region R with piecewise smooth boundary C oriented positively and for a vector field $v(x, y)$ with continuous first-order derivatives in R

(13) $\displaystyle \int_C \mathbf{v} \cdot \mathbf{n}\, ds = \int\int_R \operatorname{div} \mathbf{v}\, dx\, dy.$ ◀

Proof.   We obtain the Divergence Theorem from Green's Theorem by using the expression $\int_C (p\, dy - q\, dx)$ for $\int_C \mathbf{v} \cdot \mathbf{n}\, ds$. By equation (1) with $f = p$ and equation (2) with $f = q$, we have

$$\int_C \mathbf{v} \cdot \mathbf{n}\, ds = \int_C (p\, dy - q\, dx) = \int\int_R \left(\frac{\partial p}{\partial x} + \frac{\partial q}{\partial y} \right) dx\, dy$$

$$= \int\int_R \operatorname{div} \mathbf{v}\, dx\, dy. \quad \text{Q.E.D.}$$

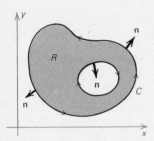

FIGURE 16.39   Unit exterior normal vectors *n* to a region *R*

Because the boundary of the region is oriented positively, the vector $\mathbf{n}$ is the *unit exterior normal* to the boundary C (Figure 16.39), and, in the case of a velocity field, the flux $\int_C \mathbf{v} \cdot \mathbf{n}\, ds$ is the (net) rate of flow out of the region. Thus, Theorem 16.7 states that the rate of flow *out* of the region is the integral of the divergence over the region. This is the origin of the term "divergence."

If the divergence is positive in the region, then there is a net flow out of the region and the fluid is expanding or there are "sources" that add fluid inside the region. If the divergence is negative, there is a net flow into the region and the fluid is contracting or there are "sinks" that remove fluid from inside the region. A fluid that does not expand or contract is called "incompressible." Thus, one result of the Divergence Theorem is that the vector field of an incompressible fluid with no sources or sinks has zero divergence.

EXAMPLE 5   What is the rate of flow of a fluid with velocity field $\mathbf{v}(x, y) = \langle 8x - y^2, x - 5y \rangle$ feet/second out of a region R of area 11 square feet?

Solution.   We need to know only the area of the region because the divergence is constant:

$$\operatorname{div} \mathbf{v} = \operatorname{div}\langle 8x - y^2, x - 5y \rangle = \frac{\partial}{\partial x}(8x - y^2) + \frac{\partial}{\partial y}(x - 5y)$$

$$= 8 - 5 = 3.$$

The rate of flow is

$$\int \int_R \text{div } \mathbf{v} \, dx \, dy = 3[\text{Area of } R] = 33 \text{ square feet/second.} \quad \square$$

Stokes' Theorem

Stokes' Theorem in the plane expresses the circulation of a vector field around the boundary of a region as the double integral of a function known as the *scalar curl* of the vector field.

Definition 16.9   **The scalar curl of a vector field $v = pi + qj$ is**

(14) $\text{curl } v = -\dfrac{\partial p}{\partial y}(x, y) + \dfrac{\partial q}{\partial x}(x, y).$ ◀

We refer to (14) as the "scalar" curl because it is a scalar (number), whereas in the case of three dimensions the "curl" that appears in Stokes' Theorem is a vector (see Miscellaneous Exercise 24).

Theorem 16.8   **(Stokes' Theorem)   For a bounded region R in the xy-plane with piecewise smooth boundary C oriented positively and for a vector field $v = pi + qj$ with continuous first-order derivatives in R,**

(15) $\displaystyle\int_C \mathbf{v} \cdot \mathbf{T} \, ds = \int \int_R \text{curl } v \, dx \, dy.$ ◀

Proof.   Stokes' Theorem is a consequence of Green's Theorem and the equation $\int_C \mathbf{v} \cdot \mathbf{T} \, ds = \int_C (p \, dx + q \, dy)$. By formula (1) with $f = q$ and formula (2) with $f = p$

$$\int_C \mathbf{v} \cdot \mathbf{T} \, ds = \int_C (p \, dx + q \, dy) = \int \int_R \left(-\frac{\partial p}{\partial y} + \frac{\partial q}{\partial x} \right) dx \, dy$$

$$= \int \int_R \text{curl } \mathbf{v} \, dx \, dy. \quad \text{Q.E.D.}$$

Stokes' Theorem enables us to compute the circulation of a vector field around the boundary of a region from its curl inside the region. In particular, a vector field has zero circulation around all curves in a rectangle if and only if its curl is zero there. Such a field is called IRROTATIONAL.

Work and Stokes' Theorem

In the case of a force field $\mathbf{F}$ the line integral $\int_C \mathbf{F} \cdot \mathbf{T} \, ds$ represents work, and Stokes' Theorem states that the work done by the force field on an object traversing the boundary C of the region R is the integral of curl $\mathbf{F}$ over R:

(16) $\text{Work} = \displaystyle\int_C \mathbf{F} \cdot \mathbf{T} \, ds = \int \int_R \text{curl } \mathbf{F} \, dx \, dy.$ ◀

In particular, the work is zero for all closed curves in a rectangle if and only if curl $\mathbf{F} = -p_y + q_x$ is zero there, and we again obtain the result that the force field is conservative if and only if $p_y = q_x$.

In the next example we use Stokes' Theorem to calculate work for a nonconservative field.

EXAMPLE 6   Use Stokes' Theorem to compute the work done by the force field $\mathbf{F} = \langle 3x^2y^2 + 2y, 9x + 2x^3y \rangle$ on an object that travels counterclockwise once around the circle $C: x^2 + y^2 = 1$.

Solution.   The curl of the force field is constant:

$$\text{curl } \mathbf{F} = \text{curl } \langle 3x^2y^2 + 2y, 9x + 2x^3y \rangle$$
$$= -\frac{\partial}{\partial y}(3x^2y^2 + 2y) + \frac{\partial}{\partial x}(9x + 2x^3y)$$
$$= -(6x^2y + 2) + (9 + 6x^2y) = 7.$$

If we consider the circle as the boundary of the region $R: x^2 + y^2 \le 1$ inside it, then the positive orientation of the circle is counterclockwise. The region R has area π, so the work is

$$\int_C \mathbf{F} \cdot \mathbf{T} \, ds = \int \int_R \text{curl } \mathbf{F} \, dx \, dy = \int \int_R 7 \, dx \, dy = 7\pi. \quad \square$$

Exercises

†*Answer provided.*
‡*Outline of solution provided.*

1.   Derive formula (1) in Green's Theorem for a region of the type shown in Figures 16.33 and 16.34.

In Exercises 2 through 5 the curve C is the positively oriented boundary of a bounded region R. Use Green's Theorem to express the line integrals over C as double integrals over R.

2.‡ $\int_C (xy^3 \, dx - 2x^3y^2 \, dy)$ **3.**† $\int_C [\sin(xy) \, dx + \cos(xy) \, dy]$

4. $\int_C (xe^{3y} \, dx - ye^{-4x} \, dy)$ **5.** $\int_C [2x \sin y \, dx + x^2\cos y \, dy]$

In Exercises 6 through 9 use formula (6) or (7) to compute the area of the regions.

6.‡   The region inside the loop of Tschirnhausen's cubic $x = t^2 - 3$, $y = \frac{1}{3}t^3 - t$ (Figure 4.64)

7.†   The region inside the *eight curve* $x = \cos t, y = \sin(t)\cos(t)$, $-\pi \underset{t}{\to} \pi$ (Figure 16.40)

8.   The region inside the piriform $x = 1 + \sin t, y = \cos t(1 + \sin t)$, $-\pi/2 \underset{t}{\to} 3\pi/2$ (Figure 11.46)

9.   The region inside the ellipse $x = a \cos t, y = b \sin t, 0 \underset{t}{\to} 2\pi$

10.†   Find a vector field $\mathbf{F}(x, y)$ such that $\int_C (\sin x \, dy - \cos y \, dx) = \int_C \mathbf{F} \cdot \mathbf{T} \, ds$ for all curves C in the xy-plane.

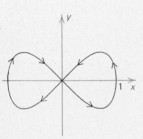

FIGURE 16.40   An eight curve

11. Find vector fields $G(x, y)$ and $H(x, y)$ such that $\int_C e^{x^2} \, dy$ equals $\int_C G \cdot T \, ds$ and $\int_C H \cdot n \, ds$ for all curves C in the xy-plane.

Compute **a.** the divergence and **b.** the scalar curl of the vector fields in Exercises 12 through 15.

12.† $\sin(xy)\mathbf{i} + \cos(x - y)\mathbf{j}$ **13.** $\langle x + y, x - y \rangle$

14. $\langle x^2y - 2xy, y^2 - xy^2 \rangle$ **15.** $7x^6y^6\mathbf{i} + 6y^7x^5\mathbf{j}$

16.† Figure 16.41 shows a vector field $\mathbf{v}(x, y)$ along a curve C. The scales on the axes are used to measure the lengths of the arrows. Which of the numbers $0, \pm 25, \pm 50, \pm 75, \pm 100$ is closest to $\int_C \mathbf{v} \cdot \mathbf{n} \, ds$?

17. Follow the instructions of Exercise 16 for the vector field and curve of Figure 16.42.

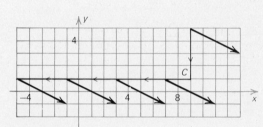

FIGURE 16.41

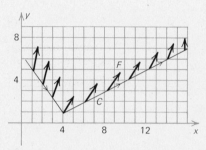

FIGURE 16.42

18.† Show that for all oriented curves C and vector fields F

$$\int_{-C} F \cdot T \, ds = -\int_C F \cdot T \, ds \text{ and } \int_{-C} F \cdot n \, ds = -\int_C F \cdot n \, ds.$$

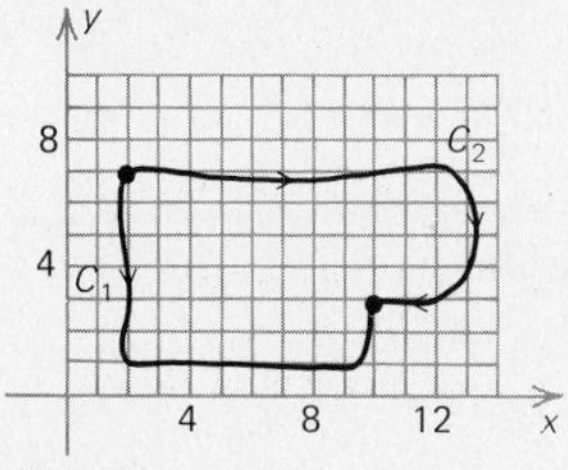

19.† The vector field F has constant curl, curl $F = \frac{1}{2}$, in the region bounded by the curves C_1 and C_2 in Figure 16.43, and $\int_{C_1} F \cdot T \, ds = 10$. Which of the numbers $0, \pm 10, \pm 20, \pm 40, \pm 50$ is closest to $\int_{C_2} F \cdot T \, ds$?

20. The vector field G has constant curl, curl $G = -2$, in the region of Figure 16.43, and $\int_{C_1} G \cdot T \, ds = -80$. Which of the numbers $0, \pm 40, \pm 80, \pm 120, \pm 160, \pm 200$ is closest to $\int_{C_2} G \cdot T \, ds$?

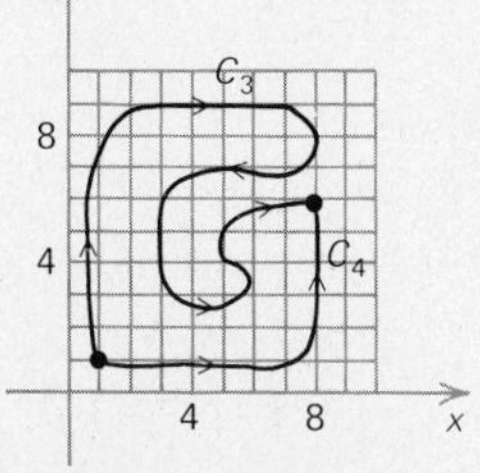

21.† The vector field A has constant divergence, div $A = -2$, in the region bounded by the curves C_3 and C_4 in Figure 16.44, and $\int_{C_3} A \cdot n \, ds = 35$. Which of the numbers $0, \pm 10, \pm 50, \pm 90, \pm 120$ is closest to $\int_{C_4} A \cdot n \, ds$?

22. The vector field B has constant divergence, div $B = 3$, in the region of Figure 16.44, and $\int_{C_3} B \cdot n \, ds = 60$. Which of the numbers $0, \pm 15, \pm 70, \pm 105, \pm 150, \pm 200$ is closest to $\int_{C_4} B \cdot n \, ds$?

FIGURE 16.43

FIGURE 16.44

FIGURE 16.45

23.‡ Does the vector field in Figure 16.45 have positive or negative divergence?

24.† Does the vector field in Figure 16.46 have positive or negative curl?

25. Are the curl and divergence of the vector field in Figure 16.47 positive or negative?

26. Derive formula (9) for the rate of flow across a curve.

FIGURE 16.46

FIGURE 16.47

Additional drill exercises   Compute the divergence and scalar curl of the vector field in each of exercises 27 through 32.

27.† $(x \sin y)\mathbf{i} - (y \cos x)\mathbf{j}$ **28.**† $xy^3\mathbf{i} + (y^2 - x^3)\mathbf{j}$

29.† $\langle x \ln y, y \ln x \rangle$ **30.**† $\langle xe^{xy} + x, ye^{xy} - y^2 \rangle$

31.† $\overrightarrow{\text{grad}}\,(x^2y^3 - x^3y^2)$ **32.**† $\dfrac{\langle y, -x \rangle}{\sqrt{x^2 + y^2}}$

MISCELLANEOUS EXERCISES

1.† Evaluate the integral of x^5 **a.** with respect to dx,   **b.** with respect to dy, and   **c.** with respect to ds over the portion $y = 1/x$, $10 \underset{x}{\longrightarrow} 1$ of a hyperbola.

2.† Compute the work done by the force field $\langle 4xy^3 + 2x, 6x^2y^2 - 3y^2 \rangle$ on an object that moves from $(1, 1)$ to $(4, 4)$.

3. Compute the work done by the force field $\langle y \cos(xy) + 1, x \cos(xy) - e^y \rangle$ on an object that moves from $(3, 2)$ to $(5, 5)$.

4.† Is the force field $\langle e^{2y} - 2x, 2xe^{2y} + 1 \rangle$ conservative? If so, find a potential function for it.

5. Is the force field $\langle \sin y + \cos x, e^{2y} + x \cos y \rangle$ conservative? If so, find a potential function for it.

6.† Which of the velocity fields $\mathbf{v}_1 = \langle ye^{xy} - x, y^2 + xe^{xy} \rangle$ and $\mathbf{v}_2 = \langle x \sin y, y \sin x + 4 \rangle$ is irrotational?

7.† Show that the velocity field $\mathbf{v}(x, y)$ of any irrotational, incompressible fluid flow with no sources or sinks in a rectangle is of the form $\mathbf{v} = \overrightarrow{\text{grad}}\, U$ where $U(x, y)$ is a solution of *Laplace's equation* $U_{xx} + U_{yy} = 0$.

8.† Suppose that an object of mass m is at $x = x(t)$, $y = y(t)$ and has nonzero velocity vector $\mathbf{v}(t)$ at time t. The quantity $\frac{1}{2} m[\,|\mathbf{v}(b)|^2 - |\mathbf{v}(a)|^2\,]$ is the increase in the object's *kinetic energy* during the time interval. **a.** Show that the work done by all forces on the object during the time interval is $m \int_a^b \mathbf{v} \cdot \dfrac{d\mathbf{v}}{dt}\, dt$. [Express the total force and the unit tangent vector in terms of $\mathbf{v}(t)$.] **b.** Show that the work is equal to the increase in kinetic energy.

9. (Exact differentials) A *differential* in the variables x and y is an expression of the form $p(x, y)\, dx + q(x, y)\, dy$. The differential of a function $U(x, y)$ is $dU = U_x(x, y)\, dx + U_y(x, y)\, dy$. **a.†** Show that $\int_C dU = U(x_1, y_1) - U(x_0, y_0)$ where (x_0, y_0) is the beginning and (x_1, y_1) the end of the curve C. **b.** A differential is called *exact* if it is the differential of a function. Show that the differential $p\, dx + q\, dy$ is exact in a rectangle if and only if $p_y = q_x$ there, and that in this case $\int_C (p\, dx + q\, dy) = U(x_1, y_1) - U(x_0, y_0)$.

10. Which of the differentials **a.†** $(1 + xy)e^{xy}\, dx + x^2 e^{xy}\, dy$, **b.** $x^2 y^3\, dx + y^2 x^3\, dy$, **c.** $x \sin y\, dx - \cos y\, dy$ are exact?

11.† a. Sketch the vector field $\mathbf{A} = \langle -y(x^2 + y^2)^{-1}, x(x^2 + y^2)^{-1} \rangle$. **b.** Show that curl $\mathbf{A} = 0$ for all $(x, y) \neq (0, 0)$ **c.** Show that $\int_C \mathbf{A} \cdot \mathbf{T}\, ds$ is a nonzero constant for all circles C with their centers at the origin and oriented counterclockwise. (This shows how Stokes' Theorem can fail when the region where the hypotheses of the theorem are valid has a "hole" in it; the vector field $\mathbf{A}$ does not have a derivative at the center $(0, 0)$ of the circles.)

12. Does the vector field in Figure 16.48 have positive or negative curl and divergence?

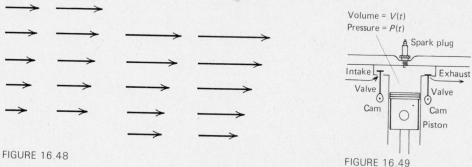

FIGURE 16.48

FIGURE 16.49

13. Figure 16.49 is a schematic sketch of a cylinder in an internal combustion engine. The piston can move up or down and the intake and exhaust valves can be open or closed. We let $P(t)$ denote the pressure inside the chamber and $V(t)$ the volume (measured to the location of the valves whether the valves are opened or closed). Figure 16.50 shows how the pressure and volume vary during one cycle of a four stroke engine. At

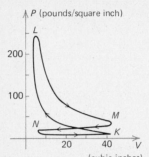

FIGURE 16.50   One cycle of a four stroke engine (Adapted from *McGraw Hill Encyclopedia of Science* Vol. 7, p. 227.)

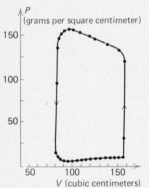

FIGURE 16.51   One cycle of a ventricle of a human heart. The dots represent times 0.02 seconds apart. (Adapted from V. B. Mountcastle, [1] *Medical Physiology*, Vol. 2., p. 905.)

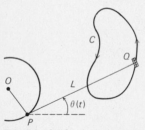

FIGURE 16.52   A planimeter

point K the cylinder has maximum volume and contains an air-gasoline mixture. The path from K to L is the compression stroke (valves closed) when the pressure rises and the volume diminishes. At point L the spark plug ignites the gasoline and the path from L to M is the expansion stroke (valves closed). The path from M to N is the exhaust stroke (exhaust valve open), and the path from N to K is the intake stroke (intake valve open).   **a.**[†]   Show that the work done on the piston during one cycle is $\int_C P\,dV$, where C is the curve in Figure 16.50.   **b.**   Show that the work is the difference of the areas inside the two loops of the curve.

14.   Figure 16.51 shows how the pressure P and volume V of one ventricle of a person's heart vary during one cycle. The work done by the heart is the line integral $-\int_C P\,dV$ with C the curve in the figure. Give an approximate value for the work by estimating either the line integral or the area inside the curve.

15.   A *planimeter* (Figure 16.52) is a draftsman's tool for measuring areas inside closed curves. The device is fastened at point O outside a circle that contains the region and has a hinge at P. A wheel at Q can turn perpendicular to the bar that holds it and a scale is marked on its circumference to measure the distance it rolls. In this problem we show that the area inside the curve C is determined by the distance the wheel rolls as it moves (turning and sliding) once counterclockwise around the curve. We let $\big(f(t), g(t)\big)$ denote the coordinates of P and suppose that the curve is traversed as t runs from a to b. We let $\theta(t)$ denote the angle shown in the figure and let L denote the length of the arm PQ.   **a.**   Use the formula $\frac{1}{2}\int_C(-y\,dx + x\,dy)$ for the area to show it equals

$$\frac{1}{2}\int_a^b\left(-g\frac{df}{dt} + f\frac{dg}{dt}\right)dt + \frac{1}{2}L^2\int_a^b\frac{d\theta}{dt}\,dt$$

$$(1) \qquad + \frac{1}{2}L\int_a^b\left(g\sin\theta\frac{d\theta}{dt} + f\cos\theta\frac{d\theta}{dt}\right)dt$$

$$+ \frac{1}{2}L\int_a^b\left(-\frac{df}{dt}\sin\theta + \frac{dg}{dt}\cos\theta\right)dt.$$

b.[†]   Show that the first two integrals in (1) are zero and that the last two are equal.   **c.**[†]   Show that the distance the wheel rolls is $\int_C \mathbf{u}\cdot\mathbf{T}\,ds$ with $\mathbf{u} = \langle-\sin\theta, \cos\theta\rangle$.   **d.**   Use the results of parts (a) and (c) to show that the area inside the curve is L times the distance rolled by the wheel.

Some basic facts concerning surface integrals and the three-dimensional versions of Green's Theorem, the Divergence Theorem, and Stokes' Theorem are presented in Exercises 16 through 30.

16.[†]   If Σ is a surface in xyz-space and $f(x, y, z)$ a function defined on Σ, the *surface integral* $\int\int_\Sigma f\,dS$ is defined as the limit of approximating sums

$$\sum_{j=1}^N f(x_j, y_j, z_j)[\text{Area of }\Sigma_j]$$

where $\Sigma_1, \Sigma_2, \Sigma_3, \ldots, \Sigma_N$ is a partition of Σ into subsurfaces and (x_j, y_j, z_j) is a point in Σ_j for each j (Figures 16.53 and 16.54). Use Rule 15.5 and the Mean Value Theorem for double integrals to show that

Σ

FIGURE 16.53   A surface Σ

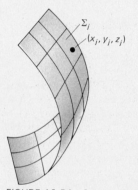

Σ_j

(x_j, y_j, z_j)

FIGURE 16.54   Subsurfaces Σ_j

(2) $$\int\int_{\Sigma} f \, dS = \int\int_{R} f(x, y, g(x, y)) \sqrt{1 + g_x(x, y)^2 + g_y(x, y)^2} \, dx \, dy$$

if Σ is the portion of the surface $z = g(x, y)$ for (x, y) in a bounded region R of the xy-plane.

17.† Express the surface integral $\int\int_{\Sigma} \sin(xyz) \, dS$ as an iterated double integral where Σ is the portion of the paraboloid $z = x^2 - y^2$ for $-1 \leq x \leq 1$, $-2 \leq y \leq 2$.

18. The three-dimensional version of *Green's Theorem* states that

(3) $$\int\int\int_{V} \frac{\partial f}{\partial x} dx \, dy \, dz = \int\int_{\Sigma} f n_1 \, dS$$

(4) $$\int\int\int_{V} \frac{\partial f}{\partial y} dx \, dy \, dz = \int\int_{\Sigma} f n_2 \, dS$$

(5) $$\int\int\int_{V} \frac{\partial f}{\partial z} dx \, dy \, dz = \int\int_{\Sigma} f n_3 \, dS$$

where V is a bounded solid in xyz-space with boundary Σ and unit exterior normal $\boldsymbol{n} = \langle n_1, n_2, n_3 \rangle$ (Figure 16.55), and $f(x, y, z)$ is a function with continuous first-order derivatives in V. Consider a solid V consisting of the points with $g(x, y) \leq z \leq h(x, y)$ and (x, y) in a bounded region of the xy-plane (Figure 16.56).   **a.** Show that

(6) $$\int\int\int_{V} \frac{\partial f}{\partial z} dx \, dy \, dz$$
$$= \int\int_{R} [f(x, y, h(x, y)) - f(x, y, g(x, y))] \, dx \, dy.$$

b.† Use (6) to derive (5) for this type of region by showing that $n_3 \, dS = dx \, dy$ on the top of the region, $n_3 \, dS = -dx \, dy$ on the bottom, and $n_3 = 0$ on the sides.   **c.** Explain why establishing formula (3), (4), or (5) for all of the subsolids in a partition of a solid V gives that result for V itself.

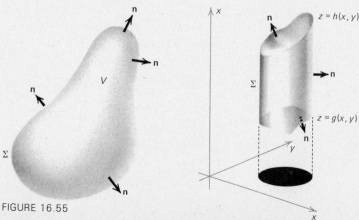

$\boldsymbol{n}$ $\boldsymbol{n}$ V $\boldsymbol{n}$ $\boldsymbol{n}$ Σ $\boldsymbol{n}$

FIGURE 16.55

x $\boldsymbol{n}$ $z = h(x, y)$ $\boldsymbol{n}$ Σ $z = g(x, y)$ $\boldsymbol{n}$ y x

FIGURE 16.56

19. The *flux* of a vector field $\mathbf{v} = p\mathbf{i} + q\mathbf{j} + r\mathbf{k}$ over a surface with unit normals $\mathbf{n}$ is the surface integral $\int\int_\Sigma \mathbf{v} \cdot \mathbf{n}\, dS$. Explain why the flux is the rate of flow across the surface in the direction of $\mathbf{n}$ if $\mathbf{v}$ is the velocity field of a fluid.

20. The *divergence* of a vector field $\mathbf{v} = p\mathbf{i} + q\mathbf{j} + r\mathbf{k}$ is div $\mathbf{v}$ $= p_x + q_y + r_z$. Compute the divergence of   **a.**[†] $\langle xyz, ye^z, x - z^2 \rangle$, **b.** $\langle xz, yz, x + y + z \rangle$,   **c.** $\overrightarrow{\text{grad}}(x^{10}y^{11}z^{12})$.

21. Use Green's Theorem (Exercise 18) to prove the three-dimensional version of the *Divergence Theorem*: for a bounded region V with boundary Σ and unit exterior normal $\mathbf{n}$ and for a vector field $\mathbf{v}$ with continuous first order derivatives in V

(7) $$\int\int_\Sigma \mathbf{v} \cdot \mathbf{n}\, dS = \int\int\int_V \text{div } \mathbf{v}\, dx\, dy\, dz.$$

22. Use Green's Theorem to prove the *gradient theorem*: under the conditions of Exercise 21

(8) $$\int\int_\Sigma f\mathbf{n}\, dS = \int\int\int_V \overrightarrow{\text{grad}} f\, dx\, dy\, dz$$

for function $f(x, y, z)$ with continuous derivatives in V.

23. The symbolic vector $\dfrac{\partial}{\partial x}\mathbf{i} + \dfrac{\partial}{\partial y}\mathbf{j} + \dfrac{\partial}{\partial z}\mathbf{k}$ is denoted $\overrightarrow{\nabla}$ (nabla). Show how $\overrightarrow{\nabla} f$ and $\overrightarrow{\nabla} \cdot \mathbf{v}$ can be interpreted to mean $\overrightarrow{\text{grad}} f$ and div $\mathbf{v}$ for a function f of three variables and a vector field $\mathbf{v}$ in space.

24. The (vector) *curl* of a three-dimensional vector field $\mathbf{v} = p\mathbf{i} + q\mathbf{j} + r\mathbf{k}$ is the vector field $\overrightarrow{\text{curl}}\, \mathbf{v} = \overrightarrow{\nabla} \times \mathbf{v}$. Compute the curls of
a.[‡] $\langle xy, y^2z^2, 2x + 3z \rangle$,   **b.**[†] $\langle ye^z, x^3 - y^2 + z^3, xyz \rangle$,
c. $\langle x^2y^3z^3, x^3y^2z^3, x^3y^3z^2 \rangle$,   **d.** $\langle x\sin(yz), y\sin z, \sin x \rangle$.

25. Show that for any function $f(x, y, z)$ with continuous second-order derivatives $\overrightarrow{\text{curl}}\, \overrightarrow{\text{grad}} f = \mathbf{O}$.

26.[†] Show that the vector field $\mathbf{F} = p\mathbf{i} + q\mathbf{j} + r\mathbf{k}$ is the gradient field of a function $U(x, y, z)$ in a box with sides parallel to the coordinate planes if and only if $\overrightarrow{\text{curl}}\, \mathbf{F} = \mathbf{O}$.

27. Use Green's Theorem to derive the *Curl Theorem*: for a vector field $\mathbf{v}(x, y, z)$ with continuous first-order derivatives in a bounded region V with boundary Σ and unit exterior normal $\mathbf{n}$

(9) $$\int\int_\Sigma \mathbf{v} \times \mathbf{n}\, dS = \int\int\int_V \overrightarrow{\text{curl}}\, \mathbf{v}\, dx\, dy\, dz.$$

FIGURE 16.57   A surface Σ with boundary curve C

28. Consider a closed curve C that is the boundary of a surface in xyz-space (Figure 16.57). We assume that Σ is *oriented*, meaning that it has a unit normal $\mathbf{n}$ at each point such that $\mathbf{n}$ is a continuous function of the coordinates of its base point. We give the boundary C *positive orientation*, which is determined by a right-hand rule (Figure 16.58): the fingers of a right hand point in the direction of the curve's orientation

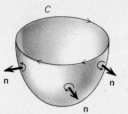

FIGURE 16.58 The boundary C oriented positively

when the thumb points in the direction of nearby normal vectors $\boldsymbol{n}$ on the surface. Stokes' Theorem in xyz-space states that for a vector field $\boldsymbol{v}$ with continuous first-order derivatives on

$$(10) \quad \int_C \boldsymbol{v} \cdot \boldsymbol{T} \, ds = \int \int_\Sigma (\overrightarrow{\operatorname{curl}} \ \boldsymbol{v}) \cdot \boldsymbol{n} \, dS.$$

Use Stokes' Theorem in the plane to prove Stokes' Theorem (10) for a surface Σ that is a region R in the xy-plane.

29. Let Σ_1 be the upper hemisphere $z = \sqrt{1 - x^2 - y^2}$ and Σ_2 the lower hemisphere $z = -\sqrt{1 - x^2 - y^2}$ both with outward pointing normal vectors. Use Stokes' Theorem to show that

$$\int \int_{\Sigma_1} (\overrightarrow{\operatorname{curl}} \ \boldsymbol{v}) \cdot \boldsymbol{n} \, dS = -\int \int_{\Sigma_2} (\overrightarrow{\operatorname{curl}} \ \boldsymbol{v}) \cdot \boldsymbol{n} \, dS$$

for any vector field $\boldsymbol{v}$ with continuous second-order derivatives.

30. Use the Divergence Theorem to obtain the result of Exercise 29.

CHAPTER 17

TAYLOR'S THEOREM, L'HOPITAL'S RULE, IMPROPER INTEGRALS

SUMMARY: This chapter deals with functions of one variable. We study the approximation of functions by Taylor polynomials (Section 17.1), l'Hopital's rule for finding limits of ratios of functions that tend to 0 or ∞ (Section 17.2), and improper integrals of unbounded functions and of functions over infinite intervals (Section 17.3).*

17-1 TAYLOR POLYNOMIALS AND TAYLOR'S THEOREM

In Section 4.10 we used approximations of graphs of functions by tangent lines to find approximate values of functions of one variable. The tangent line to the graph of $f(x)$ at $x = a$ is the graph of the first degree polynomial

$$(1) \quad P_1(x) = f(a) + f'(a)(x - a)$$

and is determined by the values $f(a)$ and $f'(a)$ of the function and its first derivative at $x = a$. In this section we show how, by using higher order derivatives, we can obtain higher degree polynomials, known as TAYLOR POLYNOMIALS,** that approximate the function more closely.

The polynomial (1) may be characterized as the first-degree polynomial that has the same value and the same first derivative as $f(x)$ at a. Similarly, the nth degree Taylor polynomial approximation of $f(x)$ has the same value and the same first n derivatives of $f(x)$ at a.

FIGURE 17.1   Brook Taylor
(1685–1731)
The Granger Collection

---

*Chapter 17, 18, and 19 may be covered any time after Chapter 8.
**Named after the British mathematician Brook Taylor (1685–1731).

Definition 17.1   Suppose that $f(x)$ has n derivatives at $x = a$ (n a positive integer). The nth degree Taylor polynomial approximation of $f(x)$ centered at a is

(2)

$$P_n(x) = f(a) + f'(a)(x - a) + \frac{1}{2}f''(a)(x - a)^2 + \frac{1}{3!}f^{(3)}(a)(x - a)^3$$

$$+ \frac{1}{4!}f^{(4)}(a)(x - a)^4 + \cdots + \frac{1}{n!}f^{(n)}(a)(x - a)^n.$$

◄

The polynomial (2) is said to be "centered" at a because it is a linear combination of powers of $x - a$. Notice that the first-degree Taylor polynomial $P_1(x)$ is the function (1) whose graph is the tangent line.

Expressed with summation notation, formula (2) reads

$$P_n(x) = \sum_{j=0}^{n} \frac{1}{j!} f^{(j)}(a)(x - a)^j.$$

◄

In these formulas $f^{(j)}(x)$ denotes the jth derivative of $f(x)$ for $j \geq 1$ and $f^{(0)}(x)$ denotes $f(x)$. The term $j!$ is called j *factorial*. It is the product $1(2)(3) \cdots (j)$ of the first j positive integers for $j \geq 1$ and is defined to be 1 for $j = 0$.

We leave as Exercise 42 the verification that $P_n(x)$ has the same value and the same first n derivatives as $f(x)$ at $x = a$.

EXAMPLE 1   Give the third degree Taylor polynomial approximation of $\ln x$ centered at $x = 1$.

Solution.   We set $f(x) = \ln x$, for which $f'(x) = x^{-1}, f''(x) = -x^{-2}$, and $f^{(3)}(x) = 2x^{-3}$. The values at $x = 1$ are $f(1) = \ln(1) = 0, f'(1) = 1$, $f''(1) = -1$, and $f^{(3)}(1) = 2$. By formula (2) with $n = 3$ and $a = 1$

$$P_3(x) = f(1) + f'(1)(x - 1) + \frac{1}{2}f''(1)(x - 1)^2$$

$$+ \frac{1}{3!}f^{(3)}(1)(x - 1)^3$$

$$= 0 + (1)(x - 1) + \tfrac{1}{2}(-1)(x - 1)^2 + \tfrac{1}{6}(2)(x - 1)^3$$
$$= (x - 1) - \tfrac{1}{2}(x - 1)^2 + \tfrac{1}{3}(x - 1)^3. \quad \square$$

Notice that the nth degree Taylor polynomial $P_n(x)$ is obtained from the $(n - 1)$st polynomial $P_{n-1}(x)$ by adding the term $\dfrac{1}{n!}f^{(n)}(a)(x - a)^n$.

Generally, higher degree Taylor polynomials give better approximations of the function. This is illustrated in Figure 17.2, where we show the graphs of $\ln x$ and its first, second, and third degree approximations centered at 1.

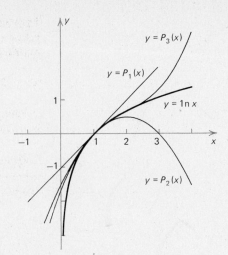

FIGURE 17.2   Taylor polynomial approximations of ln x

EXAMPLE 2   Use the fourth degree Taylor polynomial approximation of e^x centered at $x = 0$ to give an approximate value of $\sqrt{e}$.

Solution.   All the derivatives of $f(x) = e^x$ are e^x, so $f^{(j)}(x) = e^x$ and $f^{(j)}(0) = 1$ for all nonnegative integers j. The Taylor polynomial is

$$P_4(x) = 1 + x + \frac{1}{2}x^2 + \frac{1}{3!}x^3 + \frac{1}{4!}x^4.$$

The square root of e is the value of e^x at $x = \frac{1}{2}$, so we have the approximation

$$e = e^{1/2} \approx P_4\left(\frac{1}{2}\right) = 1 + \frac{1}{2} + \frac{1}{2}\left(\frac{1}{2}\right)^2 + \frac{1}{3!}\left(\frac{1}{2}\right)^3 + \frac{1}{4!}\left(\frac{1}{2}\right)^4$$

$$= 1 + \frac{1}{2} + \frac{1}{2}\left(\frac{1}{4}\right) + \frac{1}{6}\left(\frac{1}{8}\right) + \frac{1}{24}\left(\frac{1}{16}\right)$$

$$= 1 + \frac{1}{2} + \frac{1}{8} + \frac{1}{48} + \frac{1}{384}.$$

This approximation has the decimal expansion $1.64843\ldots$, which is within 0.0003 of the exact value $e^{1/2} = 1.64872\ldots$ □

Taylor's Theorem

We can determine the accuracy of the approximation of $f(x)$ by a Taylor polynomial $P_n(x)$ by studying the difference

(3) $R_n(x) = f(x) - P_n(x)$ ◄

between the function and the approximation. Equation (3) is often written in the form

(4) $f(x) = P_n(x) + R_n(x)$ ◄

and for this reason $R_n(x)$ is called the REMAINDER in the approximation. In many cases we can estimate the size and determine the sign of the remainder by using the next result, which is known as TAYLOR'S THEOREM.

Theorem 17.1 **Suppose that $f(x)$ has $n + 1$ derivatives in an open interval I and that a is a point in I. For each x in I there is at least one number c between a and x such that**

(5) $$R_n(x) = \frac{1}{(n + 1)!} f^{(n+1)}(c)(x - a)^{n+1}$$ ◀

where $R_n(x)$ is the remainder (3) in the approximation of $f(x)$ by $P_n(x)$.

When we say that c lies "between" a and x we mean that $a < c < x$ if x is greater than a and $x < c < a$ if x is less than a. We will prove Theorem 17.1 at the end of this section.

EXAMPLE 3   (a) Compute the approximation of $\ln(1.1)$ obtained by using the third-degree Taylor polynomial approximation of Example 1. (b) Use Taylor's Theorem to give upper and lower bounds on the remainder in that approximation. (c) Is the approximation greater than or less than the exact value of $\ln(1.1)$?

Solution.   (a) The Taylor polynomial is

$$P_3(x) = (x - 1) - \tfrac{1}{2}(x - 1)^2 + \tfrac{1}{3}(x - 1)^3$$

and its value at $x = 1.1$ is

(6) $$\begin{aligned} P_3(1.1) &= (1.1 - 1) - \tfrac{1}{2}(1.1 - 1)^2 + \tfrac{1}{3}(1.1 - 1)^3 \\ &= 0.1 - \tfrac{1}{2}(0.01) + \tfrac{1}{3}(0.001) = 0.095333 \ldots . \end{aligned}$$

(b) Formula (5) with $n = 3, a = 1$, and $f(x) = \ln x$, for which $f^{(4)}(x) = -6x^{-4}$, gives

$$R_3(x) = \frac{1}{4!} f^{(4)}(c)(x - 1)^4 = -\frac{1}{4c^4}(x - 1)^4.$$

At $x = 1.1$ this yields $R_3(1.1) = -\tfrac{1}{4}c^{-4}(1.1 - 1)^4 = -(0.000025)c^{-4}$, where c is a number between 1.1 and 1. The remainder is approximately -0.000025. The quantity c^{-4} must lie between $(1.1)^{-4}$ and $1^{-4} = 1$, so the remainder must satisfy $-0.000025 < R_3(1.1) < -0.000025(1.1)^{-4}$.
   (c) Because the remainder in the equation

$$\ln(1.1) = P_3(1.1) + R_3(1.1)$$

is negative, the approximation $P_3(1.1) = 0.095333 \ldots$ is greater than the exact value of $\ln(1.1)$ (which is $0.095310 \ldots$). □
   The absolute value $|f(x) - P_n(x)|$ of the difference between the values of the function f and its Taylor polynomial approximation is the *error* made in the approximation and, by formula (4), is equal to the absolute value $|R_n(x)|$ of the remainder. In Example 3 the calculations show that the error $|R_3(1.1)|$ is less than 0.000025.
   The next example deals with a polynomial $f(x) = x^4$ whose derivatives of degree 5 and higher are all zero. Because of this, the Taylor polynomials $P_n(x)$ for $n \geq 4$ are equal to x^4. The usefulness of the Taylor polynomials in this case is that they give us an easy way to

express the given polynomial x^4 as a sum of powers of $x - a$ for arbitrary constants a.

EXAMPLE 4   Use a Taylor polynomial to express x^4 as a sum of powers of $x - 1$.

Solution.   We use a Taylor polynomial centered at 1 to obtain powers of $x - 1$, and we use $P_4(x)$ because x^4 has degree 4. We have $f(x) = x^4$, $f'(x) = 4x^3, f''(x) = 12x^2, f^{(3)}(x) = 24x$, and $f^{(4)}(x) = 24$, so that $f(1) = 1, f'(1) = 4, f''(1) = 12, f^{(3)}(1) = 24, f^{(4)}(1) = 24$ and the Taylor polynomial is

$$P_4(x) = 1 + 4(x - 1) + \frac{1}{2}(12)(x - 1)^2 + \frac{1}{3!}(24)(x - 1)^3$$

$$+ \frac{1}{4!}(24)(x - 1)^4$$

$$= 1 + 4(x - 1) + 6(x - 1)^2 + 4(x - 1)^3 + (x - 1)^4.$$

This polynomial equals $f(x) = x^4$ because the corresponding remainder

$$R_4(x) = \frac{1}{5!}f^{(5)}(c)(x - 1)^5$$

involves the zero fifth derivative of x^4 and is zero. $\square$

The Generalized Mean Value Theorem

The proof of Taylor's Theorem is based on the following generalization of the Mean Value Theorem for derivatives.

Theorem 17.2   (The Generalized Mean Value Theorem for derivatives) Suppose that g and h are continuous for $a \leq x \leq b$ and have derivatives for $a < x < b$ and that $h'(x)$ is not zero. Then there is at least one number c with $a < c < b$ such that

(7) $$\frac{g(b) - g(a)}{h(b) - h(a)} = \frac{g'(c)}{h'(c)}.$$ ◄

The Mean Value Theorem (Theorem 4.8 in Section 4.12) is Theorem 17.2 in the special case of $h(x) = x$.

Proof of Theorem 17.2.   The Mean Value Theorem and the condition $h'(x) \neq 0$ show that $h(b) - h(a)$ is not zero. The function

$$F(x) = [g(b) - g(a)][h(x) - h(a)]$$
$$- [h(b) - h(a)][g(x) - g(a)]$$

is zero at a and b, is continuous for $a \leq x \leq b$, and has the derivative

(8) $$F'(x) = [g(b) - g(a)]h'(x) - [h(b) - h(a)]g'(x)$$

for $a < x < b$. Because $F(b) - F(a)$ is zero, there is at least one point c with $a < c < b$ such that $F'(c) = 0$. This equation with formula (8) yields (7).   Q.E.D.

Proof of Theorem 17.1.   We set $g(x) = R_n(x) = f(x) - P_n(x)$ and $h(x) = (x - a)^{n+1}$. Because $f(x)$ and $P_n(x)$ have the same values and the same first n derivatives at $x = a$, the value and first n derivatives of $g(x)$ at $x = a$ are zero. A short calculation shows that $h(x)$ and its first n derivatives are zero at $x = a$ and at no other points.

Consider b in the interval I with $b > a$. (A similar argument would handle the case of $b < a$.) By the Generalized Mean Value Theorem there is a number c_1 with $a < c_1 < b$ such that

$$\frac{g(b)}{h(b)} = \frac{g(b) - g(a)}{h(b) - h(a)} = \frac{g'(c_1)}{h'(c_1)}.$$

Then, by the same theorem applied to the functions $g'(x)$ and $h'(x)$, there is a number c_2 with $a < c_2 < c_1$ such that

$$\frac{g(b)}{h(b)} = \frac{g'(c_1)}{h'(c_1)} = \frac{g'(c_1) - g'(a)}{h'(c_1) - h'(a)} = \frac{g''(c_2)}{h''(c_2)}.$$

We continue in this fashion, applying the Generalized Mean Value Theorem to successive derivatives of $g(x)$ and $h(x)$ until we obtain the equation

$$(9) \qquad \frac{g(b)}{h(b)} = \frac{g^{(n+1)}(c_{n+1})}{h^{(n+1)}(c_{n+1})}$$

where $a < c_{n+1} < c_n < \cdots < c_2 < c_1 < b$.

Because $P_n(x)$ is a polynomial of degree n, its $(n + 1)$st derivative is zero and $g^{(n+1)}(c_{n+1}) = f^{(n+1)}(c_{n+1})$. The $(n + 1)$st derivative of $h(x)$ is the constant $(n + 1)!$. Therefore, when we make the substitutions $g(b) = R_n(b)$ and $h(b) = (b - a)^{n+1}$ in (9), we obtain

$$(10) \qquad \frac{R_n(b)}{(b - a)^{n+1}} = \frac{1}{(n + 1)!} f^{(n+1)}(c_{n+1})$$

which, if we write c for c_{n+1}, yields (5) to prove the theorem.   Q.E.D.

Exercises

†*Answer provided.*
‡*Outline of solution provided.*

In each of Exercises 1 through 20 give the nth degree Taylor polynomial of the function centered at $x = a$.

1.† e^x; $n = 4$; $a = 0$

2. $\cos x$; $n = 6$; $a = 0$

3.‡ $\dfrac{1}{1 - x}$; $n = 3$; $a = 0$

4. $\arctan x$; $n = 4$; $a = 0$

5.† $\sqrt{x}$; $n = 2$; $a = 9$

6. $\ln x$; $n = 3$; $a = e$

7.† e^{-3x}; $n = 4$; $a = 1$

8. $\arcsin x$; $n = 2$; $a = 0$

✳ 9.‡ $\ln(\cos x)$; $n = 4$; $a = 0$

10. $\sin x$; $n = 9$; $a = 0$

11.† $(1 + x)^{1/3}$; $n = 3$; $a = 0$

12. $x^{1/3}$; $n = 3$; $a = 1$

13.† $e^{\sin x}$; $n = 3$; $a = 0$

14. $\tan x$; $n = 3$; $a = \pi/4$

15.† $\cos(6x)$; $n = 4$; $a = 1$

16. $(1 + e^x)^{-1}$; $n = 3$; $a = 0$

17.† $\sinh x;\ n = 5;\ a = 0$ **18.** $e^{x^2};\ n = 3;\ a = 1$

19.† $x^5 - x^4;\ n = 3;\ a = 1$ **20.** $x^4;\ n = 4;\ a = 10$

In each of Exercises 21 through 30 **a.** give the nth degree Taylor polynomial approximation of the function centered at a, **b.** use the polynomial to approximate the designated value of the function, and **c.** use Taylor's Theorem to determine whether the approximation is less than or greater than the exact value.

21.‡ **a.** $e^x;\ n = 4;\ a = 0$ **b.** $e^{1/2}$

22.† **a.** $\sin x;\ n = 5;\ a = 0$ **b.** $\sin(\pi/10)$

23.† **a.** $\arcsin x;\ n = 1;\ a = \frac{1}{2}$ **b.** $\arcsin(0.48)$

24. **a.** $x^{2/3};\ n = 2;\ a = 1$ **b.** $(1.03)^{2/3}$

25.† **a.** $\sinh x;\ n = 5;\ a = 0$ **b.** $\sinh(-1)$

26. **a.** $\sqrt{1 + x};\ n = 2;\ a = 0$ **b.** $\sqrt{1.02}$

27.† **a.** $\arctan x;\ n = 1;\ a = 1$ **b.** $\arctan(0.9)$

28. **a.** $x^{1/2};\ n = 2;\ a = 25$ **b.** $\sqrt{24}$

29. **a.** $\cos x;\ n = 5;\ a = \pi/2$ **b.** $\cos(9\pi/16)$

30.† **a.** $\tan x;\ n = 2;\ a = \pi/3$ **b.** $\tan(4\pi/15)$

In each of Exercises 31 through 40 **a.** give the nth degree Taylor polynomial approximation of the function centered at $x = a$, **b.** use the polynomial to approximate the indicated value of the function, and **c.** use Taylor's Theorem to give bounds on the error made in this approximation.

31. **a.** $e^x;\ n = 4;\ a = 0$ **b.** $e^1 = e$ (Use $1 < e < 3$.)

32.† **a.** $\cos x;\ n = 3;\ a = \pi/4$ **b.** $\cos(7\pi/20)$ (Use $\pi/4 < 7\pi/20 < \pi/2$.)

33. **a.** $\ln(1 + x);\ n = 3;\ a = 0$ **b.** $\ln(2)$

34. **a.** $x^{1/2};\ n = 3;\ a = 1$ **b.** 2 (Use $1 < 2 < 4$.)

35.† **a.** $1/(1 + x);\ n = 3;\ a = 0$ **b.** $1/1.01$ (Use $0 < 0.01 < 1$.)

36. **a.** $\sin x;\ n = 5;\ a = 0$ **b.** $\sin(1)$ (Use $|\sin x| \le 1$.)

37. **a.** $\sin x;\ n = 6;\ a = 0$ **b.** $\sin(1)$ (Use $|\cos x| \le 1$.)

38.† **a.** $\cos x;\ n = 2;\ a = \pi/3$ **b.** $\cos(7\pi/24)$ (Use $\pi/4 < 7\pi/24 < \pi/3$.)

39. **a.** $x^{1/3};\ n = 1;\ a = 8$ **b.** $\sqrt[3]{7.994}$ (Use $1 < 7.994 < 8$.)

40.† **a.** $x^{60};\ n = 2;\ a = 1$ **b.** $(0.99)^{60}$ (Use $0 < 0.99 < 1$.)

41. Match the functions $\sin x$, $\cos x$, $\sinh x$, $\cosh x$, e^x, 2^x, $\arctan x$, and $\ln(1 + x)$ to the Taylor polynomials (a) through (h).

a.‡ $\displaystyle\sum_{j=0}^{n} \frac{1}{j!} x^j$ **b.**‡ $\displaystyle\sum_{j=0}^{n} \frac{(-1)^j}{(2j)!} x^{2j}$ **c.** $\displaystyle\sum_{j=0}^{n} \frac{(-1)^j}{(2j + 1)!} x^{2j+1}$

d.[†] $\displaystyle\sum_{j=0}^{n} \frac{1}{(2j+1)!} x^{2j+1}$ **e.** $\displaystyle\sum_{j=0}^{n} \frac{1}{(2j)!} x^{2j}$

f.[†] $\displaystyle\sum_{j=0}^{n} \frac{1}{j!} (x \ln 2)^j$ **g.** $\displaystyle\sum_{j=1}^{n} \frac{(-1)^{j+1}}{j} x^{j}$ **h.** $\displaystyle\sum_{j=0}^{n} \frac{(-1)^j}{2j+1} x^{2j+1}$

42. Show that $f(x)$ and its nth degree Taylor polynomial $P_n(x)$ centered at a have the same value and the same first n derivatives at $x = a$.

Exercises 43 through 46 deal with the Taylor polynomials $P_n(x)$ for $\sin x$ centered at 0 and the corresponding remainders $R_n(x)$. Use the reference table of values of $1/j!$ in Exercises 45 and 46.

43.[†] Sketch the graphs of $\sin x$, $P_1(x)$, and $P_3(x)$ in one diagram.

44. Show that $P_{2j}(x)$ is the same as $P_{2j-1}(x)$ for each positive integer j, so that $R_{2j}(x)$ may be used in estimating the errors made in using $P_{2j-1}(x)$.

45.[†] **a.** Compute $P_7(3)$.   **b.** Estimate the error $R_8(3)$ made in the approximation $P_7(3)$ of $\sin(3)$.   **c.** Use the value $\sin(3) \approx 0.1411$ to compute the error in the approximation of part (a).

46. Use R_{14} and R_{24} to estimate the errors in approximating
a. $\sin(5)$ by $P_{13}(5)$,   **b.** $\sin(6)$ by $P_{13}(6)$,   **c.** $\sin(9)$ by $P_{23}(9)$,
and   **d.** $\sin(10)$ by $P_{23}(10)$.   **e.** Figure 17.3 shows the graphs of $\sin x$, $P_{13}(x)$, and $P_{23}(x)$. Use the figure to estimate the errors in parts (a) through (d).

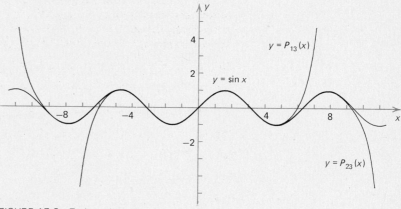

FIGURE 17.3   Taylor polynomial approximations of $\sin x$

17-2 L'HOPITAL'S RULE

The ratio $\dfrac{f(x)}{g(x)}$ is called an INDETERMINATE FORM of type $0/0$ as x tends to a if $f(x)$ and $g(x)$ both tend to 0 as x tends to a. The ratio is an indeterminate form of type ∞/∞ if $f(x)$ and $g(x)$ both tend to infinity. In this section we use the derivative

$$f'(a) = \lim_{x \to a} \frac{f(x) - f(a)}{x - a}$$

which is the limit of an indeterminate form of type $0/0$, to find limits of a variety of indeterminate forms of types $0/0$ and ∞/∞. Our calculations use the following result, which is known as L'HOPITAL'S RULE after Marquis G. F. A. de l'Hopital (1661–1704).

Theorem 17.3   Suppose that $f(x)$ and $g(x)$ are defined and that $f'(x)$ and $g'(x)$ exist with $g'(x) \neq 0$ for all $x \neq a$ in an open interval I containing a. Suppose that either $f(x)$ and $g(x)$ both tend to 0 as x tends to a, or that $f(x)$ and $g(x)$ tend to ∞ or $-\infty$. If the ratio $\dfrac{f'(x)}{g'(x)}$ has a limit as x tends to a, then

$$(1) \qquad \lim_{x \to a} \frac{f(x)}{g(x)} = \lim_{x \to a} \frac{f'(x)}{g'(x)}. \qquad\qquad\blacktriangleleft$$

The limit (1) may be a number or $\pm\infty$. The condition $x \to a$ may be replaced by $x \to a^+$ with I an open interval $a < x < b$, by $x \to a^-$ with I an open interval $b < x < a$, by $x \to \infty$ with I an open half line $x > b$, or by $x \to -\infty$ with I an open half line $x < b$.

Historical Note.   L'Hopital's rule was proved by Jean Bernoulli while he was being paid a regular salary by the marquis in return for the rights to his mathematical discoveries. L'Hopital published the result in his *Analyse des infiniments petits* (1696), which was the first textbook on differential calculus.   □

Proof for the case of 0/0.   Consider first the case of one-sided limits as $x \to a^+$ with I the interval $a < x < b$. Define $f(a)$ and $g(a)$ to be 0 so f and g are continuous for $a \leq x < b$. By the Generalized Mean Value Theorem there is, for each x with $a < x < b$, a number c with $a < c < x$ such that

$$\frac{f(x)}{g(x)} = \frac{f(x) - f(a)}{g(x) - g(a)} = \frac{f'(c)}{g'(c)}.$$

As x tends to a from the right, the number c also tends to a from the right, so the ratio $\dfrac{f(x)}{g(x)}$ has the same limit as $\dfrac{f'(x)}{g'(x)}$, as stated in the theorem. A similar argument yields (1) for limits as $x \to a^-$, and the cases of one-sided limits give (1) for limits as $x \to a$.

To handle the case of limits as $x \to \infty$, we set $x = 1/t$, so that $x \to \infty$ as $t \to 0^+$, and we define $F(t) = f(1/t)$, $G(t) = g(1/t)$. Then $F(t)$ and $G(t)$ tend to 0 as $t \to 0^+$, and $F'(t) = -t^{-2}f'(1/t)$, $G'(t) = -t^{-2}g'(1/t)$. By l'Hopital's rule (1) for the case of $t \to 0^+$, we obtain

$$\lim_{x \to \infty} \frac{f(x)}{g(x)} = \lim_{t \to 0^+} \frac{F(t)}{G(t)} = \lim_{t \to 0^+} \frac{F'(t)}{G'(t)}$$

$$= \lim_{t \to 0^+} \frac{f'\!\left(\dfrac{1}{t}\right)(-t^{-2})}{g'\!\left(\dfrac{1}{t}\right)(-t^{-2})} = \lim_{t \to 0^+} \frac{f'\!\left(\dfrac{1}{t}\right)}{g'\!\left(\dfrac{1}{t}\right)} = \lim_{x \to \infty} \frac{f'(x)}{g'(x)}$$

which is formula (1) in this case. Replacing x by $-x$ gives (1) for limits as $x \to -\infty$. Q.E.D.

The proof of l'Hopital's rule for indeterminate forms of the type ∞/∞ requires the $\epsilon\delta$-definition of limits (see Miscellaneous Exercise 22).

EXAMPLE 1 Determine

$$\lim_{x \to 0} \frac{\sin(6x)}{\sin x}.$$

Solution. Because $\sin(6x)$ and $\sin x$ both tend to 0 as x tends to 0, the ratio $\sin(6x)/\sin x$ is an indeterminate form of type 0/0 as x tends to 0. By rule (1) we have

$$\lim_{x \to 0} \frac{\sin(6x)}{\sin x} = \lim_{x \to 0} \frac{\dfrac{d}{dx}\sin(6x)}{\dfrac{d}{dx}\sin x} = \lim_{x \to 0} \frac{6\cos(6x)}{\cos x} = 6.$$

The last limit is 6 because $\cos(6x)$ and $\cos x$ both tend to 1 as x tends to 0. □

EXAMPLE 2 What is the limit of e^x/x as $x \to \infty$?

Solution. The expression e^x/x is an indeterminate form of type ∞/∞ as x tends to ∞. We have

$$\lim_{x \to \infty} \frac{e^x}{x} = \lim_{x \to \infty} \frac{\dfrac{d}{dx}e^x}{\dfrac{d}{dx}x} = \lim_{x \to \infty} \frac{e^x}{1} = \infty. \quad □$$

Because of the result of the last example, we say that e^x "grows more rapidly" than x as x tends to ∞. In fact, e^x grows more rapidly than any power x^n of x (see Exercise 34).

EXAMPLE 3 Show that $\dfrac{\ln x}{\sqrt[n]{x}}$ tends to 0 as x tends to ∞ for arbitrary positive integers n.

Solution. The quotient $(\ln x)/\sqrt[n]{x}$ is of indeterminate type ∞/∞ because both $\ln x$ and $\sqrt[n]{x} = x^{1/n}$ tend to ∞ as x tends to ∞. We have

$$\lim_{x \to \infty} \frac{\ln x}{x^{1/n}} = \lim_{x \to \infty} \frac{\dfrac{d}{dx}\ln x}{\dfrac{d}{dx}x^{1/n}} = \lim_{x \to \infty} \frac{x^{-1}}{\dfrac{1}{n}x^{-1+1/n}}$$

$$= \lim_{x \to \infty} \frac{n}{x^{1/n}} = 0.$$

We refer to the result of this exercise by stating that $\ln x$ "grows more slowly" as $x \to \infty$ than every root $\sqrt[n]{x}$ of x.   □

In the next example we have to apply l'Hopital's rule twice because the ratio $\dfrac{f'(x)}{g'(x)}$ is also an indeterminate form.

EXAMPLE 4   Find the limit

$$\lim_{x \to \pi} \frac{\sin^2 x}{\cos(3x) + 1}.$$

Solution.   The ratio $\sin^2 x / [\cos(3x) + 1]$ is an indeterminate form of type $0/0$ as $x \to \pi$. We have

$$\lim_{x \to \pi} \frac{\sin^2 x}{\cos(3x) + 1} = \lim_{x \to \pi} \frac{\dfrac{d}{dx} \sin^2 x}{\dfrac{d}{dx}[\cos(3x) + 1]}$$

(2)

$$= \lim_{x \to \pi} \frac{2 \sin x \cos x}{-3 \sin(3x)}.$$

The new ratio $2 \sin x \cos x / [-3 \sin(3x)]$ is also an indeterminate form of type $0/0$ as x tends to π. We apply l'Hopital's rule again:

$$\lim_{x \to \pi} \frac{2 \sin x \cos x}{-3 \sin(3x)} = \lim_{x \to \pi} \frac{\dfrac{d}{dx}[2 \sin x \cos x]}{\dfrac{d}{dx}[-3 \sin(3x)]}$$

(3)

$$= \lim_{x \to \pi} \frac{2 \cos^2 x - 2 \sin^2 x}{-9 \cos(3x)} = \frac{2(-1)^2 - 2(0)^2}{-9(-1)} = \frac{2}{9}.$$

Combining (2) and (3) shows that the limit of the original ratio is $\frac{2}{9}$.   □

Other types of indeterminate forms

The product $f(x)g(x)$ is an indeterminate form of type 0∞ as x tends to a if $f(x)$ tends to 0 and $g(x)$ tends to ∞. We can find the limits of many such indeterminate forms by putting the reciprocal of one of the functions in the denominator to obtain an indeterminate form of type $0/0$ or ∞/∞. The expression $f(x)^{g(x)}$ is an indeterminate form of type 0^0 as x tends to a if $f(x)$ and $g(x)$ both tend to 0. It is of the type 1^∞ if $f(x)$ tends to 1 and $g(x)$ to ∞. It is of type ∞^0 if $f(x)$ tends to ∞ and $g(x)$ to 0. The limits of such indeterminate forms can often be found by using logarithms.

EXAMPLE 5   What is the limit of $x \ln x$ as $x \to 0^+$?

Solution.   The expression $x \ln x$ is an indeterminate form of type 0∞ as x tends to 0^+ because x tends to 0 and $\ln x$ tends to $-\infty$. We have

$$\lim_{x \to 0^+} [x \ln x] = \lim_{x \to 0^+} \frac{\ln x}{\dfrac{1}{x}} = \lim_{x \to 0^+} \frac{\dfrac{d}{dx} \ln x}{\dfrac{d}{dx}\left(\dfrac{1}{x}\right)}$$

$$= \lim_{x \to 0^+} \frac{x^{-1}}{-x^{-2}} = \lim_{x \to 0^+} [-x] = 0. \quad \square$$

EXAMPLE 6 Find the limit of $f(x) = \left(1 + \dfrac{1}{x}\right)^x$ as $x \to \infty$.

Solution. The function $f(x)$ is an indeterminate form of type 1^∞ as $x \to \infty$. We know from our study of logarithms in Chapter 7 that the limit is the number e. To find it by using l'Hopital's rule we first find the limit of $\ln[f(x)]$:

$$\lim_{x \to \infty} \ln[f(x)] = \lim_{x \to \infty} \ln\left[\left(1 + \frac{1}{x}\right)^x\right] = \lim_{x \to \infty}\left[x \ln\left(1 + \frac{1}{x}\right)\right]$$

$$= \lim_{x \to \infty} \frac{\ln\left(1 + \dfrac{1}{x}\right)}{\dfrac{1}{x}} = \lim_{x \to \infty} \frac{\dfrac{d}{dx}\ln\left(1 + \dfrac{1}{x}\right)}{\dfrac{d}{dx}\left(\dfrac{1}{x}\right)}$$

$$= \lim_{x \to \infty} \frac{(-x^{-2}) \Big/ \left(1 + \dfrac{1}{x}\right)}{-x^{-2}} = \lim_{x \to \infty} \frac{1}{1 + \dfrac{1}{x}} = 1$$

Because the limit of $\ln[f(x)]$ is 1, the limit of $f(x)$ is e. $\quad \square$

EXAMPLE 7 What is the limit of $\left(\dfrac{1}{x}\right)^x$ as $x \to 0^+$?

Solution. The expression $(1/x)^x$ is of type ∞^0 as $x \to 0^+$. Taking logarithms gives

$$\lim_{x \to 0^+} \ln\left[\left(\frac{1}{x}\right)^x\right] = \lim_{x \to 0^+}\left[x \ln\left(\frac{1}{x}\right)\right] = \lim_{x \to 0^+} \frac{-\ln x}{x^{-1}}$$

$$= \lim_{x \to 0^+} \frac{\dfrac{d}{dx}(-\ln x)}{\dfrac{d}{dx}(x^{-1})} = \lim_{x \to 0^+} \frac{-x^{-1}}{-x^{-2}} = \lim_{x \to 0^+} x = 0.$$

Because its logarithm tends to 0, the function $(1/x)^x$ tends to 1. $\quad \square$

In the next example we do not (and cannot) apply l'Hopital's rule to find the limit because the expression is not indeterminate. Using l'Hopital's rule would yield the wrong result.

EXAMPLE 8   Find the limit

$$\lim_{x \to 0} \frac{\sin(3x)}{1 + \sin(4x)}.$$

Solution.   The limit is zero because the numerator, $\sin(3x)$, tends to zero and the denominator, $1 + \sin(4x)$, tends to the nonzero number 1 as x tends to zero. $\square$

If we were to make the mistake of using l'Hopital's rule in Example 8, we would think that the limit would be

$$\lim_{x \to 0} \frac{\dfrac{d}{dx}[\sin(3x)]}{\dfrac{d}{dx}[1 + \sin(4x)]} = \lim_{x \to 0} \frac{3\cos(3x)}{4\cos(4x)} = \tfrac{3}{4}.$$

Exercises   Determine the limits in Exercises 1 through 30.

1.† $\displaystyle\lim_{x \to 0} \frac{\sin(3x)}{\sin(2x)}$

2. $\displaystyle\lim_{t \to 1} \frac{t^3 + t^2 - 2}{\ln t}$

3.† $\displaystyle\lim_{x \to 2} \frac{\tan(4 - 2x)}{x^3 - 8}$

4. $\displaystyle\lim_{x \to \infty} \frac{\ln x}{x}$

5.† $\displaystyle\lim_{x \to \infty} \frac{1 - 6x^2}{5x^2 + 4x}$

6.† $\displaystyle\lim_{x \to 2} \frac{x - 2}{6x^2 - 10x - 4}$

7. $\displaystyle\lim_{x \to \pi/4} \frac{\ln(\tan x)}{\sin x - \cos x}$

8.† $\displaystyle\lim_{x \to \infty} x^2 e^{-x}$

9. $\displaystyle\lim_{x \to 2} \frac{x^2 - 4}{\sin x}$

10.† $\displaystyle\lim_{x \to 0} \frac{\ln x}{x^2 - 1}$

11. $\displaystyle\lim_{x \to 1} \frac{\ln x}{x - 1}$

12.† $\displaystyle\lim_{x \to 0} \frac{x^2}{1 - e^x}$

13.‡ $\displaystyle\lim_{x \to 0} \frac{1 - e^{x^2}}{x \sin x}$

14.† $\displaystyle\lim_{x \to 0} \frac{x^3}{\sin x - x}$

15. $\displaystyle\lim_{x \to 0} \frac{x - \sin x}{\sin(x^3)}$

16.‡ $\displaystyle\lim_{x \to \infty} \frac{4x^2 + 15}{x + e^x}$

17. $\displaystyle\lim_{x \to \pi} \frac{\ln(\sin(x/2))}{\cos(x/2)}$

18.‡ $\displaystyle\lim_{x \to 0^+} x^x$

19. $\displaystyle\lim_{t \to 0} \frac{1 + \frac{1}{3}t - (1 + t)^{1/3}}{t^2}$

20.† $\displaystyle\lim_{x \to 0} \left[\frac{1}{x} - \frac{1}{\sin x}\right]$ $\left(\text{Take a common denominator.}\right)$

21. $\displaystyle\lim_{t \to \pi} \frac{1 + \cos t}{(t - \pi)^2}$

22. $\displaystyle\lim_{x \to 0} \frac{1 - e^{2x}}{\sin(7x)}$

23.† $\displaystyle\lim_{x \to \infty} \left[x\left(\frac{\pi}{2} - \arctan x\right)\right]$

24. $\displaystyle\lim_{x \to 0} (1 - 3x)^{1/x}$

25. $\displaystyle\lim_{x \to 2} \frac{2 - \sqrt{x + 2}}{2x^2 - 10x + 8}$

26.‡ $\displaystyle\lim_{x \to \infty} \frac{\arctan(3x)}{\arctan(-x)}$

27. $\displaystyle\lim_{x \to 1} \frac{x^2 + 2x - 3}{1 - e^x}$

28.† $\displaystyle\lim_{t \to 0} \frac{\sin(t^2)}{5t^2 + t^3}$

29.[†] $\displaystyle\lim_{y \to 0} \frac{\cos(6y)}{\cos(5y)}$

30. $\displaystyle\lim_{t \to 1} \frac{2t^3 - 3t + 1}{\ln t}$

31.[†] Sketch the graphs of   **a.** $(\sin x)/x$ and   **b.** $(\cos x)/x$ by plotting the points at $x = k\,\pi/2$, $k = \pm 1, \pm 2, \ldots, \pm 6$.

32. Sketch the graph of $(\tan x)/x$ for $0 < |x| < \pi/2$.

33.[†] Sketch the graph of x^x (defined for $x > 0$). (Use the first derivative test.)

34. Use mathematical induction to show that for every positive integer n, e^x tends to ∞ faster than x^n as $x \to \infty$. (Show that e^x/x^n tends to ∞.)

35. Show that e^{-x} tends to 0 faster than x^{-n} as $x \to \infty$ for every positive integer n. (Show that $x^n e^{-x}$ tends to 0.)

36. Show that for every positive integer n, $|\ln x|$ tends to ∞ slower than $x^{-1/n}$ as $x \to 0^+$. (Show that $x^{1/n} \ln x \to 0$.)

37. Use logarithms and l'Hopital's rule to show that $\left(1 + \dfrac{a}{x}\right)^x$ tends to e^a as x tends to ∞ for arbitrary constants a.

38. What is the limit of $\left(1 - \dfrac{a}{x}\right)^{-x}$ as x tends to ∞?

17-3   IMPROPER INTEGRALS

In Chapter 5 we defined definite integrals only for bounded functions over finite intervals. In this section we show how certain integrals of unbounded functions or of functions over infinite intervals can be defined as IMPROPER INTEGRALS. We start with the case of infinite intervals of integration.

Definition 17.2   Suppose that $f(x)$ is bounded in the interval $a \leq x \leq X$ and the integral of f from a to X is defined for each $X > a$. Then

(1) $\displaystyle\int_a^\infty f(x)\,dx = \lim_{X \to \infty} \int_a^X f(x)\,dx$ ◄

provided this limit exists. If $f(x)$ is bounded in the interval $X \leq x \leq b$ and the integral of f from X to b is defined for each $X < b$, then

(2) $\displaystyle\int_{-\infty}^b f(x)\,dx = \lim_{X \to -\infty} \int_X^b f(x)\,dx$ ◄

provided the limit exists.

If the limit (1) or (2) exists and is finite, we say that the improper integral CONVERGES. If the limit does not exist or is infinite, we say the integral DIVERGES.

EXAMPLE 1   Does the improper integral

$$\int_0^\infty \frac{1}{1 + x^2} \, dx$$

converge? If so, what is its value?

Solution.   The integral converges because we have

$$\int_0^\infty \frac{1}{1 + x^2} \, dx = \lim_{X \to \infty} \int_0^X \frac{1}{1 + x^2} \, dx = \lim_{X \to \infty} \left\{ [\arctan x]_0^X \right\}$$

$$= \lim_{X \to \infty} [\arctan X] = \frac{\pi}{2}. \quad \square$$

EXAMPLE 2   Does the improper integral $\int_0^\infty \frac{x^2}{2 + x^3} \, dx$ converge?

Solution.   For $X > 0$ we have

$$\int_0^X \frac{x^2}{2 + x^3} \, dx = [\tfrac{1}{3}\ln(2 + x^3)]_0^X = \tfrac{1}{3}[\ln(2 + X^3) - \ln(2)].$$

Because $\ln(2 + X^3)$ tends to ∞ as X tends to ∞, the improper integral diverges. $\square$

EXAMPLE 3   Does the improper integral $\int_0^\infty \sin x \, dx$ converge?

Solution.   We have

$$\int_0^X \sin x \, dx = [-\cos x]_0^X = 1 - \cos X.$$

Because $\cos X$ oscillates between -1 and 1 and has no limit as $X \to \infty$, the improper integral diverges. $\square$

To find an improper integral from $-\infty$ to ∞, we first find the improper integrals from $-\infty$ to a and from a to ∞ for an arbitrary number a:

Definition 17.3   If for some number a both improper integrals

(3) $$\int_{-\infty}^a f(x) \, dx \quad \text{and} \quad \int_a^\infty f(x) \, dx$$

exist and are finite, then the improper integral

(4) $$\int_{-\infty}^\infty f(x) \, dx$$

is their sum. The integral (4) is ∞ if one of the integrals (3) is finite and the other is ∞ or both are ∞, is $-\infty$ if one of the integrals (3) is finite and the other is $-\infty$ or both are $-\infty$, and is not defined if one integral (3) is ∞ and the other is $-\infty$.

EXAMPLE 4   Does the improper integral $\int_{-\infty}^{\infty} \dfrac{1}{25 + x^2}\, dx$ converge? If so, what is its limit?

Solution.   Because $\arctan\left(\dfrac{X}{5}\right)$ tends to $\frac{1}{2}\pi$ as X tends to ∞ and tends to $-\frac{1}{2}\pi$ as X tends to $-\infty$, we have

$$\int_{0}^{\infty} \frac{1}{25 + x^2}\, dx = \lim_{X \to \infty} \int_{0}^{X} \frac{1}{25 + x^2}\, dx = \lim_{X \to \infty}\left[\frac{1}{5}\arctan\left(\frac{x}{5}\right)\right]_{0}^{X}$$

$$= \lim_{X \to \infty}\left[\frac{1}{5}\arctan\left(\frac{X}{5}\right)\right] = \frac{1}{5}\left(\frac{1}{2}\pi\right) = \frac{1}{10}\pi$$

and

$$\int_{-\infty}^{0} \frac{1}{25 + x^2}\, dx = \lim_{X \to -\infty} \int_{X}^{0} \frac{1}{25 + x^2}\, dx = \lim_{X \to -\infty}\left[\frac{1}{5}\arctan\left(\frac{x}{5}\right)\right]_{X}^{0}$$

$$= \lim_{X \to -\infty}\left[-\frac{1}{5}\arctan\left(\frac{X}{5}\right)\right] = -\frac{1}{5}\left(-\frac{1}{2}\pi\right) = \frac{1}{10}\pi.$$

Therefore, the improper integral from $-\infty$ to ∞ exists and is the sum $\frac{1}{10}\pi + \frac{1}{10}\pi = \frac{1}{5}\pi$ of the improper integrals from 0 to ∞ and from $-\infty$ to 0. $\square$

We now turn to the case where the function is not bounded at one end of a finite interval of integration.

Definition 17.4   **If for each X with $a < X < b$, $f(x)$ is bounded in the interval $a \leq x \leq X$ and the integral of f from a to X is defined, then**

$$(5) \qquad \int_{a}^{b} f(x)\, dx = \lim_{X \to b^-} \int_{a}^{X} f(x)\, dx$$

provided the limit exists. If for each X with $a < X < b$, $f(x)$ is bounded in the interval $X \leq x \leq b$ and the integral of f from X to b is defined, then

$$(6) \qquad \int_{a}^{b} f(x)\, dx = \lim_{X \to a^+} \int_{X}^{b} f(x)\, dx$$

provided the limit exists.

We say that the improper integral converges if the limit (5) or (6) exists and is finite; otherwise, we say the integral diverges.

EXAMPLE 5   Evaluate the improper integral $\int_{0}^{8} (8 - x)^{-1/3}\, dx$.

Solution.   We are dealing with an improper integral because $(8 - x)^{-1/3}$ tends to ∞ as $x \to 8^-$. We make the substitution $u = 8 - x$, $du = -dx$ to evaluate the indefinite integral

$$\int (8 - x)^{-1/3}\, dx = -\int u^{-1/3}\, du = -\frac{3}{2}u^{2/3} + C$$

$$= -\frac{3}{2}(8 - x)^{2/3} + C.$$

Then for $0 < X < 8$, we have

$$\int_0^X (8-x)^{-1/3}\, dx = [-\tfrac{3}{2}(8-x)^{2/3}]_0^X$$
$$= -\tfrac{3}{2}(8-X)^{2/3} + \tfrac{3}{2}(8)^{2/3}$$

and hence

$$\int_0^8 (8-x)^{-1/3}\, dx = \lim_{X \to 8^-} \int_0^X (8-x)^{-1/3}\, dx$$
$$= \lim_{X \to 8^-} [-\tfrac{3}{2}(8-X)^{2/3} + \tfrac{3}{2}(8)^{2/3}]$$
$$= \tfrac{3}{2}(8)^{2/3} = \tfrac{3}{2}(4) = 6. \quad \square$$

EXAMPLE 6   For which values of the constant k does the improper integral

$$(7) \qquad \int_0^1 \frac{1}{x^k}\, dx$$

converge, and what is its value in those cases?

Solution.   For $k = 1$ we have

$$\int_0^1 \frac{1}{x^k}\, dx = \int_0^1 \frac{1}{x}\, dx = \lim_{X \to 0^+} \int_X^1 \frac{1}{x}\, dx$$
$$= \lim_{X \to 0^+} [\ln(1) - \ln(X)] = \lim_{X \to 0^+} [-\ln(X)] = \infty$$

and the integral diverges.

For $k \neq 1$ we have

$$\int_0^1 \frac{1}{x^k}\, dx = \int_0^1 x^{-k}\, dx = \lim_{X \to 0^+} \int_X^1 x^{-k}\, dx$$
$$(8)$$
$$= \lim_{X \to 0^+} \left[\frac{1}{1-k} - \frac{1}{1-k} X^{1-k} \right].$$

For $k > 1$, $1 - k$ is negative and X^{1-k} tends to ∞ as $X \to 0^+$, so the limit (8) is ∞ and the integral (7) diverges. For $k < 1$, $1 - k$ is positive and X^{1-k} tends to 0 as $X \to 0^+$; the integral converges and has the value

$$\int_0^1 \frac{1}{x^k}\, dx = \lim_{X \to 0^+} \left[\frac{1}{1-k} - \frac{1}{1-k} X^{1-k} \right] = \frac{1}{1-k}. \quad \square$$

Areas of infinite regions

If a region in the xy-plane extends for all $x \geq a$ and if the portion of the region for $x \leq X$ has finite area $A(X)$ for each $X > a$, we define the area of the entire region to be the limit of $A(X)$ as $X \to \infty$, provided that limit exists.

EXAMPLE 7   What is the area of the infinite region bounded by the x-axis, the y-axis, and the graph of the function e^{-x}?

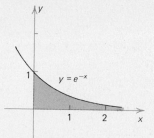

FIGURE 17.5   Infinite region of area $\int_0^\infty e^{-x}\, dx$

Solution.   The region is shown in Figure 17.5. The area of the portion out to $x = X$ is

$$A(X) = \int_0^X e^{-x}\, dx = [-e^{-x}]_0^X = 1 - e^{-X}.$$

The area of the infinite region is the limit

$$\lim_{X \to \infty} A(X) = \lim_{X \to \infty} [1 - e^{-X}] = 1. \quad \square$$

Exercises

In Exercises 1 through 25, evaluate the integrals that are defined.

†*Answer provided.*
‡*Outline of solution provided.*

1.† $\displaystyle\int_1^\infty e^{-10x}\, dx$

2. $\displaystyle\int_{-\infty}^0 x e^{-x^2}\, dx$

3.† $\displaystyle\int_0^\infty \frac{x}{1 + x^2}\, dx$

4.‡ $\displaystyle\int_5^\infty \frac{1}{x^2 - x}\, dx$

5.‡ $\displaystyle\int_\pi^\infty \sin x\, dx$

6. $\displaystyle\int_{-\infty}^1 \frac{1}{x^2 + 1}\, dx$

7.† $\displaystyle\int_0^{27} x^{-1/3}\, dx$

8.† $\displaystyle\int_0^{\pi/4} \tan x\, dx$

9.† $\displaystyle\int_{-\infty}^\infty x^3\, dx$

10. $\displaystyle\int_0^1 \frac{x}{\sqrt{x^2 - 1}}\, dx$

11.† $\displaystyle\int_{-32}^0 x^{-4/5}\, dx$

12. $\displaystyle\int_0^1 \frac{1}{\sqrt[3]{1 - x}}\, dx$

13.† $\displaystyle\int_0^4 (4 - x)^{-2}\, dx$

14. $\displaystyle\int_0^1 (1 - x^2)^{-1/2}\, dx$

15.† $\displaystyle\int_0^4 \frac{1}{\sqrt{16 - x^2}}\, dx$

16. $\displaystyle\int_{-\infty}^\infty \frac{1}{x^2 + 25}\, dx$

17.‡ $\displaystyle\int_0^1 \frac{1}{x^2 - 1}\, dx$

18. $\displaystyle\int_{10}^\infty (x - 2)^{-4/3}\, dx$

19. $\displaystyle\int_0^2 (x - 2)^{-4/3}\, dx$

20.† $\displaystyle\int_{-\infty}^0 \frac{e^x}{1 + e^x}\, dx$

21.‡ $\displaystyle\int_0^2 \frac{x}{x^2 - 5x + 6}\, dx$

22. $\displaystyle\int_{10}^\infty \frac{1}{x^2 + 4x + 4}\, dx$

23.† $\displaystyle\int_{10}^\infty \frac{2}{x^2 - 1}\, dx$

24. $\displaystyle\int_{-\infty}^\infty \frac{x}{x^4 + 1}\, dx$

25.† $\displaystyle\int_{\pi/4}^{\pi/2} \frac{\sin x}{\sqrt{\cos x}}\, dx$

Compute the areas of the regions in Exercises 26 through 33.

26.† The region bounded by $y = 0$ and $y = x^{-3}$ and to the right of $x = 2$

27.† The region bounded by $y = 0$ and $y = x^{-3}$ and between $x = 0, x = 2$

28. The region between the curves $y = -x^{-1/2}$ and $y = x^{-1/2}$ and to the right of $x = 9$

29.† The region bounded by $y = 0$ and $y = (4 - x)^{-1}$ and between $x = 0$, $x = 4$

30. The region bounded by $y = 0$ and $y = (x^2 + 16)^{-1}$

31.† The region bounded by $y = 0$ and $y = x^{-2/3}$ and to the left of $x = -8$

32. The region bounded by $y = (x - 1)^{-1}$ and $y = x^{-1}$ and to the right of $x = 5$

33. Let R be the region bounded by $y = 0$ and $y = 1/x$ and to the right of $x = 1$. Show that the solid obtained by rotating R about the x-axis has **a.**[†] infinite surface area but **b.** finite volume. **c.** The results of parts (a) and (b) suggest that you could fill the solid with paint but not paint its surface. Explain.

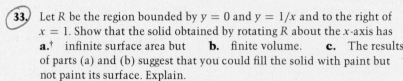

MISCELLANEOUS EXERCISES

[†]*Answer provided.*
[‡]*Outline of solution provided.*

1.[†] **a.** Use the Taylor polynomial $P_2(x)$ for $(1 + x)^{1/5}$ centered at 0 to give an approximate value of $(1.02)^{1/5}$. **b.** Use Taylor's theorem to determine whether the approximation is too large or too small and to estimate the error.

2.[†] What degree Taylor polynomial for e^x centered at 0 is needed to give e with an error of less than 10^{-6}? (Use $1 < e < 3$ and the reference tables of factorials and logarithms.)

3. What degree Taylor polynomial for $\sin x$ centered at 0 is needed to give $\sin(2)$ with an error less than 10^{-2}?

4.[†] What degree Taylor polynomial for $\ln(1 + x)$ centered at 0 is needed to give **a.** $\ln(\frac{3}{2})$ with an error less than 0.01 and **b.** $\ln(1.9)$ with an error less than 0.1?

5.[‡] Use Taylor's Theorem to express x^5 as a linear combination of powers of $x - 1$.

6. Use Taylor's Theorem to express $x^3 + x^2$ as a linear combination of powers of $x - 10$.

7. Use Taylor's Theorem to express $(1 + x)^5$ as a linear combination of powers of x.

8.[‡] Use the Taylor polynomial $P_5(x)$ for $\sin x$ centered at 0 to give an approximate value of the integral $\int_0^1 \frac{\sin x}{x} \, dx$. Estimate the error.

9. Use the Taylor polynomial $P_6(x)$ for $\cos x$ to give an approximate value of $\int_0^1 \frac{\cos x - 1}{x^2} \, dx$. Estimate the error.

10.[‡] Use the Taylor polynomial $P_3(x)$ for e^x centered at 0 to give an approximate value of $\int_0^1 e^{-x^2} \, dx$. Estimate the error.

11. Use the Taylor polynomial $P_2(x)$ for $1/(2 + x)$ to give an approximate value of $\int_0^1 \frac{1}{2 + t^3} \, dt$.

12.† Show by integration by parts that for a function $f(x)$ with $n + 1$ continuous derivatives in an interval and for x and a in that interval $f(x) = P_n(x) + R_n(x)$ with $P_n(x)$ centered at a and

(1) $R_n(x) = \dfrac{1}{n!} \displaystyle\int_a^x (x - t)^n f^{(n+1)}(t)\, dt.$

This is known as the *integral form of the remainder*.

13. The *Generalized Mean Value Theorem for integrals* states that if $g(x)$ and $h(x)$ are continuous and $h(x)$ is nonnegative for $a \le x \le b$ (or $b \le x \le a$), then there is at least one number c with $a < c < b$ (or $b < c < a$) such that

$$\int_a^b g(x) h(x)\, dx = g(c) \int_a^b h(x)\, dx.$$

Use this result to derive the derivative form of the remainder (formula (5) in Section 17.1) from formula (1) in Exercise 12.

14. Use Taylor's Theorem to prove that e is irrational. (Show that $e < 3$ and then that $(n!)e$ cannot be an integer for any positive integer n because

$$0 < (n!)e - \sum_{j=0}^{n} \frac{n!}{j!} < \tfrac{3}{4} \text{ for all integers } n > 3.)$$

15.† Let Δx be a nonzero number, and for functions $f(x)$ define $\Delta f(x) = f(x + \Delta x) - f(x).$ **a.** Compute $\Delta^2 f(x).$ **b.** Use Taylor's Theorem to show that for $f(x)$ with a continuous second derivative

$$\frac{d^2 f}{dx^2}(x) = \lim_{\Delta x \to 0} \frac{\Delta^2 f(x)}{\Delta x^2}.$$

(The corresponding calculation with infinitesimals was the origin of Leibniz notation for second derivatives.)

16.† (Circles of curvature) Suppose $f(x)$ has continuous first and second order derivatives and that $f(0) = 0, f'(0) = 0,$ and $f''(0) \ne 0.$ Let C be the circle through the points at $x = 0, x = x_1,$ and $x = x_2$ on the graph of $f(x_1 < 0 < x_2).$ Show that as x_1 and x_2 tend to 0, the circle C approaches the *circle of curvature*, which is tangent to the graph at $(0, 0)$ and has its center at $\left(0, 1/f''(0)\right).$

17.† Find the limit of $\left(\dfrac{x + 1}{x - 1}\right)^x$ as $x \to \infty.$

18. Show that $x^k \ln x \to 0$ as $x \to 0^+,$ for all $k > 0.$

19. **a.** Show that for any constant $k, (x - k \ln x) \to \infty$ as $x \to \infty.$
b. Use the result of part (a) to show that for any $k, x^k e^{-x} \to 0$ as $x \to \infty.$

20. **a.** Use the result of Exercise 19 to show that $x^{-n} e^{-1/x^2} \to 0$ as $x \to 0$ for all nonnegative integers $n.$ **b.** Show that $f(x)$ defined to be 0 for $x = 0$ and e^{-1/x^2} for all $x \ne 0$ has derivatives of all orders at $x = 0$ and that these derivatives are all 0.   **c.** What are the Taylor polynomial approximations of $f(x)$ centered at 0?

21. Find the formula for the integral of $1/x$ from a to b $(0 < a < b)$ by letting $k \to -1$ in the formula

$$\int_a^b x^k \, dx = \frac{1}{k+1} (b^{k+1} - a^{k+1}).$$

22. **a.**[†]   Use the $\epsilon\delta$-definition of the limit (Section 2.9) to show that if

$$|f(x)| \to \infty, |g(x)| \to \infty, \text{ and } \frac{f'(x)}{g'(x)} \to L \text{ (a number) as } x \to a+,$$

then $\dfrac{f(x)}{g(x)} \to L$ as $x \to a+$.   **b.**   Prove the result of part (a) for the case of $x \to \infty$.

23.[†] If you have studied parametric equations for curves (Chapter 11), express the Generalized Mean Value Theorem for derivatives as a result concerning secant and tangent lines on the curve $x = h(t), y = g(t)$.

24. Show that the integral $\int_1^\infty [(x + 10)^{-1/2} - x^{-1/2}] \, dx$ converges and find its value.

25. Show by considering areas that the improper integrals

$$\int_0^\infty \frac{1}{1 + x^2} \, dx \quad \text{and} \quad \int_0^1 \sqrt{\frac{1 - y}{y}} \, dy$$

are equal.

26. A rocket ship weighs 1000 pounds on the surface of the earth and the force of gravity on it is $16{,}000 \, r^{-2}$ pounds when it is r thousand miles above the center of the earth $(r \geq 4)$. How much work must be done against gravity for the rocket ship to "escape" the earth's gravity, that is, to be lifted an infinite distance from the surface of the earth? (Take 4 thousand miles as the radius of the earth.)

27.[†] Show that $\lim\limits_{X \to \infty} \int_{-X}^X \sin x \, dx = 0$ but that the improper integral $\int_{-\infty}^\infty \sin x \, dx$ is not defined.

28. Show that $\int_0^1 x^k \ln x \, dx$ converges for $k > -1$ but diverges for $k \leq -1$.

29. Show that $\int_1^\infty \dfrac{\ln x}{x^k} \, dx$ converges for $k > 1$ but diverges for $k \leq 1$.

30. Show that the change of variables $u = x^{1/2}$ transforms the improper integral $\int_0^4 x^{-1/2} \, dx$ into a proper integral. Evaluate both integrals.

31. If you have access to a computer or electronic calculator, compare the values of the following functions and their indicated Taylor polynomials centered at $x = 0$.   **a.**[†] e^x and $P_5(x)$ at $x = 2$ **b.** $\cos x$ and $P_6(x)$ at $x = \pi/5$ **c.**[†] $\ln(1 + x)$ and $P_7(x)$ at $x = 0.9$ **d.** $\arctan x$ and $P_9(x)$ at $x = \frac{1}{2}$ **e.**[†] $(1 + x)^e$ and $P_5(x)$ at $x = 1$ **f.** $\sinh x$ and $P_7(x)$ at $x = 1.5$

CHAPTER 18

INFINITE SEQUENCES AND SERIES

SUMMARY: This chapter deals with infinite sequences and series and criteria for determining whether they converge or diverge. In Section 18.1 we present examples of convergent and divergent infinite sequences. The ϵN-definition of convergence is given in Section 18.2 and Section 18.3 deals with finite geometric series. In Section 18.4 we introduce infinite series and give the comparison test and limit comparison test for series with nonnegative terms. The integral and ratio tests are discussed in Section 18.5, absolute convergence and alternating series in Section 18.6, and power series in Sections 18.7 and 18.8.

18-1   INFINITE SEQUENCES

We can think of an INFINITE SEQUENCE as an infinite string of numbers, such as the infinite sequence $\frac{1}{1}, \frac{1}{2}, \frac{1}{3}, \frac{1}{4}, \ldots$ of the reciprocals of the positive integers or the infinite sequence $1, 3, 5, 7, \ldots$ of the odd positive integers. We denote the infinite sequence $a_1, a_2, a_3, \ldots$ by $\{a_n\}_{n=1}^{\infty}$. Thus, the sequence of the reciprocals of the positive integers may be written $\left\{\dfrac{1}{n}\right\}_{n=1}^{\infty}$ and the sequence of odd positive integers may be written $\{2n + 1\}_{n=0}^{\infty}$.

Formally, we define an infinite sequence $\{a_n\}_{n=n_0}^{\infty}$ to be a real-valued function on the set of integers $n \geq n_0$. The value of the function at the integer n is the number a_n. We say that the infinite sequence *converges* if the numbers a_n have a finite limit as n tends to ∞. If $a_n \to \infty$ as $n \to \infty$, we say the sequence *diverges to infinity*. If $a_n \to -\infty$, we say the sequence *diverges to minus infinity*.

EXAMPLE 1   Does the sequence $\left\{\dfrac{12n}{3n + 1}\right\}_{n=1}^{\infty}$ converge? If so, what is its limit?

Solution.   We divide the numerator and denominator of $\dfrac{12n}{3n+1}$ by n to obtain

$$\lim_{n\to\infty}\frac{12n}{3n+1}=\lim_{n\to\infty}\frac{12}{3+1/n}=4.$$

The sequence converges and its limit is 4.   □

EXAMPLE 2   Does the sequence $\{5n\}_{n=3}^{\infty}$ converge?

Solution.   We have

$$\lim_{n\to\infty}5n=\infty$$

and the sequence diverges to infinity.   □

EXAMPLE 3   Does the sequence $\{(-1)^n\}_{n=0}^{\infty}$ converge?

Solution.   The number $(-1)^n$ is 1 when n is an even integer and -1 when n is an odd integer, so we are dealing with the sequence 1, -1, 1, -1, ... The sequence diverges because it has no limit.   □

EXAMPLE 4   What is the limit of the sequence $\left\{\left(1+\dfrac{1}{n}\right)^n\right\}_{n=1}^{\infty}$?

Solution.   We saw in Section 7.2 that the limit is the number e.   □

In the next example we use a convergent infinite sequence to analyze one of the paradoxes of the Greek philosopher Zeno of Elea (c. 450 B.C.) (See C. B. Boyer, [1] *A History of Mathematics*, pp. 81–84).

EXAMPLE 5   In Zeno's paradox known as "The Achilles," it is asserted that if Achilles gives a tortoise a head start in a race, he can never catch up to it. No matter how fast he runs, when he reaches the tortoise's starting position, the tortoise will have moved ahead to a second position, and when he reaches that position, the tortoise will have moved ahead, and so forth. What is wrong with this reasoning?

Solution.   We let $\{t_n\}_{n=1}^{\infty}$ be the infinite sequence of times described in the statement of the paradox: $t_1=0$ is the starting time of the race; t_2 is the time Achilles reaches the tortoise's starting position; t_3 is the time Achilles reaches the location the tortoise had at time t_2; and so forth. For each integer $n\geq 1$, Achilles reaches at time t_{n+1} the position the tortoise had at time t_n.

It is true that Achilles is behind the tortoise at all the times t_n and if he ran slower than or at the same speed as the tortoise, the numbers t_n would tend to ∞ and he would never catch up. However, if he runs faster than the tortoise, then the numbers t_n have a finite limit T and that is the instant when he catches up to the tortoise. If the tortoise had a 99 meter head start and ran $\frac{1}{10}$ meter per second while Achilles ran 10 meters per second, then Achilles would run $10 - \frac{1}{10} = \frac{99}{10}$ meters per second faster than the tortoise. Achilles would make up the 99 meter head start and catch up to the tortoise after $T = 10$ seconds. We cannot say that the tortoise will have moved ahead when Achilles reaches that point, because he and the tortoise arrive there at the same time. (See also Exercise 39.)   □

Exercises

†*Answer provided.*
‡*Outline of solution provided.*

In Exercises 1 through 4 write out the first ten terms in the sequences. (Do not do any difficult arithmetic.)

1.† $\left\{ \dfrac{k}{2^k + 1} \right\}_{k=0}^{\infty}$ **2.** $\left\{ \left(\dfrac{n-2}{n+2} \right)^n \right\}_{n=0}^{\infty}$

3.† $\left\{ \dfrac{j^j}{j!} \right\}_{j=10}^{\infty}$ **4.** $\left\{ (-1)^n \dfrac{n}{2n-1} \right\}_{n=-1}^{\infty}$

In each of Exercises 5 through 25 find the limit (if it exists) of the sequence and state whether the sequence converges or diverges. Use l'Hopital's rule in Exercises 13 through 21 and the analytic definition of the integral in Exercises 22 through 25.

5.† $\left\{ \dfrac{n - n^3}{2n^3 + 1} \right\}_{n=1}^{\infty}$ **6.** $\left\{ \dfrac{n^2 + 1}{5n} \right\}_{n=8}^{\infty}$

7.† $\left\{ \dfrac{(n+1)(n+2)(n+3)}{n + 2n^2 + 3n^3} \right\}_{n=1}^{\infty}$

8. $\left\{ \dfrac{n^3}{15 - n^{15}} \right\}_{n=3}^{\infty}$ **9.**† $\left\{ \dfrac{(-1)^n}{n} \right\}_{n=1}^{\infty}$ **10.** $\left\{ \dfrac{n^3}{7 - 4n^2} \right\}_{n=4}^{\infty}$

11.† $\{ (-n)^n \}_{n=1}^{\infty}$ **12.** $\{ n^n \}_{n=1}^{\infty}$ **13.**‡ $\left\{ n \sin\left(\dfrac{1}{n} \right) \right\}_{n=1}^{\infty}$

14.† $\left\{ n \left[\cos\left(1 + \dfrac{1}{n} \right) - \cos(1) \right] \right\}_{n=1}^{\infty}$

15.† $\left\{ n^2 \sin\left(\dfrac{1}{n} \right) \right\}_{n=1}^{\infty}$

16. $\{ ne^{-n} \}_{n=10}^{\infty}$ **17.**† $\left\{ \dfrac{\ln n}{n} \right\}_{n=1}^{\infty}$ **18.** $\left\{ n \sin\left(\dfrac{5}{n} \right) \right\}_{n=5}^{\infty}$

19.† $\left\{ \dfrac{e^n}{n^2} \right\}_{n=1}^{\infty}$ **20.** $\left\{ \dfrac{\sin^2(1/n)}{\tan(1/n)} \right\}_{n=1}^{\infty}$

21. $\left\{ \dfrac{1}{n} - \dfrac{1}{n+1} \right\}_{n=1}^{\infty}$ (Take a common denominator.)

22.‡ $\left\{ \dfrac{1}{n} \left[\left(\dfrac{1}{n} \right)^3 + \left(\dfrac{2}{n} \right)^3 + \cdots + \left(\dfrac{n-1}{n} \right)^3 + 1 \right] \right\}_{n=1}^{\infty}$

23.† $\left\{ \dfrac{1}{n} \left[\sin\left(\dfrac{1}{n} \right) + \sin\left(\dfrac{2}{n} \right) + \cdots + \sin\left(\dfrac{4n-1}{n} \right) + \sin(4) \right] \right\}_{n=1}^{\infty}$

24.† $\{ n^{-3/2} [\sqrt{1} + \sqrt{2} + \sqrt{3} + \cdots + \sqrt{n-1} + \sqrt{n}] \}_{n=1}^{\infty}$

25. $\left\{ \dfrac{1}{n} \left[\dfrac{1}{1 + \dfrac{1}{n}} + \dfrac{1}{1 + \dfrac{2}{n}} + \cdots + \dfrac{1}{1 + \dfrac{n-1}{n}} + \dfrac{1}{2} \right] \right\}_{n=1}^{\infty}$

26. **a.**[†] What is the perimeter P_n of a polygon with n equal sides that is inscribed in a circle of radius r? **b.** Show that the limit of the sequence $\{ P_n \}_{n=3}^{\infty}$ is the circumference of the circle.

27.[†] Show that the sequence $\{ r^n \}_{n=0}^{\infty}$ **a.** converges to 0 for $-1 < r < 1$, **b.** diverges to ∞ for $r > 1$, and **c.** diverges without having a limit for $r \le -1$. **d.** What is its limit for $r = 1$? (Study $\ln |r^n|$ for parts (a), (b), and (c).)

Find the limits of the sequences in Exercises 28 through 37.

28.[‡] $\{ n^{1/n} \}_{n=1}^{\infty}$ **29.** $\left\{ \left(\dfrac{1}{n} \right)^n \right\}_{n=1}^{\infty}$ **30.**[†] $\{ 5^{1/n} \}_{n=1}^{\infty}$

31. $\left\{ \left(\dfrac{1}{5} \right)^{1/n} \right\}_{n=1}^{\infty}$ **32.**[†] $\left\{ \left(1 - \dfrac{2}{n} \right)^n \right\}_{n=1}^{\infty}$ **33.** $\left\{ \left(2 - \dfrac{1}{n} \right)^n \right\}_{n=1}^{\infty}$

34.[†] $\{ (1 + n)^{1/n} \}_{n=1}^{\infty}$ **35.** $\left\{ \left(\dfrac{1}{n} \right)^{1/n} \right\}_{n=1}^{\infty}$ **36.**[‡] $\left\{ \dfrac{(2n)!}{(n!)^2} \right\}_{n=1}^{\infty}$

37.[†] $\left\{ \dfrac{(n!)^2}{(n^2)!} \right\}_{n=1}^{\infty}$

38. Show that the sequence $\{ n^k r^n \}_{n=1}^{\infty}$ tends to zero for any positive constant k and any r with $0 < r < 1$. (Use logarithms.)

39. Suppose that Achilles and a tortoise start a race at time $t = 0$ (seconds) with Achilles at $s = 0$ (meters) and the tortoise 99 meters ahead at $s = 99$. Suppose that Achilles runs 10 meters per second and the tortoise $\frac{1}{10}$ meter per second. Show that the sequence $\{ t_n \}_{n=1}$ described in the solution of Example 5 is given by the conditions that $t_1 = 0$ and $t_{n+1} = \frac{1}{100} t_n + 9.9$ for $n \ge 1$. Show by mathematical induction that $t_n = 10[1 - (\frac{1}{100})^{n-1}]$. When does Achilles catch up to the tortoise and at what point?

18-2   THE ϵN-DEFINITION OF THE LIMIT OF A SEQUENCE

In previous chapters we have been able to rely on our geometric or algebraic intuition in studying limits because in most instances we have been able to recognize the limit of an expression once we have put it in a suitable form through techniques of algebra or calculus. In this chapter we want to determine whether limits exist in situations where we cannot find the values of the limits. For this purpose we use criteria that are based on the ϵN-definition of the limit of a sequence, which we discuss in this section.

Finite limits of sequences

Definition 18.1   **The sequence $\{a_n\}_{n=1}^{\infty}$ converges and the number L is its limit if for each $\epsilon > 0$, there is a number N such that**

(1) $|a_n - L| < \epsilon$ **for $n > N$.**   ◄

If we think of the sequence $\{a_n\}$ as a function of the index n, then its graph in an ny-plane is a collection of points (Figure 18.1), and if the limit of the sequence is a number L, then the points approach the line $y = L$ as $n \to \infty$. Condition (1) means that all of the points to the right of $n = N$ lie in a strip of width 2ϵ centered on the line $y = L$ (Figure 18.2).

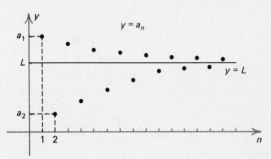

FIGURE 18.1

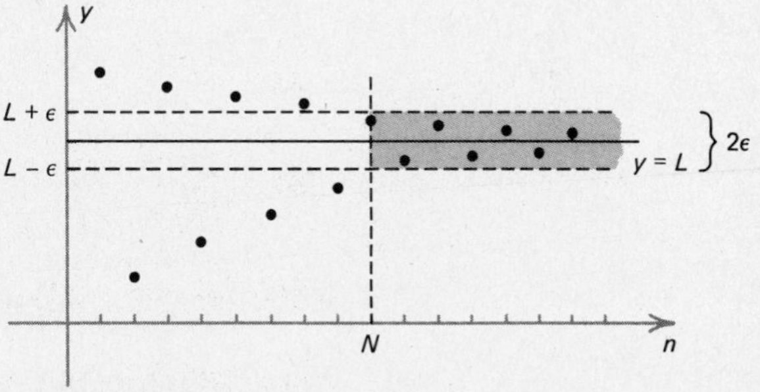

FIGURE 18.2

We can paraphrase Definition 18.1 as follows: the number a_n can be made arbitrarily close to L (within ϵ of L) by taking n sufficiently large (greater than N). We can show that a limit satisfies Definition 18.1 by giving a rule that prescribes a suitable number N for each positive number ϵ.

EXAMPLE 1   Show that by Definition 18.1

(2) $\displaystyle \lim_{n \to \infty} \frac{2n}{n+1} = 2$.

Solution.   We set $a_n = 2n/(n+1)$ and $L = 2$. Then for $n \geq 1$

(3) $|a_n - L| = \left| \dfrac{2n}{n+1} - 2 \right| = \left| \dfrac{-2}{n+1} \right| = \dfrac{2}{n+1} < \dfrac{2}{n}$.

Let ϵ be an arbitrary positive number and set $N = 2/\epsilon$. Then for $n > N$, we have by (3)

$$|a_n - L| < \frac{2}{n} < \frac{2}{N} = \epsilon$$

and Definition 18.1 is satisfied.   □

EXAMPLE 2   Use the Mean Value Theorem for derivatives to show that, according to Definition 18.1,

$$\lim_{n \to \infty} \sin\left(1 + \frac{1}{n}\right) = \sin(1).$$

Solution. The derivative of $\sin x$ is $\cos x$. By the Mean Value Theorem there is, for each n, a number c with $1 < c < 1 + \dfrac{1}{n}$ such that $\sin\!\left(1 + \dfrac{1}{n}\right) - \sin(1) = \cos(c)\!\left(\dfrac{1}{n}\right)$. Since $|\cos(c)| \leq 1$, this gives

$$\left| \sin\!\left(1 + \frac{1}{n}\right) - \sin(1) \right| \leq \frac{1}{n}.$$

Given $\epsilon > 0$, we set $N = 1/\epsilon$. Then for $n > N$, we have

$$\left| \sin\!\left(1 + \frac{1}{n}\right) - \sin(1) \right| \leq \frac{1}{n} < \frac{1}{N} = \epsilon$$

and the definition is satisfied. $\square$

The *sum* of infinite sequences $\{a_n\}_{n=n_0}$ and $\{b_n\}_{n=n_0}$ is the sequence $\{a_n + b_n\}_{n=n_0}$ and the *product* is the sequence $\{a_n b_n\}_{n=n_0}$. The *quotient* $\{a_n/b_n\}_{n=n_0}$ is defined if none of the numbers b_n are zero. The ϵN-definition of the limit of a sequence (Definition 18.1) is used in the proof of the following result, which we leave as Exercise 36.

Proposition 18.1   **If the sequence $\{a_n\}_{n=n_0}$ converges and has limit L and the sequence $\{b_n\}_{n=n_0}$ converges and has limit M, then the sum of the sequences converges and has limit $L + M$; the product converges and has limit LM; and if the quotient sequence is defined and M is not zero, the quotient sequence converges and has limit L/M.**

Infinite limits
of sequences

Definition 18.2   **The sequence $\{a_n\}_{n=1}^{\infty}$ diverges to ∞ if for each $M > 0$ there is an $N > 0$ such that**

(4) $a_n > M$ for $n > N$.

The sequence diverges to $-\infty$ if for each $M < 0$, there is an $N > 0$ such that

(5) $a_n < M$ for $n > N$.

The first half of Definition 18.2 states that the sequence diverges to ∞ if we can make a_n arbitrarily large (greater than M) by taking n sufficiently large (greater than N).

EXAMPLE 3   Use logarithms and Definition 18.2 to show that $r^n \to \infty$ as $n \to \infty$ for any constant $r > 1$.

Solution.   Because $\ln x$ is an increasing function, we can make r^n greater than M by making $\ln(r^n) = n \ln(r)$ greater than $\ln(M)$. Given $M > 0$, we set $N = \ln(M)/\ln(r)$. Then for $n > N$ we have

$$\ln(r^n) = n \ln(r) > N \ln(r) = \left[\frac{\ln(M)}{\ln(r)} \right] \ln(r) = \ln(M)$$

so that $r^n > M$ and the definition is satisfied. $\square$

Monotone sequences

A sequence $\{a_n\}_{n=1}^{\infty}$ is said to be *increasing* if $a_{n+1} > a_n$ for all $n \geq 1$. The sequence is *decreasing* if $a_{n+1} < a_n$, *nondecreasing* if $a_{n+1} \geq a_n$, and *nonincreasing* if $a_{n+1} \leq a_n$ for all $n \geq 1$. A sequence that is either nondecreasing or nonincreasing is said to be MONOTONE (Figure 18.3).

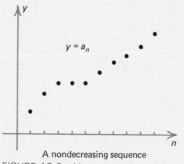

A nondecreasing sequence   A nonincreasing sequence

FIGURE 18.3   Monotone sequences

EXAMPLE 4   Is the sequence $\left\{\dfrac{n}{n+5}\right\}_{n=1}^{\infty}$ monotone?

Solution.   We set $a_n = n/(n+5)$ and $f(x) = x/(x+5)$. Then $a_n = f(n)$, and the derivative of $f(x)$

$$\frac{df}{dx}(x) = \frac{(x+5)\dfrac{d}{dx}(x) - x\dfrac{d}{dx}(x+5)}{(x+5)^2} = \frac{5}{(x+5)^2}$$

is positive for $x \geq 1$. Therefore, $f(x)$ is an increasing function and for each n, $a_n = f(n) < f(n+1) = a_{n+1}$. The sequence is increasing and hence monotone. $\square$

Bounded sequences

A sequence $\{a_n\}_{n=1}^{\infty}$ is said to be *bounded* if there is a constant M such that $|a_n| \leq M$ for all $n \geq 1$. A constant M such that $a_n \leq M$ for all $n \geq 1$ is said to be an *upper bound* for the sequence. The constant M is a *lower bound* if $a_n \geq M$ for all $n \geq 1$.

Least upper and greatest lower bounds

We will base our discussion of the theory of infinite sequences and series on the following axiom of the real numbers.

Axiom 18.1   Every infinite sequence that has an upper bound has a least upper bound. Every infinite sequence that has a lower bound has a greatest lower bound.

As the terms suggest, the least upper bound is an upper bound that is less than all other upper bounds and the greatest lower bound is a lower bound that is greater than all other lower bounds.

The next theorem shows that we can determine whether a monotone sequence converges by determining whether it is bounded.

Theorem 18.1   If a nondecreasing sequence is bounded, then it converges and its limit is its least upper bound. If a nondecreasing sequence is unbounded, then it diverges to ∞. If a nonincreasing sequence is bounded, then it converges and its limit is its greatest lower bound. If a nonincreasing sequence is unbounded, then it diverges to −∞.

Proof.   We will give the proof for nondecreasing sequences. The proof for nonincreasing sequences is similar.

Suppose that $\{a_n\}_{n=1}^{\infty}$ is nondecreasing and bounded. By Axiom 18.1 the sequence has a least upper bound L. Let ϵ be an arbitrary positive number. Because L is an upper bound, we have $a_n \leq L$ for $n \geq 1$. Because L is the least upper bound, $L - \epsilon$ is not an upper bound and we have $a_N > L - \epsilon$ for some N. Since the sequence is nondecreasing, we then have $a_n > L - \epsilon$ for $n \geq N$. The inequalities $L - \epsilon < a_n \leq L$ for $n \geq N$ imply that $|a_n - L| < \epsilon$ for $n > N$, so L is the limit of the sequence.

Suppose, on the other hand, that the nondecreasing sequence $\{a_n\}_{n=1}^{\infty}$ has no upper bound. Let M be an arbitrary positive number. M is not an upper bound for the sequence, so there is a positive integer N such that $a_N > M$. Because the sequence is nondecreasing, we then have $a_n > M$ for $n \geq N$. Definition 18.2 is verified and the sequence diverges to ∞.   Q.E.D.

EXAMPLE 5   Find the least upper bound, greatest lower bound, and limit of the nondecreasing sequence $\left\{ \dfrac{n}{n + 5} \right\}_{n=1}^{\infty}$

Solution.   We saw in Example 4 that the sequence is nondecreasing. It is bounded because for any $n \geq 1$ we have

$$(6) \qquad \frac{n}{n + 5} = \frac{1}{1 + 5/n} < 1.$$

By Axiom 18.1 the sequence has a least upper bound, and by Theorem 18.1 the least upper bound is its limit, which is

$$\lim_{n \to \infty} \frac{n}{n + 5} = \lim_{n \to \infty} \frac{1}{1 + 5/n} = 1.$$

The greatest lower bound of the sequence is its first term $1/(1 + 5) = \frac{1}{6}$. □

Exercises

Verify Definition 18.1 for the limits in Exercises 1 through 12. In Exercises 7 through 12 use the Mean Value Theorem for derivatives.

1.† $\displaystyle \lim_{n \to \infty} \frac{1}{n^2} = 0$

2. $\displaystyle \lim_{n \to \infty} \frac{3n + 1}{9n + 2} = \frac{1}{3}$

3.† $\displaystyle \lim_{n \to \infty} \sqrt{9 - \frac{5}{n}} = 3$

4. $\displaystyle \lim_{n \to \infty} \frac{1}{\sqrt{n}} = 0$

5.† $\lim_{n \to \infty} \dfrac{n^2}{n^2 + 4} = 1$

6. $\lim_{n \to \infty} \sqrt[3]{\dfrac{8n}{15 - n}} = -2$

7.† $\lim_{n \to \infty} e^{1/n} = 1$

8. $\lim_{n \to \infty} \left(1 - \dfrac{1}{n}\right)^{100} = 1$

9.† $\lim_{n \to \infty} \ln\left(\dfrac{1 + n}{n}\right) = 0$

10. $\lim_{n \to \infty} \cos\left(\dfrac{10}{n}\right) = 1$

11.† $\lim_{n \to \infty} \left[\dfrac{8n - 8}{n}\right]^{2/3} = 4$

12. $\lim_{n \to \infty} \arctan\left(\dfrac{1}{5n}\right) = 0$

Verify Definition 18.2 for the limits in Exercises 13 through 16.

13.† $\lim_{n \to \infty} \dfrac{n^2}{n + 4} = \infty$

14. $\lim_{n \to \infty} - \sqrt{n} = -\infty$

15.† $\lim_{n \to \infty} n^3 = \infty$

16. $\lim_{n \to \infty} -ne^{1/n} = -\infty$

17. Use logarithms and Definition 18.1 to show that r^n tends to 0 as n tends to infinity for any r with $-1 < r < 1$.

18. Use logarithms and Definitions 18.1 and 18.2 to prove that **a.** $n^k \to 0$ as $n \to \infty$ for $k < 0$ and **b.** $n^k \to \infty$ as $n \to \infty$ for $k > 0$.

19. For what positive integer values of n are the sequences **a.** $\{n \ln(\frac{1}{n})\}$ and **b** $\{\frac{1}{n} \ln(n)\}$ decreasing?

20. a. Use the Mean Value Theorem to show that $\tan x > x$ for $0 < x < \frac{1}{2}\pi$. **b.** Show that the sequence $\{n \sin(\frac{1}{n})\}$ is decreasing for $n \geq 1$.

In Exercises 21 through 35 determine whether the sequences are monotone, bounded above, and bounded below. Find the least upper bounds of those that are bounded above and the greatest lower bounds of those that are bounded below.

21.‡ $\left\{3 + \dfrac{1}{n}\right\}_{n=1}^{\infty}$

22. $\left\{-2 - \dfrac{3}{n}\right\}_{n=2}^{\infty}$

23.‡ $\left\{\left(1 + \dfrac{1}{n}\right)^n\right\}_{n=1}^{\infty}$

24. $\{n + (-1)^n\}_{n=0}^{\infty}$

25.† $\left\{n - \dfrac{1}{n}\right\}_{n=1}^{\infty}$

26. $\{(-1)^n + 1\}_{n=1}^{\infty}$

27.† $\left\{\dfrac{1}{n}[1 + (-1)^n]\right\}_{n=2}^{\infty}$

28.† $\left\{\dfrac{\cos(n\pi)}{n}\right\}_{n=1}^{\infty}$

29.† $\{n[1 + (-1)^{n+1}]\}_{n=0}^{\infty}$

30. $\left\{(-1)^n - \dfrac{1}{n}\right\}_{n=1}^{\infty}$

31.† $\{n(-1)^n\}_{n=0}^{\infty}$

32. $\left\{\cos\left(\dfrac{1}{n}\right)\right\}_{n=2}^{\infty}$

33.† $\{ne^{-n}\}_{n=0}^{\infty}$

34. $\left\{\sin\left[\dfrac{(-1)^n}{n}\right]\right\}_{n=1}^{\infty}$

35.† $\{\arctan(n)\}_{n=1}^{\infty}$

36. Use Definition 18.1 to prove Proposition 18.1. (Use the proof of Theorem 2.1 in Section 2.9 as a model.)

18-3   FINITE GEOMETRIC SERIES

Whereas the terms "sequence" and "series" are often interchangeable in everyday English, they have distinct meanings in calculus: we have seen that a sequence is a string of numbers; a SERIES is a sum of numbers.

In this section we discuss FINITE GEOMETRIC SERIES, which are sums of the form

$$(1) \qquad 1 + r + r^2 + r^3 + \cdots + r^n$$

with r a constant and n a positive integer. Our study of these series is based on the summation formula

$$(2) \qquad 1 + r + r^2 + r^3 + \cdots + r^n = \frac{1 - r^{n+1}}{1 - r} \qquad \blacktriangleleft$$

which is valid for $r \neq 1$. This formula can be derived for any positive integer by dividing the polynomial $1 - r$ into the polynomial 1 (see Exercise 21). The calculation for an arbitrary positive integer is indicated schematically below.

$$
\begin{array}{r}
1 + r + r^2 + \cdots + r^n \\
1 - r \overline{)\, 1 } \\
\underline{1 - r} \\
r \\
\underline{r - r^2} \\
r^2 \\
\cdot \\
\cdot \\
\underline{} \\
r^n \\
\underline{r^n - r^{n+1}} \\
r^{n+1}
\end{array}
$$

According to this calculation, when we divide $1 - r$ into 1, we obtain the sum $1 + r + r^2 + \cdots + r^n$ and a remainder of r^{n+1}. Consequently, we obtain

$$\frac{1}{1 - r} = 1 + r + r^2 + \cdots + r^n + \frac{r^{n+1}}{1 - r}.$$

Subtracting $r^{n+1}/(1 - r)$ from both sides of this equation gives formula (2).

EXAMPLE 1   Use the finite geometric series formula (2) and the value 16,777,216 of 2^{24} to compute the sum

$$1 + 2 + 2^2 + 2^3 + \cdots + 2^{23}.$$

Solution.   By (2) with $r = 2$ and $n = 23$, we have

$$1 + 2 + 2^2 + 2^3 + \cdots + 2^{23} = \frac{1 - 2^{24}}{1 - 2} = \frac{-16{,}777{,}215}{-1}$$
$$= 16{,}777{,}215. \quad \square$$

Summation notation

When we express formula (2) with summation notation, it reads

(3) $\displaystyle\sum_{j=0}^{n} r^j = \frac{1 - r^{n+1}}{1 - r}$ for $r \neq 1$.   ◀

EXAMPLE 2   Use formula (3) and the approximate value $(0.9)^{51} \approx 0.004638$ to give an approximate value of

(4) $\displaystyle\sum_{j=0}^{50} (0.9)^j.$

Solution.   By (3) with $n = 50$ and $r = 0.9$, we have

$$\sum_{j=0}^{50} (0.9)^j = \frac{1 - (0.9)^{51}}{1 - 0.9} \approx \frac{1 - 0.004638}{0.1} = 9.95362. \quad \square$$

Exercises

†*Answer provided.*
‡*Outline of solution provided.*

Use Table 18.1 and the summation formula for the finite geometric series to give approximate values of the sums in Exercises 1 through 18.

1.† $1 + \frac{1}{2} + (\frac{1}{2})^2 + (\frac{1}{2})^3 + \cdots + (\frac{1}{2})^{12}$

2.‡ $1 - \frac{1}{4} + (\frac{1}{4})^2 - (\frac{1}{4})^3 + \cdots - (\frac{1}{4})^7 + (\frac{1}{4})^8$

3.‡ $\frac{3}{4} + (\frac{3}{4})^2 + (\frac{3}{4})^3 + \cdots + (\frac{3}{4})^{31}$

4. $5^2 + 5^3 + 5^4 + \cdots + 5^{10}$

5.† $6 + 0.6 + 0.06 + 0.006 + 0.0006 + 0.00006$

TABLE 18.1

$(0.5)^{13} \approx 1.22 \times 10^{-4}$
$5^{11} = 48{,}828{,}125$
$(0.99)^{1024} \approx 3.39 \times 10^{-5}$
$(1.02)^{128} \approx 12.61$
$2^{24} = 16{,}777{,}216$
$(0.25)^9 \approx 3.82 \times 10^{-6}$
$(0.9)^{128} \approx 1.39 \times 10^{-6}$
$(1.1)^{35} \approx 28.10$
$(1.02)^{512} \approx 25{,}309$
$11^8 = 214{,}358{,}881$
$(0.75)^{32} \approx 1.00 \times 10^{-4}$
$(0.99)^{512} \approx 5.82 \times 10^{-3}$
$(1.1)^{128} \approx 198{,}730$
$(0.8)^{101} \approx 1.63 \times 10^{-10}$

6. $\displaystyle\sum_{j=0}^{127} (\tfrac{9}{10})^j$

7. $\displaystyle\sum_{j=1}^{511} (\tfrac{99}{100})^j$

8.† $\displaystyle\sum_{j=0}^{1023} (-1)^j (0.99)^j$

9.† $\displaystyle\sum_{j=0}^{100} 1^j$

10.† $\displaystyle\sum_{j=3}^{34} (1.1)^j$

11. $\displaystyle\sum_{j=0}^{127} (-1.1)^j$

12. $\displaystyle\sum_{j=2}^{127} (1.02)^j$

13.† $\displaystyle\sum_{j=0}^{100} 5(0.8)^j$

14. $\displaystyle\sum_{j=0}^{23} 2^j$

15. $\displaystyle\sum_{j=0}^{23} (-2)^j$

16. $\displaystyle\sum_{j=0}^{400} (-1)^j$

17. $\displaystyle\sum_{j=2}^{7} 11^j$

18. $\displaystyle\sum_{j=32}^{127} (1.1)^j$

19. Let B_n denote the balance due just before the nth monthly payment on a loan of B_0 dollars with the first payment occurring one month after the loan is made and with the interest calculated just before each (equal) payment at the rate of $100r\%$ per month.   **a.**   Show that

$$B_n = B_0(1 + r)^n - P \sum_{j=1}^{n-1} (1 + r)^{j \cdot} \quad \text{for } n > 1$$

where P is the monthly payment   **b.**†   Use the finite geometric series formula to find the size of payment P such that the loan is paid off in N months.

20. "As I was going to St. Ives, I met a man with seven wives. Every wife had seven sacks; every sack had seven cats; every cat had seven kits . . ." Kits, cats, sacks, wives, how many were coming from St. Ives? (Count the man. Use a finite geometric series and the value $7^5 = 16{,}807$.)

21. Use long division to derive summation formula (2) for   **a.** $n = 3$,   **b.** $n = 5$, and   **c.** $n = 7$.

22. Verify the summation formula (2) for $n = 4$ by multiplying both sides by $1 - r$.

18-4   INFINITE SERIES AND THE COMPARISON TEST FOR SERIES WITH NONNEGATIVE TERMS

If you take half of something, then take half of what remains (or one-fourth), then half of the remainder (or one-eighth), and repeat this process n times, you will have taken the fraction

(1) $\frac{1}{2} + \frac{1}{4} + \frac{1}{8} + \cdots + (\frac{1}{2})^n$

of the whole item. The formula for the sum of a finite geometric series

(2) $\displaystyle\sum_{j=0}^{n} r^j = 1 + r + r^2 + r^3 + \cdots + r^n = \frac{1 - r^{n+1}}{1 - r}$ $(r \neq 1)$

with $r = \frac{1}{2}$ and n replaced by $n - 1$ enables us to express the sum (1) in a concise form. You will have taken the fraction

(3)
$$\frac{1}{2} + \frac{1}{4} + \frac{1}{8} + \cdots + (\tfrac{1}{2})^n = \tfrac{1}{2}[1 + \tfrac{1}{2} + \tfrac{1}{4} + \cdots + (\tfrac{1}{2})^{n-1}]$$
$$= \tfrac{1}{2}\left\{ \frac{1 - (\tfrac{1}{2})^n}{1 - \tfrac{1}{2}} \right\} = 1 - (\tfrac{1}{2})^n.$$

Now suppose that the process of taking at each step one-half of the remaining amount is continued indefinitely so that the integer n tends to ∞. The quantity $(\tfrac{1}{2})^n$ tends to zero and the fraction (3) taken after n steps tends to 1. We say that the *sum* of the amounts taken in the infinite number of steps is 1 and we write

$$\tfrac{1}{2} + \tfrac{1}{4} + \tfrac{1}{8} + \cdots + (\tfrac{1}{2})^n + \cdots = 1.$$

To express this using summation notation we write

$$(4) \qquad \sum_{j=1}^{\infty} (\tfrac{1}{2})^j = 1.$$

The sum on the left of (4) is an example of a convergent INFINITE SERIES. In general, an infinite series is an expression of the form

$$(5) \qquad \sum_{j=j_0}^{\infty} a_j$$

where a_j is a real number for each integer $j \geq j_0$. We define the *sum* of the infinite series by considering the finite PARTIAL SUMS

$$(6) \qquad s_n = \sum_{j=j_0}^{n} a_j \qquad (n \geq j_0).$$

Definition 18.3   **If the partial sums (6) have a limit as n tends to ∞, we say that the limit is the sum of the infinite series (5) and we write**

$$(7) \qquad \sum_{j=j_0}^{\infty} a_j = \lim_{n \to \infty} s_n = \lim_{n \to \infty} \sum_{j=j_0}^{n} a_j.$$

We say that the infinite series *converges* if the limit (7) exists and is finite, that the series *diverges to* ∞ if the limit is ∞, that the series *diverges to* $-\infty$ if the limit is $-\infty$, and that it *diverges without having a limit* if the limit in (7) does not exist. Notice that deleting a finite number of terms from an infinite series or adding a finite number of new terms to it does not affect its convergence. In particular, for any integer $J > j_0$, the series $\sum_{j=j_0}^{\infty} a_j$ converges if and only if the series $\sum_{j=J}^{\infty} a_j$ converges.

Infinite geometric series

The summation formula

$$(2) \qquad \sum_{j=0}^{n} r^j = \frac{1 - r^{n+1}}{1 - r} \qquad (r \neq 1)$$

for the finite geometric series enables us to study the INFINITE GEOMETRIC SERIES

$$(8) \qquad \sum_{j=0}^{\infty} r^j.$$

Theorem 18.2   **For $-1 < r < 1$ the infinite geometric series (8) converges and has the value**

$$(9) \qquad \sum_{j=0}^{\infty} r^j = \frac{1}{1 - r}.$$

◀

For $r \geq 1$ the series (8) diverges to ∞ and for $r \leq -1$ it diverges without having a limit.

Proof.   The sequence $\{r^{n+1}\}_{n=0}^{\infty}$ converges to 0 for $-1 < r < 1$, diverges to ∞ for $r > 1$, and diverges without having a limit for $r \leq -1$ (see Exercise 27 of Section 18.1). Therefore, formula (2) shows that the geometric series diverges to ∞ for $r > 1$, diverges without having a limit for $r \leq -1$, and converges for $-1 < r < 1$ with the value

$$\sum_{j=0}^{\infty} r^j = \lim_{n \to \infty} \sum_{j=0}^{n} r^j = \lim \frac{1 - r^{n+1}}{1 - r} = \frac{1}{1 - r}.$$

For $r = 1$ we cannot apply formula (2), but all of the terms in the series are 1, so the partial sums

$$\sum_{j=0}^{n} r^j = \sum_{j=0}^{n} 1^j = n + 1$$

tend to infinity and the series diverges to infinity.   Q.E.D.

EXAMPLE 1   Does the infinite series

$$\sum_{j=0}^{\infty} (-\tfrac{3}{4})^j$$

converge? If so, what is its value?

Solution.   The series is the geometric series (9) with $r = -\tfrac{3}{4}$. By Theorem 18.2 the series converges and has the value

$$\sum_{j=0}^{\infty} (-\tfrac{3}{4})^j = \frac{1}{1 - (-\tfrac{3}{4})} = \frac{1}{1 + \tfrac{3}{4}} = \tfrac{4}{7}. \quad \square$$

The next example shows how repeating decimals can be expressed in terms of convergent geometric series.

EXAMPLE 2   Express the repeating decimal 0.232323 . . . as a fraction.

Solution.   We factor out the repeating digits 23 and the greatest remaining power of 10 and then use formula (9) with $r = \tfrac{1}{100}$ to obtain

$$0.232323\ldots = 23(0.010101\ldots)$$

$$= 23(10^{-2} + 10^{-4} + 10^{-6} + \cdots)$$

$$= \tfrac{23}{100}(1 + 10^{-2} + 10^{-4} + 10^{-6} + \cdots)$$

$$= \tfrac{23}{100}[1 + \tfrac{1}{100} + (\tfrac{1}{100})^2 + (\tfrac{1}{100})^3 + \cdots]$$

$$= \tfrac{23}{100} \frac{1}{1 - \tfrac{1}{100}} = \tfrac{23}{99}. \quad \square$$

Telescoping series   The next example deals with an infinite series whose sum we can determine because we can express it as a "telescoping sum," each of whose partial sums simplifies to be the sum of only two terms.

EXAMPLE 3   Use the partial fraction decomposition of the rational function $\dfrac{1}{x(x+1)}$ to find the sum of the infinite series

$$(10) \quad \sum_{j=1}^{\infty} \frac{1}{j(j+1)} = \frac{1}{1(2)} + \frac{1}{2(3)} + \frac{1}{3(4)} + \cdots$$

Solution.   To find the partial fraction decomposition we write

$$\frac{1}{x(x+1)} = \frac{A}{x} + \frac{B}{x+1} = \frac{A(x+1) + Bx}{x(x+1)}$$

which gives the equation $1 = A(x+1) + Bx$ to determine the coefficients A and B. Setting $x = 0$ shows that $A = 1$ and setting $x = -1$ gives $B = -1$. Thus, we have

$$\frac{1}{x(x+1)} = \frac{1}{x} - \frac{1}{x+1}$$

so that

$$(11) \quad \sum_{j=1}^{n} \frac{1}{j(j+1)} = \sum_{j=1}^{n} \left(\frac{1}{j} - \frac{1}{j+1} \right)$$

$$= (1 - \tfrac{1}{2}) + (\tfrac{1}{2} - \tfrac{1}{3}) + (\tfrac{1}{3} - \tfrac{1}{4}) + \cdots + \left(\frac{1}{n} - \frac{1}{n+1} \right).$$

This partial sum is said to "telescope" because the second term inside each pair of parentheses cancels with the first term in the next pair, leaving only the first term in the first pair of parentheses and the second term in the last pair. The partial sum (11) equals $1 - 1/(n+1)$ and consequently

$$\sum_{j=1}^{\infty} \frac{1}{j(j+1)} = \lim_{n \to \infty} \sum_{j=1}^{n} \frac{1}{j(j+1)} = \lim_{n \to \infty} \left(1 - \frac{1}{n+1} \right) = 1. \quad \square$$

A test for divergence   Much of the theory of infinite series is devoted to developing means to determine whether infinite series converge or diverge in cases where we cannot employ geometric series or Taylor's Theorem. The next result can be used to show that certain series diverge.

Theorem 18.3   If the infinite series

$$\sum_{j=j_0}^{\infty} a_j$$

converges, then its terms a_j tend to 0 as j tends to ∞. Therefore, if the terms a_j do not tend to zero, then the series diverges.

Proof. Suppose that the series converges. Then its value is a number L, which is the limit of the partial sums s_j as j tends to ∞. We can write $a_j = s_j - s_{j-1}$ [see formula (6)]. Hence

$$\lim_{j \to \infty} a_j = \lim_{j \to \infty} [s_j - s_{j-1}]$$
$$= \lim_{j \to \infty} s_j - \lim_{j \to \infty} s_{j-1} = L - L = 0.$$

Thus, the terms of any convergent series tend to 0, and if the terms of a series do not tend to 0, it does not converge.   Q.E.D.

EXAMPLE 4   Show that the infinite series

$$\sum_{j=1}^{\infty} (-1)^j \frac{j^2}{4j^2 - 1}$$

diverges.

Solution.   The quantity

$$\frac{j^2}{4j^2 - 1} = \frac{1}{4 - 1/j^2}$$

tends to $\frac{1}{4}$ as j tends to ∞. Therefore, the terms $(-1)^j j^2/(4j^2 - 1)$ in the infinite series do not tend to 0, and the series diverges. $\square$

The converse of Theorem 18.3 is not valid. Many infinite series diverge even though their terms tend to zero. They diverge because the terms do not tend to zero rapidly enough.

EXAMPLE 5   The infinite series

$$\sum_{j=1}^{\infty} \frac{1}{j} = 1 + \tfrac{1}{2} + \tfrac{1}{3} + \tfrac{1}{4} + \cdots$$

is known as the *harmonic series*. Show that it diverges to ∞.

Solution.   Theorem 18.3 does not apply because the terms $a_j = 1/j$ tend to zero as j tends to ∞. We have to examine the series more closely to show that it diverges.

To illustrate the idea behind the solution, let us consider the partial sum of the first 16 terms of the infinite series, grouped as follows:

$$\sum_{j=1}^{16} \tfrac{1}{j} = 1 + \tfrac{1}{2} + [\tfrac{1}{3} + \tfrac{1}{4}] + [\tfrac{1}{5} + \tfrac{1}{6} + \tfrac{1}{7} + \tfrac{1}{8}]$$
$$+ [\tfrac{1}{9} + \tfrac{1}{10} + \cdots + \tfrac{1}{16}].$$

The two terms in the first pair of square brackets are both $\geq \tfrac{1}{4}$; the four terms in the second pair are all $\geq \tfrac{1}{8}$; and the eight in the third pair are all $\geq \tfrac{1}{16}$. Therefore, each sum in square brackets is greater than $\tfrac{1}{2}$ and the entire sum of $16 = 2^4$ terms is greater than $1 + 4(\tfrac{1}{2}) = 3$.

Similarly, the next sixteen terms in a partial sum add up to more than $\frac{1}{2}$, as do the next thirty-two and the following sixty-four. In general the sum

$$\sum_{j=1}^{2^n} \frac{1}{j}$$

of the first 2^n terms, with an arbitrary positive integer n, is greater than $1 + n(\frac{1}{2})$, which tends to ∞ as n tends to ∞. Consequently, the partial sums tend to ∞ and the infinite series diverges.   □

Series with nonnegative terms

If all of the terms a_n in an infinite series

$$\sum_{j=j_0}^{\infty} a_j$$

are nonnegative, then we obtain the partial sum s_{n+1} from the partial sum s_n by adding the nonnegative term a_{n+1}. Consequently, the sequence of partial sums $\{s_n\}$ is nondecreasing. Applying Theorem 18.1 to this sequence yields the following result.

Theorem 18.4   An infinite series with nonnegative terms either converges or diverges to infinity. It converges if and only if its partial sums are bounded.

The Comparison Test

Theorem 18.4 is the basis of several tests for determining whether an infinite series converges in cases where we cannot determine the value of the series. The first of these is the COMPARISON TEST, which is given in the next theorem.

Theorem 18.5   (The Comparison Test for series with nonnegative terms)   If the infinite series $\sum\limits_{j=j_0}^{\infty} b_j$ with nonnegative terms converges and if there are constants M and J such that

(12) $0 \le a_j \le M b_j$ for $j \ge J$,

then the infinite series $\sum\limits_{j=j_0}^{\infty} a_j$ converges also.

If the inequalities (12) hold and the series $\sum\limits_{j=j_0}^{\infty} a_j$ diverges to ∞,

then the series $\sum\limits_{j=j_0}^{\infty} b_j$ diverges to ∞ also.

Proof of Theorem 18.5.   Inequalities (12) yield to the inequality

$$(13) \quad \sum_{j=J}^{n} a_j \leq M \sum_{j=J}^{n} b_j$$

for every $n \geq J$. If the series $\sum_{j=j_0}^{\infty} b_j$ converges, then its partial sums are

bounded, so that by (13) the partial sums of $\sum_{j=j_0}^{\infty} a_j$ are bounded.

Theorem 18.4 shows that the series $\sum_{j=j_0}^{\infty} a_j$ converges.

On the other hand, if the series $\sum_{j=j_0}^{\infty} a_j$ diverges, then its partial sums

tend to ∞. Inequality (13) shows that the partial sums of $\sum_{j=j_0}^{\infty} b_j$ tend to

∞, so that series also diverges to ∞.   Q.E.D.

EXAMPLE 6   Use Theorem 18.5 to show that the infinite series

$$(14) \quad \sum_{j=0}^{\infty} \frac{\sin^2 j}{2^j}$$

converges.

Solution.   Because the terms in the series are squares, they are nonnegative. Because $\sin^2 j \leq 1$, we have

$$\frac{\sin^2 j}{2^j} \leq \left(\frac{1}{2}\right)^j \quad \text{for } j \geq 0.$$

The geometric series $\sum_{j=0}^{\infty} (\tfrac{1}{2})^j$ converges, so the series (14) converges by

Theorem 18.5. $\square$

EXAMPLE 7   Use the Comparison Test to determine whether the infinite series

$$(15) \quad \sum_{j=1}^{\infty} \frac{j+2}{5j^2}$$

converges or diverges.

Solution.   We have for each $j \geq 1$

$$\frac{j+2}{5j^2} \geq \frac{j}{5j^2} = \frac{1}{5}\left(\frac{1}{j}\right)$$

and the harmonic series $\sum_{j=1}^{\infty} \frac{1}{j}$ diverges to ∞. Theorem 18.5 implies that the series (15) diverges to ∞ also. $\square$

The Limit Comparison Test

The next result is a refinement of the Comparison Test which is often easier to apply.

Theorem 18.6   (The Limit Comparison Test for series with nonnegative terms)   Suppose that the numbers a_j are nonnegative and the numbers b_j positive and that the limit

$$\lim_{j \to \infty} \frac{a_j}{b_j} = L$$

exists, with L either a nonnegative number or ∞.

(a) If L is finite and the series $\sum_{j=j_0}^{\infty} b_j$ converges, then the series $\sum_{j=j_0}^{\infty} a_j$ converges also.

(b) If L is a positive number or ∞ and $\sum_{j=j_0}^{\infty} b_j$ diverges to ∞, then $\sum_{j=j_0}^{\infty} a_j$ diverges to ∞ also.

Proof.   Suppose first that the limit L is 0 and that the series $\sum_{j=j_0}^{\infty} b_j$ converges. By Definition 18.1 with $\epsilon = 1$ there is a number J such that $(a_j/b_j) < 1$ for $j \geq J$. We then have $a_j < b_j$ for $j \geq J$, and the Comparison Test shows that the series $\sum_{j=j_0}^{\infty} a_j$ converges.

Next, suppose that the limit L is nonzero and finite. By Definition 18.1 with $\epsilon = L/2$, there is a number J such that

$$\left| \frac{a_j}{b_j} - L \right| < \frac{1}{2} L \quad \text{for } j \geq J.$$

Solving this inequality for a_j yields

(17) $\frac{1}{2} L b_j < a_j < \frac{3}{2} L b_j \quad \text{for } j \geq J.$

Inequalities (17) and the Comparison Test show that either both series $\sum_{j=j_0}^{\infty} a_j$ and $\sum_{j=j_0}^{\infty} b_j$ converge or they both diverge to ∞. This gives the two statements of the theorem for the case of nonzero finite L.

If L is ∞, Definition 18.2 with $M = 1$ shows that there is a number J such that $(a_j/b_j) > 1$ for $j \geq J$. Hence $a_j > b_j$ for $j \geq J$ and if $\sum\limits_{j=j_0}^{\infty} b_j$ diverges to ∞, then so does $\sum\limits_{j=j_0}^{\infty} a_j$ by the Comparison Test.   Q.E.D.

EXAMPLE 8   Does the series

$$(18) \qquad \sum_{j=1}^{\infty} \left(\frac{1 + j}{2j} \right)^j$$

converge or diverge?

Solution.   Because $(1 + j)/2j$ tends to $\frac{1}{2}$ as j tends to ∞, we compare (18) with the geometric series $\sum\limits_{j=1}^{\infty} \left(\frac{1}{2} \right)^j$. We set $a_j = \left(\frac{1 + j}{2j} \right)^j$ and $b_j = \left(\frac{1}{2} \right)^j$. Then we have

$$\lim_{j \to \infty} \frac{a_j}{b_j} = \lim_{j \to \infty} \frac{[(1 + j)/2j]^j}{(1/2)^j} = \lim_{j \to \infty} \left(\frac{1 + j}{j} \right)^j$$

$$= \lim_{j \to \infty} \left(1 + \frac{1}{j} \right)^j = e.$$

Because the limit is finite and the geometric series converges, the series (18) also converges by the Limit Comparison Test.   □

EXAMPLE 9   Does the infinite series $\sum\limits_{j=2}^{\infty} \frac{j^2 + 6j}{4j^3 - 7}$ converge or diverge?

Solution.   Because the degree of the numerator of $(j^2 + 6j)/(4j^3 - 7)$ is one less than the degree of the denominator, we compare with the harmonic series $\sum (1/j)$:

$$\lim_{j \to \infty} \left\{ \frac{\dfrac{j^2 + 6j}{4j^3 - 7}}{\dfrac{1}{j}} \right\} = \lim_{j \to \infty} \frac{j^3 + 6j^2}{4j^3 - 7}$$

$$= \lim_{j \to \infty} \frac{1 + (6/j)}{4 - (7/j^3)} = \frac{1}{4}.$$

Because the limit is finite and the harmonic series diverges to ∞, the Limit Comparison Test shows that the given series diverges to ∞ also.   □

Linear combinations of infinite series

If the infinite series $\sum_{j=j_0}^{\infty} a_j$ and $\sum_{j=j_0}^{\infty} b_j$ both converge, then so does the series $\sum_{j=j_0}^{\infty} (Aa_j + Bb_j)$ for any constants A and B, and we have

$$\sum_{j=j_0}^{\infty} (Aa_j + Bb_j) = A \sum_{j=j_0}^{\infty} a_j + B \sum_{j=j_0}^{\infty} b_j.$$

◀

Exercises

†Answer provided.
‡Outline of solution provided.

Determine whether the geometric series in Exercises 1 through 8 converge, diverge to infinity, or diverge without having limits. Give the values of those that converge.

1.† $\displaystyle\sum_{j=0}^{\infty} \left(\frac{1}{7}\right)^j$

2. $\displaystyle\sum_{j=0}^{\infty} (1.001)^j$

3.‡ $\displaystyle\sum_{j=2}^{\infty} (0.99)^j$

4. $\displaystyle\sum_{j=4}^{\infty} \left(-\frac{3}{4}\right)^j$

5.‡ $\displaystyle\sum_{j=5}^{\infty} 4\left(\frac{10}{11}\right)^j$

6. $\displaystyle\sum_{j=0}^{\infty} (-2)^j$

7.† $\displaystyle\sum_{j=10}^{\infty} \frac{1}{100}\left(\frac{9}{8}\right)^j$

8. $\displaystyle\sum_{j=10}^{\infty} 100\left(\frac{8}{9}\right)^j$

Use geometric series to express the repeating decimals in Exercises 9 through 12 as fractions.

9.‡ 3.42121212 . . .

10.† 5.8999 . . .

11. 0.15343434 . . .

12. 4.123123123 . . .

13. For which numbers x does the infinite geometric series

$$\sum_{j=2}^{\infty} 74\left(\frac{x}{5}\right)^j$$

converge and what are its values?

Determine whether the infinite series in Exercises 14 through 29 converge or diverge.

14.‡ $\displaystyle\sum_{n=2}^{\infty} \left(\frac{1}{2}\right)^n \left(50 + \frac{2}{n}\right)$

15. $\displaystyle\sum_{n=0}^{\infty} \frac{2 + \cos n}{3^n}$

16.† $\displaystyle\sum_{j=1}^{\infty} \left(1 + \frac{1}{j}\right)^j$

17. $\displaystyle\sum_{j=0}^{\infty} e^{j-10}$

18.† $\displaystyle\sum_{n=10}^{\infty} \frac{2n - 5}{10 + n^2}$

19. $\displaystyle\sum_{k=5}^{\infty} \left(\frac{20 + k}{3k + 10}\right)^k$

20.† $\displaystyle\sum_{j=0}^{\infty} \frac{1}{3^j + 1}$

21. $\displaystyle\sum_{k=8}^{\infty} \frac{2^k + 3}{4^k + 7}$

22.‡ $\displaystyle\sum_{j=1}^{\infty} \frac{1}{j^j}$ **23.** $\displaystyle\sum_{j=1}^{\infty} \frac{1}{j!}$

24.† $\displaystyle\sum_{n=1}^{\infty} \frac{n^{1/2}}{n^{3/2} + 5}$ **25.** $\displaystyle\sum_{n=4}^{\infty} \frac{1}{n7^n}$

26.† $\displaystyle\sum_{j=1}^{\infty} \frac{1}{\sqrt{j}}$ **27.** $\displaystyle\sum_{j=2}^{\infty} (-1)^j e^{-j}$

28.† $\displaystyle\sum_{j=0}^{\infty} \sin\left(\frac{j\pi}{4}\right)$ **29.** $\displaystyle\sum_{k=1}^{\infty} \sqrt{\frac{k^2 + 2}{k^4 + 4}}$

30. What are the values of the series in Exercises 17 and 27?

31. Suppose that the reserve requirement for banks is $100r$%, so that banks have to keep $100r$% of each deposit in reserve and cannot loan it out. Use a geometric series to determine the least upper bound on the amount of capital that could be created from $1000 of newly printed money that the government spends and then is repeatedly deposited in banks and loaned out. (The money is deposited and the bank loans out the maximum amount allowed, the loan is deposited in another bank that loans out the maximum allowed, and so forth.)

Find the sums of the infinite series in Exercises 32 through 34 by expressing them as telescoping series.

32.† $\displaystyle\sum_{j=2}^{\infty} \frac{2j + 1}{j^2(j + 1)^2}$ (Use partial fractions.)

33.† $\displaystyle\sum_{k=3}^{\infty} \ln\left\{\frac{(k + 1)^2}{k(k + 2)}\right\}$ $\left(\text{Write } \dfrac{(k + 1)^2}{k(k + 2)} = \left(\dfrac{k + 1}{k}\right)\left(\dfrac{k + 2}{k + 1}\right)^{-1}.\right)$

34. $\displaystyle\sum_{k=4}^{\infty} \frac{2}{k(k + 1)(k + 2)}$ $\left(\text{Use } \dfrac{2}{k(k + 2)} = \dfrac{1}{k} - \dfrac{1}{k + 2}.\right)$

18-5   THE INTEGRAL TEST AND THE RATIO TEST

In this section we present the INTEGRAL TEST and the RATIO TEST for determining the convergence or divergence of infinite series with nonnegative terms.

The Integral Test   In the Integral Test we determine whether an infinite series converges absolutely by studying an improper integral of a function whose values at the integers are the values a_j of the terms in the series.

Theorem 18.7   (The Integral Test for series with nonnegative terms)   Suppose that $a_j = f(j)$ for $j \geq J$ where $f(x)$ is a positive, nonincreasing function for $x \geq J$ (J a positive integer). Then the infinite series

(1) $\displaystyle\sum_{j=j_0}^{\infty} a_j$

converges if and only if the improper integral

(2) $\displaystyle\int_{J}^{\infty} f(x)\, dx$

converges.

Recall that the improper integral (2) is said to converge if the limit

$$\int_{J}^{\infty} f(x)\, dx = \lim_{X \to \infty} \int_{J}^{X} f(x)\, dx$$

is finite.

Proof of Theorem 18.7.   Consider an integer $n > J$. Because $f(x)$ is positive, the integral

(3) $\displaystyle\int_{J}^{n} f(x)\, dx$

FIGURE 18.4

Area $= \displaystyle\int_{J}^{n} f(x)\, dx$

is equal to the area of the region under the graph of f that is shown in Figure 18.4. Each of the rectangles in Figures 18.5 and 18.6 has its base on the x-axis, has width 1, and has height and area equal to $f(j)$ for some integer j. The region in Figure 18.5 has area

(4) $\displaystyle\sum_{j=J+1}^{n} f(j) = \sum_{j=J+1}^{n} a_j$

and the region in Figure 18.6 has area

(5) $\displaystyle\sum_{j=J}^{n-1} f(j) = \sum_{j=J}^{n-1} a_j.$

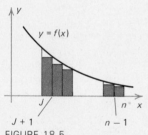

FIGURE 18.5

Area $= \displaystyle\sum_{j=J+1}^{n} f(j) = \sum_{j=J+1}^{n} a_j$

Because $f(x)$ is nonincreasing, the region of Figure 18.4 contains the region of Figure 18.5 and is contained in the region of Figure 18.6. Accordingly, the areas (3), (4), and (5) of the regions are related by the inequalities

(6) $\displaystyle\sum_{j=J+1}^{n} a_j \leq \int_{J}^{n} f(x)\, dx \leq \sum_{j=J}^{n-1} a_j.$

If the improper integral converges and has the value L, then the integral in (6) is $\leq L$ for all n, so the first of the inequalities (6) shows that the partial sums for $\displaystyle\sum_{j=j_0}^{\infty} a_j$ are bounded. Therefore, the series

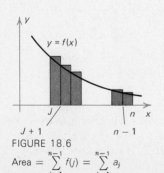

FIGURE 18.6

Area $= \displaystyle\sum_{j=J}^{n-1} f(j) = \sum_{j=J}^{n-1} a_j$

$\displaystyle\sum_{j=j_0}^{\infty} a_j$ converges.

If, on the other hand, the improper integral diverges, then the integral in (6) tends to ∞ as n tends to ∞, so that the series $\sum_{j=j_0}^{\infty} a_j$ diverges. Q.E.D.

EXAMPLE 1   Use the Integral Test to determine whether the infinite series

(7) $$\sum_{j=2}^{\infty} \frac{1}{j \ln(j)}$$

converges or diverges.

Solution.   We set $f(x) = 1/(x \ln x)$ so that $f(j)$ is the jth term in the series (7). For $x > 1$, $f(x)$ is positive and its derivative

$$\frac{df}{dx} = \frac{d}{dx}(x \ln x)^{-1} = -(x \ln x)^{-2}(1 + \ln x)$$

is negative. The function is decreasing and we can apply the Integral Test. We have

$$\int_2^{\infty} \frac{1}{x \ln x} \, dx = \lim_{n \to \infty} \int_2^n \frac{1}{x \ln x} \, dx = \lim_{n \to \infty} \left[\ln|\ln x|\right]_2^n$$
$$= \lim_{n \to \infty} \left[\ln|\ln(n)| - \ln|\ln(2)|\right] = \infty.$$

The improper integral and series (7) diverge.   □

The geometric series $\sum_{j=0}^{\infty} r^j$ is the sum of increasing powers of a fixed constant r. We have seen that it converges for $|r| < 1$ and diverges for $|r| \geq 1$. The next theorem is based on the Integral Test and deals with the sum $\sum_{j=1}^{\infty} j^{-s}$ of a fixed power of the positive integers.

Theorem 18.8   The infinite series

(8) $$\sum_{j=1}^{\infty} \frac{1}{j^s}$$

converges for $s > 1$ and diverges to ∞ for $s \leq 1$.

Proof.   The series diverges for $s \leq 0$ because in this case the terms j^{-s} do not tend to 0 as j tends to ∞. For $s > 0$ the function x^{-s} is positive and decreasing, and we can apply the Integral Test.

In the case of $s = 1$ the series (8) is the harmonic series $\sum_{j=1}^{\infty} \frac{1}{j}$ which we know is divergent from Example 5 of the last section. The Integral Test gives another proof of this result because the integral

$$\int_1^n \frac{1}{x}\,dx = \ln(n)$$

tends to ∞ as n tends to ∞. The improper integral $\int_1^\infty \frac{1}{x}\,dx$ and the harmonic series both diverge to ∞.

For $s > 1$, the integral

$$\int_1^n \frac{1}{x^s}\,dx = \int_1^n x^{-s}\,dx = \frac{n^{1-s}-1}{1-s} = \frac{1-n^{1-s}}{s-1}$$

converges to $1/(s-1)$ as n tends to ∞. Therefore, the improper integral $\int_1^\infty \frac{1}{x^s}\,dx$ and the infinite series (8) both converge.

To show that the series (8) diverges to ∞ for $0 < s < 1$, we can either use the Comparison Test with the harmonic series, since in this case x^{-s} is greater than x^{-1} for $x > 1$, or we can use the Integral Test, since for $s < 1$ the integral tends to ∞ as n tends to ∞.   Q.E.D.

In the next example we use Theorem 18.8 and the Limit Comparison Test.

EXAMPLE 2   Does the infinite series

(9) $$\sum_{j=5}^\infty \frac{4j}{j^3-10}$$

converge or diverge?

Solution.   For large j the terms in the series are close to $4j/j^3 = 4/j^2$, so we compare the series (9) with $\sum_{j=5}^\infty (1/j^2)$. We set $a_j = 4j/(j^3-10)$ and $b_j = 1/j^2$, for which $a_j > 0$, $b_j > 0$, and

(10) $$\frac{a_j}{b_j} = \frac{\dfrac{4j}{j^3-10}}{\dfrac{1}{j^2}} = \frac{4j^3}{j^3-10}.$$

The ratio (10) tends to 4 as j tends to ∞, and by Theorem 18.8 the series $\sum_{j=5}^\infty (1/j^2)$ converges. The Limit Comparison Test shows that the series (9) also converges. $\square$

The Ratio Test   The ratio $|a_{j+1}|/|a_j|$ of the absolute values of successive terms in the geometric series $\sum_{j=0}^\infty r^j$ is the constant $|r^{j+1}|/|r^j| = |r|$, and the series converges when that constant is less than 1. This result is generalized in the Ratio Test, where we study the limit of the ratio $|a_{j+1}|/|a_j|$ for other types of series. We deal here with the case of positive terms a_j.

Theorem 18.9 **(The Ratio Test for series with positive terms)**
Suppose that the numbers a_j are positive and the limit

(11) $$\lim_{j \to \infty} \frac{a_{j+1}}{a_j} = \rho$$

exists, where ρ (rho) may be a nonnegative number or ∞. If ρ
is less than 1, then the infinite series

$$\sum_{j=j_0}^{\infty} a_j$$

converges. If ρ is greater than 1 or is ∞, the series diverges
to ∞.

If ρ is equal to 1, the ratio test fails; the series may converge or
diverge.

Proof of Theorem 18.9.   Suppose first that $\rho < 1$. Let r be any number
with $\rho < r < 1$. By Definition 18.1 with $\epsilon = r - \rho$ there is a number J
such that

$$\left| \frac{a_{j+1}}{a_j} - \rho \right| < r - \rho \quad \text{for } j \geq J.$$

In particular, we have $a_{j+1}/a_j - \rho < r - \rho$, from which we obtain
$a_{j+1}/a_j < r$ and then

(12) $a_{j+1} < r a_j \quad \text{for } j \geq J.$

Applying inequality (12) repeatedly gives $a_{J+1} < r a_J,$
$a_{J+2} < r a_{J+1} < r^2 a_J, a_{J+3} < r a_{J+2} < r^3 a_J,$ and so forth. We obtain

(13) $a_{J+k} < r^k a_J \quad \text{for } k \geq 0.$

Because r is a constant with $0 < r < 1$, the geometric series $\sum_{j=0}^{\infty} r^j$
converges. The Comparison Test and inequality (13) show that the
series $\sum_{j=j_0}^{\infty} a_j$ converges.

Now suppose that ρ is a number greater than 1. Let r be a number
with $1 < r < \rho$. By Definition 18.1 with $\epsilon = \rho - r$, there is a number J
such that $|a_{j+1}/a_j - \rho| < \rho - r$ for $j \geq J$. By calculations similar to
those in the first part of the proof (see Exercise 34), this implies that

(14) $a_{j+1} > a_j \quad \text{for } j \geq J$

so that the terms in the series do not tend to 0 and the series diverges.
We leave the proof for the case of $\rho = \infty$ as Exercise 35.   Q.E.D.

EXAMPLE 3   Use the Ratio Test to determine whether the infinite series

$$(15) \quad \sum_{j=0}^{\infty} \frac{5^j}{j!}$$

converges or diverges.

Solution.   We set $a_j = 5^j/j!$ Then

$$\frac{a_{j+1}}{a_j} = \frac{5^{j+1}/(j+1)!}{5^j/j!} = \frac{5^{j+1} j!}{5^j(j+1)!} = \frac{5}{j+1}.$$

This ratio tends to 0 as $j \to \infty$, so by the Ratio Test with $\rho = 0$, the series (15) converges.   □

Exercises

†Answer provided.
‡Outline of solution provided.

In Exercises 1 through 33 determine whether the infinite series converge or diverge.

1.† $\displaystyle\sum_{j=2}^{\infty} \frac{1}{j(\ln j)^2}$ **2.** $\displaystyle\sum_{j=5}^{\infty} \frac{1}{j\sqrt{\ln j}}$ **3.‡** $\displaystyle\sum_{k=2}^{\infty} \frac{1}{\ln k}$

4.† $\displaystyle\sum_{j=50}^{\infty} \ln(j)$ **5.†** $\displaystyle\sum_{n=0}^{\infty} \frac{3n}{n^2+4}$

6. $\displaystyle\sum_{n=2}^{\infty} \frac{n}{n^3-3}$

7.† $\displaystyle\sum_{j=0}^{\infty} j\left(\frac{1}{2}\right)^j$ **8.** $\displaystyle\sum_{j=2}^{\infty} \frac{2^j}{4^j-10}$ **9.†** $\displaystyle\sum_{k=0}^{\infty} \frac{2^k}{(2k)!}$

10. $\displaystyle\sum_{j=-10}^{\infty} \frac{1}{j^2+4}$ **11.†** $\displaystyle\sum_{k=1}^{\infty} \frac{k!}{k^k}$ **12.** $\displaystyle\sum_{k=1}^{\infty} \frac{k^k}{k!}$

13.† $\displaystyle\sum_{j=4}^{\infty} \frac{\sin^2 j}{j^{3/2}}$ **14.** $\displaystyle\sum_{n=1}^{\infty} \frac{1}{n^2}(2^n)$ **15.‡** $\displaystyle\sum_{j=1}^{\infty} \frac{(j!)^2}{(j^2)!}$

16. $\displaystyle\sum_{m=1}^{\infty} me^{-m}$

17.† $\displaystyle\sum_{j=1}^{\infty} \frac{j^{1/3}}{1+j^{1/2}+j+j^{3/2}}$

18. $\displaystyle\sum_{j=0}^{\infty} \frac{\pi^j}{j^5+6}$

19.† $\displaystyle\sum_{j=75}^{\infty} \frac{(100)^j}{j!}$ **20.** $\displaystyle\sum_{k=3}^{\infty} \frac{5}{\sqrt{k^2-5}}$ **21.** $\displaystyle\sum_{j=1}^{\infty} \frac{j^j}{2^j}$

22. $\displaystyle\sum_{n=1}^{\infty} \frac{n}{\sqrt{n^4+n^2-1}}$

23.‡ $\displaystyle\sum_{k=1}^{\infty} \frac{1}{k^2}\ln k$ **24.** $\displaystyle\sum_{n=1}^{\infty} \sqrt{n}$

25.† $\displaystyle\sum_{j=2}^{\infty} \frac{1}{j^3} e^{\sin(j)}$ **26.** $\displaystyle\sum_{j=1}^{\infty} \frac{\ln(e^j)}{j}$ **27.**† $\displaystyle\sum_{k=2}^{\infty} \frac{1}{k-\sqrt{k}}$

28.‡ $\displaystyle\sum_{n=0}^{\infty} \frac{5^n + n}{6^n - n}$

29.† $\displaystyle\sum_{k=10}^{\infty} \frac{2\cdot 4\cdot 6\cdots(2k)}{1\cdot 3\cdot 5\cdots(2k-1)}$

30. $\displaystyle\sum_{n=1}^{\infty} \frac{e^n}{n^2}$

31.† $\displaystyle\sum_{n=0}^{\infty} \frac{\arctan(n)}{n^2+1}$ **32.** $\displaystyle\sum_{n=1}^{\infty} \frac{e^n}{n!}$ **33.**† $\displaystyle\sum_{k=1}^{\infty} \frac{1}{k^2 + \sin k}$

34. Complete the derivation of inequalities (14) in the proof of Theorem 18.9.

35. Prove Theorem 18.9 for the case of $\rho = \infty$.

18-6 ABSOLUTE CONVERGENCE AND THE ALTERNATING SERIES TEST

Most of the tests for convergence and divergence of infinite series that we have studied so far apply only to series with nonnegative terms.

The two exceptions are the results that the geometric series $\displaystyle\sum_{j=0}^{\infty} r^j$ converges for $|r| < 1$ whether r is positive or negative, and that any series diverges if its terms do not tend to zero.

In this section we look more closely at series which involve negative as well as positive terms. We first show how the Comparison Test, the Limit Comparison Test, the Integral Test, and the Ratio Test can be applied to such series. Then we prove the Alternating Series Test, which can be applied to certain series whose terms alternate in sign.

Absolute Convergence

The Comparison Tests, the Integral Test, and the Ratio Test for general infinite series $\displaystyle\sum_{j=j_0}^{\infty} a_j$ are based on the next theorem, which deals with the series $\displaystyle\sum_{j=j_0}^{\infty} |a_j|$ with nonnegative terms that is obtained by taking the absolute values of the terms in the original series.

Theorem 18.10   If the infinite series

(1) $\displaystyle\sum_{j=j_0}^{\infty} |a_j|$

converges, then so does the series $\displaystyle\sum_{j=j_0}^{\infty} a_j.$

Proof.   Suppose that (1) converges. Set

$$b_j = \begin{cases} a_j & \text{if } a_j \text{ is } \geq 0 \\ 0 & \text{if } a_j \text{ is } < 0 \end{cases} \quad \text{and} \quad c_j = \begin{cases} 0 & \text{if } a_j \text{ is } \geq 0 \\ -a_j & \text{if } a_j \text{ is } < 0. \end{cases}$$

The numbers b_j and c_j are nonnegative and satisfy the inequalities $b_j \leq |a_j|$ and $c_j \leq |a_j|$ for $j \geq j_0$. Hence, the Comparison Test for series with nonnegative terms (Theorem 18.5) implies that the series $\displaystyle\sum_{j=j_0}^{\infty} b_j$ and $\displaystyle\sum_{j=j_0}^{\infty} c_j$ converge. Furthermore, we have $a_j = b_j - c_j$, so that

$$\sum_{j=j_0}^{n} a_j = \sum_{j=j_0}^{n} b_j - \sum_{j=j_0}^{n} c_j.$$

The partial sums on the right of this equation have finite limits as $n \to \infty$, so the partial sum on the left has a finite limit and the series $\displaystyle\sum_{j=j_0}^{\infty} a_j$ converges.   Q.E.D.

If the series $\displaystyle\sum_{j=j_0}^{\infty} |a_j|$ converges, then the series $\displaystyle\sum_{j=j_0}^{\infty} a_j$ is said to CONVERGE ABSOLUTELY. If the series $\displaystyle\sum_{j=j_0}^{\infty} a_j$ converges but the series $\displaystyle\sum_{j=j_0}^{\infty} |a_j|$ diverges, then the series $\displaystyle\sum_{j=j_0}^{\infty} a_j$ is said to CONVERGE CONDITIONALLY.

The Comparison Test   Combining Theorems 18.5 and 18.10 yields the general statement of the Comparison Test.

Theorem 18.11   (The Comparison Test)   Suppose that $|a_j| \leq M b_j$ **for all** $j \geq J$, **where** M **and** J **are positive constants and the series** $\displaystyle\sum_{j=j_0}^{\infty} b_j$ **with nonnegative terms converges. Then the series** $\displaystyle\sum_{j=j_0}^{\infty} a_j$ **converges absolutely. If the series** $\displaystyle\sum_{j=j_0}^{\infty} a_j$ **diverges, then the series** $\displaystyle\sum_{j=j_0}^{\infty} b_j$ **diverges to** ∞.

It is important to notice that Theorem 18.11 does not apply unless the series $\sum_{j=j_0}^{\infty} b_j$ has nonnegative terms.

EXAMPLE 1   Show that the series $\sum_{j=1}^{\infty} \dfrac{(-1)^j}{j^2 + 1}$ converges absolutely.

Solution.   For $j \geq 1$ we have

$$\left| \frac{(-1)^j}{j^2 + 1} \right| < \frac{1}{j^2}$$

Because the series $\sum_{j=1}^{\infty} \dfrac{1}{j^2}$ converges (Theorem 18.8), we can apply the Comparison Test with $b_j = 1/j^2$ to show that the given series converges absolutely.   □

The Limit Comparison Test

We obtain the general statement of the Limit Comparison Test by combining Theorems 18.6 and 18.10.

Theorem 18.12   (The Limit Comparison Test)   Suppose that the limit

$$\lim_{j \to \infty} \frac{|a_j|}{b_j} = L$$

exists, with the b_j positive and L either a nonnegative number or ∞. (a) If L is finite and the series $\sum_{j=j_0}^{\infty} b_j$ converges, then the series $\sum_{j=j_0}^{\infty} a_j$ converges absolutely. (b) If L is a positive number or ∞ and $\sum_{j=j_0}^{\infty} b_j$ diverges, then $\sum_{j=j_0}^{\infty} |a_j|$ diverges.

Note.   In case (b) Theorem 18.12 gives us no information about the convergence of $\sum_{j=j_0}^{\infty} a_j$, which may converge even though the series

$\sum_{j=j_0}^{\infty} b_j$ and $\sum_{j=j_0}^{\infty} |a_j|$ both diverge.   □

EXAMPLE 2   Show that the series $\sum_{j=1}^{\infty} \dfrac{10 + (-2)^j}{3^j}$ converges absolutely.

Solution.   We use the Limit Comparison Test with $b_j = \left(\frac{2}{3}\right)^j$:

$$\lim_{j \to \infty} \left| \frac{\dfrac{10 + (-2)^j}{3^j}}{\left(\dfrac{2}{3}\right)^j} \right| = \lim_{j \to \infty} \left(\frac{3}{2}\right)^j \left| \frac{10 + (-2)^j}{3^j} \right|$$

$$= \lim_{j \to \infty} \left| \frac{10}{2^j} + (-1)^j \right| = 1.$$

The given series converges absolutely because the limit is finite and the geometric series $\displaystyle\sum_{j=1}^{\infty} \left(\frac{2}{3}\right)^j$ converges. □

The Integral and Ratio Tests

When we combine Theorem 18.10 with the Integral Test for series with nonnegative terms (Theorem 18.7), we obtain the general statement of the Integral Test.

Theorem 18.13   (The Integral Test)   Suppose that $|a_j| = f(j)$ for $j \geq J$ where $f(x)$ is a positive, nonincreasing function for $x \geq J$ (J a positive integer). Then the infinite series

$$\sum_{j=j_0}^{\infty} a_j$$

converges absolutely if and only if the improper integral

$$\int_J^{\infty} f(x)\, dx$$

converges.

The Ratio Test for general series reads as follows.

Theorem 18.14   (The Ratio Test)   Suppose that the limit

$$\lim_{j \to \infty} \frac{|a_{j+1}|}{|a_j|} = \rho$$

exists, where ρ (rho) may be a nonnegative number or ∞. If ρ is less than 1, then the infinite series

$$\sum_{j=j_0}^{\infty} a_j$$

converges absolutely. If ρ is greater than 1 or is ∞, the series diverges.

The portion of this theorem dealing with the case of $0 < \rho < 1$ follows from Theorems 18.9 and 18.10. We could show that the series diverges if ρ is greater than 1 by using the argument in the proof of Theorem 18.9 to show that the terms do not tend to zero.

EXAMPLE 3   Show that the series $\displaystyle\sum_{j=1}^{\infty} \frac{(-2)^j}{j^2}$ diverges.

Solution.   We set $a_j = (-2)^j/j^2$. Then $a_{j+1} = (-2)^{j+1}/(j+1)^2$ and

$$\frac{|a_{j+1}|}{|a_j|} = \frac{\dfrac{2^{j+1}}{(j+1)^2}}{\dfrac{2^j}{j^2}} = \frac{j^2}{2^j}\left\{\frac{2^{j+1}}{(j+1)^2}\right\} = \frac{2}{\left(1 + \dfrac{1}{j}\right)^2}.$$

The limit $\rho = 2$ of this ratio as $j \to \infty$ is greater than 1, so the series diverges by the Ratio Test. $\square$

Alternating series

All of the convergent series we have encountered up to this point have terms that tend to zero so rapidly that the series converge absolutely. We will now study series of the form

$$(2) \qquad \sum_{j=j_0}^{\infty} (-1)^j b_j$$

with positive numbers b_j. These are called ALTERNATING SERIES because their terms are alternately positive and negative. As the next theorem shows, alternating series can converge without converging absolutely. In such cases the convergence is due to cancellation between positive and negative terms in the series.

Theorem 18.15   Suppose that the positive numbers b_j $(j \geq J)$ form a nonincreasing sequence that tends to 0 as j tends to ∞; that is, suppose that $b_{j+1} \leq b_j$ for $j \geq J$ and that $b_j \to 0$ as $j \to \infty$. Then the alternating series (2) converges.

Proof.   To simplify the notation, let us suppose that $j_0 = 0$ and $J = 0$. We imagine that the partial sums

$$(3) \qquad s_n = \sum_{j=0}^{n} (-1)^j b_j = b_0 - b_1 + b_2 - \cdots + (-1)^n b_n$$

are plotted on an s-axis. The partial sum $s_0 = b_0$ is positive (Figure 18.7), and because $0 < b_1 \leq b_0$, the partial sum $s_1 = b_0 - b_1$ lies between 0 and s_0 (Figure 18.8). The partial sum $s_2 = b_0 - b_1 + b_2$ lies between s_0 and s_1 because $0 < b_2 \leq b_1$ (Figure 18.9), and the next partial sum $s_3 = b_0 - b_1 + b_2 - b_3$ lies between s_1 and s_2 because $0 < b_3 \leq b_2$ (Figure 18.10). Thus, each partial sum lies between the previous two, and the distance b_n between successive partial sums s_{n-1} and s_n tends to 0 as n tends to ∞.

To verify that s_n has a limit, we consider the sequences $\{s_{2k}\}_{k=0}^{\infty}$ and $\{s_{2k+1}\}_{k=0}^{\infty}$ of partial sums. The sequence $\{s_{2k}\}_{k=0}^{\infty}$ of those s_j's with even indices j is nonincreasing because

$$s_{2(k+1)} = s_{2k+2} = s_{2k} - b_{2k+1} + b_{2k+2}$$
$$= s_{2k} - (b_{2k+1} - b_{2k+2}) \leq s_{2k}.$$

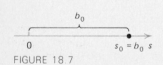

b_0

0    $s_0 = b_0$ s

FIGURE 18.7

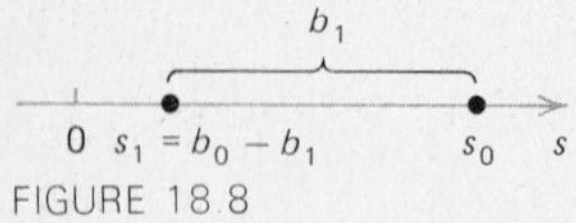

b_1

0 $s_1 = b_0 - b_1$ s_0 s

FIGURE 18.8

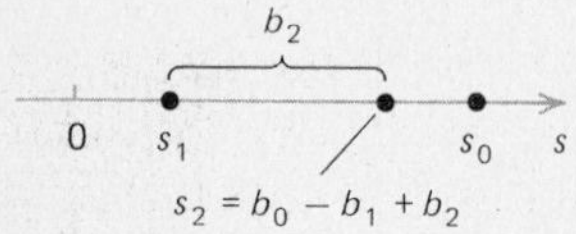

b_2

0   s_1 s_0 s

$s_2 = b_0 - b_1 + b_2$

FIGURE 18.9

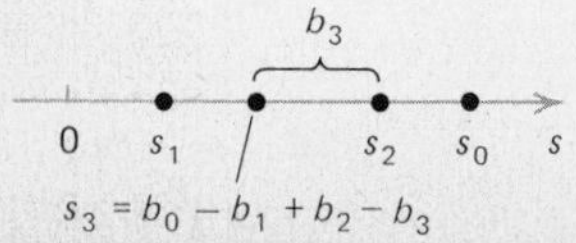

b_3

0   s_1 s_2 s_0 s

$s_3 = b_0 - b_1 + b_2 - b_3$

FIGURE 18.10

Here we use the fact that $b_{2k+1} - b_{2k+2}$ is ≥ 0. Similarly, the sequence $\{s_{2k+1}\}_{k=0}$ with odd indices is nondecreasing: we have

$$s_{2(k+1)+1} = s_{2k+3} = s_{2k+1} + (b_{2k+2} - b_{2k+3}) \leq s_{2k+1}.$$

The monotone sequences $\{s_{2k}\}_{k=0}^{\infty}$ and $\{s_{2k+1}\}_{k=0}^{\infty}$ are bounded because they lie in the interval $0 \leq s \leq b_0$, so they both have limits. The limits must be the same number L because $s_{2k+1} - s_{2k} = b_{2k+1}$ tends to 0 as $k \to \infty$. Thus, all the partial sums tend to L and the infinite series (2) converges.   Q.E.D.

EXAMPLE 4   Show that the infinite series

(4) $$\sum_{j=1}^{\infty} (-1)^j \frac{1}{j}$$

converges conditionally.

Solution.   We set $b_j = 1/j$. The conditions of Theorem 18.15 are satisfied because $b_{j+1} = 1/(j+1) < 1/j = b_j$ and $b_j = 1/j$ tends to 0 as j tends to ∞. Hence, the series (4) converges. It does not converge absolutely because the series

$$\sum_{j=1}^{\infty} \left| (-1)^j \frac{1}{j} \right| = \sum_{j=1}^{\infty} \frac{1}{j}$$

is the harmonic series which diverges.   □

An error estimate   Under the conditions of Theorem 18.15 the sum L of the infinite series lies between the partial sums s_n and s_{n+1} which are b_{n+1} units apart. Therefore, the nth partial sum s_n differs from the sum L of the infinite series by no more than the absolute value b_{n+1} of the next term:

(5) $$\left| \sum_{j=0}^{\infty} (-1)^j b_j - \sum_{j=0}^{n} (-1)^j b_j \right| \leq b_{n+1}.$$

For example, the sum of the first 10 terms of the alternating series (4) differs from the sum of the entire series by no more than $\frac{1}{10}$.

Exercises

†Answer provided.
‡Outline of solution provided.

Determine whether the infinite series in Exercises 1 through 22 converge absolutely, converge conditionally, or diverge.

1.† $\displaystyle\sum_{j=0}^{\infty} \frac{\sin j}{j^3 + 1}$ **2.** $\displaystyle\sum_{k=0}^{\infty} (-1)^k e^{-k}$ **3.‡** $\displaystyle\sum_{n=2}^{\infty} \frac{(-1)^n}{n \ln n}$

4. $\displaystyle\sum_{n=2}^{\infty} \frac{(-1)^n}{n^2 \ln n}$ **5.‡** $\displaystyle\sum_{n=45}^{\infty} (-1)^{n+1} \frac{1}{\ln(\ln n)}$

6. $\displaystyle\sum_{n=3}^{\infty} \frac{(-1)^n}{n(\ln n)^2}$ **7.†** $\displaystyle\sum_{j=1}^{\infty} \frac{(-j)^j}{(2j)!}$ **8.** $\displaystyle\sum_{j=3}^{\infty} \frac{(-3)^j}{j^3+4}$

9.† $\displaystyle\sum_{n=1}^{\infty} \frac{\cos(n\pi)}{\sqrt{n\pi}}$ **10.** $\displaystyle\sum_{k=8}^{\infty} \frac{\cos(e^k)}{k^{3/2}}$ **11.†** $\displaystyle\sum_{n=1}^{\infty} \frac{(-n)^n}{n!}$

12. $\displaystyle\sum_{n=4}^{\infty} (-1)^{n-1}\frac{1}{\sqrt{n^2-7}}$ **13.†** $\displaystyle\sum_{k=0}^{\infty} (-1)^{k+1}\arctan k$

14. $\displaystyle\sum_{p=1}^{\infty} \frac{(-1)^p}{\arctan p}$ **15.‡** $\displaystyle\sum_{j=0}^{\infty} (-1)^j\left[\frac{\pi}{2}-\arctan j\right]$

16. $\displaystyle\sum_{j=1}^{\infty} (-1)^j\frac{j}{j+1}$ **17.†** $\displaystyle\sum_{j=1}^{\infty} (-1)^j\sin\left(\frac{j\pi}{10}\right)$

18. $\displaystyle\sum_{j=1}^{\infty} (-1)^j\frac{1}{2+e^{-j}}$ **19.†** $\displaystyle\sum_{n=2}^{\infty} (-1)^n\sin\left(\frac{1}{n}\right)$

20. $\displaystyle\sum_{n=1}^{\infty} \sin\left[\frac{(-1)^n}{n}\right]$ **21.†** $\displaystyle\sum_{j=1}^{\infty} (-1)^j\frac{\ln j}{j}$

22. $\displaystyle\sum_{k=1}^{\infty} (-1)^k\cos\left(\frac{1}{k}\right)$

In Exercises 23 through 32 determine the values of x for which the infinite series converge absolutely, converge conditionally, and diverge.

23.‡ $\displaystyle\sum_{j=1}^{\infty} \frac{jx^j}{5^j}$ **24.** $\displaystyle\sum_{n=1}^{\infty} (-1)^n\frac{x^n}{n}$ **25.†** $\displaystyle\sum_{n=4}^{\infty} \frac{x^n}{\sqrt{n}}$

26. $\displaystyle\sum_{p=1}^{\infty} (-1)^p\frac{x^p}{p^3}$ **27.†** $\displaystyle\sum_{n=0}^{\infty} \frac{4^n}{x^n}$ **28.** $\displaystyle\sum_{n=1}^{\infty} \frac{1}{n!\,x^n}$

29.† $\displaystyle\sum_{n=1}^{\infty} \frac{1}{n^2+n+1}(10x)^n$ **30.** $\displaystyle\sum_{n=1}^{\infty} \frac{1}{n}\left(1+\frac{1}{x}\right)^n$

31.† $\displaystyle\sum_{n=1}^{\infty} n!\,x^n$ **32.** $\displaystyle\sum_{n=1}^{\infty} \frac{(-1)^n}{\sqrt{n}}n^x$

33.† Use inequality (5) to determine how many terms of the series
   a.   in Exercise 5 and   **b.**   in Exercise 19 it would take to approximate the sum of the entire series within $\frac{1}{5}$.

18-7 POWER SERIES AND TAYLOR SERIES

Up to now we have considered primarily infinite series $\sum_{j=1}^{\infty} a_j$ where the terms were constants a_j. We will now study a type of infinite series where the terms are functions of x, specifically constant multiples of nonnegative powers $(x - a)^j$ of $x - a$ for some constant a. Such series have the form

$$(1) \qquad \sum_{j=0}^{\infty} c_j(x - a)^j$$

and are known as POWER SERIES. Because the power series (1) involves powers of $x - a$ we say that it is *centered* at a.

Power series are important because we can use them to represent a wide variety of functions in terms of powers $(x - a)^j$ of $x - a$, thereby enabling us to use properties of $(x - a)^j$ in studying the other functions.

The geometric series

$$\sum_{j=0}^{\infty} x^j$$

is a power series centered at $x = 0$. We have seen that it converges for $|x| < 1$ and diverges for $|x| \geq 1$. Thus, the interval $-1 < x < 1$ where the series converges has its center at the point $x = 0$ where the series is centered. The next theorem shows that for any constant a, every power series centered at a converges only at $x = a$, or for all x, or in an interval whose center is at a. In the last case, the series might converge or diverge at either endpoint of the interval.

Theorem 18.16 **Either (a) the power series (1) converges only for $x = a$, or (b) it converges absolutely for all x, or (c) there is a positive number R such that it converges absolutely for $|x - a| < R$ and diverges for $|x - a| > R$.**

In case (c) of the theorem the series may converge absolutely, converge conditionally, or diverge at $x = a \pm R$. The number R is called the RADIUS OF CONVERGENCE of the power series. In case (a) we say that the radius of convergence is 0 and in case (b) that it is ∞.

The proof of Theorem 18.16 will be based on the following lemma.

Lemma 18.1 **If the power series (1) converges for some $x_0 \neq a$, then there is a constant M such that**

$$(2) \qquad |c_j(x - a)^j| \leq M \left| \frac{x - a}{x_0 - a} \right|^j \quad \text{for } j \geq 0$$

and the series converges absolutely for $|x - a| < |x_0 - a|$.

Proof. If the power series (1) converges at $x = x_0$, then by Theorem 18.3 the terms $c_j(x_0 - a)^j$ in the series tend to 0 as $j \to \infty$. This implies that the terms are bounded and there is a constant M with $|c_j(x_0 - a)^j| \leq M$ for all $j \geq 0$. Then we have

$$|c_j(x - a)^j| = |c_j(x_0 - a)^j| \left| \frac{x - a}{x_0 - a} \right|^j \leq M \left| \frac{x - a}{x_0 - a} \right|^j$$

which is inequality (2).

For $|x - a| < |x_0 - a|$, inequality (2) reads $|c_j(x - a)^j| \leq M \, r^j$ with $r = |x - a| / |x_0 - a|$ satisfying $0 \leq r < 1$. The power series converges absolutely at such a value of x by comparison with the convergent geometric series $\sum_{j=0}^{\infty} r^j$. Q.E.D.

Proof of Theorem 18.16. The power series converges at least for $x = a$ since at that value of x the terms $c_j(x - a)^j$ are zero for $j \geq 1$. Lemma 18.1 shows that if the series converges for all x, then it converges absolutely for all x.

Suppose that the series converges for some $x \neq a$ and diverges for others, and let I be the set of numbers x for which it converges. Lemma 18.1 shows that we must have $|x - a| \leq |x_1 - a|$ for all x in I if x_1 is a value for which the series diverges. Hence I is bounded. We let R denote the least upper bound on the quantity $|x - a|$ for x in I, and we consider x with $|x - a| < R$. Because R is defined as a least upper bound, there is a number x_0 in I with $|x - a| < |x_0 - a|$ and the series converges absolutely at x by the lemma. The series diverges at x with $|x - a| > R$ since R is an upper bound on the numbers $|x - a|$ such that the series converges at x. Q.E.D.

EXAMPLE 1 Determine the radius of convergence and interval of convergence of the power series

(3) $$\sum_{j=1}^{\infty} \frac{1}{j} \left(\frac{x}{3} \right)^j.$$

Solution. We use the Ratio Test. We set $a_j = \frac{1}{j} \left(\frac{x}{3} \right)^j$ with x a fixed but arbitrary nonzero number. The ratio

$$\frac{|a_{j+1}|}{|a_j|} = \frac{\left| \frac{1}{j+1} \left(\frac{x}{3} \right)^{j+1} \right|}{\left| \frac{1}{j} \left(\frac{x}{3} \right)^j \right|} = \frac{1}{1 + \frac{1}{j}} \left| \frac{x}{3} \right|$$

tends to $\rho = \frac{1}{3} |x|$ as j tends to ∞. For $|x| < 3$, ρ is less than 1 and the power series (3) converges absolutely. For $|x| > 3$, ρ is greater than 1 and the power series diverges. The radius of convergence is 3.

We still have to study the convergence of the power series at the endpoints $x = \pm 3$ of the interval of convergence. For $x = 3$, the power series (3) is the harmonic series

$$\sum_{j=1}^{\infty} \frac{1}{j}\left(\frac{3}{3}\right)^3 = \sum_{j=1}^{\infty} \frac{1}{j}$$

and diverges. For $x = -1$, it is the series

$$\sum_{j=1}^{\infty} \frac{1}{j}\left(\frac{-3}{3}\right)^j = \sum_{j=1}^{\infty} \frac{1}{j}(-1)^j$$

which converges by the Alternating Series Test. The interval of convergence of the power series (3) is $-3 \leq x < 3$. $\square$

Taylor series

If a function $f(x)$ has derivatives of all orders in an interval I and if a is a point in I, then we can form its Taylor polynomial approximations centered at a

$$(4) \qquad P_n(x) = \sum_{j=0}^{n} \frac{1}{j!} f^{(j)}(a)(x - a)^j$$

for all positive integers n. Furthermore, for all x in I such that the remainder $R_n(x) = f(x) - P_n(x)$ tends to 0 as n tends to ∞, $f(x)$ is equal to the power series

$$(5) \qquad f(x) = \sum_{j=0}^{\infty} \frac{1}{j!} f^{(j)}(a)(x - a)^j \qquad \blacktriangleleft$$

which is called the infinite TAYLOR SERIES for $f(x)$ centered at $x = a$. This formula for $f(x)$ might be valid for all x, for x in an interval, or only at $x = a$.

In the case of $a = 0$, the series (5) is also known as the MACLAURIN SERIES for $f(x)$, after the British mathematician Colin Maclaurin (1698–1746).

By Theorem 17.1 there is, for each n and each x in the interval I, at least one number c between x and a such that

$$(6) \qquad R_n(x) = \frac{1}{(n+1)!} f^{(n+1)}(c)(x - a)^{n+1}.$$

In the next example we use expression (6) for the remainder to study the Maclaurin series for e^x.

EXAMPLE 2   Find the Maclaurin series for e^x and show that it converges and has the value e^x for all x.

Solution.   In the case of $f(x) = e^x$ we have $f^{(j)}(x) = e^x$ for all x and j and $f^{(j)}(0) = 1$ for all j. The Taylor polynomial for e^x centered at $x = 0$ is

FIGURE 18.11
Colin Maclaurin (1698–1746)
The Granger Collection

$$P_n(x) = \sum_{j=0}^{n} \frac{1}{j!} x^j.$$

The remainder is

$$R_n(x) = \frac{1}{(n+1)!} e^c x^{n+1}$$

where c is a number between x and 0. The quantity e^c is less than $e^{0} = 1$ if x is negative and is less than e^x if x is positive. The quantity $x^{n+1}/(n+1)!$ tends to 0 for each x as n tends to ∞ (see Exercise 31). Therefore $R_n(x) \to 0$ and the equation $f(x) = P_n(x) + R_n(x)$ gives

$$e^x = \lim_{n \to \infty} P_n(x) = \lim_{n \to \infty} \sum_{j=0}^{n} \frac{1}{j!} x^j$$

$$= \sum_{j=0}^{\infty} \frac{1}{j!} x^j.$$

This equation is valid for all x and the radius of convergence of the power series is ∞.   □

Exercises

†Answer provided.
‡Outline of solution provided.

In Exercises 1 through 15 determine the values of x for which the power series converge absolutely, converge conditionally, and diverge.

1.‡ $\displaystyle\sum_{j=1}^{\infty} \frac{1}{j} x^j$

2.† $\displaystyle\sum_{j=0}^{\infty} \sqrt{j}\, x^j$

3.‡ $\displaystyle\sum_{j=0}^{\infty} \frac{j+3}{j^2+1} \left(\frac{x-3}{2}\right)^j$

4. $\displaystyle\sum_{j=4}^{\infty} \left(\frac{x}{j}\right)^j$

5.† $\displaystyle\sum_{j=1}^{\infty} (jx)^j$

6. $\displaystyle\sum_{j=1}^{\infty} \frac{1}{j^3\, 3^j} x^j$

7.† $\displaystyle\sum_{j=10}^{\infty} (-1)^j (10x)^j$

8. $\displaystyle\sum_{j=1}^{\infty} \frac{j!}{j^{2j}} x^{2j+1}$

9. $\displaystyle\sum_{j=0}^{\infty} (j+1)!\, x^{j+1}$

10. $\displaystyle\sum_{j=1}^{\infty} \frac{1}{\sqrt{j}}\, 5^j x^j$

11.† $\displaystyle\sum_{j=1}^{\infty} \frac{1}{2j} x^{2j}$

12. $\displaystyle\sum_{j=1}^{\infty} \frac{(-1)^j}{j\, 4^j} x^{2j}$

13. $\displaystyle\sum_{j=0}^{\infty} \frac{1}{(j!)^2} x^{2j+1}$

14.† $\displaystyle\sum_{j=2}^{\infty} \frac{x^j}{\ln(j)}$

15. $\displaystyle\sum_{j=1}^{\infty} \sqrt{\frac{j}{j^4+1}}\, x^{2j}$

Give the Maclaurin series for the functions in Exercises 16 through 23. Show that the series converge and are equal to the functions for all x.

16.‡ e^{2x} **17.**† $\sin x$ **18.** $\cos x$

19.† $\sinh x$ **20.** $\cosh x$ **21.**† $(1 + x)^5$

22. $\cos(x + 1)$ **23.** $\sin(2 - 3x)$

Give the Maclaurin series for the functions in Exercises 24 through 26 and determine their intervals of convergence.

24.† $\ln(1 + x)$ **25.**† $\dfrac{1}{2 - x}$ **26.** $(1 - x)^{-1/2}$

27.† Use the Maclaurin series for e^x to give the Maclaurin series of
    a. e^{-x^2},   **b.** $x^3 e^{-x}$, and   **c.** $\cosh x$.

28. Show that the Maclaurin series for $(1 - x)^{-1}$ is the geometric series.

29. Use the Taylor series for e^x centered at $x = 0$ and at $x = a$ to prove that $e^x = e^a e^{x-a}$.

30. The function e^{ix} with i the symbol introduced to serve as a square root of -1 can be defined by replacing x by ix in the Maclaurin series for e^x. Show that $e^{ix} = \cos x + i \sin x$.

31. Use the Ratio Test and the Maclaurin series for e^x to show that for any x, $x^j/j!$ tends to 0 as j tends to ∞.

32. *(The binomial series)*   **a.** Show that for any constant s, the Maclaurin series for $(1 + x)^s$ is

$$(1)\qquad (1 + x)^s = \sum_{j=0}^{\infty} \binom{s}{j} x^j$$

where $\binom{s}{j}$ is the *binomial coefficient*

$$\binom{s}{j} = \frac{s(s - 1)(s - 2)(s - 3) \cdots (s - j + 1)}{j!}.$$

    b. Show that the binomial series (1) has a finite number of terms if s is a nonnegative integer and has radius of convergence 1 otherwise.
    c. Use the first three terms of the binomial series (1) with $s = 1/10$ to compute an approximate value of $2^{1/10}$.

18-8 OPERATIONS WITH POWER SERIES

Much of the usefulness of power series relies on the fact that in the intervals of convergence they can be added, multiplied, divided, composed, differentiated, and integrated by performing the operations as if they were polynomials (finite sums) rather than infinite series. In this section we describe the key results concerning these operations with power series and we illustrate how they are applied. To simplify the discussions, we will deal only with power series centered at $x = 0$ (Maclaurin series) and we will omit the proofs, which can be found in books on advanced calculus.

Many of the examples and exercises of this section use the following Maclaurin series.

(1) $e^x = \sum_{j=0}^{\infty} \frac{1}{j!} x^j = 1 + x + \frac{1}{2} x^2 + \frac{1}{3!} x^3 + \frac{1}{4!} x^4 + \cdots$

(2) $\sin x = \sum_{j=0}^{\infty} \frac{(-1)^j}{(2j+1)!} x^{2j+1} = x - \frac{1}{3!} x^3 + \frac{1}{5!} x^5 - \cdots$

(3) $\cos x = \sum_{j=0}^{\infty} \frac{(-1)^j}{(2j)!} x^{2j} = 1 - \frac{1}{2} x^2 + \frac{1}{4!} x^4 - \cdots$

(4) $\sinh x = \sum_{j=0}^{\infty} \frac{1}{(2j+1)!} x^{2j+1} = x + \frac{1}{3!} x^3 + \frac{1}{5!} x^5 + \cdots$

(5) $\cosh x = \sum_{j=0}^{\infty} \frac{1}{(2j)!} x^{2j} = 1 + \frac{1}{2} x^2 + \frac{1}{4!} x^4 + \cdots$

(6) $\frac{1}{1-x} = \sum_{j=0}^{\infty} x^j = 1 + x + x^2 + x^3 + \cdots$ $\qquad (-1 < x < 1)$

(7) $\ln(1 + x) = \sum_{j=0}^{\infty} \frac{(-1)^{j+1}}{j} x^j = x - \frac{1}{2} x^2 + \frac{1}{3} x^3 - \frac{1}{4} x^4 + \cdots$

$$(-1 < x \leq 1)$$

The series for e^x, $\sin x$, $\cos x$, $\sinh x$, and $\cosh x$ converge for all x.

Sums, products, and quotients

Theorem 18.17   Suppose that the functions $f(x)$ and $g(x)$ are equal to the power series

(8) $f(x) = \sum_{j=0}^{\infty} a_j x^j$ **and** $g(x) = \sum_{j=0}^{\infty} b_j x^j$

in the open interval $|x| < R$ where R is a positive number or ∞. Then the functions $f(x) + g(x)$ and $f(x)g(x)$ have power series representations for $|x| < R$ which may be computed by adding and multiplying the power series (8) term by term. If $g(x)$ is not zero for $|x| < R$, then the quotient $f(x)/g(x)$ also has a power series representation for $|x| < R$, any number of whose terms can be computed by long division.

EXAMPLE 1   What is the Maclaurin series of $\sin x + \sin(3x)$?

Solution.   For all x we have

$$\sin x = \sum_{j=0}^{\infty} (-1)^j \frac{1}{(2j + 1)!} x^{2j+1} = x - \frac{1}{3!} x^3 + \frac{1}{5!} x^5 - \cdots$$

$$\sin(3x) = \sum_{j=0}^{\infty} (-1)^j \frac{1}{(2j + 1)!} (3x)^{2j+1}$$

$$= \sum_{j=0}^{\infty} (-1)^j \frac{3^{2j+1}}{(2j + 1)!} x^{2j+1}$$

$$= 3x - \frac{3^3}{3!} x^3 + \frac{3^5}{5!} x^5 - \cdots$$

and consequently

$$\sin x + \sin(3x) = \sum_{j=0}^{\infty} (-1)^j \left\{ \frac{1 + 3^{2j+1}}{(2j + 1)!} \right\} x^{2j+1}$$

$$= 4x - \left[\frac{1 + 3^3}{3!} \right] x^3 + \left[\frac{1 + 3^5}{5!} \right] x^5 - \cdots . \quad \square$$

EXAMPLE 2   Use the Maclaurin series (7) for $\ln(1 + x)$ to give the Maclaurin series of $x^2 \ln(1 + x)$.

Solution.   Multiplying the series (7) by x^2 gives

$$x^2 \ln(1 + x) = \sum_{j=1}^{\infty} \frac{(-1)^{j+1}}{j} x^{j+2} = x^3 - \tfrac{1}{2} x^4 + \tfrac{1}{3} x^5 - \tfrac{1}{4} x^6 + \cdots$$

for $-1 < x \le 1$. $\square$

EXAMPLE 3   Give the terms up to degree 5 in the Maclaurin series of $\sin x \cos x$.

Solution.   We have for all x

$$\sin x = x - \frac{1}{3!} x^3 \frac{1}{5!} x^5 - \cdots = x - \tfrac{1}{6} x^3 + \tfrac{1}{120} x^5 - \cdots$$

$$\cos x = 1 - \frac{1}{2} x^2 + \frac{1}{4!} x^4 - \cdots = 1 - \tfrac{1}{2} x^2 + \tfrac{1}{24} x^4 - \cdots$$

so that

$$\sin x \cos x = (x - \tfrac{1}{6} x^3 + \tfrac{1}{120} x^5 - \cdots)(1 - \tfrac{1}{2} x^2 + \tfrac{1}{24} x^4 - \cdots)$$

$$= x(1 - \tfrac{1}{2} x^2 + \tfrac{1}{24} x^4 - \cdots) - \tfrac{1}{6} x^3(1 - \tfrac{1}{2} x^2 + \tfrac{1}{24} x^4 - \cdots)$$

$$+ \tfrac{1}{120} x^5(1 - \tfrac{1}{2} x^2 + \tfrac{1}{24} x^4 - \cdots) - \cdots$$

$$= x - \tfrac{1}{2} x^3 + \tfrac{1}{24} x^5 - \cdots - \tfrac{1}{6} x^3 + \tfrac{1}{12} x^5 - \cdots + \tfrac{1}{120} x^5 - \cdots$$

$$= x - \tfrac{2}{3} x^3 + \tfrac{2}{15} x^5 - \cdots . \quad \square$$

EXAMPLE 4   Give the terms up to degree 6 in the Maclaurin series for $\dfrac{x^2}{\cos x}$. What is its radius of convergence?

Solution.   We divide x^2 by the power series

$$\cos x = 1 - \frac{1}{2}x^2 + \frac{1}{4!}x^4 - \cdots = 1 - \tfrac{1}{2}x^2 + \tfrac{1}{24}x^4 - \cdots$$

keeping track of only those terms that contribute terms of degree ≤ 6 in the dividend:

$$
\begin{array}{r}
x^2 + \tfrac{1}{2}x^4 + \tfrac{5}{24}x^6 + \cdots \\[2pt]
\hline
1 - \tfrac{1}{2}x^2 + \tfrac{1}{24}x^4 - \cdots \overline{\smash{\big)}\, x^2 \phantom{+ \tfrac{1}{2}x^4}} \\[2pt]
x^2 - \tfrac{1}{2}x^4 + \tfrac{1}{24}x^6 - \cdots \\[2pt]
\hline
\tfrac{1}{2}x^4 - \tfrac{1}{24}x^6 + \cdots \\[2pt]
\tfrac{1}{2}x^4 - \tfrac{1}{4}x^6 + \cdots \\[2pt]
\hline
\tfrac{5}{24}x^6 - \cdots \\[2pt]
\tfrac{5}{24}x^6 - \cdots \\[2pt]
\hline
+ \cdots
\end{array}
$$

This calculation shows that $x^2/(\cos x) = x^2 + \tfrac{1}{2}x^4 + \tfrac{5}{24}x^6 + \cdots$. The radius of convergence of this power series is $\tfrac{1}{2}\pi$ because the series for $\cos x$ converges for all x and $\cos x$ is zero at $x = \pm\tfrac{1}{2}\pi$. $\quad\square$

Composition

Theorem 18.18   Suppose that $u(x)$ has a convergent Maclaurin series for $|x| < R$ and $G(u)$ has a convergent power series in an open interval containing the values of $u(x)$ for $|x| < R$. Then $G(u(x))$ has a convergent Maclaurin series for $|x| < R$, whose terms can be computed from the power series for $u(x)$ and $G(u)$.

EXAMPLE 5   Give the Maclaurin series for e^{x^3}.

Solution.   When we set $u = x^3$ in the Maclaurin series

$$e^u = \sum_{j=0}^{\infty} \frac{1}{j!} u^j = 1 + u + \frac{1}{2}u^2 + \frac{1}{3!}u^3 + \frac{1}{4!}u^4 + \cdots$$

we obtain

$$e^{x^3} = \sum_{j=0}^{\infty} \frac{1}{j!} x^{3j} = 1 + x^3 + \frac{1}{2}x^6 + \frac{1}{3!}x^9 + \frac{1}{4!}x^{12} + \cdots. \quad\square$$

EXAMPLE 6   Give the terms of degree ≤ 4 in the MacLaurin series for $\sinh[\ln(1 - x)]$.

Solution.   We use the Maclaurin series

$$\sinh u = u + \frac{1}{3!}u^3 + \frac{1}{5!}u^5 + \cdots = u + \tfrac{1}{6}u^3 + \tfrac{1}{120}u^5 + \cdots$$

$$\ln(1 - x) = -x - \tfrac{1}{2}x^2 - \tfrac{1}{3}x^3 - \tfrac{1}{4}x^4 - \cdots$$

the first of which converges for all u and the second for $-1 \leq x < 1$. We obtain for $-1 < x < 1$

$$
\begin{aligned}
\sinh[\ln(1 - x)] &= \ln(1 - x) \\
&\quad + \tfrac{1}{6}[\ln(1 - x)]^3 + \tfrac{1}{120}[\ln(1 - x)]^5 + \cdots \\
&= (-x - \tfrac{1}{2}x^2 - \tfrac{1}{3}x^3 - \tfrac{1}{4}x^4 - \cdots) \\
&\quad + \tfrac{1}{6}(-x - \tfrac{1}{2}x^2 - \tfrac{1}{3}x^3 - \cdots)^3 \\
&\quad + \tfrac{1}{120}(-x - \tfrac{1}{2}x^2 - \cdots)^5 + \cdots \\
&= -x - \tfrac{1}{2}x^2 - \tfrac{1}{3}x^3 - \tfrac{1}{4}x^4 - \cdots \\
&\quad - \tfrac{1}{6}(x^3 - \tfrac{3}{2}x^4 + \cdots) + \cdots \\
&= -x - \tfrac{1}{2}x^2 - \tfrac{1}{2}x^3 + [\text{Terms of degree} \geq 5].
\end{aligned}
$$

In the next to last step of this calculation we used the formula $(a + b)^3 = a^3 + 3a^2b + 3ab^2 + b^3$.   □

Differentiation and integration

Theorem 18.19   **If a power series**

(9) $$f(x) = \sum_{j=0}^{\infty} a_j x^j$$

has a nonzero radius of convergence R, then the derivatives and antiderivatives of $f(x)$ may be computed for $|x| < R$ by differentiating and integrating the power series term by term:

(10) $$\frac{df}{dx}(x) = \sum_{j=0}^{\infty} \frac{d}{dx}(a_j x^j) = \sum_{j=1}^{\infty} ja_j x^{j-1}$$

(11) $$\int_0^x f(t)\,dt = \sum_{j=0}^{\infty} \int_0^x a_j t^j\,dt = \sum_{j=0}^{\infty} \frac{1}{j+1} a_j x^{j+1}.$$

Moreover, the three series (9), (10), and (11) have the same radius of convergence.

If the radius of convergence R is finite, then all three series in Theorem 18.19 converge absolutely for $|x| < R$, but one might converge at an endpoint $x = \pm R$ and another diverge there.

We can also integrate the power series (9) term by term over any closed finite interval contained in the interior of the interval of convergence: for $-R < a < b < R$ we have

$$\int_a^b f(x)\,dx = \sum_{j=0}^{\infty} \int_a^b a_j x^j\,dx = \sum_{j=0}^{\infty} \frac{1}{j+1} a_j(b^{j+1} - a^{j+1}).$$

EXAMPLE 7   Use the geometric series (6) to find the Maclaurin series for $(1 - x)^{-2}$. What is its radius of convergence?

Solution.   We write

$$(1 - x)^{-2} = \frac{d}{dx}(1 - x)^{-1} = \frac{d}{dx}\sum_{j=0}^{\infty} x^j$$

(12)
$$= \sum_{j=0}^{\infty} \frac{d}{dx}(x^j) = \sum_{j=1}^{\infty} jx^{j-1}$$

$$= 1 + 2x + 3x^2 + 4x^3 + \cdots$$

We start the summation at $j = 1$ in the last sum because the term for $j = 0$ is zero.

The series (12), like the geometric series, has radius of convergence 1. They both converge for $|x| < 1$.   □

EXAMPLE 8   Use the geometric series (6) to find the Maclaurin series (7) of $\ln(1 + x)$.

Solution.   Because the derivative of $\ln(1 + x)$ is $1/(1 + x)$, we have for $x > -1$

$$\int_0^x \frac{1}{1 + t}\, dt = [\ln(1 + t)]_{t=0}^{t=x} = \ln(1 + x) - \ln(1) = \ln(1 + x).$$

We obtain the Maclaurin series for $1/(1 + t)$ by replacing x by $-t$ in the geometric series (6), and then integrating that series term by term gives

$$\ln(1 + x) = \int_0^x \frac{1}{1 + t}\, dt = \sum_{j=0}^{\infty} \int_0^x (-1)^j t^j\, dt$$

$$= \sum_{j=0}^{\infty} \frac{(-1)^j}{j + 1} x^{j+1}   □$$

Computing derivatives at $x = 0$ If j and k are nonnegative integers, then the kth derivative of x^j is zero at $x = 0$ unless $j = k$ when the derivative is $k!$:

(13) $$\left[\frac{d^k}{dx^k} x^j\right]_{x=0} = \begin{cases} 0 \text{ for } 0 \le j < k \text{ and } j > k \\ k! \text{ for } j = k. \end{cases}$$

For example, if j is 3, then x^3 and its first two derivatives, $\dfrac{d}{dx}x^3 = 3x^2$

and $\dfrac{d^2}{dx^2}x^3 = 6x$, are zero at $x = 0$; the third derivative $\dfrac{d^3}{dx^3}x^3 = 6$ is

equal to the constant $6 = 3!$; and all higher derivatives are zero for all x.

Therefore, if the power series

(9) $f(x) = \displaystyle\sum_{j=0}^{\infty} a_j x^j$

has a nonzero radius of convergence, Theorem 18.19 shows that for each nonnegative integer k

(14) $f^{(k)}(0) = \displaystyle\sum_{j=0}^{\infty} a_j \left[\dfrac{d^k}{dx^k} x^j\right]_{x=0} = a_k k!.$

Thus, if we know the Maclaurin series of a function we can easily compute its derivatives at $x = 0$.

EXAMPLE 9   In Exercise 25 we will obtain the Maclaurin series

(15) $\arctan x = \displaystyle\sum_{j=0}^{\infty} \dfrac{(-1)^j}{2j+1} x^{2j+1}$

with radius of convergence 1. Use it to compute the ninth derivative of $\arctan x$ at $x = 0$.

Solution.   Formula (14) shows that $f^{(9)}(0) = 9!a_9$ where a_9 is the coefficient of x^9 in the power series for f. We obtain the power x^9 in the series (15) by taking $j = 4$. Therefore, $a_9 = (-1)^4/[2(4) + 1] = 1/9$ and

$$\left[\dfrac{d^9}{dx^9} \arctan x\right]_{x=0} = 9!a_9 = 9!/9 = 8!. \quad \square$$

Exercises

†*Answer provided*
‡*Outline of solution provided*

Use the power series (1) through (7) to give the Maclaurin series of the functions in Exercises 1 through 12.

1.† $\sin x + 5x^4 + 3$

2.† $3e^{2x} + 4e^{5x}$

3. $\cos x + \cosh x$

4.† $\ln[(1 + x)/(1 - x)]$

5. $x^2 e^{-x}$

6.† $(1 + x)/(1 - x)$

7. $\sin(x^2)$

8.† $x^3 \sinh(x^3)$

9.† $(1 - x)^{-3}$

10. $\dfrac{d^3}{dx^3} e^{x^2}$

11.† $\displaystyle\int_0^x e^{t^2}\, dt$

12. $\displaystyle\int_0^x \ln(1 - t)\, dt$

Use the power series (1) through (7) to give the first four nonzero terms in the Maclaurin series of the functions in Exercises 13 through 24.

13.† $e^x \sin x$

14.† $[\ln(1 - x)]^2$

15. $\cosh^2 x$

16. $\sin x \cos(3x)$

17.† $\dfrac{\sin x}{1 - x}$

18.† $\dfrac{e^x}{x^2 + x + 1}$

19.[†] $\sec x$ **20.**[†] e^{x+x^2}

21. $\dfrac{\ln(1 + x)}{\cos x}$ **22.**[†] $\tanh x$

23.[†] $e^{\sin x}$ **24.** $(1 + \sin x)^{-1}$

25. Integrate the geometric series (6) with x replaced by $-t^2$ to obtain the Maclaurin series (15) for $\arctan x$.

26.[†] **a.** Use the series (15) with $x = 1$ to express π as an alternating series. **b.** Give approximate values of the partial sums of the first five and six terms of the series of part (a). **c.** Use the first three terms of (15) and the equation $\frac{1}{4}\pi = \arctan(\frac{1}{2}) + \arctan(\frac{1}{3})$ to give an approximate value of π.

27. **a.** Use three terms of the series (7) to give an approximate value of $\ln(\frac{11}{9})$.   **b.** Use three terms of the power series of Exercise 4 with $x = 1/10$ to give an approximate value of $\ln(\frac{11}{9})$ $(\ln(\frac{11}{9}) = 0.20067069 \ldots)$.

What functions have the power series given in Exercises 28 through 31?

28.[†] $\displaystyle\sum_{j=1}^{\infty} j(j - 1)x^j$ **29.**[†] $\displaystyle\sum_{j=0}^{\infty} \frac{1}{(2j)!}x^{6j}$

30.[†] $\displaystyle\sum_{j=1}^{\infty} \frac{(-1)^{j+1}}{j(j + 1)}x^{j+1}$ **31.** $\displaystyle\sum_{j=0}^{\infty} \frac{1}{j!}x^{2j+1}$

If the functions in Exercises 32 through 35 are given suitable values at $x = 0$, then they have Maclaurin series with nonzero radii of convergence. Find the series.

32.[†] $\dfrac{\sin x}{x}$ **33.** $\dfrac{\ln(1 + x^2)}{x^2}$

34.[†] $\dfrac{\cos x - 1}{x^2}$ **35.** $\dfrac{e^{2x} - 1 - 2x - 2x^2}{x^3}$

36.[†] Use the power series of Exercise 5 to compute the twentieth derivative of $x^2 e^{-x}$ at $x = 0$.

37. Use the power series of Exercise 7 to compute the sixteenth derivative of $\sin(x^2)$ at $x = 0$.

MISCELLANEOUS EXERCISES

[†]*Answer provided.*
[‡]*Outline of solution provided.*

1. Show that   **a.** $(1 + n^{-1/2})^n$ tends to ∞ and   **b.** $(1 + n^{-2})^n$ tends to 1 as $n \to \infty$.

2. Are the following sequences increasing or decreasing for $n > 1$? (Study the ratios of successive terms.)

   a.[†] $\left\{\dfrac{n!}{n^n}\right\}$ **b.** $\left\{\dfrac{2^n}{n}\right\}$ **c.** $\left\{\dfrac{(2n)!}{(n!)^2}\right\}$

Determine whether the infinite series in Exercises 3 through 9 converge or diverge. State the convergence or divergence tests that you use.

3.† $\displaystyle\sum_{j=0}^{\infty} e^{j-j^2}$
  4.† $\displaystyle\sum_{j=1}^{\infty} \ln\left(1 + \frac{1}{j}\right)$

5. $\displaystyle\sum_{j=1}^{\infty} \ln\left(1 + \frac{1}{j^2}\right)$
  6. $\displaystyle\sum_{j=2}^{\infty} \left[\ln\left(\frac{1}{j}\right)\right]^2$

7. $\displaystyle\sum_{j=1}^{\infty} a_j$ where $a_j = 1/j$ if j is the square of an integer and $a_j = 0$ otherwise.

8. $\displaystyle\sum_{j=2}^{\infty} [\ln(j)]^{-j}$
  9. $\displaystyle\sum_{j=2}^{\infty} [\ln(j)]^{-\ln(j)}$

(Use the equation $\ln(j) = e^{\ln[\ln(j)]}$ to show that $[\ln(j)]^{\ln(j)} = j^{\ln[\ln(j)]}$.)

10. Show that the infinite series

$$\sum_{j=1}^{\infty} [\tfrac{1}{2}\pi - \arctan(j^2)]$$

converges. (First express $\tfrac{1}{2}\pi - \arctan x$ as an improper integral and then use the inequality $(1 + t^2)^{-1} < t^{-2}$.)

11.† Use the partial fraction decomposition of $1/(x^2 + x)$ to show that

$$\sum_{j=1}^{\infty} \frac{1}{j^2 + j} = 1.$$

12. Use the partial fraction decomposition of $1/(x^2 - 4)$ to show that

$$\sum_{j=3}^{\infty} \frac{1}{j^2 - 4} = \frac{25}{48}.$$

13. Show that

$$\sum_{j=0}^{\infty} \frac{2^j + 3^j}{6^j} = \frac{7}{2}.$$

14. Show that

$$\sum_{n=1}^{\infty} \ln\left(\frac{n}{n + 1}\right) = -\infty.$$

15. If, on the average, one car crosses a bridge every two minutes, then the probability that exactly j cars cross in an interval of length x minutes is $Q_j(x) = \dfrac{1}{j!} (\tfrac{1}{2}x)^j e^{-x/2}$. Compute the sum $\displaystyle\sum_{j=0}^{\infty} Q_j(x)$ and interpret your result.

16.† A man deposits \$100 at the beginning of each month in a savings account that pays 6% annual interest compounded monthly. There is no money in the account to start with and he makes no withdrawals. How much is in his account after 10 years?

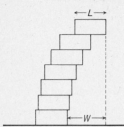

FIGURE 18.12

17. **a.** Show that

$$\ln(n+1) < 1 + \tfrac{1}{2} + \tfrac{1}{3} + \cdots + \frac{1}{n} < 1 + \ln(n)$$

for any positive integer n.   **b.** Show that it would take more than 8000 terms in a partial sum for the harmonic series for the partial sum to be greater than 10.

18. **a.** Show that for any positive numbers W, and L, identical bricks of length L could be stacked as in Figure 18.12 so that the stack would reach W to the right of the bottom brick.

19. A ball is dropped from a height of 100 feet and it bounces to a height of 80 feet. With each subsequent bounce it reaches 80% of the height of the previous bounce. What is the limit of the total distance it travels?

20.† **a.** Give the first four terms in the binomial series for $(1-s)^{-1/2}$.
    b. Make the substitution $s = t^2$ and integrate to obtain the first four terms in the Maclaurin series for $\arcsin x$.

21. Use the Maclaurin series for $(\sin t)/t$ to obtain the Maclaurin series for $\int_0^x \frac{\sin t}{t}\,dt$, where $(\sin t)/t$ is defined to be 1 at $t = 0$.

22. Use Maclaurin series to find the limits as $x \to 0$ of
    a.† $[\cos(2x) - 1 + 2x^2]/x^4$,   **b.** $\sin(6x)/\sin(7x)$, and
    c. $\ln(1 - 2x)/x$.

23. Use the Maclaurin series for $\cos x$ and the identity $\sin^2 x = \tfrac{1}{2}[1 - \cos(2x)]$ to obtain the Maclaurin series for $\sin^2 x$.

24. Give the first four nonzero terms in the Maclaurin series for
    a.‡ $e^x \sin x$,   **b.** $3\sin x + 4\cos x$,   **c.** $\sec x = 1/\cos x$,
    d. $\sin^2 x + \cos^2 x$,   **e.†** $\ln(1 + x)/(1 + x^2)$, and
    f. $x^{10}/(2 + \sin x)$.

In Exercises 25 through 29 use integrals to give upper bounds on the errors that are made when the infinite series are approximated by the designated partial sums.

25.‡ $\displaystyle\sum_{j=1}^{\infty} \frac{1}{j^3}$ by $\displaystyle\sum_{j=1}^{100} \frac{1}{j^3}$

26.† $\displaystyle\sum_{j=1}^{\infty} je^{-j^2}$ by $\displaystyle\sum_{j=1}^{20} je^{-j^2}$

27. $\displaystyle\sum_{j=-100}^{\infty} \frac{1}{1 + j^2}$ by $\displaystyle\sum_{j=-100}^{10} \frac{1}{1 + j^2}$

28.† $\displaystyle\sum_{j=0}^{\infty} \frac{j}{(1 + j^2)^3}$ by $\displaystyle\sum_{j=0}^{50} \frac{j}{(1 + j^2)^3}$

29. $\displaystyle\sum_{j=10}^{\infty} \frac{1}{j(\ln j)^2}$ by $\displaystyle\sum_{j=10}^{1000} \frac{1}{j(\ln j)^2}$

30.[†] Complete the following sentence: "The statement $\lim\limits_{n \to \infty} a_n = L$ is false if and only if there exists a positive number ϵ such that for any positive number N . . . ."

31.[†] Use the result of Exercise 30 to prove that the statement

$$\lim_{n \to \infty} \left(1 + \frac{1}{n}\right) = 1.0001$$

is false.

32. Use the result of Exercise 30 to prove that the statement $\lim\limits_{n \to \infty} (-1)^n = 1$ is false.

33. Give examples of functions $f(x)$ that are positive and continuous for $x \geq 1$ **a.**[†]   such that $\int_1^\infty f(x)\,dx$ converges but $f(j) = j^{-1}$ so the series $\sum\limits_{j=1}^\infty f(j)$ diverges and   **b.**   such that $\int_1^\infty f(x)\,dx$ diverges but $f(j) = j^{-2}$ so that the series $\sum\limits_{j=1}^\infty f(j)$ converges. (These examples show that the condition that $f(x)$ be decreasing is necessary in Theorem 18.9.)

34. Give examples of sequences $\{b_j\}_{j=0}$ of positive numbers   **a.**[†]   such that $b_j \to 0$ as $j \to \infty$ but $\sum\limits_{j=0}^\infty (-1)^j b_j$ diverges, and   **b.**   such that $b_{j+1} \leq b_j$ for each j and yet $\sum\limits_{j=0}^\infty (-1)^j b_j$ diverges. (These examples show that both conditions on the b_j in Theorem 18.12 are necessary.)

35. (*The root test*)   **a.**[†]   Show that if $\sqrt[j]{|a_j|} < r < 1$ for all $j \geq 1$, then the series $\sum\limits_{j=1}^\infty a_j$ converges absolutely.   **b.**   Show that the series converges absolutely if $\lim\limits_{j \to \infty} \sqrt[j]{|a_j|} = r < 1$.   **c.**   Show that the series diverges if $\lim\limits_{j \to \infty} \sqrt[j]{|a_j|} = r > 1$.

36. Consider a triangle ABC inscribed in a parabola as in Figure 18.13 with the vertex C at the point where the tangent line to the parabola is parallel to AB. Construct the triangles ACD and CBE as in Figure 18.14 with the vertex D at the point where the tangent line is parallel to AC and the vertex E at the point where the tangent line is parallel to BC. Show that the combined area of triangles ACD and CBE is one-fourth the area of triangle ABC. (Introduce coordinates so the parabola has the equation $y = ax^2$. Let A and B have coordinates (x_0, ax_0^2) and (x_1, ax_1^2), respectively. Find the coordinates of C, and use trapezoids to compute the areas of the triangles.)

37. Archimedes used the finite geometric series formula and the result of Exercise 36 (proved through techniques of plane geometry rather than analytic geometry) to show that the region R of Figure 18.15 bounded by the parabola and the line AB has area equal to four-thirds the area of the triangle ABC in Figure 18.13. He used an approximation procedure

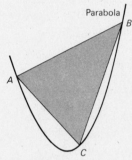

FIGURE 18.13

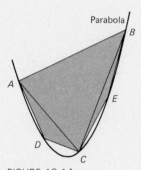

FIGURE 18.14

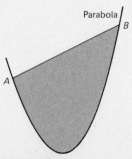

FIGURE 18.15

starting with the triangle ABC as the first approximation of R. The second approximation is obtained by adding the triangles ACD and CBE (Figure 18.14); the third approximation is obtained by adding four triangles constructed on the sides AD, DC, CE, and EB of the previous two triangles as those two triangles were constructed on the sides of triangle ABC; in the fourth approximation eight triangles are added; and so forth. Derive Archimedes' result by finding the limit of the areas of the approximations. [Archimedes did not refer specifically to limits. He completed his demonstrations with arguments by contradiction, a line of reasoning known as the *method of exhaustion* and attributed to Eudoxus (ca. 408–355 B.C.). (See pages 98–105 of Boyer, [1] *A History of Mathematics*.)]

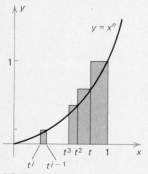

FIGURE 18.16

38. One technique Fermat employed to compute the areas of regions bounded by the curves $y = x^n$ for rational $n \neq -1$ was to approximate the region by infinite collections of rectangles. To determine the area of the region bounded by $y = x^n$ (n positive), by the x-axis, and by the line $x = 1$, for example, he would consider for each t with $0 < t < 1$ the approximation by rectangles with lower right corners at $x = 1, t, t^2, t^3, \dots, t^{j-1}, t^j, \dots$ and with upper right corners on the curve (Figure 18.16).   **a.**   Show that the area of the approximation is $(1 - t)/(1 - t^{n+1})$.   **b.**   Show by dividing the polynomial $1 - t$ into $1 - t^{n+1}$ that the limit of the area of the approximation as $t \to 1^-$ is $1/(n + 1)$ for positive integers n.   **c.**   Show that the limit is $1/(n + 1)$ for positive rational $n = p/q$ by making the substitution $s = t^{1/q}$ and dividing $1 - s$ into the numerator and denominator of the expression in part (a).

39. Use areas to show that the sequence $s_n = \left[1 + \frac{1}{2} + \frac{1}{3} + \cdots + \frac{1}{n} \right] - \ln(n)$

is bounded and decreasing. Its limit is called *Euler's constant* and has the decimal expansion $\gamma = 0.57721 \dots$.

40.† If you have studied derivatives of functions of two variables (Chapter 13), use Taylor's Theorem for functions of one variable applied to $F(t) = f(tx, ty)$ with x and y fixed to derive the special case

$$
\begin{aligned}
f(x, y) = {} & f(0, 0) + f_x(0, 0)x + f_y(0, 0)y \\
& + \tfrac{1}{2}[f_{xx}(0, 0)x^2 + 2f_{xy}(0, 0)xy + f_{yy}(0, 0)y^2] \\
& + \tfrac{1}{6}[f_{xxx}(X, Y)x^3 + 3f_{xxy}(X, Y)x^2y + 3f_{xyy}(X, Y)xy^2 + f_{yyy}(X, Y)y^3]
\end{aligned}
$$

of Taylor's Theorem for functions of two variables. Here $f(x, y)$ is assumed to have continuous third-order derivatives and (X, Y) is a point on the line segment between $(0, 0)$ and (x, y).

If you have access to a computer or electronic calculator, use the partial sums in Exercises 41 through 44 to compute approximate decimal values of the infinite series and compare the approximations with the exact values.

41.† $\displaystyle \sum_{j=1}^{8} \frac{1}{j^2}$ to approximate $\displaystyle \sum_{j=1}^{\infty} \frac{1}{j^2} = \tfrac{1}{6}\pi^2$

42. $\displaystyle\sum_{j=0}^{6} (-1)^j \frac{1}{(2j+1)^3}$ to approximate $\displaystyle\sum_{j=0}^{\infty} (-1)^j \frac{1}{(2j+1)^3} = \frac{1}{32}\pi^3$

43.† $\displaystyle\sum_{j=0}^{10} (-1)^j \frac{1}{2j+1}$ to approximate $\displaystyle\sum_{j=0}^{\infty} (-1)^j \frac{1}{2j+1} = \frac{1}{4}\pi$

44. $\displaystyle\sum_{j=1}^{10} (-1)^{j+1} \frac{1}{j}$ to approximate $\displaystyle\sum_{j=1}^{\infty} (-1)^{j+1} \frac{1}{j} = \ln(2)$

45. **a.** Use the power series

(1) $\ln(1+x) = \displaystyle\sum_{j=1}^{\infty} (-1)^{j+1}\frac{1}{j}x^j \quad (|x| < 1)$

to derive the power series

(2) $\ln\left(\dfrac{1+y}{1-y}\right) = 2 \displaystyle\sum_{j=0}^{\infty} \frac{1}{2j+1} y^{2j+1} \quad (|y| < 1).$

b. If you have access to a computer or electronic calculator use three nonzero terms of (1) and of (2) to give approximate decimal values of $\ln(1.5)$ and compare the approximations with the exact value.

CHAPTER 19

FURTHER TOPICS IN DIFFERENTIAL EQUATIONS

SUMMARY: In this chapter we resume the study of differential equations begun in Chapter 9. We study first-order linear equations in Section 19.1, first-order exact equations in Section 19.2, second-order linear equations with constant coefficients in Section 19.3, applications of second-order equations in Section 19.4, and power series solutions in Section 19.5.* Higher order linear equations with constant coefficients are discussed in the Miscellaneous Exercises.

19-1 FIRST-ORDER LINEAR DIFFERENTIAL EQUATIONS

FIRST-ORDER LINEAR DIFFERENTIAL EQUATIONS are those of the form

(1) $\dfrac{dy}{dx} + P(x)y = Q(x)$ ◀

where $y = y(x)$ is the unknown function and $P(x)$ and $Q(x)$ are functions of x. To solve equation (1) we multiply both sides by an INTEGRATING FACTOR, which transforms the left side into the derivative of a product so that both sides can be integrated. The next rule describes how to find the integrating factor.

Rule 19.1 Multiplying both sides of equation (1) by the integrating factor

(2) $I(x) = e^{\int P(x)\,dx}$

---

*Section 19.2 should be omitted if this chapter is covered before Chapter 13.

puts the equation in the form

(3) $\dfrac{d}{dx}[I(x)y(x)] = I(x)Q(x)$ ◄

which may be solved by integrating both sides with respect to x.

Derivation of Rule 19.1.   The derivative of the integrating factor (2) is

$$\frac{d}{dx}I(x) = \frac{d}{dx}e^{\int P(x)\,dx} = e^{\int P(x)\,dx}\frac{d}{dx}\int P(x)\,dx$$

$$= e^{\int P(x)\,dx}P(x) = I(x)P(x).$$

Therefore, by the product rule the left side of (3) is

$$\frac{d}{dx}[I(x)y(x)] = I(x)\frac{d}{dx}y(x) + y(x)\frac{d}{dx}I(x)$$

$$= I(x)\frac{dy}{dx}(x) + I(x)P(x)y(x)$$

$$= I(x)\left[\frac{dy}{dx}(x) + P(x)y(x)\right]$$

which is the left side of (1) multiplied by $I(x)$, as stated in the rule.   □

EXAMPLE 1   Solve the differential equation

(4) $\dfrac{dy}{dx} - 2xy = 3x.$

Solution.   The coefficient of y in the equation is $P(x) = -2x$ and has the integral $\int P(x)\,dx = -x^2 + C$. We need only one integrating factor, so we take $C = 0$ and obtain

$$I(x) = e^{\int P(x)\,dx} = e^{-x^2}.$$

Multiplying both sides of (4) by this integrating factor gives the equation

(5) $e^{-x^2}\dfrac{dy}{dx} - 2x\,e^{-x^2}y = 3x\,e^{-x^2}$

where the left side is the derivative of the product of the integrating factor e^{-x^2} and the unknown function y. We use this fact to rewrite (5) in the form

$$\frac{d}{dx}[e^{-x^2}y] = 3x\,e^{-x^2}$$

and we integrate with respect to x to obtain

$$e^{-x^2}y = \int 3x\,e^{-x^2}\,dx.$$

We then evaluate the integral by making the substitutions $u = -x^2$, $du = -2x\, dx$ and obtain $e^{-x^2}y = -\frac{3}{2}e^{-x^2} + C$, which gives the solution $y = -\frac{3}{2} + C\, e^{x^2}$. $\square$

In Chapter 9 we solved mixing problems that involved first-order separable differential equations. The following similar example leads to a linear differential equation.

EXAMPLE 2   A large tank contains 10 gallons of a solution with 5 pounds of sugar dissolved in it. A solution containing 4 pounds of sugar per gallon is then added at the rate of 3 gallons per minute. The solutions mix instantly and the mixture is drained off at the rate of 2 gallons per minute. How much sugar is there in the tank t minutes later?

Solution.   We let t denote the time (minutes) with $t = 0$ the initial time, and we let $y(t)$ (pounds) be the amount of sugar in the tank at time t. Liquid is added at the rate of 3 gallons per minute, and it is drained off at the rate of 2 gallons per minute, so that at time t there are $10 + t$ gallons in the tank and the sugar concentration is $\dfrac{y(t) \text{ pounds}}{10 + t \text{ gallons}}$.

The derivative dy/dt is the rate at which sugar is added, minus the rate at which it is removed or

$$\frac{dy}{dt} = [\text{Rate in}] - [\text{Rate out}]$$

$$= \left[4\,\frac{\text{pounds}}{\text{gallon}}\right]\left[3\,\frac{\text{gallons}}{\text{minute}}\right] - \left[\frac{y}{10 + t}\,\frac{\text{pounds}}{\text{gallon}}\right]\left[2\,\frac{\text{gallons}}{\text{minute}}\right]$$

$$= 12 - \frac{2y}{10 + t}\,\frac{\text{gallons}}{\text{minute}}.$$

We rewrite this differential equation in the form

(6) $\dfrac{dy}{dt} + \left(\dfrac{2}{10 + t}\right)y = 12.$

The coefficient $P(t)$ is $2/(10 + t)$ and has the indefinite integral

$$\int \frac{2}{10 + t}\, dt = 2\ln|10 + t| + C = \ln[(10 + t)^2] + C.$$

The integrating factor (with $C = 0$) is $I = e^{\ln[(10+t)^2]} = (10 + t)^2$. When we multiply (6) by I, we obtain

$$(10 + t)^2\frac{dy}{dt} + 2(10 + t)y = 12(10 + t)^2$$

or

$$\frac{d}{dt}[(10 + t)^2 y] = 12(10 + t)^2.$$

Integrating both sides with respect to t yields

$$(10 + t)^2 y = 4(10 + t)^3 + C$$

and setting $t = 0$, $y = 5$ in this equation shows that $C = -3500$. We solve for y to obtain

$$(7) \quad y = 4(10 + t) - 3500(10 + t)^{-2}. \quad \square$$

Exercises

Solve the differential equations and initial value problems in Exercises 1 through 13.

1.† $\dfrac{dy}{dx} + (\sin x)y = -\sin x$

2. $\dfrac{dy}{dx} - (x^{-2})y = 6x^{-2}$

3.‡ $\dfrac{dy}{dx} - \dfrac{x}{x^2 + 1}y = (x^2 + 1)^{3/2}$

4. $\dfrac{dy}{dx} + \dfrac{\sin x}{\cos x}y = \cos^2 x$

5. $\dfrac{dy}{dx} - \dfrac{2}{x + 1}y = (x + 1)^2$

6.† $\dfrac{dy}{dx} - 3x^2 y = 6x^2, \quad y(0) = 0$

7. $\dfrac{dy}{dx} - \dfrac{y}{2x} = x^{3/2}, \quad y(1) = 2$

8.† $\dfrac{dy}{dx} + \dfrac{10}{2x + 1}y = 1, \quad y(0) = 1$

9. $\dfrac{dy}{dx} - \dfrac{(x + 1)y}{x^2 + 2x + 4} = \sqrt{x^2 + 2x + 4}, \quad y(0) = 2$

10.‡ $\dfrac{dy}{dx} + 2y = \cos x, \quad y(\pi) = 0$

11. $\dfrac{dy}{dx} - y = x, \quad y(0) = 10$

12.† $\dfrac{dy}{dx} + \dfrac{2y \cos x}{\sin x} = \sin x, \quad y\!\left(\dfrac{\pi}{2}\right) = 4$

13. $\dfrac{dy}{dx} + 2y = \sin(3x), \quad y(0) = 1$ (Use the reference table of integrals.)

14.† **a.** What is the concentration of the sugar solution of Example 2 at time t? **b.** What is the limit of the concentration as $t \to \infty$?

15.† At time $t = 0$ (minutes) a tank contains 8 gallons of a solution in which 5 pounds of salt are dissolved. A solution with $\frac{1}{2}$ pound of salt per gallon is added at the rate of 4 gallons per minute. The solutions mix instantly and the mixture is drained off at the rate of 1 gallon per minute.   **a.** How much salt is there in the tank and   **b.** what is the concentration of the solution in the tank at time t?

16. Answer the questions from Exercise 15 in the case where the tank contains at $t = 0$, 1 gallon of a solution with 2 pounds of salt dissolved in it, where a solution containing 3 pounds of salt per gallon is added at the rate of 10 gallons per minute, and where the mixture is drained off at the rate of 3 gallons per minute.

17.[‡] At $t = 0$ (hours) a tank holds 9 gallons of a solution containing $\frac{2}{3}$ pounds of sugar per gallon. A solution with 2 pounds of sugar per gallon is added at the rate of one gallon per hour, and the mixture is drained off at the rate of 3 gallons per hour.   **a.** At what time t_0 is the tank empty? **b.** How much sugar is in the tank for $0 \leq t \leq t_0$?   **c.** What is the limit of the concentration of the solution in the tank as $t \to t_0{}^-$?

18. Answer the questions from Exercise 17 in the case where the tank holds 1 gallon of solution containing 3 pounds of sugar at time $t = 0$, where a solution with 1 pound of sugar per gallon is added at the rate of 2 gallons per hour, and where the mixture is drained off at the rate of 5 gallons per hour.

19.[†] Find the function $y(x)$ whose graph passes through the origin and has tangent line of slope $x - y$ at (x, y).

20. Find the family of orthogonal trajectories of the curves $x^2 - y^2 = cy$. (*Hint.* Set $u = x^2$.)

21. Find the orthogonal trajectories of the curves $y - x + Ce^{-y} = 1$.

19-2   EXACT DIFFERENTIAL EQUATIONS

A first-order differential equation

$$(1) \qquad p(x, y)\, dx + q(x, y)\, dy = 0$$

is called EXACT in a rectangle with sides parallel to the coordinates axes if the first derivatives of p and q are continuous and satisfy the condition $p_y = q_x$ in the rectangle. In this section we show how to solve exact differential equations.

Recall that the differentials dx and dy at a point on a curve are corresponding changes in x and y along the tangent line at that point and that we consider a curve to be a solution of the differential equation (1) if its differentials satisfy the equation at each point (x, y) on the curve (Figure 19.1). The next rule will enable us to find the curves that are the solutions of exact differential equations.

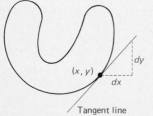

FIGURE 19.1   Differentials dx and dy at a point (x, y) on a curve

Rule 19.2   The solution of an exact differential equation (1) in a rectangle are the level curves $U(x, y) = c$ of the functions $U(x, y)$ such that

$$(2) \qquad U_x = p, \quad U_y = q. \qquad \blacktriangleleft$$

All functions $U(x, y)$ satisfying (2) have the same level curves because any two of them differ by a constant. Theorem 16.2 shows that there exist such $U(x, y)$ for every exact differential equation in a rectangle. The procedure for finding them is described in Example 1 at the end of this section.

The functions $y = y(x)$ of x that are solutions of the differential equation are defined implicitly by the equations $U(x, y) = c$ for the level curves.

Derivation of Rule 19.2.   We will derive the rule for the case where the level curve $U(x, y) = c$ is the graph of a differentiable function $y = y(x)$. Then we have $U(x, y(x)) = c$ for all x in the domain of $y(x)$ and when we use the chain rule to differentiate this equation with respect to x, we obtain

(3) $U_x(x, y(x)) + U_y(x, y(x)) y'(x) = 0.$

If dx is a nonzero change in x along the tangent line to the curve at $(x, y) = (x, y(x))$ and if dy is the corresponding change in y, then the ratio dy/dx is the derivative $y'(x)$ and multiplying (3) by dx gives

$$U_x(x, y) \, dx + U_y(x, y) \, dy = 0.$$

Since $U_x = p$ and $U_y = q$, this gives equation (1).   □

EXAMPLE 1   Show that the differential equation

(4) $(2xy - 1) \, dx + (x^2 + 1) \, dy = 0$

is exact. Then find its solution that passes through the point $(1, 2)$.

Solution.   The differential equation (4) is exact because we have $p = 2xy - 1$ and $q = x^2 + 1$ for which $p_y = 2x$ and $q_x = 2x$ are equal. By equations (2) the solutions are the level curves of a function $U(x, y)$ such that

(5) $U_x = 2xy - 1, \quad U_y = x^2 + 1.$

Integrating the first of equations (5) with respect to x with y fixed gives

(6) $U = \int (2xy - 1) \, dx = x^2y - x + \phi(y)$

where $\phi(y)$ (the constant of integration for fixed y) is a function of y that we have to determine.

We differentiate equation (6) with respect to y to obtain

(7) $U_y = \dfrac{\partial}{\partial y}[x^2y - x + \phi(y)] = x^2y + \phi'(y).$

Then comparing equation (7) with the second of equations (5) shows that we must have $\phi'(y) = 1$. Hence, $\phi(y) = y + C$, and we substitute this formula in equation (6) to see that the functions U are $U(x, y) = x^2y - x + y + C$. The level curves

(8) $x^2y - x + y = c$

of these functions are the solutions of the differential equation.

To find the curve that passes through $(1, 2)$, we set $x = 1$ and $y = 2$ in (8). We conclude that the constant c is 3 and the solution is the curve $x^2y - x + y = 3$.   □

Note. The equation $x^2y - x + y = c$ for the curves that are the solutions of the differential equation of Example 1 can be solved for y to yield the formula $y = (x + c)/(x^2 + 1)$ for the functions of x that satisfy the differential equation. □

Exercises

†*Answer provided.*
‡*Outline of solution provided.*

In each of Exercises 1 through 9 determine whether the differential equation is exact and, if so, give an equation for the curves that satisfy it.

1.‡ $(2x - 6y) \, dx + (4y^3 - 6x) \, dy = 0$

2. $(2xe^y - e^{-x}) \, dx + (x^2e^y - 2y) \, dy = 0$

3.† $(5x^3y^4 - 2y) \, dx + (3x^2y^5 + x) \, dy = 0$

4. $(\cos x)(\sin y) \, dx + (\sin x)(\cos y) \, dy = 0$

5.† $(y \cos(xy) + y \sin x) \, dx + (x \cos(xy) - \cos x) \, dy = 0$

6. $(x^2 - 3y^2) \, dx + (2xy + 4x^3) \, dy = 0$

7. $\left(5x^4 + \dfrac{1}{x - y}\right) dx - \left(6y^5 + \dfrac{1}{x - y}\right) dy = 0$

8.† $(3x^2y^2 + 2x) \, dx + (2x^3y - 3y^2) \, dy = 0$

9. $(18x - 2xy^2) \, dx + (8y - 2x^2y) \, dy = 0$

In each of Exercises 10 through 15 determine whether the differential equation is exact and, if so, give an explicit formula for the function $y(x)$ that satisfies the initial value problem.

10.‡ $(2x^3y^2 - (\sin x)(\cos x)) \, dx + (x^4y + 2y) \, dy = 0, \quad y(0) = 3$

11.† $(2xy - 2x) \, dx + (x^2 + 1) \, dy = 0, \quad y(0) = 2$

12. $(y^{1/3} - \cos x) \, dx + \tfrac{1}{3}xy^{-2/3} \, dy = 0, \quad y(\pi) = 0$

13. $(e^x - y \sin x) \, dx + (3 + \cos x) \, dy = 0, \quad y(0) = 1$

14.† $(x^2y - xy^2) dx + (xy^3 - x^3y) \, dy = 0, \quad y(1) = 2$

15. $y \sin(xy) \, dx - x \cos(xy) \, dy = 0, \quad y(\pi) = 4$

19-3 SECOND-ORDER LINEAR EQUATIONS WITH CONSTANT COEFFICIENTS

In this section we study SECOND-ORDER LINEAR DIFFERENTIAL EQUATIONS WITH CONSTANT COEFFICIENTS, which are differential equations of the form

(1) $ay''(x) + by'(x) + cy(x) = f(x)$

with a, b, and c constants ($a \neq 0$).

The homogeneous case

If $f(x)$ is the zero function, the differential equation (1) is said to be HOMOGENEOUS and has the form

(2) $ay''(x) + by'(x) + cy(x) = 0.$

Our study of this type of differential equation is based on the following theorem.

Theorem 19.1   Suppose that $y_1(x)$ and $y_2(x)$ are solutions of (2) and that neither is a constant multiple of the other. Then (a) every linear combination

(3) $y(x) = Ay_1(x) + By_2(x)$ ◄

of $y_1(x)$ and $y_2(x)$ is also a solution; (b) every solution of (2) may be expressed in the form (3); and (c) a unique solution is determined by specifying as initial conditions the values $y(a)$ and $y'(a)$ of the function and its first derivative at one point.

Expression (3) is called the GENERAL SOLUTION of the differential equation.

Proof of Theorem 19.1(a).   We consider arbitrary constants A and B and define $y(x)$ by (3). Then we have

$$ay''(x) + by'(x) + cy(x)$$
$$= a[Ay_1''(x) + By_2''(x)] + b[Ay_1'(x) + By_2'(x)]$$
$$+ c[y_1(x) + By_2(x)]$$
$$= A[ay_1''(x) + by_1'(x) + cy_1(x)]$$
$$+ B[ay_2''(x) + by_2'(x) + cy_2(x)]$$
$$= A[0] + B[0] = 0$$

which shows that $y(x)$ is a solution of (2).   Q.E.D.

The proofs of parts (b) and (c) of Theorem 19.1 may be found in textbooks on differential equations.

EXAMPLE 1   Show that $y_1(x) = e^{2x}$ and $y_2(x) = e^{-x}$ are solutions of

(4) $y''(x) - y'(x) - 2y(x) = 0$

and give the general solution of this differential equation.

Solution.   We have $y_1' = \dfrac{d}{dx} e^{2x} = 2e^{2x}$ and $y_1'' = \dfrac{d}{dx}(2e^{2x}) = 4e^{2x}$,

so that $y_1'' - y_1' - 2y_1 = 4e^{2x} - 2e^{2x} - 2e^{2x} = 0$ and y_1 is a solution.

Similarly, we have $y_2' = \dfrac{d}{dx} e^{-x} = -e^{-x}$ and $y_2'' = \dfrac{d}{dx}(-e^{-x}) = e^{-x}$,

so that $y_2'' - y_2' - 2y_2 = e^{-x} - (-e^{-x}) - 2e^{-x} = 0$ and y_2 is also a solution. Neither e^{2x} nor e^{-x} is a constant times the other, so by Theorem 19.1 the general solution of the differential equation is $y(x) = Ay_1(x) + By_2(x) = Ae^{2x} + Be^{-x}$.   □

EXAMPLE 2   Find the solution of differential equation (4) that satisfies the initial conditions $y(0) = -2$ and $y'(0) = 5$.

Solution.   The general solution $y(x) = Ae^{2x} + Be^{-x}$ has the value $y(0) = A + B$ at $x = 0$ and its derivative $y'(x) = 2Ae^{2x} - Be^{-x}$ has the value $y'(0) = 2A - B$. The initial conditions lead to the simultaneous equations $A + B = -2$, $2A - B = 5$. Adding the equations gives $3A = 3$ to show that $A = 1$. Then by using either equation we determine that $B = -3$. The solution is $y(x) = e^{2x} - 3e^{-x}$. $\square$

Finding the general solution   We find the general solution of the differential equation

(2) $ay''(x) + by'(x) + cy(x) = 0$

by studying the corresponding CHARACTERISTIC EQUATION

(5) $ar^2 + br + c = 0$ ◄

which is obtained from the differential equation by replacing $y''(x)$ with r^2, $y'(x)$ with r, and $y(x)$ with the constant 1. The solutions r of the characteristic equation are given by the quadratic formula

(6) $r = \dfrac{-b \pm \sqrt{b^2 - 4ac}}{2a}$.   ◄

Formula (6) gives two distinct real solutions r_1 and r_2 of the characteristic equation if the discriminant $b^2 - 4ac$ is positive, gives one real solution r if $b^2 - 4ac$ is zero, and gives two distinct complex solutions $\tau \pm i\omega$ if $b^2 - 4ac$ is negative.* The form of the general solution of the differential equation is different in these three cases, as is stated in the next theorem.

Theorem 19.2   (a) If $b^2 - 4ac$ is positive, then the general solution of the differential equation (2) is

(7) $y(x) = Ae^{r_1 x} + Be^{r_2 x}$ ◄

where r_1 and r_2 are the solutions of the characteristic equation.
    (b) If $b^2 - 4ac$ is zero, the general solution of the differential equation is

(8) $y(x) = (A + Bx)e^{rx}$ ◄

where r is the solution of the characteristic equation.
    (c) If $b^2 - 4ac$ is negative, the general solution of the differential equation is

(9) $y(x) = [A \cos(\omega x) + B \sin(\omega x)]e^{\tau x}$ ◄

where $\tau \pm i\omega$ are the solutions of the characteristic equation.

---

*A complex number is an expression of the form $\tau + i\omega$ where τ and ω are real numbers and i is the symbol that is introduced to serve as a square root of -1.

Proof. For the function $y = e^{rx}$ we have $y' = re^{rx}$ and $y'' = r^2e^{rx}$, so that $ay'' + by' + cy = ar^2e^{rx} + bre^{rx} + ce^{rx} = (ar^2 + br + c)e^{rx}$. Consequently, e^{rx} is a solution of the differential equation if r is a solution of the characteristic equation. This shows that e^{r_1x} and e^{r_2x} are solutions of the differential equation in case (a) and that e^{rx} is a solution in case (b). Short calculations show that xe^{rx} is also a solution in case (b) and that $\cos(\omega x)e^{\tau x}$ and $\sin(\omega x)e^{\tau x}$ are solutions in case (c) (see Exercises 24 and 25.) Theorem 19.1 then implies that the general solution is in the corresponding form (7), (8), or (9).   Q.E.D.

EXAMPLE 3   Give the general solution of the differential equation

$$y''(x) + y'(x) - 6y(x) = 0.$$

Solution.   The characteristic equation is $r^2 + r - 6 = 0$, and has the solutions

$$r = \frac{-1 \pm \sqrt{1^2 - 4(1)(-6)}}{2} = \frac{-1 \pm 5}{2} = -3 \text{ and } 2.$$

By Theorem 19.2(a) the general solution is $y(x) = Ae^{-3x} + Be^{2x}$.   □

EXAMPLE 4   Solve the initial value problem

$$\frac{d^2y}{dx^2} + 6\frac{dy}{dx} + 9y = 0, \quad y(0) = 3, \quad \frac{dy}{dx}(0) = 15.$$

Solution.   The characteristic equation is $r^2 + 6r + 9 = 0$, which may be written $(r + 3)^2 = 0$, and has the one real solution $r = -3$. By Theorem 19.2(b) the general solution of the differential equation is $y = (A + Bx)e^{-3x}$ and has the derivative

$$\frac{dy}{dx} = (A + Bx)\frac{d}{dx}e^{-3x} + \left[\frac{d}{dx}(A + Bx)\right]e^{-3x}$$
$$= (A + Bx)(-3)e^{-3x} + Be^{-3x} = (B - 3A - 3Bx)e^{-3x}.$$

Therefore, $y(0) = A$ and $\dfrac{dy}{dx}(0) = B - 3A$. To satisfy the initial conditions we must have $A = 3, B - 3A = 15$ and hence $A = 3$, $B = 24$. The solution is $y = (3 + 24x)e^{-3x}$.   □

The inhomogeneous case

The differential equation

(1) $ay''(x) + by'(x) + cy(x) = f(x)$

is called INHOMOGENEOUS if $f(x)$ is not the zero function. We base our study of this type of equation on the following result.

Theorem 19.3 **If $y_p(x)$ is one solution of the differential equation (1), then the general solution is**

(10) $y(x) = y_p(x) + Ay_1(x) + By_2(x)$ ◀

where $Ay_1(x) + By_2(x)$ is the general solution of the homogeneous equation $ay'' + by' + cy = 0$.

We refer to $y_p(x)$ as a PARTICULAR SOLUTION of (1).

Proof of Theorem 19.3. Let $y_h(x) = Ay_1(x) + By_2(x)$ be the general solution of the homogeneous differential equation. The sum (10) of $y_p(x)$ and $y_h(x)$ is a solution of the inhomogeneous equation because we have

$$ay'' + by' + cy = a(y_p'' + y_h'') + b(y_p' + y_h') + c(y_p + y_h)$$
$$= (ay_p'' + by_p' + cy_p) + (ay_h'' + by_h' + cy_h)$$
$$= f(x) + 0 = f(x).$$

Conversely, every solution $y(x)$ of (1) is of the form (10) for certain constants A and B because

$$a(y'' - y_p'') + b(y' - y_p') + c(y - y_p)$$
$$= (ay'' + by' + cy) - (ay_p'' + by_p' + cy_p)$$
$$= f(x) - f(x) = 0$$

so $y - y_p$ is a solution of the homogeneous equation.   Q.E.D.

EXAMPLE 5   The function $y_p(x) = x + \frac{1}{6}$ is a particular solution of the differential equation

(11) $y''(x) + y'(x) - 6y(x) = -6x.$

What is the general solution?

Solution.   We saw in Example 3 that the general solution of the homogeneous equation is $y_h(x) = Ae^{-3x} + Be^{2x}$. The general solution of (11) is $y(x) = y_p(x) + y_h(x) = x + \frac{1}{6} + Ae^{-3x} + Be^{2x}$.   □

The method of undetermined coefficients

We now describe the method of UNDETERMINED COEFFICIENTS for finding particular solutions of the homogeneous equation

(1) $ay''(x) + by'(x) + cy(x) = f(x)$

for functions $f(x)$ that are products or sums of products of functions of the forms

(12) x^n, $\sin(\alpha x + \beta)$, $\cos(\alpha x + \beta)$, $e^{\alpha x}$

with positive integers n and constants α and β.

The particular solution is found as a linear combination

(13) $y_p(x) = c_1 z_1(x) + c_2 z_2(x) + \cdots + c_k z_k(x)$

of functions $z_1(x), z_2(x), \ldots, z_k(x)$ in what we call a *UC* SET for f. We construct the *UC* set as follows.

We choose functions $w_1(x)$, $w_2(x)$, . . . , $w_k(x)$ such that $f(x)$ and each of its derivatives can be expressed as linear combinations of them. If none of the functions $w_j(x)$ are solutions of the homogeneous equation $ay'' + by' + cy = 0$, we take the $z_j(x)$ to be the $w_j(x)$. If any of the $w_j(x)$ are solutions of the homogeneous equation, we multiply *all* of the $w_j(x)$ by the lowest positive integer power x^n of x such that none of the resulting functions $x^n w_j(x)$ are solutions, and we take $z_j(x) = x^n w_j(x)$ ($j = 1, 2, . . . , k$).

Rule 19.3   A particular solution of equation (1) with $f(x)$ a sum of products of functions of the forms (12) may be found as a linear combination (13) of the functions in a *UC* set for $f(x)$ by substituting (13) in the differential equation and solving for the coefficients $c_1, c_2, . . . , c_k$.

The derivation of this rule can be found in textbooks on differential equations.

EXAMPLE 6   Find a *UC* set for $2xe^x$ in the differential equation

$$(14) \quad y'' + y = 2xe^x.$$

Solution.   The characteristic equation of the homogeneous equation $y'' + y = 0$ is $r^2 + 1 = 0$ and its solutions are $r = \pm i$. By Theorem 19.2(c) the general solution of the homogeneous equation is $y_h(x) = A \cos x + B \sin x$. The function $2xe^x$ and all its derivatives can be expressed as linear combinations of $w_1(x) = xe^x$ and $w_2(x) = e^x$. Neither of these is a solution of the homogeneous equation, so $z_1(x) = w_1(x) = xe^x$ and $z_2(x) = w_2(x) = e^x$ form a *UC* set.   □

EXAMPLE 7   Find a *UC* set for $\sin x$ in the differential equation

$$(15) \quad y'' + y = \sin x.$$

Solution.   We saw in Example 6 that the general solution of the homogeneous equation is $y_h(x) = A \cos x + B \sin x$. The function $\sin x$ and all of its derivatives can be expressed as linear combinations of $w_1(x) = \cos x$ and $w_2(x) = \sin x$. These are solutions of the homogeneous equation, but $x \cos x$ and $x \sin x$ are not, so the functions $z_1(x) = x w_1(x) = x \cos x$ and $z_2(x) = x w_2(x) = x \sin x$ form a *UC* set.   □

EXAMPLE 8   Find a particular solution of differential equation (14).

Solution.   We saw in Example 6 that xe^x and e^x form a *UC* set for the differential equation, so there is a particular solution in the form

$$(16) \quad y_p(x) = c_1 xe^x + c_2 e^x.$$

For this function we have

$$y_p' = \frac{d}{dx}(c_1 x e^x + c_2 e^x) = c_1 x e^x + (c_1 + c_2)e^x$$

$$y_p'' = \frac{d}{dx}[c_1 x e^x + (c_1 + c_2)e^x] = c_1 x e^x + (2c_1 + c_2)e^x$$

and consequently

$$y_p'' + y_p = 2c_1 x e^x + (2c_1 + 2c_2)e^x.$$

To have y_p satisfy the differential equation we need to have $2c_1 x e^x + (2c_1 + 2c_2)e^x = 2x e^x$. The coefficients of $x e^x$ and of e^x on the two sides of this equation must be equal, so we obtain the simultaneous equations $2c_1 = 2$, $2c_1 + 2c_2 = 0$. Hence $c_1 = 1$, $c_2 = -1$, and formula (16) gives the solution $y_p = x e^x - e^x$. □

EXAMPLE 9   Solve the initial value problem

(17) $y'' + y = 2x e^x$, $y(0) = 1$, $y'(0) = -5$.

Solution.   We found the particular solution $y_p = x e^x - e^x$ of the inhomogeneous equation in Example 8, and the general solution of the homogeneous equation is $y_h = A \cos x + B \sin x$. Therefore, the general solution of the inhomogeneous equation is $y = y_p + y_h$ or

(18) $y = x e^x - e^x + A \cos x + B \sin x$

for which $y(0) = -1 + A$. The derivative of (18) is

$$y' = x e^x - A \sin x + B \cos x$$

and has the value $y'(0) = B$. To satisfy the initial conditions we must have $-1 + A = 1$ and $B = -5$. We solve for A and B to find that $A = 2$, $B = -5$ and the solution (18) is $y = x e^x - e^x + 2 \cos x - 5 \sin x$. □

We describe in the exercises the method of VARIATION OF PARAMETERS, which may be used to solve the inhomogeneous equation (1) for functions $f(x)$ to which the method of undetermined coefficients cannot be applied.

Exercises

†*Answer provided.*
‡*Outline of solution provided.*

Solve the initial value problems in Exercises 1 through 16.

1.‡ $y'' - 3y' + 2y = 0$, $y(0) = 1$, $y'(0) = 0$

2.† $y'' - y' - 6y = 0$, $y(0) = 1$, $y'(0) = 8$

3. $y'' + 2y' - 3y = 0$, $y(0) = 4$, $y'(0) = 8$

4.† $y'' - 4y' + 4y = 0$, $y(1) = 3$, $y'(1) = -5$

5. $y'' - 2y' + y = 0$, $y(2) = 3$, $y'(2) = 4$

6.‡ $y'' - 2y' + 2y = 0$, $y(0) = 0$, $y'(0) = 1$

7.† $y'' + 4y' + 5y = 0$, $y(0) = -1$, $y'(0) = 0$

8. $y'' - 2y' + 5y = 0$, $y(0) = 1$, $y'(0) = -1$

9.‡ $y'' - y = x$, $y(0) = 0$, $y'(0) = 1$

10.† $y'' + 4y = 5e^x$, $y(0) = 0$, $y'(0) = 3$

11. $y'' - y = 2 \sin x$, $y(0) = 5$, $y'(0) = 1$

12.‡ $y'' + y = 2 \cos x$, $y(0) = 0$, $y'(0) = 1$

13.† $y'' - y = 2e^x$, $y(0) = -1$, $y'(0) = 0$

14. $y'' + 2y' = 6$, $y(0) = 4$, $y'(0) = 3$

15.† $y'' + 4y' + 4y = 2e^{-2x}$, $y(0) = 0$, $y'(0) = 1$

16. $y'' - 2y' + y = 2e^x$, $y(0) = 0$, $y'(0) = 0$

17.† (*Variation of parameters*) Suppose that $y_1(x)$ and $y_2(x)$ are solutions of the homogeneous equation $ay'' + by' + cy = 0$ and that neither is a constant multiple of the other. Show that

(19) $y(x) = c_1(x)y_1(x) + c_2(x)y_2(x)$

is a solution of the inhomogeneous equation $ay'' + by' + cy = f$ if the derivatives of the functions $c_1(x)$ and $c_2(x)$ are the solutions of the simultaneous equations

(20) $\begin{aligned} c_1'(x)y_1(x) + c_2'(x)y_2(x) &= 0 \\ c_1'(x)y_1'(x) + c_2'(x)y_2'(x) &= \frac{1}{a}f(x). \end{aligned}$

Use variation of parameters (Exercise 17) to find the general solutions of the differential equations in Exercises 18 through 21.

18.‡ $y'' + y = \sec^3 x$

19.† $y'' - 2y' + y = \dfrac{e^x}{x}$

20. $y'' - 4y' + 4y = x^{-2}e^{2x}$

21. $y'' - 2y' + y = 4e^x \ln x$ $(x > 0)$

22. Use variation of parameters to find the solution of the initial value problem $y'' + y = \csc x$ $(0 < x < \pi)$, $y(\pi/2) = 0$, $y'(\pi/2) = 1$.

23. Use initial value problems for the differential equation $y'' + y \doteq 0$ to derive the trigonometric identities   **a.†** $\sin(\alpha + \beta)$ $= \sin \alpha \cos \beta + \cos \alpha \sin \beta$ and   **b.** $\cos(\alpha + \beta) = \cos \alpha \cos \beta - \sin \alpha \sin \beta$.

24. Show that $y = xe^{rx}$ is a solution of $ay'' + by' + cy = 0$ if r is the solution of the characteristic equation $ar^2 + br + c = 0$ with $b^2 - 4ac = 0$.

25. Show that $y = e^{\tau x} \cos(\omega x)$ and $y = e^{\tau x} \sin(\omega x)$ are solutions of $ay'' + by' + cy = 0$ if $r = \tau \pm i\omega$ are the solutions of the characteristic equation $ar^2 + br + c = 0$ with $b^2 - 4ac < 0$.

19-4   APPLICATIONS OF SECOND-ORDER EQUATIONS

In this section we discuss the use of second-order linear differential equations with constant coefficients

(1) $ay''(x) + by'(x) + cy(x) = f(x)$

in the study of vibrating springs and electrical circuits.

Vibrating springs   We consider a horizontal spring that is fixed at its left end and fastened at its right end to an object of mass M (slugs) that moves horizontally along a y-axis with its scale measured in feet (Figure 19.2). (We ignore the weight of the spring and the force of gravity.) The positive side of the y-axis is on the right and its origin is at the *rest position* of the right end of the spring (its location when the spring exerts no force to the right or to the left).

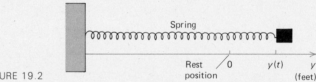

FIGURE 19.2

, We let $y(t)$ denote the y-coordinate of the right end of the spring at time t (seconds). Then $y'(t)$ is its velocity toward the right and $y''(t)$ is acceleration toward the right. We assume that the spring satisfies Hooke's law so that the force it exerts is proportional to the displacement of its end from the rest position. We assume that the object moves in a void which exerts no force on it or that it moves in a resisting medium, such as air or water, which exerts a force proportional to the object's velocity. Then the force of the spring is $-ky$ pounds toward the right when the end of the spring is at y, and the force of resistance is $-Ry'$ pounds toward the right when the object's velocity is y' feet per second. Here k is a positive constant and R is zero if there is no resisting medium and is a positive constant otherwise. We let $f(t)$ (pounds) denote the sum of any other (external) forces (toward the right) on the object. According to Newton's law the sum of the forces equals the object's mass multiplied by its acceleration y'', and we obtain the equation $-ky - Ry' + f = My''$ or

(2) $My''(t) + Ry'(t) + ky(t) = f(t)$

which is of the form (1) with a replaced by M, b by R, and c by k.

Free undamped motion   If there is no resisting medium and no external force, then R and $f(t)$ are zero. Equation (2) reads

(3) $My''(t) + ky(t) = 0$

and has the general solution

(4) $y(t) = A \cos(\omega t) + B \sin(\omega t)$

with $\omega = \sqrt{k/M}$. We can write the general solution in an alternate form by setting $C = \sqrt{A^2 + B^2}$. Then the point $(B/C, A/C)$ is on the unit circle and there is an angle ψ such that

(5) $\cos(\psi) = \dfrac{B}{C}$ and $\sin(\psi) = \dfrac{A}{C}.$

With these substitutions and the trigonometric identity

(6) $\sin(\alpha) \cos(\beta) + \cos(\alpha) \sin(\beta) = \sin(\alpha + \beta)$

formula (4) for the general solution takes the form

(7) $y(t) = C \sin(\omega t + \psi).$

Because the maximum value of (7) is C and the minimum is $-C$, the constant C is called the AMPLITUDE of the solution. The constant ψ is called the PHASE SHIFT. Formula (7) shows that the graph of the solution is a modified sine curve and that the solution has period $T = 2\pi/\omega$. The reciprocal $\omega/2\pi$ of the period is called the FREQUENCY of the solution and is the number of times the object oscillates each second. Figure 19.3 shows the graph of the solution for an object of mass $M = 1$ and a spring with $k = 5$ under the initial conditions $y(0) = 1, y'(0) = -4$.

FIGURE 19.3   Free undamped motion ($M = 1$, $R = 0$, $k = 5$, $y(0) = 1$, $y'(0) = -4$)

Free damped motion

If there is no external force but there is a resisting medium, then the differential equation (2) is

(8) $My'' + Ry' + ky = 0.$

The characteristic equation $Mr^2 + Rr + k = 0$ has the solutions

(9) $r = \dfrac{-R \pm \sqrt{R^2 - 4Mk}}{2M}.$

If the resistance is relatively small in comparison with the mass M of the object and the spring constant k, so that $R^2 - 4Mk$ is negative, then the numbers (9) are complex numbers of the form $r = \tau \pm i\omega$ with $\tau = -R/2M$ a negative number and $\omega = \sqrt{(4Mk - R^2)}/2M$ positive. The general solution of the differential equation can be written in either of the forms

(10) $y(t) = [A \cos(\omega t) + B \sin(\omega t)]e^{\tau t}$

(11) $y(t) = C \sin(\omega t + \psi) e^{\tau t}.$

The second form of the solution shows that it oscillates with frequency $\omega/2\pi$ and that, because τ is negative, the motion dies down with time. Figure 19.4 shows the graph of the solution with the same mass, spring constant, and initial conditions as in Figure 19.3 but with $R = \frac{1}{5}$, for which $R^2 - 4Mk = -19\frac{24}{25}$. The motion is called *underdamped* because the resistance is not great enough to prevent the oscillation. In the case of Figure 19.5 the resistance is stronger ($R = 2$) but the motion is still underdamped ($R^2 - 4Mk = -16$). (Note the different scales in Figures 19.4 and 19.5.)

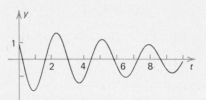

FIGURE 19.4   Free underdamped motion ($M = 1$, $R = \frac{1}{5}$, $k = 5$, $y(0) = 1$, $y'(0) = -4$)

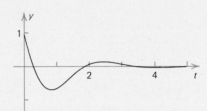

FIGURE 19.5   Free underdamped motion ($M = 1$, $R = 2$, $k = 5$, $y(0) = 1$, $y'(0) = -4$)

If the resistance is large enough, so that $R^2 - 4Mk$ is positive, then the solutions (9) of the characteristic equation are distinct negative numbers r_1 and r_2. The general solution of the differential equation is

(12) $y(t) = Ae^{r_1 t} + Be^{r_2 t}$.

Because the constants r_1 and r_2 are negative, the solution tends to 0 as t tends to ∞. The solution is zero for at most one value of t (see Exercise 26), so the spring does not oscillate and is said to be *overdamped*. This is the case in Figure 19.6 where the mass, spring constant, and initial conditions are the same as in Figures 19.3, 19.4, and 19.5 but $R = 5$ and $R^2 - 4Mk = 5$. In this case the object passes through the rest position once for $t > 0$. In the case of Figure 19.7 the resistance is so large ($R = 6$) that the object does not pass through the rest position after moving it at $t = 0$.

If $R^2 - 4Mk$ is zero, then the characteristic equation has one negative solution $r = -R/2M$ and the general solution of the differential equation is

(13) $y = (A + Bt)e^{rt}$.

The motion is similar to that in the overdamped case. The motion dies down with time and the object passes through the rest position at most once. The spring is said to be *critically damped* in this case because any further decrease in the resistance (as in a car shock absorber that is wearing out) would lead to oscillations.

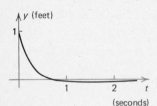

FIGURE 19.6   Free overdamped motion ($M = 1$, $R = 5$, $k = 5$, $y(0) = 1$, $y'(0) = -4$)

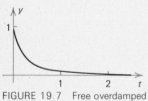

FIGURE 19.7   Free overdamped motion ($M = 1$, $R = 6$, $k = 5$, $y(0) = 1$, $y'(0) = -4$)

Electrical circuits

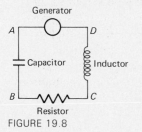

Generator

A — D

Capacitor   Inductor

B — *C*

Resistor

FIGURE 19.8

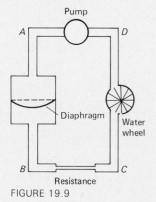

Pump

A — *D*

Diaphragm

Water
wheel

B — *C*

Resistance

FIGURE 19.9

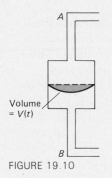

A

Volume
= *V(t)*

B

FIGURE 19.10

We will now see how second-order differential equations arise in the study of electrical circuits such as shown in the schematic sketch in Figure 19.8. The circuit contains a generator, a resistor, a capacitor, and an inductor. We think of the capacitor as parallel plates across which no current can flow and think of the inductor as a coil of wire. We suppose that the wires connecting the elements of the circuit have no resistance so that all the resistance in the circuit is contained in the resistor.

Current flow in the circuit of Figure 19.8 is analogous to water flow in the system shown in Figure 19.9, where a pump, a diaphragm, a water wheel, and a narrow pipe that resists the flow of water are connected by pipes, which we assume have no resistance. The pump produces pressure differences between points *A* and *D* in Figure 19.9; the generator produces voltage differences between points *A* and *D* in Figure 19.8. When the pressure is not the same at points *A* and *B* in Figure 19.9, the diaphragm stretches in one direction or the other; when the voltage is not the same at points *A* and *B* in Figure 19.8, charge accumulates on one side or the other of the capacitor. The narrow pipe in Figure 19.9 resists the flow of water; the resistor of Figure 19.8 resists the flow of current. The water wheel in Figure 19.9 resists changes in the velocity of the water flow; the inductor in Figure 19.8 resists changes in the current.

To derive a differential equation for the water circuit, we let $V(t)$ denote the volume by which the diaphragm is stretched below its horizontal equilibrium position at time t (Figure 19.10).

We let P_A, P_B, P_C, and P_D denote the water pressure at the points A, B, C, and D in Figure 19.9. We assume that the volume by which the diaphragm is stretched is proportional to the pressure difference between A and B:

$$(14) \quad P_A - P_B = \frac{1}{C} V(t).$$

(We use $1/C$ as the constant of proportionality so that we will have standard notation for the case of an electrical circuit.)

Water flowing through the narrow pipe causes a pressure difference between points B and C which we assume is proportional to the rate of flow $V'(t)$ of the water:

$$(15) \quad P_B - P_C = R \, V'(t).$$

We assume that the water wheel is frictionless and does not resist a constant flow of water. Then there are differences in pressure between points C and D when the wheel is accelerating or decelerating, and we assume that the difference in pressure across the water wheel is proportional to the acceleration $V''(t)$ of the water:

$$(16) \quad P_C - P_D = L \, V''(t).$$

We let $f(t)$ denote the difference in water pressure across the pump:

(17) $P_A - P_D = f(t)$.

Then adding equations (14) through (17) gives the differential equation

(18) $LV''(t) + RV'(t) + \dfrac{1}{C} V(t) = f(t)$

which governs the motion of the water in the system.

In the case of the electrical circuit we let $q(t)$ denote the charge on the lower side of the capacitor at time t. Under the assumptions that a difference in voltage across the capacitor is proportional to $q(t)$, that a difference in voltage across the resistor is proportional to the current $i(t) = q'(t)$, and that a difference in voltage across the inductor is proportional to the rate of change $v'(t) = q''(t)$ of the current, we obtain the differential equation

(19) $Lq''(t) + Rq'(t) + \dfrac{1}{C} q(t) = f(t)$

where $f(t)$ is the voltage difference $V_A - V_D$ created by the generator. In equation (19) the time is measured in seconds, the charge in *coulombs*, and the voltage in *volts*. Current is measured in *amperes*. The constant L in (19) is called the *inductance* and is measured in *henries*; the constant R is called the *resistance* and is measured in *ohms*; the constant C is the *capacitance* and is measured in *farads*. A differential equation for the current may be obtained by differentiating (19) with respect to t.

Exercises

Each of Exercises 1 through 15 deals with a spring that exerts a force of $-ky$ pounds toward the right when its free end is y feet to the right of its rest position. An object of mass M slugs is fastened to the free end and a resisting medium exerts a force of $-R\dfrac{dy}{dt}$ toward the right when the object's velocity is $\dfrac{dy}{dt}$. An external force of $f(t)$ pounds toward the right is applied to the object at time t (seconds). **a.** Determine whether the motion is free or forced and whether it is undamped, underdamped, critically damped, or overdamped. **b.** Pick the graph of the solution of the initial value problem from the curves shown in Figures 19.11 through 19.25. (The tic marks are at integer values of y and t.) **c.** Solve the initial value problem. (Ignore the weight of the spring and the force of gravity.)

1.‡ $M = 2, R = 0, k = 18, f = 0, y(0) = 1, y'(0) = 3$

2.‡ $M = 2, R = 2, k = 5, f = 0, y(0) = 2, y'(0) = -1$

3. $M = 1, R = 0, k = 25, f = 0, y(0) = -2, y'(0) = 5$

4.† $M = 2, R = 6, k = 4, f = 0, y(0) = -1, y'(0) = 4$

5. $M = 4, R = 8, k = 40, f = 0, y(0) = 0, y'(0) = 3$

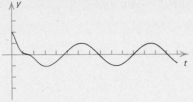

FIGURE 19.11

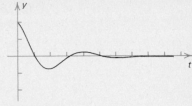

FIGURE 19.12

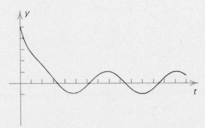

FIGURE 19.13

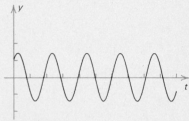

FIGURE 19.14

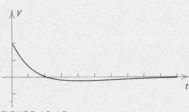

FIGURE 19.15

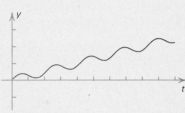

FIGURE 19.16

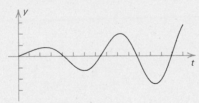

FIGURE 19.17

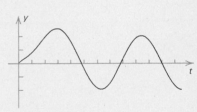

FIGURE 19.18

FIGURE 19.19

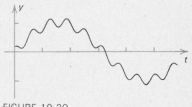

FIGURE 19.20

FIGURE 19.21

FIGURE 19.22

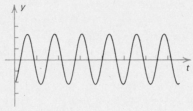

FIGURE 19.23

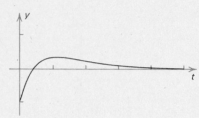

FIGURE 19.24

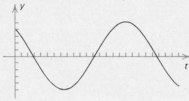

FIGURE 19.25

6.† $M = 4, R = 4, k = 1, f = 0, y(0) = 2, y'(0) = -2$

7. $M = 9, R = 0, k = 1, f = 0, y(0) = 4, y'(0) = -1$

8. $M = 2, R = 3, k = 1, f = 0, y(0) = 0, y'(0) = 1$

9.† $M = 5, R = 10, k = 10, f = 0, y(0) = -1, y'(0) = 0$

10.‡ $M = 1, R = 0, k = 1, f = \frac{2}{5}\sin t, y(0) = 0, y'(0) = \frac{2}{5}$

11. $M = 1, R = 0, k = 100, f = 99 \sin t, y(0) = \frac{1}{6}, y'(0) = 1$

12.† $M = 1, R = 2, k = 1, f = 4 \sin t, y(0) = 0, y'(0) = 1$

13. $M = 1, R = 2, k = 10, f = 9 \cos t - 2 \sin t, y(0) = 2, y'(0) = -1$

14. $M = 1, R = 3, k = 2, f = \sin t + 3 \cos t, y(0) = 5, y'(0) = -4$

15. $M = 4, R = 0, k = 36, f = 9t, y(0) = 0, y'(0) = 1$

16. What are the amplitude, period, and frequency of the motion in
    a.‡ Exercise 1,   **b.†** Exercise 3, and   **c.** Exercise 7?

17. A spring that satisfies Hooke's law is fixed at its upper end and an object that is fastened to its lower end moves vertically, subject to the force of gravity, the force of the spring, and resistance that is proportional to the object's velocity. Show that the motion is governed by the differential equation $My'' + Ry' + ky = f(t)$, where $y = 0$ at the equilibrium position of the weight and $f(t)$ is the external force at time t.

18.[†] When an object weighing 2 pounds is fastened at the lower end of a dangling spring, the spring stretches $\frac{1}{2}$ foot. The spring is stretched another $\frac{1}{2}$ foot and released. Describe the subsequent motion of the object under the assumptions that the spring satisfies Hooke's law and that there is no air resistance or other forces on the object.

19.[†] Show that the motion of a damped spring subject to an external force $\sin(at)$ approaches the *steady state motion* given by $y = C\sin(at + \psi)$ with constants C and ψ that are independent of the initial conditions.

20.   Describe the steady state motion in   **a.**[†] Exercise 12,   **b.** Exercise 13, and   **c.** Exercise 14.

21.[†] A circuit as in Figure 19.8 but with no generator has inductance of 2 henries, resistance of 4 ohms, and capacitance of $\frac{1}{10}$ farad. At time $t = 0$ there is a charge of 5 coulombs on the lower side of the capacitor and no current flow. Find the charge $q(t)$ on the capacitor as a function of t.

22.   A circuit as in Figure 19.8 has inductance of 3 henries, resistance of 6 ohms, and capacitance of $\frac{1}{3}$ farad. The generator produces a voltage difference $f(t) = \sin t$ at time t. At $t = 0$ there is no charge on the capacitor and the current is 1 ampere. Find the charge $q(t)$ and the current $i(t)$ as functions of t.

23.   Show that the current in the circuit of Figure 19.8 satisfies a differential equation of the form $Li'' + Ri' + \dfrac{1}{C}i = g(t)$.

24.   Give differential equations for the current in a circuit as in Figure 19.8 but   **a.**[†] with no capacitor and   **b.** with no inductor.

25.   Consider a circuit as in Figure 19.8 with inductance of 1 henry, resistance of 1 ohm, and capacitance of $1/k$ farads ($k > \frac{1}{2}$), and where the generator produces a voltage $-\dfrac{1}{a}\cos(at)$ at time t.   **a.**[†] Find the steady state current (see Exercise 19).   **b.** Find the amplitude of the steady state current and show that it is a maximum for $a = \sqrt{k - \frac{1}{2}}$. (This illustrates the principle used in tuning a radio. The incoming signals act like generators and the capacitance of the circuit can be adjusted so that the signals at the desired frequency produce the greatest current.)

26.   Show that for   **a.** a free overdamped spring and   **b.** a free critically damped spring the solution is zero for, at most, one nonnegative value of t.

19-5 POWER SERIES SOLUTIONS

Linear differential equations with variable coefficients arise in many branches of mathematical physics, engineering and other fields, and often their study begins with the construction of POWER SERIES SOLUTIONS of the homogeneous differential equations. In this section we show how such solutions are constructed for second order equations

$$(1) \quad a(x)\, y''(x) + b(x)\, y'(x) + c(x)\, y(x) = 0$$

with polynomial coefficients. We will restrict our attention to power series centered at $x = 0$

$$(2) \quad y(x) = a_0 + a_1 x + a_2 x^2 + \cdots = \sum_{j=0}^{\infty} a_j x^j$$

under the assumption that the coefficient $a(x)$ of $y''(x)$ in the differential equation is not zero at $x = 0$. As is illustrated in the next example, we find the coefficients a_j by substituting the power series (3) into the differential equation and equating coefficients of powers of x.

EXAMPLE 1   Construct the power series centered at $x = 0$ for the solution of the initial value problem

$$(3) \quad y''(x) + y(x) = 0, \quad y(0) = 1, \quad y'(0) = 0.$$

Solution.   We know from the results of Section 19.3 that the solution of the initial value problem (3) is $y = \cos x$, so our calculations should lead us to the power series

$$(4) \quad y = 1 - \frac{1}{2!} x^2 + \frac{1}{4!} x^4 - \cdots = \sum_{j=0}^{\infty} (-1)^j \frac{1}{(2j)!} x^{2j}$$

for that function.

We can compute the derivatives of the power series (2) by differentiating under the summation sign:

$$(5) \quad y'(x) = a_1 + 2a_2 x + 3a_3 x^2 + \cdots = \sum_{j=1}^{\infty} j a_j x^{j-1}$$

$$(6) \quad y''(x) = 2a_2 + 3(2)a_3 x + 4(3)a_4 x^2 + \cdots = \sum_{j=2}^{\infty} j(j-1)a_j x^{j-2}.$$

The sum in (5) begins at $j = 1$ because the term $0a_0$ for $j = 0$ is zero. The sum in (6) begins at $j = 2$ because the terms for $j = 0$ and 1 are zero. With formulas (2) and (6) the differential equation $y''(x) + y(x) = 0$ reads

(7) $\displaystyle\sum_{j=2}^{\infty} j(j-1)a_j x^{j-2} + \sum_{j=0}^{\infty} a_j x^j = 0.$

Because the power x^{j-2} appears in the first sum of (7) and the power x^j appears in the second, we make the substitutions $k = j - 2$, $j = k + 2$ in the first sum, which then begins at $k = 0$. To have the notation be the same in the two sums, we also replace j by k in the second sum and obtain

$$\sum_{k=0}^{\infty} (k+2)(k+1)a_{k+2}x^k + \sum_{k=0}^{\infty} a_k x^k = 0$$

or

(8) $\displaystyle\sum_{k=0}^{\infty} [(k+2)(k+1)a_{k+2} + a_k]x^k = 0.$

In order for equation (8) to be satisfied the coefficient of x^k must be zero for each k. This gives the equation $(k+2)(k+1)a_{k+2} + a_k = 0$ or

(9) $a_{k+2} = -\dfrac{1}{(k+2)(k+1)} a_k$ for $k = 0, 1, 2, \ldots$.

The initial condition $y(0) = 1$ with formula (2) shows that $a_0 = 1$. The condition $y'(0) = 0$ with formula (5) implies that $a_1 = 0$. Formula (9) then shows that $a_k = 0$ for all odd k and that $a_2 = -\dfrac{1}{(2)(1)} a_0$
$= -\dfrac{1}{2!}, a_4 = -\dfrac{1}{(4)(3)} a_2 = \dfrac{1}{4!}, a_6 = -\dfrac{1}{(6)(5)} a_4 = -\dfrac{1}{6!},$ and that in general for $j = 0, 1, 2, \ldots$.

(10) $a_{2j} = (-1)^j \dfrac{1}{(2j)!}.$

Formulas (2) and (10) yield the anticipated power series (4) for the solution. $\square$

In the next example we find a power series solution of a differential equation that we could not solve by other methods studied in this chapter.

EXAMPLE 2   Use power series centered at $x = 0$ to find the general solution of the differential equation

(11) $y''(x) - xy(x) = 0.$

Solution.   We set

(2) $y(x) = \displaystyle\sum_{j=0}^{\infty} a_j x^j.$

Then

$$(12) \quad y''(x) = \sum_{j=2}^{\infty} j(j-1)a_j x^{j-2}$$

$$(13) \quad xy(x) = \sum_{j=0}^{\infty} a_j x^{j+1}.$$

We make the substitutions $k = j - 2, j = k + 2$ in sum (12) and $k = j + 1, j = k - 1$ in sum (13). Then the differential equation $y'' - xy = 0$ reads

$$(14) \quad \sum_{k=0}^{\infty} (k+2)(k+1)a_{k+2} x^k - \sum_{k=1}^{\infty} a_{k-1} x^k = 0.$$

Setting the combined coefficient of x^k equal to 0 for $k = 0, 1, 2,$ . . . gives the equation

$$(15) \quad (0+2)(0+1)a_{0+2} = 0$$

for the case of $k = 0$ and

$$(16) \quad (k+2)(k+1)a_{k+2} - a_{k-1} = 0 \quad \text{for } k = 1, 2, 3, \ldots.$$

Equation (15) implies that $a_2 = 0$, and equation (16) yields the formula

$$(17) \quad a_{k+2} = \frac{1}{(k+2)(k+1)} a_{k-1} \quad \text{for } k = 1, 2, 3, \ldots.$$

Formula (17) enables us to express the coefficients a_j for $j \geq 3$ in terms of a_0 and a_1, which are arbitrary and would be determined by initial conditions. We have $a_3 = \dfrac{1}{(3)(2)} a_0 = \dfrac{1}{3!} a_0, a_4 = \dfrac{1}{(4)(3)} a_1$

$= \dfrac{2}{4!} a_1, a_5 = \dfrac{1}{(5)(4)} a_2 = 0, a_6 = \dfrac{1}{(6)(5)} a_3 = \dfrac{(1)(4)}{6!} a_0, a_7 = \dfrac{1}{(7)(6)} a_4$

$= \dfrac{(2)(5)}{7!} a_1, a_8 = \dfrac{1}{(8)(7)} a_5 = 0$, and so forth. These calculations lead to the formulas

$$a_{3k} = \frac{(1)(4)(7) \cdots (3k-2)}{(3k)!} a_0$$

$$a_{3k+1} = \frac{(2)(5)(8) \cdots (3k-1)}{(3k+1)!} a_1$$

for $k \geq 1$ and

$$a_{3k+2} = 0$$

which are valid for $k \leq 0$ give the expression

$$y(x) = a_0 \left[1 + \sum_{k=1}^{\infty} \frac{(1)(4)(7) \cdots (3k-2)}{(3k)!} x^{3k} \right]$$

(18)

$$\left[x + a_1 \sum_{k=0}^{\infty} \frac{(2)(5)(8) \cdots (3k-1)}{(3k+1)!} x^{3k+1} \right]$$

for the general solution of the differential equation. The Ratio Test can be used to show that the power series in (18) converges for all x (see Exercise 16). The functions defined by (18) are called *Airy functions* and are important in mathematical physics.   □

Exercises

†*Answer provided.*
†*Outline of solution provided.*

Use power series centered at $x = 0$ to find the solutions of the initial value problems in Exercises 1 through 11.

1.‡ $y'' + 2y' + y = 0$, $y(0) = 0$, $y'(0) = 1$

2. $y'' - y = 0$, $y(0) = 1$, $y'(0) = -1$

3.† $y'' - y' - 2y = 0$, $y(0) = 3$, $y'(0) = 6$

4. $y'' - 3y' = 0$, $y(0) = 3$, $y'(0) = 3$

5.† $y' - 2xy = 0$, $y(0) = 1$

6. $y' + 4y = 0$, $y(0) = 3$

7.† $y'' - xy' - y = 0$, $y(0) = 1$, $y'(0) = 0$

8. $(x - 1)y' - y = 0$, $y(0) = 1$

9.‡ $y' - y = -x^2$, $y(0) = 2$

10. $y' - y = x - 1$, $y(0) = 1$

11. $y' - 2xy = 2x$, $y(0) = 0$

Use Taylor's formula to give the terms through $a_4 x^4$ in the power series centered at $x = 0$ for the solutions of the initial value problems in Exercises 12 through 15.

12.‡ $y' = x^2 + y^2$, $y(0) = 1$

13. $y' = \cos y$, $y(0) = 0$

14.† $y'' = y^2$, $y(0) = 1$, $y'(0) = 2$

15. $y'' = \sqrt{y+1}$, $y(0) = 0$, $y'(0) = 2$

16. Use the Ratio Test to show that the power series (18) for the Airy function converges for all x.

Use power series centered at $x = 0$ to find the solutions of the initial value problems in Exercises 17 through 26.

17.† $y'' - 2xy' - 2y = 0$, $y(0) = 1$, $y'(0) = 0$

18.† $(x - 1)y'' + (2 - x)y' + 2y = 0$, $y(0) = 7$, $y'(0) = -6$

19.† $y'' - 2xy' + 8y = 0$, $y(0) = 3$, $y'(0) = 0$ (A *Hermite polynomial*)

20.† $(1 - x^2)y'' - 4xy' + 18y = 0$, $y(0) = 0$, $y'(0) = 3$

21. $(x + 1)y'' + y' - xy = 0$, $y(0) = 1$, $y'(0) = -1$

22.[†] $xy'' + y' + xy = 0, y(0) = 1, y'(0) = 0$ (A *Bessel function*)

23. $xy'' + (1 - x)y' + 3y = 0, y(0) = 6, y'(0) = -18$ (A *Laguerre polynomial*)

24.[†] $xy'' - (x^2 + 2)y' + xy = 0, y(0) = 1, y'(0) = 0$

25. $xy'' + y' + 2y = 0, y(0) = 1, y'(0) = -2$

26.[†] $xy'' - (1 + x)y' + y = 0, y(0) = 1, y'(0) = 1$

27. The differential equations in Exercises 22 through 26 are not of the type discussed in this section because the coefficients of y'' are zero at $x = 0$. Because of this we cannot prescribe both $y(0)$ and $y'(0)$ arbitrarily. Explain.

MISCELLANEOUS EXERCISES

[†]*Answer provided.*
[‡]*Outline of solution provided.*

The differential equation $\dfrac{dy}{dx} = \dfrac{A(x, y)}{B(x, y)}$ is called *homogeneous of degree k* if $A(tx, ty) = t^k A(x, y)$ and $B(tx, ty) = t^k B(x, y)$ for all nonzero x, y, and t. In this case the differential equation can be transformed into a separable equation by making the substitutions $y = xv, \dfrac{dy}{dx} = x\dfrac{dv}{dx} + v$. Use this technique to solve the initial value problems in Exercises 1 through 6.

1.[‡] $\dfrac{dy}{dx} = \dfrac{x^2 + y^2}{xy}, \quad y(1) = 2$

2.[†] $\dfrac{dy}{dx} = \dfrac{y}{x + y}, \quad y(1) = 1$

3. $\dfrac{dy}{dx} = \dfrac{3y^3 + x^2y + x^3}{3xy^2 + x^3}, \quad y(1) = 2$

4.[†] $\dfrac{dy}{dx} = \dfrac{-x}{y + \sqrt{x^2 + y^2}}, \quad y(1) = 0$

5. $\dfrac{dy}{dx} = \dfrac{y - \sqrt{x^2 - y^2}}{x}, \quad y(1) = 0$

6. $\dfrac{dy}{dx} = \dfrac{y^2 + x^2 + xy}{x^2}, \quad y(1) = 0$

7. a.[†] Show that if y is a solution of $x\dfrac{dy}{dx} + y = x$ and if its graph crosses the y-axis, then the graph passes through the origin. Sketch the graphs of the solutions that satisfy the initial conditions **b.** $y(4) = 1$, **c.** $y(1) = 1$, and **d.** $y(2) = 1$.

8. Find a function $y(x)$ such that the tangent line to its graph at $(x_0, y(x_0))$ has x-intercept $\frac{1}{2}x_0$ for all x_0.

9.[†] Solve the initial value problem $xy'' - y' = x^4, \quad y(1) = 0, y'(1) = 1$ by solving first for $v = y'$.

A differential equation of the form $\dfrac{dy}{dx} + P(x)y = Q(x)y^n$ is linear if n is

0 or 1. Otherwise it is called a *Bernoulli equation* and can be transformed into a linear equation by the substitution $v = y^{1-n}$. Use this technique to solve the initial value problems in Exercises 10 and 11.

10.† $\dfrac{dy}{dx} + xy = y^3x$, $\quad y(0) = \frac{1}{2}$ $\qquad$ **11.** $\dfrac{dy}{dx} + y = xy^2$, $\quad y(0) = \frac{1}{2}$

A differential equation $p(x, y)\, dx + q(x, y)\, dy = 0$ that is not exact can be transformed into an exact differential equation by multiplying it by a suitable integrating factor $I(x, y)$. Use this technique to solve the differential equations in Exercises 12 through 15.

12.‡ $(3xy^4 + 2y)\, dx + (4x^2y^3 + x)\, dy = 0; I = x$

13.† $y^2\, dx + (5xy - 4)\, dy = 0; I = y^3$

14. $3x^2\, dx + 2x^3y\, dy = 0; I = e^{y^2}$

15. $(9y + 15xy^2)\, dx + (8x + 14x^2y)\, dy = 0; I = x^{1/2}\, y^{1/3}$

16. The linear differential equation $\dfrac{dy}{dx} + P(x)y = Q(x)$ may be written

$[Q(x) - P(x)y]\, dx - dy = 0$. Show that it has the integrating factor $e^{\int P\, dx}$.

A differential equation of the form $ax^2y'' + bxy' + cy = 0$ with constants $a, b,$ and c is transformed into a linear equation with constant coefficients by the substitution $u = \ln x$. Use this technique with the formulas

$$\frac{d}{dx} = \frac{du}{dx}\frac{d}{du}, \quad \frac{d^2}{dx^2} = \frac{d^2u}{dx^2}\frac{d}{du} + \left(\frac{du}{dx}\right)^2\frac{d^2}{du^2}$$

to solve the differential equations in Exercises 17 through 21.

17.‡ $x^2\dfrac{d^2y}{dx^2} + 4x\dfrac{dy}{dx} + 2y = 0$ $\qquad$ **18.†** $x^2\dfrac{d^2y}{dx^2} - x\dfrac{dy}{dx} + y = 0$

19.† $x^2\dfrac{d^2y}{dx^2} + x\dfrac{dy}{dx} + y = 0$ $\qquad$ **20.** $x^2\dfrac{d^2y}{dx^2} + 3x\dfrac{dy}{dx} + 2y = 0$

21. $x^2\dfrac{d^2y}{dx^2} - 2x\dfrac{dy}{dx} - 4y = 0$

22. The *kinetic energy* of an object of mass M with velocity y' is $\frac{1}{2}M(y')^2$. The *potential energy* of a spring with spring constant k is $\frac{1}{2}ky^2$ when the free end of the spring is compressed or stretched a distance y from its rest position. Show that for a free, undamped spring the sum of the kinetic and potential energies is constant. (Multiply the differential equation $My'' + ky = 0$ by y' and integrate.)

The general solution of an nth order linear homogeneous differential equation with constant coefficients

$$a_ny^{(n)}(x) + a_{n-1}y^{(n-1)}(x) + \cdots + a_0y(x) = 0$$

is determined by the factorization of its characteristic polynomial

$$a_nr^n + a_{n-1}r^{n-1} + \cdots + a_0.$$

If the power $(r - r_j)^k$ of a linear factor appears in the factoring, then a term of the form

$$[A_0 + A_1x + A_2x^2 + \cdots + A_{k-1}x^{k-1}]e^{r_jx}$$

appears in the general solution. If the power $(ar^2 + br + c)^k$ of an irreducible quadratic polynomial appears in the factoring, then a term of the form

$$[A_0 + A_1x + \cdots + A_{k-1}x^{k-1}][B\cos(\omega x) + C\sin(\omega x)]e^{\tau x}$$

appears in the solution, where $r = \tau \pm i\omega$ are the solutions of $ar^2 + br + c = 0$. Use these facts to find the general solutions of the differential equations in Exercises 23 through 31.

23.‡ $y^{(3)} + y' = 0$ **24.** $y^{(3)} - y'' = 0$

25.† $y^{(3)} - 2y'' + y' = 0$ **26.**† $y^{(4)} - y'' = 0$

27. $y^{(5)} + y^{(3)} = 0$ **28.** $y^{(3)} - 2y'' - 3y' = 0$

29.† $y^{(4)} - y = 0$ **30.**† $y^{(4)} + 2y'' + y = 0$

31. $y^{(4)} + 2y^{(3)} + y'' = 0$

32. Use a power series centered at $x = 0$ to find the solution of the initial value problem $y^{(3)} - y = 0$, $y(0) = 1$, $y'(0) = 1$, $y''(0) = 1$.

33.† Find the first four nonzero terms in the power series centered at $x = 0$ for the solution of the initial value problem $y'' - xy' - y = e^x$, $y(0) = 0$, $y'(0) = 0$.

34. The solution of the initial value problem $xy'' + y' + xy = 0$, $y(0) = 1$ is called a *Bessel function*. Find its power series centered at $x = 0$.

35. What are the orthogonal trajectories of the family of circles tangent to the y-axis at the origin? (See the remarks before Exercise 1.)

(Euler's method) *Euler's method* of finding approximate solutions of an initial value problem $y'(x) = f(x, y)$, $y(a) = k$ for $a \le x \le b$ gives an approximate solution $y_N(x)$ for each positive integer N. We consider the partition $a = x_0 < x_1 < x_2 < \cdots < x_N = b$ of the interval $a \le x \le b$ into N equal subintervals. The graph of $y_N(x)$ is a polygonal line. The portion for $a \le x \le x_1$ starts at the point (a, k) and has slope $f(a, k)$; the portion for $x_1 \le x \le x_2$ starts at $(x_1, y_N(x_1))$ and has slope $f(x_1, y_N(x_1))$; and, in general, the portion for $x_{j-1} \le x \le x_j$ starts at $(x_{j-1}, y_N(x_{j-1}))$ and has slope $f(x_{j-1}, y_N(x_{j-1}))$. In Exercises 36 through 40   **a.** sketch the graph of the approximate solution $y_N(x)$ for $N = 4$. If you have access to a computer or electronic calculator, sketch the graphs for   **b.** $N = 8$ and   **c.** $N = 20$.

36.† $y'(x) = \frac{1}{2}(x^2 + y^2)$, $y(-1) = -1$; $-1 \le x \le 3$

37.† $y'(x) = \sin(\pi x) + \cos(\pi y)$, $y(0) = -\frac{1}{2}$; $0 \le x \le 2$

38.† $y'(x) = \sin(\pi x - \pi y)$, $y(0) = \frac{1}{2}$; $0 \le x \le 2$

39. $y'(x) = \frac{1}{2}(x^2 - y^2)$, $y(-2) = 0$; $-2 \le x \le 2$

40. $y'(x) = \cos(\pi xy)$, $y(0) = 0$; $0 \le x \le 4$

APPENDIX
REFERENCE TABLES

TABLE I   Squares, Cubes, Square
Roots, and Cube Roots

n	n^2	n^3	$\sqrt{n}$	$\sqrt{10n}$	$\sqrt[3]{n}$	$\sqrt[3]{10n}$	$\sqrt[3]{100n}$
1	1	1	1.000 000	3.162 278	1.000 000	2.154 435	4.641 589
2	4	8	1.414 214	4.472 136	1.259 921	2.714 418	5.848 035
3	9	27	1.732 051	5.477 226	1.442 250	3.107 233	6.694 330
4	16	64	2.000 000	6.324 555	1.587 401	3.419 952	7.368 063
5	25	125	2.236 068	7.071 068	1.709 976	3.684 031	7.937 005
6	36	216	2.449 490	7.745 967	1.817 121	3.914 868	8.434 327
7	49	343	2.645 751	8.366 600	1.912 931	4.121 285	8.879 040
8	64	512	2.828 427	8.944 272	2.000 000	4.308 869	9.283 178
9	81	729	3.000 000	9.486 833	2.080 084	4.481 405	9.654 894
10	100	1 000	3.162 278	10.000 00	2.154 435	4.641 589	10.000 00
11	121	1 331	3.316 625	10.488 09	2.223 980	4.791 420	10.322 80
12	144	1 728	3.464 102	10.954 45	2.289 428	4.932 424	10.626 59
13	169	2 197	3.605 551	11.401 75	2.351 335	5.065 797	10.913 93
14	196	2 744	3.741 657	11.832 16	2.410 142	5.192 494	11.186 89
15	225	3 375	3.872 983	12.247 45	2.466 212	5.313 293	11.447 14
16	256	4 096	4.000 000	12.649 11	2.519 842	5.428 835	11.696 07
17	289	4 913	4.123 106	13.038 40	2.571 282	5.539 658	11.934 83
18	324	5 832	4.242 641	13.416 41	2.620 741	5.646 216	12.164 40
19	361	6 859	4.358 899	13.784 05	2.668 402	5.748 897	12.385 62
20	400	8 000	4.472 136	14.142 14	2.714 418	5.848 035	12.599 21
21	441	9 261	4.582 576	14.491 38	2.758 924	5.943 922	12.805 79
22	484	10 648	4.690 416	14.832 40	2.802 039	6.036 811	13.005 91
23	529	12 167	4.795 832	15.165 75	2.843 867	6.126 926	13.200 06
24	576	13 824	4.898 979	15.491 93	2.884 499	6.214 465	13.388 66
25	625	15 625	5.000 000	15.811 39	2.924 018	6.299 605	13.572 09
26	676	17 576	5.099 020	16.124 52	2.962 496	6.382 504	13.750 69
27	729	19 683	5.196 152	16.431 68	3.000 000	6.463 304	13.924 77
28	784	21 952	5.291 503	16.733 20	3.036 589	6.542 133	14.094 60
29	841	24 389	5.385 165	17.029 39	3.072 317	6.619 106	14.260 43
30	900	27 000	5.477 226	17.320 51	3.107 233	6.694 330	14.422 50
31	961	29 791	5.567 764	17.606 82	3.141 381	6.767 899	14.581 00
32	1 024	32 768	5.656 854	17.888 54	3.174 802	6.839 904	14.736 13
33	1 089	35 937	5.744 563	18.165 90	3.207 534	6.910 423	14.888 06
34	1 156	39 304	5.830 952	18.439 09	3.239 612	6.979 532	15.036 95
35	1 225	42 875	5.916 080	18.708 29	3.271 066	7.047 299	15.182 94
36	1 296	46 656	6.000 000	18.973 67	3.301 927	7.113 787	15.326 19
37	1 369	50 653	6.082 763	19.235 38	3.332 222	7.179 054	15.466 80
38	1 444	54 872	6.164 414	19.493 59	3.361 975	7.243 156	15.604 91
39	1 521	59 319	6.244 998	19.748 42	3.391 211	7.306 144	15.740 61
40	1 600	64 000	6.324 555	20.000 00	3.419 952	7.368 063	15.874 01
41	1 681	68 921	6.403 124	20.248 46	3.448 217	7.428 959	16.005 21
42	1 764	74 088	6.480 741	20.493 90	3.476 027	7.488 872	16.134 29
43	1 849	79 507	6.557 439	20.736 44	3.503 398	7.547 842	16.261 33
44	1 936	85 184	6.633 250	20.976 18	3.530 348	7.605 905	16.386 43
45	2 025	91 125	6.708 204	21.213 20	3.556 893	7.663 094	16.509 64
46	2 116	97 336	6.782 330	21.447 61	3.583 048	7.719 443	16.631 03
47	2 209	103 823	6.855 655	21.679 48	3.608 826	7.774 980	16.750 69
48	2 304	110 592	6.928 203	21.908 90	3.634 241	7.829 735	16.868 65
49	2 401	117 649	7.000 000	22.135 94	3.659 306	7.883 735	16.984 99

TABLE I Squares, Cubes, Square
Roots, and Cube Roots

n	n^2	n^3	$\sqrt{n}$	$\sqrt{10n}$	$\sqrt[3]{n}$	$\sqrt[3]{10n}$	$\sqrt[3]{100n}$
50	2 500	125 000	7.071 068	22.360 68	3.684 031	7.937 005	17.099 76
51	2 601	132 651	7.141 428	22.583 18	3.708 430	7.989 570	17.213 01
52	2 704	140 608	7.211 103	22.803 51	3.732 511	8.041 452	17.324 78
53	2 809	148 877	7.280 110	23.021 73	3.756 286	8.092 672	17.435 13
54	2 916	157 464	7.348 469	23.237 90	3.779 763	8.143 253	17.544 11
55	3 025	166 375	7.416 198	23.452 08	3.802 952	8.193 213	17.651 74
56	3 136	175 616	7.483 315	23.664 32	3.825 862	8.242 571	17.758 08
57	3 249	185 193	7.549 834	23.874 67	3.848 501	8.291 344	17.863 16
58	3 364	195 112	7.615 773	24.083 19	3.870 877	8.339 551	17.967 02
59	3 481	205 379	7.681 146	24.289 92	3.892 996	8.387 207	18.069 69
60	3 600	216 000	7.745 967	24.494 90	3.914 868	8.434 327	18.171 21
61	3 721	226 981	7.810 250	24.698 18	3.936 497	8.480 926	18.271 60
62	3 844	238 328	7.874 008	24.899 80	3.957 892	8.527 019	18.370 91
63	3 969	250 047	7.937 254	25.099 80	3.979 057	8.572 619	18.469 15
64	4 096	262 144	8.000 000	25.298 22	4.000 000	8.617 739	18.566 36
65	4 225	274 625	8.062 258	25.495 10	4.020 726	8.662 391	18.662 56
66	4 356	287 496	8.124 038	25.690 47	4.041 240	8.706 588	18.757 77
67	4 489	300 763	8.185 353	25.884 36	4.061 548	8.750 340	18.852 04
68	4 624	314 432	8.246 211	26.076 81	4.081 655	8.793 659	18.945 36
69	4 761	328 509	8.306 624	26.267 85	4.101 566	8.836 556	19.037 78
70	4 900	343 000	8.366 600	26.457 51	4.121 285	8.879 040	19.129 31
71	5 041	357 911	8.426 150	26.645 83	4.140 818	8.921 121	19.219 97
72	5 184	373 248	8.485 281	26.832 82	4.160 168	8.962 809	19.309 79
73	5 329	389 017	8.544 004	27.018 51	4.179 339	9.004 113	19.398 77
74	5 476	405 224	8.602 325	27.202 94	4.198 336	9.045 042	19.486 95
75	5 625	421 875	8.660 254	27.386 13	4.217 163	9.085 603	19.574 34
76	5 776	438 976	8.717 798	27.568 10	4.235 824	9.125 805	19.660 95
77	5 929	456 533	8.774 964	27.748 87	4.254 321	9.165 656	19.746 81
78	6 084	474 552	8.831 761	27.928 48	4.272 659	9.205 164	19.831 92
79	6 241	493 039	8.888 194	28.106 94	4.290 840	9.244 335	19.916 32
80	6 400	512 000	8.944 272	28.284 27	4.308 869	9.283 178	20.000 00
81	6 561	531 441	9.000 000	28.460 50	4.326 749	9.321 698	20.082 99
82	6 724	551 368	9.055 385	28.635 64	4.344 481	9.359 902	20.165 30
83	6 889	571 787	9.110 434	28.809 72	4.362 071	9.397 796	20.246 94
84	7 056	592 704	9.165 151	28.982 75	4.379 519	9.435 388	20.327 93
85	7 225	614 125	9.219 544	29.154 76	4.396 830	9.472 682	20.408 28
86	7 396	636 056	9.273 618	29.325 76	4.414 005	9.509 685	20.488 00
87	7 569	658 503	9.327 379	29.495 76	4.431 048	9.546 403	20.567 10
88	7 744	681 472	9.380 832	29.664 79	4.447 960	9.582 840	20.645 60
89	7 921	704 969	9.433 981	29.832 87	4.464 745	9.619 002	20.723 51
90	8 100	729 000	9.486 833	30.000 00	4.481 405	9.654 894	20.800 84
91	8 281	753 571	9.539 392	30.166 21	4.497 941	9.690 521	20.877 59
92	8 464	778 688	9.591 663	30.331 50	4.514 357	9.725 888	20.953 79
93	8 649	804 357	9.643 651	30.495 90	4.530 655	9.761 000	21.029 44
94	8 836	830 584	9.695 360	30.659 42	4.546 836	9.795 861	21.104 54
95	9 025	857 375	9.746 794	30.822 07	4.562 903	9.830 476	21.179 12
96	9 216	884 736	9.797 959	30.983 87	4.578 857	9.864 848	21.253 17
97	9 409	912 673	9.848 858	31.144 82	4.594 701	9.898 983	21.326 71
98	9 604	941 192	9.899 495	31.304 95	4.610 436	9.932 884	21.399 75
99	9 801	970 299	9.949 874	31.464 27	4.626 065	9.966 555	21.472 29

TABLE II   Factorials

n	$n!$	n	$n!$	n	$n!$
0	(0)1.00000 00000	15	(12)1.30767 43680	30	(32)2.65252 85981
1	(0)1.00000 00000	16	(13)2.09227 89888	31	(33)8.22283 86542
2	(0)2.00000 00000	17	(14)3.55687 42810	32	(35)2.63130 83693
3	(0)6.00000 00000	18	(15)6.40237 37057	33	(36)8.68331 76188
4	(1)2.40000 00000	19	(17)1.21645 10041	34	(38)2.95232 79904
5	(2)1.20000 00000	20	(18)2.43290 20082	35	(40)1.03331 47966
6	(2)7.20000 00000	21	(19)5.10909 42172	36	(41)3.71993 32679
7	(3)5.04000 00000	22	(21)1.12400 07278	37	(43)1.37637 53091
8	(4)4.03200 00000	23	(22)2.58520 16739	38	(44)5.23022 61747
9	(5)3.62880 00000	24	(23)6.20448 40173	39	(46)2.03978 82081
10	(6)3.62880 00000	25	(25)1.55112 10043	40	(47)8.15915 28325
11	(7)3.99168 00000	26	(26)4.03291 46113	41	(49)3.34525 26613
12	(8)4.79001 60000	27	(28)1.08888 69450	42	(51)1.40500 61178
13	(9)6.22702 08000	28	(29)3.04888 34461	43	(52)6.04152 63063
14	(10)8.71782 91200	29	(30)8.84176 19937	44	(54)2.65827 15748

Note. Values are given in scientific notation with the exponent in parentheses: (12)1.30767   43680 denotes $1.3076743680 \times 10^{12}$.

45	(56)1.19622 22087
46	(57)5.50262 21598
47	(59)2.58623 24151
48	(61)1.24139 15593
49	(62)6.08281 86403
50	(64)3.04140 93202

TABLE III   Degrees, Minutes, and Seconds to Radians

Deg	Rad	Min	Rad	Sec	Rad
1	0.01745 33	1	0.00029 09	1	0.00000 48
2	0.03490 66	2	0.00058 18	2	0.00000 97
3	0.05235 99	3	0.00087 27	3	0.00001 45
4	0.06981 32	4	0.00116 36	4	0.00001 94
5	0.08726 65	5	0.00145 44	5	0.00002 42
6	0.10471 98	6	0.00174 53	6	0.00002 91
7	0.12217 30	7	0.00203 62	7	0.00003 39
8	0.13962 63	8	0.00232 71	8	0.00003 88
9	0.15707 96	9	0.00261 80	9	0.00004 36
10	0.17453 29	10	0.00290 89	10	0.00004 85
20	0.34906 59	20	0.00581 78	20	0.00009 70
30	0.52359 88	30	0.00872 66	30	0.00014 54
40	0.69813 17	40	0.01163 55	40	0.00019 39
50	0.87266 46	50	0.01454 44	50	0.00024 24
60	1.04719 76	60	0.01745 33	60	0.00029 09
70	1.22173 05				
80	1.39626 34				
90	1.57079 63				

TABLE IV   Trigonometric
Functions in Radian Measure

Rad	sin	tan	ctn	cos	Rad	sin	tan	ctn	cos
.00	.0000	.0000	......	1.0000	.50	.4794	.5463	1.830	.8776
.01	.0100	.0100	99.997	1.0000	.51	.4882	.5594	1.788	.8727
.02	.0200	.0200	49.993	.9998	.52	.4969	.5726	1.747	.8678
.03	.0300	.0300	33.323	.9996	.53	.5055	.5859	1.707	.8628
.04	.0400	.0400	24.987	.9992	.54	.5141	.5994	1.668	.8577
.05	.0500	.0500	19.983	.9988	.55	.5227	.6131	1.631	.8525
.06	.0600	.0601	16.647	.9982	.56	.5312	.6269	1.595	.8473
.07	.0699	.0701	14.262	.9976	.57	.5396	.6410	1.560	.8419
.08	.0799	.0802	12.473	.9968	.58	.5480	.6552	1.526	.8365
.09	.0899	.0902	11.081	.9960	.59	.5564	.6696	1.494	.8309
.10	.0998	.1003	9.967	.9950	.60	.5646	.6841	1.462	.8253
.11	.1098	.1104	9.054	.9940	.61	.5729	.6989	1.431	.8196
.12	.1197	.1206	8.293	.9928	.62	.5810	.7139	1.401	.8139
.13	.1296	.1307	7.649	.9916	.63	.5891	.7291	1.372	.8080
.14	.1395	.1409	7.096	.9902	.64	.5972	.7445	1.343	.8021
.15	.1494	.1511	6.617	.9888	.65	.6052	.7602	1.315	.7961
.16	.1593	.1614	6.197	.9872	.66	.6131	.7761	1.288	.7900
.17	.1692	.1717	5.826	.9856	.67	.6210	.7923	1.262	.7838
.18	.1790	.1820	5.495	.9838	.68	.6288	.8087	1.237	.7776
.19	.1889	.1923	5.200	.9820	.69	.6365	.8253	1.212	.7712
.20	.1987	.2027	4.933	.9801	.70	.6442	.8423	1.187	.7648
.21	.2085	.2131	4.692	.9780	.71	.6518	.8595	1.163	.7584
.22	.2182	.2236	4.472	.9759	.72	.6594	.8771	1.140	.7518
.23	.2280	.2341	4.271	.9737	.73	.6669	.8949	1.117	.7452
.24	.2377	.2447	4.086	.9713	.74	.6743	.9131	1.095	.7385
.25	.2474	.2553	3.916	.9689	.75	.6816	.9316	1.073	.7317
.26	.2571	.2660	3.759	.9664	.76	.6889	.9505	1.052	.7248
.27	.2667	.2768	3.613	.9638	.77	.6961	.9697	1.031	.7179
.28	.2764	.2876	3.478	.9611	.78	.7033	.9893	1.011	.7109
.29	.2860	.2984	3.351	.9582	.79	.7104	1.009	.9908	.7038
.30	.2955	.3093	3.233	.9553	.80	.7174	1.030	.9712	.6967
.31	.3051	.3203	3.122	.9523	.81	.7243	1.050	.9520	.6895
.32	.3146	.3314	3.018	.9492	.82	.7311	1.072	.9331	.6822
.33	.3240	.3425	2.920	.9460	.83	.7379	1.093	.9146	.6749
.34	.3335	.3537	2.827	.9428	.84	.7446	1.116	.8964	.6675
.35	.3429	.3650	2.740	.9394	.85	.7513	1.138	.8785	.6600
.36	.3523	.3764	2.657	.9359	.86	.7578	1.162	.8609	.6524
.37	.3616	.3879	2.578	.9323	.87	.7643	1.185	.8437	.6448
.38	.3709	.3994	2.504	.9287	.88	.7707	1.210	.8267	.6372
.39	.3802	.4111	2.433	.9249	.89	.7771	1.235	.8100	.6294
.40	.3894	.4228	2.365	.9211	.90	.7833	1.260	.7936	.6216
.41	.3986	.4346	2.301	.9171	.91	.7895	1.286	.7774	.6137
.42	.4078	.4466	2.239	.9131	.92	.7956	1.313	.7615	.6058
.43	.4169	.4586	2.180	.9090	.93	.8016	1.341	.7458	.5978
.44	.4259	.4708	2.124	.9048	.94	.8076	1.369	.7303	.5898
.45	.4350	.4831	2.070	.9004	.95	.8134	1.398	.7151	.5817
.46	.4439	.4954	2.018	.8961	.96	.8192	1.428	.7001	.5735
.47	.4529	.5080	1.969	.8916	.97	.8249	1.459	.6853	.5653
.48	.4618	.5206	1.921	.8870	.98	.8305	1.491	.6707	.5570
.49	.4706	.5334	1.875	.8823	.99	.8360	1.524	.6563	.5487

TABLE IV Trigonometric
Functions in Radian Measure

Rad	sin	tan	ctn	cos	Rad	sin	tan	ctn	cos
1.00	.8415	1.557	.6421	.5403	1.30	.9636	3.602	.2776	.2675
1.01	.8468	1.592	.6281	.5319	1.31	.9662	3.747	.2669	.2579
1.02	.8521	1.628	.6142	.5234	1.32	.9687	3.903	.2562	.2482
1.03	.8573	1.665	.6005	.5148	1.33	.9711	4.072	.2456	.2385
1.04	.8624	1.704	.5870	.5062	1.34	.9735	4.256	.2350	.2288
1.05	.8674	1.743	.5736	.4976	1.35	.9757	4.455	.2245	.2190
1.06	.8724	1.784	.5604	.4889	1.36	.9779	4.673	.2140	.2092
1.07	.8772	1.827	.5473	.4801	1.37	.9799	4.913	.2035	.1994
1.08	.8820	1.871	.5344	.4713	1.38	.9819	5.177	.1931	.1896
1.09	.8866	1.917	.5216	.4625	1.39	.9837	5.471	.1828	.1798
1.10	.8912	1.965	.5090	.4536	1.40	.9854	5.798	.1725	.1700
1.11	.8957	2.014	.4964	.4447	1.41	.9871	6.165	.1622	.1601
1.12	.9001	2.066	.4840	.4357	1.42	.9887	6.581	.1519	.1502
1.13	.9044	2.120	.4718	.4267	1.43	.9901	7.055	.1417	.1403
1.14	.9086	2.176	.4596	.4176	1.44	.9915	7.602	.1315	.1304
1.15	.9128	2.234	.4475	.4085	1.45	.9927	8.238	.1214	.1205
1.16	.9168	2.296	.4356	.3993	1.46	.9939	8.989	.1113	.1106
1.17	.9208	2.360	.4237	.3902	1.47	.9949	9.887	.1011	.1006
1.18	.9246	2.427	.4120	.3809	1.48	.9959	10.983	.0910	.0907
1.19	.9284	2.498	.4003	.3717	1.49	.9967	12.350	.0810	.0807
1.20	.9320	2.572	.3888	.3624	1.50	.9975	14.101	.0709	.0707
1.21	.9356	2.650	.3773	.3530	1.51	.9982	16.428	.0609	.0608
1.22	.9391	2.733	.3659	.3436	1.52	.9987	19.670	.0508	.0508
1.23	.9425	2.820	.3546	.3342	1.53	.9992	24.498	.0408	.0408
1.24	.9458	2.912	.3434	.3248	1.54	.9995	32.461	.0308	.0308
1.25	.9490	3.010	.3323	.3153	1.55	.9998	48.078	.0208	.0208
1.26	.9521	3.113	.3212	.3058	1.56	.9999	92.620	.0108	.0108
1.27	.9551	3.224	.3102	.2963	1.57	1.0000	1255.8	.0008	.0008
1.28	.9580	3.341	.2993	.2867					
1.29	.9608	3.467	.2884	.2771					

$\pi = 3.14159265 \ldots \frac{1}{2}\pi = 1.57079632 \ldots \frac{3}{2}\pi = 4.71238898 \ldots 2\pi = 6.28318530 \ldots$

TABLE V   Common Logarithms
$(\log_{10}x)$

x	0	1	2	3	4	5	6	7	8	9
10	0000	0043	0086	0128	0170	0212	0253	0294	0334	0374
11	0414	0453	0492	0531	0569	0607	0645	0682	0719	0755
12	0792	0828	0864	0899	0934	0969	1004	1038	1072	1106
13	1139	1173	1206	1239	1271	1303	1335	1367	1399	1430
14	1461	1492	1523	1553	1584	1614	1644	1673	1703	1732
15	1761	1790	1818	1847	1875	1903	1931	1959	1987	2014
16	2041	2068	2095	2122	2148	2175	2201	2227	2253	2279
17	2304	2330	2355	2380	2405	2430	2455	2480	2504	2529
18	2553	2577	2601	2625	2648	2672	2695	2718	2742	2765
19	2788	2810	2833	2856	2878	2900	2923	2945	2967	2989
20	3010	3032	3054	3075	3096	3118	3139	3160	3181	3201
21	3222	3243	3263	3284	3304	3324	3345	3365	3385	3404
22	3424	3444	3464	3483	3502	3522	3541	3560	3579	3598
23	3617	3636	3655	3674	3692	3711	3729	3747	3766	3784
24	3802	3820	3838	3856	3874	3892	3909	3927	3945	3962
25	3979	3997	4014	4031	4048	4065	4082	4099	4116	4133
26	4150	4166	4183	4200	4216	4232	4249	4265	4281	4298
27	4314	4330	4346	4362	4378	4393	4409	4425	4440	4456
28	4472	4487	4502	4518	4533	4548	4564	4579	4594	4609
29	4624	4639	4654	4669	4683	4698	4713	4728	4742	4757
30	4771	4786	4800	4814	4829	4843	4857	4871	4886	4900
31	4914	4928	4942	4955	4969	4983	4997	5011	5024	5038
32	5051	5065	5079	5092	5105	5119	5132	5145	5159	5172
33	5185	5198	5211	5224	5237	5250	5263	5276	5289	5302
34	5315	5328	5340	5353	5366	5378	5391	5403	5416	5428
35	5441	5453	5465	5478	5490	5502	5514	5527	5539	5551
36	5563	5575	5587	5599	5611	5623	5635	5647	5658	5670
37	5682	5694	5705	5717	5729	5740	5752	5763	5775	5786
38	5798	5809	5821	5832	5843	5855	5866	5877	5888	5899
39	5911	5922	5933	5944	5955	5966	5977	5988	5999	6010
40	6021	6031	6042	6053	6064	6075	6085	6096	6107	6117
41	6128	6138	6149	6160	6170	6180	6191	6201	6212	6222
42	6232	6243	6253	6263	6274	6284	6294	6304	6314	6325
43	6335	6345	6355	6365	6375	6385	6395	6405	6415	6425
44	6435	6444	6454	6464	6474	6484	6493	6503	6513	6522
45	6532	6542	6551	6561	6571	6580	6590	6599	6609	6618
46	6628	6637	6646	6656	6665	6675	6684	6693	6702	6712
47	6721	6730	6739	6749	6758	6767	6776	6785	6794	6803
48	6812	6821	6830	6839	6848	6857	6866	6875	6884	6893
49	6902	6911	6920	6928	6937	6946	6955	6964	6972	6981

Note. Decimal points are omitted in this table. The entries

	0	1	2
10	0000	0043	0086

mean that $\log_{10}(1.00) = 0.0000$, $\log_{10}(1.01) = 0.0043$, and $\log_{10}(1.02) = 0.0086$ (to four decimal place accuracy).

TABLE V   Common Logarithms
($\log_{10} x$)

x	0	1	2	3	4	5	6	7	8	9
50	6990	6998	7007	7016	7024	7033	7042	7050	7059	7067
51	7076	7084	7093	7101	7110	7118	7126	7135	7143	7152
52	7160	7168	7177	7185	7193	7202	7210	7218	7226	7235
53	7243	7251	7259	7267	7275	7284	7292	7300	7308	7316
54	7324	7332	7340	7348	7356	7364	7372	7380	7388	7396
55	7404	7412	7419	7427	7435	7443	7451	7459	7466	7474
56	7482	7490	7497	7505	7513	7520	7528	7536	7543	7551
57	7559	7566	7574	7582	7589	7597	7604	7612	7619	7627
58	7634	7642	7649	7657	7664	7672	7679	7686	7694	7701
59	7709	7716	7723	7731	7738	7745	7752	7760	7767	7774
60	7782	7789	7796	7803	7810	7818	7825	7832	7839	7846
61	7853	7860	7868	7875	7882	7889	7896	7903	7910	7917
62	7924	7931	7938	7945	7952	7959	7966	7973	7980	7987
63	7993	8000	8007	8014	8021	8028	8035	8041	8048	8055
64	8062	8069	8075	8082	8089	8096	8102	8109	8116	8122
65	8129	8136	8142	8149	8156	8162	8169	8176	8182	8189
66	8195	8202	8209	8215	8222	8228	8235	8241	8248	8254
67	8261	8267	8274	8280	8287	8293	8299	8306	8312	8319
68	8325	8331	8338	8344	8351	8357	8363	8370	8376	8382
69	8388	8395	8401	8407	8414	8420	8426	8432	8439	8445
70	8451	8457	8463	8470	8476	8482	8488	8494	8500	8506
71	8513	8519	8525	8531	8537	8543	8549	8555	8561	8567
72	8573	8579	8585	8591	8597	8603	8609	8615	8621	8627
73	8633	8639	8645	8651	8657	8663	8669	8675	8681	8686
74	8692	8698	8704	8710	8716	8722	8727	8733	8739	8745
75	8751	8756	8762	8768	8774	8779	8785	8791	8797	8802
76	8808	8814	8820	8825	8831	8837	8842	8848	8854	8859
77	8865	8871	8876	8882	8887	8893	8899	8904	8910	8915
78	8921	8927	8932	8938	8943	8949	8954	8960	8965	8971
79	8976	8982	8987	8993	8998	9004	9009	9015	9020	9025
80	9031	9036	9042	9047	9053	9058	9063	9069	9074	9079
81	9085	9090	9096	9101	9106	9112	9117	9122	9128	9133
82	9138	9143	9149	9154	9159	9165	9170	9175	9180	9186
83	9191	9196	9201	9206	9212	9217	9222	9227	9232	9238
84	9243	9248	9253	9258	9263	9269	9274	9279	9284	9289
85	9294	9299	9304	9309	9315	9320	9325	9330	9335	9340
86	9345	9350	9355	9360	9365	9370	9375	9380	9385	9390
87	9395	9400	9405	9410	9415	9420	9425	9430	9435	9440
88	9445	9450	9455	9460	9465	9469	9474	9479	9484	9489
89	9494	9499	9504	9509	9513	9518	9523	9528	9533	9538
90	9542	9547	9552	9557	9562	9566	9571	9576	9581	9586
91	9590	9595	9600	9605	9609	9614	9619	9624	9628	9633
92	9638	9643	9647	9652	9657	9661	9666	9671	9675	9680
93	9685	9689	9694	9699	9703	9708	9713	9717	9722	9727
94	9731	9736	9741	9745	9750	9754	9759	9763	9768	9773
95	9777	9782	9786	9791	9795	9800	9805	9809	9814	9818
96	9823	9827	9832	9836	9841	9845	9850	9854	9859	9863
97	9868	9872	9877	9881	9886	9890	9894	9899	9903	9908
98	9912	9917	9921	9926	9930	9934	9939	9943	9948	9952
99	9956	9961	9965	9969	9974	9978	9983	9987	9991	9996

TABLE VI   Natural Logarithms
(ln x)

x		0	1	2	3	4	5	6	7	8	9
1.0	0.0	0000	0995	1980	2956	3922	4879	5827	6766	7696	8618
1.1		9531	*0436	*1333	*2222	*3103	*3976	*4842	*5700	*6551	*7395
1.2	0.1	8232	9062	9885	*0701	*1511	*2314	*3111	*3902	*4686	*5464
1.3	0.2	6236	7003	7763	8518	9267	*0010	*0748	*1481	*2208	*2930
1.4	0.3	3647	4359	5066	5767	6464	7156	7844	8526	9204	9878
1.5	0.4	0547	1211	1871	2527	3178	3825	4469	5108	5742	6373
1.6		7000	7623	8243	8858	9470	*0078	*0682	*1282	*1879	*2473
1.7	0.5	3063	3649	4232	4812	5389	5962	6531	7098	7661	8222
1.8		8779	9333	9884	*0432	*0977	*1519	*2058	*2594	*3127	*3658
1.9	0.6	4185	4710	5233	5752	6269	6783	7294	7803	8310	8813
2.0		9315	9813	*0310	*0804	*1295	*1784	*2271	*2755	*3237	*3716
2.1	0.7	4194	4669	5142	5612	6081	6547	7011	7473	7932	8390
2.2		8846	9299	9751	*0200	*0648	*1093	*1536	*1978	*2418	*2855
2.3	0.8	3291	3725	4157	4587	5015	5442	5866	6289	6710	7129
2.4		7547	7963	8377	8789	9200	9609	*0016	*0422	*0826	*1228
2.5	0.9	1629	2028	2426	2822	3216	3609	4001	4391	4779	5166
2.6		5551	5935	6317	6698	7078	7456	7833	8208	8582	8954
2.7		9325	9695	*0063	*0430	*0796	*1160	*1523	*1885	*2245	*2604
2.8	1.0	2962	3318	3674	4028	4380	4732	5082	5431	5779	6126
2.9		6471	6815	7158	7500	7841	8181	8519	8856	9192	9527
3.0		9861	*0194	*0526	*0856	*1186	*1514	*1841	*2168	*2493	*2817
3.1	1.1	3140	3462	3783	4103	4422	4740	5057	5373	5688	6002
3.2		6315	6627	6938	7248	7557	7865	8173	8479	8784	9089
3.3		9392	9695	9996	*0297	*0597	*0896	*1194	*1491	*1788	*2083
3.4	1.2	2378	2671	2964	3256	3547	3837	4127	4415	4703	4990
3.5		5276	5562	5846	6130	6413	6695	6976	7257	7536	7815
3.6		8093	8371	8647	8923	9198	9473	9746	*0019	*0291	*0563
3.7	1.3	0833	1103	1372	1641	1909	2176	2442	2708	2972	3237
3.8		3500	3763	4025	4286	4547	4807	5067	5325	5584	5841
3.9		6098	6354	6609	6864	7118	7372	7624	7877	8128	8379
4.0		8629	8879	9128	9377	9624	9872	*0118	*0364	*0610	*0854
4.1	1.4	1099	1342	1585	1828	2070	2311	2552	2792	3031	3270
4.2		3508	3746	3984	4220	4456	4692	4927	5161	5395	5629
4.3		5862	6094	6326	6557	6787	7018	7247	7476	7705	7933
4.4		8160	8387	8614	8840	9065	9290	9515	9739	9962	*0185
4.5	1.5	0408	0630	0851	1072	1293	1513	1732	1951	2170	2388
4.6		2606	2823	3039	3256	3471	3687	3902	4116	4330	4543
4.7		4756	4969	5181	5393	5604	5814	6025	6235	6444	6653
4.8		6862	7070	7277	7485	7691	7898	8104	8309	8515	8719
4.9		8924	9127	9331	9534	9737	9939	*0141	*0342	*0543	*0744
5.0	1.6	0944	1144	1343	1542	1741	1939	2137	2334	2531	2728
5.1		2924	3120	3315	3511	3705	3900	4094	4287	4481	4673
5.2		4866	5058	5250	5441	5632	5823	6013	6203	6393	6582
5.3		6771	6959	7147	7335	7523	7710	7896	8083	8269	8455
5.4		8640	8825	9010	9194	9378	9562	9745	9928	*0111	*0293

ln(10) = 2.30258 . . . ln(100) = 4.60517 . . . ln(1000) = 6.90775 . . .

Note. The asterisk (*) indicates that the first two digits are those at the beginning of the next row.

TABLE VI Natural Logarithms
(ln x)

x		0	1	2	3	4	5	6	7	8	9
5.5	1.7	0475	0656	0838	1019	1199	1380	1560	1740	1919	2098
5.6		2277	2455	2633	2811	2988	3166	3342	3519	3695	3871
5.7		4047	4222	4397	4572	4746	4920	5094	5267	5440	5613
5.8		5786	5958	6130	6302	6473	6644	6815	6985	7156	7326
5.9		7495	7665	7834	8002	8171	8339	8507	8675	8842	9009
6.0	1.7	9176	9342	9509	9675	9840	*0006	*0171	*0336	*0500	*0665
6.1	1.8	0829	0993	1156	1319	1482	1645	1808	1970	2132	2294
6.2		2455	2616	2777	2938	3098	3258	3418	3578	3737	3896
6.3		4055	4214	4372	4530	4688	4845	5003	5160	5317	5473
6.4		5630	5786	5942	6097	6253	6408	6563	6718	6872	7026
6.5		7180	7334	7487	7641	7794	7947	8099	8251	8403	8555
6.6		8707	8858	9010	9160	9311	9462	9612	9762	9912	*0061
6.7	1.9	0211	0360	0509	0658	0806	0954	1102	1250	1398	1545
6.8		1692	1839	1986	2132	2279	2425	2571	2716	2862	3007
6.9		3152	3297	3442	3586	3730	3874	4018	4162	4305	4448
7.0		4591	4734	4876	5019	5161	5303	5445	5586	5727	5869
7.1		6009	6150	6291	6431	6571	6711	6851	6991	7130	7269
7.2		7408	7547	7685	7824	7962	8100	8238	8376	8513	8650
7.3		8787	8924	9061	9198	9334	9470	9606	9742	9877	*0013
7.4	2.0	0148	0283	0418	0553	0687	0821	0956	1089	1223	1357
7.5		1490	1624	1757	1890	2022	2155	2287	2419	2551	2683
7.6		2815	2946	3078	3209	3340	3471	3601	3732	3862	3992
7.7		4122	4252	4381	4511	4640	4769	4898	5027	5156	5284
7.8		5412	5540	5668	5796	5924	6051	6179	6306	6433	6560
7.9		6686	6813	6939	7065	7191	7317	7443	7568	7694	7819
8.0		7944	8069	8194	8318	8443	8567	8691	8815	8939	9063
8.1		9186	9310	9433	9556	9679	9802	9924	*0047	*0169	*0291
8.2	2.1	0413	0535	0657	0779	0900	1021	1142	1263	1384	1505
8.3		1626	1746	1866	1986	2106	2226	2346	2465	2585	2704
8.4		2823	2942	3061	3180	3298	3417	3535	3653	3771	3889
8.5		4007	4124	4242	4359	4476	4593	4710	4827	4943	5060
8.6		5176	5292	5409	5524	5640	5756	5871	5987	6102	6217
8.7		6332	6447	6562	6677	6791	6905	7020	7134	7248	7361
8.8		7475	7589	7702	7816	7929	8042	8155	8267	8380	8493
8.9		8605	8717	8830	8942	9054	9165	9277	9389	9500	9611
9.0		9722	9834	9944	*0055	*0166	*0276	*0387	*0497	*0607	*0717
9.1	2.2	0827	0937	1047	1157	1266	1375	1485	1594	1703	1812
9.2		1920	2029	2138	2246	2354	2462	2570	2678	2786	2894
9.3		3001	3109	3216	3324	3431	3538	3645	3751	3858	3965
9.4		4071	4177	4284	4390	4496	4601	4707	4813	4918	5024
9.5		5129	5234	5339	5444	5549	5654	5759	5863	5968	6072
9.6		6176	6280	6384	6488	6592	6696	6799	6903	7006	7109
9.7		7213	7316	7419	7521	7624	7727	7829	7932	8034	8136
9.8		8238	8340	8442	8544	8646	8747	8849	8950	9051	9152
9.9		9253	9354	9455	9556	9657	9757	9858	9958	*0058	*0158
10.0	2.3	0259	0358	0458	0558	0658	0757	0857	0956	1055	1154

TABLE VII   Exponential and
Hyperbolic Functions

x	e^x	e^{-x}	$\sinh x$	$\cosh x$	$\tanh x$
0.00	1.0000	1.00000	0.0000	1.0000	.00000
0.01	1.0101	0.99005	0.0100	1.0001	.01000
0.02	1.0202	.98020	0.0200	1.0002	.02000
0.03	1.0305	.97045	0.0300	1.0005	.02999
0.04	1.0408	.96079	0.0400	1.0008	.03998
0.05	1.0513	.95123	0.0500	1.0013	.04996
0.06	1.0618	.94176	0.0600	1.0018	.05993
0.07	1.0725	.93239	0.0701	1.0025	.06989
0.08	1.0833	.92312	0.0801	1.0032	.07983
0.09	1.0942	.91393	0.0901	1.0041	.08976
0.10	1.1052	.90484	0.1002	1.0050	.09967
0.11	1.1163	.89583	0.1102	1.0061	.10956
0.12	1.1275	.88692	0.1203	1.0072	.11943
0.13	1.1388	.87809	0.1304	1.0085	.12927
0.14	1.1503	.86936	0.1405	1.0098	.13909
0.15	1.1618	.86071	0.1506	1.0113	.14889
0.16	1.1735	.85214	0.1607	1.0128	.15865
0.17	1.1853	.84366	0.1708	1.0145	.16838
0.18	1.1972	.83527	0.1810	1.0162	.17808
0.19	1.2092	.82696	0.1911	1.0181	.18775
0.20	1.2214	.81873	0.2013	1.0201	.19738
0.21	1.2337	.81058	0.2115	1.0221	.20697
0.22	1.2461	.80252	0.2218	1.0243	.21652
0.23	1.2586	.79453	0.2320	1.0266	.22603
0.24	1.2712	.78663	0.2423	1.0289	.23550
0.25	1.2840	.77880	0.2526	1.0314	.24492
0.26	1.2969	.77105	0.2629	1.0340	.25430
0.27	1.3100	.76338	0.2733	1.0367	.26362
0.28	1.3231	.75578	0.2837	1.0395	.27291
0.29	1.3364	.74826	0.2941	1.0423	.28213
0.30	1.3499	.74082	0.3045	1.0453	.29131
0.31	1.3634	.73345	0.3150	1.0484	.30044
0.32	1.3771	.72615	0.3255	1.0516	.30951
0.33	1.3910	.71892	0.3360	1.0549	.31852
0.34	1.4049	.71177	0.3466	1.0584	.32748
0.35	1.4191	.70469	0.3572	1.0619	.33638
0.36	1.4333	.69768	0.3678	1.0655	.34521
0.37	1.4477	.69073	0.3785	1.0692	.35399
0.38	1.4623	.68386	0.3892	1.0731	.36271
0.39	1.4770	.67706	0.4000	1.0770	.37136
0.40	1.4918	.67032	0.4108	1.0811	.37995
0.41	1.5068	.66365	0.4216	1.0852	.38847
0.42	1.5220	.65705	0.4325	1.0895	.39693
0.43	1.5373	.65051	0.4434	1.0939	.40532
0.44	1.5527	.64404	0.4543	1.0984	.41364
0.45	1.5683	.63763	0.4653	1.1030	.42190
0.46	1.5841	.63128	0.4764	1.1077	.43008
0.47	1.6000	.62500	0.4875	1.1125	.43820
0.48	1.6161	.61878	0.4986	1.1174	.44624
0.49	1.6323	.61263	0.5098	1.1225	.45422

TABLE VII   Exponential and
Hyperbolic Functions

x	e^x	e^{-x}	$\sinh x$	$\cosh x$	$\tanh x$
0.50	1.6487	.60653	0.5211	1.1276	.46212
0.51	1.6653	.60050	0.5324	1.1329	.46995
0.52	1.6820	.59452	0.5438	1.1383	.47770
0.53	1.6989	.58860	0.5552	1.1438	.48538
0.54	1.7160	.58275	0.5666	1.1494	.49299
0.55	1.7333	.57695	0.5782	1.1551	.50052
0.56	1.7507	.57121	0.5897	1.1609	.50798
0.57	1.7683	.56553	0.6014	1.1669	.51536
0.58	1.7860	.55990	0.6131	1.1730	.52267
0.59	1.8040	.55433	0.6248	1.1792	.52990
0.60	1.8221	.54881	0.6367	1.1855	.53705
0.61	1.8404	.54335	0.6485	1.1919	.54413
0.62	1.8589	.53794	0.6605	1.1984	.55113
0.63	1.8776	.53259	0.6725	1.2051	.55805
0.64	1.8965	.52729	0.6846	1.2119	.56490
0.65	1.9155	.52205	0.6967	1.2188	.57167
0.66	1.9348	.51685	0.7090	1.2258	.57836
0.67	1.9542	.51171	0.7213	1.2330	.58498
0.68	1.9739	.50662	0.7336	1.2402	.59152
0.69	1.9937	.50158	0.7461	1.2476	.59798
0.70	2.0138	.49659	0.7586	1.2552	.60437
0.71	2.0340	.49164	0.7712	1.2628	.61068
0.72	2.0544	.48675	0.7838	1.2706	.61691
0.73	2.0751	.48191	0.7966	1.2785	.62307
0.74	2.0959	.47711	0.8094	1.2865	.62915
0.75	2.1170	.47237	0.8223	1.2947	.63515
0.76	2.1383	.46767	0.8353	1.3030	.64108
0.77	2.1598	.46301	0.8484	1.3114	.64693
0.78	2.1815	.45841	0.8615	1.3199	.65271
0.79	2.2034	.45384	0.8748	1.3286	.65841
0.80	2.2255	.44933	0.8881	1.3374	.66404
0.81	2.2479	.44486	0.9015	1.3464	.66959
0.82	2.2705	.44043	0.9150	1.3555	.67507
0.83	2.2933	.43605	0.9286	1.3647	.68048
0.84	2.3164	.43171	0.9423	1.3740	.68581
0.85	2.3396	.42741	0.9561	1.3835	.69107
0.86	2.3632	.42316	0.9700	1.3932	.69626
0.87	2.3869	.41895	0.9840	1.4029	.70137
0.88	2.4109	.41478	0.9981	1.4128	.70642
0.89	2.4351	.41066	1.0122	1.4229	.71139
0.90	2.4596	.40657	1.0265	1.4331	.71630
0.91	2.4843	.40252	1.0409	1.4434	.72113
0.92	2.5093	.39852	1.0554	1.4539	.72590
0.93	2.5345	.39455	1.0700	1.4645	.73059
0.94	2.5600	.39063	1.0847	1.4753	.73522
0.95	2.5857	.38674	1.0995	1.4862	.73978
0.96	2.6117	.38289	1.1144	1.4973	.74428
0.97	2.6379	.37908	1.1294	1.5085	.74870
0.98	2.6645	.37531	1.1446	1.5199	.75307
0.99	2.6912	.37158	1.1598	1.5314	.75736

TABLE VII   Exponential and
Hyperbolic Functions

x	e^x	e^{-x}	$\sinh x$	$\cosh x$	$\tanh x$
1.00	2.7183	.36788	1.1752	1.5431	.76159
1.10	3.0042	.33287	1.3356	1.6685	.80050
1.20	3.3201	.30119	1.5095	1.8107	.83365
1.30	3.6693	.27253	1.6984	1.9709	.86172
1.40	4.0552	.24660	1.9043	2.1509	.88535
1.50	4.4817	.22313	2.1293	2.3524	.90515
1.60	4.9530	.20190	2.3756	2.5775	.92167
1.70	5.4739	.18268	2.6456	2.8283	.93541
1.80	6.0496	.16530	2.9422	3.1075	.94681
1.90	6.6859	.14957	3.2682	3.4177	.95624
2.00	7.3891	.13534	3.6269	3.7622	.96403
2.10	8.1662	.12246	4.0219	4.1443	.97045
2.20	9.0250	.11080	4.4571	4.5679	.97574
2.30	9.9742	.10026	4.9370	5.0372	.98010
2.40	11.023	.09072	5.4662	5.5569	.98367
2.50	12.182	.08208	6.0502	6.1323	.98661
2.60	13.464	.07427	6.6947	6.7690	.98903
2.70	14.880	.06721	7.4063	7.4735	.99101
2.80	16.445	.06081	8.1919	8.2527	.99263
2.90	18.174	.05502	9.0596	9.1146	.99396
3.00	20.086	.04979	10.018	10.068	.99505
3.10	22.198	.04505	11.076	11.122	.99595
3.20	24.533	.04076	12.246	12.287	.99668
3.30	27.113	.03688	13.538	13.575	.99728
3.40	29.964	.03337	14.965	14.999	.99777
3.50	33.115	.03020	16.543	16.573	.99818
3.60	36.598	.02732	18.286	18.313	.99851
3.70	40.447	.02472	20.211	20.236	.99878
3.80	44.701	.02237	22.339	22.362	.99900
3.90	49.402	.02024	24.691	24.711	.99918
4.00	54.598	.01832	27.290	27.308	.99933
4.10	60.340	.01657	30.162	30.178	.99945
4.20	66.686	.01500	33.336	33.351	.99955
4.30	73.700	.01357	36.843	36.857	.99963
4.40	81.451	.01227	40.719	40.732	.99970
4.50	90.017	.01111	45.003	45.014	.99975
4.60	99.484	.01005	49.737	49.747	.99980
4.70	109.95	.00910	54.969	54.978	.99983
4.80	121.51	.00823	60.751	60.759	.99986
4.90	134.29	.00745	67.141	67.149	.99989
5.00	148.41	.00674	74.203	74.210	.99991
5.10	164.02	.00610	82.008	82.014	.99993
5.20	181.27	.00552	90.633	90.639	.99994
5.30	200.34	.00499	100.17	100.17	.99995
5.40	221.41	.00452	110.70	110.71	.99996
5.50	244.69	.00409	122.34	122.35	.99997
5.60	270.43	.00370	135.21	135.22	.99997
5.70	298.87	.00335	149.43	149.44	.99998
5.80	330.30	.00303	165.15	165.15	.99998
5.90	365.04	.00274	182.52	182.52	.99998

TABLE VII   Exponential and
Hyperbolic Functions

x	e^x	e^{-x}	$\sinh x$	$\cosh x$	$\tanh x$
6.00	403.43	.00248	201.71	201.72	.99999
6.25	518.01	.00193	259.01	259.01	.99999
6.50	665.14	.00150	332.57	332.57	1.0000
6.75	854.06	.00117	427.03	427.03	1.0000
7.00	1096.6	.00091	548.32	548.32	1.0000
7.50	1808.0	.00055	904.02	904.02	1.0000
8.00	2981.0	.00034	1490.5	1490.5	1.0000
8.50	4914.8	.00020	2457.4	2457.4	1.0000
9.00	8103.1	.00012	4051.5	4051.5	1.0000
9.50	13360.	.00007	6679.9	6679.9	1.0000
10.00	22026.	.00005	11013.	11013.	1.0000

TABLE VIII   A Table of Integrals

(1) $\displaystyle\int x^n\,dx = \frac{1}{n+1}x^{n+1} + C$ for $n \neq -1$.

(2) $\displaystyle\int \frac{1}{x}\,dx = \ln|x| + C$.

(3) $\displaystyle\int \sin x\,dx = -\cos x + C$.

(4) $\displaystyle\int \cos x\,dx = \sin x + C$.

(5) $\displaystyle\int e^x\,dx = e^x + C$.

(6) $\displaystyle\int b^x\,dx = \frac{1}{\ln b}b^x + C$ for $b > 0$.

(7) $\displaystyle\int \sec^2 x\,dx = \tan x + C$.

(8) $\displaystyle\int \csc^2 x\,dx = -\cot x + C$.

(9) $\displaystyle\int \sec x \tan x\,dx = \sec x + C$.

(10) $\displaystyle\int \csc x \cot x\,dx = -\csc x + C$.

(11) $\displaystyle\int \tan x\,dx = -\ln|\cos x| + C$.

(12) $\displaystyle\int \cot x\,dx = \ln|\sin x| + C$.

TABLE VIII   A Table of Integrals

(13a) $\displaystyle\int \sec x \, dx = \ln |\sec x + \tan x| + C = \ln \left| \tan\left(\frac{x}{2} + \frac{\pi}{4}\right) \right| + C.$

(13b) $\displaystyle\int \csc x \, dx = -\ln |\csc x + \cot x| + C = \ln \left| \tan\left(\frac{x}{2}\right) \right| + C.$

(14) $\displaystyle\int \frac{1}{x^2 + a^2} \, dx = \frac{1}{a} \arctan\left(\frac{x}{a}\right) + C.$

(15) $\displaystyle\int \frac{1}{x^2 - a^2} \, dx = \frac{1}{2a} \ln \left| \frac{x - a}{x + a} \right| + C.$

(16) $\displaystyle\int \frac{1}{\sqrt{a^2 - x^2}} \, dx = \arcsin\left(\frac{x}{a}\right) + C \quad \text{for } |x| < |a|.$

(17) $\displaystyle\int \frac{1}{\sqrt{x^2 + a^2}} \, dx = \ln(x + \sqrt{x^2 + a^2}) + C.$

(18) $\displaystyle\int \frac{1}{\sqrt{x^2 - a^2}} \, dx = \ln |x + \sqrt{x^2 - a^2}| + C \quad \text{for } |x| > |a|.$

(19) $\displaystyle\int \frac{1}{x(ax + b)} \, dx = \frac{1}{b} \ln \left| \frac{x}{ax + b} \right| + C.$

(20) $\displaystyle\int \frac{1}{x^2(ax + b)} \, dx = -\frac{1}{bx} + \frac{a}{b^2} \ln \left| \frac{ax + b}{x} \right| + C.$

(21) $\displaystyle\int x \sqrt{ax + b} \, dx = \frac{6ax - 4b}{15a^2} (ax + b)^{3/2} + C.$

(22) $\displaystyle\int \frac{1}{\sqrt{ax + b}} \, dx = \frac{2}{a} \sqrt{ax + b} + C.$

(23) $\displaystyle\int \frac{x}{\sqrt{ax + b}} \, dx = \frac{2ax - 4b}{3a^2} \sqrt{ax + b} + C.$

(24) $\displaystyle\int \sqrt{a^2 - x^2} \, dx = \frac{1}{2} x \sqrt{a^2 - x^2} + \frac{1}{2} a^2 \arcsin\left(\frac{x}{a}\right) + C.$

(25) $\displaystyle\int \sqrt{x^2 \pm a^2} \, dx = \tfrac{1}{2} x \sqrt{x^2 \pm a^2} \pm \tfrac{1}{2} a^2 \ln(x + \sqrt{x^2 \pm a^2}) + C.$

(26) $\displaystyle\int \frac{1}{(a^2 - x^2)^{3/2}} \, dx = \frac{x}{a^2 \sqrt{a^2 - x^2}} + C.$

TABLE VIII   A Table of Integrals

(27) $\displaystyle\int \frac{1}{(ax+b)(cx+d)}\,dx = \frac{1}{bc-ad}\ln\left|\frac{cx+d}{ax+b}\right| + C$ for $bc-ad \neq 0$.

(28) $\displaystyle\int \cos^2 x\,dx = \frac{1}{2}(x + \sin x \cos x) + C = \frac{x}{2} + \frac{1}{4}\sin(2x) + C$.

(29) $\displaystyle\int \sin^2 x\,dx = \frac{1}{2}(x - \sin x \cos x) + C = \frac{x}{2} - \frac{1}{4}\sin(2x) + C$.

(30) $\displaystyle\int \sin^2 x \cos^2 x\,dx = \frac{1}{8}(x - \sin x \cos^3 x + \sin^3 x \cos x) + C$.

(31) $\displaystyle\int \sin^3 x \cos^2 x\,dx = -\frac{1}{3}\cos^3 x + \frac{1}{5}\cos^5 x + C$.

(32) $\displaystyle\int \sin^3 x\,dx = -\cos x + \frac{1}{3}\cos^3 x + C$.

(33) $\displaystyle\int \sin^3 x \cos^4 x\,dx = -\frac{1}{5}\cos^5 x + \frac{1}{7}\cos^7 x + C$.

(34) $\displaystyle\int \tan^2 x\,dx = \tan x - x + C$.

(35) $\displaystyle\int \cot^2 x\,dx = -\cot x - x + C$.

(36) $\displaystyle\int \cos^m x \sin^n x\,dx = \frac{\cos^{m-1} x \sin^{n+1} x}{m+n} + \frac{m-1}{m+n}\int \cos^{m-2} x \sin^n x\,dx$

for integers $n \geq 0$ and $m \geq 2$.

(37) $\displaystyle\int \sin(ax)\sin(bx)\,dx = \frac{\sin(a-b)x}{2(a-b)} - \frac{\sin(a+b)x}{2(a+b)} + C$ for $|a| \neq |b|$.

(38) $\displaystyle\int \sin(ax)\cos(bx)\,dx = -\frac{\cos(a-b)x}{2(a-b)} - \frac{\cos(a+b)x}{2(a+b)} + C$ for $|a| \neq |b|$.

(39) $\displaystyle\int \cos(ax)\cos(bx)\,dx = \frac{\sin(a-b)x}{2(a-b)} + \frac{\sin(a+b)x}{2(a+b)} + C$ for $|a| \neq |b|$.

(40) $\displaystyle\int x\sin(ax)\,dx = \frac{1}{a^2}\sin(ax) - \frac{x}{a}\cos(ax) + C$.

(41) $\displaystyle\int x\cos(ax)\,dx = \frac{1}{a^2}\cos(ax) + \frac{x}{a}\sin(ax) + C$.

(42) $\displaystyle\int x^n \sin(ax)\,dx = -\frac{1}{a}x^n \cos(ax) + \frac{n}{a}\int x^{n-1}\cos(ax)\,dx$ for positive integers n.

TABLE VIII   A Table of Integrals

(43) $\displaystyle\int x^n \cos(ax)\, dx = \frac{1}{a} x^n \sin(ax) - \frac{n}{a} \int x^{n-1} \sin(ax)\, dx$ for positive integers n.

(44) $\displaystyle\int x e^{ax}\, dx = e^{ax}\left(\frac{x}{a} - \frac{1}{a^2}\right) + C.$

(45) $\displaystyle\int x^n e^{ax}\, dx = \frac{1}{a} x^n e^{ax} - \frac{n}{a} \int x^{n-1} e^{ax}\, dx$ for positive integers n.

(46) $\displaystyle\int e^{ax} \sin(bx)\, dx = \frac{e^{ax}}{a^2 + b^2} [a \sin(bx) - b \cos(bx)] + C.$

(47) $\displaystyle\int e^{ax} \cos(bx)\, dx = \frac{e^{ax}}{a^2 + b^2} [a \cos(bx) + b \sin(bx)] + C.$

(48) $\displaystyle\int \ln x\, dx = x \ln x - x + C.$

(49) $\displaystyle\int x \ln x\, dx = \frac{x^2}{2} \ln x - \frac{x^2}{4} + C.$

(50) $\displaystyle\int \sinh x\, dx = \cosh x + C.$

(51) $\displaystyle\int \cosh x\, dx = \sinh x + C.$

(52) $\displaystyle\int \arcsin(ax)\, dx = x \arcsin(ax) + \frac{1}{a} \sqrt{1 - (ax)^2} + C.$

(53) $\displaystyle\int \arctan(ax)\, dx = x \arctan(ax) - \frac{1}{2a} \ln\left(1 + (ax)^2\right) + C.$

(54) $\displaystyle\int \frac{1}{ax^2 + bx + c} = \frac{1}{\sqrt{b^2 - 4ac}} \ln\left|\frac{2ax + b - \sqrt{b^2 - 4ac}}{2ax + b + \sqrt{b^2 - 4ac}}\right| + C$ for $b^2 > 4ac,$

(55) $\displaystyle\int \frac{1}{ax^2 + bx + c}\, dx = \frac{2}{\sqrt{4ac - b^2}} \arctan\left[\frac{2ax + b}{\sqrt{4ac - b^2}}\right] + C$ for $b^2 < 4ac,$

(56) $\displaystyle\int \frac{1}{ax^2 + bx + c} = -\frac{2}{2ax + b} + C$ for $b^2 = 4ac.$

(57) $\displaystyle\int \frac{1}{(x^2 + a^2)^n}\, dx = \frac{1}{2a^2(n-1)} \frac{x}{(x^2 + a^2)^{n-1}}$

$$+ \frac{2n - 3}{2a^2(n-1)} \int \frac{1}{(x^2 + a^2)^{n-1}}\, dx \, (n = 2, 3, 4, \ldots).$$

ANSWERS AND OUTLINES OF SOLUTIONS

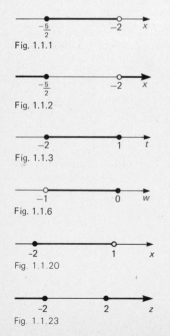

Fig. 1.1.1

$-\frac{5}{2}$ -2 x

Fig. 1.1.2

$-\frac{5}{2}$ -2 x

Section 1.1

1. Fig. 1.1.1 **2.** Fig. 1.1.2 **3.** Fig 1.1.3 **4.** Fig. 1.1.4 **5.** Fig. 1.1.5 **6.** Fig. 1.1.6
11. Fig. 1.1.11 **13.** $x - (-4) = x + 4$ **15.**‡ Midpoint $= (4 + 10)/2 = 7$;
Half-length $= (10 - 4)/2 = 3$; $|x - 7| < 3$ **16.**‡ Midpoint $= 3$; Half-
length $= 4$; $|t - 3| \geq 4$ **18a.** $-10 < x < 5$ **18c.** $x > -50$
18e. $-2 < x < 2$ **19.** Fig. 1.1.19 **20.** Fig. 1.1.20 **21.** Fig. 1.1.21 **22.** Fig. 1.1.22
23. Fig. 1.1.23 **24.** Fig. 1.1.24 **25a.** 7 **25b.** $\frac{19}{15}$ **26.** $\frac{7}{2}, \frac{1}{2}$ **27.** $-4, 0, 4$
28. $\pm 3, \pm 2$ **29.** $3, \frac{1}{2}$ **30.** $x \geq 0$ **31.** $1, -1$

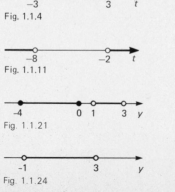

Fig. 1.1.3 | Fig. 1.1.4 | Fig. 1.1.5

Fig. 1.1.6 | Fig. 1.1.11 | Fig. 1.1.19

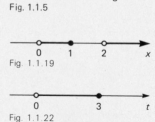

Fig. 1.1.20 | Fig. 1.1.21 | Fig. 1.1.22

Fig. 1.1.23 | Fig. 1.1.24

Section 1.2

1. $-3 \le x < 1$; Fig. 1.2.1   **3.** $-12 \le w \le 20$; Fig. 1.2.3   **4.**[‡] $5 - 2z \ge 7$ or $5 - 2z \le -7$; $2z \le -2$ or $2z \ge 12$; $z \le -1$ or $z \ge 6$; Fig. 1.2.4   **5.** $y > \frac{4}{3}$ or $y \le -2$; Fig. 1.2.5   **7.** $x > -\frac{7}{5}$; Fig. 1.2.7   **9.** $x < -\frac{1}{2}$ or $x > \frac{17}{6}$; Fig. 1.2.9 **11.** $v \ge 2$ or $v \le 1$; Fig. 1.2.11   **13.**[‡] $0 < 3x + 1$ and $3x + 1 \le 4x - 6$; $-1 < 3x$ and $7 \le x$; $x > -\frac{1}{3}$ and $x \ge 7$; $x \ge 7$; Fig. 1.2.13   **14.** $t > 2$ or $t \le -5$; Fig. 1.2.14   **16.** $-1 \le x \le 1$; Fig. 1.2.16   **17.** $y > \frac{7}{2}$ or $y < -\frac{1}{2}$; Fig. 1.2.17   **18.** $-3 < y \le 0$; Fig. 1.2.18   **19.** $t \ne 1$; Fig. 1.2.19   **20.** $t \le 9$; Fig. 1.2.20   **21.** $3 < x < 4$; Fig. 1.2.21

Fig. 1.2.1

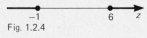
Fig. 1.2.3

Fig. 1.2.4

Fig. 1.2.5

Fig. 1.2.7

Fig. 1.2.9

Fig. 1.2.11

Fig. 1.2.13

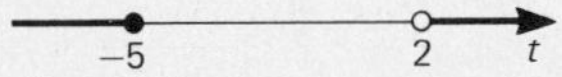

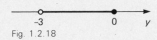
Fig. 1.2.14

Fig. 1.2.16

Fig. 1.2.17

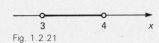
Fig. 1.2.18

Fig. 1.2.19

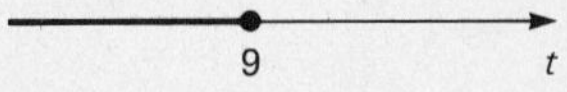

Fig. 1.2.20

Fig. 1.2.21

Section 1.3

1. Distance $= \sqrt{97}$; Fig. 1.3.1   **3.** Distance $= \sqrt{164}$; Fig. 1.3.3   **5.** $(\frac{3}{2}, -\frac{3}{2})$; Fig. 1.3.5   **7.** $Q(3, -4)$ **9.** $(4,3)$; $(4,-3)$ **11.** $\overline{PQ} = 9\sqrt{2}$; $\overline{QR} = 5\sqrt{5}$; $\overline{PR} = 5\sqrt{5}$; Isosceles; Fig. 1.3.11   **13.** $\overline{AB} = 5$; $\overline{BC} = \sqrt{61}$; $\overline{AC} = \sqrt{26}$; Not a right triangle; Fig. 1.3.13   **14.**[‡] $\overline{AB}^2 + \overline{BC}^2 = \overline{AC}^2$; $(3 - 0)^2 + (1 - 0)^2 + (2 - 3)^2 + (k - 1)^2 = (k - 0)^2 + (2 - 0)^2$; $k = 4$ **15.** $k = 2$; Rhombus

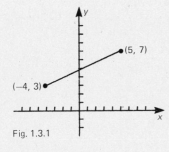
Fig. 1.3.1

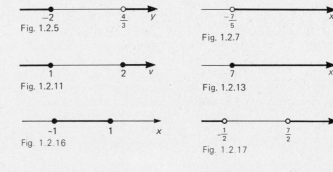

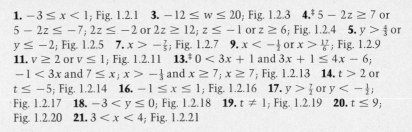

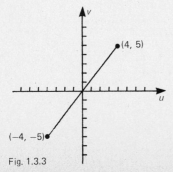
Fig. 1.3.3

Fig. 1.3.5

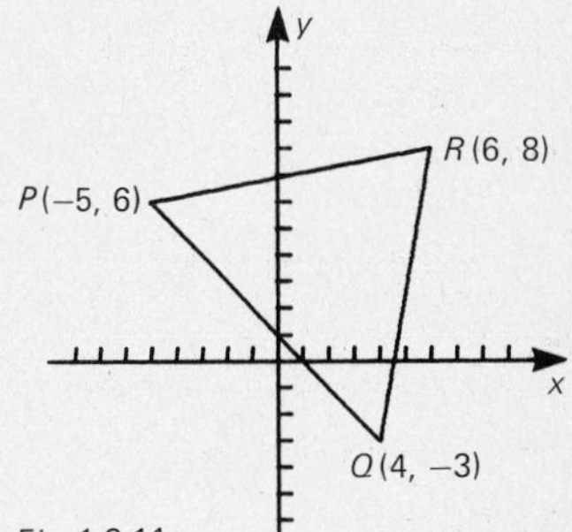

Fig. 1.3.11

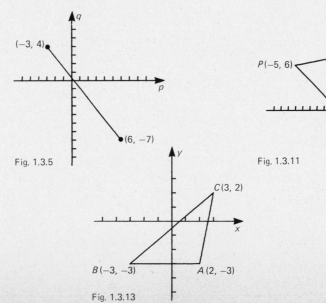
Fig. 1.3.13

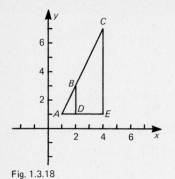

Fig. 1.3.18

17. $\overline{AB} = \sqrt{5}$; $\overline{BC} = 2\sqrt{5}$; $\overline{AC} = 3\sqrt{5}$; $\overline{AB} + \overline{BC} = \overline{AC}$; B lies on the line segment AC. **18.** Let $D = (2,1)$, $E = (4,1)$; $\overline{AE}/\overline{AD} = \overline{CE}/\overline{BD} = 3$; ΔCAE and ΔBAD are similar; $\langle CAE = \langle BAD$; B lies on AC; Fig. 1.3.18
19. M = Midpoint of hypotenuse = $(\frac{1}{2}a, \frac{1}{2}b)$; $\overline{MA} = \overline{MB} = \overline{MC} = \frac{1}{2}\sqrt{a^2 + b^2}$
23. $\overline{QR}^2 + \overline{PR}^2 = \overline{PQ}^2$; $(\frac{3}{2}, \frac{7}{2})$ **24.** $\overline{PQ}^2 + \overline{QR}^2 = \overline{PR}^2$; $(1, \frac{11}{2})$ **25.** $\overline{PQ}^2 + \overline{PR}^2 = \overline{QR}^2$; $(4,3)$

Section 1.4

1. Fig. 1.4.1 **3.**$\ddagger$ $y \geq 0$ and $x^2 + y^2 = 4^2$; Semi-circle of radius 4; Fig. 1.4.3
4. Fig. 1.4.4 **5.** Fig. 1.4.5 **7.** Fig. 1.4.7 **9.** Fig. 1.4.9 **11.** Fig. 1.4.11 **13.** No intercepts; Symmetric about the y-axis; Fig. 1.4.13 **15.** x-intercept = 0; y-intercept = 0; Symmetric about the origin; Fig. 1.4.15 **17.** No x-intercept; y-intercept = 1; Symmetric about the y-axis; Fig. 1.4.17 **20.** Fig. 1.4.20
24. x-intercepts = ± 1; y-intercepts = ± 2; Symmetric about the origin
25. x-intercept = -9; y-intercepts = ± 3; Symmetric about the x-axis
26. x-intercepts = ± 1; y-intercepts = $\pm \sqrt{3}$; Symmetric about the x-axis, the y-axis and the origin **27.** x-intercept = -2; y-intercepts = $\pm 2\sqrt{2}$; Symmetric about the x-axis **28.** x-intercepts = ± 2; No y-intercept; Symmetric about the x-axis, the y-axis and the origin

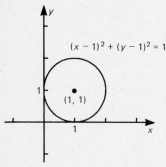

Fig. 1.4.1

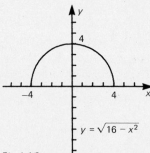

Fig. 1.4.3

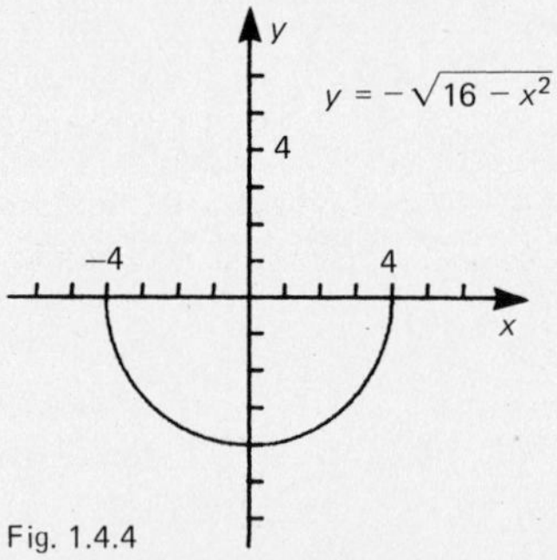

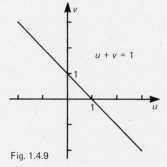

Fig. 1.4.4

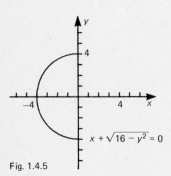

Fig. 1.4.5

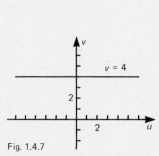

Fig. 1.4.7

Fig. 1.4.9

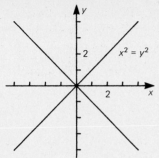

Fig. 1.4.11

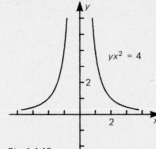

Fig. 1.4.13

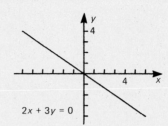

Fig. 1.4.15

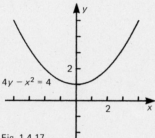

Fig. 1.4.17

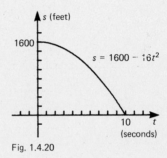

Fig. 1.4.20

Section 1.5

1. -3 **3.** 12 **5.**‡ Yes, since Slope PQ = Slope $QR = -\frac{2}{3}$ **7.** No   **9.**‡ Right triangle since Slope $PR = 6$ and Slope $QR = -\frac{1}{6}$; Area $= \frac{37}{2}$; Fig. 1.5.9
11. Right triangle; Area $= 25$; Fig. 1.5.11   **14.** Slope AD = Slope $BC = 5/k$; Slope AB = Slope $CD = \frac{1}{2}$; Rectangle for $k = -\frac{5}{2}$ **16.** $-\frac{1}{2}$ **18.** $(0,-6)$, $(1,-3)$; Slope $= 3$

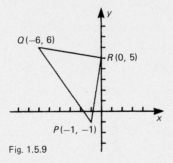

Fig. 1.5.9

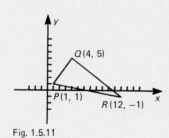

Fig. 1.5.11

Section 1.6

1. $y = -4x + 17$; Fig. 1.6.1   **3.** $y = -\frac{1}{10}x + \frac{3}{2}$; Fig. 1.6.3   **5.** $y = 3$; Fig. 1.6.5
7. $y = -\frac{5}{7}x + 5$; Slope $= -\frac{5}{7}$; Fig. 1.6.7   **9.** $y = 0x - 5$; Slope $= 0$; Fig. 1.6.9
11. $3x - 2y = 5$ **13.** $x/5 - y/6 = 1$; Slope $= \frac{6}{5}$ **15a.** $y = 7$ **15b.** $x = -4$
17.‡ Solve for y in $2x + y = 1$: $y = 1 - 2x$; Substitute in $x + y = 4$:
$x + (1 - 2x) = 4$; $x = -3, y = 7$ **18a.** (0,0), (6,3), (−3,6)   **18b.** $y = \frac{1}{2}x$
and $y = -2x$ are perpendicular.   **18c.** $\frac{45}{2}$ **21.** $P = 62.4\,h$; Fig. 1.6.21

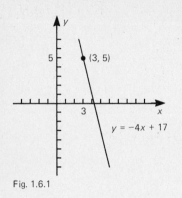
Fig. 1.6.1

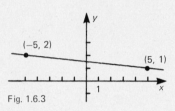
Fig. 1.6.3

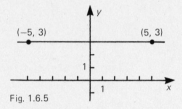
Fig. 1.6.5

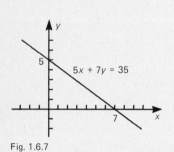
Fig. 1.6.7

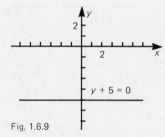
Fig. 1.6.9

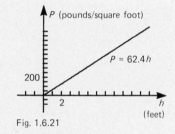
Fig. 1.6.21

Section 1.7

1.‡ Parabola that opens upward with y-intercept -3; Fig. 1.7.1   **2.** Fig. 1.7.2
4.‡ v is always positive and is large for small u and small for large u; Fig. 1.7.4
5. Fig. 1.7.5   **7.** Fig. 1.7.7   **9.**‡ Parabola that opens downward with s-intercept
16; Fig. 1.7.9   **10.** Fig. 1.7.10

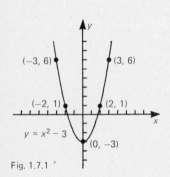
Fig. 1.7.1

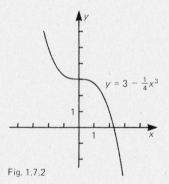
Fig. 1.7.2

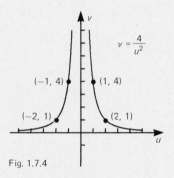
Fig. 1.7.4

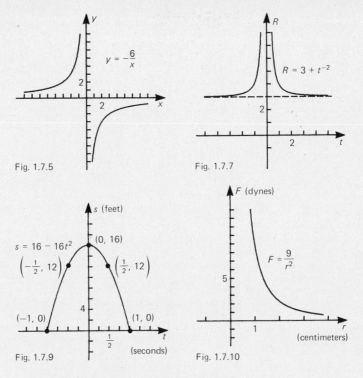

Fig. 1.7.5

Fig. 1.7.7

Fig. 1.7.9

Fig. 1.7.10

Appendix to Chapter 1

1a. $-1, -\frac{1}{9}, -27$ **1b.** $1, \frac{1}{9}, -27$ **1c.** $1, 3, -2$ **1d.** $-1, -1, -64$ **1e.** $1, \frac{1}{9}$, -8 **3a.** $x^{21/5}$ **3b.** $x^{-1/12}$ **3c.** $x^{-181/60}$ **3d.** $x^{-35/3}$ **4.** (b), (h) **5.** (b), (c), (g) **7.** (b), (c), (d), (g), (h) **11a.**‡ $(x^2 + x^{1/2} + 1)x^3 - (x^2 + x^{1/2} + 1)x^{-3}$ $= x^5 + x^{7/2} + x^3 - x^{-1} - x^{-5/2} - x^{-3}$ **11b.** $x^6 + 2x^3 + 1$ **11c.** $x^3 - 3$ $+ 3x^{-3} - x^{-6}$ **11g.** $24x^3 + 34x^2 + x - 5$ **12a.** At each stage of the division the remainder is one of the q numbers $0, 1, \ldots, q - 1$ and the new dividend is obtained by adding the next digit in the original dividend to the remainder. The new dividend must repeat within q steps after the decimal point because the original dividend has only zeros after the decimal point **12b.** Write the decimal as $x = 10^k(a_m a_{m-1} \cdots a_1 . b_1 b_2 \cdots b_n b_1 b_2 \cdots b_n \cdots)$ with k an integer and m and n positive integers and where $b_1 b_2 \cdots b_n$ is the string of digits that repeat. Then $x = 10^k[a_m a_{m-1} \cdots + (b_1 b_2 \cdots b_n/99 \cdots 9)]$ with n 9's.
13. *Step 1* P_1 reads $1 = 1^2$ and is valid; *Step 2* If P_k is true with k a positive integer, then $1 + 3 + \cdots + (2k - 1) = k^2$; Adding $2(k + 1) - 1 = 2k + 1$ to both sides gives $1 + 3 + \cdots + (2k - 1) + (2k + 1) = k^2 + 2k + 1$, which is P_{k+1} because $(k + 1)^2 = k^2 + 2k + 1$ **14.** *Step 1* P_1 reads $1^2 = \frac{1}{6}(1)(1 + 1)[2(1) + 1]$ and is correct; *Step 2* If P_k is true with k a positive integer, then $1^2 + 2^2 + \cdots + k^2 = \frac{1}{6}(k)(k + 1)(2k + 1)$; Adding $(k + 1)^2$ to both sides gives $1^2 + 2^2 + \cdots + k^2 + (k + 1)^2$ $= \frac{1}{6}(k)(k + 1)(2k + 1) + (k + 1)^2$, which is P_{k+1} because $\frac{1}{6}(k)(k + 1)(2k + 1) + (k + 1)^2 = (k + 1)[\frac{1}{6}(k)(2k + 1) + (k + 1)]$ $= \frac{1}{6}(k + 1)[k(2k + 1) + 6(k + 1)] = \frac{1}{6}(k + 1)(2k^2 + 7k + 6)$ $= \frac{1}{6}(k + 1)(k + 2)(2k + 3)$ **15.** *Step 1* P_1 reads $1 \cdot 2 = \frac{1}{3}(1)(1 + 1)(1 + 2)$

and is valid; *Step 2* If P_k is valid for a positive integer k, then $1 \cdot 2 + 2 \cdot 3 + 3 \cdot 4 + \cdots + k(k + 1) = \frac{1}{3}k(k + 1)(k + 2)$; We add $(k + 1)(k + 2)$ to both sides to obtain $1 \cdot 2 + 2 \cdot 3 + 3 \cdot 4 + \cdots + k(k + 1) + (k + 1)(k + 2) = \frac{1}{3}k(k + 1)(k + 2) + (k + 1)(k + 2)$, which is P_{k+1} because $\frac{1}{3}k(k + 1)(k + 2) + (k + 1)(k + 2) = \frac{1}{3}(k + 1)[k(k + 2) + 3(k + 2)] = \frac{1}{3}(k + 1)(k^2 + 5k + 6) = \frac{1}{3}(k + 1)(k + 2)(k + 3)$ **16.** *Step 1* P_1 reads $\dfrac{1}{1 \cdot 2} = \dfrac{1}{1 + 1}$ and is true; *Step 2* If P_k is true with k a positive integer, then

$$\frac{1}{1 \cdot 2} + \frac{1}{2 \cdot 3} + \frac{1}{3 \cdot 4} + \cdots + \frac{1}{k(k + 1)} = \frac{k}{k + 1}; \text{ Adding } \frac{1}{(k + 1)(k + 2)}$$

to both sides gives $\dfrac{1}{1 \cdot 2} + \dfrac{1}{2 \cdot 3} + \dfrac{1}{3 \cdot 4} + \cdots + \dfrac{1}{k(k + 1)} + \dfrac{1}{(k + 1)(k + 2)}$

$= \dfrac{k}{k + 1} + \dfrac{1}{(k + 1)(k + 2)}$; This is P_{k+1} because $\dfrac{k}{k + 1} + \dfrac{1}{(k + 1)(k + 2)}$

$= \dfrac{1}{k + 1}\left(k + \dfrac{1}{k + 2}\right) = \dfrac{1}{k + 1}\left(\dfrac{k^2 + 2k + 1}{k + 2}\right) = \dfrac{(k + 1)^2}{(k + 1)(k + 2)}$

$= \dfrac{k + 1}{(k + 1) + 1}$

Miscellaneous Exercises to Chapter 1

1a. $2, \frac{9}{4}, \frac{625}{256}$, undefined, $4, \frac{256}{81}$ **1b.** $256, 64, 32, 1, 4, 8$ **2.** $1, 16, -1$

3. $\dfrac{pq - p - q}{pq + p + q}$

4. $0, 4, -2, -12$ **5.**‡ $2 - 3x = 4$ or $2 - 3x = -4$; $x = -\frac{2}{3}, 2$ **6.** $t = -\frac{8}{3}, -\frac{8}{5}$

8.‡ $x + 1 = 2x - 1$ or $x + 1 = -(2x - 1)$; $x = 2, x = 0$ **10.** $z = 0$

12. $|x| = |x - y + y| \leq |x - y| + |y| \Rightarrow |x| - |y| \leq |x - y|$;

$|y| = |y - x + x| \leq |y - x| + |x| \Rightarrow |y| - |x| \leq |x - y|$ **13.** $-2 < x \leq 5$;

Fig. 1.M.13   **14.** $y \neq -\frac{6}{5}$; Fig. 1.M.14   **16.**‡ For $x > 0, 3 + x < 2x \Rightarrow x > 3$;

For $x < 0, 3 + x > 2x \Rightarrow x < 3$; $x > 3$ or $x < 0$; Fig. 1.M.16   **17.** $x > 1$ or

$x \leq -2$; Fig. 1.M.17   **19.**‡ $x^2 - x - 2 = (x - 2)(x + 1)$; Either $x - 2 < 0$

and $x + 1 > 0$ or $x - 2 > 0$ and $x + 1 < 0$; $-1 < x < 2$; Fig. 1.M.19

20. $u \geq 1$ or $u = 0$; Fig. 1.M.20   **22.** $\overline{BC} = \sqrt{65}$; $\overline{AB} = \sqrt{13}$; $\overline{AC} = \sqrt{52}$;

Yes; Fig. 1.M.22   **24.** $k = 1$ **27.** Fig. 1.M.27   **29.** The line $y = 10$; Fig. 1.M.29

31. Symmetric about the y-axis   **33.** Symmetric about the origin

35. $\frac{2}{5} < m < \frac{5}{2}$ **37.** The vertices are $A(-a, 0), B(x, \sqrt{a^2 - x^2}), C(a, 0)$;

(Slope AB)(Slope BC) $= -1$; Fig. 1.M.37   **39.** $y = x, x = 1, y = x + 4$,

$x = -2$ **41.**‡ $y - 4 = m(x - 2)$; x-intercept $= 2 - 4/m$; y-intercept

$= 4 - 2m$; Area $= \frac{1}{2}(2 - 4/m)(4 - 2m) = 8 - 2m - 8/m$;

Area $= 18 \Rightarrow m^2 + 5m + 4 = 0$; $m = -1, -4$; Vertices: $(3, 0)$ and $(0, 12)$

or $(6, 0)$ and $(0, 6)$ **42.** $y = -x - 6, y = \frac{1}{2}x$ **44.**‡ Slope $AB = -1$ and

midpoint $AB = (2, 2) \Rightarrow$ center C is on $y = x$; $\overline{CA} = \sqrt{10} \Rightarrow C = (x, x)$

with $(x - 1)^2 + (x - 3)^2 = 10$; Centers: $(0, 0), (4, 4)$

45a. $x = b, \; y = \left(\dfrac{a - b}{c}\right)x, \; y = -\dfrac{b}{c}(x - a)$

45b. The altitudes intersect at $\left(b, \left(\dfrac{a - b}{c}\right)b\right)$.

46. Fig. 1.M.46   **48.** Fig. 1.M.48

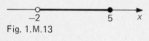

−2   5   x
Fig. 1.M.13

$-\frac{6}{5}$ y
Fig. 1.M. 14

0   3   x
Fig. 1.M. 16

−2   1   x
Fig. 1.M. 17

−1   2   x
Fig. 1.M.19

0   1   u
Fig. 1.M.20

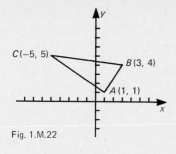

Fig. 1.M.22

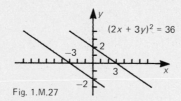

Fig. 1.M.27

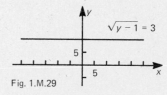

Fig. 1.M.29

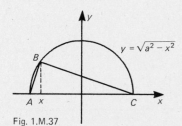

Fig. 1.M.37

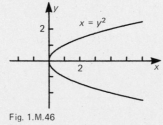

Fig. 1.M.46

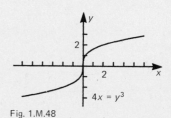

Fig. 1.M.48

Section 2.1

1. $-\frac{5}{52}, -\frac{3}{13}, 0, \frac{3}{13}, \frac{5}{52}$ **3.** $6, \sqrt{3}, 9$ **5a.** $x \neq -2, 1, -5$ **5b.** $-\frac{3}{4}$ **5c.** $\frac{6}{5}$ **5d.** 0
7a. \$1040.40 **7b.** \$1050.625 **9a**[‡] Top corners: $(\pm \sqrt{y}, y)$ (Fig. 2.1.9);
Area $= \frac{1}{2}(b_1 + b_2)h = (2 + \sqrt{y})y$ **9b.** $y > 0$ **10.** $A = 2w\sqrt{100 - \frac{1}{4}w^2}$
11a.[‡] Let $x = $ width of removed squares (ft.); Box width $= 5 - 2x$;
Length $= 7 - 2x$; Height $= x$ (Fig. 2.1.11a); Volume $= x(5 - 2x)(7 - 2x)$
cu. ft. (Fig. 2.1.11b) **11b.** $0 < x < \frac{5}{2}$ **12a.** Area $= \frac{1}{2}L(100 - L)$ square feet
12b. $0 < L < 100$ **13.** $\sqrt{(x + 2)^2 + (3x^2 - 7)^2}$ **14a.** $\sqrt{121 - L^2}$
14b. $0 < L < 11$

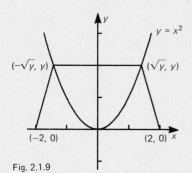

Fig. 2.1.9

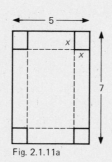

Fig. 2.1.11a

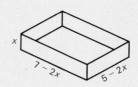

Fig. 2.1.11b

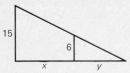

Fig. 2.1.15

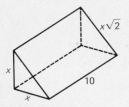

Fig. 2.1.17

15.‡ Let y = length of shadow (ft.) and x = distance from post (Fig. 2.1.15);

By similar triangles, $\dfrac{y + x}{15} = \dfrac{y}{6}$; $y = \dfrac{2}{3}x$

16. $\frac{1}{27}\pi h^3$ **17.**‡ Let x = length of legs (in.) (Fig. 2.1.17); Each end has area $\frac{1}{2}x^2$; Two sides each have area $10x$; One side has area $10x\sqrt{2}$; Total area = $x^2 + 20x + 10x\sqrt{2}$ square inches   **18a.**‡ Volume = V $= \frac{1}{3}\pi r^2 h$ (Fig. 2.1.18); $(2r)^2 = h^2 + r^2$; $r = h/\sqrt{3}$; $V = \frac{1}{9}\pi h^3$.   **18b.**‡ Solve $V = \frac{1}{9}\pi h^3$ for $h = (9V/\pi)^{1/3}$.

19.‡ (Fig. 2.1.19) Time = $\dfrac{\sqrt{x^2 + 4}\text{ mi.}}{3/2\text{ mi./hr.}} + \dfrac{(3 - x)\text{ mi.}}{6\text{ mi./hr.}}$

$= \frac{2}{3}\sqrt{x^2 + 4} + \frac{1}{6}(3 - x)$ hours

20. $\sqrt{13}\,t$ miles   **21.**‡ Set w = width (in.), h = height (Fig. 2.1.21); Volume = $w^2 h = 1000 \Rightarrow h = 1000/w^2$; Cost = $(2w^2$ sq. in.$)$ (3 cents/sq. in.) $+ (4hw$ sq. in.$)$ (2 cents/sq. in.) $= 6w^2 + 8hw$ cents $= 6w^2 + 8000/w$ cents

22a. $5x - 1000 - 20x^{2/3}$ dollars

22b. $5 - \dfrac{1000}{x} - 20x^{-1/3}$ dollars

22c. The domain of the profit function is the set of nonnegative integers; the domain of the average profit function is the set of positive integers.

23. $x \geq -1$ **24.** All x **25.** $x \neq -1$ **26.** $x \neq -3, 2, -4$ **27.** $x \neq 0, 4$

28. $-2 < x < 3$ **29.** $x < 2, 2 < x \leq 10$ **30.** $x < 2$

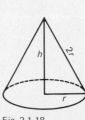

Fig. 2.1.18

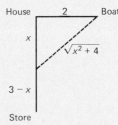

Fig. 2.1.19

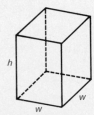

Fig. 2.1.21

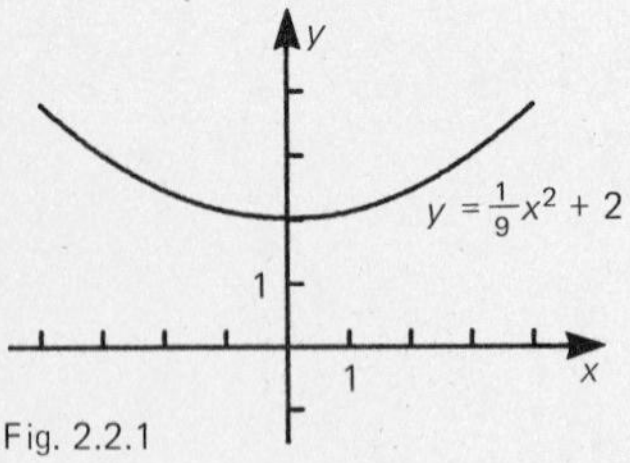

Section 2.2

1. Fig. 2.2.1   **3.** Fig. 2.2.3   **5.** Fig. 2.2.5   **7.** Fig. 2.2.7   **9.** Fig. 2.2.9   **11.** $H(t) = 0$ for $0 \leq t < 10$; $H(t) = 1$ for $10 \leq t < 25$; $H(t) = 5$ for $25 \leq t < 55$; $H(t) = 8$ for $55 \leq t \leq 90$; Fig. 2.2.11 (Any other combination of $<$ and $\leq$ is suitable provided $H(t)$ has one and only one value for each t with $0 \leq t \leq 90$.)   **13.** Fig. 2.2.13; 80 feet

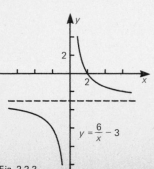

Fig. 2.2.1

Fig. 2.2.3

$y = \frac{1}{9}x^2 + 2$

$y = \frac{6}{x} - 3$

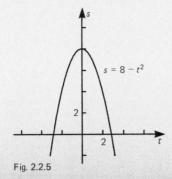

Fig. 2.2.5

$s = 8 - t^2$

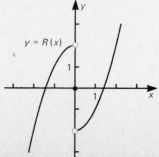

Fig. 2.2.7

$y = R(x)$

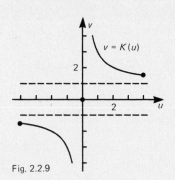

Fig. 2.2.9

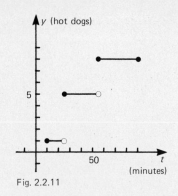

Fig. 2.2.11

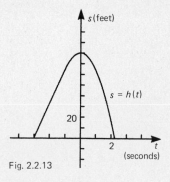

Fig. 2.2.13

Section 2.3

1. 2   **3.** 3   **5.** −1   **7a.** 1   **7b.** 0; Fig. 2.3.7   **9a.** 4   **9b.** 4   **9c.** 4; Fig. 2.3.9
11a. 4   **11b.** 1   **11c.** No limit; Fig. 2.3.11   **13a.–d.** Table 2.3.13.
13e. $\frac{1}{9}$.   **13f.** −1   **14a.** −1   **14b.** 1   **14c.** Does not exist   **14d.** −1   **14e.** 1
15a. −2   **15b.** 0   **15c.** Does not exist   **15d.** −2   **15e.** 2   **16a.** $\sqrt{5}$ **16b.** 1
16c. Does not exist   **16d.** $\sqrt{6}$ **16e.** 1   **17a.** 0   **17b.** 0   **17c.** 0   **17d.** −7
17e. 36   **18a.** $\sqrt{2}$ **18b.** $\sqrt{2}$ **18c.** $\sqrt{2}$ **18d.** $\sqrt{2}$ **18e.** $2\sqrt{2}$

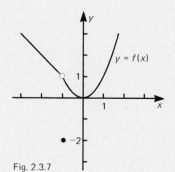

Fig. 2.3.7

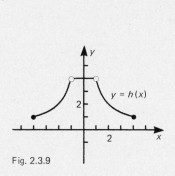

Fig. 2.3.9

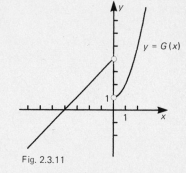

Fig. 2.3.11

TABLE 2.3.13

x	a.	b.	c.	d.
0.9	1.9	−0.7880197 . . .	5.6511815 . . .	12.216664 . . .
0.99	1.99	−0.7537561 . . .	5.9650124 . . .	14.694813 . . .
0.999	1.999	−0.7503750 . . .	5.9964996 . . .	14.969191 . . .
0.9999	1.9999	−0.7500375 . . .	5.9996440 . . .	14.996700 . . .
1.1	2.1	−0.7132083 . . .	6.3513074 . . .	18.416492 . . .
1.01	2.01	−0.7462563 . . .	6.0350126 . . .	15.311509 . . .
1.001	2.001	−0.7496250 . . .	6.0035008 . . .	15.030856 . . .
1.0001	2.0001	−0.7499625 . . .	6.0003200 . . .	15.003150 . . .
Guess of the limit	2	$-\frac{3}{4}$	6	15

Section 2.4

1. $\lim\left[\dfrac{x + 4/x}{x - 3}\right] = \dfrac{\lim(x + 4/x)}{\lim(x - 3)} = \dfrac{\lim(x) + \lim(4/x)}{\lim(x) - \lim(3)} = \dfrac{2 + 4/\lim(x)}{2 - 3}$

$= -(2 + 4/2) = -4$ where "lim" means limit as $x \to 2$

3. $\lim[f(x)(3g(x) - x)] = [\lim f(x)][\lim(3g(x) - x)]$

$= -5[\lim(3g(x)) + \lim(-x)] = -5[3\lim(g(x)) - \lim(x)]$

$= -5[3(4) - 10] = -10$ where "lim" means limit as $x \to 10^-$ **5.** $\frac{27}{25}$

7. $\frac{1}{4}$ **9.** $\frac{1}{54}$

11.‡ $\lim\limits_{x \to -1}\left[\dfrac{1 + 1/x}{x^2 - 1}\right] = \lim\limits_{x \to -1}\left[\dfrac{x + 1}{x(x^2 - 1)}\right] = \lim\limits_{x \to -1}\left[\dfrac{x + 1}{x(x + 1)(x - 1)}\right]$

$= \lim\limits_{x \to -1}\left[\dfrac{1}{x(x - 1)}\right] = \dfrac{1}{2}$

12. $-\frac{25}{2}$

14.‡ $\dfrac{(3 + h)^2 - 9}{h} = \dfrac{9 + 6h + h^2 - 9}{h} = \dfrac{6h + h^2}{h} = 6 + h \to 6$ as $h \to 0$

15. $-\frac{1}{14}$ **17.** Limit $= 1$; Fig. 2.4.17   **19.** $-\frac{1}{16}$ **20.** $\frac{1}{6}$ **21.** 4   **22.** -2 **23.** $\frac{3}{4}$

24. -2 **25.** 0   **26.** 0   **27.** 0   **28.** $\frac{2}{3}$

$y = \dfrac{x^2 + 3x + 2}{x + 1}$

Fig. 2.4.17

Section 2.5

1a. 2   **1b.** ∞ **1c.** ∞ **1d.** ∞ **1e.** 2; Fig. 2.5.1   **3a.** $-\infty$ **3b.** 5   **3c.** 5

3d. 5   **3e.** $-\infty$; Fig. 2.5.3   **5a.** -4 **5b.** $-\infty$ **5c.** ∞ **5d.** No limit   **5e.** -4;

Fig. 2.5.5   **7a.** 3   **7b.** $-\infty$ **7c.** ∞ **7d.** No limit   **7e.** -3; Fig. 2.5.7

9a. $-\infty$ **9b.** 0   **9c.** 0   **9d.** 0   **9e.** ∞; Fig. 2.5.9

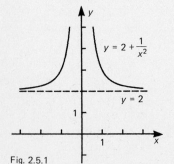

$y = 2 + \dfrac{1}{x^2}$

$y = 2$

Fig. 2.5.1

Section 2.6

1a. $-\infty$ **1b.** ∞

1c. $1 - 2x^2 - 4x^3 = x^3\left(\dfrac{1}{x^3} - \dfrac{2}{x} - 4\right) \approx -4x^3$ for large $|x|$

3a. $-\frac{2}{5}$ **3b.** $-\frac{2}{5}$

3c. $\dfrac{2x - 3}{4 - 5x} = \dfrac{x(2 - 3/x)}{x(-5 + 4/x)} = \dfrac{2 - 3/x}{-5 - 4/x} \to -\dfrac{2}{5}$ as $x \to \pm\infty$

5.‡ $(1 - x)(2 - x)(3 - x)$ has the same limits as $x \to \pm\infty$ as

$(-x)(-x)(-x) = -x^3$ **5a.** $-\infty$ **5b.** ∞

5c. $(1 - x)(2 - x)(3 - x)$

$= x^3\left(\dfrac{1}{x} - 1\right)\left(\dfrac{2}{x} - 1\right)\left(\dfrac{3}{x} - 1\right) \approx -x^3$ for large $|x|$

6.‡ $\dfrac{(x^2 + 1)(3x - 4)}{1 + 2x + 3x^2}$ has the same limits as $x \to \pm\infty$ as $\dfrac{(x^2)(3x)}{3x^2} = x$

6a. ∞ **6b.** $-\infty$

6c. $\dfrac{(x^2 + 1)(3x - 4)}{1 + 2x + 3x^2} = \dfrac{x^3(1 + 1/x^2)(3 - 4/x)}{x^2(3 + 2/x + 1/x^2)}$

$= \dfrac{x(1 + 1/x^2)(3 - 4/x)}{3 + 2/x + 1/x^2} \approx x$ for large $|x|$

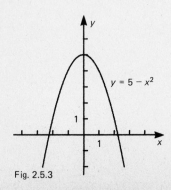

$y = 5 - x^2$

Fig. 2.5.3

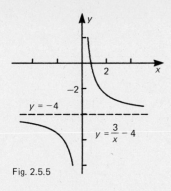

Fig. 2.5.5

8a. ∞ **8b.** $-\infty$ **10a.** $-\frac{6}{5}$ **10b.** $-\frac{6}{5}$

12.‡ $\dfrac{x^2}{x+1} - \dfrac{x^2}{x+3} = \dfrac{2x^2}{(x+1)(x+3)}$ has the same limits as $x \to \pm\infty$ as

$(2x^2)/x^2 = 2$

12a. 2   **12b.** 2   **13a.** $-\frac{3}{2}$ **13b.** $-\frac{3}{2}$ **15a.** $-\infty$ **15b.** ∞ **16a.** $\frac{3}{2}$ **16b.** $\frac{3}{2}$
17a. ∞ **17b.** ∞ **18a.** $\frac{3}{4}$ **18b.** $\frac{3}{4}$ **19a.** 0   **19b.** 0   **20a.** ∞ **20b.** ∞ **21a.** 2
21b. 2   **22a.** $-\infty$ **22b.** $-\infty$ **22c.** $-\frac{33}{16}$ **23a.** $\frac{1}{2}$ **23b.** $-\infty$ **23c.** 4
24a. ∞ **24b.** $-\infty$ **24c.** $-\infty$ **24d.** Does not exist   **25a.** $\frac{1}{2}$ **25b.** $\frac{1}{2}$
25c. $-\infty$ **25d.** ∞ **25e.** $(16)^{-1/3}$ **25f.** $(16)^{-1/3}$

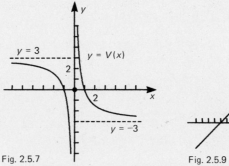

Fig. 2.5.7

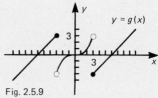

Fig. 2.5.9

Section 2.7

1. $x = 2, 5$; Fig. 2.7.1   **3.** Continuous everywhere; Fig. 2.7.3   **5.** $x \neq 0, 2$
7. $x \neq 0, -1, 1$ **9.** All x **11.‡** $f(x) = x + 3$ for $x \neq 3$, so $f(x) \to 6$ as
$x \to 3$; $k = 6$ **12.** $k = -\frac{1}{2}$ **14.** $a = 2, b = 1$ **16.** $x \neq 1$ **17.** $x \neq 0$
18. $x \neq \pm 1$ **19.** All x

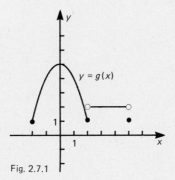

Fig. 2.7.1

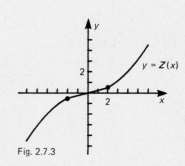

Fig. 2.7.3

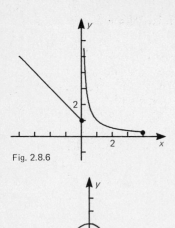

Fig. 2.8.6

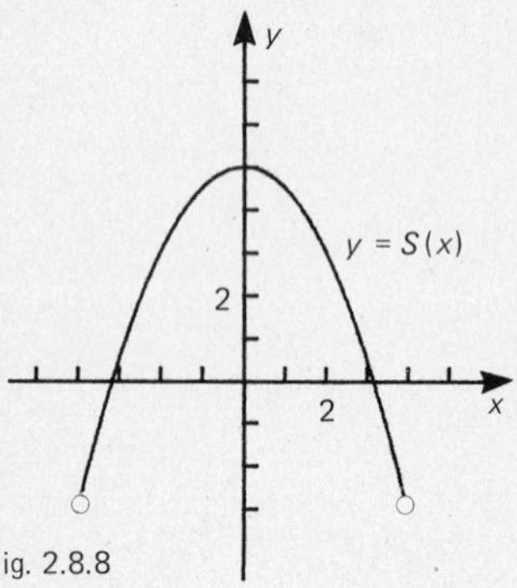

Fig. 2.8.8

Section 2.8

1. $f(x) = (x^3 - 2x)/(x + 3)$ is continuous for $-2 \leq x \leq 2$; $f(-2) = -4$, $f(2) = \frac{4}{5}$, and -3 lies between -4 and $\frac{4}{5}$, so there is a solution of $f(x) = -3$ with $-2 \leq x \leq 2$.   **3.** Set $v(x) =$ speed x feet from the start. Assume $v(x)$ is continuous. Let $2a =$ circumference of the track. $f(x) = v(x + a) - v(x)$ is continuous in the closed finite interval $0 \leq x \leq a$. If $v(a) = 0$, then $f(a) = 0$. If $v(a) \neq 0$, then $f(0) = v(a)$ and $f(a) = -v(a)$ have opposite signs. In any case $f(x) = 0$ and $v(x + a) = v(a)$ for some x in the interval.   **5.** Let $f(x) = ax^3 + bx^2 + cx + d$. If $a > 0$, then $f(x) \to \infty$ as $x \to \infty$ and $f(x) \to -\infty$ as $x \to -\infty$. If $a < 0$, then $f(x) \to -\infty$ as $x \to \infty$ and $f(x) \to \infty$ as $x \to -\infty$. In either case $f(x) > 0$ for some x and $f(x) < 0$ for other x. Since $f(x)$ is continuous, $f(x) = 0$ for some x.
6. No maximum; Q is not continuous at $x = 0$; Minimum: $Q(4) = \frac{1}{4}$; Fig. 2.8.6   **8.** Maximum: $S(0) = 5$; No minimum; The interval is not closed; Fig. 2.8.8   **10a.** 5   **10b.** -8 **11a.** 5   **11b.** -2 **12a.** No maximum   **12b.** 4   **13a.** 1   **13b.** No minimum   **14a.** 1   **14b.** -1

Section 2.9

1. For each $Y > 0$ there is a $\delta > 0$ such that $f(x) > Y$ for all x with $x_0 < x < x_0 + \delta$.   **3.** For each $Y > 0$ there is a $\delta > 0$ such that $f(x) < -Y$ for all x with $0 < |x - x_0| < \delta$.   **5.** For each $\epsilon > 0$ there is an $X > 0$ such that $|f(x) - L| < \epsilon$ for all $x > X$.   **7.** For $x > 0$, $|\sqrt{x} - 2|$ $= |x - 4|/(\sqrt{x} + 2) < |x - 4|/2$. Let $\delta =$ smaller of 2ϵ and 4. Then $0 < |x - 4| < \delta \Rightarrow x > 0$ and $|\sqrt{x} - 2| < |x - 4|/2 < 2\epsilon/2 = \epsilon$.

8. For $x > 1$, $\left|2 - \dfrac{10}{x}\right| = \dfrac{2}{x}|5 - x| < 2|5 - x|$. Let $\delta =$ smaller of 4 and $\epsilon/2$.

Then $0 < |x - 5| < \delta \Rightarrow x > 1$ and $|2 - 10/x| < 2|5 - x| < 2(\epsilon/2) = \epsilon$.
11. For $x \neq 3$, $|6 - (x^2 - 9)/(x - 3)| = |x - 3|$. Let $\delta = \epsilon$. Then $0 < |x - 3| < \delta \Rightarrow x \neq 3$ and $|6 - (x^2 - 9)/(x - 3)| = |x - 3| < \delta = \epsilon$.
13. $x^{-1/2} > Y$ if $Y^2 > 1/x$. Let $\delta = Y^{-2}$. Then $0 < x < \delta \Rightarrow x^{-1/2} > \delta^{-1/2} = Y$.
15. $|4/(x + 7)| < \epsilon$ if $|x + 7| > 4/\epsilon$. Let $X = 4/\epsilon$. Then $x > X \Rightarrow |x + 7| > 4/\epsilon \Rightarrow |4/(x + 7)| < \epsilon$.   **19.** $x^2 - 16 = (x - 4)(x + 4)$. For $|x - (-4)| < 1$, $|x - 4| < 9$ and $|x^2 - 16| < 9|x + 4|$. Let $\delta =$ the smaller of $\epsilon/9$ and 1. Then $|x - (-4)| < \delta \Rightarrow |x - 4| < 9$ and $|x^2 - 16|$ $< 9|x + 4| < 9\delta \leq \epsilon$.   **21a.**[‡] The graph is between the lines $y = \frac{1}{2}$ and $y = \frac{3}{2}$ for $1 < x < 2$ but not for $\frac{1}{2} < x < 2$ (Fig. 2.9.21a); $\delta = \frac{1}{4}, \frac{1}{2}, \frac{3}{4}, 1$
21b.[‡] For $0 < x < 2$, $|f(x) - 1|$ is less than 1 but not less than $\frac{1}{2}$; $\epsilon = 1, \frac{3}{2}, 2$ (Fig. 2.9.21b)   **23.**[‡] Any ϵ with $0 < \epsilon \leq \frac{1}{4}$ because for $L \geq 0$, $|h(x) - L| > \frac{1}{4}$ if x is just to the right of 1 and for $L < 0$, $|h(x) - L| > \frac{1}{4}$ if x is just to the left of 1.   **25.** For $0 < x < 2$, $|f(x) - 5| = 3|x - 2|$. Let $\delta =$ smaller of 2 and $\epsilon/3$. Then $2 - \delta < x < 2 \Rightarrow 0 < x < 2$, and $|f(x) - 5| = 3|x - 2| < 3\delta \leq \epsilon$.   **30.** For $|x - 3| < 1$, $|9 - x^2|$ $= |x + 3||x - 3| < 7|x - 3|$, so for $|x - 3| < 10^{-7}$, $|9 - x^2| < 7(10^{-7})$ $< 10^{-6}$ and $|9.00001 - x^2| = 10^{-5} + 9 - x^2 > 10^{-5} - 10^{-6} > 10^{-6}$. Hence for $\epsilon = 10^{-6}$ there is no δ.   **31.** If $L \geq 0$, then $|x/|x| - L|$ $= |-1 - L| \geq 1$ for $x < 0$ and if $L < 0$, then $|x/|x| - L| = |1 - L| > 1$ for $x > 0$. Hence if $\epsilon = 1$ there is no δ for any L.   **33.** $\delta = -2 + 3/\sqrt{2}$

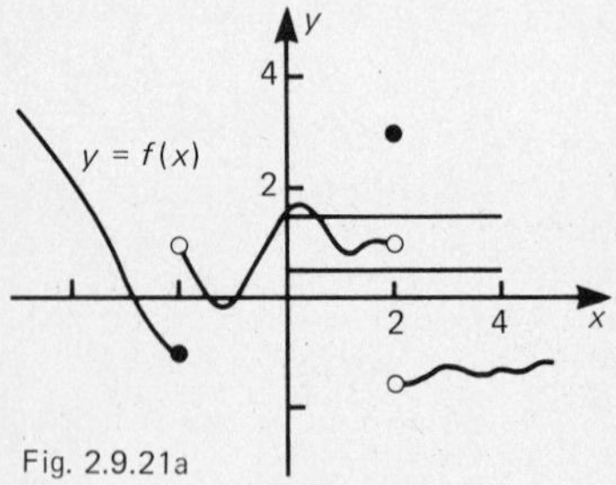

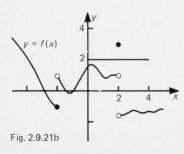

Fig. 2.9.21a

Fig. 2.9.21b

Miscellaneous Exercises to Chapter 2

1a. $W(30) = 4$ **1b.** $W(-30) = -4$ **3.** $x \neq 0, 1$ **5.** $|x| \geq 3$ **7.** $x \neq \pm 3$
9a. $\sqrt{(5-x)^2 + (3 + 6/x)^2}$ **9b.** $x \neq 0$ **11.** $A = \frac{1}{2}p^2/(2 + \sqrt{2})^2$
13. $A = (4\pi)^{1/3}(3V)^{2/3}$ **15.** $P = 5N - \frac{1}{200}N^2$ cents   **16.** $\frac{4}{3}x$ hours
17. $D(t) = 30t$ for $0 \leq t < 3$; $D(t) = \sqrt{90^2 + [30(t-3)]^2}$ for $3 \leq t < 6$;
$D(t) = \sqrt{90^2 + [30(9-t)]^2}$ for $6 \leq t < 9$; $D(t) = 30(12 - t)$
for $9 \leq t \leq 12$ **18.** Fig. 2.M.18   **20.** $W(Q) = -100 + 2Q$ for $0 \leq Q < 50$;
$W(Q) = 0$ for $50 \leq Q < 130$; $W(Q) = Q - 130$ for $130 \leq Q < 230$;
$W(Q) = 100$ for $230 \leq Q < 769$; $W(Q) = 2Q - 1438$ for $Q \geq 769$;
Fig. 2.M.20

22a.[‡] $\lim\limits_{x \to 1^-} M(x) = \lim\limits_{x \to 1^-} (3x) = 3$ **22b.**[‡] $\lim\limits_{x \to 1^+} M(x) = \lim\limits_{x \to 1^+} (x^2) = 1$

22c.[‡] $\lim\limits_{x \to 1} M(x)$ does not exist because the one-sided limits are unequal.

23a. 100   **23b.** 100   **23c.** 100   **26.** $\frac{1}{8}$ **28.** $\frac{4}{3}$ **30.** $-\frac{1}{32}$

32.[‡] For $t \neq 2$, $\dfrac{t^2 + t - 6}{t^2 + 3t - 10} = \dfrac{(t-2)(t+3)}{(t-2)(t+5)} = \dfrac{t+3}{t+5} \to \dfrac{5}{7}$ as $t \to 2$.

34. $\frac{4}{9}$ **36.**[‡] By long division, $x^4 + x^2 - 2 = (x-1)(x^3 + x^2 + 2x + 2)$;
for $x \neq 1$, $(x^4 + x^2 - 2)/(x-1) = x^3 + x^2 + 2x + 2 \to 6$ as $x \to 1$
38.[‡] For $x \neq 0$, $(x^2 + 4)/x^2 > 0$. As $x \to 0$, $x^2 + 4 \to 4$ and $x^2 \to 0$. Hence
$(x^2 + 4)/x^2 \to \infty$.   **40.**[‡] For $x < 2$, $x/(x-2) < 0$. As $x \to 2^-$, $x \to 2$ and
$x - 2 \to 0$. Hence $x/(x-2) \to -\infty$.   **42.** ∞ **44.** -2 **46.** $\frac{1}{6}$ **48.** 2.M.48
50. $x = 10$ **52.** For slowest waves, $v \approx 23$ cm/sec and $L \approx 1.8$ cm

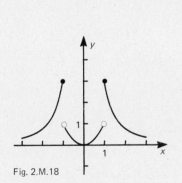

Fig. 2.M.18

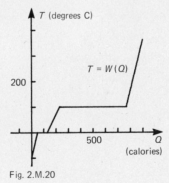

Fig. 2.M.20

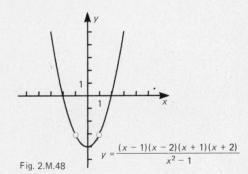

Fig. 2.M.48

Section 3.1

1. -12 **3.** 0

4.‡ $K'(2) = \lim_{t \to 2}\left[\dfrac{K(t) - K(2)}{t - 2}\right] = \lim_{t \to 2}\left[\dfrac{t^3 - 8}{t - 2}\right] = \lim_{t \to 2}[t^2 + 2t + 4] = 12$

(long division)

5.‡ $h'(25) = \lim_{t \to 25}\left[\dfrac{h(t) - h(25)}{t - 25}\right] = \lim_{t \to 25}\left[\dfrac{\sqrt{t} - 5}{t - 25}\right] = \lim_{t \to 25}\left[\dfrac{1}{\sqrt{t} + 5}\right]$

$= \dfrac{1}{10}$ (rationalization)

6.‡ $G'(4) = \lim_{u \to 4}\left[\dfrac{G(u) - G(4)}{u - 4}\right] = \lim_{u \to 4}\left[\dfrac{1/(5 - 2u) + 1/3}{u - 4}\right]$

$= \lim_{u \to 4}\left[\dfrac{8 - 2u}{3(5 - 2u)(u - 4)}\right] = \lim_{u \to 4}\left[\dfrac{-2}{3(5 - 2u)}\right] = \dfrac{2}{9}$

9.‡ Slope $= \dfrac{2^3 - 1^3}{2 - 1} = 7$; Fig. 3.1.9

10. Slope $= -1$; Fig. 3.1.10

12.‡ $y'(2) = \lim_{x \to 2}\dfrac{(4/x^2) - 1}{x - 2} = \lim_{x \to 2}\dfrac{4 - x^2}{x^2(x - 2)} = \lim_{x \to 2}\dfrac{-2 - x}{x^2} = -1$;

$y - 1 = -1(x - 2)$; $y = -x + 3$; Fig. 3.1.12

13. $y = \frac{2}{3}(x - 3)$; Fig. 3.1.13 **16.** $y = -2$; Fig. 3.1.16 **19.** $y = -\frac{1}{3}x - \frac{4}{3}$
21. $y = \frac{4}{25}x + \frac{13}{25}$ **23.** $y = -6x + 27$ **25.** Table 3.1.25 **25a.** Guess:
$f'(2) = 32$ **25c.** Guess: $f'(1) = \frac{4}{3}$ **25g.** Guess: $f'(3) = -\frac{1}{8}$ **25j.** Guess:
$f'(1) = \frac{1}{8}$ **26.** $-\frac{4}{121}$ **27.** 4 **28.** $-\frac{2}{125}$ **29.** 3 **30.** 4 **31.** $2/\sqrt{5}$

TABLE 3.1.25

Values of $\dfrac{f(x) - f(x_0)}{x - x_0}$

$x - x_0$	a.	c.	g.	j.
0.1	34.4810000 . . .	1.3550812 . . .	−0.1209835 . . .	0.1216521 . . .
0.001	32.0240000 . . .	1.333550 . . .	−0.1249583 . . .	0.1249650 . . .
0.0001	32.0023000 . . .	1.3333600 . . .	−0.1249950 . . .	0.1250000 . . .
−0.1	29.6790000 . . .	1.3105955 . . .	−0.1293294 . . .	0.1287058 . . .
−0.001	31.9760100 . . .	1.3331111 . . .	−0.1250417 . . .	0.1250350 . . .
−0.0001	31.9976000 . . .	1.3333120 . . .	−0.1250050 . . .	0.1250000 . . .

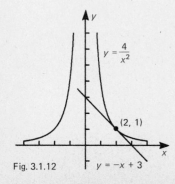

Fig. 3.1.9

Fig. 3.1.10

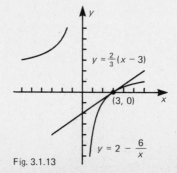

Fig. 3.1.12 $y = -x + 3$

Fig. 3.1.13

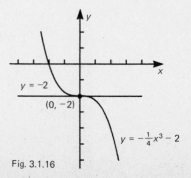

Fig. 3.1.16

Section 3.2

1. 11   **3.** $\frac{1}{6}$ **5.** $-\frac{1}{2}x$; Fig. 3.2.5a, b   **7.** 5; Fig. 3.2.7a, b   **9.** $\frac{3}{8}x^2$; Fig. 3.2.9a, b
11. $4x + 3$ **13.** $1/\sqrt{2x + 3}$ **15.** 6   **16.** $1 - x^{-2}$ **17.** $-3(3x + 1)^{-2}$
18. $-\frac{1}{2}x^{-3/2}$ **19.** $\frac{1}{2}$ **20.** 13   **21.** $\frac{1}{4}$ **22.** $(x + 1)^{-2}$

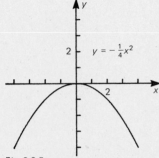

Fig. 3.2.5a

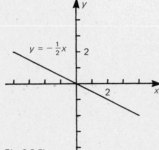

Fig. 3.2.5b

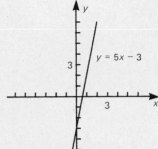

Fig. 3.2.7a

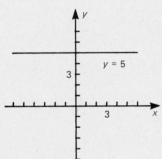

Fig. 3.2.7b

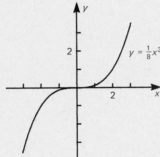

Fig. 3.2.9a

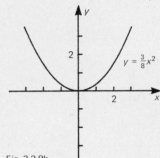

Fig. 3.2.9b

Section 3.3

1. $15x^2 - 12x + 7$; all x **3.** $-18x^{-4} + 4x^{-2}$; $x \neq 0$ **5.** $\frac{3}{2}t^{1/2} - \frac{2}{3}t^{-1/3}$;
$t > 0$

7.‡ $\dfrac{d}{dv}(\sqrt{v} - 4\sqrt[3]{v}) = \dfrac{d}{dv}(v^{1/2} - 4v^{1/3}) = \dfrac{1}{2}v^{-1/2} - \dfrac{4}{3}v^{-2/3}$; $v > 0$

8. $-1 + \frac{5}{2}s^{3/2}$; $s > 0$ **9.** $1 + \frac{3}{2}x^{-3/2}$; $x > 0$ **11.** $\frac{289}{4}$ **13.** 32   **15.** 33
17. $y = 8x - 16$ **19.** $y = -\frac{1}{70}x + \frac{4201}{70}$ **21.** $k = -4$

23. $\dfrac{d}{dx}(1) = \lim\limits_{\Delta x \to 0}\left[\dfrac{1 - 1}{\Delta x}\right] = 0$

25. Rationalize by multiplying and dividing by $\sqrt{x + \Delta x} + \sqrt{x}$:

$$\frac{d}{dx}(x^{1/2}) = \lim_{\Delta x \to 0}\left[\frac{\sqrt{x + \Delta x} - \sqrt{x}}{\Delta x}\right] = \lim_{\Delta x \to 0}\left[\frac{x + \Delta x - x}{\Delta x(\sqrt{x + \Delta x} + \sqrt{x})}\right]$$

$$= \lim_{\Delta x \to 0}\left[\frac{1}{\sqrt{x + \Delta x} + \sqrt{x}}\right] = \frac{1}{2\sqrt{x}} = \frac{1}{2}x^{-1/2}$$

27. Set $m = -n$ with n a positive rational number. As $\Delta x \to 0$,

$$\frac{1}{\Delta x}[(x + \Delta x)^m - x^m]$$

$$= \frac{1}{\Delta x}\left[\frac{1}{(x + \Delta x)^n} - \frac{1}{x^n}\right] = \left[\frac{x^n - (x + \Delta x)^n}{\Delta x}\right]\frac{1}{(x + \Delta x)^n x^n}$$

$$\to -\left(\frac{d}{dx}x^n\right)\frac{1}{x^{2n}} = -nx^{n-1}x^{-2n} = -n\,x^{-n-1} = m\,x^{m-1}$$

28. $-\frac{1}{2}x^{-3/2} + 3x^{-2}$ **29.** $2x + 6$ **30.** $-\frac{1}{2}u^{-3/2} - \frac{3}{2}u^{1/2}$ **31.** $\frac{3}{5}w^{-2/5} + \frac{5}{3}w^{2/3}$
32. $10y^4 + 12y^3 - 8y$ **33.** $6x + 2$ **34.** $-100x^{-51} - 60x^{-21} + 50x^{-11}$
35. $\frac{13}{42}x^{-29/42} - \frac{12}{35}x^{-23/35}$ **36.** $2t^{-2} - 2t^{-3}$ **37.** $\frac{1}{24}x^{-23/24}$ **38.** $a = \frac{1}{4}, b = \frac{3}{4}$
39. $a = 3, n = -\frac{1}{18}$

Section 3.4

1.‡ $\dfrac{d}{dx}[(1 + 3x - x^2)(x^2 - 5)] = (1 + 3x - x^2)\dfrac{d}{dx}(x^2 - 5)$

$$+ \left[\frac{d}{dx}(1 + 3x - x^2)\right](x^2 - 5)$$

$$= (1 + 3x - x^2)(2x) + (3 - 2x)(x^2 - 5)$$

2. $(x - x^2)'(-x^{-2} - 2x^{-3}) + (1 - 2x)(1 + x^{-1} + x^{-2})$

4.‡ $\dfrac{d}{dt}[(\sqrt{t} + \sqrt[3]{t})(\sqrt[4]{t} + \sqrt[5]{t})]$

$$= (t^{1/2} + t^{1/3})\frac{d}{dt}(t^{1/4} + t^{1/5}) + \left[\frac{d}{dt}(t^{1/2} + t^{1/3})\right](t^{1/4} + t^{1/5})$$

$$= (t^{1/2} + t^{1/3})\left(\frac{1}{4}t^{-3/4} + \frac{1}{5}t^{-4/5}\right) + \left(\frac{1}{2}t^{-1/2} + \frac{1}{3}t^{-2/3}\right)(t^{1/4} + t^{1/5})$$

5. $3(w^2 + w^{-2}) + (2w - 2w^{-3})(2 + 3w)$ **7.**‡ $F'(x) = x^2G'(x) + 2xG(x)$;
$F'(-2) = (-2)^2G'(-2) + 2(-2)G(-2) = 4(5) - 4(3) = 8$ **8.** 0

10.‡ $\dfrac{[f'(0) + g'(0)]h(0) - [f(0) + g(0)]h'(0)}{h(0)^2}$

$$= \frac{(4 + 5)(3) - (1 + 2)(6)}{3^2} = 1$$

11. $-\frac{77}{50}$

13.‡ $\dfrac{dy}{dx} = \dfrac{2(4x + 5) - (2x + 3)(4)}{(4x + 5)^2}$; For $x = 0$, $y = \dfrac{3}{5}$ and $dy/dx = -\dfrac{2}{25}$;

Tangent line: $y - \dfrac{3}{5} = -\dfrac{2}{25}x$

14. $y - \frac{1}{2} = \frac{1}{8}(x + 2)$ **17.** $(-x^4 - 3x^2 - 2x)(x^3 - 1)^{-2}$ **18.** 1
19. $(-x^{-2} + 1)(x^{-2} + 2 + x^2) + (x^{-1} + 1 + x)(-2x^{-3} + 2x)$
20. $-\frac{1}{2}[\sqrt{x}(3x - 4)]^{-1}$ **21.** $(-2x^5 - 3x^4 - 4x^3 + 2x + 1)(1 + x^4)^{-2}$
22. $-4x(x^2 - 1)^{-2}$ **23.** $1 + \frac{2}{3}x^{-1/3} - \frac{1}{3}x^{-2/3}$ **24.** $-7(x - 4)^{-2}$
25. $3x^2 - (3 - 3x^2)(x^2 + 1)^{-2}$

26. $\dfrac{d}{dx}\{f(x)[g(x)h(x)]\} = \dfrac{df}{dx}(x)g(x)h(x) + f(x)\dfrac{d}{dx}[g(x)h(x)]$

$= \dfrac{df}{dx}(x)g(x)h(x) + f(x)\left[\dfrac{dg}{dx}(x)h(x) + g(x)\dfrac{dh}{dx}(x)\right]$

$= \dfrac{df}{dx}(x)g(x)h(x) + f(x)\dfrac{dg}{dx}(x)h(x) + f(x)g(x)\dfrac{dh}{dx}(x)$

27. $2x(x^3 + 2)(x^4 + 3) + 3x^2(x^2 + 1)(x^4 + 3) + 4x^3(x^2 + 1)(x^3 + 2)$

28. $(x^{1/2} + x)(x^{-1} + x - 3) + (x + 1)(\frac{1}{2}x^{-1/2} + 1)(x^{-1} + x - 3)$
$+ (x + 1)(x^{1/2} + x)(-x^{-2} + 1)$

Section 3.5

1a. -1 ft./sec. **1b.** -2 ft./sec.; Fig. 3.5.1 **3a.** 37 mi./hr. **3b.** 37 mi./hr.;
Fig. 3.5.3 **5a.** 5 furlongs per fortnight **5b.** 0 furlongs per fortnight; Fig. 3.5.5
7a. 32 ft./sec. **7b.** $t = 1; -1 \le t < 1; 1 < t \le 3;$ **7c.** 64 ft.
8a. $-\frac{15}{2}$ mi./hr.; traveling west **8b.** 5 mi. east of the rest stop

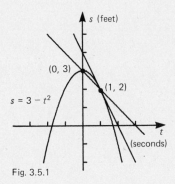

$s = 3 - t^2$

Fig. 3.5.1

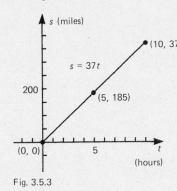

$s = 37t$

Fig. 3.5.3

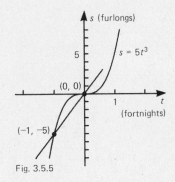

$s = 5t^3$

Fig. 3.5.5

Section 3.6

1a. $w = V^{1/3}$ **1b.** $\frac{1}{13}$ meters^{-2} **1c.** $\frac{1}{12}$ meters^{-2} **3a.** $-399/950{,}000$
$= -0.00042$ dynes/centimeter **3b.** -0.002 dynes/centimeter
5a. $3t^2 - 24t + 45$ ft./sec. **5b.** -15 ft./sec.^{2} **5c.** -12 ft./sec.^{2}
5d. 0 ft./sec.^{2} **7a.** 0.1% **7b.** 0.13% **7c.** 0.2% per hour **7d.** -0.025% per
hour **10a.** $R(x) = 0.75x$ dollars; $P(x) = \frac{1}{2}x + \frac{1}{30}x^2 - 50$ dollars
10b. Average profit $= -\frac{1}{6}$ dollar/dozen; Marginal profit $= \frac{5}{2}$ dollars/dozen
11a. $R(p) = 1000p - 300p^2 + 30p^3 - p^4$ cents **11b.** -250
13. $12t^2$ meters/second^2 **14.** $-6(t + 4)^{-3}$ centimeters/second^2
15. $90t^8 + \frac{9}{4}t^{-3/2}$ kilometers/hour^2 **16.** $\frac{88}{9}t^{5/3}$ kilometers/hour^2

Section 3.7

1.‡ $\dfrac{d}{dx}(1 - x^3)^5 = 5(1 - x^3)^4\dfrac{d}{dx}(1 - x^3) = -15x^2(1 - x^3)^4$

2. $\frac{3}{2}t^2(t^3 + 2)^{-1/2}$

4.‡ $\dfrac{d}{dx}(3x - 4)^{-1/2} = -\dfrac{1}{2}(3x - 4)^{-3/2}\dfrac{d}{dx}(3x - 4) = -\dfrac{3}{2}(3x - 4)^{-3/2}$

6.‡ $\left.\dfrac{d}{dx}\left(f(x)^{1/2}\right)\right|_{x=1} = \left[\dfrac{1}{2}f(x)^{-1/2}f'(x)\right]_{x=1} = \dfrac{1}{2}f(1)^{-1/2}f'(1)$

$= \dfrac{1}{2}(4)^{-1/2}(-5) = -\dfrac{5}{4}$

7. -320

9.‡ $\left.\dfrac{d}{dx}G(x^3)\right|_{x=2} = \left[G'(x^3)\dfrac{d}{dx}x^3\right]_{x=2} = [3x^2G'(x^3)]_{x=2} = 3(2)^2G'(8)$

$= 12(5) = 60$

10. $-\dfrac{2}{3}$

12.‡ $\left.\dfrac{d}{dt}y(x(t))\right|_{t=1} = [y'(x(t))x'(t)]_{t=1} = y'(x(1))x'(1)$

$= y'(3)x'(1) = 6(5) = 30$

13. -6

15.‡ $\left.\dfrac{d}{dx}g(f(x))\right|_{x=3} = [g'(f(x))f'(x)]_{x=3} = g'(f(3))f'(3)$

$= g'(5)f'(3) = (-1)(7) = -7$

16. -36 **18.** At $x = 2$: $y \approx 5, z \approx 3$; $dy/dx \approx 1$ and $dz/dy \approx 2$, so $dz/dx \approx 2$; At $x = 5$: $y \approx 5, z \approx 3$; $dy/dx \approx -1$ and $dz/dy \approx 2$, so $dz/dx \approx -2$ **20.** $\dfrac{20}{3}$ pounds/second **22.** $6x(6x^2 + 3)^{-1/2}$
23. $9x^{-1/2}(6\sqrt{x} + 3)^2$ **24.** $-10(x + 1)^4(x - 1)^{-6}$
25. $\dfrac{8}{3}[8x^{-2/3} + (8x + 2)^{-2/3}]$ **26.** $-8\sqrt{(5x - 2)/(3x + 2)}(5x - 2)^{-2}$
27. $-8\sqrt{(5x - 2)/(3x + 2)}(5x - 2)^{-2}$ **28.** $(5x + 4)^{1/4} + \dfrac{5}{4}x(5x + 4)^{-3/4}$
29. $\dfrac{1}{2}[1 - (2x + 1)^{1/2}]^{-1/2}[-(2x + 1)^{-1/2}]$
30. $[(x^2 + 9)(1 + \dfrac{5}{2}(5x - 2)^{-1/2}) - 2x(x + \sqrt{5x - 2})]/(x^2 + 9)^2$
31. $\dfrac{2}{3}x(x^2 + 3)^{-2/3}(x^3 + 2)^{1/2} + \dfrac{3}{2}x^2(x^2 + 3)^{1/3}(x^3 + 2)^{-1/2}$
37a. $\dfrac{1}{2}u^{-1/2}$ **37b.** $3x^2$ **37c.** $\dfrac{3}{2}x^2(x^3 + 1)^{-1/2}$ **37d.** $\sqrt{x^3 + 1}$
37e. $\dfrac{3}{2}x^2(x^3 + 1)^{-1/2}$ **38a.** $3u^2$ **38b.** $\dfrac{1}{2}x^{-1/2}$ **38c.** $\dfrac{3}{2}x^{1/2}$ **38d.** $x^{3/2} + 1$
38e. $\dfrac{3}{2}x^{1/2}$ **41.** $1 \le x < 2$ **42.** $|x - \dfrac{3}{2}| \ge \dfrac{5}{2}$

Section 3.8

1a.‡ $-60°$ is $-60(\pi/180) = -\pi/3$ radians; $\sin(-\pi/3) = -\sqrt{3}/2$;
$\cos(-\pi/3) = \dfrac{1}{2}$; Fig. 3.8.1a. **1b.‡** $135°$ is $135(\pi/180) = 3\pi/4$ radians;
$\sin(3\pi/4) = 1/\sqrt{2}$; $\cos(3\pi/4) = -1/\sqrt{2}$; Fig. 3.8.1b **1c.** $5\pi/2$ radians;
$\sin(5\pi/2) = 1$; $\cos(5\pi/2) = 0$; Fig. 3.8.1c **2a.** $\sin(7\pi) = 0$;
$\cos(7\pi) = -1$; $1260°$; Fig. 3.8.2a **2b.** $\sin(-10\pi/3) = \dfrac{1}{2}\sqrt{3}$;
$\cos(-10\pi/3) = -\dfrac{1}{2}$; $-600°$; Fig. 3.8.2b **2c.** $\sin(-3\pi/2) = 1$;
$\cos(-3\pi/2) = 0$; $-270°$; Fig. 3.8.2c

3. $\pm\dfrac{\pi}{12} + 2k\pi$ radians, k any integer

4a. $\sin(1) \approx 0.841$; $\cos(1) \approx 0.540$ **4b.** $\sin(-\dfrac{1}{5}) \approx -0.199$;
$\cos(-\dfrac{1}{5}) \approx 0.980$ **4c.** $\sin(10) \approx -0.544$; $\cos(10) \approx -0.839$
5a.‡ Table 3.8.5a; Fig. 3.8.5a **5b.** Fig. 3.8.5b

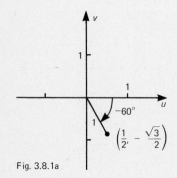

Fig. 3.8.1a

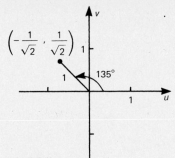

Fig. 3.8.1b

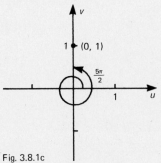

Fig. 3.8.1c

TABLE 3.8.5a

$x + \pi/6$	$\sin(x + \pi/6)$	$2\sin(x + \pi/6)$	x
0	0	0	$-\pi/6$
$\pi/2$	1	2	$\pi/3$
π	0	0	$5\pi/6$
$3\pi/2$	-1	-2	$4\pi/3$
2π	1	2	$11\pi/6$

6.‡ $160°$ is $8\pi/9$ radians; $5(8\pi/9) = 40\pi/9$ feet. **7.** $\frac{5}{2}$ ft. **8a.** 180π in./min.
8b. 180π in./min. **8c.** $\frac{9}{2}$ rev./min. **9.** 134.280 in.² **10.**‡ $\sin(5x) = -1/\sqrt{2}$
if the terminal side of the angle $5x$ is in one of the two positions which
intersect the unit circle at $v = -1/\sqrt{2}$ (Fig. 3.8.10); $5x = -(\pi/4) + 2k\pi$
or $5\pi/4 + 2k\pi$ (k an integer); $x = -\pi/20 + 2k\pi/5$ or $\pi/4 + 2k\pi/5$
(k an integer) **11.** $x = 7(2k + 1)\pi$, k any integer **12.** $(-\frac{1}{12} + k)\pi$,
$(\frac{7}{12} + k)\pi$ with k an integer **13.** $(2k + 1)\pi/4$ with k an integer
14. $\pm\sqrt{(\frac{1}{6} + 2k)\pi}$, $\pm\sqrt{(\frac{5}{6} + 2k)\pi}$ with k an integer **15.** $(2k + 1)(\frac{1}{6}\pi) - \frac{1}{3}$
with k an integer **16.** $\frac{1}{2}\pi + k\pi$ with k an integer **24.** $5\sqrt{2}\sin(-2x + \frac{3}{4}\pi)$;
Frequency $= 1/\pi$; Amplitude $= 5\sqrt{2}$ **25.** $14\sin(x - \frac{1}{3}\pi)$; Frequency
$= 1/(2\pi)$; Amplitude $= 14$ **30.** 168.4 meters **31.** 9.24 meters **32.** $\sqrt{39}$
33. $[1300 - 1200\cos(\frac{7}{8}\pi)]^{1/2} \approx 49.078$ kilometers

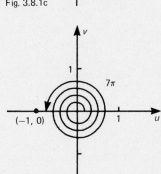

Fig. 3.8.2a

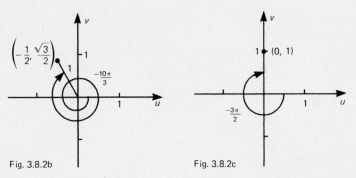

Fig. 3.8.2b

Fig. 3.8.2c

Fig. 3.8.5b

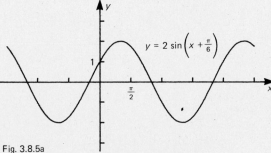

Fig. 3.8.5a

$y = 2\sin\left(x + \frac{\pi}{6}\right)$

$y = -\cos(3x)$

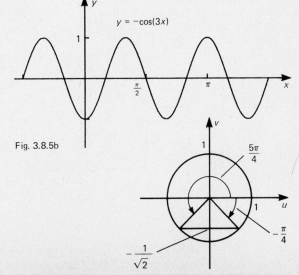

Fig. 3.8.10

Section 3.9

1. $-\sqrt{3}/2$ **3.** $3\cos(2) - 10\sin(2)$

5.‡ $\dfrac{d}{dx}(x^2\cos x) = x^2\dfrac{d}{dx}(\cos x) + \left[\dfrac{d}{dx}(x^2)\right]\cos x = -x^2\sin x + 2x\cos x$

6. $1/\cos^2 x$

8.‡ $\dfrac{d}{dt}[4\sin(6t) - 6\sin(4t)] = 4\cos(6t)\dfrac{d}{dt}(6t) - 6\cos(4t)\dfrac{d}{dt}(4t)$

$\quad = 24[\cos(6t) - \cos(4t)]$

9. $\cos(w^2) - 2w^2\sin(w^2)$

11.‡ $\dfrac{d}{dx}\sin^3(5x) = 3\sin^2(5x)\dfrac{d}{dx}\sin(5x) = 3\sin^2(5x)\cos(5x)\dfrac{d}{dx}(5x)$

$\quad = 15\sin^2(5x)\cos(5x)$

12.‡ $f(x) = \sin x$; $f(3) = \sin 3$; $f'(x) = \cos x$; $f'(3) = \cos 3$; Tangent line:
$y - \sin 3 = \cos(3)(x - 3)$ **13.** $y = -x$ **16.** $18(\cos^2 x - \sin^2 x)$

18. $10\cos x$ **19.** $(ab/c)\sin x$ **20a.** $(5\sqrt{2})\sin x$ **20b.** $(5\sqrt{2})\cos x$

28. $3\sin^2 x\cos^6 x - 5\sin^4 x\cos^4 x$ **30.** $\frac{4}{3}(x^3\cos x)^{1/3}(3x^2\cos x - x^3\sin x)$

32. $-\frac{1}{2}\sin x(\cos x)^{-1/2} - \frac{1}{2}\sin(x^{1/2})x^{-1/2}$ **33.** $(3x^2 + 1)\cos(x^3 + x)$

34. $-\sin(2x)[\cos(2x)]^{-1/2}$ **35.** $\sin(x^{1/2})\cos(x^{1/2})x^{-1/2}$

36. $[3\sin(2x)\cos(3x) - 2\sin(3x)\cos(2x)]/\sin^2(2x)$ **37.** $-\sin x[\cos(\cos x)]$

38. $5x^4(1 + \cos x)^{1/2} - \frac{1}{2}x^5\sin x(1 + \cos x)^{-1/2}$

39. $\dfrac{1}{2\cos^2 x}\sqrt{\dfrac{\cos x}{\sin x}}$

40. $3(x - \sin x^3)^2(1 - 3x^2\cos x^3)$

Section 3.11

1. The motion has two equal components, one vertical and one directed away from the focus, which form two sides of a rhombus (Figure 3.11.1). The direction of motion is along the diagonal of the rhombus, which bisects the angle.

Miscellaneous Exercises to Chapter 3

4. $y = \pm 4x$

8a. 0.0005 amperes/volt   **8b.** 0.00035 amperes/volt   **10a.** $-(3x - 4)^{-4/3}$

10b.‡ $\dfrac{d}{dx}\left[(x^2 + 1)/(1 - x^2)\right]$

$\quad = \left\{\left[\dfrac{d}{dx}(x^2 + 1)\right](1 - x^2) - (x^2 + 1)\left[\dfrac{d}{dx}(1 - x^2)\right]\right\}\Big/(1 - x^2)^2$

$\quad = 4x/(1 - x^2)^2$

10e. $-13(t^5 - 3t^2 + 1)^{-14}(5t^4 - 6t)$ **10f.** $-2/(4x + 5)^2$ **10i.** $(2u^5 + 4u^3 - 6u)/(u^2 + 1)^2$ **10j.** $(\sin x/\cos^2 x) + (\cos x/\sin^2 x)$ **10m.** $15\sin^2 5x\cos 5x$
10n. $-x\sin(x^2)/\sqrt{\cos(x^2)}$ **10q.** $(1 + \pi/4)/\sqrt{2}$ **10r.** $-\frac{3}{5}(\sqrt{3}/\pi) - \frac{18}{25}(1/\pi^2)$
11a. 26   **11b.** -6 **15a.** $\frac{3}{2}$ **15b.** 9   **15c.** -8 **15d.** 6

17. For $x > 0$, $|x| = x$ and $\dfrac{d}{dx}|x| = 1$; For $x < 0$, $|x| = -x$ and $\dfrac{d}{dx}|x|$

$\quad = -1$. The difference quotient $[|x| - |0|]/[x - 0]$ equals 1 for $x > 0$, equals -1 for $x < 0$ and has no limit as $x \to 0$.

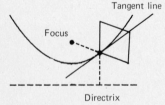

Fig. 3.11.1

22a. -2 **22b.** 1 **24a.** $\frac{9}{2}$ **24b.** 10 **24c.** $-\frac{7}{16}$ **24d.** $\frac{14}{3}4^{-5/6}$ **24e.** 281
24f. $-\frac{5}{48}$ **24g.** $-\pi$ **24h.** π **24i.** $-\frac{1}{4}\sin(\frac{1}{2})$

Section 4.1

1. Decreasing for $x < 2$; Increasing for $x > 2$; Local min at $x = 2$; Fig. 4.1.1
3.‡ $y' = x^2 + 2x = x(x + 2)$; Table 4.1.3; Increasing for $x < -2$ and for
$x > 0$; Decreasing for $-2 < x < 0$; Local max at $x = -2$; Local min at
$x = 0$; Fig. 4.1.3 **5.**‡ $y' = x^3 - 1$; Table 4.1.5; Increasing for $x > 1$;
Decreasing for $x < 1$; Local min at $x = 1$; Fig. 4.1.5 **7.** Increasing **9.** Decreasing

Table 4.1.3

$$y' > 0 \qquad y' < 0 \qquad y' > 0$$
$$\underset{-2}{\circ} \qquad \underset{0}{\circ} \qquad x$$

Table 4.1.5

$$y' < 0 \qquad y' > 0$$
$$\underset{1}{\circ} \qquad x$$

11.‡ $\left[\dfrac{d}{dx}(x\sin x)\right]_{x=\pi} = [x\cos x + \sin x]_{x=\pi} = -\pi < 0$; Decreasing

12. Decreasing

14.‡ $\left[\dfrac{d}{dx}\sin(\sqrt[3]{x})\right]_{x=1} = \left[\dfrac{1}{3}x^{-2/3}\cos(\sqrt[3]{x})\right]_{x=1} = \dfrac{1}{3}\cos(1) > 0$; Increasing

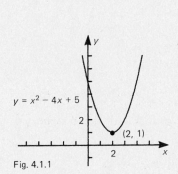

$y = x^2 - 4x + 5$

$(2, 1)$

Fig. 4.1.1

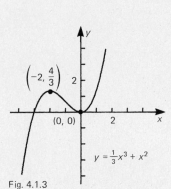

$\left(-2, \dfrac{4}{3}\right)$

$(0, 0)$

$y = \dfrac{1}{3}x^3 + x^2$

Fig. 4.1.3

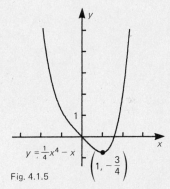

$y = \dfrac{1}{4}x^4 - x$

$\left(1, -\dfrac{3}{4}\right)$

Fig. 4.1.5

15. Increasing   **17.** Fig. 4.1.17   **18.** Fig. 4.1.18   **19.** Fig. 4.1.19   **20.** Fig. 4.1.20
21. Fig. 4.1.21   **22.** Fig. 4.1.22
23.‡ $y' = (x - 1)^2(x - 5)^8(12x - 24)$; Table 4.1.23; Local min at $x = 2$;
Fig. 4.1.23   **24.** Fig. 4.1.24   **26.** Fig. 4.1.26

Table 4.1.23

$$y' < 0 \quad y' < 0 \quad y' > 0 \quad y' > 0$$

$$\underset{1 \qquad 2 \qquad 5 \qquad x}{\circ \qquad \circ \qquad \circ}$$

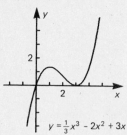

Fig. 4.1.17

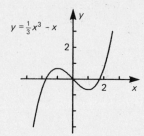

Fig. 4.1.18

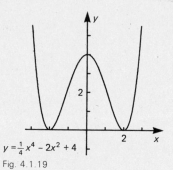

Fig. 4.1.19

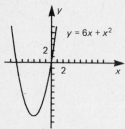

Fig. 4.1.20

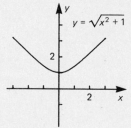

Fig. 4.1.21

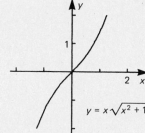

Fig. 4.1.22

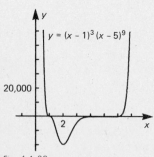

Fig. 4.1.23

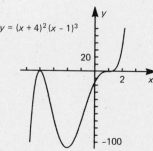

Fig. 4.1.24

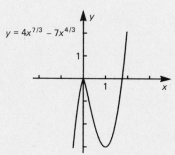

Fig. 4.1.26

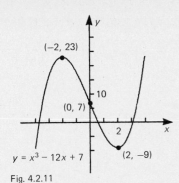

(−2, 23)

10

(0, 7)

2

$y = x^3 - 12x + 7$ (2, −9)

Fig. 4.2.11

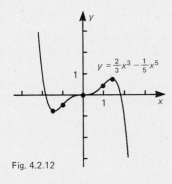

$y = \frac{2}{3}x^3 - \frac{1}{5}x^5$

1

1

Fig. 4.2.12

Section 4.2

1. $36x^2 - 12$ **3.**‡ $\dfrac{d}{dx}(x \sin x) = x \cos x + \sin x$; $\dfrac{d^2}{dx^2}(x \sin x)$

$= -x \sin x + 2 \cos x$, which equals $-\pi/2$ at $x = \pi/2$
4. $(6t^2 - 2)/(t^2 + 1)^3$ **6.** 1050

8.‡ $\dfrac{d}{dx} \cos(x^3) = -3x^2 \sin(x^3)$; $-6x \sin(x^3) - 9x^4 \cos(x^3)$

9. $- 8(\sin x)(\cos x)$ **11.** Increasing for $x < -2$ and for $x > 2$; Decreasing for $-2 < x < 2$; Concave down for $x < 0$; Concave up for $x > 0$; Local max at $x = -2$; Local min at $x = 2$; Inflection point at $x = 0$; Fig. 4.2.11
12. Decreasing for $x < -\sqrt{2}$ and $x > \sqrt{2}$; Increasing for $-\sqrt{2} < x < 0$ and $0 < x < \sqrt{2}$; Local min at $x = -\sqrt{2}$; Local max at $x = \sqrt{2}$; Concave up for $x < -1$ and $0 < x < 1$; Concave down for $-1 < x < 0$ and $x > 1$; Inflection points at $x = 0, \pm 1$; Fig. 4.2.12 **15.** Increasing for $x < 0$ and $0 < x < \frac{3}{2}$; Decreasing for $x > \frac{3}{2}$; Local max at $x = \frac{3}{2}$; Concave down for $x < 0$ and $x > 1$; Concave up for $0 < x < 1$; Inflection points at $x = 0, 1$; Fig. 4.2.15
19. Increasing for all x; Concave down for $x < 0$; Concave up for $x > 0$; Inflection point at $x = 0$; Figure 4.2.19 **21.**‡ $f'(x) = (1 - x^2)/(x^2 + 1)^2$; $f''(x) = (2x^5 - 4x^3 - 6x)/(x^2 + 1)^4$; $f'(1) = 0$ and $f''(1) = -\frac{1}{2} < 0$; Local max **22.** Local min **25.** Local min

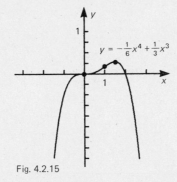

1

$y = -\frac{1}{6}x^4 + \frac{1}{3}x^3$

1

Fig. 4.2.15

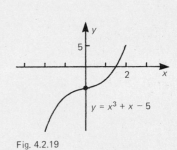

5

2

$y = x^3 + x - 5$

Fig. 4.2.19

Section 4.3

1.‡ $y' = (x^2 + 4x)/(x + 2)^2$; $y'' = 8/(x + 2)^3$; Table 4.3.1;
Increasing for $x < -4$ and $x > 0$; Decreasing for $-4 < x < -2$ and $-2 < x < 0$; Local max at $x = -4$; Local min at $x = 0$; Concave down for $x < -2$; Concave up for $x > -2$; Vertical asymptote $x = -2$; Fig. 4.3.1 **2.** Increasing for $x < 0$; Decreasing for $x > 0$; Local max at $x = 0$; Concave up for $x < -1/\sqrt{3}$ and $x > 1/\sqrt{3}$; Concave down for $-1/\sqrt{3} < x < 1/\sqrt{3}$; Inflection points at $x = \pm 1/\sqrt{3}$; Horizontal asymptote $y = 0$; Fig. 4.3.2 **6.**‡ $y' = -4x/(x^2 - 1)^2$; $y'' = (12x^2 + 4)/(x^2 - 1)^3$; Table 4.3.6; Increasing for $x < -1$ and $-1 < x < 0$; Decreasing for $0 < x < 1$ and $x > 1$; Local max at $x = 0$; Concave up for $x < -1$ and $x > 1$; Concave down for $-1 < x < 1$; Vertical asymptotes $x = \pm 1$; Horizontal asymptote $y = 1$; Fig. 4.3.6

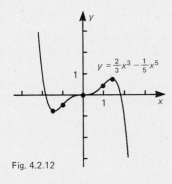

$y = \dfrac{x^2}{x + 2}$

2

2

(−4, −8)

Fig. 4.3.1

Table 4.3.1

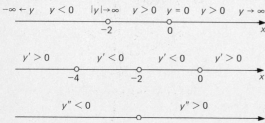

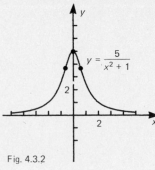

Fig. 4.3.2

Table 4.3.6

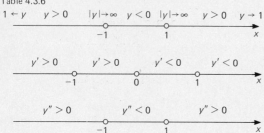

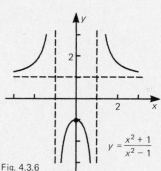

Fig. 4.3.6

7. Increasing for $x < -1$ and $x > 1$; Decreasing for $-1 < x < 0$ and $0 < x < 1$; Local max at $x = -1$; Local min at $x = 1$; Concave down for $x < 0$; Concave up for $x > 0$; Vertical asymptote $x = 0$; Fig. 4.3.7
10. Decreasing for $x < 0$ and $0 < x < \frac{1}{2}$; Increasing for $\frac{1}{2} < x < 1$ and $x > 1$; Local min at $x = \frac{1}{2}$; Vertical asymptotes $x = 0$ and $x = 1$; Horizontal asymptote $y = 0$; Fig. 4.3.10 **12.** Decreasing for $x < -2$, $-2 < x < 2$, and $x > 2$; Vertical asymptotes $x = \pm 2$; Horizontal asymptote $y = 0$; Fig. 4.3.12 **14.** Fig. 4.3.14 **16.** Fig. 4.3.16 **18.** Fig. 4.3.18 **20.** Fig. 4.3.20

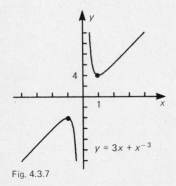

Fig. 4.3.7

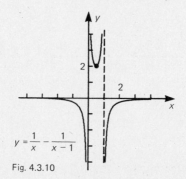

Fig. 4.3.10

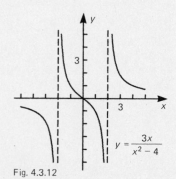

Fig. 4.3.12

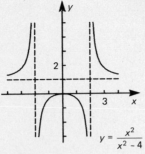

Fig. 4.3.14

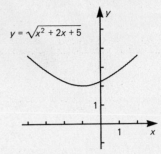

Fig. 4.3.16

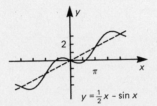

Fig. 4.3.18

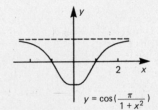

Fig. 4.3.20

Section 4.4

1.‡ $y = x^{2/3}$; $y' = \frac{2}{3}x^{-1/3}$; $y'' = -\frac{2}{9}x^{-4/3}$; Decreasing for $x < 0$;
Increasing for $x > 0$; Local min at $x = 0$; Concave down for $x < 0$ and $x > 0$;
Fig. 4.4.1   **2.** Fig. 4.4.2   **4.** Fig. 4.4.4   **6.**‡ Defined for $x \leq -3$ and $x \geq 3$;
$y' = x(x^2 - 9)^{-1/2}$; $y'' = -9(x^2 - 9)^{-3/2}$; Decreasing for $x < -3$;
Increasing for $x > 3$; Concave down for $x < -3$ and $x > 3$; Fig. 4.4.6

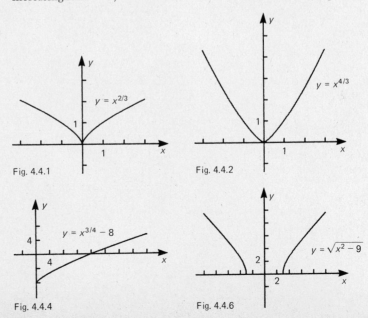

Fig. 4.4.1

Fig. 4.4.2

Fig. 4.4.4

Fig. 4.4.6

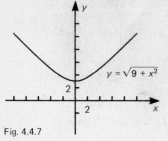

Fig. 4.4.7

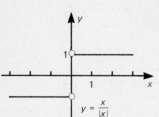

Fig. 4.4.9

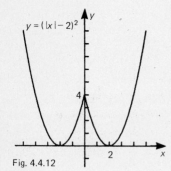

Fig. 4.4.12

7. Fig. 4.4.7   **9.** Fig. 4.4.9   **12.** Fig. 4.4.12   **14.**‡ $y' = \frac{2}{3}x^{-2/3}(x^{1/3} - 1)$;
$y'' = -\frac{2}{9}x^{-5/3}(x^{1/3} - 2)$; Table 4.4.14; Decreasing for $x < 0$ and $0 < x < 1$;
Increasing for $x > 1$; Local min at $x = 1$; Concave down for $x < 0$ and
$x > 8$; Concave up for $0 < x < 8$; Inflection point at $x = 8$ (not at $x = 0$ because we require $y'' = 0$ in our definition of inflection point);
Vertical tangent line $y = 0$; Fig. 4.4.14   **16a.** Figure 4.30   **16b.** Figure 4.33
16c. Figure 4.32   **16d.** Figure 4.29   **16e.** Figure 4.31   **21.** Fig. 4.4.21
22. Fig. 4.4.22   **23.** Fig. 4.4.23   **25.** Fig. 4.4.25   **27.** Fig. 4.4.27

Table 4.4.14

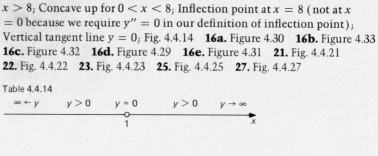

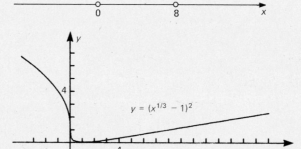

Fig. 4.4.14

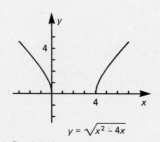

Fig. 4.4.21

$y = 2x^{1/2} - x$

Fig. 4.4.22

$y = \sqrt{x^2 - 4x}$

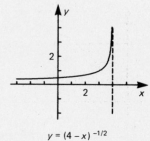

$y = (4 - x)^{-1/2}$

Fig. 4.4.23

$y = (16 - x^4)^{1/4}$

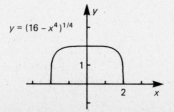

Fig. 4.4.25

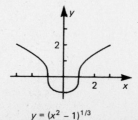

$y = (x^2 - 1)^{1/3}$

Fig. 4.4.27

Section 4.5

1.‡ $x^2f' + 2xf = 6 + 3f^2f'$; Set $x = 4$ and $f = 2$: $16f' + 16 = 6 + 12f'$;
$f'(4) = -\frac{5}{2}$ **2.** $-\frac{1}{6}$ **4.**‡ $y^{3/2} + \frac{3}{2}xy^{1/2}y' = 2 + y'$; Set $x = 0$ and $y = 4$:
$8 = 2 + y'$; $y'(0) = 6$ **5.** $-\frac{1}{6}$ **7.** 0 **8.** $-\frac{43}{13}$ **11.**‡ $\frac{2}{3}x^{-1/3} + \frac{2}{3}y^{-1/3}y' = 0$; Set
$x = 8, y = 1$: $y'(8) = -\frac{1}{2}$; Tangent line: $y = -\frac{1}{2}x + 5$ **12.** $y = -\frac{3}{16}x + \frac{35}{16}$
14.‡ $1 = -y'(\sin y)$; $y'(\frac{1}{2}) = -2/\sqrt{3}$; $y = -(2/\sqrt{3})x + (1/\sqrt{3}) + (\pi/3)$
15. $y = x$ **17a.**‡ Fig. 4.5.17; The graph of f cannot extend farther than
shown in Fig. 4.5.17 because it cannot have any vertical tangent lines or
two points with the same x-coordinate. **17b.** $x > 0.7$ **17c.** $5f^4f' - f' - 2x$
$= 0$; Set $x = 1, f = 1$: $f'(1) = \frac{1}{2}$; Tangent line: $y = \frac{1}{2}x + \frac{1}{2}$ **18a.** Fig. 4.5.18
18b. $0.7 < x < 1.2$ **18c.** $y = -2x + 2$ **20.** Fig. 4.5.20; $y = -\frac{2}{11}x + \frac{15}{11}$
22a. $y = \pm(1 - x^{2/3})^{3/2}$ **22b.** $dy/dx = -(y/x)^{1/3}$
$= \mp(1 - x^{2/3})^{1/2}/x^{1/3}$ **22c.** $dy/dx = \pm\frac{3}{2}(1 - x^{2/3})^{1/2}(-\frac{2}{3})x^{-1/3}$
$= \mp(1 - x^{2/3})^{1/2}/x^{1/3}$ **23.** $dy/dx = (2 - 2x)/4y^3$
24. $dy/dx = (2x + 5y)/(5x - 2y)$ **25.** $dy/dx = (y^4 + 1)/(1 - 4xy^3)$
26. $dy/dx = (y - 2x)/(2y - x)$
27. $dy/dx = -(2xy^3 + 3y^2x^2 + 1)/(3x^2y^2 + 2yx^3)$
28. $dy/dx = (\cos y)/(\cos y + x \sin y)$

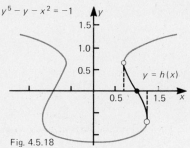

$y^5 - y - x^2 = -1$

Fig. 4.5.17

$y^5 - y - x^2 = -1$

$y = f(x)$

$y = h(x)$

Fig. 4.5.18

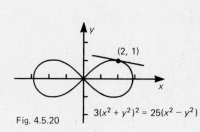

(2, 1)

$3(x^2 + y^2)^2 = 25(x^2 - y^2)$

Fig. 4.5.20

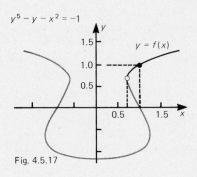

$\begin{cases} y = f(x) \\ x = g(y) \end{cases}$

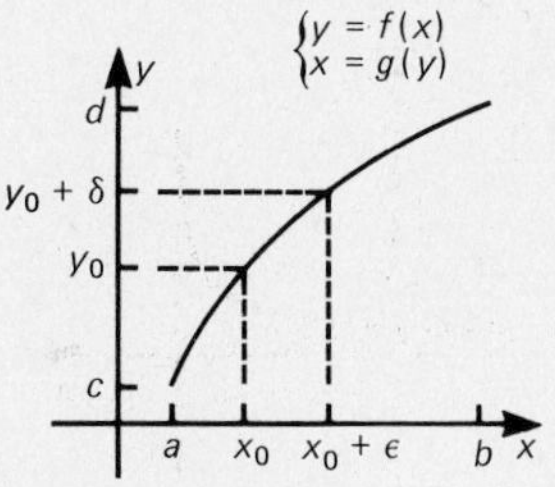

Fig. 4.6.14

Section 4.6

1. $x = \frac{1}{2}(y + 3)$ **3.** $x = \frac{1}{2}[(y - 3)^{1/3} - 3]$ **5.** $x = (y - 1)^{4/3} - 4$
7. $x = 3 + 1/(y - 3)$ **9a.** $-1 \le y \le 1$ **9b.** f is continuous for
$-\frac{1}{2}\pi \le x \le \frac{1}{2}\pi$ and $f' = \cos x$ is positive for $-\frac{1}{2}\pi < x < \frac{1}{2}\pi$, so f is increasing
in the interval $-\frac{1}{2}\pi \le x \le \frac{1}{2}\pi$; $f(\frac{1}{6}\pi) = \frac{1}{2}$, so $g(\frac{1}{2}) = \frac{1}{6}\pi$ **9c.** $2/\sqrt{3}$
10a. $y > 0$ **10b.** $f' = -2x^{-3} - 3x^{-4}$ is negative and f is continuous and
decreasing for $x > 0$; $f(1) = 2$, so $g(2) = 1$ **10c.** $-\frac{1}{5}$ **11a.** $f' = 4 - \pi \cos(\pi x)$
is positive and f is continuous and increasing for all x; $f(3) = 4(3)$
$+ \cos(3\pi) = 11$, so $g(11) = 3$ **11b.** $1/(4 + \pi)$ **11c.** All y
13.‡ $x = (y - 3)^2 + 2$; $x - 2 = (y - 3)^2$; $y = 3 \pm \sqrt{x - 2}$
15.‡ $x = (y^2 + 7)^{-1/2}$; $x^2(y^2 + 7) = 1$; $y = \pm\sqrt{x^{-2} - 7}$
16. $y = \frac{1}{2}(\pm x^{3/2} + 3)$ **18.** $y = 2 + x^{-2}$ **19.** $y = 3 + (x - 2)^{1/5}$
20. $y = \frac{1}{2}[1 + (x + 2)^3]$ **22.** $y = (x^{1/3} - 1)^3$
24. Consider $c \le y_0 < d$ and set $x_0 = g(y_0), x = g(y)$ (or equivalently y_0
$= f(x_0), y = f(x)$). For $0 < \epsilon < b - x_0$ set $\delta = f(x_0 + \epsilon) - f(x_0)$ (Fig. 4.6.14).
Then $y_0 < y < y_0 + \delta \Leftrightarrow x_0 < x < x_0 + \epsilon$ or equivalently $g(y_0) < g(y)$
$< g(y_0) + \epsilon$. Hence $g(y) \to g(y_0)$ as $y \to y_0^+$. Similarly, $g(y) \to g(y_0)$ as $y \to y_0^-$
for $c < y_0 \le d$.

Section 4.7

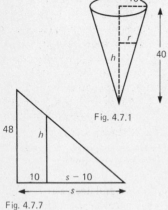

Fig. 4.7.1

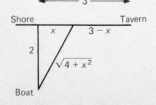

Fig. 4.7.7

1.‡ Let r = radius of the surface of the water (ft.) and h = depth of the water at time t (min.). $V = \frac{1}{3}\pi r^2 h$; By similar triangles, $r = h/4$ (Fig. 4.7.1); $V = \frac{1}{48}\pi h^3$; $dV/dt = \frac{1}{16}\pi h^2(dh/dt)$; $h = 8$ and $dV/dt = 7 \Rightarrow dh/dt = 7/4\pi$ ft./min.   **2.** $6\sqrt{3}$ cu.ft./min.   **4.**‡ $V = \frac{1}{3}\pi r^2 h$; $dV/dt = \frac{1}{3}\pi r^2(dh/dt)$ $+ \frac{2}{3}\pi r h(dr/dt)$; $r = 4, dr/dt = 5$, and $dh/dt = -6 \Rightarrow dV/dt = 8\pi$ cu. in./sec.; Increasing   **5.** Decreasing at $\frac{6}{5}$ min./hr.   **7.**‡ Let t = time (sec.), h = height of ball (ft.), s = distance from the pole to the tip of the shadow (ft.). By similar triangles (Fig. 4.7.7) $s/48 = (s - 10)/h$; $48s - 480 = sh$; $48(ds/dt)$ $= h(ds/dt) + s(dh/dt)$; At $t = 1, h = 32, dh/dt = -32$, and $s = 30$; ds/dt $= -60$; Shadow's tip has speed 60 ft./sec.   **8.** $\frac{4}{3}$ ft./sec.

10.‡ $w = 100(1 + r/4000)^{-2}$; $dw/dt = -200(1 + r/4000)^{-3}(\frac{1}{4000})(dr/dt)$; $r = 400, dr/dt = 15 \Rightarrow dw/dt = -0.561$; w is decreasing 0.561 pounds/sec.   **11.** ≈ 0.170 degrees/min.   **13.** 450π cu.in./min.   **15a.** $25\sqrt{3}$ sq.in./sec.   **15b.** 15 in./sec.   **17.** Increasing $\frac{3}{4}$ sq.ft./sec.   **18.** 64/125 candles/sq.ft.-sec.   **19a.** 50 pounds/sq.in.   **19b.** -10 pounds/sq.in.-sec.   **22.** 5   **23.** $\frac{42}{5}$ **24.** $-\frac{41}{2}$

Section 4.8

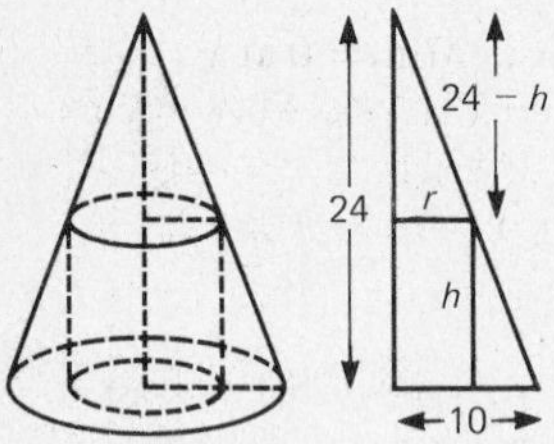

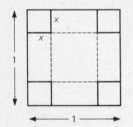

Fig. 4.8.12

Fig. 4.8.15a

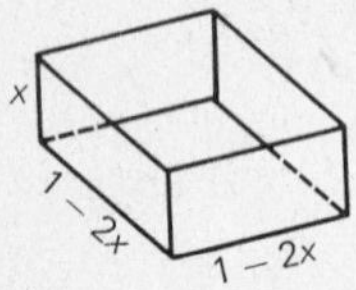

Fig. 4.8.15b

Fig. 4.8.21

1. Max $= 7$; Min $= 3$ **3.** Max $= 0$; Min $= -\frac{1}{2}$ **5.**‡ $f(x) = \frac{1}{2}x + \sin x$; $f'(x)$ $= \frac{1}{2} + \cos x$; Critical points: $x = \pm\frac{2}{3}\pi + 2k\pi$ (k any integer); $f(0) = 0$, $f(2\pi/3) = \frac{1}{3}\pi + \frac{1}{2}\sqrt{3}$; $f(\pi) = \pi/2$; Max $= (\pi/3) + \sqrt{3}/2$; Min $= 0$

7.‡ $f(x) = 5x^{2/3} - 2x^{5/3}$; $f'(x) = \frac{10}{3}x^{-1/3} - \frac{10}{3}x^{2/3} = \frac{10}{3}x^{-1/3}(1 - x)$; Critical points: $x = 0, 1$; $f(-1) = 7, f(0) = 0, f(1) = 3, f(8) = -44$; Max $= 7$; Min $= -44$ **9.**‡ Let y = length of the side that costs \$3 per foot and x = width; Cost $= [(2x + y) \text{ ft.}][2 \text{ dollars/ft.}] + [y \text{ ft.}][3 \text{ dollars/ft.}] = 4x + 5y$ dollars; $4x + 5y = 120$; $A = xy = x(120 - 4x)/5 = 24x - \frac{4}{5}x^2$ $(0 \le x \le 30)$; $A' = 24 - \frac{8}{5}x$; Critical point: $x = 15$; $A(0) = 0, A(15) = 180$, $A(30) = 0$; Max occurs for $x = 15, y = 12$.   **10.** Length $= 28$; Height $=$ width $= 14$

12.‡ Let h = height, r = radius (in.); By similar triangles, $\dfrac{r}{10} = \dfrac{24 - h}{24}$

(Fig. 4.8.12); $h = 24 - \frac{12}{5}r$; $V = \pi r^2 h = \pi r^2(24 - \frac{12}{5}r), 0 \le r \le 10$; $V' = 48\pi r - \frac{36}{5}\pi r^2$; Critical points $r = 0, \frac{20}{3}$; $V(0) = 0, V(\frac{20}{3}) > 0, V(10) = 0$; Max volume for $r = \frac{20}{3}, h = 8$ **13.** A square of side $r\sqrt{2}$ **15.**‡ Let x = width of the squares that are cut out (Fig. 4.8.15); $V = x(1 - 2x)^2, 0 \le x \le \frac{1}{2}$; $V' = 12x^2 - 8x + 1$; Critical points: $x = \frac{1}{2}, \frac{1}{6}$; $V(0) = 0, V(\frac{1}{6}) > 0, V(\frac{1}{2}) = 0$; Max volume for $x = \frac{1}{6}$ **17.** $13\frac{1}{3}$ inches long, ends 10 inches square   **18.** 100 ft. long (the length of the stable), 40 feet deep   **21.**‡ Define x (mi.) as in Fig. 4.8.21; Distance from boat to shore $= \sqrt{4 + x^2}$; Distance along shore $= 3 - x$; Time $= T = \frac{1}{5}\sqrt{4 + x^2} + \frac{1}{13}(3 - x), 0 \le x \le 3$; $T' = \frac{1}{5}x(4 + x^2)^{-1/2} - \frac{1}{13}$; Critical point: $x = \frac{5}{6}$; $T(0) = \frac{41}{65} \approx 0.63$ hours, $T(\frac{5}{6}) = 0.6$ hours, $T(3) = \frac{1}{5}\sqrt{13} \approx 0.72$ hours; The time is minimized with $x = \frac{5}{6}$ miles   **22.** 10 ft. from the point on the fence opposite P **24.** 0.56 radians   **26.** Max $= 5$ at $x = 1$; Min $= 4$ at $x = 2$ **27.** Max $= \sqrt{5}$ at $x = 0$; Min $= 1$ at $x = 2$ **28.** Max $= \sin(1)$ at $x = \pm 1$; Min $= 0$ at $x = 0$ **29.** Max $= 3$ at $x = 1$; Min $= -3$ at $x = -1$ **30.** Max $= 2^{10} - 20$ at $x = 2$; Min $= -9$ at $x = 1$ **31.** Max $= 2$ at $x = 4$; Min $= -2$ at $x = 1$ **32.** Max $= 2^{-4}$ at $x = -1$; Min $= 10^{-4}$ at $x = -3$ **34.** Max $= \frac{1}{6}\pi\sqrt{3}$ at $x = \frac{1}{3}\pi$; Min $= 0$ at $x = 0$

Section 4.9

1. Max $= \frac{1}{8}$, Min $= -1$ **3.** Max $= 1$, Min $= 0$ **5.**‡ Profit $= P = 400x$
$- (x^3 + 100x + 1500) = -x^3 + 300x - 1500$ dollars, $x \geq 0$;
$P' = -3x^2 + 300$; Critical point: $x = 10$; P increasing for $x < 10$, decreasing
for $x > 10$. **5a.** 10 gallons **5b.** 500 dollars **8.**‡ Let $w =$ width (ft.), h
$=$ height (ft.); Cost $= C = [w^2 \text{ sq.ft.}][3 \text{ dollars/sq.ft.}] + [4hw \text{ sq.ft.}] \cdot$
$[2 \text{ dollars/sq.ft.}] = 3w^2 + 8hw$ dollars; Volume $= hw^2 = 6 \Rightarrow h = 6w^{-2}$ and
$C = 3w^2 + 48w^{-1}$, $w > 0$; $dC/dw = 6w - 48w^{-2}$; Critical point: $w = 2$;
C is decreasing for $w < 2$ and increasing for $w > 2$; Minimum cost for $w = 2$,
$h = \frac{3}{2}$ **9.** $r = 8^{1/6}$ ft., $h = 8^{2/3}$ ft. **11a.** $\frac{15}{13}$ hours **11b.** $200/\sqrt{13}$ miles
14. 44 trees/acre **15.**‡ Width of top $= 6 + 12 \cos x$; Depth $= 6 \sin x$; Cross
sectional area $= A = \frac{1}{2}[6 + (6 + 12 \cos x)][6 \sin x] = 36 \sin x$
$+ 36(\sin x)(\cos x)$, $0 \leq x \leq \pi/2$; $A' = 36(\cos x + \cos^2 x - \sin^2 x)$
$= 36(2 \cos^2 x + \cos x - 1)$; Critical points where $\cos x = \frac{1}{2}$ or -1;
Critical point: $x = \pi/3$; A is increasing for $0 < x < \pi/3$ and decreasing for
$\pi/3 < x < \pi/2$; Max for $x = \pi/3$ **16.** $2\pi(1 - \sqrt{\frac{2}{3}})$ **17.** 2 dollars **18.** $1.60
26. $r = (25/\pi)^{1/3}$, $h = 4(25/\pi)^{1/3}$ **29.** Max $= 9$ at $x = 2$; Min does not exist
30. Max $= 1$ at $x = 0$; Min $= \frac{1}{17}$ at $x = 4$ **31.** Max does not exist; Min $= 0$ at
$x = 0$ **32.** Max $= 1$ at $x = 1$; Min does not exist **33.** Max $= \frac{1}{6}$ at $x = 3$;
Min $= -\frac{1}{6}$ at $x = -3$ **34.** Max $= 1$ at $x = (n + \frac{1}{2})\pi$; Min $= 0$ at $x = n\pi$, n
an integer **35.** Max $= -\frac{3}{8}$ at $x = -3$; Min does not exist **36.** Max $= 1$ for
$x > 0$; Min $= -1$ for $x < 0$ **37.** Max $= 1$ at $x = 1$; Min does not exist
38. Max $= 16/(3\sqrt{3})$ at $x = \frac{8}{3}$; The min does not exist

Section 4.10

1.‡ $f(x) = \sin x$; $f'(x) = \cos x$; $f(-\pi/3) = -\frac{1}{2}\sqrt{3}$; $f'(-\pi/3) = \frac{1}{2}$; Tangent
line: $y = -\frac{1}{2}\sqrt{3} + \frac{1}{2}(x + \pi/3)$; $\sin(-1) \approx -\frac{1}{2}\sqrt{3} + \frac{1}{2}(-1 + \pi/3)$
$\approx -\frac{1}{2}(1.7321) - \frac{1}{2} + \frac{1}{6}(3.1416) = -0.84245$; The approximation is less than
the actual value because $f''(x) = -\sin x$ is positive near $x = -\pi/3$
2. 31.75; Less **4.**‡ $f(x) = x^{1/2}$; $f'(x) = \frac{1}{2}x^{-1/2}$; $f(1) = 1$; $f'(1) = \frac{1}{2}$; Tangent
line: $y = 1 + \frac{1}{2}(x - 1)$; $\sqrt{1.03} \approx 1 + \frac{1}{2}(0.03) = 1.015$; $f''(x) = -\frac{1}{4}x^{-3/2}$ is
negative and the approximation is greater than the exact value **5.** 0.97; Less
7.‡ $f(x) = x^{1/4}$; $f'(x) = \frac{1}{4}x^{-3/4}$; $f(10000) = 10$; $f'(10000) = 0.00025$;
Tangent line: $y = 10 + 0.00025(x - 10000)$; $\sqrt[4]{10006} \approx 10 + 0.00025(6)$
$= 10.0015$; The approximation is greater than the exact value because $f''(x)$
$= -\frac{3}{16}x^{-7/4}$ is negative and the graph is concave down **8.** $25 - \frac{4}{15}$
$= 24.7333 \ldots$; greater **10a.** $4dx - 3dy = 0$ **10b.** $3dx + 4dy = 0$
10c. $dx = 0$ **10d.** $dy = 0$; Fig. 4.10.10 **12.** $dx + 2dy = 0$ **12b.** $dx + 2dy$
$= 0$; Fig. 4.10.12 **14.** $1/\sqrt{2} + \pi\sqrt{6}/30 \approx 0.9636$ **15.** -1.38 **16.** 1.5
17. 0.00417 **18.** $\pi\sqrt{3}(0.001) \approx 0.0054$ **19.** $\frac{1}{64}[-1/\sqrt{2} - 3\pi/(4\sqrt{2})]$
≈ -0.0371 **20.** $3y^2 \, dy = -3x^2 \, dx$ **21.** $(2xy^2 - y) \, dx = (x - 2x^2 y) \, dy$
22. $\cos y \, dy = dx$ **24.** $(2x - 8y + 1) \, dx = (8x - 2y^2 + 1) \, dy$

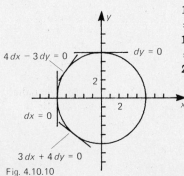

$4 \, dx - 3 \, dy = 0$

$dy = 0$

$dx = 0$

$3 \, dx + 4 \, dy = 0$

Fig. 4.10.10

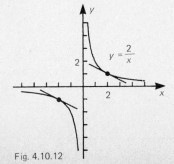

$y = \dfrac{2}{x}$

Fig. 4.10.12

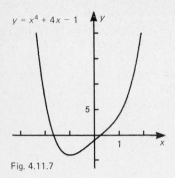

$y = x^4 + 4x - 1$

Fig. 4.11.7

Section 4.11

1.‡ $f(x) = x^3 - 2$; $f'(x) = 3x^2$; $x_{n+1} = x_n - ((x_n^3 - 2)/3x_n^2)$; $x_0 = 1$; $x_1 = \frac{4}{3} \approx 1.333$; $x_2 = \frac{91}{72} \approx 1.264$; $x_3 \approx 1.260$ **2.** 8.944
4.‡ $f(x) = \sin x + \cos x$; $f'(x) = \cos x - \sin x$; x_{n+1}
$= x_n - (\sin x_n + \cos x_n)/(\cos x_n - \sin x_n)$; $x_0 = -1$; $x_1 \approx -0.78204190$;
$x_2 \approx -0.78539818$; $x_3 \approx -0.78539816$ **5.** 0.755 **7.** $-1.663, 0.249$; Fig. 4.11.7
11. $x_3 = 2.125$; Compare with 2.238: $f(2.125) = -.4844$, $f(2.238) = 0.0086$
12a.‡ $f(x) = \sin x - x/2$; $x_0 = 3$; $x_1 = 2.0879954 \ldots$; $x_2 = 1.9122292 \ldots$;
$x_3 = 1.8956526 \ldots$; $x_4 = 1.8954942 \ldots$; $x_5 = 1.8954942 \ldots$;
$x_6 = 1.8954942 \ldots$; $x_7 = 1.8954942 \ldots$ **12b.** $x_7 = 1.1655611 \ldots$

Section 4.12

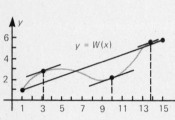

$y = W(x)$

Fig. 4.12.7

1.‡ $f(x) = \sqrt{x}$; $[f(9) - f(1)]/(9 - 1) = \frac{1}{4}$; $f'(c) = \frac{1}{2}c^{-1/2}$ equals $\frac{1}{4}$ and
$1 < c < 9$ for $c = 4$ **2.** $-\sqrt{3}$ **4.**‡ $f(x) = \cos(2x)$; $[f(\pi) - f(0)]/(\pi - 0)$
$= 0$; $f'(c) = -2\sin(2c)$ is 0 and $0 < c < \pi$ for $c = \pi/2$ **5.** $\sqrt{3}$ **7.**‡ $x = 3.5$,
$x = 10$, and $x = 13.5$ where the tangent line is parallel to the secant line
through $(1, W(1))$ and $(15, W(15))$; Fig. 4.12.7 **9.** $f(x) = x^{1/4}$;
$(x + 1)^{1/4} - x^{1/4} = [f(x + 1) - f(x)]/[(x + 1) - x] = f'(c)$ with
$x < c < x + 1$; $f'(c) = \frac{1}{4}c^{-3/4} \to 0$ as $c \to \infty$ and $c \to \infty$ as $x \to \infty$

Section 4.13

1. $x^3 + 6x - 7$

3.‡ $\dfrac{d}{dx}\left(-3\cos x - \dfrac{2}{5}x^5\right) = \dfrac{d}{dx}(-\cos x) - 2\dfrac{d}{dx}\left(\dfrac{1}{5}x^5\right) = 3\sin x - 2x^4$,

so the antiderivatives of $3\sin x - 2x^4$ are $-3\cos x - \dfrac{2}{5}x^5 + C$

4. $10x^{1/2} + 30x^{1/3} + C$ **6a.** $f(x)g(x) + C$ **6b.** $f(x)/g(x) + C$
7. $-\cos(x^3) + C$ **8.** $3x^{4/3} - 38$ **9.** $AF(x) + BG(x) + C$

Miscellaneous Exercises to Chapter 4

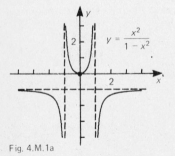

$y = \dfrac{x^2}{1 - x^2}$

Fig. 4.M.1a

1a. Fig. 4.M.1a **1e.**‡ $y' = 4x^3 - 12x - 8$ has $x - 2$ as a factor because it is
zero at $x = 2$; Dividing $x - 2$ into $4x^3 - 12x - 8$ yields
$y' = 4(x - 2)(x + 1)^2$; $y'' = 12(x + 1)(x - 1)$; Decreasing for $x < -1$ and
$-1 < x < 2$; increasing for $x > 2$; Local min at $x = 2$; Concave up for
$x < -1$ and $x > 1$; Concave down for $-1 < x < 1$; Inflection points at
$x = \pm 1$; Fig. 4.M.1e **2a.** Max $= 135$; Min $= -1$ **2c.**‡ $f'(x)$
$= x^4 + 2x^3 - 3x^2 = x^2(x + 3)(x - 1)$; Critical points: $x = 0, 1, -3$;
$f(-4) = -12.8$; $f(-3) = 18.9$; $f(0) = 0$; $f(1) = -0.3$; $f(2) = 6.4$;
Min $= -12.8$; Max $= 18.9$ **2e.**‡ $f'(x) = 4x^3 - 12x^2 + 16$;
Dividing $4x^3 - 12x^2 + 16$ by $x + 1$, $y' = 4(x + 1)(x - 2)^2$; Critical points:
$x = -1, 2$; $f(-2) = 18$; $f(-1) = -9$; $f(1) = 15$; Max $= 18$; Min $= -9$
2f. Max $= \frac{15}{4}$; Min $= 0$ **2h.**‡ By the product rule, $f'(x) = 2(x + 1)(x - 3)^3$
$+ 3(x + 1)^2(x - 3)^2 = (x + 1)(x - 3)^2[2(x - 3) + 3(x + 1)]$
$= (x + 1)(x - 3)^2(5x - 3)$; Critical points: $x = -1, \frac{3}{5}, 3$; $f(0) = -27$;
$f(\frac{3}{5}) = -(1.6)^2(2.4)^3 \approx -35.39$; $f(3) = 0$; $f(4) = 25$; Min $= -(1.6)^2(2.4)^3$;

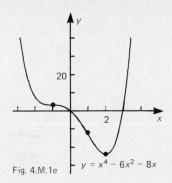

Fig. 4.M.1e

$y = x^4 - 6x^2 - 8x$

Max $= 25$ **2i.** Max $= 16$; Min $= -(1.8)^3(1.2)^2 \approx -8.4$ **3a.** Fig. 4.M.3a

3c. Fig. 4.M.3c   **3f.** Fig. 4.M.3f   **3h.** Fig. 4.M.3h   **3j.** Fig. 4.M.3j   **4a.** Fig. 4.M.4

5.‡ (*) $3y^2y' + 8yy' = 22x$; Set $x = 1$, $y = 1$: $y'(1) = 2$; Differentiate (*):

$6y(y')^2 + 3y^2y'' + 8(y')^2 + 8yy'' = 22$; Set $x = 1$, $y = 1$, $y' = 2$: $y''(1) = -\frac{34}{11}$

6. $-\frac{36}{343}$ **8.** Height $= 7/(4 + \pi)$; Width $=$ diameter $= 14/(4 + \pi)$

11a. $(96/v) + 2v^3$ dollars/mile   **11b.** 2 miles/hour

13.‡ $V'(T) = -10^{-5}[(0.02037)T^2 - (1.702)T + 6.42]$; By the quadratic

formula, $V'(T) = 0$ for

$$T = \frac{1.702 \pm \sqrt{(1.702)^2 - 4(0.02037)(6.42)}}{2(0.02037)}$$

$= T_1 \; (\approx 3.96)$ or $T_2 \; (\approx 79.59)$; $V''(T) > 0$ for $0 \le T \le 30$,

   so the minimum occurs at $T_1 \approx 3.96$ degrees

14. $D = \frac{1}{2}C$ **17.** $p = \sqrt{b/c}$ **23b.** 2.197

24a. $T_1 = \dfrac{L}{C_1}$, $T_2 = \dfrac{2h \sec \theta}{C_1} + \dfrac{L - 2h \tan \theta}{C_2}$; $T_3 = \dfrac{\sqrt{L^2 + 4h^2}}{C_1}$

24c. $C_1 = 2$ Km sec; $h = \dfrac{1}{4\sqrt{5}}$

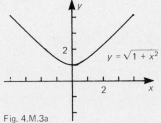

Fig. 4.M.3a

$y = \sqrt{1 + x^2}$

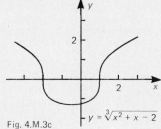

Fig. 4.M.3c

$y = \sqrt[3]{x^2 + x - 2}$

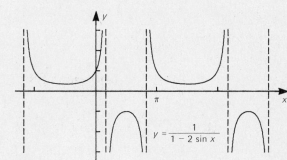

Fig. 4.M.3f

$y = \dfrac{1}{1 - 2 \sin x}$

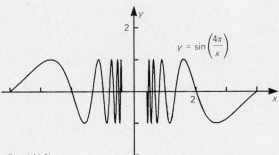

Fig. 4.M.3h

$y = \sin\left(\dfrac{4\pi}{x}\right)$

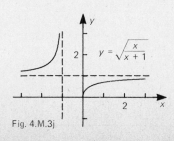

Fig. 4.M.3j

$y = \sqrt{\dfrac{x}{x + 1}}$

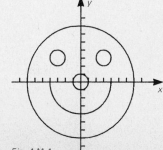

Fig. 4.M.4

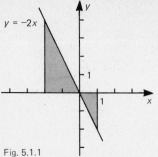

$y = -2x$

Fig. 5.1.1

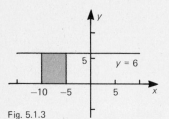

$y = 6$

-10 -5 5

Fig. 5.1.3

Section 5.1

1. 3; Fig. 5.1.1   **3.** 30; Fig. 5.1.3   **5.** $\frac{1}{4}\pi$; Fig. 5.1.5   **7.**‡ $\int_0^{-4} -4\,dx = -\int_{-4}^0 -4\,dx$
$= -(-16) = 16$; Fig. 5.1.7   **9.**‡ Fig. 5.1.9; $-[\text{Area I}] + [\text{Area II}] + [\text{Area III}]$
$= 2$ **12.**‡ Fig. 5.1.12; [Area I] $-$ [Area II] $+$ [Area III] $\approx \frac{12}{4} - \frac{11}{4} + \frac{23}{4} = 6$
(counting squares; each square has area $\frac{1}{4}$)   **13.** -10 **15a.** The area of the
region above the x-axis and below $y = f(x)$ where $f(x)$ is positive and
$a \le x \le c$ is the sum of those areas for $a \le x \le b$ and for $b \le x \le c$. The area
of the region below the x-axis and above $y = f(x)$ where $f(x)$ is negative and
$a \le x \le c$ is the sum of those areas for $a \le x \le b$ and for $b \le x \le c$.
15b. $\int_a^b f(x)\,dx + \int_b^c f(x)\,dx = \int_a^b f(x)\,dx - \int_c^b f(x)\,dx$ (by Definition 5.2)
$= \int_a^b f(x)\,dx - \left[\int_c^a f(x)\,dx + \int_a^b f(x)\,dx\right]$ (by part (a) since $c < a < b$)
$= -\int_c^a f(x)\,dx = \int_a^c f(x)\,dx$ (by Definition 5.2)

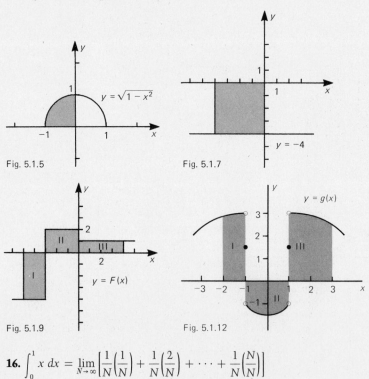

Fig. 5.1.5

$y = \sqrt{1 - x^2}$

-1 1

Fig. 5.1.7

$y = -4$

Fig. 5.1.9

$y = F(x)$

Fig. 5.1.12

$y = g(x)$

16. $\displaystyle\int_0^1 x\,dx = \lim_{N \to \infty} \left[\frac{1}{N}\left(\frac{1}{N}\right) + \frac{1}{N}\left(\frac{2}{N}\right) + \cdots + \frac{1}{N}\left(\frac{N}{N}\right)\right]$

$\displaystyle = \lim_{N \to \infty} \frac{1}{N^2}(1 + 2 + \cdots + N) = \lim_{N \to \infty} \frac{1}{N^2}\left[\frac{1}{2}N(N + 1)\right]$

$\displaystyle = \lim_{N \to \infty} \left[\frac{1}{2} + \frac{1}{2N}\right] = \frac{1}{2}$

17. $\displaystyle\int_0^1 x^3\,dx = \lim_{N \to \infty} \left[\frac{1}{N}\left(\frac{1}{N}\right)^3 + \frac{1}{N}\left(\frac{2}{N}\right)^3 + \cdots + \frac{1}{N}\left(\frac{N}{N}\right)^3\right]$

$\displaystyle = \lim_{N \to \infty} \frac{1}{N^4}(1^3 + 2^3 + \cdots + N^3) = \lim_{N \to \infty} \frac{1}{N^4}\left[\frac{1}{4}N^2(N + 1)^2\right]$

$\displaystyle = \lim_{N \to \infty} \left[\frac{1}{4} + \frac{1}{2N} + \frac{1}{4N^2}\right] = \frac{1}{4}$

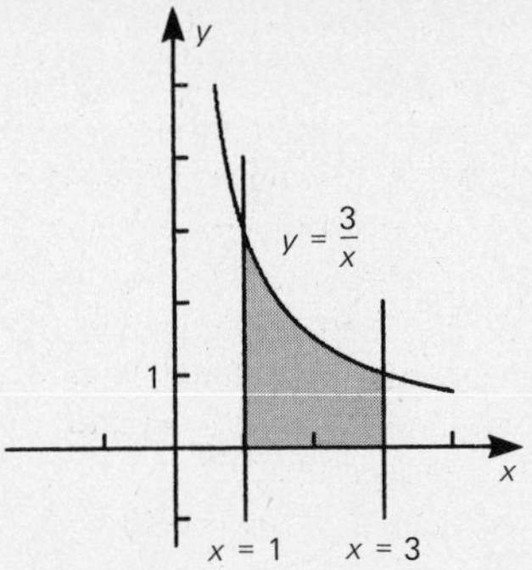

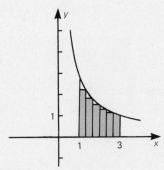

Fig. 5.2.13a

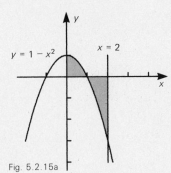

Fig. 5.2.13b

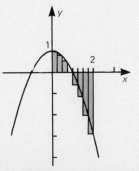

Fig. 5.2.15a

Fig. 5.2.15b

19. $\frac{1}{144}(1 + 4 + 9 + 16 + 25)$ **20.** $144(1 + \frac{1}{4} + \frac{1}{9} + \frac{1}{16} + \frac{1}{25})$
21. $\frac{1}{25}(1 + 2 + 3 + 4 + 5)$ **22.** $4 + 9 + 16 + 25$
23. $\frac{1}{100}(\sin(\frac{1}{10}) + 2\sin(\frac{2}{10}) + 3\sin(\frac{3}{10}) + 4\sin(\frac{4}{10}) + 5\sin(\frac{5}{10}))$
25. $3 - 5 + 7 - 9$

Section 5.2

1.[‡] $1^2(-1)^1 + 2^2(-1)^2 + 3^2(-1)^3 + 4^2(-1)^4 + 5^2(-1)^5 + 6^2(-1)^6$
$= -1 + 4 - 9 + 16 - 25 + 36$ **2.** $\frac{7}{5} + \frac{9}{7} + \frac{11}{9} + \frac{13}{11} + \frac{15}{13}$
4. $\sum_{j=1}^{20}(2x_j^3 - 3x_j^2)(x_j - x_{j-1})$ **6.** $\sum_{j=1}^{N}\sin(t_j)(t_j - t_{j-1})$

8. The width of each subinterval is $\dfrac{1}{10}$; The right endpoint of the jth

subinterval is $j/10$ for $j = 1, 2, \ldots, 50$; $\displaystyle\sum_{j=1}^{50}\cos\left(\frac{j}{10}\right)\left(\frac{1}{10}\right)$

9. $\displaystyle\sum_{j=1}^{100}\left(\frac{j}{100}\right)^3\left(\frac{1}{100}\right)$ **11.** $\displaystyle\sum_{j=6}^{20}\sqrt{\frac{j}{5}}\left(\frac{1}{5}\right)$

13a. Fig. 5.2.13a **13b.** Fig. 5.2.13b **13c.** $\frac{3}{4} + \frac{3}{5} + \frac{1}{2} + \frac{3}{7} + \frac{3}{8} + \frac{1}{3} \approx 2.99$
15a. Fig. 5.2.15a **15b.** Fig. 5.2.15b **15c.** $\frac{15}{64} + \frac{3}{16} + \frac{7}{64} + 0 - \frac{9}{64} - \frac{5}{16} - \frac{33}{64} - \frac{3}{4}$
≈ -1.19 **17.** $\displaystyle\int_0^3 \sin x\, dx$

19. $\displaystyle\int_{-5}^{-2}\frac{6}{1-t}dt$

21. $\displaystyle\int_0^4 [5 - (x^2 + 2)^2]\, dx$

24. $\sin\left(\dfrac{1}{2N}\right)\sin\left(\dfrac{j}{N}\right) = \dfrac{1}{2}\left[\cos\left(\dfrac{2j-1}{N}\right) - \cos\left(\dfrac{2j+1}{N}\right)\right]$, so $\displaystyle\sum_{j=1}^{N}\sin\left(\frac{j}{N}\right)$

$= \dfrac{1}{2\sin[1/(2N)]}\displaystyle\sum_{j=1}^{N}\left[\cos\left(\frac{2j-1}{N}\right) - \cos\left(\frac{2j+1}{N}\right)\right]$

$= \dfrac{\cos[1/(2N)] - \cos[1 + 1(2N)]}{2\sin[1/(2N)]}$

25. $\displaystyle\int_0^1 \sin x\, dx = \lim_{N\to\infty}\frac{1}{N}\sum_{j=1}^{N}\sin\left(\frac{j}{N}\right)$

$= \displaystyle\lim_{N\to\infty}\left\{\frac{\cos[1/(2N)] - \cos[1 + 1/(2N)]}{2N\sin[1/(2N)]}\right\}$

$= 1 - \cos(1)$ since $\cos[1/(2N)]$ tends to $\cos(0) = 1$, $\cos[1 + 1/(2N)]$

tends to $\cos(1)$, and $\dfrac{1}{2N\sin[1/(2N)]} = \dfrac{1/(2N)}{\sin[1/(2N)]}$ tends to 1 by Lemma 3.1

26. $\sin\left(\dfrac{a}{2N}\right)\sin\left(\dfrac{aj}{N}\right) = \dfrac{1}{2}\left[\cos\left(\dfrac{2aj-a}{2N}\right) - \cos\left(\dfrac{2aj+a}{2N}\right)\right]$;

$\displaystyle\int_0^a \sin x\, dx = \lim_{N\to\infty}\frac{a}{N}\sum_{j=1}^{N}\sin\left(\frac{aj}{N}\right)$

$= \displaystyle\lim_{N\to\infty}\frac{a}{2N\sin[a/(2N)]}\sum_{j=1}^{N}\left[\cos\left(\frac{2aj-a}{2N}\right) - \cos\left(\frac{2aj+a}{2N}\right)\right]$

$= \displaystyle\lim_{N\to\infty}\frac{a}{2N\sin[a/(2N)]}\left[\cos\left(\frac{a}{2N}\right) - \cos\left(\frac{2aN+a}{2N}\right)\right]$

$= \cos(0) - \cos(a)$ by Lemma 3.1

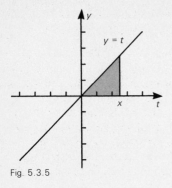

Fig. 5.3.5

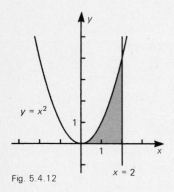

Fig. 5.4.12

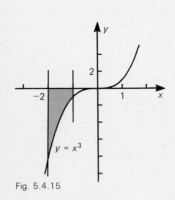

Fig. 5.4.15

31. Let I_1 = the integral from a to b, I_2 = the integral from b to c, and I_3 = the integral from a to c. If $b = x_j$ for some j in the partition $a = x_0 < x_1 < x_2 < \cdots < x_N = c$, then each corresponding Riemann sum for I_3 is a sum of Riemann sums for I_1 and I_2. If $x_{j-1} < b < x_j$ for some j, then we can change a Riemann sum for I_3 into a sum of Riemann sums for I_1 and I_2 by replacing $f(c_j)(x_j - x_{j-1})$ by $f(b)(b - x_{j-1}) + f(b)(x_j - b)$. This change is less than $2M(x_j - x_{j-1})$, where $|f(x)| \leq M$ for $a \leq x \leq c$. Hence any R.S. for I_3 equals a R.S. for I_1 plus a R.S. for I_2 plus a quantity that tends to zero, and in the limit, we obtain equation (1) in Theorem 5.5

35. $-70 \leq \int_{-8}^{27} x^{1/3} \, dx \leq 105$ **37.** Greatest = (a); Least = (c)

38. $\int_0^1 x \sin x \, dx$ **39.** $\int_0^2 \cos x \, dx$ **40.** $\int_0^1 \sqrt{1 + x^4} \, dx$ **41.** $\int_0^4 (x + x^2) \, dx$

Section 5.3

1. $\sqrt{x + 3}$ **3.** $y/(y^2 + 4)$ **5a.**‡ For $x \geq 0$, the integral is the area of the triangle bounded by the t-axis and the lines $t = x$ and $x = t$ (Fig. 5.3.5); $A = \frac{1}{2}bh = \frac{1}{2}x^2$. For $x < 0$, the integral is the area of the triangle bounded by the t-axis and the lines $y = t$ and $t = x$; $A = \frac{1}{2}bh = \frac{1}{2}(-x)(-x) = \frac{1}{2}x^2$; $\frac{1}{2}x^2$

5b.‡ $\dfrac{d}{dx}\left(\dfrac{1}{2}x^2\right) = \dfrac{d}{dx}\int_0^x t \, dt = x \Rightarrow \dfrac{1}{2}x^2$ is an anti-derivative of x; $\dfrac{1}{2}x^2 + C$

6a. $\frac{3}{2}x^2 + 6x$ **6b.** $\frac{3}{2}x^2 + 6x + C$ **6c.**‡ $\int_0^x (3t + 6) \, dt$

$= \int_0^{-2} (3t + 6) \, dt + \int_{-2}^x (3t + 6) \, dt = \frac{3}{2}x^2 + 6x$ **8.** $1/\sqrt{2}$ **10.** $\sin(81)$

12. 910   **14.** $\sqrt{3}/2$ **15.** $\frac{1}{10}$ **16.** $\frac{1}{5}(1 - 2^{-1/2})$ **17.** 21   **19.** $\frac{1}{3}\sin(1)$

21a. $\sqrt{1 + x^4}$ **21b.** $\sqrt{1 + x^4}$ **22a.** $5x$ **22b.** $10(x - 1)$ **22c.** $-2x$

Section 5.4

1. $\frac{3}{4}x^{4/3} + C$

3.‡ $\int x^{1/2} \, dx = \dfrac{1}{3/2}x^{3/2} + C = \dfrac{2}{3}x^{3/2} + C$

5. 21   **7.**‡ $\int_1^9 x^{-1/2} \, dx = [2x^{1/2}]_1^9 = 6 - 2 = 4$ **8.**‡ $\int_{96}^{97} x^0 \, dx = [x]_{96}^{97}$

$= 97 - 96 = 1$ **10.** $-\sqrt{2}$ **12.**‡ $\int_0^2 x^2 \, dx = [\frac{1}{3}x^3]_0^2 = \frac{1}{3}(2)^3 - \frac{1}{3}(0)^3 = \frac{8}{3}$;

Fig. 5.4.12   **13.** 2   **15.**‡ $-\int_{-2}^{-1} x^3 \, dx = [-\frac{1}{4}x^4]_{-2}^{-1}$

$= [-\frac{1}{4}(-1)^4] - [-\frac{1}{4}(-2)^4] = \frac{15}{4}$; Fig. 5.4.15   **17.** $\frac{1}{4}(10^4 - 1)$ **18.** $\frac{3}{11}$

19. 0   **20.** $-\frac{1}{2} + \frac{1}{2}\sqrt{2}$ **21.** $\frac{64}{5}$ **22.** $\frac{2}{201}$ **23.** 0   **24.** $\frac{7}{12}$

Section 5.5

1.‡ $3(\frac{1}{3}x^3) - 6(\frac{1}{11}x^{11}) + C = x^3 - \frac{6}{11}x^{11} + C$ **2.** $\frac{5}{6}x^6 + \frac{6}{5}x^{-5} + C$

4.‡ $[x - \frac{6}{8}x^8]_0^1 = \frac{1}{4}$ **5.** $-\frac{3}{4}\pi + 1/\sqrt{2}$

7.‡ $\left[\dfrac{7}{12/7}x^{12/7} + \dfrac{5}{12/5}x^{12/5}\right]_0^1 = \left[\dfrac{49}{12}x^{12/7} + \dfrac{25}{12}x^{12/5}\right]_0^1 = \dfrac{37}{6}$

8. $-\frac{264}{5}$ **10.** $3x + 2x^2 - \frac{5}{3}x^3 + C$ **11.** $\frac{3}{4}x^{4/3} + \frac{4}{5}x^{5/4} + \frac{5}{6}x^{6/5} + C$

12.‡ $\int (x - x^{1/2} + x^{1/3}) \, dx = \dfrac{1}{2}x^2 - \dfrac{1}{3/2}x^{3/2} + \dfrac{1}{4/3}x^{4/3} + C$

$= \dfrac{1}{2}x^2 - \dfrac{2}{3}x^{3/2} + \dfrac{3}{4}x^{4/3} + C$

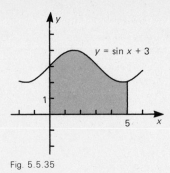

Fig. 5.5.35

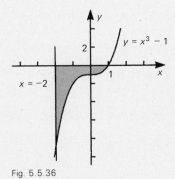

Fig. 5.5.36

13. $-\frac{123}{64}$ **15.**‡ $\int_0^1 (1 + 6t^2 + 9t^4)\, dt = [t + 2t^3 + \frac{9}{5}t^5]_0^1 = \frac{24}{5}$

17.‡ $\int_2^1 (w^{-3} - w^{-2})\, dw = [-\frac{1}{2}w^{-2} + w^{-1}]_2^1 = \cdot\frac{1}{8}$ **19.** $-\frac{4}{3}$

21.‡ $\int_{-5}^0 \sin x\, dx + \int_0^5 (1 - \cos x)\, dx = [-\cos x]_{-5}^0 + [x - \sin x]_0^5$
$= [-1] - [-\cos(-5)] + [5 - \sin(5)] - [0] = 4 + \cos(-5) - \sin(5)$
22. $2^4 + 3^5 - 2^5 = 227$ **24.** Positive because the integrand is positive in the
interval of integration and $0 < 20$ **25.** Negative because the integrand is
negative in the interval of integration and $-75 < 0$ **27.** Zero because the
integrand is odd $(f(-x) = -f(x))$ and the interval of integration is symmetric
about $x = 0$ **28.** Negative because $-5 < 0$ and the integrand is negative
30. Negative because the integrand is ≥ 0 and $4 > 0$ **31.** Positive because the
integrand is negative and $0 > -1$ **32.** Zero because the integrand is odd and
the interval of integration is symmetric about $x = 0$ **35.**‡ $\int_0^5 (\sin x + 3)\, dx$
$= [-\cos x + 3x]_0^5 = [-\cos(5) + 3(5)] - [-1 + 0] = -\cos(5) + 16$;
Fig. 5.5.35   **36.** $\frac{27}{4}$; Fig. 5.5.36

Section 5.6

1.‡ Let $u = x^3 - 1$; $du = 3x^2\, dx$; $\int (x^3 - 1)^{1/2} x^2\, dx$

$= \frac{1}{3} \int (x^3 - 1)^{1/2}\, (3x^2\, dx) = \frac{1}{3} \int u^{1/2}\, du = \frac{1}{3}\left(\frac{1}{3/2}\right)u^{3/2} + C$

$= \frac{2}{9}(x^3 - 1)^{3/2} + C$

2. $\frac{2}{15}(5u + 1)^{3/2} + C$ **4.** $-\frac{1}{3}\cos(3t + 1) + C$ **5.**‡ Let $u = \sin x$;
$du = \cos x\, dx$; $\int \sin^3 x \cos x\, dx = \int u^3\, du = \frac{1}{4}u^4 + C = \frac{1}{4}\sin^4 x + C$

7.‡ Let $u = x^2 + 9$; $du = 2x\, dx$; $\int x\sqrt{x^2 + 9}\, dx = \frac{1}{2} \int (x^2 + 9)^{1/2}\, (2x\, dx)$

$= \frac{1}{2} \int u^{1/2}\, du = \frac{1}{2}\left(\frac{1}{3/2}\right)u^{3/2} + C = \frac{1}{3}(x^2 + 9)^{3/2} + C; \left[\frac{1}{3}(x^2 + 9)^{3/2}\right]_0^1$

$= \frac{1}{3}10^{3/2} - 9$

8. $\frac{1}{3}[(21)^{3/2} - (3)^{3/2}]$

10.‡ Let $u = t + 10$, $du = dt$, $t = u - 10$; $\int \frac{t}{\sqrt{t + 10}}\, dt = \int (u - 10)\, u^{-1/2}\, du$

$= \frac{2}{3}(t + 10)^{3/2} - 20(t + 10)^{1/2} + C$

11. $-\frac{1}{9}(3y - 4)^{-1} - \frac{2}{9}(3y - 4)^{-2} + C$ **13.**‡ Let $u = x^{1/2}$, $du = \frac{1}{2}x^{-1/2}\, dx$;
$\int x^{-1/2} \sin(x^{1/2})\, dx = 2 \int \sin u\, du = -2 \cos u + C = -2 \cos\sqrt{x} + C$ **15.** $\frac{1}{13}$
17.‡ Let $u = \sin(2x)$, $du = 2 \cos(2x)\, dx$; $\int \sin(2x) \cos(2x)\, dx = \frac{1}{2} \int u\, du$
$= \frac{1}{4}u^2 + C = \frac{1}{4}\sin^2(2x) + C$

18. $-\dfrac{1}{3 \sin 3x} + C$ **20.** $\frac{3}{16}(4x + 1)^{4/3} + C$ **21.** $-\sin(3 - x) + C$

22. $-\frac{3}{16}(\cos(4x))^{4/3} + C$ **23.** $\frac{2}{3}(1 - x)^{3/2} - 2(1 - x)^{1/2} + C$
24. $\frac{1}{9}(x - 1)^9 + \frac{1}{4}(x - 1)^8 + C$ **25.** $\frac{3}{20}(x^2 + 2x + 5)^{10/3} + C$
26. $\frac{1}{2}\cos(x^{-2}) + C$ **27.** $[\cos(4) - \cos(2)]x + C$ **28.** $\frac{1}{3}[1 + \cos(2)]^{-3} - \frac{1}{24}$
29. $\frac{9}{40}$ **30.** $\frac{1}{9}$ **32.** $x + C$

Chapter 5: Appendix

1. See the answer to Exercise 27 of Section 5.2. Choose $\delta_1 > 0$ so that $|I_1 - [\text{R.S. for } I_1]| \leq \epsilon/3$ for all partitions of $a \leq x \leq b$ into subintervals of length $\leq \delta_1$. Choose $\delta_2 > 0$ so that $I_2 - [\text{R.S. for } I_2] \leq \epsilon/3$ for all partitions of $b \leq x \leq c$ into subintervals of length $\leq \delta_2$. Let δ be the smaller of δ_1, δ_2, and $\epsilon/6M$. Let S_3 be any R.S. for I_3 for a partition of $a \leq x \leq c$ into subintervals of length $\leq \delta$, and let S_1 and S_2 be Riemann sums for I_1 and I_2 constructed as in the solution of Exercise 27 of Section 5.2. Then $|S_3 - (I_1 + I_2)|$ $\leq |S_1 - I_1| + |S_2 - I_2| + |S_3 - (S_1 + S_2)| \leq \epsilon/3 + \epsilon/3 + 2M\delta \leq \epsilon$.

3a. Consider c with $x_{j-1} < c < x_j$. Let $M_1 = $ max of $f(x)$ for $x_{j-1} \leq x \leq c$, $M_2 = $ max of $f(x)$ for $c \leq x \leq x_j$, and $M_3 = $ max of $f(x)$ for $x_{j-1} \leq x \leq x_j$. If c is added to the partition, the new upper sum is obtained by replacing $M_3(x_j - x_{j-1})$ with $M_1(c - x_{j-1}) + M_2(x_j - c)$. This can only decrease the upper sum because $M_3 \geq M_1$, $M_3 \geq M_2$, and $M_3(x_j - x_{j-1})$ $= M_3(c - x_{j-1}) + M_3(x_j - c) \geq M_1(x - x_{j-1}) + M_2(x_j - c)$.

3c. Let P_1 and P_2 be partitions and P_3 the partition obtained by combining the division points of P_1 and P_2. By parts (a) and (b), [Lower sum for P_1] $\leq$ [Lower sum for P_3] $\leq$ [Upper sum for P_3] $\leq$ [Upper sum for P_2].

Miscellaneous Exercises to Chapter 5

1. Fig. 5.M.1; [Area of triangle I] $-$ [Area of triangle II] $= 16 - 1 = 15$

3. The region above the x-axis and below $y = f(x)$ where $f(x) > 0$ is the mirror image of the region below the x-axis and above $y = -f(x)$ where $-f(x) < 0$; the region below the x-axis and above $y = f(x)$ where $f(x) < 0$ is the mirror image of the region above the x-axis and below $y = -f(x)$ where $-f(x) > 0$.

6. $1 + \frac{2}{3} + \frac{4}{5} + \frac{8}{7} + \frac{16}{9}$ **8.**‡ We can obtain the powers of $\frac{1}{10}$ by taking $(\frac{1}{10})^j$ for $j = 0, 1, \ldots, 6$; With this choice of j, the alternating signs are given by $(-1)^j$; $\sum_{j=0}^{6} (-1)^j (\frac{1}{10})^j$

9.‡ The factors $1, \frac{1}{2}, \frac{1}{3}, \ldots, \frac{1}{6}$ are given by $\frac{1}{j}$ for $j = 1, 2, \ldots, 6$. With this choice of j, the exponents are given by $2j - 1$ and the alternating signs by

$(-1)^{j+1}$; $\sum_{j=1}^{6} (-1)^{j+1} \left(\frac{1}{j}\right)(10)^{2j-1}$ **10.** $\sum_{j=1}^{8} (-1)^{j-1} \frac{1}{j^2}$ **11.** $\sum_{j=1}^{6} \frac{\sqrt{2j-1}}{2j+1}$

14. $\sum_{j=1}^{N} x_{j-1}^4 (x_j - x_{j-1})$

16. $\frac{1}{50} \sum_{j=0}^{149} \frac{j/50}{1 + j/50}$

18. $\sum_{j=1}^{N} (c_j^5 - c_j)(x_j - x_{j-1})$ with $c_j = \frac{1}{2}(x_{j-1} + x_j)$ **20.** $\int_1^7 \frac{x^2}{2x + 4} dx$

22. If $0 = x_0 < x_1 < x_2 < \cdots < x_N = a$ is a partition of $0 \leq x \leq a$, then $-a = -x_N < -x_{N-1} < \cdots < -x_1 < -x_0 = 0$ is a partition of $-a \leq x \leq 0$, and the Riemann sums $\sum_{j=1}^{N} f(c_j)(x_j - x_{j-1})$ and $\sum_{j=1}^{N} f(-c_j) \cdot$ $(-x_{j-1} - (-x_j))$ for the integrals from 0 to a and from $-a$ to 0 are equal.

25.‡ Set $G(u) = \int_1^u \sqrt{t + 1}\, dt$; $G'(u) = \sqrt{u + 1}$; $\frac{d}{dx} G(x^2) = G'(x^2) \frac{d}{dx}(x^2)$

$= 2x\sqrt{x^2 + 1}$

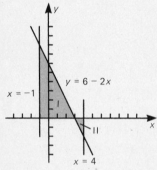

y

$y = 6 - 2x$

$x = -1$

I

II

$x = 4$

x

Fig. 5.M.1

26. $3 \sin^5 (3x)$

28.‡ Set $G(u) = \displaystyle\int_{10}^{u} \sqrt[3]{\cos t}\, dt;\ G'(u) = \sqrt[3]{\cos u};\ \dfrac{d}{dx}[-G(x^3)]$

$= -G'(x^3) \dfrac{d}{dx}(x^3) = -3x^2 \sqrt[3]{\cos(x^3)}$

30.‡ Set $G(u) = \displaystyle\int_{0}^{u} (1 - t^2)^{1/3}\, dt;\ \int_{x^2}^{x^3} (1 - t^2)^{1/3}\, dt = G(x^3) - G(x^2);$

$\dfrac{d}{dx}[G(x^3) - G(x^2)] = G'(x^3)\dfrac{d}{dx}(x^3) - G'(x^2)\dfrac{d}{dx}(x^2)$

$= 3x^2(1 - x^6)^{1/3} - 2x(1 - x^4)^{1/3}$

31. $(25x) \sin(5x) - (16x) \sin(4x)$

35. 3 **36.** $-\frac{3}{8}$ **37.** Not defined; $x^{1/4}$ is not defined for $x < 0$ **38.** 0

43. $\sqrt{2} - 1$ **45.** $-\frac{270}{7}$ **46.** Not defined because $\sin x / \sqrt{\cos x}$ tends to ∞ as $x \to 2\pi^-$ **48.**‡ $\int (x^4 + 2x^3 + 3x^2 + 2x + 1)\, dx$

$= \frac{1}{5}x^5 + \frac{1}{2}x^4 + x^3 + x^2 + x + C$ **49.** $\frac{1}{7}x^7 + \frac{2}{5}x^5 + x^3 + x + C$

51. $\frac{1}{2}x^2 + \frac{4}{3}x^{3/2} + x + C$ **53.** $\sin(\sin x) + C$ **55.** $\sqrt{8} - 2$ **56.** $\frac{51}{4}$

58. The addition indicated in Fig. 5.M.58 gives twice the desired sum. It gives $N(N + 1)$ since it gives N terms each equal to $N + 1$.

$$
\begin{array}{cccccccc}
1 + & 2 & + & 3 & + \cdots + & (N-2) + & (N-1) + & N \\
+ N + & (N-1) + & (N-2) + & \cdots + & 3 & + & 2 & + 1 \\
\hline
(N+1) + & (N+1) + & (N+1) + & \cdots + & (N+1) + & (N+1) + & (N+1)
\end{array}
$$

Fig. 5.M.58

60a.‡ $\frac{5}{2}[(\frac{5}{2})^2 + 5^2 + (\frac{15}{2})^2 + 10^2] = \frac{1875}{4} = 464.75$
60b.‡ $(1)(1^2 + 2^2 + \cdots + 10^2) = 385$

60c.‡ $\frac{1}{2}[(\frac{1}{2})^2 + 1^2 + (\frac{3}{2})^2 + \cdots + (\frac{19}{2})^2 + 10^2] = 358.75$ **60d.**‡ $\int_{0}^{10} x^2\, dx$

$= \frac{1}{3}(10)^3 = 333.33 \ldots$ **61a.** 0.5625 **61b.** $0.501388 \ldots$ **61c.** $0.480686 \ldots$
61d. $0.459697 \ldots$ **62a.** 6.146 **62b.** $5.684074 \ldots$ **62c.** $5.515572 \ldots$
62d. $5.333333 \ldots$

67. Let $f(x) = \left(\dfrac{j}{N}\right)^n$ for $\dfrac{j-1}{N} \leq x \leq \dfrac{j}{N}$ and $g(x) = \left(\dfrac{j}{N+1}\right)^n$ for

$\dfrac{j}{N+1} \leq x \leq \dfrac{j+1}{N+1}$. Then $g(x) \leq x^n \leq f(x)$ for $x \geq 0$, and

$\displaystyle\int_{0}^{1} g(x)\, dx < \int_{0}^{1} x^n\, dx < \int_{0}^{1} f(x)\, dx$, which gives the inequalities since

$\displaystyle\int_{0}^{1} g(x)\, dx = \dfrac{1}{N+1} \sum_{j=1}^{N} \left(\dfrac{1}{N+1}\right)^n = \dfrac{1}{(N+1)^{n+1}} \sum_{j=1}^{N} j^n,\ \int_{0}^{1} x^n\, dx = \dfrac{1}{n+1},$ and

$\displaystyle\int_{0}^{1} f(x)\, dx = \dfrac{1}{N} \sum_{j=1}^{N} \left(\dfrac{j}{N}\right)^n = \dfrac{1}{N^{n+1}} \sum_{j=1}^{N} j^n.$

70. The sum is the Riemann sum $\dfrac{1}{N} \displaystyle\sum_{j=1}^{N} \sqrt{\dfrac{j}{N}}$ for $\int_{0}^{1} \sqrt{x}\, dx;\ \dfrac{2}{3}$ **72.** $-\frac{170}{3}$

Section 6.1

1.‡ Fig. 6.1.1; Equate expressions for y in $y = 3 - x^2$ and $y = -2x$ to obtain an equation for the x-coordinates of the intersections; $x^2 - 2x - 3 = 0$; $x = -1, 3$; $\int_{-1}^{3} [(3 - x^2) - (-2x)] \, dx = \frac{32}{3}$ **2.** Fig. 6.1.2; $\frac{64}{3}$ **4.**‡ Fig. 6.1.4; Solve $y = x^3, y = x^2$ for x; $x^3 = x^2$; $x = 0, 1$; $x^3 < x^2$ for $0 < x < 1$; $\int_{0}^{1} (x^2 - x^3) \, dx = \frac{1}{12}$ **5.** Fig. 6.1.5; $\frac{32}{3}$ **7.**‡ Fig. 6.1.7; $y = x^3 + 1$ and $y = 4x + 1 \Rightarrow x = 0, -2, 2$; $\int_{-2}^{0} [(x^3 + 1) - (4x + 1)] \, dx + \int_{0}^{2} [(4x + 1) - (x^3 + 1)] \, dx$ or by symmetry $2 \int_{0}^{2} [(4x + 1) - (x^3 + 1)] \, dx$; 8   **8.** Fig. 6.1.8; 6   **10.**‡ Fig. 6.1.10; $y = 2 \sin x$ and $y = -3 \sin x \Rightarrow x = k\pi$, integers k; $\int_{0}^{\pi} [(2 \sin x) - (-3 \sin x)] \, dx + \int_{\pi}^{2\pi} [(-3 \sin x) - (2 \sin x)] \, dx$ or by symmetry $2 \int_{0}^{\pi} [(2 \sin x) - (-3 \sin x)] \, dx$; 20   **11.** Fig. 6.1.11; 8

13.‡ Fig. 6.1.13; $x = y^2$ and $x = 9 \Rightarrow y = \pm 3$; $\int_{-3}^{3} (9 - y^2) \, dy = 36$

14. Fig. 6.1.14; $\frac{64}{3}$ **16.**‡ Fig. 6.1.16; $x = y^3$ and $y = x^3 \Rightarrow y = y^9 \Rightarrow y = 0, \pm 1$; $\int_{-1}^{0} (x^3 - x^{1/3}) \, dx + \int_{0}^{1} (x^{1/3} - x^3) \, dx$ or $2 \int_{0}^{1} (x^{1/3} - x^3) \, dx$; 1

17. Fig. 6.1.17; $\frac{1}{3}$ **18.**‡ Fig. 6.1.18; $\int_{0}^{9} (9 - x)^{1/2} \, dx = [-\frac{2}{3}(9 - x)^{3/2}]_{0}^{9} = 18$

19. Fig. 6.1.19; $\pi + \frac{1}{3}$ **21.** Fig. 6.1.21; 8   **23.**‡ Fig. 6.1.23; $2 \int_{6}^{2} [(0) - (x^3 - 4x)] \, dx = 8$ **25.**‡ Fig. 6.1.4; See Exercise 4 for the x-integral; $x = y^{1/3}$ and $x = y^{1/2} \Rightarrow y = 0, 1$; $\int_{0}^{1} (y^{1/3} - y^{1/2}) \, dy = \frac{1}{12}$

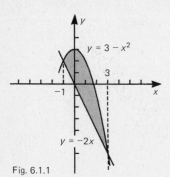

Fig. 6.1.1

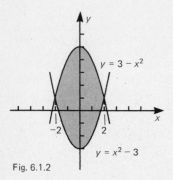

Fig. 6.1.2

Fig. 6.1.4

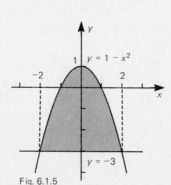

Fig. 6.1.5

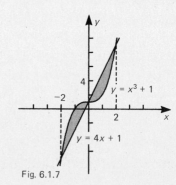

Fig. 6.1.7

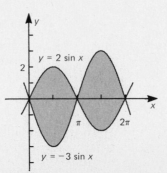

Fig. 6.1.10

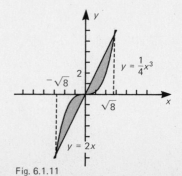

Fig. 6.1.11

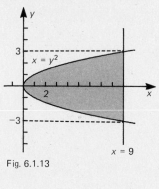

Fig. 6.1.13

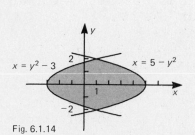

Fig. 6.1.14

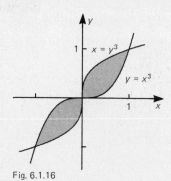

Fig. 6.1.16

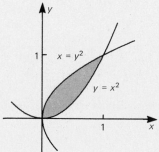

Fig. 6.1.17

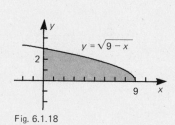

Fig. 6.1.18

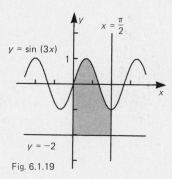

Fig. 6.1.19

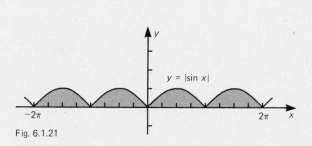

Fig. 6.1.21

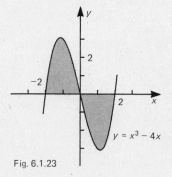

Fig. 6.1.23

The cross section at y

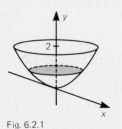

Fig. 6.2.1

The cross section at x

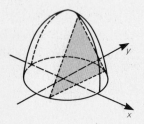

Fig. 6.2.2a

Section 6.2

1.‡ Fig. 6.2.1; The radius of the cross section at y is $x = \sqrt{2y}$;
$A(y) = \pi(\sqrt{2y})^2 = 2\pi y$; $\int_0^2 2\pi y\, dy = 4\pi$ **2.**‡ Fig. 6.2.2a; Area of equilateral
triangle of side s is $\frac{1}{4}s^2\sqrt{3}$; The cross section at x has $s = 2\sqrt{1-x^2}$
(Fig. 6.2.2b); $A(x) = \sqrt{3}(1-x^2)$; $\int_{-1}^1 \sqrt{3}(1-x^2)\, dx = \frac{4}{3}\sqrt{3}$

4.‡ Fig. 6.2.4; If $A(x) = $ area of cross section x units below the vertex, then

$$A(x) = (x/h)^2 A \text{ for } 0 \le x \le h \text{ and } V = \int_0^h \left(\frac{x}{h}\right)^2 A\, dx = \frac{1}{3}Ah$$

5. $5\sqrt{3}$ **7.**‡ Fig. 6.2.7a; The cross sections parallel to the axes of both cylinders
are squares; The cross section h units above the axes has sides of length
$s = 2\sqrt{16-h^2}$ (Fig. 6.2.7b); $\int_{-4}^4 4(16-h^2)\, dh = \frac{1024}{3}$ **8.** $\frac{4}{3}\pi$

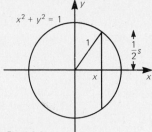

Fig. 6.2.2b

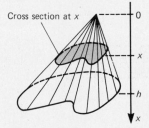

Cross section at x

Fig. 6.2.4

Cross section at h

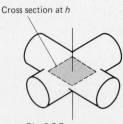

Fig. 6.2.7a

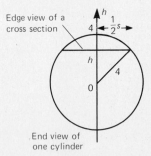

Edge view of a
cross section

End view of
one cylinder

Fig. 6.2.7b

Section 6.3

1.‡ Fig. 6.3.1; The radius of the cross-section at x is $(1-x^2) - (-3)$;
$A(x) = \pi(4-x^2)^2$; $\int_{-2}^2 \pi(4-x^2)^2\, dx = \int_{-2}^2 \pi(16 - 8x^2 + x^4)\, dx = \frac{512}{15}\pi$
2.‡ Fig. 6.3.2; The radius at x is $1 - \frac{1}{2}x^2$; $y = \frac{1}{2}x^2$ and $y = 1 \Rightarrow x = \pm\sqrt{2}$;
$\int_{-\sqrt{2}}^{\sqrt{2}} \pi(1 - \frac{1}{2}x^2)^2\, dx = \frac{16}{15}\sqrt{2}\,\pi$ **3.** Fig. 6.3.3; 3π **5.**‡ Fig. 6.3.5; The radius at y
is $(y^2 + 2) - (1)$; $\int_{-1}^1 \pi(y^2+1)^2\, dy = \frac{56}{15}\pi$ **6.**‡ Fig. 6.3.6; The radius at y is
$1 - y^{1/3}$; $\int_{-8}^1 \pi(1 - y^{1/3})^2\, dy = \frac{513}{10}\pi$ **7.** Fig. 6.3.7; 12π **9.**‡ Fig. 6.3.9; Outer
radius at x is $3x$; Inner radius at x is x^2; $A(x) = \pi(3x)^2 - \pi(x^2)^2$; $y = 3x$ and
$y = x^2 \Rightarrow x = 0, 3$; $\int_0^3 \pi(9x^2 - x^4)\, dx = \frac{162}{5}\pi$ **10.**‡ Fig. 6.3.10; Outer radius at
x is $2 - (-1)$; Inner radius at x is $2 - x^3$; $\int_{-1}^1 \pi[3^2 - (2-x^3)^2]\, dx = \frac{68}{7}\pi$

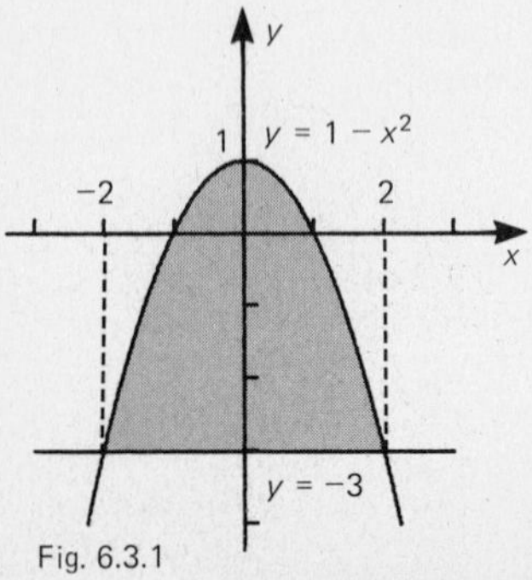

$y = 1 - x^2$

$y = -3$

Fig. 6.3.1

12.‡ Fig. 6.3.12; Outer radius at y is $3y$; Inner radius is y^2; $\int_0^3 \pi(9y^2 - y^4)\,dy$
$= \frac{162}{5}\pi$ **13.**‡ Fig. 6.3.13; Outer radius at y is $1 - (y-2)$; Inner radius at y is
$1 - (-y^2)$; $\int_{-2}^1 \pi[(3-y)^2 - (1+y^2)^2]\,dy = \frac{117}{5}\pi$ **14.** Fig. 6.3.14; 8π

16. $\int_0^\pi \pi[(\sin x + 2)^2 - 1]\,dx$

18. $\int_{-1}^0 \pi[(2 - x^{1/3})^2 - (2 - x^3)^2]\,dx + \int_0^1 \pi[(2 - x^3)^2 - (2 - x^{1/3})^2]\,dx$
$= 4\pi$; Fig. 6.3.18   **20.** Length $= 18^{2/3}$ in.; Radius $= 18^{1/6}$ in.

22. $\frac{1}{9}\pi$ **23.** $\frac{64}{3}\pi$ **24.** 20π **25.** 40π **26.** $\frac{1944}{5}\pi$ **27.** $\frac{7}{18}\pi$ **28.** $\frac{1}{3}\pi r^2 h$; Cone
29. $\pi r^2 h$; Cylinder   **30.** 36π **31.** $\frac{1}{3}\pi(2R^3 - 3R^2h + h^3)$

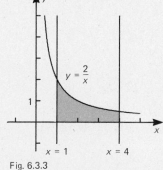

Fig. 6.3.3

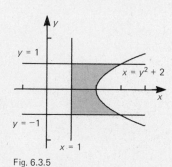

Fig. 6.3.5

Fig. 6.3.2

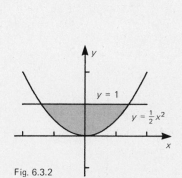

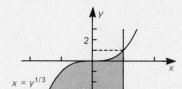

Fig. 6.3.6

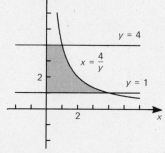

Fig. 6.3.7

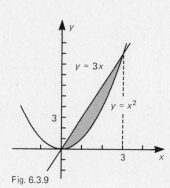

Fig. 6.3.9

Fig. 6.3.10

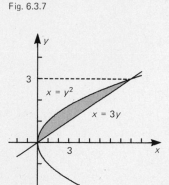

Fig. 6.3.12

Fig. 6.3.13

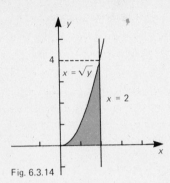

Fig. 6.3.14

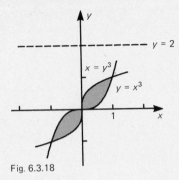

Fig. 6.3.18

Section 6.4

1a. $x - 2$ **1b.** $-x - 3$ **2a.**‡ Fig. 6.4.2; Radius of cylinder at x is x; Height of cylinder at x is $2x^{-1/2}$; $A(x) = 2\pi rh = 4\pi x^{1/2}$; $\int_1^2 4\pi x^{1/2}\, dx = \frac{8}{3}\pi(2^{3/2} - 1)$

2b.‡ Radius of the cylinder at x is $3 - x$; Height of the cylinder at x is $2x^{-1/2}$; $\int_1^2 2\pi(3 - x)(2x^{-1/2})\, dx = \frac{8}{3}\pi(7\sqrt{2} - 8)$ **2c.**‡ Outer radius at x is $x^{-1/2} - (-2)$; Inner radius at x is $-x^{-1/2} - (-2)$; $\int_1^2 \pi[(x^{-1/2} + 2)^2 - (2 - x^{-1/2})^2]\, dx = 16\pi(\sqrt{2} - 1)$ **3a.** Fig. 6.4.3; $\frac{459}{10}\pi$

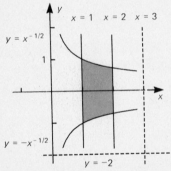

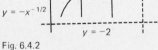

Fig. 6.4.2

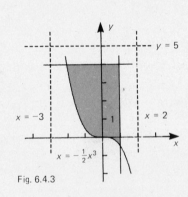

Fig. 6.4.3

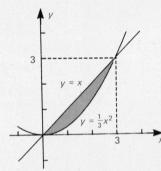

Fig. 6.4.5

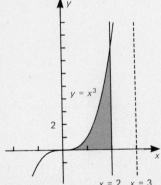

Fig. 6.4.6

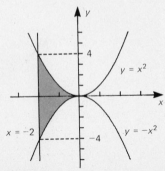

Fig. 6.4.8

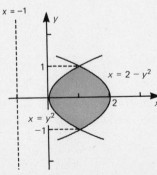

Fig. 6.4.9

3b. $\frac{405}{7}\pi$ **3c.** $\frac{1107}{20}\pi$ **5.‡** Fig. 6.4.5; $\int_0^3 2\pi x\,(x - \frac{1}{3}x^2)\,dx$
$= \int_0^3 \pi\,[(y\sqrt 3)^2 - y^2]\,dy = \frac{9}{2}\pi$ **6.** Fig. 6.4.6; $\frac{56}{5}\pi$ **8.‡** Fig. 6.4.8; The region is
symmetric about the x-axis and the axis of revolution is vertical; Inner radius
at y is $\sqrt y$ for $0 \le y \le 4$; Outer radius is 2; $2\int_0^4 \pi[2^2 - (\sqrt y)^2]\,dy$; Or: the
height of the cylinder at x is $2x^2$ and its radius is $-x$; $\int_{-2}^0 2\pi(-x)(2x^2)\,dx$; 16π
9. Fig. 6.4.9; $\frac{32}{3}\pi$ **11.** The cross-section is a ring of area = [Area of outer circle]
− [Area of inner circle] = π[Outer radius]^2 − π[Inner radius]^2; Volume
= [Area of cross-section][Height] **12a.** By the Mean Value Theorem for
derivatives, $|G(d_j) - G(c_j)| \le M_1|d_j - c_j| \le M_1|x_j - x_{j-1}|$,
where $|G'(x)| \le M_1$ for $a \le x \le b$.
$|\sum_{j=1}^N F(c_j)G(d_j)(x_j - x_{j-1}) - \sum_{j=1}^N F(c_j)G(c_j)(x_j - x_{j-1})|$
$= |\sum_{j=1}^N F(c_j)[G(d_j) - G(c_j)](x_j - x_{j-1})| \le M_1 M_2(b - a)\delta$ where
$|F(x)| \le M_2$ for $a \le x \le b$ and $|x_j - x_{j-1}| \le \delta$ for $j = 1, 2, \ldots, n$. Thus,
the sum in the statement of the exercise differs from a Riemann sum by a term
that tends to zero as the lengths of the subintervals tend to zero, so it has the
same limit, the integral. **13.** $\frac{8}{3}\pi$ **14.** $\frac{27}{2}\pi$ **15.** 36π **16.** $\frac{26}{5}\pi$ **17.** π **18.** $\frac{5}{14}\pi$
19. 4π **20.** $\frac{16}{3}\pi$

Section 6.5

1. $\frac{50}{3}$ **3.** $\int_0^\pi \sqrt{1 + \cos^2 x}\,dx$ **5.** $4\int_{-1}^1 (4 - x^2)^{-1/2}\,dx$ (both arcs)
7. $\int_{-3\sqrt 3}^{3\sqrt 3} \sqrt{(324 - 8x^2)/(324 - 9x^2)}\,dx$ **9.** $\frac{8}{27}(37)^{3/2} - \frac{1}{27}(13)^{3/2}$
11. $\frac{2}{3}(5\sqrt 5 - 1)$ **12.** 15

Section 6.6

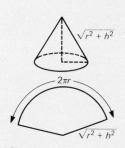

Fig. 6.6.1

1. Fig. 6.6.1; The lateral height of the cone is $\sqrt{r^2 + h^2}$, and the circumference
of its base is $2\pi r$; When the cone is cut from vertex to base and flattened it
forms the fraction $2\pi r/(2\pi\sqrt{r^2 + h^2})$ of a circle of radius $\sqrt{r^2 + h^2}$;
$\pi r\sqrt{r^2 + h^2}$ **3.** $\frac{1}{27}\pi\,[(145)^{3/2} - 1]$ **4.** $2\pi\int_{-1}^1 x^2\sqrt{1 + 4x^2}\,dx$
6.‡ Use formula (1) with x replaced by y; $x = y^{1/3}$;
$\frac{dx}{dy} = \frac{1}{3}x^{-2/3}$; $2\pi\int_1^8 (y^{1/3} + 1)\sqrt{1 + \frac{1}{9}y^{-4/3}}\,dy$
7. $2\pi\int_{-2}^2 (2 + y^2)\sqrt{1 + 4y^2}\,dy$ **9.** $\int_1^2 2\pi(x + x^{-1})\sqrt{1 + (1 - x^{-2})^2}\,dx$
10. $\int_{\frac{1}{2}\pi}^{\frac{3}{2}\pi} 2\pi\left(\frac{\cos x}{x}\right)\sqrt{1 + [(\sin x)/x + (\cos x)/x^2]^2}\,dx$
11. $\int_{-3}^3 2\pi(4)\,dx = 48\pi$

Section 6.7

1.‡ $s(t) = \int (t^3 + 4)\,dt = \frac{1}{4}t^4 + 4t + C$; $s(0) = 5 \Rightarrow C = 5$; Or:
$s(t) = s(0) + \int_0^t v(u)\,du = 5 + \int_0^t (u^3 + 4)\,du$; $\frac{1}{4}t^4 + 4t + 5$
2. $5t + t^{-1} - 6$ **4.‡** $v(t) = \int t\,dt = \frac{1}{2}t^2 + C$; $v(0) = 3 \Rightarrow C = 3$; $s(t)$
$= \int (\frac{1}{2}t^2 + 3)\,dt = \frac{1}{6}t^3 + 3t + C$; $s(0) = -2 \Rightarrow C = -2$; $\frac{1}{6}t^3 + 3t - 2$
5. $-\frac{1}{12}t^4 + t^2 + 15t + 3$ **7.‡** $s(8) = s(1) + \int_1^8 v(t)\,dt = \int_1^8 6t^{2/3}\,dt = \frac{558}{5}$
meters **8.** \$166.66

10.‡ $dV/dt = [\text{Rate in}] - [\text{Rate out}] = 10t^4 + 100t + 4$ cu.ft./day;

$$V(1) = V(0) + \int_0^1 \frac{dV}{dt}(t)\, dt = 81 \text{ cu.ft.}$$

11a. $t = 0, s = 10; t = 1, s = \frac{59}{6}$ **11b.** $t = \frac{3}{2}$ **12.** $\int_{144}^{168} g(t)\, dt$ inches
15a.‡ Use a vertical s-axis with its origin on the ground; $v(t) = -32t + 96$;
$s(t) = -16t^2 + 96t$ **15b.**‡ $s = $ max when $v = 0$; $t = 3$; $s = 144$ feet
15c.‡ $s = -16t^2 + 96 = -16t(t - 6)$; $t = 6$; $v(6) = -96$ feet/second;
speed $= 96$ feet per second   **16.** 144 ft.   **18.**‡ $v = t(t - 2)(t + 2)$ is negative
for $t < -2$ and $0 < t < 2$ and is positive for $-2 < t < 0$ and $t > 2$;
$\int_{-3}^{-2} (4t - t^3)\, dt + \int_{-2}^{0} (t^3 - 4t)\, dt + \int_0^1 (4t - t^3)\, dt = 12$ ft.
20.‡ $v = 3t^2 - 48$ is positive for $t < -4$ and $t > 4$ and negative for
$-4 < t < 4$; $\int_0^4 (48 - 3t^2)\, dt + \int_4^5 (3t^2 - 48)\, dt = 141$ centimeters
22a. $-35{,}000$ ft./sec.   **22b.** 7800 pounds   **24.** 2200 ft.
27. 33.075 meters   **28a.** 1 second   **28b.** -48 feet/second   **29a.** 29.4 meters
29b. 30.625 meters   **31a.** $(10)^{1/4}$ seconds   **31b.** $4(10)^{3/4}$ meters/second
32a. 0 kilometers   **32b.** $\frac{32}{3}$ kilometers   **32c.** 3 kilometers/second at
$t = 3$ seconds   **34a.** $t = 1$ second or $t = 2$ seconds
34b. $s = \frac{5}{4}$ meters at $t = \frac{3}{2}$ seconds

Section 6.8

1. 1000 yard-pounds   **3.**‡ Force toward the right by the spring (Figure 6.46)
$= -ks$; $-6 = -k(1) \Rightarrow k = 6$; Force toward the left on the spring $= -6s$;
$s = -1$ when the length is 14; $s = -3$ when the length is 12; $\int_{-3}^{-1} -6s\, ds$
$= 24$ inch-ounces   **4.** 48 foot-pounds   **6.** 20 dyne-centimeters   **8.**‡ $s = 2t$;
$t = \frac{1}{2}s$; $F = 100 - 3t = 100 - \frac{3}{2}s$ pounds; $\int_0^{20} (100 - \frac{3}{2}s)\, ds$
$= 1700$ foot-pounds   **9a.** 140.4π foot-pounds   **9b.** $140.4\pi/55$ seconds

Section 6.9

1. 499.2 pounds   **3.** 135.2 pounds   **4a.** Volume of can $= \frac{5}{54}\pi$ cu.ft.;
(110 pounds/cu.ft.)$(\frac{5}{54}\pi$ cu.ft.$) = \frac{275}{27}\pi$ pounds   **4b.** $\int_0^{5/6} 110(\frac{2}{3}\pi)h\, dh$
$= \frac{1375}{54}\pi$ pounds   **5.** 546.1 pounds

7.‡ Introduce a vertical h-axis with its zero at the top of the plate as in
Fig. 6.9.7; The portion of the plate between $h = h_{j-1}$ and $h = h_j$ has area
$(20/\sqrt{3})(h_j - h_{j-1})$, and the force on it is approximately

$$62.4(15 + h_j)(20/\sqrt{3})(h_j - h_{j-1}); \quad 62.4\left(\frac{20}{\sqrt{3}}\right)\int_0^{5\sqrt{3}} (15 + h)\, dh$$

$= 93{,}600 + 46{,}800/\sqrt{3}$ pounds
8. $312\sqrt{3} - 3.9\sqrt{2}$ pounds   **10.** 332.8 pounds   **12.** $\frac{26}{7}$ pounds

Section 6.10

1a. $(\frac{1}{2}, 0)$

1c.‡ $\overline{x} = \dfrac{6(2) + 3(1) + 1(3)}{6 + 3 + 1} = \dfrac{9}{5}$; $\overline{y} = \dfrac{6(-2) + 3(1) + 1(4)}{6 + 3 + 1} = -\dfrac{1}{2}$

1d. $(\frac{2}{7}, \frac{2}{7})$ **2a**‡ Divide the region into three rectangles as in Fig. 6.10.2; the
centroid is the center of gravity of three points, one at the centroid of each
rectangle and of weight equal to the area of the rectangle, i.e., of a point at
$(0, \frac{3}{2})$ weighing 4, a point at $(\frac{3}{2}, 0)$ weighing 2, and a point at $(\frac{1}{2}, -\frac{3}{2})$ weighing
3; $(\frac{1}{2}, \frac{1}{6})$

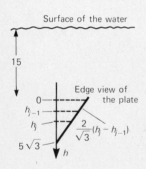

Fig. 6.9.7

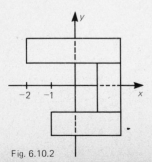

Fig. 6.10.2

2b. $\left(2, \dfrac{27\pi - 16}{18\pi + 32}\right)$ **3a.** 36 pounds; $\overline{x} = 0$ **6a.** $\frac{4}{63}$ pounds; $(\frac{7}{11}, \frac{21}{55})$

6c. $\frac{32}{3}$ pounds; $(\frac{6}{35}, 0)$ **7a.** $(\frac{12}{5}, \frac{3}{4})$ **7c.** $(0, 0)$ **12a.** $(0, 4r/3\pi)$

Section 6.11

1a. 2   **1c.** $2/\pi$ **2a.** $c = -\sqrt{3}$ **5a.** $f(x)$ is discontinuous at $x = 4$ but, nevertheless, does satisfy the conclusion of the theorem: $\frac{1}{5}\int_0^5 f(x)\,dx = \frac{7}{5} = f(\frac{7}{5})$ **6.** $\frac{5555}{2}$ **8.** 58 ft./sec.   **10.** $\frac{1}{9}\int_{-8}^1 x^{2/3}\,dx = \frac{11}{5}$ equals $c^{2/3}$ at $c = -(\frac{11}{5})^{3/2}$ and this number is between -8 and 1

Miscellaneous Exercises to Chapter 6

1.‡ Fig. 6.M.1; $\int_{-1}^4 [(x^2 + 3x + 4) - 2x^2]\,dx = \frac{125}{6}$ **3.** Fig. 6.M.3; $\frac{4}{3}$
5.‡ Fig. 6.M.5; $\int_0^1 (1 - \sqrt{x})^2\,dx = \frac{1}{6}$ **7.‡** Fig. 6.M.7; $\int_{-1}^1 (x^4 - 2x^2 + 1)\,dx = \frac{16}{15}$
9.‡ Fig. 6.M.9; $16 + 2\int_1^3 (9x^{-2} + 1)\,dx$ or $\int_1^9 6y^{-1/2}\,dy$; 24   **12.‡** Fig. 6.M.12; $\int_{\pi/6}^{5\pi/6} (\sin x - \frac{1}{2})\,dx = \sqrt{3} - \pi/3$ **13.** Fig. 6.M.13; $2^{1/2} - \pi 2^{-3/2}$

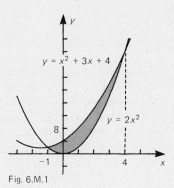

Fig. 6.M.1

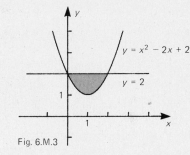

Fig. 6.M.3

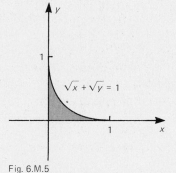

Fig. 6.M.5

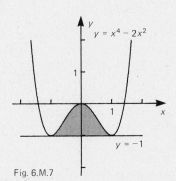

Fig. 6.M.7

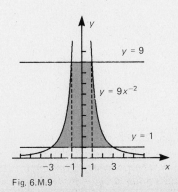

Fig. 6.M.9

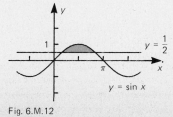

Fig. 6.M.12

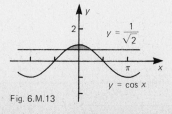

Fig. 6.M.13

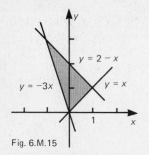

$y = 2 - x$

$y = -3x$

$y = x$

Fig. 6.M.15

0

10

$\sqrt{100 - x^2}$

Fig. 6.M.28

15. Fig. 6.M.15; 2   **17.**‡ Fig. 6.M.17; $\int_{-1}^{2} (y + 2 - y^2) \, dy$ or

$\int_0^1 2\sqrt{x} \, dx + \int_1^4 (\sqrt{x} - x + 2) \, dx$; $\frac{9}{2}$ **18.** Fig. 6.M.18; 1   **20a.** $\frac{3}{2}\pi$

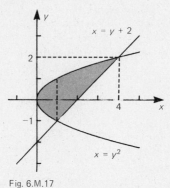

Fig. 6.M.17

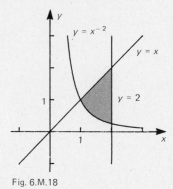

Fig. 6.M.18

20b. $\frac{1}{2}\pi$ **20c.** $\frac{1}{3}\pi$ **20d.** $\frac{5}{3}\pi$ **21c.** $\int_0^8 2\pi x \, (4 - x^{2/3}) \, dx = 64\pi$ **22a.** 64π
22b. $\frac{128}{15}\pi$ **22c.** 8π **22d.** 64π **24.** $\frac{16}{5}\pi$ **26.** π **28.** Introduce an x-axis as in
Fig. 6.M.28; The radius of the horizontal cross-section at x is $\sqrt{100 - x^2}$;
$\int_4^{10} \pi (100 - x^2) \, dx = 288\pi$ cu.in.   **29.** $25/(84\pi)$ in./min.   **31.** $2\pi^2 a^2 b$
32. $166.4\,\pi/27$ pounds   **33.** $f(x) = x^2 + 1$ **34c.** Length of the portion for
$x \geq 0, y \geq 0$ is $\lim[\frac{3}{2}(b^{2/3} - a^{2/3})] = \frac{3}{2}$; 3   **37.** 64 ft.   **38.** 144 ft.; 28 ft./sec.

40. $\left| \int_a^b v(t) \, dt \right| \leq \int_a^b |v(t)| \, dt$ **42.** $1146\frac{2}{3}$ dollars   **44.** $\int_a^b [1 + |f'(x)|] \, dx$
46.‡ $F = (1.6 \times 10^9) r^{-2}$ pounds at a distance r miles from the center of the
earth; $\int_{4000}^{4300} (1.6 \times 10^9) r^{-2} \, dr = 28,000$ mile-pounds   **48.** $\frac{13}{2}$ in.
50. 100 foot-pounds   **52.** $62.4(wHL + \frac{1}{2} wL^2 \cos \alpha)$;
Max $= 62.4(wHL + \frac{1}{2} wL^2)$; Min $= 62.4 \, wHL$ **53a.** $P(y) = \int_y^Y \rho(t) \, dt$
53b. $dP/dy = -0.00012 \, P$ **54.** Upper approximation
$= \sum_{j=1}^N (62.4) \, D(y_{j-1}) W(c_j)(y_j - y_{j-1})$ where $W(c_j) = $ max of $W(y)$ for
$y_{j-1} \leq y \leq y_j$; Lower approximation $= \sum_{j=1}^N (62.4) \, D(y_j) W(c_j)(y_j - y_{j-1})$
where $W(c_j) = $ min of $W(y)$ for $y_{j-1} \leq y \leq y_j$ **56a.** $1875\pi/2$ kilograms
56b. $\frac{8}{3}$ meters above the base on the axis   **59.** $(\bar{x}, 0)$ where
$\bar{x} = \left[\int_a^b x \, f(x)^2 \, dx \right] / \left[\int_a^b f(x)^2 \, dx \right]$ **61.** $\int_a^b x^2 \rho(x) \, dx$ **62.** 272.25 ft.
63.‡ Let $h = $ depth of the bottom of the boat; Volume of displaced water
$= 12 \int_0^h \sqrt{y} \, dy = 8h^{3/2}$; $(62.4)(8h^{3/2}) = 200 \Rightarrow h = (25/62.4)^{2/3} \approx 0.54$ ft.
65. $\int_{p_0}^{p_1} D(p) \, dp$ (Approximate the region under the graph of $D(p)$ by thin
horizontal rectangles. The area of the portion of each rectangle to the right of
p_0 is the amount saved by the customers corresponding to that rectangle.)

Section 7.1

1a. $\frac{1}{2}$ **1b.** 2   **1c.** 1   **2a.** 10^5 **2c.** $3\sqrt{10}$ **2e.** 10^{-4} **2h.** 10^4 or 10^{-4} **2j.** $\frac{1}{4} 6^{1/5}$
3a. 0.6590   **3c.** 2.5740   **3e.** $-3 + 0.5315$ **5a.**‡ Write log for $\log_{10}$;
$\log(2.16) \approx 0.3345$; $\log(57.3) = \log(5.73 \times 10) = 1 + \log(5.73) \approx 1.7582$;
$\log(2.16/57.3) = \log(2.16) - \log(57.3) \approx 0.3345 - 1.7582 = -2 + 0.5763$
$\approx \log(3.77 \times 10^{-2})$; $3.77 \times 10^{-2} = 0.0377$ **5b.** 3.06×10^4 **5d.** 1.54

5f. 1.33×10^{35} **5h.**‡ Write log for $\log_{10}$; $\log(0.0123) \approx -2 + 0.0899$; $\log(\sqrt[4]{0.0123}) = \frac{1}{4}\log(0.0123) \approx \frac{1}{4}(-4 + 2.0899) \approx -1 + 0.5225 \approx \log(0.333)$; 0.333 **5j.** 2.59 **5l.**‡ Write log for $\log_{10}$; $\log[\log(5^{\sqrt{10}})] = \log[\sqrt{10}\log(5)]$ $= \log(\sqrt{10}) + \log[\log(5)] \approx \frac{1}{2}\log(10) + \log(0.6990) \approx \frac{1}{2} + (-1 + 0.8445)$ $= 0.3445 \approx \log(2.21) \approx \log[\log(162)]$; 162 **6a.**‡ Write log for $\log_{10}$; $\sqrt{x}\log(3) = \log(7)$; $\log(\sqrt{x}) + \log[\log(3)] = \log[\log(7)]$; $\frac{1}{2}\log x \approx 0.2484$; $\log x \approx 0.4968 \approx \log(3.14)$; 3.14 **6b.** -0.79

7. Set $y = \sqrt{x}$; As $x \to \infty$, $y \to \infty$ and $\dfrac{1}{\sqrt{x}}\log_{10}x = \dfrac{2}{\sqrt{x}}\log_{10}(\sqrt{x})$

$= \dfrac{2}{y}\log_{10}y \to 0$

10a. 5.2×10^5 kilowatt hours **10b.** 1.6×10^7 kilowatt hours
12. $I_{\text{Sirius}}/I_{\text{Betelgeuse}} = 10$ **15a.** $f(x) = Cb^x$ $(C > 0, b > 0)$
15b. $f(x) = Cx^k$ $(C > 0)$ **16.** -1 **17.** 0 **18.** $\pm\sqrt{2}$ **19.** $-\frac{2}{3}$ **20.** $\frac{1}{2}$ **21.** -4
22. $0, 2$ **23.** $1, 2$ **24.** $-1, 2$ **25.** All x **26.** $\frac{1}{3}\log_2(7)$ **27.** $\frac{1}{4}[\log_2(5)]^2$
28. $2\log_2(9)$ **29.** $-\log_2(5)$ **30.** $-\frac{1}{2}\log_2(10)$ **31.** $\pm\sqrt{\log_2(7)/\log_2(3)}$
32. $-\frac{1}{3}\log_{10}(2)$ **33.** $\sqrt[3]{\log_{10}(6)/\log_{10}(5)}$ **34.** $1, 2$ **35.** $\pm\sqrt{2}$ **36.** $\frac{1}{4}\log_{10}(75)$
37. $0, \log_{10}(7)$ **38.** $4^{-6/5}$ **39.** $\frac{8}{5}, \frac{1}{40}$ **40.** $0, \frac{1}{8}$ **41.** 1 **42.** $10^{10/3}$ **43.** $-\frac{15}{16}$

Section 7.2

1. $\frac{9}{4} = 2.25$; $\frac{64}{27} = 2.370\ldots$; $\frac{625}{256} = 2.441\ldots$ **3a.** 1.60944 **3b.** -1.60944
3c. 13.5477 **5a.**‡ $\ln(1.89)/\ln(5) \approx 0.63658/1.60944 \doteq 0.395\ldots$
5b. 3.440 **6.** $\frac{1}{300}$ **7.** $1/[100\ln(10)]$

9.‡ $\dfrac{1}{\sin x}\dfrac{d}{dx}\sin x = \dfrac{\cos x}{\sin x}$

10.‡ $\cos(\ln x)\dfrac{d}{dx}\ln x = \dfrac{1}{x}\cos(\ln x)$

11. $[1 + (\ln x)^{-1/2}]/2x$ **12.** $\ln(3) + 1$ **15.** $1/[4y\sqrt{\ln(\sqrt{y})}]$
17. $10(\log_{10}x)^9/[x\ln(10)]$ **19.** Fig. 7.2.19; Max at $x = e$

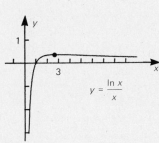

1

3

$y = \dfrac{\ln x}{x}$

Fig. 7.2.19

22. $\lim\limits_{n \to \infty}\left(1 + \dfrac{1}{n}\right)^k = \left[\lim\limits_{n \to \infty}\left(1 + \dfrac{1}{n}\right)\right]^k = 1^k = 1$; $\lim\limits_{n \to \infty}\left(1 + \dfrac{1}{k}\right)^n = \lim\limits_{n \to \infty}c^n = \infty$

since $c = 1 + \dfrac{1}{k} > 1$

24. The area under $y = \dfrac{1}{t}$ for $1 \leq t \leq x_1$ equals that for $x_2 \leq t \leq x_1x_2$;

The Riemann sum $\sum\limits_{j=1}^{N}\dfrac{1}{t_j}(t_j - t_{j-1})$ for the first integral and corresponding to

the partition $1 = t_0 < t_1 < \cdots < t_N = x_1$ equals the Riemann sum

$\sum\limits_{j=1}^{N}\dfrac{1}{x_2t_j}(x_2t_j - x_2t_{j-1})$ for the second integral corresponding to the partition

$x_2 = x_2t_0 < x_2t_1 < \cdots < x_2t_N = x_2x_N$ **26.** $\sqrt{e}$ **29.** ± 2 **30.** $0, \pm 2$
31. $e^{34/3}$ **32.** $\pm e^{25/2}$ **33.** $[\frac{1}{5}\ln(6)]^{1/3}$ **34.** $\frac{1}{7}\ln(5)$ **35.** $\sqrt[4]{3}$ **37.** $[\ln(2)]^2$
39. $x > 0$; $\dfrac{1}{x}[\ln(x^3)]^{-2/3}$ **40.** $x \leq -1$ or $x \geq 1$; $\dfrac{1}{x}[\ln(x^2)]^{-1/2}$
41. All x; $-(\sin x)/(2 + \cos x)$ **42.** $x > 1$; $[x\ln(x)\ln(10)]^{-1}$
43. $x > 1$; $[x\ln(10)\log_{10}(x)]^{-1}$ **45.** $x > 0$; $(\ln x + 1)\cos(x\ln x)$

Section 7.3

1. $\ln(6)$

3.‡ $u = x^2 + 4, du = 2x\,dx; \int \dfrac{1}{x^2 + 4}\,dx = \dfrac{1}{2}\int \dfrac{1}{u}\,du = \dfrac{1}{2}\ln|u| + C$

$= \dfrac{1}{2}\ln|x^2 + 4| + C; \left[\dfrac{1}{2}\ln|x^2 + 4|\right]_0^5 = \dfrac{1}{2}[\ln(29) - \ln(4)]$

4. $\ln(4)$ **6.**‡ $u = \ln x, du = \dfrac{1}{x}\,dx; \int \dfrac{1}{u}\,du = \ln|\ln x| + C$

7. $\dfrac{2}{3}(\ln x)^{3/2} + C$ **9.** $\dfrac{3}{4}\ln(\dfrac{17}{2})$ **11.** Fig. 7.3.11; $\dfrac{3}{2} - 2\ln(2)$

13.‡ Fig. 7.3.13; $\displaystyle\int_0^{2/3}\left[3 - \dfrac{1}{1-x}\right]dx = [3x + \ln|1-x|]_0^{2/3} = 2 - \ln(3)$

14. Fig. 7.3.14; $6\pi + 2\pi\ln(4)$ **17a.** $2x/(x^2 - 9)$ **20.** $\pi[\ln(2) - \frac{1}{2}]$
21. $3/[\ln(4)]$ **22.** $-7 - 4\ln(3)$

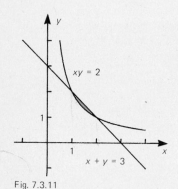

Fig. 7.3.11

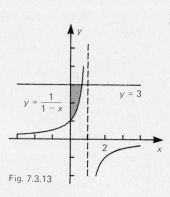

Fig. 7.3.13

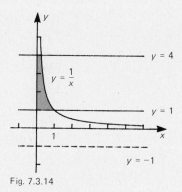

Fig. 7.3.14

Section 7.4

1. $\dfrac{(2x + 3)(3x + 4)}{5x + 6}\left[\dfrac{2}{2x + 3} + \dfrac{3}{3x + 4} - \dfrac{5}{5x + 6}\right]$

3. $\left[\dfrac{x^2 + 1}{x^2 - 1}\right]^{1/2}\left[\dfrac{x}{x^2 + 1} - \dfrac{x}{x^2 - 1}\right]$

5. $(1 + \ln x)\,x^x$

7.‡ $\ln|F(x)| = 10\ln|3 - 2x| + 20\ln|6 - 5x| + 30\ln|9 - 8x|$;

$\dfrac{F'(x)}{F(x)} = \dfrac{-20}{3 - 2x} - \dfrac{100}{6 - 5x} - \dfrac{240}{9 - 8x}$; -360

9. $x^{1/7}(2x + 1)^{1/6}(x + 1)^{10}\left[\dfrac{1}{7x} + \dfrac{1}{3(2x + 1)} + \dfrac{10}{x + 1}\right]$

10. $\sqrt{(x + 1)(x - 3)(x + 2)(x - 4)}\,\dfrac{1}{2}\left[\dfrac{1}{x + 1} + \dfrac{1}{x - 3} + \dfrac{1}{x + 2} + \dfrac{1}{x - 4}\right]$

11. $x(\sin x)(\cos x)(\ln x)\left[\dfrac{1}{x} + \dfrac{\cos x}{\sin x} - \dfrac{\sin x}{\cos x} + \dfrac{1}{x \ln x}\right]$

13. $\dfrac{x \sin x}{(x^2 + 1)^5 \cos x}\left(\dfrac{1}{x} + \dfrac{\cos x}{\sin x} - \dfrac{10x}{x^2 + 1} + \dfrac{\sin x}{\cos x}\right)$

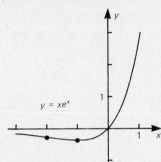

Fig. 7.5.25a

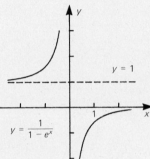

Fig. 7.5.25c

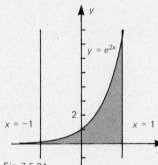

Fig. 7.5.34

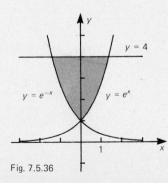

Fig. 7.5.36

Section 7.5

1a.‡ $(e^{\ln x})^3 = x^3$ **1b.** $-x^2$

2a.‡ $\ln(e^{x/2}) = \ln(4); \dfrac{x}{2} = \ln(4); 2\ln(4)$

2b. $e^5 - 1$ **3.** $6e^{6x}$ **5.** $e^{-x}(1 - x)$ **7.** $-e^x/(1 - e^x))$ **9.** $e^x(x + 2)/(x + 3)^2$

11.‡ $5^{-x} = \exp[\ln(5^{-x})] = e^{-x\ln(5)}$, where $\exp(u) = e^u$;

$$e^{-x\ln(5)}\frac{d}{dx}[-x\ln(5)] = -5^{-x}\ln(5)$$

13. $e^x + ex^{e-1}$

15. $(\sin x)^x\left[\ln(\sin x) + \dfrac{x\cos x}{\sin x}\right]$

17. $(3 + x)^{3-x}\left[-\ln(3 + x) + \dfrac{3 - x}{3 + x}\right]$

19. $3e^{\sin(3x)}\cos(3x)$ **22.** $-1.2\,e^{-0.12}$ kg./cu.m.   **23.** $y - e^2 = 4e^2(x - 1)$

25a. Fig. 7.5.25a; Local min at $x = -1$; Inflection point at $x = -2$; Asymptote $x = 0$ **25c.** Fig. 7.5.25c; Asymptotes $y = 0, y = 1, x = 0$

26. $\frac{1}{3}(1 - e^{-6})$ **28.** $\sin(e^t) + C$

30. $\dfrac{1}{\pi + 1}(\pi^{\pi + 1} - 1) - \dfrac{\pi}{\ln(\pi)}(\pi^{\pi - 1} - 1)$

32. $\frac{1}{7}(e - e^{\cos(7)})$ **34.** Fig. 7.5.34; $\frac{1}{2}(e^2 - e^{-2})$ **36.** Fig. 7.5.36; $8\ln(4) - 6$

38. $-\ln(1.99)$ or $\ln(100)$ **41a.** e^{-2}

41b. $\left(\dfrac{1}{2}\right)^8 \dfrac{1}{8!}e^{-1/2}$ **44.** $x \neq 0; (1 - e^x + xe^x)(1 - e^x)^{-2}$

45. All $x; \cos(x - e^x)(1 - e^x)$ **46.** $x \leq 0; -\frac{3}{2}e^{3x}(1 - e^{3x})^{-1/2}$

47. All $x; 2\pi x(1 + x^2)^{(\pi - 1)}$ **48.** $x > 0; \dfrac{\ln(10)}{x}10^{\ln x} + \dfrac{10}{x}(\ln x)^9$

50. $x > 0; \sqrt{2}x^{\sqrt{2}-1}$ **52.** $\frac{1}{2}(e^4 - 1)$ **53.** $\dfrac{1}{\ln(3)}3^x + \frac{1}{4}x^4 + C$

54. $x - e^{-x} + C$ **56.** $\frac{1}{6}e^{6x} + \frac{2}{3}e^{3x} + x + C$

Section 7.6

1. $y = -3e^{-4t}$ **2.** $z = \frac{15}{16}4^x$ **5.**‡ $dV/dt = -0.05\,V; V = Ce^{-0.05t}$; $V(t + 10)/V(t) = e^{-1/2} = 0.6065\ldots$ **7.** $\frac{1}{70}\ln(45) \approx 0.0224$

8. $5700[\ln(1.25)/\ln(2)] \approx 1835$ years   **10.**‡ $P(t) = 5000e^{rt}; P(t + 10)$ $= 2P(t) \Rightarrow e^{10r} = 2; r = \frac{1}{10}\ln(2)$ and $P(t) = 5000\exp[\frac{1}{10}t\ln(2)]$; Or: $e^r = 2^{1/10}; P(t) = (5000)\,2^{t/10}; P(14) = (5000)\,2^{7/5} \approx 13,200$ **11.** 10,000

13. $1 - e^{-1/2} \approx 0.393$ **19.** $70 - \frac{5}{2}\sqrt{2} \approx 66.46$ degrees   **20.** $135\ln(\frac{27}{47})$ ≈ -74.83 degrees/minute

Section 7.7

1. $100(1.06)^{10} \approx 179$ dollars   **3.**‡ A deposit of C dollars yields $1.05\,C$ dollars after one year compounded annually at 5% and yields Ce^r compounded continuously at $100r\%$; $1.05 = e^r; r = \ln(1.05) \approx 0.04879$ **5.** 7.5 years

7. In 25 years   **8.‡** Consider $0 = t_0 < t_1 < \cdots < t_N = T$. Income for $t_{j-1} \le t \le t_j$ will be about $R(t_j)(t_j - t_{j-1})$, and it would take about $e^{-rt_j} R(t_j)(t_j - t_{j-1})$ invested at $t = 0$ to earn the same amount. The present value is approximately $\sum_{j=1}^{N} e^{-rt_j} R(t_j)(t_j - t_{j-1})$, which is a Riemann sum for the integral in (4).   **11.** $\$15{,}000(1 + 0.035)^{-4} \approx \$13{,}071.63$

Section 7.8

1a. $-1/\sqrt{2}$ **1c.** 0 **1d.** -2 **1f.** $-\sqrt{2}$ **2a.‡** $\tan(\frac{5}{7}\pi) = \tan(-\frac{2}{7}\pi)$ $= -\tan(\frac{2}{7}\pi) \approx -\tan(0.90) \approx -1.26$ **2b.** 1.70 **2d.‡** $\csc(5) = 1/\sin(5)$ $= -1/\sin(2\pi - 5) \approx -1/\sin(1.28) \approx -1.04$ **2e.** -2.57 **2g.** 2.07
3a.‡ $\cos(3x) = 1/\sqrt{2}$; $3x = \pm\frac{1}{4}\pi + 2k\pi$ (integers k); $x = \pm\frac{1}{12}\pi + \frac{2}{3}k\pi$
3b. $-\frac{1}{6}\pi + 2k\pi$ or $-\frac{5}{6}\pi + 2k\pi$ (integers k)   **3d.‡** $\frac{1}{5}x = \frac{1}{6}\pi + k\pi$; $\frac{5}{6}\pi + 5k\pi$ (integers k)   **3e.** $1 - \frac{3}{4}\pi + k\pi$ (integers k)   **3g.** $\frac{1}{30}\pi + \frac{1}{10}k\pi$ (integers k)   **4a.‡** $\tan(1.16) \approx 2.3$; Angles with the same tangents differ by multiples of π; $1.16 + k\pi$ (integers k)   **4b.** $\pm 1.32 + (2k + 1)\pi$ (integers k)
4d. $-0.125 + \frac{1}{2}k\pi$ (integers k)

5. $\dfrac{d}{dx}\left(\dfrac{\sin x}{\cos x}\right) = \left[\cos x\left(\dfrac{d}{dx}\sin x\right) - \sin x\left(\dfrac{d}{dx}\cos x\right)\right] \Big/ \cos^2 x$

$= (\cos^2 x + \sin^2 x)/\cos^2 x = 1/\cos^2 x = \sec^2 x;\ \dfrac{d}{dx}(\cos x)^{-1}$

$= -(\cos x)^{-2}\dfrac{d}{dx}\cos x = \dfrac{1}{\cos x}\dfrac{\sin x}{\cos x} = \sec x \tan x$

7. -2 **10.** $4\sec^2(4t)$ **12.** $-2\cot y \csc^2 y$ **14.** $8\tan x \sec^2 x\,(1 - \tan^2 x)$

16.‡ $3\tan^2(6x)\dfrac{d}{dx}\tan(6x) = 3\tan^2(6x)\sec^2(6x)\dfrac{d}{dx}(6x)$

$= 18\tan^2(6x)\sec^2(6x)$

18. $y - \sqrt{2} = 3\sqrt{2}(x - \frac{1}{12}\pi)$ **20.** $\frac{1}{2}\sec(2x) + C$ **22.** $\frac{2}{3}(\tan x)^{3/2} + C$
24.‡ $u = \sec x,\ du = \sec x \tan x;\ \int \sec^4 x \sec x \tan x\, dx = \int u^4\, du$
$= \frac{1}{5}u^5 + C = \frac{1}{5}\sec^5 x + C$ **26.** 1 **28.‡** $u = 2x,\ du = 2\,dx;\ \frac{1}{2}\int (\sin u)^{-2}\, du$
$= \frac{1}{2}\int \csc^2 u\, du = -\frac{1}{2}\cot(2x) + C;\ [-\frac{1}{2}\cot(2x)]_{5\pi/8}^{7\pi/8}$
$= [-\frac{1}{2}\cot(\frac{7}{4}\pi)] - [-\frac{1}{2}\cot(\frac{5}{4}\pi)] = 1$ **30.‡** Fig. 7.8.30; $8\sin x$
$= \csc^2 x \Rightarrow \sin^3 x = \frac{1}{8} \Rightarrow \sin x = \frac{1}{2}$; Intersections at $x = \frac{1}{6}\pi + 2k\pi$ and $x = \frac{5}{6}\pi + 2k\pi$ (integers k); $\int_{\pi/6}^{5\pi/6} (8\sin x - \csc^2 x)\, dx = 6\sqrt{3}$

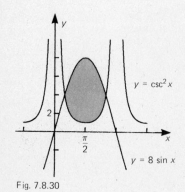

Fig. 7.8.30

$y = \csc^2 x$

$y = 8 \sin x$

32.‡ Use an x-axis as in Fig. 7.8.32; $x = 15\tan\Theta$; $\dfrac{d\Theta}{dt} = 4\pi$ radians per minute

(t measured in minutes); $\dfrac{dx}{dt} = 15\sec^2\Theta\dfrac{d\Theta}{dt}$; When $x = 20$, $\sec\Theta = \dfrac{5}{3}$;

$\dfrac{500}{3}\pi$ miles/minute

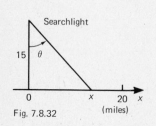

Searchlight

15 θ

0 x 20 x
(miles)

Fig. 7.8.32

34. Max $= 2$; Min $= -2$ **36.** $5/\sqrt{26}$ **37.‡** $-(1 + y')\csc^2(x + y) = 0$; Set $x = 0$ and $y = \pi/6$: $y'(0) = -1$ **39a.** $32[\cos^2(4x) - \sin^2(4x)]$
39c. $2\sin x + 4x\cos x - x^2\sin x$ **40a.‡** $\sin x \approx \sin(\frac{1}{4}\pi) + \cos(\frac{1}{4}\pi)(x - \frac{1}{4}\pi)$ $= (1/\sqrt{2}) + (1/\sqrt{2})(x - \frac{1}{4}\pi);\ (1/\sqrt{2}) + (1/\sqrt{2})(\frac{1}{16}\pi) \approx 0.846$
40b. $2 + 2\sqrt{3}(\frac{1}{9}\pi) \approx 3.21$ **41.** $600/\pi$ coulombs

43. $\frac{1}{3}n\pi$, n an integer   **44.** $(\frac{1}{4} + \frac{1}{2}n)\pi$, n an integer   **45.** $(\frac{1}{4} + \frac{1}{2}n)\pi$, n an integer
46. $(\frac{1}{4} + \frac{1}{2}n)\pi$, n an integer   **47.** $(\pm\frac{1}{15} + \frac{1}{5}n)\pi$, n an integer
48. $(\frac{2}{3} + 2n)\pi$, $(\frac{4}{3} + 2n)\pi$, n an integer   **49.** No solutions

51. $(\frac{1}{3} + 2n)\pi$, $(\frac{5}{3} + 2n)\pi$, $2n\pi$, n an integer   **53.** $\dfrac{1}{x}\sec^2(\ln x)$

54. $(\sin x \cos x)^{-1}$ **55.** $6(\sin x)(\cos x)\sec^3(\sin^2 x)\tan(\sin^2 x)$
57. 0 **59.** $\frac{1}{4}\tan^4 x + C$ **60.** $\frac{1}{3}\sec^3 x + C$ **61.** $\ln|1 + \tan x| + C$
63. $-x^{-1} - \cot x + C$ **65.** $\arctan(\frac{8}{15}) \approx 0.490$ radians
66. $y - 1 = (-2 \pm \frac{5}{3}\sqrt{3})(x - 1)$ **68.** 0.322 radians

Section 7.9

1a.‡ The angles with the same tangent as $\frac{4}{5}\pi$ are $\frac{4}{5}\pi + k\pi$ (integers k); The one between $-\frac{1}{2}\pi$ and $\frac{1}{2}\pi$ is $-\frac{1}{5}\pi$ **1b.** $\frac{4}{5}\pi$ **1d.**‡ $\Theta = \arctan(5)$ is the angle in Fig. 7.9.1; $\sin\Theta = 5/\sqrt{26}$ **1e.** $-1/\sqrt{15}$ **1g.** -5 **1i.** $\pi/3$ **1k.** $2\pi/3$
2a.‡ $\sin(0.33) \approx 0.32$ and $-\frac{1}{2}\pi \le 0.33 \le \frac{1}{2}\pi$, so $\arcsin(0.32) \approx 0.33$ **2b.** 1.38
2d.‡ $\tan(1.16) \approx 2.3$ so $\tan(-1.6) \approx -2.3$; $-\frac{1}{2}\pi < -1.16 < \frac{1}{2}\pi$;
$\arctan(-2.3) \approx -1.16$ **3.** $2x(1 - x^4)^{-1/2}$ **5.** $3/[1 + (3t - 4)^2]$

7.‡ $\dfrac{1}{3}[\arctan(6x^2)]^{-2/3}\dfrac{d}{dx}\arctan(6x^2) = \dfrac{1}{3}[\arctan(6x^2)]^{-2/3}\dfrac{\dfrac{d}{dx}(6x^2)}{1 + (6x^2)^2}$

$= 4x[\arctan(6x^2)]^{-2/3}[1 + 36x^4]^{-1}$

9. $7\sec z\tan z/(1 + 49\sec^2 z)$ **11.** $\arcsin x + x(1 - x^2)^{-1/2}$ **13.**‡ $u = x^2$,
$du = 2x\,dx$; $\frac{3}{2}\int(1 - u^2)^{-1/2}\,du = \frac{3}{2}\arcsin u + C = \frac{3}{2}\arcsin(x^2) + C$
14. $\frac{1}{3}\arctan(x^3) + C$ **17.** $\pi/6$ **19.** $\frac{1}{2}(\arctan x)^2 + C$ **21.**‡ $u = 2x$; $du = 2\,dx$;
$\int(25 - 4x^2)^{-1/2}\,dx = \frac{1}{2}\int(25 - u^2)^{-1/2}\,du = \frac{1}{2}\arcsin(\frac{2}{5}x) + C$;
$[\frac{1}{2}\arcsin(\frac{2}{5}x)]^2_{-1} = \frac{1}{2}\arcsin(\frac{4}{5}) - \frac{1}{2}\arcsin(-\frac{2}{5})$ **24.** $y - \frac{1}{6}\pi = (2/\sqrt{3})(x - \frac{1}{2})$
26. Fig. 7.9.26; $2\arctan(3)$ **28.**‡ Use an h-axis and let Θ be the angle shown in Fig. 7.9.28; $\Theta = \arctan(h/30)$; At $h = 40$, $\Theta = \arctan(\frac{4}{3}) \approx 0.92$ radians

28b.‡ $\dfrac{d\Theta}{dt} = \left[1 + \dfrac{1}{900}h^2\right]^{-1}\dfrac{1}{30}\dfrac{dh}{dt}$; Set $h = 40$ and $\dfrac{dh}{dt} = 10$: $\dfrac{3}{25}$ radians/sec.

29.‡ Let x = distance from the wall and define angles α and β as in Fig. 7.9.29;

The angle subtended $= \Theta = \alpha - \beta = \arctan\left(\dfrac{8}{x}\right) - \arctan\left(\dfrac{5}{x}\right)$;

$\dfrac{d\Theta}{dx} = \dfrac{-8}{x^2 + 64} + \dfrac{5}{x^2 + 25}$; $\dfrac{d\Theta}{dx} = 0$ for $x = \pm\sqrt{40}$; $\sqrt{40}$

32. $\Theta = \arctan(\mu)$

36. $u = \dfrac{x}{a}$, $du = \dfrac{1}{a}\,dx$; $\displaystyle\int\dfrac{1}{\sqrt{a^2 - x^2}}\,dx = \dfrac{1}{a}\int\dfrac{1}{\sqrt{1 - (x/a)^2}}\,dx$

$= \displaystyle\int\dfrac{1}{\sqrt{1 - u^2}}\,du = \arcsin u + C = \arcsin\left(\dfrac{x}{a}\right) + C$; $\displaystyle\int\dfrac{1}{a^2 + x^2}\,dx$

$= \dfrac{1}{a^2}\displaystyle\int\dfrac{1}{1 + (x/a)^2}\,dx = \dfrac{1}{a}\int\dfrac{1}{1 + u^2}\,du = \dfrac{1}{a}\arctan u + C = \dfrac{1}{a}\arctan\left(\dfrac{x}{a}\right) + C$
37a. $1/(3\sqrt{8})$ **37b.** $1/(3\sqrt{8})$ **38.** $\mathrm{arcsec}(4) - \mathrm{arcsec}(2)$
39. $\mathrm{arcsec}(-100) - \mathrm{arcsec}(-10) = \mathrm{arcsec}(10) - \mathrm{arcsec}(100)$ since $\mathrm{arcsec}(-x) = \pi - \mathrm{arcsec}\,x$

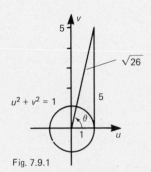

Fig. 7.9.1

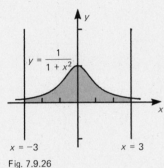

$y = \dfrac{1}{1 + x^2}$

$x = -3$ $x = 3$

Fig. 7.9.26

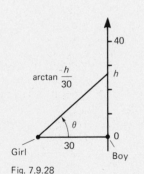

$\arctan\dfrac{h}{30}$

Girl   Boy

Fig. 7.9.28

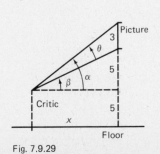

Picture

Critic

Floor

Fig. 7.9.29

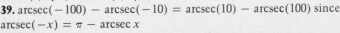

Section 7.10

1. $\dfrac{d}{dx}\sinh x = \dfrac{d}{dx}\left[\dfrac{1}{2}(e^x - e^{-x})\right] = \dfrac{1}{2}(e^x + e^{-x}) = \cosh x; \dfrac{d}{dx}\cosh x$

$= \dfrac{d}{dx}\left[\dfrac{1}{2}(e^x + e^{-x})\right] = \dfrac{1}{2}(e^x - e^{-x}) = \sinh x$

2. $6\cosh(6x)$ **4.** $x^{-2/3}\cosh^2(x^{1/3})\sinh(x^{1/3})$ **6.** $\tanh x$ **8.** $\sinh x$
10. $\cosh(\ln x)/x$ **12.** $\frac{1}{37}\sinh(3700)$ **14.** $2\cosh(\sqrt{x}) + C$
16. $\ln|1 + \sinh x| + C$

17.‡ $u = \cosh x, du = \sinh x\, dx; \displaystyle\int \dfrac{\sinh x}{\cosh x}\, dx = \int \dfrac{1}{u}\, du = \ln|\cosh x| + C$

20. $\sinh[\ln(10)] = \frac{99}{20}$ **22a.** 1.1752 **22c.** -0.4621 **23a.** $\cosh^2 x - \sinh^2 x$
$= [\frac{1}{2}(e^x + e^{-x})]^2 - [\frac{1}{2}(e^x - e^{-x})]^2$
$= \frac{1}{4}(e^{2x} + 2 + e^{-2x}) - \frac{1}{4}(e^{2x} - 2 + e^{-2x}) = 1$ **23b.** $1 - \tanh^2 x$
$= (\cosh^2 x - \sinh^2 x)/\cosh^2 x = 1/\cosh^2 x = \operatorname{sech}^2 x$ **23d.** $\cosh x \cosh y$
$= \frac{1}{4}(e^x + e^{-x})(e^y + e^{-y}) = \frac{1}{4}(e^{x+y} + e^{x-y} + e^{y-x} + e^{-x-y});$
$\sinh x \sinh y = \frac{1}{4}(e^x - e^{-x})(e^y - e^{-y}) = \frac{1}{4}(e^{x+y} - e^{x-y} - e^{y-x} + e^{-x-y});$
$\cosh x \cosh y + \sinh x \sinh y = \frac{1}{2}(e^{x+y} + e^{-x-y}) = \cosh(x + y);$ Replace y
by $-y$ to obtain the other identity   **24a.**‡ $\cosh x = \sqrt{1 + \sinh^2 x}$ since
$\cosh x > 0; \sqrt{1 + 4^2} = \sqrt{17}$ **24b.** $\sqrt{8}$

25a. $\dfrac{d}{dx}\tanh x = \dfrac{d}{dx}\dfrac{\sinh x}{\cosh x} = \dfrac{\cosh^2 x - \sinh^2 x}{\cosh^2 x} = \operatorname{sech}^2 x$

25b. $\dfrac{d}{dx}\coth x = \dfrac{d}{dx}\dfrac{\cosh x}{\sinh x} = \dfrac{\sinh^2 x - \cosh^2 x}{\sinh^2 x} = -\operatorname{csch}^2 x$

25c. $\dfrac{d}{dx}\operatorname{sech} x = \dfrac{d}{dx}\dfrac{1}{\cosh x} = -\dfrac{\sinh x}{\cosh^2 x} = -\operatorname{sech} x \tanh x$

25d. $\dfrac{d}{dx}\operatorname{csch} x = \dfrac{d}{dx}\dfrac{1}{\sinh x} = -\dfrac{\cosh x}{\sinh^2 x} = -\operatorname{csch} x \coth x$

28. As $x \to \infty, e^{-2x} \to 0$ and $\tanh x = (1 - e^{-2x})/(1 + e^{-2x}) \to 1$; As
$x \to -\infty, e^{2x} \to 0$ and $\tanh x = (e^{2x} - 1)/(e^{2x} + 1) \to -1$
30. $y = (H/\rho)[\cosh(\rho x/H) - \cosh(\rho/H)]$ **32a.** Set $u = e^y; 2x = u + 1/u;$
$u = x + \sqrt{x^2 - 1}$ with the plus sign because $u \geq 1; \ln(x + \sqrt{x^2 - 1})$

33a.‡ $\sinh(\operatorname{arcsinh} x) = x; \cosh(\operatorname{arcsinh} x)\dfrac{d}{dx}\operatorname{arcsinh} x = 1;$

$\sqrt{1 + x^2}\dfrac{d}{dx}\operatorname{arcsinh} x = 1; 1/\sqrt{1 + x^2}$

33b. $1/\sqrt{x^2 - 1}$ **35.**‡ Set $f(x) = \sinh x + 2\cosh x; f'(x) = \cosh x + 2\sinh x$
$= \sqrt{1 + \sinh^2 x} + 2\sinh x; f''(x) = \sinh x + 2\cosh x$
$= \sinh x + 2\sqrt{1 + \sinh^2 x} > 0$ for all x; The minimum occurs where
$f'(x) = 0; \sinh x = -\sqrt{\frac{1}{3}}; \cosh x = \sqrt{\frac{4}{3}};$ Min $= \sqrt{3}$.   **37.** $\bar{x} = 0;$
$\bar{y} = \frac{1}{2}[\operatorname{csch}(1) + \cosh(1)]$ **39.** $\pi(b - a) + \frac{1}{2}\pi[\sinh(2b) - \sinh(2a)]$
40c. $f^2 \approx gh/L^2$ **40d.** $f^2 \approx g/(2\pi L)$

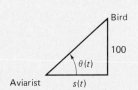

Fig. 7.M.39

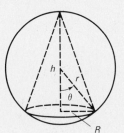

Fig. 7.M.43

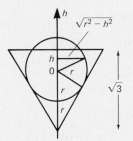

Fig. 7.M.45

Miscellaneous Exercises to Chapter 7

1.‡ $\sin(3x) \geq 0$; $2k\pi \leq 3x \leq (2k+1)\pi$; $\frac{2}{3}k\pi \leq x \leq \frac{1}{3}(2k+1)\pi$ (integers k);

$$\frac{d}{dx}[\sin(3x)]^{1/2} = \frac{1}{2}[\sin(3x)]^{-1/2}\cos(3x)\frac{d}{dx}(3x) = \frac{3}{2}[\sin(3x)]^{-1/2}\cos(3x)$$

2. $\sqrt{(2k-\frac{1}{2})\pi} \leq |x| \leq \sqrt{(2k+\frac{1}{2})\pi}$ for $k = 1, 2, 3, \ldots$ or $|x| < \sqrt{\frac{1}{2}\pi}$;
$-2x\tan(x^2)$ **5.** $(2k-\frac{1}{2})\pi \leq x \leq (2k+\frac{1}{2})\pi$, integers k; $-\pi(\sin x)(\cos x)^{\pi-1}$

7. $x > 0$; $\frac{2\ln x}{x}x^{\ln x}$

8.‡ $(x-2)(x+1) > 0$; $x < -1$ or $x > 2$; $(2x-1)/(x^2-x-2)$
10. $x \neq \frac{1}{12}k\pi$ for integers k; 0 **12.** $x \neq (k+\frac{1}{2})\pi$ for integers k;
$(1+x^2)^{\tan x}[2x(1+x^2)^{-1}\tan x + (\sec^2 x)\ln(1+x^2)]$ **14.** The domain
consists of the isolated points $x = k\pi$ for integers k and the function has no
derivative.   **16.** $x \neq (k+\frac{1}{2})\pi$ for integers k; 1

18. $\frac{1}{e+1}(\pi^{e+1} - e^{e+1}) + \frac{1}{\pi+1}(\pi^{\pi+1} - e^{\pi+1}) + (e^\pi - e^e) + \frac{1}{\ln\pi}(\pi^\pi - \pi^e)$

20. $\ln(\frac{2}{3})$ **22.** $\int_0^{\pi/4}(2^{-1/2} - \sin x)\,dx + \int_{\pi/4}^2(\sin x - 2^{-1/2})\,dx$
$= \pi 2^{-3/2} - 1 - \cos(2)$ **24a.** Figure 7.46   **24b.** Figure 7.48   **24c.** Figure 7.47
24d. Figure 7.44   **24e.** Figure 7.45   **26a.** Figure 7.54   **26b.** Figure 7.56
26c. Figure 7.57   **26d.** Figure 7.58   **26e.** Figure 7.55   **30.** $k = \pm\sqrt{2}$
32. $\arctan(\frac{4}{3}) \approx 0.93$ radians   **33.** $2\pi\ln(3)$ **35.** $\pi\tan(\frac{1}{9}\pi^2)$ **37.** π; The infinite
region between the x-axis and the curve $y = (1+x^2)^{-1}$ has finite area π.
39.‡ Let $s(t) = $ horizontal distance of bird and $\Theta(t) = $ angle with the
horizontal (Fig. 7.M.39); $s = 100\cot\Theta$; $ds/dt = -100\csc^2\Theta\,(d\Theta/dt)$;
$\Theta = \pi/3$, $d\Theta/dt = -\frac{1}{4}$: $33\frac{1}{3}$ feet/second   **41.** $400/\sqrt{34}$ feet/second
42. 137 ± 3 feet   **43.**‡ Let $R = $ radius of the base, $h = $ height (Fig. 7.M.43);
$V = \frac{1}{3}\pi R^2 h$; $(h-r)^2 + R^2 = r^2$; $V = \frac{1}{3}\pi(2h^2 r - h^3)$;
$dV/dh = \frac{1}{3}\pi(4rh - 3h^2)$; Or: $R = r\sin\Theta$, $h = r + r\cos\Theta$ (Fig. 7.M.43);
$V = \frac{1}{3}\pi r^3\sin^2\Theta(1 + \cos\Theta)$; $dV/d\Theta = \frac{1}{3}\pi r^3\sin\Theta[3\cos^2\Theta + 2\cos\Theta - 1]$;
$\cos\Theta = \frac{1}{3}$; $h = \frac{4}{3}r$, $R = (\frac{2}{3}\sqrt{2})r$ **44.** $(250\pi)/(9\sqrt{3})$ **45.** Fig. 7.M.45;
Radius $= \frac{1}{2}\sqrt{3}$ inches   **47.** Radius $= \frac{5}{2}$ inches   **48.** Area of cross-section
$= 5r - r^2\cos(5/r)\sin(5/r)$; $5 - 2r\cos(5/r)\sin(5/r)$
$+ 5\cos^2(5/r) - 5\sin^2(5/r) = 0$ **49a.** Fig. 7.M.49a   **49b.** Fig. 7.M.49b

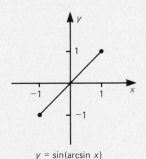

$y = \sin(\arcsin x)$

Fig. 7.M.49a

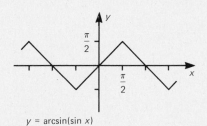

$y = \arcsin(\sin x)$

Fig. 7.M.49b

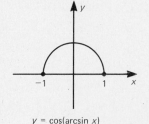

$y = \cos(\arcsin x)$

Fig. 7.M.49c

49c. Fig. 7.M.49c   **50a.** $20(\pi - \Theta) + 100\sqrt{2 - 2\cos\Theta}$ **50b.** $\Theta = 0$
51. $\pi/4$ **52.** Find the maximum of $h = 50 - 10\csc\Theta + 5\cot\Theta$;
$dh/d\Theta = 10\csc\Theta\cot\Theta - 5\csc^2\Theta$; $\cos\Theta = \frac{1}{2}$; $\Theta = \pi/3$; The angle at the ring
is $2\pi/3$ **53.** $(5^{2/3} + 6^{2/3})^{3/2}$ **55a.** Minute hand: 2π radians/hour;
Hour hand: $\frac{1}{6}\pi$ radians/hour   **55b.** $(605\,\pi\sqrt{3})/(6\sqrt{331})$ inches/hour

57. $q_0 e^{-t/RC}$ **59.** One consequence: An object with zero initial velocity in mid-air would never fall. **60.** $[I(y + \Delta y) - I(y)] = -k\,I(y)\,\Delta y$; $dI/dy = -kI$; $I = I_0 e^{-ky}$ **62.** $1600\pi \ln(2)$ inch-pounds **66.** $s(t) = t - 1 - \ln t$ **67.** Set $f(x) = (\ln x)/x$; Max of f is $f(e) = 1/e$; $f(5) = (\ln 5)/5$ so $(\ln 5)/5 < 1/e$; $f(1) = 0$, so by the Intermediate Value Theorem $f(x_0) = (\ln 5)/5$ for some $1 < x_0 < e$; $x = x_0$ and $x = 5$ are the only two solutions because f is increasing for $x < e$ and decreasing for $x > e$. **70.** $y = \cosh x$ **71.** $2/(\ln 3)$ **72a.** 0 **72b.** ∞ **73a.** $1/e$ **73b.** $|x^2 \ln x| = x|x \ln x| < x/e$ for $0 < x < 1$ **73c.** $|x^a \ln x| = (2/a)|y^2 \ln y|$ with $y = x^{a/2}$, and $y \to 0^+$ as $x \to 0^+$ **83a.** The integral is the present value of the income up to time T and the other term is the present value of the income from the sale. **83b.** $f(T) + S'(T) - rS(T) = 0$ **87a.** $\int_0^T R(t)C(t)\,dt$

Section 8.1

1a. $u = x^2 + 6$; (1) **1b.** $\frac{1}{3}(x^2 + 6)^{3/2} + C$ **3a.** $u = \sin x$; (1) **3b.** $\frac{2}{3}(\sin x)^{3/2} + C$ **5a.** $u = \tan x$; (1) **5b.** $\frac{1}{3}\tan^3 x + C$ **7a.** $u = x + \sin x$; (1) **7b.** $\ln|x + \sin x| + C$ **9a.** $u = \sin 3x$; (1) **9b.** $\frac{1}{3}\ln|\sin 3x| + C$ **11a.** $u = \ln x$; (1) **11b.** $2\sqrt{\ln x} + C$ **13a.** $u = x^2$; (8) **13b.** $\frac{1}{2}\arcsin(x^2/3) + C$ **15a.** $u = \arcsin x$; (1) **15b.** $\frac{1}{2}(\arcsin x)^2 + C$ **17a.** $u = e^{-x}$; (3) **17b.** $\cos(e^{-x}) + C$ **19a.** $u = \csc(6x) + \cot(6x)$; (1) **19b.** $-\frac{1}{6}\ln|\csc(6x) + \cot(6x)| + C$ **21a.** $u = x^4$; (8) **21b.** $\frac{1}{24}\pi$ **23a.** $u = 4 - x^2$; (1) **23b.** $2 - \sqrt{3}$ **25a.** $u = \cos x$; (9) **25b.** $-\arctan(\cos x) + C$ **27a.** $u = 1 - x^3$; (1) **27b.** $-\frac{2}{3}\ln(\frac{26}{7})$ **29a.** $u = \sqrt{x}$; (6) **29b.** $2\sec(\sqrt{x}) + C$ **31a.** $u = \sin x$; (1) **31b.** $\frac{1}{2}\sin^2(2)$ **33a.** $u = x^3$; (9) **33b.** $\frac{1}{9}\arctan(x^3/3) + C$ **35a.** $u = \cos x$; (1) **35b.** $2 - 2\sqrt{\cos(1)}$ **37a.** $u = 6 + e^{3x}$; (1) **37b.** $\frac{1}{24}(6 + e^{3x})^8 + C$ **39a.** $u = \arctan x$; (1) **39b.** $\frac{1}{2}(\arctan x)^2 + C$ **41a.** $u = 6 + x^3$; (1) **41b.** $\frac{1}{3}\ln(\frac{7}{6})$ **43a.** $u = \ln x$; (1) **43b.** $\frac{1}{2}\ln^2 x + C$ **45a.** $u = \tan x$; (8) **45b.** $\arcsin(\tan x) + C$ **47a.** $u = \sin x$; (1), (11)

47b. $\dfrac{1}{\ln \pi}\pi^{\sin x} + \dfrac{1}{\pi + 1}(\sin x)^{\pi + 1} + C$

49a.‡ $u = 1 - x$, $x = 1 - u$, $du = -dx$; (1) **49b.**‡ $\int [(u - 1)/\sqrt{u}]\,du$ $= \frac{2}{3}u^{3/2} - 2u^{1/2} + C = \frac{2}{3}(1 - x)^{3/2} - 2(1 - x)^{1/2} + C$ **51a.** $u = 1 - 2x$; (1) **51b.** $\frac{1}{62}$ **53a.** $u = 16 + 2x$; (1) **53b.** 20

57.‡ $u = x^{1/2}$, $du = \dfrac{1}{2}x^{-1/2}\,dx$; $\displaystyle\int \dfrac{1}{(3 + \sqrt{x})\sqrt{x}}\,dx = \int \dfrac{2}{3 + u}\,du$

$= 2\ln|3 + u| + C = 2\ln(3 + \sqrt{x}) + C$

58. $-\frac{20}{3}\ln|2 - 3x^{1/4}| + C$ **60.**‡ $u = \sqrt{\sin x}$, $\cos x\,dx = 2u\,du$; $\int [2u/(3u^2 + 4u)]\,du = \frac{2}{3}\ln|3u + 4| + C = \frac{2}{3}\ln(3\sqrt{\sin x} + 4) + C$ **61.** $\frac{2}{3}(e^x - 2)^{3/2} + 4(e^x - 2)^{1/2} + C$ **63.** $3\arctan(2)$

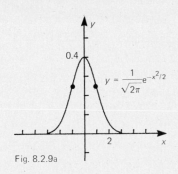

$y = \dfrac{1}{\sqrt{2\pi}} e^{-x^2/2}$

Fig. 8.2.9a

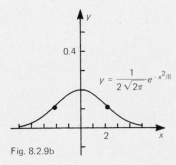

$y = \dfrac{1}{2\sqrt{2\pi}} e^{-x^2/8}$

Fig. 8.2.9b

Section 8.2

1. $\int_0^4 2u\, e^u\, du$ **3.** $2\int_2^4 (u^{11} - u^{10})\, du$ **5.** $\frac{1}{3}\int_3^9 \ln u\, du$

7. $\int_4^8 \dfrac{u^{1/2}}{u-3}\, du$

9. The second derivative is $[(x-k)^2 - \sigma^2]$ multiplied by a constant and by a nonzero exponential, so it changes sign at $x = k \pm \sigma$ and its inflection points are 2σ units apart; Figs. 8.2.9a, b   **10.** Set $u = x/\sigma$;
$(\sigma\sqrt{2\pi})^{-1} \int_{-X}^{X} \exp[-x^2/(2\sigma^2)]\, dx = (2\pi)^{-1/2} \int_{-X/\sigma}^{X/\sigma} \exp[-u^2/2]\, du \to 1$ as
$X \to \infty$ **11a.** The probability that the bullet will hit the wall somewhere is 1.
11b. $\frac{1}{2}$

Section 8.3

1. $\frac{1}{5}x \sin(5x) + \frac{1}{25}\cos(5x) + C$ **3.**‡ $u = x^2, du = 2x\, dx; \frac{1}{2}\int \sin u\, du$
$= -\frac{1}{2}\cos u + C = -\frac{1}{2}\cos(x^2) + C$ **5.** $\frac{1}{2}x^2 \ln x - \frac{1}{4}x^2 + C$
7. $\ln|\ln x| + C$ **9.** $\frac{1}{4}[\arcsin(2x)]^2 + C$ **11.** $\frac{1}{2}e^x(\sin x - \cos x) + C$
13.‡ First $u = \sin(2x), dv/dx = \sin x$; Then $u = \cos(2x), dv/dx = \cos x$;
$\int \sin(2x) \sin x\, dx = -\sin(2x)\cos x + 2\int \cos(2x)\cos x\, dx$
$= -\sin(2x)\cos x + 2\cos(2x)\sin x + 4\int \sin(2x)\sin x\, dx$;
$\frac{1}{3}\sin(2x)\cos x - \frac{2}{3}\cos(2x)\sin x + C$ **15.** $\frac{1}{5}x^5 \ln x - \frac{1}{25}x^5 + C$
17. $\frac{1}{2}e^{x^2} + C$ **19.** $\frac{4}{7}\sin(3x)\sin(4x) + \frac{3}{7}\cos(3x)\cos(4x) + C$ **21.** $\pi - 2$
23. $\frac{1}{6}\sin^2(3x) + C$ **25.**‡ First $u = \sin x; dv/dx = \sin x$; Then $\cos^2 x$
$= 1 - \sin^2 x; \int \sin^2 x\, dx = -\sin x \cos x + \int \cos^2 x\, dx = -\sin x \cos x$
$+ \int (1 - \sin^2 x)\, dx; \int \sin^2 x\, dx = -\frac{1}{2}\sin x \cos x + \frac{1}{2}x + C; \frac{1}{4}\pi$
27. $2^x[x(\ln 2)^{-1} - (\ln 2)^{-2}] + C$ **29.** $x \cosh x - \sinh x + C$
31. $\frac{1}{2}[x \sin(\ln x) + x \cos(\ln x)] + C$ **33.** $-x^2(1 - x^2)^{1/2} - \frac{2}{3}(1 - x^2)^{3/2}$
$+ C$ **35.** $11 \ln(11) - 10$ **37a.** $1 - e^{-3}$ gm.   **37b.** $(1 - 4e^{-3})/(1 - e^{-3})$ cm.
39. $4\pi e^3$ **41.**‡ $e^{ax}\sin(bx) = (aA - bB)e^{ax}\sin(bx) + (bA + aB)e^{ax}\cos(bx)$;
$aA - bB = 1, bA + aB = 0; A = a/(a^2 + b^2), B = -b/(a^2 + b^2)$
43. $A = b/(a^2 - b^2), B = -a/(a^2 - b^2)$

Section 8.4

1. $-\frac{1}{4}\cos(4x) + \frac{1}{12}\cos^3(4x) + C$ **3.**‡ $\cos^2 x = 1 - \sin^2 x$ and $u = \sin x$;
$\int u^{\sqrt{2}}(1 - u^2)\, du = (1 + \sqrt{2})^{-1}(\sin x)^{1+\sqrt{2}} - (3 + \sqrt{2})^{-1}(\sin x)^{3+\sqrt{2}} + C$
5. $\frac{3}{8}x - \frac{1}{4}\sin(2x) + \frac{1}{32}\sin(4x) + C$ **7.**‡ $\frac{1}{2}\int [\sin(4x) + \sin(2x)]\, dx$
$= -\frac{1}{8}\cos(4x) - \frac{1}{4}\cos(2x) + C$ **9.** $\frac{1}{2}\sin x + \frac{1}{30}\sin(15x) + C$
11.‡ $\int \tan^2(6x)\, dx = \int [\sec^2(6x) - 1]\, dx = \frac{1}{6}\tan(6x) - x + C$
13. $-\frac{1}{2}x^2 - \frac{1}{2}\cot(x^2) + C$ **15.** $\tan x + \frac{1}{3}\tan^3 x + C$ **17.** $2^{-1/2} + \frac{1}{2}\ln(\sqrt{2} + 1)$
22. First $u = \sin^{m-1}x, dv/dx = \cos^n x \sin x$, then $\cos^{n+2}x$

$= \cos^n x(1 - \sin^2 x); \int \sin^m x \cos^n x\, dx = -\dfrac{1}{n+1}\sin^{m-1}x \cos^{n+1}x$

$+ \dfrac{1}{n+1}\int (m-1)\sin^{m-2}x \cos^{n+2}x\, dx = -\dfrac{1}{n+1}\sin^{m-1}x \cos^{n+1}x$

$+ \dfrac{m-1}{n+1}\int \sin^{m-2}x \cos^n x\, dx - \dfrac{m-1}{n+1}\int \sin^m x \cos^n x\, dx$;

$\int \sin^m x \cos^n x\, dx = -\dfrac{1}{m+n}\sin^{m-1}x \cos^{n+1}x + \dfrac{m-1}{m+n}\int \sin^{m-2}x \cos^n x\, dx$

26.[‡] Use formula (10) with $n = 6$ and $n = 4$; $\frac{1}{5}\sec^4 x \tan x + \frac{4}{5}\int \sec^4 x\,dx$
$= \frac{1}{5}\sec^4 x \tan x + \frac{4}{15}\sec^2 x \tan x + \frac{8}{15}\tan x + C$ **28.** $\frac{1}{3}\tan^3 x - \tan x +$
$+ x + C$

Section 8.5

1.[‡] $x = 4\tan\Theta$, $dx = 4\sec^2\Theta\,d\Theta$; $\sqrt{x^2 + 16} = 4\sec\Theta$; $\frac{1}{16}\int \cot\Theta\,d\Theta$
$= \frac{1}{16}\ln|\sin\Theta| + C = \frac{1}{16}\ln[x(x^2 + 16)^{-1/2}] + C$ **3.**[‡] $x = 3\sin\Theta$,
$dx = 3\cos\Theta\,d\Theta$, $\sqrt{9 - x^2} = 3\cos\Theta$; $\frac{1}{3}\int \csc\Theta\,d\Theta = -\frac{1}{3}\ln|\csc\Theta + \cot\Theta| + C$
$= -\frac{1}{3}\ln|(3/x) + \sqrt{9 - x^2}/x| + C$ **5.** $\frac{625}{8}\arcsin(\frac{1}{5}x) - \frac{1}{8}x(25 - x^2)^{3/2}$
$+ \frac{1}{8}x^3(25 - x^2)^{1/2} + C$ **7.** $\frac{1}{7}(16 - x^2)^{7/2} - \frac{16}{5}(16 - x^2)^{5/2} + C$
9. $\frac{1}{3}(1 + x^2)^{3/2} - (1 + x^2)^{1/2} + C$ **11.** $2\arcsin(x/2) + \frac{1}{2}x\sqrt{4 - x^2} + C$
13. $\frac{1}{2}\ln|1 + x| - \frac{1}{2}\ln|1 - x| + C$ **15.** $x(1 - x^2)^{-1/2} - \arcsin x + C$

17.[‡] Complete the square: $\int \dfrac{1}{(x - 3)^2 + 16}\,dx$; $x - 3 = 4\tan\Theta$;

$\dfrac{1}{4}\arctan[(x - 3)/4] + C$

19.[‡] Complete the square: $\int \dfrac{1}{\sqrt{100 - (x + 4)^2}}\,dx$; $x + 4 = 10\sin\Theta$;

$\arcsin[(x + 4)/10] + C$

21. $\dfrac{x - 1}{\sqrt{1 - (x - 1)^2}} + C$

Section 8.6

1.[‡] $\dfrac{13x + 6}{x(x + 2)(x + 3)} = \dfrac{A}{x} + \dfrac{B}{x + 2} + \dfrac{C}{x + 3}$; $13x + 6 = A(x + 2)(x + 3)$
$+ Bx(x + 3) + Cx(x + 2)$; $x = 0$: $A = 1$; $x = -2$: $B = 10$; $x = -3$:
$C = -11$; $\dfrac{1}{x} + \dfrac{10}{x + 2} - \dfrac{11}{x + 3}$

2. $2 + \dfrac{4}{x + 2} - \dfrac{3}{x - 2}$

3.[‡] $\dfrac{x - 2}{(x + 1)^2} = \dfrac{A}{x + 1} + \dfrac{B}{(x + 1)^2}$; $x - 2 = A(x + 1) + B$; $x = -1$: $B = -3$;

$x - 2 = Ax + A - 3 \Rightarrow A = 1$; $\dfrac{1}{x + 1} - \dfrac{3}{(x + 1)^2}$

5. $x^3 - 16x + \dfrac{256x}{x^2 + 16}$ **7.** $\dfrac{1}{(x + 1)^2} - \dfrac{1}{(x + 1)^3}$

9.[‡] $\dfrac{x^2 + x + 1}{x(x - 1)(2x + 1)} = \dfrac{A}{x} + \dfrac{B}{x - 1} + \dfrac{C}{2x + 1}$;

$x^2 + x + 1 = A(x - 1)(2x + 1) + Bx(2x + 1) + Cx(x - 1)$;

$x = 0$: $A = -1$; $x = 1$: $B = 1$; $x = -\dfrac{1}{2}$: $C = 1$; $-\dfrac{1}{x} + \dfrac{1}{x - 1} + \dfrac{1}{2x + 1}$

11.‡ $\dfrac{x^3}{(x^2 + 1)^2} = \dfrac{Ax + B}{x^2 + 1} + \dfrac{Cx + D}{(x^2 + 1)^2}$; $x^3 = (Ax + B)(x^2 + 1) + Cx + D$;

$x^3 = Ax^3 + Bx^2 + (A + C)x + (B + D)$; $1 = A, 0 = B, 0 = A + C$;

$0 = B + D$; $C = -1, D = 0$; $\dfrac{x}{x^2 + 1} - \dfrac{x}{(x^2 + 1)^2}$

13. $x + 1 + \dfrac{1}{5}\left[\dfrac{-36x - 27}{x^2 + 2x + 5} + \dfrac{1}{x}\right]$ **15.** $\dfrac{3}{x + 2} - \dfrac{6}{(x + 2)^2}$ **17.** $\dfrac{1}{x^2} - \dfrac{1}{(x + 1)^2}$

19. $\dfrac{3}{x^2 + 4} + \dfrac{12x - 10}{(x^2 + 4)^2}$ **21.** $-\dfrac{1}{x} + \dfrac{2}{x - 1} - \dfrac{3}{x - 2} + \dfrac{4}{x - 3}$

23. $-\ln|x| + 6\ln|x + 1| + C$ **25.** $\ln|x^2 + 3x + 2| + C$

27.‡ $\dfrac{1}{2}\displaystyle\int \dfrac{2x}{x^2 + 4}dx + 3\int \dfrac{1}{x^2 + 4}dx = \dfrac{1}{2}\ln(x^2 + 4) + \dfrac{3}{2}\arctan\left(\dfrac{1}{2}x\right) + C$

28. $x + \frac{13}{2}\ln(x^2 + 7) - \sqrt{7}\arctan(x/\sqrt{7}) + C$ **29.** $\ln|x| - 1/(x + 1) + C$

31.‡ $u = x^2 + 1$; $du = 2x\,dx$; $\displaystyle\int \dfrac{3}{u^3}du = -\dfrac{3}{2}u^{-2} + C = -\dfrac{3}{2}(x^2 + 1)^{-2} + C$

33. $-3\ln(5) + 2\ln(26) - 2\ln(2)$ **35.** $\frac{1}{2}\ln|x^2 + 2x + 10| + C$

37.‡ $\dfrac{1}{2}\left(\dfrac{2x + 2}{x^2 + 2x + 5}\right) - \dfrac{1}{x^2 + 2x + 5}$; Set $u = x^2 + 2x + 5$ in the first

expression and complete the square in the second; $\dfrac{1}{2}\ln|x^2 + 2x + 5|$

$-\dfrac{1}{2}\arctan[(x + 1)/2] + C$

39. $-4\ln(3) + 3\ln(2)$ **41.** $\arctan x + (x^2 + 1)^{-1} + C$
43. $10 - 2\arctan(10)$ **45.** $-\frac{3}{4} - \ln(40)$

48. $x^2 + \left(\dfrac{b}{a}\right)x = -\dfrac{c}{a}$; $x^2 + \left(\dfrac{b}{a}\right)x + \dfrac{b^2}{4a^2} = \dfrac{b^2}{4a^2} - \dfrac{c}{a}$;

$\left(x + \dfrac{b}{2a}\right)^2 = \dfrac{b^2 - 4ac}{4a^2}$; $x = \dfrac{-b \pm \sqrt{b^2 - 4ac}}{2a}$

Section 8.7

1.‡ $-2\displaystyle\int \dfrac{u^2 - 1}{(u^2 + 1)(u^2 + 2u - 1)}du$; Partial fractions:

$\dfrac{1}{2}\ln|u^2 + 2u - 1| - \dfrac{1}{2}\ln|u^2 + 1| - \arctan u + C$

$= \dfrac{1}{2}\ln|\sin x - \cos x| - \dfrac{1}{2}x + C$

3.‡ $\displaystyle\int \dfrac{1}{u^2 + 4}du = \dfrac{1}{2}\arctan\left[\dfrac{1}{2}\tan\left(\dfrac{1}{2}x\right)\right] + C$

5. $-\frac{1}{4}[\tan(\frac{1}{2}x)]^{-2} - \frac{1}{2}\ln|\tan(\frac{1}{2}x)| + C$ **7.** $-\frac{1}{9}\pi + \frac{5}{6}\arctan(2/\sqrt{3})$

Section 8.8

1.‡ Formula (24) with $a = 13$; $[\frac{1}{2} x(169 - x^2)^{1/2} + \frac{169}{2} \arcsin(\frac{1}{13}x)]_0^{12}$
$= 30 + \frac{169}{2} \arcsin(\frac{12}{13})$

2. $\dfrac{1}{4} \ln \left| \dfrac{x + 2}{x - 2} \right| + C$

4.‡ $u = x^2$; $\frac{1}{2} \int \sqrt{1 - u^2}\, du$; Formula (24): $\frac{1}{4} u \sqrt{1 - u^2} + \frac{1}{4} \arcsin u + C$
$= \frac{1}{4} x^2 \sqrt{1 - x^4} + \frac{1}{4} \arcsin(x^2) + C$

5.‡ $u = e^x$: $\int \dfrac{du}{1 - u^2}$; Formula (15): $\dfrac{1}{2} \ln \left| \dfrac{u + 1}{u - 1} \right| + C = \dfrac{1}{2} \ln \left| \dfrac{e^x + 1}{e^x - 1} \right| + C$

6. $10\sqrt{61} + 72 \ln(\frac{5}{6} + \frac{1}{6}\sqrt{61})$ **9.** $-\frac{1}{2}\cos(x^2) + \frac{1}{6}\cos^3(x^2) + C$
11. $\frac{1}{8}\cos(4x) - \frac{1}{12}\cos(6x) + C$

12.‡ $u = 2x + 5$; $du = 2\, dx$; $\displaystyle\int \dfrac{1}{9 + (2x + 5)^2}\, dx = \dfrac{1}{2} \int \dfrac{1}{9 + u^2}\, du$

$= \dfrac{1}{2}\left(\dfrac{1}{3}\right)\arctan(u/3) + C = \dfrac{1}{6} \arctan\left(\dfrac{2x + 5}{3}\right) + C$ by formula (14);

$\displaystyle\int_0^{10} \dfrac{1}{9 + (2x + 5)^2}\, dx = \left[\dfrac{1}{6} \arctan\left(\dfrac{2x + 5}{3}\right)\right]_0^{10} = \dfrac{1}{6}\left[\arctan\left(\dfrac{25}{3}\right) - \arctan\left(\dfrac{5}{3}\right)\right]$

14. $\frac{1}{20}\arctan(10)$ **15.** $\frac{1}{4}\ln(101)$

16. $\ln \left| \dfrac{\cos x + 1}{\cos x} \right| + C$

20. $\ln|x + e^x| + C$

23.‡ Use formula (57) with $n = 2$, $a = 1$; $\dfrac{1}{2}\left(\dfrac{x}{x^2 + 1}\right) + \dfrac{1}{2} \displaystyle\int \dfrac{1}{x^2 + 1}\, dx$

$= \dfrac{1}{2}\left(\dfrac{x}{x^2 + 1}\right) + \dfrac{1}{2}\arctan x + C$

24. $\dfrac{1}{2\ln(3)}\, 3^{x^2} + C$

25. $e^x \arcsin(e^x) + \sqrt{1 - e^{2x}} + C$ **26.** 0 **28.** $-x^2 \cos x + 2 \cos x$
$+ 2x \sin x + C$ **29.** $-2\cot(\sqrt{x}) + C$ **34.** $\sin x \ln|\sin x| - \sin x + C$

37.‡ $u = 5 - 10x$; $du = -10\, dx$; $-\dfrac{1}{10} \displaystyle\int \dfrac{1}{u}\, du = -\dfrac{1}{10} \ln|u| + C$

$= -\dfrac{1}{10} \ln|5 - 10x| + C$ by formula (2); $-\dfrac{1}{10}(\ln 25 - \ln 5)$

38. $\dfrac{1}{7} \ln \left| \dfrac{x - 3}{x + 4} \right| + C$

39. $\frac{1}{3}\arctan[(x - 2)/3] + C$ **42.** $1 + \frac{1}{2}\sin(2)\cos(2)$ **43.** $\ln(\frac{3}{2})$
45. $\frac{3}{2}\sqrt{37} + \frac{1}{4}\ln(6 + \sqrt{37})$ **48.** Fig. 8.8.48; $A = \frac{1}{2}\ln(3)$
50. $60 \arctan(60) - \frac{1}{2}\ln(3601)$

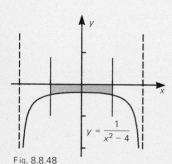

$y = \dfrac{1}{x^2 - 4}$

Fig. 8.8.48

Section 8.9

1a.‡ $\Delta x = \frac{1}{4}$; $\frac{1}{4}(e^{1/8} + e^{3/8} + e^{5/8} + e^{7/8}) \approx \frac{1}{4}(e^{0.12} + e^{0.38} + e^{0.62} + e^{0.88})$
$\approx \frac{1}{4}(1.1275 + 1.4623 + 1.8589 + 2.4109) = \frac{1}{4}(6.8596) \approx 1.7150$
1b.‡ $\frac{1}{4}(\frac{1}{2})(e^0 + 2e^{1/4} + 2e^{1/2} + 2e^{3/4} + e^1) \approx \frac{1}{8}[1 + 2(1.2840) + 2(1.6487)$
$+ 2(2.1170) + 2.7183] = \frac{1}{8}(13.8177) \approx 1.7272$ **1c.**‡ $\frac{1}{3}(\frac{1}{4})(e^0 + 4e^{1/4}$
$+ 2e^{1/2} + 4e^{3/4} + e^1) \approx \frac{1}{12}[1 + 4(1.2840) + 2(1.6487) + 4(2.1170) + 2.7183]$
$= \frac{1}{12}(20.6197) \approx 1.7183$ **3a.** 1.4309 **3b.** 1.3865 **3c.** 1.4166 **5a.** 25.2436
5b. 28.4651 **5c.** 26.4389 **7a.** 0.43407 **7b.** 0.43157 **7c.** 0.43184
9a. 53.88206 **9b.** 58.92795 **9c.** 54.73342 **11.** If the graph of $g(x)$ is the line
through $(x_{j-1}, f(x_{j-1}))$ and $(x_j, f(x_j))$, then the integral of $g(x)$ from x_{j-1}
to x_j is $\frac{1}{2}\Delta x\,[f(x_{j-1}) + f(x_j)]$. The sum of these expressions for $j = 1, 2, \ldots,$
N is the expression in formula (5). **12.** For each x_0 and Δx, the integral of
$ax^2 + bx + c$ from $x_0 - \Delta x$ to $x_0 + \Delta x$ is $[\frac{1}{3}ax^3 + \frac{1}{2}bx^2 + cx]_{x_0 - \Delta x}^{x_0 + \Delta x}$
$= \frac{1}{3}\Delta x(6ax_0^2 + 6bx_0 + 2a(\Delta x)^2 + 6c)$; If $f(x)$ equals $ax^2 + bx + c$ at x
$= x_0 - \Delta x, x = x_0$, and $x = x_0 + \Delta x$, then $f(x_0 - \Delta x) + 4f(x_0) + f(x_0 + \Delta x)$
$= 6ax_0^2 + 6bx_0 + 2a(\Delta x)^2 + 6c$; Therefore, if the graph of $g(x)$ is the parabola
passing through $(x_{2k-1}, f(x_{2k-1})), (x_{2k}, f(x_{2k}))$, and $(x_{2k+1}, f(x_{2k}))$, then the
integral of $g(x)$ from x_{2k-1} to x_{2k+1} is $\frac{1}{3}\Delta x\,[f(x_{2k-1}) + 4f(x_{2k}) + f(x_{2k+1})]$;
The sum of these expressions for $k = 1, 2, \ldots, K$, yields formula (6).
14a. 0.3697220 . . . **14b.** 0.3697224 . . . **16a.** 0.8049852 . . .
16b. 0.8161624 . . . **18a.** 0.1779646 . . . **18b.** 0.1778720 . . .

Miscellaneous Exercises to Chapter 8 *(Part I)*

1. Partial fractions **3.** $u = \ln x$ **5.** Integration by parts **7.** Integration by
parts **9.** $u = 4\cos x$ **11.** $u = x^2 + 1$ **13.** $\cos^2(4x) = \frac{1}{2}[1 + \cos(8x)]$ or
integration by parts and $\sin^2(4x) = 1 - \cos^2(4x)$ **15.** $u = \sin x$ and
$\cos^2 x = 1 - \sin^2 x$ **17.** Integration by parts **19.** $u = x^2$ and $v = \cos u$
21. $u = \sin x$ **23.** $u = \sin(5x)$ and $\cos^2(5x) = 1 - \sin^2(5x)$ **25.** Integration
by parts **27.** $u = 1 + x^2$ **29.** $x = \sin \Theta$ **31.** Partial fractions **33.** $u = x^3$
35. $u = x^2$

37. $\dfrac{x}{x^2 - 2x + 17} = \dfrac{x - 1}{x^2 - 2x + 17} + \dfrac{1}{(x - 1)^2 + 16}$; Use $u = x^2 - 2x + 17$

in the first integral and $v = x - 1$ in the second.

39. $u = 9 - x^2$ **41.** $x = \sin \Theta$ **43.** $u = x^{1/2}$ **45.** $u = x + 1$ **47.** $u = \cos x$
49. $u = 1 - 3x$

Miscellaneous Exercises to Chapter 8 *(Part II)*

1.‡ $x^3 - 8 = (x - 2)(x^2 + 2x + 4)$; $\dfrac{A}{x - 2} + \dfrac{Bx + C}{x^2 + 2x + 4} = \dfrac{1}{x^3 - 8}$;

$A + B = 0, 2A - 2B + C = 0, 4A - 2C = 1$;

$\displaystyle\int\left[\dfrac{1}{12(x - 2)} - \dfrac{x + 4}{12(x^2 + 2x + 4)}\right]dx = \dfrac{1}{12}\ln|x - 2| - \dfrac{1}{24}\ln|x^2 + 2x + 4|$
$- [1/(4\sqrt{3})]\arctan[(x + 1)/\sqrt{3}] + C$

3. $\frac{1}{4}(\ln x)^4 + C$ **5.** $-\frac{1}{3}x^2\cos(3x) + \frac{2}{9}x\sin(3x) + \frac{2}{27}\cos(3x) + C$
7. $x\arcsin(3x) + \frac{1}{3}\sqrt{1 - 9x^2} + C$ **9.** $-\frac{1}{4}e^{4\cos x} + C$

11.‡ $\displaystyle\int(x^3 - x)\,dx + \int\dfrac{x}{x^2 + 1}\,dx = \dfrac{1}{4}x^4 - \dfrac{1}{2}x^2 + \dfrac{1}{2}\ln(x^2 + 1) + C$

(Use $u = x^2 + 1$)

13. $\frac{1}{2}x + \frac{1}{16}\sin(8x) + C$ **15.** $\frac{1}{6}\sin^6 x - \frac{1}{8}\sin^8 x + C$

17.‡ $(x - 1)\arctan(x - 1) - \displaystyle\int \frac{x - 1}{(x - 1)^2 + 1}\, dx$

$$= (x - 1)\arctan(x - 1) - \frac{1}{2}\ln[1 + (x - 1)^2] + C$$

19. $-\frac{1}{2}\ln|\cos x^2| + C$ **21.** $\arctan(\sin x) + C$ **23.** $\frac{1}{5}\sin(5x) - \frac{1}{15}\sin^3(5x) + C$
25. $-\frac{1}{3}x\cos(3x) + \frac{1}{9}\sin(3x) + C$ **27.** $-\frac{1}{2}(1 + x^2)^{-1} + C$ **29.**‡ $x = \sin\Theta$,
$dx = \cos\Theta\, d\Theta,\ \sqrt{1 - x^2} = \cos\Theta;\ \int\sin^3\Theta\cos^6\Theta\, d\Theta = \int(u^8 - u^6)\, du$
$= \frac{1}{9}\cos^9\Theta - \frac{1}{7}\cos^7\Theta + C = \frac{1}{9}(1 - x^2)^{9/2} - \frac{1}{7}(1 - x^2)^{7/2} + C$ with $u = \cos\Theta$
31. $\ln|x| - \frac{1}{2}(x^2 + 1)^{-1} + C$ **33.** $-\frac{1}{3}\cot(x^3) + C$ **35.** $(5^{x^2})/[2\ln(5)] + C$

37.‡ $\displaystyle\int \frac{x - 1}{x^2 - 2x + 17}\, dx + \int \frac{1}{x^2 - 2x + 17}\, dx$

$$= \frac{1}{2}\ln|x^2 - 2x + 17| + \frac{1}{4}\arctan[(x - 1)/4] + C$$

39. $-\frac{5}{14}(9 - x^2)^{7/5} + C$ **41.** $\frac{1}{2}(\arcsin x - x\sqrt{1 - x^2}) + C$ **43.** $2\,e^{\sqrt{x}} + C$
45. $\frac{2}{7}(x + 1)^{7/2} - \frac{4}{5}(x + 1)^{5/2} + \frac{2}{3}(x + 1)^{3/2} + C$ **47.** $-\arctan(\cos x) + C$
49. $-\frac{1}{3}\sinh(1 - 3x) + C$ **50a.** $4(x + \frac{3}{4} - \frac{1}{4}\sqrt{17})(x + \frac{3}{4} + \frac{1}{4}\sqrt{17})$
50b.‡ Because the polynomial is zero at $x = -2$, it has $x + 2$ as a factor; Fig.
8.M.50b; $(x + 2)(x^2 + x + 4)$; The quadratic factor is irreducible because
$b^2 - 4ac = 1^2 - 4(1)(4)$ is negative.
50c. $(x + 1)(x - \frac{7}{2} + \frac{1}{2}\sqrt{65})(x - \frac{7}{2} - \frac{1}{2}\sqrt{65})$ **50d.**‡ The polynomial has
$x + 2$ and $x - 1$ as factors; Fig. 8.M.50d; $(x + 2)(x - 1)(x^2 - x + 3)$
50e. $7x^4(x - \frac{2}{7} + \frac{4}{7}\sqrt{2})(x - \frac{2}{7} - \frac{4}{7}\sqrt{2})$ **50f.** $2(x + 1 + \frac{1}{2}\sqrt{2})(x + 1 - \frac{1}{2}\sqrt{2})$
50i. $(x + \sqrt{2})^2(x - \sqrt{2})^2$ **50j.** $(x - 2)(x + 2)(x^2 + 4)$

51. $4\ln|x| + 2\ln|x + 1| + \dfrac{1}{x + 1} + C$

53. $\ln|x| + \ln|x - 2| + \dfrac{3}{x - 2} + C$

55.‡ $\dfrac{2x^2 + x + 2}{(x^2 + 1)^2} = \dfrac{Ax + B}{x^2 + 1} + \dfrac{Cx + D}{(x^2 + 1)^2};\ 2x^2 + x + 2$

$$= Ax^3 + Bx^2 + (A + C)x + (B + D);\ A = 0, B = 2, A + C = 1,$$

$B + D = 2;\ \displaystyle\int \frac{2}{x^2 + 1}\, dx + \int \frac{x}{(x^2 + 1)^2}\, dx = 2\arctan x - \frac{1}{2}(x^2 + 1)^{-1} + C$

57. $\dfrac{1}{2}\ln|x^2 - 2x + 10| + \dfrac{1}{3}\arctan\!\left(\dfrac{x - 1}{3}\right) + C$

59. $3\ln|x + 2| + 4\ln|x + 5| + C$

61. $\dfrac{1}{3}x^3 + x^2 + \dfrac{4}{3}\arctan\!\left(\dfrac{x - 2}{3}\right) + C$

63. $\frac{1}{3}x^3 + x + \ln|x + 1| + 2\ln|x - 3| + C$

65.‡ $u = \sqrt{x}, dx = 2u\, du;\ -2\displaystyle\int \frac{u}{u - 3}\, dx = -2\int\left(1 + \frac{3}{u - 3}\right) du$

$= -2u - 6\ln|u - 3| + C = -2\sqrt{x} - 6\ln|\sqrt{x} - 3| + C$

66.‡ $u = x^{1/3}, dx = 3u^2\, du;\ 3\displaystyle\int \frac{du}{u(1 + u)} = 3\int\left(\frac{1}{u} - \frac{1}{1 + u}\right) du$

$= 3\ln(x^{1/3}) - 3\ln(1 + x^{1/3}) + C$

$$
\begin{array}{r}
x^2 + x\ \ + 4 \\
x + 2\ \overline{\smash{\big)}\ x^3 + 3x^2 + 6x + 8} \\
\underline{x^3 + 2x^2} \\
x^2 + 6x \\
\underline{x^2 + 2x} \\
4x + 8 \\
\underline{4x + 8}
\end{array}
$$

Fig. 8.M.50b

$$
\begin{array}{r}
x^2\ -\ x\ \ + 3 \\
x^2 + x - 2\ \overline{\smash{\big)}\ x^4 + 0x^3 + 0x^2 + 5x - 6} \\
\underline{x^4 + x^3 - 2x^2} \\
-\,x^3 + 2x^2 + 5x \\
\underline{-\,x^3 - x^2 + 2x} \\
3x^2 + 3x - 6 \\
\underline{3x^2 + 3x - 6}
\end{array}
$$

Fig. 8.M.50d

67. $u = (1 - e^x)^{1/2}, x = \ln(1 - u^2), dx = [2u/(u^2 - 1)]\,du;$

$$2\int \frac{u^2}{u^2 - 1}\,du = 2\int \left(1 + \frac{1}{u^2 - 1}\right)du$$

$$= 2\sqrt{1 - e^x} + \ln|1 - \sqrt{1 - e^x}| - \ln|1 + \sqrt{1 - e^x}| + C.$$

69. $u = e^x, dx = du/u;\ \displaystyle\int \frac{u + 2}{u(2 - u)}\,du = \int \left(\frac{1}{u} + \frac{2}{2 - u}\right)du$

$$= x - 2\ln|2 - e^x| + C$$

71. $4\ln|1 - x^{1/4}| - \frac{4}{5}x^{5/4} - x - \frac{4}{3}x^{3/4} - 2x^{1/2} - 4x^{1/4} + C$

73. $2x^{1/2} - 4x^{1/4} + 4\ln|1 + x^{1/4}| + C$ **75.** $5/\ln(2) - 5$ **76.** 800π pounds

77. $12\pi^2$ **78.** $5\sqrt{101} + \frac{1}{2}\ln(10 + \sqrt{101})$ **79.** $\frac{232}{15}$ **81.** $(V_{max})(i_{max})/2$

Section 9.1

1. $y = e^{1 - \cos t}$

3.‡ $\dfrac{df}{dx} = -3f^4;\ f^{-4}\,df = -3\,dx;\ -\dfrac{1}{3}f^{-3} = -3x + C_1;\ f = (9x + C)^{-1/3}$

5. $y = (\arctan x + C)^{-1}$ **6.** $y = \sin(x + C)$ **7.** $y = \arctan(x + C) + n\pi$
with integers n **9.** $K(x) = \frac{1}{9}(x^{3/2} + 2)^2$ **11.** $F = 2\tan(2x + \frac{1}{4}\pi)$ **13.** $y = (\frac{1}{2} + \ln x)^{-1}$

15.‡ $\dfrac{dP}{dt} = P(10 - P);\ \dfrac{dP}{P(10 - P)} = dt;$ Partial fractions:

$$\frac{1}{10}\left[\frac{1}{P} + \frac{1}{10 - P}\right]dP = dt;\ \ln\left[\frac{P}{10 - P}\right] = 10t + C;\ P(0) = 2 \Rightarrow C = \ln(1/4);$$

$$\frac{P}{10 - P} = \frac{1}{4}e^{10t};\ P(t) = \frac{10e^{10t}}{4 + e^{10t}}$$

16. $f = \dfrac{1 + 3e^{10t}}{1 + e^{10t}}$

18. $x + \ln|x| + y - \ln|y| = C$ **19.** $\arctan x - \frac{1}{2}\ln(y^2 + 1) = C$
21a. $y = 4/(\cos x + 1)$ **21b.** $y = 4/(\cos x - 3)$ **21c.** Fig. 9.1.21
22. $\frac{1}{4}x^4 - 3x^2 + y^2 = 9$; Fig. 9.1.22 **24.** $y = 27/(19 - x^3)$; Fig. 9.1.24

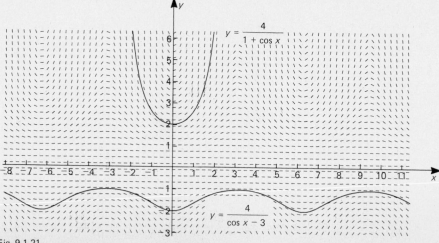

$$y = \frac{4}{1 + \cos x}$$

$$y = \frac{4}{\cos x - 3}$$

Fig. 9.1.21

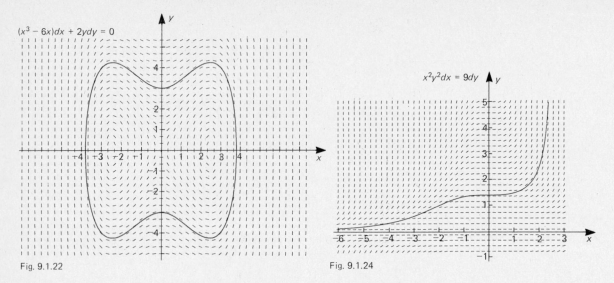

$(x^3 - 6x)dx + 2ydy = 0$

$x^2y^2dx = 9dy$

Fig. 9.1.22

Fig. 9.1.24

Section 9.2

1.‡ Let $v(t)$ = velocity at time t (seconds); $F = ma = m\dfrac{dv}{dt}$;

$m = 6560/32 = 205$ slugs; $-kv^2 = 205\dfrac{dv}{dt}$; $-\dfrac{205}{v} = -kt + C$; Set $t = 0$

and $v = 10$: $C = -\dfrac{205}{10}$; $205\left(\dfrac{1}{10} - \dfrac{1}{v}\right) = -kt$; Set $t = 20$ and $v = 5$:

$k = 41/40 = 1.025$

3. $P = (\frac{1}{2}t + 20)^2$ **5.** $P = \frac{1}{2}L$ **6.** $dP/dt = kP(L - P)$ with $L \approx 675$ and $k \approx 7.7 \times 10^{-4}$ **7.** $D = 8.7\ln(P/0.0083)$

9a.‡ Every molecule of KI_3 that is formed uses one molecule of $C_2H_4Br_2$ and three molecules of KI; $[C_2H_4Br_2] = a - y$; $[KI] = b - 3y$;

$$\dfrac{dy}{(a - y)(b - 3y)} = 0.3\,dt;$$ Partial fractions:

$$\dfrac{1}{b - 3a}\left(\dfrac{1}{a - y} - \dfrac{3}{b - 3y}\right)dy = 0.3\,dt;\ (*)\ \ln\left[(b - 3y)/(a - y)\right]$$

$\doteq 0.3(b - 3a)t + \ln(b/a)$

9b.‡ For $b > 3a$, the right side of $(*) \to \infty$, so $(b - 3y)/(a - y) \to \infty$; $y \to a$
9c.‡ For $b < 3a$, the right side of $(*) \to -\infty$, so $(b - 3y)/(a - y) \to 0$; $y \to b/3$ **10.** $y = 2a - (10^{10}t + \frac{1}{4}a^{-2})^{-1/2}$ **12.** $y = (\frac{1}{2}kt + 10)^{-2}$
13a. $y = a(\frac{1}{2} + \sqrt{0.00253/0.160})^{-1} \approx 1.6a$ **13b.** Fig. 9.2.13; Along both curves y approaches the equilibrium concentration as $t \to \infty$ **16.** 40 months
17a. $y = C/x$ **17e.** $4x^2 + y^2 = C$ **17g.** $y^{-2} = x^{-2} + C$
17j. $y^2\ln y - \frac{1}{2}y^2 + x^2 = C$

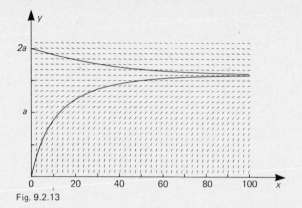

Fig. 9.2.13

Miscellaneous Exercises to Chapter 9

1a. $y = x^x$ **1c.** $y^2 + 1 = e^{-\arctan x}$ **3.** $T = 70 + 30\left(\frac{2}{3}\right)^{t/60}$

4a.‡ Let $s(t) =$ pounds of salt in the tank at time t (minutes);

$$\frac{ds}{dt} = [\text{Rate in}] - [\text{Rate out}] = \left[\frac{1}{2}\frac{\text{pound}}{\text{gallon}}\right]\left[6\frac{\text{gallons}}{\text{minute}}\right]$$

$$- \left[\frac{s(t)}{120}\frac{\text{pounds}}{\text{gallon}}\right]\left[6\frac{\text{gallons}}{\text{minute}}\right]; \int \frac{20}{60 - s}\,ds = \int dt; \; -\ln|60 - s| = \frac{1}{20}t + C;$$

Set $t = 0, s = 20$: $C = -\ln(40)$; $s = 60 - 40\,e^{-t/20}$ pounds

4b.‡ Concentration $= s(t)/120 \to \frac{1}{2}$ pound per gallon as $t \to \infty$

6. $H = k\ln(M) +$ How happy he would be with one dollar

7. $dV/dt = -k\,V^{2/3}$; $V = (V_0^{1/3} - \frac{1}{3}kt)^3$; It disappears

10a.‡ $da/dt = -0.001\,a + 0.005\,b = 0.005 - 0.006\,a$

10b.‡ $-\dfrac{1}{6}\ln|5 - 6a| = 0.001\,t + C; C = -\dfrac{1}{6}\ln|5 - 6a(0)|$;

$$\frac{5 - 6a(0)}{5 - 6a(t)} = e^{0.006\,t}; \; a(t) = \frac{5}{6} - \left[\frac{5}{6} - a(0)\right]e^{-0.006\,t}$$

10c.‡ At equilibrium $da/dt = 0$; $a = \frac{5}{6}$, $b = \frac{1}{6}$ **12b.** L is the amount of the substance in the solution when equilibrium is reached; $x(0)$ is greater than L if the solution is initially supersaturated.   **16.** $y = [H_2]$ is defined implicitly by the equation $kt = -2(m - 2)\,y^{-1/2} - \frac{4}{3}ay^{-3/2} - (8 - 6m)/(3\sqrt{a})$ and the condition $y(0) = a$ **17.**‡ Set $m(x) = f'(x)$; $m' = (\rho/k)\sqrt{1 + m^2}$ and $m(0) = 0$; $m = \tan\Theta$; $\ln|\sqrt{1 + m^2} + m| = (\rho x/k) + C$; $C = 0$; $m(x) = \sinh(\rho x/k)$; $f(x) = (k/\rho)\cosh(\rho x/k) + C$; $C = b - k/\rho$; $f(x) = (k/\rho)[\cosh(\rho x/k) - 1] + b$ **18.** $dF/dt = k(1 - F)$; $F(0) = 0$; $F(t) = 1 - e^{-kt}$ **19.** For $0 \le t \le \frac{5}{16}\pi$, $v = 40\tan(\frac{1}{4}\pi - \frac{4}{5}t)$; For $t \ge \frac{5}{16}\pi$, $v = 40\tanh(\frac{1}{4}\pi - \frac{4}{5}t)$

Section 10.1

1.‡ $p = 2$; Vertex $(0, 2)$; $y - 2 = (x - 0)^2/[4(2)]$; $y = \frac{1}{8}x^2 + 2$; Fig. 10.1.1
2. $x - 2 = \frac{1}{4}(y - 3)^2$; Fig. 10.1.2   **4.** $x + 2 = \frac{1}{24}y^2$; Fig. 10.1.4

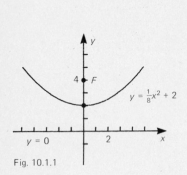

Fig. 10.1.1

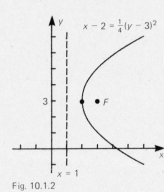

Fig. 10.1.2

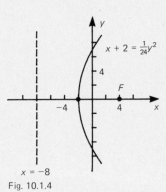

Fig. 10.1.4

6.‡ If the directrix is $x = 2k$, the vertex is $(k, 0)$; $x - k = -\dfrac{1}{4k}y^2$;

Set $x = 3, y = 4$: $k^2 - 3k - 4 = 0$; $k = 4$ or -1; $x - 4 = -\dfrac{1}{16}y^2$ or

$x + 1 = \dfrac{1}{4}y^2$; Fig. 10.1.6

7. $y = \frac{1}{9}x^2$; Fig. 10.1.7   **9.** $y = 3 + x^2$; Fig. 10.1.9   **11.** Fig. 10.1.11
13. Fig. 10.1.13   **15.** Fig. 10.1.15   **17.** Fig. 10.1.17   **19.** Fig. 10.1.19   **22.**‡ The
distance from (x, y) to $y = x$ is $|x - y|/\sqrt{2}$; $(y - 3)^2 + (x - 0)^2$
$= \frac{1}{2}(x - y)^2$; $x^2 + 2xy + y^2 - 12y + 18 = 0$; Fig. 10.1.22
23. $4x^2 - 4xy + y^2 + 8x + 16y - 16 = 0$; Fig. 10.1.23
25. $4x^2 + 4xy + y^2 + 56x - 72y - 304 = 0$; Fig. 10.1.25   **27.** $\frac{27}{2}\pi$ cubic
inches

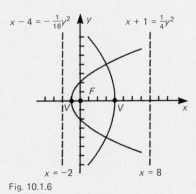

Fig. 10.1.6

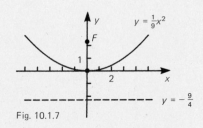

Fig. 10.1.7

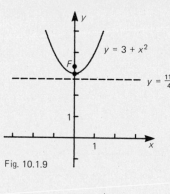

Fig. 10.1.9

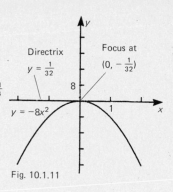

Fig. 10.1.11

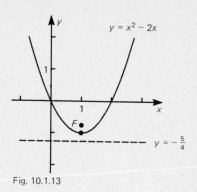

Fig. 10.1.13

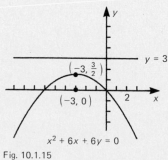

Fig. 10.1.15

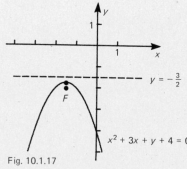

Fig. 10.1.17

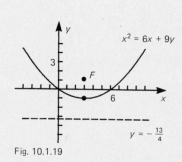

Fig. 10.1.19

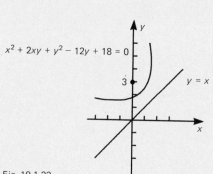

Fig. 10.1.22

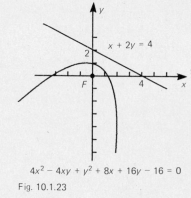

Fig. 10.1.23

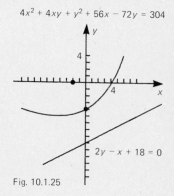

Fig. 10.1.25

Section 10.2

1. Fig. 10.2.1 **3.** Fig. 10.2.3 **5.** Fig. 10.2.5 **7.** Fig. 10.2.7 **9.** $\frac{1}{25}x^2 + \frac{1}{16}y^2 = 1$
11.‡ The major axis is on the x-axis and the center is at $(12, 0)$; $a = 13$;
$c = 12$; $b = 5$; $\frac{1}{169}(x - 12)^2 + \frac{1}{25}y^2 = 1$ **12.** $(x - 1)^2 + \frac{1}{2}(y - 2)^2 = 1$
14. $\frac{1}{16}(x - 1)^2 + (y - 10)^2 = 1$ **16.** $\frac{1}{16}(x + 4)^2 + \frac{1}{9}(y - 3)^2 = 1$ **18.**‡ The
major axis is on the line $x = 2$ and the center is at $(2, 1)$; $a = 6, c = 2$,
$b = \sqrt{32}$; $\frac{1}{32}(x - 2)^2 + \frac{1}{36}(y - 1)^2 = 1$
20.‡ $\sqrt{x^2 + y^2} + \sqrt{(x - 1)^2 + (y - 1)^2} = 2$; $(x - 1)^2 + (y - 1)^2$
$= 4 - 4\sqrt{x^2 + y^2} + x^2 + y^2$; $4(x^2 + y^2) = (1 + x + y)^2$;
$3x^2 - 2xy + 3y^2 - 2x - 2y - 1 = 0$

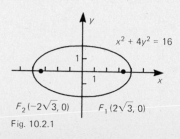

Fig. 10.2.1

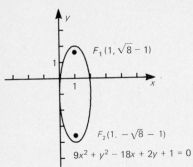

$9x^2 + y^2 - 18x + 2y + 1 = 0$

Fig. 10.2.3

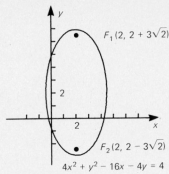

$4x^2 + y^2 - 16x - 4y = 4$

Fig. 10.2.5

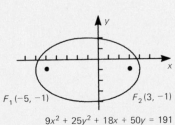

$9x^2 + 25y^2 + 18x + 50y = 191$

Fig. 10.2.7

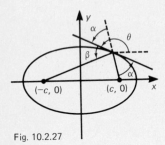

Fig. 10.2.27

21. $99x^2 - 6xy + 91y^2 + 108x - 276y - 2016 = 0$ **23.** πab **25.** Oblate
spheroid: $V = \frac{4}{3}\pi a^2 b$; Prolate spheroid: $V = \frac{4}{3}\pi ab^2$ **27.‡** $m_1 = \tan\Theta$ = Slope
of tangent line $= -b^2 x_0/(a^2 y_0)$ (Fig. 10.2.27); m_2 = Slope of line from
(x_0, y_0) to $(c, 0) = \tan(\Theta - \alpha) = y_0/(x_0 - c)$; m_3 = Slope of line from
(x_0, y_0) to $(-c, 0) = \tan(\Theta + \beta) = y_0/(x_0 + c)$; Use $\tan(A - B)$
$= [\tan(A) - \tan(B)]/[1 + \tan(A)\tan(B)]$; $\tan\alpha = \tan[\Theta - (\Theta - \alpha)]$
$= (m_1 - m_2)/(1 + m_1 m_2) = b^2/(y_0 c)$; $\tan\beta = \tan[(\Theta + \beta) - \Theta]$
$= (m_3 - m_1)/(1 + m_3 m_1) = b^2/(y_0 c)$; $\tan\alpha = \tan\beta \Rightarrow \alpha = \beta$ since
$0 < \alpha, \beta < \pi/2$

Section 10.3

1. Fig. 10.3.1   **3.** Fig. 10.3.3   **5.** Fig. 10.3.5   **7.** Fig. 10.3.7   **9.** Fig. 10.3.9
11. $\frac{1}{3}x^2 - \frac{1}{7}y^2 = 1$ **13.** $\frac{1}{16}(x - 5)^2 - \frac{1}{9}(y - 3)^2 = 1$ **15.‡** The center is $(0, 0)$
and the axis is on the x-axis; $a = 1, b = 2$; $x^2 - \frac{1}{4}y^2 = 1$ **16.** $\frac{2}{9}x^2 - \frac{2}{9}y^2 = 1$
18. $(x - 1)^2 - \frac{1}{8}(y - 5)^2 = 1$ **20.** $\frac{1}{9}(x - 1)^2 - \frac{1}{16}(y - 1)^2 = 1$
22.‡ $(x^2 + y^2)^{1/2} - [(x - 9)^2 + (y - 12)^2]^{1/2} = \pm 5$; $(x - 9)^2 + (y - 12)^2$
$= 25 \pm 10(x^2 + y^2)^{1/2} + x^2 + y^2$; $(200 - 18x - 24y)^2 = 100(x^2 + y^2)$;
$56x^2 + 216xy + 119y^2 - 1800x - 2400y + 10000 = 0$;   **23.** $xy = 8$
25a.‡ $y^2 + (5x)y + (4x^2 - 10) = 0$; Quadratic formula
$y = \frac{1}{2}[-5x \pm \sqrt{(5x)^2 - 4(4x^2 - 10)}] = \frac{1}{2}[-5x \pm 3x\sqrt{1 + 40/(9x^2)}]$
$\approx \frac{1}{2}[-5x \pm 3x]$ as $x \to \pm\infty$ since $0 < \sqrt{1 + t} - 1 = t/[1 + \sqrt{1 + t}] < t$

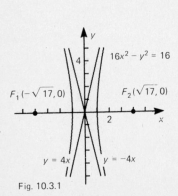

$y = 4x$ $y = -4x$

Fig. 10.3.1

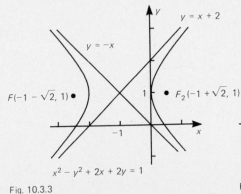

$x^2 - y^2 + 2x + 2y = 1$

Fig. 10.3.3

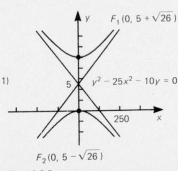

$y^2 - 25x^2 - 10y = 0$

Fig. 10.3.5

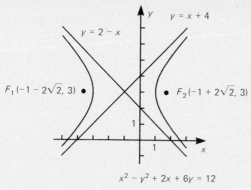

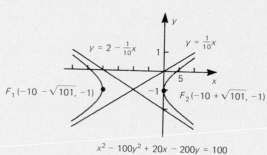

Fig. 10.3.7

$$x^2 - y^2 + 2x + 6y = 12$$

$$x^2 - 100y^2 + 20x - 200y = 100$$

Fig. 10.3.9

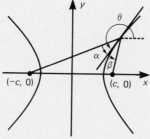

Fig. 10.3.27

for $t > 0$; $y = -x$, $y = -4x$ **25b.** $y = x$, $y = 5x$ **27.**‡ Fig. 10.3.27; m_1 $= \tan \Theta =$ Slope of the tangent line $= b^2 x_0/(a^2 y_0)$; $m_2 =$ Slope of the line from (x_0, y_0) to $(c, 0) = \tan(\Theta + \beta) = y_0/(x_0 - c)$; $m_3 =$ Slope of the line from (x_0, y_0) to $(-c, 0) = \tan(\Theta - \alpha) = y_0/(x_0 + c)$; $\tan \beta$ $= \tan[(\Theta + \beta) - \Theta] = (m_2 - m_1)/(1 + m_1 m_2) = b^2/(y_0 c)$; $\tan \alpha$ $= \tan[\Theta - (\Theta - \alpha)] = (m_1 - m_3)/(1 + m_1 m_3) = b^2/(y_0 c)$; $\tan \alpha = \tan \beta$; $\alpha = \beta$ because $0 < \alpha$, $\beta < \pi/2$

30. For (x, y) on the hyperbola with $|x|$ large, $y = \pm \dfrac{b}{a} x \sqrt{1 - (a^2/x^2)}$ $\approx \pm \dfrac{b}{a} x$ since $\sqrt{1 + t} - 1 < t$ for $t > 0$ (see the answer to Exercise 25a)

Section 10.4

1. Hyperbola   **3.** Parabola   **5.** Parabola   **7.**‡ $\cot(2\alpha) = (A - C)/B$ $= -1/\sqrt{3}$; $\alpha = -\frac{1}{6}\pi$; $x = \frac{1}{2}\sqrt{3}x' + \frac{1}{2}y'$, $y = -\frac{1}{2}x' + \frac{1}{2}\sqrt{3}y'$; $(x')^2 - (y')^2 = 4$; Fig. 10.4.7; Vertices: $x = \sqrt{3}$, $y = -1$ and $x = -\sqrt{3}$, $y = 1$; Asymptotes: $y = \left[\dfrac{1 + \sqrt{3}}{1 - \sqrt{3}}\right]x$ and $y = \left[\dfrac{\sqrt{3} - 1}{\sqrt{3} + 1}\right]x$ **8.** Fig. 10.4.8

11.‡ $\cot(2\alpha) = 0$; $\alpha = \pi/4$; $x = (x' - y')/\sqrt{2}$, $y = (x' + y')/\sqrt{2}$; $(x')^2 + 2x' - (y')^2 - 2y' = 1$; $(x' + 1)^2 - (y' + 1)^2 = 1$; The center is at $x' = -1$, $y' = -1$ or $x = 0$, $y = -\sqrt{2}$; The axis is $y' = -1$ or

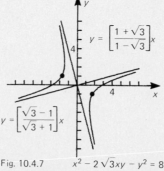

$$y = \left[\dfrac{1 + \sqrt{3}}{1 - \sqrt{3}}\right]x$$

$$y = \left[\dfrac{\sqrt{3} - 1}{\sqrt{3} + 1}\right]x$$

Fig. 10.4.7 $x^2 - 2\sqrt{3}xy - y^2 = 8$

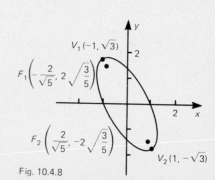

Fig. 10.4.8

$y = x - \sqrt{2}$; Fig. 10.4.11 **12.** Fig. 10.4.12 **14.** Fig. 10.4.14
22.‡ $\tan \alpha = \frac{4}{3} + \sqrt{(\frac{4}{3})^2 + 1} = 3$; $\cos \alpha = 1/\sqrt{10}$, $\sin \alpha = 3/\sqrt{10}$
(Fig. 10.4.22a); $x = (x' - 3y')/\sqrt{10}$, $y = (3x' + y')/\sqrt{10}$; $(y')^2 - (x')^2 = 1$;
Fig. 10.4.22b **24.** Fig. 10.4.24 **26.** Fig. 10.4.26 **27.** Fig. 10.4.27 **28.** Fig. 10.4.28
29. Fig. 10.4.29 **30.** Fig. 10.4.30 **31.** Fig. 10.4.31 **35.** Fig. 10.4.35
36. Fig. 10.4.36 **37.** Fig. 10.4.37 **38.** Fig. 10.4.38

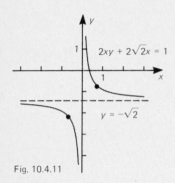

Fig. 10.4.11

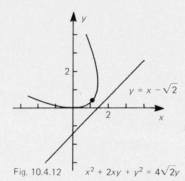

Fig. 10.4.12 $x^2 + 2xy + y^2 = 4\sqrt{2}y$

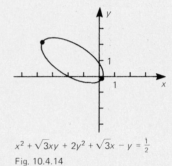

$x^2 + \sqrt{3}xy + 2y^2 + \sqrt{3}x - y = \frac{1}{2}$
Fig. 10.4.14

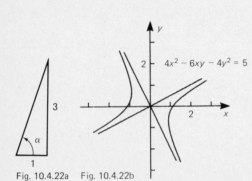

Fig. 10.4.22a   Fig. 10.4.22b

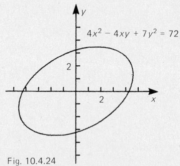

Fig. 10.4.24

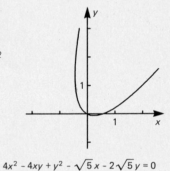

$4x^2 - 4xy + y^2 - \sqrt{5}x - 2\sqrt{5}y = 0$
Fig. 10.4.26

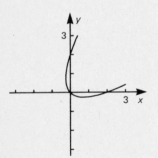

$x^2 - 2xy + y^2 - 2x - 2y = 0$
Fig. 10.4.27

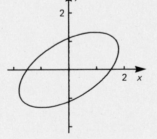

$7x^2 - 6\sqrt{3}xy + 13y^2 = 16$
Fig. 10.4.28

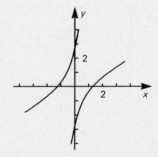

$32x^2 - 60xy + 7y^2 = 52$
Fig. 10.4.29

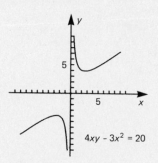

$4xy - 3x^2 = 20$

Fig. 10.4.30

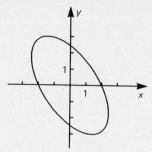

$3x^2 + 2\sqrt{2}\,xy + 2y^2 = 12$

Fig. 10.4.31

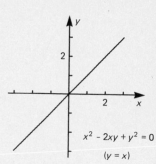

$x^2 - 2xy + y^2 = 0$

$(y = x)$

Fig. 10.4.35

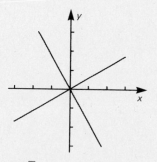

$\sqrt{3}\,x^2 - 2xy - \sqrt{3}\,y^2 = 0$

$(x = \sqrt{3}\,y,\, y = -\sqrt{3}\,x)$

Fig. 10.4.36

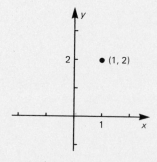

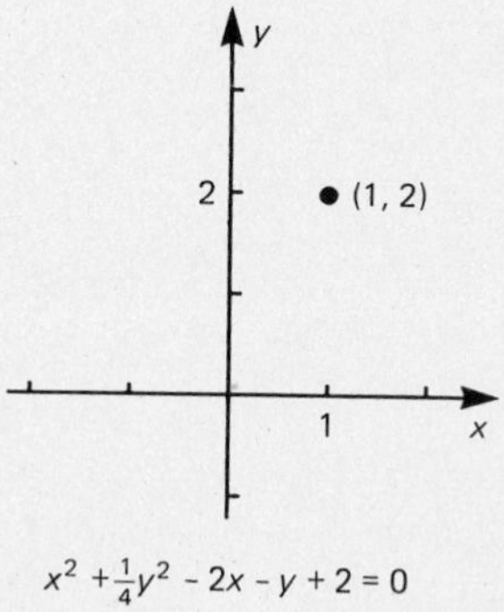

$x^2 + \tfrac{1}{4}y^2 - 2x - y + 2 = 0$

Fig. 10.4.37

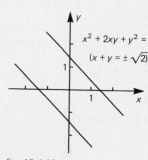

$x^2 + 2xy + y^2 = 2$

$(x + y = \pm\sqrt{2})$

Fig. 10.4.38

Section 10.5

1a. $\tfrac{1}{9}x^2 + \tfrac{1}{8}y^2 = 1$; Fig. 10.5.1a   **1b.** $9x^2 - \tfrac{9}{8}y^2 = 1$; Fig. 10.5.1b   **2a.** 0
2b. $\tfrac{2}{3}\sqrt{2}$ **2c.** $\sqrt{1.1}$

4. For $0 < e < 1$, the conic section is an ellipse whose focus at

$\left(\dfrac{-2e^2}{1 - e^2}, 0\right)$ and directrix at $x = \dfrac{-1 - e^2}{1 - e^2}$ move off to the left as $e \to 1^-$.

At $e = 1$, the second focus and directrix are gone and the conic section is a

parabola. For $e > 1$, the focus at $\left(\dfrac{2e^2}{e^2 - 1}, 0\right)$ and directrix $x = \dfrac{e^2 + 1}{e^2 - 1}$

are on the right. As $e \to \infty$, the focus moves to the left toward the point $(2, 0)$
and the directrix moves to the left toward the fixed directrix $x = 1$.

8a. Fig. 10.5.8a   **8b.** Fig. 10.5.8b
9. $e = \tfrac{3}{4}\sqrt{2}$; Foci $= (\pm 3, 0)$; Directrices:

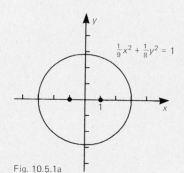

$\tfrac{1}{9}x^2 + \tfrac{1}{8}y^2 = 1$

Fig. 10.5.1a

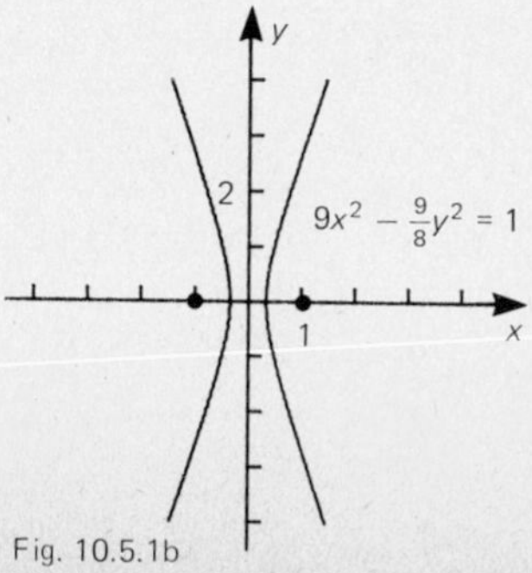

$9x^2 - \tfrac{9}{8}y^2 = 1$

Fig. 10.5.1b

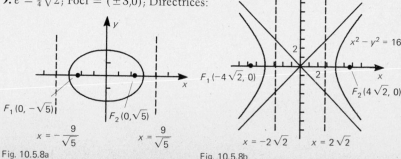

$F_1(0, -\sqrt{5})$ $F_2(0, \sqrt{5})$

$x = -\dfrac{9}{\sqrt{5}}$ $x = \dfrac{9}{\sqrt{5}}$

Fig. 10.5.8a

$x^2 - y^2 = 16$

$F_1(-4\sqrt{2}, 0)$ $F_2(4\sqrt{2}, 0)$

$x = -2\sqrt{2}$ $x = 2\sqrt{2}$

Fig. 10.5.8b

$x = \pm \frac{8}{3}$ **10.** $e = \frac{2}{3}\sqrt{3}$; Foci $= (0, \pm 2)$; Directrices: $y = \pm \frac{3}{2}$
11. $e = \frac{2}{3}$; Foci $= (\pm 2, 0)$; Directrices: $x = \pm \frac{9}{2}$ **13.** $e = \frac{1}{2}\sqrt{2}$; Foci $= (\pm 1, 2)$;
Directrices: $x = \pm 2$ **16.** Circle; $e = 0$; Focus $= (0,0)$

Section 10.6

1a. $(-3/\sqrt{2}, 3/\sqrt{2})$; Fig. 10.6.1a   **1c.** $(-\frac{1}{4}\sqrt{3}, -\frac{1}{4})$; Fig. 10.6.1c
1d. $(0,0)$; Fig. 10.6.1d   **1g.** $(2\cos(3), 2\sin(3)) \approx (-1.98, 0.28)$; Fig. 10.6.1g
1i. $(-4\cos(-10); -4\sin(-10)) \approx (3.36, -2.18)$; Fig. 10.6.1i

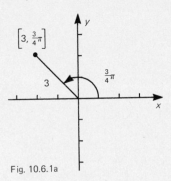

Fig. 10.6.1a

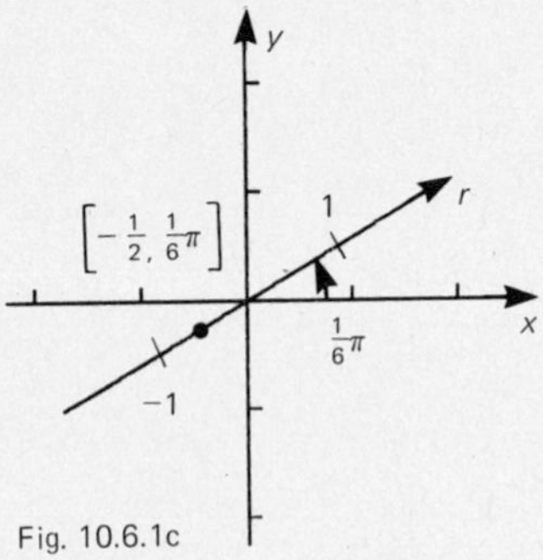

Fig. 10.6.1c

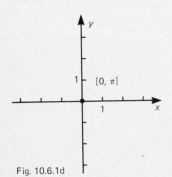

Fig. 10.6.1d

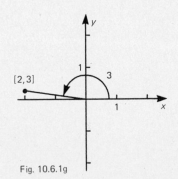

Fig. 10.6.1g

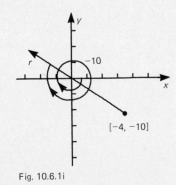

Fig. 10.6.1i

2a. $(3\sqrt{2}, \frac{5}{4}\pi + 2n\pi)$ or $(-3\sqrt{2}, \frac{1}{4}\pi + 2n\pi)$ with n an integer
2c. $(7, 2n\pi)$ or $(-7, (2n + 1)\pi)$ with n an integer   **2e.** $(2, \frac{2}{3}\pi + 2n\pi)$ or
$(-2, -\frac{1}{3}\pi + 2n\pi)$ with n an integer   **2g.**‡ $r = (3^2 + 4^2)^{1/2} = 5$,
$\Theta = \arctan(\frac{4}{3}) + 2n\pi \approx 0.927 + 2n(3.142)$ or $r = -5$,
$\Theta = \arctan(\frac{4}{3}) + (2n + 1)\pi \approx 0.927 + (2n + 1)(3.142)$ with n an integer
2i. $[3\sqrt{5}, \arctan(2) + (2n + 1)\pi] \approx [6.708, 1.107 + (2n + 1)(3.142)]$ or
$[-3\sqrt{5}, \arctan(2) + 2n\pi] \approx [-6.708, 1.107 + 2n(3.124)]$ with integers n
3. $(1/\sqrt{2}, -1/\sqrt{2})$ **4.** $(-\pi, 0)$ **5.** $(3\cos(3), 3\sin(3))$ **6.** $(-2, -2\sqrt{3})$
7. $[0, \Theta]$, any Θ **8.** $[\sqrt{146}, \arctan(\frac{5}{11}) + 2k\pi]$,
$[-\sqrt{146}, \arctan(\frac{5}{11}) + (2k + 1)\pi]$ with integers k **9.** $[2, (\frac{1}{2} + 2k)\pi]$,
$[-2, (\frac{3}{2} + 2k)\pi]$, with integers k **10.** $[\sqrt{12}, (\frac{7}{6} + 2k)\pi], [-\sqrt{12}, (\frac{1}{6} + 2k)\pi]$
with integers k **11.** $[4\sqrt{2}, (\frac{5}{4} + 2k)\pi], [-4\sqrt{2}, (\frac{1}{4} + 2k)\pi]$ with integers k
12. $[\sqrt{101}, \arctan(-10) + (2k + 1)\pi], [-\sqrt{101}, \arctan(-10) + 2k\pi]$ with
integers k **13.** $[\sqrt{101}, \arctan(10) + (2k + 1)\pi], [-\sqrt{101}, \arctan(10) + 2k\pi]$
with integers k **14.** $[\sqrt{101}, \arctan(-1/10) + 2k\pi]$,
$[-\sqrt{101}, \arctan(-1/10) + (2k + 1)\pi]$ with integers k

Section 10.7

1. $r = 2\sec\Theta$ **3.** $\Theta = \arctan\left(\frac{1}{3}\right)$ **5.** $r = \pm\sqrt{\sec 2\Theta}$ **7.** $r = -2\cos\Theta$
9. $r = 2(\cos\Theta \pm 1)$ **11.** Fig. 10.7.11   **13.** Fig. 10.7.13   **15.** Fig. 10.7.15
17. Fig. 10.7.17   **19.** Fig. 10.7.19   **23.** Fig. 10.7.23   **24.** Fig. 10.7.24
29. $r^2 = 8\cos\Theta\sin\Theta$ **30.** $r^2 = 7$ **31.** $r^2\sin^2\Theta = 6r\cos\Theta + 9$

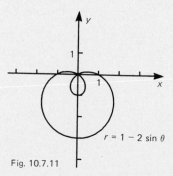

Fig. 10.7.11

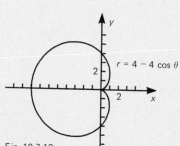

Fig. 10.7.13

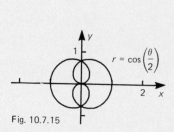

Fig. 10.7.15

Fig. 10.7.17

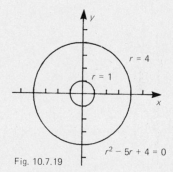

Fig. 10.7.19

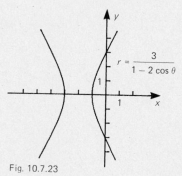

Fig. 10.7.23

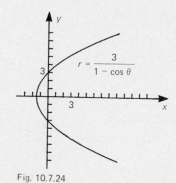

Fig. 10.7.24

32. $r = 4 \cos \Theta$ **33.** $\tan \Theta = 2$ **35.** $r = -5 \csc \Theta$ **37.** Fig. 10.7.37
38. Fig. 10.7.38   **39.** Fig. 10.7.39   **41.** Fig. 10.7.41   **43.** Fig. 10.7.43
44. Fig. 10.7.44

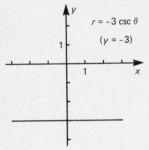

$r = -3 \csc \theta$

$(y = -3)$

Fig. 10.7.37

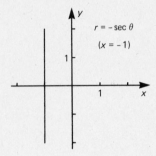

$r = -\sec \theta$

$(x = -1)$

Fig. 10.7.38

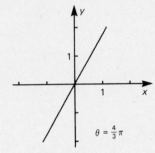

$\theta = \frac{4}{3}\pi$

Fig. 10.7.39

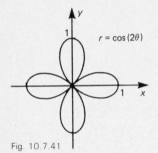

$r = \cos(2\theta)$

Fig. 10.7.41

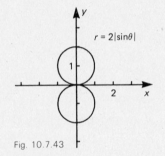

$r = 2|\sin\theta|$

Fig. 10.7.43

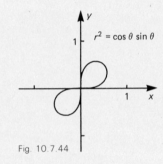

$r^2 = \cos \theta \sin \theta$

Fig. 10.7.44

Section 10.8

1. 9π **3.** $\frac{1}{3}\pi$ **5.** $\pi + 3\sqrt{3}$ **7.** $\frac{1}{3}\pi - \frac{1}{4}\sqrt{3}$ **9.** 1   **11.** $\frac{9}{2}\pi$ **12.** 2   **13.** $\frac{9}{4}\pi$

Miscellaneous Exercises to Chapter 10

1. $-2x^2 + 4x$ **6a.** 94.45 million miles   **6b.** 91.35 million miles
11a. Fig. 10.M.11a   **14.** Fig. 10.M.14   **17a.**‡ For $x \neq 0$, $y = -\frac{3}{2} \pm \frac{1}{2}\sqrt{5x^2 + 20}$
$= -\frac{3}{2} \pm \frac{1}{2}\sqrt{5}x \sqrt{1 + 20/x^2}$; $y = -\frac{3}{2} \pm \frac{1}{2}\sqrt{5}x$ **17b.** $y = 0$ and $y = x$

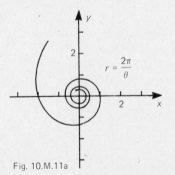

$r = \dfrac{2\pi}{\theta}$

Fig. 10.M.11a

$r = 1 + 2 \sin\left(\dfrac{\theta}{2}\right)$

Fig. 10.M.14

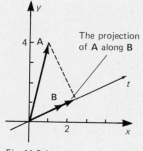

Fig. 11.1.5

Section 11.1

1a. $\langle 2, -8 \rangle$ **1b.** $\langle -7, -5 \rangle$ **2a.** $\langle 5, 8 \rangle$; $\sqrt{89}$; $\langle 5/\sqrt{89}, 8/\sqrt{89} \rangle$
2b. $\langle 2, -4 \rangle$; $\sqrt{20}$; $\langle 2/\sqrt{20}, -4/\sqrt{20} \rangle$ **3.‡** The line has slope 2, so the vector
$\mathbf{i} + 2\mathbf{j}$ is parallel to it; $(1/\sqrt{5})\mathbf{i} + (2/\sqrt{5})\mathbf{j}$ and $-(1/\sqrt{5})\mathbf{i} - (2/\sqrt{5})\mathbf{j}$
5a.‡ Fig. 11.1.5; $\overrightarrow{OS} = \overrightarrow{OP} + \overrightarrow{PS} = \overrightarrow{OP} + \overrightarrow{QR} = \langle 3, 4 \rangle + \langle 4, 1 \rangle = \langle 7, 5 \rangle$;
$S = (7, 5)$ **6a.** $\langle -\frac{5}{2}, \frac{5}{2}\sqrt{3} \rangle$ **7a.‡** $2s + 3t = 6$ and $5s + 2t = -7$; $s = -3$,
$t = 4$ **8.‡** Set $Q = (x_0, y_0)$, $R = (x_1, y_1)$, $P = (x, y)$;
$\overrightarrow{OP} = \frac{1}{4}\overrightarrow{QR} = \frac{1}{4}\langle x_1 - x_0, y_1 - y_0 \rangle$; $\overrightarrow{OP} = \overrightarrow{OQ} + \overrightarrow{QP}$
$= \langle \frac{3}{4}x_0 + \frac{1}{4}x_1, \frac{3}{4}y_0 + \frac{1}{4}y_1 \rangle$; $P = (\frac{3}{4}x_0 + \frac{1}{4}x_1, \frac{3}{4}y_0 + \frac{1}{4}y_1)$ **9a.** $(\frac{10}{3}, \frac{11}{3})$
10.‡ Fig. 11.1.10; $\mathbf{A} + \mathbf{B} = \mathbf{C} + \mathbf{D} \Rightarrow \frac{1}{2}(\mathbf{A} + \mathbf{B}) = \frac{1}{2}(\mathbf{C} + \mathbf{D})$ **11.** Fig. 11.1.11;
$\overrightarrow{MN} = \overrightarrow{MA} + \overrightarrow{AN} = \frac{1}{2}\overrightarrow{BA} + \frac{1}{2}\overrightarrow{AC} = \frac{1}{2}(\overrightarrow{BA} + \overrightarrow{AC}) = \frac{1}{2}\overrightarrow{BC}$ **13a.** $60\mathbf{i} + 10\mathbf{j}$
13b. $\sqrt{3700}$ **16a.** $8\mathbf{i} - \mathbf{j}$ **16b.** $-8\mathbf{i} + \mathbf{j}$ **18a.** $\langle -1, 7 \rangle$ **18b.** $\langle 5, -2 \rangle$
18c. $\langle 0, 12 \rangle$ **19a.** $\langle \frac{12}{5}, -\frac{19}{5} \rangle$ **19b.** 7 **19c.** $\sqrt{68}$ **20.** $(-1, -2)$, $(5, 2)$, $(3, 6)$
21a. $\langle -5, -\frac{19}{2} \rangle$ **21b.** $\langle -\frac{5}{2}, -\frac{9}{2} \rangle$ **21c.** $\langle 1, -\frac{9}{13} \rangle$ **22a.** $\frac{1}{4}\pi$ **22b.** $\frac{3}{4}\pi$
22c. $\frac{7}{4}\pi$ **22d.** $\frac{1}{6}\pi$ **22e.** $2\pi - \arctan(2)$ **23.** $\langle 49/\sqrt{2}, 51/\sqrt{2} \rangle$ miles
24. $\langle -2, -10 \rangle$ knots **25.** $\langle -70\sqrt{2}, 20\sqrt{2} \rangle$ pounds

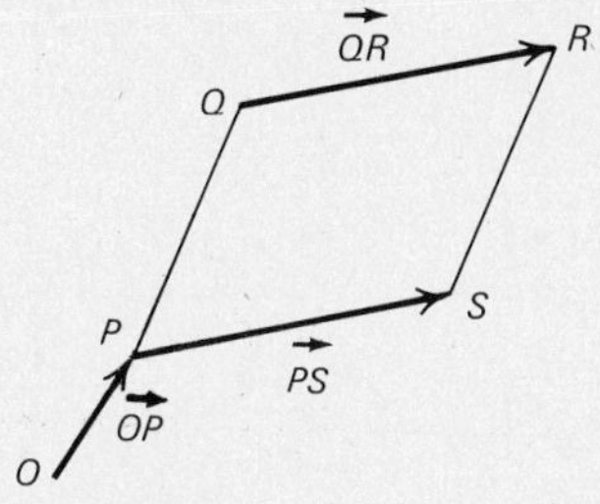

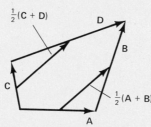

Fig. 11.1.10

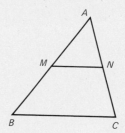

Fig. 11.1.11

Section 11.2

1a. -16 **1c.** -1 **2a.** 0.28 radians **2c.** 2.82 radians **3a.** $\pm \langle 1/\sqrt{5}, -2/\sqrt{5} \rangle$
3c. $\pm \langle 7/\sqrt{58}, 3/\sqrt{58} \rangle$ **4a.** $6/\sqrt{5}$; $\langle \frac{12}{5}, \frac{6}{5} \rangle$; Fig. 11.2.4a **4c.** $-9/\sqrt{2}$;
$\langle -\frac{9}{2}, -\frac{9}{2} \rangle$; Fig. 11.2.4c **5a.‡** Fig. 11.2.5; 0.8 **5b.** 0 **6.** $500\sqrt{2}$ foot-pounds
7.‡ If $\mathbf{i} + b\mathbf{j}$ makes an angle of $45°$ with $3\mathbf{i} + 2\mathbf{j}$, then $(\mathbf{i} + b\mathbf{j}) \cdot (3\mathbf{i} + 2\mathbf{j})$
$= |\mathbf{i} + b\mathbf{j}| |3\mathbf{i} + 2\mathbf{j}| (1/\sqrt{2})$; $5b^2 - 24b - 5 = 0$; $b = 5$ or $-\frac{1}{5}$;
$\langle 1/\sqrt{26}, 5/\sqrt{26} \rangle$ or $\langle 5/\sqrt{26}, -1/\sqrt{26} \rangle$ **9.** Fig. 11.2.9; $\overrightarrow{AC} \cdot \overrightarrow{DB}$
$= (\overrightarrow{AB} - \overrightarrow{DA}) \cdot (\overrightarrow{DA} + \overrightarrow{AB}) = |\overrightarrow{AB}|^2 - |\overrightarrow{DA}|^2$; $\overrightarrow{AC} \cdot \overrightarrow{DB} = 0 \Leftrightarrow |\overrightarrow{AB}|$
$= |\overrightarrow{DA}|$ **12.** $\langle A, B \rangle$ is perpendicular to the line; Let $P(x_1, y_1)$ be any point on
the line; $\overrightarrow{PQ} = \langle x_0 - x_1, y_0 - y_1 \rangle$ runs from P to $Q(x_0, y_0)$ and the distance
from Q to the line is the absolute value of the component of $\overrightarrow{PQ}$ in the
direction of $\langle A, B \rangle$ (Fig. 11.2.12); $|\overrightarrow{PQ} \cdot \langle A, B \rangle|/|\langle A, B \rangle|$
$= |A(x_0 - x_1) + B(y_0 - y_1)|/\sqrt{A^2 + B^2} = |Ax_0 + By_0 + C|/\sqrt{A^2 + B^2}$
because $Ax_1 + By_1 = -C$ **13a.** $\frac{16}{5}$ **13b.** $7/\sqrt{26}$ **15a.** 16 **15b.** 8 **15c.** -14

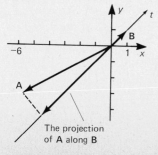

Fig. 11.2.4a

Fig. 11.2.4c

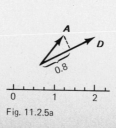

Fig. 11.2.5a

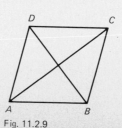

Fig. 11.2.9

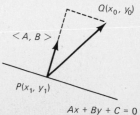

Fig. 11.2.12

16a. $\langle x, \frac{2}{5}x \rangle$ **16b.** $\langle x, 0 \rangle$ **16c.** $\langle 0, x \rangle$ **17.** $\langle \frac{4}{3}, -2 \rangle$ **18a.** $\frac{3}{2}$ **18b.** $-\frac{8}{3}$ **19.** 0
20. $\langle \frac{2}{5}\sqrt{5}, \frac{1}{5}\sqrt{5} \rangle, \langle -\frac{2}{5}\sqrt{5}, -\frac{1}{5}\sqrt{5} \rangle$ **21a.** $-\frac{1}{5}\sqrt{5}$ **21b.** 0 **21c.** $-1/\sqrt{10001}$
22. $\langle 3, 15 \rangle$ **23a.** $\langle -\frac{27}{5}, -\frac{36}{5} \rangle$ **23b.** $\langle 0, 0 \rangle$ **23c.** $\langle \frac{4}{5}, -\frac{2}{5} \rangle$ **25a.** 2 foot-pounds
25b. -12 foot-pounds **25c.** 0 foot-pounds **26.** $10 \cos(\frac{5}{12}\pi)$ knots

Section 11.3

1.‡ Figs. 11.3.1a, b, c **3.**‡ $\frac{1}{9}x^2 - \frac{1}{25}y^2 = \cosh^2 t - \sinh^2 t = 1$, so the curve is on the hyperbola $\frac{1}{9}x^2 - \frac{1}{25}y^2 = 1$; It is half the hyperbola because $x = 3 \cosh t$ takes on all values ≥ 3; Fig. 11.3.3 **4.** $\frac{1}{9}x^2 + \frac{1}{25}y^2 = \cos^2 t + \sin^2 t = 1$
6. Fig. 11.3.6 **8.** $\frac{1}{3}(5^{3/2} - 1)$ **11.** Fig. 11.3.11; $\sqrt{37}(e^\pi - 1) \approx 135$
12a. $\sqrt{13}$ meters/minute **12c.** $\sqrt{\frac{1}{16}\pi^2 + 1}$ meters/minute
13a. $4x^{-2} + 4y^{-2} = \cos^2 t + \sin^2 t = 1$ **13b.** $(2\sqrt{2}, -2\sqrt{2})$ **13c.** 4
15a. $x = k, y = 1/(1 + k^2)$ **15b.** $y = 1/(1 + x^2)$ **17.** $x = 7\cos(s/7)$, $y = 7\sin(s/7), 0 \leq s \leq \frac{7}{2}\pi$

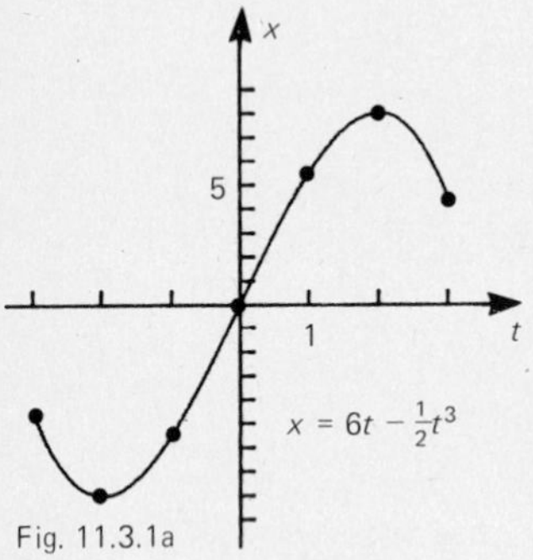

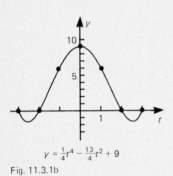

Fig. 11.3.1a

$x = 6t - \frac{1}{2}t^3$

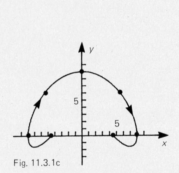

Fig. 11.3.1c

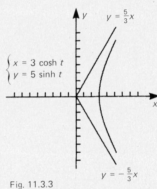

$\begin{cases} x = 3\cosh t \\ y = 5\sinh t \end{cases}$

$y = \frac{5}{3}x$

$y = -\frac{5}{3}x$

Fig. 11.3.3

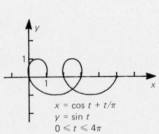

$x = \cos t + t/\pi$
$y = \sin t$
$0 \leq t \leq 4\pi$

Fig. 11.3.6

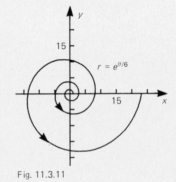

$r = e^{\theta/6}$

Fig. 11.3.11

$y = \frac{1}{4}t^4 - \frac{13}{4}t^2 + 9$

Fig. 11.3.1b

Section 11.4

1a. Fig. 11.4.1a **1c.** Fig. 11.4.1c **1e.** Fig. 11.4.1e **2a.** $R'(t)$ $= (t\cos t + \sin t)\mathbf{i} - [(t\sin t + \cos t)/t^2]\mathbf{j}; R'(\frac{1}{2}\pi) = \mathbf{i} - (2/\pi)\mathbf{j};$ $T(\frac{1}{2}\pi) = (\pi^2 + 4)^{-1/2}\langle \pi, -2 \rangle$; Tangent line: $y = -2x/\pi + 1$
2c. $\langle 1, 0 \rangle$; $y = 0$ **3a.**‡ $5t^4\mathbf{i} - 7t^6\mathbf{j}; 20t^3\mathbf{i} - 42t^5\mathbf{j}$; **3c.** $(1/t)\mathbf{i}$ $+ [3/(1 + 9t^2)]\mathbf{j}; -(1/t^2)\mathbf{i} - [54t/(1 + 9t^2)^2]\mathbf{j}$ **4a.** $(\frac{1}{2}t^2 + \frac{1}{2})\mathbf{i}$ $+ (\frac{1}{3} - \frac{1}{3}t^3)\mathbf{j}$ **4c.** $(\frac{1}{3}e^{3t} + \frac{14}{3})\mathbf{i} + (\frac{1}{4}e^{-4t} + \frac{19}{4})\mathbf{j}$ **5a.** $\langle -2, -7 \rangle$
5b. $\langle 0, -9 \rangle$ **5c.** Fig. 11.4.5 **7a.** 8π **7b.** $-6\mathbf{i} + 7\mathbf{j}$ **7c.** $6\mathbf{i} + 7\mathbf{j}$
8.‡ $x(1) \approx -2, x(2) \approx -1, x(3) \approx -3, x(4) \approx 0, x(5) \approx 3, x(6) \approx 1,$

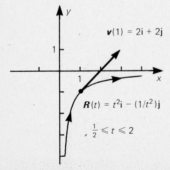

$\mathbf{v}(1) = 2\mathbf{i} + 2\mathbf{j}$

$R(t) = t^2\mathbf{i} - (1/t^2)\mathbf{j}$

$\frac{1}{2} \leq t \leq 2$

Fig. 11.4.1a

$x(7) \approx 2; y(1) \approx -2, y(2) \approx 0, y(3) \approx 3, y(4) \approx 5, y(5) \approx 3, y(6) \approx 0,$
$y(7) \approx -2;$ Fig. 11.4.8a; Draw plausible tangent lines to the graphs of $x(t)$
and $y(t)$ (Figs. 11.4.8b, c); $x'(\frac{7}{2}) \approx 4, y'(\frac{7}{2}) \approx 1, \mathbf{v}(\frac{7}{2}) \approx 4\mathbf{i} + \mathbf{j};$
$x'(5) \approx 0, y'(5) \approx -4, \mathbf{v}(5) \approx -4\mathbf{j};$ Fig. 11.4.8a

12. $\dfrac{d}{du}\mathbf{F}(t(u)) = \left[\dfrac{d}{du}g(t(u))\right]\mathbf{i} + \left[\dfrac{d}{du}h(t(u))\right]\mathbf{j}$

$= g'(t(u))\,t'(u)\mathbf{i} + h'(t(u))\,t'(u)\mathbf{j} = \mathbf{F}'(t(u))\,t'(u)$

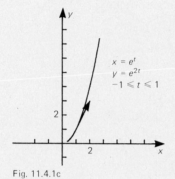

$x = e^t$
$y = e^{2t}$
$-1 \leqslant t \leqslant 1$

Fig. 11.4.1c

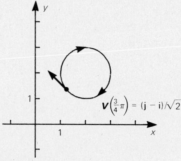

$\mathbf{v}\left(\dfrac{3}{4}\pi\right) = (\mathbf{j} - \mathbf{i})/\sqrt{2}$

Fig. 11.4.1e

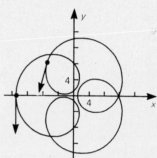

Fig. 11.4.5

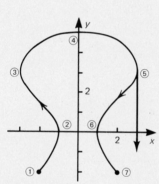

Fig. 11.4.8a

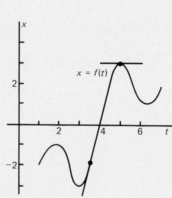

$x = f(t)$

Fig. 11.4.8b

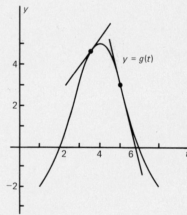

$y = g(t)$

Fig. 11.4.8c

Section 11.5

1a. $\kappa = -1, \rho = 1$ **1b.** $\kappa = 2, \rho = \frac{1}{2}$ **3.** $\kappa = -1/(10\sqrt{5}), \rho = 10\sqrt{5}$
5a.‡ $\mathbf{R}(x) = x\mathbf{i} + (1/x)\mathbf{j}; \mathbf{R}'(1) = \mathbf{i} - \mathbf{j}; \mathbf{T}(1) = (\mathbf{i} - \mathbf{j})/\sqrt{2};$
$\mathbf{N}(1) = (\mathbf{i} + \mathbf{j})/\sqrt{2}; \rho(1) = \sqrt{2};$ The position vector of the center of
curvature $= \mathbf{R}(1) + \rho(1)\mathbf{N}(1) = \langle 2, 2\rangle;$ Center of curvature $= (2, 2)$
5b. $(-2, -2)$ **5c.** Fig. 11.5.5 **6.** Fig. 11.5.6 **8.** $2/(3\sqrt{3})$; Fig. 11.5.8
10a.‡ $\rho \approx \frac{1}{2};$ The center of curvature is to the right of $\mathbf{T}; \kappa \approx -2$ **10b.**‡ $\rho \approx 1;$
The center of curvature is to the left of $\mathbf{T}; \kappa \approx 1$ **10c.**‡ $\rho \approx \frac{3}{2};$ The center of
curvature is to the right of $\mathbf{T}; \kappa \approx -\frac{2}{3}$ **10d.**‡ Fig. 11.5.10 **12a.** $\mathbf{T}(t)$
$= \langle -\sin t, \cos t\rangle; \mathbf{N}(t) = \langle -\cos t, -\sin t\rangle$

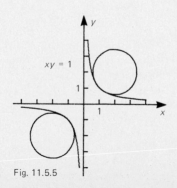

$xy = 1$

Fig. 11.5.5

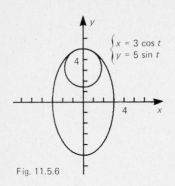

Fig. 11.5.6

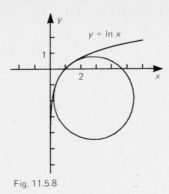

Fig. 11.5.8

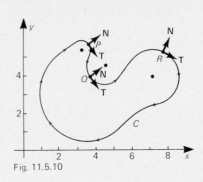

Fig. 11.5.10

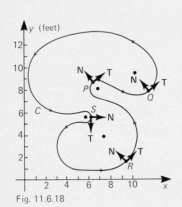

Fig. 11.6.18

Section 11.6

1a.[‡] $\mathbf{v}(t) = \dfrac{d}{dt}[\sin(2t)\mathbf{i} + 3\cos(4t)\mathbf{j}] = 2\cos(2t)\mathbf{i} - 12\sin(4t)\mathbf{j}$

millimeters/minute; $\mathbf{a}(t) = \dfrac{d}{dt}[2\cos(2t)\mathbf{i} - 12\sin(4t)\mathbf{j}]$

$= -4\sin(2t)\mathbf{i} - 48\cos(4t)\mathbf{j}$ millimeters/minute^2

1b. $(4e^{2t} + 4te^{2t})\mathbf{i} - 9e^{-3t}\mathbf{j}$ yards/year^2 **2a.** $x = Mt\cos\psi$,
$y = Mt\sin\psi - 16t^2$ **3.** $\frac{1}{12}\pi, \frac{5}{12}\pi$ **4.** No   **7.** $576[\text{Sine of }5°] \approx 50.2$ feet
8.[‡] $d^2s/dt^2 = 0$ so $\mathbf{a} = \kappa(ds/dt)^2\mathbf{N}$; $\kappa = 1/300$; $ds/dt = 200$; $\mathbf{F} = m\mathbf{a}$;
$m = 100$; All of the force is perpendicular to the track and is due to friction;
$|\mathbf{F}| = 100|\mathbf{a}| = 13{,}333\frac{1}{3}$ pounds   **9.** $26{,}400$ feet/second   **12.** $\frac{9}{8}$ pounds
14.[‡] To find the curvature at $(0, -2)$, give the ellipse the parametric equations
$x = 3\cos u, y = 2\sin u$; $u = \frac{3}{2}\pi$ at $(0, -2)$; $x'(\frac{3}{2}\pi) = 3, y'(\frac{3}{2}\pi) = 0$,
$x''(\frac{3}{2}\pi) = 0, y''(\frac{3}{2}\pi) = 2$; $\kappa = \frac{3}{8}$; $\mathbf{T} = \mathbf{i}, \mathbf{N} = \mathbf{j}$; $3\mathbf{i} + 5\mathbf{j} = \mathbf{a} = (d^2s/dt^2)\mathbf{i}$
$+ \frac{3}{8}(ds/dt)^2\mathbf{j}$; $d^2s/dt^2 = 3, ds/dt = \sqrt{45/2}$; The object has speed $\sqrt{45/2}$
feet/second and is speeding up at the rate of 3 feet/second^2 **15.** $-\frac{1}{8}$
18. Curvature at $R \approx \frac{1}{3}$; Curvature at $P \approx -\frac{4}{3}$; Curvature at $O \approx \frac{1}{2}$;
Curvature at $S \approx -2$ (Fig. 11.6.18)   **19.**[‡] $\mathbf{a} \approx \mathbf{T} - 4\mathbf{N}$ and $\kappa \approx -\frac{4}{3}$
$\Rightarrow d^2s/dt^2 \approx 1$ and $ds/dt \approx \sqrt{3}$; The object's speed is approximately $\sqrt{3}$
feet/second, and it is speeding up approximately 1 foot/second^2
21. $\mathbf{a} \approx -\frac{2}{4}\mathbf{i} + \frac{2}{4}\mathbf{j}$ feet/second^2 **23.** $(2t + 3)\mathbf{i} + (-16t^2 - 3t + 4)\mathbf{j}$
24. $(-5t + 10)\mathbf{i} + (-16t^2 + 15t + 3)\mathbf{j}$ **26.** $40\mathbf{i} + (-16t^2 + 16t + 50)\mathbf{j}$
27. $(3t + 10)\mathbf{i} + (2t - 4.9t^2)\mathbf{j}$ **29.** $(20t)\mathbf{i} + (-4.9t^2 + 24.5t)\mathbf{j}$
30. $(10\sqrt{3}t)\mathbf{i} + (-4.9t^2 + 10t)\mathbf{j}$ **31.** $(\frac{4}{5}t^{5/2} + 3t + \frac{31}{5})\mathbf{i} + (-\ln t - 10)\mathbf{j}$
32. It approaches $(0, \frac{1}{2}\pi)$

Section 11.7

1. $(4\mathbf{u}_r + 3\pi\mathbf{u}_\theta)/\sqrt{16 + 9\pi^2} = [(-4 - 3\pi)\mathbf{i} + (4 - 3\pi)\mathbf{j}]/\sqrt{32 + 18\pi^2}$
3. 5   **6.** $|a\mathbf{A} + b\mathbf{B}|^2 = (a\mathbf{A} + b\mathbf{B})\cdot(a\mathbf{A} + b\mathbf{B}) = a^2\mathbf{A}\cdot\mathbf{A} + 2ab\mathbf{A}\cdot\mathbf{B}$
$+ b^2\mathbf{B}\cdot\mathbf{B} = a^2 + b^2$ **11.** Because $r(\Theta_0) = 0$ and $r'(\Theta_0) \neq 0$, $r(\Theta)$ is not
zero for $\Theta \neq \Theta_0$ near Θ_0. For such Θ the vector $\mathbf{u}_r(\Theta)$ is parallel to the secant
line through $(0, 0)$ and $(r(\Theta)\cos\Theta, r(\Theta)\sin\Theta)$ and approaches $u_r(\Theta_0)$ as
$\Theta \to \Theta_0$. Hence $\mathbf{u}_r(\Theta_0)$ is parallel to the tangent line, which is the line
$\Theta = \Theta_0$.

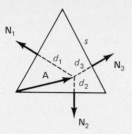

Fig. 11.M.2

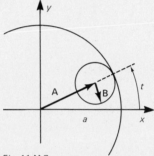

Fig. 11.M.7

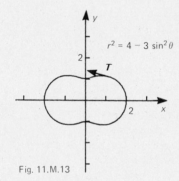

Fig. 11.M.13

Miscellaneous Exercises to Chapter 11

1. The forces do not act at the same point.   **2.** Let $\mathbf{N}_1$, $\mathbf{N}_2$, and $\mathbf{N}_3$ be unit normal vectors to the sides of the triangle and d_1, d_2, and d_3 the distances from the point to the sides; $\mathbf{N}_1 + \mathbf{N}_2 + \mathbf{N}_3 = \mathbf{O}$; Let $\mathbf{A}$ be the vector from one vertex to the point (Fig. 11.M.2); $d_1 = -\mathbf{A} \cdot \mathbf{N}_1$, $d_2 = -\mathbf{A} \cdot \mathbf{N}_2$, $d_3 = \frac{1}{2}s\sqrt{3} - \mathbf{A} \cdot \mathbf{N}_3$, where s is the length of the side of the triangle and $\frac{1}{2}s\sqrt{3}$ is its height; $d_1 + d_2 + d_3 = \frac{1}{2}s\sqrt{3} - \mathbf{A} \cdot (\mathbf{N}_1 + \mathbf{N}_2 + \mathbf{N}_3)$ $= \frac{1}{2}s\sqrt{3}$ **7a.**‡ Let t be the angle of inclination of the position vector $\mathbf{A}$ of the center of the smaller circle; $\mathbf{A} = (\frac{3}{4}a\cos t)\mathbf{i} + (\frac{3}{4}a\sin t)\mathbf{j}$; Let $\mathbf{B}$ be the vector from the center of the smaller circle to the point on it (Fig. 11.M.7); $\mathbf{B} = [\frac{1}{4}a\cos(-3t)]\mathbf{i} + [\frac{1}{4}a\sin(-3t)]\mathbf{j}$; $x = \frac{1}{4}a\cos(3t) + \frac{3}{4}a\cos t$, $y = -\frac{1}{4}a\sin(3t) + \frac{3}{4}a\sin t$ **7b.**‡ $x = a\cos^3 t$, $y = a\sin^3 t$; $x^{2/3} + y^{2/3} = a^{2/3}$ **9.** $x = \operatorname{arcsinh} s$, $y = \cosh(\operatorname{arcsinh} s) = \sqrt{1 + s^2}$

11a.‡ $\dfrac{dy}{dx} = \dfrac{dy/dt}{dx/dt} = -(\cos t)/(e^{-t} + 3t^2)$; $\dfrac{d^2 y}{dx^2} = \left[\dfrac{d}{dt}\left(\dfrac{dy}{dx}\right)\right] \Big/ \left(\dfrac{dx}{dt}\right)$

$= [-(e^{-t} + 3t^2)\sin t + (e^{-t} - 6t)\cos t]/(e^{-t} + 3t^2)^3$.

11b. $5t^4/(1 + \ln t)$; $(15t^3 + 20t^3 \ln t)/(1 + \ln t)^3$
13a. $-3\sin\Theta\cos\Theta/\sqrt{4 - 3\sin^2\Theta}$ **13b.** $dx/d\Theta = [-7\sin\Theta + 6\sin^3\Theta]$ $/\sqrt{4 - 3\sin^2\Theta}$; $dy/d\Theta = [-2\cos\Theta + 6\cos^3\Theta]/\sqrt{4 - 3\sin^2\Theta}$; $\langle -4/\sqrt{17}, 1/\sqrt{17}\rangle$; Fig. 11.M.13   **15.**‡ Use u as parameter for C_2 and equate the expressions for x and y: $t^3 + 6 = u^2 - 3$, $3t + 4 = 3u - 5$; $u = t + 3$; $t^3 + t^2 - 6t = 0$; $t = 0, 3,$ or -2; $(6, 4)$, $(33, 13)$ and $(-2, -2)$
17. $x = a(\cos t + t\sin t)$, $y = a(\sin t - t\cos t)$ **19.** $x = t + \sin t$, $y = 1 - \cos t$; Or, with $t + \pi$ replacing t: $x = (t - \sin t) - \pi$, $y = (\cos t - 1) + 2$; The evolute is obtained by shifting the cycloid 2 units up and π units to the left.   **20.** He should swim upstream at an angle of $\pi/3$ with the bank of the river.   **22.** $24\sqrt{880}$ feet/second ≈ 712 feet/second

25a.‡ $\dfrac{d}{dt}(e^{kt/m}\mathbf{v}) = -ge^{kt/m}\mathbf{j}$; $e^{kt/m}\mathbf{v} = -(gm/k)e^{kt/m}\mathbf{j} + \mathbf{C}$;

Set $t = 0$, $\mathbf{v} = \mathbf{v}_0$: $\mathbf{C} = \mathbf{v}_0 + (gm/k)\mathbf{j}$; $\mathbf{v} = e^{-kt/m}\mathbf{v}_0 - (gm/k)(1 - e^{-kt/m})\mathbf{j}$

25b.‡ $\mathbf{v} \to -(mg/k)\mathbf{j}$ (the terminal velocity)   **26a.** The tangential component of the acceleration vector is $d^2\Theta/dt^2$ since $s = \Theta$, and the tangential component of the force is $-mg\sin\Theta$; The tangential components of the equation $\mathbf{F} = m\mathbf{a}$ give the differential equation   **26b.** $d\Theta/dt = \pm\sqrt{2g(\cos\Theta - \cos\Theta_0)}$ **26c.** $\sqrt{2g(1 - \cos\Theta_0)}$
26d. When Θ is increasing, $dt/d\Theta = [2g(\cos\Theta - \cos\Theta_0)]^{-1/2}$; Period $= 4\int_0^{\Theta_0} [2g(\cos\Theta - \cos\Theta_0)]^{-1/2} d\Theta$ **27.** $\frac{1}{4}\pi$
30a. Let $\mathbf{R}_e(t)$ and $\mathbf{R}_m(t)$ be the position vectors of the earth and the sun;

$$\mathbf{R}_e(t) = -93\left[\cos\left(\frac{2\pi t}{365}\right)\mathbf{i} + \sin\left(\frac{2\pi t}{365}\right)\mathbf{j}\right],$$

$$\mathbf{R}_m(t) = 36\left[\cos\left(\frac{2\pi t}{88}\right)\mathbf{i} + \sin\left(\frac{2\pi t}{88}\right)\mathbf{j}\right]$$

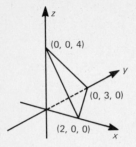

Fig. 12.1.1

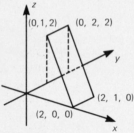

Fig. 12.1.3

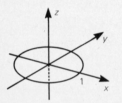

Fig. 12.1.5

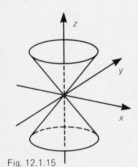

Fig. 12.1.15

Section 12.1

1. Fig. 12.1.1   **3.** Fig. 12.1.3   **5.** Fig. 12.1.5   **7.** Fig. 12.1.7   **9.** Fig. 12.1.9
12. Fig. 12.1.12   **14a.** 5   **14b.** $\sqrt{17}$ **14c.** $\sqrt{10}$ **14d.** 1   **14e.** 4   **14f.** 3
14g. $\sqrt{26}$ **15.** Fig. 12.1.15; $\frac{1}{4}\pi$ **17a.** $\langle 0, 19, 10 \rangle$ **17b.** $\langle -2, 31, 18 \rangle$
17c. $\sqrt{14}$ **17d.** $\sqrt{53}$ **17e.** -39 **17f.** $(\mathbf{A} + \mathbf{B}) \cdot (\mathbf{A} - \mathbf{B}) = \mathbf{A} \cdot \mathbf{A} - \mathbf{B} \cdot \mathbf{B}$
$= |\mathbf{A}|^2 - |\mathbf{B}|^2$ **19.** $k = 2$ a rectangle but not a square   **20.** 0.33; 0.19; 2.62
21. $(7, -3, -12)$ **22.** As is the case for vectors in the plane, $\mathbf{A} \cdot \mathbf{B} = \mathbf{A} \cdot \mathbf{C}$
with $\mathbf{A} \neq 0$ implies only that $\mathbf{B} - \mathbf{C}$ is perpendicular to $\mathbf{A}$ **23.** $\mathbf{A}$ is parallel
to $\mathbf{C}$ and perpendicular to $\mathbf{B}$; $\mathbf{B}$ is perpendicular to $\mathbf{C}$ **24a.**‡ Suppose that
edges of the cube are along the coordinate axes, that one corner is at $(0,0,0)$,
and the opposite one is at (s,s,s); The angle α between the diagonal and an
edge is the angle between $\langle s, s, s \rangle$ and $\langle s, 0, 0 \rangle$; $\cos \alpha = 1/\sqrt{3}$; the acute
angle α is $\arccos(1/\sqrt{3}) \approx 0.96$ radians   **27a.** $1/\sqrt{26}$ **27b.** $\langle -\frac{1}{26}, 0, \frac{5}{26} \rangle$
28a. $\cos\alpha = 3/\sqrt{22}$; $\alpha \approx 0.88$ radians; $\cos\beta = -2/\sqrt{22}$; $\beta \approx 2.01$ radians;
$\cos\gamma = -3/\sqrt{22}$; $\gamma \approx 2.26$ radians   **28b.** $\cos\alpha = 1/\sqrt{21}$; $\alpha \approx 1.35$ radians;
$\cos\beta = -2/\sqrt{21}$; $\beta \approx 2.02$ radians; $\cos\gamma = 4/\sqrt{21}$; $\gamma \approx 0.51$ radians
29. $\langle 1/\sqrt{3}, 1/\sqrt{3}, 1/\sqrt{3} \rangle$ or $\langle -1/\sqrt{3}, -1/\sqrt{3}, -1/\sqrt{3} \rangle$
31.‡ $\mathbf{F} \cdot \overrightarrow{PQ} = \langle 3, -4, 1 \rangle \cdot \langle 1, -1, 6 \rangle = 13$ foot-pounds   **32.** The vectors
making an angle α with $\mathbf{i}$ lie along a half cone with the x-axis as its axis; The
vectors making an angle β with $\mathbf{j}$ lie along a half cone with the y-axis as its
axis; The cones intersect along two lines if $\alpha + \beta > \pi/2$, intersect along one
line if $\alpha + \beta = \pi/2$, and intersect only at their vertices if $\alpha + \beta < \pi/2$

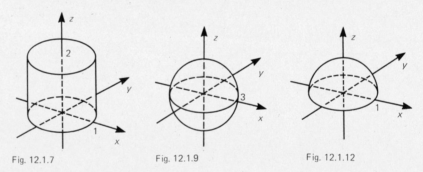

Fig. 12.1.7   Fig. 12.1.9   Fig. 12.1.12

33. $3\sqrt{2}$ **34.** $(2, -1, -1)$ **35a.** $\frac{23}{30}\sqrt{30}$ **35b.** $-\frac{11}{2}\sqrt{2}$ **35c.** $-\sqrt{11}$
36a. $\langle \frac{5}{2}, \frac{5}{2}, 5 \rangle$ **36b.** $\langle 0, -\frac{24}{37}, -\frac{4}{37} \rangle$ **36c.** $\langle 0, 0, 0 \rangle$ **37.** $\langle 1, -3, 7 \rangle$
38. 20 foot-pounds   **39a.** $-\frac{11}{3}$ dynes   **39b.** -11 dyne-centimeters
40. $\langle 2, -\sqrt{95}, -1 \rangle$ kilometers/hour

Section 12.2

1. -39 **2.** 76   **4.** $\langle -4, -14, -10 \rangle$ **6.** $\langle -16, 10, 7 \rangle$ **8.** $2\sqrt{5}$
10. $D = x_1(y_2 - y_3) - y_1(x_2 - x_3) + (x_2 y_3 - x_3 y_2) = (x_1 - x_3)(y_2 - y_3)$
$- (y_1 - y_3)(x_2 - x_3)$, so with $\mathbf{A} = \langle x_1 - x_3, y_1 - y_3, 0 \rangle$ and
$\mathbf{B} = \langle x_2 - x_3, y_2 - y_3, 0 \rangle$, $\mathbf{A} \times \mathbf{B} = D\mathbf{k}$ and the area is $\frac{1}{2}|\mathbf{A} \times \mathbf{B}|$
11. $|\mathbf{A} \times \mathbf{B}|^2 = |\mathbf{A}|^2|\mathbf{B}|^2 \sin^2\Theta = |\mathbf{A}|^2|\mathbf{B}|^2 - (\mathbf{A} \cdot \mathbf{B})^2$
12. $\pm (1/\sqrt{179})\langle -1, -3, 13 \rangle$ **14.** $\pm(1/\sqrt{19})\langle -3, -1, 3 \rangle$ **16.**‡ Fig. 12.2.16;
$\overrightarrow{OT} = \overrightarrow{OP} + \overrightarrow{PS} + \overrightarrow{PQ} + \overrightarrow{PR} = \langle 9, 4, 5 \rangle$; $(9, 4, 5)$ **17.** $(2, 2, -5)$
20a.‡ Volume $= |\overrightarrow{PQ} \cdot (\overrightarrow{PR} \times \overrightarrow{PS})| = 2$ **20b.** 91
21. The area of the base $= \frac{1}{2}|\mathbf{A} \times \mathbf{B}|$; The height is the absolute value of the
component of $\mathbf{C}$ in the direction of $\mathbf{A} \times \mathbf{B}$; The volume
$= \frac{1}{3}[\text{Area of the base}][\text{Height}] = \frac{1}{6}|\mathbf{A} \times \mathbf{B}||\mathbf{C} \cdot (\mathbf{A} \times \mathbf{B})|/|\mathbf{A} \times \mathbf{B}|$

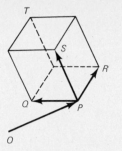

O

Fig. 12.2.16

$= \frac{1}{6} |(\mathbf{A} \times \mathbf{B}) \cdot \mathbf{C}|$ **22a.** $\frac{3}{2}$ **23.** Since $\overrightarrow{PR} \times \overrightarrow{PS}$ is perpendicular to the plane through P, R, and S, Q is in the plane if and only if $\overrightarrow{PQ}$ is perpendicular to $\overrightarrow{PR} \times \overrightarrow{PS}$ **24.** Yes **26.** If the angle α between $\mathbf{A} \times \mathbf{B}$ and $\mathbf{C}$ satisfies $0 \le \alpha \le \pi/2$, then the angle between $\mathbf{A}$ and $\mathbf{B} \times \mathbf{C}$ satisfies the same conditions, and $(\mathbf{A} \times \mathbf{B}) \cdot \mathbf{C}$ and $\mathbf{A} \cdot (\mathbf{B} \times \mathbf{C})$ are both the volume of the parallelopiped with sides formed by $\mathbf{A}$, $\mathbf{B}$, and $\mathbf{C}$; If $(\mathbf{A} \times \mathbf{B}) \cdot \mathbf{C}$ is negative, then it and $\mathbf{A} \cdot (\mathbf{B} \times \mathbf{C})$ are the negative of the volume.

27. $(x_1\mathbf{i} + y_1\mathbf{j} + z_1\mathbf{k}) \times (x_2\mathbf{i} + y_2\mathbf{j} + z_2\mathbf{k}) = x_1x_2(\mathbf{i} \times \mathbf{i})$
$+ y_1x_2(\mathbf{j} \times \mathbf{i}) + z_1x_2(\mathbf{k} \times \mathbf{i}) + x_1y_2(\mathbf{i} \times \mathbf{j}) + y_1y_2(\mathbf{j} \times \mathbf{j})$
$+ z_1y_2(\mathbf{k} \times \mathbf{j}) + x_1z_2(\mathbf{i} \times \mathbf{k}) + y_1z_2(\mathbf{j} \times \mathbf{k}) + z_1z_2(\mathbf{k} \times \mathbf{k})$
$= (y_1z_2 - z_1y_2)\mathbf{i} - (x_1z_2 - z_1x_2)\mathbf{j} + (x_1y_2 - y_1x_2)\mathbf{k}$

30. 9 **31.** 0 **32.** 0 **33.** $\langle 1, 0, -1 \rangle$ **34.** $\langle -10, 22, 2 \rangle$ **35.** $\langle -7, 1, -6 \rangle$
36. $\langle 1, -11, -4 \rangle$ **37.** $\frac{1}{2}\sqrt{109}$ **38.** $\frac{1}{2}\sqrt{21}$ **39.** $\frac{1}{2}\sqrt{514}$ **40.** $\frac{1}{2}\sqrt{3}$ **41.** 4 **42.** $\frac{1}{6}$

Section 12.3

1a. True **1b.** False **1c.** True **1d.** False **1e.** False **1f.** True **1g.** False
1h. True **1i.** False **1j.** True **1k.** True **2.** $x = 2 + 4t, y = 4, z = 3 - 7t$
4. $x = -3, y = -2 + 3t, z = 5t$ **6.**‡ The coefficients of x, y, and z in $6x - 4y = 0$ are 6, -4, and 0, so $\langle 6, -4, 0 \rangle$ is perpendicular to the plane and parallel to the line; $x = 3 + 6t, y = -4 - 4t, z = 5$ **8.** $x = 6 + t$, $y = 2 - 2t, z = 3 - 13t$ **10.**‡ $\mathbf{A} = \langle 2, -4, 0 \rangle$ is perpendicular to the plane $2x - 4y = 7$; $\mathbf{B} = \langle 0, 3, 5 \rangle$ is perpendicular to the plane $3y + 5z = 0$; $\mathbf{A} \times \mathbf{B} = \langle -20, -10, 6 \rangle$ is parallel to the line as is $\langle -10, -5, 3 \rangle$; $x = 6 - 10t, y = -5t, z = -3 + 3t$ **12.**‡ $\mathbf{A} = \langle 1, 1, 1 \rangle$ and $\mathbf{B} = \langle 1, -2, -1 \rangle$ are perpendicular to the planes; $\mathbf{A} \times \mathbf{B} = \langle 1, 2, -3 \rangle$ is parallel to the line of intersection; To find a point on the intersection, set $z = 0$ in the two equations: $x + y = 1, x - 2y = 1 \Rightarrow y = 0, x = 1$; The point $(1,0,0)$ is on the intersection; $x = 1 + t, y = 2t, z = -3t$
13. $x = 2t, y = -3 + 4t, z = -\frac{19}{4} + 3t$ **15.** $4x + 5z = 52$
17. $3x - y + z = 0$ **18.**‡ $\langle 2, -1, 1 \rangle$ is perpendicular to the planes; $(-3, -2, 6)$ is on the line at $t = 0$; $2(x + 3) - (y + 2) + (z - 6) = 0$
20.‡ $(0, -4, -3)$ is on the intersection at $x = 0$; $(4, 0, -7)$ is on the intersection at $y = 0$; $\langle 0, -4, -3 \rangle \times \langle 4, 0, -7 \rangle = \langle 28, -12, 16 \rangle$ is perpendicular to the desired plane; $7x - 3y + 4z = 0$ **22.** $x - y - z = 8$
24.‡ Set $\mathbf{P} = (1, -1, 2), \mathbf{Q} = (4, -1, 5), \mathbf{R} = (1, 2, -1); \overrightarrow{PQ} = \langle 3, 0, 3 \rangle$
and $\overrightarrow{PR} = \langle 0, 3, -3 \rangle$ are parallel to the plane and $\overrightarrow{PQ} \times \overrightarrow{PR}$
$= \langle -9, 9, 9 \rangle$ is perpendicular to it; $-(x - 1) + (y + 1) + (z - 2) = 0$;
$-x + y + z = 0$ **26.** $\frac{1}{2}x - \frac{1}{3}y + \frac{1}{5}z = 1$ **27.** $\frac{1}{5}y - \frac{1}{2}z = 1$
29. Fig. 12.3.29 **31.** Fig. 12.3.31 **33.** Fig. 12.3.33 **35.** $x = 0$,
$y = 0, z = 0, \frac{1}{3}x - \frac{1}{4}y + \frac{1}{5}z = 1$ **36.** $16/\sqrt{10}$ **38.** $\sqrt{6}$
42. No; $(300, 1680, 560)$ **43.** $x = 13t, y = -20t, z = 2t$

Section 12.4

1. For each partition $a = t_0 < t_1 < t_2 < \cdots < t_N = b$ the polygonal line joining $(x(t_j), y(t_j), z(t_j))$ for $j = 0, 1, 2, \ldots, N$ has length

$$\sum_{j=1}^{N} \sqrt{[x(t_j) - x(t_{j-1})]^2 + [y(t_j) - y(t_{j-1})]^2 + [z(t_j) - z(t_{j-1})]^2}$$

$$= \sum_{j=1}^{N} \sqrt{\left[\frac{dx}{dt}(c_j)\right]^2 + \left[\frac{dy}{dt}(d_j)\right]^2 + \left[\frac{dz}{dt}(e_j)\right]^2} \, (t_j - t_{j-1}),$$

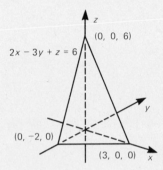

$2x - 3y + z = 6$

$(0, 0, 6)$

$(0, -2, 0)$

$(3, 0, 0)$

Fig. 12.3.29

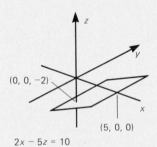

$(0, 0, -2)$

$(5, 0, 0)$

$2x - 5z = 10$

Fig. 12.3.31

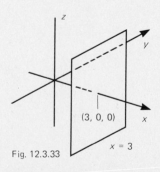

$(3, 0, 0)$

$x = 3$

Fig. 12.3.33

where by the Mean Value Theorem for derivatives, the numbers c_j, d_j, and e_j lie between t_{j-1} and t_j. This is an approximating sum for integral (3).

2. Speed $= \dfrac{ds}{dt}(t) = \dfrac{d}{dt}\displaystyle\int_a^t \sqrt{[x'(u)]^2 + [y'(u)]^2 + [z'(u)]^2}\, du$

$= \sqrt{[x'(t)]^2 + [y'(t)]^2 + [z'(t)]^2}$ by the Fundamental Theorem for derivatives of integrals.

3. $\langle 6t, \cos t, -\sin 2t \rangle$, $\sqrt{36t^2 + \cos^2 t + 4\cos^2 t \sin^2 t}$, $\langle 6, -\sin t, -2\cos 2t \rangle$

4. $\langle 2\cos 2t, 3\cos 3t, 3\cos 3t \rangle$, $\sqrt{4\cos^2 2t + 18\cos^2 3t}$

$\langle -4\sin 2t, -9\sin 3t, -9\sin 3t \rangle$ **6.** $\displaystyle\int_0^{10} \sqrt{\cos^2 t + (e^t + te^t)^2 + (3t^2 - 3)^2}\, dt$

8a. $\dfrac{d}{dt}[\mathbf{A}(t) \cdot \mathbf{B}(t)] = \displaystyle\lim_{\Delta t \to 0} \dfrac{1}{\Delta t}[\mathbf{A}(t + \Delta t) \cdot \mathbf{B}(t + \Delta t) - \mathbf{A}(t) \cdot \mathbf{B}(t)]$

$= \displaystyle\lim_{\Delta t \to 0} \dfrac{1}{\Delta t}\{\mathbf{A}(t + \Delta t) \cdot [\mathbf{B}(t + \Delta t) - \mathbf{B}(t)] + [\mathbf{A}(t + \Delta t) - \mathbf{A}(t)] \cdot \mathbf{B}(t)\}$

$= \mathbf{A}(t) \cdot \dfrac{d\mathbf{B}}{dt}(t) + \dfrac{d\mathbf{A}}{dt}(t) \cdot \mathbf{B}(t)$

9. If $|\mathbf{A}(t)|$ is constant, then $\dfrac{d}{dt}|\mathbf{A}(t)|^2 = \dfrac{d}{dt}[\mathbf{A}(t) \cdot \mathbf{A}(t)] = 2\mathbf{A}(t) \cdot \dfrac{d\mathbf{A}}{dt}(t)$

is zero

11. $\dfrac{d}{dt}[\mathbf{v}(t) \times \mathbf{R}(t)] = \mathbf{a}(t) \times \mathbf{R}(t) + \mathbf{v}(t) \times \mathbf{v}(t) = 0$

since $\mathbf{a}(t)$ is parallel to $\mathbf{R}(t)$; $\mathbf{v}(t) \times \mathbf{R}(t) = \mathbf{C}$; $\mathbf{R}(t)$ is perpendicular to the constant vector $\mathbf{C}$; The motion is in the plane through the origin perpendicular to $\mathbf{C}$

12. $\dfrac{d}{dt}[e^{-kt}\mathbf{R}(t)] = e^{-kt}[\mathbf{v}(t) - k\mathbf{R}(t)] = 0$; $\mathbf{R}(t) = e^{kt}\mathbf{R}(0)$;

The motion is on the line through the origin and parallel to $\mathbf{R}(0)$

14. $\frac{1}{2}\sqrt{3}$ **16.** $(\frac{1}{30}t^6 + t + 7)\mathbf{i} + (\frac{1}{20}t^5 + 2t + 6)\mathbf{j} + (\frac{1}{12}t^4 + 3t + 5)\mathbf{k}$

17. $(-\frac{1}{4}\sin(2t) + \frac{1}{2}t + 5)\mathbf{i} + (5t)\mathbf{j} + (\frac{1}{4}\sin(2t) - \frac{1}{2}t)\mathbf{k}$ **18.** $\mathbf{i} + (t - 10)\mathbf{j}$

$+ (\frac{1}{2}t^2 - 10t + 50)\mathbf{k}$ **20.** $(5t + 4)\mathbf{i} + (6t + 3)\mathbf{j} + (-16t^2 + 10t + 2)\mathbf{k}$

21. $(10t + 20)\mathbf{i} + 40\mathbf{j} + (-16t^2 + 96t + 56)\mathbf{k}$ **22.** $(50t + 50)\mathbf{i}$

$+ (60t + 100)\mathbf{j} + (-16t^2 + 160t + 200)\mathbf{k}$ **23.** $(6t)\mathbf{i} + (-8t)\mathbf{j}$

$+ (-16t^2 + 37t)\mathbf{k}$ **25.** $(10t)\mathbf{i} + (-20t)\mathbf{j} + (-4.9t^2 + 30.5t)\mathbf{k}$

26. $(-\frac{35}{29}\sqrt{58}t)\mathbf{i} + (-\frac{70}{29}\sqrt{58}t)\mathbf{j} + (-4.9t^2 + 10)\mathbf{k}$

Section 12.5

1a. Replacing z by $-z$ gives an equivalent equation   **1b.** Replacing y by $-y$ gives an equivalent equation   **1c.** Replacing x by $-x$ gives an equivalent equation   **1d.** Replacing x by $-x$ and y by $-y$ gives an equivalent equation   **1e.** Replacing x by $-x$ and z by $-z$ gives an equivalent equation   **1f.** Replacing y by $-y$ and z by $-z$ gives an equivalent equation   **1g.** Replacing x by $-x$, y by $-y$, and z by $-z$ gives an equivalent equation   **2.** Circular cylinder; $x^2 = 1$, two lines; $x^2 + z^2 = 1$, circle; $z^2 = 1$, two lines; Symmetric about the origin, all three axes, and all three coordinate planes; Fig. 12.5.3   **3.** Parabolic cylinder; $x = y^2$, parabola; $x = 0$, line; $y^2 = 0$, line; Symmetric about the x-axis, the xy-plane, and the xz-plane; Fig. 12.5.3   **4.** Elliptic

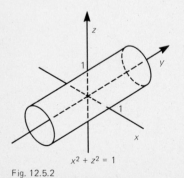

$x^2 + z^2 = 1$

Fig. 12.5.2

cylinder; $4x^2 + y^2 = 4$, ellipse; $4x^2 = 4$, two lines; $y^2 = 4$, two lines; Symmetric about the origin, all three axes, and all three coordinate planes; Fig. 12.5.4 **6.** Cylinder; $x^4 - 2x^2 = 0$, three lines; $z = x^4 - 2x^2$ $z = 0$, line; Symmetric about the z-axis, the yz-plane, the xz-plane; Fig. 12.5.6 **8.** Circular paraboloid; $x^2 + y^2 = 1$, circle; $z = 1 - x^2$, parabola; $z = 1 - y^2$, parabola; Symmetric about the z-axis, the yz-plane, and the xz-plane; Fig. 12.5.8 **10.** Hyperbolic paraboloid; $xy = 0$, two lines; $x = 0$, line; $y = 0$, line; Symmetric about the z-axis; Fig. 12.5.10 **12.** Hyperboloid of two sheets; $x^2 + y^2 = -1$, no solution; $z^2 - x^2 = 1$, hyperbola; $z^2 - y^2 = 1$, hyperbola; Symmetric about the origin, all three axes, all three coordinate planes; Fig. 12.5.12 **13.** Hyperboloid of one sheet; $x^2 + y^2 = 1$, circle; $x^2 - z^2 = 1$, hyperbola; $y^2 - z^2 = 1$, hyperbola; Symmetric about the origin, all three axes, all three coordinate planes; Fig. 12.5.13 **15.** Circular cone; $x^2 + y^2 = 0$, point; $z^2 = x^2$, two lines; $z^2 = y^2$, two lines; Symmetric about the origin, all three axes, all three coordinate planes; Fig. 12.5.15 **17.** Hemisphere; $x^2 + y^2 = 16$, circle; $z = \sqrt{16 - x^2}$, half-circle; $z = \sqrt{16 - y^2}$, half-circle; Symmetric about the z-axis, the xz-plane, and the yz-plane; Fig. 12.5.17 **19.** Half of a circular cone; $x^2 + y^2 = 0$, point; $z = -|x|$, inverted V-shaped curve; $z = -|y|$, inverted V-shaped curve; Symmetric about the z-axis, the xz-plane, the yz-plane; Fig. 12.5.19

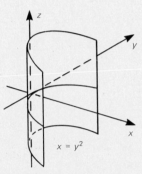

$x = y^2$

Fig. 12.5.3

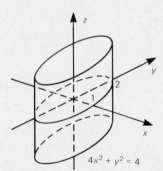

$4x^2 + y^2 = 4$

Fig. 12.5.4

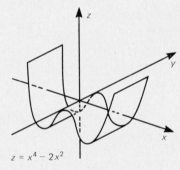

$z = x^4 - 2x^2$

Fig. 12.5.6

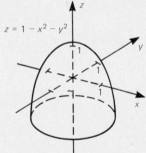

$z = 1 - x^2 - y^2$

Fig. 12.5.8

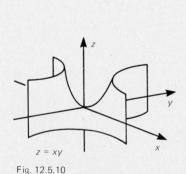

$z = xy$

Fig. 12.5.10

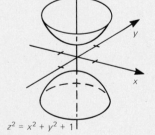

$z^2 = x^2 + y^2 + 1$

Fig. 12.5.12

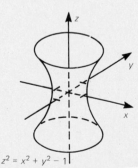

$z^2 = x^2 + y^2 - 1$

Fig. 12.5.13

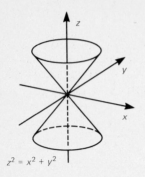

$z^2 = x^2 + y^2$

Fig. 12.5.15

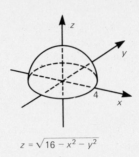

$z = \sqrt{16 - x^2 - y^2}$

Fig. 12.5.17

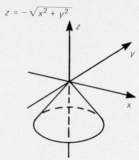

$z = -\sqrt{x^2 + y^2}$

Fig. 12.5.19

Appendix to Chapter 12

1.‡ $2\begin{vmatrix} 1 & -4 \\ 0 & 4 \end{vmatrix} - 4\begin{vmatrix} 0 & -4 \\ -3 & 4 \end{vmatrix} + 2\begin{vmatrix} 0 & 1 \\ -3 & 0 \end{vmatrix} = 62$

3. $\frac{73}{8}$ **5.** 63

7.‡ $\begin{vmatrix} 1 & -1 & 1 \\ 1 & 0 & 2 \\ 2 & -1 & 1 \end{vmatrix} = -2; \ x = \dfrac{1}{-2}\begin{vmatrix} 4 & -1 & 1 \\ 0 & 0 & 2 \\ 6 & -1 & 1 \end{vmatrix} = 2; \ y = \dfrac{1}{-2}\begin{vmatrix} 1 & 4 & 1 \\ 1 & 0 & 2 \\ 2 & 6 & 1 \end{vmatrix} = -3;$

$z = \dfrac{1}{-2}\begin{vmatrix} 1 & -1 & 4 \\ 1 & 0 & 0 \\ 2 & -1 & 6 \end{vmatrix} = -1$

9. $x = 1, y = 0, z = 1$ **11.** $x = 1, y = -1, z = 2$ **12.** $x = 4, y = 0, z = 3$
13. $x = \frac{1}{2}, y = \frac{3}{2}, z = -1$ **14.** $x = -\frac{1}{5}, y = \frac{2}{5}, z = 1$

$z = x^2 + y^2 - 2x - 2y + 2$

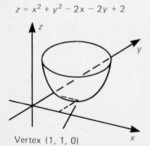

Vertex (1, 1, 0)

Fig. 12.M.26a

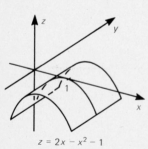

$z = 2x - x^2 - 1$

Fig. 12.M.26c

Miscellaneous Exercises to Chapter 12

6a. $2/\sqrt{6}$ **6b.** $4/\sqrt{14}$ **14.**‡ Solve $2t = u + 4, -t + 3 = 2u + 1, 4t - 2 = 6$ for t and u: $t = 2, u = 0$; $(4, 1, 6)$ **15.** $\sqrt{\frac{10}{7}}$ **16a.** The normal vector to Π_1 is at right angles to the intersection of Π_1 and Π_3; The normal vector to Π_2 is at right angles to the intersection of Π_2 and Π_3 **16b.** $\arccos(7/\sqrt{60}) \approx 0.44$
17a. $\frac{1}{2}\pi - $ [Angle between **A** and **n**] **17b.** $\frac{1}{2}\pi - \arccos(2/\sqrt{17}) \approx 0.51$
20. Divide by $|\mathbf{A}|$ and choose new coordinates so that $\mathbf{A} = \mathbf{i}$; Set $\mathbf{B}$ $= b_1\mathbf{i} + b_2\mathbf{j} + b_3\mathbf{k}$ and $\mathbf{C} = c_1\mathbf{i} + c_2\mathbf{j} + c_3\mathbf{k}$; $(\mathbf{C} \cdot \mathbf{A})\mathbf{B} - (\mathbf{B} \cdot \mathbf{A})\mathbf{C}$ $= (\mathbf{C} \cdot \mathbf{i})\mathbf{B} - (\mathbf{B} \cdot \mathbf{i})\mathbf{C} = c_1\mathbf{B} - b_1\mathbf{C} = (c_1 b_2 - b_1 c_2)\mathbf{j} + (c_1 b_3 - c_3 b_1)\mathbf{k}$ $= \mathbf{i} \times (\mathbf{B} \times \mathbf{C}) = \mathbf{A} \times (\mathbf{B} + \mathbf{C})$ **21.** Reduce to the case of $\mathbf{A} = \mathbf{i}$ as in Exercise 20; Set $\mathbf{B} = b_1\mathbf{i} + b_2\mathbf{j} + b_3\mathbf{k}, \mathbf{C} = c_1\mathbf{i} + c_2\mathbf{j} + c_3\mathbf{k}$,
$\mathbf{D} = d_1\mathbf{i} + d_2\mathbf{j} + d_3\mathbf{k}; \mathbf{A} \times \mathbf{B} = \mathbf{i} \times \mathbf{B} = -b_3\mathbf{j} + b_2\mathbf{k}; (\mathbf{A} \times \mathbf{B}) \cdot (\mathbf{C} \times \mathbf{D})$
$= b_3(c_1 d_3 - c_3 d_1) + b_2(c_1 d_2 - c_2 d_1) = b_3 c_1 d_3 - b_3 c_3 d_1 + b_2 c_1 d_2 - b_2 c_2 d_1$
$= c_1(b_1 d_1 + b_2 d_2 + b_3 d_3) - d_1(b_1 c_1 + b_2 c_2 + b_3 c_3)$
$= (\mathbf{A} \cdot \mathbf{C})(\mathbf{B} \cdot \mathbf{D}) - (\mathbf{A} \cdot \mathbf{D})(\mathbf{B} \cdot \mathbf{C})$ **22.** Use Exercise 20 **25.** $(3, -2, 1)$,
$\sqrt{14}$ **26a.**‡ $z = (x^2 - 2x + 1) + (y^2 - 2y + 1) = (x - 1)^2 + (y - 1)^2$;
Circular paraboloid; Fig. 12.M.26a **26c.** Parabolic cylinder; Fig. 12.M.26c
27. $x = -\frac{44}{13} + 25t, y = -\frac{53}{26} + 22t, z = 13t$

Section 13.1

1a. 108 **1b.** 1000 **2.** $\frac{1}{2}\sqrt{3}\,e^4$ **5.**‡ Because we cannot take square roots of negative numbers or divide by zero, $(x-y)^{-1/2}$ is defined for $x > y$; Fig. 13.1.5 **7.** Fig. 13.1.7 **8.** Fig. 13.1.8 **10.** Fig. 13.1.10 **13.** Fig. 13.1.13 **14.** Fig. 13.1.14 **15.** Fig. 13.1.15 **17.** Fig. 13.1.17 **21a.** Figure 13.23 **21b.** Figure 13.21

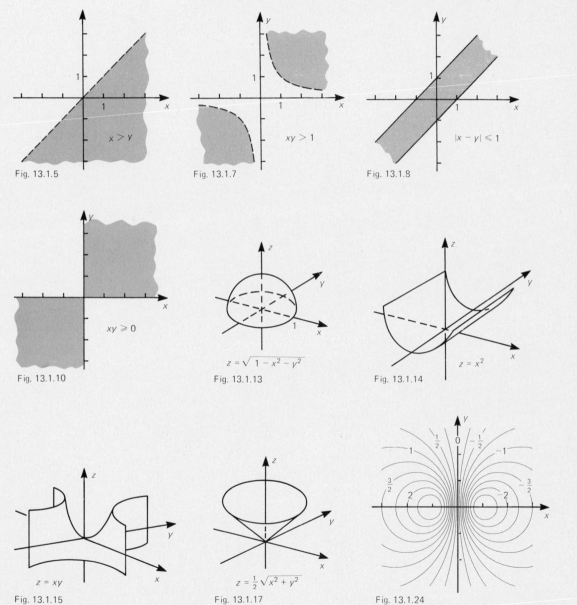

$x > y$

Fig. 13.1.5

$xy > 1$

Fig. 13.1.7

$|x - y| \leqslant 1$

Fig. 13.1.8

$xy \geqslant 0$

Fig. 13.1.10

$z = \sqrt{1 - x^2 - y^2}$

Fig. 13.1.13

$z = x^2$

Fig. 13.1.14

$z = xy$

Fig. 13.1.15

$z = \frac{1}{2}\sqrt{x^2 + y^2}$

Fig. 13.1.17

Fig. 13.1.24

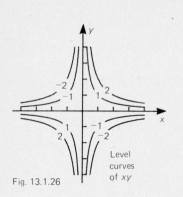

Level curves of xy

Fig. 13.1.26

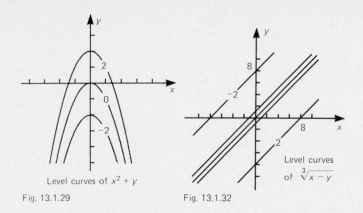

Level curves of $x^2 + y$

Fig. 13.1.29

Level curves of $\sqrt[3]{x - y}$

Fig. 13.1.32

21d. Figure 13.25   **22a.** Figure 13.26   **22b.** Figure 13.31   **22d.** Figure 13.30
24a. Fig. 13.1.24   **26.** Fig. 13.1.26   **29.** Fig. 13.1.29   **32.** Fig. 13.1.32
35.‡ $y \sin x = c \Rightarrow y = c \csc x$; Fig. 13.1.35   **38.** The value of the function at P
is the length of time it takes for the wave to get to P.

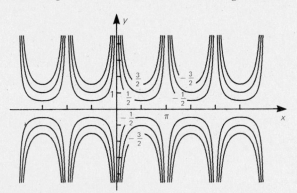

Level curves of $y \sin x$

Fig. 13.1.35

Section 13.2

1. $y^5 + 24x^3y^7$ **2.** $x^3 \cos(xy) + 1$ **5.** $\partial f/\partial x = 2xy^4 \cos(x^2y^4)$; $\partial f/\partial y$
$= 4y^3x^2 \cos(x^2y^4)$ **6.** $G_x = -y/(1 - xy)$; $G_y = -x/(1 - xy)$ **8.** $p_u(1, 2)$
$= 2e \cos(4)$; $p_v(1, 2) = -4e \sin(4)$ **10.** $\ln(xy) + 1$ **12.** $-2s/(r - s)^2$
14. $f_{xx} = 12x^2y^5$; $f_{yy} = 20x^4y^3$; $f_{xy} = 20x^3y^4 = f_{yx}$ **15.** $f_{xx} = -4/(2x - 3y)^2$;
$f_{xy} = f_{yx} = 6/(2x - 3y)^2$; $f_{yy} = -9/(2x - 3y)^2$ **16.** $f_{xx} = 90y^2(1 + xy)^8$;
$f_{yy} = 90x^2(1 + xy)^8$; $f_{xy} = f_{yx} = 10(1 + xy)^9 + 90xy(1 + xy)^8$
20. $P_{rr} = [-r^2s^2 \sin(rs) - 2rs \cos(rs) + 2 \sin(rs)]/(r^3s)$;
$P_{ss} = [-r^2s^2 \sin(rs) - 2rs \cos(rs) + 2 \sin(rs)]/(s^3r)$; $P_{sr} = P_{rs}$
$= [\sin(rs) - r^2s^2 \sin(rs) - rs \cos(rs)]/(r^2s^2)$ **22a.** 120,000   **22b.** 120,000
22c. 48,000   **25.** $\frac{2}{3}\pi rh$ **27.** $\partial\rho/\partial T = \frac{760}{273} \rho_0/P$; $\partial\rho/\partial P = -760(1 + \frac{1}{273} T) \rho_0/P^2$
28a. 890 gram-calories   **28b.** -5 gram-calories/degree   **28c.** 130
gram-calories/month   **29a.** $Y_t(1000, 20)$ is the rate of growth (cubic feet per
year) of an acre of trees with 1000 trees on it 20 years after they are planted.

33. $f_x = \dfrac{1}{2}x^{-1/2}y^{1/4} + 2xy^4$; $f_y = \dfrac{1}{4}x^{1/2}y^{-3/4} + 4x^2y^3$ **34.** $f_x = \dfrac{2x^2 + y^2}{\sqrt{x^2 + y^2}}$;

$f_y = \dfrac{xy}{\sqrt{x^2 + y^2}}$ **35.** $f_x = \dfrac{1}{y}\cos\!\left(\dfrac{x}{y}\right) - y\sin(xy)$; $f_y = -\dfrac{x}{y^2}\cos\!\left(\dfrac{x}{y}\right) - x\sin(xy)$

36. $f_x = -2y(x - y)^{-2}$; $f_y = 2x(x - y)^{-2}$ **37.** $g_x = \dfrac{-y}{x^2 + y^2}$; $g_y = \dfrac{x}{x^2 + y^2}$

38. $g_x = ye^{x-y} + xye^{x-y}$; $g_y = xe^{x-y} - xye^{x-y}$

39. $g_x = (-2x^5y^3 - 2xy^8)(x^4 - y^5)^{-2}$; $g_y = (3x^6y^2 + 2x^2y^7)(x^4 - y^5)^{-2}$

40. $g_x = \dfrac{\sec x \tan x}{\sec x + \tan y}$; $g_y = \dfrac{\sec^2 y}{\sec x + \tan y}$ **41.** $f_{xx} = 2y^3 - 6y^2x$;

$f_{yy} = 6x^2y - 2x^3$; $f_{xy} = f_{yx} = 6xy^2 - 6yx^2$ **42.** $f_{xx} = -2y^2(x + y)^{-3}$;
$f_{yy} = -2x^2(x + y)^{-3}$; $f_{xy} = f_{yx} = 2xy(x + y)^{-3}$ **43.** $f_{xx} = -y^2\sin(xy)$;
$f_{yy} = -x^2\sin(xy)$; $f_{xy} = f_{yx} = \cos(xy) - xy\sin(xy)$

44. $g_{xx} = -\dfrac{1}{4}x^{-3/2}(2 - y)^{1/3}$; $g_{yy} = \dfrac{2}{9}x^{1/2}(2 - y)^{-5/3}$; $g_{xy} = g_{yx} =$

$-\dfrac{1}{6}x^{-1/2}(2 - y)^{-2/3}$ **45.** $g_{xx} = -(x + y)^{-2}$;

$g_{yy} = (-2y^2 - x^2 - 2xy)(xy + y^2)^{-2}$; $g_{xy} = g_{yx} = -(x + y)^{-2}$
46. $g_{xx} = (-4 + y^2)(4 - x^2 - y^2)^{-3/2}$; $g_{yy} = (-4 + x^2)(4 - x^2 - y^2)^{-3/2}$;
$g_{xy} = g_{yx} = (-xy)(4 - x^2 - y^2)^{-3/2}$

Section 13.3

1. $-\frac{375}{22}$ **3.** 1   **5.** 0   **7.** $[(\sin t)e^{t^3\sin t}]/t^3$ **9.** $P(u, v) = ue^v\ln(u^3v^3e^v + 1)$
11a.‡ $g(2) = f(8, 16) = 3$

11b.‡ $\dfrac{dg}{dt}(t) = f_x(t^3, t^4)\dfrac{d}{dt}(t^3) + f_y(t^3, t^4)\dfrac{d}{dt}(t^4)$

$= f_x(t^3, t^4)(3t^2) + f_y(t^3, t^4)(4t^3)$; $\dfrac{dg}{dt}(2) = 12 f_x(8, 16) + 32 f_y(8, 16)$

$= 12(5) + 32(-7) = -164$

12a. 0   **12b.** -44 **15a.**‡ $F(1, 2) = f(8, 17) = 10$ **15b.**‡ $F_u(u, v)$
$= f_x(u^2v^3, 3u + 7v)(2uv^3) + f_y(u^2v^3, 3u + 7v)(3)$; $F_u(1, 2)$
$= f_x(8, 17)(16) + f_y(8, 17)(3) = 77$ **15c.**‡ $F_v(u, v)$
$= f_x(u^2v^3, 3u + 7v)(3u^2v^2) + f_y(u^2v^3, 3u + 7v)(7)$; $F_v(1, 2)$
$= f_x(8, 17)(12) + f_y(8, 17)(7) = 53$ **16a.** 4   **16b.** $20 + 2\sin(2)$ **16c.** 0
18a.‡ $P(3, 2) = R(x(3, 2), y(3, 2)) = R(1, 0) = 8$ **18b.**‡ $P_u(3, 2)$
$= R_x(x(3, 2), y(3, 2))x_u(3, 2) + R_y(x(3, 2), y(3, 2))y_u(3, 2)$
$= R_x(1, 0)x_u(3, 2) + R_y(1, 0)y_u(3, 2) = (9)(5) + (10)(7) = 115$
18c.‡ $P_v(3, 2) = R_x(x(3, 2), y(3, 2))x_v(3, 2) + R_y(x(3, 2), y(3, 2))y_v(3, 2)$
$= R_x(1, 0)x_v(3, 2) + R_y(1, 0)y_v(3, 2) = (9)(6) + (10)(4) = 94$
21a. $E(x(90), y(90))$ calories   **21b.** $E_x(x(90), y(90))x'(90)$
$+ E_y(x(90), y(90))y'(90)$ calories/hour   **22a.** $y'(x)$
$= -f_x(x, y(x))/f_y(x, y(x))$ **22b.** $y''(x)$
$= [-f_{yy}f_x^2/f_y^3] + [2f_{xy}f_x/f_y^2] - [f_{xx}/f_y]$ **23.** $f_x = F'(x + y) + F'(x - y)$;
$f_y = F'(x + y) - F'(x - y)$; $f_{xx} = F''(x + y) + F''(x - y) = f_{yy}$
26. $F(1) - F(0) = F'(c)(1 - 0) = F'(c)$ with $0 < c < 1$; $F(1) = f(x_1, y_1)$;
$F(0) = f(x_0, y_0)$; $F'(c) = f_x(X, Y)(x_1 - x_0) + f_y(X, Y)(y_1 - y_0)$ with
$X = x_0 + c(x_1 - x_0)$ and $Y = y_0 + c(y_1 - y_0)$

30. $\lim \dfrac{x^3 y - y}{2x + y + 1} = \dfrac{\lim(x^3 y - y)}{\lim(2x + y + 1)} = \dfrac{\lim(x^3)\lim(y) - \lim(y)}{\lim(2x) + \lim(y) + 1}$

$= \dfrac{0(3) - 3}{0 + 3 + 1} = -\dfrac{3}{4}$ where "lim" denotes $\lim\limits_{(x,y) \to (0,3)}$ or $\lim\limits_{x \to 0}$ or $\lim\limits_{y \to 3}$

31. $\lim \dfrac{(x + y)(x - 2y)}{6x + 7y} = \dfrac{\lim(x + y)\lim(x - 2y)}{\lim(6x) + \lim(7y)}$

$= \dfrac{[\lim(x) + \lim(y)][\lim(x) - \lim(2y)]}{6(1) + 7(2)} = \dfrac{(1 + 2)(1 - 4)}{6 + 14} = -\dfrac{9}{20}$ where

"lim" denotes $\lim\limits_{(x,y) \to (1,2)}$ or $\lim\limits_{x \to 1}$ or $\lim\limits_{y \to 2}$ **35.** $y \ne e^x$; -1 **36.** $y \ge x$; 0
37. $x \ne 0$; The limit does not exist because y/x equals m for $y = mx$ with $x \ne 0$, so y/x has different limits as (x, y) approaches $(0, 0)$ along different lines through the origin **38.** The first and third quadrants, excluding the origin but including the rest of the x- and y-axes; The limit does not exist because the function equals $m^{1/2}(1 + m^2)^{-1/2}$ for $y = mx, x \ne 0$ ($m \ge 0$) and has different limits along different lines through the origin

40. $x - 1 \le y \le x + 1$; $\frac{1}{2}\pi$ **41.** All x; 0

Section 13.4

1a. $\langle 108, -48 \rangle$ **1b.** $\frac{516}{5}$ **3a.**‡ $\overrightarrow{\text{grad}}\, f(x, y) = \langle 2xe^{2y}, 2x^2 e^{2y} \rangle$; $\langle 8e^6, 32e^6 \rangle$
3b.‡ $\langle 8e^6, 32e^6 \rangle \cdot \langle 2/\sqrt{13}, -3/\sqrt{13} \rangle = -80e^6/\sqrt{13}$ **4a.** $\langle \frac{1}{2}, -\frac{1}{2} \rangle$ **4b.** 0
6a.‡ $\overrightarrow{\text{grad}}\, f(x, y) = \langle \cos(xy) - xy\sin(xy), -x^2\sin(xy) \rangle$;
$\langle \cos 6 - 6\sin 6, -4\sin 6 \rangle$ **6b.**‡ $\mathbf{u} = \langle -2/\sqrt{13}, -3/\sqrt{13} \rangle$;
$(24\sin 6 - 2\cos 6)/\sqrt{13}$ **8.** Fig. 13.4.8 **10a.**‡ At $(-1, 1)$,
$\overrightarrow{\text{grad}}\,(x^5 y^{20}) = \langle 5, -20 \rangle$; The maximum directional derivative is the length $5\sqrt{17}$ of the gradient and is in the direction of the gradient; The unit vector in that direction is $\langle 1/\sqrt{17}, -4/\sqrt{17} \rangle$ **10b.**‡ The minimum directional derivative is the negative $-5\sqrt{17}$ of the length of the gradient and is in the direction opposite that of the gradient; $\langle -1/\sqrt{17}, 4/\sqrt{17} \rangle$ **12a.** $\frac{1}{3}\sqrt{37}$;
$\langle 6/\sqrt{37}, -1/\sqrt{37} \rangle$ **12b.** $-\frac{1}{3}\sqrt{37}$; $\langle -6/\sqrt{37}, 1/\sqrt{37} \rangle$ **14.**‡ $\overrightarrow{\text{grad}}\, f(x, y)$
$= (3x^2 y^3 - y)\mathbf{i} + (3x^3 y^2 - x)\mathbf{j}$; The vectors are perpendicular to
$\overrightarrow{\text{grad}}\, f(10, 10) = 299{,}990(\mathbf{i} + \mathbf{j})$; $(1/\sqrt{2})(\mathbf{i} - \mathbf{j})$ and $-(1/\sqrt{2})(\mathbf{i} - \mathbf{j})$
15. $(1/\sqrt{13})(3\mathbf{i} + 2\mathbf{j})$ and $-(1/\sqrt{13})(3\mathbf{i} + 2\mathbf{j})$ **17.**‡ set $f(x, y)$
$= xy^3 + 6x^2 y$; $\overrightarrow{\text{grad}}\, f = \langle y^3 + 12xy, 3xy^2 + 6x^2 \rangle$; $\overrightarrow{\text{grad}}\, f(1, 1) = \langle -13, 9 \rangle$;
$\pm \langle -13, 9 \rangle/\sqrt{250}$ **19.**‡ $\langle f_x, f_y \rangle \cdot \langle 1, -1 \rangle/\sqrt{2} = 7/\sqrt{2}$; $\langle f_x, f_y \rangle \cdot \langle \frac{4}{5}, \frac{3}{5} \rangle = 0$;
$f_x - f_y = 7$ and $4f_x + 3f_y = 0$; $f_x(1, 2) = 3, f_y(1, 2) = -4$ **20.** $P_x(0, 0)$
$= -2\sqrt{3}$; $P_y(0, 0) = 10$

22.‡ Along an s-axis in the direction $\mathbf{u}$ and through the point $(3, 1)$, the level

curves $f = 2$ and $f = 4$ are approximately $\dfrac{5}{4}$ units apart, and $f = 4$ is on the

positive side of $f = 2$; $\Delta f = 2$ and $\Delta s \approx \dfrac{5}{4}$; $D_{\mathbf{u}} f(3, 1) \approx \dfrac{\Delta f}{\Delta s} \approx \dfrac{2}{5/4} = \dfrac{8}{5}$;

Fig. 13.4.22 **23.** -5

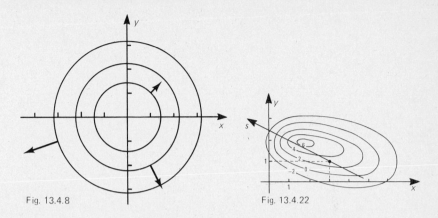

Fig. 13.4.8   Fig. 13.4.22

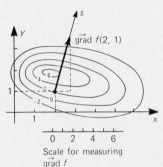

grad $f(2, 1)$

0 2 4 6
Scale for measuring
grad f

Fig. 13.4.25

25.‡ On an s-axis perpendicular to the imagined level curve through

$(2, 1)$, $\Delta f = 2$ and $\Delta s \approx \dfrac{3}{8}$. For **u** in the direction of greatest increase

$|\overrightarrow{\text{grad}}\, f(2, 1)| = D_{\mathbf{u}}f(2, 1) \approx \dfrac{\Delta f}{\Delta s} = \dfrac{2}{3/8} = 5\dfrac{1}{3}$; From the sketch of $\overrightarrow{\text{grad}}\, f(2, 1)$ in

Fig. 13.4.25, we see that $\overrightarrow{\text{grad}}\, f(2, 1) \approx \dfrac{3}{2}\mathbf{i} + 5\mathbf{j}$. (We used a separate scale for

measuring the length of the gradient vector in the figure so the arrow would fit
in the picture.)

26. $-10\mathbf{i} - 8\mathbf{j}$ **28a.**‡ $F_x(5, -1) = x$-component of $\overrightarrow{\text{grad}}\, F(5, -1) \approx -3$
28b. -1 **28c.** 0 **28e.** $1/\sqrt{2} \approx 0.71$

Section 13.5

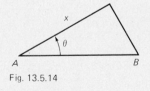

Fig. 13.5.14

1.‡ Set $f(x, y) = x^2y^{-3}$; $f(2, 1) = 4$; $f_x(2, 1) = 4$; $f_y(2, 1) = -12$; Normal
vector: $\langle 4, -12, -1 \rangle$; Tangent plane: $z = 4 + 4(x - 2) - 12(y - 1)$ or
$z = 4x - 12y + 8$ **2.** $\langle 1, -3, -1 \rangle$; $z = x - 3y$ **5.** $\langle -\frac{1}{4}\sqrt{2}, -\frac{1}{4}\sqrt{6}, -1 \rangle$;
$z = -\frac{1}{4}\sqrt{2} - \frac{1}{4}\sqrt{2}(x - \frac{1}{4}\pi) + \frac{1}{4}\sqrt{6}(y - \frac{2}{3}\pi)$ **7.** $\langle \frac{3}{25}, \frac{4}{25}, -1 \rangle$;
$z = \ln 5 + \frac{3}{25}(x - 3) + \frac{4}{25}(y - 4)$ **9.**‡ Set $f(x, y) = (x^2 + y^3 + 10)^{1/3}$;
$f(3, 2) = 3$; $f_x(3, 2) = \frac{2}{9}$; $f_y(3, 2) = \frac{4}{9}$; Tangent plane:
$z = 3 + \frac{2}{9}(x - 3) + \frac{4}{9}(y - 2)$; $f(3.02, 1.97) \approx 3 + \frac{2}{9}(3.02 - 3) + \frac{4}{9}(1.97 - 2)$
$= 2.9911$ **10.** 1.5006 **12.** -0.05 **14.**‡ Fig. 13.5.14; Area $= A(x, \Theta)$
$= 5x \sin \Theta$; $A_x(8, \pi/6) = \frac{5}{2}$; $A_\Theta(8, \pi/6) = 20\sqrt{3}$; $dA = \frac{5}{2}\,dx + 20\sqrt{3}\,d\Theta$;
Maximum error $\approx \frac{5}{2}(0.001) + 20\sqrt{3}(0.02) \approx 0.695$ square inches   **15.** 0.002 rad.
17. $df = 2xy^5\,dx + 5x^2y^4\,dy$ **18.** $dG = (u - 3v)^{-1}\,du - 3(u - 3v)^{-1}\,dv$
19. $dH = (2r + 4s)\,dr + (-3s^2 + 4r)\,ds$ **21.** 0.19 square meters
22. 0.08π square inches   **23.** 1.03 square centimeters   **24.** 0.0025

Section 13.6

2a. $-6x\,e^{-3y}\sin(4z)$ **2b.** $-12x^2\,e^{-3y}\cos(4z)$ **2c.** $8x\,e^{-3y}\cos(4z)$
4. $(\sin x)(\cos y)e^z$ **7.** $z_x = (x - y\cos\Theta)/z$, $z_y = (y - x\cos\Theta)/z$,
$z_\Theta = (xy\sin\Theta)/z$ **9a.**‡ $\overrightarrow{\text{grad}}\,(x^4y^{-2}z^3) = \langle 4x^3y^{-2}z^3, -2x^4y^{-3}z^3, 3x^4y^{-2}z^2 \rangle$;
$\langle 27, -\frac{27}{4}, \frac{27}{4} \rangle$ **9b.**‡ $\langle 27, -\frac{27}{4}, \frac{27}{4} \rangle \cdot \langle \frac{3}{13}, -\frac{4}{13}, -\frac{12}{13} \rangle = \frac{27}{13}$ **11a.** $\langle \frac{1}{5}, \frac{1}{2}, \frac{1}{6} \rangle$

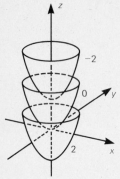

Fig. 13.6.15

11b. $\frac{7}{45}\sqrt{3}$　**13a.** $\langle 0, 9, -9 \rangle$　**13b.** 0　**15.** Fig. 13.6.15

17a.‡ grad $(x^3y^2 + y^3z^2) = \langle 3x^2y^2, 2x^3y + 3y^2z^2, 2y^3z \rangle$; The gradient $\langle 12, 112, 48 \rangle$ at $(1, 2, 3)$ is a normal vector to the level surface　**17b.**‡ The tangent plane has normal vector $\langle 12, 112, 48 \rangle$ and passes through $(1, 2, 3)$; $12(x - 1) + 112(y - 2) + 48(z - 3) = 0$ or $3x + 28y + 12z = 95$

19a. $\langle 1, 0, 25 \rangle$　**19b.** $x + 25z = 10$　**20a.** $\langle 10, 5, 2 \rangle$

20b. $10x + 5y + 2z = 30$　**23.** $x = 2 + 3t, y = 1 + 10t, z = -1 - 14t$

25.‡ Set $f(x, y, z) = x^2y^4(1 + z^2)^{-1}$; $f(5, 1, 2) = 5$; $f_x(5, 1, 2) = 2$; $f_y(5, 1, 2) = 20$; $f_z(5, 1, 2) = -4$; At $(5, 1, 2)$, $df = 2\,dx + 20\,dy - 4\,dz$; $f(4.996, 1.003, 1.995) \approx f(5, 1, 2) + df$ for $dx = -0.004, dy = 0.003$, and $dz = -0.005$; $5 + 2(-0.004) + 20(0.003) - 4(-0.005) = 5.072$　**27.** 6.07

29. $f_x = \frac{1}{2}x^{-1/2}y^{1/4}$; $f_y = \frac{1}{4}x^{1/2}y^{-3/4} + \frac{1}{5}y^{-4/5}z^{1/6}$; $f_z = \frac{1}{6}y^{1/5}z^{-5/6}$

30. $f_x = (x + 3y - 4z)^{-1}$; $f_y = 3(x + 3y - 4z)^{-1}$; $f_z = -4(x + 3y - 4z)^{-1}$

31. $f_x = \sin y$; $f_y = x \cos y - \cos z$; $f_z = y \sin z$

32. $g_x = (y^2z + z^2y)(x + y + z)^{-2}$; $g_y = (x^2z + xz^2)(x + y + z)^{-2}$; $g_z = (x^2y + xy^2)(x + y + z)^{-2}$　**33.** $g_x = y \arctan(yz)$; $g_y = x \arctan(yz) + (xyz)(1 + y^2z^2)^{-1}$; $g_z = (xy^2)(1 + y^2z^2)^{-1}$

34. $g_x = ye^{x^2z} + 2x^2yze^{x^2z}$; $g_y = xe^{x^2z}$; $g_z = x^3ye^{x^2z}$　**35.** $f_{xx} = 2y$; $f_{yy} = -6yz^2$; $f_{zz} = -2y^3$; $f_{xy} = f_{yx} = 2x$; $f_{xz} = f_{zx} = 1$; $f_{yz} = f_{zy} = -6y^2z$

36. $f_{xx} = 20(x^2 - y^3 + z^4)^9 + 360x^2(x^2 - y^3 + z^4)^8$; $f_{yy} = -60y(x^2 - y^3 + z^4)^9 + 810y^4(x^2 - y^3 + z^4)^8$; $f_{zz} = 120z^2(x^2 - y^3 + z^4)^9 + 1440z^6(x^2 - y^3 + z^4)^8$; $f_{xy} = f_{yx} = -540xy^2(x^2 - y^3 + z^4)^8$; $f_{xz} = f_{zx} = 720xz^3(x^2 - y^3 + z^4)^8$; $f_{yz} = f_{zy} = -1080y^2z^3(x^2 - y^3 + z^4)^8$　**37.** $g_{xx} = -y^2z^2 \sin(xyz)$; $g_{yy} = -x^2z^2 \sin(xyz)$; $g_{zz} = -x^2y^2 \sin(xyz)$; $g_{xy} = g_{yx} = z \cos(xyz) - xyz^2 \sin(xyz)$; $g_{xz} = g_{zx} = y \cos(xyz) - xy^2z \sin(xyz)$; $g_{yz} = g_{zy} = x \cos(xyz) - x^2yz \sin(xyz)$

38. $g_{xx} = 2y^3 \tan z$; $g_{yy} = 6x^2y \tan z$; $g_{zz} = 2x^2y^3 \sec^2 z \tan z$; $g_{xy} = g_{yx} = 6xy^2 \tan z$; $g_{xz} = g_{zx} = 2xy^3 \sec^2 z$; $g_{yz} = g_{zy} = 3x^2y^2 \sec^2 z$

39. $37/(12\sqrt{6})$　**40.** $-5/(8\sqrt{6})$

41. $\frac{5}{2}\sqrt{305}$; $(305)^{-1/2} \langle 2, 1, 10\sqrt{3} \rangle$　**42.** $48\sqrt{265}$; $(265)^{-1/2} \langle -6, 15, 2 \rangle$

43. $P(2) = 3$; $P'(2) = -48$　**44.** $g(1,3) = 7$; $g_x(1,3) = 3$; $g_y(1,3) = 2$

45. $H(0,0,0) = 3$; $H_x(0,0,0) = \frac{5}{6}$; $H_y(0,0,0) -\frac{2}{3}$; $H_z(0,0,0) = -\frac{1}{2}$

46. $G(1,2,3) = 12$; $G_x(1,2,3) = 2$; $G_y(1,2,3) = -30$; $G_z(1,2,3) = 30$

47. $A(1,10,100) = 0.1$; $A_u(1,10,100) = 0.2$; $A_v(1,10,100) = 0.02$; $A_w(1,10,100) = -0.003$　**48.** $K_x = 2xA'(x^2)$; $K_y = 3A^2(y)A'(y)$; $K_z = 4z^3A'(z^4)$

49. $df = [yz \sin(x + y + z) + xyz \cos(x + y + z)]\,dx + [xz \sin(x + y + z) + xyz \cos(x + y + z)]\,dy + [xy \sin(x + y + z) + xyz \cos(x + y + z)]\,dz$

50. $dg = 2xe^{3y-4z}\,dx + 3x^2e^{3y-4z}\,dy - 4x^2e^{3y-4z}\,dz$

51. $dh = \frac{\cosh x}{\sinh z}\,dx + \frac{\sinh y}{\sinh z}\,dy - (\sinh x + \cosh y)\left(\frac{\cosh z}{\sinh^2 z}\right)dz$

52. $dK = \frac{1}{5}(x^2 + y^3 - z^4)^{-4/5}(2x\,dx + 3y^2\,dy - 4z^3\,dz)$

53. $\frac{dN}{dt} = D_1\frac{d\alpha}{dt} + D_2\frac{d\beta}{dt} + D_3$

Miscellaneous Exercises to Chapter 13

1a.‡ Holding y, z, and w fixed and differentiating with respect to x gives the x-derivative $yzw \cos(xyzw)$; Holding x, z, and w fixed and differentiating with respect to y gives the y-derivative $xzw \cos(xyzw)$; The z-derivative is $xyw \cos(xyzw)$; The w-derivative is $xyz \cos(xyzw)$ **1b.** x-derivative: $2xy^2$; y-derivative: $2x^2y + 2yz^2$; z-derivative: $2y^2z + 2zu^2$; u-derivative: $2z^2u + 2uv^2$; v-derivative: $2u^2v$ **2a.** $\overrightarrow{\text{grad}}(x^7y^6z^5w^4) = \langle 7x^6y^6z^5w^4,$ $6x^7y^5z^5w^4, 5x^7y^6z^4w^4, 4x^7y^6z^5w^3 \rangle$

3. $\dfrac{\partial}{\partial x}(xy)^y = y^{y+1}x^{y-1}$; $\dfrac{\partial}{\partial y}(xy)^y = (xy)^y + (xy)^y\ln(xy)$

8a. $V = \frac{1}{3}\pi h^3 \tan^2\Theta$ **9a.** $z_x(x, y) = -f_x(x, y, z(x, y))/f_z(x, y, z(x, y))$; $z_y(x, y) = -f_y(x, y, z(x, y))/f_z(x, y, z(x, y))$ **10.** $z_x(-1, -1) = \frac{11}{26}$; $z_y(-1, -1) = \frac{11}{26}$ **15a.** 3 **16.** $[f(x, y) - f(x, y_0)] - [f(x_0, y) - f(x_0, y_0)]$ $= [f_x(c_1, y) - f_x(c_1, y_0)](x - x_0) = f_{xy}(c_1, c_2)(x - x_0)(y - y_0)$ with c_1 between x and x_0 and c_2 between y and y_0 by the Mean Value Theorem applied first to $F(x) = f(x, y) - f(x, y_0)$ and then to $G(y) = f_x(c_1, y)$; $[f(x, y) - f(x_0, y)] - [f(x, y_0) - f(x_0, y_0)] = [f_y(x, c_4) - f_y(x_0, c_4)](y - y_0)$ $= f_{yx}(c_3, c_4)(x - x_0)(y - y_0)$ with c_4 between y and y_0 and c_3 between x and x_0 by the Mean Value Theorem applied first to $H(y) = f(x, y) - f(x_0, y)$ and then to $f_y(x, c_4)$; $f_{xy}(c_1, c_2) = f_{yx}(c_3, c_4)$; (c_1, c_2) and (c_3, c_4) tend to (x_0, y_0) as (x, y) tends to (x_0, y_0), so $f_{xy}(x_0, y_0) = f_{yx}(x_0, y_0)$ since f_{xy} and f_{yx} are continuous **21a.** $y \neq 1$ **21b.** $\frac{1}{2}\pi$ **21c.** $-\frac{1}{2}\pi$

24. $\dfrac{d}{dx}G(x, a(x), b(x)) = G_x(x, a(x), b(x)) + G_u(x, a(x), b(x))a'(x)$

$+ G_v(x, a(x), b(x))b'(x)$; $G_x(x, u, v) = \displaystyle\int_u^v f_x(x, t)\, dt$; $G_u(x, u, v) = -f(x, u)$;

$G_v(x, u, v) = f(x, v)$

25a.‡ $e^{-(x^3)^2}\dfrac{d}{dx}(x^3) - e^{-(x^2)^2}\dfrac{d}{dx}(x^2) + \displaystyle\int_{x^2}^{x^3}\dfrac{\partial}{\partial x}(e^{-t^2})\, dt = 3x^2\, e^{-x^6} - 2x\, e^{-x^4}$

since e^{-t^2} does not depend on x

25d.‡ $e^{x(5x)^2}\dfrac{d}{dx}(5x) - e^{x(-x)^2}\dfrac{d}{dx}(-x) + \displaystyle\int_{-x}^{5x}\dfrac{\partial}{\partial x}(e^{xt^2})\, dt$

$= 5\, e^{25x^3} + e^{x^3} + \displaystyle\int_{-x}^{5x} t^2\, e^{xt^2}\, dt$

25e. $-\sqrt[3]{\sin(x^2)} + \displaystyle\int_x^{10} \frac{1}{3} t \cos(xt)\, [\sin(xt)]^{-2/3}\, dt$

Section 14.1

1. $(1, 2)$ **3.** $(0, n\pi)$, n any integer **5.**‡ $f_x = 4y - 4x^3$, $f_y = 4x - 4y^3$; At critical points $4y - 4x^3 = 0$ and $4x - 4y^3 = 0$; $y = x^3$ and $x = y^3$; $x = x^9$; $x = 0, 1, -1$; $(0, 0), (1, 1), (-1, -1)$ **6.** $(0, 1), (0, -1), (1, 0), (-1, 0)$ **7.**‡ $f_x = -12x^3 - 12xy$, $f_y = 6y^2 - 6x^2$; $f_y = 0 \Rightarrow x^2 = y^2$; $f_x = 0 \Rightarrow$ $x(x^2 + y) = 0 \Rightarrow x = 0$ or $y = -x^2$; If $x = 0$, then $y = 0$; If $y = -x^2$, then $y = -y^2$ so that $y = 0$ or -1; $y = -1 \Rightarrow x = \pm 1$; $(0, 0), (1, -1), (-1, -1)$ **9.** $(2, 0), (2, 3\sqrt{2}), (2, -3\sqrt{2}), (-2, 0), (-2, 3\sqrt{2}), (-2, -3\sqrt{2})$

12. $(0,0)$, $(-\frac{1}{2}, -\frac{1}{3})$ **13.** $(x, 0)$ all x **14.** The lines $x - y = n\pi$, n any integer
16.‡ $f_x = (1 - x)ye^{-x-y}$, $f_y = x(1 - y)e^{-x-y}$; Critical points: $(0, 0)$ and $(1, 1)$;
The maximum occurs at $(1, 1)$ or along one of the sides where $x = 0$ or $y = 0$;
$f = 0$ for $x = 0$ or $y = 0$, $f(1, 1) = e^{-2}$; Maximum $= e^{-2}$ **18.** $(1, 1, 1)$,
$(1, -1, -1)$, $(-1, 1, -1)$, $(-1, -1, 1)$ **20.**‡ Let x, y, and z be the length of the
sides, which are positive when the volume is a maximum; Surface area $= A$
$= 2xy + 2xz + 2yz$; $z = (A - 2xy)/(2x + 2y)$; Volume $= V = xyz$
$= (Axy - 2x^2y^2)/(2x + 2y)$; $V_x = (2Ay^2 - 8xy^3 - 4x^2y^2)/(2x + 2y)^2$;
$V_y = (2Ax^2 - 8x^3y - 4x^2y^2)/(2x + 2y)^2$; $2y^2(A - 4xy - 2x^2) = 0$ and
$2x^2(A - 4xy - 2y^2) = 0$; $2x^2 = A - 4xy = 2y^2$; $y = x$; $A = 4xy + 2x^2$
$= 6x^2$ and $A = 2x^2 + 4xz$; $z = x$ **22.**‡ Area $= 1 = 2xy + x^2\tan \Theta$;
$y = (1 - x^2\tan \Theta)/(2x)$; Perimeter $= P = 2x + 2y + 2x \sec \Theta$
$= 2x(1 + \sec \Theta) + (1 - x^2\tan \Theta)/x$; $P_\Theta = 2x \sec \Theta \tan \Theta - x \sec^2\Theta$
$= x \sec \Theta(2 \tan \Theta - \sec \Theta)$; $P_\Theta = 0 \Rightarrow 2 \tan \Theta = \sec \Theta$; $\sin \Theta = \frac{1}{2}$; $\Theta = \frac{1}{6}\pi$;
$\sec \Theta = 2/\sqrt{3}$; $\tan \Theta = 1/\sqrt{3}$; $P_x = 2(1 + \sec \Theta) - x^{-2} - \tan \Theta$;
$P_x = 0 \Rightarrow x^2 = (2 + 2\sec \Theta - \tan \Theta)^{-1}$; $x = (2 + \sqrt{3})^{-1/2}$;
$y = (1 + \sqrt{3})(6 + 3\sqrt{3})^{-1/2}$ **26.** $y = 0.474x + 2.002$ **30.** 102
31a. Minimum $= -5$ **32a.** The level curves $x^2 + xy + y^2 + 3x - 9y = c$
are conic sections with $B^2 - 4AC = 1^2 - 4(1)(1) = -3$ negative, so they are
ellipses and the graph $z = x^2 + xy + y^2 + 3x - 9y$ is an elliptic paraboloid;
The intersection of the graph with the xz-plane is the parabola $z = x^2 + 3x$,
$y = 0$, which opens upward, so the surface opens upward **32b.** -39
35. Max $= 17$ at $(3, -2)$ **37.** Max $= 1$ at $(\frac{1}{2}\pi + 2k\pi, 0)$, k an integer
38. Min $= 0$ at $(x, 0)$ **40.** Max $= (2e)^{-1/2}$ at $(2^{-1/2}, 0)$; Min $= -(2e)^{-1/2}$ at
$(-2^{-1/2}, 0)$ **41.** Min $= -2$ at $(1, 1)$ and $(-1, -1)$ **43.** Max $= e$ at
$(2k\pi, \frac{1}{2}\pi + 2n\pi)$; Min $= -e$ at $(\pi + 2k\pi, \frac{1}{2}\pi + 2n\pi)$, k and n integers
44. $x = \frac{1}{3}$, $y = 1$, $z = 2$ **45.** $a = \frac{1}{3}$, $b = 18$ **46.** $4\sqrt{2}$

Section 14.2

1a.‡ Set $f = 3xy - x^3 - y^3 + \frac{1}{8}$; $f_x = 3y - 3x^2$, $f_y = 3x - 3y^2$; Critical
points: $(0, 0)$ and $(1, 1)$; $f_{xx} = -6x$, $f_{xy} = 3$, $f_{yy} = -6y$; Table 14.2.1a;
Figure 14.17 **1b.** Figure 14.20

TABLE 14.2.1a

Critical point	$A = f_{xx}$	$B = f_{xy}$	$C = f_{yy}$	$B^2 - AC$	Type
$(0, 0)$	0	3	0	9	Saddle point
$(1, 1)$	-6	3	-6	-27	Local maximum

1d. Set $f = 2y^3 - 3x^4 - 6x^2y + \frac{1}{16}$; $f_x = -12x^3 - 12xy$, $f_y = 6y^2 - 6x^2$;
Critical points: $(0, 0)$, $(-1, -1)$, $(1, -1)$; $f_{xx} = -36x^2 - 12y$, $f_{xy} = -12x$,
$f_{yy} = 12y$; Table 14.2.1d; Figure 14.19 **2a.** Figure 14.26 **2b.** Figure 14.24
2d. Figure 14.27 **3.** Saddle points at $(0, n\pi)$ for all integers n **5.** Local
minimum at $(0, 0)$ **6.** Set $f = x^2y - x^2 - 2y^2$; $f_x = 2xy - 2x$,
$f_y = x^2 - 4y$; Critical points: $(0, 0)$, $(2, 1)$, $(-2, 1)$; $f_{xx} = 2y - 2$, $f_{xy} = 2x$,
$f_{yy} = -4$; Table 14.2.6 **9.** Local minimum at $(3, 0)$; Saddle point at $(-1, 0)$
10. Saddle points at $(0, 0)$, $(0, 4)$, $(3, 0)$; Local minimum at $(1, \frac{4}{3})$ **11.** Saddle
points at $(-1, 0)$, $(-1, -1)$, $(0, -1)$; Local minimum at $(-\frac{2}{3}, -\frac{2}{3})$

17. Critical points at $(x, 0)$, all x; Second derivative test fails   **18.** Saddle point at $(\frac{1}{2}\sqrt{2}, 0)$; Local minimum at $(-\frac{1}{2}\sqrt{2}, 0)$ **21.** Saddle point at $(1, 0)$ and $(-1, 0)$ **22.** Saddle point at $(0, 0)$; local maxima at $(1, 0)$ and $(-1, 0)$
24. Saddle point at $(0, 0)$ **25.** Local minimum at $(x, 0)$ **26.** Critical point at $(2, 5)$; Second derivative test fails

TABLE 14.2.1d

Critical point	$A = f_{xx}$	$B = f_{xy}$	$C = f_{yy}$	$B^2 - AC$	Type
$(0, 0)$	0	0	0	0	The test fails
$(-1, -1)$	-24	12	-12	-144	Local maximum
$(1, 1)$	-24	-12	-12	-144	Local maximum

TABLE 14.2.6

Critical point	$A = f_{xx}$	$B = f_{xy}$	$C = f_{yy}$	$B^2 - AC$	Type
$(0, 0)$	-2	0	-4	-8	Local maximum
$(2, 1)$	0	4	-4	16	Saddle point
$(-2, 1)$	0	-4	-4	16	Saddle point

Section 14.3

1.‡ $\overrightarrow{\text{grad}}(2x + y) = \lambda\overrightarrow{\text{grad}}(x^2 + 2y^2) \Rightarrow 2 = 2\lambda x, 1 = 4\lambda y; x = 1/\lambda$, $y = 1/(4\lambda); x^2 + 2y^2 = 18 \Rightarrow \lambda = \pm\frac{1}{4}; \lambda = \frac{1}{4} \Rightarrow x = 4, y = 1; \lambda = -\frac{1}{4} \Rightarrow x = -4, y = -1; 2x + y$ is 9 at $(4, 1)$ and is -9 at $(-4, -1)$, so the maximum is 9 at $(4, 1)$ and the minimum is -9 at $(-4, -1)$ **2.** Maximum $= 9$ at $(2, -2)$; Minimum $= -7$ at $(-2, 2)$ **4.**‡ $\overrightarrow{\text{grad}}(5 - 6x - 3y - 2z)$ $= \lambda\overrightarrow{\text{grad}}(4x^2 + 2y^2 + z^2) \Rightarrow -6 = 8\lambda x, -3 = 4\lambda y, -2 = 2\lambda z;$ $x = -3/(4\lambda), y = -3/(4\lambda), z = -1/\lambda; 4x^2 + 2y^2 + z^2 = 70 \Rightarrow \lambda = \pm\frac{1}{4};$ $\lambda = \frac{1}{4} \Rightarrow x = 3, y = 3, z = 4; \lambda = -\frac{1}{4} \Rightarrow x = -3, y = -3, z = -4;$ The maximum is 40 at $(-3, -3, -4)$ and the minimum is -30 at $(3, 3, 4)$
5. Maximum $= 33$ at $(3, 0, 6)$; Minimum $= -33$ at $(-3, 0, -6)$
6.‡ $\overrightarrow{\text{grad}}(xy) = \lambda \overrightarrow{\text{grad}}(2x^2 + 3y^2) \Rightarrow y = 4\lambda x, x = 6\lambda y; y^2 = \frac{2}{3}x^2;$ $2x^2 + 3y^2 = 12 \Rightarrow x = \pm\sqrt{3}; y = \pm\sqrt{2};$ The maximum is $\sqrt{6}$ at $(\sqrt{3}, \sqrt{2})$ and $(-\sqrt{3}, -\sqrt{2})$; The minimum is $-\sqrt{6}$ at $(-\sqrt{3}, \sqrt{2})$ and $(\sqrt{3}, -\sqrt{2})$
7. Maximum $= 15$ at $(\sqrt{5}, \sqrt{6}, \sqrt{\frac{15}{2}}), (\sqrt{5}, -\sqrt{6}, -\sqrt{\frac{15}{2}}),$ $(-\sqrt{5}, \sqrt{6}, -\sqrt{\frac{15}{2}})$, and $(-\sqrt{5}, -\sqrt{6}, \sqrt{\frac{15}{2}})$; The minimum is -15 at $(-\sqrt{5}, -\sqrt{6}, -\sqrt{\frac{15}{2}}), (-\sqrt{5}, \sqrt{6}, \sqrt{\frac{15}{2}}), (\sqrt{5}, -\sqrt{6}, \sqrt{\frac{15}{2}})$, and $(\sqrt{5}, \sqrt{6}, -\sqrt{\frac{15}{2}})$ **8.**‡ $\overrightarrow{\text{grad}}(x^2 + 2x + y^2) = \lambda \overrightarrow{\text{grad}}(3x^2 + 2y^2) \Rightarrow 2x + 2 = 6\lambda x, 2y = 4\lambda y;$ If $y \neq 0$, then $\lambda = \frac{1}{2}$ and $x = 2$ so that $3x^2 + 2y^2 = 48$ implies that $y = \pm\sqrt{18}$; If $y = 0$, then $3x^2 + 2y^2 = 48$ implies that $x = \pm4$; The maximum is 26 at $(2, \sqrt{18})$ and $(2, -\sqrt{18})$; The minimum is 8 at $(-4, 0)$
10. Maximum $= 20$ at $(-2, 4)$; Minimum $= 0$ at $(0, 0)$ **12.**‡ x^2y^2 has an absolute minimum of 0 along the x- and y-axes, where it has critical points; Since it has no other critical points, the maximum must occur somewhere on

the boundary $x^2 + 4y^2 = 24$ at points where x and y are not zero;
$\overrightarrow{\mathrm{grad}}(x^2 y^2) = \lambda \overrightarrow{\mathrm{grad}}(x^2 + 4y^2) \Rightarrow y^2 = \lambda, x^2 = 4\lambda; \lambda = 3;$ The maximum is
36 at $(\sqrt{12}, \sqrt{3}), (-\sqrt{12}, \sqrt{3}), (\sqrt{12}, -\sqrt{3}),$ and $(-\sqrt{12}, -\sqrt{3})$
13. Maximum $= 8$ at $(2, 0)$; Minimum $= -1$ at $(-1, 0)$
20. He would buy approximately 2 acre-feet of water and 16 pounds of fertilizer
for approximately $88 per acre   **21.** If $Y(x, y)$ is the yield and k the constant
selling price, then the profit is $P(x, y) = kY(x, y) - C(x, y)$;
$\overrightarrow{\mathrm{grad}}\,P = k\,\overrightarrow{\mathrm{grad}}\,Y - \overrightarrow{\mathrm{grad}}\,C$; Since $\overrightarrow{\mathrm{grad}}\,C$ is not $\mathbf{0}$, P and Y do not have the
same critical points; If the maximum of Y occurs inside the triangle (beneath
its upper boundary $20x + 3y = c$), then it occurs at a local maximum of Y,
which cannot be a local maximum of P. Because $\overrightarrow{\mathrm{grad}}\,C$ is perpendicular to the
upper boundary, $\overrightarrow{\mathrm{grad}}\,P$ is perpendicular to it whenever $\overrightarrow{\mathrm{grad}}\,Y$ is; Hence, if the
maximum of Y occurs on the upper boundary, then the maximum of P for
(x, y) on the upper boundary probably occurs at the same point, and that is
probably the maximum of P in the entire triangle.   **22.** Figure 14.3.22;
Maximum ≈ 1550; Minimum ≈ 400 **25.** f has a critical point on the
constraint curve or surface

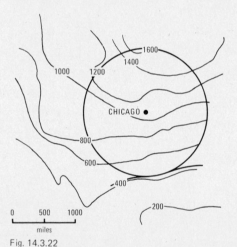

Fig. 14.3.22

Miscellaneous Exercises to Chapter 14

1. Local minima at $(0, -\frac{1}{2}\pi + 2n\pi)$; Saddle points at $(0, \frac{1}{2}\pi + 2n\pi)$
(integers n)   **5.** Maximum $= 12$; Minimum $= 0$ **8.** Fig. 14.M.8; Maximum
$= 24$; Minimum $= \frac{7}{2}$; The maximum occurs at a corner of the square where
the boundary is not a level curve of a function with nonzero gradient

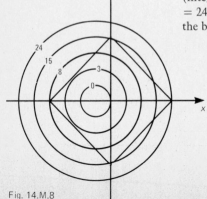

Fig. 14.M.8

12. $(1, 4)$ **14.** The tangent line to the intersection of the constraint surfaces is perpendicular to $\overrightarrow{\text{grad}}\, g$ and $\overrightarrow{\text{grad}}\, h$; At the maximum or minimum, $\overrightarrow{\text{grad}}\, f$ is perpendicular to the tangent line and hence can be expressed as a linear combination of $\overrightarrow{\text{grad}}\, g$ and $\overrightarrow{\text{grad}}\, h$.   **15.**‡ $\overrightarrow{\text{grad}}\,(x^2 + y^2 + z^2)$
$= \lambda\,\overrightarrow{\text{grad}}\,(x^2 + y^2 - z) + \mu\,\overrightarrow{\text{grad}}\,(x + 2y + 2z)$; $2x = 2\lambda x + \mu$,
$2y = 2\lambda y + 2\mu$, $2z = -\lambda + 2\mu$; $y = 2x$; The constraint conditions give
$x = -1, \frac{1}{2}$; The maximum distance is $\sqrt{30}$ at $(-1, -2, 5)$; The minimum
distance is $\sqrt{\frac{45}{16}}$ at $(\frac{1}{2}, 1, \frac{5}{4})$ **16.** Maximum $= 31$; Minimum $= 11$
17. $\overrightarrow{\text{grad}}\,(Ax^2 + 2Bxy + Cy^2) = \lambda\,\overrightarrow{\text{grad}}\,(x^2 + y^2) \Rightarrow (*)\, Ax + By = \lambda x$ and
$(**)\, Bx + Cy = \lambda y$; Eliminating x and y from $(*)$ and $(**)$ gives
$(A - \lambda)(C - \lambda) = B^2$, which is the determinant equation; Multiplying $(*)$ by
x and $(**)$ by y and adding gives $Ax^2 + 2Bxy + Cy^2 = \lambda(x^2 + y^2) = \lambda$
18. An ellipsoid of revolution (obtained by rotating an ellipse about its major
axis) with foci at the elephant and the camel

Section 15.1

1.‡ Fig. 15.1.1; $\int_0^2 \left\{ \int_{x^2}^{2x} 4x^3 y\, dy \right\} dx = \int_0^2 [2x^3 y^2]_{y=x^2}^{y=2x}\, dx$
$= \int_0^2 (8x^5 - 2x^7)\, dx = \frac{64}{3}$ **2.** $\frac{7}{6}$ **4.**‡ Fig. 15.1.4; $\int_{-3}^4 \left\{ \int_0^5 e^x \sin y\, dy \right\} dx$
$= \int_{-3}^4 [-e^x \cos y]_{y=0}^{y=5}\, dx = (1 - \cos(5))(e^4 - e^{-3})$ **5.** $\frac{1}{15} 2^{17}$ **8.**‡ Fig. 15.1.8;
$\int_{-1}^1 \left\{ \int_{y^2}^{2-y^2} y^2\, dx \right\} dy = \int_{-1}^1 [xy^2]_{x=y^2}^{x=2-y^2}\, dy = \int_{-1}^1 (2y^2 - 2y^4)\, dy = \frac{8}{15}$
9. $e^4 - 1$ **10.**‡ Fig. 15.1.10; $\int_0^1 \left\{ \int_y^{4y} x^{1/2} y^{1/2}\, dx \right\} dy = \int_0^1 [\frac{2}{3} x^{3/2} y^{1/2}]_{x=y}^{x=4y}\, dy = \frac{14}{9}$
11. $4\sin(2) - 4\sin(1)$ **15.** $\frac{1}{6}$ **16.** $2 - \sqrt{2} + 2(e^{1/\sqrt{2}} - e)$ **18.**‡ Fig. 15.1.18;
$\int_0^1 \left\{ \int_x^{3x} x^2\, dy \right\} dx + \int_1^2 \left\{ \int_x^{4-x} x^2\, dy \right\} dx = \int_0^1 2x^3\, dx + \int_1^2 (4x^2 - 2x^3)\, dx = \frac{7}{3}$

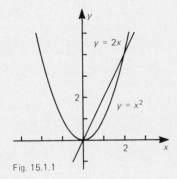

Fig. 15.1.1

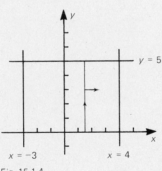

Fig. 15.1.4

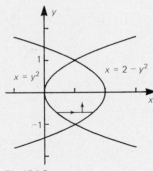

Fig. 15.1.8

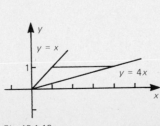

Fig. 15.1.10

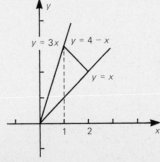

Fig. 15.1.18

20. $\int_0^1 \left\{ \int_{y/3}^{3y} f(x, y)\, dx \right\} dy + \int_1^3 \left\{ \int_{y/3}^{3/y} f(x, y)\, dx \right\} dy$ or

$\int_0^1 \left\{ \int_{x/3}^{3x} f(x, y)\, dy \right\} dx + \int_1^3 \left\{ \int_{x/3}^{3/x} f(x, y)\, dy \right\} dx$ **21.‡** Fig. 15.1.21;

$\int_0^4 \left\{ \int_{-\sqrt{y}}^{\sqrt{y}} f(x, y)\, dx \right\} dy$ **22.** $\int_{-1}^1 \left\{ \int_{y^{1/3}}^1 f(x, y)\, dx \right\} dy$ **25.** $\int_{-1}^3 \left\{ \int_x^3 f(x, y)\, dy \right\} dx$

27. $\frac{1}{2}(1 - e^{-16})$ **29.‡** Fig. 15.1.29;

$\int_0^1 \left\{ \int_{-\sqrt{y}}^{\sqrt{y}} f(x, y)\, dx \right\} dy + \int_1^4 \left\{ \int_{y-2}^{\sqrt{y}} f(x, y)\, dx \right\} dy$

30. $\int_0^1 \left\{ \int_{x/2}^x f(x, y)\, dy \right\} dx + \int_1^2 \left\{ \int_{x/2}^1 f(x, y)\, dy \right\} dx$

33. $\int_a^b \left\{ \int_c^d F(x) G(y)\, dy \right\} dx = \int_a^b \left\{ F(x) \int_c^d G(y)\, dy \right\} dx$

$= \left\{ \int_c^d G(y)\, dy \right\} \left\{ \int_a^b F(x)\, dx \right\}$ **34.** 4 **35.** $\frac{49}{2} \ln(10)$

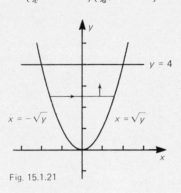

Fig. 15.1.21

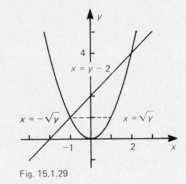

Fig. 15.1.29

Section 15.2

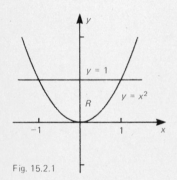

Fig. 15.2.1

1.‡ Fig. 15.2.1; $\iint_R (x^2 + y^2 + 1)\, dx\, dy = \int_{-1}^1 \left\{ \int_{x^2}^1 (x^2 + y^2 + 1)\, dy \right\} dx = \frac{76}{35}$

2. $4 + 2\pi^2$ **4.‡** $\int_0^4 \left\{ \int_0^5 (x^2 + z^2 + 3)\, dz \right\} dx = \frac{1000}{3}$ **6.‡** Fig. 15.2.6;

$\iint_R (6 - 3x - 2y)\, dx\, dy = \int_0^2 \left\{ \int_0^{3-3x/2} (6 - 3x - 2y)\, dy \right\} dx = 6$ **7.** $\frac{16}{3}$

8.‡ The plane has the equation $2x - 2y + z = 4$; Its intersection with the plane $z = -2$ has the simultaneous equations $x - y = 3, z = -2$, so the projection of the solid on the xy-plane is the triangle R in Fig. 15.2.8;

$\iint_R [(2y - 2x + 4) - (-2)]\, dx\, dy = \int_0^3 \left\{ \int_{x-3}^0 (2y - 2x + 6)\, dy \right\} dx = 9$

11.‡ $\int_0^1 \left\{ \int_0^{\sqrt{1-x^2}} xy\, dy \right\} dx = \frac{1}{8}$ **12.** $\frac{8}{3}$ **14.‡** $\bar{x} = 8 \int_0^1 \left\{ \int_0^{\sqrt{1-x^2}} x^2 y\, dy \right\} dx = \frac{8}{15}$;

$\bar{y} = \bar{x}$ by symmetry; $(\frac{8}{15}, \frac{8}{15})$ **15.** $(\frac{5}{4}, \frac{9}{16})$

17. $\int_0^{3\pi} \left\{ \int_0^{3\pi} \sqrt{1 + \cos^2 x\, \sin^2 y + \cos^2 y\, \sin^2 x}\, dy \right\} dx$

18. $\int_{-1}^1 \left\{ \int_{-\sqrt{1-x^4}}^{\sqrt{1-x^4}} \sqrt{1 + 16x^6 + 4y^2}\, dy \right\} dx$ **20a.** $2Rh \arcsin(k/R)$

21. Let (x_j, y_j) be in R_j where $R_1, R_2, \ldots, R_N$ is a partition of R; The weight of the portion of the plate in R_j is approximately $\rho(x_j, y_j)[\text{Area of } R_j]$ and its moment of inertia about (x_0, y_0) is approximately

$[(x_j - x_0)^2 + (y_j - y_0)^2] \rho(x_j, y_j)[\text{Area of } R_j]$; The sum of these quantities is the approximate moment of inertia of the plate and is an approximating sum for the integral in the statement of the problem **22.‡** Fig. 15.2.22;

$\iint_R (x^2 + y^2)(15xy^2)\, dx\, dy = 15 \int_0^1 \left\{ \int_{x^2}^x (x^3 y^2 + xy^4)\, dy \right\} dx = \frac{11}{28}$ **23.** $\frac{4}{5}$

25a. $(a^3 b + b^3 a)/3$

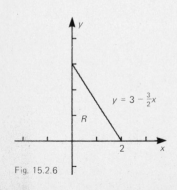

Fig. 15.2.6

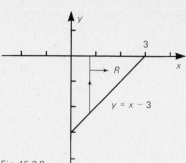

Fig. 15.2.8

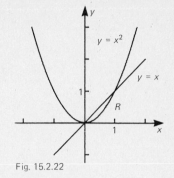

Fig. 15.2.22

Section 15.3

1.‡ $\int_0^{2\pi}\left\{\int_{\sqrt 2}^2 r^{-3}\,dr\right\}d\Theta=\tfrac14\pi$ **2.** $\pi(1-e^{-16})$ **4.**‡ Fig. 15.3.4;

$\int_{-\pi/2}^{\pi/2}\left\{\int_{2\cos\theta}^{4\cos\theta} r^3\,dr\right\}d\Theta=\tfrac{45}2\pi$ **5.**‡ Figure 15.3.5; Because we want nonnegative values for r, we consider $-\tfrac16\pi\le\Theta\le\tfrac16\pi$, which gives us one petal of the rose;

$3\int_{-\pi/6}^{\pi/6}\left\{\int_0^{\cos(3\Theta)} r\,dr\right\}d\Theta=\tfrac14\pi$ **7.** $\tfrac{16}3\pi$ **10.**‡ Fig. 15.3.10;

$\int_{-\pi/2}^{\pi/2}\left\{\int_0^{2\cos\theta} 2r\sqrt{4-r^2}\,dr\right\}d\Theta=\tfrac{16}3\pi-\tfrac{64}9$ **12.**‡ Fig. 15.3.12;

$\int_0^\pi\left\{\int_0^{2\sin\Theta}(2r\sin\Theta-r^2)r\,dr\right\}d\Theta=\tfrac12\pi$

13. $\int_1^2\left\{\int_0^{\sqrt{1-(x-1)^2}}\dfrac{\sqrt{x^2+y^2}}{1+y}\,dy\right\}dx$ or $\int_0^1\left\{\int_1^{1+\sqrt{1-y^2}}\dfrac{\sqrt{x^2+y^2}}{1+y}\,dx\right\}dy$

17. Let $k\to R^-$; $2\pi R^2$ **21.** The area of the ring between the circles $r=r_{j-1}$ and $r=r_j$ is $\pi(r_j^2-r_{j-1}^2)$; The portion for $\Theta_{k-1}\le\Theta\le\Theta_k$ is the fraction $\Delta\Theta/2\pi$ of the entire ring and has area $\tfrac12(r_j^2-r_{j-1}^2)\Delta\Theta$

$=\tfrac12(r_j+r_{j-1})(r_j-r_{j-1})\Delta\Theta=\tfrac12(r_j+r_{j-1})\Delta r\Delta\Theta$

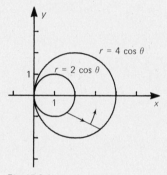

Fig. 15.3.4

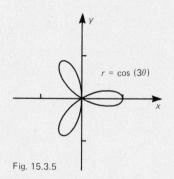

Fig. 15.3.5

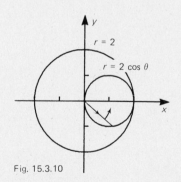

Fig. 15.3.10

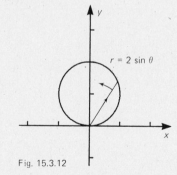

Fig. 15.3.12

Section 15.4

1.‡ $\int_{-1}^1\left\{\int_{-1}^1\left\{\int_0^{\sqrt{x^2+y^2}} z\,dz\right\}dy\right\}dx=\int_{-1}^1\left\{\int_{-1}^1\tfrac12(x^2+y^2)\,dy\right\}dx=\tfrac43$ **2.** $\tfrac{64}{21}$

4.‡ $\int_{-1}^1\left\{\int_{x^2}^1\left\{\int_0^y 15y^{1/2}z^{1/2}\,dz\right\}dy\right\}dx=\int_{-1}^1\left\{\int_{x^2}^1 10y^2\,dy\right\}dx=\tfrac{40}7$

5. $\tfrac13(4-\sin(4))(e^{18}-1)$ **7.** $\tfrac4{525}$ **8.** $\int_{a_1}^{b_1}\int_{a_2}^{b_2}\int_{a_3}^{b_3} f(x)g(y)h(z)\,dz\,dy\,dx$

$=\int_{a_1}^{b_1}\int_{a_2}^{b_2}\left\{f(x)g(y)\int_{a_3}^{b_3} h(z)\,dz\right\}dy\,dx=\left\{\int_{a_3}^{b_3} h(z)\,dz\right\}\int_{a_1}^{b_1}\int_{a_2}^{b_2} f(x)g(y)\,dy\,dx$

$=\left\{\int_{a_1}^{b_1} f(x)\,dx\right\}\left\{\int_{a_2}^{b_2} g(y)\,dy\right\}\left\{\int_{a_3}^{b_3} h(z)\,dz\right\}$ by Exercise 33 of Section 15.1

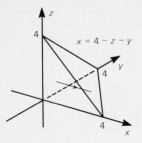

Fig. 15.4.21a

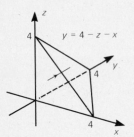

Fig. 15.4.21b

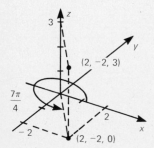

Fig. 15.5.1

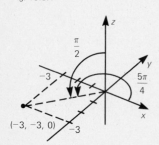

Fig. 15.5.4a

9.[‡] $\int_0^3 x\,dx \int_1^5 \sin y\,dy \int_{-3}^0 z^2\,dz = [\frac{9}{2}][\cos(1) - \cos(5)][9]$
$= \frac{81}{2}[\cos(1) - \cos(5)]$ **10.** $\frac{1}{24}(e^8 - e^{-8})^2(e^9 - e^{-9})$ **12.**[‡] (x,y) varies over the circle $x^2 + y^2 \le 1$ in the xy-plane; For each such (x,y), z varies from -5 to $x^2 - y^2$; The solid is bounded by the cylinder $x^2 + y^2 = 1$, the plane $z = -5$, and the hyperbolic paraboloid $z = x^2 - y^2$ **13.** The portion of the sphere $x^2 + y^2 + z^2 \le 4$ where x, y, and z are nonnegative **15.**[‡] (y,z) varies over the triangle bounded by the y-axis, the z-axis and the line $z = 2 - 2y$ in the yz-plane; For each such (y,z), x varies from 0 to $4 - 4y - 2z$; The line $z = 2 - 2y$ is the intersection of the yz-plane with the plane $x = 4 - 4y - 2z$; The solid is the tetrahedron bounded by $x = 4 - 4y - 2z$ and the coordinate planes **16.** The tetrahedron bounded by the coordinate planes and the plane $x + y - z = 2$ **18.**[‡] Weight $= \int_0^2 \int_{-1}^1 \int_{y^2}^1 (z + 2)\,dz\,dy\,dx = \frac{104}{15}$;
$\bar{x} = \frac{15}{104} \int_0^2 \int_{-1}^1 \int_{y^2}^1 x(z + 2)\,dz\,dy\,dx = 1$; $\bar{y} = \frac{15}{104} \int_0^2 \int_{-1}^1 \int_{y^2}^1 y(z + 2)\,dz\,dy\,dx = 0$; $\bar{z} = \frac{15}{104} \int_0^2 \int_{-1}^1 \int_{y^2}^1 z(z + 2)\,dz\,dy\,dx = \frac{57}{91}$
19. $\bar{x} = \bar{y} = \frac{11}{15}, \bar{z} = \frac{7}{12}$ **21.**[‡] Fig. 15.4.21a: $\int_0^4 \int_0^{4-z} \int_0^{4-z-y} f(x,y,z)\,dx\,dy\,dz$ or Fig. 15.4.21b: $\int_0^4 \int_0^{4-z} \int_0^{4-z-x} f(x,y,z)\,dy\,dx\,dz$ **22.** $\int_0^3 \int_{-2}^2 \int_{x^3}^8 f(x,y,z)\,dy\,dx\,dz$ or $\int_0^3 \int_{-8}^8 \int_{-2}^{\sqrt[3]{y}} f(x,y,z)\,dx\,dy\,dz$

25. $\frac{1016}{1155}$ **26.** $18(1 - e^{-4})$ **27.** $\frac{1}{60}$ **28.** 8 **29.** $\frac{1}{36}(e^{12} - 85)$ **30.** $\frac{1}{16}\ln(\frac{4}{3})$
31. $\frac{1}{12}(e - 1)$

Section 15.5

1a.[‡] $r = \sqrt{(2)^2 + (-2)^2} = 2\sqrt{2}$; Fig. 15.5.1; $\Theta = \frac{7}{4}\pi$; $[2\sqrt{2}, \frac{7}{4}\pi, 3]$
1b. $[2, \frac{7}{6}\pi, -5]$ **2a.**[‡] $x = 3\cos(\frac{5}{6}\pi) = -\frac{3}{2}\sqrt{3}$; $y = 3\sin(\frac{5}{6}\pi) = \frac{3}{2}$; $(-\frac{3}{2}\sqrt{3}, \frac{3}{2}, 4)$ **2b.** $(0, -1, -2)$ **3a.**[‡] $x = 4\sin(\frac{1}{3}\pi)\cos(\frac{3}{4}\pi)$ $= 4(\frac{1}{2}\sqrt{3})(-\frac{1}{2}\sqrt{2}) = -\sqrt{6}$; $y = 4\sin(\frac{1}{3}\pi)\sin(\frac{3}{4}\pi) = 4(\frac{1}{2}\sqrt{3})(\frac{1}{2}\sqrt{2}) = \sqrt{6}$; $z = 4\cos(\frac{1}{3}\pi) = 4(\frac{1}{2}) = 2$; $(-\sqrt{6}, \sqrt{6}, 2)$ **3b.** $(0, 0, 0)$ **3d.** $(0, 0, -5)$
3f. $(0, 0, \sqrt{2})$ **4a.**[‡] $\rho = \sqrt{(-3)^2 + (-3)^2 + (0)^2} = 3\sqrt{2}$; Fig. 15.5.4a; $[3\sqrt{2}, \frac{5}{4}\pi, \frac{1}{2}\pi]$ **4b.**[‡] $\rho = \sqrt{(\sqrt{2})^2 + (\sqrt{2})^2 + (-2)^2} = 2\sqrt{2}$; Fig. 15.5.4b; $\cos\phi = z/\rho = -1/\sqrt{2}$; $\phi = \frac{3}{4}\pi$; $\Theta = \frac{1}{4}\pi$ since x and y are positive and equal; $[2\sqrt{2}, \frac{1}{4}\pi, \frac{3}{4}\pi]$ **4c.** $[4, \frac{4}{3}\pi, \frac{1}{2}\pi]$ **4d.** $[100, \Theta, \pi]$ with any Θ **4f.** $[4, \frac{5}{6}\pi, \frac{5}{6}\pi]$
4h. $[2, \frac{1}{4}\pi, \frac{1}{3}\pi]$ **5a.** $x^2 + y^2 = 4$ **5b.** $y = x$ with $x \ge 0$ **5c.** $y = -\sqrt{3}x$ with $x \le 0$ **5e.** $x^2 + y^2 + z^2 = 4$ **5f.** $z = \sqrt{x^2 + y^2}$
5g. $z = -\sqrt{\frac{1}{3}(x^2 + y^2)}$ **6a.** $z = -r$ **6b.** $\Theta = \frac{1}{4}\pi$ **6c.** $z = -\sqrt{4 - r^2}$
6e. $r = 2\cos\Theta$ **6g.** $r = 3\csc\Theta$ **6i.** $z = \sqrt{3}r$ **7a.** $\phi = \frac{3}{4}\pi$ **7b.** $\Theta = \frac{1}{4}\pi$
7c. $\rho = 2$ and $\frac{1}{2}\pi \le \phi \le \pi$ **7e.** $\rho\sin\phi = 2\cos\Theta$ **7g.** $\rho\sin\phi = 3\csc\Theta$
7i. $\phi = \frac{1}{6}\pi$ **8.**[‡] $\int_0^{2\pi} \{ \int_0^1 \{ \int_0^r r^2\,dz \}\,dr \}\,d\Theta = \int_0^{2\pi} \{ \int_0^1 r^3\,dr \}\,d\Theta = \int_0^{2\pi} \frac{1}{4}\,d\Theta = \frac{1}{2}\pi$
9. $\frac{2}{15}\pi$ **11.**[‡] $\int_0^{2\pi} \{ \int_0^1 \{ \int_{r^2}^{\sqrt{2-r^2}} r\,dz \}\,dr \}\,d\Theta = \frac{4}{3}\pi\sqrt{2} - \frac{7}{6}\pi$ **12.** $\frac{4}{3}\pi(2\sqrt{2} - 1)$
14. $\frac{25}{6}\pi$

15. $\int_0^2 \{ \int_0^{\sqrt{4-x^2}} \{ \int_{-x^2-y^2}^4 \frac{z(x^2 + y^2)}{y + 4}\,dz \}\,dy \}\,dx$

17.[‡] $\int_0^{\pi} \{ \int_0^1 \{ \int_0^{\pi/2} (\rho\cos\phi)^2\,\rho^2\,\sin\phi\,d\phi \}\,d\rho \}\,d\Theta = \frac{1}{15}\pi$

18.[‡] $\int_0^{2\pi} \{ \int_0^2 \{ \int_0^{5\pi/6} \rho^3\sin\phi\,d\phi \}\,d\rho \}\,d\Theta = 8\pi + 4\pi\sqrt{3}$ **20.** $\frac{8}{15}\pi k R^5$

21. $\int_{-1/\sqrt{2}}^{1/\sqrt{2}} \int_0^{\sqrt{(1/2)-x^2}} \int_{-\sqrt{1-x^2-y^2}}^{-\sqrt{x^2+y^2}} x(x^2 + y^2 + z^2)\,dz\,dy\,dx$

23. $\int_0^{\pi} \int_0^{\pi/2} \int_0^5 \rho^3\sin^2\phi\sin\Theta\,d\rho\,d\phi\,d\Theta$ **25.** $(\frac{32}{35}, -\frac{32}{35}, -\frac{32}{35})$ **28.** $\frac{1}{4}\pi$
30a. See the solution of Exercise 21 of Section 15.3 with r replaced by ρ and Θ replaced by ϕ

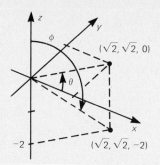

Fig. 15.5.4b

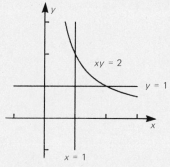

Fig. 15.M.11

30b. The vertical line through the center of gravity of the region must intersect the region; Let $[\rho_j^*, 0, \phi_n^*]$ be the spherical coordinates of any point on that line and in the region **30c.** Volume = [Area][Distance traveled by the center of gravity] = $[\frac{1}{2}(\rho_j + \rho_{j-1})\Delta\rho\Delta\phi][\rho_j^* \sin(\phi_n^*)\Delta\Theta]$ **30d.** The sum of the volumes from part (c) multiplied by $f(\rho_j \sin(\phi_n)\cos(\Theta_k),$ $\rho_j\sin(\phi_n)\sin(\Theta_k), \rho_j\cos(\phi_n))$ is an approximating sum for the integral **31.**‡ Spherical coordinates: $[4000, \frac{1}{6}\pi, \frac{1}{6}\pi]$; Rectangular coordinates: $(1000\sqrt{3}, 1000, 2000\sqrt{3})$ **32.** $(0, -2000\sqrt{2}, 2000\sqrt{2})$ **34.**‡ Timimoun has spherical coordinates $[4000, 0, \frac{1}{3}\pi]$ and Bennett has spherical coordinates $[4000, -\frac{3}{4}\pi, \frac{1}{6}\pi]$; The unit vector $\mathbf{A} = \langle \sin(\frac{1}{3}\pi)\cos(0), \sin(\frac{1}{3}\pi)\sin(0),$ $\cos(\frac{1}{3}\pi)\rangle = \langle \frac{1}{2}\sqrt{3}, 0, \frac{1}{2}\rangle$ points from the center of the earth toward Timimoun; The unit vector $\mathbf{B} = \langle \sin(\frac{1}{6}\pi)\cos(-\frac{3}{4}\pi), \sin(\frac{1}{6}\pi)\sin(-\frac{3}{4}\pi),$ $\cos(\frac{1}{6}\pi)\rangle = \frac{1}{2}\langle -1/\sqrt{2}, -1/\sqrt{2}, \sqrt{3}\rangle$ points from the center of the earth toward Bennett; The angle α between $\mathbf{A}$ and $\mathbf{B}$ is given by $\cos\alpha = \mathbf{A} \cdot \mathbf{B}$ $= \frac{1}{4}\sqrt{3} - \frac{1}{8}\sqrt{6} \approx 0.217$; $\alpha \approx 1.44$; The great circle distance is the length of the arc subtended by the angle α on a circle of radius 4000; 5760 miles **35.** 10,600 miles **36.** 9200 miles **38.** $\frac{256}{9}$ **39.** $\frac{3716}{15}\pi$ **40.** $-\frac{243}{10}$ **41.** $\frac{3}{4}$ **42.** $\frac{4}{3}\pi[\sin(216) - \sin(1)]$ **43.** $\frac{1}{8}\pi - \frac{1}{32}\pi\sin(4)$ **44.** $\frac{128}{7}\pi$

Miscellaneous Exercises to Chapter 15

1. $\int_{-\pi/4}^{\pi/4}\int_{1}^{\tan y} f(x,y)\,dx\,dy$ **4.** $\int_{0}^{4}\int_{2-\sqrt{4-y}}^{2+\sqrt{4-y}} f(x,y)\,dx\,dy$ **5.** $\frac{1}{3}\sqrt{2}$ $+ \frac{1}{6}\ln(\sqrt{2}+1) - \frac{1}{6}\ln(\sqrt{2}-1)$ **6.** 38π **8.** $(2\sqrt{2}+2)/15$ **9.** $\frac{91}{12}$ **11.**‡ Fig. 15.M.11; $\int_{1}^{2}\left\{\int_{1}^{2/x}\left\{\int_{xy}^{2} z\,dz\right\}dy\right\}dx = \frac{8}{3}\ln(2) - \frac{29}{18}$ **13.** $\frac{1}{2}\pi$ **15.** $2\pi(5e^6 + 7e^{-6})$ **18.** $2\pi(\frac{1}{3}R^3 + \frac{1}{6}k^3 - \frac{1}{2}R^2k)$

20b. $\int_{0}^{\pi}\frac{\cos\alpha}{R^2}\sin\phi\,d\phi = \int_{b-\rho}^{b+\rho}\frac{1}{R^2\rho b}[b - (\rho^2 + b - R^2)/2b]\,dR = 2b^{-2}$

23. $\frac{1}{4}\pi - \frac{1}{3}$

Section 16.1

1.‡ $dx = 3t^2\,dt$; $\int_{0}^{3}(t^3 + t^2)(3t^2\,dt) = \frac{5103}{10}$ **2.** $-\frac{4}{5}\sinh(5)$ **4a.**‡ $x = 1 + 3t,$ $y = 1 + 4t, 0 \underset{t}{\to} 1$; $\int_{0}^{1}(1+3t)(4\,dt) = 10$

4b.‡ $\int_{0}^{1}(1+3t)(1+4t)(\sqrt{9+16}\,dt) = \frac{85}{2}$ **5a.** $\frac{1}{5}\sqrt{85}(1-e^{-5})$ **5b.** $\frac{6}{7}[\cos(7) - 1]$ **7a.**‡ $\int_{1}^{-1} t^3t^4(3t^2\,dt) = 0$ **7b.**‡ $\int_{1}^{-1} t^3t^4(2t\,dt) = -\frac{4}{9}$ **7c.**‡ $\int_{-1}^{1} t^3t^4(\sqrt{9t^4 + 4t^2}\,dt) = 0$ because the integrand is odd **9a.**‡ $x = \cos t,$ $y = \sin t, \pi \underset{t}{\to} 0$; $\int_{\pi}^{0}(\cos t)(-\sin t\,dt) = 0$ **9b.**‡ $\int_{\pi}^{0}(\cos t)(\cos t\,dt) = -\frac{1}{2}\pi$ **9c.**‡ $\int_{0}^{\pi}\cos t\,dt = 0$ **11a.** $3\int_{\pi}^{0}\cos(3t)\sin[\sin t\sin(2t)]\,dt$ **11b.** $\int_{0}^{\pi} e^{\sin(3t)}\sqrt{\cos^2 t + 4\cos^2(2t) + 9\cos^2(3t)}\,dt$ **13a.**‡ $\int_{C} f\,dx \approx 8\int_{C}dx$ $= 8\{[x \text{ at the end}] - [x \text{ at the beginning}]\} \approx 8(-6 - 8) = -112$; -100 is the closest **13b.**‡ $\int_{C} f\,dy \approx 8\int_{C}dx = 8\{[y \text{ at the end}] - [y \text{ at the beginning}]\}$ $\approx 8(-2 - (-3)) = 8$; 10 is the closest **13c.**‡ $\int_{C} f\,ds \approx 8\int_{C}ds$ $= 8[\text{The length of } C] \approx 8(25) = 200$ **14a.** 20 **14b.** 10 **14c.** -400 **16a.**‡ $C = C_1 - C_2 - C_3$ **16b.**‡ $\int_{C} f\,dx = \int_{C_1} f\,dx - \int_{C_2} f\,dx - \int_{C_3} f\,dx$; 3 **16c.**‡ $\int_{C} f\,ds = \int_{C_1} f\,ds + \int_{C_2} f\,ds + \int_{C_3} f\,ds$; 10 **19.** For any partition $a = t_0 < t_1 < t_2 < \cdots < t_N = b$ the approximating sum $\sum_{j=1}^{N} f(x(t_j), y(t_j))\sqrt{[x(t_j) - x(t_{j-1})]^2 + [y(t_j) - y(t_{j-1})]^2}$ for the line

integral on the left of (15) can be expressed in the form
$\sum_{j=1}^{N} f(x(t_j), y(t_j)) \sqrt{[x'(c_j)]^2 + [y'(d_j)]^2} (t_j - t_{j-1})$ of an approximating
sum for the definite integral on the right of (15); Here, by the Mean Value
Theorem for derivatives, we have $x(t_j) - x(t_{j-1}) = x'(c_j)(t_j - t_{j-1})$ and
$y(t_j) - y(t_{j-1}) = y'(d_j)(t_j - t_{j-1})$ with $t_{j-1} < c_j, d_j < t_j$ **20.** -24
22. $\frac{3}{2}(5^{7/2} - 2^{7/2})/(5^{3/2} - 2^{3/2})$ **23.** 0 **24.** $\frac{2}{3}(e^{-1} - 1)$
25. $\frac{9}{5}[\cos(7) - \cos(2)] - \frac{5}{9}[\sin(-5) - \sin(4)]$ **26.** $\frac{1}{3}(1 - e^9) - \frac{1}{2}\sin^2(1) - \frac{1}{4}$
27. $-\frac{1}{3}$ **28.** $-5\sqrt{74}$ **29.** $\frac{1}{3}(26^{3/2} - 2^{3/2})$ **30.** $2\sqrt{5}[(\sin^2(10) + 4)^{3/2} - 8]$

Section 16.2

1. Fig. 16.2.1   **3.** Fig. 16.2.3   **5.** Fig. 16.2.5   **7.** Fig. 16.2.7   **10.** A gradient field;
$U = x \sin(3y) + \frac{1}{2}x^2 - \frac{1}{2}y^2 + C$ **12.** A gradient field; $U = \ln|1 + xy| + C$
14. A gradient field; $U = e^{xy} - x + C$ **16.**‡ Set $p = yz + y + 1$,
$q = xz + x + z, r = xy + y - 2; p_y = z + 1 = q_x, p_z = y = r_x$,
$q_z = x + 1 = r_y$; A gradient field; $U = \int (yz + y + 1) \, dx$
$= xyz + xy + x + \phi(y, z); U_y = xz + x + \phi_y(y, z); \phi_y = z$;
$\phi = \int z \, dy = yz + \psi(z); U = xyz + xy + x + yz + \psi(z)$;
$U_z = xy + y + \psi'(z); \psi' = -2; \psi = -2z + C$;
$U = xyz + xy + x + yz - 2z + C$ **17.** A gradient field;
$U = \sin(x + z) + e^{yz} + x + C$ **18.** Not a gradient field
20. Not path-independent   **22.** Path-independent; $55 - e^2 + \sin(3)$
24. Not path-independent   **26.** -25 **28.** Yes; $U = \sin(xyz); \sin(24)$ **29.** No
31a. $\frac{17}{35}$ **31b.** $\frac{1}{2}$ **31c.** $\frac{1}{3}$

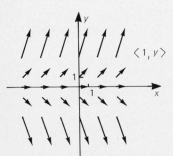

$\langle 1, y \rangle$

Fig. 16.2.1

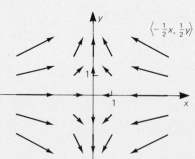

$\langle -\frac{1}{2}x, \frac{1}{2}y \rangle$

Fig. 16.2.3

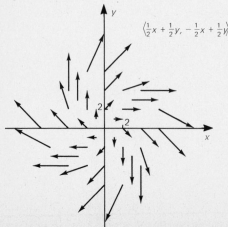

$\langle \frac{1}{2}x + \frac{1}{2}y, -\frac{1}{2}x + \frac{1}{2}y \rangle$

Fig. 16.2.5

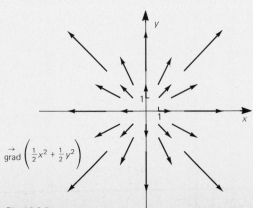

$\overrightarrow{\text{grad}}\left(\frac{1}{2}x^2 + \frac{1}{2}y^2\right)$

Fig. 16.2.7

Fig. 16.2.32a

Fig. 16.2.32b

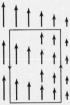

Fig. 16.3.23

32. The force fields in Figures 16.27 and 16.28; Consider closed curves as in Figs. 16.2.32a, b; The work done on an object that traverses either curve is positive   **36.** Suppose first that the line integral is path independent; Let C be any closed curve and let P and Q be two points on it; Let C_1 be the portion of C from P to Q and C_2 the portion from Q to P; $-C_2$ runs from P to Q and $C = C_1 + C_2$; $\int_C (p\,dx + q\,dy) = \left\{ \int_{C_1} + \int_{C_2} \right\}(p\,dx + q\,dy)$
$= \left\{ \int_{C_1} - \int_{-C_2} \right\}(p\,dx + q\,dy) = 0$. Suppose the line integral is zero for closed curves and C_1 and C_2 both run from P to Q; $C = C_1 - C_2$ is closed and hence $\int_{C_1}(p\,dx + q\,dy) - \int_{C_2}(p\,dx + q\,dy) = \int_C (p\,dx + q\,dy) = 0$.

Section 16.3

2.‡ $\displaystyle\iint_R \left[-\frac{\partial}{\partial y}(xy^3) + \frac{\partial}{\partial x}(-2x^3y^2) \right] dx\,dy = \iint_R (-3xy^2 - 6x^2y^2)\,dx\,dy$

3. $\iint [-x\cos(xy) - y\sin(xy)]\,dx\,dy$ **6.‡** The loop oriented counterclockwise has parametric equations $x = t^2 - 3$, $y = \frac{1}{3}t^3 - t$, $-\sqrt{3} \xrightarrow{t} \sqrt{3}$; Area $= \int_C x\,dy = \int_{-\sqrt{3}}^{\sqrt{3}} (t^2 - 3)(t^2 - 1)\,dt = \frac{8}{5}\sqrt{3}$ **7.** $\frac{4}{3}$

10. $\langle -\cos y, \sin x \rangle$ **12a.** $y\cos(xy) + \sin(x - y)$ **12b.** $-x\cos(xy)$ $- \sin(x - y)$ **16.** -50 **18.** Reversing the orientation of C replaces $\mathbf{T}$ by $-\mathbf{T}$ and $\mathbf{n}$ by $-\mathbf{n}$ and does not change ds; Or: $\int_{-C} \mathbf{F} \cdot \mathbf{T}\,ds$
$= \int_{-C}(p\,dx + q\,dy) = -\int_C (p\,dx + q\,dy) = -\int_C \mathbf{F} \cdot \mathbf{T}\,ds$ and
$\int_{-C} \mathbf{F} \cdot \mathbf{n}\,ds = \int_{-C}(-q\,dx + p\,dy) = -\int_C (-q\,dx + p\,dy) = -\int_C \mathbf{F} \cdot \mathbf{n}\,ds$ since reversing the orientation of C replaces dx by $-dx$ and dy by $-dy$
19. -20 **21.** -50 **23.‡** Positive because the flux across a closed curve as in Fig. 16.3.23 is positive   **24.** Negative
27. div $= \sin y - \cos x$; curl $= -x\cos y + y\sin x$ **28.** div $= y^3 + 2y$; curl $= -3xy^2 - 3x^2$ **29.** div $= \ln y + \ln x$; curl $= -\dfrac{x}{y} + \dfrac{y}{x}$
30. div $= 2e^{xy} + 2xye^{xy} + 1 - 2y$; curl $= -x^2e^{xy} + y^2e^{xy}$
31. div $= 2(y - x)^3$; curl $= 0$ **32.** div $= 0$; curl $= -(x^2 + y^2)^{-1/2}$

Miscellaneous Exercises to Chapter 16

1a. $\frac{1}{6}(1 - 10^6)$ **1b.** $\frac{1}{4}(10^4 - 1)$ **1c.** $\frac{1}{6}[(10^4 + 1)^{3/2} - 2^{3/2}]$ **2.** 1998
4. Yes; $U = xe^{2y} - x^2 + y + C$ **6.** $\mathbf{v}_1$ **7.** Irrotational $\Rightarrow$ curl $\mathbf{v} = 0$ $\Rightarrow \mathbf{v} = \overrightarrow{\text{grad}}\,U(x,y)$; Incompressible with no sources or sinks $\Rightarrow$ div $\mathbf{v}$ $= \text{div}[\overrightarrow{\text{grad}}\,U] = U_{xx} + U_{yy} = 0$

8a. Work $= \displaystyle\int_C \mathbf{F} \cdot \mathbf{T}\,ds = \int_C m\frac{d\mathbf{v}}{dt} \cdot \mathbf{T}\,ds = \int_a^b m\frac{d\mathbf{v}}{dt} \cdot \mathbf{T}\frac{ds}{dt}\,dt$

$= m\displaystyle\int_a^b \frac{d\mathbf{v}}{dt} \cdot \mathbf{v}\,dt$ since $\mathbf{v} = \dfrac{ds}{dt}\mathbf{T}$

8b. Work $= m\displaystyle\int_a^b \frac{1}{2}\frac{d}{dt}(\mathbf{v} \cdot \mathbf{v})\,dt = m\int_a^b \frac{1}{2}\frac{d}{dt}|\mathbf{v}|^2\,dt$

$= \dfrac{1}{2}m|\mathbf{v}(b)|^2 - \dfrac{1}{2}m|\mathbf{v}(a)|^2$

Fig. 16.M.11

9a. $\int_C dU = \int_a^b [U_x(x(t), y(t)) x'(t) + U_y(x(t), y(t)) y'(t)] dt$

$= \int_a^b \frac{d}{dt} U(x(t), y(t)) dt = U(x(b), y(b)) - U(x(a), y(a))$

$= U(x_1, y_1) - U(x_0, y_0)$ where $C: x = x(t), y = y(t), a \underset{t}{\to} b$

10a. Exact   **11a.** Fig. 16.M.11

11b. curl $\mathbf{A} = -\frac{\partial}{\partial y}[-y(x^2 + y^2)^{-1}] + \frac{\partial}{\partial x}[x(x^2 + y^2)^{-1}] = 0$ for

$(x, y) \neq (0, 0)$

11c. Give C the parametric equations $x = a \cos t, y = a \sin t, 0 \underset{t}{\to} 2\pi$;

$\int_C \mathbf{A} \cdot \mathbf{T} ds = \int_0^{2\pi} \langle -\frac{1}{a} \sin t, \frac{1}{a} \cos t \rangle \cdot \langle -\sin t, \cos t \rangle a \, dt = 2\pi$

since $ds = a \, dt$

13a. Suppose a cycle occurs in the time interval $a \leq t \leq b$; Let A be the area of the top of the piston and $y(t)$ its distance from the top of the cylinder; The force on the piston at time t is $P(t)A$ so the work done on the piston in one cycle is approximately *$\sum_{j=1}^N P(t_j) A[y(t_j) - y(t_{j-1})]$

$= \sum_{j=1}^N P(t_j)[V(t_j) - V(t_{j-1})]$ where $a = t_0 < t_1 < t_2 < \cdots < t_N = b$ is a partition of $a \leq t \leq b$ and $V(t) = Ay(t) + $ [a constant] is the volume in the cylinder at time t; The limit of the sum on the right of (*) is $\int_C P \, dV$

15b. The first integral in equation (1) is zero because P traverses the boundary of a region with zero area; The second integral is zero because $\Theta(a) = \Theta(b)$; The last two integrals are shown to be equal by an integration by parts, utilizing the facts that $f(a) = f(b)$, $g(a) = g(b)$, and $\Theta(a) = \Theta(b)$

16. Let R_j be the projection of Σ_j on the xy-plane, let (x_j, y_j) be a point in R_j, and set $z_j = g(x_j, y_j)$; $\sum_{j=1}^N f(x_j, y_j, z_j)$ [Area of Σ_j]

$= \sum_{j=1}^N f(x_j, y_j, z_j) \iint_{R_j} \sqrt{1 + g_x(x, y)^2 + g_y(x, y)^2} \, dx \, dy$

$= \sum_{j=1}^N f(x_j, y_j, g(x_j, y_j)) \sqrt{1 + g_x(X_j, Y_j)^2 + g_y(X_j, Y_j)^2}$ [Area of R_j]

is an approximating sum for the surface integral and for the double integral in equation (2); Here (X_j, Y_j) is a point in R_j for each j

17. $\int_{-2}^2 \int_{-1}^1 \sin(x^3 y - xy^3) \sqrt{1 + 4x^2 + 4y^2} \, dx \, dy$ **18b.** On the top of the solid, $\mathbf{n} = \langle -h_x, -h_y, 1 \rangle / \sqrt{h_x^2 + h_y^2 + 1}$ and $dS = \sqrt{h_x^2 + h_y^2 + 1} \, dx \, dy$, so $n_3 dS = dx \, dy$; On the bottom of the solid $\mathbf{n} = \langle g_x, g_y, -1 \rangle / \sqrt{g_x^2 + g_y^2 + 1}$ and $dS = \sqrt{g_x^2 + g_y^2 + 1} \, dx \, dy$ so $n_3 dS = -dx \, dy$; $n_3 = 0$ on the sides because $\mathbf{n}$ is horizontal there   **20a.** $yz + e^z - 2z$ **24a.**‡ Fig. 16.M.24;

$$\langle \frac{\partial}{\partial x}, \frac{\partial}{\partial y}, \frac{\partial}{\partial z} \rangle \times \langle xy, y^2z^2, 2x + 3z \rangle = \begin{vmatrix} \mathbf{i} & \mathbf{j} & \mathbf{k} \\ \frac{\partial}{\partial x} & \frac{\partial}{\partial y} & \frac{\partial}{\partial z} \\ xy & y^2z^2 & 2x + 3z \end{vmatrix}$$

$$= \mathbf{i} \begin{vmatrix} \frac{\partial}{\partial y} & \frac{\partial}{\partial z} \\ y^2z^2 & 2x + 3z \end{vmatrix} - \mathbf{j} \begin{vmatrix} \frac{\partial}{\partial x} & \frac{\partial}{\partial z} \\ xy & 2x + 3z \end{vmatrix} + \mathbf{k} \begin{vmatrix} \frac{\partial}{\partial x} & \frac{\partial}{\partial y} \\ xy & y^2z^2 \end{vmatrix}$$

$$= \left[\frac{\partial}{\partial y}(2x + 3z) - \frac{\partial}{\partial z}(y^2z^2) \right] \mathbf{i} - \left[\frac{\partial}{\partial x}(2x + 3z) - \frac{\partial}{\partial z}(xy) \right] \mathbf{j} + \left[\frac{\partial}{\partial x}(y^2z^2) - \frac{\partial}{\partial y}(xy) \right] \mathbf{k}$$

$$= -2zy^2 \mathbf{i} - 2\mathbf{j} - x\mathbf{k}$$

Fig. 16.M.24

$\langle -2zy^2, -2, -x \rangle$ **24b.** $\langle xz - 3z^2, ye^z - yz, 3x^2 - e^z \rangle$

26. $\overrightarrow{\text{curl}} \langle p, q, r \rangle = (r_y - q_z)\mathbf{i} - (r_x - p_z)\mathbf{j} + (q_x - p_y)\mathbf{k}$ is **0** if and only if $p_y = q_x$, $p_z = r_x$, and $q_z = r_y$; Use Theorem 16.3

Section 17.1

1. $1 + x + \dfrac{1}{2!}x^2 + \dfrac{1}{3!}x^3 + \dfrac{1}{4!}x^4$

3. $1 + x + x^2 + x^3$ **5.** $3 + \frac{1}{6}(x - 9) - \frac{1}{216}(x - 9)^2$

7. $e^{-3}\left[1 + (-3)(x - 1) + \dfrac{9}{2!}(x - 1)^2 + \dfrac{(-27)}{3!}(x - 1)^3 + \dfrac{81}{4!}(x - 1)^4 \right]$

9. $-\dfrac{1}{2!}x^2 - \dfrac{2}{4!}x^4$

11. $1 + \dfrac{1}{3}x + \left(-\dfrac{2}{9}\right)\dfrac{1}{2!}x^2 + \left(\dfrac{10}{27}\right)\dfrac{1}{3!}x^3$

13. $1 + x + \frac{1}{2}x^2$

15. $\cos(6) + (-6)\sin(6)(x - 1) + \dfrac{(-36)}{2!}\cos(6)(x - 1)^2$

$+ \dfrac{216}{3!}\sin(6)(x - 1)^3 + \dfrac{1296}{4!}\cos(6)(x - 1)^4$

17. $x + \dfrac{1}{3!}x^3 + \dfrac{1}{5!}x^5$

19. $(x - 1) + \dfrac{8}{2!}(x - 1)^2 + \dfrac{36}{3!}(x - 1)^3$

21a.‡ $e^x \approx P_4(x) = 1 + x + \dfrac{1}{2!}x^2 + \dfrac{1}{3!}x^3 + \dfrac{1}{4!}x^4$

21b.‡ $e^{1/2} \approx P_4(\frac{1}{2}) = \frac{633}{384} = 1.6484 \ldots$

21c.‡ $R_4\left(\dfrac{1}{2}\right) = \dfrac{1}{5!}\left(\dfrac{1}{2}\right)^5 e^c$ with $0 < c < \dfrac{1}{2}$ is positive, so $P_4\left(\dfrac{1}{2}\right) < e^{1/2}$

22a. $x - \dfrac{1}{3!}x^3 + \dfrac{1}{5!}x^5$

22b. $\frac{1}{10}\pi - \frac{1}{6000}\pi^3 + \frac{1}{12}(10^{-6})\pi^5 = 0.3090170 \ldots$ **22c.** $\sin(\frac{1}{10}\pi)$
$< P_5(\frac{1}{10}\pi)$ **23a.** $\frac{1}{6}\pi + (2/\sqrt{3})(x - \frac{1}{2})$ **23b.** $\frac{1}{6}\pi - 0.04/\sqrt{3} = 0.50050 \ldots$
23c. $\arcsin(0.48) > P_1(0.48)$

25a. $x + \dfrac{1}{3!}x^3 + \dfrac{1}{5!}x^5$

25b. $-\frac{141}{120} = -1.175$ **25c.** $\sinh(-1) < P_5(-1)$ **27a.** $\frac{1}{4}\pi + \frac{1}{2}(x - 1)$
27b. $\frac{1}{4}\pi - 0.05 = 0.735 \ldots$ **27c.** $\arctan(0.9) < P_1(0.9)$
30a. $\sqrt{3} + 4(x + \frac{1}{3}\pi) + 4\sqrt{3}(x - \frac{1}{3}\pi)^2$ **30b.** $3 - \frac{4}{15}\pi + \frac{4}{225}\sqrt{3}\pi^2$
$= 1.198 \ldots$ **30c.** $\tan(\frac{4}{15}\pi) < P_2(\frac{4}{15}\pi)$

31a.‡ $1 + x + \dfrac{1}{2!}x^2 + \dfrac{1}{3!}x^3 + \dfrac{1}{4!}x^4$

31b.‡ $1 + 1 + \frac{1}{2} + \frac{1}{6} + \frac{1}{24} = \frac{65}{24} = 2.7083 \ldots$

31c.‡ $R_4(1) = \dfrac{1}{5!}e^c$ with $0 < c < 1$; $\dfrac{1}{120} = \dfrac{1}{5!} < R_4(1) = \dfrac{1}{5!}e^c < \dfrac{1}{5!}(3) = \dfrac{1}{40}$

32a. $[1 - (x - \frac{1}{4}\pi) - \frac{1}{2}(x - \frac{1}{4}\pi)^2 + \frac{1}{6}(x - \frac{1}{4}\pi)^3]/\sqrt{2}$
32b. $[1 - \frac{1}{10}\pi - \frac{1}{200}\pi^2 + \frac{1}{6000}\pi^3]/\sqrt{2} = 0.45372 \ldots$

32c. $R_3\left(\frac{7}{20}\pi\right) = \frac{1}{4!}\cos(c)\left(\frac{1}{10}\pi\right)^4$ with $\frac{1}{4}\pi < c < \frac{1}{2}\pi$ because $\frac{7}{20}\pi < \frac{1}{2}\pi$;

$0 < R_3\left(\frac{7}{20}\pi\right) < \frac{1}{24}\left(\frac{1}{10}\pi\right)^4 \bigg/ \sqrt{2} = (2.86 \ldots)10^{-4}$

35a. $1 - x + x^2 - x^3$ **35b.** 0.990099 **35c.** $R_3(0.01) = (1 + c)^{-5}(0.01)^4$
with $0 < c < 1$; $(3.125)10^{-10} < R_3(0.01) < 10^{-8}$ **38a.** $\frac{1}{2} - \frac{1}{2}\sqrt{3}(x - \frac{1}{3}\pi)$
$-\frac{1}{4}(x - \frac{1}{3}\pi)^2$ **38b.** $\frac{1}{2} + \frac{1}{48}\sqrt{3}\pi - \frac{1}{2304}\pi^2 = 0.60907 \ldots$
38c. $R_2(\frac{7}{24}\pi) = -\frac{1}{6}\sin(c)(\frac{1}{24}\pi)^3$ with $\frac{1}{4}\pi < c < \frac{1}{3}\pi$ because $\frac{1}{4}\pi < \frac{7}{24}\pi$;
$-0.00032 \ldots = -\frac{1}{12}\sqrt{3}(\frac{1}{24}\pi)^3 < R_2(\frac{7}{24}\pi) < -\frac{1}{6}(1/\sqrt{2})(\frac{1}{24}\pi)^3$
$= -0.00026 \ldots$ **40a.** $1 + 60(x - 1) + 1770(x - 1)^2$ **40b.** 0.577
40c. $R_2(0.99) = -34220(c^{58})10^{-6}$ with $0 < c < 1$; $-0.03422 < R_2(0.99) < 0$
41a.[‡] If $f(x)$ has this Taylor polynomial, then $f(0) = f'(0) = \ldots$
$= f^{(n)}(0) = 1$; $f(x) = e^x$ **41b.**[‡] If $f(x)$ has this Taylor polynomial, then
$f^{(2j)}(0) = (-1)^j$ and $f^{(2j+1)}(0) = 0$ for all nonnegative integers j; $\cos x$
41d. $\sinh x$ **41f.** 2^x **43.** Fig. 17.1.43 **45a.** $\frac{459}{5040} = 0.09107 \ldots$

45b. $R_8(3) = \frac{1}{9!}\cos(c)(3^9)$ with $0 < c < 3$; $|R_8(3)| < \frac{1}{9!}(3^9) = 0.0542 \ldots$

45c. $\sin(3) - P_7(3) = R_8(3) = 0.050 \ldots$

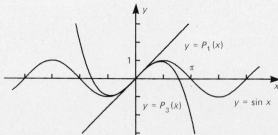

Fig. 17.1.43

Section 17.2

1. $\frac{3}{2}$ **3.** $-\frac{1}{6}$ **5.** $-\frac{6}{5}$ **6.** $\frac{1}{14}$ **8.** 0 **10.** ∞ (L'Hopital's rule does not apply)
12. 0

13.[‡] $\lim_{x \to 0} \dfrac{-2xe^{x^2}}{\sin x + x \cos x} = \lim_{x \to 0} \dfrac{-2e^{x^2} - 4x^2e^{x^2}}{2\cos x - x \sin x} = -1$ **14.** -6

16.[‡] $\lim_{x \to \infty} \dfrac{8x}{1 + e^x} = \lim_{x \to \infty} \dfrac{8}{e^x} = 0$

18. $\ln(x^x) = x \ln x$; $\lim_{x \to 0^+}(x \ln x) = \lim_{x \to 0^+} \dfrac{\ln x}{1/x} = \lim_{x \to 0^+} \dfrac{1/x}{-1/x^2}$

$= \lim_{x \to 0^+}(-x) = 0$; $\lim_{x \to 0^+} x^x = 1$
20. 0 **23.** 1 **26.**[‡] L'Hopital's rule does not apply; $\arctan(3x) \to \frac{1}{2}\pi$ and
$\arctan(-x) \to -\frac{1}{2}\pi$ as $x \to \infty$; -1 **28.** $\frac{1}{5}$ **29.** 1 **31a.** Fig. 17.2.31a
31b. Fig. 17.2.31b **33.** Fig. 17.2.33; Local minimum at $x = e^{-1}$;
$\lim_{x \to 0^+} x^x = 1$

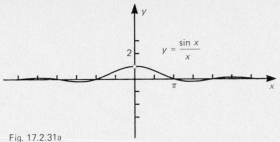

Fig. 17.2.31a

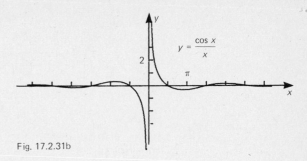

Fig. 17.2.31b

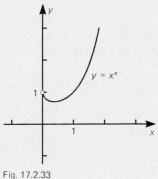

Fig. 17.2.33

Section 17.3

1. $\frac{1}{10}e^{-10}$ **3.** ∞

4.‡ $\lim\limits_{X \to \infty} \int_5^X \left(\dfrac{1}{x-1} - \dfrac{1}{x} \right) dx = \lim\limits_{X \to \infty} \left[\ln(1 - 1/X) - \ln\left(\dfrac{4}{5}\right) \right] = -\ln\left(\dfrac{4}{5}\right)$

5.‡ $\int_\pi^X \sin x\, dx = \cos(\pi) - \cos(X)$ has no limit as $X \to \infty$; the improper integral is undefined **7.** $\frac{27}{2}$ **8.** $\ln(\sqrt{2})$ **9.** Undefined **11.** 10 **13.** ∞ **15.** $\frac{1}{2}\pi$

17.‡ $\dfrac{1}{2} \int_0^X \left(\dfrac{1}{x-1} - \dfrac{1}{x+1} \right) dx = \dfrac{1}{2}(\ln|X-1| - \ln|X+1|) \to -\infty$ as

$X \to 1^-$; $-\infty$

20. $\ln(2)$

21.‡ $\int_0^X \left(\dfrac{3}{x-3} - \dfrac{2}{x-2} \right) dx = 3\ln|X-3| - 2\ln|X-2| - 3\ln(3) + 2\ln(2)$

$\to \infty$ as $X \to 2^-$; ∞

23. $\ln\left(\frac{11}{9}\right)$ **25.** $2^{3/4}$ **26.** $\frac{1}{8}$ **27.** ∞ **29.** ∞ **31.** ∞

33a. $\int_1^X \dfrac{2\pi}{x} \sqrt{1 + x^{-4}}\, dx > \int_1^X \dfrac{2\pi}{x} dx \to \infty$ as $X \to \infty$

Miscellaneous Exercises to Chapter 17

1a. 1.003968 **1b.** The approximation is less than the exact value; $0 < R_2(0.02) < (3.84 \ldots)10^{-7}$ **2.** 9 **4a.** 4 **4b.** 5 **5.**‡ $P_5(x)$ for $f(x) = x^5$ centered at 1 equals $f(x)$ because $f^{(j)}(x) = 0$ for $j \geq 6$; $x^5 = 1 + 5(x-1) + 10(x-1)^2 + 10(x-1)^3 + 5(x-1)^4 + (x-1)^5$

8.‡ $P_5(x) = x - \dfrac{1}{6}x^3 + \dfrac{1}{120}x^5$; $\int_0^1 \dfrac{1}{x}\sin x\, dx \approx \int_0^1 \dfrac{1}{x} P_5(x)\, dx$

$= \dfrac{1703}{1800} = 0.9461 \ldots$; $\left| \dfrac{1}{x} R_5(x) \right| \leq \dfrac{1}{6!}|x^5|$; $|\text{Error}| \leq \dfrac{1}{720} \int_0^1 x^5\, dx$

$= \dfrac{1}{4320} = (2.31 \ldots)10^{-4}$; Or, since $P_5(x) = P_6(x)$, $\left| \dfrac{1}{x} R_6(x) \right| \leq \dfrac{1}{7!}|x^6|$;

$|\text{Error}| \leq \dfrac{1}{5040} \int_0^1 x^6\, dx = \dfrac{1}{35280} = (2.83 \ldots)10^{-5}$

10.‡ $P_3(t) = 1 + t + \dfrac{1}{2}t^2 + \dfrac{1}{6}t^3$; $e^{-x^2} \approx P_3(-x^2) = 1 - x^2 + \dfrac{1}{2}x^4 - \dfrac{1}{6}x^6$;

$$\int_0^1 e^{-x^2}\,dx \approx \int_0^1 P_3(-x^2)\,dx = \frac{78}{105} = 0.7428; \text{ For } -1 < t < 0,$$

$$\frac{1}{4!e}t^4 < R_3(t) < \frac{1}{4!}t^4; \text{ Use } e < 3: \frac{1}{72}x^8 < R_3(-x^2) < \frac{1}{24}x^8 \text{ for } 0 < x < 1;$$

$$\frac{1}{648} < \text{Error} < \frac{1}{216}$$

12. For $n = 0$ equation (1) in the statement of the exercise states

$$f(x) = f(a) + \int_a^x f'(t)\,dt, \text{ which is the Fundamental Theorem of Calculus;}$$

We will prove (1) for all nonnegative integers n by showing that if it is valid for $n = k$ (k an arbitrary nonnegative integer), then it is valid for $n = k + 1$ (This line of reasoning is called *mathematical induction*); If (1) is true for

$$n = k, \text{ then } f(x) = P_k(x) + \frac{1}{k!}\int_a^x (x - t)^k f^{(k+1)}(t)\,dt; \text{ Integration by parts}$$

with $u = f^{(k+1)}(t)$ and $\dfrac{dv}{dt} = (x - t)^k$ gives $f(x) = P_k(x)$

$$+ \left[-\frac{1}{(k+1)!}(x - t)^{k+1}f^{(k+1)}(t) \right]_a^x + \frac{1}{(k+1)!}\int_a^x (x - t)^{k+1}f^{(k+2)}(t)\,dt$$

which gives (1) for $n = k + 1$ since $P_k(x) + \dfrac{1}{(k+1)!}(x - a)^{k+1} = P_{k+1}(x)$

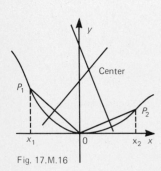

Fig. 17.M.16

15a. $f(x + 2\Delta x) - 2f(x + \Delta x) + f(x)$ **15b.** $f(x) = f(a) + f'(a)(x - a)$ $+ \frac{1}{2}f''(c)(x - a)^2$ with c between a and x; $f(a + \Delta x) = f(a) + f'(a)\Delta x$ $+ \frac{1}{2}f''(c_1)(\Delta x)^2$ and $f(a + 2\Delta x) = f(a) + 2f'(a)\Delta x + 2f''(c_2)(\Delta x)^2$ with c_1 and c_2 between a and $a + \Delta x$; $(\Delta^2 f(a))/(\Delta x)^2 = 2f''(c_2) - f''(c_1)$ $\rightarrow f''(a)$ as $\Delta x \rightarrow 0$ **16.** Consider $a \neq 0$ with $f''(x) \neq 0$ for x between 0 and a; The perpendicular bisector of the line segment from $0(0,0)$ to $A(a, f(a))$ has the equation $y = \frac{1}{2}f(a) - [a/f(a)](x - \frac{1}{2}a)$ and has y-intercept $\frac{1}{2}f(a) + \frac{1}{2}a^2/f(a)$, which by l'Hopital's rule (applied twice) tends to $1/f''(0)$ as $a \rightarrow 0$; Consider $x_1 < 0 < x_2$ with $f''(x) > 0$ for $x_1 \leq x \leq x_2$; The circle through the points $P_1(x_1, f(x_1))$, $0(0,0)$, $P_2(x_2, f(x_2))$ has its center at the intersection of the intersection of the perpendicular bisectors of the line segments $P_1 0$ and $0P_2$ (Fig. 17.M.16); Because the y-intercepts of the perpendicular bisectors tend to $1/f''(0)$, the center of the circle tends to $(0, 1/f''(0))$; A similar argument applies to the case of $f''(x) < 0$ **17.** e^2

22. Given ϵ with $0 < \epsilon < 1$, choose $b > a$ so that $(*)|[f'(x)/g'(x)] - L|$ $< \epsilon/2$ for $a < x < b$; Then, in particular $(**)|f'(x)/g'(x)| < 1 + |L|$ for

$$a < x < b; \text{ Choose } \delta \text{ with } a < a + \delta < b \text{ so } (***) \left| \frac{\dfrac{f(b)}{f(x)} - \dfrac{g(b)}{g(x)}}{1 - \dfrac{f(b)}{f(x)}} \right| < \frac{\epsilon}{2 + 2|L|}$$

for $a < x < a + \delta$; Consider x with $a < x < a + \delta$; By the Generalized Mean

Value Theorem, there is a c with $x < c < b$ such that $\dfrac{f(b) - f(x)}{g(b) - g(x)} = \dfrac{f'(c)}{g'(c)}$,

which yields $\dfrac{f(x)}{g(x)} = \left[\dfrac{1 - \dfrac{g(b)}{g(x)}}{1 - \dfrac{f(b)}{f(x)}} \right] \dfrac{f'(c)}{g'(c)}$; The equation

$\dfrac{f(x)}{g(x)} - L = \left[\dfrac{\dfrac{f(b)}{f(x)} - \dfrac{g(b)}{g(x)}}{1 - \dfrac{f(b)}{f(x)}} \right] \dfrac{f'(c)}{g'(c)} + \left[\dfrac{f'(c)}{g'(c)} - L \right]$ and inequalities

(*), (**), and (***) imply that $|[f(x)/g(x)] - L| < \epsilon$ for $a < x < a + \delta$

23. There is at least one number c with $a < c < b$ such that the slope $g'(c)/h'(c)$ of the tangent line at $(h(c),g(c))$ on the curve is equal to the slope $[g(b) - g(a)]/[h(b) - h(a)]$ of the secant line through the points $(h(a),g(a))$ and $(h(b),g(b))$ **27.** $\int_{-X}^{X} \sin x \, dx = \cos(-X) - \cos(X) = 0$ for all X, so the limit as $X \to \infty$ is 0; $\int_{0}^{X} \sin x \, dx = 1 - \cos(X)$ has no limit as $X \to \infty$ and $\int_{X}^{0} \sin x \, dx = \cos(X) - 1$ has no limit as $X \to -\infty$
31a. $e^2 = 7.389056 \ldots$; $P_5(2) = 7.266666 \ldots$ **31c.** $\ln(1.9)$ $= 0.641853 \ldots$; $P_7(0.9) = 0.671827 \ldots$ **31e.** $2^e = 6.580885 \ldots$; $P_5(1) = 6.583538 \ldots$

Section 18.1

1. 0; $1/(2^1 + 1)$; $2/(2^2 + 1)$; $3/(2^3 + 1)$; $4/(2^4 + 1)$; $5/(2^5 + 1)$; $6/(2^6 + 1)$; $7/(2^7 + 1)$; $8/(2^8 + 1)$; $9/(2^9 + 1)$ **3.** $10^{10}/10!$; $11^{11}/11!$; $12^{12}/12!$; $13^{13}/13!$; $14^{14}/14!$; $15^{15}/15!$; $16^{16}/16!$; $17^{17}/17!$; $18^{18}/18!$; $19^{19}/19!$ **5.** Converges to $-\frac{1}{2}$
7. Converges to $\frac{1}{3}$ **9.** Converges to 0 **11.** Diverges without a limit

13.‡ By l'Hopital's rule: $\lim\limits_{n \to \infty} \dfrac{\sin(1/n)}{1/n} = \lim\limits_{n \to \infty} \cos(1/n) = 1$

14. Converges to $-\sin(1)$ **15.** Diverges to ∞ **17.** Converges to 0
19. Diverges to ∞

22.‡ Converges to $\dfrac{1}{4}$ because the nth term is a Riemann sum for $\int_0^1 x^3 \, dx$

corresponding to the partition $0 < \dfrac{1}{n} < \dfrac{2}{n} < \cdots < \dfrac{n-1}{n} < 1$

23. Converges to $1 - \cos(4)$ **24.** Converges to $\frac{2}{3}$ **26a.** $2rn \sin(\pi/n)$
27a. For $-1 < r < 1$, $\ln(|r^n|) = n \ln|r| \to -\infty$ as $n \to \infty$ since $\ln(|r|) < 0$, so $|r^n| \to 0$ **27b.** For $r > 1$, $\ln(r^n) = n \ln(r) \to \infty$ as $n \to \infty$ since $\ln(r) > 0$ so $r^n \to \infty$ **27c.** For $r < -1$, $\ln(|r^n|) = n \ln(|r|) \to \infty$, so $|r^n| \to \infty$; r^n diverges without a limit because it is positive for even n and negative for odd n **27d.** 1
28.‡ 1 since $\lim\limits_{n \to \infty} \ln(n^{1/n}) = \lim\limits_{n \to \infty} [\ln(n)]/n = \lim\limits_{n \to \infty} (1/n) = 0$ by l'Hopital's rule
30. 1 **32.** e^{-2} **34.** 1

36.‡ Cancel $n!$: $\left(\dfrac{2n}{n} \right)\left(\dfrac{2n-1}{n-1} \right)\left(\dfrac{2n-2}{n-2} \right) \cdots \left(\dfrac{n+1}{1} \right) > 2^n$; ∞ **37.** 0

Section 18.2

1. Set $N = 1/\sqrt{\epsilon}$; For $n > N$, we have $n^2 > 1/\epsilon$ and $|0 - 1/n^2| < \epsilon$

3. Set $N = 5/(3\epsilon)$; For $n > N$, $|3 - \sqrt{9 - 5/n}| = \left|\dfrac{5/n}{3 + \sqrt{9 - 5/n}}\right| < \dfrac{5}{3n} < \epsilon$

5. Set $N = 2/\sqrt{\epsilon}$; For $n > N$, $\left|\dfrac{n^2}{n^2 + 4} - 1\right| = 4/(n^2 + 4) < 4/n^2 < \epsilon$

7. By the Mean Value Theorem applied to e^x, $e^{1/n} - 1 = (e^c)/n$ with $0 < c < 1/n$; Set $N = e/\epsilon$; For $n > N$, $|e^{1/n} - 1| < e/n < \epsilon$ **9.** By the Mean Value Theorem applied to $\ln x$, $\ln(1 + 1/n) = 1/(cn)$ with $1 < c < 1 + 1/n$; Set $N = 1/\epsilon$; For $n > N$, $|\ln(1 + 1/n)| < 1/n < \epsilon$ **11.** By the Mean Value Theorem applied to $x^{2/3}$, $|(8 - 8/n)^{2/3} - 4| = \frac{2}{3}c^{-1/3}(8/n)$ with $8 - 8/n < c < 8$; Set $N = 3/(16\epsilon)$; For $n > N$, $|(8 - 8/n)^{2/3} - 4| < 16/(3n) < \epsilon$ **13.** $n^2/(n + 4) = n(1 + 4/n)^{-1} > n/5$ for $n \geq 1$; Set $N = 5M$; For $n > N$, $n^2/(n + 4) > N/5 = M$ **15.** Set $N = M^{1/3}$; For $n > N$, $n^3 > M$ **21.‡** Monotone because it is decreasing; $3 + (1/n) \to 3$ as $n \to \infty$; Bounded above and bounded below; Least upper bound $= 4$; Greatest lower bound $= 3$

23.‡ For $x > 0$, the derivative of $x \ln(1 + 1/x)$ is $\ln(1 + 1/x) - \dfrac{1}{x + 1}$

$$= \int_1^{1+1/x} \left(\frac{1}{t} - \frac{x}{x + 1}\right) dt = \int_1^{1+1/x} \left(\frac{1}{t} - \frac{1}{1 + 1/x}\right) dt,$$ which is positive since

$t \leq 1 + 1/x$ in the interval of integration; Increasing, and hence monotone; $(1 + 1/n)^n \to e$ as $n \to \infty$; Bounded above and below; Least upper bound $= e$; Greatest lower bound $= 2$

25. Increasing and hence monotone; Bounded below but not above; Greatest lower bound $= 0$ **27.** Not monotone; Bounded above and below; Greatest lower bound $= 0$; Least upper bound $= 1$ **28.** Not monotone; Bounded above and below; Greatest lower bound $= -1$; Least upper bound $= \frac{1}{2}$ **29.** Not monotone; Bounded below; Greatest lower bound $= 0$ **31.** Not monotone; not bounded above or below   **33.** Not monotone (because of the first term); Bounded above and below; Greatest lower bound $= 0$; Least upper bound $= 1/e$ **35.** Increasing and hence monotone; Bounded above and below; Greatest lower bound $= \frac{1}{4}\pi$; Least upper bound $= \frac{1}{2}\pi$

Section 18.3

1. 1.999756

2.‡ $\dfrac{1 - (-1/4)^9}{1 - (-1/4)} \approx \dfrac{4}{5}(1 + 3.82 \times 10^{-6}) = 0.800003056$

3.‡ $\dfrac{3}{4}\left[\dfrac{1 - (3/4)^{31}}{1 - (3/4)}\right] \approx 2.9996$

5. 6.66666   **8.** 0.502495528   **9.** 101   **10.** 267.7   **13.** 24.999999999185 **19b.** $P = rB_0/\left(1 - (1 + r)^{-N}\right)$

Section 18.4

1. Converges; $\frac{7}{6}$ **3.**‡ Converges ($r = 0.99$); $(0.99)^2 \sum_{j=0}^{\infty} (0.99)^j$
$= (0.99)^2(1 - 0.99)^{-1} = 98.01$ **5.**‡ Converges ($r = \frac{10}{11}$); $4(\frac{10}{11})^5 \sum_{j=0}^{\infty} (\frac{10}{11})^j$
[...] **7.** Diverges to ∞ **9.**‡ $3.4 + 0.0212121 \ldots$
[...](0.212121 \ldots) = 3.4 + \frac{21}{10}(0.010101 \ldots) = 3.4 + 2.1 \sum_{j=1}^{\infty} (\frac{1}{100})^j$
[...]$2.1(\frac{1}{99}) = \frac{1129}{330}$ **10.** $\frac{59}{10}$ **14.**‡ $50 + 2/n \le 51$ for $n \ge 2$; Converges by
[Com]parison Test with $b_n = (\frac{1}{2})^n$ **16.** Diverges because $(1 + 1/j)^j \to e$
[...converg]es by the Limit Comparison Test with $b_n = 1/n$ **20.** Converges by
[...T]est with $b_j = (\frac{1}{3})^j$ **22.**‡ $(1/j^j) \le (1/2^j)$ for $j \ge 2$; Converges by
[Compa]rison Test with $b_j = (\frac{1}{2})^j$ **24.** Diverges by the Limit Comparison
[Test with $b_n] = 1/n$ **26.** Diverges by the Limit Comparison Test with $b_j = 1/j$
[...]s because $\sin(\frac{1}{4}j\pi)$ does not tend to 0   **32.** $\frac{1}{4}$ **33.** $\ln(\frac{4}{3})$

[Sectio]n 18.5

[...]s by the Integral Test
[...]$\frac{\ln k)}{/k} = \lim_{k \to \infty} \frac{k}{\ln(k)} = \lim_{k \to \infty} k = \infty$ by l'Hopital's rule; Diverges by
[...] Comparison Test with $b_k = 1/k$
[...] because $\ln(j)$ does not tend to 0   **5.** Diverges by the Integral
[...Co]nverges by the Ratio Test   **9.** Converges by the Ratio Test
[...g]es by the Ratio Test   **13.** Converges by the Comparison Test with

$$\frac{)!]^2/[(j + 1)^2]!}{j!)^2/(j^2)!} = \frac{1}{[(j + 1)^2 - 1][(j + 1)^2 - 2] \cdots [j^2 + 1]}$$

[...]rges by the Ratio Test
[...]rges by the Limit Comparison Test with $b_j = j^{-7/6}$ **19.** Converges
[...] Ratio Test

$$\frac{\ln k)/k^2}{k^{-3/2}} = \lim_{k \to \infty} \frac{\ln k}{k^{1/2}} = \lim_{k \to \infty} \frac{2}{k^{1/2}} = 0$$ by l'Hopital's Rule; Converges

by the Limit Comparison Test with $b_k = k^{-3/2}$

25. Converges by the Comparison Test with $b_j = j^{-3}$ **27.** Diverges by the
Comparison Test with $b_k = 1/k$

28.‡ $\dfrac{(5^n + n)/(6^n + n)}{5^n/6^n} = \dfrac{1 + n5^{-n}}{1 + n6^{-n}} \to 1$ as $n \to \infty$ since by l'Hopital's rule

$n5^{-n}$ and $n6^{-n} \to 0$; Converges by the Limit Comparison Test with $b_n = \left(\dfrac{5}{6}\right)^n$

29. Diverges by the Ratio Test; Or, because the terms are all > 1
31. Converges by the Comparison Test with $b_n = n^{-2}$; Or, by the Integral Test
33. Converges by the Limit Comparison Test with $b_k = k^{-2}$

Section 18.6

1. Converges absolutely by the Comparison Test with $b_j = j^{-3}$ **3.‡** Converges by the Alternating Series Test because $1/[(n + 1) \ln(n + 1)] < 1/(n \ln n)$ and $1/(n \ln n) \to 0$; Converges conditionally because the sum of $1/(n \ln n)$ diverges by the Integral Test

5.‡ $1/\ln[\ln(n + 1)] < 1/\ln(\ln n)$ and $1/\ln(\ln n) \to 0$; Converges by the Alternating Series Test; Converges conditionally because the sum of $1/\ln(\ln n)$ diverges by the Limit Comparison Test with $b_n = 1/n$: $\displaystyle \lim_{n \to \infty} \frac{1/\ln(\ln n)}{1/n}$

$$= \lim_{n \to \infty} \frac{n}{\ln(\ln n)} = \lim_{n \to \infty} (n \ln n) = \infty \text{ by l'Hopital's Rule}$$

7. Converges absolutely by the Ratio Test   **9.** Converges conditionally by the Alternating Series Test and the Comparison Test with $b_n = n^{-1/2}$
11. Diverges by the Ratio Test   **13.** Diverges because $(-1)^{k+1} \arctan k$ does not tend to 0

15.‡ $\left[\frac{1}{2}\pi - \arctan(j + 1)\right] - \left[\frac{1}{2}\pi - \arctan j\right] < 0$ and $\frac{1}{2}\pi - \arctan j \to 0$;

Converges by the Alternating Series Test; $\frac{1}{2}\pi - \arctan j$

$$= \int_j^\infty \frac{1}{x^2 + 1} dx > \int_j^\infty \frac{1}{2x^2} = \frac{1}{2j};$$ Converges conditionally by the Comparison

Test with $b_j = 1/j$

17. Diverges because $(-1)^j \sin(j\pi/10)$ does not tend to 0   **19.** Converges conditionally by the Alternating Series Test and the Limit Comparison Test with $b_n = 1/n$ **21.** Converges conditionally by the Alternating Series Test and the Comparison Test with $b_j = 1/j$ (or the Integral Test)

23.‡ $\left|\frac{(j + 1)x^{j+1}/5^{j+1}}{jx^j/5^j}\right| = \left(1 + \frac{1}{j}\right)\left|\frac{x}{5}\right| \to \left|\frac{x}{5}\right|$; By the Ratio Test the series

converges absolutely for $|x| < 5$ and diverges for $|x| > 5$; It diverges for $x = \pm 5$ because then its terms do not tend to 0

25. Converges absolutely for $|x| < 1$; Converges conditionally for $x = -1$; Diverges for $x < -1$ and for $x \geq 1$ **27.** Converges absolutely for $|x| > 4$; Diverges for $0 < |x| \leq 4$ (Not defined at $x = 0$)   **29.** Converges absolutely for $|x| \leq \frac{1}{10}$; Diverges for $|x| > \frac{1}{10}$ **31.** Converges absolutely for $x = 0$; Diverges for $x \neq 0$ **33a.** $n > e^{e^5} \approx 2.85 \times 10^{64}$ **33b.** $n + 1 > [\arcsin(\frac{1}{5})]^{-1} \approx 4.97$; $n \geq 4$

Section 18.7

1.‡ $\left|\frac{x^{j+1}/(j + 1)}{x^j/j}\right| = \frac{1}{1 + 1/j}|x| \to |x|$; Converges absolutely for $|x| < 1$ and

diverges for $|x| > 1$ by the Ratio Test; Diverges for $x = 1$ (harmonic series); Converges conditionally for $x = -1$ by the Alternating Series Test

2. Converges absolutely for $|x| < 1$; Diverges for $|x| \geq 1$

3.‡ $\left|\frac{[(j + 4)/((j + 1)^2 + 1)][(x - 3)/2]^{j+1}}{[(j + 3)/(j^2 + 1)][(x - 3)/2]^j}\right| \to \left|\frac{x - 3}{2}\right|$; Converges absolutely for $|x - 3| < 2$ and diverges for $|x - 3| > 2$ by the Ratio Test; Diverges for $x = 5$ by the Limit Comparison Test with $b_j = 1/j$; Converges Conditionally for $x = 1$ by the Alternating Series Test

5. Converges absolutely for $x = 0$; Diverges for $x \neq 0$ **7.** Converges absolutely for $|x| < \frac{1}{10}$; Diverges for $|x| \geq \frac{1}{10}$ **11.** Converges absolutely for $|x| < 1$; Diverges for $|x| \geq 1$ **14.** Converges absolutely for $|x| < 1$; Converges conditionally for $x = -1$; Diverges for $x < -1$ and for $x \geq 1$

16.‡ Set $f(x) = e^{2x}$; $f^{(j)}(x) = 2^j e^{2x}$; $f^{(j)}(0) = 2^j$; $(*) \ \sum_{j=0}^{\infty} \frac{1}{j!}(2x)^j$; Converges

absolutely for all x by the Ratio Test; By Taylor's Theorem

$$e^{2x} = \sum_{j=0}^{n} \frac{1}{j!}(2x)^j + R_n(x) \text{ where } R_n(x) = \frac{1}{(n+1)!}e^{2c}(2x)^{n+1} \text{ with } c$$

between 0 and x; $|R_n(x)| \leq \dfrac{1}{(n+1)!}e^{2|x|}|2x|^{n+1} \to 0$ because $(*)$ converges

17. $\sum_{j=1}^{\infty} \frac{1}{(2j+1)!}(-1)^j x^{2j+1}$; $\sin x = \sum_{j=1}^{n} \frac{1}{(2j+1)!}(-1)^j x^{2j+1} + R_{2n+1}(x)$;

$|R_{2n+1}(x)| = \left| \dfrac{1}{(2n+2)!}(-1)^{n+1}(\sin c) x^{2n+2} \right| \leq \dfrac{1}{(2n+2)!}|x|^{2n+2} \to 0$

19. $\sum_{j=0}^{\infty} \frac{1}{(2j+1)!} x^{2j+1}$; $\sinh x = \sum_{j=0}^{n} \frac{1}{(2j+1)!} x^{2j+1} + R_{2n+1}(x)$;

$|R_{2n+1}(x)| = \left| \dfrac{1}{(2n+2)!}(\sinh c) x^{2n+2} \right| < \dfrac{1}{(2n+2)!}|\sinh x| |x|^{2n+2} \to 0$

21. $1 + 5x + 10x^2 + 10x^3 + 5x^4 + x^5$; The sum is finite

24. $\sum_{j=1}^{\infty} (-1)^j \frac{1}{j} x^j$; $-1 < x \leq 1$ **25.** $\sum_{j=0}^{\infty} \frac{1}{2}\left(\frac{1}{2}x\right)^j$; $-1 < x < 2$

27a. $\sum_{j=0}^{\infty} \frac{1}{j!}(-1)^j x^{2j}$ **27b.** $\sum_{j=0}^{\infty} \frac{1}{j!}(-1)^j x^{j+3}$ **27c.** $\sum_{j=0}^{\infty} \frac{1}{(2j)!} x^{2j}$

Section 18.8

1. $3 + x - \frac{1}{6}x^3 + 5x^4 + \sum_{j=2}^{\infty} \frac{(-1)^j}{(2j+1)!} x^{2j+1}$

2. $\sum_{j=0}^{\infty} \left\{ \frac{3(2)^j + 4(5)^j}{j!} \right\} x^j$ **4.** $\sum_{j=1}^{\infty} \frac{2}{(2j-1)} x^{2j-1}$ **6.** $1 + \sum_{j=1}^{\infty} 2x^j$

8. $\sum_{j=0}^{\infty} \frac{1}{(2j+1)!} x^{6(j+1)}$ **9.** $\sum_{j=2}^{\infty} \frac{j(j-1)}{2} x^{j-2}$ **11.** $\sum_{j=0}^{\infty} \frac{1}{(2j+1)j!} x^{2j+1}$

13. $x + x^2 + \frac{1}{3}x^3 - \frac{1}{24}x^5 + \cdots$ **14.** $x^2 + x^3 + \frac{11}{12}x^4 + \frac{5}{6}x^5 + \cdots$
17. $x + x^2 + \frac{5}{6}x^3 + \frac{5}{6}x^4 + \cdots$ **18.** $1 - \frac{1}{2}x^2 + \frac{2}{3}x^3 - \frac{1}{8}x^4 + \cdots$
19. $1 + \frac{1}{2}x^2 + \frac{5}{24}x^4 + \frac{17}{240}x^6 + \cdots$ **20.** $1 + x - \frac{3}{2}x^2 + \frac{7}{6}x^3 + \cdots$
22. $x - \frac{1}{3}x^3 + \frac{2}{15}x^5 - \frac{17}{315}x^7 + \cdots$ **23.** $1 + x + \frac{1}{2}x^2 - \frac{1}{8}x^4 + \cdots$

26a. $\pi = 4 \sum_{j=0}^{\infty} \frac{(-1)^j}{2j+1}$ **26b.** 3.34; 2.98   **26c.** 3.1456   **28.** $2x^2(1-x)^{-3}$

29. $\cosh(x^3)$ **30.** $\int_0^x \ln(1 + t)\, dt = x\ln(1 + x) + \ln(1 + x) - x$

32. $\displaystyle\sum_{j=0}^{\infty} \frac{(-1)^j}{(2j + 1)!} x^{2j}$ **34.** $\displaystyle\sum_{j=1}^{\infty} \frac{(-1)^j}{(2j)!} x^{2j-2}$ **36.** 380

Miscellaneous Exercises to Chapter 18

2a. Decreasing   **3.** Converges by the Ratio Test   **4.** Diverges by the Limit Comparison Test with $b_j = 1/j$

11. $\dfrac{1}{x(x + 1)} = \dfrac{1}{x} - \dfrac{1}{x + 1}$; $\displaystyle\sum_{j=1}^{N}\left(\dfrac{1}{j} - \dfrac{1}{j+1}\right) = 1 - \dfrac{1}{N + 1}$ (All the terms cancel but these two); 1

16. $\$100(1.005^{121} - 1)/0.005 \approx \$16{,}569$ **20a.** $1 + \frac{1}{2}s + \frac{3}{8}s^2 + \frac{5}{16}s^3$;
20b. $x + \frac{1}{6}x^3 + \frac{3}{40}x^5 + \frac{5}{112}x^7$ **22a.** $\frac{2}{3}$ **24a.**‡ $e^x \sin x = (1 + x + \frac{1}{2}x^2$
$+ \frac{1}{6}x^3 + \frac{1}{24}x^4 + \cdots)(x - \frac{1}{6}x^3 + \frac{1}{120}x^5 - \cdots) = x + x^2 + (\frac{1}{2} - \frac{1}{6})x^3$
$+ (\frac{1}{6} - \frac{1}{6})x^4 + (\frac{1}{24} - \frac{1}{12} + \frac{1}{120})x^5 + \cdots = x + x^2 + \frac{1}{3}x^3 - \frac{1}{30}x^5 + \cdots$
25.‡ $0 < \sum_{j=101}^{\infty} j^{-3} < \int_{100}^{\infty} x^{-3}\, dx = \frac{1}{20{,}000}$ **26.** $\frac{1}{2}e^{-400} \approx 1.8(10^{-44})$
28. $\frac{1}{4}(2501)^{-2} \approx 4(10^{-8})$ **30.** "$\ldots |a_n - L| \geq \epsilon$ for at least one $n > N$."
31. Set $\epsilon = 0.00005$; For any $n > 20{,}000$, we have $1/n < 0.00005$ and $|(1 + 1/n) - 1.0001| = 0.0001 - 1/n > \epsilon$ **33a.** Let the graph of $f(x)$ consist of the line segment from $(1,1)$ to $(\frac{3}{2},0)$, and the four line segments joining $(n - 1/n, 0)$, $(n, 1/n)$, $(n + 1/n, 0)$, and $(n + 1 - 1/(n + 1), 0)$ for $n = 2, 3, 4, \ldots$; $\int_1^{\infty} f(x)\, dx = \frac{1}{2} + \sum_{n=1}^{\infty} n^{-2} < \infty$ **34a.** Let $b_j = 0$ for even j and $b_j = 1/j$ for odd j **35a.** $|a_j| < r^j$, so the series converges absolutely by the Comparison Test with $b_j = r^j$ **40.** $F(1) = F(0) + F'(0) + \frac{1}{2}F''(0)$ $+ \frac{1}{6}F^{(3)}(c)$ with $0 < c < 1$; Set $X = cx$ and $Y = cy$ **41.** Partial sum $= 1.527422 \ldots$; $\frac{1}{6}\pi^2 = 1.644934 \ldots$ **43.** Partial sum $= 0.808078 \ldots$; $\frac{1}{4}\pi = 0.785398 \ldots$

Section 19.1 $(\exp\{u\} = e^u)$

1. $y = -1 + C\,e^{\cos x}$

3.‡ $I = \exp\left\{-\displaystyle\int \frac{x}{x^2 + 1}\, dx\right\} = (x^2 + 1)^{-1/2}$; $\dfrac{d}{dx}(Iy) = x^2 + 1$

$Iy = \dfrac{1}{3}x^3 + x + C$; $y = \left(\dfrac{1}{3}x^3 + x + C\right)(x^2 + 1)^{1/2}$

6. $y = -2 + 2e^{x^3}$ **8.** $y = \frac{1}{6}x + \frac{1}{12} + \frac{11}{12}(2x + 1)^{-5}$

10.‡ $I = \exp\left\{\displaystyle\int 2\, dx\right\} = e^{2x}$; $\dfrac{d}{dx}(Iy) = e^{2x}\cos x$; $Iy = \left(\dfrac{1}{5}\sin x + \dfrac{2}{5}\cos x\right)e^{2x}$

$+ C$; $y = \dfrac{1}{5}\sin x + \dfrac{2}{5}\cos x + Ce^{-2x}$; Set $x = \pi$ and $y = 0$: $C = \dfrac{2}{5}e^{2\pi}$;

$y = \dfrac{1}{5}\sin x + \dfrac{2}{5}\cos x + \dfrac{2}{5}e^{2\pi - 2x}$

12. $y = (4 - \cos x + \frac{1}{3}\cos^3 x)/\sin^2 x$ **14a.** $4 - 3500(10 + t)^{-3}$ pounds per gallon   **14b.** 4 pounds per gallon   **15a.** $\frac{1}{2}(8 + 3t) + 2(8 + 3t)^{-1/3}$ pounds
15b. $\frac{1}{2} + 2(8 + 3t)^{-4/3}$ pounds per gallon   **17a.**‡ The tank contains $9 - 2t$ gallons up to $t = t_0$ with $t_0 = \frac{9}{2}$ (hours)

17b.[‡] Let y = pounds of sugar in the tank at time t; $\dfrac{dy}{dt} + \dfrac{3y}{9 - 2t} = 2$;

$I = (9 - 2t)^{-3/2}$; $y = 2(9 - 2t) - \dfrac{4}{9}(9 - 2t)^{3/2}$

17c.[‡] $2 - \frac{4}{9}(9 - 2t)^{1/2} \to 2$ as $t \to (\frac{9}{2})^-$ **19.** $y(x) = x - 1 + e^{-x}$

Section 19.2

1.[‡] $\dfrac{\partial}{\partial y}(2x - 6y) = -6$ and $\dfrac{\partial}{\partial x}(4y^3 - 6x) = -6$; Exact;

$U = \displaystyle\int (2x - 6y)\,dx = x^2 - 6xy + \phi(y)$; $U_y = -6x + \phi'(y)$;

$\phi'(y) = 4y^3$; $\phi = y^4 + C$; $U = x^2 - 6xy + y^4 + C$; $x^2 - 6xy + y^4 = c$

3. Not exact   **5.** Exact; $\sin(xy) - y\cos x = c$ **8.** Exact, $x^3 y^2 + x^2 - y^3 = c$

10.[‡] $\dfrac{\partial}{\partial y}[2x^3 y^2 - (\sin x)(\cos x)] = 4x^3 y$ and $\dfrac{\partial}{\partial x}(x^4 y + 2y) = 4x^3 y$;

Exact; $U = \displaystyle\int [2x^3 y^2 - (\sin x)(\cos x)]\,dx = \dfrac{1}{2}x^4 y^2 - \dfrac{1}{2}\sin^2 x + \phi(y)$;

$U_y = x^4 y + \phi'(y)$; $\phi'(y) = 2y$; $\phi(y) = y^2 + C$; $U = \dfrac{1}{2}x^4 y^2$

$- \dfrac{1}{2}\sin^2 x + y^2 + C$; $\dfrac{1}{2}x^4 y^2 - \dfrac{1}{2}\sin^2 x + y^2 = c$; Set $x = 0, y = 3$: $c = 9$;

$y = [(18 + \sin^2 x)/(2 + x^4)]^{1/2}$

11. $y = (x^2 + 2)/(x^2 + 1)$ **14.** Not exact

Section 19.3

1.[‡] $r^2 - 3r + 2 = 0$; $r = 1$ and 2; $y = Ae^x + Be^{2x}$; $y' = Ae^x + 2Be^{2x}$;
$A + B = 1$ and $A + 2B = 0$; $A = 2, B = -1$; $y = 2e^x - e^{2x}$
2. $y = 2e^{3x} - e^{-2x}$ **4.** $y = (14 - 11x)e^{2x-2}$ **6.**[‡] $r^2 - 2r + 2 = 0$;
$r = 1 \pm i$; $y = (A\cos x + B\sin x)e^x$; $A = 0$ and $A + B = 1$; $y = (\sin x)e^x$
7. $y = -(\cos x + 2\sin x)e^{-2x}$ **9.**[‡] $r^2 - 1 = 0$; $y_h = Ae^x + Be^{-x}$; The
functions 1 and x form a U.C. set; $y_p = a + bx$; $y_p'' - y_p = -a - bx$;
$a = 0, b = -1$; $y = -x + Ae^x + Be^{-x}$; $A + B = 0$ and $-1 + A - B = 1$;
$A = 1, B = -1$; $y = -x + e^x - e^{-x}$ **10.** $y = e^x - \cos(2x) + \sin(2x)$
12.[‡] $r^2 + 1 = 0$; $y_h = A\cos x + B\sin x$; $x\cos x$ and $x\sin x$ form a U.C. set;
$y_p = ax\cos x + bx\sin x$; $y_p'' + y_p = -2a\sin x + 2b\cos x$; $a = 0, b = 1$;
$y = x\sin x + A\cos x + B\sin x$; $A = 0, B = 1$; $y = x\sin x + \sin x$
13. $y = xe^x - e^x$ **15.** $y = x^2 e^{-2x} + x e^{-2x}$ **17.** If the first of equations (20)
is satisfied, then $y'(x) = [c_1(x)y_1'(x) + c_2(x)y_2'(x)]$
$+ [c_1'(x)y_1(x) + c_2'(x)y_2(x)] = c_1(x)y_1'(x) + c_2(x)y_2'(x)$, and then,
if the second equation is satisfied, we have $ay'' + by' + cy$
$= a[c_1'(x)y_1'(x) + c_2'(x)y_2'(x)] + c_1(x)[ay_1''(x) + by_1'(x) + cy_1(x)]$
$+ c_2(x)[ay_2''(x) + by_2'(x) + cy_2(x)] = f(x)$ **18.**[‡] $y_1 = \cos x$,
$y_2 = \sin x$; $c_1'(x)\cos x + c_2'(x)\sin x = 0$ and $c_1'(x)(-\sin x) + c_2'(x)\cos x$
$= \sec^3 x$; $c_1'(x) = -\sec^2 x\tan x$ and $c_2'(x) = \sec^2 x$; $c_1(x) = -\frac{1}{2}\tan^2 x + A$
and $c_2(x) = \tan x + B$; $y = (-\frac{1}{2}\tan^2 x + A)\cos x + (\tan x + B)\sin x$;

Or: $y = \frac{1}{2}(\sin x)(\tan x) + A\cos x + B\sin x$ **19.** $y = (A + Bx + x\ln|x|)e^x$
23a. $\sin(x + \beta)$ and $(\cos\beta)\sin x + (\sin\beta)\cos x$ are both solutions of the initial value problem $y'' + y = 0$, $y(0) = \sin\beta$, $y'(0) = \cos\beta$

Section 19.4

1a.[‡] The motion is free because $f = 0$; It is undamped because $R = 0$
1b.[‡] The solution is of the form $y = C\cos(\omega t + \psi)$, so its graph must be in Figure 19.14, 19.23, or 19.25; The function of Figure 19.14 satisfies the initial conditions $y(0) = 1$, $y'(0) = 3$; Figure 19.14   **1c.**[‡] Characteristic equation: $2r^2 + 18 = 0$; $r = \pm 3i$; $y = A\cos(3t) + B\sin(3t)$; $y' = -3A\sin(3t)$ $+ 3B\cos(3t)$; $y(0) = 1 \Rightarrow A = 1$; $y'(0) = 3 \Rightarrow B = 1$; $y = \cos(3t) + \sin(3t)$
2a.[‡] The motion is free because $f = 0$; It is underdamped because $R \neq 0$ and $R^2 - 4Mk = -36$ is negative   **2b.**[‡] The motion oscillates and dies down, so it is given by Figure 19.12, 19.19, or 19.21; The function of Figure 19.12 satisfies the initial conditions $y(0) = 2$, $y'(0) = -1$; Figure 19.12
2c.[‡] $2r^2 + 2r + 5 = 0$; By the quadratic formula, $r = -\frac{1}{2} \pm \frac{3}{2}i$; $y = e^{-t/2}[A\cos(\frac{3}{2}t) + B\sin(\frac{3}{2}t)]$; $y' = e^{-t/2}[-\frac{1}{2}A\cos(\frac{3}{2}t) - \frac{1}{2}B\sin(\frac{3}{2}t)$ $- \frac{3}{2}A\sin(\frac{3}{2}t) + \frac{3}{2}B\cos(\frac{3}{2}t)]$; $y(0) = 2 \Rightarrow A = 2$; $y'(0) = -1 \Rightarrow B = 0$; $y = 2e^{-t/2}\cos(\frac{3}{2}t)$ **4a.** Free, overdamped motion   **4b.** Figure 19.24
4c. $y = 2e^{-t} - 3e^{-2t}$ **6a.** Free, critically damped motion   **6b.** Figure 19.15
6c. $y = (2 - t)e^{-t/2}$ **9a.** Free, underdamped motion   **9b.** Figure 19.19
9c. $y = -e^{-t}(\cos t + \sin t)$ **10a.**[‡] Forced, undamped motion because $f \neq 0$ and $R = 0$ **10b.**[‡] By the method of undetermined coefficients, the solution is of the form $(A + c_1 t)\cos(\omega t) + (A + c_2 t)\sin(\omega t)$; The motion oscillates and grows, so it is given by Figure 19.17   **10c.**[‡] $y_p = c_1 t\cos t$ $+ c_2 t\sin t$; $y_p'' + y_p = -2c_1\sin t + 2c_2\cos t \Rightarrow c_1 = -\frac{1}{5}$ and $c_2 = 0$; $y = -\frac{1}{5}t\cos t + A\cos t + B\sin t$; $y' = -\frac{1}{5}\cos t + \frac{1}{5}t\sin t - A\sin t$ $+ B\cos t$; $y(0) = 0 \Rightarrow A = 0$; $y'(0) = \frac{2}{5} \Rightarrow B = \frac{3}{5}$; $y = -\frac{1}{5}t\cos t + \frac{3}{5}\sin t$
12a. Forced, critically damped motion   **12b.** Figure 19.18
12c. $y = (2 + 3t)e^{-t} - 2\cos t$ **16a.**[‡] $y = \cos(3t) + \sin(3t)$ with $A = B = 1$; Amplitude $= \sqrt{A^2 + B^2} = \sqrt{2}$ feet; Period $= T \Leftrightarrow 3T = 2\pi$; $T = \frac{2}{3}\pi$ seconds; Frequency $= 1/T = 3/(2\pi)$ cycles per second
16b. Amplitude $= 3$; Period $= 2\pi/\sqrt{5}$; Frequency $= \sqrt{5}/(2\pi)$
18. The object oscillates about its rest position with amplitude $\frac{1}{2}$ foot and period $\frac{1}{4}\pi$ seconds   **19.** $y = c_1\cos(at) + c_2\sin(at) + e^{\tau t}[A\cos(\omega t)$ $+ B\sin(\omega t)]$ with $\tau < 0$; The terms involving the exponential tend to zero as t tends to infinity, so the steady state motion is given by $y = c_1\cos(at) + c_2\sin(at) = C\sin(at + \psi)$; The initial conditions determine A and B, whereas the differential equation determines c_1 and c_2, so the steady state motion is independent of the initial conditions
20a. $y = -2\cos t$ **21.** $e^{-t}[5\cos(2t) + \frac{5}{2}\sin(2t)]$ **24a.** $Lq'' + Rq' = f$
25a. $i(t) = c_1\cos(at) + c_2\sin(at)$ with $c_1 = (k - a^2)/D$ and $c_2 = -a/D$ where $D = (k - a^2)^2 + a^2$

Section 19.5

1.[‡] $y = \displaystyle\sum_{j=0}^{\infty} a_j x^j$; $y'' + 2y' + y$

$= \displaystyle\sum_{j=0}^{\infty} [(j + 2)(j + 1)a_{j+2} + 2(j + 1)a_{j+1} + a_j]x^j$;

$a_{j+2} = -[2(j+1)a_{j+1} + a_j]/[(j+2)(j+1)]$ for $j \geq 0$; $y(0) = 0$
$\Rightarrow a_0 = 0$; $y'(0) = 1 \Rightarrow a_1 = 1$; $a_2 = -1$; $a_3 = 1/2!$; $a_4 = -1/3!$;

$a_j = (-1)^{j-1}/(j-1)!$ for $j \geq 1$; $y = \displaystyle\sum_{j=1}^{\infty} \frac{1}{(j-1)!}(-1)^{j-1} x^j$

3. $3 \displaystyle\sum_{j=0}^{\infty} \frac{1}{j!}(2x)^j$

5. $\displaystyle\sum_{k=0}^{\infty} \frac{1}{k!} x^{2k}$

7. $\displaystyle\sum_{k=0}^{\infty} \frac{1}{2^k k!} x^{2k}$

9.‡ $y = \sum_{j=0}^{\infty} a_j x^j$; $\sum_{j=0}^{\infty} [(j+1)a_{j+1} - a_j]x^j = -x^2$; $y(0) = 2 \Rightarrow a_0 = 2$;
$a_1 - a_0 = 0$; $2a_2 - a_1 = 0$; $3a_3 - a_2 = -1$; $a_{j+1} = a_j/(j+1)$ for
$j \geq 3$; $y = 2 + 2x + x^2$

12.‡ $y'(x) = x^2 + y(x)^2$; $y(0) = 1 \Rightarrow y'(0) = 1$; Differentiate the differential
equation: $y'' = 2x + 2yy'$; $y(0) = 1$ and $y'(0) = 1 \Rightarrow y''(0) = 2$;
Differentiate again: $y^{(3)} = 2 + 2yy'' + 2(y')^2$; $y^{(3)}(0) = 8$; Differentiate
again: $y^{(4)} = 2yy^{(3)} + 2y'y'' + 4yy''$; $y^{(4)}(0) = 28$; Use Taylor's formula:
$y = 1 + x + x^2 + \frac{4}{3}x^3 + \frac{7}{6}x^4 + \cdots$

14. $y = 1 + 2x + \frac{1}{2}x^2 + \frac{2}{3}x^3 + \frac{5}{12}x^4 + \cdots$

17. $y = e^{x^2}$ $\quad$ **18.** $y = x^2 - 6x + 7$ $\quad$ **19.** $y = 4x^4 - 12x^2 + 3$

20. $y = -7x^3 + 3x$ $\quad$ **22.** $y = \displaystyle\sum_{n=0}^{\infty} \frac{(-1)^n}{(n!)^2}\left(\frac{x}{2}\right)^{2n}$ $\quad$ **24.** $y = e^{x^2/2}$ $\quad$ **26.** $y = e^x$

Miscellaneous Exercises to Chapter 19

1.‡ $x(dv/dx) = 1/v$; $v^2 = 2\ln|x| + C$; $y^2 = 2x^2 \ln|x| + Cx^2$;
$y(1) = 2 \Rightarrow C = 4$; $y^2 = 2x^2 \ln|x| + 4x^2$ $\quad$ **2.** $y(1 + \ln|y|) = x$
4. $y + \sqrt{x^2 + y^2} = 1$

7a. $\dfrac{d}{dx}(xy) = x \Rightarrow xy = \dfrac{1}{2}x^2 + C$; $x = 0 \Rightarrow C = 0$ and then $y = 0$ at $x = 0$

9. $y = \frac{1}{15}x^5 + \frac{1}{3}x^2 - \frac{2}{5}$ $\quad$ **10.** $y = (1 + 3e^{x^2})^{-1/2}$ $\quad$ **12.**‡ $U_x = 3x^2y^4 + 2xy$;
$U = x^3y^4 + x^2y + \phi(y)$; $U_y = 4x^3y^3 + x^2 + \phi'(y)$ and $U_y = 4x^3y^3 + x^2$;
$\phi'(y) = 0$; $U = x^3y^4 + x^2y + C$; The solutions are the curves $x^3y^4 + x^2y + c$
13. $xy^5 - y^4 = c$

17.‡ $\dfrac{d^2y}{du^2} + 3\dfrac{dy}{du} + 2y = 0$; $y = Ae^{-u} + Be^{-2u}$; $y = Ax^{-1} + Bx^{-2}$

18. $y = x(A + B\ln x)$ $\quad$ **19.** $y = A\cos(\ln x) + B\sin(\ln x)$
23.‡ $r^3 + r = 0$; $r = 0, \pm i$; $y = A + B\cos x + C\sin x$
25. $y = A + (B + Cx)e^x$ $\quad$ **26.** $y = A + Bx + Ce^{-x} + De^x$
29. $y = Ae^x + Be^{-x} + C\cos x + D\sin x$
30. $y = (A + Bx)(C\cos x + D\sin x)$ $\quad$ **33.** $y = \frac{1}{2}x^2 + \frac{1}{6}x^3 + \frac{1}{6}x^4 + \frac{1}{24}x^5 + \cdots$

36. Figs. 19.M.36a, b, c, d [Fig. 19.M.36d shows the graph of $y_{200}(x)$; The graph of the exact solution has a vertical asymptote near $x = 3$]

37. Figs. 19.M.37a, b, c, d [Fig. 19.M.37d shows the graph of $y_{200}(x)$]

38. Figs. 19.M.38a, b, c, d [Fig. 19.M.38d shows the graph of $y_{200}(x)$]

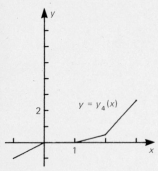

Fig. 19.M.36a

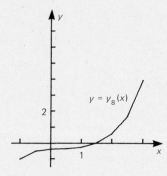

Fig. 19.M.36b

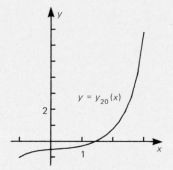

Fig. 19.M.36c

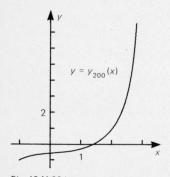

Fig. 19.M.36d

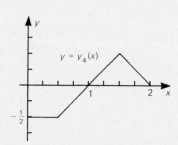

Fig. 19.M.37a

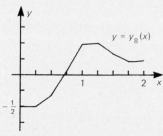

Fig. 19.M.37b

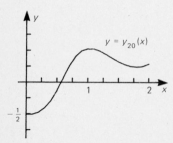

Fig. 19.M.37c

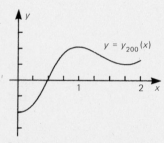

Fig. 19.M.37d

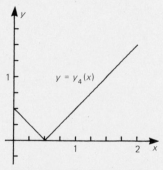

Fig. 19.M.38a

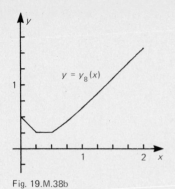

Fig. 19.M.38b

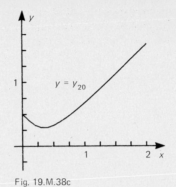

Fig. 19.M.38c

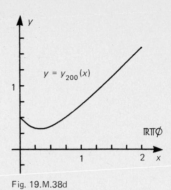

Fig. 19.M.38d

REFERENCES

Batschelet, Edward, [1] *Introduction to Mathematics for Life Scientists*, New York: Springer Verlag, 1971.

Beeton, A. M., [1] "Changes in the Environment and Biota of the Great Lakes," *Eutrophication: Causes, Consequences, Correctives*, Washington, D.C.: National Academy of Sciences, 1969.

Bell, E. T., [1] *Men of Mathematics*, New York: Simon and Schuster, 1937.

Bishop, Errett, [1] *Foundations of Constructive Analysis*, New York: McGraw Hill, 1967.

Bos, H. J. M., [1] "Differentials, Higher-Order Differentials and the Derivative in Leibnizian Calculus," *Archive for the History of the Exact Sciences*, Vol. 14, 1975.

Boyer, C. B., [1] *A History of Mathematics*, New York: John Wiley & Sons, 1968.
[2] *History of Analytic Geometry*, New York: Scripta Mathematica, 1956.
[3] *History of the Calculus*, New York: Dover, 1959.
[4] *The Concepts of the Calculus*, New York: Dover, 1959.
[5] "Pascal: The Man and the Mathematician," *Scripta Mathematica*, 26, 1963.

Brasch, F., [1] *Johann Kepler, 1571–1630, A Tercentenary Commemoration of His Life and Works*, Baltimore: Williams and Williams, 1931.

Brewster, D., [1] *Memoirs of the Life, Writings, and Discoveries of Sir Isaac Newton*, New York: Johnson Reprint Corp., 1965.

Brombacher, W. G. and M. R. Houseman, [1] "Balloon Altitude, Barometric and Photogrammetric Measurements," *The Stratosphere Flight of 1935 in the Balloon Explorer II*, Washington, D.C.: The National Geographic Society, 1938.

Byers, H. R., [1] *The Thunderstorm*, Washington, D.C.: U.S. Department of Commerce, 1949.

Cairns, Edward, [1] *Mathematics for Applied Engineering*, Englewood Cliffs, New Jersey: Prentice-Hall, Inc., 1967.

Cardano, G., [1] *The Book of My Life*, New York: Dover Publications, 1929.

Caspar, M., [1] *Kepler*, transl. by C. Doris Hellman, New York: Abelard-Schuman, 1959.

Castellan, G. W., [1] *Physical Chemistry*, Reading, Massachusetts: Addison-Wesley Publishing Company, 1964.

Cavalli-Sforza, L. L. and W. F. Bodmer, [1] *The Genetics of Human Populations*, San Francisco: W. H. Freeman and Co., 1971.

Child, J., [1] *The Early Mathematical Manuscripts of Leibniz*, Chicago: Open Court, 1920.

Critchfield, Howard, [1] *General Climatology*, Englewood Cliffs, New Jersey: Prentice-Hall, Inc., 1974.

Davies, O. L. and P. L. Goldsmith, [1] *Statistical Methods in Research and Production*, Edinburgh: Oliver and Boyd, 1972.

Davson, Hugh, [1] *A Textbook of General Physiology*, Boston: Little, Brown & Co., 1964.

De Morgan, A., [1] *Essays on the Life and Work of Newton*, Chicago: Open Court, 1914.

Dictionary of Scientific Biography, New York: Charles Scribner's Sons, 1971.

Draper, Jean and Klingman, Jane, [1] *Mathematical Analysis: Business and Economic Applications*, New York: Harper and Row, 1967.

Ekman, S., [1] *Zoogeography of the Sea*, London: Sidgwich and Jackson, 1953.

Encyclopaedia Britannica, Chicago: Encyclopaedia Britannica, 1965.

Eves, Howard, [1] *An Introduction to the History of Mathematics*, New York: Holt, Rinehart and Winston, 1969.

Feodosiev and Siniarev, [1] *Introduction to Rocket Technology*, New York: Academic Press, 1959.

Flammarion, C., [1] *The Flammarion Book of Astronomy*, New York: Simon and Schuster, 1964.

Frost, Percival, [1] *An Elementary Treatise on Curve Tracing*, New York: Chelsea Publishing Company, 1872.

Gingerich, O., [1] *Frontiers in Astronomy*, San Francisco: W. H. Freeman and Co., 1970.

Gould, S. H., [1] "The Method of Archimedes," *American Mathematical Monthly*, 62, 1955, 473–476.

Green, C. R. and Sellers, W. D., [1] *Arizona Climate*, Tucson: University of Arizona Press, 1964.

Gucker, F. T. and R. L. Seifert, [1] *Physical Chemistry*, New York: W. W. Norton and Co., First Edition, 1966.

Haldane, E. S., [1] *Descartes: His Life and Times*, London: John Murray, 1905.

Handbook of Chemistry and Physics, Cleveland, Ohio: Chemical Rubber Publishing Company, 1960.

Handbook of Physiology, Washington, D.C.: American Physiological Society, 1962.

Heath, T. L., [1] *Apollonius of Perga, Treatise on Conic Sections*, New York: Barnes and Noble, 1961. [2] *A History of Greek Mathematics*, 2 vols., Oxford: Clarendon Press, 1921.

Herrick, C., [1] *Mathematics for Electronics*, Columbus, Ohio: Charles E. Merrill Books, Inc., 1967.

Hey, J. S., [1] *The Radio Universe*, New York: The Pergamon Press, 1971.

Hoerner, S., [1] *Fluid Dynamic Drag*, Second Edition, published by the author, 1958.

Kennedy, Edward, [1] "The History of Trigonometry," *Historical Notes for the Mathematics Classroom*, NCTM Yearbook XXXV, Washington, D.C., 1969.

Kinsman, B., [1] *Wind Waves*, Englewood Cliffs, N.J.: Prentice-Hall, 1965.

Krebs, C. J., [1] *Ecology; The Experimental Analysis of Distribution and Abundance*, New York: Harper & Row, 1972.

Lawrence, J. D., [1] *A Catalog of Special Plane Curves*, New York: Dover, 1972.

Marchaj, C. A., [1] *Sailing Theory and Practice*, New York: Dodd, Mead, and Company, 1964.

May, Kenneth, [1] *Bibliography and Research Manual of the History of Mathematics*, Toronto: University of Toronto Press, 1973.

McGraw-Hill Encyclopedia of Science and Technology, New York: McGraw-Hill, 1971.

More, W. J., [1] *Physical Chemistry*, Englewood Cliffs, New Jersey: Prentice-Hall, Inc., Third Edition, 1962.

More, L. T., [1] *Isaac Newton, A Biography*, New York: Dover, 1962.

Mountcastle, V. B., [1] *Medical Physiology*, Saint Louis: C. V. Mosby Co., 1974.

Neiburger, M., [1] "The role of meteorology in the study and control of air pollution," *Bulletin of the American Meteorological Society*, 50, December, 1966.

Noble, B., [1] *Applications of Undergraduate Mathematics in Engineering*, New York: Macmillan, 1967.

Perrin, Charles, [1] *Mathematics for Chemists*, New York: Wiley-Interscience, 1970.

Read, C. B., [1] "Articles on the History of Mathematics: A Bibliography of Articles Appearing in Six Periodicals," *School Science and Mathematics*, 1959.

Richards, C. E., [1] "Analog Simulation in Fish Population Studies," *Analog/Hybrid Computer Educational Users Group Transactions*, July 1970.

Richmond, D. E., [1] "Inverse square laws: a simple treatment," *Selected Papers on Calculus*, Buffalo: Mathematical Association of America, 1969.

Ritter, R., [1] *François Viète*, Paris, 1895.

Robinson, A., [1] *Non-Standard Analysis*, Amsterdam: North Holland Pub. Co., 1966.

Sabra, A. I., *Theory of Light From Descartes to Newton*, London: Oldbourne, 1967.

Schlichting, H., [1] *Boundary Layer Theory*, New York: McGraw-Hill, 1968.

Shelton, R. W. and S. R. Kerr, [1] "Population Density of Monsters in Loch Ness," *Limnology and Oceanography*, September 1972.

Shepard, F. P. and R. F. Dill, [1] *Submarine Canyons and Other Sea Valleys*, Chicago: Rand McNally, 1966.

Shephard, R., [1] *Alive Man: The Physiology of Physical Activity*, Springfield, Illinois: Charles C Thomas Publishers, 1972.

Smith, David E., [1] *History of Mathematics*, New York: Ginn and Company, 1951.
[2] *A Source Book in Mathematics*, New York: Dover, 1954.

Spencer, P. W. "The Turning of the Leaves," *Natural History*, October, 1973.

Strahler, A. N., [1] *Physical Geography*, New York: Wiley, 1969.

Strahler, A. N. and A. H. Strahler, [1] *Environmental Geoscience*, Santa Barbara: Hamilton Pub. Co., 1973.

Struik, D., [1] *A Source Book in Mathematics*, 1200–1800, Cambridge: Harvard Univ. Press, 1969.

Truesdell, C., "Leonard Euler, Supreme Geometer (1701–1783)," *Studies in Eighteenth-Century Culture*, Cleveland: The Press of Case Western Reserve University, 1972.

Turnbull, H. W., [1] *The Great Mathematicians*, New York: New York University Press, 1961.

Van Allan, J. A., [1] "Radiation Belts Around the Earth," *Scientific American*, March 1959.

Verhoogen, J., [1] *The Earth*, New York: Holt, Rinehart, and Winston, 1970.

Vrooman, J. R., [1] *Descartes: A Bibliography*, New York: Putnam, 1970.

Williamson, S., [1] *Fundamentals of Air Pollution*, Reading, Massachusetts: Addison-Wesley Publishing Company, 1973.

Woepcke, F., [1] *Extrait du Fakhrî, Traité d'Algèbre, par Aboù Bekr Mohammed Ben Alhaçan Alkarkhî*, Paris: l'Imprimerie Impériale, 1853.

Wykes, A., [1] *Doctor Cardano, Physician Extraordinary*, London: Muller, 1969.

INDEX

A

Abscissa, 7
Absolute convergence, 813
Absolute value, 3
Acceleration, 114
   due to gravity, 115
   vector, 529, 570
      normal and tengential components, 531
Agnesi, Maria Gaetana (1718–1799), 516
Air pressure, 320, 353
Air resistance, 125, 437, 451
Airy function, 861
Alcohol in blood, 116
D'Alembert, J. L. (1717–1783), 147
Alternating current, 365, 429, 545
Alternating series, 816
Alternating series test, 816
Altitudes of a triangle, 38
Amplitude, 132, 851
Analytic geometry, 1
Angles between curves, 314
Angle between tangent lines, 384
Angle of inclination of a line, 18
Antiderivatives, 206
Appollonius (247–222 B.C.), 29, 143, 452, 495
Approximate integration, 423
    midpoint rule, 424
    Simpson's rule, 424
    trapezoid rule, 424
Archimedes (287–212 B.C.), 29, 143, 247, 248, 275, 318, 452, 495, 728
Archimedean spiral, 143, 474, 541
Arclength, 135
   as a parameter, 513
Arccosecant, 369
Arccosine, 367
Arccotangent, 369
Arcsecant, 369
Arcsine, 366
Arctangent, 368
Areas
   computed by x-integration, 261
   computed by y-integration, 264
   in polar coordinates, 491
Aristotle (384–322 B.C.), 145
Arithmetic mean, 312
Astroid, 172, 318, 542, 543
Asymptotes, 58, 464
Atmospheric pressure, 347
Average cost, 59
Average power, 429
Average rate of change, 112, 605

Average value
   geometric interpretation, 314
   of $f(x)$ in an interval, 313
   of $f(x,y)$ in a region, 698
   of $f(x,y)$ or $f(x,y,z)$ on a curve, 731
Average velocity, 107

B

Barometric pressure, 184
Barrow, Isaac (1630–1677), 145, 249
Berkeley, George (1685–1753), 147
Bernoulli, Daniel (1700–1782), 381
Bernoulli, Jacques (1654–1705), 381, 511
Bernoulli, Jean (1677–1748), 381, 511, 772
Bernoulli's differential equation, 863
Bernoulli's equation of physical and moral wealth, 449
Bernoulli's law, 184
Bessel function, 862
Bicorn, 172
Binomial coefficient, 823
Binomial series, 823
Binormal to a curve, 572
Biological growth, 351
Blood flow from the heart, 389
Bolzano, Bernhard (1781–1848), 147
Boundary of a set, 657
Bounded function, 224
Bounded interval, 656
Bounded sequence, 791
Bowditch curve, 523
Boyle's law, 387
Brachistochrone problem, 511
Brahe, Tycho (1546–1601), 538
Briggs, Henry (1561–1639), 380
Bürgi, Jobst (1552–1632), 380

C

Carbon dioxide in the atmosphere, 149
Carbon-14 dating, 353
Cardano, Gerolamo (1501–1576), 29
Cardioid, 199, 488, 490, 510, 542
Catalog of curves, 25
   in polar coordinates, 437
Catalog of surfaces, 573
Catenary, 376, 378, 542
Catenoid, 378
Cauchy, Augustin (1789–1857), 147, 249
Cavalieri, Bonaventura (1598–1647), 247
Center of curvature, 527
Center of gravity, 305
   of a flat plate, 308, 698

   of a rod, 306
   of a solid, 712
Centroid, 310
Chain rule
   for $G(s(t))$, 118
   for $F(x(u,v))$, 617
   for $f(x(t),y(t))$, 618
   for $f(x(u,v),y(u,v))$, 620
   for $f(x(t),y(t),z(t))$, 645
   for $f(x(u,v),y(u,v),z(u,v))$, 645
Changing variables in definite integrals, 394
Characteristic equation, 844
Characteristic of common logarithms, 327
Chemical reaction rates, 353, 441, 446, 447, 450
Chlorophyl, 86
Circle of curvature, 527, 783
Circular paraboloid, 576
Circulation around a curve, 753
Cissoid, 494
Closed curve, 746
Closed interval, 656
Common logarithms, 324
   table, 871–872
Comparison test, 801, 813
Component of one vector in the direction of another, 506, 551
Composition of functions, 118, 615
Compound interest, 354
Computer/calculator exercises
   approximate integration, 426
   derivatives, 93
   Euler's method, 864
   infinite series, 797
   limits, 52
   Newton's method, 201
   Riemann sums, 259
   Taylor's formula, 784
Concavity, 157
Conchoid of Nicomedes, 494
Conditional convergence, 813
Cones, 452, 574
   elements of, 452
   elliptic, 578
   vertex of, 452
Conic sections, 452
   degenerate, 475
   directrices and eccentricity, 477
   discriminant, 474
   in polar coordinates, 489
   rotation of axes, 470
Conservative force field, 746
Constants of integration, 236
Constraint curve, 676
Consumer surplus, 321

Continuity
  ϵ-δ definition, 79, 179
  of $f(x)$, 66
  of $f(x,y)$, 615
  of $f(x,y,z)$, 644
  of polynomials and rational functions, 67
  of sums, products and quotients, 67
Continuously compounded interest, 355
Contour curves, 590
Convergence
  absolute, 813
  conditional, 813
  of improper integrals, 777
Coordinate line, 1
Coordinate planes, 7
Coordinate surfaces, 714, 718
Coordinates
  cylindrical, 714
  polar, 481
  rectangular, 7, 546
  spherical, 714, 717
Cosecant function, 358
Cosh x, 373
Cosine function, 127
Cosines, law of, 133, 505
Cotangent function, 358
Coth x, 375
Coulomb's law, 28, 300
Crab nebula, 629
Cramer's rule, 580
Critical points
  of $f(x)$, 154
  of $f(x,y)$, 659
Cross curve, 515
Cross product, 555
  and determinants, 558
Cross section, 575
  of $f(x,y)$, 583, 625
  of $f(x,y,z)$, 646
Csch x, 375
Cubic curves, 26
Curl
  scalar, 755
  vector, 762
Curl theorem, 762
Curtate cycloid, 516
Curvature, 524, 570
  center of, 527
  circles of, 527, 783
  radius of, 527
  zero, 527
Curves
  a brief catalog of, 25, 487
  angles between, 314
  average values, 731
  closed, 746
  constraint, 676
  cross, 515
  cubic, 26
  curvature of, 524
  length, 282, 511, 731
  negative of, 732
  orientation, 509, 733
  parametric equation of, 509
  perpendicular, 103
  sum of two, 733
  tangent, 103
Curves in three dimensions, 569
  acceleration vector, 570
  binormal, 572
  curvature, 570
  length, 569
  orientation, 762

principal normal, 572
  unit tangent vector, 569
  velocity vector, 569
Cycloid, 510, 534
Cylinders, 573
  parabolic, 576
Cylindrical coordinates, 714
  integrals in, 715

D

Dandelin, G. P. (1794–1847), 477
Decibels, 328, 651
Decimal expansions of real numbers, 32
Declination of a compass, 683
Decreasing function, 153
Definite integrals, 235
Degenerate conic sections, 475
Degrees to radians, table, 868
Deltoid, 543
Demand functions, 117, 210, 321, 450
Density, 305, 308, 698, 711
  of air, 42, 150
  of water, 209
Dependent variable, 171
Derivative
  as a function, 95
  definition, 88
  directional, 625
  one-sided, 206
  partial, 586, 603
Descartes, René (1596–1650), 29, 144, 381, 453, 495
Descartes' folium, 169
Determinants, 557
  computing cross products by, 558
  of systems of equations, 580
Difference quotient, 100
Differential, 146, 195, 434, 640
  at a point on a curve, 195
  of $f(x)$, 195
  of $f(x,y)$, 640
  of $f(x,y,z)$, 647
  total, 640, 647
Differential equation, 349, 430
  Bernoulli's, 863
  Euler's, 652
  exact, 759, 840, 863
  first-order linear, 836
  general solution, 843
  homogeneous, 843
    of degree k, 862
  inhomogeneous, 845
  Laplace's, 611, 759
  logistic, 440
  nth order linear with constant
    coefficients, 863
  particular solution, 846
  power series solution, 858
  second-order linear with constant
    coefficients, 842
  separable, 430
Differentiation operator, 97
Diffusion, 449
Dimension analysis, 120, 620
Direction angles, 551
Direction cosines, 551
Direction fields, 432
Directional derivatives, 625, 646
Directrices of conic sections, 477
Discontinuous functions, 65
Discriminant of a conic section, 474, 577

Discriminant of a quadratic polynomial, 412
Displacement vector, 497, 549
Dissolution, 450
Distance
  between parallel planes, 565
  in a coordinate plane, 8
  in space, 548
  on a coordinate line, 4
  to a line, 508
  to a plane, 584
  total and net, 290
Divergence theorem
  in space, 762
  in the plane, 753, 754
Domain
  of $f(x)$, 39
  of $f(x,y)$, 586
  of $f(x,y,z)$, 643
Doppler effect, 651
Dot product
  in two dimensions, 504
  in three dimensions, 550
Double integrals, 685
  existence, 688
  in polar coordinates, 703
Drug sensitivity, 210
Dummy variable, 214

E

$e = 2.718281828\ldots\ldots$, 325
Earthquakes, 327, 682
Eccentricity of conic sections, 477
Elasticity of demand, 210, 450
Electric current in a light bulb, 149
Electrical circuits, 853
Electroencephalograms, 654
Elements of a cone, 452
Ellipse, 452, 458
  axes, 460
  center, 458
  foci, 458
  major axis, 460
  minor axis, 460
  optical property, 462
  vertices, 460
Ellipsoids, 318, 463, 574, 578
Elliptic cone, 578
Elliptic integrals, 463, 515
Elliptic paraboloid, 576
Epicycle, 545
Epitrochoid, 522
ϵ-δ-definition of a definite integral, 250
ϵ-δ-definition of limits and continuity, 74, 179
$\epsilon$$N$-definition of the limit of a sequence, 789
Equations of lines in planes, 21
  point-slope form, 22
  slope-intercept form, 23
  two-intercept form, 23
  two-point form, 22
Equations of lines in space, 562, 585
Euclid (ca. 300 B.C.), 29, 248, 452
Eudoxus (ca. 408–355 B.C.), 834
Euler, Leonhard (1707–1783), 147, 381
Euler's constant, 834
Euler's differential equation, 652
Euler's method, 864
Even and odd functions, 166
Evolute, 543
Exact differentials, 759
Exact differential equation, 840
Exhaustion, method of, 834

Exponents, 31
  irrational, 323
Exponential functions, 322
  e^x, 342
  table of, 875–878
Extreme value theorem, 71, 658

F

Factorials, 765
  table of, 868
Falling bodies, 437
Fermat, Pierre (1601–1665), 29, 145, 247, 453,
  495, 834
Fermat's principle, 190
First derivative test for $f(x)$, 152
  at a point, 153
  in an arbitrary interval, 204
  in an open interval, 153
First derivative test for $f(x, y)$, 656
First-order linear differential equation, 836
Flux, 752, 762
Focal length of a lens, 184
Foci
  of ellipses, 458
  of hyperbolas, 463
  of parabolas, 453
Folium of Descartes, 169
Force field, 746
Force vector, 497
Four-petaled rose, 486
Free damped and undamped motion, 851
Freeth's nephroid, 494
Frequency, 132, 852
Friction, 372
Functions of one variable, 39
  bounded, 224
  continuous, 65
  composite, 118
  decreasing, 153
  discontinuous, 65
  domain of, 39
  even, 166
  graphs of, 44
  increasing, 153
  inverse, 173
  monotone, 250
  nondecreasing and nonincreasing, 250
  odd, 166
  periodic, 130, 166
  piecewise continuous, 253
  piecewise monotone, 250
  product of two, 67
  quotient of two, 67
  range, 173
  rational, 62
  sum of two, 67
  unbounded, 254
  uniformly continuous, 253
Functons of two variables, 586
  continuity, 615
  critical points, 659
  cross sections, 583, 625
  domains, 586
  graphs, 587
  homogeneous, 652
  level curves, 590
  limits and continuity, 613, 615
  local maxima and minima, 658
Functions of three variables, 643
  directional derivatives and gradients, 646
  limits, continuity and partial deriva-
    tives, 644

Fundamental theorem of algebra, 413
Fundamental theorem of calculus, 229
  for derivatives of integrals, 230
  for integrals of derivatives, 232

G

Galilei, Galileo (1564–1642), 144, 145, 247,
  249, 453
Generalized mean value theorem
  for derivatives, 768
  for integrals, 783
Genetics, 449
Geometric mean, 643
Geometric series
  finite, 794
  infinite, 797
Gradient, 629, 646
  field, 739
  theorem, 762
  vector, 629, 646
Graph
  of equations in two variables, 12
  of $f(x)$, 44
  of $f(x,y)$, 587
Gravity, 84
  acceleration due to, 115
  center of, 305, 698
Great circle, 722
  distance, 722, 725
Greatest lower bound, 791
Green's theorem
  in space, 761
  in the plane, 749
Gregory, James (1638–1675), 145

H

Half-life, 350
Halley, Edmund, 147, 539
Hanging cable, 376, 378, 451, 544
Harmonic series, 800
Heart
    work done by, 760
    rate of blood flow, 389
Heat, 112
Heat current density, 636
Heat flow equation, 653
Helix, 569
Heron's formula, 585
Hipparchus (ca. 180–125 B.C.), 379
Hippopede, 543
Historical notes, 28, 143, 247, 379, 511, 772
Homogeneous differential equation, 843, 863
Homogeneous function, 652
Hooke's law, 297
de l'Hopital, Marquis G. F. A. (1661–1704), 772
l'Hopital's rule, 772
l'Huilier, S. (1750–1840), 147
Huygens, Christiaan (1629–1695), 511
Hydrostatic pressure, 301
Hyperbola, 26, 452, 463
    asymptotes, 464
    axis, 463
    center, 463
    foci, 463
    in navigation, 468
    optical property, 467
    rectangular, 467
    vertices, 463
Hyperbolic functions, 373
    cosh x, 373

  coth x, 375
  csch x, 375
  graphs, 374, 375
  inverses, 376
  sech x, 375
  sinh x, 373
  tanh x, 375
  table of, 875–878
Hyperbolic functions and hyperbolas, 374
Hyperbolic paraboloid, 577
Hyperbolic spiral, 493, 494
Hyperboloids, 574
  of one sheet, 578
  of two sheets, 578
Hypocycloid, 542
Hypotrochoid, 515

I

$i = \sqrt{-1}$, 382, 844
Icarus (astroid), 494
Ice formation, 445
Implicit differentiation, 168
Implicitly defined functions, 168
Improper integrals, 777
Increasing function, 153
Indefinite integral, 235
Independent variable, 171
Indeterminate forms, 771
Indifference curve, 684
Indivisibles, 247, 728
Induction, mathematical, 32
Inequalities, 2
Inertia, 320, 702
Infinite sequence, 785
  bounded, 791
  decreasing, 791
  limits, 789, 790
  nonincreasing, 791
  increasing, 791
  monotone, 791
  nondecreasing, 791
Infinite series, 797
  with nonnegative terms, 801
Infinitesimals, 145
Inflection points, 157
Inhomogeneous differential equation, 845
Initial conditions, 350, 431
Initial value problem, 350, 431
Instantaneous velocity, 109
Integral, 213
  algebra of, 239
  analytic definition, 220, 224
  changing variables in, 394
  definite, 235
  double, 686
  elliptic, 463, 515
  $\varepsilon\delta$-definition, 250
  geometric definition, 213
  improper, 777
  in cylindrical coordinates, 715
  in spherical coordinates, 718
  indefinite, 235
  iterated, 689, 708
  line, 729
  of products and powers of trigonometric
    functions, 402
  of rates of change, 288
  of rational functions, 412, 416
  of rational functions of sine and
    cosine, 419
  sign, 214
  surface, 760
  triple, 707

Integral form of the remainder in Taylor's formula, 783
Integral tables, 390, 410, 878–881
  use of, 421
Integral test, 806, 815
Integrand, 214
Integrating factor, 836
Integration
  approximate methods of, 423
  constant of, 236
  interval of, 214
  by inverse trigonometric substitution, 407
  by partial fractions, 411
  by parts, 397
  by reduction formulas, 405
  by substitution, 242, 390
  by undetermined coefficients, 401, 402
  limits of, 214
  reversing the order of, 692
Intercepts, 566
  of curves, 15
Intercept equations
  of a line, 23
  of a plane, 566
Interest
  compound, 354
  compounded continuously, 355
Intermediate value theorem, 70
Internal combustion engine, 759
Intervals, 2, 656
  partitions of, 221
Inverse functions, 173
Inverse function theorem, 175
Inverse square law, 539
Inverse hyperbolic functions, 376
Inverse trigonometric functions, 366
  arccos x, 367
  arccot x, 369
  arccsc x, 369
  arcsec x, 369
  arcsin x, 366
  arctan x, 368
  derivatives, 367, 369
Inverse trigonometric substitution, 370, 407
Involute, 543
Irrational exponents, 323
Irrational numbers, 31
Irreducible quadratic polynomials, 412
Irrotational, incompressible fluid flow, 755
Iterated integrals, 689, 708

K

Kepler, Johann (1571–1630), 248, 453, 538
Kepler's laws, 538
Kinetic energy, 863

L

Lagrange, J. L. (1736–1813), 674
Lagrange multipliers, 674
  with two constraints, 684
Lambert's law of absorption, 387
Laplace's differential equation, 611, 759
Law of cosines, 133
Law of sines, 133
Laws of arithmetic, 30
Least squares method, 194, 664
Least upper bound, 791
Leibniz, Gottfried (1646–1716), 94, 146, 248, 381
Leibniz notation for derivatives, 94

Lemniscate, 172, 493, 494
Level curves, 590
Level surfaces, 643
Lift of an airplane, 116
Light intensity, 348
Limaçon, 488, 490, 493, 541, 706
Limit comparison test, 803, 814
Limits of $f(x)$
  $\varepsilon\delta$-definition, 74
  existence, 51, 61
  finite, 48, 74
  infinite, 57, 78
  of polynomials, 62
  of rational functions, 62
  one-sided, 50
  two-sided, 50, 51
Limits of $f(x,y)$, 613
Limits of $f(x,y,z)$, 644
Limit theorems, 53, 80
Limits of integration, 214
Lines
  equations of, 21
  parametric equations of, 562
  slope of, 17
  symmetric equations, 585
Line integrals, 729
  path-independent, 742
Linear combination of two functions, 67
Linear density, 305
Linear relations, 25
Lissajous curve, 523
Loblolly pine trees, 610
Local maxima and minima
  of $f(x)$, 154
  of $f(x,y)$, 658
Loch Ness monsters, 86
Logarithm
  base b, 324, 334
  common (base 10), 324
    table of, 871–872
  derivatives of, 330, 334
  graphs of, 330, 335
  natural (base e), 326, 329
    table of, 873–874
Logarithmic differentiation, 341
Logarithmic spiral, 494, 515, 541, 542, 543
Logistic differential equation, 440
LORAN (Long Range Navigation), 453, 468
Lotka, A. J. (1880–1949), 442
Lotka-Volterra equations, 443
Lower bound of an infinite sequence, 791

M

Maclaurin, Colin (1698–1746), 821
Maclaurin series, 821
Magnitude
  of an earthquake, 327
  of a star, 328
Mantissa of common logarithms, 327
Marginal cost, profit and revenue, 115, 193, 210, 318
  with two variables, 684
Mass and weight, 291, 529
Mathematical induction, 32
Maxima and minima
  of $f(x)$, 154
  of $f(x,y)$, 658
  of $f(x,y,z)$, 658
Max-min problems, 186, 190
  with integer variables, 191
Mean value theorem
  for derivatives of $f(x)$, 202

for derivatives of $f(x,y)$, 624
for double integrals, 699
for integrals, 314
  geometric interpretation, 315
generalized, 768, 783
Medians of a triangle, 11, 37, 502
Menaechmus (ca. 360 B.C.), 29, 452
Mercury, orbit of, 544
Merton rule, 145, 148
Merton College, Oxford, 145
Midpoint
  formula, 9
  method, 201
  rule for integration, 424
Minimal surface, 611
  differential equation of, 611
  Scherck's, 611
Mixing problems, 448, 449, 838
Moles (6×10^{23} molecules), 441
Moments, 306
  about a line, 308
  about a plane, 712
  about a point, 306
  of flat plates, 308, 698
  of inertia, 320, 702
  of solids, 712
Monkey saddle, 601, 702
Monotone functions, 250
Monotone sequences, 791
Multiple integrals, 686

N

Nabla, 762
Napier, John (1550–1617), 380
Natural logarithms, 326, 329
  table of, 873–874
Negative of a curve, 732
Nephroid, 494
Newton, Isaac (1642–1727), 30, 94, 146, 249, 381, 453, 539
Newton's law
  $F = ma$, 291, 437, 529
  inverse square, 539
  of cooling, 354, 448
  of gravity, 28, 210, 319, 727
Newton's method, 199
Nicomedes' conchoid, 494
Nile river, 293
Normal component of acceleration vector, 531
Normal line, 93
Normal probability density, 396
Normal vector
  to a curve, 570
  to a level surface of $f(x,y,z)$, 647
  to a plane, 563
  to the graph of $f(x,y)$, 639
  unit, 526

O

Oblate spheroid, 463
Ocean waves, 379
Odd function, 166
One-sided derivative, 206
One-sided limit, 50
Open interval, 656
Optics, 184, 189, 190
  ellipses and, 462
  hyperbolas and, 467
  parabolas and, 456
Oresme, Nicole (ca. 1323–1382), 248

Ordered pairs, 7
Ordinate, 7
Oriented curve, 509, 733
Orthogonal trajectories, 444, 448, 451, 840
Oxygen consumption, 125
Oxygen intake, 321

P

Pappus (ca. AD 300), 311, 452, 495
Pappus' first and second theorems, 311
Parabola, 26, 148, 452, 453
    axis, 453
    directrices, 453
    focal radius, 456
    focus, 453
    optical property, 456
    vertex, 453
Parabolic cylinder, 576
Parabolic trajectories, 456
Paraboloids, 574, 575
    circular, 576
    elliptic, 576
    hyperbolic, 577
Parameters, 509
    variation of, 849
Parametric equations, 509
    in polar coordinates, 510
    of curves, 509
    of lines, 562
Partial derivatives, 586, 603
    higher order, 607
    of $f(x,y)$, 586, 603
    of $f(x,y,z)$, 644
    of $f(x_1,x_2, \ldots x_n)$, 651
    mixed, 607
        equality of, 607, 652
Partial sums, 797
Partitions of intervals, 221
Pascal, Blaise (1623–1682), 247
Path-independent line integrals, 742
Path of steepest ascent, 636
Pendulum, 544
Penicillin concentration, 665
Periodic functions, 130, 166
Perpendicular curves, 103
Perpendicular vectors, 505
pH, 327
Phase shift, 132, 851
Piecewise continuous function, 253, 688
Piecewise monotone function, 250
Piecewise smooth function, 686
Piriform, 523
Planar density, 308, 698
Planes
    angle between a line and, 584
    angle between two, 584
    coordinate, 7
    distance between parallel, 565
    distance from a point to, 568, 584
    equations of, 563
    intercept equations of, 566
    normal vector to, 563
    tangent, 637, 647
Planetary motion, 536
Planimeter, 760
Point-slope equation of a line, 22
Polar coordinates, 481
    a brief catalog of curves in, 487
    area in, 491
    conic sections and, 489
    double integrals in, 703
    graphs of equations in, 484, 485

parametric equations in, 510
    speed in, 537
    symmetry in, 484
    vectors in, 501, 536
Pollution potential, 665
Polynomials, 62
    continuity of, 67
Population growth, 352, 446, 448, 451
    in a crowded environment, 439
Position vector, 497, 548
Potential energy, 863
Potential function, 747
Poiseuille's law, 493
Power series, 819
    differentiation and integration of, 827
    multiplication and division of, 824
    radius of convergence, 819
    solutions of differential equations, 858
Predator/prey, 442, 448
Present value, 356, 388
Pressure, 301
    air, 320, 353
    atmospheric, 347
    barometric, 184
    hydrostatic, 301
Prime notation for derivatives, 94
Prime numbers, 337
Prime number theorem, 337
Principia Mathematica, 539
Principle normal and binormal, 572
Prism, 43
Probability, 348, 396, 451, 831
    normal density, 396
Product rule, 104
Products of functions, 67
Projectiles, 386, 530
Projection of a vector on a plane, 507
Projection of one vector along another, 507, 551
Prolate spheroid, 463
Ptolemy (second century AD), 379
    theory of the solar system, 538
Pythagorean identities, 130, 362
Pythagorean theorem, 8
    generalized, 542
    in three dimensions, 547

Q

Quadrants, 8
Quadratic formula, 412, 419
Quadric surfaces, 574
Quotient rule, 105
Quotients of functions, 67

R

Radians, 127
    table of conversion to degrees, 868
Radioactive decay, 350
Radius of convergence, 819
Radius of curvature, 527
Range of $f(x)$, 173
Rate of change, 112
    average rates of change, 112, 605
    instantaneous rates of change, 112
    integrals of, 288
Rate of blood flow from the heart, 389
Ratio test, 810, 815
Rational function, 62
    continuity, 67
    graphing, 160
Rational numbers, 31
Rationalization, 55
Reactions to drugs, 210

Real numbers, 1
    algebra of, 30
    decimal expansion, 32
Real-valued functions of a real variable, 39
Rectangular coordinates
    in two dimensions, 7
    in three dimensions, 546
Reference tables, 865
Related rate problems, 179
Remainder, 766
    integral form, 783
Reserve requirement for banks, 806
Resistors, 450
Richter scale, 327
Riemann, G. F. B. (1826–1866), 223, 249
Riemann sums, 223
Right-hand rule, 555, 762
Roberval, de Gilles Persone (1602–1675), 144, 247
Root mean square value of a function, 321
Root test, 833
Rotation of axes, 470
Rumor spreading, 449

S

Saddle point, 667
St. Ives, 796
Satellite, 494, 534
Scalar curl, 755
Scalar product (*see* dot product)
Scalar triple product, 560
Scherck's minimal surface, 611
Secant function, 358
Secant line, 88
Sech x, 375
Second derivative test for $f(x)$
    concavity and, 157
    for a local maximum or minimum, 159
    in an open interval, 157
Second derivative test for $f(x,y)$, 666
Second-order linear differential equation, 842
Seismic exploration, 211
Separable differential equation, 430
Sequences (*see* infinite sequences)
Series
    alternating, 816
    binomial, 823
    geometric, 794, 797
    harmonic, 800
    infinite, 797
    Maclaurin, 821
    power, 819
    Taylor, 821
    telescoping, 799
Simpson's rule for integration, 424
Sine function, 127
Sines, law of, 133
Sinh x, 373
Skew lines, 565
Slope
    of a line, 17
    of parallel lines, 19
    of perpendicular lines, 19
Slope-intercept equation of a line, 23
Snell's law, 190
Solar radiation, 609
Sound barrier, 126
Specific gravity, 602
Speed, 290, 497, 512, 569
    in polar coordinates, 537
Spherical coordinates, 714, 717
    in geography, 722
    integrals in, 718

Spheroid, 463
Spirals
    Archimedian, 143, 474, 541
    hyperbolic, 493, 494
    logarithmic, 494, 515, 541, 542, 543
Springs, 296, 387, 850
Squares, cubes, square roots and cube roots,
    tables of, 866–867
Standard deviation, 396
Star magnitude, 328
Steady-state motion, 857
Stimulus and sensation, 440, 446
Stirling's formula, 337
Stokes' law, 595
Stokes' theorem
    in space, 763
    in the plane, 755
Subtropical inversion, 113
Sums of functions, 67
Surfaces, 573
    catalog of, 573
    quadric, 574
Surface area, 700
Surface integrals, 760
Surfaces of revolution, areas of, 284, 727
Survival function, 321
Symmetry of a curve, 13
    about a line, 14
    about a plane, 13
    about the origin, 13
    about the x-axis, 14
    about the y-axis, 14
    in polar coordinates, 484
Symmetry of surfaces, 580
Symmetric equations of a line in space, 585
System of equations, 580

T

Tangent curves, 103
Tangent function, 358
Tangent line, 87
    angles between, 384
    equation for, 91
    vertical, 92, 164
Tangent line approximations, 194
Tangent plane
    to level surfaces of $f(x,y,z)$, 647
    to the graph of $f(x,y)$, 637
Tangent vector, unit, 520, 569
Tangential component of acceleration
    vector, 531
Tanh x, 375
Tautochrone problem, 511
Taylor, Brook (1685–1731), 764
Taylor polynomial, 764
Taylor series, 821
Taylor's theorem
    for $f(x)$, 766
    for $f(x,y)$, 835
Telescoping series, 799
Thermal conductivity, 636
Thin lens equation, 184
Topographical maps, 594
Torricelli, Evangelista (1608–1647),
    144, 247
Torus, 317

Total differential
    of $f(x,y)$, 640
    of $f(x,y,z)$, 647
Traces, 579
Transcendental functions, 322
Translation of axes, 455, 460, 465
Translation of vectors, 498
Trapezoid rule for integration, 424
Triangle inequality, 36
Trigonometric functions, 127, 357
    differentiation formulas, 135, 358
    graphs of, 361
    integration formulas, 237, 359, 360
    table of, 869–870
Triple integrals, 707
Triple product $A \cdot (B \times C)$, 560
Tschirnhausen's cubic, 197, 515
Tsunami waves, 602
Two-intercept equation of a line, 23
Two-point equation of a line, 22
Two-sided limit, 50, 51

U

U.C. set, 846
Unbounded function, 254
Unbounded interval, 656
Undetermined coefficients
    integration by, 401, 402
    method of, 846
Unit circle, 128
Unit normal vector, 526
Unit tangent vector, 520, 569
Unit vector, 498
Uniformly continuous, 253
Upper and lower approximate solutions, 299
Upper bound of an infinite sequence, 791
Uranium-238 dating, 353
Utility function, 684

V

V-2 rocket, 117
Value of a function, 39
Van Allen belts, 644
Variable, 39
    dependent, 171
    dummy, 214
    independent, 171
    of a function, 39
Variation of parameters, 849
Vectors, 496
    acceleration, 498, 529
    arithmetic of, 498
    component of one vector in the direction of
        another, 506, 551
    components of, 499
    direction angles, 551
    direction cosines, 551
    displacement, 497, 549
    force, 497
    in three dimensions, 546, 548
    length of, 496, 500
    normal, 526, 563, 570, 639
    perpendicular, 505
    polar form, 501, 536
    position, 497, 548

projection of one vector along
    another, 507, 551
translation of, 498
unit, 498
unit normal, 526
unit tangent, 520, 569
velocity, 497, 518, 569
zero, 496
Vector analysis, 729
Vector curl, 762
Vector fields, 738
    force vector fields, 738
    gradient fields, 739
    velocity vector fields, 739, 752
Vector-valued functions, 516
    antiderivatives, 521
    derivatives, 517, 520
    limits of, 517
Velocity
    average, 108
    instantaneous, 109
    vector, 497, 518, 569
    vector field, 739, 752
Vertex of a cone, 452
Vertical tangent lines, 92, 164
Vibrating springs, 850
    amplitude, 851
    critically damped, 852
    free undamped, 850
    overdamped, 852
    steady state motion, 857
    underdamped, 852
Viète, François (1540–1603), 29, 380
Voltaire, 248
Volterra, Vito (1860–1940), 442
Volume
    as a triple integral, 711
    by double integrals, 696
    computed by slicing, 266
Volumes of solids of revolution, 269, 276
    method of cylinders, 276, 280
    method of disks, 269, 272
    method of washers, 273

W

Wallis, John (1616–1703), 381
Water pressure, 25
Water resistance, 445
Water waves, 86, 379
Wave equation, 653
Weber's law, 440, 446
Weight and density, 305, 698, 711
Weight and mass, 291
    of a flat plate, 308, 698
Well tempered scale of music, 328
Wind chill, 664
Witch of Agnesi, 515
Work, 507, 508
    and Stokes' theorem, 755
    as a line integral, 745
    end of, 1005

Z

Zeno's paradox, 786, 788
Zero curvature, 527
Zero vector, 496

Basic differentiation formulas

$$\frac{d}{dx} x^n = nx^{n-1}$$

$$\frac{d}{dx} u^n = nu^{n-1}\frac{du}{dx}$$

$$\frac{d}{dx}\ln|x| = \frac{1}{x}$$

$$\frac{d}{dx}\ln|u| = \frac{1}{u}\frac{du}{dx}$$

$$\frac{d}{dx}\sin x = \cos x$$

$$\frac{d}{dx}\sin u = \cos u\frac{du}{dx}$$

$$\frac{d}{dx}\cos x = -\sin x$$

$$\frac{d}{dx}\cos u = -\sin u\frac{du}{dx}$$

$$\frac{d}{dx}\tan x = \sec^2 x$$

$$\frac{d}{dx}\tan u = \sec^2 u\frac{du}{dx}$$

$$\frac{d}{dx}\cot x = -\csc^2 x$$

$$\frac{d}{dx}\cot u = -\csc^2 u\frac{du}{dx}$$

$$\frac{d}{dx}\sec x = \sec x \tan x$$

$$\frac{d}{dx}\sec u = \sec u \tan u\frac{du}{dx}$$

$$\frac{d}{dx}\csc x = -\csc x \cot x$$

$$\frac{d}{dx}\csc u = -\csc u \cot u\frac{du}{dx}$$

$$\frac{d}{dx}\arcsin x = \frac{1}{\sqrt{1-x^2}}$$

$$\frac{d}{dx}\arcsin u = \frac{1}{\sqrt{1-u^2}}\frac{du}{dx}$$

$$\frac{d}{dx}\arctan x = \frac{1}{x^2+1}$$

$$\frac{d}{dx}\arctan u = \frac{1}{u^2+1}\frac{du}{dx}$$

$$\frac{d}{dx}e^x = e^x$$

$$\frac{d}{dx}e^u = e^u\frac{du}{dx}$$

$$\frac{d}{dx}b^x = (\ln b)b^x$$

$$\frac{d}{dx}b^u = (\ln b)b^u\frac{du}{dx}$$

$$\frac{d}{dx}\sinh x = \cosh x$$

$$\frac{d}{dx}\sinh u = \cosh u\frac{du}{dx}$$

$$\frac{d}{dx}\cosh x = \sinh x$$

$$\frac{d}{dx}\cosh u = \sinh u\frac{du}{dx}$$

The Product Rule $$\frac{d}{dx}(uv) = u\frac{dv}{dx} + v\frac{du}{dx}$$

The Quotient Rule $$\frac{d}{dx}\left(\frac{u}{v}\right) = \frac{v\dfrac{du}{dx} - u\dfrac{dv}{dx}}{v^2}$$

The Chain Rule $$\frac{d}{dt}G(s(t)) = \frac{dG}{ds}(s(t))\frac{ds}{dt}(t)$$